Problem Books in Mathematics

Series Editors:
Peter Winkler
Department of Mathematics
Dartmouth College
Hanover, NH 03755
USA

For further volumes:
http://www.springer.com/series/714

Vladimir V. Tkachuk

A Cp-Theory Problem Book

Special Features of Function Spaces

 Springer

Vladimir V. Tkachuk
Departamento de Matematicas
Universidad Autonoma Metropolitana-Iztapalapa
San Rafael Atlixco, Mexico City, Mexico

ISSN 0941-3502
ISBN 978-3-319-37794-0   ISBN 978-3-319-04747-8 (eBook)
DOI 10.1007/978-3-319-04747-8
Springer Cham Heidelberg New York Dordrecht London

Mathematics Subject Classification (2010): 54C35

Printed on acid-free paper

Springer is part of Springer Science+Business Media (www.springer.com)

Preface

This is the second volume of the series of books of problems in C_p-theory entitled *A C_p-Theory Problem Book*, i.e., this book is a continuation of the first volume subtitled *Topological and Function Spaces*. The series was conceived as an introduction to C_p-theory with the hope that each volume will also be used as a reference guide for specialists.

The first volume provides a self-contained introduction to general topology and C_p-theory and contains some highly nontrivial state-of-the-art results. For example, Sect. 1.4 presents Shapirovsky's theorem on the existence of a point-countable π-base in any compact space of countable tightness and Sect. 1.5 brings the reader to the frontier of the modern knowledge about realcompactness in the context of function spaces.

This present volume introduces quite a few topics from scratch but dealing with topology and C_p-theory is already a professional endeavour. The objective is to study the behaviour of general topological properties in function spaces and establish the results on duality of cardinal functions and classes with respect to the C_p-functor. The respective background includes a considerable amount of top-notch results both in topology and set theory; the author's obsession with keeping this work self-contained implied that an introduction to advanced set theory had to be provided in Sect. 1.1. The methods developed in this section made it possible to present a very difficult example of Todorčević of a compact strong S-space.

Of course, it was impossible to omit the famous Baturov's theorem on coincidence of the Lindelöf number and extent in subspaces of $C_p(X)$ for any Lindelöf Σ-space X and the result of Christensen on σ-compactness of X provided that $C_p(X)$ is analytic. The self-containment policy of the author made it obligatory for him to give a thorough introduction to Lindelöf Σ-spaces in Sect. 1.3 and to the descriptive set theory in Sect. 1.4.

We use all topological methods developed in the first volume, so we refer to its problems and solutions when necessary. Of course, the author did his best to keep *every* solution as independent as possible, so a short argument could be repeated several times in different places.

The author wants to emphasize that if a postgraduate student mastered the material of the first volume, it will be more than sufficient to understand every problem and solution of this book. However, for a concrete topic, much less might be needed. Finally, let me outline some points which show the potential usefulness of the present work:

- *The only background needed is some knowledge of set theory and real numbers; any reasonable course in calculus covers everything needed to understand this book.*
- *The student can learn all of general topology required without recurring to any textbook or papers; the amount of general topology is strictly minimal and is presented in such a way that the student works with the spaces $C_p(X)$ from the very beginning.*
- *What is said in the previous paragraph is true as well if a mathematician working outside of topology (e.g., in functional analysis) wants to use results or methods of C_p-theory; he (or she) will find them easily in a concentrated form or with full proofs if there is such a need.*
- *The material we present here is up to date and brings the reader to the frontier of knowledge in a reasonable number of important areas of C_p-theory.*
- *This book seems to be the first self-contained introduction to C_p-theory. Although there is an excellent textbook written by Arhangel'skii (1992a), it heavily depends on the reader's good knowledge of general topology.*

Mexico City, Mexico   Vladimir V. Tkachuk

Contents

Detailed Summary of Exercises

1.1. Tampering with Additional Axioms: Some Hereditary Properties

1.2. Monolithity, Stability and Their Generalizations

1.3. Whyburn Spaces, Calibers and Lindelöf Σ-Property

1.4. A Glimpse of Descriptive Set Theory

1.5. Additivity of Properties: Mappings Between Function Spaces

Introduction

The term "C_p-theory" was invented to abbreviate the phrase "The theory of function spaces endowed with the topology of pointwise convergence". The credit for the creation of C_p-theory must undoubtedly be given to Alexander Vladimirovich Arhangel'skii. The author is proud to say that Arhangel'skii also was the person who taught him general topology and directed his Ph.D. thesis. Arhangel'skii was the first to understand the need to unify and classify a bulk of heterogeneous results from topological algebra, functional analysis and general topology. He was the first to obtain crucial results that made this unification possible. He was also the first to formulate a critical mass of open problems which showed this theory's huge potential for development.

Later, many mathematicians worked hard to give C_p-theory the elegance and beauty it boasts nowadays. The author hopes that the work he presents for the reader's judgement will help to attract more people to this area of mathematics.

The main text of this volume consists of 500 statements formulated as problems; it constitutes Chap. 1. These statements provide a gradual development of many popular topics of C_p-theory to bring the reader to the frontier of the present-day knowledge. A complete solution is given to every problem of the main text.

The material of Chap. 1 is divided into five sections with 100 problems in each one. The sections start with an introductory part where the definitions and concepts to be used are given. The introductory part of any section *never exceeds two pages and covers everything that was not defined previously.* Whenever possible, we try to save the reader the effort of ploughing through various sections, chapters and volumes, so we give the relevant definitions in the current section not caring much about possible repetitions.

Chapter 1 ends with some bibliographical notes to give the most important references related to its results. The selection of references is made according to the author's preferences and by no means can be considered complete. However, a complete list of contributors to the material of Chap. 1 can be found in our bibliography of 300 items. It is my pleasant duty to acknowledge that I consulted the paper of Arhangel'skii (1998a) to include quite a few of its 375 references in my bibliography.

Sometimes, as we formulate a problem, we use without reference definitions and constructions introduced in other problems. The general rule is to try to find the relevant definition *not more than ten problems before*.

The first section of Chap. 1 deals with hereditary properties in $C_p(X)$. To understand the respective results, the reader needs a topological background including the ability to manage additional axioms of ZFC and apply strong and difficult methods of set theory. The pursuit of self-containment obliged the author to give an introduction to advanced set theory. In this section the reader can find the applications of continuum hypothesis, Martin's axiom, Jensen's axiom, Souslin trees and Luzin spaces.

The non-C_p material presented in Chap. 1 also includes an introduction to descriptive set theory and Lindelöf Σ-spaces. This helped to keep this work self-contained when we gave the proofs of Baturov's theorem on $C_p(X)$ for a Lindelöf Σ-space X and Christensen's theorem on σ-compactness of X provided that $C_p(X)$ is analytic. There are many topics in Chap. 1 which are developed up to the frontier of the present-day knowledge. In particular, Sect. 1.5 includes the famous Gerlits–Pytkeev theorem about coincidence of the Fréchet–Urysohn property and k-property in any space $C_p(X)$.

The complete solutions of all problems of Chap. 1 are given in Chap. 2. Chapter 3 begins with a selection of 100 statements which were proved as auxiliary facts in the solutions of the problems of the main text. This material is split into six sections to classify the respective results and make them easier to find. Chapter 4 consists of 100 open problems presented in ten sections with the same idea: to classify this bulk of problems and make the reader's work easier.

Chapter 4 also witnesses an essential difference between the organization of our text and the book of Arhangel'skii and Ponomarev (1974): *we never put unsolved problems in the main text as is done in their book*. All problems formulated in Chap. 1 are given complete solutions in Chap. 2 and the unsolved ones are presented in Chap. 4.

There is little to explain about how to use this book as a reference guide. In this case the methodology is not that important and the only thing the reader wants is to find the results he (or she) needs as fast as possible. To help with this, the titles of chapters and sections give the first approximation. To better see the material of a chapter, one can consult the second part of the Contents section where a detailed summary is given; it is supposed to cover all topics presented in each section. Besides, the index can also be used to find necessary material.

To sum up the main text, I believe that the coverage of C_p-theory will be reasonably complete and many of the topics can be used by postgraduate students who want to specialize in C_p-theory. Formally, this book can also be used as an introduction to general topology. However, it would be a somewhat biased introduction, because the emphasis is always given to C_p-spaces and the topics are only developed when they have some applications in C_p-theory.

To conclude, let me quote an old saying which states that the best way for one to learn a theorem is to prove it oneself. This text provides a possibility to do this. If the reader's wish is to read the proofs, then they are concentrated immediately after the main text.

Chapter 1
Duality Theorems and Properties
of Function Spaces

This chapter presents some fundamental aspects of set theory, descriptive set theory, general topology and C_p-theory.

Section 1.1 introduces some advanced concepts of set theory. We give the statements and applications of the continuum hypothesis, Martin's axiom and Jensen's axiom. The next thing under the study is the behavior of spread, hereditary Lindelöf number and hereditary density in function spaces. The most important results of this section are the duality theorems for s^*, hd^* and hl^* (Problems 025–030) and Todorcevic's example of a strong S-space (Problem 098).

In Sect. 1.2 we deal with monolithity, stability and their generalizations. The principal results are presented as several generic theorems on duality between $\eta(\kappa)$-monolithity and $\theta(\kappa)$-stability, formulated in Problems 146–151.

Section 1.3 starts with Whyburn spaces and their properties. Next, we introduce Lindelöf Σ-spaces and their most important characterizations. The rest of the section is devoted to calibers, precalibers and small diagonals. The most important results include Baturov's theorem on extent in subspaces of $C_p(X)$ for a Lindelöf Σ-space X (Problem 269) and Gruenhage's theorem on Lindelöf Σ-spaces with a small diagonal (Problem 300).

In Sect. 1.4 we introduce the basic notions of descriptive set theory and give their applications to C_p-theory. This section features three main results: Christensen's theorem on analyticity of $C_p(X)$ (Problem 366), Fremlin's theorem on K-analytic spaces whose compact subspaces are metrizable (Problem 395) and Pytkeev's theorem on condensations of Borel sets (Problem 354).

The first part of Sect. 1.5 comprises some results on decompositions of $C_p(X)$ into a finite or countable union of subspaces with "nice" properties. The second part is devoted to the study of the existence of good mappings between $C_p(X)$ and $C_p(Y)$ and the simplest implications this has for the spaces X and Y. We also have two main results in this section: Gerlits–Pytkeev theorem on k-property in $C_p(X)$ (Problem 465) and Tkachuk's theorem on discreteness of X if $C_p(X)$ is homeomorphic to a retract of a $G_{\delta\sigma}$-subspace of $\mathbb{R}^X$ (Problem 500).

V.V. Tkachuk, *A Cp-Theory Problem Book: Special Features of Function Spaces*,
Problem Books in Mathematics, DOI 10.1007/978-3-319-04747-8_1,
© Springer International Publishing Switzerland 2014

1.1   Some Additional Axioms and Hereditary Properties

A space X is *left-separated (right-separated)* if there exists a well-order $<$ on X such that the set $\{y \in X : y < x\}$ is closed (open) in X for any $x \in X$. A space X is *scattered* if any subspace $Y \subset X$ has an isolated point. Recall that $\varphi^*(X) = \sup\{\varphi(X^n) : n \in \mathbb{N}\}$ and $h\varphi(X) = \sup\{\varphi(Y) : Y \subset X\}$ for any cardinal invariant φ. All results of this book are proved assuming that ZFC axioms hold. The abbreviation ZFC stands for Zermelo–Fraenkel–Choice. This axiomatic system is the most accepted one at the present moment. We won't need to have the knowledge of what the axioms of ZFC exactly say. It is sufficient to know that all they do is to postulate some very natural properties of sets. For the reader who wants to learn more, the book of Kunen (1980) is an excellent introduction to this subject.

In the twentieth century topologists and set-theorists discovered that there were some very natural problems which could not be solved using ZFC axioms only; to fix this, quite a few additional axioms have been created. Practically all of those axioms are proved to be *consistent with ZFC* which means that if ZFC has no contradiction, then ZFC, together with the axiom in question, does not have one. In this section we formulate the most popular additional axioms and their applications. All results of this book are proved in ZFC if no additional assumptions are formulated explicitly; however, we sometimes emphasize this.

The statement CH (called *Continuum Hypothesis*) says that the first uncountable ordinal is equal to the continuum, i.e., $\omega_1 = \mathfrak{c}$. The statement "$\kappa^+ = 2^\kappa$ for any infinite cardinal κ" is called *Generalized Continuum Hypothesis (GCH)*.

A partial order on a set $\mathcal{P}$ is a relation $\leq$ on $\mathcal{P}$ with the following properties:

(PO1) $p \leq p$ for any $p \in \mathcal{P}$;

(PO2) $p \leq q$ and $q \leq r$ imply $p \leq r$;

(PO3) $p \leq q$ and $q \leq p$ imply $p = q$.

The pair $(\mathcal{P}, \leq)$ is called *a partially ordered set*. If the order is clear, we will write $\mathcal{P}$ instead of $(\mathcal{P}, \leq)$. Let $(\mathcal{P}, \leq)$ be a partially ordered set. The elements $p, q \in P$ are called *compatible* if there is $r \in P$ such that $r \leq p$ and $r \leq q$. If p and q are not compatible, they are called *incompatible*. A set $A \subset \mathcal{P}$ is *an antichain* if the elements of A are pairwise incompatible. We say that $(\mathcal{P}, \leq)$ *has the property ccc* if any antichain of $\mathcal{P}$ is countable. A set $D \subset \mathcal{P}$ is called *dense* in $\mathcal{P}$ if, for every $p \in \mathcal{P}$, there is $q \in D$ such that $q \leq p$.

A non-empty set $F \subset \mathcal{P}$ is *a filter* if it has the following properties:

(F1)   for any $p, q \in F$, there is $r \in F$ such that $r \leq p$ and $r \leq q$;

(F2)   if $p \in F$ and $p \leq q$, then $q \in F$.

Given an infinite cardinal κ, we denote by $MA(\kappa)$ the following statement: for any ccc partial order $\mathcal{P}$ and any family $\mathcal{D}$ of dense subsets of $\mathcal{P}$ with $|\mathcal{D}| \leq \kappa$, there is a filter $F \subset \mathcal{P}$ such that $F \cap D \neq \emptyset$ for any $D \in \mathcal{D}$. Now, *Martin's axiom, MA*, says that $MA(\kappa)$ holds for any infinite $\kappa < \mathfrak{c}$.

A subset $C \subset \omega_1$ is called *club* ($\equiv$*closed and unbounded*) if C is uncountable and closed in the order topology on ω_1. A set $S \subset \omega_1$ is *stationary* if $S \cap C \neq \emptyset$

for any club $C \subset \omega_1$. *Jensen's axiom* $\diamondsuit$ is the statement: for each $\alpha < \omega_1$, there is a set $A_\alpha \subset \alpha$ such that, for any $A \subset \omega_1$, the set $\{\alpha \in \omega_1 : A \cap \alpha = A_\alpha\}$ is stationary. The principle $\diamondsuit^+$ is the following statement: for each $\alpha \in \omega_1$, there is a countable family $\mathcal{A}_\alpha \subset \exp(\alpha)$ such that, for any $A \subset \omega_1$, there is a club $C \subset \omega_1$ for which $A \cap \alpha \in \mathcal{A}_\alpha$ and $C \cap \alpha \in \mathcal{A}_\alpha$ for any $\alpha \in C$. The sequence $\{\mathcal{A}_\alpha : \alpha < \omega_1\}$ is called *a $\diamondsuit^+$-sequence*.

A space X is called *zero-dimensional* if X has a base consisting of clopen sets. A point $x \in X$ is called *a P-point* if any countable intersection of neighborhoods of x is a neighborhood of x. An uncountable dense-in-itself space X is called *Luzin* (also written *Lusin*) if any nowhere dense subspace of X is countable. Say that X is an *L-space* if $hl(X) = \omega < d(X)$; if $hd(X) = \omega < l(X)$, then X is called *an S-space*. The axiom SA says that there are no S-spaces, i.e., that every regular hereditarily separable space is Lindelöf. Furthermore, X is *a strong S-space* if $hd^*(X) = \omega < l(X)$; if $hl^*(X) = \omega < d(X)$, then X is called *strong L-space*.

A *tree* is a partially ordered set $(\mathcal{T}, \leq)$ such that, for every $x \in \mathcal{T}$, the set $L_x = \{y \in \mathcal{T} : y < x\}$ is well ordered by $\leq$. We will often write $\mathcal{T}$ instead of $(\mathcal{T}, \leq)$. If $\mathcal{T}$ is a tree and $x \in \mathcal{T}$, then *the height* of x in $\mathcal{T}$ or $\mathrm{ht}(x, \mathcal{T})$ is the ordinal isomorphic to L_x. For each ordinal α, the *α-th level* of $\mathcal{T}$ or $\mathrm{Lev}_\alpha(\mathcal{T})$ is the set $\{x \in \mathcal{T} : \mathrm{ht}(x, \mathcal{T}) = \alpha\}$. The height $\mathrm{ht}(\mathcal{T})$ of the tree $\mathcal{T}$ is the least α such that $\mathrm{Lev}_\alpha(\mathcal{T}) = \emptyset$. A subset $\mathcal{T}' \subset \mathcal{T}$ is called *a subtree* of $\mathcal{T}$ if $L_x \subset \mathcal{T}'$ for every $x \in \mathcal{T}'$. A subset $C \in \mathcal{T}$ is called *a chain* if C is linearly ordered by $\leq$, i.e., every two elements of C are comparable. *An antichain* of $\mathcal{T}$ is a set $A \subset \mathcal{T}$ such that $x, y \in A$ and $x \neq y$ implies $x \not\leq y$ and $y \not\leq x$. For every infinite cardinal κ, a *κ-Souslin tree* is a tree $\mathcal{T}$ such that $|\mathcal{T}| = \kappa$ and every chain and every antichain have cardinality $< \kappa$. An ω_1-Souslin tree is called *Souslin tree*. If κ is a regular cardinal, *a κ-tree* is a tree of height κ with levels of cardinality $< \kappa$. A *κ-Aronszajn* tree is a κ-tree with no chains of cardinality κ. An ω_1-Aronszajn tree is called *Aronszajn tree*.

If $f : X \to Y$ and $Z \subset X$, we denote the restriction of f to Z by $f|_Z$ or $f|Z$. If we have maps $f, g : X \to Y$, then $f \approx g$ if the set $\{x \in X : f(x) \neq g(x)\}$ is finite. Given functions $f : X \to Y$ and $g : X_1 \to Y_1$, we say that $f \subset g$ if $X \subset X_1$, $Y \subset Y_1$ and $g|X = f$. Now, ω^α is the set of all maps from α to ω and $\omega^{<\omega_1} = \bigcup\{\omega^\alpha : \alpha < \omega_1\}$. Any ω_1-sequence $\{s_\alpha : \alpha < \omega_1\} \subset \omega^{<\omega_1}$ such that $s_\alpha \in \omega^\alpha$ is an injective map and $s_\beta|\alpha \approx s_\alpha$ for all $\alpha < \beta < \omega_1$ is called *Aronszajn coding*. Denote by **P** the set of all monotonically increasing functions from ω^ω, i.e., $\mathbf{P} = \{f \in \omega^\omega : f(i) < f(j)$ whenever $i < j\}$. Given $f, g \in \omega^\omega$, we say that $f <^* g$ if there exists $m \in \omega$ such that $f(n) < g(n)$ for all $n \geq m$. A sequence $\{f_\alpha : \alpha < \gamma\} \subset \omega^\omega$ is called *strictly $<^*$-increasing* if $f_\alpha <^* f_\beta$ for all $\alpha < \beta < \gamma$. A set $S \subset \omega^\omega$ is *$<^*$-cofinal* in ω^ω if, for any $f \in \omega^\omega$, we have $f <^* g$ for some $g \in S$.

001. Given an infinite cardinal κ prove that the following properties are equivalent for any space X:

  (i) $hl(X) \leq \kappa$;
  (ii) $l(X) \leq \kappa$ and every $U \in \tau(X)$ is a union of $\leq \kappa$-many closed subsets of X;
  (iii) $l(X) \leq \kappa$ and every closed $F \subset X$ is a G_κ-set in X;
  (iv) $l(U) \leq \kappa$ for any open $U \subset X$.

  In particular, a space X is hereditarily Lindelöf if and only if it is Lindelöf and perfect.

002. Prove that a space X is hereditarily normal if and only if any open subspace of X is normal.

003. Prove that if X is perfectly normal, then any $Y \subset X$ is also perfectly normal.

004. Let X be any space. Prove that $hd(X) = \sup\{|A| : A$ is a left-separated subspace of $X\}$. In particular, the space X is hereditarily separable if and only if every left-separated subspace of X is countable.

005. Let X be any space. Prove that $hl(X) = \sup\{|A| : A$ is a right-separated subspace of $X\}$. In particular, the space X is hereditarily Lindelöf if and only if every right-separated subspace of X is countable.

006. Prove that a space is right-separated if and only if it is scattered.

007. Let X be a left-separated space. Prove that $hl(X) \leq s(X)$. In particular, any left-separated space of countable spread is hereditarily Lindelöf.

008. Let X be a right-separated space. Prove that $hd(X) \leq s(X)$. In particular, any right-separated space of countable spread is hereditarily separable.

009. Prove that any space has a dense left-separated subspace.

010. Suppose that $s(X) = \omega$. Prove that X has a dense hereditarily Lindelöf subspace.

011. Prove that for any space X, we have $hl^*(X) = hl(X^\omega)$. In particular, if all finite powers of X are hereditarily Lindelöf, then X^ω is hereditarily Lindelöf.

012. Prove that for any space X, we have $hd^*(X) = hd(X^\omega)$. In particular, if all finite powers of X are hereditarily separable, then X^ω is hereditarily separable.

013. Prove that for any space X, we have $s^*(X) = s(X^\omega)$.

014. Suppose that $s(X \times X) \leq \kappa$. Prove that $hl(X) \leq \kappa$ or $hd(X) \leq \kappa$. In particular, if $s(X \times X) = \omega$, then X is hereditarily separable or hereditarily Lindelöf.

015. Prove that $|X| \leq 2^{hl(X)}$ for any space X. In particular, any hereditarily Lindelöf space has cardinality $\leq \mathfrak{c}$.

016. Prove that $s(X \times X) \leq s(C_p(X)) \leq s^*(X)$ for any space X.

017. Prove that $hd(X \times X) \leq hl(C_p(X)) \leq hd^*(X)$ for any space X.

018. Prove that $hl(X \times X) \leq hd(C_p(X)) \leq hl^*(X)$ for any space X.

019. For an arbitrary $n \in \mathbb{N}$, let $J_n = J(n)$ be the hedgehog with n spines. Prove that $s(X^n) \leq s(C_p(X, J_n)) \leq s(C_p(X) \times C_p(X))$ for any space X.

020. For an arbitrary $n \in \mathbb{N}$, let $J_n = J(n)$ be the hedgehog with n spines. Prove that $hd(X^n) \leq hl(C_p(X, J_n)) \leq hl(C_p(X) \times C_p(X))$ for any space X.

021. For an arbitrary $n \in \mathbb{N}$, let $J_n = J(n)$ be the hedgehog with n spines. Prove that $hl(X^n) \leq hd(C_p(X, J_n)) \leq hd(C_p(X) \times C_p(X))$ for any space X.

022. For any space X prove that $s(C_p(X) \times C_p(X)) = s^*(C_p(X))$.

023. For any space X prove that $hl(C_p(X) \times C_p(X)) = hl^*(C_p(X))$.

024. For any space X prove that $hd(C_p(X) \times C_p(X)) = hd^*(C_p(X))$.

025. Prove that $s^*(X) = s^*(C_p(X))$ for any space X.

026. Prove that $hl^*(X) = hd^*(C_p(X))$ for any space X.

027. Prove that $hd^*(X) = hl^*(C_p(X))$ for any space X.

028. For an infinite cardinal κ, suppose that $s(C_p(X)) = \kappa$ and $\Delta(X) \leq \kappa$. Prove that $s^*(C_p(X)) \leq \kappa$ and hence $s^*(C_p(X)) = s(C_p(X))$. In particular, if X is a space with a G_δ-diagonal, then $s^*(C_p(X)) = s(C_p(X)) = s^*(X)$.

029. For an infinite cardinal κ, suppose that $hl(C_p(X)) = \kappa$ and $\Delta(X) \leq \kappa$. Prove that $hl^*(C_p(X)) \leq \kappa$ and hence $hl^*(C_p(X)) = hl(C_p(X))$. In particular, if X is a space with a G_δ-diagonal, then $hl^*(C_p(X)) = hl(C_p(X)) = hd^*(X)$.

030. (Velichko's theorem) Prove that $hd^*(C_p(X)) = hd(C_p(X)) = hl^*(X)$ for any space X.

031. Prove that $s(C_p(C_p(X))) = s^*(X)$ for any space X.

032. Prove that $hd(C_p(C_p(X))) = hd^*(X)$ for any space X.

033. Prove that $hl(C_p(C_p(X))) = hl^*(X)$ for any space X.

034. Prove that for a zero-dimensional space X, we have $s^*(X) = s(C_p(X))$.

035. Prove that $hd^*(X) = hl(C_p(X))$ for any zero-dimensional space X.

036. Prove that under SA, the following conditions are equivalent:

  (i) $s(C_p(X)) = \omega$;
  (ii) $hl((C_p(X))^\omega) = \omega$;
  (iii) $hd((C_p(X))^\omega) = \omega$.

  In particular, if SA holds, then $hl(C_p(X)) = \omega$ implies $hl((C_p(X))^\omega) = \omega$ and $s(C_p(X)) = \omega$ implies $s((C_p(X))^\omega) = \omega$.

037. Prove that the following statements are equivalent (remember that all spaces are assumed to be Tychonoff):

  (i)   there is a space X with $s(X) = \omega$ and $d(X) > \omega$;
  (ii) there is a space X with $hl(X) = \omega$ and $d(X) > \omega$;
  (iii) there is a left-separated space X with $s(X) = \omega$ and $|X| = \omega_1$.

038. (Δ-system lemma) Prove that for any regular uncountable cardinal κ, if $\mathcal{U}$ is a family of finite sets with $|\mathcal{U}| = \kappa$, then there exists a set F (called the Δ-root for $\mathcal{U}$) and a family $\mathcal{V} \subset \mathcal{U}$ (called the Δ-system for $\mathcal{U}$) such that $|\mathcal{V}| = \kappa$ and $A \cap B = F$ for any distinct $A, B \in \mathcal{V}$.

039. Prove that under CH, there exists a hereditarily Lindelöf non-separable dense subspace X of the space $\{0, 1\}^{\omega_1}$. In particular, L-spaces exist under CH.

040. Prove that under CH, there exists a hereditarily separable non-Lindelöf dense subspace X of the space $\{0, 1\}^{\omega_1}$. Thus, S-spaces exist under CH.

041. Prove that under CH, any sequential compact space has points of countable character.

042. Prove that under CH, there is a P-point in $\beta\omega \setminus \omega$.

043. Let X be a Luzin space. Prove that $hl(X) = \omega$ and $hd(X) \leq \omega_1$.

044. Prove that if a Luzin space X is separable, then all closed subsets of X are separable.

045. Prove that no Luzin space can be condensed onto a compact space.

046. Prove that under CH, there is a Luzin second countable space as well as a Luzin non-separable space.

047. Prove that $MA(\omega)$ holds in ZFC (and hence CH implies MA), while $MA(\mathfrak{c})$ is false in ZFC.

048. Prove that $MA(\kappa)$ is equivalent to $MA(\kappa)$ restricted to ccc partially ordered sets of cardinality $\leq \kappa$, i.e., if $MA(\kappa)$ is true for all ccc partial orders of cardinality $\leq \kappa$, then it is true for all ccc partial orders.

049. Let $(\mathcal{P}, \leq)$ be a partially ordered set. Call a subset $A \subset \mathcal{P}$ *centered* if for any $n \in \mathbb{N}$ and any $p_1, \dots, p_n \in A$, there is $r \in \mathcal{P}$ such that $r \leq p_i$ for all $i \leq n$. Assume $MA + \neg CH$ and take any ccc partially ordered set $\mathcal{P}$. Prove that for any uncountable $R \subset \mathcal{P}$, there exists an uncountable centered $Q \subset R$. In particular, all elements of Q are pairwise compatible.

050. Assume $MA + \neg CH$. Let X_t be a space with $c(X_t) = \omega$ for every $t \in T$. Prove that $c(X) = \omega$, where $X = \prod\{X_t : t \in T\}$.

051. Given families $\mathcal{A}, \mathcal{B} \subset \exp(\omega)$ such that $|\mathcal{A}| \leq \kappa$, $|\mathcal{B}| \leq \kappa$ and $\kappa < \mathfrak{c}$, suppose that $B \backslash (\bigcup \mathcal{A}')$ is infinite for every $B \in \mathcal{B}$ and any finite family $\mathcal{A}' \subset \mathcal{A}$. Prove that $MA(\kappa)$ implies that there exists $M \subset \omega$ such that $B \backslash M$ is infinite for any $B \in \mathcal{B}$ while $A \backslash M$ is finite for any $A \in \mathcal{A}$.

052. (Booth lemma) Let $\mathcal{C} \subset \exp(\omega)$ be a family such that $|\mathcal{C}| = \kappa < \mathfrak{c}$ and $\bigcap \mathcal{C}'$ is infinite for every finite $\mathcal{C}' \subset \mathcal{C}$. Prove that $MA(\kappa)$ implies that there exists an infinite $L \subset \omega$ such that $L \backslash C$ is finite for any $C \in \mathcal{C}$.

053. Let $\mathcal{A} \subset \exp(\omega)$ be an almost disjoint family ($\equiv$ all elements of $\mathcal{A}$ are infinite while $A \cap B$ is finite whenever A and B are distinct elements of $\mathcal{A}$). Suppose that κ is an infinite cardinal and $|\mathcal{A}| = \kappa < \mathfrak{c}$. Prove that $MA(\kappa)$ implies that $\mathcal{A}$ is not maximal.

054. Assume $MA + \neg CH$. Let X be a space such that $\chi(x, X) < \mathfrak{c}$ for some $x \in X$. Prove that for any countable $A \subset X$ with $x \in \overline{A}$, there exists a sequence $\{a_n\}_{n \in \omega} \subset A$ such that $a_n \to x$.

055. Let X be a second countable space of cardinality $< \mathfrak{c}$. Prove that under MA, any subset of X is a G_δ in X.

056. Prove that MA implies $2^\kappa = \mathfrak{c}$ for any infinite $\kappa < \mathfrak{c}$.

057. Let X be a second countable dense-in-itself space. Given a cardinal κ such that $0 < \kappa < \mathfrak{c}$, suppose that N_α is a nowhere dense subset of the space X for any $\alpha < \kappa$. Prove that under MA, the set $\bigcup\{N_\alpha : \alpha < \kappa\}$ is of first category in X.

058. Prove that Martin's axiom is equivalent to the following statement: "Given a compact space X such that $c(X) = \omega$, for any family γ of nowhere dense subsets of X with $|\gamma| < \mathfrak{c}$, we have $\bigcup \gamma \neq X$".

059. Show that under $MA + \neg CH$, if $s^*(X) = \omega$, then $hd(X^\omega) = hl(X^\omega) = \omega$. In particular, neither strong S-spaces nor strong L-spaces exist under $MA + \neg CH$.

060. Prove that under MA+¬CH, every compact space of countable spread is hereditarily separable.

061. Prove that under MA+¬CH, every compact space of countable spread is perfectly normal.

062. Suppose that MA+¬CH hold. Prove that every compact space X with $s(X \times X) \leq \omega$ is metrizable.

063. Prove that MA+¬CH implies that there are no Luzin spaces.

064. Let $C_n \subset \omega_1$ be a club for any natural n. Prove that $\bigcap\{C_n : n \in \omega\}$ is a club.

065. Prove that

    (i) every stationary subset of ω_1 is uncountable;

    (ii) not all uncountable subsets of ω_1 are stationary;

    (iii) if $A \subset \omega_1$ contains a stationary set, then A is stationary;

    (iv) any intersection of a stationary subset of ω_1 with a club is stationary;

    (v) if $A_n \subset \omega_1$ for each $n \in \omega$ and $\bigcup\{A_n : n \in \omega\}$ is stationary, then A_n is stationary for some $n \in \omega$.

066. Let A be a stationary subset of ω_1. Prove that there exists a disjoint family of stationary subsets $\{A_\alpha : \alpha < \omega_1\}$ such that $A_\alpha \subset A$ for each $\alpha \in \omega_1$.

067. (Fodor's Lemma, also called Pressing-Down Lemma) Let $A \subset \omega_1$ be a stationary subset of ω_1. Suppose that $f : A \to \omega_1$ is a map such that $f(\alpha) < \alpha$ for any $\alpha \in A$. Prove that there is $\alpha_0 \in \omega_1$ such that the set $f^{-1}(\alpha_0)$ is stationary (and, in particular, $|f^{-1}(\alpha_0)| = \omega_1$).

068. Given any ordinal $\alpha < \omega_1$, let $S_\alpha = \{f \in \omega^\alpha : f$ is an injection$\}$. In the set $S = \bigcup_{\alpha<\omega_1} S_\alpha$, consider the following partial order: $f \leq g$ if and only if $f \subset g$. Observe that $(S, \leq)$ is a tree with all its chains countable. Prove that there exists an ω_1-sequence $\{s_\alpha : \alpha < \omega_1\} \subset S$ such that $s_\alpha \in S_\alpha$ and $s_\beta|\alpha \approx s_\alpha$ for all $\alpha < \beta < \omega_1$. Deduce from this fact that the subtree $T = \bigcup_{\alpha<\omega_1}\{s \in S_\alpha : s \approx s_\alpha\}$ of the tree S is an Aronszajn tree. Hence Aronszajn codings and trees exist in ZFC.

069. Observe that Jensen's axiom implies CH; prove that it is equivalent to any of the following statements:

    (i) for every $\alpha < \omega_1$, there exists a countable family $\mathcal{A}_\alpha \subset \exp(\alpha)$ such that for any $A \subset \omega_1$, the set $\{\alpha : A \cap \alpha \in \mathcal{A}_\alpha\}$ is stationary;

    (ii) for any $\alpha < \omega_1$, there is a set $B_\alpha \subset \alpha \times \alpha$ such that for any $B \subset \omega_1 \times \omega_1$, the set $\{\alpha : B \cap (\alpha \times \alpha) = B_\alpha\}$ is stationary;

    (iii) for any $\alpha < \omega_1$, there is a function $f_\alpha : \alpha \to \alpha$ such that for any $f : \omega_1 \to \omega_1$, the set $\{\alpha : f|_\alpha = f_\alpha\}$ is stationary;

    (iv) for any $\alpha < \omega_1$, there is a function $g_\alpha : \alpha \to \omega_1$ such that for any $g : \omega_1 \to \omega_1$, the set $\{\alpha : g|_\alpha = g_\alpha\}$ is stationary;

    (v) there exists a set S of cardinality ω_1 and a set of functions $\{h_\alpha : \alpha < \omega_1\}$ such that $h_\alpha : \alpha \to S$ for all $\alpha < \omega_1$ and, for any $h : \omega_1 \to S$, the set $\{\alpha : h|_\alpha = h_\alpha\}$ is stationary;

    (vi) for any set T of cardinality ω_1, there exists a family $\{k_\alpha : \alpha < \omega_1\}$ of functions such that $k_\alpha : \alpha \to T$ for all $\alpha < \omega_1$, and, for any function $k : \omega_1 \to T$, the set $\{\alpha : k|_\alpha = k_\alpha\}$ is stationary.

070. Prove that if Jensen's axiom holds, then there is a Souslin tree.

071. Prove that MA+¬CH implies there are no Souslin trees.

072. Given two topologies τ and μ on the same set X, say that τ is *weaker* than μ if $\tau \subset \mu$. If $\tau \subset \mu$ and $\tau \neq \mu$, then τ is said to be *strictly weaker than* μ. If $(X, \leq)$ is a linearly ordered set and $Y \subset X$, then $\leq_Y$ is the order $\leq$ considered only on the points of Y. Let $(L, \leq)$ be any linearly ordered space.

   (i) Prove that for any $M \subset L$, the topology $\tau(\leq_M)$ on M generated by the order $\leq_M$ is weaker than the topology τ_M^L of the subspace on M induced from L.

   (ii) Show that $\tau(\leq_M)$ can be strictly weaker than τ_M^L even if M is a dense or a clopen subspace of L.

   (iii) Assume that the set M is *order dense in L*, i.e., for any $a, b \in L$, if $a < b$, then $a \leq p < q \leq b$ for some $p, q \in M$. Prove that M is dense in L and $\tau(\leq_M) = \tau_M^L$.

   (iv) Prove that there exists a compact linearly ordered space $(K, \preceq)$ such that for some order dense $N \subset K$, there exists an order isomorphism between $(N, \preceq_N)$ and $(L, \leq)$. In particular, $\tau(\preceq_N) = \tau_N^K$, the space N is dense in K and $(L, \leq)$ is order isomorphic to $(N, \preceq_N)$.

073. Prove that under $\diamond$, there exists a linearly ordered hereditarily Lindelöf non-separable compact space.

074. Prove that a linearly ordered compact L-space exists if and only if there exists a Souslin tree.

075. Let L be a non-separable linearly ordered space such that $c(L) \leq \omega$. Prove that $c(L \times L) > \omega$. In particular, if X is a perfectly normal non-separable linearly ordered compact space, then $c(X) = \omega$ but $c(X \times X) > \omega$.

076. Let X be a linearly ordered hereditarily Lindelöf non-separable compact space. Prove that $C_p(X)$ is not Lindelöf.

077. Suppose that $s^*(X) = \omega$. Prove that X condenses onto a hereditarily separable space.

078. Suppose that $C_p(X)$ has countable spread. Prove that it can be condensed onto a hereditarily separable space.

079. Prove that under Jensen's axiom, there is a space X of countable spread which does not condense onto a hereditarily separable space.

080. For an arbitrary space X, assume that Y is a second countable space such that the space $C_p(X, Y)$ is dense in Y^X. Fix any base $\mathcal{B} \subset \tau^*(Y)$ in the space Y; an open set $U \subset C_p(X, Y)$ is called $\mathcal{B}$-*standard (or standard with respect to X, Y and $\mathcal{B}$)* if there exist $n \in \mathbb{N}$, points $x_1, \ldots, x_n \in X$ and sets $O_1, \ldots, O_n \in \mathcal{B}$ such that $U = \{f \in C_p(X, Y) : f(x_i) \in O_i$ for all $i = 1, \ldots, n\}$. Prove that $C_p(X, Y)$ is perfectly normal if and only if any open subset of $C_p(X, Y)$ is a union of countably many $\mathcal{B}$-standard open subsets of $C_p(X, Y)$.

081. Suppose that $C_p(X)$ is perfectly normal. Prove that all closed subsets of $X \times X$ are separable.

082. Let X be a compact space with $C_p(X)$ perfectly normal. Prove that $X \times X$ is hereditarily separable.

083. Let X be a compact space with $C_p(X)$ perfectly normal. Prove that under MA+¬CH, the space X is metrizable.

084. Prove that the following properties are equivalent for any space X:

   (i) $C_p(X) \times C_p(X)$ is perfectly normal;

   (ii) $(C_p(X))^n$ is perfectly normal for any natural $n \geq 1$;

   (iii) $(C_p(X))^\omega$ is perfectly normal;

   (iv) $C_p(X, Y)$ is perfectly normal for any second countable space Y;

   (v) for every $n \in \mathbb{N}$, all closed subsets of X^n are separable.

085. Prove that for any compact space X, the space $C_p(X) \times C_p(X)$ is perfectly normal if and only if $(C_p(X))^\omega$ is hereditarily Lindelöf.

086. Prove that under SA, if $C_p(X)$ is perfectly normal, then $(C_p(X))^\omega$ is hereditarily Lindelöf.

087. Let X be a space with a G_δ-diagonal. Prove that $C_p(X)$ is perfectly normal if and only if $(C_p(X))^\omega$ is perfectly normal.

088. Prove that under MA+¬CH, all closed subspaces of $C_p(X)$ are separable if and only if $(C_p(X))^\omega$ is hereditarily separable.

089. Prove that under CH, there exists a subspace X of $\{0, 1\}^{\omega_1}$ such that for all $n \in \mathbb{N}$, all closed subsets of X^n are separable and X is not hereditarily separable. Therefore, $C_p(X)$ is a perfectly normal non-Lindelöf space.

090. Prove that a compact space X is metrizable if and only if X^3 is hereditarily normal.

091. Prove that $w(X) = \Delta(X)$ for any infinite compact space X. Deduce from this fact that a compact space X is metrizable if and only if the diagonal of X is a G_δ-subspace of $X \times X$.

092. Prove that a countably compact space X is metrizable if and only if every subspace $Y \subset X$ with $|Y| \leq \omega_1$ is metrizable.

093. Give an example of a non-metrizable pseudocompact space P such that every $Y \subset P$ with $|Y| \leq \omega_1$ is metrizable.

094. Let X be a non-metrizable compact space. Prove that there exists a continuous map of X onto a (non-metrizable compact) space of weight ω_1.

095. Let X be a perfectly normal compact space. Prove that $d(X) \leq \omega_1$.

096. Let Y be a subspace of a perfectly normal compact space X. Prove that $nw(Y) = w(Y)$.

097. Prove that under Continuum Hypothesis, there exists a strictly $<^*$-increasing ω_1-sequence $S = \{f_\alpha : \alpha < \omega_1\} \subset \mathbf{P}$ which is $<^*$-cofinal in ω^ω.

098. Assuming CH, prove that there exists a space Y with the following properties:

   (i) $|Y| = \omega_1$ and Y is scattered;

   (ii) Y is locally compact and locally countable, i.e., every point of Y has a countable neighborhood (and hence Y is not Lindelöf);

   (iii) Y^k is hereditarily separable for every $k \in \mathbb{N}$.

   In particular, strong S-spaces exist under CH.

099. Prove that under CH, there exists a scattered compact space X which is not first countable (and hence not metrizable), while X^ω is hereditarily separable, and hence, $(C_p(X))^\omega$ is hereditarily Lindelöf. This implies that under CH, there exist strong L-spaces and strong compact S-spaces. Observe that under MA+¬CH, any compact space X is metrizable whenever $(C_p(X))^\omega$ is hereditarily Lindelöf.

100. Prove that under CH, there exists a non-normal X such that the space $(C_p(X))^\omega$ is hereditarily Lindelöf.

1.2 Monolithity, Stability and Their Generalizations

As usual, all spaces are assumed to be Tychonoff. Given a cardinal κ, a space X is κ-*monolithic* if $A \subset X$ and $|A| \leq \kappa$ imply $nw(\overline{A}) \leq \kappa$. A space is *monolithic* if it is κ-monolithic for any infinite cardinal κ. A space X is κ-*stable* if for any continuous onto map $f : X \to Y$, we have $nw(Y) \leq \kappa$ whenever $iw(Y) \leq \kappa$. A space is called *stable* if it is κ-stable for any infinite cardinal κ. A space X is said to be *strongly* κ-*monolithic* if for any $A \subset X$ with $|A| \leq \kappa$, we have $w(\overline{A}) \leq \kappa$. If a space is strongly κ-monolithic for any infinite cardinal κ, it is called *strongly monolithic*. A space X is κ-*scattered* if $|\overline{A}| \leq \kappa$ for any $A \subset X$ with $|A| \leq \kappa$. The space X is κ-*simple* if for any continuous onto map $f : X \to Y$, we have $|Y| \leq \kappa$ if $w(Y) \leq \kappa$. The space X is *simple* if it is κ-simple for any infinite cardinal κ.

Let $\mathcal{P}$ be a topological property. A space X is called $\mathcal{P}(\kappa)$-*monolithic* if for any set $A \subset X$ with $|A| \leq \kappa$, the subspace $\overline{A}$ has the property $\mathcal{P}$. For example, a space X is *normal*(ω)-*monolithic* if the closure of any countable subset of X is normal. Analogously, if θ is a cardinal function, then a space X is called $\theta(\kappa)$-*monolithic* if for any $A \subset X$ with $|A| \leq \kappa$, we have $\theta(\overline{A}) \leq \kappa$. For example, a space X is said to be $s^*(\omega)$-*monolithic or spread*$^*(\omega)$-*monolithic* if for any countable $A \subset X$, we have $s^*(\overline{A}) \leq \omega$ (recall that this means that spread of all finite powers of $\overline{A}$ is countable). Note that a space X is κ-monolithic if and only if it is $nw(\kappa)$-monolithic and X is strongly κ-monolithic if and only if it is $w(\kappa)$-monolithic.

Let $\mathcal{P}$ be a topological property. A space X is called $\mathcal{P}(\kappa)$-*stable* if for any continuous onto map $f : X \to Y$, the space Y has $\mathcal{P}$ whenever $iw(Y) \leq \kappa$. Analogously, if θ is a cardinal function, a space X is called $\theta(\kappa)$-*stable* if for any continuous onto map $f : X \to Y$, we have $\theta(Y) \leq \kappa$ whenever $iw(Y) \leq \kappa$. For example, a space X is *Fréchet–Urysohn*(ω)-*stable* if for any continuous onto map $f : X \to Y$, the space Y is Fréchet–Urysohn if it can be condensed onto a second countable space. Note that in this notation, a space is κ-stable if and only if it is $nw(\kappa)$-stable.

Now, if $\mathcal{F}$ is a class of continuous maps, a space X is called $\mathcal{P}(\kappa)$-$\mathcal{F}$-*stable* if for every map $f : X \to Y$ with $f \in \mathcal{F}$, the space Y has the property $\mathcal{P}$ in case it can be condensed onto a space of weight κ. In an analogous way, a space X is said to be $\theta(\kappa)$-$\mathcal{F}$-*stable* if for every continuous onto map $f : X \to Y$ with $f \in \mathcal{F}$, we have $\theta(Y) \leq \kappa$ if Y can be condensed onto a space of weight κ. For example, a space X is *tightness*$^*(\kappa)$-$\mathbb{R}$-*quotient-stable,* if for any $\mathbb{R}$-quotient map $f : X \to Y$, all finite powers of Y have tightness $\leq \kappa$ whenever $iw(Y) \leq \kappa$. In all the above definitions of monolithity and stability, if the cardinal κ is omitted, then the respective property is defined to be fulfilled for all infinite κ. For example, a space X is χ-*open-stable* if for any infinite cardinal κ and any open continuous onto map $f : X \to Y$, the space Y has character $\leq \kappa$ if it can be condensed onto a space of weight $\leq \kappa$.

All linear spaces in this book are considered over the space $\mathbb{R}$ of the real numbers. A linear space L, equipped with a topology τ, is called *linear topological space* if the linear operations $(x, y) \to x + y$ and $(t, x) \to tx$ are continuous with respect to τ. A subset A of a linear space L is called *convex* if $x, y \in A$ implies

$tx + (1 - t)y \in A$ for any number $t \in [0, 1]$. The *convex hull* $\mathrm{conv}(A)$ of a subset A of a linear space L is the set $\{t_1 x_1 + \cdots + t_n x_n : n \in \mathbb{N}, x_1, \ldots, x_n \in A, t_1, \ldots, t_n \in [0, 1], t_1 + \cdots + t_n = 1\}$. A linear topological space L is called *locally convex* if it has a base which consists of convex sets.

A space X is called *a P-space* if every G_δ-subset of X is open in X. Given an infinite cardinal κ, a *κ-modification of a space X* is the space with the same underlying set, whose topology is generated by all G_κ-subsets of X. A subset $\{x_\alpha : \alpha < \kappa\}$ of a space X is called *free sequence* if for any $\beta < \kappa$, we have $\overline{\{x_\alpha : \alpha < \beta\}} \cap \overline{\{x_\alpha : \alpha \geq \beta\}} = \emptyset$. The *point-finite cellularity* $p(X)$ of a space X is the supremum of cardinalities of point-finite families of non-empty open subsets of X. If X is a space, then $a(X) = \sup\{\kappa : A(\kappa) \text{ embeds in } X\}$. The cardinal function a is called *the Alexandroff number*. Note that $a(X)$ makes sense only if X has a nontrivial convergent sequence. A space X is *Hurewicz* if for any sequence $\{\mathcal{U}_n : n \in \omega\}$ of open covers of X, we can choose, for each $n \in \omega$, a finite $\mathcal{V}_n \subset \mathcal{U}_n$ such that $\bigcup\{\mathcal{V}_n : n \in \omega\}$ is a cover of X.

Given a product $X = \prod\{X_t : t \in T\}$ of the spaces X_t, and a point $x \in X$, let $\Sigma(X, x) = \{y \in X : |\{t \in T : y(t) \neq x(t)\}| \leq \omega\}$. The space $\Sigma(X, x)$ is called the *Σ-product of spaces $\{X_t : t \in T\}$ with the center x*. If X is the same product, then the set $\sigma(X, x) = \{y \in X : |\{t \in T : y(t) \neq x(t)\}| < \omega\}$ is called the *σ-product of spaces $\{X_t : t \in T\}$ with the center x*. If some statement about Σ-products or σ-products is made with no center specified, then this statement holds for an arbitrary center. The symbols $\Sigma(A)$ and $\sigma(A)$ are reserved for the respective Σ- and σ-products of real lines with the center zero, i.e., $\Sigma(A) = \{x \in \mathbb{R}^A : |\{a \in A : x(a) \neq 0\}| \leq \omega\}$ and $\sigma(A) = \{x \in \mathbb{R}^A : |\{a \in A : x(a) \neq 0\}| < \omega\}$.

A space X is *a k-space* if for any non-closed $A \subset X$, there exists a compact subspace $K \subset X$ such that $A \cap K$ is not closed in K. A subspace Y of a space X is *a retract* of X if there exists a continuous map $r : X \to Y$ (called *a retraction*) such that $r(y) = y$ for all $y \in Y$. If X is a space and $Y \subset X$, then $C_p(Y|X) = \pi_Y(C_p(X)) = \{f \in C_p(Y) : \text{there exists } g \in C_p(X) \text{ such that } g|Y = f\} \subset C_p(Y)$.

Given a class $\mathcal{F}$ of mappings and a class $\mathcal{X}$ of spaces, we might say that a space X is an $\mathcal{F}$-preimage of a space from $\mathcal{X}$. This means that there exists a surjective map $f : X \to Y$ such that $f \in \mathcal{F}$ and $Y \in \mathcal{X}$. For example, X is a perfect preimage of a second countable space if there is a second countable space Y and a perfect map of X onto Y. The reader could also have noticed that we often say phrases like "Y is an open continuous image of X" or "open continuous onto map". Why say so if we defined an open map as a continuous and surjective one with some additional property? Well, the answer is that in many books, the continuity and surjectivity are not always assumed for open, closed and even perfect maps. Since this book is also supposed to be a reference guide for specialists (who almost never read definitions), the author rather prefers some excess in assumptions than a possible misunderstanding.

101. Prove that any metrizable space is strongly monolithic.

102. Prove that if X is a metrizable space with $w(X) \leq \mathfrak{c}$, then X condenses onto a second countable space.

103. Suppose that A is a proper closed subset of a metric space (X, d) and let $d(x, A) = \inf\{d(x, a) : a \in A\}$ for any point $x \in X$. Prove that there exists a family $\{U_s, a_s : s \in S\}$ (called a Dugundji system for $X \setminus A$) such that

   (i) $U_s \subset X \setminus A$ and $a_s \in A$ for any $s \in S$;
   (ii) $\{U_s : s \in S\}$ is an open locally finite (in $X \setminus A$) cover of $X \setminus A$;
   (iii) $d(x, a_s) \leq 2d(x, A)$ for any $s \in S$ and any $x \in U_s$.

104. Let A be a closed subspace of a metrizable space X. Suppose that $f : A \to L$ is a continuous map of A into a locally convex linear topological space L. Prove that there exists a continuous map $F : X \to L$ such that $F|A = f$ and $F(X) \subset \text{conv}(f(A))$.

105. Prove that every metrizable space of uncountable weight can be mapped continuously onto a metrizable space of weight ω_1.

106. Prove that a metrizable space is ω-stable if and only if it is separable.

107. Prove that any Σ-product of spaces with a countable network is monolithic. In particular, any Σ-product of second countable spaces is monolithic.

108. Prove that any space X is κ-stable for any $\kappa \geq nw(X)$. In particular, any space with a countable network is stable.

109. Prove that any product of spaces with a countable network is stable. In particular, any product of second countable spaces is stable.

110. Prove that any Σ-product of spaces with countable network is stable. In particular, any Σ-product of second countable spaces is stable.

111. Prove that any σ-product of spaces with countable network is stable. In particular, any σ-product of second countable spaces is stable.

112. Prove that any σ-product of Lindelöf P-spaces is ω-stable.

113. Prove that for any cardinal κ, if X is κ-monolithic, then any subspace $Y \subset X$ is κ-monolithic.

114. Let κ be an infinite cardinal. Suppose that X_α is a κ-monolithic space for each $\alpha < \kappa$. Prove that $\prod\{X_\alpha : \alpha < \kappa\}$ is also κ-monolithic. In particular, any finite product of κ-monolithic spaces is κ-monolithic.

115. Prove that if a space X is covered with a locally finite family of closed monolithic subspaces, then X is monolithic.

116. Give an example of a space which is not ω-monolithic being a union of two monolithic subspaces.

117. Show that $\mathbb{R}^{\omega_1}$ is not ω-monolithic. Hence the product of uncountably many monolithic spaces can fail to be ω-monolithic.

118. Prove that any compact space is stable.

119. Prove that any pseudocompact space is ω-stable.

120. Show that any compact ω-monolithic space of countable tightness must be Fréchet–Urysohn. Give an example of a monolithic compact space of uncountable tightness.

121. Let $f : X \to Y$ be a closed continuous onto map. Prove that if the space X is κ-monolithic, then so is Y.

122. Give an example showing that a continuous image of a monolithic space is not necessarily ω-monolithic.

123. Given an infinite cardinal κ, prove that any continuous image of a κ-stable space is κ-stable. In particular, any retract of a κ-stable space is κ-stable.

124. Suppose that $X = \bigcup\{X_\alpha : \alpha < \kappa\}$ and X_α is a κ-stable space for any $\alpha < \kappa$. Prove that X is κ-stable. In particular, any space which is a countable union of stable subspaces is stable.

125. Give an example of an ω-stable space X such that some closed subspace of X is not ω-stable.

126. Give an example of an ω-stable space whose square is not ω-stable.

127. Prove that any Lindelöf P-space is ω-simple.

128. Prove that for any scattered space X with $l(X) \leq \kappa$, the Lindelöf degree of the κ-modification of X does not exceed κ. In particular, if X is a Lindelöf (or compact!) scattered space, then the ω-modification of X is Lindelöf.

129. Prove that any Lindelöf scattered space (and, in particular, any compact scattered space) is a simple space.

130. Give an example of a pseudocompact scattered non-simple space.

131. Prove that the following are equivalent for any space X:

  (i) the space X is weight(ω)-stable;
  (ii) the space X is character(ω)-stable;
  (iii) the space X is Fréchet–Urysohn(ω)-stable;
  (iv) the space X is sequential(ω)-stable;
  (v) The space X is k-property(ω)-stable.
  (vi) the space X is $\pi w(\omega)$-stable;
  (vii) the space X is $\pi\chi(\omega)$-stable;
  (viii) the space X is pseudocompact.

132. Prove that the following conditions are equivalent for any space X:

  (i) for any countable $A \subset C_p(X)$ the space $\overline{A}$ is σ-compact;
  (ii) for any countable $A \subset C_p(X)$ the space $\overline{A}$ is σ-countably compact;
  (iii) for any countable $A \subset C_p(X)$ the space $\overline{A}$ is σ-pseudocompact;
  (iv) for any countable $A \subset C_p(X)$ the space $\overline{A}$ is locally compact;
  (v) $C_p(X)$ is a Hurewicz space;
  (vi) $C_p(X)$ is a σ-locally compact space;
  (vii) for any countable $A \subset C_p(X)$, the space $\overline{A}$ is Hurewicz;
  (viii) for any countable $A \subset C_p(X)$, the space $\overline{A}$ is σ-locally compact;
  (ix) X is finite.

133. Prove that any non-scattered countably compact space can be continuously mapped onto $\mathbb{I}$.

134. Prove that the following conditions are equivalent for any compact X:

  (i) $C_p(X)$ is a Fréchet–Urysohn space;

(ii) $\overline{A}$ is a Fréchet–Urysohn space for any countable $A \subset C_p(X)$;

(iii) X is scattered.

135. Prove that for any Lindelöf P-space X, the space $C_p(X)$ must be strongly ω-monolithic and Fréchet–Urysohn.

136. Prove that for any Lindelöf scattered space X, the space $C_p(X)$ is a strongly monolithic Fréchet–Urysohn space.

137. Let X be a Lindelöf P-space. Prove that the set $\overline{A}$ is Čech-complete for any countable $A \subset C_p(X)$.

138. Let X be a pseudocompact space. Suppose that $\overline{A}$ is Čech-complete for any countable $A \subset C_p(X)$. Prove that X is finite.

139. Suppose that $\overline{A}$ is normal for any countable $A \subset C_p(X)$. Prove that $\overline{A}$ is collectionwise-normal for any countable $A \subset C_p(X)$, i.e., if the space $C_p(X)$ is normal(ω)-monolithic, then it is collectionwise-normal(ω)-monolithic.

140. Prove that under MA+¬CH, there exists a normal(ω)-monolithic space which is not collectionwise-normal(ω)-monolithic.

141. Suppose that $\overline{A}$ is normal for any countable $A \subset C_p(X)$. Prove that $\overline{A}$ is countably paracompact for any countable $A \subset C_p(X)$.

142. Suppose that $\overline{A}$ is hereditarily normal for any countable $A \subset C_p(X)$. Prove that $\overline{A}$ is perfectly normal for any countable $A \subset C_p(X)$.

143. Given an infinite cardinal κ, prove that $C_p(X, \mathbb{I})$ is κ-stable if and only if $C_p(X)$ is κ-stable.

144. Given an infinite cardinal κ, prove that $C_p(X, \mathbb{I})$ is κ-monolithic if and only if $C_p(X)$ is κ-monolithic.

145. Let θ be any cardinal function. Prove that the Hewitt realcompactification υX of a space X is $\theta(\omega)$-stable if and only if X is $\theta(\omega)$-stable. In particular, υX is ω-stable if and only if X is ω-stable.

146. Let η and θ be cardinal functions such that $\theta(Z) = \eta(C_p(Z))$ for any space Z. Suppose that η is hereditary, i.e., for any space Z and any subspace $Y \subset Z$, we have the inequality $\eta(Y) \leq \eta(Z)$. Prove that for any infinite cardinal κ, a space X is $\theta(\kappa)$-stable if and only if $C_p(X)$ is $\eta(\kappa)$-monolithic.

147. Let η and θ be cardinal functions such that $\theta(Z) = \eta(C_p(Z))$ for any space Z. Suppose that the function η is closed-hereditary, i.e., for any space Z and any closed $Y \subset Z$, we have $\eta(Y) \leq \eta(Z)$. Prove that for any infinite cardinal κ, a space X is $\theta(\kappa)$-$\mathbb{R}$-quotient-stable if and only if $C_p(X)$ is $\eta(\kappa)$-monolithic.

148. Suppose that θ and η are cardinal functions such that

(i) for any space Z, if a space Y is a continuous image of Z, then $\eta(Y) \leq \eta(Z)$;

(ii) for any space Z and any closed $A \subset Z$, we have $\theta(A) = \eta(C_p(A|Z))$.

Prove that for an arbitrary space X, the space $C_p(X)$ is $\eta(\kappa)$-stable if and only if X is $\theta(\kappa)$-monolithic.

149. Suppose that θ and η are cardinal functions such that

(i) for any space Z, if a space Y is a quotient image of Z, then $\eta(Y) \leq \eta(Z)$;

(ii) for any space Z and any closed $A \subset Z$ we have $\theta(A) = \eta(C_p(A|Z))$.

Prove that for an arbitrary space X, the space $C_p(X)$ is $\eta(\kappa)$-quotient-stable if and only if X is $\theta(\kappa)$-monolithic.

150. Suppose that θ and η are cardinal functions such that

(i) for any space Z, if a space Y is an $\mathbb{R}$-quotient continuous image of Z, then $\eta(Y) \leq \eta(Z)$;

(ii) for any space Z and any closed $A \subset Z$ we have $\theta(A) = \eta(C_p(A|Z))$.

Prove that for an arbitrary space X, the space $C_p(X)$ is $\eta(\kappa)$-$\mathbb{R}$-quotient-stable if and only if X is $\theta(\kappa)$-monolithic.

151. Suppose that θ and η are cardinal functions such that

(i) for any space Z, if Y is an open continuous image of Z, then $\eta(Y) \leq \eta(Z)$;

(ii) for any space Z and any closed $A \subset Z$, we have $\theta(A) = \eta(C_p(A|Z))$.

Prove that for an arbitrary space X, the space $C_p(X)$ is $\eta(\kappa)$-open-stable if and only if X is $\theta(\kappa)$-monolithic.

152. Given a space X and an infinite cardinal κ prove that X is κ-monolithic if and only if $C_p(X)$ is κ-stable. In particular, X is monolithic if and only if $C_p(X)$ is stable.

153. Prove that if $C_p(X)$ is a stable space, then $(C_p(X))^\kappa$ is also a stable space for any cardinal κ.

154. Suppose that X is an arbitrary space and κ is an infinite cardinal. Prove that $C_p(X)$ is κ-monolithic if and only if X is κ-stable. In particular, $C_p(X)$ is monolithic if and only if X is stable.

155. Prove that X is a monolithic space if and only if so is $C_pC_p(X)$.

156. Prove that X is a stable space if and only if so is $C_pC_p(X)$.

157. Prove that X is κ-simple if and only if $C_p(X)$ is strongly κ-monolithic.

158. Prove that the following properties are equivalent for any space X:

(i) $C_p(X)$ is strongly κ-monolithic;

(ii) $C_p(X)$ is $\pi w(\kappa)$-monolithic;

(iii) $C_p(X)$ is $\pi\chi(\kappa)$-monolithic;

(iv) $C_p(X)$ is $\chi(\kappa)$-monolithic.

159. Given a space X and an infinite cardinal κ prove that X is $s^*(\kappa)$-monolithic if and only if $C_p(X)$ is $s^*(\kappa)$-stable.

160. Given an arbitrary space X and an infinite cardinal κ prove that if $C_p(X)$ is spread(κ)-stable, then X is spread(κ)-monolithic.

161. Give an example of a spread-monolithic space X such that $C_p(X)$ is not spread(ω)-stable.

162. Given a space X and an infinite cardinal κ prove that X is $s^*(\kappa)$-stable if and only if $C_p(X)$ is $s^*(\kappa)$-monolithic.

163. Prove that if $C_p(X)$ is spread(κ)-monolithic, then X is spread(κ)-stable.

164. Show that there exists a spread-stable space X such that $C_p(X)$ is not spread(ω)-monolithic.

165. Given a space X and an infinite cardinal κ prove that X is $hd^*(\kappa)$-monolithic if and only if $C_p(X)$ is $hl^*(\kappa)$-stable.

166. Given an arbitrary space X and an infinite cardinal κ prove that if $C_p(X)$ is $hl(\kappa)$-stable, then X is $hd(\kappa)$-monolithic.

167. Show that there exists an hd-monolithic space X such that $C_p(X)$ is not $hl(\omega)$-stable.

168. Given a space X and an infinite cardinal κ prove that X is $hd^*(\kappa)$-stable if and only if $C_p(X)$ is $hl^*(\kappa)$-monolithic.

169. Given an arbitrary space X and an infinite cardinal κ prove that if $C_p(X)$ is $hl(\kappa)$-monolithic, then X is $hd(\kappa)$-stable.

170. Give an example of an hd-stable space X such that the space $C_p(X)$ is not $hl(\omega)$-monolithic.

171. Given a space X and an infinite cardinal κ prove that X is $hl^*(\kappa)$-stable if and only if $C_p(X)$ is $hd(\kappa)$-monolithic.

172. Give an example of an hl-stable space X such that the space $C_p(X)$ is not $hd(\omega)$-monolithic.

173. Given a space X and an infinite cardinal ν prove that X is $hl^*(\nu)$-monolithic if and only if $C_p(X)$ is $hd(\nu)$-stable.

174. Show that there exists an hl-monolithic space X such that $C_p(X)$ is not $hd(\omega)$-stable.

175. Given a space X and an infinite cardinal κ prove that X is $p(\kappa)$-stable if and only if $C_p(X)$ is $a(\kappa)$-monolithic.

176. Given a space X and an infinite cardinal κ prove that X is $l^*(\kappa)$-stable if and only if $C_p(X)$ is $t(\kappa)$-monolithic.

177. Given a space X and an infinite cardinal κ prove that X is $d(\kappa)$-stable if and only if $C_p(X)$ is $iw(\kappa)$-monolithic.

178. Given an arbitrary space X and an infinite cardinal κ prove that the following conditions are equivalent:

   (i) $C_p(X)$ is $iw(\kappa)$-monolithic;
   (ii) $C_p(X)$ is $\Delta(\kappa)$-monolithic;
   (iii) $C_p(X)$ is $\psi(\kappa)$-monolithic.

179. Give an example of a space X which is pseudocharacter-monolithic but not diagonal-number-monolithic.

180. Give an example of a space X which is diagonal-number-monolithic but not i-weight-monolithic.

181. Suppose that X is an arbitrary space and κ is an infinite cardinal. Prove that X is $t_m(\kappa)$-$\mathbb{R}$-quotient-stable if and only if $C_p(X)$ is $q(\kappa)$-monolithic.

182. Give an example of a space X for which $C_p(X)$ is q-monolithic and X is not $t_m(\omega)$-stable.

183. Given an infinite cardinal κ, prove that X is $l^*(\kappa)$-monolithic if and only if $C_p(X)$ is $t(\kappa)$-quotient-stable.

184. Given an infinite cardinal κ, suppose that $C_p(X)$ is $l(\kappa)$-monolithic. Prove that the space X is tightness*(κ)-$\mathbb{R}$-quotient-stable.

185. Given an arbitrary space X, suppose that $C_p(X)$ is $t(\kappa)$-monolithic for some infinite cardinal κ. Prove that it is $t^*(\kappa)$-monolithic.

186. Suppose that $C_p(X)$ is Fréchet–Urysohn(κ)-monolithic for some infinite cardinal κ. Prove that it is Fréchet–Urysohn*(κ)-monolithic.

187. Given an infinite cardinal κ, prove that X is κ-scattered if and only if $C_p(X)$ is $w(\kappa)$-open-stable.

188. Let X be a $d(\omega)$-stable space such that X^n is Hurewicz for all $n \in \mathbb{N}$. Prove that for any $A \subset C_p(X)$ and any $f \in \overline{A}\backslash A$, there is a discrete $D \subset A$ such that f is the only accumulation point of D.

189. Let X be an $l(\omega)$-monolithic space of countable spread. Prove that X is Lindelöf.

190. Let X be an $hl(\omega)$-monolithic space of countable spread. Prove that X is hereditarily Lindelöf. In particular, if X is an ω-monolithic space of countable spread, then X is hereditarily Lindelöf.

191. Let X be an ω-monolithic space such that $C_p(X)$ has countable spread. Prove that $hl^*(X) = \omega$ and hence $hd^*(C_p(X)) = \omega$.

192. Suppose that a space X is ω-monolithic, ω-stable and $s(C_p(X)) = \omega$. Prove that $nw(X) = \omega$.

193. Assume SA. Prove that for any space X, if $C_p(X)$ is $s(\omega)$-monolithic, then it is $hl^*(\omega)$-monolithic. In particular, if the space $C_p(X)$ is $hl(\omega)$-monolithic, then it is $hl^*(\omega)$-monolithic.

194. Give an example of an $hl(\omega)$-monolithic non-$hl^*(\omega)$-monolithic space.

195. Assume SA. Let X be a ω-stable space such that $s(C_p(X)) = \omega$. Prove that X has a countable network.

196. Assume SA. Prove that if $s(C_p(X)) = \omega$, then for any $Y \subset X$ we have $s(C_p(Y)) = \omega$.

197. Assume MA+¬CH. Let X be an ω-monolithic space such that the spread of $C_p(X)$ is countable. Prove that X has a countable network.

198. Assume MA+¬CH and suppose that $C_p(X)$ contains no uncountable free sequences. Prove that for any $Y \subset X$, the space Y is hereditarily Lindelöf if and only if it is hereditarily separable.

199. Assume MA+¬CH and let X be an ω-monolithic space with $l^*(X) = \omega$. Suppose that $C_p(X)$ is Lindelöf and $Y \subset X$ has countable spread. Prove that Y has a countable network.

200. Assuming MA+¬CH prove that if $C_p(X)$ is hereditarily stable, then X has a countable network.

1.3 Whyburn Spaces, Calibers and Lindelöf Σ-Property

All spaces are assumed to be Tychonoff. Given two families $\mathcal{A}$ and $\mathcal{B}$ of subsets of a space X, say that $\mathcal{A}$ *is a network with respect to* $\mathcal{B}$ if for any $B \in \mathcal{B}$ and any open $U \supset B$, there is $A \in \mathcal{A}$ such that $B \subset A \subset U$. A space X is *a Σ-space,* if X has a closed cover $\mathcal{C}$ such that all elements of $\mathcal{C}$ are countably compact and there exists a σ-discrete family $\mathcal{F}$ which is a network with respect to $\mathcal{C}$. A *Lindelöf Σ-space* is a Σ-space which has the Lindelöf property. A space X is *locally compact* if every $x \in X$ has a compact neighborhood.

A space X is called *a p-space* if there is a sequence $\{\mathcal{U}_n : n \in \omega\}$ of open (in βX) covers of X such that for every $x \in X$, the set $\bigcap\{\mathrm{St}(x, \mathcal{U}_n) : n \in \omega\}$ is contained in X. A space X is *disconnected* if there exist $U, V \in \tau^*(X)$ such that $U \cap V = \emptyset$ and $U \cup V = X$. A space is *connected* if it is not disconnected. A space X is *a k-space* if for any non-closed $A \subset X$, there exists a compact subspace $K \subset X$ such that $A \cap K$ is not closed. Recall that $\mathrm{ext}(X) = \sup\{|D| : D$ is a closed discrete subspace of $X\}$. The cardinal $\mathrm{ext}(X)$ is called *the extent* of the space X. Say that a space X is $K_{\sigma\delta}$ if there exists a space Y such that $X \subset Y$ and $X = \bigcap\{Y_n : n \in \omega\}$ where each Y_n is a σ-compact subset of Y. A compact space K is called *a compactification* of a space X if X is homeomorphic to a dense subspace of K.

A subset F of a space X is called *almost closed* if $\overline{F}\backslash F$ is a singleton. A space X is *Whyburn or a Whyburn space* if for any non-closed $A \subset X$ and any $x \in \overline{A}\backslash A$, there exists an almost closed $F \subset A$ with $x \in \overline{F}$. The space X is *weakly Whyburn or a weakly Whyburn space* if for any non-closed $A \subset X$, there exists $x \in \overline{A}\backslash A$ and an almost closed $F \subset A$ such that $x \in \overline{F}$. A space X is called *submaximal* if it has no isolated points and every dense subspace of X is open in X. A space X is *Hurewicz* if for any sequence $\{\mathcal{U}_n : n \in \omega\}$ of open covers of X, we can choose, for each $n \in \omega$, a finite $\mathcal{V}_n \subset \mathcal{U}_n$ such that $\bigcup\{\mathcal{V}_n : n \in \omega\}$ is a cover of X. A space X is *radial* if for any $A \subset X$ and any $x \in \overline{A}\backslash A$, there exist a regular cardinal κ and a transfinite sequence $S = \{x_\alpha : \alpha < \kappa\} \subset A$ such that $S \to x$, i.e., for any open $U \ni x$, there is $\alpha < \kappa$ such that for each $\beta \geq \alpha$, we have $x_\beta \in U$. The space X is *pseudoradial* if $A \subset X$ and $A \neq \overline{A}$ imply that there is a regular cardinal κ and a transfinite sequence $S = \{x_\alpha : \alpha < \kappa\} \subset A$ such that $S \to x \notin A$.

Let X be a space with $Y \subset X$ and $Z \subset X$. We say that *a family* $\mathcal{F} \subset \exp(X)$ *separates Y from Z* if for any $y \in Y$ and any $z \in Z$, there exists $F \in \mathcal{F}$ such that $y \in F$ and $z \notin F$. If $Z = \emptyset$, then the agreement is that $\mathcal{F}$ separates Y from Z if and only if $Y \subset \bigcup \mathcal{F}$. Given spaces X and Y, a map $p : X \to \exp(Y)$ is called *compact-valued* if $p(x)$ is a compact subset of Y for each $x \in X$. We will often write $p : X \to Y$ instead of $p : X \to \exp(Y)$. A compact-valued map $p : X \to Y$ is called *upper semicontinuous* if for any open $U \subset Y$, the set $p^{-1}(U) = \{x \in X : p(x) \subset U\}$ is open in X; say that p is *an onto map,* if $\bigcup\{p(x) : x \in X\} = Y$.

An uncountable regular cardinal κ is *a caliber* of a space X if for any indexed family $\mathcal{U} = \{U_\alpha : \alpha < \kappa\} \subset \tau^*(X)$, there exists a set $A \subset \kappa$ such that $|A| = \kappa$ and $\bigcap\{U_\alpha : \alpha \in A\} \neq \emptyset$. An uncountable regular cardinal κ is called *a precaliber* of a space X if for any indexed family $\mathcal{U} = \{U_\alpha : \alpha < \kappa\} \subset \tau^*(X)$, there exists $A \subset \kappa$ such that $|A| = \kappa$ and the family $\mathcal{U}' = \{U_\alpha : \alpha \in A\}$ has finite intersection property, i.e., $\bigcap \mathcal{V} \neq \emptyset$ for any finite $\mathcal{V} \subset \mathcal{U}'$. The diagonal $\Delta = \{(x, x) : x \in X\} \subset X^2$ of a space X is κ-*small* if for any $A \subset X^2 \backslash \Delta$ with $|A| = \kappa$, there exists $B \subset A$ such that $|B| = \kappa$ and $\overline{B} \cap \Delta = \emptyset$. The ω_1-small diagonal is called *small*. A continuous onto map $f : X \to Y$ is called *pseudo-open or hereditarily quotient* if for any $y \in Y$, we have $y \in \text{Int}_Y(f(U))$ whenever $f^{-1}(y) \subset U \in \tau(X)$.

For any uncountable cardinal κ, let $L(\kappa) = \kappa \cup \{a\}$, where $a \notin \kappa$. If $x \in \kappa$, let $\mathcal{B}_x = \{\{x\}\}$. If $x = a$, then $\mathcal{B}_x = \{\{a\} \cup (\kappa \backslash B) : B$ is a countable subset of $\kappa\}$. The families $\{\mathcal{B}_x : x \in L(\kappa)\}$ generate a topology $\tau(L(\kappa))$ for which $\mathcal{B}_x$ is a local base at x for any $x \in L(\kappa)$. We call $L(\kappa)$ *the Lindelöfication of the discrete space of cardinality* κ. A class of spaces $\mathcal{P}$ is *invariant* under some operation if the result of this operation is in $\mathcal{P}$ when applied to elements of $\mathcal{P}$. For example, $\mathcal{P}$ is invariant under finite products if for any $X_1, \ldots, X_n \in \mathcal{P}$, we have $X_1 \times \cdots \times X_n \in \mathcal{P}$. Analogously, $\mathcal{P}$ is invariant with respect to continuous images (closed subspaces) if for any $X \in \mathcal{P}$, any continuous image (any closed subspace) of X belongs to $\mathcal{P}$.

Suppose that X is a set without topology and $[\cdot]$ is an operator on $\exp(X)$ such that $[\emptyset] = \emptyset$, $[A \cup B] = [A] \cup [B]$, $A \subset [A]$ and $[[A]] = [A]$ for all $A, B \subset X$. Then there exists a unique topology τ on X such that $[A] = \text{cl}_\tau(A)$ for any $A \subset X$. We will say that τ *is generated by the closure operator* $[\cdot]$. Now, if $\langle \cdot \rangle$ is an operator on $\exp(X)$ such that $\langle \emptyset \rangle = \emptyset$, $\langle A \cap B \rangle = \langle A \rangle \cap \langle B \rangle$, $\langle A \rangle \subset A$ and $\langle \langle A \rangle \rangle = \langle A \rangle$ for all $A, B \subset X$, then there exists a unique topology τ on X such that $\langle A \rangle = \text{Int}_\tau(A)$ for any $A \subset X$. We will say that τ *is generated by the interior operator* $\langle \cdot \rangle$.

Assume that X is a set without topology and $\mathcal{B}$ is a family of subsets of X such that $\bigcup \mathcal{B} = X$ and for any $U, V \in \mathcal{B}$, if $x \in U \cap V$ then there exists $W \in \mathcal{B}$ such that $x \in W \subset U \cap V$. Then there exists a unique topology τ on the set X such that $\mathcal{B}$ is a base of (X, τ). The topology τ is said to be *generated by the family $\mathcal{B}$ as a base*. If $\mathcal{S}$ is a family of subsets of X such that $\bigcup \mathcal{S} = X$, then there exists a unique topology τ on X such that $\mathcal{S}$ is a subbase of (X, τ); we will say that the topology τ is *generated by $\mathcal{S}$ as a subbase*. Next, suppose that $\mathcal{B}_x \subset \exp(X)$ for any $x \in X$ and the family $\{\mathcal{B}_x : x \in X\}$ has the following properties:

(LB1) $\mathcal{B}_x \neq \emptyset$ and $\bigcap \mathcal{B}_x \ni x$ for every $x \in X$;

(LB2)   if $x \in X$ and $U, V \in \mathcal{B}_x$, then there is $W \in \mathcal{B}_x$ such that $W \subset U \cap V$;

(LB3)   if $x \in U \in \mathcal{B}_y$, then there is $V \in \mathcal{B}_x$ such that $V \subset U$.

Then there exists a unique topology τ on X such that $\mathcal{B}_x$ is a local base of (X, τ) at x for any $x \in X$. We will call τ *the topology generated by the families* $\{\mathcal{B}_x : x \in X\}$ *as local bases*. Finally assume that $\mathcal{F}$ is a family of maps defined on X such that every $f \in \mathcal{F}$ maps X into a topological space Y_f. There exists a unique topology τ on the set X such that the family $\{f^{-1}(U) : f \in \mathcal{F}$ and $U \in \tau(Y_f)\}$ is a subbase of (X, τ); we will say that the topology τ is *generated by the family* $\mathcal{F}$.

201. Prove that any continuous image of a hereditarily normal compact space is hereditarily normal.

202. Let X be a compact space for which X^2 is hereditarily normal. Prove that X is perfectly normal and hence first countable.

203. Given a compact space X, let $\Delta = \{(x, x) : x \in X\} \subset X^2$ be the diagonal of X. Prove that if $X^2 \backslash \Delta$ is paracompact, then X is metrizable.

204. Observe that any Fréchet–Urysohn space must be Whyburn. Prove that any countably compact Whyburn space is Fréchet–Urysohn.

205. Give an example of a pseudocompact Whyburn space which fails to be Fréchet–Urysohn.

206. Observe that a continuous image of a Whyburn space need not be Whyburn. Prove that any image of a Whyburn space under a closed map is a Whyburn space. Prove that the same is true for weakly Whyburn spaces.

207. Prove that every space with a unique non-isolated point is Whyburn. In particular, there exist Whyburn spaces of uncountable tightness.

208. Prove that any submaximal space is Whyburn.

209. Prove that any radial space is weakly Whyburn.

210. Prove that every Whyburn k-space is Fréchet–Urysohn. In particular, any sequential Whyburn space as well as any Čech-complete Whyburn space must be Fréchet–Urysohn.

211. Prove that every compact weakly Whyburn space is pseudoradial but not necessarily sequential.

212. Give an example of a Whyburn space which is not pseudoradial.

213. Prove that any scattered space is weakly Whyburn.

214. Observe that any sequential space is a k-space. Prove that any hereditarily k-space (and hence any hereditarily sequential space) is Fréchet–Urysohn.

215. Prove that there exist hereditarily weakly Whyburn spaces which are not Whyburn.

216. Prove that if X is $d(\omega)$-stable and X^n is a Hurewicz space for each natural n, then $C_p(X)$ is a Whyburn space. In particular, if X is a σ-compact space, then $C_p(X)$ is a Whyburn space.

217. Prove that for a paracompact space X, if $C_p(X)$ is Whyburn, then X is a Hurewicz space. In particular, if X is metrizable and $C_p(X)$ is Whyburn, then X is separable.

218. Given a space X such that $C_p C_p(X)$ is Whyburn, prove that X has to be finite if it is either countably compact or has a countable network.

219. Prove that there exists a separable metrizable space X such that $C_p(X)$ is not weakly Whyburn.

220. Prove that there exists a compact weakly Whyburn space which fails to be hereditarily weakly Whyburn.

221. Prove that any metrizable space is a p-space and a Σ-space at the same time.

222. Prove that $C_p(X)$ is a p-space if and only if X is countable.

223. Prove that every Lindelöf p-space is a Lindelöf Σ-space. Give an example of a p-space which is not a Σ-space.

224. Prove that

(i) Any closed subspace of a Σ-space is a Σ-space. In particular, any closed subspace of a Lindelöf Σ-space is a Lindelöf Σ-space.

(ii) Any closed subspace of a p-space is a p-space. In particular, any closed subspace of a Lindelöf p-space is a Lindelöf p-space.

225. Prove that X is a Lindelöf Σ-space if and only if X has a countable network with respect to a compact cover $\mathcal{C}$.

226. Suppose that $X = \bigcup\{X_n : n \in \omega\}$ where X_n is countably compact and closed in X for any $n \in \omega$. Prove that X is a Σ-space. As a consequence, every σ-compact space is a Lindelöf Σ-space.

227. Prove that the Sorgenfrey line and the space $L(\omega_1)$ are examples of Lindelöf spaces which are not Lindelöf Σ.

228. Prove that any space with a σ-discrete network is a Σ-space. In particular, if $nw(X) \leq \omega$, then X is a Lindelöf Σ-space.

229. Let X be a metrizable space. Prove that $C_p(X)$ is a Σ-space if and only if X is second countable.

230. Prove that any p-space is a k-space. Give an example of a Lindelöf Σ-space which is not a k-space.

231. Give an example of a countable space which is not a p-space. Note that this example shows that not every Lindelöf Σ-space is a p-space.

232. Prove that any Čech-complete space is a p-space. Give an example of a p-space which is not Čech-complete.

233. Prove that the following conditions are equivalent for any space X:

(i) for an arbitrary compactification bX of the space X, there exists a countable family of compact subspaces of bX which separates X from $bX \backslash X$;

(ii) there exists a compactification bX of the space X and a countable family of compact subspaces of bX which separates X from $bX \backslash X$;

(iii) there exists a compactification bX of the space X and a countable family of Lindelöf Σ-subspaces of bX which separates X from $bX \backslash X$;

(iv) there exists a space Z such that X is a subspace of Z and there is a countable family of compact subspaces of Z which separates X from $Z \backslash X$;

(v) X is a Lindelöf Σ-space.

234. Let X be a space of countable tightness such that $C_p(X)$ is a Σ-space. Prove that if $C_p(X)$ is normal, then it is Lindelöf.

235. Let X be a Σ-space with a G_δ-diagonal. Prove that X has a σ-discrete network.

236. Prove that for any Lindelöf Σ-space X with $\psi(X) \leq \omega$, we have $|X| \leq \mathfrak{c}$.

237. Prove that under CH, there exists a hereditarily separable compact space X such that $C_p(X)$ does not have a dense Σ-subspace.

238. Prove that under CH, the space $C_p(\beta\omega)$ is not a Σ-space.

239. Prove that for any metrizable X, the space $C_p(X)$ has a dense Lindelöf Σ-subspace.

240. Let $p : X \to Y$ be compact-valued upper semicontinuous onto map. Prove that $l(Y) \le l(X)$.

241. Let $p : X \to Y$ be compact-valued upper semicontinuous onto map. Prove that if X is compact, then so is Y.

242. Let $p : X \to Y$ be compact-valued upper semicontinuous onto map. Prove that if X is a Lindelöf Σ-space, then so is Y.

243. Prove that

   (i) any continuous image of a Lindelöf Σ-space is a Lindelöf Σ-space;

   (ii) any perfect preimage of a Lindelöf Σ-space is a Lindelöf Σ-space.

244. Prove that $w(X) = nw(X) = iw(X)$ for any Lindelöf p-space X. In particular, any Lindelöf p-space with a countable network has a countable base.

245. Prove that any perfect image and any perfect preimage of a Lindelöf p-space is a Lindelöf p-space. Give an example of a closed continuous onto map $f :$ $X \to Y$ such that X is a Lindelöf p-space and Y is not a p-space.

246. Suppose that $C_p(X)$ is a closed continuous image of a Lindelöf p-space. Prove that X is countable.

247. Show that an open continuous image of a p-space need not be a p-space. Supposing that $C_p(X)$ is an open continuous image of a p-space, prove that X is countable (and hence $C_p(X)$ is a p-space).

248. Prove that a space X has the Lindelöf Σ-property if and only if there exist a second countable space M and a compact K such that X is a continuous image of a closed subspace of $K \times M$.

249. Prove that the following properties are equivalent for any space X:

   (i) there exist a second countable space M and a space Y such that Y can be mapped perfectly onto M and continuously onto X;

   (ii) there exists an upper semicontinuous compact-valued onto map $\varphi :$ $M \to X$ for some second countable space M;

   (iii) X is a Lindelöf Σ-space.

250. Give an example of a space X which embeds into $C_p(Y)$ for some Lindelöf p-space Y and is not embeddable into $C_p(Z)$ for any $K_{\sigma\delta}$-space Z.

251. Give an example of a p-space Y and a pseudo-open (and hence quotient) map $\varphi : Y \to C_p(X)$ of Y onto $C_p(X)$ for an uncountable space X.

252. Prove that X is a Lindelöf p-space if and only if there is a perfect map of X onto a second countable space.

253. Prove that X is a Lindelöf Σ-space if and only if it is a continuous image of a Lindelöf p-space.

254. Prove that the class of Lindelöf Σ-spaces is the smallest one which contains all compact spaces, all second countable spaces, and is invariant with respect to finite products, continuous images and closed subspaces.

255. Suppose that X_n is a Lindelöf p-space for each $n \in \omega$. Prove that the product $\prod\{X_n : n \in \omega\}$ is a Lindelöf p-space.

256. Suppose that X_n is a Lindelöf Σ-space for each $n \in \omega$. Prove that the product $\prod\{X_n : n \in \omega\}$ is a Lindelöf Σ-space.

257. Let X be a space such that $X = \bigcup\{X_n : n \in \omega\}$, where each X_n is a Lindelöf Σ-space. Prove that X is a Lindelöf Σ-space.

258. Suppose that Z is a space and $X_n \subset Z$ is Lindelöf Σ for each $n \in \omega$. Prove that $X = \bigcap\{X_n : n \in \omega\}$ is a Lindelöf Σ-space.

259. Let X be a Lindelöf Σ-space such that each compact subset of X is finite. Prove that X is countable.

260. Give an example of a Lindelöf p-space X such that $nw(X) > \omega$ and all compact subsets of X are countable.

261. Prove that any $K_{\sigma\delta}$-space is Lindelöf Σ. Show that there exists a $K_{\sigma\delta}$-space which is not Lindelöf p.

262. Let X be a $K_{\sigma\delta}$-space such that all compact subsets of X are countable and $\psi(X) \leq \omega$. Prove that X is countable.

263. Suppose that X is a Lindelöf Σ-space and $C_p(X)$ has the Baire property. Prove that X is countable. In particular, if X is a space with a countable network and $C_p(X)$ is Baire, then X is countable.

264. Prove that there exists an uncountable Lindelöf space X for which $C_p(X)$ has the Baire property.

265. Suppose that $C_p(X)$ is a Lindelöf Σ-space and has the Baire property. Prove that X is countable.

266. Prove that every Lindelöf Σ-space is stable, and hence, for every Lindelöf Σ-space X, the space $C_p(X)$ is monolithic.

267. Prove that if υX is a Lindelöf Σ-space, then X is ω-stable.

268. Prove that any product and any σ-product of Lindelöf Σ-spaces is stable. Show that any Σ-product of Lindelöf Σ-spaces is ω-stable.

269. (Baturov's theorem). Let X be a Lindelöf Σ-space. Prove that for any set $Y \subset C_p(X)$, we have $\text{ext}(Y) = l(Y)$.

270. Prove that every subspace of X is a Lindelöf Σ-space if and only if X has a countable network.

271. Prove that every subspace of X is a Lindelöf p-space if and only if X is second countable.

272. Observe first that there exist hereditarily Čech-complete non-metrizable spaces. Therefore a hereditarily p-space need not be metrizable. Prove that any hereditarily Čech-complete space is scattered.

273. Prove that $\omega_1 + 1$ is a scattered compact space which is not hereditarily Čech-complete.

274. Prove that every subspace of X is σ-compact if and only if X is countable.

275. Prove that

    (i) If an uncountable regular cardinal κ is a caliber of a space X, then κ is a precaliber of X.

(ii) if an infinite successor cardinal κ is a precaliber of a space X, then $c(X) < \kappa$. In particular, if ω_1 is a precaliber of X, then $c(X) = \omega$.

276. Let κ be an uncountable regular cardinal. Prove that if κ is a precaliber (caliber) of X_n for every $n \in \omega$, then κ is a precaliber (caliber) of $\bigcup\{X_n : n \in \omega\}$.

277. Let κ be an uncountable regular cardinal. Prove that if κ is a precaliber (caliber) of X, then κ is a precaliber (caliber) of every continuous image of X.

278. Suppose that κ is an uncountable regular cardinal and Y is a dense subspace of X. Prove that

(i) if κ is a caliber of Y, then it is a caliber of X;
(ii) κ is a precaliber of Y if and only if it is a precaliber of X.

279. Show that an uncountable regular cardinal κ is a caliber of a compact space X if and only if it is a precaliber of X.

280. Let κ be an uncountable regular cardinal. Prove that if κ is a precaliber of X_t for every $t \in T$, then κ is a precaliber of $\prod\{X_t : t \in T\}$.

281. Let κ be an uncountable regular cardinal. Prove that if κ is a caliber of X_t for every $t \in T$, then κ is a caliber of $\prod\{X_t : t \in T\}$.

282. Prove that any product X of separable spaces satisfies the *Shanin condition*, i.e., every uncountable regular cardinal is a caliber of X.

283. Prove that any uncountable regular cardinal is a precaliber of $C_p(X)$ for any space X.

284. Prove that there exists a space X such that ω_1 is a precaliber of X while the point-finite cellularity of X is uncountable. Observe that if ω_1 is a caliber of X, then $p(X) = \omega$.

285. Let X be a metrizable space. Prove that any regular uncountable cardinal is a caliber of $C_p(X)$.

286. Prove that an uncountable regular cardinal κ is a precaliber of X if and only if it is a caliber of βX.

287. Suppose that X is a compact space of countable tightness. Prove that if ω_1 is a caliber of X, then X is separable. Give an example of a non-separable compact space X such that ω_1 is a caliber of X.

288. Assuming MA+$\neg$CH, prove that ω_1 is a precaliber of any space which has the Souslin property.

289. Assume the axiom of Jensen ($\diamond$). Prove that there exists a space X with $c(X) = \omega$ while ω_1 is not a precaliber of X.

290. Prove that for any uncountable regular cardinal κ, the diagonal of $C_p(X)$ is κ-small if and only if κ is a caliber of X. In particular, ω_1 is a caliber of X if and only if $C_p(X)$ has a small diagonal.

291. Prove that an uncountable regular cardinal κ is a caliber of X if and only if it is a caliber of $C_p(C_p(X))$.

292. Suppose that an uncountable regular cardinal κ is a caliber of $C_p(X)$. Prove that for any $Y \subset X$ the cardinal κ is a caliber of $C_p(Y)$.

293. Let κ be an uncountable regular cardinal. Prove that if κ is a caliber of $C_p(X)$, then the diagonal of X is κ-small. In particular, if ω_1 is a caliber of $C_p(X)$, then the diagonal of X is small.

294. Let κ be an uncountable regular cardinal. Prove that if all finite powers of X are Lindelöf and X has a κ-small diagonal, then κ is a caliber of $C_p(X)$. As a consequence, if $l^*(X) = \omega$, then X has a κ-small diagonal if and only if κ is a caliber of $C_p(X)$. In particular, if X is compact, then the diagonal of X is small if and only if ω_1 is a caliber of $C_p(X)$.

295. Prove that any compact space of weight $\leq \omega_1$ with a small diagonal is metrizable.

296. Let X be a compact space with a small diagonal. Prove that if X is ω-monolithic and has countable tightness, then it is metrizable.

297. Let X be a compact space with a small diagonal. Prove that if X is monolithic, then it is metrizable.

298. Prove that under CH, any compact space with a small diagonal is metrizable.

299. Assume that $2^{\omega_1} = \omega_2$. Prove that any compact X, with ω_1 and ω_2 calibers of $C_p(X)$, is metrizable.

300. Note that every Lindelöf Σ-space with a diagonal G_δ must have a countable network. Prove that under CH, any Lindelöf Σ-space with a small diagonal has a countable network.

1.4   A Glimpse of Descriptive Set Theory

All spaces are assumed to be Tychonoff. A space X is called *zero-dimensional* if it has a base which consists of clopen sets. The space X is *strongly zero-dimensional* if any finite open cover of X has a disjoint open refinement. We say that *the large inductive dimension of X is equal to zero* (denoting it by $\operatorname{Ind} X = 0$) if for any closed $F \subset X$ and open $U \supset F$, there is a clopen $V \subset X$ such that $F \subset V \subset U$. The symbol $\mathbb{P}$ stands for the space of the irrational numbers with its topology inherited from $\mathbb{R}$. A space X is called *submetrizable* if it can be condensed onto a metrizable space.

Given spaces X and Y, the expression $X \simeq Y$ says that X is homeomorphic to Y; a map $\varphi : X \to \exp(Y)$ is called *lower semicontinuous* if for any open $U \subset Y$, the set $\varphi_l^{-1}(U) = \{x \in X : \varphi(x) \cap U \neq \emptyset\}$ is open in X. The map φ is called *upper semicontinuous* if for any open $U \subset Y$, the set $\varphi_u^{-1}(U) = \{x \in X : \varphi(x) \subset U\}$ is open in X. If $\varphi(x)$ is a compact subset of Y for each $x \in X$, the map φ is called *compact-valued*. Given a set X, an arbitrary function $d : X \times X \to \mathbb{R}$ is called *a pseudometric on X* if $d(x,x) = 0$, $d(x,y) \geq 0$ for any $x, y \in X$ and $d(x,z) \leq d(x,y) + d(y,z)$ for any $x, y, z \in X$.

Denote by A the set of numeric sequences $\alpha = \{\alpha_i : i \in \mathbb{N}\}$ such that $\alpha_i = 0$ or $\alpha_i = 2$ for all $i \in \mathbb{N}$. Given $\alpha = \{\alpha_i : i \in \mathbb{N}\} \in A$, let $x(\alpha) = \sum_{i=1}^{\infty} 3^{-i} \cdot \alpha_i$. Let $\mathbb{K} = \{x(\alpha) : \alpha \in A\}$. The set $\mathbb{K}$ is called *the Cantor perfect set* or simply *the Cantor set*. If we refer to the Cantor set as a topological space, its topology is considered to be inherited from $\mathbb{R}$. The space $\mathbb{D}$ is the set $\{0, 1\}$ with the discrete topology. A *Tychonoff cube* is the space $\mathbb{I}^\kappa$ for some infinite cardinal κ. Analogously, *a Cantor cube* is the space $\mathbb{D}^\kappa$ for some infinite cardinal κ. Recall that a space X is said to be *pseudocompact* if $C(X) = C^*(X)$, i.e., every continuous function is bounded on X.

Given a space X, denote by $\Sigma_0^0(X)$ the family of all open subsets of X. If, for some countable ordinal ξ, we have $\Sigma_\xi^0(X)$, let $\Pi_\xi^0(X) = \{X \backslash U : U \in \Sigma_\xi^0\}$. In particular, $\Pi_0^0(X)$ is the family of all closed subsets of X. Now, if ξ is a countable ordinal and we have $\Sigma_\eta^0(X)$ for all $\eta < \xi$, we let $\Sigma_\xi^0(X) = \{U \subset X : \text{there exists a sequence } \{\xi_n : n \in \omega\} \subset \xi \text{ such that } U = \bigcup \{U_n : n \in \omega\}, \text{ where } U_n \in \Pi_{\xi_n}^0(X) \text{ for each } n \in \omega\}$. This makes it possible to define the families $\Sigma_\xi^0(X)$ and $\Pi_\xi^0(X)$ for all countable ordinals ξ. We will also need the families $\Delta_\xi^0(X) = \Sigma_\xi^0(X) \cap \Pi_\xi^0(X)$ for all countable ordinals ξ.

The family $\mathbb{B}(X) = \bigcup \{\Sigma_\xi^0(X) \cup \Pi_\xi^0(X) : \xi < \omega_1\}$ is called *the family of Borel subsets of the space X*. For any ordinal $\xi < \omega_1$, the collection $\Sigma_\xi^0(X)$ is called *the family of Borel subsets of X of additive class ξ* and $\Pi_\xi^0(X)$ is *the family of all Borel subsets of X of multiplicative class ξ*. In the classical terminology the elements of the family $\Sigma_1^0(X)$ are called F_σ-sets, the elements of $\Pi_1^0(X)$ are G_δ-sets and the family $\Sigma_2^0(X)$ consists of $G_{\delta\sigma}$-sets; the sets that belong to $\Pi_2^0(X)$ are called $F_{\sigma\delta}$-subsets of X, the family $\Sigma_3^0(X)$ consists of $F_{\sigma\delta\sigma}$-sets, the family $\Pi_3^0(X)$ is referred to as the family of $G_{\delta\sigma\delta}$-sets and so on.

A space is called *Polish* if it is second countable and completely metrizable. A second countable space is called *a Borel set* if it is homeomorphic to a Borel subspace of some Polish space. A space is *analytic* if it is a continuous image of $\mathbb{P}$. A metrizable space X is an absolute Borel set of additive class $\xi > 0$, if X, when embedded into an arbitrary metric space M, belongs to $\Sigma_\xi^0(M)$. Analogously, X is an absolute Borel set of multiplicative class ξ, if, for any embedding φ of X into a metric space M, the set $\varphi(X)$ belongs to $\Pi_\xi^0(M)$. In particular, X is an absolute G_δ if X is a G_δ-subset whenever it is embedded into a metric space; X is an absolute $F_{\sigma\delta}$ if for any metric space M, any subspace of M homeomorphic to X is $F_{\sigma\delta}$-set in M.

A space X is called *a $K_{\sigma\delta}$-space* if it is homeomorphic to a subspace Y of some space Z such that Y is a countable intersection of σ-compact subspaces of Z. A space is called *K-analytic* if it is a continuous image of a $K_{\sigma\delta}$-space. Given a family $\mathcal{U}$ of topologies on a set X, the family $\sup \mathcal{U}$ is the topology whose subbase is the family $\bigcup \mathcal{U}$.

A sequence $\{\mathcal{U}_n : n \in \omega\}$ of covers of a space X is called *complete* if, for any filter $\mathcal{F}$ on X such that $\mathcal{F} \cap \mathcal{U}_n \neq \emptyset$ for all $n \in \omega$, we have $\bigcap \{\overline{F} : F \in \mathcal{F}\} \neq \emptyset$. If X is a space, let $B_1(X) = \{f \in \mathbb{R}^X : f_n \to f$ for some sequence $\{f_n : n \in \omega\} \subset C_p(X)\}$. We will consider the set $B_1(X)$ with the topology inherited from $\mathbb{R}^X$. A compact space K is called *Rosenthal compact* if K is homeomorphic to a subspace of $B_1(M)$ for some Polish space M.

We denote by $\omega^{<\omega}$ the set of all finite sequences of elements of ω, including the empty one. In other words, $\omega^0 = \{\emptyset\}$ and $\omega^{<\omega} = \bigcup \{\omega^n : n \in \omega\}$. If $f \in \omega^\omega$ and $n \in \mathbb{N}$, then $f|n = f|\{0, \ldots, n-1\}$; this function is sometimes identified with the sequence $(f(0), \ldots, f(n-1))$. If $n = 0$, then $f|n = \emptyset$. Given two functions $f, g \in \omega^\omega$, we write $f \leq g$, if $f(n) \leq g(n)$ for every $n \in \omega$.

A family $\mathcal{B} \subset \exp(X)$ is $\mathbb{P}$-*directed* if $\mathcal{B} = \{B_f : f \in \omega^\omega\}$, and, for any $f, g \in \omega^\omega$, we have $B_f \subset B_g$ whenever $f \leq g$. Thus, *a $\mathbb{P}$-directed compact cover of a space X* is a $\mathbb{P}$-directed family $\mathcal{B}$ such that the elements of $\mathcal{B}$ are compact subsets of X and $\bigcup \mathcal{B} = X$.

If f is a function, then $\operatorname{dom}(f)$ is its domain; given another function g, the expression $f \subset g$ says that $\operatorname{dom}(f) \subset \operatorname{dom}(g)$ and $g|\operatorname{dom}(f) = f$. If we have a set of functions $\{f_i : i \in I\}$ such that $f_i|(\operatorname{dom}(f_i) \cap \operatorname{dom}(f_j)) = f_j|(\operatorname{dom}(f_i) \cap \operatorname{dom}(f_j))$ for any $i, j \in I$, then we can define a function f with $\operatorname{dom}(f) = \bigcup_{i \in I} \operatorname{dom}(f_i)$ as follows: given any $x \in \operatorname{dom}(f)$, find any $i \in I$ with $x \in \operatorname{dom}(f_i)$ and let $f(x) = f_i(x)$. It is easy to check that the value of f at x does not depend on the choice of i so we have consistently defined a function f which will be denoted by $\bigcup \{f_i : i \in I\}$.

Given two linearly ordered sets $(A, \prec_A)$ and $(B, \prec_B)$ *the lexicographic order* $<$ on the set $A \times B$ is introduced as follows: given distinct points $p = (a, b)$ and $q = (c, d)$ of the set $A \times B$, we let $p < q$ if $a \prec_A c$; if $c \prec_A a$, then $q < p$. Now, if $a = c$ and $b \prec_B d$, then $p < q$; if $d \prec_B b$, then $q < p$. It is easy to see that if $\prec_A$ and $\prec_B$ are well-orders, then $<$ is also a well-order.

301. Let X be a zero-dimensional space. Prove that any subspace of X is also zero-dimensional.

302. Prove that any product of zero-dimensional spaces is a zero-dimensional space.

303. Given a cardinal κ and an infinite space X with $w(X) \leq \kappa$, prove that X is zero-dimensional if and only if it can be embedded in $\mathbb{D}^\kappa$.

304. Prove that any space X is a perfect image of a zero-dimensional space Y such that $w(Y) \leq w(X)$.

305. Prove that any non-zero-dimensional space can be continuously mapped onto $\mathbb{I}$.

306. Prove that any Lindelöf space is zero-dimensional if and only if it is strongly zero-dimensional. In particular, any compact and any second countable zero-dimensional space is strongly zero-dimensional.

307. Let X be a space with $|X| < \mathfrak{c}$. Prove that X is zero-dimensional. In particular, any countable space is strongly zero-dimensional.

308. For an arbitrary space X, prove that X is strongly zero-dimensional if and only if $\text{Ind} X = 0$. Observe that as a consequence, any strongly zero-dimensional space is normal.

309. Prove that any strongly zero-dimensional space is zero-dimensional. Give an example of a normal zero-dimensional space which is not strongly zero-dimensional.

310. Prove that a closed subspace of a strongly zero-dimensional space is strongly zero-dimensional. Give an example of a strongly zero-dimensional space X such that some $Y \subset X$ is not strongly zero-dimensional.

311. Let X be a normal space such that $X = \bigcup \{X_n : n \in \omega\}$, where each X_n is strongly zero-dimensional and closed in X. Prove that X is strongly zero-dimensional.

312. Prove that there exists a space X which is not zero-dimensional and we have the equality $X = \bigcup \{X_n : n \in \omega\}$, where each X_n is strongly zero-dimensional and closed in X.

313. Prove that the space $\mathbb{P}$ of the irrationals is homeomorphic to ω^ω and hence $\mathbb{P}$ is zero-dimensional.

314. Let X be a paracompact space. Prove that X is strongly zero-dimensional if and only if every open cover of X has a disjoint open refinement.

315. Let P be a strongly zero-dimensional paracompact space and suppose that M is a completely metrizable space. Denote by $CL(M)$ the set of all closed non-empty subsets of M and let $\varphi : P \rightarrow CL(M)$ be a lower semicontinuous map. Prove that φ has a continuous selection, i.e., there exists a continuous map $f : P \rightarrow M$ such that $f(x) \in \varphi(x)$ for any $x \in P$.

316. Let M be a strongly zero-dimensional completely metrizable space. Prove that any closed non-empty $F \subset M$ is a closed retract of M, i.e., there exists a closed continuous map $f : M \rightarrow F$ such that $f(x) = x$ for any $x \in F$.

317. Let F be a closed subspace of $\mathbb{P}$ and suppose that a space X is a continuous image of F. Prove that X is also a continuous image of $\mathbb{P}$.

318. Given a second countable space X and an arbitrary ordinal $\xi < \omega_1$, prove that there exists a set $U \in \Sigma_\xi^0(X \times \mathbb{K})$ such that $\Sigma_\xi^0(X) = \{U[y] : y \in \mathbb{K}\}$, where $U[y] = \{x \in X : (x, y) \in U\}$ for any $y \in \mathbb{K}$. Observe that as an easy consequence, for any second countable space X and every countable ordinal ξ, there exists a set $V \in \Pi_\xi^0(X \times \mathbb{K})$ such that $\Pi_\xi^0(X) = \{V[y] : y \in \mathbb{K}\}$.

319. Prove that for any uncountable Polish space X and every countable ordinal ξ, the classes $\Sigma_\xi^0(X)$ and $\Pi_\xi^0(X)$ do not coincide.

320. Let X be a second countable space. Given countable ordinals ξ and $\eta > \xi$, prove that $\Sigma_\xi^0(X) \cup \Pi_\xi^0(X) \subset \Delta_\eta^0(X)$. Show that if X is an uncountable Polish space, then $\Sigma_\xi^0(X) \cup \Pi_\xi^0(X) \neq \Delta_\eta^0(X)$.

321. Suppose that X is a second countable space. Prove that for every countable limit ordinal η, we have $\bigcup\{\Sigma_\xi^0(X) : \xi < \eta\} \subset \Delta_\eta^0(X)$. Show that if X is uncountable and Polish, then the inclusion is strict, i.e., $\bigcup\{\Sigma_\xi^0(X) : \xi < \eta\} \neq \Delta_\eta^0(X)$.

322. Prove that there exists a countable space which cannot be embedded into $C_p(B)$ for any Borel set B.

323. Prove that a second countable space X is an absolute F_σ if and only if X is σ-compact.

324. Prove that a second countable space X is an absolute G_δ if and only if X is Čech-complete.

325. Suppose that X is a Polish space and $f : X \to Y$ is a perfect map. Prove that Y is Polish (remember that any perfect map is onto).

326. Let X be a Polish space. Suppose that $f : X \to Y$ is a continuous surjective open map. Prove that there is a closed $F \subset X$ such that $f(F) = Y$ and $f|F$ is a perfect map.

327. Prove that any open continuous image of a Polish space is a Polish space.

328. Prove that X is a Polish space if and only if it is an open continuous image of $\mathbb{P}$.

329. Prove that a second countable space is Polish if and only if it is a closed continuous image of $\mathbb{P}$. Show that a closed continuous image of $\mathbb{P}$ is not necessarily first countable.

330. Prove that X is homeomorphic to a Borel subset of some Polish space if and only if it is homeomorphic to a Borel subset of $\mathbb{R}^\omega$.

331. Let X be a Borel set. Prove that every $Y \in \mathbb{B}(X)$ is also a Borel set. In particular, any closed and any open subspace of a Borel set is a Borel set.

332. Given second countable spaces X and Y and a continuous map $f : X \to Y$, prove that for every Borel subset A of the space Y, the set $f^{-1}(A)$ is Borel in X.

333. Prove that any countable product of Borel sets is a Borel set. Show that for any second countable space X, if $X = \bigcup\{X_i : i \in \omega\}$ and each X_i is a Borel set, then X is also a Borel set.

334. Prove that every Borel set is an analytic space.

335. Prove that

(i) any closed subspace of an analytic space is an analytic space;

(ii) any open subspace of an analytic space is an analytic space;

(iii) any countable product of analytic spaces is an analytic space.

336. Assume that Y is a space and $X_i \subset Y$ is an analytic space for all $i \in \omega$. Prove that $X = \bigcap\{X_i : i \in \omega\}$ is also an analytic space.

337. Assume that $X = \bigcup\{X_i : i \in \omega\}$ and X_i is an analytic space for every $i \in \omega$. Prove that X is an analytic space.

338. Let X and Y be Polish spaces. Suppose that $f : X \to Y$ is a continuous map. Prove that for any analytic set $B \subset Y$, the set $f^{-1}(B)$ is also analytic.

339. Let A and B be two disjoint analytic subsets of a Polish space M. Prove that there exist Borel subsets A' and B' of the space M such that $A \subset A'$, $B \subset B'$ and $A' \cap B' = \emptyset$.

340. Let X be a subspace of a Polish space M. Prove that X is Borel if and only if X and $M \setminus X$ are analytic.

341. Prove that X is a Borel set if and only if there exists a closed subspace of $\mathbb{P}$ which condenses onto X. As a consequence, if a Borel set X can be condensed onto a second countable space Y, then Y is a also Borel set.

342. Show that there exists a subspace $X \subset \mathbb{R}$ which is not analytic (and hence not Borel).

343. Prove that

(i) any closed subset of a K-analytic space is a K-analytic space;

(ii) any countable product of K-analytic spaces is a K-analytic space.

344. Assume that Y is a space and $X_i \subset Y$ is a K-analytic space for all $i \in \omega$. Prove that $X = \bigcap\{X_i : i \in \omega\}$ is also a K-analytic space.

345. Assume that $X = \bigcup\{X_i : i \in \omega\}$ and X_i is a K-analytic space for all $i \in \omega$. Prove that X is a K-analytic space.

346. Observe that there exist K-analytic non-analytic spaces. Show that any analytic space is a K-analytic space. Prove that for any space X with a countable network, X is analytic if and only if it is K-analytic.

347. Prove that a non-empty Polish space X is homeomorphic to $\mathbb{P}$ if and only if X is zero-dimensional and any compact subspace of X has empty interior.

348. Prove that a metrizable compact X is homeomorphic to the Cantor set if and only if X is zero-dimensional and has no isolated points.

349. Prove that a countable metrizable space X is homeomorphic to $\mathbb{Q}$ if and only if X has no isolated points.

350. Prove that every countable metrizable space is homeomorphic to a closed subspace of $\mathbb{Q}$.

351. Let X be a second countable σ-compact space. Prove that X is not Polish if and only if it contains a closed subspace homeomorphic to $\mathbb{Q}$.

352. Let X be an analytic non-σ-compact space. Prove that some closed subspace of X is homeomorphic to $\mathbb{P}$.

353. Prove that any uncountable analytic space contains a closed subspace which is homeomorphic to the Cantor set.

354. Prove that any non-σ-compact Borel set can be condensed onto $\mathbb{I}^\omega$ as well as onto $\mathbb{R}^\omega$.

355. Give an example of a non-σ-compact subspace of the real line $\mathbb{R}$ which cannot be condensed onto a compact space.

356. Prove that $\mathbb{Q}$ cannot be condensed onto a compact space.

357. Prove that for any metrizable compact X, the space $C_p(X)$ condenses onto a compact space.

358. Prove that a Polish space X is dense-in-itself if and only if $\mathbb{P}$ condenses onto X.

359. Prove that for any metrizable compact X, the space $\mathbb{P}$ condenses onto $C_p(X)$.

360. Prove that $C_p(X)$ is analytic if and only if $\mathbb{R}^\omega$ maps continuously onto $C_p(X)$. Observe that not every analytic space is a continuous image of $\mathbb{R}^\omega$.

361. Suppose that X is an infinite space such that $C_p(X)$ is analytic. Prove that $C_p(X)$ can be continuously mapped onto $\mathbb{R}^\omega$. Deduce from this fact that if X and Y are infinite spaces such that $C_p(X)$ and $C_p(Y)$ are analytic, then each of the spaces $C_p(X)$ and $C_p(Y)$ maps continuously onto the other one.

362. Prove that for any second countable σ-compact space X, the space $C_p(X)$ is $K_{\sigma\delta}$.

363. Let X be a space with a countable network. Prove that X is analytic if and only if every second countable continuous image of X is analytic.

364. Let X be a space with a countable network. Prove that X is σ-compact if and only if every second countable continuous image of X is σ-compact.

365. Prove that a second countable space M is Polish if and only if there exists a map $f : \mathbb{P} \to \exp(M)$ with the following properties:

(a) $f(s)$ is compact for any $s \in \mathbb{P}$;
(b) for any $s, t \in \mathbb{P}$, if $s \leq t$, then $f(s) \subset f(t)$;
(c) for any compact $K \subset M$, there is $s \in \mathbb{P}$ such that $K \subset f(s)$.

366. Prove that if $C_p(X)$ is analytic, then X is σ-compact.

367. Prove that the following are equivalent for an arbitrary second countable space X:

(i) $C_p(X)$ is analytic;
(ii) $C_p(X)$ is a $K_{\sigma\delta}$-space;
(iii) X is σ-compact.

Observe that, as a consequence, the spaces $C_p(\mathbb{P})$ and $C_p(\mathbb{R}^\omega)$ are not analytic.

368. For a second countable space X let A be a countable dense subspace of X. Prove that the following conditions are equivalent:

(i) $C_p(A|X)$ is a Borel set;
(ii) $C_p(A|X)$ is analytic;
(iii) X is σ-compact.

369. Given a countable space X prove that $C_p(X)$ is analytic if and only if $C_p(X, \mathbb{I})$ is analytic.

370. Prove that a countable space X embeds into $C_p(\mathbb{P})$ if and only if $C_p(X)$ is analytic.

371. Take any point $\xi \in \beta\omega\backslash\omega$ and consider the space $X = \omega \cup \{\xi\}$ with the topology inherited from $\beta\omega$. Prove that neither $C_p(X)$ nor $\mathbb{R}^X\backslash C_p(X)$ is analytic. As a consequence, X cannot be embedded into $C_p(\mathbb{P})$.

372. Prove that if $\alpha < \omega_1$, then there exists a countable space X with a unique non-isolated point such that $C_p(X) \in \mathbb{B}(\mathbb{R}^X)\backslash(\bigcup_{\beta<\alpha} \Sigma_\beta^0(\mathbb{R}^X))$, i.e., the space $C_p(X)$ can have an arbitrarily high Borel complexity for a countable space X with a unique non-isolated point.

373. Prove that the following are equivalent for any metrizable space X:

   (i) X is an absolute $F_{\sigma\delta}$;
   (ii) there is a completely metrizable space M such that X is an $F_{\sigma\delta}$-subset of M;
   (iii) X has a complete sequence of σ-discrete closed covers.

374. Prove that $C_p(X)$ is an absolute $F_{\sigma\delta}$ for any countable metrizable X.

375. Let K be a compact space. Given a countable $X \subset C_p(K)$, prove that $C_p(X)$ is an absolute $F_{\sigma\delta}$.

376. Prove that any second countable space embeds into $C_p(\mathbb{K})$, where $\mathbb{K}$ is the Cantor set.

377. Give an example of a second countable X such that for any compact K, the space X cannot be embedded in $C_p(K)$ as a closed subspace.

378. Prove that any countable second countable space embeds into $C_p(\mathbb{K})$ as a closed subspace.

379. Given a function $f : X \to \mathbb{R}$, consider the following conditions:

   (i) for any open $U \subset \mathbb{R}$ the set $f^{-1}(U)$ is an F_σ-set in X;
   (ii) there exists a sequence $\{f_n : n \in \omega\} \subset C_p(X)$ which converges to f.

   Prove that (ii)$\Longrightarrow$(i) for any space X. Show that if X is second countable, then also (i)$\Longrightarrow$(ii) and hence (i) $\Longleftrightarrow$ (ii).

380. Prove that if $X = \mathbb{R}$, then $B_1(X) \neq \mathbb{R}^X$.

381. Prove that a compact space X is countable if and only if $B_1(X) = \mathbb{R}^X$.

382. Prove that under MA+¬CH, there exists an uncountable $X \subset \mathbb{R}$ such that $B_1(X) = \mathbb{R}^X$.

383. Prove that the two arrows space is Rosenthal compact.

384. Prove that every Rosenthal compact space is Fréchet–Urysohn.

385. Let X be a separable compact space. Prove that X is Rosenthal compact if and only if for any dense countable $A \subset X$, the space $C_p(A|X)$ is analytic.

386. Let X be a compact space. Assume that A and B are dense countable subsets of X such that $C_p(A|X)$ is analytic and $C_p(B|X)$ is not. Prove that X contains a subspace homeomorphic to $\beta\omega$.

387. Suppose that X is a compact space and A is a dense countable subset of X such that $C_p(A|X)$ is analytic. Prove that X is Rosenthal compact or else $\beta\omega$ embeds in X.

388. Prove that a space X is K-analytic if and only if X is an image of $\mathbb{P}$ under a compact-valued upper semicontinuous map, i.e., when there exists a compact-valued upper semicontinuous $\varphi : \mathbb{P} \to X$ such that $\bigcup\{\varphi(p) : p \in \mathbb{P}\} = X$.

389. Prove that an arbitrary space X is K-analytic if and only if there exists a family $\mathcal{K} = \{K_f : f \in \omega^\omega\}$ of compact subsets of X with the following properties:

  (i) the family $\mathcal{K}$ is a cover of X, i.e., $\bigcup \mathcal{K} = X$;
  (ii) if a sequence $f_n \in \omega^\omega$ converges to $f \in \omega^\omega$ and $x_n \in K_{f_n}$ for all $n \in \omega$, then the sequence $\{x_n\}$ has an accumulation point which belongs to K_f.

390. Prove that an arbitrary space X is K-analytic if and only if there exists a space Y which maps perfectly onto $\mathbb{P}$ and continuously onto X.

391. Prove that an arbitrary space X is K-analytic if and only if X is realcompact and has a $\mathbb{P}$-directed compact cover.

392. Suppose that X can be condensed onto a metrizable space. Prove that X is analytic if and only if it has a $\mathbb{P}$-directed compact cover.

393. Let X be a compact space. Prove that $C_p(X)$ is K-analytic if and only if it has a $\mathbb{P}$-directed compact cover.

394. Give an example of a non-K-analytic space which has a $\mathbb{P}$-directed compact cover.

395. Assuming MA+¬CH, prove that if X is a K-analytic space such that every compact subspace of X is metrizable, then X has a countable network. Observe that if every compact subspace of an analytic space X is countable, then X is countable.

396. Suppose that X is a compact space and some outer base $\mathcal{B}$ of its diagonal $\Delta = \{(x, x) : x \in X\}$ in $X \times X$ can be indexed as $\mathcal{B} = \{O_p : p \in \omega^\omega\}$ in such a way that $O_p \subset O_q$ whenever $p, q \in \omega^\omega$ and $q \leq p$. Prove that X is metrizable.

397. Suppose that $C_p(X)$ is K-analytic and X is separable. Prove that $C_p(X)$ is analytic.

398. Let X be a compact space such that $C_p(X)$ is K-analytic. Prove that X is a Fréchet–Urysohn space.

399. Prove that the following conditions are equivalent for any space X:

  (i) $C_p(C_p(X))$ is K-analytic;
  (ii) $C_p(C_p(X))$ is analytic;
  (iii) X is finite.

400. Prove that the following properties are equivalent for any space X:

  (i) X is hereditarily K-analytic;
  (ii) X is hereditarily analytic;
  (iii) X is countable.

1.5   Additivity of Properties: Mappings Between Function Spaces

All spaces are assumed to be Tychonoff. A topological property $\mathcal{P}$ will be called *finitely additive* if a space X has $\mathcal{P}$ provided that X is a finite union of its subspaces with the property $\mathcal{P}$. We say that $\mathcal{P}$ is *countably additive* if a space X has $\mathcal{P}$ given that $X = \bigcup\{X_n : n \in \omega\}$ and $X_n \vdash \mathcal{P}$ (this is read: "X_n has $\mathcal{P}$") for every $n \in \omega$. The finite and countable additivity of a cardinal function are defined analogously, i.e., a cardinal function φ is *finitely additive* if the conditions $X = X_1 \cup \ldots \cup X_n$ and $\varphi(X_i) \leq \kappa$ for each $i \leq n$ imply $\varphi(X) \leq \kappa$. We say that φ is *countably additive* if the conditions $X = \bigcup\{X_n : n \in \omega\}$ and $\varphi(X_n) \leq \kappa$ for every $n \in \omega$ imply $\varphi(X) \leq \kappa$. A cardinal function φ is called *completely additive* if for any infinite cardinal κ and any space X, we have $\varphi(X) \leq \kappa$ whenever $X = \bigcup\{X_\alpha : \alpha < \kappa\}$ and $\varphi(X_\alpha) \leq \kappa$ for each $\alpha < \kappa$.

A property $\mathcal{P}$ is called finitely (countably) additive in some class $\mathcal{C}$ of spaces if for any $X \in \mathcal{C}$, the space X has $\mathcal{P}$ whenever X is a finite (countable) union of its subspaces with the property $\mathcal{P}$. Analogously, a cardinal function φ is called *finitely additive in a class $\mathcal{C}$* if for any space $X \in \mathcal{C}$, the conditions $X = X_1 \cup \ldots \cup X_n$ and $\varphi(X_i) \leq \kappa$ for each $i \leq n$ imply $\varphi(X) \leq \kappa$. We say that φ is *countably additive in a class $\mathcal{C}$* if for any $X \in \mathcal{C}$, we have $\varphi(X) \leq \kappa$ whenever $X = \bigcup\{X_n : n \in \omega\}$ and $\varphi(X_n) \leq \kappa$ for every $n \in \omega$. A cardinal function φ is called *completely additive in a class $\mathcal{C}$* if for any infinite cardinal κ and any $X \in \mathcal{C}$, we have $\varphi(X) \leq \kappa$ whenever $X = \bigcup\{X_\alpha : \alpha < \kappa\}$ and $\varphi(X_\alpha) \leq \kappa$ for each $\alpha < \kappa$.

If X is a space and $\mathcal{C}_n \subset \tau(X)$ for each $n \in \omega$, the sequence $\{\mathcal{C}_n : n \in \omega\}$ is called *pseudocomplete* if for any family $\{U_n : n \in \omega\}$ such that $\overline{U}_{n+1} \subset U_n$ and $U_n \in \mathcal{C}_n$ for every $n \in \omega$, we have $\bigcap\{U_n : n \in \omega\} \neq \emptyset$. A space X is called *pseudocomplete* if there is a pseudocomplete sequence $\{\mathcal{B}_n : n \in \omega\}$ of π-bases in X. The space X is *homogeneous* if for any $x, y \in X$, there is a homeomorphism $h : X \to X$ such that $h(x) = y$. A subset F of a space X is called *a zero-set in X* if there exists a function $f \in C(X)$ such that $F = f^{-1}(0)$. Say that $U \subset X$ is *a cozero-set in X* if $X \backslash U$ is a zero-set in X.

Let $\mathcal{M}$ be an infinite maximal almost disjoint family in ω and $M = \omega \cup \mathcal{M}$. If $x \in \omega$, let $\mathcal{B}_x = \{x\}$. Given $x \in \mathcal{M}$, define $\mathcal{B}_x = \{\{x\} \cup (x \backslash A) : A$ is a finite subset of $x\}$ (remember that for any $x \in \mathcal{M}$, we can consider x to be a point of M or a subset of ω). The families $\{\mathcal{B}_x : x \in M\}$ generate a topology τ_M on M as local bases and the space (M, τ_M) is called *a Mrowka space*. Given a cardinal κ, let $A(\kappa) = \kappa \cup \{a\}$, where $a \notin \kappa$. If $x \in \kappa$, let $\mathcal{B}_x = \{\{x\}\}$. If $x = a$, then $\mathcal{B}_x = \{\{a\} \cup (\kappa \backslash B) : B$ is a finite subset of $\kappa\}$. The families $\{\mathcal{B}_x : x \in A(\kappa)\}$ generate a topology $\tau(A(\kappa))$ for which $\mathcal{B}_x$ is a local base at x for any $x \in A(\kappa)$. The space $A(\kappa)$ is called *Alexandroff one-point compactification* of the discrete space of cardinality κ. For any uncountable cardinal κ, let $L(\kappa) = \kappa \cup \{a\}$, where $a \notin \kappa$. If $x \in \kappa$, let $\mathcal{B}_x = \{\{x\}\}$. If $x = a$, then $\mathcal{B}_x = \{\{a\} \cup (\kappa \backslash B) : B$ is a countable subset of $\kappa\}$. The families $\{\mathcal{B}_x : x \in L(\kappa)\}$ generate a topology $\tau(L(\kappa))$ for which $\mathcal{B}_x$ is a local base at x for any $x \in L(\kappa)$. The space $L(\kappa)$ is called *the Lindelöfication of the discrete space of cardinality* κ.

Given a space X and $A \subset C(X)$, denote by $\overline{A}^u$ the set $\{f \in C(X) :$ there exists a sequence $\{f_n : n \in \omega\} \subset A$ such that $f_n \rightrightarrows f\}$. The closure operator $A \to \overline{A}^u$ generates a topology τ_u (called *the uniform convergence topology*) on the space $C(X)$, such that $\overline{A}^u = \mathrm{cl}_{\tau_u}(A)$ for every $A \subset C(X)$. The space $(C(X), \tau_u)$ will be denoted $C_u(X)$. If O is a subset of $\mathbb{R}$, then $C_u(X, O)$ is the set $C(X, O)$ with the topology inherited from $C_u(X)$. If the set $C^*(X)$ of continuous bounded functions on X is considered with the topology inherited from $C_u(X)$, it is denoted by $C_u^*(X)$. A set $A \subset C_p(X)$ is *strongly dense (or uniformly dense)* in $C_p(X)$ if $\overline{A}^u = C_p(X)$. We will identify any space X with the subspace $E(X) = \{e_x : x \in X\} \subset C_p(C_p(X))$, where $e_x(f) = f(x)$ for any $x \in X$.

As usual, the expression $X \simeq Y$ says that the spaces X and Y are homeomorphic. A space X is *metacompact* if every open cover of X has a point-finite open refinement. A space X is called *Dieudonné complete* if it embeds as a closed subspace into a product of metrizable spaces. Now, X is *realcompact* if it embeds as a closed subspace into a product of real lines. Given a space X, the extent of X is defined as follows: $\mathrm{ext}(X) = \sup\{|D| : D$ is a closed discrete subspace of $X\}$. A space X is *Fréchet–Urysohn* if for any $A \subset X$ and any $x \in \overline{A}$, there is a sequence $\{a_n : n \in \omega\} \subset A$ such that $a_n \to x$. A space X is *sequential* if for any non-closed $A \subset X$, there is a sequence $(a_n) \subset A$ which converges to some point of $X \backslash A$. A space X is *a k-space* if for any non-closed $A \subset X$, there exists a compact subspace $K \subset X$ such that $A \cap K$ is not closed. We say that X is *κ-scattered* if $|\overline{A}| \leq \kappa$ for any $A \subset X$ with $|A| \leq \kappa$.

Given a sequence $S = \{A_n : n \in \omega\}$ of subsets of a space X, consider the set $\lim S = \bigcup\{\bigcap\{A_m : m \geq n\} : n \in \omega\}$. The set $\lim S$ is called *the limit of the sequence* S. The fact that $A = \lim S$ will also be expressed as $S \to A$ or $A_n \to A$. Recall that a set X is *ω-covered* by a family $\mathcal{U}$ if for any finite $A \subset X$, there is $U \in \mathcal{U}$ such that $A \subset U$. *The Gerlits property* φ is formulated as follows: X has φ if for any family $\mathcal{U} = \bigcup\{\mathcal{U}_n : n \in \omega\} \subset \tau(X)$ with $\mathcal{U}_n \subset \mathcal{U}_{n+1}$ for each $n \in \omega$, which ω-covers the space X, there exists a sequence $\{X_n : n \in \omega\}$ of subsets of X such that $X_n \to X$ and X_n is ω-covered by $\mathcal{U}_n$.

A map $f : X \to Y$ is *finite-to-one* if the set $f^{-1}(y)$ is finite for any $y \in Y$. Let X be a space. Given an infinite cardinal κ, a function $f : X \to \mathbb{R}$ is called *strictly κ-continuous* if for any $A \subset X$ with $|A| \leq \kappa$, there exists $g \in C(X)$ with $f|_A = g|_A$. Let $t_m(X) = \min\{\kappa :$ any strictly κ-continuous function on X is continuous$\}$. The cardinal $t_m(X)$ is called *weak functional tightness of the space* X. A subspace Y of a space X is called *κ-placed* in X if for any $x \in X \backslash Y$, there exists a G_κ-set H in X such that $x \in H \subset X \backslash Y$. Let $q(X) = \min\{\kappa : X$ is κ-placed in $\beta X\}$. The cardinal $q(X)$ is called *the Hewitt–Nachbin number of the space* X.

401. Observe first that the space $A(\omega_1)$ is the union of two discrete (and hence metrizable) subspaces of countable i-weight. Deduce from this observation that the first axiom of countability, metrizability, i-weight, P-property and pseudocharacter are not finitely additive.

402. Representing $L(\omega_1)$ as a union of two metrizable subspaces, observe that sequentiality, π-character, the Fréchet–Urysohn property, Čech-completeness and k-property are not finitely additive.

403. Let $\xi \in \beta\omega\backslash\omega$ and observe that the space $\omega \cup \{\xi\}$ is a union of two second countable spaces while $w(\omega \cup \{\xi\}) > \omega$. Therefore weight is not finitely additive.

404. Give an example of a non-realcompact space which is the union of two hereditarily realcompact subspaces.

405. Prove that if φ is a cardinal function and $\varphi \in$ {network weight, spread, Lindelöf number, hereditary Lindelöf number, density, hereditary density, extent, Souslin number, point-finite cellularity}, then φ is completely additive and hence countably additive.

406. Prove that pseudocompleteness, π-weight and the Baire property are finitely additive but not countably additive.

407. Considering any Mrowka space, prove that normality is not finitely additive.

408. Suppose that $X \times X = Y \cup Z$, where Y and Z are first countable. Prove that X is also first countable.

409. Suppose that $X \times X = Y \cup Z$, where Y and Z have countable pseudocharacter. Prove that $\psi(X) \leq \omega$.

410. Suppose that $X \times X = Y \cup Z$, where Y and Z have countable tightness. Prove that $t(X) \leq \omega$.

411. Suppose that $X \times X = Y \cup Z$, where Y and Z have countable weight. Prove that $w(X) \leq \omega$.

412. Suppose that X is a separable space such that $X \times X = Y \cup Z$, where Y and Z are metrizable. Prove that X is metrizable.

413. Suppose that X is a compact space such that $X \times X = Y \cup Z$, where Y and Z are metrizable. Prove that X is metrizable.

414. Give an example of a non-metrizable space X such that $X \times X$ is the union of two metrizable subspaces.

415. Suppose that $X^\omega = \bigcup\{X_n : n \in \omega\}$. Prove that for some $n \in \omega$, there is $Y \subset X_n$ such that there exists an open continuous map of Y onto X^ω, and hence, there exists an open continuous map of Y onto X. As a consequence, if X^ω is the countable union of first countable subspaces, then X is first countable.

416. Given an arbitrary space X, suppose that X^ω is the finite union of metrizable subspaces. Prove that X is metrizable.

417. Given a countably compact space X, suppose that X^ω is a countable union of metrizable subspaces. Prove that X is metrizable.

418. Give an example of a non-metrizable space X such that X^ω is a countable union of its metrizable subspaces.

419. For an arbitrary space X and any $f, g \in C^*(X)$, let

$$d(f, g) = \sup\{|f(x) - g(x)| : x \in X\}.$$

Prove that d is a complete metric on the set $C^*(X)$ and the topology, generated by d, coincides with the topology of $C_u^*(X)$.

420. Let $\mathcal{P}$ be an F_σ-hereditary property, i.e., $X \vdash \mathcal{P}$ implies $Y \vdash \mathcal{P}$ whenever Y is an F_σ-subspace of X. Suppose that $C_p(X)$ is a finite union of subspaces which have the property $\mathcal{P}$. Prove that $C_p(X)$ is a finite union of *dense* subspaces, each one of which has the property $\mathcal{P}$.

421. Let $\mathcal{P}$ be a hereditary property. Suppose that $C_p(X)$ is a finite union of subspaces which have the property $\mathcal{P}$. Prove that there is $n \in \mathbb{N}$ and $\varepsilon > 0$ such that $C_p(X, (-\varepsilon, \varepsilon)) = Y_1 \cup \ldots \cup Y_n$, where $Y_i \vdash \mathcal{P}$ and $\overline{Y_i}^u \supset C(X, (-\varepsilon, \varepsilon))$ for each $i \in \{1, \ldots, n\}$.

422. Suppose that $C_p(X)$ is a finite union of its paracompact (not necessarily closed) subspaces. Prove that $C_p(X)$ is Lindelöf and hence paracompact.

423. Suppose that $C_p(X) = Y_1 \cup \ldots \cup Y_n$, where Y_i is realcompact for each $i \leq n$. Prove that $C_p(X)$ is realcompact.

424. Suppose that $C_p(X) = Y_1 \cup \ldots \cup Y_n$, where Y_i is homeomorphic to $\mathbb{R}^{\kappa_i}$ for each $i \leq n$. Prove that X is discrete.

425. Suppose that $C_p(X) = Y_1 \cup \ldots \cup Y_n$, where Y_i is hereditarily realcompact for each $i \leq n$. Prove that $iw(C_p(X)) = \psi(C_p(X)) = \omega$ and hence $C_p(X)$ is hereditarily realcompact.

426. Given an infinite cardinal κ, suppose that the space $C_p(X)$ is a finite union of its κ-monolithic (not necessarily closed) subspaces. Prove that $C_p(X)$ must be κ-monolithic.

427. Given an infinite cardinal κ suppose that the space $C_p(X)$ is a finite union of its spread(κ)-monolithic (not necessarily closed) subspaces. Prove that $C_p(X)$ is spread(κ)-monolithic.

428. Given an infinite cardinal κ suppose that the space $C_p(X)$ is a finite union of its $hd(\kappa)$-monolithic (not necessarily closed) subspaces. Prove that $C_p(X)$ must be $hd(\kappa)$-monolithic.

429. Given an infinite cardinal κ suppose that the space $C_p(X)$ is a finite union of its $hl(\kappa)$-monolithic (not necessarily closed) subspaces. Prove that $C_p(X)$ must be $hl(\kappa)$-monolithic.

430. Suppose that $C_p(X)$ is a finite union of its Dieudonné complete subspaces. Prove that $C_p(X)$ is realcompact and hence Dieudonné complete.

431. Let X be an arbitrary space. Suppose that $C_p(X) = \bigcup\{Z_n : n \in \omega\}$. Prove that there exists a function $f \in C_p(X)$ and $\varepsilon > 0$ such that for some $n \in \omega$, the set $(Z_n + f) \cap C(X, (-\varepsilon, \varepsilon))$ is dense in $C_u(X, (-\varepsilon, \varepsilon))$ and hence also in $C_p(X, (-\varepsilon, \varepsilon))$.

432. Suppose that $C_p(X) = \bigcup\{Z_n : n \in \omega\}$ and each Z_n is closed in $C_p(X)$. Prove that some Z_n contains a homeomorphic copy of $C_p(X)$.

433. Let $\mathcal{P}$ be a hereditary property. Suppose that $C_p(X) = \bigcup\{Z_n : n \in \omega\}$, where each Z_n is closed in $C_p(X)$ and has $\mathcal{P}$. Prove that $C_p(X)$ also has $\mathcal{P}$.

434. Suppose that $C_p(X) = \bigcup\{Z_n : n \in \omega\}$, where each Z_n is locally compact. Prove that X is finite.

435. Suppose that $C_p(X) = \bigcup\{Z_n : n \in \omega\}$ and each Z_n is locally pseudocompact. Prove that $C_p(X)$ is σ-pseudocompact.

436. Suppose that $C_p(X) = \bigcup\{Z_n : n \in \omega\}$, where each Z_n is realcompact and closed in $C_p(X)$. Prove that $C_p(X)$ is realcompact.

437. Prove that any metacompact collectionwise-normal space is paracompact.

438. Prove that if $C_p(X)$ is normal and metacompact, then it is Lindelöf.

439. Prove that $C_p(\beta\omega)$ is not metacompact.

440. Prove that $C_p(L(\kappa))$ is not metacompact for any uncountable κ.

441. Prove that neither the Baire property nor pseudocompleteness is countably additive in spaces $C_p(X)$.

442. Prove that π-weight and π-character are not countably additive in spaces $C_p(X)$.

443. Suppose that the space $C_p(X)$ is a countable union of its Čech-complete (not necessarily closed) subspaces. Prove that X is countable and discrete (and hence $C_p(X)$ is Čech-complete).

444. Given an infinite cardinal κ suppose that $C_p(X)$ is a union of countably many (not necessarily closed) subspaces of character $\leq \kappa$. Prove that $\chi(C_p(X)) \leq \kappa$ and hence $|X| \leq \kappa$.

445. Prove that weight is countably additive in spaces $C_p(X)$.

446. Prove that metrizability is countably additive in spaces $C_p(X)$.

447. Prove that tightness is countably additive in spaces $C_p(X)$.

448. Prove that pseudocharacter is countably additive in spaces $C_p(X)$.

449. Prove that i-weight and diagonal number are countably additive in spaces $C_p(X)$.

450. Prove that the Fréchet–Urysohn property is countably additive in spaces $C_p(X)$.

451. Suppose that X is a metrizable space and $C_p(X) = \bigcup\{Y_i : i \in \omega\}$, where Y_i is hereditarily realcompact (not necessarily closed) for every $i \in \omega$. Prove that $nw(C_p(X)) = iw(C_p(X)) = \omega$ and hence $C_p(X)$ is hereditarily realcompact.

452. Suppose that X is a pseudocompact space and $C_p(X) = \bigcup\{Z_n : n \in \omega\}$, where each Z_n is paracompact and closed in $C_p(X)$. Prove that $C_p(X)$ is Lindelöf.

453. Give an example of a non-normal space which is a countable union of its closed normal subspaces.

454. Let X be a compact space. Suppose that $C_p(X) = \bigcup\{Z_n : n \in \omega\}$, where each Z_n is normal and closed in $C_p(X)$. Prove that $C_p(X)$ is Lindelöf.

455. Let X be a metrizable space. Suppose that $C_p(X)$ is a countable union of its (not necessarily closed) normal subspaces. Prove that X is second countable and hence $C_p(X)$ is normal.

456. Let X be an arbitrary space. Given a uniformly dense $Y \subset C_p(X)$, prove that $t(X) \leq l(Y)$.

457. For an arbitrary space X and a uniformly dense $Y \subset C_p(X)$ prove that $nw(Y) = nw(C_p(X))$ and $d(Y) = d(C_p(X))$.

458. For an arbitrary space X and a uniformly dense $Y \subset C_p(X)$ prove that $hd(Y) = hd(C_p(X))$, $hl(Y) = hl(C_p(X))$ and $s(Y) = s(C_p(X))$.

459. Suppose that X is a space and $Y \subset C_p(X)$ is uniformly dense in $C_p(X)$. Prove that if Y is a Lindelöf Σ-space, then $C_p(X)$ is also Lindelöf Σ.

460. Suppose that X is a space and $Y \subset C_p(X)$ is uniformly dense in $C_p(X)$. Prove that

   (i) if Y is K-analytic, then $C_p(X)$ is K-analytic;
   (ii) if Y is analytic, then $C_p(X)$ is analytic.

461. Given a space X and a uniformly dense subspace $Y \subset C_p(X)$, prove that $t(Y) = t(C_p(X))$.

462. Let X be a space with $\text{ext}^*(X) \leq \kappa$. Prove that $t(Y) \leq \kappa$ for any compact $Y \subset C_p(X)$.

463. Suppose that X has the Gerlits property φ. Prove that all continuous images and all closed subspaces of X have φ.

464. Prove that $C_p(X)$ is a Fréchet–Urysohn space if and only if X has the Gerlits property φ and $t(C_p(X)) = \omega$.

465. (Gerlits–Pytkeev theorem) Prove that the following conditions must be equivalent for any space X:

   (i) $C_p(X)$ is a Fréchet–Urysohn space;
   (ii) $C_p(X)$ is a sequential space;
   (iii) $C_p(X)$ is a k-space.

466. Suppose that X is a σ-compact space such that $C_p(X) = \bigcup_{n \in \omega} Y_n$ where Y_n is a k-space for every $n \in \omega$. Prove that $C_p(X)$ is a Fréchet–Urysohn space. In particular, if X is σ-compact and $C_p(X)$ is a countable union of sequential spaces, then $C_p(X)$ is a Fréchet–Urysohn space.

467. Given spaces X and Y suppose that $\varphi : X \to Y$ is a continuous map and let $\varphi^*(f) = f \circ \varphi$ for any $f \in C_p(Y)$; then $\varphi^* : C_p(Y) \to R = \varphi^*(C_p(Y)) \subset C_p(X)$. We can define a map $r_\varphi : C_p(C_p(X)) \to C_p(C_p(Y))$ by the equality $r_\varphi(\delta) = (\delta|R) \circ \varphi^*$ for any $\delta \in C_p(C_p(X))$. Prove that r_φ is a continuous ring homomorphism such that $r_\varphi|_X = \varphi$ (here we identify X and Y with their canonical copies in $C_p(C_p(X))$ and $C_p(C_p(Y))$, respectively). Prove that φ is the unique continuous ring homomorphism with this property, i.e., if $s : C_p(C_p(X)) \to C_p(C_p(Y))$ is a continuous ring homomorphism such that $s|_X = \varphi$, then $s = r_\varphi$.

468. Suppose that X is an ω-monolithic compact space. Prove that for every surjective continuous map $\varphi : X \to Y$, the map $r_\varphi : C_p(C_p(X)) \to C_p(C_p(Y))$ is surjective.

469. Given spaces X and Y let $\varphi : X \to Y$ be a continuous onto map. Prove that the mapping $r_\varphi : C_p(C_p(X)) \to r_\varphi(C_p(C_p(X))) \subset C_p(C_p(Y))$ is open if and only if φ is $\mathbb{R}$-quotient.

470. Suppose that there exists a continuous map of $C_p(X)$ onto $C_p(Y)$. Prove that $nw(Y) \leq nw(X)$.

471. Suppose that there exists a continuous map of $C_p(X)$ onto $C_p(Y)$. Prove that $iw(Y) \leq iw(X)$.

472. Suppose that there exists a continuous map of $C_p(X)$ onto $C_p(Y)$. Prove that $s^*(Y) \leq s^*(X)$, $hl^*(Y) \leq hl^*(X)$ and $hd^*(Y) \leq hd^*(X)$.

473. Suppose that there exists a continuous map of $C_p(X)$ onto $C_p(Y)$. Prove that if X is κ-monolithic, then Y is also κ-monolithic.

474. Suppose that there exists a quotient map of $C_p(X)$ onto $C_p(Y)$. Prove that $l^*(Y) \leq l^*(X)$ and $q(Y) \leq q(X)$.

475. Suppose that there exists a quotient map of $C_p(X)$ onto $C_p(Y)$. Prove that if X is $l^*(\kappa)$-monolithic, then Y is also $l^*(\kappa)$-monolithic.

476. Suppose that there exists a continuous open map of $C_p(X)$ onto $C_p(Y)$. Prove that $|Y| \leq |X|$.

477. Suppose that there exists a continuous open map of $C_p(X)$ onto $C_p(Y)$. Prove that if X is κ-scattered, then Y is also κ-scattered.

478. Suppose that there exists a continuous closed map of $C_p(X)$ onto $C_p(Y)$. Prove that if X is κ-stable, then Y is also κ-stable.

479. Give an example of spaces X and Y for which there is a continuous map of $C_p(X)$ onto $C_p(Y)$ while $|Y| > |X|$.

480. Give an example of spaces X and Y for which there is a continuous map of $C_p(X)$ onto $C_p(Y)$ while $l^*(Y) > l^*(X)$.

481. Give an example of spaces X and Y for which there is a continuous map of $C_p(X)$ onto $C_p(Y)$ while $q(Y) > q(X)$.

482. Give an example of spaces X and Y for which there is a continuous map of $C_p(X)$ onto $C_p(Y)$ while X is compact and Y is not σ-compact.

483. Give an example of spaces X and Y for which there is an open continuous map of $C_p(X)$ onto $C_p(Y)$ while $d(Y) > d(X)$.

484. Give an example of spaces X and Y for which there is an open continuous map of $C_p(X)$ onto $C_p(Y)$ while $t_m(Y) > t_m(X)$.

485. Give an example of spaces X and Y for which there is an open continuous map of $C_p(X)$ onto $C_p(Y)$ while $c(Y) > c(X)$ and $p(Y) > p(X)$.

486. Give an example of spaces X and Y for which there is a continuous map of $C_p(X)$ onto $C_p(Y)$ while X is discrete and Y is not discrete.

487. Suppose that there exists a perfect map of $C_p(X)$ onto $C_p(Y)$. Prove that $d(X) = d(Y)$.

488. Suppose that there exists a perfect map of $C_p(X)$ onto $C_p(Y)$. Prove that $nw(X) = nw(Y)$.

489. Suppose that there exists a perfect map of $C_p(X)$ onto $C_p(Y)$. Prove that $|X| = |Y|$.

490. Suppose that there exists a perfect map of $C_p(X)$ onto $C_p(Y)$. Prove that $hd^*(X) = hd^*(Y)$.

491. Can $C_p(\beta\omega \setminus \omega)$ be mapped continuously onto $C_p(\omega_1)$?

492. Suppose that there exists a perfect irreducible map $\varphi : C_p(X) \to \mathbb{R}^\kappa$ for some cardinal κ. Prove that X is discrete.

493. Let $M = \prod_{t \in T} M_t$ where M_t is a metrizable space for all $t \in T$; assume that $\varphi : M \to C_p(X)$ is a closed continuous onto map. Prove that for every $t \in T$, we can choose a closed separable $N_t \subset M_t$ in such a way that $\varphi(\prod_{t \in T} N_t) = C_p(X)$. In particular, a space $C_p(X)$ is a closed continuous image of a product of (completely) metrizable spaces if and only if it is a closed continuous image of a product of separable (completely) metrizable spaces.

494. Let M be a product of completely metrizable spaces. Suppose that there exists a continuous closed onto map $\varphi : M \to C_p(X)$. Prove that X is discrete. In particular, X is discrete if $C_p(X)$ is a closed continuous image of $\mathbb{R}^\kappa$ for some cardinal κ.

495. Let X be a pseudocompact space. Suppose that $C_p(X)$ contains a dense subspace which is a continuous image of a product of separable spaces. Prove that X is compact and metrizable.

496. Let X be a pseudocompact space. Suppose that $C_p(X)$ is a closed continuous image of a product of metrizable spaces. Prove that X is countable.

497. Let X be a pseudocompact space. Suppose that $C_p(X)$ is an open continuous image of a product of separable metrizable spaces. Prove that X is countable.

498. Let M be a product of separable completely metrizable spaces. Assuming that there is a finite-to-one open map $\varphi : M \to C_p(X)$, prove that X is discrete.

499. Let M be a product of separable completely metrizable spaces. Assuming that there is a finite-to-one open map $\varphi : C_p(X) \to M$ prove that X is discrete.

500. Let H be a $G_{\delta\sigma}$-subspace of $\mathbb{R}^\kappa$ for some κ. Prove that if $C_p(X)$ is homeomorphic to a retract of H, then X is discrete. In particular, if $C_p(X)$ is homeomorphic to a retract of $\mathbb{R}^\kappa$ for some cardinal κ, then X is discrete.

1.6   Bibliographic Notes

The main text of Chap. 1 consists of problems of the following types:

(i)   textbook statements which give a gradual development of some topic;
(ii)   folkloric statements that might not be published but are known by specialists;
(iii)   famous theorems cited in textbooks and well-known surveys;
(iv)   comparatively recent results which have practically no presence in textbooks.

We will almost never cite original papers for the results of the first three types. We will cite them for a very small selection of results of the fourth type. This selection is made according to the preferences of the author and *does not mean that all statements of the fourth type will be mentioned.* I bring my apologies to readers who might think that I did not select something more important than what is selected. The point is that such a selection has to be subjective because it is impossible to mention all contributors. As a consequence, *there are many statements which are published as results in papers, but this fact is never mentioned in this book.* There are statements of the main text which constitute published or unpublished results of the author. However, they are treated exactly like the results of others: some are mentioned and some aren't. On the other hand, the bibliography contains (to the best knowledge of the author) the papers and books of *all contributors to the material of this book.*

Section 1.1 contains quite a few nontrivial facts from set theory. This material is best covered in Kunen's book (1980). The general topology part is a group of results (sometimes quite difficult ones) which are present in all major surveys on cardinal invariants. See, e.g., Arhangel'skii (1978), Arhangel'skii and Ponomarev (1974) and Juhász (1980). The C_p-theory part is concerned with the behavior of spread, hereditary Lindelöf number and hereditary density. The main duality theorems (Problems 025–030) were proved in Velichko (1981) and Zenor (1980). The results on perfect normality of $C_p(X)$ (Problems 081–089) were published in Tkachuk (1995). The multiplicativity of countable spread and countable hereditary Lindelöf number under SA was proved in Arhangel'skii (1989b). A highly nontrivial example of a compact strong S-space (Problem 099) was constructed in Todorcevic (1989).

Section 1.2 basically consists of textbook results of Arhangel'skii on monolithity and stability and their development in Tkachuk (1991). The books of Arhangel'skii (1992a) and Bessaga and Pelczinski (1975) give all background material and more. The SA results in Problems 195 and 196 were published by Arhangel'skii in 1996b.

Section 1.3 contains an introduction to Lindelöf Σ-spaces. Most results on this class are folkloric and their formulations are dispersed in many papers and books. However, the author knows no source where they would all be presented with complete proofs and systematized. The introductory material culminates in a famous theorem of Baturov (Problem 269) proved in 1987. We postpone the applications of Baturov's theorem until the next chapter. The investigations of small diagonals were triggered by a paper of Hušek (1977). One of the most important results here is Juhász' theorem (Problem 298) published in Juhász (1992).

Section 1.4 constitutes a brief introduction to descriptive set theory and its applications in C_p-theory. The textbook results can be found in Kuratowski (1966). Problem 354 is, in fact, a very deep result of Pytkeev (1976). The statement of Problem 366 was proved by Christensen (1974). The examples from Problem 372 were constructed in Lutzer et al. (1985). The statement of Problem 395 was proved in Fremlin (1977).

In Sect. 1.5, the results on countable additivity of properties in $C_p(X)$ were proved in Tkachuk (1994). The equivalencies in Problem 465 were proved in Gerlits (1983) and Pytkeev (1992a). The result of Problem 500 was established by Tkachuk in the paper (1994).

Chapter 2
Solutions of Problems 001–500

This chapter brings the reader to the cutting edge of many areas of C_p-theory, so the treatment of topology and C_p-theory is already professional. When you read a solution of a problem of the main text, it has more or less the same level of exposition as a published paper on a similar topic. Recall that we already gave 500 solutions of the problems of the main text and proved a total of more than 200 statements inside the solutions. These inside statements (called facts) are sometimes quite difficult and are often a particular case or the whole of some famous theorems. A reader who mastered this material is more than prepared for reading the most advanced papers in C_p-theory.

The author hopes, however, that reading our solutions is more helpful than ploughing through the proofs in published papers; the reason is that we are not so constrained by the amount of the available space as a journal contributor, so we take much more care about all details of the proof. It is also easier to work with the references in our solutions than with those in research papers because in a paper the author does not need to bother about whether the reference is accessible for the reader whereas we only refer to what we have proved in this book apart from some very simple facts of calculus and set theory.

Another important difference between this chapter and the first one is that we use the textbook facts from general topology without giving a reference to them. This book is self-contained so all necessary results are proved in the first chapter, but the references to standard things have to stop sometime. This makes it difficult for a beginner to read the second chapter results without reading the first one. However, a reader who mastered the material of the first four chapters of Engelking book (1977) will have no problem with this.

We also stopped giving references to some very simple facts of C_p-theory. The reader can easily find the respective proofs using the index. Our reference omission rule can be expressed as follows: we omit references to textbook results from topology *proved in the first volume*. We omit references to some very simple and fundamental facts of C_p-theory also proved in the first volume. We denote the first

V.V. Tkachuk, *A Cp-Theory Problem Book: Special Features of Function Spaces*, Problem Books in Mathematics, DOI 10.1007/978-3-319-04747-8_2, © Springer International Publishing Switzerland 2014

volume by [TFS], so a reference "Problem 225 of [TFS]" says that we refer to the statement of Problem 225 of "A C_p-Theory Problem Book. Topological and Function Spaces".

When we refer to a solution, please, keep in mind that the solutions of this volume are denoted by the letter T. Therefore, "T.337" means "the solution of Problem 337 of this book. The reference "S.118" stands for "solution of Problem 118 of [TFS]".

There are quite a few phrases like "it is easy to see" or "it is an easy exercise"; the reader should trust the author's word and experience that the statements like that are *really* easy to prove as soon as one has the necessary background. On the other hand, the highest percentage of errors comes exactly from omissions of all kinds, so my recommendation is that, even though you should trust the author's claim that the statement is easy to prove or disprove, you shouldn't take just his word for the truthfulness of *any* statement. Verify it yourself and if you find any errors communicate them to me to correct the respective parts.

T.001. *Given an infinite cardinal κ prove that the following properties are equivalent for any space X:*

(i) $hl(X) \le \kappa$;
(ii) $l(X) \le \kappa$ and every $U \in \tau(X)$ is a union of $\le \kappa$-many closed subsets of X;
(iii) $l(X) \le \kappa$ and every closed $F \subset X$ is a G_κ-set in X;
(iv) $l(U) \le \kappa$ for any open $U \subset X$.

In particular, a space X is hereditarily Lindelöf if and only if it is Lindelöf and perfect.

Solution. Say that $H \subset X$ is an F_κ-set if H is a union of $\le \kappa$-many closed subsets of X.

(i)$\Longrightarrow$(ii). Assume that $hl(X) \le \kappa$ and take any $U \in \tau(X)$. For any point $x \in U$ there is $U_x \in \tau(x, X)$ such that $\overline{U}_x \subset U$. Since $l(U) \le \kappa$, the open cover $\{U_x : x \in U\}$ of the space U has a subcover of cardinality $\le \kappa$, i.e., there is $A \subset U$ such that $|A| \le \kappa$ and $\bigcup\{U_x : x \in A\} = U$. Therefore $\bigcup\{\overline{U}_x : x \in A\} = U$ and hence U is an F_κ-set.

It is a matter of passing to complements to see that (ii) $\Longleftrightarrow$ (iii).

(ii)$\Longrightarrow$(iv). If (ii) holds, then take any $U \in \tau(X)$; then U is an F_κ-set, i.e., $U = \bigcup \mathcal{F}$ where $|\mathcal{F}| \le \kappa$ and each $F \in \mathcal{F}$ is closed in X; thus $l(F) \le \kappa$ for any $F \in \mathcal{F}$. It is an easy exercise to see that a union $\le \kappa$-many spaces of Lindelöf number $\le \kappa$ has the Lindelöf number $\le \kappa$ so $l(U) \le \kappa$.

(iv)$\Longrightarrow$(i). Suppose that $l(U) \le \kappa$ for each open $U \subset X$ and take an arbitrary set $Y \subset X$; given any $\gamma \subset \tau(Y)$ with $\bigcup \gamma = Y$, for each $U \in \gamma$ choose $O_U \in \tau(X)$ such that $O_U \cap Y = U$. The set $O = \bigcup\{O_U : U \in \gamma\}$ is open in X and hence $l(O) \le \kappa$; since $\{O_U : U \in \gamma\}$ is an open cover of O, there is $\gamma' \subset \gamma$ such that $|\gamma'| \le \kappa$ and $\bigcup\{O_U : U \in \gamma'\} = O$. It is immediate that $\bigcup \gamma' = Y$ which proves that $l(Y) \le \kappa$.

T.002. *Prove that a space X is hereditarily normal if and only if any open subspace of X is normal.*

Solution. We must only prove that X is hereditarily normal if every open subspace of X is normal. Assume that every $W \in \tau(X)$ is normal; take any $Y \subset X$ and any disjoint sets $F, G \subset Y$ which are closed in Y. It is easy to see that $P = \overline{F} \cap \overline{G} \subset X \backslash Y$ (the bar denotes the closure in X). Therefore the open set $U = X \backslash P$ contains Y. The space U is normal; it is immediate that $F' = \overline{F} \cap U$ and $G' = \overline{G} \cap U$ are disjoint closed subsets of U. Choose any open (in U and hence in X) sets O_1 and O_2 such that $F' \subset O_1$, $G' \subset O_2$ and $O_1 \cap O_2 = \emptyset$. Finally observe that $O_1' = O_1 \cap Y$ and $O_2' = O_2 \cap Y$ are disjoint open subsets of Y such that $F \subset O_1$ and $G \subset O_2$, so Y is normal.

T.003. *Prove that if X is perfectly normal, then any $Y \subset X$ is also perfectly normal.*

Solution. Any open $U \subset X$ must be an F_σ-set in X because X is perfect. Since any F_σ-subset of a normal space is normal (Fact 1 of S.289), the space U is normal. It turns out that any open subset of X is normal. Hence any $Y \subset X$ is normal by Problem 002. Since it is trivial that any subspace of X is perfect, we showed that any $Y \subset X$ is perfectly normal.

T.004. *Let X be any space. Prove that $hd(X) = \sup\{|A| : A$ is a left-separated subspace of $X\}$. In particular, the space X is hereditarily separable if and only if every left-separated subspace of X is countable.*

Solution. If β is an ordinal and $Y = \{y_\alpha : \alpha < \beta\}$ is a set indexed by the set β, say that a well-order $<$ on Y is *induced by the indexation* if $y_\alpha < y_{\alpha'}$ is equivalent to $\alpha < \alpha'$ for any $\alpha, \alpha' < \beta$. We say that a well-order $<$ on a space Y *witnesses left-separation of Y (or left-separates Y, or Y is left-separated by $<$)* if the set $L_x^Y = \{y \in Y : y < x\}$ is closed in Y for any $x \in Y$.

Fact 1. Suppose that Y is a left-separated space with its left-separation witnessed by a well-order $<$. Then any subspace $Z \subset Y$ is also left-separated by the order restricted to Z from Y.

Proof. Denote by $\prec$ the restriction of the order $<$ to Z. Then, for any $x \in Z$, we have $L_x^Z = \{z \in Z : z \prec x\} = L_x^Y \cap Z$. Therefore L_x^Z is closed in Z for any $x \in Z$, so Z is left-separated by $\prec$. Fact 1 is proved.   □

Fact 2. Let κ be a regular infinite cardinal. If Y is a space which has a left-separated subspace of cardinality κ, then there exists a subspace $Z = \{z_\alpha : \alpha < \kappa\} \subset Y$ such that Z is left-separated by its indexation.

Proof. Take any $T \subset Y$ which is left-separated by an order $<$ and $|T| = \kappa$. There is $T' \subset T$ such that $|T'| = \kappa$ and the cardinality of the set $L_x^{T'}$ is less than κ for each $x \in T'$. Indeed, if $|L_x^T| < \kappa$ for each $x \in T$, we can take $T' = T$. If this is not true, then let $x_0 = \min\{y \in T : |L_y^T| = \kappa\}$. It is evident that $T' = L_{x_0}^T$ is as promised.

Now pick $z_0 \in T'$ arbitrarily; assume that $\beta < \kappa$ and we have constructed a set $Z_\beta = \{z_\alpha : \alpha < \beta\} \subset T'$ such that

(*) $\alpha < \alpha'$ implies $z_\alpha < z_{\alpha'}$ whenever $\alpha < \alpha' < \beta$.

If Z_β is cofinal in T', then $T' = \bigcup\{L_x^{T'} : x \in Z_\beta\}$ which is impossible because $|L_x^{T'}| < \kappa$ for each $x \in Z_\beta$ and the cardinal κ is regular. Thus there is $z_\beta \in T'$ such that $z_\alpha < z_\beta$ for all $\alpha < \beta$. It is immediate that $(*)$ holds for all $\alpha \leq \beta$ and hence our inductive construction can be continued to provide a set $Z = \{z_\alpha : \alpha < \kappa\} \subset T'$ with the property $(*)$ for all $\alpha < \kappa$. It is evident that the order $<$ on Z is induced by the indexation of Z. Since the order $<$ left-separates Z by Fact 1, the space Z is left-separated by its indexation, so Fact 2 is proved. $\square$

Returning to our solution assume that $hd(X) \leq \kappa$ and X has a left-separated subspace of cardinality κ^+; by Fact 2 there is $Y = \{x_\alpha : \alpha < \kappa^+\} \subset X$ which is left-separated by its indexation. Since $d(Y) \leq \kappa$, there is $A \subset Y$ with $|A| \leq \kappa$ and $Y \subset \overline{A}$ (the bar denotes the closure in X). There exists an ordinal $\beta < \kappa$ such that $A \subset X_\beta = \{x_\alpha : \alpha < \beta\}$; the set X_β being closed in Y, we have $x_\beta \notin \overline{X}_\beta$, whereas $x_\beta \in \overline{A} \subset \overline{X}_\beta$. This contradiction proves that the cardinality of any left-separated $Y \subset X$ does not exceed $hd(X)$.

Now suppose that $|A| \leq \kappa$ for any left-separated $A \subset X$. Assume that there is $Y \subset X$ such that $d(Y) \geq \kappa^+$ and pick any $y_0 \in Y$. Suppose that $\beta < \kappa^+$ and we have chosen points $\{y_\alpha : \alpha < \beta\}$ in the set Y so that $y_\gamma \notin \overline{\{y_\alpha : \alpha < \gamma\}}$ for each $\gamma < \beta$. Since $d(Y) > \kappa$, the set $\{y_\alpha : \alpha < \beta\}$ cannot be dense in the space Y; choose any $y_\beta \in Y \backslash \overline{\{y_\alpha : \alpha < \beta\}}$. It is clear that this inductive construction gives us a set $A = \{y_\alpha : \alpha < \kappa^+\} \subset X$ such that $y_\beta \notin \overline{\{y_\alpha : \alpha < \beta\}}$ for each $\beta < \kappa^+$. Therefore A is a left-separated (by its indexation) subspace of X with $|A| > \kappa$ which is a contradiction. This proves that $hd(X) \leq \kappa$, so our solution is complete.

T.005. *Let X be any space. Prove that $hl(X) = \sup\{|A| : A$ is a right-separated subspace of $X\}$. In particular, the space X is hereditarily Lindelöf if and only if every right-separated subspace of X is countable.*

Solution. If β is an ordinal and $Y = \{y_\alpha : \alpha < \beta\}$ is a set indexed by the set β, say that a well-order $<$ on Y is *induced by the indexation* if $y_\alpha < y_{\alpha'}$ is equivalent to $\alpha < \alpha'$ for any $\alpha, \alpha' < \beta$. We say that a well-order $<$ on a space Y *witnesses right-separation of Y (or right-separates Y, or Y is right-separated by $<$)* if the set $L_x^Y = \{y \in Y : y < x\}$ is open in Y for any $x \in Y$.

Fact 1. Suppose that Y is a right-separated space with its right-separation witnessed by a well-order $<$. Then any subspace $Z \subset Y$ is also right-separated by the order restricted to Z from Y.

Proof. Denote by $\prec$ the restriction of the order $<$ to Z. Then, for any $x \in Z$, we have $L_x^Z = \{z \in Z : z \prec x\} = L_x^Y \cap Z$. Therefore L_x^Z is open in Z for any $x \in Z$, so Z is right-separated by $\prec$. Fact 1 is proved. $\square$

Fact 2. Let κ be a regular infinite cardinal. If a space Y has a right-separated subspace of cardinality κ, then there exists a subspace $Z = \{z_\alpha : \alpha < \kappa\} \subset Y$ such that Z is right-separated by its indexation.

Proof. Take any $T \subset Y$ which is right-separated by an order $<$ and $|T| = \kappa$. There is $T' \subset T$ such that $|T'| = \kappa$ and the cardinality of the set $L_x^{T'}$ is less than κ for each $x \in T'$. Indeed, if $|L_x^T| < \kappa$ for each $x \in T$, we can take $T' = T$. If this is not true, then let $x_0 = \min\{y \in T : |L_y^T| = \kappa\}$. It is evident that $T' = L_{x_0}^T$ is as promised.

Now pick $z_0 \in T'$ arbitrarily; assume that $\beta < \kappa$ and we have constructed a set $Z_\beta = \{z_\alpha : \alpha < \beta\} \subset T'$ such that

(*) $\alpha < \alpha'$ implies $z_\alpha < z_{\alpha'}$ whenever $\alpha < \alpha' < \beta$.

If Z_β is cofinal in T', then $T' = \bigcup\{L_x^{T'} : x \in Z_\beta\}$ which is impossible because $|L_x^{T'}| < \kappa$ for each $x \in Z_\beta$ and the cardinal κ is regular. Thus there is $z_\beta \in T'$ such that $z_\alpha < z_\beta$ for all $\alpha < \beta$. It is immediate that (*) holds for all $\alpha \leq \beta$ and hence our inductive construction can be continued to provide a set $Z = \{z_\alpha : \alpha < \kappa\} \subset T'$ with the property (*) for all $\alpha < \kappa$. It is evident that the order $<$ on Z is induced by the indexation of Z. Since the order $<$ right-separates Z by Fact 1, the space Z is right-separated by its indexation, so Fact 2 is proved. □

Returning to our solution assume that $hl(X) \leq \kappa$ and X has a right-separated subspace of cardinality κ^+. By Fact 2 we can assume, without loss of generality, that there exists a subset $Y = \{x_\alpha : \alpha < \kappa^+\} \subset X$ right-separated by its indexation. If $U_\alpha = \{x_\beta : \beta < \alpha\}$, then $\mathcal{U} = \{U_\alpha : \alpha < \kappa^+\}$ is an open cover of the space Y. Since $l(Y) \leq \kappa$, there is $A \subset \kappa^+$ with $|A| \leq \kappa$ such that $Y = \bigcup\{U_\alpha : \alpha \in A\}$. There exists an ordinal $\beta < \kappa$ such that $A \subset \beta$; it is immediate that $U_\alpha \subset U_\beta \subset Y\backslash\{y_\beta\}$ for all $\alpha \in A$. Therefore $x_\beta \notin \bigcup\{U_\alpha : \alpha \in A\} \subset U_\beta$ which is a contradiction. This contradiction proves that the cardinality of any right-separated $Y \subset X$ does not exceed $hl(X)$.

Now suppose that $|A| \leq \kappa$ for any right-separated $A \subset X$. Assume that there is $Y \subset X$ such that $l(Y) \geq \kappa^+$ and take any open cover $\mathcal{U}$ of the space Y which has no subcover of cardinality $\leq \kappa$. Pick any $y_0 \in Y$ and any $U_0 \in \mathcal{U}$ for which $y_0 \in U_0$. Suppose that $\beta < \kappa^+$ and we have chosen a set $\{y_\alpha : \alpha < \beta\} \subset Y$ and a family $\{U_\alpha : \alpha < \beta\} \subset \mathcal{U}$ such that $y_\gamma \in U_\gamma\backslash\bigcup\{U_\alpha : \alpha < \gamma\}$ for each $\gamma < \beta$. Since $l(Y) > \kappa$, the set $\{U_\alpha : \alpha < \beta\}$ cannot be a cover of Y; choose any $y_\beta \in Y\backslash\bigcup\{U_\alpha : \alpha < \beta\}$ and any $U_\beta \in \mathcal{U}$ with $y_\beta \in U_\beta$. It is clear that this inductive construction gives us a set $A = \{y_\alpha : \alpha < \kappa^+\} \subset X$ and a family $\{U_\alpha : \alpha < \kappa^+\} \subset \mathcal{U}$ such that $y_\beta \in U_\beta\backslash\bigcup\{U_\alpha : \alpha < \beta\}$ for each $\beta < \kappa^+$. It is easy to see that the set $\{y_\alpha : \alpha < \beta\}$ is open in A for any $\beta < \kappa^+$ and therefore A is a right-separated (by its indexation) subspace of X with $|A| > \kappa$ which is a contradiction. This proves that $hl(X) \leq \kappa$, so our solution is complete.

T.006. *Prove that a space is right-separated if and only if it is scattered.*

Solution. If X is a right-separated space, take any well-order $<$ on X which witnesses this property. Given any $Y \subset X$, let $y \in Y$ be the $<$-minimal point of Y. If $y \in \overline{Y\backslash\{y\}}$, then $Y\backslash\{y\} \neq \emptyset$ so we can choose a $<$-minimal element $x \in Y\backslash\{y\}$. The set $U = \{z \in X : z < x\}$ is open in X and $U \cap Y = \{y\}$ which contradicts the fact that $y \in \overline{Y\backslash\{y\}}$. This proves that y is an isolated point of Y and hence X is scattered.

Given a space Z we denote by $IP(Z)$ the set of isolated points of Z. It is clear that the set $Z \setminus IP(Z)$ is closed in Z. Now assume that X is a non-empty scattered space and let $U_0 = \emptyset$. Suppose that $\beta < \kappa = |X|^+$ and we have chosen $U_\alpha \in \tau(X)$ for all $\alpha < \beta$ so that the following conditions are satisfied:

(1) $U_\alpha \subset U_{\alpha'}$ if $\alpha < \alpha' < \beta$;
(2) if $\gamma < \beta$ is a limit ordinal, then $U_\gamma = \bigcup \{U_\alpha : \alpha < \gamma\}$;
(3) if $\alpha + 1 < \beta$, then $U_{\alpha+1} = U_\alpha \cup IP(X \setminus U_\alpha)$.

If $\bigcup \{U_\alpha : \alpha < \beta\} = X$, then our inductive construction stops. If not, we let $U_\beta = \bigcup \{U_\alpha : \alpha < \beta\}$ if β is a limit ordinal; if $\beta = \gamma + 1$, then let $U_\beta = U_\gamma \cup IP(X \setminus U_\gamma)$. It is worth mentioning that, in case of a successor β, the set U_β is open in X because the set $X \setminus U_\beta = (X \setminus U_\gamma) \setminus IP(X \setminus U_\gamma)$ is closed in $X \setminus U_\gamma$ and hence in X. Observe also that X is scattered so it follows from (3) that $U_{\alpha+1} \setminus U_\alpha \neq \emptyset$ whenever $\alpha + 1 \leq \beta$.

It is evident that the properties (1)–(3) are now satisfied for all $\alpha \leq \beta$ so our inductive construction can be continued as long as the union of chosen sets is not equal to X. The family $\mathcal{U}_\beta = \{U_{\alpha+1} \setminus U_\alpha : \alpha < \beta\}$ is disjoint for any limit ordinal β. Since $\mathcal{U}_\beta$ consists of non-empty sets, our construction cannot involve κ steps because $|X| < \kappa$. Therefore $\bigcup \{U_\alpha : \alpha < \beta\} = X$ for some $\beta < \kappa$.

Given any $x \in X$, let $\alpha(x) = \min\{\alpha < \beta : x \in U_\alpha\}$. It follows from (2) that $\alpha(x)$ is a successor ordinal for all $x \in X$. The properties (1) and (2) imply that the family $\{U_{\alpha+1} \setminus U_\alpha : \alpha + 1 \leq \beta\}$ is a disjoint cover of X; fix a well-order $<_\alpha$ on the set $U_{\alpha+1} \setminus U_\alpha$ for each α with $\alpha + 1 \leq \beta$.

We are now in position to define the required well-order on X. For any distinct $x, y \in X$, let $x < y$ if $\alpha(x) < \alpha(y)$; if $\alpha(x) > \alpha(y)$, then $y < x$. Now, if we have the equality $\alpha(x) = \alpha(y) = \alpha + 1$, then $\{x, y\} \subset A_{\alpha+1} \setminus A_\alpha$; we let $x < y$ if $x <_\alpha y$ and $y < x$ if $y <_\alpha x$. It is routine to verify that $<$ is indeed a well-order on X. Let us show that it witnesses the fact that X is right-separated.

Given any point $x \in X$, there is $\alpha < \beta$ such that $\alpha + 1 = \alpha(x)$. Observe that $W = \{y \in X : y < x\} = U_\alpha \cup P$ where $P = \{y \in U_{\alpha+1} \setminus U_\alpha : y <_\alpha x\}$. Therefore $X \setminus W = (X \setminus U_\alpha) \setminus P$ is a closed set in $X \setminus U_\alpha$ because P consists of isolated points of $X \setminus U_\alpha$ by (3). Any closed subset of $X \setminus U_\alpha$ is closed in X so W is open in X. This proves that X is right-separated.

T.007. *Let X be a left-separated space. Prove that $hl(X) \leq s(X)$. In particular, any left-separated space of countable spread is hereditarily Lindelöf.*

Solution. The following fact is very useful in any consideration where spread is involved.

Fact 1. Given a space Z with $s(Z) \leq \kappa$ and an open cover $\mathcal{U}$ of the space Z, there exists a discrete $D \subset Z$ and a family $\mathcal{U}' \subset \mathcal{U}$ such that $|\mathcal{U}'| \leq \kappa$ and $Z = \overline{D} \cup (\bigcup \mathcal{U}')$.

Proof. Take any $z_0 \in Z$ and any $U_0 \in \mathcal{U}$ with $z_0 \in U_0$; suppose that $\beta < \kappa^+$ and we have chosen a set $\{z_\alpha : \alpha < \beta\} \subset Z$ and a family $\{U_\alpha : \alpha < \beta\} \subset \mathcal{U}$ such that

(*) $z_\gamma \in U_\gamma \setminus (\overline{\{z_\alpha : \alpha < \gamma\}} \cup (\bigcup \{U_\alpha : \alpha < \gamma\}))$ for all $\gamma < \beta$.

Observe that $(*)$ implies that the set $D_\beta = \{z_\alpha : \alpha < \beta\}$ is discrete because $U_\gamma \cap D_\beta = \{z_\gamma\}$ for each $\gamma < \beta$. If $Z_\beta = \overline{\{z_\alpha : \alpha < \gamma\}} \cup (\bigcup\{U_\alpha : \alpha < \gamma\}) \neq Z$, then take any $z_\beta \in Z \backslash Z_\beta$ and $U_\beta \in \mathcal{U}$ with $z_\beta \in U_\beta$. It is immediate that $(*)$ holds for all $\gamma \leq \beta$, so our inductive construction can go on as long as $Z_\beta \neq Z$. If we carry it out for all $\beta < \kappa^+$, then we obtain a discrete set $D_{\kappa^+} \subset Z$ of cardinality κ^+ which contradicts $s(Z) \leq \kappa$. Thus we have $Z_\beta = Z$ for some $\beta < \kappa^+$, so $D = D_\beta$ and $\mathcal{U}' = \{U_\alpha : \alpha < \beta\}$ are as promised. Fact 1 is proved. $\square$

Returning to our solution let $s(X) = \kappa$; fix any order $\prec$ on X which witnesses that X is left-separated. Given any $Z \subset X$ and $z \in Z$, let $L_z^Z = \{y \in Z : y \prec z\}$. To prove that $hl(X) \leq \kappa$ assume the contrary; then $\mathcal{H} = \{H \subset X : l(H) > \kappa\} \neq \emptyset$. We claim that there is $Y \in \mathcal{H}$ such that $l(L_x^Y) \leq \kappa$ for any $x \in Y$. Indeed, take any $H \in \mathcal{H}$ and consider $A = \{z \in H : l(L_z^H) > \kappa\}$. If $A = \emptyset$, then we can take $Y = H$. If $A \neq \emptyset$, then, for the point $y = \min(A)$, the set $Y = L_y^H$ is as promised.

Take any open cover $\mathcal{U}$ of the space Y which has no subcover of cardinality $\leq \kappa$. It follows from our choice of Y that for any $x \in Y$, there is a family $\mathcal{U}_x \subset \mathcal{U}$ such that $|\mathcal{U}_x| \leq \kappa$ and $L_x^Y \subset \bigcup \mathcal{U}_x$. Apply Fact 1 to find a family $\mathcal{U}' \subset \mathcal{U}$ and a discrete set $D \subset Y$ such that $|\mathcal{U}'| \leq \kappa$ and $Y = \bigcup \mathcal{U}' \cup \overline{D}$ (the bar denotes the closure in Y). We have $|D| \leq \kappa$ because $s(Y) \leq s(X) \leq \kappa$. The set D cannot be cofinal in Y; for otherwise, the family $V = \bigcup\{\mathcal{U}_x : x \in D\}$ is a subfamily of $\mathcal{U}$ of cardinality $\leq \kappa$ which covers Y. Thus there is $z \in Y$ such that $D \subset L_z^Y$. Since the order $\prec$ left-separates Y (Fact 1 of T.004), the set L_z^Y is closed in Y so $\overline{D} \subset L_z^Y$. As a consequence, $Y \backslash L_z^Y \subset \bigcup \mathcal{U}'$ and therefore $\mathcal{U}_z \cup \mathcal{U}' \subset \mathcal{U}$ is a subcover of Y of cardinality $\leq \kappa$ which is a contradiction. This shows that $\mathcal{H} = \emptyset$, so our solution is complete.

T.008. *Let X be a right-separated space. Prove that $hd(X) \leq s(X)$. In particular, any right-separated space of countable spread is hereditarily separable.*

Solution. Let $s(X) = \kappa$; suppose that $hd(X) > \kappa$. Then there is a left-separated $Y \subset X$ with $|Y| = \kappa^+$ by Problem 004. Since $s(Y) \leq s(X) = \kappa$ and Y is left-separated, we can apply Problem 007 to conclude that $hl(Y) \leq \kappa$. The space Y is a subspace of a right-separated space X, so Y is right-separated by Fact 1 of T.005; since $|Y| = \kappa^+$, we have $hl(Y) = \kappa^+$ by Problem 005. This contradiction shows that $hd(X) \leq \kappa = s(X)$.

T.009. *Prove that any space has a dense left-separated subspace.*

Solution. Take any space X and fix a well-order $<$ on X. For any $U \in \tau^*(X)$ let x_U be the $<$-minimal point of U. We claim that $Y = \{x_U : U \in \tau^*(X)\}$ is a dense left-separated subspace of X. The density of Y follows from the fact that $x_U \in Y \cap U$ for any $U \in \tau^*(X)$. Now pick any $y \in Y$ and $U \in \tau^*(X)$ such that $y = x_U$. To prove that the set $L_y = \{z \in Y : z < y\}$ is closed in Y take any $x \in Y \backslash L_y$; then $x = x_W$ for some $W \in \tau^*(X)$. Given any $z \in L_y$, we have $z < y \leq x$, so z cannot belong to W because x is the $<$-minimal point of W. This shows that W

is a neighborhood of x with $W \cap L_y = \emptyset$ so $x \notin \overline{L}_y$. The point $x \in Y \backslash L_y$ has been chosen arbitrarily, so L_y is closed in Y for any $y \in Y$. Therefore Y is a dense left-separated subspace of X.

T.010. *Suppose that $s(X) = \omega$. Prove that X has a dense hereditarily Lindelöf subspace.*

Solution. There is a dense left-separated $Y \subset X$. Since $s(Y) \le s(X) \le \omega$, we can apply Problem 007 to conclude that $hl(Y) \le s(Y) = \omega$ and therefore Y is a dense hereditarily Lindelöf subspace of X.

T.011. *Prove that for any space X, we have $hl^*(X) = hl(X^\omega)$. In particular, if all finite powers of X are hereditarily Lindelöf, then X^ω is hereditarily Lindelöf.*

Solution. It is evident that X^n embeds in X^ω for all $n \in \mathbb{N}$, so it suffices to show that $hl(X^\omega) \le hl^*(X)$. For each $n \in \mathbb{N}$ let $\pi_n : X^\omega \to X^n$ be the natural projection defined by $\pi_n(x) = x|n$ for all $x \in X^\omega$ (recall that we identify any ordinal with the set of its predecessors and, in particular, $n = \{0, \dots, n-1\}$). Note that the family $\mathcal{B}_n = \{U_0 \times \cdots \times U_{n-1} \subset X^n : U_i \in \tau(X) : \text{for all } i < n\}$ is a base in X^n for all $n \in \mathbb{N}$. Let $\mathcal{C}_n = \{\pi_n^{-1}(U) : U \in \mathcal{B}_n\}$ for all $n \in \mathbb{N}$; then $\mathcal{C} = \bigcup\{\mathcal{C}_n : n \in \mathbb{N}\}$ is a base in X^ω. $\quad \bullet$

Let $hl^*(X) = \kappa$; suppose that $hl(X^\omega) > \kappa$. Then there is a right-separated subspace of X^ω of cardinality κ^+ (see Problem 005). Apply Fact 2 of T.005 to find a subspace $Y = \{y_\alpha : \alpha < \kappa^+\} \subset X^\omega$ which is right-separated by its indexation. The set $L_\alpha = \{y_\beta : \beta < \alpha + 1\}$ is open in Y for all $\alpha < \kappa^+$; since $\mathcal{C}$ is a base in X^ω, there is $n_\alpha \in \mathbb{N}$ and $U_\alpha \in \mathcal{C}_{n_\alpha}$ such that $y_\alpha \in U_\alpha \cap Y \subset L_\alpha$ for each $\alpha < \kappa^+$. The cardinal κ^+ is regular, so there is $m \in \mathbb{N}$ such that the set $A = \{\alpha < \kappa^+ : n_\alpha = m\}$ has cardinality κ^+; let $Y' = \{y_\alpha : \alpha \in A\}$.

Consider the set $Z = \pi_m(Y') = \{z_\alpha = \pi_m(y_\alpha) : \alpha \in A\} \subset X^m$. For each $\alpha \in A$ take $V_\alpha \in \mathcal{B}_m$ such that $\pi_m^{-1}(V_\alpha) = U_\alpha$. Observe that $z_\alpha \in V_\alpha$ for all $\alpha \in A$; besides, if $\beta > \alpha$ and $\beta \in A$, then $y_\beta \notin U_\alpha$. Since $U_\alpha = \pi_m^{-1}(V_\alpha)$, we have $z_\beta \notin V_\alpha$ for each $\beta > \alpha$ with $\beta \in A$. In particular, the map $\pi_m : Y' \to Z$ is a bijection.

Thus we can define a well-order $\prec$ on Z letting $z_\alpha \prec z_\beta$ if and only if $\alpha < \beta$. Given an arbitrary $z \in Z$, observe that $z = z_\beta$ for some $\beta \in A$ and therefore $Q_z = \{z' \in Z : z' \prec z\} = \{z_\alpha \in Z : \alpha \in A \text{ and } \alpha < \beta\}$. Since $z_\beta \notin V_\alpha$ for each $\alpha < \beta$ with $\alpha \in A$, for the set $G_\beta = \bigcup\{V_\alpha : \alpha \in A \text{ and } \alpha < \beta\}$, we have $G_\beta \cap Z = \{z_\alpha : \alpha \in A \text{ and } \alpha < \beta\} = Q_z$ and hence the set Q_z is open in Z for each $z \in Z$, i.e., Z is a right-separated subspace of X^m. Since $|Z| = \kappa^+$, this contradicts $hl(X^m) \le hl^*(X) = \kappa$ (see Problem 005), so we proved that $hl(X^\omega) \le \kappa = hl^*(X)$ and therefore $hl(X^\omega) = hl^*(X)$.

T.012. *Prove that for any space X, we have $hd^*(X) = hd(X^\omega)$. In particular, if all finite powers of X are hereditarily separable, then X^ω is hereditarily separable.*

Solution. It is evident that X^n embeds in X^ω for all $n \in \mathbb{N}$, so it suffices to show that $hd(X^\omega) \le hd^*(X)$. For each $n \in \mathbb{N}$ let $\pi_n : X^\omega \to X^n$ be the natural projection defined by $\pi_n(x) = x|n$ for all $x \in X^\omega$ (recall that we identify any ordinal with the set of its predecessors and, in particular, $n = \{0, \ldots, n-1\}$). Note that the family $\mathcal{B}_n = \{U_0 \times \cdots \times U_{n-1} \subset X^n : U_i \in \tau(X) : \text{for all } i < n\}$ is a base in X^n for all $n \in \mathbb{N}$. Let $\mathcal{C}_n = \{\pi_n^{-1}(U) : U \in \mathcal{B}_n\}$ for all $n \in \mathbb{N}$; then $\mathcal{C} = \bigcup\{\mathcal{C}_n : n \in \mathbb{N}\}$ is a base in X^ω.

Let $hd^*(X) = \kappa$; suppose that $hd(X^\omega) > \kappa$. Then there is a left-separated subspace of X^ω of cardinality κ^+ (see Problem 004). Apply Fact 2 of T.004 to find a subspace $Y = \{y_\alpha : \alpha < \kappa^+\} \subset X^\omega$ which is left-separated by its indexation. The set $L_\alpha = \{y_\beta : \beta < \alpha\}$ is closed in Y for all $\alpha < \kappa^+$; since $\mathcal{C}$ is a base in X^ω, there is $n_\alpha \in \mathbb{N}$ and $U_\alpha \in \mathcal{C}_{n_\alpha}$ such that $y_\alpha \in U_\alpha \cap Y \subset Y \backslash L_\alpha$ for each $\alpha < \kappa^+$. The cardinal κ^+ is regular, so there is $m \in \mathbb{N}$ such that the set $A = \{\alpha < \kappa^+ : n_\alpha = m\}$ has cardinality κ^+; let $Y' = \{y_\alpha : \alpha \in A\}$.

Consider the set $Z = \pi_m(Y') = \{z_\alpha = \pi_m(y_\alpha) : \alpha \in A\} \subset X^m$. For each $\alpha \in A$ take $V_\alpha \in \mathcal{B}_m$ such that $\pi_m^{-1}(V_\alpha) = U_\alpha$. Observe that $z_\alpha \in V_\alpha$ for all $\alpha \in A$; besides, if $\beta > \alpha$ and $\beta \in A$, then $y_\beta \notin U_\beta$. Since $U_\beta = \pi_m^{-1}(V_\beta)$, we have $z_\alpha \notin V_\beta$ for each $\beta > \alpha$ with $\beta \in A$. In particular, the map $\pi_m : Y' \to Z$ is a bijection.

Thus we can define a well-order $\prec$ on Z letting $z_\alpha \prec z_\beta$ if and only if $\alpha < \beta$. Given an arbitrary $z \in Z$, observe that $z = z_\beta$ for some $\beta \in A$ and therefore $Q_z = \{z' \in Z : z' \prec z\} = \{z_\alpha \in Z : \alpha \in A \text{ and } \alpha < \beta\}$. Since $z_\alpha \notin V_\beta$ for each $\alpha < \beta$ with $\alpha \in A$, for the set $G_\beta = \bigcup\{V_\alpha : \alpha \in A \text{ and } \alpha \geq \beta\}$, we have $G_\beta \cap Z = \{z_\alpha : \alpha \in A \text{ and } \alpha \geq \beta\} = Y \backslash Q_z$, and hence the set $Y \backslash Q_z$ is open in Z for each $z \in Z$. As a consequence, Q_z is closed in Z for all $z \in Z$, so Z is a left-separated subspace of X^m. Since $|Z| = \kappa^+$, this contradicts $hd(X^m) \leq hd^*(X) = \kappa$ (see Problem 004), so we proved that $hd(X^\omega) \leq \kappa = hd^*(X)$ and therefore $hd(X^\omega) = hd^*(X)$.

T.013. *Prove that for any space X, we have $s^*(X) = s(X^\omega)$.*

Solution. It is evident that X^n embeds in X^ω for all $n \in \mathbb{N}$, so it suffices to show that $s(X^\omega) \leq s^*(X)$. For each $n \in \mathbb{N}$ let $\pi_n : X^\omega \to X^n$ be the natural projection defined by $\pi_n(x) = x|n$ for all $x \in X^\omega$ (recall that we identify any ordinal with the set of its predecessors and, in particular, $n = \{0, \ldots, n-1\}$). Note that the family $\mathcal{B}_n = \{U_0 \times \cdots \times U_{n-1} \subset X^n : U_i \in \tau(X) : \text{for all } i < n\}$ is a base in X^n for all $n \in \mathbb{N}$. Let $\mathcal{C}_n = \{\pi_n^{-1}(U) : U \in \mathcal{B}_n\}$ for all $n \in \mathbb{N}$; then $\mathcal{C} = \bigcup\{\mathcal{C}_n : n \in \mathbb{N}\}$ is a base in X^ω.

Let $s^*(X) = \kappa$; suppose that $s(X^\omega) > \kappa$. Then there is a discrete subspace $Y = \{y_\alpha : \alpha < \kappa^+\} \subset X^\omega$. Since $\mathcal{C}$ is a base in X^ω, there is $n_\alpha \in \mathbb{N}$ and $U_\alpha \in \mathcal{C}_{n_\alpha}$ such that $U_\alpha \cap Y = \{y_\alpha\}$ for each $\alpha < \kappa^+$. The cardinal κ^+ is regular, so there is $m \in \mathbb{N}$ such that the set $A = \{\alpha < \kappa^+ : n_\alpha = m\}$ has cardinality κ^+; let $Y' = \{y_\alpha : \alpha \in A\}$.

Consider the set $Z = \pi_m(Y') = \{z_\alpha = \pi_m(y_\alpha) : \alpha \in A\} \subset X^m$. For each $\alpha \in A$ take $V_\alpha \in \mathcal{B}_m$ such that $\pi_m^{-1}(V_\alpha) = U_\alpha$. Observe that $z_\alpha \in V_\alpha$ for all $\alpha \in A$; besides, if $\beta \neq \alpha$ and $\beta \in A$, then $y_\beta \notin U_\alpha$. Since $U_\alpha = \pi_m^{-1}(V_\alpha)$, we have $z_\beta \notin V_\alpha$ for each $\beta \neq \alpha$ with $\beta \in A$. In particular, the map $\pi_m : Y' \to Z$ is a bijection. Besides, $z_\beta \notin V_\alpha$ for each $\beta \neq \alpha$ implies that $V_\alpha \cap Z = \{z_\alpha\}$ for each $\alpha \in A$ and hence Z is a discrete subspace of X^m. Since $|Z| = \kappa^+$, this contradicts $s(X^m) \leq s^*(X) = \kappa$, so we proved that $s(X^\omega) \leq \kappa = s^*(X)$ and therefore $s(X^\omega) = s^*(X)$.

T.014. *Suppose that* $s(X \times X) \leq \kappa$. *Prove that* $hl(X) \leq \kappa$ *or* $hd(X) \leq \kappa$. *In particular, if* $s(X \times X) = \omega$, *then* X *is hereditarily separable or hereditarily Lindelöf.*

Solution. Assume that $hd(X) > \kappa$ and $hl(X) > \kappa$. It follows from $hd(X) > \kappa$ that there exists a left-separated subspace of X of cardinality κ^+ (see Problem 004). Apply Fact 2 of T.004 to find a subspace $Y = \{y_\alpha : \alpha < \kappa^+\} \subset X$ which is left-separated by its indexation. Analogously, $hl(X) > \kappa$ implies that there exists a subspace $Z = \{z_\alpha : \alpha < \kappa^+\} \subset X$ which is right-separated by its indexation (see Fact 2 of T.005).

Let $D = \{(y_\alpha, z_\alpha) : \alpha < \kappa\} \subset X \times X$; for any $\alpha < \kappa^+$ the set $Y_\alpha = \{y_\beta : \beta < \alpha\}$ is closed in Y, so there exists $U_\alpha \in \tau(y_\alpha, X)$ such that $U_\alpha \cap Y_\alpha = \emptyset$. Analogously, the set $\{z_\beta : \beta < \alpha + 1\}$ is open in Z so the set $Z_\alpha = \{z_\beta : \alpha < \beta\}$ is closed in Z and therefore there exists $V_\alpha \in \tau(z_\alpha, X)$ such that $V_\alpha \cap Z_\alpha = \emptyset$. As a consequence, $W_\alpha = U_\alpha \times V_\alpha \in \tau((y_\alpha, z_\alpha), X \times X)$ and $W_\alpha \cap D = \{(y_\alpha, z_\alpha)\}$ for each $\alpha < \kappa^+$. This shows that D is a discrete subspace of $X \times X$; since $|D| = \kappa^+$, this is a contradiction with $s(X \times X) \leq \kappa$. Thus it is impossible that $hd(X) > \kappa$ and $hl(X) > \kappa$, so we have $hd(X) \leq \kappa$ or $hl(X) \leq \kappa$.

T.015. *Prove that* $|X| \leq 2^{hl(X)}$ *for any space* X. *In particular, any hereditarily Lindelöf space has cardinality* $\leq \mathfrak{c}$.

Solution. Given any set A, denote by $P(A, \kappa)$ the family of all subsets of A of cardinality $\leq \kappa$. Then $P(P(A, \kappa), \kappa)$ is the collection of all families of cardinality $\leq \kappa$ of subsets of A of cardinality $\leq \kappa$ each. If Z is an arbitrary space and we have families $\mathcal{V} \subset \tau(Z)$ and $\mathcal{D} \subset \exp(Z)$, say that the pair $(\mathcal{V}, \mathcal{D})$ is *incomplete* if $O(\mathcal{V}, \mathcal{D}) = (\bigcup \mathcal{V}) \cup (\bigcup \{\overline{D} : D \in \mathcal{D}\}) \neq Z$.

Fact 1. For any space Z, we have $|Z| \leq 2^{\psi(Z) \cdot s(Z)}$.

Proof. Let $\psi(Z) \cdot s(Z) = \kappa$. For any $z \in Z$ fix a family $\mathcal{V}_z \subset \tau(z, Z)$ such that $|\mathcal{V}_z| \leq \kappa$ and $\bigcap \mathcal{V}_z = \{z\}$. Take any $z_0 \in Z$ and let $A_0 = \{z_0\}$. Suppose that $\beta < \kappa^+$ and we have constructed a family $\{A_\alpha : \alpha < \beta\}$ of subsets of Z with the following properties:

(1) $|A_\alpha| \leq 2^\kappa$ for each $\alpha < \beta$;
(2) $A_\alpha \subset A_\gamma$ whenever $\alpha < \gamma < \beta$;
(3) for any $\gamma < \beta$, if $A'_\gamma = \bigcup \{A_\alpha : \alpha < \gamma\}$ and $\mathcal{C}_\gamma = \bigcup \{\mathcal{V}_z : z \in A'_\gamma\}$, then for any families $\mathcal{D} \in P(P(A'_\gamma, \kappa), \kappa)$ and $\mathcal{V} \in P(\mathcal{C}_\gamma, \kappa)$, such that the pair $(\mathcal{V}, \mathcal{D})$ is incomplete, we have $A_\gamma \backslash O(\mathcal{V}, \mathcal{D}) \neq \emptyset$.

To construct A_β, let $A'_\beta = \bigcup \{A_\alpha : \alpha < \beta\}$ and $\mathcal{C}_\beta = \bigcup \{\mathcal{V}_z : z \in A'_\beta\}$. Observe that we have $|P(P(A'_\beta, \kappa), \kappa)| \leq 2^\kappa$ and $|\mathcal{C}_\beta| \leq 2^\kappa$ because $|P(B, \kappa)| = 2^\kappa$ for any set B of cardinality 2^κ. For any family $\mathcal{V} \in P(\mathcal{C}_\beta, \kappa)$ and any $\mathcal{D} \in P(P(A'_\beta, \kappa), \kappa)$ such that the pair $(\mathcal{V}, \mathcal{D})$ is incomplete, choose a point $p(\mathcal{V}, \mathcal{D}) \in Z \backslash O(\mathcal{V}, \mathcal{D})$. Now, let $A_\beta = A'_\beta \cup \{p(\mathcal{V}, \mathcal{D}) : \mathcal{D} \in P(P(A'_\beta, \kappa), \kappa), \mathcal{V} \in P(\mathcal{C}_\beta, \kappa)$ and the pair $(\mathcal{V}, \mathcal{D})$ is incomplete$\}$. It is immediate that the properties (1)–(3) hold for the family $\{A_\alpha : \alpha \leq \beta\}$, so we can continue our inductive construction to obtain a family $\mathcal{A} = \{A_\alpha : \alpha < \kappa^+\}$ with the properties (1)–(3).

Assume first that $A = \bigcup \mathcal{A} \neq Z$ and fix any $t \in Z \backslash A$. For any $U \in \mathcal{V}_t$, and any $z \in A \backslash U$, choose $V_z \in \mathcal{V}_z$ with $t \notin V_z$; the family $\mathcal{H} = \bigcup \{V_z : z \in A \backslash U\}$ is an open cover of $A \backslash U$, so we can apply Fact 1 of T.007 to find a family $\mathcal{H}_U \subset \mathcal{H}$ and a set $D_U \subset A \backslash U$ such that $|\mathcal{H}_U| \leq \kappa$, the set D_U is discrete (and hence $|D_U| \leq \kappa$ as well) and $(\bigcup \mathcal{H}_U) \cup \overline{D}_U \supset A \backslash U$. Since $A = \bigcup \{A \backslash U : U \in \mathcal{V}_t\}$, we have $A \subset O(\mathcal{V}, \mathcal{D})$ where $\mathcal{V} = \bigcup \{\mathcal{H}_U : U \in \mathcal{V}_t\}$ and $\mathcal{D} = \{D_U : U \in \mathcal{V}_t\}$. Let $\mathcal{W} = \bigcup \{V_z : z \in A\}$; since $|\mathcal{V}| \leq \kappa$, $|\mathcal{D}| \leq \kappa$ and $\mathcal{V} \subset \mathcal{W}$, there exists $\beta < \kappa$ such that $\mathcal{D} \subset P(P(A'_\beta, \kappa), \kappa)$ and $\mathcal{V} \in P(C_\beta, \kappa)$. Observe that $\overline{D}_U \subset Z \backslash U$ for each $U \in \mathcal{V}_t$ and therefore $t \notin \overline{D}_U$ for all $U \in \mathcal{V}_t$. In addition, $t \notin \bigcup \mathcal{H}_U$ for each $U \in \mathcal{V}_t$ which implies that $t \notin \bigcup \mathcal{V}$. Thus $t \notin (\bigcup \mathcal{V}) \cup (\bigcup \{\overline{D}_U : U \in \mathcal{V}_t\}) = O(\mathcal{V}, \mathcal{D})$ which shows that the pair $(\mathcal{V}, \mathcal{D})$ is incomplete and therefore $p = p(\mathcal{V}, \mathcal{D}) \in A_\beta$. However, $A_\beta \subset A \subset O(\mathcal{V}, \mathcal{D})$ so $p \in Z \backslash O(\mathcal{V}, \mathcal{D}) \subset Z \backslash A$; this contradiction shows that $A = Z$ and hence $|Z| \leq \kappa^+ \cdot 2^\kappa = 2^\kappa = 2^{\psi(X) \cdot s(X)}$, so Fact 1 is proved. $\square$

Finally, observe that $s(X) \leq hl(X)$ and $\psi(X) \leq hl(X)$, so we can apply Fact 1 to conclude that $|X| \leq 2^{s(X) \cdot \psi(X)} \leq 2^{hl(X) \cdot hl(X)} = 2^{hl(X)}$, so our solution is complete.

T.016. *Prove that* $s(X \times X) \leq s(C_p(X)) \leq s^*(X)$ *for any space* X.

Solution. Given points $x_1, \ldots, x_n \in X$ and $O_1, \ldots, O_n \in \tau^*(\mathbb{R})$, recall that the set $[x_1, \ldots, x_n; O_1, \ldots, O_n] = \{f \in C_p(X) : f(x_i) \in O_i \text{ for all } i \leq n\}$ is called *a standard open subset of* $C_p(X)$. Standard open sets $[x_1, \ldots, x_n; O_1, \ldots, O_n]$ where $n \in \mathbb{N}$, $x_1, \ldots, x_n \in X$ and $O_1, \ldots, O_n$ are rational intervals form a base in $C_p(X)$ (see Problem 056 of [TFS]).

Let $s(C_p(X)) = \kappa$; we will prove first that $s(X) \leq \kappa$. Assuming the contrary we can find a discrete subspace $D \subset X$ with $|D| = \kappa^+$. For each $d \in D$ choose $U_d \in \tau(d, X)$ such that $U \cap D = \{d\}$ and a function $f_d \in C(X, [0, 1])$ such that $f_d(d) = 1$ and $f_d|(X \backslash U_d) \equiv 0$. Let $O_d = \{f \in C_p(X) : f(d) > 0\}$; it is clear that O_d is an open subset of $C_p(X)$ and $f_d \in O_d$ for each $d \in D$. The set $F = \{f_d : d \in D\}$ is discrete; to see this, take any $a \in D \backslash \{d\}$. Then $d \notin U_a$ and hence $f_a(d) = 0$ which shows that $f_a \notin O_d$. This implies $O_d \cap F = \{f_d\}$ for each $d \in D$ and therefore the correspondence $d \to f_d$ is a bijection between D and F. As a consequence, F is a discrete subspace of $C_p(X)$ of cardinality κ^+; this contradiction with $s(C_p(X)) \leq \kappa$ shows that $s(X) \leq \kappa$.

Now assume that $s(X \times X) > \kappa$ and fix a discrete subspace $E \subset X \times X$ with $|E| = \kappa^+$. The set $\Delta = \{(x, x) : x \in X\}$ is a subspace of $X \times X$ homeomorphic to X so $s(\Delta) \leq \kappa$. Thus $|E \cap \Delta| \leq \kappa$ and hence $|E \backslash \Delta| = \kappa^+$, i.e., we have a discrete subspace $D = E \backslash \Delta \subset (X \times X) \backslash \Delta$ of cardinality κ^+. For each $d = (x, y) \in D$ there exist $U_d \in \tau(x, X)$ and $V_d \in \tau(y, X)$ such that $(U_d \times V_d) \cap D = \{d\}$. Since $x \neq y$, we lose no generality assuming that $U_d \cap V_d = \emptyset$ for each $d \in D$. Take any functions $g_d, h_d \in C(X, [0, 1])$ such that $g_d(x) = h_d(y) = 1$, $g_d|(X \backslash U_d) \equiv 0$, $h_d|(X \backslash V_d) \equiv 0$ and let $f_d = g_d - h_d$. Then $O_d = \{f \in C_p(X) : f(x) > 0$ and $f(y) < 0\}$ is an open subset of $C_p(X)$ and $f_d \in O_d$ for all $d \in D$.

We claim that the set $F = \{f_d : d \in D\}$ is discrete; to see this, assume that $d = (x, y) \in D$ and $a = (s, t) \in D\backslash\{d\}$. Since $d \notin U_a \times V_a$, we have $x \notin U_a$ or $y \notin V_a$. In the first case we have $f_a(x) \leq 0$ and in the second one we obtain $f_a(y) \geq 0$; thus, in both cases $f_a \notin O_d$ and therefore $O_d \cap D = \{d\}$ and, in particular, the correspondence $d \to f_d$ is a bijection between D and F. As a consequence, F is a discrete subspace of $C_p(X)$ of cardinality κ^+; this contradiction with $s(C_p(X)) \leq \kappa$ shows that $s(X \times X) \leq \kappa$ and therefore $s(X \times X) \leq s(C_p(X))$.

To prove that $s(C_p(X)) \leq s^*(X)$ assume that $s^*(X) \leq \kappa < s(C_p(X))$ and fix a discrete subspace $F \subset C_p(X)$ with $|F| = \kappa^+$. For each $f \in F$, there exists open standard set $O_f \in \tau(f, C_p(X))$ such that $O_f \cap F = \{f\}$. We have $O_f = [x_1^f, \ldots, x_{n_f}^f; O_1^f, \ldots, O_{n_f}^f]$ where $x_1^f, \ldots, x_{n_f}^f$ are distinct points of X and $O_1^f \ldots, O_{n_f}^f$ are rational intervals. Since there are only countably many of all possible n-tuples of rational intervals for any $n \in \mathbb{N}$, there exist $n \in \mathbb{N}$, a set $G \subset F$ and rational intervals $O_1, \ldots, O_n$ such that $|G| = \kappa^+$, $n_f = n$, $O_i^f = O_i$, $i = 1, \ldots, n$ for each $f \in G$. Let $x_f = (x_1^f, \ldots, x_n^f) \in X^n$ for all $f \in G$. For the set $O(f) = f^{-1}(O_1) \times \cdots \times f^{-1}(O_n)$ we have $f \in O(f)$ for all $f \in G$. We claim that the set $D = \{x_f : f \in G\} \subset X^n$ is discrete.

To show this, pick any $f \in G$; if $g \in G\backslash\{f\}$, then $f \notin O_g$ so $f(x_i^g) \notin O_i$ for some $i \leq n$. This implies $x_i^g \notin f^{-1}(O_i)$ and therefore $x_g \notin O(f)$. An evident consequence is that $O(f) \cap D = \{x_f\}$. Consequently, the subspace D is discrete and the correspondence $f \to x_f$ is a bijection between G and D. This gives us a discrete $D \subset X^n$ of cardinality κ^+ which contradicts $s(X^n) \leq s^*(X) = \kappa$. We established that $s(C_p(X)) \leq \kappa = s^*(X)$, so our solution is complete.

T.017. *Prove that $hd(X \times X) \leq hl(C_p(X)) \leq hd^*(X)$ for any space X.*

Solution. Given points $x_1, \ldots, x_n \in X$ and sets $O_1, \ldots, O_n \in \tau^*(\mathbb{R})$, the set

$$[x_1, \ldots, x_n; O_1, \ldots, O_n] = \{f \in C_p(X) : f(x_i) \in O_i \text{ for all } i \leq n\}$$

is called *a standard open subset of $C_p(X)$*. Standard open sets $[x_1, \ldots, x_n; O_1, \ldots, O_n]$ where $n \in \mathbb{N}$, $x_1, \ldots, x_n \in X$ and $O_1, \ldots, O_n$ are rational intervals form a base in $C_p(X)$ (see Problem 056 of [TFS]).

Let $hl(C_p(X)) = \kappa$; we will prove first that $hd(X) \leq \kappa$. Assuming the contrary we can find a left-separated subspace $D \subset X$ with $|D| = \kappa^+$ (see Problem 004). Denote by $<$ an order that left-separates D; for any point $d \in D$, consider the set $D_d = \{z \in D : z < d\}$. For each $d \in D$ the set D_d is closed in D and $d \notin D_d$, so we can choose a set $U_d \in \tau(d, X)$ with $U_d \cap D_d = \emptyset$ and a function $f_d \in C(X, [0, 1])$ such that $f_d(d) = 1$ and $f_d|(X\backslash U_d) \equiv 0$. Let $O_d = \{f \in C_p(X) : f(d) > 0\}$; it is clear that O_d is an open subset of $C_p(X)$ and $f_d \in O_d$ for each $d \in D$. Take any $a \in D\backslash\{d\}$; if $a < d$, then $a \notin U_d$ so $f_d(a) = 0 \neq f_a(a) = 1$. Therefore $f_d \neq f_a$. If $d < a$, then $d \notin U_a$, and hence $f_a(d) = 0 \neq f_d(d) = 1$ which shows again that $f_a \neq f_d$. Consequently, the correspondence $d \to f_d$ is a bijection, so we can define a well-order $\prec$ on the set $F = \{f_d : d \in D\}$ by letting $f_d \prec f_a$ if and only if $d < a$.

We claim that the well-order $\prec$ right-separates F; to see this, take any $f \in F$. To prove that the set $F_f = \{g \in F : g \prec f\}$ is open in F, take any $g \prec f$. There is $a \in D$ such that $g = f_a$; given any $f' \in F \backslash F_f$, we have $f' = f_d$ for some $d \in D$. Since $f' \notin F_f$, we have $a < d$. It follows from $a \notin U_d$ that we have $f_d(a) = 0$ whence $f' = f_d \notin O_a$. The point $f' \in F \backslash F_f$ was taken arbitrarily, so we established that $O_a \cap F \subset F_f$. It is evident that $g \in O_a$, so every $g = f_a \in F_f$ has a neighborhood $W(g) = O_a$ such that $W(g) \cap F \subset F_f$. Thus F_f is open in F for each $f \in F$ and therefore F is a right-separated subspace of $C_p(X)$. Since $|F| = |D| = \kappa^+$, this is a contradiction with $hl(C_p(X)) \leq \kappa$ (see Problem 005) which shows that $hd(X) \leq \kappa$.

Now assume that $hd(X \times X) > \kappa$ and fix a left-separated subspace $E \subset X \times X$ with $|E| = \kappa^+$. The set $\Delta = \{(x, x) : x \in X\}$ is a subspace of $X \times X$ homeomorphic to X so $hd(\Delta) \leq \kappa$. Thus $|E \cap \Delta| \leq \kappa$ and hence $|E \backslash \Delta| = \kappa^+$, i.e., we have a left-separated subspace $D = E \backslash \Delta \subset (X \times X) \backslash \Delta$ of cardinality κ^+. Denote by $<$ an order that left-separates D and let $D_z = \{d \in D : d < z\}$ for any $z \in D$. For each point $d = (x, y) \in D$ there exist $U_d \in \tau(x, X)$ and $V_d \in \tau(y, X)$ such that $(U_d \times V_d) \cap D_d = \emptyset$. Since $x \neq y$, we lose no generality assuming that $U_d \cap V_d = \emptyset$ for every point $d \in D$. Take any functions $g_d, h_d \in C(X, [0, 1])$ such that $g_d(x) = h_d(y) = 1$, $g_d|(X \backslash U_d) \equiv 0$, $h_d|(X \backslash V_d) \equiv 0$ and let $f_d = g_d - h_d$. Then $O_d = \{f \in C_p(X) : f(x) > 0$ and $f(y) < 0\}$ is an open subset of $C_p(X)$ and $f_d \in O_d$ for all $d \in D$.

If $a, d \in D$ and $a \neq d$, we can assume, without loss of generality that $a < d$. If $a = (x', y')$ and $d = (x, y)$, then $a \notin U_d \times V_d$ which implies $x' \notin U_d$ or $y' \notin V_d$. In the first case we have $f_d(x') \leq 0$ while $f_a(x') = 1 > 0$; in the second case we obtain $f_d(y') \geq 0$ while $f_a(y') = -1 < 0$. Therefore in both cases we have $f_a \neq f_d$ which proves that the correspondence $d \to f_d$ is a bijection. This makes it possible to define a well-order $\prec$ on the set $F = \{f_d : d \in D\}$ by letting $f_a \prec f_d$ if and only if $a < d$.

We prove next that the well-order $\prec$ right-separates F. Take any $f \in F$; to prove that the set $F_f = \{g \in F : g \prec f\}$ is open in F, take any $g \prec f$. There is $a = (x, y) \in D$ such that $g = f_a$; given any $f' \in F \backslash F_f$, we have $f' = f_d$ for some $d = (x', y') \in D$ with $a < d$. It follows from $a \notin U_d \times V_d$ that $x \notin U_d$ or $y \notin V_d$; in the first case we have $f_d(x) \leq 0$ and in the second one we obtain $f_d(y) \geq 0$ so in both cases $f' = f_d \notin O_a$.

The point $f' \in F \backslash F_f$ was taken arbitrarily, so we established that $O_a \cap F \subset F_f$ for any $g = f_a \in F_f$. Thus any $g = f_a \in F_f$ has a neighborhood $W(g) = O_a$ such that $W(g) \cap F \subset F_f$; this proves that F_f is open in F for each $f \in F$ and therefore F is a right-separated subspace of $C_p(X)$. Since $|F| = |D| = \kappa^+$, this is a contradiction with $hl(C_p(X)) \leq \kappa$ (see Problem 005) which shows that $hd(X \times X) \leq \kappa = hl(C_p(X))$.

To prove that $hl(C_p(X)) \leq hd^*(X)$ assume that $hd^*(X) \leq \kappa < hl(C_p(X))$ and fix a right-separated subspace $F \subset C_p(X)$ with $|F| = \kappa^+$. Let $<$ be the well-order that right-separates F. For each $g \in F$, the set $F_g = \{f \in F : g < f\}$ is closed in F. This is evident if $F_g = \emptyset$; if not, then for the function $g' = \min F_g$ we have $F_g = F \backslash \{f \in F : f < g'\}$, so F_g is closed in F being the complement of the set $\{f \in F : f < g'\}$ which is open in F because F is right-separated.

Therefore, for each $f \in F$, there exists a standard open set $O_f \in \tau(f, C_p(X))$ such that $O_f \cap F_f = \emptyset$. We have $O_f = [x_1^f, \ldots, x_{n_f}^f; O_1^f, \ldots, O_{n_f}^f]$ where $x_1^f, \ldots, x_{n_f}^f$ are distinct points of X and $O_1^f, \ldots, O_{n_f}^f$ are rational intervals. Since there are only countably many of all possible n-tuples of rational intervals for any $n \in \mathbb{N}$ and κ^+ is a regular uncountable cardinal, there exist $n \in \mathbb{N}$, a set $G \subset F$ and rational intervals $O_1, \ldots, O_n$ such that $|G| = \kappa^+$, $n_f = n$, $O_i^f = O_i$, $i = 1, \ldots, n$ for each $f \in G$. Let $x_f = (x_1^f, \ldots, x_n^f) \in X^n$ for all $f \in G$. For the set $O(f) = f^{-1}(O_1) \times \cdots \times f^{-1}(O_n)$, we have $x_f \in O(f)$ for all $f \in G$. We claim that the set $D = \{x_f : f \in G\} \subset X^n$ is left-separated.

To show this, pick any $f, g \in G$ such that $g < f$. Since $f \in F_g$, we have $f \notin O_g$, so $f(x_i^g) \notin O_i$ for some $i \leq n$. This implies $x_i^g \notin f^{-1}(O_i)$ and therefore $x_g \notin O(f)$. Thus we have proved that

(*) for any $f, g \in G$ such that $g < f$, we have $x_f \in O(f)$ and $x_g \notin O(f)$.

An evident consequence is that $x_f \neq x_g$ for any distinct $f, g \in G$, so the correspondence $f \to x_f$ is a bijection between G and D. This makes it possible to well-order the set D by letting $x_f \prec x_g$ if and only if $f < g$. Fix $x \in D$; we must prove that the set $D_x = \{y \in D : y \prec x\}$ is closed in D. Take any $y \in D_x$; there are $f, g \in G$ such that $x = x_f$ and $y = x_g$. Since $y \prec x$, we have $g < f$ and hence $x_g \notin O(f)$ by (*). This shows that every $x = x_f \in D$ has a neighborhood $W_x = O(f)$ such that $W_x \cap D_x = \emptyset$; an easy consequence is that D_x is closed in D for any $x \in D$, so D is left-separated by the well-order $\prec$. However, $|D| = |G| = \kappa^+$ and $G \subset X^n$ which contradicts $hd(X^n) \leq hd^*(X) \leq \kappa$. Thus $hl(C_p(X)) \leq \kappa = hd^*(X)$, so our solution is complete.

T.018. *Prove that $hl(X \times X) \leq hd(C_p(X)) \leq hl^*(X)$ for any space X.*

Solution. Given points $x_1, \ldots, x_n \in X$ and sets $O_1, \ldots, O_n \in \tau^*(\mathbb{R})$, the set

$$[x_1, \ldots, x_n; O_1, \ldots, O_n] = \{f \in C_p(X) : f(x_i) \in O_i \text{ for all } i \leq n\}$$

is called *a standard open subset of* $C_p(X)$. Standard open sets $[x_1, \ldots, x_n; O_1, \ldots, O_n]$ where $n \in \mathbb{N}$, $x_1, \ldots, x_n \in X$ and $O_1, \ldots, O_n$ are rational intervals form a base in $C_p(X)$ (see Problem 056 of [TFS]).

Let $hd(C_p(X)) = \kappa$; we will prove first that $hl(X) \leq \kappa$. Assuming the contrary we can find a right-separated subspace $D \subset X$ with $|D| = \kappa^+$ (see Problem 005). Denote by $<$ an order that right-separates D; for any $d \in D$, let $D_d = \{z \in D : d < z\}$. For each $d \in D$ the set D_d is closed in D. This is evident if $D_d = \emptyset$; if not, then $d' = \min D_d$ is well-defined and the set $D_d = D \setminus \{z \in D : z < d'\}$ is closed being the complement of the set $\{z \in D : z < d'\}$ which is open in D because D is right-separated.

Since $d \notin D_d$ we can choose a set $U_d \in \tau(d, X)$ with $U_d \cap D_d = \emptyset$ and a function $f_d \in C(X, [0, 1])$ such that $f_d(d) = 1$ and $f_d|(X \setminus U_d) \equiv 0$. Consider the set $O_d = \{f \in C_p(X) : f(d) > 0\}$; it is clear that O_d is an open subset of $C_p(X)$ and $f_d \in O_d$ for each $d \in D$. Take any distinct points $a, d \in D$; without loss of generality we can assume that $a < d$. Then $d \notin U_a$ so $f_a(d) = 0 \neq f_d(d) = 1$. This proves that $f_d \neq f_a$ and therefore the correspondence $d \to f_d$ is a bijection. Thus we can define a well-order $\prec$ on the set $F = \{f_d : d \in D\}$ by letting $f_d \prec f_a$ if and only if $d < a$. Given any $f \in F$, take any $g \prec f$. There are $a, d \in D$ such that $g = f_a$ and $f = f_d$. Since $g \prec f$, we have $a < d$. It follows from $d \notin U_a$ that we have $f_a(d) = 0$ whence $g = f_a \notin O_d$. This proves that $g \notin O_d$ for any $g \prec f$ and therefore $O_d \cap F_f = \emptyset$ where $F_f = \{g \in F : g \prec f\}$. As a consequence, each $f = f_d \in D$ has an open neighborhood $O(f) = O_d$ such that $O(f) \cap F_f = \emptyset$; this implies that F_f is closed in F for any $f \in F$ and hence F is a left-separated subspace of $C_p(X)$ with the well-order $\prec$ witnessing this. Since $|F| = |D| = \kappa^+$, this is a contradiction with $hd(C_p(X)) \leq \kappa$ (see Problem 004) which shows that $hl(X) \leq \kappa$.

Now assume that $hl(X \times X) > \kappa$ and fix a right-separated subspace $E \subset X \times X$ with $|E| = \kappa^+$. The set $\Delta = \{(x,x) : x \in X\}$ is a subspace of $X \times X$ homeomorphic to X so $hl(\Delta) \leq \kappa$. Thus $|E \cap \Delta| \leq \kappa$ and hence $|E \backslash \Delta| = \kappa^+$, i.e., we have a right-separated subspace $D = E \backslash \Delta \subset (X \times X) \backslash \Delta$ of cardinality κ^+. Denote by $<$ an order that right-separates D and let $D_d = \{z \in D : d < z\}$ for any $d \in D$. Since every D_d is closed in D, for each $d = (x, y) \in D$ there exist $U_d \in \tau(x, X)$ and $V_d \in \tau(y, X)$ such that $(U_d \times V_d) \cap D_d = \emptyset$. Since $x \neq y$, we lose no generality assuming that $U_d \cap V_d = \emptyset$ for each $d \in D$. Take any functions $g_d, h_d \in C(X, [0, 1])$ such that $g_d(x) = h_d(y) = 1$, $g_d|(X \backslash U_d) \equiv 0$, $h_d|(X \backslash V_d) \equiv 0$ and let $f_d = g_d - h_d$. Then $O_d = \{f \in C_p(X) : f(x) > 0$ and $f(y) < 0\}$ is an open subset of $C_p(X)$ and $f_d \in O_d$ for all $d \in D$.

If $a, d \in D$ and $a \neq d$, we can assume without loss of generality that $d < a$. If $a = (x', y')$ and $d = (x, y)$, then $a \notin U_d \times V_d$ which implies $x' \notin U_d$ or $y' \notin V_d$. In the first case we have $f_d(x') \leq 0$ while $f_a(x') = 1 > 0$; in the second case we obtain $f_d(y') \geq 0$ while $f_a(y') = -1 < 0$. Since in both cases we have $f_a \neq f_d$, the correspondence $d \to f_d$ is a bijection. This makes it possible to define a well-order $\prec$ on the set $F = \{f_d : d \in D\}$ by letting $f_d \prec f_a$ if and only if $d < a$.

We prove next that the well-order $\prec$ left-separates F. Take any $f \in F$; to prove that the set $F_f = \{g \in F : g \prec f\}$ is closed in F, take any $g \prec f$. There is a point $a = (x', y') \in D$ such that $g = f_a$; we have $f = f_d$ for some $d = (x, y) \in D$ with $a < d$. It follows from $d \notin U_a \times V_a$ that $x \notin U_a$ or $y \notin V_a$; in the first case we have $f_a(x) \leq 0$ and in the second one we obtain $f_a(y) \geq 0$, so in both cases $g = f_a \notin O_d$. The point $f \in F$ was taken arbitrarily, so we established that $O_d \cap F_f = \emptyset$ for any $f = f_d \in F$. This proves that F_f is closed in F for each $f \in F$ and therefore F is a left-separated subspace of $C_p(X)$. Since $|F| = |D| = \kappa^+$, this is a contradiction with $hd(C_p(X)) \leq \kappa$ (see Problem 004) which shows that $hl(X \times X) \leq \kappa = hd(C_p(X))$.

To prove that $hd(C_p(X)) \leq hl^*(X)$ assume that $hl^*(X) \leq \kappa < hd(C_p(X))$ and fix a left-separated subspace $F \subset C_p(X)$ with $|F| = \kappa^+$. Let $<$ be a well-order that right-separates F. For each $f \in F$, the set $F_f = \{g \in F : g < f\}$ is closed in F. Therefore, for each $f \in F$, there exists a standard open set $O_f \in \tau(f, C_p(X))$ such that $O_f \cap F_f = \emptyset$. We have $O_f = [x_1^f, \ldots, x_{n_f}^f; O_1^f, \ldots, O_{n_f}^f]$ where $x_1^f, \ldots, x_{n_f}^f$ are distinct points of X and $O_1^f, \ldots, O_{n_f}^f$ are rational intervals. Since there are only countably many of all possible n-tuples of rational intervals for any $n \in \mathbb{N}$ and κ^+ is a regular uncountable cardinal, there exist $n \in \mathbb{N}$, a set $G \subset F$ and rational intervals $O_1, \ldots, O_n$ such that

$$|G| = \kappa^+, \ n_f = n, \ O_i^f = O_i, \ i = 1, \ldots, n$$

for each $f \in G$. Let $x_f = (x_1^f, \ldots, x_n^f) \in X^n$ for all $f \in G$. For the set $O(f) = f^{-1}(O_1) \times \cdots \times f^{-1}(O_n)$, we have $x_f \in O(f)$ for all $f \in G$. We claim that the set $D = \{x_f : f \in G\} \subset X^n$ is right-separated.

To show this, pick any $f, g \in G$ such that $g < f$. Since $g \in F_f$, we have $g \notin O_f$ so $g(x_i^f) \notin O_i$ for some $i \leq n$. This implies $x_i^f \notin g^{-1}(O_i)$ and therefore $x_f \notin O(g)$. Thus we have proved that

(*) for any $f, g \in G$ such that $g < f$, we have $x_f \in O(f)$ and $x_f \notin O(g)$.

An evident consequence is that $x_f \neq x_g$ for any distinct $f, g \in G$, so the correspondence $f \to x_f$ is a bijection between G and D. This makes it possible to well-order the set D by letting $x_f \prec x_g$ if and only if $f < g$. Fix $x \in D$; we must prove that the set $D_x = \{y \in D : y \prec x\}$ is open in D. Take any $y \in D_x$ and any $x' \in D \backslash D_x$; there are $f, g \in G$ such that $x' = x_f$ and $y = x_g$. Since $y \prec x'$, we have $g < f$ and hence $x' = x_f \notin O(g)$ by $(*)$. This shows that every $y = x_g \in D_x$ has a neighborhood $W_y = O(g)$ such that $W_y \cap D \subset D_x$; an easy consequence is that D_x is open in D for any $x \in D$ so D is right-separated by the well-order $\prec$. However, $|D| = |G| = \kappa^+$ and $G \subset X^n$ which contradicts $hl(X^n) \leq hl^*(X) \leq \kappa$. Thus $hd(C_p(X)) \leq \kappa = hl^*(X)$, so our solution is complete.

T.019. *For an arbitrary $n \in \mathbb{N}$, let $J_n = J(n)$ be the hedgehog with n spines. Prove that $s(X^n) \leq s(C_p(X, J_n)) \leq s(C_p(X) \times C_p(X))$ for any space X.*

Solution. For an arbitrary $n \in \mathbb{N}$, let $M_n = \{1, \ldots, n\}$. If Z is a space and $n \geq 2$, let $\Delta_{ij}^n(Z) = \{z = (z_1, \ldots, z_n) \in Z^n : z_i = z_j\}$ for any distinct $i, j \in M_n$. The set $\Delta_n(Z) = \bigcup\{\Delta_{ij}^n(Z) : 1 \leq i < j \leq n\}$ is called the n-*diagonal* of Z.

Fact 0. For any $n \in \mathbb{N}$, $n \geq 2$ and any space Z, we have $s(Z^n) = s(Z^n \backslash \Delta_n(Z))$.

Proof. Since $Z^n \backslash \Delta_n(Z) \subset Z^n$, we have $s(Z^n \backslash \Delta_n(Z)) \leq s(Z^n)$. We prove the inverse inequality by induction on n; to simplify the notation, let $\Delta_n = \Delta_n(Z)$ and $\Delta = \Delta_2$. If $p_1 : Z \times Z \to Z$ is the natural projection onto the first coordinate, then $p_1(Z^2 \backslash \Delta) = Z$ (we consider Z to be infinite), so $s(Z^2 \backslash \Delta) \geq s(Z)$ because continuous maps do not increase the spread. Since Δ is homeomorphic to Z, we have $s(\Delta) = s(Z) \leq s(Z^2 \backslash \Delta)$. It is an easy exercise that a finite union of spaces of spread $\leq \kappa$ also has spread $\leq \kappa$, so if $\kappa = s(Z^2 \backslash \Delta)$, it follows from $Z^2 = \Delta \cup (Z^2 \backslash \Delta)$ that $s(Z^2) \leq \kappa$ and hence $s(Z^2) = s(Z^2 \backslash \Delta)$.

Assume that we have proved, for all $n \leq m$, that we have $s(Z^n) \leq s(Z^n \backslash \Delta_n)$. Let $s(Z^{m+1} \backslash \Delta_{m+1}) = \kappa$; denote by $p_m : Z^{m+1} \to Z^m$ the natural projection onto the face determined by the first m coordinates. Since $p_m(Z^{m+1}\backslash\Delta_{m+1}) = Z^m\backslash\Delta_m$, we have $s(Z^m) \leq s(Z^m\backslash\Delta_m) \leq \kappa$ by the induction hypothesis and the fact that continuous maps do not increase the spread.

Observe also that $\Delta_{ij}^{m+1}(Z)$ is homeomorphic to $\Delta \times Z^{m-1}$; since Δ is homeomorphic to Z, the space $\Delta_{ij}^{m+1}(Z)$ is homeomorphic to Z^m for any distinct $i, j \in M_n$ and hence Δ_{m+1} is a finite union of spaces homeomorphic to Z^m. As a consequence, $s(\Delta_{m+1}) \leq s(Z^m) \leq s(Z^m\backslash\Delta_m) \leq \kappa$. Now it follows from $s(\Delta_{m+1}) \leq \kappa$ and $Z^{m+1} = \Delta_{m+1} \cup (Z^{m+1}\backslash\Delta_{m+1})$ that $s(Z^{m+1}) \leq \kappa$. As we accomplished the required inductive step, we can conclude that $s(Z^n) = s(Z^n\backslash\Delta_n)$ for all $n \in \mathbb{N}$, $n \geq 2$, so Fact 0 is proved. $\square$

For notational convenience we consider that $J_n = \{0\} \cup I_1 \cup \cdots \cup I_n$ where $I_k = (0, 1] \times \{k\}$ for each $k \in M_n$. Assume that $s(C_p(X, J_n)) = \kappa$ and we have a discrete $D \subset X^n$ with $|D| = \kappa^+$. By Fact 0 we can assume, without loss of generality, that $D \subset X^n\backslash\Delta_n(X)$. Let $\{d_\alpha : \alpha < \kappa^+\}$ be a faithful enumeration of the set D where $d_\alpha = (x_1^\alpha, \ldots, x_n^\alpha)$ for each $\alpha < \kappa^+$. For each $\alpha < \kappa$ choose sets $U_1^\alpha, \ldots, U_n^\alpha$ such that

(1) $U_i^\alpha \in \tau(x_i^\alpha, X)$ for each $i \in M_n$;
(2) if $i, j \in M_n$ and $i \neq j$, then $\text{cl}_X(U_i^\alpha) \cap \text{cl}_X(U_j^\alpha) = \emptyset$;
(3) $W_\alpha \cap D = \{d_\alpha\}$ where $W_\alpha = U_1^\alpha \times \cdots \times U_n^\alpha$ for each $\alpha < \kappa^+$.

For all $\alpha < \kappa^+$ let $V_\alpha = \bigcup\{U_i^\alpha : i \in M_n\}$ and take a function $f_i^\alpha \in C(X, [0, 1])$ such that $f_i^\alpha(x_i^\alpha) = 1$ and $f_i^\alpha|(X\backslash U_i^\alpha) \equiv 0$ for all $i \in M_n$. Let $g_\alpha(x) = 0$ for all $x \in X\backslash V_\alpha$; if $x \in V_\alpha$, then there is a unique $i \in M_n$ such that $x \in U_i^\alpha$, so let $g_\alpha(x) = (f_i^\alpha(x), i)$. This gives us a map $g_\alpha : X \to J_n$ for all $\alpha < \kappa^+$.

Observe that the map $\psi_k : [0, 1] \to I_k \cup \{0\} \subset J_n$ defined by $\psi_k(0) = 0$ and $\psi_k(t) = (t, k)$ for all $t \in (0, 1]$ is continuous. Therefore the map $\psi_k \circ f_k^\alpha : X \to J_n$ is continuous for all $k \in M_n$. Now, to see that g_α is continuous, take any $x \in X$; the family $\mathcal{U}_\alpha = \{U_1^\alpha, \ldots, U_n^\alpha\}$ is discrete by (2), so there is $U \in \tau(x, X)$ such that U intersects at most one element of $\mathcal{U}_\alpha$, say, U_k^α. It is evident that $g_\alpha|U = (\psi_k \circ f_k^\alpha)|U$, so $g_\alpha|U$ is a continuous map. Therefore we can apply Fact 1 of S.472 to conclude that the mapping g_α is continuous, i.e., $g_\alpha \in C_p(X, J_n)$ for each $\alpha < \kappa^+$.

If $\alpha, \beta < \kappa^+$ and $\alpha \neq \beta$, then $d_\alpha \notin W_\beta$ by (3), so there is $k \in M_n$ such that $x_k^\alpha \notin U_k^\beta$. This implies $g_\alpha(x_k^\alpha) = (1, k) \in I_k$ while $g_\beta(x_k^\alpha) \notin I_k$ because $g_\beta^{-1}(I_k) \subset U_k^\beta$. This shows that $g_\alpha \neq g_\beta$ and therefore the correspondence $\alpha \to g_\alpha$ is a bijection between the sets D and $E = \{g_\alpha : \alpha < \kappa^+\} \subset C_p(X, J_n)$.

For each $\alpha < \kappa^+$, the set $O_\alpha = \{f \in C_p(X, J_n) : f(x_k^\alpha) \in I_k$ for all $k \in M_n\}$ is open in $C_p(X, J_n)$ and $g_\alpha \in O_\alpha$ for all $\alpha < \kappa^+$. If $\alpha, \beta < \kappa^+$ and $\alpha \neq \beta$, then $d_\alpha \notin W_\beta$ by (3), so there is $k \in M_n$ such that $x_k^\alpha \notin U_k^\beta$. This implies $g_\beta(x_k^\alpha) \notin I_k$ because $g_\beta^{-1}(I_k) \subset U_k^\beta$; as a consequence, $g_\beta \notin O_\alpha$ which proves that $O_\alpha \cap E = \{g_\alpha\}$. Thus E is a discrete subspace of $C_p(X, J_n)$ and $|E| = |D| = \kappa^+$ which contradicts $s(C_p(X, J_n)) \leq \kappa$. This contradiction proves that $s(X^n) \leq \kappa = s(C_p(X, J_n))$ for each $n \in \mathbb{N}$.

Fact 1. The space $J(n)$ embeds in $\mathbb{R}^2$ for each $n \in \mathbb{N}$.

Proof. Given any points $z, z' \in \mathbb{R}^2$ such that $z = (x, y)$ and $z' = (x', y')$, let $d(z, z') = \sqrt{(x - x')^2 + (y - y')^2}$. Then d is a metric which generates the usual topology on $\mathbb{R}^2$ (see Problems 130 and 205 of [TFS]). Denote by ρ the usual metric on $J(n)$ (see Problem 222 of [TFS]). Let $\varphi(0) = w = (0, 0) \in \mathbb{R}^2$ and, for any $k \in M_n$ and any $x = (t, k) \in I_k$, let $\varphi(x) = (t, kt) \in \mathbb{R}^2$. Observe that for the map $\varphi : J(n) \to \mathbb{R}^2$, we have

(∗) for any $x, y \in J(n)$, we have $d(\varphi(x), \varphi(y)) \leq (n + 1) \cdot \rho(x, y)$.

Indeed, if $x = (t, k) \in I_k \cup \{0\}$ and $y = (s, m) \in I_m \cup \{0\}$ for some $k \neq m$, then $d(\varphi(x), \varphi(y)) \leq d(\varphi(x), w) + d(w, \varphi(y)) = t\sqrt{1 + k^2} + s\sqrt{1 + m^2}$. It follows from $k \leq n$ and $m \leq n$ that we have $\sqrt{1 + k^2} < n + 1$ and $\sqrt{1 + m^2} < n + 1$, so $d(\varphi(x), \varphi(y)) \leq (s + t)(n + 1) = (n + 1) \cdot \rho(x, y)$. If $x = (t, k) \in I_k \cup \{0\}$ and $y = (s, k) \in I_k \cup \{0\}$, then $d(\varphi(x), \varphi(y)) = |t - s| \cdot \sqrt{1 + k^2} \leq (n + 1) \cdot |t - s| = (n + 1) \cdot \rho(x, y)$.

An immediate consequence of (∗) is that the map $\varphi : J(n) \to \mathbb{R}^2$ is continuous; it is clear that $J(n)$ is compact and φ is injective, so φ is a homeomorphism between $J(n)$ and $T_n = \varphi(J(n)) \subset \mathbb{R}^2$. Fact 1 is proved.   □

Returning to our solution note that $C_p(X, J_n) \subset C_p(X, \mathbb{R}^2)$ by Fact 1 and Problem 089 of [TFS]. Thus $s(C_p(X, J_n)) \leq s(C_p(X, \mathbb{R}^2)) = s(C_p(X) \times C_p(X))$ because the spaces $C_p(X, \mathbb{R}^2)$ and $C_p(X) \times C_p(X)$ are homeomorphic by Problem 112 of [TFS]. This proves that $s(C_p(X, J_n)) \leq s(C_p(X) \times C_p(X))$, so our solution is complete.

T.020. *For an arbitrary $n \in \mathbb{N}$, let $J_n = J(n)$ be the hedgehog with n spines. Prove that $hd(X^n) \leq hl(C_p(X, J_n)) \leq hl(C_p(X) \times C_p(X))$ for any space X.*

Solution. For an arbitrary $n \in \mathbb{N}$, let $M_n = \{1, \ldots, n\}$. If Z is a space and $n \geq 2$, let $\Delta_{ij}^n(Z) = \{z = (z_1, \ldots, z_n) \in Z^n : z_i = z_j\}$ for any distinct $i, j \in M_n$. The set $\Delta_n(Z) = \bigcup\{\Delta_{ij}^n(Z) : 1 \leq i < j \leq n\}$ is called the *n-diagonal* of Z.

Fact 0. For any $n \in \mathbb{N}$, $n \geq 2$ and any space Z, we have $hd(Z^n) = hd(Z^n \backslash \Delta_n(Z))$.

Proof. Since $Z^n \backslash \Delta_n(Z) \subset Z^n$, we have $hd(Z^n \backslash \Delta_n(Z)) \leq hd(Z^n)$. We prove the inverse inequality by induction on n; to simplify the notation, let $\Delta_n = \Delta_n(Z)$ and $\Delta = \Delta_2$. If $p_1 : Z \times Z \to Z$ is the natural projection onto the first coordinate, then $p_1(Z^2 \backslash \Delta) = Z$ (we consider Z to be infinite), so $hd(Z^2 \backslash \Delta) \geq hd(Z)$ because continuous maps do not increase the hereditary density. Since Δ is homeomorphic to Z, we have $hd(\Delta) = hd(Z) \leq hd(Z^2 \backslash \Delta)$. It is an easy exercise that a finite union of spaces of hereditary density $\leq \kappa$ also has hereditary density $\leq \kappa$, so if $\kappa = hd(Z^2 \backslash \Delta)$, it follows from $Z^2 = \Delta \cup (Z^2 \backslash \Delta)$ that $hd(Z^2) \leq \kappa$ and hence $hd(Z^2) = hd(Z^2 \backslash \Delta)$.

Assume that we have proved, for all $n \leq m$, that we have $hd(Z^n) \leq hd(Z^n \backslash \Delta_n)$. Let $hd(Z^{m+1} \backslash \Delta_{m+1}) = \kappa$; denote by $p_m : Z^{m+1} \to Z^m$ the natural projection onto the face determined by the first m coordinates. Since $p_m(Z^{m+1}\backslash\Delta_{m+1}) = Z^m\backslash\Delta_m$, we have $hd(Z^m) \leq hd(Z^m\backslash\Delta_m) \leq \kappa$ by the induction hypothesis and the fact that continuous maps do not increase the hereditary density.

Observe also that $\Delta_{ij}^{m+1}(Z)$ is homeomorphic to $\Delta \times Z^{m-1}$; since Δ is homeomorphic to Z, the space $\Delta_{ij}^{m+1}(Z)$ is homeomorphic to Z^m for any distinct $i, j \in M_n$, and hence Δ_{m+1} is a finite union of spaces homeomorphic to Z^m. As a consequence, $hd(\Delta_{m+1}) \leq hd(Z^m) \leq hd(Z^m\backslash\Delta_m) \leq \kappa$. Now it follows from $hd(\Delta_{m+1}) \leq \kappa$ and $Z^{m+1} = \Delta_{m+1} \cup (Z^{m+1}\backslash\Delta_{m+1})$ that $hd(Z^{m+1}) \leq \kappa$. As we accomplished the required inductive step, we can conclude that $hd(Z^n) = hd(Z^n\backslash\Delta_n)$ for all $n \in \mathbb{N}$, $n \geq 2$ so Fact 0 is proved.   □

For notational convenience we consider that $J_n = \{0\} \cup I_1 \cup \cdots \cup I_n$ where $I_k = (0, 1] \times \{k\}$ for each $k \in M_n$. Assume that $hl(C_p(X, J_n)) = \kappa$ and $hd(X^n) > \kappa$. Then there is a subspace $D = \{d_\alpha : \alpha < \kappa^+\} \subset X^n$ which is left-separated by its indexation (see Problem 004 and Fact 2 of T.004); by Fact 0, we can assume, without loss of generality, that $D \subset X^n\backslash\Delta_n(X)$. Let $d_\alpha = (x_1^\alpha, \ldots, x_n^\alpha)$ for each $\alpha < \kappa^+$. For every $\alpha < \kappa$ choose sets $U_1^\alpha, \ldots, U_n^\alpha$ such that:

(1) $U_i^\alpha \in \tau(x_i^\alpha, X)$ for each $i \in M_n$;
(2) if $i, j \in M_n$ and $i \neq j$, then $\mathrm{cl}_X(U_i^\alpha) \cap \mathrm{cl}_X(U_j^\alpha) = \emptyset$;
(3) $W_\alpha \cap D_\alpha = \emptyset$ where $D_\alpha = \{d_\beta : \beta < \alpha\}$ and $W_\alpha = U_1^\alpha \times \cdots \times U_n^\alpha$ for each $\alpha < \kappa^+$.

For all $\alpha < \kappa^+$ let $V_\alpha = \bigcup\{U_i^\alpha : i \in M_n\}$ and take a function $f_i^\alpha \in C(X, [0, 1])$ such that $f_i^\alpha(x_i^\alpha) = 1$ and $f_i^\alpha|(X\backslash U_i^\alpha) \equiv 0$ for all $i \in M_n$. Let $g_\alpha(x) = 0$ for all $x \in X\backslash V_\alpha$; if $x \in V_\alpha$, then there is a unique $i \in M_n$ such that $x \in U_i^\alpha$ so let $g_\alpha(x) = (f_i^\alpha(x), i)$. This gives us a map $g_\alpha : X \to J_n$ for all $\alpha < \kappa^+$.

Observe that the map $\psi_k : [0, 1] \to I_k \cup \{0\} \subset J_n$ defined by $\psi_k(0) = 0$ and $\psi_k(t) = (t, k)$ for all $t \in (0, 1]$ is continuous. Therefore the map $\psi_k \circ f_k^\alpha : X \to J_n$ is continuous for all $k \in M_n$. Now, to see that g_α is continuous, take any $x \in X$; the family $\mathcal{U}_\alpha = \{U_1^\alpha, \ldots, U_n^\alpha\}$ is discrete by (2), so there is $U \in \tau(x, X)$ such that U intersects at most one element of $\mathcal{U}_\alpha$, say, U_k^α. It is evident that $g_\alpha|U = (\psi_k \circ f_k^\alpha)|U$, so $g_\alpha|U$ is a continuous map. Therefore we can apply Fact 1 of S.472 to conclude that the mapping g_α is continuous, i.e., $g_\alpha \in C_p(X, J_n)$ for each $\alpha < \kappa^+$.

If $\alpha < \beta < \kappa^+$, then $d_\alpha \notin W_\beta$ by (3), so there is $k \in M_n$ such that $x_k^\alpha \notin U_k^\beta$. This implies $g_\alpha(x_k^\alpha) = (1, k) \in I_k$ while $g_\beta(x_k^\alpha) \notin I_k$ because $g_\beta^{-1}(I_k) \subset U_k^\beta$. This shows that $g_\alpha \neq g_\beta$ and therefore the correspondence $\alpha \to g_\alpha$ is a bijection between the sets D and $E = \{g_\alpha : \alpha < \kappa^+\} \subset C_p(X, J_n)$.

For each $\alpha < \kappa^+$ the set $O_\alpha = \{f \in C_p(X, J_n) : f(x_k^\alpha) \in I_k$ for all $k \in M_n\}$ is open in $C_p(X, J_n)$ and $g_\alpha \in O_\alpha$ for all $\alpha < \kappa^+$. If $\alpha < \beta < \kappa^+$, then $d_\alpha \notin W_\beta$ by (3), so there is $k \in M_n$ such that $x_k^\alpha \notin U_k^\beta$. This implies $g_\beta(x_k^\alpha) \notin I_k$ because $g_\beta^{-1}(I_k) \subset U_k^\beta$, so we have the following property:

(*) $g_\beta \notin O_\alpha$ whenever $\alpha < \beta < \kappa^+$.

Take any $\beta < \kappa^+$; given any $\alpha < \beta$, we have $\alpha < \beta'$ for any $\beta' \geq \beta$. An immediate consequence of (∗) is that $g_{\beta'} \notin O_\alpha$. Therefore $O_\alpha \cap E \subset E_\beta$ where $E_\beta = \{g_\gamma : \gamma < \beta\}$. This shows that the set E_β is open in E for each $\beta < \kappa^+$, and hence E is a right-separated (by its indexation) subspace of $C_p(X, J_n)$. Since $|E| = |D| = \kappa^+$, this contradicts $hl(C_p(X, J_n)) \leq \kappa$; thus $hd(X^n) \leq \kappa = hl(C_p(X, J_n))$ for each $n \in \mathbb{N}$.

Finally note that $C_p(X, J_n) \subset C_p(X, \mathbb{R}^2)$ by Problem 089 of [TFS] and Fact 1 of T.019. Thus $hl(C_p(X, J_n)) \leq hl(C_p(X, \mathbb{R}^2)) = hl(C_p(X) \times C_p(X))$ because the spaces $C_p(X, \mathbb{R}^2)$ and $C_p(X) \times C_p(X)$ are homeomorphic by Problem 112 of [TFS]. This proves that $hl(C_p(X, J_n)) \leq hl(C_p(X) \times C_p(X))$, so our solution is complete.

T.021. *For an arbitrary $n \in \mathbb{N}$, let $J_n = J(n)$ be the hedgehog with n spines. Prove that $hl(X^n) \leq hd(C_p(X, J_n)) \leq hd(C_p(X) \times C_p(X))$ for any space X.*

Solution. For an arbitrary $n \in \mathbb{N}$, let $M_n = \{1, \ldots, n\}$. If Z is a space and $n \geq 2$, let $\Delta_{ij}^n(Z) = \{z = (z_1, \ldots, z_n) \in Z^n : z_i = z_j\}$ for any distinct $i, j \in M_n$. The set $\Delta_n(Z) = \bigcup\{\Delta_{ij}^n(Z) : 1 \leq i < j \leq n\}$ is called the *n-diagonal* of Z.

Fact 0. For any $n \in \mathbb{N}$, $n \geq 2$ and any space Z, we have $hl(Z^n) = hl(Z^n \backslash \Delta_n(Z))$.

Proof. Since $Z^n \backslash \Delta_n(Z) \subset Z^n$, we have $hl(Z^n \backslash \Delta_n(Z)) \leq hl(Z^n)$. We prove the inverse inequality by induction on n; to simplify the notation, let $\Delta_n = \Delta_n(Z)$ and $\Delta = \Delta_2$. If $p_1 : Z \times Z \to Z$ is the natural projection onto the first coordinate, then $p_1(Z^2 \backslash \Delta) = Z$ (we consider Z to be infinite), so $hl(Z^2 \backslash \Delta) \geq hl(Z)$ because continuous maps do not increase the hereditary Lindelöf number. Since Δ is homeomorphic to Z, we have $hl(\Delta) = hl(Z) \leq hl(Z^2 \backslash \Delta)$. It is an easy exercise that a finite union of spaces with hereditary Lindelöf number $\leq \kappa$ also has hereditary Lindelöf number $\leq \kappa$, so if $\kappa = hl(Z^2 \backslash \Delta)$, it follows from $Z^2 = \Delta \cup (Z^2 \backslash \Delta)$ that $hl(Z^2) \leq \kappa$ and hence $hl(Z^2) = hl(Z^2 \backslash \Delta)$.

Assume that we have proved, for all $n \leq m$, that we have $hl(Z^n) \leq hl(Z^n \backslash \Delta_n)$. Let $hl(Z^{m+1} \backslash \Delta_{m+1}) = \kappa$; denote by $p_m : Z^{m+1} \to Z^m$ the natural projection onto the face determined by the first m coordinates. Since $p_m(Z^{m+1} \backslash \Delta_{m+1}) = Z^m \backslash \Delta_m$, we have $hl(Z^m) \leq hl(Z^m \backslash \Delta_m) \leq \kappa$ by the induction hypothesis and the fact that continuous maps do not increase the hereditary density.

Observe also that $\Delta_{ij}^{m+1}(Z)$ is homeomorphic to $\Delta \times Z^{m-1}$; since Δ is homeomorphic to Z, the space $\Delta_{ij}^{m+1}(Z)$ is homeomorphic to Z^m for any distinct $i, j \in M_n$, and hence Δ_{m+1} is a finite union of spaces homeomorphic to Z^m. As a consequence, $hl(\Delta_{m+1}) \leq hl(Z^m) \leq hl(Z^m \backslash \Delta_m) \leq \kappa$. Now it follows from $hl(\Delta_{m+1}) \leq \kappa$ and $Z^{m+1} = \Delta_{m+1} \cup (Z^{m+1} \backslash \Delta_{m+1})$ that $hl(Z^{m+1}) \leq \kappa$. As we accomplished the required inductive step, we can conclude that $hl(Z^n) = hl(Z^n \backslash \Delta_n)$ for all $n \in \mathbb{N}$, $n \geq 2$, so Fact 0 is proved. □

For notational convenience we consider that $J_n = \{0\} \cup I_1 \cup \cdots \cup I_n$ where $I_k = (0, 1] \times \{k\}$ for each $k \in M_n$. Assume that $hd(C_p(X, J_n)) = \kappa$ and $hl(X^n) > \kappa$. Then there is a subspace $D = \{d_\alpha : \alpha < \kappa^+\} \subset X^n$ which is right-separated by its indexation (see Problem 005 and Fact 2 of T.005); by Fact 0, we can assume, without loss of generality, that $D \subset X^n \backslash \Delta_n(X)$. Let $d_\alpha = (x_1^\alpha, \ldots, x_n^\alpha)$ for each $\alpha < \kappa^+$. For every $\alpha < \kappa$ choose sets $U_1^\alpha, \ldots, U_n^\alpha$ such that

(1) $U_i^\alpha \in \tau(x_i^\alpha, X)$ for each $i \in M_n$;
(2) if $i, j \in M_n$ and $i \neq j$, then $\mathrm{cl}_X(U_i^\alpha) \cap \mathrm{cl}_X(U_j^\alpha) = \emptyset$;
(3) $W_\alpha \cap D \subset D_\alpha$ where $D_\alpha = \{d_\beta : \beta < \alpha + 1\}$ and $W_\alpha = U_1^\alpha \times \cdots \times U_n^\alpha$ for each $\alpha < \kappa^+$.

For all $\alpha < \kappa^+$ let $V_\alpha = \bigcup \{U_i^\alpha : i \in M_n\}$ and take a function $f_i^\alpha \in C(X, [0, 1])$ such that $f_i^\alpha(x_i^\alpha) = 1$ and $f_i^\alpha|(X \backslash U_i^\alpha) \equiv 0$ for all $i \in M_n$. Let $g_\alpha(x) = 0$ for all $x \in X \backslash V_\alpha$; if $x \in V_\alpha$, then there is a unique $i \in M_n$ such that $x \in U_i^\alpha$ so let $g_\alpha(x) = (f_i^\alpha(x), i)$. This gives us a map $g_\alpha : X \to J_n$ for all $\alpha < \kappa^+$.

Observe that the map $\psi_k : [0, 1] \to I_k \cup \{0\} \subset J_n$ defined by $\psi_k(0) = 0$ and $\psi_k(t) = (t, k)$ for all $t \in (0, 1]$ is continuous. Therefore the map $\psi_k \circ f_k^\alpha : X \to J_n$ is continuous for all $k \in M_n$. Now, to see that g_α is continuous, take any $x \in X$; the family $\mathcal{U}_\alpha = \{U_1^\alpha, \ldots, U_n^\alpha\}$ is discrete by (2), so there is $U \in \tau(x, X)$ such that U intersects at most one element of $\mathcal{U}_\alpha$, say, U_k^α. It is evident that $g_\alpha|U = (\psi_k \circ f_k^\alpha)|U$, so $g_\alpha|U$ is a continuous map. Therefore we can apply Fact 1 of S.472 to conclude that the mapping g_α is continuous, i.e., $g_\alpha \in C_p(X, J_n)$ for each $\alpha < \kappa^+$.

If $\alpha < \beta < \kappa^+$, then $d_\beta \notin W_\alpha$ by (3), so there is $k \in M_n$ such that $x_k^\beta \notin U_k^\alpha$. We have $g_\beta(x_k^\beta) = (1, k) \in I_k$ while $g_\alpha(x_k^\beta) \notin I_k$ because $g_\alpha^{-1}(I_k) \subset U_k^\alpha$. This shows that $g_\alpha \neq g_\beta$ and therefore the correspondence $\alpha \to g_\alpha$ is a bijection between the sets D and $E = \{g_\alpha : \alpha < \kappa^+\} \subset C_p(X, J_n)$.

For each $\alpha < \kappa^+$ the set $O_\alpha = \{f \in C_p(X, J_n) : f(x_k^\alpha) \in I_k$ for all $k \in M_n\}$ is open in $C_p(X, J_n)$ and $g_\alpha \in O_\alpha$ for all $\alpha < \kappa^+$. If $\alpha < \beta < \kappa^+$, then $d_\beta \notin W_\alpha$ by (3), so there is $k \in M_n$ such that $x_k^\beta \notin U_k^\alpha$. This implies $g_\alpha(x_k^\beta) \notin I_k$ because $g_\alpha^{-1}(I_k) \subset U_k^\alpha$, so we have the following property:

(*) $g_\alpha \notin O_\beta$ whenever $\alpha < \beta < \kappa^+$.

Take any $\beta < \kappa^+$; given any $\alpha < \beta$, we have $g_\alpha \notin O_\beta$ by (*). Therefore $O_\beta \cap E_\beta = \emptyset$ where $E_\beta = \{g_\alpha : \alpha < \beta\}$. This shows that the set E_β is closed in E for each $\beta < \kappa^+$, and hence E is a left-separated (by its indexation) subspace of $C_p(X, J_n)$. Since $|E| = |D| = \kappa^+$, this contradicts $hd(C_p(X, J_n)) \leq \kappa$; thus $hl(X^n) \leq \kappa = hd(C_p(X, J_n))$ for each $n \in \mathbb{N}$.

Finally note that $C_p(X, J_n) \subset C_p(X, \mathbb{R}^2)$ by Problem 089 of [TFS] and Fact 1 of T.019. Thus $hd(C_p(X, J_n)) \leq hd(C_p(X, \mathbb{R}^2)) = hd(C_p(X) \times C_p(X))$ because the spaces $C_p(X, \mathbb{R}^2)$ and $C_p(X) \times C_p(X)$ are homeomorphic by Problem 112 of [TFS]. This proves that $hd(C_p(X, J_n)) \leq hd(C_p(X) \times C_p(X))$, so our solution is complete.

T.022. *For any space X prove that $s(C_p(X) \times C_p(X)) = s^*(C_p(X))$.*

Solution. It is evident that $s(C_p(X) \times C_p(X)) \leq s^*(C_p(X))$, so we must only prove that $s^*(C_p(X)) \leq s(C_p(X) \times C_p(X))$. Let X_i be a homeomorphic copy of X for all $i \in \mathbb{N}$. Then $(C_p(X))^n$ is homeomorphic to the space $C_p(X_1 \oplus \cdots \oplus X_n)$ for all $n \in \mathbb{N}$

(Problem 114 of [TFS]). Let $Y_n = X_1 \oplus \cdots \oplus X_n$; then Y_n^k is a finite union of spaces homeomorphic to X^k. Since finite unions do not increase spread, we have $s(Y_n^k) \leq s(X^k) \leq s^*(X)$ for all $k \in \mathbb{N}$. Consequently, $s^*(Y_n) \leq s^*(X)$ for all $n \in \mathbb{N}$. Apply Problem 016 to conclude that $s((C_p(X))^n) = s(C_p(Y_n)) \leq s^*(Y_n) \leq s^*(X)$ for all $n \in \mathbb{N}$, i.e., $s^*(C_p(X)) \leq s^*(X)$. Since $s(X^n) \leq s(C_p(X) \times C_p(X))$ for all $n \in \mathbb{N}$ by Problem 019, we have $s^*(X) \leq s((C_p(X))^2)$. Finally, $s^*(C_p(X)) \leq s^*(X) \leq s(C_p(X) \times C_p(X))$, so $s^*(C_p(X)) = s(C_p(X) \times C_p(X))$.

T.023. *For any space X prove that $hl(C_p(X) \times C_p(X)) = hl^*(C_p(X))$.*

Solution. It is evident that $hl(C_p(X) \times C_p(X)) \leq hl^*(C_p(X))$, so we must only prove that $hl^*(C_p(X)) \leq hl(C_p(X) \times C_p(X))$. Let X_i be a homeomorphic copy of X for all $i \in \mathbb{N}$. Then $(C_p(X))^n$ is homeomorphic to the space $C_p(X_1 \oplus \cdots \oplus X_n)$ for all $n \in \mathbb{N}$ (see Problem 114 of [TFS]). Let $Y_n = X_1 \oplus \cdots \oplus X_n$; then Y_n^k is a finite union of spaces homeomorphic to X^k. Since finite unions do not increase the hereditary density, we have $hd(Y_n^k) \leq hd(X^k) \leq hd^*(X)$ for all $k \in \mathbb{N}$. Consequently, $hd^*(Y_n) \leq hd^*(X)$ for all $n \in \mathbb{N}$. Apply Problem 017 to conclude that $hl((C_p(X))^n) = hl(C_p(Y_n)) \leq hd^*(Y_n) \leq hd^*(X)$ for all $n \in \mathbb{N}$, i.e., $hl^*(C_p(X)) \leq hd^*(X)$. Since $hd(X^n) \leq hl(C_p(X) \times C_p(X))$ for all $n \in \mathbb{N}$ by Problem 020, we have $hd^*(X) \leq hl((C_p(X))^2)$. Finally, $hl^*(C_p(X)) \leq hd^*(X) \leq hl(C_p(X) \times C_p(X))$ so $hl^*(C_p(X)) = hl(C_p(X) \times C_p(X))$.

T.024. *For any space X prove that $hd(C_p(X) \times C_p(X)) = hd^*(C_p(X))$.*

Solution. It is evident that $hd(C_p(X) \times C_p(X)) \leq hd^*(C_p(X))$, so we must only prove that $hd^*(C_p(X)) \leq hd(C_p(X) \times C_p(X))$. Let X_i be a homeomorphic copy of the space X for all $i \in \mathbb{N}$. Then $(C_p(X))^n$ is homeomorphic to the space $C_p(X_1 \oplus \cdots \oplus X_n)$ for all $n \in \mathbb{N}$ (see Problem 114 of [TFS]). Let $Y_n = X_1 \oplus \cdots \oplus X_n$; then Y_n^k is a finite union of spaces homeomorphic to X^k. Since finite unions do not increase the hereditary Lindelöf number, we have $hl(Y_n^k) \leq hl(X^k) \leq hl^*(X)$ for all $k \in \mathbb{N}$. Consequently, $hl^*(Y_n) \leq hl^*(X)$ for all $n \in \mathbb{N}$. Apply Problem 018 to conclude that $hd((C_p(X))^n) = hd(C_p(Y_n)) \leq hl^*(Y_n) \leq hl^*(X)$ for all $n \in \mathbb{N}$ and hence $hd^*(C_p(X)) \leq hl^*(X)$. Since $hl(X^n) \leq hd(C_p(X) \times C_p(X))$ for all $n \in \mathbb{N}$ by Problem 021, we conclude that $hl^*(X) \leq hd((C_p(X))^2)$. Finally, observe that the inequalities $hd^*(C_p(X)) \leq hl^*(X) \leq hd(C_p(X) \times C_p(X))$ show that $hd^*(C_p(X)) = hd(C_p(X) \times C_p(X))$.

T.025. *Prove that $s^*(X) = s^*(C_p(X))$ for any space X.*

Solution. Let X_i be a homeomorphic copy of X for all $i \in \mathbb{N}$. Then $(C_p(X))^n$ is homeomorphic to the space $C_p(X_1 \oplus \cdots \oplus X_n)$ for all $n \in \mathbb{N}$ by Problem 114 of [TFS]. Let $Y_n = X_1 \oplus \cdots \oplus X_n$; then Y_n^k is a finite union of spaces homeomorphic to X^k. Since finite unions do not increase spread, we have $s(Y_n^k) \leq s(X^k) \leq s^*(X)$ for all $k \in \mathbb{N}$. Consequently, $s^*(Y_n) \leq s^*(X)$ for all $n \in \mathbb{N}$. Apply Problem 016 to conclude that $s((C_p(X))^n) = s(C_p(Y_n)) \leq s^*(Y_n) \leq s^*(X)$ for all $n \in \mathbb{N}$; this implies that $s^*(C_p(X)) \leq s^*(X)$. Since $s(X^n) \leq s(C_p(X) \times C_p(X))$ for all $n \in \mathbb{N}$ by Problem 019, we have $s^*(X) \leq s((C_p(X))^2) \leq s^*(C_p(X))$ so $s^*(X) = s^*(C_p(X))$.

T.026. *Prove that* $hl^*(X) = hd^*(C_p(X))$ *for any space* X.

Solution. Let X_i be a homeomorphic copy of X for all $i \in \mathbb{N}$. Then $(C_p(X))^n$ is homeomorphic to the space $C_p(X_1 \oplus \cdots \oplus X_n)$ for all $n \in \mathbb{N}$ by Problem 114 of [TFS]. Consider the space $Y_n = X_1 \oplus \cdots \oplus X_n$; then Y_n^k is a finite union of spaces homeomorphic to X^k. It is evident that finite unions do not increase hereditary Lindelöf number, so we have $hl(Y_n^k) \le hl(X^k) \le hl^*(X)$ for all $k \in \mathbb{N}$. Consequently, $hl^*(Y_n) \le hl^*(X)$ for all $n \in \mathbb{N}$. We have $hd((C_p(X))^n) = hd(C_p(Y_n)) \le hl^*(Y_n) \le hl^*(X)$ for all $n \in \mathbb{N}$ (see Problem 018); this implies that $hd^*(C_p(X)) \le hl^*(X)$. Since $hl(X^n) \le hd(C_p(X) \times C_p(X))$ for all $n \in \mathbb{N}$ by Problem 021, we have $hl^*(X) \le hd((C_p(X))^2) \le hd^*(C_p(X))$ whence $hl^*(X) = hd^*(C_p(X))$.

T.027. *Prove that* $hd^*(X) = hl^*(C_p(X))$ *for any space* X.

Solution. Let X_i be a homeomorphic copy of X for all $i \in \mathbb{N}$. Then $(C_p(X))^n$ is homeomorphic to the space $C_p(X_1 \oplus \cdots \oplus X_n)$ for all $n \in \mathbb{N}$ by Problem 114 of [TFS]. Consider the space $Y_n = X_1 \oplus \cdots \oplus X_n$; then Y_n^k is a finite union of spaces homeomorphic to X^k. It is evident that finite unions do not increase hereditary density, so we have $hd(Y_n^k) \le hd(X^k) \le hd^*(X)$ for all $k \in \mathbb{N}$. Consequently, $hd^*(Y_n) \le hd^*(X)$ for all $n \in \mathbb{N}$. We have $hl((C_p(X))^n) = hl(C_p(Y_n)) \le hd^*(Y_n) \le hd^*(X)$ for all $n \in \mathbb{N}$ (see Problem 017); this implies that $hl^*(C_p(X)) \le hd^*(X)$. Since $hd(X^n) \le hl(C_p(X) \times C_p(X))$ for all $n \in \mathbb{N}$ by Problem 020, we have $hd^*(X) \le hl((C_p(X))^2) \le hl^*(C_p(X))$ whence $hd^*(X) = hl^*(C_p(X))$.

T.028. *For an infinite cardinal* κ, *suppose that* $s(C_p(X)) = \kappa$ *and* $\Delta(X) \le \kappa$. *Prove that* $s^*(C_p(X)) \le \kappa$ *and hence* $s^*(C_p(X)) = s(C_p(X))$. *In particular, if* X *is a space with a* G_δ-*diagonal, then* $s^*(C_p(X)) = s(C_p(X)) = s^*(X)$.

Solution. For an arbitrary $n \in \mathbb{N}$, let $M_n = \{1, \ldots, n\}$; we will also need the set $S_n = \{\sigma : \sigma$ is a bijection and $\sigma : M_n \to M_n\}$. If Z is a space and $n \ge 2$, let $\Delta_{ij}^n(Z) = \{z = (z_1, \ldots, z_n) \in Z^n : z_i = z_j\}$ for any distinct $i, j \in M_n$. The set $\Delta_n(Z) = \bigcup\{\Delta_{ij}^n(Z) : 1 \le i < j \le n\}$ is called the n-*diagonal* of the space Z. If the space Z is clear, we write Δ_n instead of $\Delta_n(Z)$.

Fact 1. Given a space Z and a natural number $n \ge 2$, we have $s(Z^n \backslash U) \le s(C_p(Z))$ for any $U \in \tau(\Delta_n(Z), Z^n)$.

Proof. Assume that $s(C_p(Z)) = \lambda$; if we are given a point $z = (z_1, \ldots, z_n) \in Z^n$, then $\text{supp}(z) = \{z_1, \ldots, z_n\}$; let $\Delta_n = \Delta_n(Z)$. For any point $z \in Z^n \backslash \Delta_n$, consider the set $P_z = \{z' \in Z^n \backslash \Delta_n : \text{supp}(z') = \text{supp}(z)\}$. It is evident that $|P_z| = n!$ for any $z \in Z^n \backslash \Delta_n$. It is also clear that either $P_z \cap P_y = \emptyset$ or $P_z = P_y$ for any $y, z \in Z^n \backslash \Delta_n$. Therefore the family $\mathcal{P} = \{P_z : z \in Z^n \backslash \Delta_n\}$ is disjoint; if we choose an element $z_P \in P$ for each $P \in \mathcal{P}$, we obtain a set $C = \{z_P : P \in \mathcal{P}\}$ such that $C \subset Z^n \backslash \Delta_n$ and $|C \cap P_z| = 1$ for every $z \in Z^n \backslash \Delta_n$.

Given any permutation $\sigma \in S_n$, consider the map $p_\sigma : Z^n \to Z^n$ defined by $p_\sigma(z) = (z_{\sigma(1)}, \ldots, z_{\sigma(n)})$ for any $z = (z_1, \ldots, z_n) \in Z^n$. It is immediate that each p_σ is a homeomorphism such that $p_\sigma(\Delta_n) = \Delta_n$. As a consequence, the set $W = \bigcap \{p_\sigma(U) : \sigma \in S_n\}$ is an open neighborhood of Δ_n such that $W \subset U$ and $P_z \subset Z^n \backslash W$ for any $z \in Z^n \backslash W$. Since $Z^n \backslash U \subset Z^n \backslash W$, it suffices to prove that $s(Z^n \backslash W) \leq \lambda$. Suppose that there is a discrete $D \subset Z^n \backslash W$ of cardinality λ^+. Since P_z is finite for any $z \in Z^n \backslash \Delta_n$, the set $R = \{z \in C : P_z \cap D \neq \emptyset\}$ has cardinality λ^+. For each $z \in R$ choose $d_z \in P_z \cap D$; then $D' = \{d_z : z \in R\}$ is again a discrete subspace of $Z^n \backslash W$ of cardinality λ^+ such that

(*) $|D' \cap P_z| = 1$ for each $z \in D'$.

The set $F = \overline{D'} \backslash D'$ is a closed subset of Z^n and $F \subset Z^n \backslash W$ (the bar denotes the closure in Z^n). Let $T = Z^n \backslash (W \cup F)$; there exist $E \subset D'$ and $m \in \mathbb{N}$ such that $|E| = \lambda^+$ and $|P_d \cap T| = m$ for every $d \in E$. For each $d = (d_1, \ldots, d_n) \in E$ and $i \leq n$, choose a set $O_i^d \in \tau(d_i, Z)$ such that

(1) $O_i^d \cap O_j^d = \emptyset$ if $i \neq j$;
(2) $O^d \cap D' = \{d\}$ where $O^d = O_1^d \times \cdots \times O_n^d$;
(3) if $y = (d_{j_1}, \ldots, d_{j_n}) \in \Delta_n$, then $O_{j_1}^d \times \cdots \times O_{j_n}^d \subset W$;
(4) if $z = (d_{j_1}, \ldots, d_{j_n}) \in (P_d \backslash \{d\}) \cap T$, then $(O_{j_1}^d \times \cdots \times O_{j_n}^d) \cap (E \cup F) = \emptyset$.

For each $d \in E$ and any $i \in M_n$ take a function $f_i^d \in C(Z, [0, 1])$ such that $f_i^d(d_i) = 1$ and $f_i^d(Z \backslash O_i^d) \subset \{0\}$; let $f_d = f_1^d + \cdots + f_n^d$. If $d = (d_1, \ldots, d_n) \in E$, then the set $U_d = \{f \in C_p(Z) : f(d_i) > 0 \text{ for all } i \in M_n\}$ is open and $f_d \in U_d$. Take any distinct $a, d \in E$ with $a = (a_1, \ldots, a_n)$ and $d = (d_1, \ldots, d_n)$. If $f_d \in U_a$, then for all $i \leq n$, there is $j_i \in M_n$ such that $f_{j_i}^d(a_i) > 0$ and hence $a_i \in O_{j_i}^d$. An immediate consequence is that $a \in O_{j_1}^d \times \cdots \times O_{j_n}^d$. If there exist $i, k \leq n$ such that $i \neq k$ and $j_i = j_k$, then $y = (d_{j_1}, \ldots, d_{j_n}) \in \Delta_n$ and $a \in O_{j_1}^d \times \cdots \times O_{j_n}^d \cap W$ by (3) which contradicts $E \cap W = \emptyset$.

Thus $j_i \neq j_k$ if $i \neq k$, i.e., the map $\sigma : M_n \to M_n$ defined by $\sigma(i) = j_i$ is a bijection; let $v = \sigma^{-1}$. Given any $z = (d_{k_1}, \ldots, d_{k_n}) \in P_a \cap T$, the point $a(z) = (a_{v(k_1)}, \ldots, a_{v(k_n)})$ belongs to $P_a \cap T$ because $a(z) \in (O_{k_1}^d \times \cdots \times O_{k_n}^d) \backslash W$ and $(O_{k_1}^d \times \cdots \times O_{k_n}^d) \cap F = \emptyset$ by the property (4). This shows that the set $A = \{a\} \cup \{a(z) : z \in P_a \cap T\}$ is contained in $P_a \cap T$ which contradicts the fact that $P_a \cap T$ has m elements by the choice of E while $|A| = m + 1$ and $A \subset P_a \cap T$.

This contradiction shows that $f_d \notin U_a$ whenever $d \neq a$ which implies that $U_a \cap E = \{f_a\}$ for each $a \in E$ and, in particular, $f_d \neq f_a$ for distinct d and a, i.e., the correspondence $d \to f_d$ is a bijection. Therefore $\{f_a : a \in E\}$ is a discrete subspace of $C_p(Z)$ of cardinality λ^+ which is a contradiction with $s(C_p(Z)) \leq \lambda$. Thus $s(Z^n \backslash U) \leq \lambda$ and Fact 1 is proved.   □

Returning to our solution, fix $n \in \mathbb{N}$ with $n \geq 2$; take an arbitrary family $\mathcal{U} \subset \tau(X \times X)$ such that $|\mathcal{U}| \leq \kappa$ and $\bigcap \mathcal{U} = \Delta = \Delta_2(X)$. Given distinct $i, j \in M_n$, let $q_{ij} : X^n \to X \times X$ be the natural projection onto the face defined by i and j, i.e., for any $x = (x_1, \ldots, x_n) \in X^n$ we have $q_{ij}(x) = (x_i, x_j) \in X \times X$. It is clear that

$\Delta_{ij}^n(X) = q_{ij}^{-1}(\Delta)$ and therefore $\Delta_{ij}^n(X) = \bigcap \mathcal{U}_{ij}$ where $\mathcal{U}_{ij} = \{q_{ij}^{-1}(U) : U \in \mathcal{U}\}$. If $B_n = \{(i, j) \in M_n \times M_n : i < j\}$, then the family $\mathcal{V} = \{U = \bigcup\{U_{ij} : (i, j) \in B_n\} : U_{ij} \in \mathcal{U}_{ij}$ for all $(i, j) \in B_n\}$ consists of open subsets of X^n and $\bigcap \mathcal{V} = \Delta_n(X)$. It is evident that $|\mathcal{V}| \leq \kappa$ and $X^n \backslash \Delta_n(X) = \bigcup\{X^n \backslash V : V \in \mathcal{V}\}$. For any set $F \in \mathcal{F} = \{X^n \backslash V : V \in \mathcal{V}\}$ we have $s(F) \leq \kappa$ by Fact 1. It follows from $|\mathcal{F}| \leq \kappa$ and $\bigcup \mathcal{F} = X^n \backslash \Delta_n(X)$ that $s(X^n \backslash \Delta_n(X)) \leq \kappa$ (it is an easy exercise to show that a union of $\leq \kappa$-many spaces of spread $\leq \kappa$ each has spread $\leq \kappa$). Let $\Delta_n = \Delta_n(X)$; we proved that:

(**) $s(X^n \backslash \Delta_n) \leq \kappa$ for any $n \in \mathbb{N}$ with $n \geq 2$.

Now we can prove that $s^*(X) \leq \kappa$ by induction on the power of X. If $n = 2$, then $s(X^n) \leq \kappa$ by Problem 016. Assume that we proved that $s(X^m) \leq \kappa$; then, for $n = m+1$, we have $X^n = \Delta_n \cup (X^n \backslash \Delta_n)$. Observe that Δ_{ij}^n is homeomorphic to $\Delta \times X^{n-2}$ which in turn is homeomorphic to $X^{n-1} = X^m$ because Δ is homeomorphic to X. As a consequence, Δ_n is a finite union of spaces homeomorphic to X^m so $s(\Delta_n) \leq \kappa$ by the induction hypothesis. Since $s(X^n \backslash \Delta_n) \leq \kappa$ by (**), we have $s(X^n) \leq \kappa$, so our inductive step is accomplished. This proves that $s^*(X) \leq \kappa$ and hence $s^*(C_p(X)) = s^*(X) \leq \kappa$ (see Problem 025), so our solution is complete.

T.029. *For an infinite cardinal κ, suppose that $hl(C_p(X)) = \kappa$ and $\Delta(X) \leq \kappa$. Prove that $hl^*(C_p(X)) \leq \kappa$ and hence $hl^*(C_p(X)) = hl(C_p(X))$. In particular, if X is a space with a G_δ-diagonal, then $hl^*(C_p(X)) = hl(C_p(X)) = hd^*(X)$.*

Solution. For an arbitrary $n \in \mathbb{N}$, let $M_n = \{1, \ldots, n\}$. If Z is a space and $n \geq 2$, let $\Delta_{ij}^n(Z) = \{z = (z_1, \ldots, z_n) \in Z^n : z_i = z_j\}$ for any distinct $i, j \in M_n$. The set $\Delta_n(Z) = \bigcup\{\Delta_{ij}^n(Z) : 1 \leq i < j \leq n\}$ is called the *n-diagonal* of the space Z. If the space Z is clear, we write Δ_n instead of $\Delta_n(Z)$. Call a set $U \in \tau(Z^n)$ *marked* if $U = U_1 \times \cdots \times U_n$ where $\{U_i : i \in M_n\} \subset \tau(Z)$ and $U_i \cap U_j = \emptyset$ for any distinct $i, j \in M_n$.

Fact 1. Given $n \in \mathbb{N}$, for any space Z and any marked set $U \subset Z^n$, we have $hd(U) \leq hl(C_p(Z))$.

Proof. Assume that $hl(C_p(Z)) = \lambda$; we have $U = U_1 \times \cdots \times U_n$ where the family $\{U_i : i \in M_n\}$ consists of open subsets of Z and $U_i \cap U_j = \emptyset$ for any distinct $i, j \in M_n$. If $hd(U) > \lambda$, then there exists a set $P = \{z_\alpha : \alpha < \lambda^+\} \subset U$ which is left-separated by its indexation (see Problem 004 and Fact 2 of T.004). For any $\alpha < \lambda^+$, we have $z_\alpha = (z_\alpha^1, \ldots, z_\alpha^n)$ where $z_\alpha^i \in U_i$ for any $i \in M_n$.

For each $\alpha < \lambda^+$ there exist $O_\alpha^1, \ldots, O_\alpha^n \in \tau(Z)$ with the following properties:

(1) $z_\alpha^i \in O_\alpha^i$ for all $i \in M_n$;
(2) $O_\alpha \cap P_\alpha = \emptyset$ where $O_\alpha = O_\alpha^1 \times \cdots \times O_\alpha^n$ and $P_\alpha = \{z_\beta : \beta < \alpha\}$;
(3) $O_\alpha^i \subset U_i$ for all $i \in M_n$.

For any $\alpha < \lambda^+$ and $i \in M_n$ take a function $f_\alpha^i \in C(Z, [0, 1])$ such that $f_\alpha^i(z_\alpha^i) = 1$ and $f_\alpha^i(Z \backslash O_\alpha^i) \subset \{0\}$; let $f_\alpha = f_\alpha^1 + \cdots + f_\alpha^n$. We will prove that the set $E = \{f_\alpha : \alpha < \lambda^+\}$ is right-separated by its indexation.

Let $U_\alpha = \{f \in C_p(Z) : f(z_\alpha^i) > 0$ for all $i \in M_n\}$. It is clear that U_α is an open subset of $C_p(Z)$ and $f_\alpha \in U_\alpha$ for all $\alpha < \lambda^+$. Assume that $\beta < \alpha$ and $f_\alpha \in U_\beta$. Then, for any $i \in M_n$, we have $f_\alpha(z_\beta^i) > 0$ and hence $z_\beta^i \in O_\alpha^j$ for some $j \in M_n$. However, $z_\beta^i \in U_i$ and $O_\alpha^j \cap U_i = \emptyset$ for any $j \neq i$. Therefore $z_\beta^i \in O_\alpha^i$ for all $i \in M_n$ whence $z_\beta \in O_\alpha$ which contradicts (2). This contradiction shows that we have

(*) $f_\alpha \notin U_\beta$ for any $\beta < \alpha < \lambda^+$ and, in particular, the map $z_\alpha \to f_\alpha$ is a bijection between P and E.

Take any $\alpha < \lambda^+$; given any $\beta < \alpha$, we have $\beta < \alpha'$ for any $\alpha' \geq \alpha$. An immediate consequence of (*) is that $f_{\alpha'} \notin U_\beta$. Therefore $U_\beta \cap E \subset E_\alpha$ where $E_\alpha = \{f_\gamma : \gamma < \alpha\}$. This shows that the set E_α is open in E for each $\alpha < \lambda^+$ and hence E is a right-separated (by its indexation) subspace of $C_p(Z)$. Since $|E| = \lambda^+$ by (*), we have a contradiction with $hl(C_p(Z)) \leq \lambda$ (see Problem 005). Fact 1 is proved.   □

Fact 2. Given a space Z and $n \geq 2$, we have $hd(Z^n \backslash U) \leq hl(C_p(Z))$ for any $U \in \tau(\Delta_n(Z), Z^n)$.

Proof. Let $hl(C_p(Z)) = \lambda$; we have $s(Z^n \backslash U) \leq s(C_p(Z)) \leq hl(C_p(Z)) = \lambda$ by Fact 1 of T.028. If $hd(Z^n \backslash U) > \lambda$, then there exists a set $D = \{z_\alpha : \alpha < \lambda^+\} \subset Z^n \backslash U$ which is left-separated by its indexation (see Problem 004 and Fact 2 of T.004). For any $\alpha < \lambda^+$ there is a marked set $U \in \tau(z_\alpha, Z^n)$. We have $hd(U) \leq \lambda$ by Fact 1. Thus $U_\alpha = U \cap D$ cannot have λ^+-many elements, i.e., $|U_\alpha| \leq \lambda$. The family $\mathcal{U} = \{U_\alpha : \alpha < \lambda^+\}$ is an open cover of D and $s(D) \leq \lambda$. Besides, the space D is left-separated, so we can apply Problem 007 to conclude that $hl(D) \leq s(D) \leq \lambda$. In particular, there is $\mathcal{U}' \subset \mathcal{U}$ such that $|\mathcal{U}'| \leq \lambda$ and $\bigcup \mathcal{U}' = D$. Since each $U \in \mathcal{U}'$ has cardinality $\leq \lambda$, we have $|D| \leq \lambda$ which is a contradiction. Therefore $hd(Z^n \backslash U) \leq \lambda$ and Fact 2 is proved.   □

Returning to our solution, fix $n \in \mathbb{N}$ with $n \geq 2$; take an arbitrary family $\mathcal{U} \subset \tau(X \times X)$ such that $|\mathcal{U}| \leq \kappa$ and $\bigcap \mathcal{U} = \Delta = \Delta_2(X)$. Given distinct $i, j \in M_n$, let $q_{ij} : X^n \to X \times X$ be the natural projection onto the face defined by i and j, i.e., for any $x = (x_1, \ldots, x_n) \in X^n$, we have $q_{ij}(x) = (x_i, x_j) \in X \times X$. It is clear that $\Delta_{ij}^n(X) = q_{ij}^{-1}(\Delta)$ and therefore $\Delta_{ij}^n(X) = \bigcap \mathcal{U}_{ij}$ where $\mathcal{U}_{ij} = \{q_{ij}^{-1}(U) : U \in \mathcal{U}\}$. If $B_n = \{(i, j) \in M_n \times M_n : i < j\}$, then the family $\mathcal{V} = \{U = \bigcup \{U_{ij} : (i, j) \in B_n\} : U_{ij} \in \mathcal{U}_{ij}$ for all $(i, j) \in B_n\}$ consists of open subsets of X^n and $\bigcap \mathcal{V} = \Delta_n(X)$. It is evident that $|\mathcal{V}| \leq \kappa$ and $X^n \backslash \Delta_n(X) = \bigcup \{X^n \backslash V : V \in \mathcal{V}\}$. For any set $F \in \mathcal{F} = \{X^n \backslash V : V \in \mathcal{V}\}$ we have $hd(F) \leq \kappa$ by Fact 2. It follows from $|\mathcal{F}| \leq \kappa$ and $\bigcup \mathcal{F} = X^n \backslash \Delta_n(X)$ that $hd(X^n \backslash \Delta_n(X)) \leq \kappa$ (it is an easy exercise to show that a union of $\leq \kappa$-many spaces of hereditary density $\leq \kappa$ each has hereditary density $\leq \kappa$). Let $\Delta_n = \Delta_n(X)$; we proved that

(**) $hd(X^n \setminus \Delta_n) \leq \kappa$ for any $n \in \mathbb{N}$ with $n \geq 2$.

Now we can prove that $hd^*(X) \leq \kappa$ by induction on the power of X. If $n = 2$, then $hd(X^n) \leq \kappa$ by Problem 017. Assume that we established that $hd(X^m) \leq \kappa$; then, for $n = m + 1$, we have $X^n = \Delta_n \cup (X^n \setminus \Delta_n)$. Observe that the space Δ_{ij}^n is homeomorphic to $\Delta \times X^{n-2}$ which is homeomorphic to $X^{n-1} = X^m$ because Δ is homeomorphic to X. As a consequence, Δ_n is a finite union of spaces homeomorphic to X^m so $hd(\Delta_n) \leq \kappa$ by the induction hypothesis. Since $hd(X^n \setminus \Delta_n) \leq \kappa$ by (**), we have $hd(X^n) \leq \kappa$, so our inductive step is accomplished. This proves that $hd^*(X) \leq \kappa$ and hence $hl^*(C_p(X)) = hd^*(X) \leq \kappa$ (see Problem 027), so our solution is complete.

T.030 (Velichko's theorem). *Prove that $hd^*(C_p(X)) = hd(C_p(X)) = hl^*(X)$ for any space X.*

Solution. For an arbitrary $n \in \mathbb{N}$, let $M_n = \{1, \ldots, n\}$. If Z is a space and $n \geq 2$, let $\Delta_{ij}^n(Z) = \{z = (z_1, \ldots, z_n) \in Z^n : z_i = z_j\}$ for any distinct $i, j \in M_n$. The set $\Delta_n(Z) = \bigcup \{\Delta_{ij}^n(Z) : 1 \leq i < j \leq n\}$ is called the *n-diagonal* of the space Z. If the space Z is clear, we write Δ_n instead of $\Delta_n(Z)$. Call a set $U \in \tau(Z^n)$ *marked* if $U = U_1 \times \cdots \times U_n$ where $\{U_i : i \in M_n\} \subset \tau(Z)$ and $U_i \cap U_j = \emptyset$ for any distinct $i, j \in M_n$.

Fact 1. Given $n \in \mathbb{N}$, for any space Z and any marked set $U \subset Z^n$, we have $hl(U) \leq hd(C_p(Z))$.

Proof. Assume that $hd(C_p(Z)) = \lambda$; we have $U = U_1 \times \cdots \times U_n$ where the family $\{U_i : i \in M_n\}$ consists of open subsets of Z and $U_i \cap U_j = \emptyset$ for any distinct $i, j \in M_n$. If $hl(U) > \lambda$, then there exists a set $P = \{z_\alpha : \alpha < \lambda^+\} \subset U$ which is right-separated by its indexation (see Problem 005 and Fact 2 of T.005). For any $\alpha < \lambda^+$, we have $z_\alpha = (z_\alpha^1, \ldots, z_\alpha^n)$ where $z_\alpha^i \in U_i$ for any $i \in M_n$.

For each $\alpha < \lambda^+$ there exist $O_\alpha^1, \ldots, O_\alpha^n \in \tau(Z)$ with the following properties:

(1) $z_\alpha^i \in O_\alpha^i$ for all $i \in M_n$;
(2) $O_\alpha \cap P_\alpha = \emptyset$ where $O_\alpha = O_\alpha^1 \times \cdots \times O_\alpha^n$ and $P_\alpha = \{z_\beta : \alpha < \beta\}$;
(3) $O_\alpha^i \subset U_i$ for all $i \in M_n$.

For any $\alpha < \lambda^+$ and $i \in M_n$ take a function $f_\alpha^i \in C(Z, [0, 1])$ such that $f_\alpha^i(z_\alpha^i) = 1$ and $f_\alpha^i(Z \setminus O_\alpha^i) \subset \{0\}$; let $f_\alpha = f_\alpha^1 + \cdots + f_\alpha^n$. We will prove that the set $E = \{f_\alpha : \alpha < \lambda^+\}$ is left-separated by its indexation.

Let $U_\alpha = \{f \in C_p(Z) : f(z_\alpha^i) > 0 \text{ for all } i \in M_n\}$. It is clear that U_α is an open subset of $C_p(Z)$ and $f_\alpha \in U_\alpha$ for all $\alpha < \lambda^+$. Assume that $\alpha < \beta$ and $f_\alpha \in U_\beta$. Then, for any $i \in M_n$, we have $f_\alpha(z_\beta^i) > 0$ and hence $z_\beta^i \in O_\alpha^j$ for some $j \in M_n$. However, $z_\beta^i \in U_i$ and $O_\alpha^j \cap U_i = \emptyset$ for any $j \neq i$. Therefore $z_\beta^i \in O_\alpha^i$ for all $i \in M_n$ whence $z_\beta \in O_\alpha$ which contradicts (2). This contradiction shows that we have

(*) $f_\alpha \notin U_\beta$ for any $\alpha < \beta < \lambda^+$ and, in particular, the map $z_\alpha \to f_\alpha$ is a bijection
between P and E.

Take any $\beta < \lambda^+$; given any $\alpha < \beta$, we have $f_\alpha \notin U_\beta$ by (*). Therefore
$U_\beta \cap E_\beta = \emptyset$ where $E_\beta = \{f_\gamma : \gamma < \beta\}$. This shows that the set E_β is closed in
E for each $\beta < \lambda^+$ and hence E is a left-separated (by its indexation) subspace of
$C_p(Z)$. Since $|E| = \lambda^+$ by (*), we have a contradiction with $hd(C_p(Z)) \le \lambda$ (see
Problem 004). Fact 1 is proved.   □

Fact 2. Given a space Z and $n \ge 2$, we have $hl(Z^n \backslash U) \le hd(C_p(Z))$ for any
$U \in \tau(\Delta_n(Z), Z^n)$.

Proof. Let $hd(C_p(Z)) = \lambda$; we have $s(Z^n \backslash U) \le s(C_p(Z)) \le hd(C_p(Z)) = \lambda$
by Fact 1 of T.028. Observe that $t(C_p(Z)) \le hd(C_p(Z)) = \lambda$ and therefore
$l^*(Z) \le \lambda$ by Problem 149 of [TFS]. The set $Z^n \backslash U$ is closed in Z^n so
$l(Z^n \backslash U) \le \lambda$. For every $z \in F = Z^n \backslash U$ there exists a marked set $U_z \in \tau(z, Z^n)$.
The family $\mathcal{U} = \{U_z : z \in F\}$ is an open cover of the space F, so there is
$\mathcal{U}' \subset \mathcal{U}$ such that $|\mathcal{U}'| \le \lambda$ and $\bigcup \mathcal{U}' \supset F$. We have $hl(V) \le \lambda$ for each $V \in \mathcal{U}'$
so $hl(\bigcup \mathcal{U}') \le \lambda$ (it is an easy exercise that a union of $\le \lambda$-many spaces with
hereditary Lindelöf number $\le \lambda$ has hereditary Lindelöf number $\le \lambda$). Therefore
$hl(Z^n \backslash U) \le hl(\bigcup \mathcal{U}') \le \lambda$ and Fact 2 is proved.   □

Returning to our solution, let $hd(C_p(X)) = \kappa$; observe that $hl(X \times X) \le \kappa$ by
Problem 018. Let $\Delta = \Delta_2(X)$; for any $z \in (X \times X) \backslash \Delta$, pick any $U_z \in \tau(z, X \times X)$
such that $\overline{U}_z \cap \Delta = \emptyset$. It follows immediately from $hl(X^2) \le \kappa$ that the open cover
$\mathcal{U} = \{U_z : z \in (X \times X) \backslash \Delta\}$ of the space $(X \times X) \backslash \Delta$ has a subcover $\mathcal{U}'$ such that
$|\mathcal{U}'| \le \kappa$. Therefore $(X \times X) \backslash \Delta = \bigcup\{\overline{U} : U \in \mathcal{U}'\}$, i.e., $(X \times X) \backslash \Delta$ is a union of
$\le \kappa$-many closed sets and therefore $\Delta(X) \le \kappa$.

Fix $n \in \mathbb{N}$ with $n \ge 2$; take an arbitrary family $\mathcal{U} \subset \tau(X \times X)$ such that $|\mathcal{U}| \le \kappa$
and $\bigcap \mathcal{U} = \Delta$. Given distinct $i, j \in M_n$, let $q_{ij} : X^n \to X \times X$ be the natural
projection onto the face defined by i and j, i.e., for any $x = (x_1, \ldots, x_n) \in X^n$ we
have $q_{ij}(x) = (x_i, x_j) \in X \times X$. It is clear that $\Delta^n_{ij}(X) = q_{ij}^{-1}(\Delta)$ and therefore
$\Delta^n_{ij}(X) = \bigcap \mathcal{U}_{ij}$ where $\mathcal{U}_{ij} = \{q_{ij}^{-1}(U) : U \in \mathcal{U}\}$. If $B_n = \{(i, j) \in M_n \times M_n : i <$
$j\}$, then the family $\mathcal{V} = \{U = \bigcup\{U_{ij} : (i, j) \in B_n\} : U_{ij} \in \mathcal{U}_{ij}$ for all $(i, j) \in B_n\}$
consists of open subsets of X^n and $\bigcap \mathcal{V} = \Delta_n(X)$. It is evident that $|\mathcal{V}| \le \kappa$ and
$X^n \backslash \Delta_n(X) = \bigcup\{X^n \backslash V : V \in \mathcal{V}\}$. For any set $F \in \mathcal{F} = \{X^n \backslash V : V \in \mathcal{V}\}$ we
have $hl(F) \le \kappa$ by Fact 2. It follows from $|\mathcal{F}| \le \kappa$ and $\bigcup \mathcal{F} = X^n \backslash \Delta_n(X)$ that
$hl(X^n \backslash \Delta_n(X)) \le \kappa$. Let $\Delta_n = \Delta_n(X)$; we proved that

(**) $hl(X^n \backslash \Delta_n) \le \kappa$ for any $n \in \mathbb{N}$ with $n \ge 2$.

Now we can prove that $hl^*(X) \le \kappa$ by induction on the power of X. If
$n = 2$, then $hl(X^n) \le \kappa$ by Problem 018. Assume that we established that
$hl(X^m) \le \kappa$; then, for $n = m + 1$, we have $X^n = \Delta_n \cup (X^n \backslash \Delta_n)$. Observe
that the space Δ^n_{ij} is homeomorphic to $\Delta \times X^{n-2}$ which is homeomorphic to
$X^{n-1} = X^m$ because Δ is homeomorphic to X. As a consequence, Δ_n is a finite
union of spaces homeomorphic to X^m, so $hl(\Delta_n) \le \kappa$ by the induction hypothesis.

Since $hl(X^n \setminus \Delta_n) \leq \kappa$ by $(**)$, we have $hl(X^n) \leq \kappa$ so our inductive step is accomplished. This proves that $hl^*(X) \leq \kappa$ and hence $hd^*(C_p(X)) = hl^*(X) \leq \kappa$ (see Problem 026), so our solution is complete.

T.031. *Prove that* $s(C_p(C_p(X))) = s^*(X)$ *for any space* X.

Solution. If X is empty, there is nothing to prove. If $X \neq \emptyset$, then we have the equalities $s^*(X) = s^*(C_p(X)) = s^*(C_p(C_p(X)))$ by Problem 025. Apply Problem 182 of [TFS] to conclude that there exists a space Y such that $C_p(X)$ is homeomorphic to $Y \times \mathbb{R}$. It follows from Problem 177 of [TFS] that the space $C_p(C_p(X))$ is homeomorphic to $(C_p(C_p(X)))^\omega$ and therefore $s^*(C_p(C_p(X))) = s(C_p(C_p(X))) = s^*(X)$.

T.032. *Prove that* $hd(C_p(C_p(X))) = hd^*(X)$ *for any space* X.

Solution. If X is empty, there is nothing to prove. If $X \neq \emptyset$, then we have the equalities $hd^*(X) = hl^*(C_p(X)) = hd^*(C_p(C_p(X)))$ by Problem 026 and Problem 027. Now apply Problem 030 to conclude that $hd^*(C_p(C_p(X))) = hd(C_p(C_p(X))) = hd^*(X)$.

T.033. *Prove that* $hl(C_p(C_p(X))) = hl^*(X)$ *for any space* X.

Solution. If X is empty, there is nothing to prove. If $X \neq \emptyset$, then we have the equalities $hl^*(X) = hd^*(C_p(X)) = hl^*(C_p(C_p(X)))$ by Problem 026 and Problem 027. Apply Problem 182 of [TFS] to conclude that there exists a space Y such that $C_p(X)$ is homeomorphic to $Y \times \mathbb{R}$. It follows from Problem 177 of [TFS] that the space $C_p(C_p(X))$ is homeomorphic to $(C_p(C_p(X)))^\omega$ and therefore $hl^*(C_p(C_p(X))) = hl(C_p(C_p(X))) = hl^*(X)$.

T.034. *Prove that for a zero-dimensional space* X, *we have* $s^*(X) = s(C_p(X))$.

Solution. For an arbitrary $n \in \mathbb{N}$, let $M_n = \{1, \ldots, n\}$. If Z is a space and $n \geq 2$, let $\Delta_{ij}^n(Z) = \{z = (z_1, \ldots, z_n) \in Z^n : z_i = z_j\}$ for any distinct $i, j \in M_n$. The set $\Delta_n(Z) = \bigcup \{\Delta_{ij}^n(Z) : 1 \leq i < j \leq n\}$ is called the n-diagonal of Z.

Let $\Delta_n = \Delta_n(X)$ for any $n \geq 2$; assume that for some infinite cardinal κ, we have $s(C_p(X)) = \kappa$ and $s(X^n) > \kappa$ for some $n \in \mathbb{N}$. Then $n > 2$ by Problem 016; by Fact 0 of T.019 there exists a discrete faithfully indexed $D = \{d_\alpha : \alpha < \kappa^+\} \subset X^n \setminus \Delta_n$. We have $d_\alpha = (x_1^\alpha, \ldots, x_n^\alpha)$ for each $\alpha < \kappa^+$. Given any $\alpha < \kappa^+$, choose sets $U_1^\alpha, \ldots, U_n^\alpha$ such that

(1) U_i^α is a clopen subset of X and $x_i^\alpha \in U_i^\alpha$ for each $i \in M_n$;
(2) if $i, j \in M_n$ and $i \neq j$, then $U_i^\alpha \cap U_j^\alpha = \emptyset$;
(3) $W_\alpha \cap D = \{d_\alpha\}$ where $W_\alpha = U_1^\alpha \times \cdots \times U_n^\alpha$ for each $\alpha < \kappa^+$.

For all $\alpha < \kappa^+$ let $V_\alpha = \bigcup \{U_i^\alpha : i \in M_n\}$ and take a function f_i^α such that $f_i^\alpha(U_i^\alpha) = \{i\}$ and $f_i^\alpha|(X \setminus U_i^\alpha) \equiv 0$ for all $i \in M_n$. It is evident that f_i^α is a continuous function for all $i \in M_n$, so the function $g_\alpha = f_1^\alpha + \cdots + f_n^\alpha$ is also continuous for any $\alpha < \kappa^+$. Let $O_\alpha = \{f \in C_p(X) : |f(x_i^\alpha) - g_\alpha(x_i^\alpha)| < 1/3$ for all $i \in M_n\}$. It is clear that O_α is open in $C_p(X)$ and $g_\alpha \in O_\alpha$ for all $\alpha < \kappa^+$.

If $\alpha \neq \beta$, then $d_\alpha \notin W_\beta$ by (3) and hence there is $i \in M_n$ such that $x_i^\alpha \notin U_i^\beta$. Therefore $g_\beta(x_i^\alpha) \in (\{0\} \cup M_n)\backslash\{i\}$ which implies $|g_\beta(x_i^\alpha) - g_\alpha(x_i^\alpha)| \geq 1 > \frac{1}{3}$ which shows in turn that $g_\beta \notin O_\alpha$. As a consequence, for the set $E = \{g_\alpha : \alpha < \kappa^+\}$ we have $O_\alpha \cap E = \{g_\alpha\}$ for each $\alpha < \kappa^+$. Thus E is a discrete subset of $C_p(X)$ of cardinality $\kappa^+ > \kappa = s(C_p(X))$ which is a contradiction. This contradiction shows that $s(X^n) \leq s(C_p(X))$ for all natural numbers n, i.e., $s^*(X) \leq s(C_p(X))$. However, $s(C_p(X)) \leq s^*(C_p(X)) = s^*(X)$ by Problem 025, so we have $s(C_p(X)) = s^*(X)$.

T.035. *Prove that $hd^*(X) = hl(C_p(X))$ for any zero-dimensional space X.*

Solution. For an arbitrary $n \in \mathbb{N}$, let $M_n = \{1, \dots, n\}$. If Z is a space and $n \geq 2$, let $\Delta_{ij}^n(Z) = \{z = (z_1, \dots, z_n) \in Z^n : z_i = z_j\}$ for any distinct $i, j \in M_n$. The set $\Delta_n(Z) = \bigcup\{\Delta_{ij}^n(Z) : 1 \leq i < j \leq n\}$ is called the n-*diagonal* of Z.

Let $\Delta_n = \Delta_n(X)$ for any $n \geq 2$; assume that for some infinite cardinal κ, we have $hl(C_p(X)) = \kappa < hd(X^n)$ for some $n \in \mathbb{N}$. Then $n > 2$ by Problem 017; by Fact 0 of T.020 and Fact 2 of T.004, there exists a set $D = \{d_\alpha : \alpha < \kappa^+\} \subset X^n\backslash\Delta_n$ which is left-separated by its indexation. We have $d_\alpha = (x_1^\alpha, \dots, x_n^\alpha)$ for each $\alpha < \kappa^+$. Given any $\alpha < \kappa^+$, choose sets $U_1^\alpha, \dots, U_n^\alpha$ such that

(1) U_i^α is a clopen subset of X and $x_i^\alpha \in U_i^\alpha$ for each $i \in M_n$;
(2) if $i, j \in M_n$ and $i \neq j$, then $U_i^\alpha \cap U_j^\alpha = \emptyset$;
(3) $W_\alpha \cap D \subset \{d_\beta : \beta \geq \alpha\}$ where $W_\alpha = U_1^\alpha \times \cdots \times U_n^\alpha$ for each $\alpha < \kappa^+$.

For all $\alpha < \kappa^+$ let $V_\alpha = \bigcup\{U_i^\alpha : i \in M_n\}$ and take a function f_i^α such that $f_i^\alpha(U_i^\alpha) = \{i\}$ and $f_i^\alpha|(X\backslash U_i^\alpha) \equiv 0$ for all $i \in M_n$. It is evident that f_i^α is a continuous function for all $i \in M_n$, so the function $g_\alpha = f_1^\alpha + \cdots + f_n^\alpha$ is also continuous for any $\alpha < \kappa^+$. Let $O_\alpha = \{f \in C_p(X) : |f(x_i^\alpha) - g_\alpha(x_i^\alpha)| < 1/3$ for all $i \in M_n\}$. It is clear that O_α is open in $C_p(X)$ and $g_\alpha \in O_\alpha$ for all $\alpha < \kappa^+$.

If $\alpha < \beta$, then $d_\alpha \notin W_\beta$ by (3), and hence there is $i \in M_n$ such that $x_i^\alpha \notin U_i^\beta$. Therefore $g_\beta(x_i^\alpha) \in (\{0\} \cup M_n)\backslash\{i\}$ which implies $|g_\beta(x_i^\alpha) - g_\alpha(x_i^\alpha)| \geq 1 > \frac{1}{3}$ which shows in turn that $g_\beta \notin O_\alpha$. As a consequence, for the set $E = \{g_\alpha : \alpha < \kappa^+\}$ we have $O_\alpha \cap E \subset \{g_\beta : \beta \leq \alpha\}$ for each $\alpha < \kappa^+$. Thus E is a right-separated (by its indexation) subset of $C_p(X)$ of cardinality $\kappa^+ > \kappa = hl(C_p(X))$ which is a contradiction. This contradiction shows that $hd(X^n) \leq hl(C_p(X))$ for all natural n, i.e., $hd^*(X) \leq hl(C_p(X))$. However, $hl(C_p(X)) \leq hl^*(C_p(X)) = hd^*(X)$ by Problem 027, so we have $hl(C_p(X)) = hd^*(X)$.

T.036. *Prove that, under SA, the following conditions are equivalent:*

(i) $s(C_p(X)) = \omega$;
(ii) $hl((C_p(X))^\omega) = \omega$;
(iii) $hd((C_p(X))^\omega) = \omega$.

In particular, if SA holds, then $hl(C_p(X)) = \omega$ implies $hl((C_p(X))^\omega) = \omega$ and $s(C_p(X)) = \omega$ implies $s((C_p(X))^\omega) = \omega$.

Solution. It is evident that (ii)$\Longrightarrow$(i). The axiom SA says that (iii)$\Longrightarrow$(ii), so we only must prove that (i)$\Longrightarrow$(iii). Assume that $s(C_p(X)) = \omega$.

Fact 1. Suppose that Y and Z are spaces such that $s(Y \times Z) \leq \kappa$ for some infinite cardinal κ. Then either $hd(Y) \leq \kappa$ or $hl(Z) \leq \kappa$.

Proof. Assume that $hd(Y) > \kappa$ and $hl(Z) > \kappa$. It follows from $hd(Y) > \kappa$ that there exists a left-separated subspace of Y of cardinality κ^+ (see Problem 004). Apply Fact 2 of T.004 to find a subspace $D = \{y_\alpha : \alpha < \kappa^+\} \subset Y$ which is left-separated by its indexation. Analogously, $hl(Z) > \kappa$ implies that there exists a subspace $E = \{z_\alpha : \alpha < \kappa^+\} \subset Z$ which is right-separated by its indexation (see Fact 2 of T.005).

Consider the set $H = \{(y_\alpha, z_\alpha) : \alpha < \kappa^+\} \subset Y \times Z$; for any ordinal $\alpha < \kappa^+$ the set $D_\alpha = \{y_\beta : \beta < \alpha\}$ is closed in D, so there exists $U_\alpha \in \tau(y_\alpha, Y)$ such that $U_\alpha \cap D_\alpha = \emptyset$. Analogously, the set $\{z_\beta : \beta < \alpha + 1\}$ is open in E, so the set $E_\alpha = \{z_\beta : \alpha < \beta\}$ is closed in E, and therefore there exists $V_\alpha \in \tau(z_\alpha, Z)$ such that $V_\alpha \cap E_\alpha = \emptyset$. As a consequence, $W_\alpha = U_\alpha \times V_\alpha \in \tau((y_\alpha, z_\alpha), Y \times Z)$ and $W_\alpha \cap H = \{(y_\alpha, z_\alpha)\}$ for each $\alpha < \kappa^+$. This shows that H is a discrete subspace of $Y \times Z$; since $|H| = \kappa^+$, this is a contradiction with $s(Y \times Z) \leq \kappa$. Thus it is impossible that $hd(Y) > \kappa$ and $hl(Z) > \kappa$, so we have $hd(Y) \leq \kappa$ or $hl(Z) \leq \kappa$. Fact 1 is proved. $\qquad\square$

Observe that $s(X \times X) \leq \omega$ by Problem 016. Hence $hl(X) \leq \omega$ or $hd(X) \leq \omega$ by Fact 1; if $hd(X) \leq \omega$, then apply SA to conclude that $hl(X) \leq \omega$ anyway.

Assume first that for every point $x \in X$, there exists a set $U_x \in \tau(x, X)$ such that $hl(C_p(\overline{U}_x)) \leq \omega$. We have $hd(\overline{U}_x \times \overline{U}_x) \leq \omega$ by Problem 017 and therefore $hl(\overline{U}_x \times \overline{U}_x) \leq \omega$ by SA. Since the space X is Lindelöf, there exists a countable $A \subset X$ such that $\bigcup\{U_x : x \in A\} = X$. The set $W = \bigcup\{U_x \times U_x : x \in A\}$ is open in $X \times X$ and $\Delta = \{(x, x) : x \in X\} \subset W$. Any countable union of hereditarily Lindelöf spaces is hereditarily Lindelöf, so $hl(W) = \omega$ which shows that Δ is a G_δ-set in W and hence in $X \times X$.

Now assume that there is a point $x \in X$ such that $hl(C_p(\overline{U})) > \omega$ for any set $U \in \tau(x, X)$. For the space $Y = X \backslash \{x\}$, we have $Y = \bigcup\{Y_n : n \in \omega\}$ where the subspaces Y_n can be chosen closed in X with $Y_n \subset Y_{n+1}$ for all $n \in \omega$. For each $n \in \omega$, take any $V_n \in \tau(x, X)$ such that $\overline{V}_n \cap Y_n = \emptyset$. The set $\overline{V}_n \cup Y_n$ is a closed subspace of the (Lindelöf and hence normal) space X, so the restriction maps $C_p(X)$ continuously onto $C_p(Y_n \cup \overline{V}_n)$. Since $Y_n \cup \overline{V}_n$ is homeomorphic to $Y_n \oplus \overline{V}_n$, the space $C_p(Y_n \cup \overline{V}_n)$ is homeomorphic to $C_p(Y_n) \times C_p(\overline{V}_n)$. Thus $s(C_p(Y_n) \times C_p(\overline{V}_n)) = \omega$; since $hl(C_p(\overline{V}_n)) > \omega$, we have $hd(C_p(Y_n)) \leq \omega$ by Fact 1. As a consequence, we obtain $hl(Y_n \times Y_n) \leq \omega$. As $Y \times Y = \bigcup\{Y_n \times Y_n : n \in \omega\}$, we have $hl(Y \times Y) = \omega$ and therefore $hl(X \times X) \leq \omega$. This implies again that Δ is a G_δ-subset of X.

We proved that in both cases $s(C_p(X)) = \omega$ implies, under SA, that $\Delta(X) = \omega$. Apply Problem 028 to conclude that $s^*(X) = \omega$. As before, for any natural n, the inequality $s(X^n \times X^n) \leq \omega$ implies $hl(X^n) \leq \omega$ by Fact 1 and SA. This proves that $hl^*(X) = \omega$ and hence $hd((C_p(X))^\omega) = hd^*(C_p(X)) = hl^*(X) = \omega$ (see Problem 012), so (i)$\Longrightarrow$(iii) and our proof is complete.

T.037. *Prove that the following statements are equivalent (remember that all spaces are assumed to be Tychonoff):*

(i)   *there is a space X with $s(X) = \omega$ and $d(X) > \omega$;*
(ii) *there is a space X with $hl(X) = \omega$ and $d(X) > \omega$;*
(iii) *there is a left-separated space X with $s(X) = \omega$ and $|X| = \omega_1$.*

Solution. If X is a space with $s(X) = \omega < d(X)$, then there is a left-separated uncountable subspace $Y \subset X$ (see Problem 004). It is an easy exercise that no uncountable left-separated space is separable, so $d(Y) > \omega$. Besides, Y has to be hereditarily Lindelöf by Problem 007 which proves that (i)$\Longrightarrow$(ii).

If X is as in (ii), then there is an uncountable left-separated $Y \subset X$. Take any $Z \subset Y$ with $|Z| = \omega_1$. Then Z is a left-separated space with $|Z| = \omega_1$ and $s(Z) = \omega$; this shows that the implication (ii)$\Longrightarrow$(iii) also holds.

Finally, if X is as in (iii), then X cannot be separable being uncountable and left-separated. Hence X also satisfies (i) and the implication (iii)$\Longrightarrow$(i) is established.

T.038 (Δ-system lemma).*Prove that for any regular uncountable cardinal κ, if $\mathcal{U}$ is a family of finite sets with $|\mathcal{U}| = \kappa$, then there exists a set F (called the Δ-root for $\mathcal{U}$) and a family $\mathcal{V} \subset \mathcal{U}$ (called the Δ-system for $\mathcal{U}$) such that $|\mathcal{V}| = \kappa$ and $A \cap B = F$ for any distinct $A, B \in \mathcal{V}$.*

Solution. It follows from regularity of κ that there is $n \in \omega$ and a subfamily $\mathcal{U}' \subset \mathcal{U}$ such that $|\mathcal{U}'| = \kappa$ and $|U| = n$ for any $U \in \mathcal{U}'$. Thus, it is sufficient to prove our statement for $\mathcal{U}'$. This shows that we can assume that there is $n \in \omega$ such that $|U| = n$ for any $U \in \mathcal{U}$. Our proof will be by induction on n.

If $n = 0$, then all elements of $\mathcal{U}$ are empty, so our statement holds vacuously. Assume that we have proved our Delta-lemma for all $n < m$ and take any family $\mathcal{U}$ of cardinality κ such that $|U| = m$ for all $U \in \mathcal{U}$. Let $X = \bigcup \mathcal{U}$; we must consider two cases:

Case 1. $|\{U \in \mathcal{U} : x \in U\}| < \kappa$ for any $x \in X$.

Let $\mathcal{V}$ be a maximal disjoint subfamily of $\mathcal{U}$; if $|\mathcal{V}| < \kappa$, then $|\bigcup \mathcal{V}| < \kappa$. If $Y = \bigcup \mathcal{V}$, then the family $\mathcal{U}' = \{U \in \mathcal{U} : U \cap Y \neq \emptyset\}$ has cardinality $< \kappa$ because κ is regular and each point of Y belongs to $< \kappa$ elements of $\mathcal{U}$. However, if $U \cap Y = \emptyset$, then the family $\mathcal{V} \cup \{U\} \subset \mathcal{U}$ is disjoint and strictly larger than $\mathcal{V}$ which contradicts maximality of $\mathcal{V}$. Therefore $U \cap Y \neq \emptyset$ for all $U \in \mathcal{U}$, i.e., $\mathcal{U} = \mathcal{U}'$ and hence $|\mathcal{U}| = |\mathcal{U}'| < \kappa$ which is a contradiction with $|\mathcal{U}| = \kappa$.

This contradiction shows that $\mathcal{V}$ is a disjoint subfamily of $\mathcal{U}$ of cardinality κ, so we can let $F = \emptyset$ finishing our proof in this case.

Case 2.   There is $x \in X$ such that $|\{U \in \mathcal{U} : x \in U\}| = \kappa$.

Let $\mathcal{U}' = \{U \in \mathcal{U} : x \in U\}$; the family $\mathcal{W} = \{U \backslash \{x\} : U \in \mathcal{U}'\}$ has cardinality κ and $|W| = m - 1$ for each $W \in \mathcal{W}$. Therefore we can apply the induction hypothesis to find a set F' and a subfamily $\mathcal{W}' \subset \mathcal{W}$ such that $|\mathcal{W}'| = \kappa$ and $V \cap W = F'$ for any distinct $V, W \in \mathcal{W}'$. Now, letting $F = F' \cup \{x\}$ and

$V = \{W \cup \{x\} : W \in W'\}$ we obtain a family $V \subset U$ such that $|V| = \kappa$ and $A \cap B = F$ for any distinct $A, B \in V$. Thus our statement holds in this case as well, so our proof is complete.

T.039. *Prove that, under CH, there exists a hereditarily Lindelöf non-separable dense subspace X of the space $\{0, 1\}^{\omega_1}$. In particular, L-spaces exist under CH.*

Solution. Denote by $\mathbb{D}$ the discrete two-point space $\{0, 1\}$. We will need the space $\Sigma = \{x \in \mathbb{D}^{\omega_1} : |x^{-1}(1)| \leq \omega\}$. It was proved in Fact 3 of S.307 that Σ is dense in $\mathbb{D}^{\omega_1}$, countably compact and non-compact while $\overline{A}$ is compact for any countable $A \subset \Sigma$ (the bar denotes the closure in Σ). An immediate consequence is that Σ is a dense countably compact non-separable subspace of $\mathbb{D}^{\omega_1}$. Fix an arbitrary base $\mathcal{B} = \{B_\alpha : \alpha < \omega_1\}$ in the space Σ such that $B_\alpha \neq \emptyset$ for each $\alpha < \omega_1$.

Fact 1. Let Z be a space such that $c(Z) \leq \omega$ and $w(Z) \leq \omega_1$. If CH holds, then there exists a family $\mathcal{N}$ of nowhere dense closed subspaces of Z such that $|\mathcal{N}| \leq \omega_1$ and, for any nowhere dense $F \subset Z$, there is $N \in \mathcal{N}$ such that $F \subset N$. We will say that the family $\mathcal{N}$ *is cofinal* in the family of all nowhere dense subsets of Z.

Proof. Take any base U in the space Z with $|U| \leq \omega_1$. Since CH holds, the family of all countable subfamilies of U has cardinality at most ω_1. Therefore the family $\mathcal{N} = \{Z \backslash (\bigcup V) : V \text{ is a countable disjoint subfamily of } U \text{ and } \bigcup V \text{ is dense in } Z\}$ consists of closed nowhere dense subsets of Z and $|\mathcal{N}| \leq \omega_1$.

To see that $\mathcal{N}$ is as promised, take any nowhere dense $F \subset Z$. Let V be a maximal disjoint subfamily of U such that $V \cap F = \emptyset$ for any $V \in V$. Since Z has the Souslin property, the family V is countable; it follows from maximality of V that $\bigcup V$ is dense in Z whence $N = Z \backslash (\bigcup V) \in \mathcal{N}$. It is immediate that $F \subset N$, so Fact 1 is proved. □

Returning to our solution, apply Fact 1 to the space Σ to find a cofinal family $\mathcal{N} = \{N_\alpha : \alpha < \omega_1\}$ of closed nowhere dense subsets of Σ. The space Σ has the Baire property (see Problem 274(ii) of [TFS]) and hence so does B_α for all $\alpha < \omega_1$ (see Problem 275 of [TFS]). The set N_0 being nowhere dense, we can choose a point $x_0 \in B_0 \backslash N_0$. Assume that $\alpha < \omega_1$ and we have chosen points $\{x_\beta : \beta < \alpha\}$ so that

(*) $\quad x_\beta \in B_\beta \backslash (\bigcup \{N_\gamma : \gamma \leq \beta\})$

for all $\beta < \alpha$. Since B_α has the Baire property, it cannot be covered by the first category set $N = \bigcup \{N_\beta : \beta \leq \alpha\}$, so we can choose a point $x_\alpha \in B_\alpha \backslash N$.

This shows that our inductive construction can go on giving us a subspace $X = \{x_\beta : \beta < \omega_1\} \subset \Sigma$ such that (*) holds for all $\beta < \omega_1$. We claim that X is the required space. Observe first that X is dense in Σ and hence in $\mathbb{D}^{\omega_1}$ because $X \cap B_\alpha \neq \emptyset$ for all $\alpha < \omega_1$. As a consequence, X cannot be separable because a countable dense subset of X would be a countable dense subset of Σ which is not separable, a contradiction.

The space X is dense in Σ and hence it has no isolated points. Now, if A is a right-separated subspace of X, then A is scattered (see Problem 006) and hence there is a discrete subspace D of X such that $A \subset \overline{D}$. It is an easy exercise that every discrete subspace of a space without isolated points is nowhere dense in this space so $A \subset \overline{D}$ is nowhere dense in X and hence in Σ. By Fact 1, there is $\alpha < \omega_1$ such that $A \subset N_\alpha$. By (∗), we have $X \cap N_\alpha \subset \{x_\beta : \beta < \alpha\}$ and therefore $A \subset \{x_\beta : \beta < \alpha\}$ which shows that A is countable. Applying Problem 005 we conclude that X is hereditarily Lindelöf. Therefore $hd(X) \geq d(X) \geq \omega_1 > hl(X) = \omega \geq s(X)$ and our solution is complete.

T.040. *Prove that, under CH, there exists a hereditarily separable non-Lindelöf dense subspace X of the space $\{0, 1\}^{\omega_1}$. Thus, S-spaces exist under CH.*

Solution. Denote by $\mathbb{D}$ the discrete two-point space $\{0, 1\}$. If f is a function, then $\mathrm{dom}(f)$ is its domain. For any set A, let $\mathrm{Fn}(A)$ be the set of all finite functions from A to $\mathbb{D}$, i.e., $\mathrm{Fn}(A) = \{s : s \in \mathbb{D}^B \text{ for some finite } B \subset A\}$. In this definition the set B can be empty, i.e., the empty function is considered to be an element of $\mathrm{Fn}(A)$. Any $s \in \mathrm{Fn}(A)$ determines a standard open subset $[s]$ of $\mathbb{D}^A$ by the formula $[s] = \{f \in \mathbb{D}^A : f|\mathrm{dom}(s) = s\}$. It is a standard practice to identify functions and their graphs so the fact that $f|\mathrm{dom}(s) = s$ can be written as $s \subset f$. We will use this agreement here for the sake of brevity. Given any set $A \subset \omega_1$, let $\pi_A : \mathbb{D}^{\omega_1} \to \mathbb{D}^A$ be the natural projection of $\mathbb{D}^{\omega_1}$ onto the face $\mathbb{D}^A$. Call a subset Z of the space $\mathbb{D}^{\omega_1}$ *finally dense* in $\mathbb{D}^{\omega_1}$ if there is $\alpha < \omega_1$ such that $\pi_{\omega_1 \setminus \alpha}(Z)$ is dense in $\mathbb{D}^{\omega_1 \setminus \alpha}$. A set $Z \subset \mathbb{D}^{\omega_1}$ is called *hereditarily finally dense (or HFD)* if any infinite $Y \subset Z$ is finally dense in $\mathbb{D}^{\omega_1}$.

Fact 1. Assume that $\mathcal{U} = \{U_n : n \in \omega\}$ is a family of infinite sets. Then there exists a disjoint family $\mathcal{V} = \{V_n : n \in \omega\}$ such that V_n is infinite and $V_n \subset U_n$ for all $n \in \omega$. We will say that the family $\mathcal{V}$ is a (disjoint) π-net for $\mathcal{U}$.

Proof. Let $\{n_k : k \in \omega\}$ be an enumeration of ω in which each $n \in \omega$ occurs infinitely often (see Fact 3 of S.286). Take $x_0 \in U_{n_0}$ arbitrarily; if we have chosen $x_0, \ldots, x_k$, we can choose $x_{k+1} \in U_{n_{k+1}} \setminus \{x_0, \ldots, x_k\}$ because the set $U_{n_{k+1}}$ is infinite. This inductive construction gives us a set $\{x_i : i \in \omega\}$ such that the correspondence $i \mapsto x_i$ is an injection. If $A_n = \{i \in \omega : n_i = n\}$, then the family $\{A_n : n \in \omega\}$ is disjoint and consists of infinite sets. Therefore the set $V_n = \{x_i : i \in A_n\}$ is infinite for each $n \in \omega$; the family $\{V_n : n \in \omega\}$ is clearly disjoint, consists of infinite sets and $V_n \subset U_n$ for each $n \in \omega$. Fact 1 is proved. □

Fact 2. Under CH there exists an HFD space $Z = \{z_\beta : \beta < \omega_1\} \subset \mathbb{D}^{\omega_1}$.

Proof. Apply CH to fix an enumeration $\{A_\beta : \beta < \omega_1\}$ of all countably infinite subsets of ω_1. For each $\beta < \omega_1$, let $\delta_\beta = \min\{\alpha \in \omega_1 : A_\beta \subset \alpha\}$; it is evident that $\delta_\beta \geq \omega$ for all $\beta < \omega_1$. We will define, by induction on α, the element $z_\beta(\alpha)$ for all $\beta < \omega_1$. To start the procedure, let $z_\beta(n) = 0$ for all $n \in \omega$ and $\beta < \omega_1$. Suppose that $\alpha < \omega_1$ and we defined $z_\beta(\gamma)$ for all $\beta < \omega_1$ and $\gamma < \alpha$ in such a way that

(i) $z_\beta(\gamma) = 0$ if $\beta \geq \gamma$;
(ii) if $\beta < \gamma < \alpha$ and $\gamma \geq \delta_\beta$, then, for any $s \in \mathrm{Fn}(\gamma \setminus \delta_\beta)$ (including $s = \emptyset$), the sets $\{\xi : \xi \in A_\beta, s \subset z_\xi \text{ and } z_\xi(\gamma) = 1\}$ and $\{\xi : \xi \in A_\beta, s \subset z_\xi \text{ and } z_\xi(\gamma) = 0\}$ are infinite.

Observe that is (i) trivially fulfilled for $\alpha = \omega$ while (ii) holds vacuously. Assume first that $\beta < \alpha$ and $\delta_\beta < \alpha$. Given any non-empty function $s \in \text{Fn}(\alpha \backslash \delta_\beta)$, we have $\text{dom}(s) = \{\gamma_1, \ldots, \gamma_n\}$ where $\gamma_1 < \cdots < \gamma_n < \alpha$. If $s' = s|\{\gamma_1, \ldots, \gamma_{n-1}\}$, then the induction hypothesis can be applied to the function s' and ordinals β and $\gamma = \gamma_n$ to conclude that the sets $\{\xi : \xi \in A_\beta, \ s' \subset z_\xi$ and $z_\xi(\gamma_n) = 1\}$ and $\{\xi : \xi \in A_\beta, \ s' \subset z_\xi$ and $z_\xi(\gamma_n) = 0\}$ are infinite. As a consequence, the set $\{\xi : \xi \in A_\beta, \ s' \subset z_\xi$ and $z_\xi(\gamma_n) = s(\gamma_n)\}$ is infinite and coincides with the set $C_{\beta,s} = \{\xi : \xi \in A_\beta$ and $s \subset z_\xi\}$. Therefore the family $C = \{C_{\beta,s} : \beta < \alpha, \ \delta_\beta < \alpha$ and $s \in \text{Fn}(\alpha \backslash \delta_\beta)\} \cup \{A_\beta : \delta_\beta = \alpha\}$ is countable and consists of countable infinite sets. Apply Fact 1 to find a π-net $\mathcal{N} = \{N_k : k \in \omega\}$ for the family C. Find an infinite $M_k \subset N_k$ such that $N_k \backslash M_k$ is also infinite for all $k \in \omega$. Given any $\beta < \alpha$, let $z_\beta(\alpha) = 1$ if $\beta \in \bigcup\{M_k : k \in \omega\}$ and $z_\beta(\alpha) = 0$ otherwise. Letting $z_\beta(\alpha) = 0$ for all $\beta \geq \alpha$, we finish our inductive construction.

It is now trivial that (i) holds for all $\beta \leq \alpha$; let us prove that (ii) is also fulfilled for all $\gamma \leq \alpha$. Since the statement of (ii) holds and does not depend on α when $\gamma < \alpha$, it suffices to show that (ii) is true for $\gamma = \alpha$. In case when $\delta_\beta = \gamma' < \alpha$ we proved that the set $C_{\beta,s}$ is infinite for any $s \in \text{Fn}(\alpha \backslash \delta_\beta)$. By our construction, there is $k \in \omega$ such that $N_k \subset C_{\beta,s}$. Observe that the set $\{\xi \in A_\beta : s \subset z_\xi$ and $z_\xi(\alpha) = 1\}$ coincides with the set $P = C_{\beta,s} \cap \{\xi : z_\xi(\alpha) = 1\}$ and $M_k \subset P$ by our construction. Since M_k is infinite, the set P is also infinite. Analogously, the set $\{\xi \in A_\beta : s \subset z_\xi$ and $z_\xi(\alpha) = 0\}$ coincides with the set $Q = C_{\beta,s} \cap \{\xi : z_\xi(\alpha) = 0\}$ and $N_k \backslash M_k \subset Q$ by our construction. Since $N_k \backslash M_k$ is infinite, the set Q is also infinite. This shows that (ii) holds for all $\beta < \alpha$ such that $\delta_\beta < \alpha$.

If $\delta_\beta = \alpha$, then the only element of $\text{Fn}(\alpha \backslash \delta_\beta)$ is the empty function, so the set $\{\xi \in A_\beta : s \subset z_\xi$ and $z_\xi(\alpha) = 1\}$ coincides with $Q = A_\beta \cap \{\xi : z_\xi(\alpha) = 1\}$. There is $l \in \omega$ such that $N_l \subset A_\beta$; by our construction we have $M_l \subset Q$ and hence Q is infinite. Analogously, the set $\{\xi \in A_\beta : s \subset z_\xi$ and $z_\xi(\alpha) = 0\}$ coincides with $P = A_\beta \cap \{\xi : z_\xi(\alpha) = 0\}$. By our construction we have $N_l \backslash M_l \subset P$ and hence P is infinite.

We have finally shown that (i) and (ii) are fulfilled for all $\gamma \leq \alpha$ and therefore our inductive construction can be continued giving us a set $Z = \{z_\beta : \beta < \omega_1\} \subset \mathbb{D}^{\omega_1}$ with the properties (i) and (ii). To see that Z is an HFD, take any infinite $Y \subset Z$. There exists $\beta < \omega_1$ such that $\{z_\xi : \xi \in A_\beta\} \subset Y$. We claim that Y is dense beyond δ_β. Indeed, it suffices to prove that for any $s \in \text{Fn}(\omega_1 \backslash \delta_\beta)$, there is $y \in Y$ such that $s \subset y$. If $s = \emptyset$, there is nothing to prove. If $\text{dom}(s) = \{\gamma_1, \ldots, \gamma_n\}$ where $\gamma_1 < \cdots < \gamma_n$, then we can apply (ii) to β and $\gamma = \gamma_n + 1 > \delta_\beta$ to see that there exists $\xi \in A_\beta$ such that $z_\xi(\gamma) = 1$ and $s \subset z_\xi$. Therefore $y = z_\xi \in Y$ and $s \subset y$. Thus any infinite $Y \subset Z$ is finally dense, so Z is an HFD and Fact 2 is proved. $\square$

If $x, y \in \mathbb{D}^{\omega_1}$ and the set $\{\alpha < \omega_1 : x(\alpha) \neq y(\alpha)\}$ is countable, we will call the points x and y *similar*.

Fact 3. Suppose that $Z = \{z_\beta : \beta < \omega_1\} \subset \mathbb{D}^{\omega_1}$ is an HFD. If a point $y_\alpha \in \mathbb{D}^{\omega_1}$ is similar to z_α for each $\alpha < \omega_1$, then the set $Y = \{y_\beta : \beta < \omega_1\}$ is also an HFD.

Proof. Take any infinite $Y' \subset Y$; there is a countably infinite set $A \subset \omega_1$ such that $P = \{y_\alpha : \alpha \in A\} \subset Y'$. The set $Z' = \{z_\alpha : \alpha \in A\}$ is finally dense, so we can take $\beta < \omega_1$ such that Z' is dense beyond β. Since each y_α is similar to z_α, there is a countable $B \subset \omega_1$ such that $\pi_{\omega_1 \setminus B}(y_\alpha) = \pi_{\omega_1 \setminus B}(z_\alpha)$ for all $\alpha \in A$. Take any $\gamma < \omega_1$ such that $\beta \cup B \subset \gamma$; it is immediate that $\pi_{\omega_1 \setminus \gamma}(P) = \pi_{\omega_1 \setminus \gamma}(Z')$. It follows from density of $\pi_{\omega_1 \setminus \beta}(Z')$ in $\mathbb{D}^{\omega_1 \setminus \beta}$ that $\pi_{\omega_1 \setminus \gamma}(Y') \supset \pi_{\omega_1 \setminus \gamma}(P) = \pi_{\omega_1 \setminus \gamma}(Z')$ is dense in $\mathbb{D}^{\omega_1 \setminus \gamma}$. Therefore Y' is finally dense in $\mathbb{D}^{\omega_1}$ and Fact 3 is proved.   □

Fact 4. Any HFD space $Z = \{z_\beta : \beta < \omega_1\} \subset \mathbb{D}^{\omega_1}$ is hereditarily separable.

Proof. If Z is not hereditarily separable, then there exists $Y = \{y_\alpha : \alpha < \omega_1\} \subset Z$ such that $y_\alpha \notin \overline{\{y_\beta : \beta < \alpha\}}$ for all $\alpha < \omega_1$. Choose a finite function $s_\alpha \in \text{Fn}(\omega_1)$ such that

(*) $y_\alpha \in [s_\alpha]$ and $[s_\alpha] \cap \{y_\beta : \beta < \alpha\} = \emptyset$

for all $\alpha < \omega_1$. Apply Delta-lemma (see Problem 038) to find an uncountable $A \subset \omega_1$ and a finite set $F \subset \omega_1$ such that $\text{dom}(s_\alpha) \cap \text{dom}(s_\beta) = F$ for any distinct $\alpha, \beta \in A$. Since the set $\mathbb{D}^F$ is finite, there is an uncountable $A' \subset A$ such that $y_\alpha | F = y_\beta | F$ for all $\alpha, \beta \in A'$. Take any countably infinite set $Y' \subset \{y_\alpha : \alpha \in A'\}$. Since Y is an HFD, there exists $\gamma < \omega_1$ such that Y' is dense beyond γ. Increasing γ if necessary, we can consider that $\{\beta : y_\beta \in Y'\} \cup F \subset \gamma$.

The family $\mathcal{S} = \{\text{dom}(s_\alpha) \setminus F : \alpha < \omega_1\}$ is disjoint, so only countably many elements of $\mathcal{S}$ can intersect the set γ. As a consequence, there exists $\beta_0 \in A'$ such that $\text{dom}(s_{\beta_0}) \setminus F \subset \omega_1 \setminus \gamma$ and $\beta < \beta_0$ whenever $y_\beta \in Y'$. The set Y' being dense beyond γ there is $\beta < \omega_1$ such that $y_\beta \in Y'$ and $y_\beta | (\text{dom}(s_{\beta_0}) \setminus F = y_{\beta_0} | \text{dom}(s_{\beta_0}) \setminus F$. We saw already that $y_\beta | F = y_{\beta_0} | F$ and therefore $y_\beta | \text{dom}(s_{\beta_0}) = y_{\beta_0} | \text{dom}(s_{\beta_0})$ which shows that $y_\beta \in [s_{\beta_0}]$ contradicting (*). This contradiction shows that Z is hereditarily separable, so Fact 4 is proved.   □

Returning to our solution, take any HFD space $Z = \{z_\alpha : \alpha < \omega_1\} \subset \mathbb{D}^{\omega_1}$ (Fact 2). For each $\alpha < \omega_1$, let $y_\alpha(\beta) = 0$ for all $\beta \leq \alpha$ and $y_\alpha(\beta) = z_\alpha(\beta)$ for each $\beta > \alpha$. Then y_α is similar to z_α for each $\alpha < \omega_1$, so the set $\{y_\alpha : \alpha < \omega_1\}$ is also an HFD (as a matter of fact, if we take the HFD space $Z = \{z_\alpha : \alpha < \omega_1\}$ constructed in Fact 2, then $y_\alpha = z_\alpha$ for all $\alpha < \omega_1$).

Observe that the family $\mathcal{B} = \{[s] : s \in \text{Fn}(\omega_1)\}$ is a base in $\mathbb{D}^{\omega_1}$ and $|\mathcal{B}| = \omega_1$. Choose any enumeration $\{B_\alpha : \alpha < \omega_1\}$ of $\mathcal{B}$ in which every $U \in \mathcal{B}$ occurs ω_1 times (see Fact 3 of S.286). We have $B_\alpha = [s_\alpha]$ for some $s_\alpha \in \text{Fn}(\omega_1)$; let $D_\alpha = \text{dom}(s_\alpha)$ for each $\alpha < \omega_1$. Let $x_\alpha(\beta) = y_\alpha(\beta)$ for all $\beta \in \omega_1 \setminus D_\alpha$ and $x_\alpha(\beta) = s_\alpha(\beta)$ for all $\beta \in D_\alpha$. Since we changed each y_α at finitely many coordinates, the space $X = \{x_\alpha : \alpha < \omega_1\}$ is still an HFD by Fact 3. The set X intersects each B_α because $x_\alpha \in B_\alpha \cap X$; this shows that X is dense in $\mathbb{D}^{\omega_1}$. Any infinite subset of an HFD is trivially an HFD, so we can consider that the point u whose all coordinates are zeros does not belong to X (for otherwise we just throw it away from X).

Let $W_\alpha = \{x \in \mathbb{D}^{\omega_1} : x(\alpha) = 1\}$. Then W_α is an open subset of $\mathbb{D}^{\omega_1}$ and the family $\mathcal{W} = \{W_\alpha : \alpha < \omega_1\}$ covers X because $u \notin X$. However the cover $\mathcal{W}$ does not have a countable subcover. Indeed, if $A \subset \omega_1$ is countable, then take any

$v \in \omega_1$ with $\beta < v$ for all $\beta \in A$. The function $s : \{v\} \to \mathbb{D}$ defined by $s(v) = 1$ occurs ω_1 times in the enumeration $\{s_\alpha : \alpha < \omega_1\}$, so there exists $\alpha > v$ such that $s_\alpha = s$. Observe that $D_\alpha = \{v\}$, so when we changed y_α to x_α we only affected the vth coordinate of y_α. Therefore $x_\alpha(\beta) = y_\alpha(\beta) = 0$ for all $\beta \in A$ and hence $x_\alpha \notin \bigcup\{W_\beta : \beta \in A\}$. As a consequence, the cover $\mathcal{W}$ of the space X has no countable subcover, so X is not Lindelöf. Finally, X is hereditarily separable by Fact 4, so our solution is complete.

T.041. *Prove that, under CH, any sequential compact space has points of countable character.*

Solution. Given a space X and a set $A \subset X$ denote by $\mathrm{Seq}(A)$ the set of all limits of sequences lying in A. It is clear that $A \subset \mathrm{Seq}(A) \subset \overline{A}$. Let $S_0(A) = A$ and, if we have sets $\{S_\beta(A) : \beta < \alpha\}$ for some $\alpha < \omega_1$, let $S_\alpha(A) = \mathrm{Seq}(\bigcup\{A_\beta : \beta < \alpha\})$. Observe that the family $\{S_\alpha(A) : \alpha < \omega_1\}$ has the following properties:

(1) $A \subset S_\alpha(A) \subset \overline{A}$ for all $\alpha < \omega_1$;
(2) $S_\alpha(A) \subset S_\beta(A)$ if $\alpha < \beta < \omega_1$

and therefore the set $S(A) = \bigcup\{S_\alpha(A) : \alpha < \omega_1\}$ also lies between A and the closure of A.

Fact 1. A space X is sequential if and only if $S(A) = \overline{A}$ for any $A \subset X$.

Proof. Suppose that $S(A) = \overline{A}$ for any $A \subset X$. If the space X is not sequential, then there is a non-closed $A \subset X$ such that $\mathrm{Seq}(A) = A$ and hence $S(A) = A \neq \overline{A}$, a contradiction which proves sufficiency.

Now assume that X is a sequential space and take any $A \subset X$. Observe that $\mathrm{Seq}(S(A)) = S(A)$; indeed, if B is a convergent sequence from $S(A)$, then $B \subset S_\alpha(A)$ for some $\alpha < \omega_1$, and hence the limit of B belongs to $S_{\alpha+1}(A) \subset S(A)$. Since X is sequential, the set $S(A)$ has to be closed and hence $S(A) \subset \overline{A} \subset \overline{S(A)} = S(A)$ and Fact 1 is proved.   □

Fact 2. If X is a sequential space and $|A| \leq \mathfrak{c}$ for some $A \subset X$, then $|\overline{A}| \leq \mathfrak{c}$.

Proof. Given any set B, let $P_\omega(B) = \{C \subset B : |C| \leq \omega\}$. Observe first that if $|B| \leq \mathfrak{c}$, then $|P_\omega(B)| \leq \mathfrak{c}$. Now, if $B \subset X$ then the set of all convergent sequences lying in B is a subfamily of $P_\omega(B)$; therefore $|\mathrm{Seq}(B)| \leq \mathfrak{c}$ whenever $|B| \leq \mathfrak{c}$.

We have $|S_0(A)| = |A| \leq \mathfrak{c}$. Assume that $\beta < \omega_1$ and we have proved that $|S_\alpha(A)| \leq \mathfrak{c}$ for all $\alpha < \beta$. Then $|\bigcup\{S_\alpha(A) : \alpha < \beta\}| \leq \omega \cdot \mathfrak{c} = \mathfrak{c}$ and therefore $|S_\beta(A)| = |\mathrm{Seq}(\bigcup\{S_\alpha(A) : \alpha < \beta\})| \leq \mathfrak{c}$ as well which proves that $|S_\beta(A)| \leq \mathfrak{c}$ for all $\beta < \omega_1$. Consequently, $|S(A)| \leq \omega_1 \cdot \mathfrak{c} = \mathfrak{c}$. Finally, apply Fact 1 to conclude that $|\overline{A}| = |S(A)| \leq \mathfrak{c}$ finishing the proof of Fact 2.   □

Fact 3. Any sequential space has countable tightness.

Proof. Let X be a sequential space. It suffices to prove that for any non-closed $A \subset X$, there is a countable $B \subset A$ such that $\overline{B} \backslash A \neq \emptyset$ (see Lemma of S.162). Since X is sequential, there is a convergent sequence $S \subset A$ (taken without its limit) such that $S \to x \in X \backslash A$. It is clear that the set $B = S$ is countable and $x \in \overline{B} \backslash A$, so $t(X) \leq \omega$ and Fact 3 is proved.   □

Fact 4. Let X be a compact space of countable tightness. Then there exists a countable set $A \subset X$ and a non-empty G_δ-set $H \subset X$ such that $H \subset \overline{A}$.

Proof. Recall that a set $\{x_\alpha : \alpha < \omega_1\} \subset X$ is called *a free sequence (of length ω_1)* if $\overline{\{x_\alpha : \alpha < \beta\}} \cap \overline{\{x_\alpha : \alpha \geq \beta\}} = \emptyset$ for any $\beta < \omega_1$. Assume that for any countable $A \subset X$, no non-empty G_δ-subset of X is contained in the closure of A. Choose a point $x_0 \in X$ and let $F_0 = X$. Assume that $\beta < \omega_1$ and we have chosen points $\{x_\alpha : \alpha < \beta\}$ and non-empty closed G_δ-sets $\{F_\alpha : \alpha < \beta\}$ such that

(1) $x_\alpha \in F_\alpha \subset F_\gamma$ whenever $\gamma < \alpha < \beta$;
(2) $\overline{\{x_\alpha : \alpha < \gamma\}} \cap F_\gamma = \emptyset$ for all $\gamma < \beta$.

The set $F = \bigcap \{F_\alpha : \alpha < \beta\}$ is a non-empty G_δ-subset of X; since the set $A = \{x_\alpha : \alpha < \beta\}$ is countable, we have $H' = F \backslash \overline{A} \neq \emptyset$. Evidently, H is a non-empty G_δ-subset of X, so we can find a closed non-empty G_δ-subset $F_\beta \subset H$ (see Fact 2 of S.328). Picking any point $x_\beta \in F_\beta$ we complete our inductive construction of the set $S = \{x_\alpha : \alpha < \omega_1\}$ and the family $\{F_\alpha : \alpha < \omega_1\}$ with the properties (1) and (2).

We claim that the set S is a free sequence in X. Indeed, if $\beta < \omega_1$, then we have $\{x_\alpha : \alpha \geq \beta\} \subset F_\beta$ by (1); thus $\overline{\{x_\alpha : \alpha \geq \beta\}} \subset \overline{F_\beta} = F_\beta$ and therefore $\overline{\{x_\alpha : \alpha < \beta\}} \cap \overline{\{x_\alpha : \alpha \geq \beta\}} \subset \overline{\{x_\alpha : \alpha < \beta\}} \cap F_\beta = \emptyset$ for any $\beta < \omega_1$ by (2). As a consequence, the compact space X has an uncountable free sequence which contradicts $t(X) = \omega$ (see Problem 328 of [TFS]). Fact 4 is proved.   □

Returning to our solution, assume CH and take any compact sequential space X. The space X has countable tightness by Fact 3 and therefore there exists a countable $A \subset X$ and a non-empty G_δ-set $H \subset X$ such that $H \subset \overline{A}$ by Fact 4. We have $|H| \leq |\overline{A}| \leq \mathfrak{c}$ by Fact 2. Apply again Fact 2 of S.328 to find a non-empty closed G_δ-set $P \subset H$; then $|P| \leq |H| \leq \mathfrak{c} = \omega_1$. If $\chi(x, P) \geq \omega_1$ for any $x \in P$, then we can apply Problem 330 of [TFS] to conclude that $|P| \geq 2^{\omega_1} > \omega_1$ which is a contradiction. Therefore $\chi(x, P) \leq \omega$ for some $x \in P$. Since P is a G_δ-set in X and $\{x\}$ is a G_δ-set in P, the point x is a G_δ-set in X (see Fact 2 of S.358). As a consequence, $\chi(x, X) = \psi(x, X) = \omega$ (see Problem 327 of [TFS]), i.e., x is a point of countable character in X and our solution is complete.

T.042. *Prove that, under CH, there is a P-point in $\beta\omega \backslash \omega$.*

Solution. Let $\omega^* = \beta\omega \backslash \omega$; given an infinite set $A \subset \omega$, let $[A] = \overline{A} \cap \omega^*$ (the closure is taken in $\beta\omega$). Then the family $\mathcal{B} = \{[A] : A$ is an infinite subset of $\omega\}$ is a base in ω^* (see Fact 2 of S.370). Every element of $\mathcal{B}$ is clopen in ω^* (Fact 1 of S.370) and, under CH, the cardinality of $\mathcal{B}$ is equal to ω_1. If U is an arbitrary clopen subset of ω^*, then $U = \bigcup \mathcal{B}'$ for some $\mathcal{B}' \subset \mathcal{B}$; since U is compact, we can consider that $\mathcal{B}'$ is finite. Since there are at most ω_1 finite subfamilies of $\mathcal{B}$, the family $\mathcal{C}$ of all clopen subsets of ω^* also has cardinality ω_1.

Fact 1. No non-empty open subset of ω^* is a union of $\leq \omega_1$ of nowhere dense subsets of ω^*.

Proof. Assume that, on the contrary, we have $U \subset \bigcup\{F_\alpha : \alpha < \omega_1\}$ where F_α is a nowhere dense subset of ω^*. Since the closure of a nowhere dense subset is nowhere dense, we can assume that F_α is closed in ω^* for each $\alpha < \omega_1$. Since $\mathcal{B}$ is a base in ω^*, we can find $U_0 \in \mathcal{B}$ such that $U_0 \subset U \backslash F_0$. Suppose that $\beta < \omega_1$ and we have sets $\{U_\alpha : \alpha < \beta\}$ with the following properties:

(1) $U_\alpha \in \mathcal{B}$ and $U_\alpha \cap F_\alpha = \emptyset$ for all $\alpha < \beta$;
(2) $U_\alpha \subset U_\gamma$ whenever $\gamma < \alpha < \beta$.

By compactness of ω^*, the set $P = \bigcap\{U_\alpha : \alpha < \beta\}$ is closed and non-empty; being a G_δ-subset of ω^* it has a non-empty interior by Problem 370 of [TFS]; choose any $U_\beta \in \mathcal{B}$ with $U_\beta \subset \mathrm{Int}(P)\backslash F_\beta$. It is evident that (1) and (2) hold for all $\alpha \leq \beta$ and therefore our inductive construction can be continued giving us a family $\{U_\alpha : \alpha < \omega_1\}$ with the properties (1) and (2). The property (2) and compactness of ω^* imply that the set $Q = \bigcap\{U_\alpha : \alpha < \omega_1\}$ is non-empty. If $x \in Q$, then $x \in U$ and $x \notin F_\alpha$ for any $\alpha < \omega_1$ by (1). Therefore $x \in U\backslash\bigcup\{F_\alpha : \alpha < \omega_1\}$; this contradiction finishes the proof of Fact 1. □

Returning to our solution observe that the family $\mathcal{U}$ of all countable unions of elements of $\mathcal{C}$ has cardinality $\leq \omega_1^\omega = \mathfrak{c} = \omega_1$; given any $U \in \mathcal{U}$, the set $\overline{U}\backslash U$ is nowhere dense in ω^*. Since $|\mathcal{U}| \leq \omega_1$, we can apply Fact 1 to conclude that there exists a point $x \in \omega^*\backslash\bigcup\{\overline{U}\backslash U : U \in \mathcal{U}\}$. We claim that x is a P-point of the space ω^*. To prove it, take any family $\{V_n : n \in \omega\} \subset \tau(x, \omega^*)$. Since $\mathcal{B}$ is a base in ω^*, there exists a family $\{W_n : n \in \omega\} \subset \mathcal{B}$ such that $W_{n+1} \subset W_n$ and $x \in W_n \subset V_n$ for all $n \in \omega$. It suffices to show that $x \in \mathrm{Int}(W)$ where $W = \bigcap\{W_n : n \in \omega\}$. Now, if this is not true, then $x \in \overline{U}$ where $U = \omega^*\backslash W$ is open and $U = \bigcup\{\omega^*\backslash W_n : n \in \omega\}$. Observe that $\omega^*\backslash W_n \in \mathcal{C}$ for each $n \in \omega$ and therefore $U \in \mathcal{U}$. As a consequence, $x \in \overline{U}\backslash U$ for some $U \in \mathcal{U}$ which is a contradiction with our choice of the point x. Thus x is a P-point of ω^*, so our solution is complete.

T.043. *Let X be a Luzin space. Prove that $hl(X) = \omega$ and $hd(X) \leq \omega_1$.*

Solution. Take any right-separated $A \subset X$; then A is scattered (see Problem 006) and hence it has a dense discrete subspace D. It is easy to see that every discrete subspace is nowhere dense in a space without isolated points. Therefore $A \subset \mathrm{cl}_X(D)$ is also nowhere dense in X and hence countable because X is a Luzin space. We proved that every right-separated $A \subset X$ is countable and hence X is hereditarily Lindelöf by Problem 005.

To show that $hd(X) \leq \omega_1$, we will need several observations. Take any $Y \subset X$; if $U = \mathrm{Int}(\overline{Y})$, then $Y\backslash U$ is nowhere dense in X and hence countable. The set $Y' = Y \cap U$ is dense in U and hence it is dense-in-itself (unless it is empty, of course). Any nowhere dense subset of Y' is, evidently, nowhere dense in X and hence countable. This shows that Y' is a Luzin space (or an empty space). As a consequence, any $Y \subset X$ is a union of a Luzin space Y' (which may be empty) and a countable set. Therefore, to prove that $d(Y) \leq \omega_1$ for every $Y \subset X$, it suffices to establish that $d(Z) \leq \omega_1$ for any Luzin space Z.

We will prove first that any non-empty open subset of Z contains a non-empty open subset of density $\leq \omega_1$. Take any $U \in \tau^*(Z)$; there is a left-separated $L \subset U$ which is dense in U (see Problem 009). Let $<$ be the well-order that left-separates L. For any $x \in L$, let $L_x = \{y \in L : y < x\}$. It is an exercise for the reader that if $|L_x| \leq \omega$ for all $x \in L$, then $|L| \leq \omega_1$. If $|L| \leq \omega_1$, there is nothing to prove because then $d(U) \leq \omega_1$; if $|L| > \omega_1$, then the set L_x is uncountable for some $x \in L$; let $z = \min\{y \in L : |L_y| > \omega\}$. The set L_z has cardinality ω_1 and hence it must be dense in some open set $V \subset U$. It is evident that $d(V) \leq \omega_1$ and hence we established that for every $U \in \tau^*(Z)$, there is $V \in \tau^*(Z)$ such that $V \subset U$ and $d(V) \leq \omega_1$.

Now let $\mathcal{U}$ be a maximal disjoint family of non-empty open subsets of Z of density $\leq \omega_1$. Since every non-empty open $U \subset Z$ contains an element of $\mathcal{U}$, the set $\bigcup \mathcal{U}$ is dense in Z. We have $c(Z) \leq hl(Z) = \omega$, so $\mathcal{U}$ is countable. It is an easy exercise that a countable union of spaces of density $\leq \omega_1$ has density $\leq \omega_1$, so $d(\bigcup \mathcal{U}) \leq \omega_1$. Any dense subset of $\bigcup \mathcal{U}$ is also dense in Z, so $d(Z) \leq \omega_1$ and our solution is complete.

T.044. *Prove that if a Luzin space X is separable, then all closed subsets of X are separable.*

Solution. Fix any dense countable $A \subset X$; if F is a closed subspace of X, then $P = F \setminus \text{Int}(F)$ is nowhere dense in X and hence countable. The set $Q = A \cap \text{Int}(F)$ is countable and dense in $\text{Int}(F)$ whence the set $P \cup Q$ is a countable dense subset of the space F.

T.045. *Prove that no Luzin space can be condensed onto a compact space.*

Solution. Denote by $\mathbb{D}$ the discrete two-point space $\{0, 1\}$; we let $\mathbb{D}^0 = \{\emptyset\}$. If f is a function, then $\text{dom}(f)$ is its domain. It is a standard practice to identify functions and their graphs, so the fact that for functions f and g, we have $f|\text{dom}(g) = g$ can be written as $g \subset f$. We will use this agreement here for the sake of brevity. As usual, any ordinal is identified with the set of its predecessors; in particular, $n = \{0, \ldots, n-1\}$ for any $n \in \omega$. If $s \in \mathbb{D}^k$ for some $k \in \omega$, then, for any $i \in \mathbb{D}$, we denote by $s^\frown i$ the function $t \in \mathbb{D}^{k+1}$ such that $t|k = s$ and $t(k) = i$.

Fact 1. If Z is a Fréchet–Urysohn space without isolated points, then there exists a closed separable dense-in-itself subspace $Y \subset Z$.

Proof. Since no point of Z is isolated, it follows from the Fréchet–Urysohn property of Z that, for each $a \in Z$, there is a sequence $S_a \subset Z \setminus \{a\}$ which converges to a. Take any countably infinite $A_0 \subset Z$; if we have countable sets $A_0 \subset \cdots \subset A_n \subset Z$, let $A_{n+1} = A_n \cup (\bigcup\{S_a : a \in A_n\})$. It is clear that this construction gives us a sequence $\{A_n : n \in \omega\}$ of countable subsets of Z such that every point $a \in A_n$ is a limit of a sequence from $A_{n+1} \setminus \{a\}$. Consequently, the set $A = \bigcup\{A_n : n \in \omega\}$ is countable and has no isolated points. Therefore the space $Y = \overline{A}$ is the promised dense-in-itself separable closed subspace of Z. Fact 1 is proved.   □

Fact 2. Let Z be a compact dense-in-itself space; given a family $\{F_n : n \in \omega\}$ of closed nowhere dense subspaces of Z, there exists an uncountable closed subspace $K \subset Z \setminus \bigcup \{F_n : n \in \omega\}$.

Proof. Since F_0 is nowhere dense in Z, we can choose $U_\emptyset \in \tau^*(Z)$ with $\overline{U}_\emptyset \cap F_0 = \emptyset$. Suppose that $n \geq 1$ and, for any $k < n$ and every $s \in \mathbb{D}^k$, we have a non-empty open subset U_s of the space Z such that

(1) if $s \in \mathbb{D}^k$, then $\overline{U}_s \cap F_k = \emptyset$;
(2) if $s, t \in \mathbb{D}^k$ and $s \neq t$, then $\overline{U}_s \cap \overline{U}_t = \emptyset$;
(3) if $s \in \mathbb{D}^k$, $t \in \mathbb{D}^m$ and $t \subset s$, then $\overline{U}_s \subset U_t$.

Take any $t \in \mathbb{D}^{n-1}$; since $U_t \in \tau^*(Z)$ and F_n is nowhere dense in Z, we can choose $V, W \in \tau^*(Z)$ such that $\overline{V} \cup \overline{W} \subset U_t$, $\overline{V} \cap \overline{W} = \emptyset$ and $(\overline{V} \cup \overline{W}) \cap F_n = \emptyset$. Let $U_{t \frown 0} = V$ and $U_{t \frown 1} = W$. After carrying out this procedure for all $t \in \mathbb{D}^{n-1}$ we will obtain a non-empty open set U_s for every $s \in \mathbb{D}^n$. The property (1), evidently, holds for $k = n$. Given distinct $s, t \in \mathbb{D}^n$, we have two cases:

(1) $s|(n-1) = t|(n-1)$; then, by our construction, $U_s = V$ and $U_t = W$ for some $V, W \in \tau^*(Z)$ such that $\overline{V} \cap \overline{W} = \emptyset$. Therefore $\overline{U}_s \cap \overline{U}_t = \overline{V} \cap \overline{W} = \emptyset$, so the property (2) also holds for $k = n$.
(2) $s' = s|(n-1) \neq t' = t|(n-1)$; then $\overline{U}_s \cap \overline{U}_t \subset \overline{U}_{s'} \cap \overline{U}_{t'} = \emptyset$ (the last equality is valid by the property (2) for $k = n-1$), and hence (2) is fulfilled in this case as well.

To check the property (3) observe that it suffices to show that it holds for $k = n$ and any $m < n$. By our construction we have $\overline{U}_{s'} \subset U_s$ where $s' = s|(n-1)$. The property (3) applied to $k = n-1$ and $m \leq k$ shows that $\overline{U}_t \subset \overline{U}_{s'} \subset U_s$, and hence (3) is also true when $k = n$. Thus our inductive construction can go on providing us, for any $n \in \omega$, a non-empty open set U_s for each $s \in \mathbb{D}^n$ in such a way that the properties (1)–(3) are fulfilled for all $k \in \omega$.

The set $K = \bigcap \{\bigcup \{\overline{U}_s : s \in \mathbb{D}^n\} : n \in \omega\}$ is compact being the intersection of closed subspaces of Z. If $z \in K$, then there is a sequence $\{s_n : n \in \omega\}$ such that $s_n \in \mathbb{D}^n$ and $z \in \bigcap \{\overline{U}_{s_n} : n \in \omega\}$. It follows from (1) that $\overline{U}_{s_n} \cap F_n = \emptyset$ for all $n \in \omega$ and therefore $z \notin F_n$ for all $n \in \omega$. This shows that $K \subset Z \setminus \bigcup \{F_n : n \in \omega\}$. Finally, to see that the set K is uncountable, observe that for any $s \in \mathbb{D}^\omega$, the set $K_s = \bigcap \{\overline{U}_{s|n} : n \in \omega\}$ is non-empty because the family $\{\overline{U}_{s|n} : n \in \omega\}$ is decreasing by (3) and consists of compact sets.

Given distinct $s, t \in \mathbb{D}^\omega$, there is a number $n \in \omega$ such that $s|n \neq t|n$ and hence $K_s \cap K_t \subset \overline{U}_{s|n} \cap \overline{U}_{t|n} = \emptyset$ by (2). As a consequence, if we choose a point $y_s \in K_s$ for all $s \in \mathbb{D}^\omega$, then $s \mapsto y_s$ is an injection of $\mathbb{D}^\omega$ to K and hence $|K| \geq |\mathbb{D}^\omega| = \mathfrak{c} > \omega$. Fact 2 is proved. $\square$

Fact 3. If Z is a first countable dense-in-itself compact space, then there exists an uncountable disjoint family $\mathcal{K}$ of closed uncountable subsets of Z.

Proof. Any first countable space is Fréchet–Urysohn, so we can apply Fact 1 to find a separable closed dense-in-itself $Y \subset Z$. Of course, it suffices to construct the promised family in Y, so we can consider that Z is separable; fix any countable dense $D \subset Z$. The family $\{\{d\} : d \in D\}$ is countable and consists of nowhere dense subsets of Z. This makes it possible to apply Fact 2 to find an uncountable compact subset $K_0 \subset Z \backslash D$. It follows from density of D that K_0 is nowhere dense in Z.

Suppose that $\alpha < \omega_1$ and we have a family $\{K_\beta : \beta < \alpha\}$ of uncountable nowhere dense subsets of Z with the following properties:

(4) $K_\beta \subset Z \backslash D$ for all $\beta < \alpha$;
(5) $K_\beta \cap K_\gamma = \emptyset$ for any distinct $\beta, \gamma < \alpha$.

The family $\mathcal{F} = \{\{d\} : d \in D\} \cup \{F_\beta : \beta < \alpha\}$ is countable and consists of nowhere dense subsets of Z. Therefore we can apply Fact 2 to find an uncountable dense-in-itself compact set $K_\alpha \subset Z \backslash \bigcup \mathcal{F}$. It follows from $K_\alpha \subset Z \backslash D$ that K_α is nowhere dense in Z, so our inductive construction can be continued, giving us a family $\mathcal{K} = \{K_\alpha : \alpha < \omega_1\}$ with properties (4) and (5). An immediate consequence of (5) is that the family $\mathcal{K}$ is disjoint; since its consists of uncountable compact subsets of Z, the proof of Fact 3 is complete. $\square$

Now it is easy to finish our solution. Suppose that there is a condensation $f : L \to K$ of a Luzin space L onto a compact space K. It is immediate that any continuous image of a hereditarily Lindelöf space is hereditarily Lindelöf, so $hl(K) \leq \omega$ because $hl(L) \leq \omega$ (see Problem 043). Therefore K is perfect (see Problem 001) and hence $\chi(K) = \psi(K) = \omega$ (see Problem 327 of [TFS]). If x is an isolated point of K, then $f^{-1}(x)$ is an isolated point of L; this contradiction shows that K is dense-in-itself. By Fact 3, there is an uncountable disjoint family $\mathcal{K}$ of uncountable closed subsets of K. The family $\mathcal{L} = \{f^{-1}(F) : F \in \mathcal{K}\}$ is also disjoint and consists of uncountable closed subsets of L. Since L is a Luzin space, every $P \in \mathcal{L}$ has non-empty interior in L. The interiors of all elements of $\mathcal{L}$ form a disjoint uncountable family of non-empty open subsets of L which is a contradiction with $c(L) \leq s(L) \leq hl(L) \leq \omega$. This contradiction shows that no Luzin space can condense onto a compact space.

T.046. *Prove that, under CH, there is a Luzin second countable space as well as a Luzin non-separable space.*

Solution. Denote by $\mathbb{D}$ the discrete two-point space $\{0, 1\}$. We will also need the space $\Sigma = \{x \in \mathbb{D}^{\omega_1} : |x^{-1}(1)| \leq \omega\}$.

Fact 1. Let X be a dense-in-itself space with the Baire property such that $c(X) \leq \omega$ and $w(X) \leq \omega_1$. Then, under CH, there is a dense Luzin subspace in the space X.

Proof. Fix an arbitrary base $\mathcal{B} = \{B_\alpha : \alpha < \omega_1\}$ in the space X such that $B_\alpha \neq \emptyset$ for each $\alpha < \omega_1$ (repetitions are allowed in this enumeration to cover the case when X has a countable base). Since $c(X) = \omega$ and $w(X) \leq \omega_1$, we can apply Fact 1 of T.039 to conclude that there exists a family $\mathcal{N} = \{N_\alpha : \alpha < \omega_1\}$ of closed nowhere dense subsets of X which is *cofinal* in the family of all nowhere dense subsets of X, i.e., for any nowhere dense subset N of the space X, there is $\alpha < \omega_1$ such that

$N \subset N_\alpha$. Since the space X has the Baire property, so does B_α for all $\alpha < \omega_1$ (see Problem 275 of [TFS]). The set N_0 being nowhere dense, we can choose a point $x_0 \in B_0 \backslash N_0$. Assume that $\alpha < \omega_1$ and we have chosen points $\{x_\beta : \beta < \alpha\}$ so that

(*) $x_\beta \in B_\beta \backslash (\bigcup \{N_\gamma : \gamma \leq \beta\} \cup \{x_\gamma : \gamma < \beta\})$

for all $\beta < \alpha$. Since the subspace B_α has the Baire property, it cannot be covered by a first category set $P = (\bigcup \{N_\beta : \beta \leq \alpha\}) \cup \{x_\beta : \beta < \alpha\}$, so we can choose a point $x_\alpha \in B_\alpha \backslash P$.

This shows that our inductive construction can go on giving us a subspace $L = \{x_\beta : \beta < \omega_1\} \subset X$ such that (*) holds for all $\beta < \omega_1$. We claim that L is a Luzin space. It follows from (*) that $x_\alpha \neq x_\beta$ whenever $\alpha \neq \beta$ and therefore L is uncountable. Furthermore, the set L is dense in X because $L \cap B_\alpha \neq \emptyset$ for all $\alpha < \omega_1$; thus the space L has no isolated points. Now, if A is a nowhere dense subspace of L, then A is nowhere dense in X; since $\mathcal{N}$ is cofinal in the family of all nowhere dense subsets of X, there is $\alpha < \omega_1$ such that $A \subset N_\alpha$. By (*), we have $L \cap N_\alpha \subset \{x_\beta : \beta < \alpha\}$ and therefore $A \subset \{x_\beta : \beta < \alpha\}$ which shows that A is countable. This proves that L is a Luzin space and finishes the proof of Fact 1. □

It was proved in Fact 3 of S.307 that Σ is dense in $\mathbb{D}^{\omega_1}$, countably compact and non-compact, while $\overline{A}$ is compact for any countable $A \subset \Sigma$ (the bar denotes the closure in Σ). An immediate consequence is that Σ is a dense countably compact non-separable subspace of $\mathbb{D}^{\omega_1}$. Any countably compact space has the Baire property (see Problem 274 of [TFS]); besides, $c(\Sigma) = \omega$ and $w(\Sigma) = \omega_1$, so, under CH, we can apply Fact 1 to conclude that there is dense Luzin subspace $L \subset \Sigma$. The space L cannot be separable because any countable dense subset of L would be dense in Σ which is not separable. This shows that CH implies existence of a non-separable Luzin space.

Finally, if $X = \mathbb{R}$ and CH holds, then X is an uncountable dense-in-itself Baire space with $c(X) = w(X) = \omega$. Therefore we can apply Fact 1 again to conclude that there is a dense Luzin subspace M of the space $\mathbb{R}$. Hence M is an example of a second countable Luzin space, so our solution is complete.

T.047. *Prove that $MA(\omega)$ holds in ZFC (and hence CH implies MA), while $MA(\mathfrak{c})$ is false in ZFC.*

Solution. Given any function f the set $\text{dom}(f)$ is the domain of f. If f, g are functions and $f | \text{dom}(g) = g$, i.e., the function f extends g, we say that $g \subset f$. Take any non-empty partially ordered set $(\mathcal{P}, \leq)$; suppose that D_n is dense in $\mathcal{P}$ for each $n \in \omega$. Pick any $d_0 \in D_0$; assume that we have $d_0 \geq \ldots \geq d_n$ such that $d_i \in D_i$ for all $i \leq n$. Since the set D_{n+1} is dense in $\mathcal{P}$, we can choose $d_{n+1} \in D_{n+1}$ such that $d_{n+1} \leq d_n$. This shows that our inductive construction can be continued to obtain a set $\{d_n : n \in \omega\}$ such that $d_{n+1} \leq d_n$ and $d_n \in D_n$ for all $n \in \omega$.

Now let $\mathcal{F} = \{p \in \mathcal{P} : \text{there is } n \in \omega \text{ such that } d_n \leq p\}$. We leave to the reader the trivial verification of the fact that $\mathcal{F}$ is a filter; since $\mathcal{F} \cap D_n \ni d_n$, the filter $\mathcal{F}$ intersects every D_n and hence we proved $MA(\omega)$ for any (not necessarily ccc) partially ordered set. Since CH says that the only infinite cardinal smaller than $\mathfrak{c}$ is ω, this proves that CH implies the Martin's axiom.

To show that $MA(\mathfrak{c})$ is false, let $\mathcal{P} = \bigcup \{ \mathbb{D}^B : B$ is a finite subset of $\omega \}$; here $\mathbb{D} = \{0, 1\}$. Let $f \leq g$ if $g \subset f$; we omit the trivial verification of the fact that $\leq$ is a partial order on $\mathcal{P}$. The set $\mathcal{P}$ is countable and hence ccc. Given any $h \in \mathbb{D}^\omega$, let $D_h = \{ f \in \mathcal{P} :$ there is $n \in \omega$ such that $n \in \mathrm{dom}(f)$ and $f(n) \neq h(n) \}$. Observe that the set D_h is dense in $\mathcal{P}$ for any $h \in \mathbb{D}^\omega$. Indeed, for any $g \in \mathcal{P}$, we can take any $n \in \omega \backslash \mathrm{dom}(g)$ and define $f \in \mathcal{P}$ by $f|\mathrm{dom}(g) = g$ and $f(n) = 1 - h(n)$. It is clear that $f \in D_h$ and $f \leq g$.

For each $n \in \omega$, let $D_n = \{ f \in \mathcal{P} : n \in \mathrm{dom}(f) \}$. The set D_n is also dense in $\mathcal{P}$ for every $n \in \omega$. To see it, take any $g \in \mathcal{P}$; if $n \in \mathrm{dom}(g)$, then $g \in D_n$. If not, then define $f \in \mathcal{P}$ by $f|\mathrm{dom}(g) = g$ and $f(n) = 0$. It is clear that $f \in D_n$ and $f \leq g$.

The family $\mathcal{D} = \{ D_n : n \in \omega \} \cup \{ D_h : h \in \mathbb{D}^\omega \}$ consists of dense subsets of $\mathcal{P}$ and $|\mathcal{D}| = \mathfrak{c}$. We claim that there is no filter in $\mathcal{P}$ which intersects all elements of $\mathcal{D}$. To see this, assume that $\mathcal{F}$ is a filter in $\mathcal{P}$ with $\mathcal{F} \cap D \neq \emptyset$ for all $D \in \mathcal{D}$. Since $\mathcal{F} \cap D_n \neq \emptyset$ for each $n \in \omega$, there is $f_n \in \mathcal{F}$ such that $n \in \mathrm{dom}(f_n)$. Note that we have the following property of $\mathcal{F}$:

(*) if $f, g \in \mathcal{F}$, then $f(n) = g(n)$ for any $n \in \mathrm{dom}(f) \cap \mathrm{dom}(g)$;

because there exists $h \in \mathcal{F}$ such that $f, g \subset h$ and hence $f(n) = h(n) = g(n)$. As a consequence, we can define a function $h \in \mathbb{D}^\omega$ by $h(n) = f_n(n)$ for each $n \in \omega$. An immediate consequence of (*) is that $f \subset h$ for any $f \in \mathcal{F}$. However, $\mathcal{F} \cap D_h \neq \emptyset$ and hence there is $f \in \mathcal{F}$ such that $f \in D_h$ which means $f(n) \neq h(n)$ for some $n \in \mathrm{dom}(f)$ and therefore $f \not\subset h$; this contradiction shows that $MA(\mathfrak{c})$ is false.

T.048. *Prove that $MA(\kappa)$ is equivalent to $MA(\kappa)$ restricted to ccc partially ordered sets of cardinality $\leq \kappa$, i.e., if $MA(\kappa)$ is true for all ccc partial orders of cardinality $\leq \kappa$, then it is true for all ccc partial orders.*

Solution. Assume that $MA(\kappa)$ holds for all ccc partial orders of cardinality $\leq \kappa$ and take a ccc partially ordered set $(\mathcal{P}, \leq)$ of arbitrary cardinality. Fix any family $\{ D_\alpha : \alpha < \kappa \}$ of dense subsets of $\mathcal{P}$. For each $\alpha < \kappa$ define a function $f_\alpha : \mathcal{P} \to \mathcal{P}$ as follows: given any $p \in \mathcal{P}$, there is $q \in D_\alpha$ such that $q \leq p$; let $f_\alpha(p) = q$. Besides, for any compatible $p, q \in \mathcal{P}$, choose $b(p, q) \in \mathcal{P}$ such that $b(p, q) \leq p$ and $b(p, q) \leq q$ (if p and q are incompatible, we do not choose anything for the pair (p, q)). Thus $b : \mathcal{C} \to \mathcal{P}$ where $\mathcal{C} = \{ (p, q) \in \mathcal{P} \times \mathcal{P} : p$ and q are compatible$\}$.

Take any $p_0 \in \mathcal{P}$ and let $Q_0 = \{ p_0 \}$; assume that $m < \omega$ and we have sets $Q_0 \subset \cdots \subset Q_m \subset \mathcal{P}$ such that

(*) $|Q_n| \leq \kappa$ for each $n \leq m$ and $b((Q_n \times Q_n) \cap \mathcal{C}) \cup (\bigcup \{ f_\alpha(Q_n) : \alpha < \kappa \}) \subset Q_{n+1}$

for any $n < m$. Since $|Q_m| \leq \kappa$, the set $Q = b((Q_m \times Q_m) \cap \mathcal{C}) \cup (\bigcup \{ f_\alpha(Q_m) : \alpha < \kappa \})$ also has cardinality $\leq \kappa$; if we let $Q_{m+1} = Q_m \cup Q$, the property (*) also holds for $n = m + 1$ and hence we can construct an increasing sequence

$\{Q_m : m \in \omega\}$ of subsets of $\mathcal{P}$ such that $(*)$ holds for all $n \in \omega$. We claim that the set $\mathcal{Q} = \bigcup \{Q_m : m \in \omega\}$ has the following properties:

(1) $E_\alpha = D_\alpha \cap \mathcal{Q}$ is dense in $(\mathcal{Q}, \leq)$ for each $\alpha < \kappa$;
(2) the set $\mathcal{Q}$ with the order $\leq$ induced from $\mathcal{P}$ is ccc.

Take any $q \in \mathcal{Q}$. Then $q \in Q_n$ for some $n \in \omega$; the property $(*)$ implies that $p = f_\alpha(q) \in D_\alpha \cap \mathcal{Q}$ and $p \leq q$. This proves (1). To establish (2), take any uncountable $A \subset \mathcal{Q}$. Since $\mathcal{P}$ is ccc, the set A cannot be an antichain in $\mathcal{P}$. Thus there are $p, q \in A$ which are compatible in $\mathcal{P}$. There is $n \in \omega$ for which $p, q \in Q_n$; since $(p, q) \in \mathcal{C}$, the point $r = b(p, q)$ belongs to Q_{n+1} and hence to $\mathcal{Q}$. As a consequence, we found $r \in \mathcal{Q}$ such that $r \leq p$ and $r \leq q$ which shows that p and q are compatible in $\mathcal{Q}$, i.e., A is not an antichain in $\mathcal{Q}$.

Thus we can apply our reduced form of Martin's axiom to the set $\mathcal{Q}$ to find a filter $\mathcal{F}'$ in $\mathcal{Q}$ such that $\mathcal{F}' \cap E_\alpha \neq \emptyset$ for all $\alpha < \kappa$. The set $\mathcal{F}'$ may not be a filter in $\mathcal{P}$; however the set $\mathcal{F} = \{p \in \mathcal{P} : \text{there is } q \in \mathcal{F}' \text{ such that } q \leq p\}$ is a filter in $\mathcal{P}$ (check it, please) and $\mathcal{F} \cap D_\alpha \neq \emptyset$ for all $\alpha < \kappa$.

T.049. *Let* $(\mathcal{P}, \leq)$ *be a partially ordered set. Call a subset* $A \subset \mathcal{P}$ *centered if for any* $n \in \mathbb{N}$ *and any* $p_1, \ldots, p_n \in A$, *there is* $r \in \mathcal{P}$ *such that* $r \leq p_i$ *for all* $i \leq n$. *Assume* MA$+\neg$CH *and take any ccc partially ordered set* $\mathcal{P}$. *Prove that for any uncountable* $R \subset \mathcal{P}$, *there exists an uncountable centered* $Q \subset R$. *In particular, all elements of* Q *are pairwise compatible.*

Solution. Passing to smaller uncountable set R if necessary, we can assume that $R = \{r_\alpha : \alpha < \omega_1\}$ where $r_\alpha \neq r_\beta$ if $\alpha \neq \beta$. Given an arbitrary $p \in \mathcal{P}$, let $O(p) = \{r \in \mathcal{P} : r \leq p\}$; if $A \subset \mathcal{P}$, then $O(A) = \bigcup\{O(p) : p \in A\}$. We will also need the sets $R_\alpha = \{r_\beta : \beta > \alpha\}$ and $T_\alpha = \{p \in \mathcal{P} : O(p) \cap O(R_\alpha) \neq \emptyset\}$ for each $\alpha < \omega_1$. It is clear that $T_\beta \subset T_\alpha$ if $\beta \geq \alpha$; however, we have the following property:

$(*)$ there exists $\alpha < \omega_1$ such that $T_\beta = T_\alpha$ for all $\beta \geq \alpha$.

If $(*)$ is not true, then there exist sets $\{\alpha_\gamma : \gamma < \omega_1\} \subset \omega_1$ and $\{p_\gamma : \gamma < \omega_1\} \subset \mathcal{P}$ for which $\alpha_\gamma < \alpha_\delta$ whenever $\gamma < \delta$ and $p_\gamma \in T_{\alpha_\gamma} \setminus T_{\alpha_{\gamma+1}}$ for all $\gamma < \omega_1$. Choose $s_\gamma \in S_\gamma = \{r_\alpha : \alpha_\gamma < \alpha \leq \alpha_{\gamma+1}\}$ with $O(p_\gamma) \cap O(s_\gamma) \neq \emptyset$ for all $\gamma < \omega_1$; this makes it possible to take $q_\gamma \in O(p_\gamma) \cap O(s_\gamma)$ for all $\gamma < \omega_1$. Now, if $\gamma < \delta$, then $O(q_\gamma) \subset O(p_\gamma)$ and hence $O(q_\gamma) \cap O(R_{\alpha_\delta}) = \emptyset$. Besides, $S_\delta \subset R_{\alpha_\delta}$ and therefore $O(q_\delta) \subset O(s_\delta) \subset O(S_\delta) \subset O(R_{\alpha_\delta})$ which shows that $O(q_\gamma) \cap O(q_\delta) = \emptyset$, i.e., q_γ is incompatible with q_δ. Thus $\{q_\gamma : \gamma < \omega_1\}$ is an uncountable antichain in $\mathcal{P}$ which contradicts the ccc property of $\mathcal{P}$ and proves $(*)$.

Fix $\alpha < \omega_1$ for which $(*)$ is fulfilled and consider the partially ordered set $\mathcal{Q} = O(R_\alpha)$ with the order $\leq$ induced from $\mathcal{P}$. We first show that $\mathcal{Q}$ is ccc. Indeed, if $A \subset \mathcal{Q}$ is uncountable, then A cannot be an antichain in $\mathcal{P}$, so there are $p, q \in A$ such that $r \in O(p) \cap O(q)$ for some $r \in \mathcal{P}$. By definition of $\mathcal{Q}$, there is $p' \in R_\alpha$ such that $p \leq p'$; this shows that $r \leq p'$ and hence $r \in \mathcal{Q}$; consequently, p and q are also compatible in $\mathcal{Q}$, i.e., A is not an antichain in $\mathcal{Q}$, so $\mathcal{Q}$ is also ccc.

The set $D_\gamma = O(R_\gamma)$ is dense in Q for all $\gamma \geq \alpha$; to see this take any $r \in Q$. It is evident that $\emptyset \neq O(r) \cap Q = O(r) \cap O(R_\alpha)$, so we can apply $(*)$ for $\beta = \gamma$ to conclude that $O(r) \cap O(R_\gamma) \neq \emptyset$ which shows that there is $s \in O(R_\gamma) = D_\gamma$ with $s \leq r$. Now applying $MA(\omega_1)$ to the ccc partially ordered set Q and the family $\{D_\gamma : \gamma < \omega_1\}$ of its dense subsets, we obtain a filter $\mathcal{F}$ in Q such that $\mathcal{F} \cap D_\gamma \neq \emptyset$ for all $\gamma < \omega_1$. Finally, let $Q = \{r \in R : p \leq r \text{ for some } p \in \mathcal{F}\}$. The set Q is centered for if $q_1, \ldots, q_n \in Q$, then there are $s_1, \ldots, s_n \in \mathcal{F}$ such that $s_i \leq q_i$ for all $i \leq n$. It is an easy consequence of the property (F1) of the definition of the filter that there exists $s \in \mathcal{F}$ such that $s \leq s_i$ for all $i \leq n$. Therefore $s \leq q_i$ for all $i \leq n$ and hence Q is centered. To see that $|Q| = \omega_1$ observe that for any $\gamma < \omega_1$, there is $r \in \mathcal{F} \cap D_\gamma$ which implies existence of $\beta > \gamma$ such that $r \in O(r_\beta)$, i.e., $r \leq r_\beta$ whence $r_\beta \in Q$. As a consequence, the set $\{\beta < \omega_1 : r_\beta \in Q\}$ is cofinal in ω_1 and hence $|Q| = \omega_1$.

T.050. *Assume $MA + \neg CH$. Let X_t be a space with $c(X_t) = \omega$ for every $t \in T$. Prove that $c(X) = \omega$, where $X = \prod\{X_t : t \in T\}$.*

Solution. Say that ω_1 *is a precaliber of a space* Z if for any uncountable family $\mathcal{U} \subset \tau^*(Z)$, we can find an uncountable centered $\mathcal{U}' \subset \mathcal{U}$.

Fact 1. Let Z be a space with $c(Z) = \omega$. If $MA + \neg CH$ holds, then ω_1 is a precaliber of the space Z.

Proof. Let $\mathcal{P} = \tau^*(Z)$ where $U \leq V$ if $U \subset V$. It is clear that $\leq$ is a partial order on $\mathcal{P}$; it is immediate that $U, V \in \mathcal{P}$ are compatible if and only if $U \cap V \neq \emptyset$. An immediate consequence is that $\mathcal{P}$ is ccc because $c(Z) = \omega$. Observe also that a set $A \subset \mathcal{P}$ is centered in the sense of Problem 049 if and only if the family A is centered as a family of subsets of Z. Therefore we can apply Problem 049 to conclude that any uncountable family $\mathcal{U} \subset \tau^*(Z)$ contains an uncountable centered subfamily, i.e., ω_1 is a precaliber of Z. Fact 1 is proved. □

Fact 2. Given an infinite cardinal κ, suppose that $c(Z_s) \leq \kappa$ for all $s \in S$. Assume additionally that $c(\prod_{s \in A} Z_s) \leq \kappa$ for each finite $A \subset S$. Then $c(\prod_{s \in S} Z_s) \leq \kappa$.

Proof. Let $Z = \prod\{Z_s : s \in S\}$ and $Z_A = \prod\{Z_s : s \in A\}$ for each $A \subset S$. If $c(Z) > \kappa$, then there is a disjoint family $\{U_\alpha : \alpha < \kappa^+\} \subset \tau^*(Z)$. Shrinking each U_α if necessary, we can assume that every U_α is a standard open subset of Z, i.e., $U_\alpha = (\prod_{s \in A_\alpha} U_\alpha^s) \times (\prod_{s \in S \setminus A_\alpha} Z_s)$ where A_α is a finite subset of S and $U_\alpha^s \in \tau^*(Z_s)$ for each $\alpha < \kappa^+$ and $s \in A_\alpha$. Since κ^+ is an uncountable regular cardinal, we can apply Delta-lemma (see Problem 038) to find a set $E \subset \kappa^+$ and a finite $A \subset S$ such that $|E| = \kappa^+$ and $A_\alpha \cap A_\beta = A$ for any distinct $\alpha, \beta \in E$. For any $\alpha \in E$ let $V_\alpha = \prod_{s \in A} U_\alpha^s$; then $\mathcal{V} = \{V_\alpha : \alpha \in E\} \subset \tau^*(Z_A)$. As $c(Z_A) \leq \kappa$ and $|\mathcal{V}| > \kappa$, there exist distinct $\alpha, \beta \in E$ such that $V_\alpha \cap V_\beta \neq \emptyset$. Take any $y \in V_\alpha \cap V_\beta$ and define a function $z : S \to \bigcup_{s \in S} Z_s$ as follows: $z(s) = y(s)$ for all $s \in A$; let $z(s) \in U_\alpha^s$ for all $s \in A_\alpha \setminus A$ and $z(s) \in U_\beta^s$ for all $s \in A_\beta \setminus A$. This definition is consistent because $A_\alpha \setminus A$ is disjoint from $A_\beta \setminus A$. Finally, define $z(s)$ arbitrarily for all $s \in S \setminus (A_\alpha \cup A_\beta)$. It is immediate that $z \in U_\alpha \cap U_\beta$; this contradiction finishes the proof of Fact 2. □

Returning to our solution, apply Fact 1 to conclude that ω_1 is a precaliber of X_t for all $t \in T$. By Fact 2, it suffices to show that every finite product of spaces with precaliber ω_1 has the Souslin property. Observe that it is evident that any space with precaliber ω_1 has the Souslin property, so it is sufficient to show that if ω_1 is a precaliber of Y and Z, then it is a precaliber of $Y \times Z$ for then a trivial induction shows that any finite product of spaces with precaliber ω_1 also has precaliber ω_1 and hence the Souslin property.

So assume that ω_1 is a precaliber of Y and Z. Given any uncountable family $\mathcal{U} \subset \tau^*(Y \times Z)$ we can assume that $\mathcal{U} = \{U_\alpha : \alpha < \omega_1\}$ and, for any $\alpha < \omega_1$, we have $U_\alpha = V_\alpha \times W_\alpha$ for some $V_\alpha \in \tau^*(Y)$ and $W_\alpha \in \tau^*(Z)$. Since ω_1 is a precaliber of Y, there is an uncountable $A \subset \omega_1$ such that the family $\mathcal{V} = \{V_\alpha : \alpha \in A\}$ is centered; apply also the fact that ω_1 is a precaliber of Z to choose an uncountable $B \subset A$ such that the family $\mathcal{W} = \{W_\alpha : \alpha \in B\}$ is also centered. We claim that the family $\mathcal{U}' = \{U_\alpha : \alpha \in B\}$ is centered as well. To see this, take any $U_{\alpha_1}, \ldots, U_{\alpha_n} \in \mathcal{U}'$; since $\mathcal{V}$ is centered, we can find a point $y \in V_{\alpha_1} \cap \cdots \cap V_{\alpha_n}$. The family $\mathcal{W}$ being centered, there is a point $z \in W_{\alpha_1} \cap \cdots \cap W_{\alpha_n}$. Hence $w = (y, z) \in U_{\alpha_1} \cap \cdots \cap U_{\alpha_n}$, so the family $\mathcal{U}' \subset \mathcal{U}$ is centered and uncountable. This shows that precaliber ω_1 is preserved by finite products and finishes our solution.

T.051. *Given families $A, B \subset \exp \omega$ such that $|A| \leq \kappa$, $|B| \leq \kappa$ and $\kappa < \mathfrak{c}$, suppose that $B \backslash (\bigcup A')$ is infinite for every $B \in \mathcal{B}$ and any finite family $A' \subset A$. Prove that $MA(\kappa)$ implies that there exists $M \subset \omega$ such that $B \backslash M$ is infinite for any $B \in \mathcal{B}$ while $A \backslash M$ is finite for any $A \in \mathcal{A}$.*

Solution. Consider the set $\mathcal{P} = \{(K, \mathcal{C}) : K$ is a finite subset of ω and $\mathcal{C}$ is a finite subfamily of $\mathcal{A}\}$. We will introduce a partial order $\leq$ on the set $\mathcal{P}$ as follows: given $P, P' \in \mathcal{P}$ with $P = (K, \mathcal{C})$ and $P' = (K', \mathcal{C}')$, we let $P \leq P'$ if $P' \subset P$, $\mathcal{C}' \subset \mathcal{C}$ and $(K \backslash K') \cap (\bigcup \mathcal{C}') = \emptyset$. It is straightforward that $\leq$ is indeed a partial order on $\mathcal{P}$. Besides, $(\mathcal{P}, \leq)$ is ccc. To see this, observe that for any finite $K \subset \omega$, if $P = (K, \mathcal{C}) \in \mathcal{P}$ and $P' = (K, \mathcal{C}') \in \mathcal{P}$, then P and P' are compatible because $Q = (K, \mathcal{C} \cup \mathcal{C}')$ is a common extension of P and P', i.e., $Q \leq P$ and $Q \leq P'$. Now, if $\mathcal{P}'$ is an uncountable subset of $\mathcal{P}$, then there are distinct $P, P' \in \mathcal{P}'$ such that $P = (K, \mathcal{C})$ and $P' = (K, \mathcal{C}')$ for some finite $K \subset \omega$ (because there are only countably may finite subsets of ω). Therefore P and P' are compatible and hence $\mathcal{P}'$ is not antichain.

Given any $A \in \mathcal{A}$ let $D_A = \{P = (K, \mathcal{C}) \in \mathcal{P} : A \in \mathcal{C}\}$. The set D_A is dense in $\mathcal{P}$ for any $A \in \mathcal{A}$ for if $P = (K, \mathcal{C}) \in \mathcal{P}$, then $Q = (K, \mathcal{C} \cup \{A\})$ belongs to D_A and is an extension of P. Observe also that the set $D(B, n) = \{P = (K, \mathcal{C}) \in \mathcal{P} : K \cap B$ is not contained in $n = \{0, \ldots, n - 1\}\}$ is dense in $\mathcal{P}$ for any $B \in \mathcal{B}$ and $n \in \omega$. Indeed, for any $P = (K, \mathcal{C}) \in \mathcal{P}$ the set $B \backslash (\bigcup \mathcal{C})$ is infinite, so there is $m \in B \backslash (\bigcup \mathcal{C})$ with $m \geq n$. It is immediate that $Q = (K \cup \{m\}, \mathcal{C})$ is an extension of P and $Q \in D(B, n)$.

The family $\mathcal{D} = \{D_A : A \in \mathcal{A}\} \cup \{D(B, n) : B \in \mathcal{B}$ and $n \in \omega\}$ consists of dense subsets of $\mathcal{P}$ and has cardinality $\leq \kappa$, so $MA(\kappa)$ is applicable to obtain a filter $\mathcal{F} \subset \mathcal{P}$ such that $\mathcal{F} \cap D \neq \emptyset$ for any $D \in \mathcal{D}$. Let $L = \bigcup \{K : (K, \mathcal{C}) \in \mathcal{F}$ for some (finite) $\mathcal{C} \subset \mathcal{A}\}$. We claim that

(1) the set $L \cap B$ is infinite for any $B \in \mathcal{B}$.

Indeed, if $L \cap B$ is finite for some $B \in \mathcal{B}$, then $L \cap B \subset n$ for some $n \in \omega$. However, there is some $F = (K, \mathcal{C}) \in \mathcal{F} \cap D(B, n)$ and therefore $K \subset L$ and $K \cap B \not\subset n$ which is a contradiction with $L \cap B \supset K \cap B$. We will also need the following property:

(2) $L \cap A$ is finite for any $A \in \mathcal{A}$.

To show that the property (2) holds take any $F = (K, \mathcal{C}) \in \mathcal{F} \cap D_A$. Given any $F' = (K', \mathcal{C}') \in \mathcal{F}$ the elements F and F' are compatible because $\mathcal{F}$ is a filter. Thus there is $F'' = (K'', \mathcal{C}'') \in \mathcal{F}$ with $F'' \le F$ and $F'' \le F'$. This implies $K'' \supset K$, $K'' \supset K'$ and $(K'' \backslash K) \cap (\bigcup \mathcal{C}) = \emptyset$. Since $A \in \mathcal{C}$, we have $(K'' \backslash K) \cap A = \emptyset$ and therefore $K' \cap A \subset K'' \cap A \subset K \cap A$. Since F' was taken arbitrarily, we have $L \cap A = \bigcup \{K' \cap A : (K', \mathcal{C}') \in \mathcal{F}$ for some $\mathcal{C}' \subset \mathcal{A}\} \subset K \cap A$ and therefore $L \cap A \subset K \cap A$ is a finite set, i.e., (2) is proved.

Finally, observe that it is an immediate consequence of (1) and (2) that the set $M = \omega \backslash L$ is as promised, so our solution is complete.

T.052 (Booth lemma). *Let $\mathcal{C} \subset \exp \omega$ be a family such that $|\mathcal{C}| = \kappa < \mathfrak{c}$ and $\bigcap \mathcal{C}'$ is infinite for every finite $\mathcal{C}' \subset \mathcal{C}$. Prove that $MA(\kappa)$ implies that there exists an infinite $L \subset \omega$ such that $L \backslash C$ is finite for any $C \in \mathcal{C}$.*

Solution. Let $\mathcal{A} = \{\omega \backslash C : C \in \mathcal{C}\}$ and $\mathcal{B} = \{\omega\}$. Then, for any finite $\mathcal{A}' \subset \mathcal{A}$, we have $\omega \backslash (\bigcup \mathcal{A}') = \bigcap \mathcal{C}'$ where $\mathcal{C}' = \{\omega \backslash A : A \in \mathcal{A}'\}$ is a finite subfamily of $\mathcal{C}$. Thus $B \backslash (\bigcup \mathcal{A}')$ is infinite for any (and the unique) $B \in \mathcal{B}$. Therefore we can apply Problem 051 to the families $\mathcal{A}$ and $\mathcal{B}$ to conclude that there is $M \subset \omega$ such that $\omega \backslash M$ is infinite and $A \backslash M$ is finite for any $A \in \mathcal{A}$. Consider the set $L = \omega \backslash M$; we already saw that L is infinite. Now if $C \in \mathcal{C}$, then $(\omega \backslash C) \in \mathcal{A}$, and hence $(\omega \backslash C) \backslash (\omega \backslash L) = L \backslash C$ is finite, so C is as promised.

T.053. *Let $\mathcal{A} \subset \exp \omega$ be an almost disjoint family ($\equiv$ all elements of $\mathcal{A}$ are infinite while $A \cap B$ is finite whenever A and B are distinct elements of $\mathcal{A}$). Suppose that κ is an infinite cardinal and $|\mathcal{A}| = \kappa < \mathfrak{c}$. Prove that $MA(\kappa)$ implies that $\mathcal{A}$ is not maximal.*

Solution. Assume first that the set $D = \omega \backslash (\bigcup \mathcal{A}')$ is finite for some finite $\mathcal{A}' \subset \mathcal{A}$. Since $\mathcal{A}$ is infinite, there is $A \in \mathcal{A} \backslash \mathcal{A}'$; we have $A \backslash D \subset \bigcup \mathcal{A}'$ and hence the infinite set $A \backslash D$ is a union of a finite family $\{A' \cap A : A' \in \mathcal{A}'\}$ of finite sets which is a contradiction. This shows that $\omega \backslash (\bigcup \mathcal{A}')$ is infinite for any finite $\mathcal{A}' \subset \mathcal{A}$, so if $\mathcal{B} = \{\omega\}$, then the families $\mathcal{A}$ and $\mathcal{B}$ satisfy the premises of Problem 051. Therefore there exists $M \subset \omega$ such that $\omega \backslash M$ is infinite and $A \backslash M$ is finite for any $A \in \mathcal{A}$. Finally, if $L = \omega \backslash M$, then L is an infinite subset of ω such that $L \cap A$ is finite for any $A \in \mathcal{A}$. Hence $\mathcal{A}$ is not maximal because the family $\mathcal{A} \cup \{L\}$ is almost disjoint and is strictly larger than $\mathcal{A}$.

T.054. *Assume $MA + \neg CH$. Let X be a space such that $\chi(x, X) < \mathfrak{c}$ for some $x \in X$. Prove that for any countable $A \subset X$ with $x \in \overline{A}$, there exists a sequence $\{a_n\}_{n \in \omega} \subset A$ such that $a_n \to x$.*

Solution. If $x \in A$, then our statement is trivially true, so let us assume that $x \notin A$. Then the set $U \cap A$ is infinite for any $U \in \tau(x, X)$. Consider the family $C = \{U \cap A : U \in \mathcal{U}\}$ where $\mathcal{U}$ is some base at the point x with $|\mathcal{U}| < \mathfrak{c}$. Note that $\bigcap C'$ is infinite for any finite $C' \subset C$ because this set is also the intersection of A with a neighborhood of x. Our idea is to identify A with ω and apply the Booth lemma (see Problem 052). Formally, take any bijection $\varphi : A \to \omega$; the family $\mathcal{B} = \{\varphi(P) : P \in \mathcal{A}\}$ has cardinality $< \mathfrak{c}$ and consists of infinite subsets of ω such that $\bigcap \mathcal{B}'$ is infinite for every finite $\mathcal{B}' \subset \mathcal{B}$. Therefore we can apply Problem 052 to the family $\mathcal{B}$ to find an infinite set $M \subset \omega$ such that $M \backslash P$ is finite for any $P \in \mathcal{B}$. Since φ is a bijection, we have the same property for the set $L = \varphi^{-1}(M)$, i.e., L is infinite and $L \backslash C$ is finite for any $C \in C$. Take any enumeration $\{a_n : n \in \omega\}$ of the set L. Then $\{a_n : n \in \omega\} = L \subset A$ and, for any $W \in \tau(x, X)$, there is $U \in \mathcal{U}$ such that $x \in U \subset W$. Thus $C = U \cap A \in C$, so $L \backslash C$ is finite. Since $C \subset W$, the set $L \backslash W$ is also finite and hence there is $m \in \omega$ such that $a_n \in W$ for all $n \geq m$. This proves that $a_n \to x$.

T.055. *Let X be a second countable space of cardinality $< \mathfrak{c}$. Prove that, under MA, any subset of X is a G_δ in X.*

Solution. We need some easy facts about second countable spaces.

Fact 1. For any second countable space Z there exists a metric d on Z and a base $\mathcal{B} = \{U_n : n \in \omega\}$ in the space Z such that $\tau(d) = \tau(Z)$ and $\text{diam}_d(U_n) \to 0$ when $n \to \infty$.

Proof. We can consider that Z is a subset of $\mathbb{I}^\omega$ (see Problem 209 of [TFS]); take any metric ρ on $\mathbb{I}^\omega$ which generates the topology of $\mathbb{I}^\omega$. It is straightforward that the metric ρ as well as the restriction d of the metric ρ to $Z \times Z$ is totally bounded. Therefore, for each $n \in \mathbb{N}$, there exists a finite set $A_n \subset Z$ such that $\bigcup \{B_d(a, \frac{1}{n}) : a \in A_n\} = Z$. Let $\mathcal{B} = \{B_d(a, \frac{1}{n}) : n \in \mathbb{N} \text{ and } a \in A_n\}$; if $\varepsilon > 0$, then there is $m \in \mathbb{N}$ such that $\frac{1}{n} < \frac{\varepsilon}{3}$ for all $n \geq m$. If $\mathcal{B}' = \{B_d(a, \frac{1}{n}) : n < m, a \in A_n\}$, then $\mathcal{B}'$ is a finite subfamily of $\mathcal{B}$ such that $\text{diam}_d(U) < \varepsilon$ for any $U \in \mathcal{B} \backslash \mathcal{B}'$. This shows that if we take any faithful enumeration $\{U_n : n \in \omega\}$ of the family $\mathcal{B}$, then $\text{diam}_d(U_n) \to 0$ if $n \to \infty$.

To prove that $\mathcal{B}$ is a base in Z, take any $z \in Z$ and any $U \in \tau(z, Z)$; there is $n \in \mathbb{N}$ such that $B_d(z, \frac{2}{n}) \subset U$. Take any $a \in A_n$ such that $z \in B_d(a, \frac{1}{n})$; it is immediate that $B_d(a, \frac{1}{n}) \subset B_d(z, \frac{2}{n}) \subset U$ which proves that $\mathcal{B}$ is a base in Z, so Fact 1 is proved. $\square$

Now apply Fact 1 to find a metric d which generates the topology of X and a base $\mathcal{U} = \{U_n : n \in \omega\}$ in X such that $\text{diam}_d(U_n) \to 0$ when $n \to \infty$. It is evident that $\mathcal{U}$ has the following property:

(*)   for any infinite $\mathcal{U}' \subset \mathcal{U}$, either $\bigcap \mathcal{U}' = \emptyset$ or $|\bigcap \mathcal{U}'| = 1$.

For the cardinal $\kappa = |X| < \mathfrak{c}$ choose any enumeration $\{x_\alpha : \alpha < \kappa\}$ of the set X and take an arbitrary $Y \subset X$. Let $P_\alpha = \{n \in \omega : x_\alpha \in U_n\}$ for each $\alpha < \kappa$ and consider the families $\mathcal{B} = \{P_\alpha : x_\alpha \in Y\}$ and $\mathcal{A} = \{P_\alpha : x_\alpha \in X \backslash Y\}$. It is clear that

$\mathcal{A}, \mathcal{B} \subset \exp \omega$ and $|\mathcal{A}| \leq \kappa$, $|\mathcal{B}| \leq \kappa$. Besides, if $\mathcal{A}'$ is a finite subfamily of $\mathcal{A}$ and $P_\alpha \in \mathcal{B}$, then $P_\alpha \backslash (\bigcup \mathcal{A}')$ cannot be finite; for otherwise, there is $P_\beta \in \mathcal{A}'$ such that $P = P_\beta \cap P_\alpha$ is infinite and hence the infinite intersection $\bigcap \{U_n : n \in P\}$ contains the two-point set $\{x_\alpha, x_\beta\}$ which is impossible by $(*)$ (observe that our enumeration of X need not be faithful; however, the points x_α and x_β are distinct because $x_\alpha \in Y$ and $x_\beta \in X \backslash Y$).

Consequently, we can apply Problem 051 to the families $\mathcal{A}$ and $\mathcal{B}$ to find an infinite set $M \subset \omega$ such that $B \backslash M$ is infinite for any $B \in \mathcal{B}$ and $A \backslash M$ is finite for any $A \in \mathcal{A}$. For the set $L = \omega \backslash M$, the set $A \cap L$ is finite for any $A \in \mathcal{A}$ and $B \cap L$ is infinite for any $B \in \mathcal{B}$. Now let $W_n = \bigcup \{U_m : m \in L \text{ and } m \geq n\}$ for all $n \in \omega$. Given any $\alpha < \kappa$ such that $x_\alpha \in Y$ and any $n \in \omega$ note that $L \cap P_\alpha$ is infinite and therefore there is $m \geq n$ such that $m \in P_\alpha \cap L$ whence $x_\alpha \in U_m \subset W_n$. This proves that $Y \subset W_n$ for all $n \in \omega$ and hence $Y \subset H = \bigcap \{W_n : n \in \omega\}$.

On the other hand no point of $X \backslash Y$ can belong to H; indeed, if $x_\alpha \in X \backslash Y$, then $L \cap P_\alpha$ is finite and hence there is $n \in \omega$ such that $m \notin L \cap P_\alpha$ for any $m \geq n$. As a consequence $x_\alpha \notin W_n$ and hence $x_\alpha \notin H$. This proves that $Y = \bigcap \{W_n : n \in \omega\}$, so Y is a G_δ-set in X and our solution is complete.

T.056. *Prove that MA implies $2^\kappa = \mathfrak{c}$ for any infinite $\kappa < \mathfrak{c}$.*

Solution. It is evident that $2^\kappa \geq 2^\omega = \mathfrak{c}$, so we must only prove that $2^\kappa \leq \mathfrak{c}$. Since $\kappa < \mathfrak{c}$, we can take a set $X \subset \mathbb{R}$ with $|X| = \kappa$. Fix any countable base $\mathcal{B}$ in the space X; since every open subset of X is a union of a subfamily of $\mathcal{B}$, the number of all open subsets of X does not exceed $|\exp \mathcal{B}| \leq \mathfrak{c}$. If $\mathcal{G}$ is the family of all G_δ-subsets of X, then $|\mathcal{G}| \leq |\tau(X)|^\omega \leq \mathfrak{c}^\omega = \mathfrak{c}$. Now apply Problem 055 to conclude that every subset of X is a G_δ-set in X, so we have $2^\kappa = |\exp(X)| = |\mathcal{G}| \leq \mathfrak{c}$ and therefore $2^\kappa = \mathfrak{c}$.

T.057. *Let X be a second countable dense-in-itself space. Given a cardinal κ such that $0 < \kappa < \mathfrak{c}$, suppose that N_α is a nowhere dense subset of the space X for any $\alpha < \kappa$. Prove that, under MA, the set $\bigcup \{N_\alpha : \alpha < \kappa\}$ is of first category in X.*

Solution. Take any base $\mathcal{U} = \{U_n : n \in \omega\} \subset \tau^*(X)$ of the space X and let $B_j = \{i \in \omega : U_i \subset U_j\}$ for each $j \in \omega$. Since X has no isolated points, the set B_j is infinite for all $j \in \omega$. Let $\mathcal{B} = \{B_j : j \in \omega\}$; consider also the set $A_\alpha = \{i \in \omega : U_i \cap N_\alpha \neq \emptyset\}$ for every $\alpha < \kappa$. The families $\mathcal{B}$ and $\mathcal{A} = \{A_\alpha : \alpha < \kappa\}$ have cardinality strictly less than $\mathfrak{c}$. If $k, j \in \omega$ and $\mathcal{A}' = \{A_{\alpha_1}, \ldots, A_{\alpha_k}\} \subset \mathcal{A}$, then the set $D = N_{\alpha_1} \cup \cdots \cup N_{\alpha_k}$ is nowhere dense in X, so there is $W \in \tau^*(X)$ such that $W \subset U_j \backslash D$. Since X is dense-in-itself, the set $P = \{i \in \omega : U_i \subset W\}$ is infinite and $P \subset B_j \backslash (\bigcup \mathcal{A}')$. Therefore $B_j \backslash (\bigcup \mathcal{A}')$ is infinite for any finite $\mathcal{A}' \subset \mathcal{A}$ and any $j \in \omega$. Consequently, we can apply Problem 051 to families $\mathcal{A}$ and $\mathcal{B}$ to obtain a set $M \subset \omega$ such that $A \backslash M$ is finite for every $A \in \mathcal{A}$ and $B \backslash M$ is infinite for each $B \in \mathcal{B}$. Hence, for $L = \omega \backslash M$, the set $L \cap A$ is finite for all $A \in \mathcal{A}$ and $L \cap B$ is infinite for each $B \in \mathcal{B}$. The set $W_n = \bigcup \{U_k : k \in L \text{ and } k \geq n\}$ is open and dense in X for all $n \in \omega$. Of course, we must only check density of W_n.

To do it, fix any $j \in \omega$; since there are infinitely many $k \in L \cap B_j$, we can find $k \geq n$ with $k \in L \cap B_j$. By definitions of L, W_n and B_j we have $U_k \subset U_j \cap W_n$. This shows that $W_n \cap U_j \neq \emptyset$ for each $j \in \omega$ and hence W_n is dense in X. Consequently, the set $F_n = X \setminus W_n$ is nowhere dense in X, so the set $F = \bigcup \{F_n : n \in \omega\}$ is of first category.

Take any $\alpha < \kappa$; since the set $L \cap A_\alpha$ is finite, there is $n \in \omega$ such that $L \cap A_\alpha \subset n$ and therefore $W_n \cap N_\alpha = \emptyset$. An immediate consequence is that $N_\alpha \subset F_n \subset F$. The ordinal α was taken arbitrarily, so we proved that $N = \bigcup \{N_\alpha : \alpha < \kappa\} \subset F$ and hence the set N is of first category in X.

T.058. *Prove that Martin's axiom is equivalent to the following statement: "Given a compact space X such that $c(X) = \omega$, for any family γ of nowhere dense subsets of X with $|\gamma| < \mathfrak{c}$, we have $\bigcup \gamma \neq X$".*

Solution. The following statement:

"Given a compact space X such that $c(X) = \omega$, for any family γ of nowhere dense subsets of X with $|\gamma| < \mathfrak{c}$, we have $\bigcup \gamma \neq X$"

will be called the topological version of Martin's axiom. We must prove the equivalence of Martin's axiom and its topological version.

Take any compact space X with $c(X) = \omega$ and assume that γ is a family of nowhere dense subsets of X with $|\gamma| < \mathfrak{c}$. Let $\mathcal{P} = \tau^*(X)$; introduce a partial order on $\mathcal{P}$ by $U \leq V$ if and only if $U \subset V$. It is clear that any $U, V \in \mathcal{P}$ are incompatible if and only if $U \cap V = \emptyset$. The Souslin property of X implies that any antichain in $\mathcal{P}$ is countable being a disjoint family of non-empty open subsets of X. Hence $\mathcal{P}$ is ccc. For any $N \in \gamma$ let $D_N = \{U \in \mathcal{P} : \overline{U} \cap N = \emptyset\}$. The set N is nowhere dense, so for any $V \in \mathcal{P}$, there is $U \in \mathcal{P}$ such that $U \subset V$ and $\overline{U} \cap N = \emptyset$ (we used regularity of X here: remember that all our spaces are assumed Tychonoff if the opposite is not stated explicitly). This proves that D_N is a dense subset of $\mathcal{P}$ and therefore the family $\mathcal{D} = \{D_N : N \in \gamma\}$ is a family of dense subsets of $\mathcal{P}$ such that $|\mathcal{D}| < \mathfrak{c}$. If MA holds, then there is a filter $\mathcal{F}$ in $\mathcal{P}$ such that $\mathcal{F} \cap D_N \neq \emptyset$ for all $N \in \gamma$. It is evident that $\mathcal{F}$ is a centered family of subsets of X, so by compactness of X, we have $P = \bigcap \{\overline{F} : F \in \mathcal{F}\} \neq \emptyset$. Pick any $x \in P$; given any $N \in \gamma$, there is $F \in \mathcal{F} \cap D_N$ and hence $\overline{F} \cap N = \emptyset$. Since $x \in \overline{F}$, this shows that $x \notin N$ for any $N \in \gamma$ and therefore $x \notin \bigcup \gamma$. Thus Martin's axiom implies the topological version of Martin's axiom.

Fact 1. Let $(\mathcal{P}, \leq)$ be an arbitrary partially ordered set. Then, for any $p, q \in \mathcal{P}$, the set $D(p, q) = \{r \in \mathcal{P} : r \leq p \text{ and } r \leq q\} \cup \{r \in \mathcal{P} : r \perp q\} \cup \{r \in \mathcal{P} : r \perp p\}$ is dense in $\mathcal{P}$.

Proof. Take any $r \in \mathcal{P}$; if there exists some $s \leq r$ which is incompatible either with p or with q, then $s \in D(p, q)$ and we are done. Hence we can assume that each $s \leq r$ is compatible with both p and q. In particular, r is compatible with p, so there is $t \in \mathcal{P}$ with $t \leq r$ and $t \leq p$. By our assumption t has to be compatible with q, so there is $s \in \mathcal{P}$ such that $s \leq t$ and $s \leq q$. Since $s \leq t \leq r$ and $s \leq t \leq p$, we have $s \leq p$, so $s \in D(p, q)$ and Fact 1 is proved. $\square$

Now assume that the topological version of Martin's axiom is true and take any cardinal $\kappa < \mathfrak{c}$; we must prove $MA(\kappa)$ for an arbitrary non-empty ccc partially ordered set $(\mathcal{P}, \leq)$. For any $p \in \mathcal{P}$ let $N_p = \{q \in \mathcal{P} : q \leq p\}$. It is immediate that $\bigcup \{N_p : p \in \mathcal{P}\} = \mathcal{P}$; if $r \in N_p \cap N_q$, then $N_r \subset N_p \cap N_q$, so the family $\{N_p : p \in \mathcal{P}\}$ is a base of some topology τ on the set $\mathcal{P}$ (see Problem 006 of [TFS]). Denote by P the space $(\mathcal{P}, \tau)$. Observe that P need not even be a T_1-space.

Let $\mathcal{B} = \{U \in \tau^* = \tau \backslash \{e\} : U = \mathrm{Int}(\overline{U})\}$ (the bar denotes the closure in P and the interior is taken in P as well). The elements of $\mathcal{B}$ will be called *regular* open sets of P. Let us formulate some simple properties of regular open sets (we leave their easy proofs to the reader):

(1) $\mathrm{Int}(\overline{W}) \in \mathcal{B}$ for any $W \in \tau^*$;
(2) if $W \in \tau$ and $\overline{W} \neq P$, then $P \backslash \overline{W} \in \mathcal{B}$;
(3) if $U, V \in \mathcal{B}$ and $U \cap V \neq \emptyset$, then $U \cap V \in \mathcal{B}$.

To apply our topological version of the Martin's axiom, we are going to use the family $\mathcal{B}$ for constructing a compact Hausdorff space X. This construction (called *the Stone space of a Boolean algebra*) can be carried out in a very general situation; however our concrete case is sufficient for our solution.

A set $\mathcal{F} \subset \mathcal{B}$ is called *a filter* if it has the following properties:

(4) $U \cap V \in \mathcal{F}$ for any $U, V \in \mathcal{F}$;
(5) $U \in \mathcal{F}$ and $U \subset V \in \mathcal{B}$ implies $V \in \mathcal{F}$.

A filter $\mathcal{F} \subset \mathcal{B}$ is called *an ultrafilter on* $\mathcal{B}$ if it is maximal with respect to inclusion, i.e., if $\mathcal{G} \subset \mathcal{B}$ is a filter and $\mathcal{F} \subset \mathcal{G}$, then $\mathcal{F} = \mathcal{G}$. An easy application of the Zorn's lemma shows that any filter on $\mathcal{B}$ is contained in an ultrafilter on $\mathcal{B}$.

Fact 2. The following properties are equivalent for any $\mathcal{F} \subset \mathcal{B}$:

(6) $\mathcal{F}$ is an ultrafilter on $\mathcal{B}$;
(7) $\mathcal{F}$ is a maximal centered subfamily of $\mathcal{B}$;
(8) $\mathcal{F}$ is centered and, for any $U \in \mathcal{B}$, either $U \in \mathcal{F}$ or $P \backslash \overline{U} \in \mathcal{F}$.

Proof. Let us first show that any centered family $\mathcal{C}$ on $\mathcal{B}$ is contained in a filter on $\mathcal{B}$. Let $\wedge \mathcal{C}$ be the family of all finite intersections of the elements from $\mathcal{C}$. It is immediate that $U \cap V \in \wedge \mathcal{C}$ for any $U, V \in \wedge \mathcal{C}$. If $\mathcal{C}' = \{U \in \mathcal{B} : U \supset V$ for some $V \in \wedge \mathcal{C}\}$, then $\mathcal{C} \subset \wedge \mathcal{C} \subset \mathcal{C}'$ and $\mathcal{C}'$ is a filter: we leave the trivial verification of this fact to the reader.

Now, if $\mathcal{F}$ is an ultrafilter, then it is centered; this is proved by an easy induction using (4). Assume that there is a strictly larger centered family $\mathcal{C} \supset \mathcal{F}$. By the observation above, there is a filter $\mathcal{C}' \supset \mathcal{C}$. Since $\mathcal{F}$ is an ultrafilter, we have $\mathcal{F} = \mathcal{C}'$ and therefore $\mathcal{C} = \mathcal{F} = \mathcal{C}'$ which is a contradiction showing that (6)$\Longrightarrow$(7).

Now, if $\mathcal{F}$ is a maximal centered family on $\mathcal{B}$, then, again, take a filter $\mathcal{F}' \supset \mathcal{F}$. Since $\mathcal{F}'$ is also centered and $\mathcal{F}$ is maximal centered, we have $\mathcal{F} = \mathcal{F}'$, i.e., $\mathcal{F}$ is a filter. Furthermore, no filter on $\mathcal{B}$ can be larger than $\mathcal{F}$ because any filter is centered and $\mathcal{F}$ is maximal centered. As a consequence, $\mathcal{F}$ is an ultrafilter and we established that (6) $\Longleftrightarrow$ (7).

If $\mathcal{F}$ is a maximal centered family on $\mathcal{B}$ take any $U \in \mathcal{B}$. If $U \notin \mathcal{F}$, then, by maximality of $\mathcal{F}$, there are $U_1, \ldots, U_n \in \mathcal{F}$ such that $U \cap U_1 \cap \cdots \cap U_n = \emptyset$. Analogously, if $V = P \backslash \overline{U} \notin \mathcal{F}$, then we can choose sets $V_1 \ldots, V_k \in \mathcal{F}$ such that $V \cap V_1 \ldots \cap V_k = \emptyset$. Since $\mathcal{F}$ is centered, the set $W = (\bigcap_{i \leq n} U_i) \cap (\bigcap_{i \leq k} V_i)$ is non-empty. It is clear that $W \cap (U \cup V) = \emptyset$ which contradicts the fact that $U \cup V$ is dense in P. This proves that any maximal centered subfamily of $\mathcal{B}$ satisfies (8) and hence (7)$\Longrightarrow$(8).

Finally, take any $\mathcal{F}$ as in (8). If $\mathcal{F}$ is not maximal, then there is $U \in \mathcal{B}$ such that $U \notin \mathcal{F}$ and $\mathcal{F}' = \mathcal{F} \cup \{U\}$ is still centered. The property (8) says that $V = P \backslash \overline{U} \in \mathcal{F}$ and hence the sets U and V belong to the centered family $\mathcal{F}'$. However, $U \cap V = \emptyset$; this contradiction shows that (8)$\Longrightarrow$(7), so Fact 1 is proved. $\square$

Returning to our solution let $X = \{\mathcal{F} : \mathcal{F}$ is an ultrafilter on $\mathcal{B}\}$. Given any $U \in \mathcal{B}$, let $O_U = \{x \in X : U \in x\}$. Let us check that the family $\mathcal{O} = \{O_U : U \in \mathcal{B}\}$ generates a topology on X as a base. Since no $x \in X$ is empty, there is $U \in \mathcal{B}$ with $U \in x$. Therefore $x \in O_U$ which proves that $\bigcup \mathcal{O} = X$. Now, if $x \in O_U \cap O_V$, then $U, V \in x$ and hence $W = U \cap V \in x$. Thus $x \in O_W \subset O_U \cap O_V$, so we can apply Problem 006 of [TFS] to see that $\mathcal{O}$ is a base of a topology μ on X. Next we prove that

(9) (X, μ) is a Hausdorff compact space.

In the sequel we will write X instead of (X, μ). By Problem 118 of [TFS], to prove compactness of X, it suffices to show that for every $\mathcal{U} \subset \mathcal{O}$ such that $\bigcup \mathcal{U} = X$, there is a finite $\mathcal{U}' \subset \mathcal{U}$ with $\bigcup \mathcal{U}' = X$. Fix a family $\mathcal{B}' \subset \mathcal{B}$ such that $\mathcal{U} = \{O_U : U \in \mathcal{B}'\}$. If the cover $\mathcal{U}$ has no finite subcover, then for any $U_1, \ldots, U_n \in \mathcal{B}'$ there is $x \in X$ such that $x \notin \bigcup_{i \leq n} O_{U_i}$ which is equivalent to $U_i \notin x$ for all $i \leq n$. Therefore $P \backslash \overline{U}_i \in x$ for all $i \leq n$ (Fact 2) and hence $\bigcap \{P \backslash \overline{U}_i : i \leq n\} \neq \emptyset$. This shows that the family $\gamma = \{P \backslash \overline{U} : U \in \mathcal{B}'\}$ is centered; if x is any ultrafilter on $\mathcal{B}$ with $\gamma \subset x$, then $P \backslash \overline{U} \in x$ for all $U \in \mathcal{B}'$ and therefore $U \notin x$ for all $U \in \mathcal{B}'$. Thus $x \notin O_U$ for all $U \in \mathcal{B}'$ whence $x \notin \bigcup \{O_U : U \in \mathcal{B}'\} = \bigcup \mathcal{U} = X$ which is a contradiction. Therefore $\mathcal{U}$ has a finite subcover, so X is compact.

To see that X is Hausdorff, take distinct $x, y \in X$. There is $U \in \mathcal{B}$ such that $U \in x$ but $U \notin y$. Thus $V = P \backslash \overline{U} \in y$ by Fact 2; it is immediate that $x \in O_U$, $y \in O_V$ and $O_U \cap O_V = \emptyset$ so X is Hausdorff and hence Tychonoff (see Problem 124 of [TFS]).

Let us establish next that X has the Souslin property. It suffices to show that no uncountable subfamily of $\mathcal{O}$ can be disjoint. Suppose that $\mathcal{B}' \subset \mathcal{B}$ is an uncountable family. Observe that $O_U \cap O_V = O_{U \cap V}$, so the family $\{O_U : U \in \mathcal{B}'\}$ is disjoint if and only if $\mathcal{B}'$ is disjoint. Now if $\mathcal{B}'$ is disjoint, then for each $U \in \mathcal{B}'$ there is $p(U) \in P$ such that $N_{p(U)} \subset U$. It turns out that the family $\{N_{p(U)} : U \in \mathcal{B}'\}$ is disjoint, i.e., the set $\{p(U) : U \in \mathcal{B}'\}$ is an uncountable antichain. This contradiction with ccc property of P shows that $c(X) \leq \omega$.

Now assume that D_α is a dense subset of P for all $\alpha < \kappa$. For each $p \in P$, let $U_p = \text{Int}(\overline{N}_p)$. Consider the set $W_\alpha = \bigcup \{O_{U_p} : p \in D_\alpha\}$. We claim that W_α is an open dense subspace of X. Since the openness is clear, let us prove that W_α is dense in X. Given any $U \in \mathcal{B}$ take any $p \in \mathcal{P}$ with $N_p \subset U$. The set D_α is dense in $\mathcal{P}$, so there is $q \in D_\alpha$ with $q \leq p$. As a consequence, $N_q \subset N_p \subset U$ and therefore $U_q = \text{Int}(\overline{N}_q) \subset \text{Int}\overline{U} = U$ which shows that $O_{U_q} \subset O_U \cap W_\alpha$. We proved that W_α intersects every element of $\mathcal{O}$, so W_α is dense in X for all $\alpha < \kappa$.

The set $F_\alpha = X \backslash W_\alpha$ is nowhere dense in X, so by the topological version of Martin's axiom, there is $x \in X \backslash \bigcup \{F_\alpha : \alpha < \kappa\} = \bigcap \{W_\alpha : \alpha < \kappa\}$. Recall that x is an ultrafilter on $\mathcal{B}$. We claim that the family $\mathcal{C} = \{p \in \mathcal{P} : U_p \in x\}$ is centered (the term "centered" is used in the sense of Problem 049 here; we hope that its meaning is always clear from the context). Indeed, if $p_1, \ldots, p_n \in \mathcal{F}$, then $U_{p_1}, \ldots, U_{p_n} \in x$ and hence $U = U_{p_1} \cap \cdots \cap U_{p_n} \neq \emptyset$. Observe that N_{p_i} is dense in U_{p_i} and hence in U for every $i \leq n$. Therefore $N_{p_1} \cap U \neq \emptyset$; if $k < n$ and we proved that $G_k = N_{p_1} \cap \cdots \cap N_{p_k} \cap U \neq \emptyset$, then G_k is a non-empty open subset of U, so $N_{p_{k+1}} \cap G_k = N_{p_1} \cap \cdots N_{p_{k+1}} \cap U \neq \emptyset$. This shows that our inductive proof can go on to finally establish that $G_n = N_{p_1} \cap \cdots \cap N_{p_n} \cap U = N_{p_1} \cap \cdots \cap N_{p_n} \neq \emptyset$. If $r \in N_{p_1} \cap \cdots \cap N_{p_n}$, then $r \leq p_i$ for all $i \leq n$, so $p_1, \ldots, p_n$ are compatible and hence $\mathcal{C}$ is centered. Observe also that $p \in \mathcal{C}$ and $p \leq q$ implies $q \in \mathcal{C}$ because $U_q \supset U_p \in x$ implies $U_q \in x$, i.e., $q \in \mathcal{C}$. Note also that $\mathcal{C} \cap D_\alpha \neq \emptyset$ for each $\alpha < \kappa$ because, by definition of W_α, there exists $p \in D_\alpha$ such that $x \in O_{U_p}$, i.e., $U_p \in x$ and therefore $p \in \mathcal{C} \cap D_\alpha$.

To sum up, we proved that the topological version of Martin's axiom implies that for any ccc partially ordered set $\mathcal{P}$ if $\mathcal{D}$ is a family of dense subsets of $\mathcal{P}$ with $|\mathcal{D}| \leq \kappa$, then there is a centered $\mathcal{C} \subset \mathcal{P}$ such that $\mathcal{C} \cap D \neq \emptyset$ for any $D \in \mathcal{D}$ and $\mathcal{C}$ is backwards closed, i.e., $p \in \mathcal{C}$ and $p \leq q$ implies $q \in \mathcal{C}$.

Now take any ccc partially ordered set $\mathcal{P}$ of cardinality $\leq \kappa$. Given any family $\mathcal{D}$ of dense subsets of $\mathcal{P}$ with $|\mathcal{D}| \leq \kappa$, let $\mathcal{D}' = \mathcal{D} \cup \{D(p,q) : p, q \in \mathcal{P}\}$ where $D(p,q) = \{r \in \mathcal{P} : r \leq p$ and $r \leq q\} \cup \{r \in \mathcal{P} : r \perp q\} \cup \{r \in \mathcal{P} : r \perp p\}$ for all $p, q \in \mathcal{P}$. Each $D(p,q)$ is dense in $\mathcal{P}$ by Fact 2, so $\mathcal{D}'$ is a family of dense subsets of $\mathcal{P}$ and $|\mathcal{D}'| \leq \kappa$. We showed that there must exist a centered backwards closed set $\mathcal{F}$ such that $\mathcal{F} \cap D \neq \emptyset$ for any $D \in \mathcal{D}'$. We claim that $\mathcal{F}$ is a filter. Indeed, take any $p, q \in \mathcal{F}$; since $\mathcal{F} \cap D(p,q) \neq \emptyset$, there is $r \in \mathcal{F} \cap D(p,q)$. The family $\mathcal{F}$ is centered, so it is impossible that $r \perp q$ or $r \perp p$. Thus $r \leq p$ and $r \leq q$ which shows that $\mathcal{F}$ is a filter such that $\mathcal{F} \cap D \neq \emptyset$ for any $D \in \mathcal{D}$ and hence $MA(\kappa)$ is established for all ccc partially ordered sets of cardinality κ. As a consequence $MA(\kappa)$ holds in any ccc partially ordered set by Problem 048. Since the cardinal $\kappa < \mathfrak{c}$ was chosen arbitrarily, we established that the Martin's axiom holds in its full generality whenever its topological version is true.

T.059. *Show that, under MA+¬CH, if $s^*(X) = \omega$, then $hd(X^\omega) = hl(X^\omega) = \omega$. In particular, neither strong S-spaces nor strong L-spaces exist under MA+¬CH.*

Solution. Let us prove first that for any space Z, under MA+¬CH, it follows from $s^*(Z) = \omega$ that $hl(Z) = \omega$. If this is not true, then there exists a space Z such that $s^*(Z) = \omega$ and there is a right-separated $R \subset Z$ with $|R| = \omega_1$ (see Problem 005). Since $s^*(R) = \omega$ and $hl(R) > \omega$, we can assume that $Z = R$; furthermore, without loss of generality, we can assume that $Z = \{x_\alpha : \alpha < \omega_1\}$ and Z is right-separated by its indexation (see Fact 2 of T.005).

The set $Z_\alpha = \{x_\beta : \beta \leq \alpha\}$ is an open neighborhood of x_α, so we can choose $U_\alpha \in \tau(x_\alpha, Z)$ such that $\overline{U}_\alpha \subset Z_\alpha$ for all $\alpha < \omega_1$. A subspace $Y \subset Z$ is called *special* if $x_\alpha \in Y$ implies that $Y \cap U_\alpha$ is finite. It is clear that every special subspace of Z is discrete.

Consider the set $\mathcal{P} = \{p : p \text{ is a finite subset of } Z\}$. Given $p, q \in \mathcal{P}$, let $p \leq q$ if $p \supset q$ and $(p \backslash q) \cap (\bigcup_{x_\alpha \in q} U_\alpha) = \emptyset$. We omit a straightforward verification that $(\mathcal{P}, \leq)$ is a partially ordered set. If $p \in \mathcal{P}$, let $i_-(p) = \min\{\alpha < \omega_1 : x_\alpha \in p\}$ and $i_+(p) = \max\{\alpha < \omega_1 : x_\alpha \in p\}$. Let us show that $\mathcal{P}$ is ccc. If there is an uncountable antichain $\mathcal{C} \subset \mathcal{P}$, then we can apply the Delta-lemma (see Problem 038) to find an uncountable $\mathcal{C}' \subset \mathcal{C}$ and a finite set $r \subset Z$ such that $p \cap q = r$ for any distinct $p, q \in \mathcal{C}'$. The family $\mathcal{C}_0 = \{p \backslash r : p \in \mathcal{C}'\}$ is disjoint and consists of finite sets, so, for any $\alpha < \omega_1$, the family $\{d \in \mathcal{C}_0 : d \cap \alpha \neq \emptyset\}$ is countable. This makes it possible to choose an uncountable $\mathcal{C}'' \subset \mathcal{C}'$ with the following properties:

(1) $i_-(p \backslash r) > i_+(r)$ for any $p \in \mathcal{C}''$;
(2) $\mathcal{C}'' = \{p_\alpha : \alpha < \omega_1\}$ and $i_+(p_\alpha \backslash r) < i_-(p_\beta \backslash r)$ for all $\alpha < \beta < \omega_1$;
(3) there is $n \in \mathbb{N}$ such that $|p \backslash r| = n$ for all $p \in \mathcal{C}''$.

If $p_\alpha \backslash r = \{y_1^\alpha, \ldots, y_n^\alpha\}$ (the order of the elements of $p_\alpha \backslash r$ is fixed but taken arbitrarily), then let $z_\alpha = (y_1^\alpha, \ldots, y_n^\alpha) \in Z^n$ for each $\alpha < \omega_1$. It follows from (2) and the definition of the family $\{U_\gamma : \gamma < \omega_1\}$ that

(4) for any $\alpha < \beta < \omega_1$, we have $(p_\beta \backslash r) \cap (\bigcup_{x_\gamma \in p_\alpha} \overline{U}_\gamma) = \emptyset$

because $\bigcup_{x_\gamma \in p_\alpha} \overline{U}_\gamma \subset \{x_\nu : \nu \leq i_+(p_\alpha)\}$. Thus, if $\alpha < \beta$, then for incompatibility of p_α and p_β it is necessary that $(p_\alpha \backslash r) \cap W_\beta \neq \emptyset$ where $W_\beta = \bigcup_{x_\gamma \in p_\beta \backslash r} U_\gamma$. Now consider the set $O_\beta = \{z = (z_1, \ldots, z_n) \in Z^n : z_i \in W_\beta \text{ for some } i \leq n\}$. It is clear that O_β is an open subset of Z^n; our last observation shows that $z_\alpha \in O_\beta$ for each $\alpha \leq \beta$. On the other hand, it follows from (4) that $z_{\beta+1} \notin \overline{O}_\beta$; therefore $G_\beta = O_{\beta+1} \backslash \overline{O}_\beta$ is an open subset of Z^n such that $G_\beta \cap (\{z_{\alpha+1} : \alpha < \omega_1\}) = \{z_{\beta+1}\}$ for each $\beta < \omega_1$. Consequently, the set $\{z_{\alpha+1} : \alpha < \omega_1\}$ is an uncountable discrete subspace of Z^n which contradicts $s(Z^n) \leq s^*(Z) = \omega$. This contradiction proves that $\mathcal{P}$ is ccc.

For each $\alpha < \omega_1$ let $D_\alpha = \{p \in \mathcal{P} : i_+(p) > \alpha\}$; the set D_α is dense in $\mathcal{P}$ for each $\alpha < \omega_1$. Indeed, if $p \in \mathcal{P}$, then let $\beta = \max\{i_+(p) + 1, \alpha + 1\}$; then $q = p \cup \{x_\beta\} \leq p$ and $q \in D_\alpha$. If MA+¬CH holds, then the family $\mathcal{D} = \{D_\alpha : \alpha < \omega_1\}$ has cardinality $< \mathfrak{c}$ and consists of dense subsets of $\mathcal{P}$. Choose any filter $\mathcal{F} \subset \mathcal{P}$ such that $\mathcal{F} \cap D_\alpha \neq \emptyset$ for all $\alpha < \omega_1$. We claim that the set $Y = \bigcup \mathcal{F}$ is special. To see this, take any $x_\alpha \in Y$ and any $p \in \mathcal{F}$ such that $x_\alpha \in p$. If q is an arbitrary element of $\mathcal{F}$, then q is compatible with p and hence there is $r \leq p$ and $r \leq q$. As a consequence, $(q \backslash p) \cap U_\alpha \subset (r \backslash p) \cap U_\alpha = \emptyset$ by the definition of the order $\leq$. Therefore $q \cap U_\alpha \subset p \cap U_\alpha$; since this is true for any $q \in \mathcal{F}$, the set $Y \cap U_\alpha \subset p \cap U_\alpha$ is finite and therefore Y is a special subset of Z. Finally, observe that Y has to be uncountable because any countable set is contained in some Z_α while $\mathcal{F} \cap D_{\alpha+1} \neq \emptyset$ which shows that there is $p \in \mathcal{F}$ with $p \backslash Z_\alpha \neq \emptyset$ and hence $Y \backslash Z_\alpha \neq \emptyset$. This final contradiction shows that $hl(Z) \leq \omega$.

Thus we have proved that, under MA+¬CH, any space Z with $s^*(Z) = \omega$ is hereditarily Lindelöf. Besides, if $s^*(Z) = \omega$, then $s^*(Z^n) = s^*(Z) = \omega$ for any $n \in \mathbb{N}$ and therefore $hl^*(Z) = \omega$. Furthermore, $s^*(C_p(Z)) = s^*(Z) = \omega$ (see Problem 025) and hence $hl^*(C_p(Z)) = \omega$ which shows that $hd^*(Z) = hl^*(C_p(Z)) = \omega$ (see Problem 027). Thus, for $Z = X$, it follows from $s^*(X) = \omega$ and MA+¬CH that $hd^*(X) = hl^*(X) = \omega$. Finally, apply Problem 011 and Problem 012 to conclude that, under MA+¬CH, we have $hl(X^\omega) = hd(X^\omega) = \omega$ for each space X such that $s^*(X) = \omega$.

T.060. *Prove that, under MA+¬CH, every compact space of countable spread is hereditarily separable.*

Solution. Let Z be a compact space with $s(Z) = \omega$. Observe first that $t(Z) = \omega$ because any free sequence is a discrete subset of Z, so all free sequences in Z are countable (see Problem 328 of [TFS]; note that Martin's axiom is not needed for this conclusion). Let us show first that there is a countable $A \subset Z$ such that $\overline{A}$ has non-empty interior. Assuming that this is not true we are going to construct a family $\{F_\alpha : \alpha < \omega_1\}$ with the following properties:

(1) F_α is a closed separable subset of Z (and hence $\mathrm{Int}(F_\alpha) = \emptyset$);
(2) $F_\alpha \subset F_\beta$ if $\alpha < \beta < \omega_1$;
(3) For each $\alpha < \omega_1$ there is a countable $A \subset F_{\alpha+1} \backslash F_\alpha$ such that $F_\alpha \subset \overline{A}$.

To start off, pick any $x \in Z$ and let $F_0 = \{x\}$. If $\alpha < \omega_1$ and we have a collection $\{F_\beta : \beta \leq \alpha\}$ with the properties (1)–(3), fix a dense countable set $B = \{b_n : n \in \omega\} \subset F_\alpha$; since F_α has empty interior, every b_n is in a closure of a countable set $A_n \subset Z \backslash F_\alpha$. If $A = \bigcup_{n \in \omega} A_n$ and $F_{\alpha+1} = \overline{F_\alpha \cup A}$, then the conditions (1)–(3) are fulfilled for the family $\{F_\beta : \beta \leq \alpha + 1\}$. If α is a limit ordinal and we have a family $\{F_\beta : \beta < \alpha\}$ with (1)–(3), let $F_\alpha = \overline{\bigcup\{F_\beta : \beta < \alpha\}}$. It is immediate that (1)–(3) are also fulfilled for the collection $\{F_\beta : \beta \leq \alpha\}$. Thus our inductive construction can be continued providing us a family $\{F_\alpha : \alpha < \omega_1\}$ with the properties (1)–(3). Let $F = \overline{\bigcup\{F_\alpha : \alpha < \omega_1\}}$. It is an easy exercise to see that $t(Z) = \omega$ implies that F is a closed subspace of Z and hence compact. Note also that $c(F) \leq s(F) \leq s(Z) = \omega$. Besides, (3) implies that F_α is nowhere dense in $F_{\alpha+1}$ and hence in F. Therefore the compact space F is a union of ω_1-many (and hence $< \mathfrak{c}$-many) nowhere dense subsets which is a contradiction with the topological version of Martin's axiom (see Problem 058).

This proves that, under Martin's axiom, every compact space of countable spread has a non-empty separable open set. Now, given any compact space Z with $s(Z) = \omega$, consider a maximal disjoint family $\mathcal{U}$ of non-empty separable open subsets of Z. Observe that $\mathcal{U}$ is countable because $c(Z) \leq s(Z) = \omega$. Besides, $\bigcup \mathcal{U}$ has to be dense in Z; for otherwise, there is $W \in \tau^*(Z)$ such that $P = \overline{W} \subset Z \backslash (\bigcup \mathcal{U})$. Since P is a compact space of countable spread, it has a non-empty open separable $V \subset P$. The set $U = V \cap W$ is separable, non-empty and open in Z; it is evident that $U \notin \mathcal{U}$ and the family $\mathcal{U} \cup \{U\}$ is disjoint which contradicts maximality of $\mathcal{U}$. This contradiction shows that $\mathcal{U}$ is dense in Z. Being a countable union of separable subspaces, the space $\mathcal{U}$ is also separable. If we take any countable $A \subset \bigcup \mathcal{U}$ which is dense in $\bigcup \mathcal{U}$, then A is dense in Z, so Z is separable. As a consequence, every compact space of countable spread is separable under MA+¬CH. Thus, if Z is compact and $s(Z) = \omega$, then all closed subsets of Z are separable under MA+¬CH because the previous statement is also true for all closed subspaces of Z.

Finally, take any $Y \subset Z$; since $\overline{Y}$ is separable, we can take a countable $A \subset \overline{Y}$ with $\overline{A} = \overline{Y}$. It follows from $t(Z) = \omega$ that, for any $a \in A$, there is a countable $B_a \subset Y$ such that $a \in \overline{B_a}$. The set $B = \bigcup \{B_a : a \in A\} \subset Y$ is countable and $\overline{B} \supset \overline{A} = \overline{Y} \supset Y$. Therefore B is a countable dense subset of Y and we proved that every subspace of Z is separable. Thus MA+¬CH implies that every compact space of countable spread is hereditarily separable.

T.061. *Prove that, under MA+¬CH, every compact space of countable spread is perfectly normal.*

Solution. To show that every compact space of countable spread is hereditarily Lindelöf under MA+¬CH, take any compact X with $s(X) = \omega$. If X is not hereditarily Lindelöf, then there is $Y = \{y_\alpha : \alpha < \omega_1\} \subset X$ which is right-separated by its indexation (see Fact 2 of T.005). The space $\overline{Y}$ is also a counterexample to our statement, so to obtain a contradiction, there is no loss of generality in assuming that $\overline{Y} = X$.

The set $Y_\alpha = \{y_\beta : \beta \leq \alpha\}$ is an open neighborhood of y_α in Y, so we can choose $U_\alpha \in \tau(y_\alpha, X)$ such that $\overline{U}_\alpha \cap Y \subset Y_\alpha$ for all $\alpha < \omega_1$. A subspace $Z \subset Y$ is called *special* if $y_\alpha \in Z$ implies that $Z \cap U_\alpha$ is finite. It is clear that every special subspace of Y is discrete. Given any $A \subset \omega_1$, let $Y_A = \{y_\alpha : \alpha \in A\}$.

Consider the set $\mathcal{P} = \{p : p$ is a non-empty finite subset of $Y\}$, and for any $n \in \mathbb{N}$, let $\mathcal{P}_n = \{p \in \mathcal{P} : |p| = n\}$. Given $p, q \in \mathcal{P}$, let $p \preceq q$ if $p \supset q$ and $(p \backslash q) \cap (\bigcup_{y_\alpha \in q} U_\alpha) = \varnothing$. We omit a straightforward verification that $(\mathcal{P}, \preceq)$ is a partially ordered set. Given any $p \in \mathcal{P}$, let $i_-(p) = \min\{\alpha < \omega_1 : y_\alpha \in p\}$ and $i_+(p) = \max\{\alpha < \omega_1 : y_\alpha \in p\}$. To proceed by contradiction, assume that $\mathcal{P}$ is ccc. The set $D_\alpha = \{p \in \mathcal{P} : i_+(p) > \alpha\}$ is dense in $\mathcal{P}$ for each $\alpha < \omega_1$. Indeed, if $p \in \mathcal{P}$, let $\beta = \max\{i_+(p) + 1, \alpha + 1\}$; then $q = p \cup \{y_\beta\} \preceq p$ and $q \in D_\alpha$. Thus the family $\mathcal{D} = \{D_\alpha : \alpha < \omega_1\}$ consists of dense subsets of $\mathcal{P}$. If MA+¬CH holds, then $\mathcal{D}$ has cardinality $< \mathfrak{c}$, so we can choose a filter $\mathcal{F} \subset \mathcal{P}$ such that $\mathcal{F} \cap D_\alpha \neq \varnothing$ for all $\alpha < \omega_1$. We claim that the set $Z = \bigcup \mathcal{F}$ is special. To see this, take any $y_\alpha \in Z$ and any $p \in \mathcal{F}$ such that $y_\alpha \in p$. If q is an arbitrary element of $\mathcal{F}$, then q is compatible with p and hence there is $r \in \mathcal{F}$ such that $r \preceq p$ and $r \preceq q$. As a consequence, $(q \backslash p) \cap U_\alpha \subset (r \backslash p) \cap U_\alpha = \varnothing$ by the definition of the order $\preceq$. Therefore $q \cap U_\alpha \subset p \cap U_\alpha$; since this is true for any $q \in \mathcal{F}$, the set $Z \cap U_\alpha \subset p \cap U_\alpha$ is finite and therefore Z is a special subset of Y. Finally, observe that Z has to be uncountable because any countable set is contained in some Y_α while $\mathcal{F} \cap D_{\alpha+1} \neq \varnothing$ which shows that there is $p \in \mathcal{F}$ with $p \backslash Y_\alpha \neq \varnothing$ and hence $Z \backslash Y_\alpha \neq \varnothing$. Thus Z is an uncountable discrete subspace of X which contradicts $s(X) = \omega$. This contradiction shows that $\mathcal{P}$ cannot be ccc.

Let $R_\alpha = Y \setminus Y_\alpha$ for all $\alpha < \omega_1$; we will also need the set $R = \bigcap \{\overline{R}_\alpha : \alpha < \omega_1\}$. By compactness of X, the set R is compact and non-empty. It is an immediate consequence of the fact that Y_α is open in Y that $\overline{R}_\alpha \cap Y_\alpha = \emptyset$ for each $\alpha < \omega_1$ and therefore $R \cap Y = \emptyset$. Given $p, q \in \mathcal{P}$, we say that $p < q$ if $i_+(p) < i_-(q)$.

Say that a family $\mathcal{B}$ of subsets of Y is *adequate* if it has the following properties:

(4) every $B \in \mathcal{B}$ is countable and $\overline{B} \cap R = \emptyset$;
(5) there is $n \in \mathbb{N}$, a set $\{q_\alpha : \alpha < \omega_1\} \subset \mathcal{P}_n$ and a family $\{B_\alpha : \alpha < \omega_1\} \subset \mathcal{B}$ such that $q_\alpha < q_\beta$ and $B_\beta \cap q_\alpha \neq \emptyset$ whenever $\alpha < \beta < \omega_1$.

We proved that $\mathcal{P}$ cannot be ccc; therefore there is an uncountable antichain $\mathcal{C} \subset \mathcal{P}$; applying the Delta-lemma (see Problem 038) we can find an uncountable $\mathcal{C}' \subset \mathcal{C}$ and a finite set $r \subset Y$ such that $p \cap q = r$ for any distinct $p, q \in \mathcal{C}'$. The family $\mathcal{C}_0 = \{p \setminus r : p \in \mathcal{C}'\}$ is disjoint and consists of finite sets, so, for any $\alpha < \omega_1$, the family $\{d \in \mathcal{C}_0 : d \cap \alpha \neq \emptyset\}$ is countable. This makes it possible to choose an uncountable $\mathcal{C}'' \subset \mathcal{C}'$ with the following properties:

(6) $i_-(p \setminus r) > i_+(r)$ for any $p \in \mathcal{C}''$;
(7) $\mathcal{C}'' = \{p_\alpha : \alpha < \omega_1\}$ and $p_\alpha \setminus r < p_\beta \setminus r$ for all $\alpha < \beta < \omega_1$;
(8) there is $n \in \mathbb{N}$ such that $|p_\alpha \setminus r| = n$ for all $\alpha < \omega_1$.

It follows from the property (7) and the definition of the family $\{U_\alpha : \alpha < \omega_1\}$ that, for any ordinals $\alpha < \beta < \omega_1$, we have $(p_\beta \setminus r) \cap (\bigcup_{x_\gamma \in p_\alpha} \overline{U}_\gamma) = \emptyset$ because $\bigcup_{x_\gamma \in p_\alpha} \overline{U}_\gamma \subset \{x_\nu : \nu \leq i_+(p_\alpha)\}$. Thus, if $\alpha < \beta$, then for incompatibility of p_α and p_β it is necessary that

(9) $(p_\alpha \setminus r) \cap W_\beta \neq \emptyset$ where $W_\beta = \bigcup_{x_\gamma \in p_\beta \setminus r} U_\gamma$.

Now it follows from (6)–(9) that the family $\mathcal{D}$ of all finite unions of the family $\{U_\alpha \cap Y : \alpha < \omega_1\}$ is adequate because we can let $q_\alpha = p_\alpha \setminus r$ and $B_\alpha = W_\alpha$ for all $\alpha < \omega_1$.

Say that a collection $\{B_\alpha : \alpha < \omega_1\}$ of subsets of Y is *cofinally centered on a set* $Z \subset Y$ if for any uncountable $T \subset Z$ there is $\alpha < \omega_1$ such that the family $\{B_\beta \cap T : \beta \geq \alpha\}$ is centered.

Lemma. Suppose that $\mathcal{B}$ is an adequate family closed under finite unions, i.e., any finite union of elements of $\mathcal{B}$ is an element of $\mathcal{B}$. Then there exists an uncountable $A \subset \omega_1$ and a collection $\{B_\alpha : \alpha < \omega_1\} \subset \mathcal{B}$ which is cofinally centered on the set $Y_A = \{y_\alpha : \alpha \in A\}$.

Proof. Fix the smallest $n \in \mathbb{N}$ for which there exists a set $\{q_\alpha : \alpha < \omega_1\} \subset \mathcal{P}_n$ and a family $\mathcal{B}' = \{B_\alpha : \alpha < \omega_1\} \subset \mathcal{B}$ which satisfy (5). Let $A = \{i_-(q_\alpha) : \alpha < \omega_1\}$ and suppose that $\mathcal{B}'$ is not cofinally centered on Y_A. Then there exists an uncountable counterexample $C \subset A$ such that, for each $\alpha < \omega_1$, there is a finite set $b_\alpha \subset \omega_1 \setminus \alpha$ for which

(*) $Y_C \cap (\bigcap \{B_\beta : \beta \in b_\alpha\}) = \emptyset$.

The set C will be still a counterexample if we make it smaller. The property $(*)$ will still hold if we substitute each b_α with some $b_{\beta(\alpha)}$ for which $\beta(\alpha) > \alpha$. Thus we can make the necessary changes in the set C and in the collection $\{b_\alpha : \alpha < \omega_1\}$ to show that we can assume, without loss of generality, that there is an increasing enumeration $\{\delta_\alpha : \alpha < \omega_1\}$ of the set C such that

(10) $b_\alpha < \delta_\alpha$ for each $\alpha < \omega_1$;
(11) if, for each $\alpha < \omega_1$, we let $r_\alpha = q_\gamma$ for the ordinal γ defined by $i_-(q_\gamma) = \delta_\alpha$, then $\alpha < \beta < \omega_1$ implies $r_\alpha < b_\beta$.

Consider the sets $s_\alpha = r_\alpha \backslash \{\delta_\alpha\}$ and $C_\alpha = \bigcup\{B_\gamma : \gamma \in b_\alpha\}$ for each $\alpha < \omega_1$. If $\alpha < \beta$, then, by $(*)$, there is some $\gamma \in b_\beta$ such that $y_{\delta_\alpha} \notin B_\gamma$. However $B_\gamma \cap r_\alpha \neq \emptyset$ by (5) and (11), so $B_\gamma \cap s_\alpha \neq \emptyset$. Since $\gamma \in b_\beta$, we have $B_\gamma \subset C_\beta$ which implies $s_\alpha \cap C_\beta \neq \emptyset$. An easy consequence of (5) for the set $\{q_\alpha : \alpha < \omega_1\}$ is that $s_\alpha < s_\beta$ whenever $\alpha < \beta < \omega_1$. Thus the property (5) is fulfilled for the set $\{s_\alpha : \alpha < \omega_1\} \subset \mathcal{P}_{n-1}$ and the family $\{C_\alpha : \alpha < \omega_1\} \subset \mathcal{B}$ contradicting the minimality of n and finishing the proof of our lemma. $\square$

Returning to our solution, apply the lemma to our adequate family $\mathcal{D}$ to choose an uncountable $A \subset \omega_1$ and a collection $\{D_\alpha : \alpha < \omega_1\} \subset \mathcal{D}$ which is cofinally centered on Y_A. By definition of $\mathcal{D}$, the set $K_\alpha = \overline{D}_\alpha$ is compact and disjoint from R for every $\alpha < \omega_1$. Let $H_\alpha = \bigcap_{\beta > \alpha} K_\beta$ for each $\alpha < \omega_1$ and consider the set $H = \bigcup_{\alpha < \omega_1} H_\alpha$. Since $t(X) = \omega$, the set H is compact (observe that some of the sets H_α can be empty). The set H being disjoint from R, the decreasing family $\{\overline{R}_\alpha \cap H : \alpha < \omega_1\}$ of compact subsets of X has empty intersection. Therefore $\overline{R}_\delta \cap H = \emptyset$ for some $\delta < \omega_1$. For the set $C = A \backslash \delta$, find $\alpha < \omega_1$ such that the family $\{D_\beta \cap Y_C : \beta > \alpha\}$ is centered. Then the family $\{K_\beta \cap Y_C : \beta > \alpha\}$ is also centered and therefore the family of compact sets $\{K_\beta \cap \overline{R}_\delta : \beta > \alpha\}$ is centered because $Y_C = R_\delta$. By compactness of the space X, there is a point $x \in \bigcap\{K_\beta \cap \overline{R}_\delta : \beta > \alpha\} = H_\alpha \cap \overline{R}_\delta \subset H \cap \overline{R}_\delta = \emptyset$ which is a contradiction. This contradiction shows that all right-separated subspaces of X are countable and hence X is perfectly normal (see Problem 001), so our solution is complete.

T.062. *Suppose that* MA+¬CH *hold. Prove that every compact space* X *with* $s(X \times X) \leq \omega$ *is metrizable.*

Solution. If Z is a space, let $\Delta_Z = \{(z, z) : z \in Z\} \subset Z \times Z$ be the diagonal of Z. Given a cover $\mathcal{U}$ of the space Z, let $\text{St}(z, \mathcal{U}) = \bigcup\{U \in \mathcal{U} : z \in U\}$ for any $z \in Z$. A cover $\mathcal{V}$ of the space Z is *a barycentric refinement* of the cover $\mathcal{U}$ if, for any $z \in Z$, there is $U \in \mathcal{U}$ such that $\text{St}(z, \mathcal{V}) \subset U$. Any compact space Z is paracompact and hence any open cover of Z has an open barycentric refinement (see Problem 230 of [TFS]).

Fact 1. Let Z be a compact space such that $\Delta = \Delta_Z$ is a G_δ-subset of $Z \times Z$. Then Z is second countable and hence metrizable.

Proof. The diagonal Δ is closed in $Z \times Z$; if it is a G_δ-subset of $Z \times Z$, then, by paracompactness of Z, there exists a sequence $\{O_n : n \in \omega\} \subset \tau(Z \times Z)$ and a collection $\{\mathcal{U}_n : n \in \omega\}$ of open finite covers of Z such that

(1) $O_{n+1} \subset O_n$ for each $n \in \omega$;
(2) $\bigcap\{O_n : n \in \omega\} = \Delta$;
(3) $\bigcup\{U \times U : U \in \mathcal{U}_n\} \subset O_n$ for every $n \in \omega$;
(4) the cover $\mathcal{U}_{n+1}$ is a barycentric refinement of $\mathcal{U}_n$ for all $n \in \omega$.

The family $\mathcal{S} = \bigcup\{\mathcal{U}_n : n \in \omega\}$ is countable and $\bigcup \mathcal{S} = Z$, so there exists a topology τ on the set Z generated by $\mathcal{S}$ as a subbase (see Problem 008 of [TFS]). Let us prove that $Y = (Z, \tau)$ is a Hausdorff space. Take any distinct $a, b \in Z$; since $c = (a, b) \notin \Delta$, there is $n \in \omega$ such that $c \notin O_n$. We claim that the set $\{a, b\}$ is not contained in $\mathrm{St}(x, \mathcal{U}_{n+1})$ for any $x \in Z$. Indeed, if there is $x \in Z$ with $\{a, b\} \subset \mathrm{St}(x, \mathcal{U}_{n+1})$, then there is $W \in \mathcal{U}_n$ such that $\mathrm{St}(x, \mathcal{U}_{n+1}) \subset W$ and therefore $\{a, b\} \subset W$. Thus $c \in W \times W \subset O_n$ which is a contradiction.

Now, take $U, V \in \mathcal{U}_{n+1}$ such that $a \in U$ and $b \in V$; if $z \in U \cap V$, then $\{a, b\} \subset \mathrm{St}(z, \mathcal{U}_{n+1})$ which we proved not to be possible. Therefore $U \cap V = \emptyset$, so the space Y is Hausdorff. The identity map $i : Z \to Y$ is continuous because $i^{-1}(W) = W \in \tau(Z)$ for any $W \in \mathcal{S}$ (see Problem 009 of [TFS]). Being a continuous image of a compact space Z, the space Y is a compact Hausdorff and hence Tychonoff space (see Problem 124 of [TFS]). The subbase $\mathcal{S}$ of the space Y is countable; the family of all finite intersections of the elements of $\mathcal{S}$ is a countable base of Y, so Y is second countable and hence metrizable (see Problems 209 and 212 of [TFS]). As a consequence, i condenses Z onto a metrizable space Y. Therefore Z is metrizable by Problem 140 of [TFS]. Any metrizable compact space is second countable, so Fact 1 is proved.   □

Returning to our solution, observe that the inequality $s(X \times X) \leq \omega$ implies that, under MA+¬CH, the space $X \times X$ is perfectly normal (see Problem 061) and hence its diagonal Δ_X is a G_δ-set in $X \times X$. Now apply Fact 1 to conclude that X is second countable and metrizable.

T.063. *Prove that MA+¬CH implies that there are no Luzin spaces.*

Solution. Suppose to the contrary that there exists a Luzin space X and consider the family $\mathcal{U} = \{U \in \tau^*(X) : |U| \leq \omega\}$; if $X' = \bigcup \mathcal{U}$, then $\mathcal{U}$ is an open cover of X'. Since $hl(X) \leq \omega$ (see Problem 043), there is a countable $\mathcal{U}' \subset \mathcal{U}$ such that $X' = \bigcup \mathcal{U}'$. Consequently, X' is countable. If some $V \in \tau^*(X \backslash X')$ is countable, then there is $W \in \tau(X)$ such that $W \cap (X \backslash X') = V$. Therefore $W = V \cup (W \cap X') \in \tau(X)$ is countable which is a contradiction because $W \not\subset \bigcup \mathcal{U}$. Thus all non-empty open subsets of $Y = X \backslash X'$ are uncountable and, in particular, Y has no isolated points. Furthermore, any nowhere dense subset A of Y is also nowhere dense in X and hence $|A| \leq \omega$. This proves that

(*) if there is a Luzin space, then there is a Luzin space whose all non-empty open subsets are uncountable. We will call such spaces *everywhere uncountable*.

If A is any set, let $\text{Fin}(A)$ be the family of all finite subsets of A; if f is function, then $\text{dom}(f)$ is its domain. If $\mathcal{U}$ and $\mathcal{V}$ are families of subsets of a set X, then $\mathcal{V}$ is *properly inscribed in* $\mathcal{U}$ if for any $V \in \mathcal{V}$ there is $U \in \mathcal{U}$ such that V is properly contained in U, i.e., $V \subset U$ and $V \neq U$.

Fact 1. Let M be a second countable space. Suppose that Z is an uncountable subset of M. Then, under MA+¬CH, there exists a disjoint family $\{F_\alpha : \alpha < \omega_1\}$ of closed subsets of M such that $|F_\alpha \cap Z| \geq \omega_1$ for each $\alpha < \omega_1$.

Proof. Fix some countable base $\mathcal{B}$ of the space M which is closed under finite unions, i.e., any finite union of elements of $\mathcal{B}$ belongs to $\mathcal{B}$. It is easy to find a disjoint family $\{Z_\alpha : \alpha < \omega_1\} \subset \exp(Z)$ such that $|Z_\alpha| = \omega_1$ for all $\alpha < \omega_1$. Let $\mathcal{P}$ be the set of pairs $p = (s, a)$ such that

(1) s and a are functions for which there is a finite set $D(p) \subset \omega_1 \times \omega$ such that $\text{dom}(s) = \text{dom}(a) = D(p)$;
(2) $s : D(p) \to \mathcal{B}$ and $a : D(p) \to \text{Fin}(Z)$;
(3) $a(\alpha, n) \in \text{Fin}(Z_\alpha)$ for each $(\alpha, n) \in D(p)$;
(4) $a(\alpha, n) \subset s(\alpha, n)$ for any $(\alpha, n) \in D(p)$.

Given any elements $p = (s, a) \in \mathcal{P}$ and $q = (t, b) \in \mathcal{P}$, say that $p \leq q$ if $D(q) \subset D(p)$ and $p(\alpha, n) \subset q(\alpha, n)$, $b(\alpha, n) \subset a(\alpha, n)$ for each $(\alpha, n) \in D(q)$. It is immediate that $(\mathcal{P}, \leq)$ is a partially ordered set. Let us show that $\mathcal{P}$ has ccc. Assume that S is an uncountable subset of $\mathcal{P}$; by the Delta-lemma (see Problem 038) there is an uncountable $S' \subset S$ and a finite set $F \subset \omega_1 \times \omega$ such that $D(p) \cap D(q) = F$ for any distinct $p, q \in S'$. Since $\mathcal{B}$ is countable, there are only countably many functions from F to $\mathcal{B}$. Thus there is an uncountable $S'' \subset S'$ such that, for any $p = (s, a) \in S''$ and $q = (t, b) \in S''$, we have $s|F = t|F$. As a consequence, given distinct $p = (s, a) \in S''$ and $q = (t, b) \in S''$, we can define a function $w : D(p) \cup D(q) \to \mathcal{B}$ by the conditions $w(\alpha, n) = s(\alpha, n)$ for all $(\alpha, n) \in D(p)$ and $w(\alpha, n) = t(\alpha, n)$ for all $(\alpha, n) \in D(q)\backslash D(p)$. Furthermore, let $c(\alpha, n) = a(\alpha, n)$ for all points $(\alpha, n) \in D(p)\backslash D(q)$ and $c(\alpha, n) = b(\alpha, n)$ for all $(\alpha, n) \in D(q)\backslash D(p)$; besides, let $c(\alpha, n) = a(\alpha, n) \cup b(\alpha, n)$ for all $(\alpha, n) \in F$. It is evident that the point $r = (w, c) \in \mathcal{P}$ is a common extension of p and q, i.e., $r \leq p$ and $r \leq q$, so S is not an antichain which shows that $\mathcal{P}$ has ccc.

Given any $\alpha < \omega_1$ and $z \in Z_\alpha$, let $E(z, \alpha) = \{(s, a) \in \mathcal{P} : z \in a(\alpha, n)$ for some $n \in \omega\}$. The set $E(z, \alpha)$ is dense in $\mathcal{P}$ for all $\alpha < \omega_1$ and $z \in Z_\alpha$. Indeed, if $p = (s, a) \in \mathcal{P}$, then take any $n \in \omega$ such that $(\alpha, n) \notin D(p)$, any $B \in \mathcal{B}$ with $z \in B$ and let $q = (t, b) \in \mathcal{P}$ be defined by $t|D(p) = s$, $b|D(p) = a$, $b(\alpha, n) = \{z\}$ and $t(\alpha, n) = B$. It is evident that $q \leq p$ and $q \in E(z, \alpha)$.

Now, for any distinct ordinals $\alpha, \beta \in \omega_1$ and any $m, n \in \omega$, let $H(\alpha, \beta, m, n) = \{p = (s, a) \in \mathcal{P} : \overline{s(\alpha, n)} \cap \overline{s(\beta, m)} = \emptyset\}$. The set $H(\alpha, \beta, m, n)$ is also dense in $\mathcal{P}$ for all possible 4-tuples (α, β, m, n). To show this, take any point $p = (s, a) \in \mathcal{P}$. If $(\alpha, n) \notin D(p)$, then choose any $z \in Z_\alpha$ and any $U \in \mathcal{B} \cap \tau(z, M)$ and define $p' = (s', a') \in \mathcal{P}$ by $s'|D(p) = s$, $a'|D(p) = a$, $s'(\alpha, n) = U$ and $s'(\alpha, n) = \{z\}$. It is clear that $p' \leq p$, so it suffices to find $q \in H(\alpha, \beta, m, n)$ with $q \leq p'$. An identical reasoning for the case when $(\beta, m) \notin D(p)$ shows that we can assume, without loss of generality, that $Q = \{(\alpha, n), (\beta, m)\} \subset D(p)$.

The finite sets $a(\alpha, n) \subset Z_\alpha$ and $a(\beta, m) \subset Z_\beta$ are disjoint because $Z_\alpha \cap Z_\beta = \emptyset$. Since the family $\mathcal{B}$ is closed under finite unions, we can find sets $U, V \in \mathcal{B}$ such that $a(\alpha, n) \subset U$, $a(\beta, m) \subset V$ and $\overline{U} \cap \overline{V} = \emptyset$. Now let $t(\alpha, n) = U \cap s(\alpha, n)$, $t(\beta, m) = s(\beta, m) \cap V$ and $t(\gamma, k) = s(\gamma, k)$ for all $(\gamma, k) \in D(p) \backslash Q$. It is immediate that $q = (t, a) \in H(\alpha, \beta, m, n)$ and $q \leq p$ which shows that $H(\alpha, \beta, m, n)$ is dense in $\mathcal{P}$ for all distinct $\alpha, \beta \in \omega_1$ and any $m, n \in \omega$. Thus the family

$$\mathcal{D} = \{ E(z, \alpha) : \alpha < \omega_1, \, z \in Z_\alpha \} \cup \{ H(\alpha, \beta, m, n) : \alpha, \beta \in \omega_1, \, m, n \in \omega \text{ and } \alpha \neq \beta \}$$

consists of dense subsets of $\mathcal{P}$ and has cardinality $< \mathfrak{c}$. Applying the Martin's axiom we can find a filter $\mathcal{F} \subset \mathcal{P}$ such that $\mathcal{F} \cap D \neq \emptyset$ for any $D \in \mathcal{D}$. We will consider for technical purposes that for any $p = (s, a) \in \mathcal{P}$, we have $s(\alpha, n) = M$ and $a(\alpha, n) = \emptyset$ for all $(\alpha, n) \in (\omega_1 \times \omega) \backslash D(p)$. Consider the set $F_{\alpha, n} = \bigcap \{ s(\alpha, n) : (s, a) \in \mathcal{F} \text{ for some } a \}$ and let $K_\alpha = \bigcup_{n \in \omega} F_{\alpha, n}$. It is clear that the set $F_{\alpha, n}$ is closed in M for any $(\alpha, n) \in \omega_1 \times \omega$. Observe first that $Z_\alpha \subset K_\alpha$ for each $\alpha < \omega_1$; indeed, if $z \in Z_\alpha$, then pick any $p = (s, a) \in \mathcal{F} \cap E(z, \alpha)$. There is some $n \in \omega$ such that $z \in p(\alpha, n)$. Now take any $q = (t, b) \in \mathcal{F}$; there is $r = (u, c) \in \mathcal{F}$ such that $r \leq p$ and $r \leq q$. In particular, $u(\alpha, n) \subset s(\alpha, n) \cap t(\alpha, n)$ while $c(\alpha, n) \supset a(\alpha, n) \cup b(\alpha, n)$ and therefore $z \in c(\alpha, n) \subset u(\alpha, n)$. As a consequence, $z \in u(\alpha, n) \subset t(\alpha, n) \subset \overline{t(\alpha, n)}$ which shows that $z \in \overline{t(\alpha, n)}$ for all $q = (t, b) \in \mathcal{F}$, i.e., $z \in F_{\alpha, n} \subset K_\alpha$; the point $z \in Z_\alpha$ was chosen arbitrarily, so $Z_\alpha \subset K_\alpha$ for all $\alpha < \omega_1$.

Now take any distinct ordinals $\alpha, \beta < \omega_1$ and any $m, n \in \omega$. There is a point $p = (s, a) \in \mathcal{F} \cap H(\alpha, \beta, m, n)$; it is clear that $F_{\alpha, n} \cap F_{\beta, m} \subset \overline{s(\alpha, n)} \cap \overline{s(\beta, m)} = \emptyset$. An immediate consequence is that $K_\alpha \cap K_\beta = \emptyset$ for any distinct $\alpha, \beta < \omega_1$.

Since each K_α covers an uncountable set Z_α, there is $n(\alpha) \in \omega$ such that $F_{\alpha, n(\alpha)} \cap Z_\alpha$ is uncountable. The sets $F_\alpha = F_{\alpha, n(\alpha)}$ are as promised, so Fact 1 is proved. $\square$

Fact 2. Let X be a Luzin space. If M is a second countable space and MA+¬CH holds, then $f(X)$ is countable for any continuous map $f : X \to M$.

We can consider that $M = f(X)$; assume that M is uncountable. If, in the formulation of Fact 1, we let $Z = M$, then we obtain an uncountable disjoint family $\mathcal{U}$ of uncountable closed subsets of M. The family $\mathcal{V} = \{ f^{-1}(P) : P \in \mathcal{U} \}$ consists of disjoint closed uncountable subsets of X. The space X being Luzin, we have $\text{Int}_X(V) \neq \emptyset$ for any $V \in \mathcal{V}$. Therefore $\{ \text{Int}_X(V) : V \in \mathcal{V} \}$ is an uncountable disjoint family of non-empty open subsets of X which contradicts the fact that $c(X) \leq hl(X) = \omega$ (see Problem 043). Fact 2 is proved.

Fact 3. Let I be the space $[0, 1]$ with the natural topology. No Luzin space can be mapped continuously onto I. Note that no additional axioms are needed for the proof of this fact while, under MA+¬CH, it is an immediate consequence of Fact 2.

Proof. By Fact 3 of T.045, there exists a disjoint family $\{ F_\alpha : \alpha < \omega_1 \}$ of uncountable closed subsets of I. If L is a Luzin space and $f : L \to I$ is a continuous onto map, then $U_\alpha = \text{Int}_L(f^{-1}(F_\alpha)) \neq \emptyset$ for all $\alpha < \omega_1$ because

$f^{-1}(F_\alpha)$ is an uncountable closed subset of L. Thus $\{U_\alpha : \alpha < \omega_1\}$ is an uncountable disjoint family of non-empty open subsets of L which is a contradiction with $c(L) \leq hl(L) = \omega$ (see Problem 043). Fact 3 is proved. □

A space is zero-dimensional if it has a base which consists of clopen sets.

Fact 4. Let Z be an arbitrary space. If Z is not zero-dimensional, then it maps continuously onto I. No additional axioms are needed for the proof of this fact.

Proof. Since Z is not zero-dimensional, there is a point $z \in Z$ and $U \in \tau(z, Z)$ such that $U \neq Z$ and there is no clopen set W for which $z \in W \subset U$. The space Z being Tychonoff, there is a continuous function $f : Z \to I$ such that $f(z) = 1$ and $f|(Z \setminus U) \equiv 0$. If $t \in I \setminus f(Z)$, then $t \in (0, 1)$ and $W = f^{-1}((t, 1])$ is an open subset of Z with $z \in W \subset U$. Besides, $Z \setminus W = f^{-1}([0, t]) = f^{-1}([0, t))$ is again an open set, so W is clopen which is a contradiction. Thus $f(Z) = I$ and Fact 4 is proved. □

Fact 5. Any Luzin space is zero-dimensional (no additional axioms are needed here either).

Proof. If L is a Luzin space, then it cannot be continuously mapped onto I by Fact 3. Therefore L is zero-dimensional by Fact 4, so Fact 5 is proved. □

Fact 6. Let X be an everywhere uncountable Luzin space. Suppose that we have a family $\mathcal{U} = \{U(n, k) : n, k \in \omega\}$ of clopen subsets of X (some of which might be empty) such that the family $\mathcal{U}_n = \{U(n, k) : k \in \omega\}$ is disjoint and $\bigcup \mathcal{U}_n$ is dense in X for all $n \in \omega$. Then, under MA+¬CH, the set

$$H = \bigcup\{\text{Int}(\bigcap\{U(n, h(n)) : n \in \omega\}) : h \in \omega^\omega\}$$

is dense in X.

Proof. Let $E_h = \bigcap\{U(n, h(n)) : n \in \omega\}$ for each $h \in \omega^\omega$. Given any $U \in \mathcal{U}$, let $\chi_U : X \to \{0, 1\}$ be the characteristic function of U defined by $\chi_U(U) = \{1\}$ and $\chi_U(X \setminus U) = \{0\}$ (if $U = \emptyset$, then $\chi_U \equiv 0$; if $U = X$, then $\chi_U \equiv 1$). The function $f = \Delta\{\chi_U : U \in \mathcal{U}\}$ maps X onto a space $Y \subset \{0, 1\}^\mathcal{U}$. Since $\mathcal{U}$ is countable, the space Y is second countable, so $|Y| \leq \omega$ by Fact 2. Next observe that for every $h \in \omega^\omega$, the set E_h is non-empty if and only if there is a point $y_h \in Y$ such that $E_h = f^{-1}(y_h)$. Indeed, given $U(n, k) \in \mathcal{U}$, let $y_h(U(n, k)) = 1$ if and only if $k = h(n)$; otherwise let $y_h(U) = 0$. It is immediate that $y_h \in Y$ if and only if $E_h \neq \emptyset$ and $f^{-1}(y_h) = E_h$.

Now take any set $W \in \tau^*(X)$. Since we have only countably many of non-empty elements of the family $\{E_h : h \in \omega^\omega\}$ and $|W| > \omega$, there is $h \in \omega^\omega$ such that $E_h \cap W$ is uncountable. Since X is a Luzin space, there is $V \in \tau^*(X)$ such that $V \subset \overline{E_h \cap W}$. Every $U(n, k)$ is also closed, so E_h is closed as well. Thus $V \subset \overline{E_h \cap W} \subset \overline{E_h} \cap \overline{W} = E_h \cap \overline{W}$. Since $V \cap W \neq \emptyset$, we have $W_1 = V \cap W \neq \emptyset$, and hence $W_1 \subset E_h \cap W$ which shows that $H \cap W \neq \emptyset$. Since the set W was chosen arbitrarily, we have established that H is dense in X, so Fact 6 is proved. □

Returning to our solution, assume MA+¬CH and suppose that there exists a Luzin space. By $(*)$ there exists an everywhere uncountable Luzin space X. Let $T_0 = \{X\}$. Suppose that $\alpha < \omega_1$ and we constructed families $\{T_\beta : \beta < \alpha\}$ of clopen subsets of X with the following properties:

(5) T_β is countable, disjoint and $\bigcup T_\beta$ is dense in X for every $\beta < \alpha$;
(6) T_β is properly inscribed in T_γ whenever $\gamma < \beta < \alpha$.

If $\alpha = \alpha_0 + 1$, then for each $U \in T_{\alpha_0}$ choose a proper clopen $O_U \subset U$ (which is possible because X is zero-dimensional (Fact 5) and everywhere uncountable (see $(*)$). Let $T_\alpha = \{O_U : U \in T_{\alpha_0}\} \cup \{U \setminus O_U : U \in T_{\alpha_0}\}$. It is immediate that (5) and (6) still hold for the collection $\{T_\beta : \beta \leq \alpha\}$.

If α is a limit ordinal, then choose an increasing sequence $\{\beta_n : n \in \omega\}$ cofinal in α and choose an enumeration $\{U(n,k) : k \in \omega\}$ for the family T_{β_n} for all $n \in \omega$. If some T_{β_n} is finite, then enumerate its elements as $\{U(n,k) : k < m_n\}$ for some $m_n \in \omega$ and let $U(n,k) = \emptyset$ for all $k \geq m_n$. Now we can apply Fact 6 to conclude that the open set $H = \bigcup\{\text{Int}(\bigcap\{U(n,h(n)) : n \in \omega\}) : h \in \omega^\omega\}$ is dense in X. By zero-dimensionality of X (Fact 5), there exists a disjoint family $T_\alpha \subset \tau(X)$ of clopen subsets of X such that $\bigcup T_\alpha$ is dense in X and each $U \in T_\alpha$ is contained in the set $E_h = \bigcap\{U(n,h(n)) : n \in \omega\}$ for some $h \in \omega^\omega$.

The property (5) is, clearly, fulfilled for the collection $\{T_\beta : \beta \leq \alpha\}$. To see that (6) also holds, we must prove it for $\beta = \alpha$ and $\gamma < \alpha$. Take any $U \in T_\alpha$; there is $n \in \omega$ such that $\gamma < \beta_n$. There is $h \in \omega^\omega$ such that $U \subset E_h$. Therefore $U \subset U(n,h(n)) \in T_{\beta_n}$; since T_{β_n} is properly inscribed in T_γ, there is $W \in T_\gamma$ such that $U(n,h(n))$ is properly included in W. Thus U is properly included in W, so (5) and (6) hold for the collection $\{T_\beta : \beta \leq \alpha\}$. Therefore our inductive construction can go on to give us the collection $\{T_\alpha : \alpha < \omega_1\}$ such that (5) and (6) are fulfilled for all $\alpha < \omega_1$.

Observe that the family $T = \bigcup\{T_\alpha : \alpha < \omega_1\}$ has the following property:

(7) if $U, V \in T$ and $U \cap V \neq \emptyset$, then either $U \subset V$ or $V \subset U$ and in both cases the inclusion is proper.

Indeed, if $U \cap V \neq \emptyset$, then $U \in T_\alpha$ and $V \in T_\beta$ where $\alpha \neq \beta$. Assume, for example, that $\alpha < \beta$. Since T_β is properly inscribed in T_α, there is $W \in T_\alpha$ such that $V \subset W$. If $U \neq W$, then $U \cap W = \emptyset$ because T_α is disjoint and hence $U \cap V = \emptyset$, a contradiction. Thus $U = W$ and hence $V \subset U$ and the inclusion is proper by (6).

Given $U, V \in T$, let $U \leq V$ iff $U \subset V$. Then $\leq$ is a partial order on T and $(T, \leq)$ is ccc because $c(X) = \omega$ and $T \subset \tau(X)$. The family T^* of non-empty elements of T is uncountable; for otherwise, $T^* \subset \bigcup\{T_\beta : \beta < \alpha\}$ for some $\alpha < \omega_1$. However, there are non-empty elements of T_α because $\bigcup T_\alpha$ is dense in X, and by (6), no element of T_α can belong to $\bigcup\{T_\beta : \beta < \alpha\}$ which is a contradiction. It is evident that $(T^*, \leq)$ also has ccc, so, under MA+¬CH, we can apply Problem 049 to find an uncountable $S \subset T^*$ such that all elements of S are pairwise compatible. By (7), this means $U \subset V$ or $V \subset U$ for any $U, V \in S$.

Take any $U \in S$. If $U \in T_\alpha$, then, for any $\beta < \alpha$ there is at most one element of T_β which contains U. Therefore

(8) for any $U \in S$ the family $P_U = \{V \in S : U \subset V\}$ is countable.

Choose any $U_0 \in S$; suppose that we have $\{U_\alpha : \alpha < \beta\}$ for some $\beta < \omega_1$ such that U_α is properly contained in U_γ if $\gamma < \alpha < \beta$. Since $P = \bigcup\{P_{U_\alpha} : \alpha < \beta\}$ is countable, there is $U_\beta \in S \backslash P$. By (7), the set U_β is properly included in U_α for each $\alpha < \beta$. As a consequence, we can construct a collection $\{U_\alpha : \alpha < \omega_1\}$ such that $\alpha < \beta < \omega_1$ implies $U_\beta \subset U_\alpha$ and the last inclusion is proper. If $V_\alpha = U_0 \backslash U_\alpha$ for all $\alpha < \omega_1$, then $\{V_\alpha : \alpha < \omega_1\} \subset \tau^*(X)$. Since X is hereditarily Lindelöf, there is $\beta < \omega_1$ such that $H = \bigcup\{V_\alpha : \alpha < \beta\} = G = \bigcup\{V_\alpha : \alpha < \omega_1\}$. Since $V_\alpha \subset V_\beta$ for all $\alpha < \beta$, we have $H \subset V_\beta = U_0 \backslash U_\beta$. Since $U_{\beta+1}$ is properly included in U_β, there is $x \in U_\beta \backslash U_{\beta+1}$. Thus $x \in U_0 \backslash U_{\beta+1} = V_{\beta+1}$ and therefore $x \in G$. However, $H \subset U_0 \backslash U_\beta$ so $x \notin H$. This contradiction shows that the existence of our family T is contradictory, so our solution is complete.

T.064. *Let $C_n \subset \omega_1$ be a club for any natural n. Prove that $\bigcap\{C_n : n \in \omega\}$ is a club.*

Solution. It is evident that $C = \bigcap_{n \in \omega} C_n$ is closed in ω_1; to see that it is unbounded, fix any $\alpha < \omega_1$. Let $\{n_k : k \in \omega\}$ be an enumeration of ω in which every $m \in \omega$ occurs infinitely many times. Since C_{n_0} is unbounded, there is $\alpha_0 \in C_{n_0}$ such that $\alpha < \alpha_0$. Suppose that we have $\alpha_0, \ldots, \alpha_k$ such that $\alpha < \alpha_0 < \cdots < \alpha_k < \omega_1$ and $\alpha_i \in C_{n_i}$ for all $i \leq k$. Since $C_{n_{k+1}}$ is unbounded, there is $\alpha_{k+1} \in C_{n_{k+1}}$ with $\alpha_k < \alpha_{k+1}$. This proves that our inductive construction can continue to provide an increasing sequence $\{\alpha_i : i \in \omega\} \subset \omega_1$ such that $\alpha < \alpha_0$ and $\alpha_i \in C_{n_i}$ for all $i \in \omega$.

Let $\beta = \sup\{\alpha_i : i \in \omega\}$. Then the sequence $S = \{\alpha_i : i \in \omega\}$ converges to β. Furthermore, for every $m \in \omega$ the set $S_m = \{\alpha_i : n_i = m\}$ is an infinite subsequence of S and therefore $S_m \to \beta$ as well. Since $S_m \subset C_m$ and C_m is closed, the ordinal β belongs to C_m for all $m \in \omega$. Thus $\beta \in C$ and $\beta > \alpha$, so C is uncountable.

T.065. *Prove that*

  (i) *every stationary subset of ω_1 is uncountable;*
  (ii) *not all uncountable subsets of ω_1 are stationary;*
  (iii) *if $A \subset \omega_1$ contains a stationary set, then A is stationary;*
  (iv) *Any intersection of a stationary subset of ω_1 with a club is stationary;*
  (v) *if $A_n \subset \omega_1$ for each $n \in \omega$ and $\bigcup\{A_n : n \in \omega\}$ is stationary, then A_n is stationary for some $n \in \omega$.*

Solution. If a set $A \subset \omega_1$ is countable then $A \subset \alpha$ for some $\alpha < \omega_1$. However, the set $C = \omega_1 \backslash \alpha$ is a club such that $A \cap C = \emptyset$. Hence A is not stationary and (i) is proved.

Let C be the set of all limit ordinals of ω_1. Topologically, C is the set of non-isolated points of ω_1, so C is closed in ω_1. Besides, C cannot be countable because otherwise the infinite (and even uncountable) set $\omega_1 \backslash \alpha$ is closed and discrete in ω_1 for some $\alpha < \omega_1$ which is a contradiction with countable compactness of ω_1

(see Problem 314 of [TFS]). Thus C is a club; an immediate consequence is that $A = \omega_1 \setminus C$ is not stationary. Since the uncountable set $\{\alpha + 1 : \alpha < \omega\}$ is contained in A, the set A is uncountable and non-stationary so (ii) is proved.

If $A \supset A'$ and the set A' is stationary, then, for any club $C \subset \omega_1$, we have $A \cap C \supset A' \cap C \neq \emptyset$, so A is stationary and (iii) is proved.

Assume that $A \subset \omega_1$ is stationary and $C \subset \omega_1$ is a club. Given any club $D \subset \omega_1$, the set $C \cap D$ is also a club (see Problem 064) and therefore $A \cap (C \cap D) = (A \cap C) \cap D \neq \emptyset$. Thus the set $A \cap C$ intersects any club in ω_1, so $A \cap C$ is stationary and (iv) is proved.

To settle (v), assume that the set A_n is not stationary for all $n \in \omega$ and let $A = \bigcup \{A_n : n \in \omega\}$. There is a club $C_n \subset \omega_1$ such that $A_n \cap C_n = \emptyset$ for all $n \in \omega$. The set $C = \bigcap \{C_n : n \in \omega\}$ is a club by Problem 064 while $A \cap C = \emptyset$ which is a contradiction. Therefore some A_n must be stationary, so our solution is complete.

T.066. *Let A be a stationary subset of ω_1. Prove that there exists a disjoint family of stationary subsets $\{A_\alpha : \alpha < \omega_1\}$ such that $A_\alpha \subset A$ for each $\alpha \in \omega_1$.*

Solution. For any countable ordinal $\alpha > 0$ choose an onto map $f_\alpha : \omega \to \alpha$. If $A(\alpha, n) = \{\beta \in A : f_\beta(n) = \alpha\}$ for all $\alpha > 0$ and $n \in \omega$, then the family $\mathcal{A}_n = \{A(\alpha, n) : \alpha > 0\}$ is disjoint for every $n \in \omega$. Indeed, if $\alpha \neq \alpha'$ and $\beta \in A(\alpha, n) \cap A(\alpha', n)$, then $f_\beta(n) = \alpha = \alpha'$ which is a contradiction. It is easy to check that $\bigcup \{A(\alpha, n) : n \in \omega\} = (\omega_1 \setminus (\alpha + 1)) \cap A$ for every $\alpha > 0$. Therefore $\bigcup \{A(\alpha, n) : n \in \omega\}$ is a stationary set for each $\alpha > 0$ (see Problem 065) which implies by Problem 065 that $A(\alpha, n_\alpha)$ is stationary for some $n_\alpha \in \omega$. There exists $m \in \omega$ such that the set $P = \{\alpha > 0 : n_\alpha = m\}$ is uncountable. Therefore, for all ordinals $\alpha \in P$, the set $A(\alpha, n_\alpha) = A(\alpha, m)$ belongs to $\mathcal{A}_m$. As a consequence, the disjoint family $\mathcal{A}_m$ has uncountably many stationary elements and hence we can choose an uncountable subfamily $\mathcal{A}' \subset \mathcal{A}_m$ such that every $A \in \mathcal{A}'$ is stationary. Choosing any enumeration $\{A_\alpha : \alpha < \omega_1\}$ of the family $\mathcal{A}'$ we obtain the promised disjoint uncountable family of stationary subsets of A.

T.067 (Fodor's lemma; also called pressing-down lemma). *Let $A \subset \omega_1$ be a stationary subset of ω_1. Suppose that $f : A \to \omega_1$ is a map such that $f(\alpha) < \alpha$ for any $\alpha \in A$. Prove that there is $\alpha_0 \in \omega_1$ such that the set $f^{-1}(\alpha_0)$ is stationary (and, in particular, $|f^{-1}(\alpha_0)| = \omega_1$).*

Solution. Assume that $f^{-1}(\alpha)$ is not stationary (maybe empty) for any $\alpha < \omega_1$. Then we have a club $C_\alpha \subset \omega_1$ such that $f^{-1}(\alpha) \cap C_\alpha = \emptyset$ for each $\alpha < \omega_1$. Consider the set $C = \{\beta < \omega_1 : \beta \in \bigcap_{\alpha < \beta} C_\alpha\}$. We claim that C is a club.

If $\beta \in \overline{C} \setminus C$, then there is an increasing sequence $\{\beta_n : n \in \omega\} \subset C$ which converges to β. Fix any $\alpha < \beta$; there is $m \in \omega$ such that $\alpha < \beta_m$. Since $\beta_n \in C$, we have $\beta_n \in C_\alpha$ for all $n \geq m$. The set C_α is closed in ω_1 so $\beta = \lim \beta_n \in C_\alpha$. The ordinal α was chosen arbitrarily, so $\beta \in C_\alpha$ for all $\alpha < \beta$, i.e., $\beta \in C$. This contradiction shows that C is a closed subset of ω_1.

To see that C is uncountable, fix any $\nu \in \omega_1 \setminus \{0\}$. The set $D_0 = \bigcap \{C_\alpha : \alpha < \nu\}$ is a club by Problem 064, so there is $\beta_0 > \nu$ such that $\beta_0 \in D_0$. Let $\beta_{-1} = \nu$

and assume that we have ordinals $\beta_{-1} < \beta_0 < \cdots < \beta_n$ such that $\beta_i \in \bigcap \{C_\alpha : \alpha < \beta_{i-1}\}$ for all $i \in \{0, \ldots, n\}$. Since the set $D_{n+1} = \bigcap \{C_\alpha : \alpha < \beta_n\}$ is a club by Problem 064, there is $\beta_{n+1} \in D_{n+1}$ such that $\beta_n < \beta_{n+1}$. Thus our inductive construction can be carried out for all $n \in \omega$ to construct an increasing sequence $\{\beta_n : n \in \omega\}$ such that $\beta_n \in \bigcap \{C_\alpha : \alpha < \beta_{n-1}\}$ for all $n \in \omega$. Let $\beta = \sup\{\beta_n : n \in \omega\}$; given any $\gamma < \beta$, there is $m \in \omega$ such that $\gamma < \beta_m$. By the definition of β_n, we have $\beta_n \in \bigcap \{C_\alpha : \alpha < \beta_{n-1}\} \subset C_\gamma$ for all $n \geq m + 1$. Thus the sequence $S = \{\beta_n : n \geq m + 1\}$ is contained in C_γ; the set C_γ being closed, we have $\beta = \lim S \in C_\gamma$. It turns out that $\beta \in C_\gamma$ for each $\gamma < \beta$ so $\beta \in C$. Since $\beta > \nu$, we proved that C is a club.

Since the set A is stationary, there exists $\beta \in C \cap A$; if $\alpha = f(\beta) < \beta$, then $\beta \in f^{-1}(\alpha) \cap C$ and hence $\beta \in C_\alpha$ by the definition of the set C. This proves that $\beta \in f^{-1}(\alpha) \cap C_\alpha$ while $f^{-1}(\alpha) \cap C_\alpha = \emptyset$ by the choice of the set C_α. The obtained contradiction shows that our assumption is false and hence there is $\alpha_0 < \omega_1$ such that $f^{-1}(\alpha_0)$ is stationary.

T.068. *Given $\alpha < \omega_1$, let $S_\alpha = \{f \in \omega^\alpha : f$ is an injection$\}$. In the set $S = \bigcup_{\alpha < \omega_1} S_\alpha$, consider the following partial order: $f \leq g$ if and only if $f \subset g$. Observe that $(S, \leq)$ is a tree with all its chains countable. Prove that there exists an ω_1-sequence $\{s_\alpha : \alpha < \omega_1\} \subset S$ such that $s_\alpha \in S_\alpha$ and $s_\beta|\alpha \approx s_\alpha$ for all $\alpha < \beta < \omega_1$. Deduce from this fact that the subtree $T = \bigcup_{\alpha < \omega_1} \{s \in S_\alpha : s \approx s_\alpha\}$ of the tree S is an Aronszajn tree. Hence Aronszajn codings and trees exist in ZFC.*

Solution. It is evident that $\leq$ is a partial order on S. Given any $f \in \omega^\alpha$, observe that the set $L_f = \{g \in S : g < f\}$ coincides with the set $\{f|\beta : \beta < \alpha\}$. It is easy to see that the map $i : L_f \to \alpha$ defined by $i(f|\beta) = \beta$, is an order isomorphism, so L_f is well-ordered being isomorphic to α. This proves that S is a tree. If $\{t_\alpha : \alpha < \omega_1\}$ is a chain in S, then $t_\beta|\alpha = t_\alpha$ whenever $\alpha < \beta < \omega_1$. Let $t(\alpha) = t_{\alpha+1}(\alpha)$ for all $\alpha < \omega_1$. It is straightforward that $t : \omega_1 \to \omega$ and $t|\alpha = t_\alpha$ for all $\alpha < \omega_1$. An immediate consequence is that t is an injection which is a contradiction. Therefore

(*) S is a tree with no uncountable chains.

If f is a function, then $\mathrm{dom}(f)$ is its domain. Suppose that we have a set of functions $\{f_i : i \in I\}$ such that $f_i|(\mathrm{dom}(f_i) \cap \mathrm{dom}(f_j)) = f_j|(\mathrm{dom}(f_i) \cap \mathrm{dom}(f_j))$ for any $i, j \in I$. Then we can define a function f with $\mathrm{dom}(f) = \bigcup_{i \in I} \mathrm{dom}(f_i)$ as follows: given any $x \in \mathrm{dom}(f)$, find any $i \in I$ with $x \in \mathrm{dom}(f_i)$ and let $f(x) = f_i(x)$. It is easy to check that the value of f at x does not depend on the choice of i, so we have consistently defined a function f which will be denoted by $\bigcup \{f_i : i \in I\}$ (this makes sense if we identify each function with its graph). We are going to construct the promised ω_1-sequence $\{s_\alpha : \alpha < \omega_1\}$ by transfinite induction. Start off with $s_0 = \emptyset$. Assume that $\gamma < \omega_1$ and we have a set $\{s_\alpha : \alpha < \gamma\}$ with the following properties:

(1) $s_\alpha \in S_\alpha$ and $s_\beta|\alpha \approx s_\alpha$ whenever $\alpha < \beta < \gamma$;
(2) $\omega \setminus s_\alpha(\{\beta : \beta < \alpha\})$ is infinite for every $\alpha < \gamma$.

Let $\mathrm{ran}(s_\alpha) = s_\alpha(\{\beta : \beta < \alpha\})$ for all $\alpha < \gamma$. If $\gamma = \beta + 1$ for some ordinal β, then pick any $n \in \omega \backslash \mathrm{ran}(s_\beta)$; the function s_γ, defined by the conditions $s_\gamma | \beta = s_\beta$ and $s_\gamma(\beta) = n$, maps γ to ω. It is clearly injective and satisfies (1) and (2).

If γ is a limit ordinal, fix any increasing sequence $\{\gamma_n : n \in \omega\}$ which converges to γ. Let $t_0 = s_{\gamma_0}$; assume that we have functions $t_i \in S_{\gamma_i}$ for all $i \leq n$ such that

(3) $t_{i+1} | \gamma_i = t_i$ for all $i = 0, \ldots, n - 1$.
(4) $t_i \approx s_{\gamma_i}$ for all $i = 0, \ldots, n$.

Define an auxiliary function $t'_{n+1} \in \omega^{\gamma_{n+1}}$ by the conditions $t'_{n+1} | \gamma_n = t_n$ and $t'_{n+1} | (\gamma_{n+1} \backslash \gamma_n) = s_{\gamma_{n+1}} | (\gamma_{n+1} \backslash \gamma_n)$. The function t'_{n+1} might fail to be injective. However, we have $s_{\gamma_{n+1}} | \gamma_n \approx s_{\gamma_n} \approx t_n$ and hence there is a finite set $P \subset \gamma_n$ such that $s_{\gamma_{n+1}} | (\gamma_n \backslash P) = t_n | (\gamma_n \backslash P) = t'_{n+1} | (\gamma_n \backslash P)$. Thus $t'_{n+1} | (\gamma_{n+1} \backslash P)$ is injective because it coincides with the injective function $s_{\gamma_{n+1}}$. Recall that the function $t'_{n+1} | \gamma_n = t_n$ is also injective, so the injectivity of t'_{n+1} can fail only in case when the set $Q = t'_{n+1}(P) \cap s_{\gamma_{n+1}}(\gamma_{n+1} \backslash \gamma_n)$ is non-empty. If this is the case, then there exists a finite set $Q' \subset \gamma_{n+1} \backslash \gamma_n$ such that $t'_{n+1}(Q') = t_n(Q)$ and hence $t'_{n+1} | (\gamma_{n+1} \backslash Q')$ is an injection which coincides with t_n on γ_n. Since the set $\omega \backslash \mathrm{ran}(s_{\gamma_{n+1}})$ is infinite, we can define a function t_{n+1} as follows: $t_{n+1} | (\gamma_{n+1} \backslash Q') = t'_{n+1} | (\gamma_{n+1} \backslash Q')$ and $t_{n+1} | Q' : Q' \to \omega \backslash \mathrm{ran}(s_{\gamma_{n+1}})$ is defined arbitrarily with the only condition that $t_{n+1} | Q'$ be injective. It is immediate that $t_{n+1} | \gamma_n = t_n$ and $t_{n+1} \approx s_{\gamma_{n+1}}$, so our inductive construction can be continued to obtain a sequence $\{t_i : i \in \omega\}$ with the properties (3) and (4) true for all $i \in \omega$.

Observe that it follows from the fact that each t_i is injective that the function $t = \bigcup \{t_i : i \in \omega\}$ is injective as well. Besides,

(5) for each $\alpha < \gamma$, we have $t | \alpha \approx s_\alpha$

because there is some $\gamma_i > \alpha$ and hence $t | \alpha = (t | \gamma_i) | \alpha \approx s_{\gamma_i} | \alpha \approx s_\alpha$ by (1). It is possible that (2) does not hold for the function t; to correct this, note that the set $R = t(\{\gamma_i : i \in \omega\})$ is infinite because of injectivity of t and hence we can find an infinite (and faithfully indexed) set $R' = \{\beta_i : i \in \omega\} \subset R$ such that $R \backslash R'$ is also infinite. Finally, let $W = \gamma \backslash \{\gamma_i : i \in \omega\}$ and define s_γ by $s_\gamma | W = t | W$ and $s_\gamma(\gamma_i) = \beta_i$ for all $i \in \omega$. The injectivity of s_γ is evident; the property (2) holds for $\alpha = \gamma$ because $\omega \backslash s_\gamma(\{\beta : \beta < \gamma\})$ contains an infinite set $R \backslash R'$. Finally, (1) is true because for each $\alpha < \gamma$ the function $s_\gamma | \alpha$ can be obtained from $t | \alpha$ by changing the values of $t | \alpha$ on the finite set $\{\gamma_i : i \in \omega\} \cap \alpha$. Therefore $s_\gamma | \alpha \approx t | \alpha \approx s_\alpha$ by (5).

This finishes our inductive construction and shows that there exists a collection $\{s_\alpha : \alpha < \omega_1\}$ with the properties (1) and (2). Observe, furthermore, that the set $T = \bigcup_{\alpha < \omega_1} \{s \in S_\alpha : s \approx s_\alpha\}$ is a subtree of S because given any $\alpha < \omega_1$ and any $s \in T_\alpha = \{s \in S_\alpha : s \approx s_\alpha\}$ we have $s | \beta \approx s_\alpha | \beta \approx s_\beta$ for any $\beta < \alpha$ [see (1)]. It is evident that the αth level of T is the set T_α. All elements of T_α are obtained from s_α by changing its value at some finite set of points. Since there are only countably many finite subsets of α, the set T_α is countable, and hence T is a tree without uncountable chains [because even the tree $S \supset T$ does not have uncountable chains by $(*)$] with all of its levels countable. Therefore T is an Aronszajn tree, so our solution is complete.

T.069. *Observe that Jensen's axiom implies CH; prove that it is equivalent to any of the following statements:*

(i) *for every $\alpha < \omega_1$, there exists a countable family $\mathcal{A}_\alpha \subset \exp \alpha$ such that, for any $A \subset \omega_1$, the set $\{\alpha : A \cap \alpha \in \mathcal{A}_\alpha\}$ is stationary;*

(ii) *for any $\alpha < \omega_1$, there is a set $B_\alpha \subset \alpha \times \alpha$ such that, for any $B \subset \omega_1 \times \omega_1$, the set $\{\alpha : B \cap (\alpha \times \alpha) = B_\alpha\}$ is stationary;*

(iii) *for any $\alpha < \omega_1$, there is a function $f_\alpha : \alpha \to \alpha$ such that, for any $f : \omega_1 \to \omega_1$, the set $\{\alpha : f|_\alpha = f_\alpha\}$ is stationary.*

(iv) *for any $\alpha < \omega_1$, there is a function $g_\alpha : \alpha \to \omega_1$ such that, for any $g : \omega_1 \to \omega_1$, the set $\{\alpha : g|_\alpha = g_\alpha\}$ is stationary.*

(v) *there exists a set S of cardinality ω_1 and a set of functions $\{h_\alpha : \alpha < \omega_1\}$ such that, $h_\alpha : \alpha \to S$ for all $\alpha < \omega_1$ and, for any $h : \omega_1 \to S$, the set $\{\alpha : h|_\alpha = h_\alpha\}$ is stationary;*

(vi) *for any set T of cardinality ω_1, there exists a family $\{k_\alpha : \alpha < \omega_1\}$ of functions such that $k_\alpha : \alpha \to T$ for all $\alpha < \omega_1$ and, for any function $k : \omega_1 \to T$, the set $\{\alpha : k|_\alpha = k_\alpha\}$ is stationary.*

Solution. Let $\{A_\alpha : \alpha < \omega_1\}$ be a collection of sets which witnesses the Jensen's axiom. Given any $P \subset \omega$, there is $\alpha \in \omega_1 \backslash \omega$ such that $P \cap \alpha = P \cap \omega = P = A_\alpha$. Letting $f(\alpha) = P$, we obtain a surjection $f : W \to \exp(\omega)$ for some $W \subset \omega_1$. Therefore $\mathfrak{c} = |\exp \omega| \leq |W| \leq \omega_1$ and hence $\mathfrak{c} = \omega_1$ if the Jensen's axiom holds, i.e., the Jensen's axiom implies CH.

Given any set Q, a set $P \subset \omega_1$ and a function $f : P \to Q$, viewing every ordinal $\alpha \in P$ as a set $\{\gamma : \gamma < \alpha\}$ may cause confusion when we take images under f. So, our agreement will be as follows: the expression $f(\alpha)$ means *the element* of Q which is the respective image of *the point* α; this element can also be considered *a subset* of Q if $Q \subset \omega_1$, i.e., $f(\alpha)$ could be either a point of Q or the set $\{\gamma : \gamma < f(\alpha)\}$ if $Q \subset \omega_1$. However, if we want the image of the set $\alpha \subset \omega_1$, then we use square brackets, i.e., the set $f(\{\gamma : \gamma < \alpha\})$ is denoted by $f[\alpha]$.

Fact 1. Suppose that we have a continuous map $r : \omega_1 \to \omega_1$ such that $\alpha < \beta < \omega_1$ implies $r(\alpha) \leq r(\beta)$. If $Q = \bigcup\{\alpha \times r(\alpha) : \alpha < \omega_1\} \subset \omega_1 \times \omega_1$, then, for any surjective map $s : \omega_1 \to Q$, the set $C = \{\alpha < \omega_1 : s[\alpha] = \alpha \times r(\alpha)\}$ is a club.

Proof. To see that C is closed in ω_1, take any increasing sequence $\{\alpha_n : n \in \omega\} \subset C$ which converges to an ordinal α. We have $s[\alpha_n] = \alpha_n \times r(\alpha_n)$ for each $n \in \omega$. Since the mapping r is continuous and non-decreasing, we have $\bigcup_{n \in \omega} r(\alpha_n) = r(\alpha)$ and hence $\bigcup\{\alpha_n \times r(\alpha_n) : n \in \omega\} = \alpha \times r(\alpha)$. Furthermore,

$$s[\alpha] = s(\bigcup\{\alpha_n : n \in \omega\}) = \bigcup\{s[\alpha_n] : n \in \omega\} = \bigcup\{\alpha_n \times r(\alpha_n) : n \in \omega\} = \alpha \times r(\alpha),$$

which shows that $\alpha \in C$, i.e., the set C is closed in ω_1.

To show that C is unbounded, fix any $\alpha < \omega_1$ and let $\alpha_0 = \alpha$. Since $s[\alpha_0]$ is countable, there exists $\beta_0 \in \omega_1 \backslash \alpha_0$ such that $s[\alpha_0] \subset \beta_0 \times r(\beta_0)$ (the fact that r is non-decreasing was used here again).

Assume that $n \in \omega$ and we have sequences of ordinals $\{\alpha_i : i \le n\}$ and $\{\beta_i : i \le n\}$ with the following properties:

(1) $\alpha = \alpha_0$ and $\alpha_i \le \beta_i \le \alpha_{i+1}$ for any $i = 0, \ldots, n - 1$.
(2) $s[\alpha_i] \subset \beta_i \times r(\beta_i) \subset s[\alpha_{i+1}]$ for all $i = 0, \ldots, n - 1$.

The set $\beta_n \times r(\beta_n)$ is countable and the map s surjective, so there is an ordinal $\alpha_{n+1} \in \omega_1 \backslash \beta_n$ such that $\beta_n \times r(\beta_n) \subset s[\alpha_{n+1}]$. Once more use the fact that r is non-decreasing to find an ordinal $\beta_{n+1} \in \omega_1 \backslash \alpha_{n+1}$ such that $s[\alpha_{n+1}] \subset \beta_{n+1} \times r(\beta_{n+1})$. It is evident that (1) and (2) are now fulfilled for all $i \le n$ and hence we can carry out our inductive construction to obtain sequences $\{\alpha_i : i \in \omega\}$ and $\{\beta_i : i \in \omega\}$ with the properties (1) and (2) for all $i \in \omega$. Let $\gamma = \lim \alpha_n = \lim \beta_n$. Then

$$s[\gamma] = s \left(\bigcup \{\alpha_n : n \in \omega\} \right) = \bigcup \{s[\alpha_n] : n \in \omega\} \subset \bigcup \{\beta_n \times r(\beta_n) : n \in \omega\} = \gamma \times r(\gamma).$$

We used (2) to assert the inclusion; as before, the last equality holds because r is continuous and non-decreasing. On the other hand

$$s[\gamma] = s \left(\bigcup_{n \in \omega} \alpha_{n+1} \right) = \bigcup \{s[\alpha_{n+1}] : n \in \omega\} \supset \bigcup \{\beta_n \times r(\beta_n) : n \in \omega\} = \gamma \times r(\gamma),$$

which proves that $s[\gamma] = \gamma \times r(\gamma)$. Thus, for any $\alpha < \omega_1$ we found $\gamma \ge \alpha$ with $\gamma \in C$. Hence C is a club and Fact 1 is proved.   □

Fact 2. Given any $f : \omega_1 \to \omega_1$, the set $C_f = \{\alpha < \omega_1 : f[\alpha] \subset \alpha\}$ is a club.

Proof. Assume that $\{\alpha_n : n \in \omega\} \subset C_f$ is an increasing sequence with $\alpha = \lim \alpha_n$. Then

$$f[\alpha] = f \left(\bigcup_{n \in \omega} \alpha_n \right) = \bigcup_{n \in \omega} f[\alpha_n] \subset \bigcup_{n \in \omega} \alpha_n = \alpha,$$

which shows that $\alpha \in C_f$ and hence C_f is closed in ω_1.

To see that C_f is unbounded, take any $\alpha < \omega_1$ and let $\alpha_0 = \alpha$. Suppose that we have $\alpha = \alpha_0 < \cdots < \alpha_n$ such that $f[\alpha_i] \subset \alpha_{i+1}$ for all $i < n$. Since $f[\alpha_n]$ is a countable subset of ω_1, there is $\alpha_{n+1} > \alpha_n$ such that $f[\alpha_n] \subset \alpha_{n+1}$. Thus we can inductively construct a sequence $\{\alpha_i : i \in \omega\} \subset \omega_1$ such that $f[\alpha_i] \subset \alpha_{i+1}$ for all $i \in \omega$. If $\beta = \lim \alpha_i$, then

$$f[\beta] = f \left(\bigcup_{i \in \omega} \alpha_i \right) = \bigcup_{i \in \omega} f[\alpha_i] \subset \bigcup_{i \in \omega} \alpha_{i+1} = \beta,$$

which shows that $\beta \in C_f$. Thus, for any $\alpha < \omega_1$ there is $\beta > \alpha$ with $\beta \in C_f$; hence C_f is a club, so Fact 2 is proved.   □

Returning to our solution, assume that $\Diamond$ holds; fix a collection $\{A_\alpha : \alpha < \omega_1\}$ which witnesses this and let $\mathcal{A}_\alpha = \{A_\alpha\}$ for all $\alpha < \omega_1$. It is immediate that the ω_1-sequence $\{\mathcal{A}_\alpha : \alpha < \omega_1\}$ satisfies (i) so $\Diamond \Longrightarrow$(i).

Now assume that (i) is true and fix any bijection $s : \omega_1 \to \omega \times \omega_1$. Given any set $A \subset \omega_1$, let $A^* = s(A)$; if $B \subset \omega \times \omega_1$, then $B' = s^{-1}(B)$. For each $\alpha < \omega_1$, let $\mathcal{B}_\alpha = \{A^* : A \in \mathcal{A}_\alpha\}$. We claim that

(3) for any $B \subset \omega \times \omega_1$, the set $P_B = \{\alpha < \omega_1 : B \cap (\omega \times \alpha) \in \mathcal{B}_\alpha\}$ is stationary.

To prove (3) note first that the set $C = \{\alpha < \omega_1 : s[\alpha] = \omega \times \alpha\}$ is a club by Fact 1 applied to the map s and the map $r : \omega_1 \to \omega_1$ defined by $r(\alpha) = \omega$ for all $\alpha \in \omega_1$. By (i) the set $H_0 = \{\alpha : B' \cap \alpha \in \mathcal{A}_\alpha\}$ is stationary and therefore so is the set $H = H_0 \cap C$ (see Problem 065). If $\alpha \in H$, then $s[\alpha] = \omega \times \alpha$ and hence $B \cap (\omega \times \alpha) = (B' \cap \alpha)^* \in \mathcal{B}_\alpha$ because $B' \cap \alpha \in \mathcal{A}_\alpha$. As a consequence, $H \subset P_B$, so P_B is stationary and (3) is proved.

Let $\{B_\alpha^k : k \in \omega\}$ be an enumeration of $\mathcal{B}_\alpha$ for all $\alpha < \omega_1$ (repetitions are allowed and there is no loss of generality to assume that $\mathcal{B}_\alpha \neq \emptyset$ for all $\alpha < \omega_1$). We will need the sets $B_{\alpha,n}^k = \{\xi < \omega_1 : (n, \xi) \in B_\alpha^k\}$ for all $k, n \in \omega$ and $\alpha < \omega_1$. Consider the collection $\mathcal{C}_n = \{B_{\alpha,n}^n : \alpha < \omega_1\}$. We claim that there exists $n \in \omega$ for which the ω_1-sequence $\mathcal{C}_n$ witnesses $\Diamond$.

Indeed, if this is not the case, then for any $n \in \omega$ there exists a set $B_n \subset \omega_1$ such that the set $\{\alpha : B_n \cap \alpha = B_{\alpha,n}^n\}$ is non-stationary. Now, looking at the set $B = \bigcup\{\{n\} \times B_n : n \in \omega\} \subset \omega \times \omega_1$, we can see that $B \cap (\omega \times \alpha) = B_\alpha^n$ implies $B \cap (\{n\} \times \alpha) = B_{\alpha,n}^n$ and therefore $B_n \cap \alpha = B_{\alpha,n}^n$. This shows that the set $P_n = \{\alpha : B \cap (\omega \times \alpha) = B_\alpha^n\}$ is non-stationary. Any countable union of non-stationary sets is non-stationary by Problem 065, so $\bigcup_{n \in \omega} P_n = \{\alpha : B \cap (\omega \times \alpha) \in \mathcal{B}_\alpha\}$ is non-stationary which contradicts (3) and completes the proof of (i)$\Longrightarrow \Diamond$. Thus (i) $\Longleftrightarrow \Diamond$.

Take any bijection $s : \omega_1 \to \omega_1 \times \omega_1$. If $r : \omega_1 \to \omega_1$ is defined by $r(\alpha) = \alpha$ for all $\alpha \in \omega_1$, then Fact 1 can be applied to r and s to conclude that the set $C = \{\alpha \in \omega_1 : s[\alpha] = \alpha \times \alpha\}$ is a club. Assume that a collection $\{A_\alpha : \alpha < \omega_1\}$ witnesses $\Diamond$ and let $B_\alpha = s(A_\alpha)$ for all $\alpha \in C$; if $\alpha \in \omega_1 \backslash C$, then let $B_\alpha = \emptyset$. Given any set $B \subset \omega_1 \times \omega_1$, the set $Q = \{\alpha : s^{-1}(B) \cap \alpha = A_\alpha\}$ is stationary and hence so is the set $Q_0 = Q \cap C$. For any $\alpha \in Q_0$, we have

$$B_\alpha = s(A_\alpha) = s(s^{-1}(B) \cap \alpha) = B \cap s(\alpha) = B \cap (\alpha \times \alpha)$$

which shows that the set $\{\alpha : B \cap (\alpha \times \alpha) = B_\alpha\}$ is stationary because it contains the stationary set Q_0. Thus (ii) holds, i.e., we proved that $\Diamond \Longrightarrow$(ii).

Now assume that (ii) is true and fix an ω_1-sequence $\{B_\alpha : \alpha < \omega_1\}$ which witnesses this. Let $p : \omega_1 \times \omega_1 \to \omega_1$ be the natural projection to the first factor; we claim that the ω_1-sequence $\{p(B_\alpha) : \alpha < \omega_1\}$ witnesses $\Diamond$. It is evident that $A_\alpha = p(B_\alpha) \subset \alpha$ for all $\alpha < \omega_1$. Given any set $A \subset \omega_1$, consider the set $B = A \times A$. The set $R = \{\alpha : B \cap (\alpha \times \alpha) = B_\alpha\}$ is stationary; for any $\alpha \in R$ we have $A_\alpha = p(B_\alpha) = p(B \cap (\alpha \times \alpha)) = A \cap \alpha$ and hence the set $R_0 = \{\alpha : A \cap \alpha = A_\alpha\}$ is stationary because $R \subset R_0$. This settles (ii)$\Longrightarrow \Diamond$ and hence $\Diamond \Longleftrightarrow$ (ii).

To see that (iii)$\Longrightarrow$(iv), let $g_\alpha = f_\alpha$ for each $\alpha < \omega_1$. The implication (iv)$\Longrightarrow$(v) becomes evident if we let $S = \omega_1$ and $h_\alpha = g_\alpha$ for all $\alpha < \omega_1$.

Assume that the property (v) holds and take any set T of cardinality ω_1; fix any bijection $s : S \to T$ and let $k_\alpha = s \circ h_\alpha$ for all $\alpha < \omega_1$. Given any function $k : \omega_1 \to T$, let $h = s^{-1} \circ k$; since (v) holds, the set $W = \{\alpha : h|\alpha = h_\alpha\}$ is stationary. If $\alpha \in W$ and $\beta < \alpha$, then $k(\beta) = s(h(\beta)) = s(h_\alpha(\beta)) = k_\alpha(\beta)$, i.e., $k|\alpha = k_\alpha$, so the set $W' = \{\alpha : k|\alpha = k_\alpha\}$ is stationary because $W' \supset W$. This proves that (v)$\Longrightarrow$(vi). Since the implication (vi)$\Longrightarrow$(v) is evident, we have (v) $\Longleftrightarrow$ (vi). It is also immediate that (vi) is stronger than (iv); as a consequence, (iv) $\Longleftrightarrow$ (v) $\Longleftrightarrow$ (vi).

Now assume that (iv) holds and fix an ω_1-sequence $\{g_\alpha : \alpha < \omega_1\}$ that witnesses it. Let $D = \{\alpha : g_\alpha[\alpha] \subset \alpha\}$. For any $\alpha \in D$, let $f_\alpha = g_\alpha$; if $\alpha \in \omega_1 \backslash D$, then let $f_\alpha(\beta) = 0$ for all $\beta < \alpha$. We claim that the set $\{f_\alpha : \alpha < \omega_1\}$ witnesses (iii). To see this, take any function $f : \omega_1 \to \omega_1$ and apply Fact 2 to conclude that the set $C = \{\alpha : f[\alpha] \subset \alpha\}$ is a club.

The set $E = \{\alpha < \omega_1 : f|\alpha = g_\alpha\}$ is stationary by our assumption, so the set $C' = E \cap C$ is also stationary. If $\alpha \in C'$, then $f|\alpha = g_\alpha = f_\alpha$, so the set $U = \{\alpha < \omega_1 : f|\alpha = f_\alpha\}$ is stationary because $C' \subset U$. This shows that (iii) holds and proves (iv)$\Longrightarrow$(iii). Consequently, (iii) $\Longleftrightarrow$ (iv) $\Longleftrightarrow$ (v) $\Longleftrightarrow$ (vi).

We prove next that (ii)$\Longrightarrow$(iii). If $f : X \to Y$ is a function, then the set $\Gamma_f = \{(x, f(x)) : x \in X\} \subset X \times Y$ is its graph. It is easy to see that two functions $f, g : X \to Y$ coincide if and only if $\Gamma_f = \Gamma_g$. Fix an ω_1-sequence $\{B_\alpha : \alpha < \omega_1\}$ as in (ii) and let $D = \{\alpha < \omega_1 : \text{there is a function } p_\alpha : \alpha \to \alpha \text{ such that } B_\alpha = \Gamma_{p_\alpha}\}$. For each $\alpha \in D$, let $f_\alpha = p_\alpha$; if $\alpha \in \omega_1 \backslash D$, then let $f_\alpha(\beta) = 0$ for all $\beta < \alpha$. The ω_1-sequence $\{f_\alpha : \alpha < \omega_1\}$ witnesses (iii). Indeed, if $f : \omega_1 \to \omega_1$, then $\Gamma_f \subset \omega_1 \times \omega_1$, and therefore the set $E = \{\alpha : \Gamma_f \cap (\alpha \times \alpha) = B_\alpha\}$ is stationary. Given any $\alpha \in E$, the set $\Gamma_f \cap (\alpha \times \alpha)$ is a graph of some function on α and hence $B_\alpha = \Gamma_f \cap (\alpha \times \alpha)$ is also a graph of some function on α. Therefore, by our choice of D, we have $E \subset D$ and hence $\Gamma_f \cap (\alpha \times \alpha) = \Gamma_{p_\alpha} = \Gamma_{f_\alpha}$ for each $\alpha \in E$. It is an easy exercise to check that $\Gamma_f \cap (\alpha \times \alpha) = \Gamma_{f_\alpha}$ is equivalent to $f|\alpha = f_\alpha$, so the set $E' = \{\alpha : f|\alpha = f_\alpha\}$ is stationary because $E \subset E'$. Thus (iii) holds, so we proved the implication (ii)$\Longrightarrow$(iii).

Finally, assume that the set $\{f_\alpha : \alpha < \omega_1\}$ witnesses (iii) and let $W_\alpha = f_\alpha[\alpha]$ for all $\alpha < \omega_1$. The collection $\mathcal{W} = \{W_\alpha : \alpha < \omega_1\}$ is almost what we need to prove $\diamondsuit$. To see it, take any *non-empty* set $W \subset \omega_1$ and fix some $\gamma \in W$. Define a function $f : \omega_1 \to \omega_1$ by $f(\alpha) = \alpha$ if $\alpha \in W$ and $f(\alpha) = \gamma$ for all $\alpha \in \omega_1 \backslash W$. Since (iii) holds, the set $B = \{\alpha : f|\alpha = f_\alpha\}$ is stationary and hence $B' = B \backslash \gamma$ is also stationary. If $\alpha \in B'$, then $f_\alpha(\beta) = \beta$ for any $\beta \in W \cap \alpha$ and $f_\alpha(\beta) = \gamma$ for any $\beta \in \alpha \backslash W$. Consequently, $W_\alpha = f_\alpha[\alpha] = (W \cap \alpha) \cup \{\gamma\} = W \cap \alpha$ because $\gamma \in W \cap \alpha$. Hence, for stationary-many α's, we have $W \cap \alpha = W_\alpha$, i.e.,

(4) for any *non-empty* $W \subset \omega_1$, the set $\{\alpha : W \cap \alpha = W_\alpha\}$ is stationary.

Therefore the only set not captured by the collection $\mathcal{W}$ is the empty set because $W_\alpha \neq \emptyset$ for all ordinals $\alpha < \omega_1$. To correct this situation, observe that the set $E = \{\alpha < \omega_1 : W_\alpha = \{0\}\}$ is stationary [just let $W = \{0\}$ and apply (4)]. Apply

Problem 066 to find disjoint stationary sets $E_0, E_1 \subset E$ such that $E = E_0 \cup E_1$. Now let $A_\alpha = W_\alpha$ for all $\alpha \in \omega_1 \backslash E_0$ and $A_\alpha = \emptyset$ for all $\alpha \in E_0$. Then the empty set is captured by the collection $\mathcal{A} = \{A_\alpha : \alpha < \omega_1\}$ because if $A = \emptyset$, then the set $\{\alpha : A \cap \alpha = A_\alpha\}$ is stationary being at least as large as E_0. To see that we still capture all non-empty sets by the family $\mathcal{A}$, take any non-empty $A \subset \omega_1$. The set $D = \{\alpha > 0 : A \cap \alpha = W_\alpha\}$ is stationary. If $A \neq \{0\}$, then there is $\beta \in A \backslash \{0\}$ and hence the set $\{\alpha : A \cap \alpha = A_\alpha\}$ is stationary because it contains the stationary set $D \backslash \beta$. If, on the other hand, $A = \{0\}$, then the set $\{\alpha : A \cap \alpha = A_\alpha\}$ is stationary because it contains the stationary set E_1. Thus the set $\{\alpha : A \cap \alpha = A_\alpha\}$ is stationary for all sets $A \subset \omega_1$. Therefore (iii)$\Longrightarrow \diamond$ and this was the last implication we needed to show that $\diamond \Longleftrightarrow$ (i) $\Longleftrightarrow$ (ii) $\Longleftrightarrow$ (iii) $\Longleftrightarrow$ (iv) $\Longleftrightarrow$ (v) $\Longleftrightarrow$ (vi), so our solution is complete.

T.070. *Prove that if Jensen's axiom holds, then there is a Souslin tree.*

Solution. All through this solution we assume $\diamond$. Let $T = \bigcup\{\omega^\alpha : \alpha < \omega_1\}$; denote by T_α the set ω^α for all $\alpha < \omega_1$. Given $f, g \in T$, let $f \leq g$ iff $f \subset g$. It is an easy exercise that $\leq$ is a partial order on T. Any subset of T will be also considered a partially ordered set with the order induced from T. For any $g \in T$, we denote by $ht(g)$ the unique ordinal α such that $g \in T_\alpha$. Observe that the set $L_g = \{f \in T : f < g\}$ can also be represented as $\{f | \beta : \beta < \alpha = ht(g)\}$. It is immediate that the correspondence $\beta \mapsto f | \beta$ is an isomorphism of α onto L_g, so L_g is well-ordered for any $g \in T$, and hence T is a tree. If $f \in T$, $ht(f) = \alpha$ and $n \in \omega$, then $g = f \frown n$ is the function defined by $g | \alpha = f$ and $g(\alpha) = n$. It is clear that $ht(g) = \alpha + 1$.

If f is a function, then $dom(f)$ is its domain. Suppose that we have a set of functions $\{f_i : i \in I\}$ such that $f_i | (dom(f_i) \cap dom(f_j)) = f_j | (dom(f_i) \cap dom(f_j))$ for any $i, j \in I$. Then we can define a function f with $dom(f) = \bigcup_{i \in I} dom(f_i)$ as follows: given any $x \in dom(f)$, find any $i \in I$ with $x \in dom(f_i)$ and let $f(x) = f_i(x)$. It is easy to check that the value of f at x does not depend on the choice of i, so we have consistently defined a function f which will be denoted by $\bigcup\{f_i : i \in I\}$ (this makes sense if we identify each function with its graph).

Given any set Q, a set $P \subset \omega_1$ and a function $f : P \to Q$, viewing every ordinal $\alpha \in P$ as a set $\{\gamma : \gamma < \alpha\}$ may cause confusion when we take images under f. So, our agreement will be as follows: the expression $f(\alpha)$ means *the element* of Q which is the respective image of *the point* α; this element can also be considered *a subset* of Q if $Q \subset \omega_1$, i.e., $f(\alpha)$ could be either a point of Q or the set $\{\gamma : \gamma < f(\alpha)\}$ if $Q \subset \omega_1$. However, if we want the image of the set $\alpha \subset \omega_1$, then we use square brackets, i.e., the set $f(\{\gamma : \gamma < \alpha\})$ is denoted by $f[\alpha]$.

Observe that $|T_\alpha| = |\omega^\alpha| = \mathfrak{c} = \omega_1$ because we have CH under the Jensen's axiom (see Problem 069). Thus there exists a collection $\{f_\alpha : \alpha < \omega_1\}$ such that $f_\alpha : \alpha \to T$ for each $\alpha < \omega_1$ and, given any function $f : \omega_1 \to T$, the set $\{\alpha : f | \alpha = f_\alpha\}$ is stationary (see Problem 069).

We are going to construct a subtree S of the tree T by induction on $\alpha < \omega_1$. This construction will give us a set $S_\alpha \subset T_\alpha$, and the resulting tree will be the set

$S = \bigcup_{\alpha < \omega_1} S_\alpha$. To start off, let $S_0 = T_0 = \emptyset$. Suppose that $\beta < \omega_1$ and we have constructed a collection $\{S_\alpha : \alpha < \beta\}$ with the following properties:

(1) $S_\alpha \subset T_\alpha$ and $|S_\alpha| \leq \omega$ for all $\alpha < \beta$;
(2) if $s \in S_\alpha$ and $\alpha + 1 < \beta$, then $s^\frown n \in S_{\alpha+1}$ for all $n \in \omega$;
(3) if $\alpha < \gamma < \beta$ and $s \in S_\alpha$, then there is $t \in S_\gamma$ such that $s < t$;
(4) if $\alpha < \gamma < \beta$ and $t \in S_\gamma$, then $t|\alpha \in S_\alpha$;
(5) if $\gamma < \beta$ is a limit ordinal and the set $A = f_\gamma[\gamma]$ is a maximal antichain in $S(\gamma) = \bigcup_{\alpha < \gamma} S_\alpha$, then, for any $t \in S_\gamma$ there is $s \in A$ such that $s < t$.

If $\beta = \gamma + 1$ is a successor ordinal, then let $S_\beta = \{s^\frown n : s \in S_\gamma \text{ and } n \in \omega\}$. The properties (1), (2), (4) and (5) clearly hold for all $\alpha \leq \beta$. To see that (3) also holds, take any $s \in S_\alpha$ for some $\alpha < \beta$. Since (3) is true for all $\alpha < \beta$, there is $u \in S_\gamma$ such that $s < u$. Therefore $t = u^\frown 0 \in S_\beta$ and $s < t$. Thus all properties (1)–(5) hold for all $\alpha \leq \beta$.

If β is a limit ordinal, choose an increasing sequence $\{\beta_n : n \in \omega\} \subset \beta$ with $\lim \beta_n = \beta$; we must consider two cases:

(1) The set $f_\beta[\beta]$ is not a maximal antichain in $S(\beta)$. For each $s \in S(\beta)$, there is $m \in \omega$ such that $\mathrm{ht}(s) < \beta_m$ and therefore there is $s_0 \in S_{\beta_m}$ with $s < s_0$ [see (3)]. Using the property (3), we can construct, by an evident induction, a sequence $\{s_n : n \in \omega\} \subset S(\beta)$ such that $s_n \in S_{\beta_{m+n}}$ and $s_n < s_{n+1}$ for all $n \in \omega$. Letting $w(s) = \bigcup_{n \in \omega} s_n$ we obtain a function $w(s) \in T_\beta$ such that $s < w(s)$. If $S_\beta = \{w(s) : s \in S(\beta)\}$, then the conditions (1)–(5) are satisfied for all $\alpha \leq \beta$.

(2) If $A = f_\beta[\beta]$ is a maximal antichain in $S(\beta)$, then take any $s \in S(\beta)$ and observe that there is $t_s \in A$ such that $s \leq t_s$ or $t_s \leq s$ because otherwise $s \notin A$ and $A \cup \{s\}$ is an antichain in $S(\beta)$ which contradicts the maximality of A. There exists $m \in \omega$ such that $\mathrm{ht}(s \cup t_s) \leq \beta_m$; apply (3) to find $s_0 \in S_{\beta_m}$ such that $s \cup t_s \leq s_0$. As before, we can construct by induction a sequence $\{s_n : n \in \omega\} \subset S(\beta)$ such that $s_n \in S_{\beta_{m+n}}$ and $s_n < s_{n+1}$ for all $n \in \omega$. Letting $w(s) = \bigcup_{n \in \omega} s_n$ we obtain a function $w(s) \in T_\beta$ such that $s < w(s)$ and $t_s < w(s)$. If $S_\beta = \{w(s) : s \in S(\beta)\}$, then the conditions (1)–(5) are satisfied for all $\alpha \leq \beta$. Indeed, (1)–(4) are evident by our construction. As to (5) it must only be checked for the set $A = f_\beta[\beta] \subset S(\beta)$. Since every $t \in S_\beta$ is $w(u)$ for some $u \in S(\beta)$, we have $t_u < w(s) = t$, so, if $s = t_u$, then $s \in A$ and $s < t$ which shows that (5) is also fulfilled for all $\alpha \leq \beta$.

Therefore our inductive construction can be continued until we obtain a set $S = \bigcup \{S_\alpha : \alpha < \omega_1\}$. It follows from (4) that $L_s \subset S$ for any $s \in S$ and hence S is a tree. It is also immediate from our construction that all levels of S are countable, so S is an ω_1-tree. All levels of S are non-empty by (3), so $|S| = \omega_1$. Let us show that all antichains of S are countable.

Suppose that $U \subset S$ is an uncountable antichain. It is an easy exercise that every antichain is contained in a maximal antichain, so we can assume, without loss of generality, that U is a maximal antichain in S. Take any bijection $f : \omega_1 \to U$. Next observe that the set

(6) $C = \{\alpha < \omega_1 : f[\alpha] \subset S(\alpha)$ and $f[\alpha]$ is a maximal antichain in $S(\alpha)\}$ is a club.

To see this, assume that $\{\alpha_n : n \in \omega\} \subset \omega_1$ is an increasing sequence such that $\alpha_n \in C$ for all $n \in \omega$. If $\alpha = \lim \alpha_n$, then $f[\alpha] = \bigcup_{n\in\omega} f[\alpha_n] \subset \bigcup_{n\in\omega} S(\alpha_n) = S(\alpha)$ so $f[\alpha] \subset S(\alpha)$.

Now, if $f[\alpha]$ is not a maximal antichain in $S(\alpha)$, then there is $s \in S(\alpha)\backslash f[\alpha]$ such that $\{s\} \cup f[\alpha]$ is still an antichain. However, $s \in S(\alpha_n)$ for some $n \in \omega$, and hence $\{s\} \cup f[\alpha_n]$ is an antichain in $S(\alpha_n)$ strictly larger than $f[\alpha_n]$ which contradicts the maximality of $f[\alpha_n]$ in $S(\alpha_n)$. Therefore C is a closed subset of ω_1.

To see that C is unbounded, fix any $\alpha < \omega_1$ and let $\alpha_0 = \alpha$. There is $\beta_0 < \omega_1$ such that $\alpha_0 < \beta_0$ and $f[\alpha_0] \subset S(\beta_0)$. Assume that we have increasing sequences $\{\alpha_i : i \leq n\}$ and $\{\beta_i : i \leq n\}$ such that

(7) $\alpha_i < \beta_i < \alpha_{i+1}$ for all $i \leq n - 1$;
(8) if $i \leq n - 1$ and $s \in S(\beta_i)$, then some $t_s \in f[\alpha_{i+1}]$ is compatible with s;
(9) $f[\alpha_i] \subset S(\beta_i)$ for all $i \leq n$.

Since the set U is a maximal antichain, for every function $s \in S(\beta_n)$ there is $t_s \in U$ which is compatible with s. Choose any ordinal $\alpha_{n+1} > \beta_n$ for which $f[\alpha_{n+1}] \supset \{t_s : s \in S(\beta_n)\}$. Since the set $f[\alpha_{n+1}]$ is countable, there is $\beta_{n+1} > \alpha_{n+1}$ such that $f[\alpha_{n+1}] \subset S(\beta_{n+1})$. It is clear that (7)–(9) now hold for $\{\alpha_i : i \leq n + 1\}$ and $\{\beta_i : i \leq n + 1\}$, and hence our inductive construction can go ahead giving us increasing sequences $\{\alpha_i : i \in \omega\} \subset \omega_1$ and $\{\beta_i : i \in \omega\} \subset \omega_1$ such that (7)–(9) take place for all $i \in \omega$. We claim that $\beta = \lim \beta_n = \lim \alpha_n \in C$, i.e., $f[\beta] \subset S(\beta)$ and the set $f[\beta]$ is a maximal antichain in $S(\beta)$.

For the first assertion observe that

$$f[\beta] = f\left(\bigcup_{n\in\omega} \alpha_n\right) = \bigcup_{n\in\omega} f[\alpha_n] \subset \bigcup_{n\in\omega} S(\beta_n) = S(\beta)$$

[the inclusion takes place by (9)] and hence $f[\beta] \subset S(\beta)$. For the second statement assume that $s \in S(\beta)\backslash f[\beta]$ is incompatible with all elements of $f[\beta]$. Choose $n \in \omega$ such that $s \in S(\beta_n)$. By (7) there is $t_s \in f[\alpha_{n+1}] \subset f[\beta]$ which is compatible with s. It is clear that $t_s \in f[\beta]$ and hence s is compatible with an element of $f[\beta]$. This contradiction shows that $f[\beta]$ is a maximal antichain in $S(\beta)$ and hence, for any $\alpha < \omega_1$, there is $\beta > \alpha$ with $\beta \in C$. Therefore C is a club, so we finished the proof of (6).

Recalling that the collection $\{f_\alpha : \alpha < \omega_1\}$ witnesses the Jensen's axiom, we convince ourselves that the set $E = \{\alpha : f|\alpha = f_\alpha\}$ must be stationary and therefore so is the set $H = E \cap C$. Take any $\alpha \in H$; by definition of C, the set $A = f[\alpha] = f_\alpha[\alpha]$ is a maximal antichain in $S(\alpha)$.

Since the set A is countable, we can pick $s \in U\backslash A$ such that $\text{ht}(s) \geq \alpha$. Then $t = s|\alpha \in S_\alpha$ by (4); apply (5) to conclude that there is $u \in f_\alpha[\alpha]$ with $u < t$. We have $u < t < s$ for distinct $u, s \in U$ which is a contradiction. Thus S has ccc.

To finish our proof that S is a Souslin tree, assume that A is an uncountable chain in S. Passing to an appropriate uncountable subset of A if necessary, we can assume that $A = \{s_\alpha : \alpha < \omega_1\}$ where $s_\alpha < s_\beta$ whenever $\alpha < \beta$. It is immediate that $s = \bigcup_{\alpha < \omega_1} s_\alpha$ is well-defined and hence there exists a function $s : \omega_1 \to \omega$ such that $s|\alpha \in S$ for all $\alpha < \omega_1$.

Let $t_\alpha = (s|\alpha)^\frown n_\alpha$ where $n_\alpha = s(\alpha) + 1$ for all $\alpha < \omega_1$. It is clear that $t_\alpha \in S_{\alpha+1}$ and $t_\alpha \perp s|(\alpha + 1)$ for all $\alpha < \omega_1$; we claim that $B = \{t_\alpha : \alpha < \omega_1\}$ is an antichain. Assume that $\alpha < \beta$; the only possibility for t_α to be compatible with t_β is to satisfy the condition $t_\alpha < t_\beta$ because $\alpha + 1 = \mathrm{ht}(t_\alpha) < \mathrm{ht}(t_\beta) = \beta + 1$. However, $s|(\alpha + 1) \le s|\beta < t_\beta$ and therefore $s|(\alpha + 1) < t_\beta$. It turns out that the incompatible elements t_α and $s|(\alpha + 1)$ are less than t_β while L_{t_β} must be well-ordered because S is a tree. This contradiction shows that B is an uncountable antichain in S which in turn contradicts the ccc property of S we established above. Hence S is a Souslin tree and our solution is complete.

T.071. *Prove that $MA + \neg CH$ implies there are no Souslin trees.*

Solution. Assume that $MA + \neg CH$ holds and $(S, \le)$ is a Souslin tree. Since S is uncountable and has ccc, we can apply Problem 049 to conclude that there is an uncountable set $C \subset S$ such that every $s, t \in C$ are compatible. However, in any tree any compatible elements are comparable. Thus $s \le t$ or $t \le s$ for any $s, t \in C$, i.e., C is an uncountable chain in S which is a contradiction with the fact that S is a Souslin tree.

T.072. *Given two topologies τ and μ on the same set X, say that τ is weaker than μ if $\tau \subset \mu$. If $\tau \subset \mu$ and $\tau \ne \mu$, then τ is said to be strictly weaker than μ. If $(X, \le)$ is a linearly ordered set and $Y \subset X$, then $\le_Y$ is the order $\le$ considered only on the points of Y. Let $(L, \le)$ be any linearly ordered space:*

(i) *Prove that for any $M \subset L$, the topology $\tau(\le_M)$ on M generated by the order $\le_M$ is weaker than the topology τ_M^L of the subspace on M induced from L.*

(ii) *Show that $\tau(\le_M)$ can be strictly weaker than τ_M^L even if M is a dense or a clopen subspace of L.*

(iii) *Assume that M is order dense in L, i.e., for any $a, b \in L$, if $a < b$, then $a \le p < q \le b$ for some $p, q \in M$. Prove that M is dense in L and $\tau(\le_M) = \tau_M^L$.*

(iv) *Prove that there exists a compact linearly ordered space $(K, \le)$ such that, for some order dense $N \subset K$, there exists an order isomorphism between $(N, \le_N)$ and $(L, \le)$. In particular, $\tau(\le_N) = \tau_N^K$, the space N is dense in K and $(L, \le)$ is order isomorphic to $(N, \le_N)$.*

Solution. Given any points $a, b \in L$, we let $(a, b) = \{s \in L : a < s < b\}$. Besides, $(a, \to) = \{s \in L : a < s\}$ and $(\leftarrow, a) = \{s \in L : s < a\}$. If $\mathcal{B} = \{(a, b) : a, b \in L$ and $a < b\} \cup \{(a, \to) : a \in L\} \cup \{(\leftarrow, a) : a \in L\}$, then $\mathcal{B}$ is a base of the space L.

(i) For any $p, q \in M$ such that $p < q$, we let $(p, q)_M = (p, q) \cap M$. Analogously, $(p, \to)_M = (p, \to) \cap M$ and $(\leftarrow, p) = (\leftarrow, p) \cap M$ for any $p \in M$. Observe that $(p, q)_M = \{x \in M : p <_M x <_M q\}$, i.e., $(p, q)_M$ is the interval in $(M, \leq_M)$ determined by the order $\leq_M$ and points p, q. Analogously, $(p, \rightarrow)_M = \{x \in M : p <_M x\}$ and $(\leftarrow, p)_M = \{x \in M : x <_M p\}$ and therefore the family

$$\mathcal{B}_M = \{(p, q)_M : p, q \in M \text{ and } p < q\} \cup \{(p, \rightarrow)_M : p \in M\} \cup \{(\leftarrow, p)_M : p \in M\}$$

is a base of the space $(M, \tau(\leq_M))$. By our definition, for any $U \in \mathcal{B}_M$ there is $V \in \mathcal{B}$ such that $U = V \cap M$. Therefore every element of $\mathcal{B}_M$ belongs to the topology τ_M^L. Since $\mathcal{B}_M$ is a base of $\tau(\leq_M)$, every $W \in \tau(\leq_M)$ is a union of a subfamily of $\mathcal{B}_M$. Since τ_M^L is a topology, every union of a subfamily of $\mathcal{B}_M$ also belongs to τ_M^L. Consequently, $\tau(\leq_M) \subset \tau_M^L$ and hence we proved (i).

(ii) Let $L = \{-\frac{1}{n} : n \in \mathbb{N}\} \cup \{\frac{1}{n} : n \in \mathbb{N}\}$; the order on L is induced from the order $\leq$ in $\mathbb{R}$. The set $M = \{-\frac{1}{n} : n \in \mathbb{N}\} \cup \{1\}$ is clopen in L (prove it please!) and the point $s = 1$ is isolated in M if M has the subspace topology induced from L. Indeed, $\{s\} = (\frac{1}{2}, \rightarrow)$, so s is even isolated in L and hence in M. However, s is not isolated in $(M, \tau(\leq_M))$. To see this, observe that the base at s in M is given by the sets $(t, \rightarrow)_M$ where t runs over $M \setminus \{s\}$. Now, if $t \in M \setminus \{s\}$, then $t = -\frac{1}{n}$ for some $n \in \mathbb{N}$ and therefore $(t, \rightarrow)_M \supset \{-\frac{1}{m} : m > n\}$ is an infinite set. Thus every neighborhood of s in $(M, \tau(\leq_M))$ is infinite, so s is not isolated in $(M, \tau(\leq_M))$. Hence it is possible that $\tau(\leq_M) \neq \tau_M^L$ for a clopen M.

To show that the same is possible for a dense set $M \subset L$, consider the space $L = \{-\frac{1}{n} : n \in \mathbb{N}\} \cup \{0\} \cup \{1\}$ with the order induced from $\mathbb{R}$. It is evident that $0 \in \overline{\{-\frac{1}{n} : n \in \mathbb{N}\}}$, so the set $M = L \setminus \{0\}$ is dense in L. As before, the point $s = 1$ is isolated in (M, τ_M^L) because s is isolated in L. However, s is not isolated in M: we proved it in the previous case considering the same M. This shows that it is possible that $\tau(\leq_M) \neq \tau_M^L$ for a dense $M \subset L$, so (ii) is established.

(iii) Assume that M is order dense in L. To see that M is dense in L it suffices to show that $M \cap B \neq \emptyset$ for any non-empty $B \in \mathcal{B}$. We have three possibilities for the set B:

(1) If $B = (\leftarrow, a)$ for some $a \in L$, pick any $s \in B$; then $s < a$, so there are $p, q \in M$ such that $s \leq p < q \leq a$. Therefore $p \in (\leftarrow, a) \cap M$, so $M \cap B \neq \emptyset$ in this case.

(2) If $B = (a, \rightarrow)$ for some $a \in L$, then again take any $s \in B$ and observe that $a < s$ implies that there are $p, q \in M$ with $a \leq p < q \leq s$ and hence $q \in (a, \rightarrow) \cap M$ which shows that $M \cap B \neq \emptyset$ in this case as well.

(3) If $B = (a, b)$ for some $a, b \in L$ such that $a < b$, pick any $s \in (a, b)$. Again, there are $p, q \in M$ with $a \leq p < q \leq s$ which implies $q \in (a, b) \cap M$ and therefore $M \cap B \neq \emptyset$, so M is dense in L.

As to $\tau_M^L = \tau(\leq_M)$, it suffices to show that $\tau_M^L \subset \tau(\leq_M)$. Take any $U \in \tau_M^L$ and $V \in \tau(L)$ such that $U = V \cap M$. Given any $s \in U$ there exists an interval $I \in \mathcal{B}$ such that $s \in I \subset V$. We must consider the same three cases:

(1) $I = (a, b)$ for some $a, b \in L$ with $a < s < b$. By order density of M, there are $p, q, r, t \in M$ such that $a \le p < q \le s$ and $s \le r < t \le b$. Therefore $s \in (p, t)_M \subset U$.

(2) $I = (a, \to)$ for some $a \in L$ with $a < s$. By order density of M, there are $p, q \in M$ such that $a \le p < q \le s$. Consequently, $s \in (p, \to)_M \subset U$.

(3) $I = (\leftarrow, a)$ for some $a \in L$ with $s < a$. By order density of M, there are $p, q \in M$ such that $s \le p < q \le a$. Consequently, $s \in (\leftarrow, q)_M \subset U$.

Thus we proved that for any $x \in U$, there is $I \in \mathcal{B}_M$ such that $s \in I \subset U$. Therefore $\mathcal{B}_M$ is a base for τ_M^L, so $\tau_M^L \subset \tau(\le_M)$ and (iii) is proved.

(iv) Given a subset $x \subset L$, call x a *gap ray* if it satisfies the following conditions:

(1) x has no maximal element, i.e., there is no $s \in x$ such that $t \le s$ for all $t \in x$;

(2) $s \in x$ and $t \le s$ implies $t \in x$;

(3) there is no minimal element in $L \backslash x$.

Call a set $x \subset L$ *a point ray* if $x = L_s = \{t \in L : t < s\}$ for some $s \in L$. A set $x \subset L$ is a *left ray* if it satisfies (2). It is evident that any point ray is a left ray.

Let $K = \{x \subset L : x$ is a gap ray or a point ray$\}$; note that we have

(4) any left ray x without a maximal element is in K,

because if there is a minimal element s in the set $(L \backslash x)$, then $x = L_s$. If $x, y \in K$, then let $x \preceq y$ if $x \subset y$. It is evident that $\preceq$ is a partial order on K. We will also need the following property of left rays:

(5) If $x \subset L$ is a left ray and $s \in L$, then $x \subset L_s$ if and only if $s \notin x$.

Indeed, if $t \in x$ and $t \ge s$, then $s \in x$, so $s \notin x$ implies $t < s$ for all $t \in x$, i.e., $x \subset L_s$. If, on the other hand, $x \subset L_s$, then $t < s$ for any $t \in x$ and hence $s \notin x$. This proves (5).

It is immediate from the definition that if $x \subset L$ is a left ray and $s \in x$, then $L_s \subset x$. On the other hand, if $L_s \subset x$ and $x \ne L_s$, then $s \in x$; for otherwise, $x \subset L_s$ by (5) and hence $x = L_s$. Thus we proved that

(6) for any left ray $x \subset L$ if $s \in x$, then $L_s \subset x$; besides if $L_s \ne x$, then $L_s \subset x$ if and only if $s \in x$.

To establish that $\preceq$ is also a linear order, take any $x, y \in K$. If $x = y$, then there is nothing to prove. If $x \ne y$, then $x \backslash y \ne \emptyset$ or $y \backslash x \ne \emptyset$. If $x \backslash y \ne \emptyset$, then take any $s \in x \backslash y$ and apply (5) and (6) to conclude that $y \subset L_s \subset x$ and hence $y \preceq x$. Analogously, if $y \backslash x \ne \emptyset$, then $x \preceq y$ which proves that $\preceq$ is a linear order on K. If $a, b \in K$, then $(a, b)_K = \{x \in K : a \prec x \prec b\}$; analogously, $(a, \to)_K = \{x \in K : a \prec x\}$ and $(\leftarrow, b)_K = \{x \in K : x \prec b\}$. The family

$$\mathcal{B}(K) = \{(a, b)_K : a, b \in K \text{ and } a \prec b\} \cup \{(a, \to)_K : a \in K\} \cup \{(\leftarrow, b)_K : b \in K\}$$

is a base in the space K.

Let us show that $(K, \preceq)$ is compact. Take any closed non-empty $F \subset K$; observe that any $x \in F$ is a subset of L so $z = \bigcup\{x : x \in F\}$ is also a subset of L. It is immediate that z is a left ray. We prove first that $z \in K$. If z has no maximal element then $z \in K$ by (4). If we have a maximal element $t \in z$ then $t \in x$ for some $x \in F$ and hence $z \subset L_t \cup \{t\} \subset x$, i.e., $z \subset x$ and therefore $z = x$ because z is a union of all elements of F which implies $x \subset z$. Thus $z = x \in F$ so $z \in K$.

To prove that $z \in F$ assume that this is not true; since F is closed, there is $B \in \mathcal{B}(K)$ such that $z \in B \subset K \backslash F$. We have three possibilities for the set B. Assume that $B = (a, b)_K$ for some $a, b \in K$ with $a \prec z \prec b$; since no element of F can be greater than or equal to z, we have $z \in (a, \rightarrow)_K \subset K \backslash F$. The case $B = (\leftarrow, a)_K$ is impossible because $F \subset (\leftarrow, a)_K$ for any $a \succ z$ and hence $(\leftarrow, a)_K \cap F \neq \emptyset$. Thus, in all possible cases, there is $a \in K$ such that $a \prec z$ and $(a, \rightarrow)_K \cap F = \emptyset$. Since $z \neq a$, there is $s \in z \backslash a$; it follows the definition of z that there is $x \in F$ such that $s \in x$ and therefore $s \in x \backslash a$. It follows from (5) and (6) that $a \subset L_s \subset x$, so $a \preceq x$. Besides, $s \in x \backslash a$ implies $a \neq x$ and therefore $a \prec x$ which shows that $x \in (a, \rightarrow)_K \cap F = \emptyset$, a contradiction. Consequently, $z \in F$ is the maximal element of F.

To show that F also has a minimal element, let $w = \bigcap\{x : x \in F\}$. It is an easy exercise that any intersection of left rays is a left ray. It is also evident that $w \subset x$ for any $x \in F$, so if $w \in F$, then w is the minimal element of F. However, it is not even clear whether $w \in K$.

Assume for a moment that $w \in K$. If $w \notin F$, then there is $B \in \mathcal{B}(K)$ such that $w \in B \subset K \backslash F$. There are three possibilities for the set B. Note first that B cannot be $(a, \rightarrow)_K$ for any $a \in K$ because $w \in B$ implies $a \prec w$; since $w \preceq x$ for any $x \in F$, we have $a \prec x$ for any $x \in F$, i.e., $F \subset (a, \rightarrow)_K$ which contradicts the equality $(a, \rightarrow)_K \cap F = \emptyset$. Now, if $B = (a, b)_K$ for some $a, b \in K$, then $(w, b)_K \cap F = \emptyset$; if $B = (\leftarrow, b)_K$, then since there is no $x \in F$ with $x \preceq w$, we also have $(w, b)_K \cap F = \emptyset$.

To sum up, if $w \in K \backslash F$, then there is $b \in K$ such that $(w, b)_K \cap F = \emptyset$. Since $w \prec b$, there is $t \in b \backslash w$; by definition of w, there is $x \in F$ such that $t \notin x$. Apply (5) and (6) to conclude that $x \subset L_t \subset b$. Since $t \in b$, we have $L_t \neq b$ and therefore $x \preceq L_t \prec b$, so $x \in (w, b)_K \cap F$ which is a contradiction.

Thus we proved that

(7) if $w \in K$, then $w \in F$.

Now assume that $w \notin K$. By (4), the ray w has a maximal element s. The point $u = L_s \in K$ does not belong to F because $s \notin u$. Observe also that $u \subset x$ for all $x \in F$ and hence $u \prec x$ for each $x \in F$. The set F being closed in K there is $B \in \mathcal{B}(K)$ such that $u \in B \subset K \backslash F$. Again, we have three possibilities for B. If $B = (a, b)_K$, then $(u, b)_K \cap F = \emptyset$. If $B = (a, \rightarrow)_K$, then $a \prec u \prec x$ for every $x \in F$ which shows that $F \subset B$, a contradiction with $B \cap F = \emptyset$. If $B = (\leftarrow, b)_K$, then it follows from $(\leftarrow, u)_K \cap F = \emptyset$ that $(u, b)_K \cap F = \emptyset$. As a result we showed that $w \notin K$ implies $(u, b)_K \cap F = \emptyset$ for some $b \in K$ with $u \prec b$.

Now $u \prec b$ implies $u \subset b$ and $u \neq b$. Since $w \backslash u = \{s\}$ and $w \notin K$ cannot coincide with $b \in K$, there exists $t \in b \cap (L \backslash w)$. By definition of w, there is

$x \in F$ such that $t \notin x$. Apply (5) and (6) to conclude that $x \subset L_t \subset b$. Since $t \in b$, we have $L_t \neq b$ and therefore $x \preceq L_t \prec b$, so $x \in (u,b)_K \cap F$ which is a contradiction. This contradiction shows that $w \in K$ and hence $w \in F$ by (7). We have finally proved that F has both the smallest and the largest elements. The closed non-empty set $F \subset K$ was chosen arbitrarily, so K is compact by Problem 305 of [TFS].

Given any $s \in L$, let $\varphi(s) = L_s$ for all $s \in L$; this defines a map $\varphi : L \to K$. If $s, t \in L$ and $s < t$, then $L_s \subset L_t$, i.e., $\varphi(s) \preceq \varphi(t)$. Besides, $s \in L_t \setminus L_s$, so $\varphi(s) \neq \varphi(t)$ and hence $\varphi(s) \prec \varphi(t)$. This shows that $\varphi : L \to \varphi(L)$ is an order isomorphism and hence L is order isomorphic to the set $N = \varphi(L) \subset K$. Observe first that N is precisely the set of point rays in K. Let us prove that N is order dense in K. Take any $a, b \in K$ with $a \prec b$. If both a and b are point rays, then we can take $p = a$ and $q = b$ to obtain $p, q \in N$ such that $a \preceq p \prec q \preceq b$.

Now if $b = L_s$ for some $s \in L$, then $a \subset L_s$ and it follows from $a \neq L_s$ that there is some $t \in L_s \setminus a$. Apply (5) to conclude that $a \subset L_t$ and hence $a \preceq L_t$. Now, if $p = L_t$ and $q = L_s = b$, then $a \preceq p < q \preceq b$, so the order density of N is established in this case as well.

Finally, assume that b is a gap ray. Since $a \prec b$, we can choose some $s \in b \setminus a$. No gap ray has a maximal element, so there is $t \in b$ with $s < t$. Now, if $p = L_s$ and $q = L_t$, then $a \preceq p \prec q \preceq b$ (we used (5) and (6) again) which finishes the proof of order density of N in K. Now apply (iii) to conclude that N is dense in K and τ_N^K coincides with the topology generated by the order $\preceq_N$. We have now proved everything promised in (iv), so our solution is complete.

T.073. *Prove that, under $\diamondsuit$, there exists a linearly ordered hereditarily Lindelöf non-separable compact space.*

Solution. We will first prove some facts about linearly ordered spaces. Let $(L, \leq)$ be a linearly ordered space. Given any points $a, b \in L$, such that $a < b$, we let $(a,b) = \{s \in L : a < s < b\}$. If $a > b$, it is convenient to consider that $(a,b) = (b,a)$. Of course, $(a,b) = \emptyset$ if $a = b$. Furthermore, $[a,b] = (a,b) \cup \{a,b\}$ for any $a, b \in L$; besides, $(a, \rightarrow) = \{s \in L : a < s\}$ and $(\leftarrow, a) = \{s \in L : s < a\}$. By definition, the family $\mathcal{B}_L = \{(a,b) : a, b \in L\} \cup \{(a, \rightarrow) : a \in L\} \cup \{(\leftarrow, a) : a \in L\}$ is a base of the space L. The elements of $\mathcal{B}_L$ are called *open intervals* in L. We consider the whole space L to be an open interval as well.

Fact 1. Let $(K, \leq)$ be a compact linearly ordered space. Then every non-empty open $U \subset K$ is a disjoint union of open intervals.

Proof. Since K is compact, there exist $m = \min K$ and $M = \max K$ (see Problem 305 of [TFS]). Say that a set $P \subset K$ is *convex* if, for any $x, y \in P$, we have $[x, y] \subset P$. Note that every open interval is a convex set. Besides,

(1) if a family $\mathcal{C}$ consists of convex sets and $\bigcap \mathcal{C} \neq \emptyset$, then $\bigcup \mathcal{C}$ is also a convex set.

Indeed, take any $x, y \in C = \bigcup \mathcal{C}$. There are $A, B \in \mathcal{C}$ such that $x \in A$ and $y \in B$. Since $\bigcap \mathcal{C} \neq \emptyset$, we can take some $z \in A \cap B$. Then $[x, z] \subset A$ and $[z, y] \subset B$, so $[x, y] \subset [x, z] \cup [z, y] \subset A \cup B \subset C$ which shows that $[x, y] \subset C$ for any $x, y \in C$, i.e., C is a convex set.

Given $x, y \in U$, say that $x \sim y$ if $[x, y] \subset U$. It is immediate that $\sim$ is an equivalence relation on U. Let $E_x = \{y \in U : y \sim x\}$ be the equivalence class of x for all $x \in U$. The set E_x is convex because $y, z \in E_x$ implies $[y, z] \subset [y, x] \cup [x, z] \subset E_x$. Besides, E_x is open in K; indeed, if $y \in E_x$, then there is $B \in \mathcal{B}_K$ such that $y \in B \subset U$. Since B is a convex set, we have $B \subset E_y$; since $E_x = E_y$, we have $B \subset E_x$, and hence every $y \in E_x$ is contained in E_x together with an open set, i.e., E_x is open for every $x \in U$. Since K is compact, there exist $M_x = \max \overline{E}_x$ and $m_x = \min \overline{E}_x$ (see Problem 305 of [TFS]). If $m_x < z < M_x$, then the set $B = (z, \to)$ is open and contains $M_x \in \overline{E}_x$, so there is $b \in E_x$ with $z < b$. Analogously, there is $a \in E_x$ with $a < z$. Therefore $z \in [a, b] \subset E_x$ and hence $z \in E_x$. This proves that $(m_x, M_x) \subset E_x$. If $m_x \notin E_x$ and $M_x \notin E_x$, then $E_x = (m_x, M_x)$ is an open interval.

Assume now that $M_x \notin E_x$ and $m_x \in E_x$. If $m_x = m$, then $E_x = (\leftarrow, M_x)$ is an open interval. If $m < m_x$, then the set $[m, m_x)$ is closed in K being the intersection of a closed set $[m, m_x]$ with the closed set $K \backslash E_x$. If $n_x = \max[m, m_x)$ (we can take the respective maximum by Problem 305 of [TFS]), then $E_x = (n_x, M_x)$ is an open interval. This proves that E_x is an open interval if $m_x \in E_x$ and $M_x \notin E_x$. Analogously, E_x is an open interval if $m_x \notin E_x$ and $M_x \in E_x$. Finally, assume that $E_x = [m_x, M_x]$. We still have some subcases here.

(1) if $m_x = m$ and $M_x = M$, then $E_x = K$ is an open interval;
(2) if $m_x = m$ and $M_x < M$, then the set $(M_x, M]$ has a minimal element N_x and hence $E_x = (\leftarrow, N_x)$ is an open interval;
(3) if $m_x > m$ and $M_x = M$, then the set $[m, m_x)$ has a maximal element n_x, so $E_x = (n_x, \to)$ is an open interval;
(4) if $m_x > m$ and $M_x < M$, then the set $(M_x, M]$ has a minimal element N_x, so $E_x = (n_x, N_x)$ is an open interval.

This proves that E_x is an open interval for any $x \in U$. It is clear that U is a disjoint union of the sets E_x, so Fact 1 is proved. □

Fact 2. Let $(L, \leq)$ be a linearly ordered space. If $c(L) \leq \omega$, then L is hereditarily Lindelöf.

Proof. There exists a compact linearly ordered space $(K, \preceq)$ such that L is order isomorphic to a dense $N \subset K$ such that the topology of the subspace on N coincides with the topology generated by the restriction of the order $\preceq$ to N (see Problem 072). Since N is homeomorphic to L, we have $\omega = c(L) = c(N) = c(K)$, so it suffices to prove our Fact for K because if K is hereditarily Lindelöf, then so is N and hence L.

Observe that $s(K) = c(K) = \omega$ (see Fact 2 of S.304). Note also that $t(K) = \omega$ because any free sequence is a discrete subset of K, so all free sequences in K are countable (see Problem 328 of [TFS]). Therefore $\chi(K) = t(K) = \omega$ (see Problem 303 of [TFS]). For each $x \in K$ fix a local base $\{U_n^x : n \in \omega\} \subset \mathcal{B}_K$ at the point x such that $U_{n+1}^x \subset U_n^x$ for all $n \in \omega$. Take any points $a, b \in K$ such that $a < b$ and observe that the set $F_n = (a, b) \setminus (U_n^a \cup U_n^b) = [a, b] \setminus (U_n^a \cup U_n^b)$ is closed for each $n \in \omega$. Given any $x \in (a, b)$, there is $n \in \omega$ such that $x \notin U_n^b \cup U_n^a$ and hence $x \in F_n$. This shows that $\bigcup_{n \in \omega} F_n = (a, b)$, so we proved that every open interval (a, b) is an F_σ-set in K.

Now if $G_n = (\leftarrow, b) \setminus U_n^b$, then the set G_n is closed in K and $\bigcup_{n \in \omega} G_n = (\leftarrow, b)$ and therefore $(\leftarrow, b)$ is an F_σ-set for any $b \in K$. Analogously, $(a, \rightarrow)$ is an F_σ-set for any $a \in K$ so we proved that

(2) every open interval is an F_σ-set in K.

Now take any non-empty open $U \subset K$; there is a family $\mathcal{U}$ of disjoint open intervals such that $\bigcup \mathcal{U} = U$ (see Fact 1). Since $c(K) = \omega$, the family $\mathcal{U}$ is countable. Any countable union of F_σ-sets is an F_σ-set, so U is an F_σ-set by (2). It turns out that every open subset of K is an F_σ-set, so K is perfectly normal. Applying Problem 001 we conclude that K is hereditarily Lindelöf, so Fact 2 is proved.   □

We will need to introduce some notation for working with trees. Given a tree $(S, \leq)$ and any $s \in S$, the set $L(s, S) = \{t \in S : t < s\}$ is well-ordered and hence isomorphic to an ordinal α; let $\mathrm{ht}(s, S) = \alpha$. Let $S(\alpha) = \{s \in S : \mathrm{ht}(s, S) = \alpha\}$; call $S(\alpha)$ *the α-th level of S*. If $s \in S$, then $H(s, S) = \{t \in S : s \leq t\}$. Another important observation is that given any $s \in S$ with $\mathrm{ht}(s, S) = \alpha$, for any $\beta < \alpha$ there is a unique $t = s(\beta) \in S(\beta)$ with $t \leq s$.

Fact 3. Assume that there exists a Souslin tree. Then there exists a Souslin tree S with the following properties:

(3) $S(0)$ is a singleton;
(4) given any $\alpha < \beta < \omega_1$ and any $s \in S(\alpha)$, there is $t \in S(\beta)$ such that $s < t$;
(5) for any $\alpha < \omega_1$ any $s \in S(\alpha)$ and any $n \in \mathbb{N}$, there is $\beta > \alpha$ such that $|\{t \in S(\beta) : s < t\}| \geq n$.

Proof. Take any Souslin tree T and consider the set $U = \{t \in T : |H(t, T)| = \omega_1\}$. Observe that $s \leq t$ implies $H(s, T) \supset H(t, T)$ and hence $|H(t, T)| = \omega_1$ implies $|H(s, T)| = \omega_1$ for any $s < t$ which shows that $L(t, T) \subset U$ for any $t \in U$ and therefore U is a subtree of T. Of course, we haven't even proved that U is non-empty. We will show more, namely, that $|U| = \omega_1$ and hence U is also a Souslin tree and, besides, $|H(s, U)| = \omega_1$ for any $s \in U$.

To prove that $|U| = \omega_1$, note that for each $\alpha < \omega_1$, the set $T_\alpha' = T \setminus (\bigcup_{\beta < \alpha} T(\beta))$ is uncountable and $T_\alpha' = \bigcup \{H(s, T) : s \in T(\alpha)\}$. Therefore $H(s, T)$ is uncountable for some $s \in T(\alpha)$ and hence $s \in U$. It turns out that $T(\alpha) \cap U \neq \emptyset$ for each $\alpha < \omega_1$ so $|U| = \omega_1$. Furthermore, if $s \in U$, then for each $\alpha > \mathrm{ht}(s, T)$, the uncountable set $H(s, T) \cap T_\alpha'$ is covered by the countable union $\bigcup \{H(t, T) : t \in T(\alpha)\}$ which shows that $H(t, T) \cap H(s, T) \cap T_\alpha'$ is uncountable for some

$t \in T(\alpha)$ and hence $t \in U$. Besides, $H(t, T) \cap H(s, T) \neq \emptyset$ implies that $t \in H(s, T) \cap U = H(s, U)$. Thus, for any $\alpha > \text{ht}(s, T)$, we found an element of $H(s, U)$ in $T(\alpha)$ which shows that $|H(s, U)| = \omega_1$.

Now, pick an arbitrary $a \in U$ and let $S = H(a, U)$; we claim that S is a tree with the required properties. It is clear that $S(0) = \{a\}$ so (3) is fulfilled. To see that (4) is true, fix any $\alpha < \omega_1$, any $\beta > \alpha$ and any $s \in S(\alpha)$. The set $H(s, S) = H(s, U)$ is uncountable and hence there is $u \in H(s, S)$ such that $\text{ht}(u, S) > \beta$. Since $L(u, S)$ is isomorphic to some $\gamma > \beta$, there is $t < u$ such that $t \in S(\beta)$. We have $s < u$ and $t < u$; since $L(u, S)$ is well-ordered, the elements t and s are comparable and hence $s < t$. This settles (4).

To prove the property (5) by induction on n, we start off with $n = 2$; since the uncountable set $H(s, S)$ cannot be a chain, there exist $a, b \in H(s, S)$ with $H(a, S) \cap H(b, S) = \emptyset$. If $\text{ht}(a, S) = \text{ht}(b, S)$, then we are done. If not, then assume that $\gamma = \text{ht}(a, S) < \text{ht}(b, S)$. Then $b' = b(\gamma) \neq a$ and $a, b' \in S(\gamma) \cap H(s, S)$ which proves (5) for $n = 2$.

Now, if the property (5) is true for $n = k$, then fix any $\gamma > \alpha$ and distinct $t_1, \ldots, t_k \in S(\gamma) \cap H(s, S)$. Applying (5) for $n = 2$ to t_k, we can find $\beta > \gamma$ and distinct $a, b \in S(\beta) \cap H(t_k, S)$. The property (4) guarantees existence of $s_1, \ldots, s_{k-1} \in S(\beta)$ such that $s_i(\gamma) = t_i$ for all $i \leq k - 1$. It is clear that $s_1, \ldots, s_{k-1}, a, b \in H(s, S)$ are distinct and belong to $S(\beta)$, so (5) holds for $n = k + 1$. Therefore (5) holds for any $n \geq 2$ and Fact 3 is proved. □

Fact 4. If there is a Souslin tree, then there exists a linearly ordered hereditarily Lindelöf non-separable compact space.

Proof. Apply Fact 3 to fix a Souslin tree S with the properties (3)–(5) and fix any well-order $\preceq$ on S. Denote by L the set of all maximal chains of S (with respect to inclusion). Given any $x \in L$ let $\delta(x) = \sup\{\text{ht}(s, S) : s \in x\}$. Observe that for any $x \in L$ and any $s \in x$ with $\text{ht}(s, S) = \alpha$, there is $t \in S(\alpha + 1) \cap H(s, S)$ by (4). Therefore the chain x cannot have a maximal element, and hence $\delta(x)$ is a limit ordinal for any $x \in L$. Besides, if $x \in L$ and $s \in x$, then $L(s, S) \subset x$, so the set $x \cap S(\alpha)$ is a singleton for any $\alpha < \delta(x)$. Let $x[\alpha]$ be the element of $S(\alpha)$ such that $x \cap S(\alpha) = \{x[\alpha]\}$. If we have distinct points $x, y \in L$, then let $d(x, y) = \min\{\alpha : x[\alpha] \neq y[\alpha]\}$.

We define an order $\leq$ on L as follows: if $x, y \in L$ are Distinct, then $x < y$ if and only if $x[d(x, y)] \prec y[d(x, y)]$. It is evident that any two elements of L are comparable. It is immediate that $x < y$ and $y < x$ cannot hold at the same time, so $x \leq y$ and $y \leq x$ implies $x = y$. To check the transitivity, assume that $x \leq y$ and $y \leq z$ for some $x, y, z \in L$. The proof is trivial if there is some equality among x, y, z, so assume that $x < y$ and $y < z$; let $\alpha = d(x, y)$ and $\beta = d(y, z)$:

(1) If $\beta < \alpha$, then $y[\beta] \prec z[\beta]$ while $x[\beta] = y[\beta]$ and hence $x[\beta] \prec z[\beta]$. Since $x[\gamma] = y[\gamma] = z[\gamma]$ for all $\gamma < \beta$, we have $d(x, z) = \beta$ which shows that $x < z$.

(2) If $\alpha = \beta$, then $x[\alpha] \prec y[\alpha] \prec z[\alpha]$ and therefore $x[\alpha] \prec z[\alpha]$ by transitivity of $\prec$. Since $x[\gamma] = y[\gamma] = z[\gamma]$ for all $\gamma < \alpha$, we have $d(x, z) = \alpha$ which shows that $x < z$.

(3) If $\alpha < \beta$, then $y[\alpha] = z[\alpha]$ which implies that $x[\alpha] \prec y[\alpha] = z[\alpha]$. Since $x[\gamma] = y[\gamma] = z[\gamma]$ for all $\gamma < \alpha$, we have $d(x, z) = \alpha$ which shows that $x < z$.

Thus $\leq$ is a linear order on L. Considering that we endow L with the topology generated by $\leq$, let us show that $c(L) = \omega$. Assume that there is an uncountable disjoint family $\mathcal{U}$ of non-empty open subsets of L. Since $\mathcal{B}_L$ is a base in L, we can assume that $\mathcal{U}$ consists of elements of $\mathcal{B}_L$. Furthermore, given $a, b \in L$, one of the sets $(\leftarrow, a)$ and $(\leftarrow, b)$ is contained in the other which shows that $\mathcal{U}$ can contain at most one set $(\leftarrow, a)$. Analogously, $\mathcal{U}$ can contain at most one element $(b, \rightarrow)$, so throwing away at most two elements from $\mathcal{U}$, we can consider that $\mathcal{U} = \{(a_\alpha, b_\alpha) : \alpha < \omega_1\}$ where the intervals (a_α, b_α) are non-empty and disjoint. There will be no loss of generality to assume that $a_\alpha < b_\alpha$ for all $\alpha < \omega_1$. Choose any $x_\alpha \in (a_\alpha, b_\alpha)$ for each $\alpha < \omega_1$. Observe that if $\mu_\alpha = d(a_\alpha, b_\alpha)$, then $x_\alpha[\beta] = a_\alpha[\beta] = b_\alpha[\beta]$ for every $\beta < \mu_\alpha$. Consider also the ordinals $\nu_\alpha = d(x_\alpha, a_\alpha)$ and $\xi_\alpha = d(x_\alpha, b_\alpha)$. Since $a_\alpha[\mu_\alpha] \neq b_\alpha[\mu_\alpha]$, the point $x_\alpha[\mu_\alpha]$ must be distinct from either $a_\alpha[\mu_\alpha]$ or $b_\alpha[\mu_\alpha]$. Thus, $\min\{\nu_\alpha, \xi_\alpha\} = \mu_\alpha$; let $\zeta_\alpha = \max\{\nu_\alpha, \xi_\alpha\}$; observe that any chain $x \in L$ with $x_\alpha[\zeta_\alpha] \in x$ coincides with x_α on all levels under ζ_α. Since the relation $a_\alpha < x_\alpha < b_\alpha$ is determined by the points $x_\alpha[\mu_\alpha]$, $x_\alpha[\zeta_\alpha]$ and the fact that $x_\alpha[\beta] = a_\alpha[\beta] = b_\alpha[\beta]$ for all $\beta < \mu_\alpha$, we have

(6) $x \in (a_\alpha, b_\alpha)$ for any $x \in L$ such that $x_\alpha[\zeta_\alpha] \in x$.

Now, if $p_\alpha = x_\alpha[\zeta_\alpha]$, then $A = \{p_\alpha : \alpha < \omega_1\}$ is an antichain in S. To see this, assume that $\alpha \neq \beta$ and $p_\beta \in H(p_\alpha, S)$. It is easy to see that there exists a maximal chain $x \ni p_\beta$. As a consequence, $p_\alpha \in x$ and therefore we can apply (6) to conclude that $x \in (a_\alpha, b_\alpha) \cap (a_\beta, b_\beta)$ which is a contradiction. Hence A is an uncountable antichain in S; since S is s Souslin tree, this gives another contradiction which shows that $c(L) = \omega$.

Our next step is to prove that L is not separable. To check this, assume that X is a countable subset of L. Since every $x \in X$ is a countable chain of S, there is $\alpha < \omega_1$ such that $\delta(x) < \alpha$ for any $x \in X$. We have $S(\alpha) \neq \emptyset$ by (4), so pick any $s \in S(\alpha)$. Apply (5) to find $\beta > \alpha$ and distinct $t, u, v \in S(\beta)$ such that $\{t, u, v\} \subset H(s, S)$. There exist $a, b, c \in L$ with $t \in a$, $u \in b$ and $v \in c$. Since the situation is symmetric, we can assume, without loss of generality, that $a \prec b \prec c$, i.e., $b \in (a, c)$ and hence the interval (a, c) is non-empty. However, for every $x \in (a, c) \cap X$, we have $x[\gamma] = a[\gamma] = c[\gamma] = s(\gamma)$ for all $\gamma < \alpha$ which shows that $\delta(x) \geq \alpha$ which contradicts the choice of α. Therefore $X \cap (a, c) = \emptyset$ and hence X is not dense in L. This proves that L is not separable.

Finally, let $(K, \sqsubseteq)$ be a compact linearly ordered space such that $(L, \leq)$ is isomorphic to an order dense $N \subset K$. Of course, N is also homeomorphic to L so $c(N) = \omega$. Since N is dense in K by Problem 072, we have $c(K) = \omega$ and hence K is hereditarily Lindelöf by Fact 2. To see that K is not separable observe that K is perfectly normal by Problem 001 and hence $t(K) \leq \chi(K) = \psi(K) = \omega$ (see Problem 327 of [TFS]). Now, if a countable $X \subset K$ is dense in K, then $X \subset \overline{N}$ and hence, for any $x \in X$, we can choose a countable $A_x \subset N$ with $x \in \overline{A}_x$. The set $A = \bigcup \{ A_x : x \in X \} \subset N$ is countable and $x \in \overline{A}$ for all $x \in X$. Therefore $X \subset \overline{A}$ and hence $N \subset K = \overline{X} \subset \overline{A}$ which shows that a countable set A is dense in N contradicting the fact that N is homeomorphic to a non-separable space L. This contradiction shows that K is a non-separable hereditarily Lindelöf compact space, so Fact 4 is proved.   □

Now it is very easy to solve our problem. If Jensen's axiom holds, then there exists a Souslin tree by Problem 070. Applying Fact 4 we conclude that there exists a non-separable hereditarily Lindelöf linearly ordered compact space, so our solution is complete.

T.074. *Prove that a linearly ordered compact L-space exists if and only if there exists a Souslin tree.*

Solution. We have proved in Fact 4 of T.073 that if there is a Souslin tree, then there exists a compact linearly ordered hereditarily Lindelöf non-separable space. Thus we only have to establish necessity constructing a Souslin tree from a linearly ordered hereditarily Lindelöf non-separable compact space.

Let $(L, \leq)$ be any linearly ordered space. Say that *the order $\leq$ is dense in L* if for any $a, b \in L$ for which $a < b$, there is $c \in L$ such that $a < c < b$. Given any $a, b \in L$, such that $a < b$ we let $(a, b)_L = \{ s \in L : a < s < b \}$. If $a > b$, it is convenient to consider that $(a, b)_L = (b, a)_L$. Of course, $(a, b)_L = \emptyset$ if $a = b$. Furthermore, $[a, b]_L = (a, b)_L \cup \{ a, b \}$ for any $a, b \in L$; besides, $(a, \rightarrow)_L = \{ s \in L : a < s \}$ and $(\leftarrow, a)_L = \{ s \in L : s < a \}$. By definition, the family

$$\mathcal{B}_L = \{ (a, b)_L : a, b \in L \} \cup \{ (a, \rightarrow)_L : a \in L \} \cup \{ (\leftarrow, a)_L : a \in L \}$$

is a base of the space L. The elements of $\mathcal{B}_L$ are called *open intervals* in L. We consider the whole space L to be an open interval as well. We will also need the family $\mathcal{C}_L = \{ (a, b)_L : a, b \in L \}$; the elements of $\mathcal{C}_L$ are called *proper intervals*. A linearly ordered space L is called *a Souslin line* if $c(L) = \omega < d(L)$. Given any linearly ordered space $(L, \leq)$ and any $A \subset L$, a point $a \in A$ is *extreme for A* if a is either maximal or a minimal point of A. By $\leq_A$ we denote the order $\leq$ restricted to the points of A. The sets $(a, \rightarrow)_L$ are called *right rays* and the sets $(\leftarrow, b)_L$ are the *left rays* of L.

Fact 1. Suppose that there is a Souslin line. Then there exists a Souslin line $(S, \leq)$ with the following properties:

(i) S has neither minimal nor maximal element;
(ii) the order $\leq$ is dense in S;
(iii) no $U \in \tau^*(S)$ is separable.

Proof. Take any Souslin line $(X, \preceq)$. Given $x, y \in X$, let $x \sim y$ if the interval $(x, y)_X$ is separable or empty. For technical reasons we will consider empty spaces separable too. Observe that

(1) an interval $(x, y)_X$ is separable if and only if so is $[x, y]_X$.

Indeed, sufficiency holds in (1) because $(x, y)_X$ is open in $[x, y]_X$ and necessity is true because adding the set $\{a, b\}$ to a countable dense subset of $(x, y)_X$ we obtain a countable dense subset of $[x, y]_X$.

It is immediate from the definition that for any $x, y \in X$, we have $x \sim x$ and $x \sim y$ is equivalent to $y \sim x$. To see that $\sim$ is transitive, assume that $x \sim y$ and $y \sim z$. Then $(x, z)_X$ is an open subset of the separable space $[x, y]_X \cup [y, z]_X$, so $(x, z)_X$ is separable and hence $\sim$ is an equivalence relation. For any $x \in X$ consider the equivalence class $E_x = \{y \in X : y \sim x\}$ of the point x. The family $L = \{E_x : x \in X\}$ of all equivalence classes for $\sim$ is crucial for constructing our promised Souslin line. Observe that

(2) every $E_x \in L$ is a *convex set*, i.e., $y, z \in E_x$ implies $[y, z]_X \subset E_x$.

Indeed, the space $[y, z]_X$ is separable because $y \sim x \sim z$; if we take an arbitrary $t \in [y, z]_X$, then $(y, t)_X$ is separable being an open subset of a separable space $[y, z]_X$; therefore $t \sim y \sim x$ which shows that any $t \in [y, z]_X$ belongs to E_x, i.e., $[y, z]_X \subset E_x$ so (2) is proved.

It turns out that any element of E_x order represents it in the following sense:

(3) Assume that $E_x, E_y \in L$ and $E_x \neq E_y$; take any $z \in E_x$ and $t \in E_y$. Then $z \prec t$ if and only if $z' \prec t'$ for any $z' \in E_x$ and $t' \in E_y$.

Sufficiency is evident. As to necessity, assume that $t' \prec z'$ for some $z' \in E_x$ and $t' \in E_y$. Then $u \prec z'$ for all $u \in E_y$; for otherwise, $z' \in [t', u]_X$ for some $u \in E_y$ and hence $z' \in E_y$ by (2), a contradiction. In particular, $t \prec z'$ and therefore $t \in [z, z']_X$ whence $t \in E_x$ by convexity of E_x [see (2)]; this contradiction shows that (3) is true.

Now we can introduce the following linear order $\leq$ on L: given distinct points $a = E_x \in L$ and $b = E_y \in L$, let $a < b$ if and only if $x \prec y$. The property (3) shows that this definition is consistent and gives us a linear order on L. To be completely rigorous, we should choose a set $W \subset X$ such that $\bigcup\{E_w : w \in W\} = X$ and $u \neq w$ implies $E_u \cap E_w = \emptyset$ for any $u, w \in W$. The property (3) implies that for any $u, w \in W$, we have $E_u < E_w$ if and only if $u \prec w$, so $(L, \leq)$ is isomorphic to the set $(W, \preceq_W)$. Our construction of L guarantees that

(4) the order $\leq$ is dense in L.

Indeed, if $(a, b)_L = \emptyset$ for some $a, b \in L$ with $a < b$, take any $x \in a$ and $y \in b$. Since the interval $(x, y)_X$ is an open subset of a separable space $a \cup b$ (here we consider $a \cup b$ to be a subspace of X), the set $(x, y)_X$ is separable (maybe empty) and hence $x \sim y$ which implies $E_x = E_y$ and contradicts $a = E_x \neq b = E_y$. This proves (4). Our next observation is that

(5) the set E_x is separable for every $x \in X$.

Let $\mathcal{U}$ be a maximal disjoint family of non-empty intervals $(u, v)_X$ such that $u, v \in E_x$. The family $\mathcal{U}$ is countable because $c(X) = \omega$. For any $U = (u, v)_X \in \mathcal{U}$ we have $u \sim x \sim v$ and hence the space $(u, v)_X$ is separable; apply (1) to fix a countable dense set $D_U \subset [u, v]_X$ with $\{u, v\} \subset D_U$. Consider the countable set $D = \bigcup \{D_U : U \in \mathcal{U}\} \cup N$ where N is the (possibly empty) set of extreme points of E_x. The set D is dense in E_x; to show it, assume that $z \in E_x \backslash \overline{\bigcup \mathcal{U}}$ and pick any $W \in \mathcal{B}_X$ with $z \in W$ and $W \cap (\bigcup \mathcal{U}) = \emptyset$. Observe that z cannot be an extreme point of E_x, so there are $r, s \in E_x$ such that $r < z < s$. The set $(r, s)_X \cap W$ has to be a proper interval, say $(u, v)_X$. It is evident that $u, v \in E_x$; the maximality of $\mathcal{U}$ implies that there is $(p, q)_X \in \mathcal{U}$ such that $(p, q)_X \cap (u, v)_X \neq \emptyset$. The set D being dense in $(p, q)_X$, we have $D \cap (u, v)_X \cap (p, q)_X \neq \emptyset$ which contradicts $D \cap (u, v)_X = \emptyset$. This shows that the set D is dense in E_x so (5) is proved. For any $a \in L$ considered as a subspace of X, apply (5) to choose a dense countable $P_a \subset a$. Now let us prove that for the linearly ordered space $(L, \leq)$, we have the following property:

(6) no non-empty open subspace of L is separable.

It suffices to show that no non-empty $U \in \mathcal{B}_L$ is separable. If $U = (\leftarrow, a)_L$ for some $a \in L$, then take any $b \in U$. If U is separable, then $(b, a)_L$ is also separable being a non-empty open subset of $(\leftarrow, a)_L$ [see (4)]. Analogously, if some non-empty interval $(a, \rightarrow)_L$ is separable, then for any $b \in (a, \rightarrow)_L$ the interval $(a, b)_L$ is also separable. Thus to prove (6) it suffices to show that no interval $(a, b)_L$ can be separable for distinct $a, b \in L$. The family $\{E_x : x \in X\}$ is disjoint; since $c(X) = \omega$, the set $M = \{a \in L : \text{Int}_X(a) \neq \emptyset\}$ is countable.

Now assume that (6) is false and choose $a, b \in L$ with $a < b$ such that some countable $Q \subset (a, b)_L$ is dense in $(a, b)_L$. Pick any $z \in a$ and $t \in b$; we claim that the interval $(z, t)_X$ is separable. Let $M' = M \cap [a, b]_L$; it is evident that the set $D = \bigcup \{P_c : c \in Q \cup M'\} \cup P_a \cup P_b$ is countable. Assume that $w \in (z, t)_X$ and $w \notin \text{cl}_X(D)$; take any $B \in \mathcal{B}_X$ such that $w \in B$ and $B \cap D = \emptyset$. It is clear that $B' = B \cap (z, t) \in \mathcal{B}_X$ and $w \in B' \subset X \backslash D$. Besides, $B' = (s, u)_X$ for some $s, u \in [z, t]_X$. If $s \sim u$, then $c = E_s = E_u \in M'$ and hence $(s, u)_X \cap D \supset P_c \neq \emptyset$. This contradiction shows that $c = E_s \neq E_u = d$ and therefore $(c, d)_L$ is a non-empty open subset of $(a, b)_L$ [see (4)]. By density of Q in $(a, b)_L$, there is $e \in Q \cap (c, d)_L$ and hence $D \cap (s, u)_X \supset D \cap e \supset P_e$. This contradiction shows that $w \in \text{cl}_X(D)$; the point $w \in (z, t)_X$ was chosen arbitrarily, so $(z, t)_X \subset \text{cl}_X(D)$ and therefore $(z, t)_X$ is separable whence $z \sim t$ which again gives a contradiction with $a = E_z \neq b = E_t$. This finishes the proof of (6).

Our next step is to prove that $c(L) = \omega$. It suffices to show that any disjoint subfamily $\mathcal{U} \subset \tau^*(L)$ of the family $\mathcal{B}_L$ is countable. Assuming the contrary pick any uncountable $\mathcal{U} \in \tau^*(L) \cap \mathcal{B}_L$ and observe that any two distinct non-empty left rays have non-empty intersection as well as any two distinct non-empty right rays of L. This shows that throwing away at most two elements from $\mathcal{U}$, we will have a disjoint uncountable family of proper intervals of L. Thus there will be no loss of generality to assume that $\mathcal{U} = \{(a_\alpha, b_\alpha)_L : \alpha < \omega_1\}$. It follows from (4) that taking a smaller interval inside each $(a_\alpha, b_\alpha)_L$ if necessary, we can assume that the family $\{[a_\alpha, b_\alpha]_L : \alpha < \omega_1\}$ is disjoint as well.

Choose $x_\alpha \in a_\alpha$ and $y_\alpha \in b_\alpha$ for all $\alpha < \omega_1$. Observe that the interval $(x_\alpha, y_\alpha)_X$ is non-empty for each $\alpha < \omega_1$. Indeed, the property (4) implies that $V = (a_\alpha, b_\alpha)_L \neq \emptyset$; if $c \in V$, then $v \in (x_\alpha, y_\alpha)_X$ for any $v \in c$ by (3). Therefore the family $\mathcal{V} = \{(x_\alpha, y_\alpha)_X : \alpha < \omega_1\}$ is uncountable and consists of non-empty intervals of X. Besides, $\mathcal{V}$ is disjoint because $\alpha \neq \beta$ and $u \in (x_\alpha, y_\alpha)_X \cap (x_\beta, y_\beta)_X$ implies $E_u \in [a_\alpha, b_\alpha]_L \cap [a_\beta, b_\beta]_L = \emptyset$ which is a contradiction. Thus $\mathcal{V} \subset \tau^*(X)$ is disjoint which contradicts $c(X) = \omega$. This proves that $c(L) = \omega$.

Finally, let $S = L \backslash E$ where E is the (possibly empty) set of the extreme points of L. Considering the set S with the order induced from L, we will write $\leq$ instead of $\leq_S$. It follows from (4) that S is *order dense in* L in the sense of Problem 072 and hence the topology of S as a subspace of L is generated by the order $\leq$ (see Problem 072). Note first that S is a non-empty open subspace of L and therefore $c(S) = \omega$. It is an immediate consequence of (6) that no non-empty open subset of S is separable, so S is a Souslin line with the property (iii). Now, if $a \in S$, then $a \notin E$ and hence there are $b, c \in L$ such that $b < a < c$. Apply (4) to find $b', c' \in L$ such that $b < b' < a < c' < c$; it is evident that $b', c' \in S$ and hence a cannot be an extreme point of S. Therefore (i) is true for S. To see that $\leq$ is dense in S, take any $a, b \in S$ with $a < b$; by (4) there is $c \in L$ with $a < c < b$. It is clear that c cannot be an extreme point of L, so $c \in S$ and hence (ii) holds in S as well. Fact 1 is proved.   □

Returning to our solution, assume that there exists a linearly ordered hereditarily Lindelöf non-separable compact space. Any such space is a Souslin line, so we can apply Fact 1 to fix a Souslin line $(S, \leq)$ with the properties (i)–(iii). It is an easy consequence of (i) that $\mathcal{C}_S$ is a base in S. We omit the index S when using intervals in S, i.e., (a, b) is $(a, b)_S$ and $[a, b] = [a, b]_S$ for any $a, b \in S$.

Let T_0 be a maximal disjoint family of non-empty proper intervals of S. It is clear that $\bigcup T_0$ is dense in S. Assume that $\alpha < \omega_1$ and we have families $\{T_\beta : \beta < \alpha\}$ with the following properties:

(7) T_β is a disjoint family of non-empty proper intervals of S for any $\beta < \alpha$;
(8) $\bigcup T_\beta$ is dense in S for all $\beta < \alpha$;
(9) if $\beta < \gamma < \alpha$, then, for any $V \in T_\beta$ and any $W \in T_\gamma$, we have either $V \cap W = \emptyset$ or $W \subset V$ and $V \backslash \overline{W} \neq \emptyset$.

Assume first that $\alpha = \nu + 1$. For each $U \in T_\nu$ denote by P_U a maximal disjoint family of non-empty proper intervals (a, b) such that $a, b \in U$. Observe that $\overline{(a, b)} \subset [a, b]$ and therefore $U \backslash \overline{(a, b)} \supset \{a, b\}$ is a non-empty set for every $(a, b) \in P_U$. The maximality of P_U implies that $\bigcup P_U$ is dense in U for any $U \in T_\nu$. Therefore the family $T_\alpha = \bigcup \{P_U : U \in T_\nu\}$ has a union which is dense in S, i.e., (8) holds for T_α. The property (7) is evidently true for $\beta = \alpha$. To check (9) we can assume, without loss of generality, that $\gamma = \alpha$. If $V \in T_\beta$, $W \in T_\alpha$ and $V \cap W \neq \emptyset$, then, by our construction, there is $U \in T_\gamma$ such that $W \in P_U$. Since

$V \cap U \supset V \cap W \neq \emptyset$, we have $U \subset V$; since $W \subset U$, we have $W \subset V$. Besides, $V \backslash \overline{W} \supset U \backslash \overline{W} \neq \emptyset$, so $V \backslash \overline{W} \neq \emptyset$ and hence (9) also holds.

Now assume that α is a limit ordinal. Let $T'_\alpha = \bigcup \{T_\beta : \beta < \alpha\}$ and consider the family $\mathcal{U} = \{U \in \mathcal{C}_S : U \neq \emptyset,$ and for any $V \in T'_\alpha$, we have either $U \cap V = \emptyset$ or $U \subset V$ and $V \backslash \overline{U} \neq \emptyset\}$. From the definition of $\mathcal{U}$, it is not even clear whether $\mathcal{U} \neq \emptyset$. However, we will prove that

(10) if $\mathcal{W}$ is a maximal disjoint subfamily of $\mathcal{U}$, then $\bigcup \mathcal{W}$ is dense in S.

To prove (10) let us first show that

(*) for any non-empty interval $I = (p,q)$, there is $U = (a,b) \in \mathcal{U}$ such that $[a,b] \subset I$.

Since the family T'_α is countable, there is a countable set $P \subset S$ such that all endpoints of the intervals from T'_α belong to P. Since the set I is not separable by (iii), there is a non-empty interval $(a,b) \subset I$ such that $[a,b] \subset I$ and $P \cap [a,b] = \emptyset$. To see that $U = (a,b) \in \mathcal{U}$ take any $V = (c,d) \in T'_\alpha$ such that $c < d$ and $V \cap U \neq \emptyset$. The choice of U implies $c, d \notin (a,b)$; if $c > b$ or $d < a$, then $U \cap V = \emptyset$, a contradiction. Since $c \notin [a,b]$, we have $c < a$. Analogously, $d \notin [a,b]$ implies $d > b$ and hence $[a,b] \subset V$ together with $H = V \backslash \overline{U} \neq \emptyset$ because $\{a,b\} \subset H$. This proves that $U \in \mathcal{U}$ and therefore (*) holds.

Now it is easy to establish (10). Assume that $G = \bigcup \mathcal{W}$ is not dense in S and take any non-empty interval $I = (p,q)$ such that $I \cap G = \emptyset$. We proved that there exists $U = (a,b) \in \mathcal{U}$ such that $U \subset I$. Therefore the family $\mathcal{W} \cup \{U\} \subset \mathcal{U}$ is still disjoint which contradicts maximality of $\mathcal{W}$. This settles (10).

To finish our inductive step, let T_α be any maximal disjoint subfamily of $\mathcal{U}$. We have $|T_\alpha| \leq \omega$ because $c(S) = \omega$. The property (10) together with the method of construction of the family T_α imply that the properties (7)–(9) hold for all $\beta \leq \alpha$, so our inductive procedure can be continued to obtain a collection $\{T_\alpha : \alpha < \omega_1\}$ with the properties (7)–(9) fulfilled for all $\beta < \omega_1$.

Let $T = \bigcup \{T_\alpha : \alpha < \omega_1\}$; given any $U, V \in T$, let $U \preceq V$ if $V \subset U$. It is evident that $\preceq$ is a partial order on T. We are going to show that $(T, \preceq)$ is a Souslin tree.

To first establish that T is a tree, take any $U \in T$; then $U \in T_\alpha$ for some $\alpha < \omega_1$. If $V \in T_\beta$ for some $\beta > \alpha$, then it is impossible that $V \subset U$ by (9); the property (7) implies that no $V \in T_\alpha \backslash \{U\}$ is contained in U. Therefore if $V \prec U$, then $V \in T_\beta$ for some $\beta < \alpha$. By (7) the set $L_U = \{V \in T : V \prec U\}$ has at most one element in each T_β. Furthermore, the property (8) implies that for each $\beta < \alpha$, there is $V \in T_\beta$ with $V \cap U \neq \emptyset$ and hence $V \supset U$. This shows that $L_U \cap T_\beta = \{V_\beta\}$ for some $V_\beta \in T_\beta$. As a consequence, the correspondence $V_\beta \to \beta$ is an isomorphism of L_U onto α which proves that L_U is well-ordered and hence T is a tree.

The properties (8) and (9) show that T is uncountable because each T_α is non-empty and $T_\alpha \cap T_\beta = \emptyset$ if $\alpha < \beta$ by (9). If $\mathcal{A} \subset T$ is an antichain, then $U \cap V = \emptyset$ for any distinct $U, V \in \mathcal{A}$ by (7) and (9). Thus $\mathcal{A}$ is countable because $c(S) = \omega$. Finally, if $\mathcal{C}$ is an uncountable chain, then we can consider that $\mathcal{C}$ is maximal and hence $C = \{U_\alpha : \alpha < \omega_1\}$ where $\{U_\alpha\} = C \cap T_\alpha$ for each $\alpha < \omega_1$. By (9) the family $\{U_\alpha \backslash \overline{U}_{\alpha+1} : \alpha < \omega_1\}$ is uncountable and consists of non-empty disjoint open subsets of S which again contradicts $c(S) = \omega$. Hence $(T, \preceq)$ is a Souslin tree and our solution is complete.

T.075. *Let L be a non-separable linearly ordered space such that $c(L) \leq \omega$. Prove that $c(L \times L) > \omega$. In particular, if X is a perfectly normal non-separable linearly ordered compact space, then $c(X) = \omega$ but $c(X \times X) > \omega$.*

Solution. Given a linearly ordered space $(M, \leq)$ and any $a, b \in M$, such that $a < b$, we let $(a, b)_M = \{s \in M : a < s < b\}$; besides, $(a, \rightarrow)_M = \{s \in M : a < s\}$ and $(\leftarrow, a)_M = \{s \in M : s < a\}$. By definition of a linearly ordered space, the family

$$\mathcal{B}_M = \{(a, b)_M : a, b \in M\} \cup \{(a, \rightarrow)_M : a \in M\} \cup \{(\leftarrow, a)_M : a \in M\}$$

is a base of the space M.

Fact 1. Let M be an arbitrary linearly ordered space. If $U \in \tau^*(M)$ and U has no isolated points, then there are $a, b \in U$ such that $a < b$, the interval $(a, b)_M$ is non-empty and $(a, b)_M \subset U$.

Proof. Pick any $s \in U$; since $\mathcal{B}_M$ is a base in M, there is $V \in \mathcal{B}_M$ such that $s \in V \subset U$. Therefore there exists a non-empty $V \in \mathcal{B}_M$ with $V \subset U$. If $V = (a, b)_M$ for some $a, b \in U$, then there is nothing to prove. If $V = (\leftarrow, c)_M$ for some $c \in M$, then V is infinite because it has no isolated points. This makes it possible to choose $a, b, d \in V$ such that $a < d < b$. In particular, $a, b \in V \subset U$, $(a, b)_M \subset V \subset U$ and $(a, b)_M \neq \emptyset$ because $d \in (a, b)_M$. Analogously, if $V = (c, \rightarrow)_M$, then V has to be infinite because it has no isolated points. This makes it possible to choose $a, b, d \in V$ such that $a < d < b$. We have again $a, b \in V \subset U$, $(a, b)_M \subset V \subset U$ and $(a, b)_M \neq \emptyset$ because $d \in (a, b)_M$; thus Fact 1 is proved. $\qquad \square$

Returning to our solution, denote by D the set of all isolated points of L. We omit the index L in the name of the base in L and its intervals, i.e., we use $\mathcal{B}$ instead of $\mathcal{B}_L$ and (a, b) instead of $(a, b)_L$ for any $a, b \in L$.

Since $\{d\}$ is a non-empty open set for each $d \in D$ and the family $\{\{d\} : d \in D\}$ is disjoint, we have $|D| \leq \omega$ because $c(L) = \omega$. The space L is not separable, so $U = L \backslash \overline{D}$ is not separable and has no isolated points. Apply Fact 1 to choose a non-empty interval $I_0 = (a_0, c_0) \subset U$. Since I_0 contains no isolated points, it is infinite, so there is $b_0 \in (a_0, c_0)$ such that both intervals (a_0, b_0) and (b_0, c_0) are non-empty.

Assume that $\beta < \omega_1$ and we have constructed points $a_\alpha, b_\alpha, c_\alpha \in L$ with the following properties:

(1) $a_\alpha < b_\alpha < c_\alpha$ for all $\alpha < \beta$;
(2) $(a_\alpha, b_\alpha) \neq \emptyset$ and $(b_\alpha, c_\alpha) \neq \emptyset$ for each $\alpha < \beta$;
(3) $(a_\alpha, c_\alpha) \cap \{b_\xi : \xi < \alpha\} = \emptyset$ for all $\alpha < \beta$.

The set $B = \{b_\xi : \xi < \beta\}$ is countable, so $W = U \backslash \overline{B}$ is a non-empty open set without isolated points. We can apply Fact 1 again to find a non-empty interval $I_\beta = (a_\beta, c_\beta) \subset W$; since I_β contains no isolated points, it is infinite, so there is $b_\beta \in (a_\beta, c_\beta)$ such that both intervals (a_β, b_β) and (b_β, c_β) are non-empty. It is evident that the properties (1)–(3) hold for all $\alpha \leq \beta$, so our inductive construction can be continued to obtain a collection $\{a_\alpha, b_\alpha, c_\alpha : \alpha < \omega_1\}$ of points of L such that the properties (1)–(3) are fulfilled for all $\alpha < \omega_1$.

It is clear that $U_\alpha = (a_\alpha, b_\alpha) \times (b_\alpha, c_\alpha)$ is an open subset of $L \times L$. The property (2) implies $U_\alpha \neq \emptyset$ for all $\alpha < \omega_1$. If $\alpha < \beta$, then $b_\alpha \notin (a_\beta, c_\beta)$ so we have two cases.

(1) $b_\alpha \leq a_\beta$. Then $(a_\alpha, b_\alpha) \cap (a_\beta, b_\beta) = \emptyset$ and hence $U_\alpha \cap U_\beta = \emptyset$.
(2) $c_\beta \leq b_\alpha$. Then $(b_\beta, c_\beta) \cap (b_\alpha, c_\alpha) = \emptyset$ and again $U_\alpha \cap U_\beta = \emptyset$.

This shows that the family $\{U_\alpha : \alpha < \omega_1\} \subset \tau^*(L \times L)$ is disjoint and hence $c(L \times L) > \omega$. Finally observe that if X is a linearly ordered non-separable perfectly normal compact space, then X is hereditarily Lindelöf by Problem 001. As a consequence, $c(X) \leq hl(X) = \omega$ and hence the result proved for L is applicable to X. Thus $c(X \times X) > \omega$, so our solution is complete.

T.076. *Let X be a linearly ordered hereditarily Lindelöf non-separable compact space. Prove that $C_p(X)$ is not Lindelöf.*

Solution. Let $\leq$ be a linear order which generates the topology of X. Given $a, b \in X$, such that $a < b$, let $(a, b) = \{x \in X : a < x < b\}$ and $[a, b] = (a, b) \cup \{a, b\}$. We will also need the intervals $[a, b) = \{x \in X : a \leq x < b\}$ and $(a, b] = \{x \in X : a < x \leq b\}$. Assume that $C_p(X)$ is Lindelöf; then $t(X) = \omega$ by Problem 189 of [TFS]. Denote by D the set of all isolated points of X. Since $\{d\}$ is a non-empty open set for each $d \in D$ and the family $\{\{d\} : d \in D\}$ is disjoint, we have $|D| \leq \omega$ because $c(X) = \omega$. The space X is not separable, so $U = L \backslash \overline{D}$ is not separable and has no isolated points.

Recall that a family $\mathcal{F} \subset \exp(X)$ is called *point-countable* if $\{F \in \mathcal{F} : x \in F\}$ is countable for every $x \in X$. Any compact space of countable tightness has a point-countable π-base (see Problem 332 of [TFS]), so fix a point-countable π-base $\mathcal{C}$ in the space X. It is evident that the family $\mathcal{C}' = \{C \in \mathcal{C} : C \subset U\}$ is a π-base in the space U. Of course, $\mathcal{C}'$ is point-countable; besides,

(1) U has no countable π-base,

because otherwise U is separable which is false. Observe that each $C \in \mathcal{C}'$ is a non-empty open set without isolated points. This makes it possible to apply Fact 1 of T.075 to choose a non-empty interval $W_C = (a_C, b_C) \subset C$ for each $C \in \mathcal{C}'$. It is evident that $\mathcal{W} = \{W_C : C \in \mathcal{C}'\}$ is also a point-countable π-base of the space U.

For every set $W = (a, b) \in \mathcal{W}$ take any function $f_W \in C_p(X)$ such that $f_W : X \to [0, 1]$, $f_W(x) = 0$ for all $x \leq a$ and $f_W(x) = 1$ for all $x \geq b$. This choice is possible because X is a normal space and the sets $F = \{x \in X : x \leq a\}$ and $G = \{x \in X : b \leq x\}$ are closed and disjoint.

Consider the set $P = \{f_W : W \in \mathcal{W}\}$; we claim that

(*) for any $f \in C_p(X)$ there is $O_f \in \tau(f, C_p(X))$ such that the set $O_f \cap P$ is countable.

Assume first that there exists $x \in X$ such that $f(x) \notin \{0, 1\}$. Then the set $O_f = \{g \in C_p(X) : g(x) \neq 1 \text{ and } g(x) \neq 0\}$ is open in $C_p(X)$ and $f \in O_f$. Now, if $f_W \in O_f$ for some $W = (a, b) \in \mathcal{W}$, then $f_W(x) \in (0, 1)$ and hence $x \in (a, b)$. Since there are only countably many $W = (a, b) \in \mathcal{W}$ such that $x \in W$, the set $P \cap O_f$ is countable.

Now assume that $f(X) \subset \{0, 1\}$. Since X is compact, it has a minimal element m and a maximal element M. It is evident that $f_W(m) = 0$ and $f_W(M) = 1$ for every $W \in \mathcal{W}$. If $f \notin \overline{P}$, then there is $O_f \in \tau(f, C_p(X))$ such that $O_f \cap P = \emptyset$, so there is nothing to prove. Therefore we can assume that $f \in \overline{P}$ and hence $f(m) = 0$, $f(M) = 1$. Thus the set $H = f^{-1}(1)$ is closed and non-empty, so it has a minimal element $p \in H$. Observe that $[m, p) = f^{-1}(0) \cap [m, p]$ is a closed subset of X and therefore it has a maximal element q (see Problem 305 of [TFS]), so the interval (q, p) is empty. Consider the set $O_f = \{g \in C_p(X) : g(q) < 1 \text{ and } g(p) > 0\}$. It is immediate that $O_f \in \tau(f, C_p(X))$.

Now, if $f_W \in O_f$ for some $W = (a, b) \in \mathcal{W}$, then $f_W(p) > 0$ and hence $p > a$; furthermore, $f_W(q) < 1$ implies $q < b$. It is impossible that $b < p$ because then $b \in (q, p) = \emptyset$, a contradiction. Thus $p \leq b$, i.e., $p \in (a, b]$. Analogously, if $q < a$, then $a \in (q, p) = \emptyset$ which is a contradiction. Therefore $a \leq q$ and hence $q \in [a, b)$. Observe also that the equalities $p = b$ and $q = a$ cannot hold at the same time because otherwise $\emptyset = (q, p) = (a, b) \neq \emptyset$ which is a contradiction. Thus $p \in (a, b)$ or $q \in (a, b)$. The family $\mathcal{W}$ being point-countable, there are only countably many $W = (a, b)$ such that $\{p, q\} \cap W \neq \emptyset$. This proves that $O_f \cap P$ is countable in this case as well so (*) is proved.

The family $\mathcal{U} = \{O_f : f \in C_p(X)\}$ is an open cover of the Lindelöf space $C_p(X)$. If $\mathcal{U}' \subset \mathcal{U}$ is a countable subcover of $\mathcal{U}$, then $P = \bigcup\{P \cap O_f : O_f \in \mathcal{U}'\}$ is countable because $O_f \cap P$ is countable for each $O_f \in \mathcal{U}'$ (see (*)).

Now observe that for any $W = (a, b) \in \mathcal{W}$ and $W' = (a', b') \in \mathcal{W}$, if $f_W = f_{W'}$, then $W = (a, b) = f_W^{-1}((0, 1)) = f_{W'}^{-1}((0, 1)) = (a', b') = W'$. As a consequence, $\mathcal{W} = \{W : f_W \in P\}$ is a countable π-base of U which contradicts (1). Thus $C_p(X)$ cannot be Lindelöf, so our solution is complete.

T.077. *Suppose that* $s^*(X) = \omega$. *Prove that* X *condenses onto a hereditarily separable space.*

Solution. Given any $n \in \mathbb{N}$, it follows from $s(X^n \times X^n) \leq \omega$ that $hd(X^n) = \omega$ or $hl(X^n) = \omega$ (see Problem 014). If $hd(X^n) = \omega$ for some n, then there is nothing to prove because the space X itself is hereditarily separable. Therefore we can assume that $hl^*(X) = \omega$ and hence $hd^*(C_p(X)) = \omega$. In particular, $d(C_p(X)) = \omega$ and hence X condenses onto a second countable space (see Problem 174 of [TFS]) which is, of course, hereditarily separable.

T.078. *Suppose that $C_p(X)$ has countable spread. Prove that it can be condensed onto a hereditarily separable space.*

Solution. For an arbitrary $n \in \mathbb{N}$, let $M_n = \{1, \ldots, n\}$; we will also need the set $S_n = \{\sigma : \sigma \text{ is a bijection and } \sigma : M_n \to M_n\}$. If Z is a space and $n \geq 2$, let $\Delta_{ij}^n(Z) = \{z = (z_1, \ldots, z_n) \in Z^n : z_i = z_j\}$ for any distinct $i, j \in M_n$. The set $\Delta_n(Z) = \bigcup\{\Delta_{ij}^n(Z) : 1 \leq i < j \leq n\}$ is called the n-*diagonal* of the space Z.

Fact 1. If Z is a left-separated space, then Z^n is left-separated for any $n \in \mathbb{N}$.

Proof. Let us prove first that $Z \times Z$ is left-separated. Take any well-order $<$ on Z which left-separates Z; a well-order $\prec$ on $Z \times Z$ which left-separates $Z \times Z$ will be defined as follows: given $x = (a, b) \in Z \times Z$ and $y = (c, d) \in Z \times Z$, let $x \prec y$ if $\max(a, b) < \max(c, d)$; if $\max(a, b) = \max(c, d)$, then $x \prec y$ if $a < c$; finally, if $\max(a, b) = \max(c, d)$ and $a = c$, then $x \prec y$ if $c < d$.

We omit a routine verification of the fact that $\prec$ is a linear order on $Z \times Z$. Let us prove that $\prec$ well-orders $Z \times Z$. For any $x = (a, b) \in Z \times Z$, let $m(x) = \max(a, b)$, $p_1(x) = a$ and $p_2(x) = b$. Observe that we only use symbols min and max for the minimum and maximum with respect to the order $<$. Given any non-empty set $A \subset Z \times Z$, let $m_0 = \min\{m(x) : x \in A\}$ and $A_0 = \{x \in A : m(x) = m_0\}$; it is clear that A_0 is a non-empty set. Therefore, the element $r_0 = \min\{p_1(x) : x \in A_0\}$ is well-defined and hence the set $A_1 = \{x \in A_0 : p_1(x) = r_0\}$ is non-empty. Thus we have the element $r_1 = \min\{p_2(x) : x \in A_1\}$ and the set $A_2 = \{x \in A_1 : p_2(x) = r_1\}$ is non-empty. It is immediate that the set A_2 can have at most one point and this point is the $\prec$-minimal element of A.

To see that the order $\prec$ left-separates $Z \times Z$, take any $x = (a, b) \in Z \times Z$ and let $A = \{c \in Z : c < a\}$ and $B = \{c \in Z : c < b\}$. Since the order $<$ left-separates Z, the sets A and B are closed in Z. We have three cases:

(1) $a = b$; then $A = B$ and $P_x = \{z \in Z \times Z : z \prec x\} = (A \times A) \cup (\{a\} \times A) \cup (A \times \{a\})$ which shows that P_x is closed being a union of three closed sets;
(2) $a < b$; then the set $P_x = (B \times B) \cup (A \times \{b\})$ is closed being a union of two closed sets;
(3) $b < a$; then the set $P_x = (A \times A) \cup (A \times \{a\}) \cup (\{a\} \times B)$ is closed being a union of three closed sets.

Therefore the well-order $\prec$ left-separates $Z \times Z$, i.e., we proved our Fact for $n = 2$. As a consequence, the space Z^{2^k} is left-separated for each $k \in \mathbb{N}$. It is clear that a subspace of a left-separated space is left-separated; since for any $n \in \mathbb{N}$, the space Z^n is a subspace of some Z^{2^k}, Fact 1 is proved. $\square$

Fact 2. For any space Z if $\Delta_2(Z)$ is a G_κ-set in $Z \times Z$, then $\Delta_n(Z)$ is a G_κ-set in Z^n for any $n \in \mathbb{N}$ and any infinite cardinal κ.

Proof. Take an arbitrary family $\mathcal{U} \subset \tau(Z \times Z)$ such that $|\mathcal{U}| \leq \kappa$ and $\bigcap \mathcal{U} = \Delta = \Delta_2(Z)$. Given distinct $i, j \in M_n$, let $q_{ij} : Z^n \to Z \times Z$ be the natural projection onto the face defined by i and j, i.e., for any $z = (z_1, \ldots, z_n) \in Z^n$ we have $q_{ij}(z) = (z_i, z_j) \in Z \times Z$. It is clear that $\Delta_{ij}^n(Z) = q_{ij}^{-1}(\Delta)$ and therefore

$\Delta_{ij}^n(Z) = \bigcap \mathcal{U}_{ij}$ where $\mathcal{U}_{ij} = \{q_{ij}^{-1}(U) : U \in \mathcal{U}\}$. If $B_n = \{(i, j) \in M_n \times M_n : i < j\}$, then the family $\mathcal{V} = \{U = \bigcup\{U_{ij} : (i, j) \in B_n\} : U_{ij} \in \mathcal{U}_{ij}$ for all $(i, j) \in B_n\}$ consists of open subsets of Z^n and $\bigcap \mathcal{V} = \Delta_n(Z)$. Since it is evident that $|\mathcal{V}| \leq \kappa$, Fact 2 is proved.   □

Fact 3. Given any space Y, a subspace $Z \subset Y$ and a natural number $n \geq 2$, we have $s(Z^n \backslash U) \leq s(C_p(Y))$ for any $U \in \tau(\Delta_n(Z), Z^n)$.

Proof. This proof is a slight modification of the proof of Fact 1 of T.028. Assume that $s(C_p(Y)) = \lambda$; if we are given a point $z = (z_1, \ldots, z_n) \in Z^n$, then $\mathrm{supp}(z) = \{z_1, \ldots, z_n\}$; let $\Delta_n = \Delta_n(Z)$. For any point $z \in Z^n \backslash \Delta_n$, consider the set $P_z = \{z' \in Z^n \backslash \Delta_n : \mathrm{supp}(z') = \mathrm{supp}(z)\}$. It is evident that $|P_z| = n!$ for any $z \in Z^n \backslash \Delta_n$. It is also clear that either $P_z \cap P_y = \emptyset$ or $P_z = P_y$ for any $y, z \in Z^n \backslash \Delta_n$. Therefore the family $\mathcal{P} = \{P_z : z \in Z^n \backslash \Delta_n\}$ is disjoint; if we choose an element $z_P \in P$ for each $P \in \mathcal{P}$, we obtain a set $C = \{z_P : P \in \mathcal{P}\}$ such that $C \subset Z^n \backslash \Delta_n$ and $|C \cap P_z| = 1$ for every $z \in Z^n \backslash \Delta_n$.

Given any permutation $\sigma \in S_n$, consider the map $p_\sigma : Z^n \to Z^n$ defined by $p_\sigma(z) = (z_{\sigma(1)}, \ldots, z_{\sigma(n)})$ for any $z = (z_1, \ldots, z_n) \in Z^n$. It is immediate that each p_σ is a homeomorphism such that $p_\sigma(\Delta_n) = \Delta_n$. As a consequence, the set $W = \bigcap\{p_\sigma(U) : \sigma \in S_n\}$ is an open (in Z) neighborhood of Δ_n such that $W \subset U$ and $P_z \subset Z^n \backslash W$ for any $z \in Z^n \backslash W$. Since $Z^n \backslash U \subset Z^n \backslash W$, it suffices to prove that $s(Z^n \backslash W) \leq \lambda$. Suppose that there is a discrete $D \subset Z^n \backslash W$ of cardinality λ^+. Since P_z is finite for any $z \in Z^n \backslash \Delta_n$, the set $R = \{z \in C : P_z \cap D \neq \emptyset\}$ has cardinality λ^+. For each $z \in R$, choose $d_z \in P_z \cap D$; then $D' = \{d_z : z \in R\}$ is again a discrete subspace of $Z^n \backslash W$ of cardinality λ^+ such that

(*) $|D' \cap P_z| = 1$ for each $z \in D'$.

The set $F = \overline{D'} \backslash D'$ is a closed subset of Z^n and $F \subset Z^n \backslash W$ (the bar denotes the closure in Z^n). Let $T = Z^n \backslash (W \cup F)$; there exist $E \subset D'$ and $m \in \mathbb{N}$ such that $|E| = \lambda^+$ and $|P_d \cap T| = m$ for every $d \in E$. For each $d = (d_1, \ldots, d_n) \in E$ and $i \leq n$, choose a set $O_i^d \in \tau(d_i, Z)$ such that

(1) $O_i^d \cap O_j^d = \emptyset$ if $i \neq j$;
(2) $O^d \cap D' = \{d\}$ where $O^d = O_1^d \times \cdots \times O_n^d$;
(3)   if $y = (d_{j_1}, \ldots, d_{j_n}) \in \Delta_n$, then $O_{j_1}^d \times \cdots \times O_{j_n}^d \subset W$.
(4)   if $z = (d_{j_1}, \ldots, d_{j_n}) \in (P_d \backslash \{d\}) \cap T$, then $(O_{j_1}^d \times \cdots \times O_{j_n}^d) \cap (E \cup F) = \emptyset$.

For each $d \in E$ and any $i \in M_n$, we have $d_i \notin \mathrm{cl}_Y(Z \backslash O_i^d)$ and hence there exists a function $f_i^d \in C(Y, [0, 1])$ such that $f_i^d(d_i) = 1$ and $f_i^d(Z \backslash O_i^d) \subset \{0\}$; let $f_d = f_1^d + \cdots + f_n^d$. If $d = (d_1, \ldots, d_n) \in E$, then the set $U_d = \{f \in C_p(Y) : f(d_i) > 0$ for all $i \in M_n\}$ is open and $f_d \in U_d$. Take any distinct $a, d \in E$ with $a = (a_1, \ldots, a_n)$ and $d = (d_1, \ldots, d_n)$. If $f_d \in U_a$, then, for all $i \leq n$, there is $j_i \in M_n$ such that $f_{j_i}^d(a_i) > 0$ and hence $a_i \in O_{j_i}^d$. An immediate consequence is that $a \in O_{j_1}^d \times \cdots \times O_{j_n}^d$. If there exist $i, k \leq n$ such that $i \neq k$ and $j_i = j_k$, then $y = (d_{j_1}, \ldots, d_{j_n}) \in \Delta_n$ and $a \in O_{j_1}^d \times \cdots \times O_{j_n}^d \cap W$ by (3) which contradicts $E \cap W = \emptyset$.

Thus $j_i \neq j_k$ if $i \neq k$, i.e., the map $\sigma : M_n \to M_n$ defined by $\sigma(i) = j_i$, is a bijection; let $\nu = \sigma^{-1}$. Given any $z = (d_{k_1}, \ldots, d_{k_n}) \in P_a \cap T$, the point $a(z) = (a_{\nu(k_1)}, \ldots, a_{\nu(k_n)})$ belongs to $P_a \cap T$ because $a(z) \in (O_{k_1}^d \times \cdots \times O_{k_n}^d) \setminus W$ and $(O_{k_1}^d \times \cdots \times O_{k_n}^d) \cap F = \emptyset$ by the property (4). This shows that the set $A = \{a\} \cup \{a(z) : z \in P_d \cap T\}$ is contained in $P_a \cap T$ which contradicts the fact that $P_a \cap T$ has m elements by the choice of E while $|A| = m + 1$ and $A \subset P_a \cap T$.

This contradiction shows that $f_d \notin U_a$ whenever $d \neq a$ which implies that $U_a \cap E = \{f_a\}$ for each $a \in E$ and, in particular, $f_d \neq f_a$ for distinct d and a, i.e., the correspondence $d \to f_d$ is a bijection. Therefore $\{f_a : a \in E\}$ is a discrete subspace of $C_p(Y)$ of cardinality λ^+ which is a contradiction with $s(C_p(Y)) \leq \lambda$. Thus $s(Z^n \setminus U) \leq \lambda$ and Fact 3 is proved. $\square$

Fact 4. Given any space Y such that $s(C_p(Y)) \leq \kappa$ for some infinite cardinal κ, there is a dense $Z \subset Y$ such that $hl^*(Z) \leq \kappa$.

Proof. There exists a dense left-separated $Z \subset Y$ (see Problem 009). We have $s(Y \times Y) \leq \kappa$ (see Problem 016) and hence $s(Z \times Z) \leq \kappa$. Since $Z \times Z$ is also left-separated by Fact 1, we have $hl(Z \times Z) \leq \kappa$ by Problem 007 and, in particular, $\Delta(Z) \leq \kappa$. Apply Fact 2 to conclude that $\Delta_n(Z)$ is a G_κ-set in Z^n for any $n \in \mathbb{N}$, $n \geq 2$.

Fix any $n \in \mathbb{N}$, $n > 2$; since $\Delta_n(Z)$ is a G_κ-set in the space Z^n, there exists a family $\mathcal{V} \subset \tau(\Delta_n(Z), Z^n)$ such that $|\mathcal{V}| \leq \kappa$ and $\bigcap \mathcal{V} = \Delta_n(Z)$. For any set $F \in \mathcal{F} = \{Z^n \setminus V : V \in \mathcal{V}\}$ we have $s(F) \leq \kappa$ by Fact 3. It follows from $|\mathcal{F}| \leq \kappa$ that $s(\bigcup \mathcal{F}) \leq \kappa$ (it is an easy exercise to show that a union of $\leq \kappa$-many spaces of spread $\leq \kappa$ each has spread $\leq \kappa$). It is evident that $Z^n \setminus \Delta_n(Z) = \bigcup \mathcal{F}$, so we have $s(Z^n \setminus \Delta_n(Z)) \leq \kappa$. Applying Fact 0 of T.019, we convince ourselves that $s(Z^n) = s(Z^n \setminus \Delta_n(Z)) \leq \kappa$. Besides, the space Z^n is left-separated by Fact 1, so we can apply Problem 007 again to conclude that $hl(Z^n) \leq \kappa$. The number $n \in \mathbb{N}$ was chosen arbitrarily, so we showed that $hl^*(Z) \leq \kappa$. Fact 4 is proved. $\square$

Returning to our solution, apply Fact 4 to find a dense $Z \subset X$ with $hl^*(Z) \leq \omega$. Since Z is dense in X, the space $C_p(X)$ condenses onto a space $Y \subset C_p(Z)$. We have $hd(Y) \leq hd^*(C_p(Z)) = hl^*(Z) = \omega$ (see Problem 026) and therefore $C_p(X)$ condenses onto a hereditarily separable space Y. Thus our solution is complete.

T.079. *Prove that, under Jensen's axiom, there is a space X of countable spread which does not condense onto a hereditarily separable space.*

Solution. It was proved in Problem 073 that, under Jensen's axiom, there exists a hereditarily Lindelöf non-separable compact space X. We have $s(X) \leq hl(X) = \omega$ while every condensation of X is a homeomorphism because X is compact. Hence X is a space of countable spread which cannot be condensed even onto a separable space.

T.080. *For an arbitrary space X, let Y be a second countable space such that the space $C_p(X, Y)$ is dense in Y^X. Fix any base $\mathcal{B} \subset \tau^*(Y)$ in the space Y; an open set $U \subset C_p(X, Y)$ is called $\mathcal{B}$-standard (or standard with respect to X, Y and $\mathcal{B}$)*

if there exist $n \in \mathbb{N}$, points $x_1, \ldots, x_n \in X$ and sets $O_1, \ldots, O_n \in \mathcal{B}$ such that $U = \{ f \in C_p(X, Y) : f(x_i) \in O_i \text{ for all } i = 1, \ldots, n \}$. Prove that $C_p(X, Y)$ is perfectly normal if and only if any open subset of $C_p(X, Y)$ is a union of countably many $\mathcal{B}$-standard open subsets of $C_p(X, Y)$.

Solution. For technical purposes we will use an even more general concept of a standard set with respect to a base. Given spaces Z, T, a base $\mathcal{C} \subset \tau^*(T)$ of the space T and a set $P \subset C_p(Z, T)$, call a set $U \in \tau(P)$ *standard in P with respect to Z, T and $\mathcal{C}$* if there is a $\mathcal{C}$-standard open subset U' of the space $C_p(Z, T)$ such that $U = U' \cap P$. According to the definition, a set $U' \in \tau(C_p(Z, T))$ is $\mathcal{C}$-standard in $C_p(Z, T)$ if there exist $n \in \mathbb{N}$, $z_1, \ldots, z_n \in Z$ and $O_1, \ldots, O_n \in \mathcal{C}$ such that $U' = [z_1, \ldots, z_n; O_1, \ldots, O_n](Z, T, \mathcal{C}) = \{ f \in C_p(Z, T) : f(z_i) \in O_i \text{ for all } i = 1, \ldots, n \}$. Given any $A \subset Z$, let $\pi_A : C_p(Z, T) \to C_p(A, T)$ be the restriction map, i.e., $\pi_A(f) = f|A$ for all $f \in C_p(Z, T)$. Observe that π_A is continuous being the restriction of the natural projection $p_A : T^Z \to T^A$ of the product space T^Z onto its face T^A (see Problem 107 of [TFS]). A set $F \subset Z$ is called *a zero-set* if there is $f \in C_p(Z)$ such that $F = f^{-1}(0)$. A set $U \subset Z$ is *a cozero-set* if $Z \backslash U$ is a zero-set.

Fact 1. Given arbitrary spaces Z and T and any base $\mathcal{C} \subset \tau^*(T)$ of the space T, take any non-empty set $A \subset Z$ and assume that $V \subset C_A = \pi_A(C_p(Z, T)) \subset C_p(A, T)$ is a standard open subset of C_A (with respect to A, T and $\mathcal{C}$). Then $U = \pi_A^{-1}(V)$ is a $\mathcal{C}$-standard open subset of $C_p(Z, T)$.

Proof. By definition, there exist $z_1, \ldots, z_n \in A$ and $O_1, \ldots, O_n \in \mathcal{C}$ such that $V = V' \cap C_A$ where $V' = [z_1, \ldots, z_n; O_1, \ldots, O_n](A, T, \mathcal{C})$. It suffices to show that $U = \pi_A^{-1}(V) = W = [z_1, \ldots, z_n; O_1, \ldots, O_n](Z, T, \mathcal{C})$. Indeed, if we take any $f \in W$, then $f(z_i) \in O_i$ for each $i \leq n$ and hence $\pi_A(f)(z_i) = f(z_i) \in O_i$ for all $i \leq n$ which shows that $\pi_A(f) \in V$, i.e., $f \in \pi_A^{-1}(V)$. This proves that $W \subset \pi_A^{-1}(V)$. On the other hand, if $f \in \pi_A^{-1}(V)$, then $f(z_i) = \pi_A(f)(z_i) \in O_i$ for each $i \leq n$ and therefore $f \in W$. Thus $U = W$ and Fact 1 is proved. $\square$

Fact 2. A space Z is perfectly normal if and only if every closed subset of Z is a zero-set or, equivalently, if every open subset of Z is a cozero-set.

Proof. The statements with zero-sets and with cozero-sets are clearly equivalent, so necessity follows immediately from Fact 1 of S.358. Now assume that every closed $F \subset Z$ is a zero-set. Every zero-set is a G_δ-set: this is an easy exercise ,so Z is perfect and we only have to prove normality of Z. Given disjoint closed $F, G \subset Z$ fix $f, g \in C(Z)$ such that $F = f^{-1}(0)$ and $G = g^{-1}(0)$. If $h(z) = \frac{|f(z)|}{|f(z)| + |g(z)|}$ for all $z \in Z$, then h is a continuous function on Z for which we have $h(F) \subset \{0\}$ and $h(G) \subset \{1\}$, so Z is normal and Fact 2 is proved. $\square$

Fact 3. Given any space Z and a second countable space T take any base $\mathcal{C} \subset \tau^*(T)$ in the space T. If any $U \in \tau^*(C_p(Z, T))$ is a countable union of $\mathcal{C}$-standard subsets of $C_p(Z, T)$, then $C_p(Z, T)$ is perfectly normal. Observe that we are stating a stronger fact than sufficiency in Problem 080 because we do not assume that $C_p(Z, T)$ is dense in T^Z.

Proof. $\mathcal{S} = \{[z; O](Z, T, \mathcal{C}) : z \in Z \text{ and } O \in \mathcal{C}\}$. It is evident that every $\mathcal{C}$-standard open subset of $C_p(Z, T)$ is a finite intersection of elements of $\mathcal{S}$. Any $O \in \mathcal{C}$ is a cozero-set because any second countable space is perfectly normal (see Fact 2). Let $\varphi : T \to \mathbb{R}$ be a continuous function such that $F = T \backslash O = \varphi^{-1}(0)$. For any point $z \in Z$ the mapping $\pi_{\{z\}} : C_p(Z, T) \to T$ is continuous being the restriction to $C_p(Z, T)$ of the natural projection of T^Z to its zth factor. Furthermore

$$[z; O](Z, T, \mathcal{C}) = \pi_{\{z\}}^{-1}(O) = \pi_{\{z\}}^{-1}(\varphi^{-1}(\mathbb{R} \backslash \{0\})) = (\varphi \circ \pi_{\{z\}})^{-1}(\mathbb{R} \backslash \{0\}),$$

which shows that $[z; O](Z, T, \mathcal{C})$ is a cozero-set for any $z \in Z$ and $O \in \mathcal{C}$. Observe that any finite intersection of cozero-sets is a cozero-set because any finite union of zero-sets is a zero-set (see Fact 1 of S.499). As a consequence, any $\mathcal{C}$-standard open subset of $C_p(Z, T)$ is a cozero-set. By our assumption every open set is a countable union of $\mathcal{C}$-standard sets which we proved to be cozero-sets of $C_p(Z, T)$. Any countable union of cozero-sets is a cozero-set because any countable intersection of zero-sets is a zero-set (see Fact 1 of S.499). Therefore every open subset of $C_p(Z, T)$ is a cozero-set, so we can apply Fact 2 to conclude that $C_p(Z, T)$ is perfectly normal. Fact 3 is proved. □

Returning to our solution observe that if every open $U \subset C_p(X, Y)$ is a countable union of $\mathcal{B}$-standard open subsets of $C_p(X, Y)$, then Fact 3 is applicable to conclude that $C_p(X, Y)$ is perfectly normal and hence we proved sufficiency.

Now assume that the space $C_p(X, Y)$ is perfectly normal. Given any non-empty open set $U \subset C_p(X, Y)$, there exists a continuous function $\varphi : C_p(X, Y) \to \mathbb{R}$ such that $F = C_p(X, Y) \backslash U = \varphi^{-1}(0)$ (see Fact 2). Since $\mathbb{R}$ is second countable and $C_p(X, Y)$ is a dense subspace of the product Y^X of second countable spaces, we can apply Problem 299 of [TFS] to find a countable set $A \subset X$ and a continuous function $\delta : C_A = \pi_A(C_p(X, Y)) \to \mathbb{R}$ such that $\varphi = \delta \circ \pi_A$. If $G = \delta^{-1}(0)$, then $F = \pi_A^{-1}(G)$ and hence $U = \pi_A^{-1}(V)$ where $V = C_A \backslash G$. The set V is open in the second countable space C_A; since the standard open subsets of C_A (with respect to A, Y and $\mathcal{B}$) form a base in C_A, the set V is a union of these standard open sets. Any second countable space is hereditarily Lindelöf, so there is a countable family $\mathcal{V}$ of standard subsets of C_A (with respect to A, Y and $\mathcal{B}$) such that $\bigcup \mathcal{V} = V$. If $\mathcal{U} = \{\pi_A^{-1}(V) : V \in \mathcal{V}\}$, then $\mathcal{U}$ consists of $\mathcal{B}$-standard subsets of $C_p(X, Y)$ by Fact 1; since $|\mathcal{U}| \leq \omega$ and $\bigcup \mathcal{U} = U$, we represented any open $U \subset C_p(X, Y)$ as a countable union of $\mathcal{B}$-standard open subsets of $C_p(X, Y)$. This proves necessity and completes our solution.

T.081. *Suppose that $C_p(X)$ is perfectly normal. Prove that all closed subsets of $X \times X$ are separable.*

Solution. Given arbitrary spaces Z and T take any base $\mathcal{C}$ in the space T; say that a set $U \in \tau(C_p(Z, T))$ is $\mathcal{C}$-*standard in the space* $C_p(Z, T)$ if there exist $n \in \mathbb{N}$, distinct points $z_1, \ldots, z_n \in Z$ and sets $O_1, \ldots, O_n \in \mathcal{C}$ such that

$$U = [z_1, \ldots, z_n; O_1, \ldots, O_n](Z, T, \mathcal{C})$$

$$= \{f \in C_p(Z, T) : f(z_i) \in O_i \text{ for all } i = 1, \ldots, n\}.$$

Given any $n \in \mathbb{N}$, let $J_i = (0, 1] \times \{i\}$ for each $i \in M_n = \{1, \ldots, n\}$ and $H_n = \bigcup \{J_i : i \in M_n\} \cup \{\theta\}$. For any $x, y \in H_n$, $x = (t, k)$, $y = (s, l)$, let $\rho(x, y) = |t - s|$ if $k = l$. If $k \neq l$, then $\rho(x, y) = t + s$. Let $\rho(x, \theta) = t$, $\rho(\theta, y) = s$ and $\rho(\theta, \theta) = 0$. The space H_n with the topology generated by the metric ρ is called *the Kowalsky hedgehog with n spines*. It is evident that for every number $i \in M_n$, there exists a homeomorphism $h_i : [0, 1] \to J_i \cup \{\theta\}$ such that $h_i(0) = \theta$. Observe that h_i is a continuous map considered as a function from $[0, 1]$ to the space $H_n \supset J_i \cup \{\theta\}$. Observe that the family $\mathcal{B}_n = \{U \in \tau^*(H_n) :$ either $\theta \in U$ or $U \subset J_i$ for some $i \leq n\}$ is a base in the space H_n. Given any space Z, let $\Delta_n(Z) = \{z = (z_1, \ldots, z_n) : \text{there are distinct } i, j \leq n \text{ with } z_i = z_j\}$.

Fact 1. Let Z be an arbitrary space; take any distinct points $z_1, \ldots, z_k \in Z$ and a set $U_i \in \tau(z_i, Z)$ for each $i \leq k$ such that $\overline{U}_i \cap \overline{U}_j = \emptyset$ if $i \neq j$. Suppose also that $n \in \mathbb{N}$ and we are given points $t_1, \ldots, t_k \in H_n \backslash \{\theta\}$. Then there exists a function $f \in C_p(Z, H_n)$ such that $f(Z \backslash \bigcup \{U_i : i \leq k\}) \subset \{\theta\}$, and, for every $i \leq k$, we have $f(z_i) = t_i$ and $f(U_i) \subset J_{q(i)} \cup \{\theta\}$ where $q(i) \in M_n$ is the unique natural number for which $t_i \in J_{q(i)}$.

Proof. The Tychonoff property of Z implies that there exists $g_i \in C_p(Z, [0, 1])$ such that $g_i(Z \backslash U_i) \subset \{0\}$ and $g_i(z_i) = 1$ for all $i \leq k$. Let $r_i = h_{q(i)}^{-1}(t_i)$ for all $i \leq k$; for the function $f_i = r_i \cdot g_i \in C_p(Z, [0, 1])$ we have $f_i(Z \backslash U_i) \subset \{0\}$ and $f_i(z_i) = r_i$ for each $i \leq k$. Finally define a function $f : X \to H_n$ as follows: $f(z) = \theta$ if $z \in Z \backslash (\bigcup \{U_i : i \leq n\})$; if $i \leq n$ and $z \in U_i$, then let $f(z) = h_{q(i)}(f_i(z))$.

Observe first that by our definition of f, we have $f(Z \backslash \bigcup \{U_i : i \leq n\}) \subset \{\theta\}$ and $f(z_i) = h_{q(i)}(f_i(z_i)) = h_{q(i)}(r_i) = t_i$ for all $i \leq k$. Besides, $f | U_i = (h_{q(i)} \circ f_i) | U_i$ which shows that $f(U_i) = h_{q(i)}(f_i(U_i)) \subset h_{q(i)}([0, 1]) = J_{q(i)} \cup \{\theta\}$. To see that f is continuous take any $z \in Z$; since the closures of U_i's are disjoint, there is $G \in \tau(x, X)$ such that G intersects at most one of the sets $U_1, \ldots, U_k$, say U_i. Then $f | G = (h_{q(i)} \circ f_i) | G$ is a continuous function, so f is continuous by Fact 1 of S.472. Fact 1 is proved.   □

Fact 2. For any $n \in \mathbb{N}$ and any space Z the set $C_p(Z, H_n)$ is dense in H_n^Z.

Proof. The space H_n^Z is homeomorphic to the space $C_p(Z', H_n)$ where Z' is the set Z considered with the discrete topology. The family $\mathcal{D} = \tau^*(H_n)$ is, evidently, a base in the space H_n. The $\mathcal{D}$-standard subsets of the space $C_p(Z', H_n)$ form a base in $C_p(Z', H_n)$, so it suffices to show that $C_p(Z, H_n)$ intersects every $\mathcal{D}$-standard subset of $C_p(Z', H_n)$.

To establish this, take any $\mathcal{D}$-standard subset U of the space $C_p(Z', H_n)$. By definition, there exist distinct points $z_1, \ldots, z_k \in Z$ and sets $O_1, \ldots, O_k \in \tau^*(H_n)$ such that $U = \{f \in C_p(Z', H_n) : f(z_i) \in O_i$ for all $i \leq k\}$. Since $H_n \backslash \{\theta\}$ is dense in H_n, we have $O_i \backslash \{\theta\} \neq \emptyset$; pick $t_i \in O_i \backslash \{\theta\}$ for all $i \leq k$. Then $t_i \in H_n \backslash \{\theta\}$ for each $i \leq k$ and hence we can apply Fact 1 to conclude that there is $f \in C_p(Z, H_n)$ such that $f(z_i) = t_i \in O_i$ for all $i \leq k$. As a consequence $f \in U \cap C_p(Z, H_n)$, so $C_p(Z, H_n)$ is dense in $C_p(Z', H_n) = H_n^Z$ and Fact 2 is proved.   □

Fact 3. Given any space Z, let F be a non-empty closed subspace of $Z^n \setminus \Delta_n(Z)$. Let $U_F^z = [z_1, \ldots, z_n; J_1, \ldots, J_n](Z, H_n, \mathcal{B}_n)$ for every $z = (z_1, \ldots, z_n) \in F$. Assume that W is a $\mathcal{B}_n$-standard subset of $C_p(Z, H_n)$, such that $W \subset \bigcup \{U_F^z : z \in F\}$. Then $W = [z_1, \ldots, z_n, y_1, \ldots, y_k; O_1, \ldots, O_n, G_1, \ldots, G_k](Z, H_n, \mathcal{B}_n)$ for some point $z = (z_1, \ldots, z_n) \in F$ and $O_i \subset J_i$ for all $i = 1, \ldots, n$.

Proof. Since W is $\mathcal{B}_n$-standard, there exist $m \in \mathbb{N}$, distinct points $x_1, \ldots, x_m \in Z$ and sets $O_1, \ldots, O_m \in \mathcal{B}_n$ such that $W = [x_1, \ldots, x_m; O_1, \ldots, O_m](Z, H_n, \mathcal{B}_n)$. If $\theta \in O_i$, then let $q(i) = 0$; if $\theta \notin O_i$, then there is a unique $q(i) \in M_n$ such that $O_i \subset J_{q(i)}$. Choose $V_i \in \tau(x_i, Z)$ for all $i \leq m$ in such a way that the family $\{\overline{V}_i : i \leq m\}$ is disjoint. We claim that

(1) for any $l \in M_n$, there is $i \leq m$ such that $q(i) = l$.

Indeed, if this is not the case, then let $N = \{i \in M_m : q(i) \neq 0\}$ and choose any $t_i \in O_i$ for all $i \in N$. By Fact 1 there exists a function $f \in C_p(Z, H_n)$ such that $f(Z \setminus \bigcup\{V_i : i \in N\}) \subset \{\theta\}$, $f(x_i) = t_i$ and $f(V_i) \subset J_{q(i)} \cup \{\theta\}$ for all $i \in N$. Observe that if $q(i) = 0$, then $x_i \notin \bigcup\{V_i : i \in N\}$, so $f(x_i) = \theta \in O_i$ as well and hence $f \in W$. Since $q(i) \neq l$ for all $i \leq m$, we have $f(Z) \cap J_l = \emptyset$. However there is $z = (z_1, \ldots, z_n) \in F$ such that $f \in U_F^z$. The definition of U_F^z requires that $f(z_l) \in J_l$, a contradiction which shows that (1) is true.

It follows from (1) that changing the order of the points $x_1, \ldots, x_m$ and the sets $O_1, \ldots, O_m$ if necessary we can assume that there are $k_1, \ldots, k_n \in \mathbb{N}$ such that $W = [z_1^1, \ldots, z_{k_1}^1, \ldots, z_1^n, \ldots, z_{k_n}^n, y_1, \ldots, y_p;$

$$O_1^1, \ldots, O_{k_1}^1, \ldots, O_1^n, \ldots, O_{k_n}^n, Q_1, \ldots, Q_p](Z, H_n, \mathcal{B}_n),$$

where $\theta \in Q_i$ for all $i = 1, \ldots, p$ and $O_i^l \subset J_l$ for $l = 1, \ldots, n$ and $i = 1, \ldots, k_l$. For every n-tuple $j = (j_1, \ldots, j_n)$ such that $1 \leq j_i \leq k_i$, let $z^j = (z_{j_1}^1, \ldots, z_{j_n}^n)$. If some z^j is in F, then everything is proved. If not, then we can choose $Q_i^l \in \tau(z_i^l, Z)$ for every $l \in M_n$ and $i \in M_{k_l}$ such that the family $\mathcal{Q} = \{\mathrm{cl}_Z(Q_i^l) : l \in M_n, i \in M_{k_l}\}$ is disjoint, $y_1, \ldots, y_p \notin \bigcup \mathcal{Q}$ and for every n-tuple $j = (j_1, \ldots, j_n)$ with $1 \leq j_i \leq k_i$ the set $Q^j = Q_{j_1}^1 \times \cdots \times Q_{j_n}^n$ is disjoint from F. Fact 1 shows that there exists a function $g \in C_p(Z, H_n)$ such that $g(z_i^l) \in O_i^l$, $g^{-1}(J_l) \subset \bigcup\{Q_i^l : i \in M_{k_l}\}$ and $g(X \setminus \bigcup\{Q_i^l : l \in M_n, i \in M_{k_l}\}) \subset \{\theta\}$.

There is $z = (z_1 \ldots, z_n) \in F$ with $g \in U_F^z$. Then $g(z_l) \in J_l$ for each $l \in M_n$ and therefore $z_l \in \bigcup\{Q_i^l : i \in M_{k_l}\}$ i.e., $z_l \in Q_{j_l}^l$ for some $j_l \in M_{k_l}$. Now we have a contradiction for the n-tuple $j = (j_1, \ldots, j_n)$ because $z \in Q^j \cap F = \emptyset$, so Fact 3 is proved. □

Fact 4. If $n \geq 1$ and the space $C_p(Z, H_n)$ is perfectly normal, then all closed subspaces of Z^n are separable.

Proof. For the sake of brevity, call a space T a *CS-space* if all closed subspaces of T are separable. We will prove our fact by induction on n. For the case when $n = 1$, take any non-empty closed $F \subset Z$; for each $z \in F$, consider the set

$U_z = \{f \in C_p(Z, H_1) : f(z) \in J_1\}$. The set $U_F = \bigcup\{U_z : z \in F\}$ is open in $C_p(Z, H_1)$, so we can apply Problem 080 to conclude that there is a countable family $\mathcal{V}$ of $\mathcal{B}_1$-standard subsets of $C_p(Z, H_1)$ such that $\bigcup \mathcal{V} = U_F$. For each $V \in \mathcal{V}$ we have $V = [z_1, \ldots z_k; O_1, \ldots, O_k](Z, H_1, \mathcal{B}_1)$ where $z_i \in Z$ and $O_i \in \mathcal{B}_1$ for all $i \in M_k$; let $\mathrm{supp}(V) = \{z_1, \ldots, z_k\}$. The set $A = \bigcup\{\mathrm{supp}(V) : V \in \mathcal{V}\}$ is countable and hence so is the set $B = A \cap F$. It suffices to show that $\overline{B} = F$.

Assuming the contrary we can find $y \in F \backslash \overline{B}$ and a function $f \in C_p(Z, H_1)$ such that $f(B) \subset \{\theta\}$ and $f(y) \in J_1$. It is clear that $f \in U_y \subset U_F$, so there is $V = [z_1, \ldots z_k; O_1, \ldots, O_k](Z, H_1, \mathcal{B}_1) \in \mathcal{V}$ such that $f \in V$. Apply Fact 3 to conclude that there is $z = z_i \in \mathrm{supp}(V)$ for which $z \in F$ and $O_i \subset J_1$. However, $z_i \in A \cap F = B$ and $f(z_i) = \theta$ by our choice of f, so $f(z_i) \notin O_i$, a contradiction with $f \in V$. Therefore B is dense in F and hence F is separable. This proves that Z is a CS-space, so the case when $n = 1$ is settled.

Assume that $n > 1$ and our fact is proved for all $m < n$; let $\Delta_n = \Delta_n(Z)$. Observe that H_m is a subspace of H_n for all $m < n$ and hence $C_p(Z, H_m)$ is perfectly normal being a subspace of $C_p(Z, H_n)$. The inductive hypothesis implies that X^m is a CS-space for all $m < n$. The space $\Delta_{ij}^n = \{z = (z_1, \ldots, z_n) \in Z^n : z_i = z_j\}$ is homeomorphic to $Z^{M_n \backslash \{i,j\}} \times \Delta_2$ which is homeomorphic to Z^{n-1} because Δ_2 is homeomorphic to Z. Since $\Delta_n = \bigcup\{\Delta_{i,j}^n : i, j \in M_n, i \neq j\}$, the space Δ_n is a finite union of spaces homeomorphic to X^{n-1}. It is immediate that a finite union of CS-spaces is a CS-space, so the induction hypothesis implies that Δ_n is a CS-space. Therefore it suffices to show that $Z^n \backslash \Delta_n$ is a CS-space. We consider that the base $\mathcal{B}_n$ is fixed in H_n.

Let F be a closed set in $Z^n \backslash \Delta_n$. For any $z = (z_1, \ldots, z_n) \in F$ consider the set $U_z = \{f \in C_p(Z, H_n) : f(z_i) \in J_i \text{ for all } i \in M_n\}$. The set $U_F = \bigcup\{U_z : z \in F\}$ is open in $C_p(Z, H_n)$; Fact 2 makes it possible to apply Problem 080 to conclude that there exists a countable family $\mathcal{V}$ of $\mathcal{B}_n$-standard subsets of $C_p(Z, H_n)$ such that $U_F = \bigcup \mathcal{V}$. Let $A = \bigcup\{(\mathrm{supp}(V))^n : V \in \mathcal{V}\}$; it is evident that A is countable, so it suffices to prove that the countable set $B = A \cap F$ is dense in F.

If it is not so, then take any $y = (y_1, \ldots, y_n) \in F \backslash \overline{B}$; for each $i \in M_n$ there exists $W_i \in \tau(y_i, Z)$ such that the family $\{\overline{W}_i : i \in M_n\}$ is disjoint and $W \cap B = \emptyset$ where $W = W_1 \times \cdots \times W_n$. Apply Fact 1 to find $f \in C_p(Z, H_n)$ with $f(y_i) \in J_i$, $f^{-1}(J_i) \subset W_i$ for all $i \in M_n$ and $f(Z \backslash \bigcup\{W_i : i \in M_n\}) = \{\theta\}$. There exists $V \in \mathcal{V}$ such that $f \in V$. Now Fact 3 guarantees existence of distinct $z_1, \ldots, z_n \in \mathrm{supp}(V)$ with $z = (z_1, \ldots, z_n) \in F$ and $f(z_i) \in J_i$ for all $i \in M_n$. But then $z_i \in W_i$ and therefore $z \in W$; since $z \in (\mathrm{supp}(V))^n \subset A \cap F = B$, we have $z \in W \cap B$ which is a contradiction. Fact 4 is proved. $\square$

Returning to our solution, let $h(t) = h_1(t)$ for all $t \in [0, 1]$; if $t \in [-1, 0)$, then let $h(t) = h_2(-t)$. It is an easy exercise that the map $h : \mathbb{I} \to H_2$ is a homeomorphism and hence H_2 is homeomorphic to $\mathbb{I}$. As a consequence, the space $C_p(X, H_2)$ is homeomorphic to a subspace of $C_p(X)$. Thus perfect normality of $C_p(X)$ implies perfect normality of $C_p(X, H_2)$, so $X \times X$ is a CS-space by Fact 4 and hence our solution is complete.

T.082. *Let X be a compact space with $C_p(X)$ perfectly normal. Prove that $X \times X$ is hereditarily separable.*

Solution. For the sake of brevity we say that a space Z is a CS-*space* if all closed subspaces of Z are separable. Our solution will be based on two simple statements.

Fact 1. If Z is a CS-space, then $s(Z) = \omega$.

Proof. Indeed, if $D \subset Z$ is uncountable and discrete, then $F = \overline{D}$ is closed in Z, so there is a countable dense $A \subset F$. However, D is open in F, so for any $d \in D \backslash A$ (which exists because A is countable and D is not), the set $U = \{d\}$ is non-empty, open in F and $U \cap A = \emptyset$; this contradiction completes the proof of Fact 1. □

Fact 2. If K is a compact CS-space, then K is hereditarily separable.

Proof. Observe first that any free sequence in K is a discrete subset of K; we have $s(K) = \omega$ by Fact 1, so all free sequences in K are countable and hence $t(K) = \omega$ (see Problem 328 of [TFS]). Take any $Y \subset K$; since $\overline{Y}$ is separable, we can take a countable $A \subset \overline{Y}$ with $\overline{A} = \overline{Y}$. It follows from $t(K) = \omega$ that for any $a \in A$, there is a countable $B_a \subset Y$ such that $a \in \overline{B_a}$. The set $B = \bigcup\{B_a : a \in A\} \subset Y$ is countable and $\overline{B} \supset \overline{A} = \overline{Y} \supset Y$. Therefore B is a countable dense subset of Y and we proved that every subspace of K is separable. Thus K is hereditarily separable and Fact 2 is proved. □

Returning to our solution, observe that all closed subspaces of $X \times X$ are separable by Problem 081, so we can apply Fact 2 to conclude that $X \times X$ is hereditarily separable.

T.083. *Let X be a compact space with $C_p(X)$ perfectly normal. Prove that under MA+¬CH, the space X is metrizable.*

Solution. Perfect normality of $C_p(X)$ implies that all closed subspaces of $X \times X$ are separable (see Problem 081). Therefore $s(X \times X) = \omega$ by Fact 1 of T.082. Under Martin's axiom and the negation of CH the countability of the spread of $X \times X$ implies metrizability of X (see Problem 062).

T.084. *Prove that the following properties are equivalent for any space X:*

(i) $C_p(X) \times C_p(X)$ is perfectly normal;
(ii) $(C_p(X))^n$ is perfectly normal for any natural $n \geq 1$;
(iii) $(C_p(X))^\omega$ is perfectly normal;
(iv) $C_p(X, Y)$ is perfectly normal for any second countable space Y;
(v) for every $n \in \mathbb{N}$, all closed subsets of X^n are separable.

Solution. For the sake of brevity we say that a space Z is a CS-*space* if all closed subspaces of Z are separable. Given arbitrary spaces Z and T, take any base $\mathcal{C}$ in the space T which is invariant under finite intersections, i.e., the intersection of finitely many elements of $\mathcal{C}$ is still an element of $\mathcal{C}$. A set $U \in \tau(C_p(Z, T))$ is called $\mathcal{C}$-*standard in the space* $C_p(Z, T)$ if there exist $n \in \mathbb{N}$, distinct points $z_1, \ldots, z_n \in Z$ and sets $O_1, \ldots, O_n \in \mathcal{C}$ such that $U = [z_1, \ldots, z_n; O_1, \ldots, O_n] = \{f \in C_p(Z, T) : f(z_i) \in O_i$ for all $i = 1, \ldots, n\}$. Observe that if we omit the condition saying the points $z_1, \ldots, z_n$ in the definition of U are distinct, then we obtain the same concept because if $z_i = z_j$ then considering $O_i \cap O_j$ instead of O_i for the point z_i and omitting z_j and O_j we obtain the same set U. After finitely many of such transformations we obtain the same set U written as $[y_1, \ldots, y_k; W_1, \ldots, W_k]$ where $y_1, \ldots, y_k$ are distinct and $W_1, \ldots, W_k \in \mathcal{C}$. That is why, in this solution, *we do not assume that the points $z_1, \ldots, z_n$ in the definition of U are distinct*.

It is evident that (iii)$\Longrightarrow$(ii)$\Longrightarrow$(i). Since the space $(C_p(X))^\omega$ is homeomorphic to $C_p(X, \mathbb{R}^\omega)$, we have (iv)$\Longrightarrow$(iii).

Now assume that (v) holds; take any second countable space Y and any countable base $\mathcal{B} \subset \tau^*(Y)$ of the space Y which is invariant under finite intersections. Fix any $U \in \tau^*(C_p(X, Y))$; given any $n \in \mathbb{N}$ and any $B = (B_1, \ldots, B_n) \in \mathcal{B}^n$, the set

$$P(B, n) = \{x = (x_1, \ldots, x_n) \in X^n : W(x, B) = [x_1, \ldots, x_n; B_1, \ldots, B_n] \subset U\}$$

is closed in X^n. Indeed, if $y = (y_1, \ldots, y_n) \in X^n \backslash P(B, n)$, then there is a function $f \in W(y, B)$ such that $f \notin U$. The set $O_f = f^{-1}(B_1) \times \cdots \times f^{-1}(B_n)$ is open in X^n and $y \in O_f$. If $z = (z_1, \ldots, z_n) \in O_f$, then $f(z_i) \in B_i$ for all $i \leq n$ and hence $f \in W(z, B)$; since $f \notin U$, we proved that $W(z, B)$ is not contained in U and hence $z \notin P(B, n)$. The point $z \in O_f$ was chosen arbitrarily and therefore $O_f \cap P(B, n) = \emptyset$ which proves that $P(B, n)$ is closed in X^n. Since (v) is fulfilled for X, we can choose, for all $n \in \mathbb{N}$ and all $B \in \mathcal{B}^n$, a countable set $Q(B, n) \subset P(B, n)$ which is dense in $P(B, n)$. The family $\mathcal{W} = \{W(x, B) : x \in Q(B, n)$ for some $n \in \mathbb{N}$ and $B \in \mathcal{B}^n\}$ is countable and consists of $\mathcal{B}$-standard subsets of $C_p(X, Y)$.

If $g \in U$, then, by the definition of the topology of pointwise convergence, there are $n \in \mathbb{N}$, $x = (x_1, \ldots, x_n) \in X^n$ and $B = (B_1, \ldots, B_n) \in \mathcal{B}^n$ such that $g \in W(x, B) \subset U$; it follows from the inclusion $W(x, B) \subset U$ that $x \in P(B, n)$. The set $V = g^{-1}(B_1) \times \cdots \times g^{-1}(B_n)$ is an open neighborhood of x in X^n; since $Q(B, n)$ is dense in $P(B, n)$, there exists $y \in Q(B, n) \cap V$. It is clear that $g \in W(y, B) \subset U$ which shows that $W(y, B) \in \mathcal{W}$ and hence $g \in \bigcup \mathcal{W}$. The function $g \in U$ was chosen arbitrarily, so $\bigcup \mathcal{W} = U$ and we proved that every $U \in \tau^*(C_p(X, Y))$ is a countable union of $\mathcal{B}$-standard subsets of $C_p(X, Y)$. Applying Fact 3 of T.080 we conclude that $C_p(X, Y)$ is perfectly normal and hence (v)$\Longrightarrow$(iv).

Finally, assume that $C_p(X) \times C_p(X)$ is perfectly normal. Since $(C_p(X))^2$ is homeomorphic to $C_p(X, \mathbb{R}^2)$, the space $C_p(X, \mathbb{R}^2)$ is also perfectly normal. For every $n \in \mathbb{N}$ the space H_n is homeomorphic to a subset of $\mathbb{R}^2$ by Fact 1 of T.019 and hence $C_p(X, H_n)$ is homeomorphic to a subspace of $C_p(X, \mathbb{R}^2)$. Therefore $C_p(X, H_n)$ is perfectly normal for each $n \in \mathbb{N}$, so we can apply Fact 4 of T.081 to conclude that X^n is a CS-space for all $n \in \mathbb{N}$, i.e., the property (v) holds. We proved that (i)$\Longrightarrow$(v) so our solution is complete.

T.085. *Prove that for any compact space X, the space $C_p(X) \times C_p(X)$ is perfectly normal if and only if $(C_p(X))^\omega$ is hereditarily Lindelöf.*

Solution. For the sake of brevity we say that a space Z is *a CS-space* if all closed subspaces of Z are separable. If $C_p(X) \times C_p(X)$ is perfectly normal, then X^n is a CS-space for all $n \in \mathbb{N}$ by Problem 084. Since X is compact, we can apply Fact 2 of T.082 to conclude that $hd(X^n) = \omega$ for all $n \in \mathbb{N}$. Therefore $hl^*(C_p(X)) = \omega$ by Problem 027 and hence $hl((C_p(X))^\omega) = \omega$ by Problem 011.

T.086. *Prove that, under SA, if $C_p(X)$ is perfectly normal, then $(C_p(X))^\omega$ is hereditarily Lindelöf.*

Solution. If $C_p(X)$ is perfectly normal, then $\mathrm{ext}(C_p(X)) = \omega$ by Problem 295 of [TFS] (in fact, only normality suffices for this). If $D \subset C_p(X)$ is an uncountable discrete subspace, then D is open in $F = \overline{D}$; perfect normality of F implies that $D = \bigcup\{D_n : n \in \omega\}$ where each D_n is closed in F and hence in $C_p(X)$. As a consequence, some D_n is a closed uncountable subspace of $C_p(X)$, a contradiction with $\mathrm{ext}(C_p(X)) = \omega$. This proves that $s(C_p(X)) = \omega$. Now, if SA holds, we can apply Problem 036 to conclude that $(C_p(X))^\omega$ is hereditarily Lindelöf.

T.087. *Let X be a space with a G_δ-diagonal. Prove that $C_p(X)$ is perfectly normal if and only if $(C_p(X))^\omega$ is perfectly normal.*

Solution. For the sake of brevity we say that a space Z is *a CS-space* if all closed subspaces of Z are separable. For any $n \in \mathbb{N}$, let $M_n = \{1, \ldots, n\}$; denote by S_n the family of all bijections from M_n to M_n. By $\mathrm{id} : M_n \to M_n$, we denote the identity map. If Z is a space and $n \in \mathbb{N}$, then $\Delta_n(Z) = \{z = (z_1, \ldots, z_n) \in Z^n : z_i = z_j$ for some distinct $i, j \in M_n\}$. If $\sigma \in S_n$, then we have a map $\varphi_\sigma^Z : Z^n \to Z^n$ defined by $\varphi_\sigma^Z(z) = (z_{\sigma(1)}, \ldots, z_{\sigma(n)})$ for any $z = (z_1, \ldots, z_n) \in Z^n$. It is clear that φ_σ^Z is a homeomorphism of Z^n onto Z^n such that $\varphi_\sigma^Z(Z^n \backslash \Delta_n(Z)) = Z^n \backslash \Delta_n(Z)$. Recall that given a space Z and $A \subset Z$, the pseudocharacter $\psi(A, Z)$ of A in Z is the minimal cardinality of a family $\mathcal{U} \subset \tau(Z)$ such that $\bigcap \mathcal{U} = A$.

Call an open set $U \subset C_p(Z, [0, 1])$ *standard* in the space $C_p(Z, [0, 1])$ if there exist $k \in \mathbb{N}$, distinct points $z_1, \ldots, z_k \in Z$ and sets $O_1, \ldots, O_k \in \tau^*([0, 1])$ such that $U = [z_1, \ldots, z_k; O_1, \ldots, O_k] = \{f \in C_p(Z, [0, 1]) : f(x_i) \in O_i$ for all $i \in M_k\}$; let $\mathrm{supp}(U) = \{z_1, \ldots, z_k\}$.

Fact 1. If $C_p(Z)$ is perfectly normal, then $s(C_p(Z)) = \omega$. In particular, if Z is perfectly normal and has a G_δ-diagonal, then $s^*(Z) = \omega$.

Proof. We have $\mathrm{ext}(C_p(Z)) = \omega$ by Problem 295 of [TFS] (in fact, only normality suffices for this). If $D \subset C_p(Z)$ is an uncountable discrete subspace, then D is open in $F = \overline{D}$; perfect normality of F implies that $D = \bigcup\{D_n : n \in \omega\}$ where each D_n is closed in F and hence in $C_p(Z)$. As a consequence, some D_n is a closed uncountable discrete subspace of $C_p(Z)$; this contradiction with $\mathrm{ext}(C_p(X)) = \omega$ shows that $s(C_p(Z)) = \omega$. If Z has a G_δ-diagonal, then $s^*(Z) = s^*(C_p(Z)) = s(C_p(Z)) = \omega$ by Problem 028 so Fact 1 is proved. □

Fact 2. Assume that $C_p(Z)$ is perfectly normal. If $n \in \mathbb{N}$ and a set $P \subset Z^n \backslash \Delta_n(Z)$ is closed in Z^n, then P is separable.

Proof. We will first prove a weaker fact, namely, that

(*) if F is a closed subspace of Z^n such that $F \subset Z^n \backslash \Delta_n(Z)$ and $\varphi_\sigma^Z(F) \cap F = \emptyset$
for any $\sigma \in S_n \backslash \{\text{id}\}$, then F is separable.

It is clear that the space $C_p(Z, [0, 1])$ is also perfectly normal. For an arbitrary
point $z = (z_1, \ldots, z_n) \in F$, let $U_z = \{f \in C_p(Z, [0, 1]) : f(z_i) > 0$ for all
$i \in M_n\}$. We claim that the set $U_F = \bigcup\{U_z : z \in F\}$ has the following property:

(1) if $W = [z_1, \ldots, z_p; Q_1, \ldots, Q_p]$ is an arbitrary standard subset of the space
$C_p(Z, [0, 1])$ such that $W \subset U_F$, then there are numbers $i_1, \ldots, i_n \in M_p$ such
that $(z_{i_1}, \ldots, z_{i_n}) \in F$ and $O_{i_k} \subset (0, 1]$ for all $k \in M_n$.

Suppose that $W = [z_1, \ldots, z_p; Q_1, \ldots, Q_p]$ is a standard subset of $C_p(Z, [0, 1])$
for which $W \subset U_F$ and (1) does not hold. Denote by f_0 the function which is
identically zero on Z; then $f_0 \notin U_F$ because otherwise $f_0 \in U_t$ for some point
$t = (t_1, \ldots, t_n) \in F$ and hence $f_0(t_i) > 0$ for all $i \in M_n$ which is a contradiction.
However, if $0 \in Q_i$ for all $i \in M_p$, then $f_0 \in W$ which is impossible because
$W \subset U_F$. This shows that we can assume without loss of generality that there is
some $r \in M_p$ such that $Q_1, \ldots, Q_r \subset (0, 1]$ and $0 \in Q_i$ for all $i \in M_p \backslash M_r$.

Since the condition (1) is not satisfied for the set W, we have $(z_{i_1}, \ldots, z_{i_n}) \notin F$
for any $\{z_{i_1}, \ldots, z_{i_n}\} \subset \{z_1, \ldots, z_r\}$. The set F is closed in Z^n, so we can choose
$G_i \in \tau(z_i, Z)$ for all $i \in M_p$ in such a way that the family $\{G_i : i \in M_p\}$ is disjoint
and

$$(G_{i_1} \times \cdots \times G_{i_n}) \cap F = \emptyset \text{ if } \{i_1, \ldots, i_n\} \subset M_r.$$

It is easy to see that there exists $f \in C_p([0, 1])$ such that $f(z_i) \in Q_i$ for all
$i \in M_r$ and $f(Z \backslash \bigcup\{G_i : i \in M_r\}) \subset \{0\}$. It is clear that $f \in W$, so there
is $t = (t_1, \ldots, t_n) \in F$ such that $f \in U_t$. By definition of the set U_t, we have
$f(t_i) > 0$ for all $i \in M_n$; therefore $\{t_1, \ldots, t_n\} \subset \bigcup\{G_i : i \in M_r\}$ which implies
that for each $k \in M_n$, there is $i_k \in M_r$ for which $t_k \in G_{i_k}$. As a consequence
$t \in (G_{i_1} \times \cdots \times G_{i_n}) \cap F$ contradicting the choice of the sets G_i. Thus (1) is true.

Now apply Problem 080 to conclude that there is a family $\{W_k : k \in \omega\}$ of
standard subsets of $C_p(Z, [0, 1])$ such that $U_F = \bigcup\{W_k : k \in \omega\}$. Since the set
$B = \bigcup\{\text{supp}(W_k) : k \in \omega\}$ is countable, so is the set $A = B^n$. Thus it suffices to
show that $A \cap F$ is dense in F.

Assume that this is not true and fix any point $y = (y_1, \ldots, y_n) \in F \backslash \overline{A \cap F}$.
It follows from the hypothesis in (*) that $\{\varphi_\sigma^Z(y) : \sigma \in S_n\} \cap \overline{A \cap F} = \emptyset$ and
therefore there exist disjoint sets $H_1, \ldots, H_n \in \tau(Z)$ such that $y_i \in H_i$ for all
$i \in M_n$ and $(H_{j_1} \times \cdots \times H_{j_n}) \cap (A \cap F) = \emptyset$ whenever $\{j_1, \ldots, j_n\} \subset M_n$.
Choose any function $g \in C_p(Z, [0, 1])$ such that $g(Z \backslash \bigcup\{H_i : i \in M_n\}) \subset \{0\}$ and
$g(y_i) = 1$ for all $i \in M_n$. Then $g \in U_y \subset U_F$ and therefore there exists $k \in \omega$ for
which $g \in W_k$. The property (1) implies that $W_k = [z_1, \ldots, z_p; Q_1, \ldots, Q_p]$ where
$z = (z_{i_1}, \ldots, z_{i_n}) \in F$ for some $i_1, \ldots, i_n \in M_p$ and $Q_{i_k} \subset (0, 1]$ for all $k \in M_n$.

Given any $k \in M_n$, we have $g(z_{i_k}) > 0$, so there exists $j_k \in M_n$ such that $z_{i_k} \in H_{j_k}$. Consequently, $z \in (H_{j_1} \times \cdots \times H_{j_n}) \cap (A \cap F)$, and this contradiction completes the proof of $(*)$.

Now let us prove first that P is locally separable, i.e.,

(2) for each $z = (z_1, \ldots, z_n) \in P$, there is a separable $W_z \in \tau(z, P)$.

Since all points $z_1, \ldots, z_n$ are distinct, we can choose a set $V_i \in \tau(z_i, Z)$ for every $i \in M_n$ in such a way that $\overline{V}_i \cap \overline{V}_j = \emptyset$ if $i \neq j$. Observe that we have $V = V_1 \times \cdots \times V_n \in \tau(z, Z^n)$ and $\varphi_\sigma^Z(\overline{V}) \cap \overline{V} = \emptyset$ for each $\sigma \in S_n \setminus \{id\}$. Therefore the set $\overline{V} \cap P$ satisfies the conditions given in the hypothesis of $(*)$ and hence $(*)$ can be applied to conclude that $\overline{V} \cap P$ is separable. The set $W_z = V \cap P$ is open in P and $z \in W_z$. Besides, W_z is open in a separable smaller subspace $\overline{V} \cap P$, so W_z is also separable which shows that we proved (2).

Let $\mathcal{W} = \{W_z : z \in P\}$; then $\mathcal{W}$ is an open cover of P with separable subspaces of P. Observe also that $s(C_p(Z)) = \omega$ by Fact 1 and apply Fact 1 of T.028 to conclude that $s(P) = \omega$ as well. By Fact 1 of T.007 we can find a countable $\mathcal{W}' \subset \mathcal{W}$ and a discrete (and hence countable) set $D \subset P$ such that $\bigcup \mathcal{W}' \cup \overline{D} = P$. A countable union of separable spaces is separable, so there is a countable $A \subset \bigcup \mathcal{W}'$ which is dense in $\bigcup \mathcal{W}'$. Therefore $A \cup D$ is a countable dense subset of P, so P is separable and Fact 2 is proved. $\square$

Fact 3. Given an infinite cardinal κ, if Z is a space with $\psi(\Delta_2(Z), Z) \leq \kappa$, then $\psi(\Delta_n(Z), Z^n) \leq \kappa$ for any $n \geq 3$. In particular, if Z has a G_δ-diagonal, then $\Delta_n(Z)$ is a G_δ-subset of Z^n for all $n \geq 2$.

Proof. Fix $n \in \mathbb{N}$ with $n \geq 3$; take an arbitrary family $\mathcal{U} \subset \tau(Z \times Z)$ such that $|\mathcal{U}| \leq \kappa$ and $\bigcap \mathcal{U} = \Delta_2(Z)$. Given distinct $i, j \in M_n$, let $q_{ij} : Z^n \to Z \times Z$ be the natural projection onto the face defined by i and j, i.e., for any $z = (z_1, \ldots, z_n) \in Z^n$ we have $q_{ij}(z) = (z_i, z_j) \in Z \times Z$. It is an immediate consequence from the definition of q_{ij} that $\Delta_{ij}^n(Z) = \{z = (z_1, \ldots, z_n) \in Z^n : z_i = z_j\} = q_{ij}^{-1}(\Delta_2(Z))$ and therefore $\Delta_{ij}^n(Z) = \bigcap \mathcal{U}_{ij}$ where $\mathcal{U}_{ij} = \{q_{ij}^{-1}(U) : U \in \mathcal{U}\}$. If $B_n = \{(i, j) \in M_n \times M_n : i < j\}$ then the family $\mathcal{V} = \{U = \bigcup \{U_{ij} : (i, j) \in B_n\} : U_{ij} \in \mathcal{U}_{ij}$ for all $(i, j) \in B_n\}$ consists of open subsets of Z^n and $\bigcap \mathcal{V} = \Delta_n(Z)$. Since it is evident that $|\mathcal{V}| \leq \kappa$, Fact 3 is proved. $\square$

Returning to our solution, observe that we must only prove that perfect normality of $C_p(X)$ implies perfect normality of $(C_p(X))^\omega$. By Problem 084 it suffices to show that if $C_p(X)$ is perfectly normal, then X^n is a CS-space for any $n \in \mathbb{N}$. By Problem 081 this is true for $n = 2$. Assume that $n > 2$ and we proved CS-property of X^m for all $m < n$; let $\Delta_n = \Delta_n(X)$. The space $\Delta_{ij}^n = \{x = (x_1, \ldots, x_n) \in X^n : x_i = x_j\}$ is homeomorphic to $X^{M_n \setminus \{i, j\}} \times \Delta_2$ which is homeomorphic to X^{n-1} because Δ_2 is homeomorphic to X. Since $\Delta_n = \bigcup \{\Delta_{ij}^n : i, j \in M_n, i \neq j\}$, the space Δ_n is a finite union of spaces homeomorphic to X^{n-1}. It is immediate that a finite union of CS-spaces is a CS-space, so the induction hypothesis implies that Δ_n is a CS-space. Therefore it suffices to show that $X^n \setminus \Delta_n$ is a CS-space.

Take any closed $F \subset X^n \backslash \Delta_n$. Fact 3 implies that Δ_n is a G_δ-subset of X^n; let $\mathcal{U} \subset \tau(X^n)$ be a countable family with $\bigcap \mathcal{U} = \Delta_n$. For each $U \in \mathcal{U}$ the set $F_U = F \backslash U$ is separable being a closed subset of X^n (see Fact 2). It is evident that $F = \bigcup \{F_U : U \in \mathcal{U}\}$, so F is a countable union of its separable subspaces. Thus F is separable so our solution is complete.

T.088. *Prove that, under MA+¬CH, all closed subspaces of $C_p(X)$ are separable if and only if $(C_p(X))^\omega$ is hereditarily separable.*

Solution. For the sake of brevity we say that a space Z is *a CS-space* if all closed subspaces of Z are separable. It suffices to show that, under MA+¬CH, if $C_p(X)$ is a CS-space, then $(C_p(X))^\omega$ is hereditarily separable. The space $C_p(X)$ is separable being a closed subspace of itself; therefore $iw(X) = \omega$ and hence $\Delta(X) = \omega$. Another observation we need is that $s^*(X) = \omega$ by Fact 1 of T.087. Now, since MA+¬CH holds, we have $hl^*(X) = \omega$ by Problem 059. As a consequence, we can apply Problems 026 and 012 to conclude that $hd((C_p(X))^\omega) = \omega$.

T.089. *Prove that, under CH, there exists a subspace X of $\{0, 1\}^{\omega_1}$ such that for all $n \in \mathbb{N}$, all closed subsets of X^n are separable and X is not hereditarily separable. Therefore, $C_p(X)$ is a perfectly normal non-Lindelöf space.*

Solution. If Z is a space, then $A \subset Z$ is called *a meager* subset of Z if A is of first category in Z, i.e., A is a countable union of nowhere dense subsets of Z. For any $n \in \mathbb{N}$ let $\Delta_n(Z) = \{z = (z_1, \ldots, z_n) \in Z^n : z_i = z_j$ for some distinct $i, j \in M_n = \{1, \ldots, n\}\}$. Of course, $\Delta_1(Z) = \emptyset$. Let $\mathbb{D}$ be the set $\{0, 1\}$ endowed with the discrete topology. A family $\mathcal{N}$ of nowhere dense subsets of Z is called *cofinal* in Z if, for every nowhere dense $P \subset Z$, there is $N \in \mathcal{N}$ such that $P \subset N$. Say that a set $P \subset Z$ is *somewhere dense in Z* if $\text{Int}(\overline{P}) \neq \emptyset$.

Fact 1. Suppose that Z is any space and Y is a space with $\pi w(Y) = \omega$. Given a set $A \subset Z \times Y$, let $A_z = \{y \in Y : (z, y) \in A\}$. Assume that $P \subset Z \times Y$ is a nowhere dense (meager) set; then there is a meager set $M(P) \subset Z$ such that P_z is a nowhere dense (meager) set in Y for any $z \in Z \backslash M(P)$.

Proof. Fix any π-base $\mathcal{B} = \{B_n : n \in \omega\}$ in the space Y. Assume first that P is nowhere dense in $Z \times Y$ and let $A_n = \{z \in Z : \overline{P_z} \supset B_n\}$ for all $n \in \omega$. If $A = \bigcup \{A_n : n \in \omega\}$ and $z \in Z \backslash A$, then $\overline{P_z}$ contains no B_n in its closure and hence P_z is nowhere dense in Y. If some A_n is not nowhere dense, then there is some $U \in \tau^*(Z)$ with $\text{cl}_Z(A_n) \supset U$. Then $U \times B_n \subset \text{cl}_{Z \times Y}(P)$; to see it, take any point $p = (u, b) \in U \times B_n$. Given any basic neighborhood $W = G \times H \in \tau(p, Z \times Y)$, there is $z \in G \cap A_n$ because $U \subset \text{cl}_Z(A_n)$. Since $\text{cl}_Y(P_z) \supset B_n$, we have $\text{cl}_{Z \times Y}(\{z\} \times P_z) \supset \{z\} \times B_n$. Thus (z, b) is in the closure of $\{z\} \times P_z$, so $(Z \times H) \cap (\{z\} \times P_z) \neq \emptyset$ and hence there is $t \in P_z \cap H$. It is immediate that $(z, t) \in (G \times H) \cap P$ which shows that $p \in \text{cl}_{Z \times Y}(P)$. As a consequence, a non-empty open set $U \times B_n$ is in the closure of P which is a contradiction. Thus our fact is proved for a nowhere dense set P.

Now, if $P = \bigcup_{n \in \omega} P_n$ where each P_n is nowhere dense, then there is a meager set $M(P_n)$ in Z such that $(P_n)_z$ is a nowhere dense set for each $z \in Z \backslash M(P_n)$. It is clear that $M(P) = \bigcup \{M(P_n) : n \in \omega\}$ is a meager subset of Z, and if $z \in Z \backslash M(P)$, then $P_z = \bigcup \{(P_n)_z : n \in \omega\}$ is a meager set because so is $(P_n)_z$ for every $n \in \omega$. Fact 1 is proved.   □

Fact 2. Suppose that Y is a product of second countable spaces. If we have a space Z such that $c(Z \times Y) = \omega$, then for any nowhere dense (meager) set $P \subset Z \times Y$ there is a meager set $M(P) \subset Z$ such that $P_z = \{y \in Y : (z, y) \in P\}$ is a nowhere dense (meager) set in Y for any $z \in Z \backslash M(P)$.

Proof. Assume first that P is a nowhere dense subspace of $Z \times Y$. By our hypothesis $Y = \prod \{Y_t : t \in T\}$ where Y_t is second countable for any $t \in T$. Given any set $S \subset T$, let $Y_S = \prod \{Y_t : t \in S\}$ and let $q_S : Y \to Y_S$ and $\pi_S : Z \times Y \to Z \times Y_S$ be the respective natural projections.

Call an open subset W of the product $Z \times Y$ *a standard subset of* $Z \times Y$ if $W = W_0 \times U$ for some $W_0 \in \tau(Z)$ and $U = \prod_{t \in T} U_t$ such that $U_t \in \tau(Y_t)$ for all $t \in T$ and $U_t = Y_t$ for all but finitely many t's; let $\text{supp}(W) = \{t \in T : U_t \neq Y_t\}$. It is evident that the family $\mathcal{B}$ of all standard subsets of $Z \times Y$ is a base in $Z \times Y$.

Let $\mathcal{U}$ be a maximal disjoint subfamily of the collection $\{B \in \mathcal{B} : B \cap P = \emptyset\}$. Since P is nowhere dense, the set $G = \bigcup \mathcal{U}$ is dense in $Z \times Y$. Besides, $\mathcal{U}$ is countable because $c(Z \times Y) = \omega$. Therefore the set $S = \bigcup \{\text{supp}(W) : W \in \mathcal{U}\}$ is also countable. Observe that the set $Q = \pi_S(P)$ is nowhere dense in $Z \times Y_S$. Indeed, $\pi_S^{-1}(\pi_S(W)) = W$ for every $W \in \mathcal{U}$ and therefore $\pi_S(W) \cap Q = \emptyset$ for each $W \in \mathcal{U}$ which in turn implies $Q \cap \pi_S(G) = \emptyset$. The map π_S is open and continuous (see Problem 107 of [TFS]), so $\pi_S(G)$ is an open dense subset of $Z \times Y_S$ contained in $(Z \times Y_S) \backslash Q$ which shows that Q is nowhere dense in $Z \times Y_S$. Since the space Y_S is second countable, we can apply Fact 1 to conclude that there is a meager set $M(P) \subset Z$ such that Q_z is nowhere dense in Y_S for any $z \in Z \backslash M(P)$. It is an easy exercise that any inverse image of a nowhere dense set under an open map is a nowhere dense set, so $q_S^{-1}(Q_z)$ is nowhere dense in Y. On the other hand, $P_z \subset q_S^{-1}(Q_z)$ for each $z \in Z$ which shows that P_z is nowhere dense in Y for any $z \in Z \backslash M(P)$, so our fact is proved for a nowhere dense P.

Now, if $P = \bigcup_{n \in \omega} P_n$ where each P_n is nowhere dense, then there is a meager set $M(P_n)$ in Z such that $(P_n)_z$ is a nowhere dense set for each $z \in Z \backslash M(P_n)$. It is clear that $M(P) = \bigcup \{M(P_n) : n \in \omega\}$ is a meager subset of Z, and if $z \in Z \backslash M(P)$, then $P_z = \bigcup \{(P_n)_z : n \in \omega\}$ is a meager set because so is $(P_n)_z$ for every $n \in \omega$. Fact 2 is proved.   □

Fact 3. Suppose that $Z = \{z_\alpha : \alpha < \omega_1\}$ is a faithfully indexed separable space (recall that *faithfully indexed* means $z_\alpha \neq z_\beta$ whenever $\alpha \neq \beta$). Given any $n \in \mathbb{N}$ and a point $z = (z_{\beta_1}, \ldots, z_{\beta_n}) \in Z^n$, let $\max(z) = \max\{\beta_1, \ldots, \beta_n\}$ and $\min(z) = \min\{\beta_1, \ldots, \beta_n\}$. An ω_1-sequence $P = \{p_\alpha : \alpha < \omega_1\} \subset Z^n$ is called *increasing* if $\alpha < \beta$ implies $\max(p_\alpha) < \min(p_\beta)$. Assume that for any $n \in \mathbb{N}$, any increasing ω_1-sequence $P \subset Z^n \backslash \Delta_n(Z)$ is somewhere dense in Z^n. Then Z^n is a CS-space for all $n \in \mathbb{N}$.

Proof. For each ordinal $\alpha < \omega_1$, let $Z_\alpha = \{z_\beta : \beta < \alpha\}$. We will prove this fact by induction on n. If $n = 1$, then $\Delta_n(Z) = \emptyset$ and every uncountable subset of Z contains an ω_1-sequence which is increasing. Thus every uncountable set is somewhere dense and therefore every nowhere dense subset of Z is countable (observe that we cannot say that Z is Luzin because it might have isolated points). Given any closed $F \subset Z$, the set $U = \text{Int}(F)$ has a countable dense set A because Z is separable. The set $F \backslash U$ is nowhere dense and hence countable; this shows that $(F \backslash U) \cup A$ is a countable set which is dense in F. This proves that Z is a CS-space.

Now assume that $n > 2$ and we proved CS-property of Z^m for all $m < n$; let $\Delta_n = \Delta_n(Z)$. For each $i \in M_n$ we denote by $p_i : Z^n \to Z$ the natural projection of Z^n onto its ith factor. The space $\Delta_{ij}^n = \{z = (z_1, \ldots, z_n) \in Z^n : z_i = z_j\}$ is homeomorphic to $Z^{M_n \backslash \{i,j\}} \times \Delta_2$ which is homeomorphic to Z^{n-1} because Δ_2 is homeomorphic to Z. Since $\Delta_n = \bigcup \{\Delta_{ij}^n : i, j \in M_n, i \neq j\}$, the space Δ_n is a finite union of spaces homeomorphic to Z^{n-1}. It is immediate that a finite union of CS-spaces is a CS-space, so the induction hypothesis implies that Δ_n is a CS-space.

Take any closed $F \subset Z^n$ and let $\mathcal{U} = \{U \in \tau(Z^n) : U \subset \overline{P}$ for some increasing ω_1-sequence $P \subset F \backslash \Delta_n\}$. It is clear that $G = \bigcup \mathcal{U}$ is an open set in Z^n contained in F. Since Z is separable, so is Z^n and hence there is a countable $A \subset G$ with $G \subset \overline{A}$. The set $F \backslash G$ is closed in Z^n and there is no increasing ω_1-sequence $P \subset (F \backslash G) \backslash \Delta_n$ because otherwise $W = \text{Int}(\overline{P}) \neq \emptyset$ and hence $W \subset G$, a contradiction. Thus there exists $\alpha < \omega_1$ such that $(F \backslash G) \backslash \Delta_n \subset H = \bigcup \{p_i^{-1}(Z_\alpha) : i \in M_n\}$ which shows that $F \backslash G \subset H \cup \Delta_n$. It is evident that H is a countable union of spaces homeomorphic to Z^{n-1}; we observed already that Δ_n is a finite union since of spaces homeomorphic to Z^{n-1}. Since any countable union of CS-spaces is a CS-space, this proves that $H \cup \Delta_n$ is a CS-space, so the closed subspace $F \backslash G$ of the space $H \cup \Delta_n$ has a dense countable subset B. It is clear that $A \cup B$ is a countable dense subset of F, so Fact 3 is proved. $\square$

Recall that $\Sigma = \{x \in \mathbb{D}^{\omega_1} : |x^{-1}(1)| \leq \omega\}$ is a countably compact dense subspace of the space $\mathbb{D}^{\omega_1}$; besides, Σ is not separable because the closure of any countable subset of Σ is a metrizable compact space (see Fact 3 of S.307). Take any base $\mathcal{W} = \{W_\alpha : \alpha \in \omega_1 \backslash \omega\} \subset \tau^*(\Sigma)$ of the space Σ. Any countably compact space has the Baire property (see Problem 274 of [TFS]) and hence every open subset of Σ has the Baire property (see Problem 275 of [TFS]). By density of Σ in $\mathbb{D}^{\omega_1}$ any nowhere dense subspace of $\mathbb{D}^{\omega_1}$ which is contained in Σ is nowhere dense in Σ. This shows that if some W_α is meager in $\mathbb{D}^{\omega_1}$, then it is meager in itself which is a contradiction with the Baire property of W_α. Thus W_α is not a meager subset of $\mathbb{D}^{\omega_1}$ for each $\alpha < \omega_1$.

Denote by S_n the set of all bijections from M_n onto itself. Given any $\sigma \in S_n$ and any $z = (z_1, \ldots, z_n) \in (\mathbb{D}^{\omega_1})^n$, let $T_\sigma(z) = (z_{\sigma(1)}, \ldots, z_{\sigma(n)})$. It is clear that $T_\sigma : (\mathbb{D}^{\omega_1})^n \to (\mathbb{D}^{\omega_1})^n$ is a homeomorphism. Call a subset $A \subset (\mathbb{D}^{\omega_1})^n$ *symmetric* if $T_\sigma(A) = A$ for any $\sigma \in S_n$.

From now on we assume that CH holds. Observe that for each $k \in \mathbb{N}$, there exists a family $\mathcal{N}_k = \{N_\alpha^k : \alpha < \omega_1\}$ with the following Properties:

(1) N_α^k is a closed symmetric nowhere dense subset of $(\mathbb{D}^{\omega_1})^k$ for each $k \in \mathbb{N}$ and $\alpha < \omega_1$;
(2) $\mathcal{N}_k$ is cofinal in the family of all nowhere dense subsets of $(\mathbb{D}^{\omega_1})^k$ for each $k \in \mathbb{N}$.

Indeed, for any natural k, there is a cofinal family $\mathcal{N}_k'$ of cardinality ω_1 of nowhere dense closed subspaces of $(\mathbb{D}^{\omega_1})^k$ by Fact 1 of T.039. Given any $N \in \mathcal{N}_k'$, the set $N' = \bigcup \{T_\sigma(N) : \sigma \in S_k\}$ is symmetric, closed and nowhere dense in $(\mathbb{D}^{\omega_1})^k$ while $N \subset N'$. Thus the family $\mathcal{N}_k = \{N' : N \in \mathcal{N}_k'\}$ is still cofinal and consists of closed symmetric nowhere dense subsets of $(\mathbb{D}^{\omega_1})^k$; it is evident that $\mathcal{N}_k$ has cardinality ω_1. Taking any enumeration of $\mathcal{N}_k$ we obtain the promised family $\mathcal{N}_k$ with the properties (1) and (2). If $M_\alpha^k = \bigcup \{N_\beta^k : \beta < \alpha\}$ for any $k \in \mathbb{N}$ and $\alpha < \omega_1$, then let $\mathcal{M}_k = \{M_\alpha^k : \alpha \in \omega_1 \backslash \omega\}$ for each $k \in \mathbb{N}$. It is immediate that the families $\mathcal{M}_k$ have the following properties:

(3) $\mathcal{M}_k$ consists of symmetric meager subsets of $(\mathbb{D}^{\omega_1})^k$ and $M_\alpha^k \subset M_\beta^k$ whenever $\omega \leq \alpha < \beta < \omega_1$;
(4) if M is a meager subset of $(\mathbb{D}^{\omega_1})^k$, then there exists $\alpha = \mu(M, k) \in \omega_1 \backslash \omega$ such that $M \subset M_\alpha^k$.

Take any $n \in \mathbb{N}$ and any natural $k > n$. For any $z = (z_1, \ldots, z_n) \in (\mathbb{D}^{\omega_1})^k$, let $p_n^k(z) = (z_1, \ldots, z_n)$ and $q_n^k(z) = (z_{n+1}, \ldots, z_k)$. Observe that if we consider $(\mathbb{D}^{\omega_1})^k$ to be the product $(\mathbb{D}^{\omega_1})^n \times (\mathbb{D}^{\omega_1})^{k-n}$, then p_n^k and q_n^k are the natural projections onto the respective factors. In particular, we can apply Fact 2 to the product $(\mathbb{D}^{\omega_1})^n \times (\mathbb{D}^{\omega_1})^{k-n}$ and the set M_α^k to find a meager set $R(k, n, \alpha)$ in the space $(\mathbb{D}^{\omega_1})^n$ such that

(5) the set $M_\alpha^k[x] = q_n^k((p_n^k)^{-1}(x) \cap M_\alpha^k)$ is meager in the space $(\mathbb{D}^{\omega_1})^{k-n}$ for every $x \in (\mathbb{D}^{\omega_1})^n \backslash R(k, n, \alpha)$.

Observe that substituting $R(k, n, \alpha)$ with a larger set we still have (5), so we can assume, without loss of generality, that $R(k, n, \alpha) \subset R(k, n, \beta)$ whenever $\alpha < \beta$. For technical reasons, it is convenient to let $R(1, 1, \alpha) = M_\alpha^1$ for all $\alpha \in \omega_1 \backslash \omega$.

The space $\mathbb{D}^{\omega_1}$ is separable (see Problem 108 of [TFS]), so we can fix a countable faithfully indexed dense subset $\{x_n : n \in \omega\}$ of $\mathbb{D}^{\omega_1}$. Choose a point $x_\omega \in \mathbb{D}^{\omega_1}$ arbitrarily and let $\nu_\alpha = \alpha$ for all $\alpha \leq \omega$. Assume that $\omega_1 > \alpha > \omega$ and we have chosen points $\{x_\beta : \beta < \alpha\} \subset \mathbb{D}^{\omega_1}$ and ordinals $\{\nu_\beta : \beta < \alpha\} \subset \omega_1$ such that

(6) $\nu_\beta \geq \beta$ for each $\beta < \alpha$;
(7) if $\omega \leq \beta < \alpha$, then for any $n \in \mathbb{N}$ and any infinite ordinals $\gamma < \beta_1 < \cdots < \beta_n < \beta$ we have $\nu_\beta > \mu(M_\gamma^k[(x_{\beta_1}, \ldots, x_{\beta_n})], k - n)$ for all $k > n$;
(8) for any number $n \in \mathbb{N}$ and any infinite ordinals $\gamma < \beta_1 < \cdots < \beta_n < \alpha$ if $x = (x_{\beta_1}, \ldots, x_{\beta_n})$, then $M_\gamma^k[x]$ is a meager set in $(\mathbb{D}^{\omega_1})^{k-n}$ for any $k > n$;
(9) given any natural number l, we have $(x_{\beta_1}, \ldots, x_{\beta_l}) \notin M_\gamma^l$ for any infinite ordinals $\gamma < \beta_1 < \cdots < \beta_l < \alpha$.

To construct the point x_α and the ordinal ν_α, let $\delta = \sup\{\mu(M_\gamma^k[x], k - n) :$ $n, k \in \mathbb{N}, \ k > n, \ x = (x_{\beta_1}, \ldots, x_{\beta_n})$ for some infinite $\gamma < \beta_1 < \cdots < \beta_n < \alpha\}$. Take any $\nu_\alpha > \max\{\delta, \alpha\}$; the set $S = \bigcup\{R(k, 1, \nu_\alpha) : k \in \mathbb{N}\}$ is meager, so there exists a point $x_\alpha \in W_\alpha \backslash (S \cup \{x_\beta : \beta < \alpha\})$ (this choice of x_α is possible because W_α is not a meager subset of $\mathbb{D}^{\omega_1}$). We claim that the sets $\{x_\beta : \beta \le \alpha\}$ and $\{\nu_\beta : \beta \le \alpha\}$ still satisfy (6)–(9). The properties (6) and (7) hold for ν_α by our choice of ν_α. To prove (8) we can assume that $\beta_n = \alpha$. If $n = 1$, then for any $k \in \mathbb{N}$ we have $x_{\beta_n} = x_\alpha \notin R(k, 1, \gamma)$ because $\gamma < \alpha < \nu_0$ which implies $R(k, 1, \gamma) \subset R(k, 1, \nu_0) \subset S$ while we have chosen x_α outside of S. Therefore $M_\gamma^k[x]$ is a meager set in $(\mathbb{D}^{\omega_1})^{k-1}$ by the definition of $R(k, 1, \gamma)$.

Now, fix any $n > 1$ and $k > n$; let $y = (x_{\beta_1}, \ldots, x_{\beta_{n-1}})$. By the property (8) for the ordinals $\gamma < \beta_1 < \cdots < \beta_{n-1}$, the set $M_\gamma^k[y]$ is meager in $(\mathbb{D}^{\omega_1})^{k-n+1}$; besides, $M_\gamma^k[y]$ has been involved in the definition of δ and hence $\beta = \mu(M_\gamma^k[y], k - n + 1) < \delta$. By the definition of the function μ, we have $M_\gamma^k[y] \subset M_\beta^{k-n+1}$. Observe also that $x_\alpha \notin R(k - n + 1, 1, \delta)$ and hence $x_\alpha \notin R(k - n + 1, 1, \beta)$ which shows that $M_\beta^{k-n+1}[x_\alpha]$ is of first category in $(\mathbb{D}^{\omega_1})^{k-n}$. Finally note that $M_\gamma^k[x] = (M_\gamma^k[y])[x_\alpha] \subset M_\beta^{k-n+1}[x_\alpha]$ is a meager set because so is $M_\beta^{k-n+1}[x_\alpha]$. This shows that (8) is true.

To show that the property (9) holds assume that $\gamma < \beta_1 < \cdots < \beta_l < \alpha$ and $x = (x_{\beta_1}, \ldots, x_{\beta_l}) \in M_\gamma^l$. By the induction hypothesis, we can assume that $\beta_l = \alpha$. If $l = 1$, then $M_\gamma^l = R(1, 1, \gamma) \subset R(1, 1, \nu_\alpha)$ because $\nu_\alpha > \gamma$. By our choice of x_α we have $x_\alpha \notin R(1, 1, \nu_\alpha)$ and hence $x_\alpha \notin R(1, 1, \gamma)$ which is a contradiction showing that (9) is true if $l = 1$.

Now assume that $l > 1$ and consider the point $y = (x_{\beta_1}, \ldots, x_{\beta_{l-1}})$. It is evident that $x_{\beta_l} \in M_\gamma^l[y]$ and hence $M_\gamma^l[y] \ne \emptyset$. Applying the property (7) for the numbers $k = l$, $n = l - 1$ and the ordinals $\gamma < \beta_1 < \cdots < \beta_l$, we conclude that $\nu_{\beta_l} > \beta = \mu(M_\gamma^l[y], 1)$ and hence $M_\gamma^l[y] \subset M_\beta^1$. Since $\beta < \nu_{\beta_l}$, the point x_{β_l} was chosen outside of M_β^1 which is a contradiction because $x_{\beta_l} \in M_\gamma^l[y] \subset M_\beta^1$.

This proves that we have the properties (6)–(9) for the sets $\{x_\beta : \beta \le \alpha\}$ and $\{\nu_\beta : \beta \le \alpha\}$ and hence our inductive construction can be continued, giving us sets $\{x_\beta : \beta < \omega_1\}$ and $\{\nu_\beta : \beta < \omega_1\}$ with the properties (6)–(9) for all $\alpha \in \omega_1 \backslash \omega$.

Let $X = \{x_\alpha : \alpha < \omega_1\}$; we claim that the space X is as promised. The space X is separable because the countable set $A = \{x_n : n \in \omega\}$ is dense in $\mathbb{D}^{\omega_1}$ and hence in X.

Now fix any $n \in \mathbb{N}$ and assume that $S = \{p_\alpha : \alpha < \omega_1\} \subset X^n \backslash \Delta_n(X)$ is an increasing ω_1-sequence. If S is nowhere dense in X^n, then it is nowhere dense in $(\mathbb{D}^{\omega_1})^n$. Apply (4) to find $\alpha < \omega_1 \backslash \omega$ such that $S \subset M_\alpha^n$. However, the sequence S is increasing, and hence there exists $\beta > \alpha$ such that $p_\beta = (x_{\beta_1}, \ldots, x_{\beta_n})$ where $\beta_i > \alpha$ for all $i \in M_n$. Since $p_\beta \in X^n \backslash \Delta_n(X)$, all coordinates of p_β are distinct, and hence there exists a bijection $\sigma \in S_n$ such that $p = T_\sigma(p_\beta) = (x_{\beta_1'}, \ldots, x_{\beta_n'})$ where $\beta_1' < \cdots < \beta_n'$. We have $\alpha < \beta_1' < \cdots < \beta_n'$, so the property (9) is applicable to the point p to conclude that $p \notin M_\alpha^n$. The set M_α^n being symmetric, we have $p_\beta = T_{\sigma^{-1}}(p) \notin T_{\sigma^{-1}}(M_\alpha^n) = M_\alpha^n$ which is a contradiction with $p_\beta \in S \subset M_\alpha^n$.

This shows that no increasing sequence of $X^n \backslash \Delta_n(X)$ is nowhere dense in X^n and therefore we can apply Fact 3 to conclude that X^n is a CS-space for all $n \in \mathbb{N}$.

Observe finally that $X' = X \cap \Sigma \subset X$ is a dense subspace of Σ because it intersects W_α for each $\alpha \in \omega_1 \backslash \omega$; since Σ is not separable (see Fact 3 of S.307), the space X' cannot be separable, so X is not hereditarily separable and our solution is complete.

T.090. *Prove that a compact space X is metrizable if and only if X^3 is hereditarily normal.*

Solution. If X is metrizable, then X^3 is metrizable as well, so X^3 is hereditarily normal because so is every metrizable space. This proves necessity.

Fact 1. Given any spaces Y and Z, assume that Z has a countable non-closed subset and Y has a closed subspace which is not a G_δ-set in Y. Then $Y \times Z$ is not hereditarily normal.

Proof. Assume that $Y \times Z$ is hereditarily normal; fix a set $B_0 = \{z_n : n \in \omega\} \subset Z$ for which there is a point $z \in \overline{B_0} \backslash B_0$, and a closed $F \subset Y$ which is not a G_δ-set in Y. If $B = B_0 \cup \{z\}$, then the space $Y \times B$ has to be hereditarily normal and, in particular, the space $T = (Y \times B) \backslash (F \times \{z\})$ has to be normal. It is easy to see that the sets $P = (Y \backslash F) \times \{z\}$ and $Q = F \times B_0$ are closed in T and disjoint. Since T is normal, there are $U, V \in \tau(T)$ such that $P \subset U$, $Q \subset V$ and $U \cap V = \emptyset$. The sets U and V are also open in $Y \times B$ because T is open in $Y \times B$. Therefore the set $W_n = \{y \in Y : (y, z_n) \in V\}$ is open in Y for each n; it is also immediate that $F \subset W_n$ for all $n \in \omega$. As F is not a G_δ-set in Y, we can take a point $y \in (\bigcap \{W_n : n \in \omega\}) \backslash F$. Then $R = \{y\} \times B_0 \subset V$ and therefore $t = (y, z) \in \overline{R}$. The set $U \ni t$ being open in $Y \times B$ we have $U \cap R \neq \emptyset$; since $R \subset V$ we have $U \cap V \neq \emptyset$, this contradiction completes the proof of Fact 1. □

Fact 2. Any infinite compact space has a non-closed countable subspace.

Proof. Let K be an infinite compact space; take any countably infinite $A \subset K$. If A is not closed, then we are done; if it is, then A is a countable compact space. The space A cannot be discrete because no infinite discrete space is compact. Thus there is a non-isolated point a in the space A. Therefore $A \backslash \{a\}$ is a non-closed countable subset of K and Fact 2 is proved. □

Returning to our solution observe that X^3 is homeomorphic to $X^2 \times X$, so we can apply Fact 1 to the spaces $Y = X^2$ and $Z = X$ to conclude that all closed subsets of X^2 must be G_δ-sets in X^2 because $Z = X$ has a non-closed countable set by Fact 2. In particular, the diagonal $\Delta = \{(x, x) : x \in X\}$ is a G_δ-subset in X^2. Thus we can apply Fact 1 of T.062 to conclude that X is metrizable. This proves sufficiency and makes our solution complete.

T.091. *Prove that $w(X) = \Delta(X)$ for any infinite compact space X. Deduce from this fact that a compact space X is metrizable if and only if the diagonal of X is a G_δ-subspace of $X \times X$.*

Solution. Let $\Delta = \{(x, x) : x \in X\}$ be the diagonal of X. Suppose that $w(X) \leq \kappa$ and fix a base $\mathcal{B}$ in the space X with $|\mathcal{B}| \leq \kappa$. The family $\mathcal{P} = \{(U, V) : U, V \in \mathcal{B}$ and $\overline{U} \cap \overline{V} = \emptyset\}$ has cardinality at most $\kappa \cdot \kappa = \kappa$. If $(U, V) \in \mathcal{P}$, then $F(U, V) = \overline{U} \times \overline{V}$ is a closed subset of $X \times X$ such that $F(U, V) \cap \Delta = \emptyset$, i.e., $F(U, V) \subset (X \times X) \backslash \Delta$. If $(x, y) \in (X \times X) \backslash \Delta$, then $x \neq y$, so there exist sets $U', V' \in \tau(X)$ such that $x \in U'$, $y \in V'$ and $U' \cap V' = \emptyset$. Pick $U, V \in \mathcal{B}$ for which $x \in U \subset U'$ and $y \in V \subset V'$; then $(U, V) \in \mathcal{P}$ and $(x, y) \in F(U, V)$. This proves the equality $(X \times X) \backslash \Delta = \bigcup \{F(U, V) : (U, V) \in \mathcal{P}\}$; since $|\mathcal{P}| \leq \kappa$, we showed that $(X \times X) \backslash \Delta$ is an F_κ-subset of $X \times X$ and hence Δ is a G_κ-subset of $X \times X$ which implies that $\Delta(X) \leq \kappa$. Thus $\Delta(X) \leq w(X)$; observe that we did not need compactness of X to prove this inequality.

To establish that $w(X) \leq \Delta(X)$ assume that $\Delta(X) = \kappa$ and choose a family $\mathcal{H} \in \tau(\Delta, X \times X)$ such that $|\mathcal{H}| \leq \kappa$ and $\Delta = \bigcap \mathcal{H}$. For any $H \in \mathcal{H}$ take a finite cover $\mathcal{U}_H$ of the space X such that $U \times U \subset H$ for any $U \in \mathcal{U}_H$; by paracompactness of X we can find a finite open barycentric refinement $\mathcal{V}_H$ of the cover $\mathcal{U}_H$.

The family $\mathcal{S} = \bigcup \{\mathcal{V}_H : H \in \mathcal{H}\}$ has cardinality $\leq \kappa$ and $\bigcup \mathcal{S} = X$, so there exists a topology τ on the set X generated by $\mathcal{S}$ as a subbase (see Problem 008 of [TFS]). Let us prove that $Y = (X, \tau)$ is a Hausdorff space. Take any distinct $a, b \in X$; since $c = (a, b) \notin \Delta$, there is $H \in \mathcal{H}$ such that $c \notin H$. We claim that the set $\{a, b\}$ is not contained in $\mathrm{St}(x, \mathcal{V}_H)$ for any $x \in X$. Indeed, if there is $x \in X$ with $\{a, b\} \subset \mathrm{St}(x, \mathcal{V}_H)$, then we can find a set $W \in \mathcal{U}_H$ such that $\mathrm{St}(x, \mathcal{V}_H) \subset W$ and therefore $\{a, b\} \subset W$. Thus $c \in W \times W \subset H$ which is a contradiction.

Now, take any sets $U, V \in \mathcal{V}_H$ such that $a \in U$ and $b \in V$; if $z \in U \cap V$, then $\{a, b\} \subset \mathrm{St}(z, \mathcal{V}_H)$ which we proved not to be possible. Therefore $U \cap V = \emptyset$ and hence the space Y is Hausdorff. The identity map $i : X \to Y$ is continuous because $i^{-1}(W) = W \in \tau(X)$ for any $W \in \mathcal{S}$ (see Problem 009 of [TFS]). Being a continuous image of a compact space X, the space Y is a compact Hausdorff and hence Tychonoff space (see Problem 124 of [TFS]). The subbase $\mathcal{S}$ of the space Y has cardinality at most κ; the family of all finite intersections of the elements of $\mathcal{S}$ is a base of Y of cardinality $\leq \kappa$ so $w(Y) \leq \kappa$. As a consequence, i condenses X onto a space Y with $w(Y) \leq \kappa$. This condensation has to be a homeomorphism (see Problem 123 of [TFS]), so $w(X) \leq \kappa$ and hence we proved that $w(X) \leq \Delta(X)$, i.e., $w(X) = \Delta(X)$.

Finally, observe that a compact space X is metrizable if and only if it is second countable (see Problems 209 and 212 of [TFS]) which, in turn, happens if and only if $\Delta(X) = \omega$, i.e., the diagonal of X is a G_δ-subset of $X \times X$.

T.092. *Prove that a countably compact space X is metrizable if and only if every subspace $Y \subset X$ with $|Y| \leq \omega_1$ is metrizable.*

Solution. Given an arbitrary space Z, a set $Y \subset Z$ and a family $\mathcal{A} \subset \exp(Z)$, let $\mathcal{A}|Y = \{A \cap Y : A \in \mathcal{A}\}$; besides, $\mathcal{A}[z] = \{A \in \mathcal{A} : z \in A\}$ for any $z \in Z$.

Fact 1. Suppose that Z is a space and $A \subset Z$. Then $\chi(p, A) = \chi(p, \overline{A})$ for any point $p \in A$.

Proof. Since $A \subset \overline{A}$, we have $\chi(p, A) \leq \chi(p, \overline{A})$. Now assume that $\chi(p, A) = \kappa$ and take any local base $\mathcal{B}$ of the point p in the space A such that $|\mathcal{B}| \leq \kappa$. For each $U \in \mathcal{B}$ fix $O_U \in \tau(\overline{A})$ such that $O_U \cap A = U$. Then $\mathcal{B}' = \{O_U : U \in \mathcal{B}\}$ is a family of open neighborhoods of p in the space $\overline{A}$ and $|\mathcal{B}'| \leq \kappa$. To see that $\mathcal{B}'$ is a local base at p in the space $\overline{A}$, take any $W \in \tau(p, \overline{A})$. There exist $V \in \tau(p, \overline{A})$ and $U \in \mathcal{B}$ such that $\mathrm{cl}_{\overline{A}}(V) \subset W$ and $U \subset V \cap A$; we have

$$O_U \subset \mathrm{cl}_{\overline{A}}(O_U) = \mathrm{cl}_{\overline{A}}(U) \subset \mathrm{cl}_{\overline{A}}(V \cap A) \subset \mathrm{cl}_{\overline{A}}(V) \subset W$$

and hence $O_U \subset W$ which proves that $\mathcal{B}'$ is a local base at the point p in the space $\overline{A}$ whence $\chi(p, \overline{A}) \leq |\mathcal{B}'| \leq \kappa$. Consequently, $\chi(p, \overline{A}) \leq \chi(p, A)$ which implies $\chi(p, A) = \chi(p, \overline{A})$, so Fact 1 is proved.   □

Fact 2. Let Z be any space. Assume that $w(Y) \leq \omega$ for every $Y \subset Z$ with $|Y| \leq \omega_1$. Then $w(Z) = \omega$.

Proof. If there is a left-separated $Y \subset Z$ with $|Y| = \omega_1$, then $w(Y) = \omega$ and hence $hd(Y) = \omega$ which implies that there are no uncountable left-separated subspaces in Y by Problem 004. This contradiction shows that Z is hereditarily separable (see Problem 004). Take any dense countable $A \subset Z$; for any $z \in Z$, the set $A \cup \{z\}$ is countable, so $\chi(z, A \cup \{z\}) = \omega$ because $w(A \cup \{z\}) = \omega$ by our hypothesis. Now apply Fact 1 to conclude that

$$\chi(z, Z) = \chi\left(z, \overline{\{z\} \cup A}\right) = \chi(z, \{z\} \cup A) = \omega$$

for any $z \in Z$, i.e., $\chi(Z) = \omega$. Fix a countable local base $\mathcal{B}_z$ at the point z in the space Z for all $z \in Z$. Given any set $A \subset Z$, let $\mathcal{B}(A) = \bigcup\{\mathcal{B}_z : z \in A\}$.

Take any $z_0 \in Z$ and let $Y_0 = \{z_0\}$; suppose that $\alpha < \omega_1$ and we have sets $\{Y_\beta : \beta < \alpha\}$ with the following properties:

(1) $Y_\beta \subset Z$ and $|Y_\beta| \leq \omega$ for all $\beta < \alpha$;
(2) $Y_\beta \subset Y_\gamma$ whenever $\beta < \gamma < \alpha$;
(3) for any $\gamma < \alpha$ if $Y'_\gamma = \bigcup\{Y_\beta : \beta < \gamma\}$, then $\mathcal{B}(Y'_\gamma)|Y_\gamma$ is not a base in Y_γ.

The family $\mathcal{B}' = \mathcal{B}(\bigcup\{Y_\beta : \beta < \alpha\})$ is countable, so it cannot be a base for the space Z. Therefore there exists a point $z_\alpha \in Z$ and a set $U \in \tau(z_\alpha, Z)$ such that $V \backslash U \neq \emptyset$ for any $V \in \mathcal{B}'[z_\alpha]$. Choose a point $z_V \in V \backslash U$ for each $V \in \mathcal{B}'[z_\alpha]$ and let $Y_\alpha = \{z_\alpha\} \cup (\bigcup\{Y_\beta : \beta < \alpha\}) \cup \{z_V : V \in \mathcal{B}'[z_\alpha]\}$. It is immediate that the properties (1) and (2) hold for the family $\{Y_\beta : \beta \leq \alpha\}$. The property (3) also takes place because the point z_α and the set $U \cap Y_\alpha \in \tau(z_\alpha, Y_\alpha)$ witness that $\mathcal{B}'|Y_\alpha$ is not a base in Y_α. Thus our inductive construction can be continued to obtain a family $\{Y_\alpha : \alpha < \omega_1\}$ for which (1)–(3) are satisfied for all $\beta < \omega_1$. The set $Y = \bigcup\{Y_\alpha : \alpha < \omega_1\}$ has cardinality at most ω_1 so $w(Y) = \omega$.

It is evident that $\mathcal{B}(Y)|Y$ is a base in Y, so we can apply claim of S.088 to conclude that there is a countable $\mathcal{V} \subset \mathcal{B}(Y)$ such that $\mathcal{V}|Y$ is a base in Y. The property (2) implies that there is $\alpha < \omega_1$ such that $\mathcal{V} \subset \mathcal{B}(Y_\alpha)$ and hence $\mathcal{B}(Y_\alpha)|Y$

is a base in Y. An immediate consequence is that $\mathcal{B}(Y_\alpha)|Y_{\alpha+1}$ is a base in $Y_{\alpha+1}$ which contradicts (3) and finishes the proof of Fact 2. $\qquad\square$

Given a space Z and a set $Y \subset Z$, call a family $\mathcal{U} \subset \tau(Z)$ *an external base of Y in Z* if for any $y \in Y$ and any $U \in \tau(y, Z)$ there is $V \in \mathcal{U}$ such that $y \in V \subset U$.

Fact 3. Given any space Z and $Y \subset Z$ suppose that we have a collection $\mathcal{U} \subset \tau(Z)$ such that $|\mathcal{U}| \leq \kappa$ and $\mathcal{U}$ is not an external base of Y in Z. Then there exists a set $A \subset Z$ such that $|A| \leq \kappa$ and there is a point $z \in A \cap Y$ which witnesses that $\mathcal{U}|A$ is not a base in A.

Proof. Since $\mathcal{U}$ is not an external base of Y in Z, there exists $z \in Y$ and $W \in \tau(z, Z)$ such that there is no $U \in \mathcal{U}$ such that $z \in U \subset W$. For each $U \in \mathcal{U}[z]$ pick any $z_U \in U \backslash W$ and let $A = \{z\} \cup \{z_U : U \in \mathcal{U}[z]\}$. It is clear that $|A| \leq |\mathcal{U}| \leq \kappa$; evidently, the point $z \in Y \cap A$ and its neighborhood $W \cap A \in \tau(z, A)$ witness the fact that $\mathcal{U}|A$ is not a base in A, so Fact 3 is proved. $\qquad\square$

Returning to our solution, observe that if X is metrizable, then every $Y \subset X$ is metrizable so necessity is clear. Now assume that X is not metrizable while any $Y \subset X$ is metrizable if $|Y| \leq \omega_1$. Then $w(X) \geq \omega_1$, so Fact 2 is applicable to conclude that there is $E \subset X$ such that $|E| \leq \omega_1$ and $w(E) \geq \omega_1$. Take any $z \in \overline{E}$; since $\{z\} \cup E$ is metrizable, we have $\chi(z, \{z\} \cup E) = \omega$ and hence $\chi(z, \overline{E}) = \omega$ by Fact 1. The point $z \in \overline{E}$ was chosen arbitrarily so $\chi(\overline{E}) = \omega$; besides, $w(\overline{E}) \geq w(E) \geq \omega_1$. It is evident that $\overline{E}$ is a non-metrizable countably compact space such that any subspace $Y \subset \overline{E}$ is metrizable whenever $|Y| \leq \omega_1$. This shows that for obtaining a contradiction, we can assume without loss of generality that $X = \overline{E}$ and hence $\chi(X) = \omega$.

Fix a countable local base $\mathcal{U}_x$ at the point x for each $x \in X$. Given any set $A \subset X$, we will need the family $\mathcal{U}(A) = \bigcup\{\mathcal{U}_x : x \in A\}$. Given a family $\mathcal{A}$ of open sets of a space Z we will say that a point $z \in Z$ *witnesses that $\mathcal{A}$ is not a base in Z* if there exists $U \in \tau(z, Z)$ such that there is no $V \in \mathcal{A}$ for which $z \in V \subset U$.

Take a point $y_0 \in X$ and let $Y_0 = \{y_0\}$. Assume that $\alpha < \omega_1$ and we have a family $\{Y_\beta : \beta < \alpha\}$ of subsets of X with the following properties:

(4) $|Y_\beta| \leq \omega$ for all $\beta < \alpha$;
(5) if $\beta < \beta' < \alpha$, then $Y_\beta \subset Y_{\beta'}$;
(6) if $\beta < \alpha$ is a limit ordinal, then $Y_\beta = \bigcup\{Y_\gamma : \gamma < \beta\}$;
(7) if $\beta = \gamma + 1 < \alpha$, then for any finite $\mathcal{V} \subset \mathcal{U}(Y_\gamma)$ such that $\bigcup \mathcal{V} \neq X$, we have $Y_\beta \backslash (\bigcup \mathcal{V}) \neq \emptyset$;
(8) if β is a limit ordinal with $\beta + 1 < \alpha$, then there is a point $z \in \overline{Y}_\beta \cap Y_{\beta+1}$ which witnesses that $\mathcal{U}(Y_\beta)|Y_{\beta+1}$ is not a base in $Y_{\beta+1}$.

If α is a limit ordinal, then let $Y_\alpha = \bigcup\{Y_\beta : \beta < \alpha\}$; it is immediate that we still have the properties (4)–(8) for all $\beta \leq \alpha$. If $\alpha = \gamma + 1$ where γ is a non-limit ordinal, consider the family $\mathcal{W} = \{\mathcal{V} : \mathcal{V}$ is a finite subfamily of $\mathcal{U}(Y_\gamma)$ such that $\bigcup \mathcal{V} \neq X\}$. Choose a point $a(\mathcal{V}) \in X \backslash (\bigcup \mathcal{V})$ for any $\mathcal{V} \in \mathcal{W}$ and let $Y_\alpha = Y_\gamma \cup \{a(\mathcal{V}) : \mathcal{V} \in \mathcal{W}\}$; it is evident that the properties (4)–(8) now hold for all $\beta \leq \alpha$.

Now, if $\alpha = \gamma + 1$ where γ is a limit ordinal, let us prove first that $\mathcal{U}(Y_\gamma)$ is not an external base for $\overline{Y}_\gamma$. Indeed, if $\mathcal{U}(Y_\gamma)$ is an external base for $\overline{Y}_\gamma$, then $\overline{Y}_\gamma \neq X$; for otherwise, $w(X) = \omega$ which is a contradiction. Fix any $q \in X \backslash \overline{Y}_\gamma$ and consider the family $\mathcal{A} = \{U \in \mathcal{U}(Y_\gamma) : q \notin U\}$. Since $\mathcal{U}(Y_\gamma)$ is an external base of $\overline{Y}_\gamma$, the family $\mathcal{A}$ is a countable open cover of $\overline{Y}_\gamma$; the latter space being countably compact, there is a finite $\mathcal{V} \subset \mathcal{A}$ such that $\overline{Y}_\gamma \subset \bigcup \mathcal{V}$. Since γ is a limit ordinal, the property (6) implies that there exists $\beta < \gamma$ such that $\mathcal{V} \subset \mathcal{U}(Y_\beta)$. Now $\mathcal{V} \subset \mathcal{U}(Y_\beta)$, $Y_\beta \subset Y_\gamma \subset \overline{Y}_\gamma \subset \bigcup \mathcal{V}$ and $\bigcup \mathcal{V} \neq X$ which shows that there is a point $x \in Y_{\beta+1} \backslash (\bigcup \mathcal{V})$ by (7) which is a contradiction with $Y_{\beta+1} \subset Y_\gamma \subset \bigcup \mathcal{V}$.

Now that we established that $\mathcal{U}(Y_\gamma)$ is not an external base for $\overline{Y}_\gamma$, apply Fact 3 to find a countable set $Q \subset Z$ and a point $z \in \overline{Y}_\gamma \cap Q$ such that z witnesses that $\mathcal{U}(Y_\gamma)|Q$ is not a base in Q. Now consider the family $\mathcal{W} = \{\mathcal{V} : \mathcal{V}$ is a finite subfamily of $\mathcal{U}(Y_\gamma)$ such that $\bigcup \mathcal{V} \neq X\}$. Choose a point $a(\mathcal{V}) \in X \backslash (\bigcup \mathcal{V})$ for any $\mathcal{V} \in \mathcal{W}$ and let $Y_\alpha = Q \cup Y_\gamma \cup \{a(\mathcal{V}) : \mathcal{V} \in \mathcal{W}\}$; it is evident that the properties (4)–(7) now hold for all $\beta \leq \alpha$. As to (8), it holds because $z \in \overline{Y}_\gamma \cap Y_{\gamma+1}$ and z witnesses that $\mathcal{U}(Y_\gamma)|Y_{\gamma+1}$ is not a base in $Y_{\gamma+1}$ due to the fact that z does it for the family $\mathcal{U}(Y_\gamma)|Q$ and $Q \subset Y_{\gamma+1}$.

Therefore our inductive construction can be continued to obtain a collection $\{Y_\beta : \beta < \omega_1\}$ for which the properties (4)–(8) hold for all ordinals $\beta < \omega_1$. The space $Y = \bigcup \{Y_\beta : \beta < \omega_1\}$ is metrizable because $|Y| \leq \omega_1$. Take any σ-discrete base $\mathcal{B}$ in the space Y (see Problem 221 of [TFS]) and fix, for each $V \in \mathcal{B}$, a set $O_V \in \tau(Z)$ such that $O_V \cap Y = V$. Let $\mathcal{C} = \{O_V : V \in \mathcal{B}\}$ and observe that $\mathcal{C}_\alpha = \{U \in \mathcal{C} : U \cap Y_\alpha \neq \emptyset\}$ is countable for all α because Y_α is countable and each $y \in Y$ belongs to at most countably many elements of $\mathcal{C}$. Let $\mathcal{U} = \mathcal{U}(Y)$; given an ordinal $\alpha < \omega_1$, call a pair $(B, B') \in \mathcal{C}_\alpha \times \mathcal{C}_\alpha$ adequate if there is $U \in \mathcal{U}$ for which $B \cap Y \subset U \cap Y \subset B' \cap Y$. For each $\alpha < \omega_1$ and for each adequate pair $(B, B') \in \mathcal{C}_\alpha \times \mathcal{C}_\alpha$, choose $W_{B,B'} \in \mathcal{U}$ such that $B \cap Y \subset W_{B,B'} \cap Y \subset B' \cap Y$ and let $\mathcal{E}_\alpha = \{W_{B,B'} \in \mathcal{U} : (B, B') \in \mathcal{C}_\alpha \times \mathcal{C}_\alpha$ is an adequate pair$\}$. Note that $|\mathcal{E}_\alpha| \leq \omega$ and hence there exists $\beta(\alpha) > \alpha$ such that $\mathcal{E}_\alpha \subset \mathcal{U}(Y_{\beta(\alpha)})$.

Take any $\alpha_0 < \omega_1$ and inductively let $\alpha_{n+1} = \beta(\alpha_n)$ for all $n \in \omega$. For the sequence $\{\alpha_n : n \in \omega\} \subset \omega_1$ we have

(9) $\alpha_n < \alpha_{n+1}$ and $\mathcal{E}_{\alpha_n} \subset \mathcal{U}(Y_{\alpha_{n+1}})$ for all $n \in \omega$.

Observe that $\alpha = \sup\{\alpha_n : n \in \omega\}$ is a limit ordinal; the property (8) shows that there exists $z \in \overline{Y}_\alpha \cap Y_{\alpha+1}$ and $W \in \tau(z, X)$ such that

(10) $U \cap Y \not\subset W \cap Y$ for every $U \in \mathcal{U}(Y_\alpha)[z]$.

Observe that $\mathcal{C}|Y$ and $\mathcal{U}|Y$ are bases in the space Y and therefore there exist $B, B' \in \mathcal{C}$ and $U \in \mathcal{U}$ such that $z \in B \cap Y \subset U \cap Y \subset B' \cap Y \subset W \cap Y$. We have $z \in \overline{Y}_\alpha$, so $B \cap Y_\alpha \neq \emptyset$ and $B' \cap Y_\alpha \neq \emptyset$. It follows from (5) and (6) that there exists $n \in \omega$ such that $B \cap Y_{\alpha_n} \neq \emptyset$ and $B' \cap Y_{\alpha_n} \neq \emptyset$ whence $B, B' \in \mathcal{C}_{\alpha_n}$. Since the pair $(B, B') \in \mathcal{C}_{\alpha_n} \times \mathcal{C}_{\alpha_n}$ is adequate, we have at our disposal the set $W_{B,B'} \in \mathcal{E}_{\alpha_n}$ for which $z \in W_{B,B'} \cap Y \subset B' \cap Y \subset W \cap Y$. Apply the property (9) to conclude that $W_{B,B'} \in \mathcal{E}_{\alpha_n} \subset \mathcal{U}(Y_{\alpha_{n+1}}) \subset \mathcal{U}(Y_\alpha)$ and therefore $W_{B,B'} \in \mathcal{U}(Y_\alpha)$

which is a contradiction with the property (10). This contradiction shows that Y is a non-metrizable subspace of X of cardinality ω_1, so we obtained a contradiction with our hypothesis. Thus our assumption of non-metrizability of X is false, so X is metrizable and our solution is complete.

T.093. *Give an example of a non-metrizable pseudocompact space P such that every $Y \subset P$ with $|Y| \leq \omega_1$ is metrizable.*

Solution. Let $\kappa = 2^{\omega_1}$ and take any set T of cardinality κ. It is easy to find a disjoint family $\{T_\alpha : \alpha < \kappa\} \subset \exp(T)$ such that $\bigcup\{T_\alpha : \alpha < \kappa\} = T$ and $|T_\alpha| = \kappa$ for each $\alpha < \kappa$. If $Q = \bigcup\{\mathbb{I}^S : S$ is a countable subset of $T\}$, then $|Q| = |T|^\omega \cdot \mathfrak{c} = (2^{\omega_1})^\omega \cdot 2^\omega = 2^{\omega_1} \cdot 2^\omega = 2^{\omega_1}$, so we can fix an enumeration $\{q_\alpha : \alpha < \kappa\}$ of the set Q. For every $\alpha < \kappa$, let S_α be the unique countable subset of T such that $q_\alpha \in \mathbb{I}^{S_\alpha}$.

For any $\alpha < \kappa$ define $x_\alpha \in \mathbb{I}^T$ as follows: $x_\alpha(t) = q_\alpha(t)$ if $t \in S_\alpha$; if $t \in T_\alpha \backslash S_\alpha$, then $x_\alpha(t) = 1$ and $x_\alpha(t) = 0$ for any $t \in T \backslash (T_\alpha \cup S_\alpha)$. The space $P = \{x_\alpha : \alpha < \kappa\}$ is as promised.

To prove it observe first that P fills all countable faces of $\mathbb{I}^\kappa$, i.e., for any countable $S \subset T$ and any $q \in \mathbb{I}^S$, there is $p \in P$ such that $p|S = q$. Indeed, $q = q_\alpha$ and $S = S_\alpha$ for some $\alpha < \kappa$, so $x_\alpha|S = q_\alpha = q$ by the definition of x_α. This shows that the set P is dense in $\mathbb{I}^T$ and hence pseudocompact by Fact 2 of S.433. Another easy observation is that P is not compact because P is dense in $\mathbb{I}^T$ and $P \neq \mathbb{I}^T$ due to the fact that the point $u \in \mathbb{I}^T$ defined by $u(t) = 1$ for all $t \in T$ does not belong to P. As a consequence, P is not metrizable because any metrizable pseudocompact space is compact.

Finally, if a set $A \subset \kappa$ has cardinality ω_1, then $|A| \leq \omega_1 < 2^{\omega_1} = \kappa$, and hence, for any $\beta < \kappa$, the set $\bigcup\{S_\alpha : \alpha \in A\}$ cannot cover the set T_β. Take any $t \in T \backslash (\bigcup\{S_\alpha : \alpha \in A \backslash \{\beta\}\})$; then $x_\beta(t) = 1$ while $x_\alpha(t) = 0$ for all $\alpha \in A \backslash \{\beta\}$. Therefore x_β cannot be an accumulation point for the set $\{x_\alpha : \alpha \in A\}$. Since any subset of cardinality $\leq \omega_1$ in P can be written as $\{x_\alpha : \alpha \in A\}$ for some $A \subset \kappa$ with $|A| \leq \omega_1$, we proved that if $Y \subset P$ and $|Y| \leq \omega_1$ then Y has no accumulation points in P, i.e., Y is closed and discrete in P. Thus P is not metrizable while every $Y \subset P$ with $|Y| \leq \omega_1$ is metrizable being a discrete space. We finally proved that P has all the required properties, so our solution is complete.

T.094. *Let X be a non-metrizable compact space. Prove that there exists a continuous map of X onto a (non-metrizable compact) space of weight ω_1.*

Solution. The space $C_p(X)$ cannot be separable by Problem 213 of [TFS]. This implies that $hd(C_p(X)) \geq d(C_p(X)) \geq \omega_1$ and hence there is a left-separated $P \subset C_p(X)$ with $|P| = \omega_1$. For each $x \in X$ let $e_x(f) = f(x)$ for all $f \in P$. We have a map e defined by $e(x) = e_x$. Then $e : X \to C_p(P)$ and e is a continuous map (see Problem 166 of [TFS]); let $Y = e(X)$. Since $w(C_p(P)) = |P| = \omega_1$ (see Problem 169 of [TFS]), we have $w(Y) \leq w(C_p(P)) = \omega_1$ so $w(Y) \leq \omega_1$.

We also have the mapping $e^* : C_p(Y) \to C_p(X)$ defined by $e^*(f) = f \circ e$ for any $f \in Y$. The map e^* is an embedding (see Problem 163 of [TFS]) and it is easy to see that $P \subset e^*(C_p(Y))$. Thus P is an uncountable left-separated subspace of $e^*(C_p(Y))$. Since $e^*(C_p(Y))$ is homeomorphic to $C_p(Y)$, the latter has an uncountable left-separated subspace and therefore $w(Y) = nw(Y) = nw(C_p(Y)) \geq hd(C_p(Y)) \geq \omega_1$ (we applied Problem 172 of [TFS], Problem 004 and Fact 4 of S.307). Therefore $w(Y) \geq \omega_1$; we have proved already that $w(Y) \leq \omega_1$, so Y is a continuous image of the space X such that $w(Y) = \omega_1$. Of course, Y is compact and non-metrizable (see Problems 119 and 212 of [TFS]).

T.095. *Let X be a perfectly normal compact space. Prove that $d(X) \leq \omega_1$.*

Solution. For every non-empty closed set $F \subset X$ there exists a countable family $\mathcal{O}_F \subset \tau(F, X)$ such that $\bigcap \mathcal{O}_F = F$; this is possible because X is perfectly normal. Pick any $x_0 \in X$ and let $Y_0 = \{x_0\}$. Suppose that $\alpha < \omega_1$ and we have chosen sets $\{Y_\beta : \alpha < \beta\}$ with the following properties:

(1) $Y_\beta \subset X$ and Y_β is countable for all $\beta < \alpha$;
(2) $Y_\beta \subset Y_\gamma$ if $\beta < \gamma < \alpha$;
(3) let $F_\beta = \overline{Y}_\beta$ for all $\beta < \alpha$; if $\gamma < \alpha$, $\mathcal{U}_\gamma = \bigcup\{\mathcal{O}_{F_\beta} : \beta < \gamma\}$ and $\mathcal{V}$ is a finite subfamily of $\mathcal{U}_\gamma$ such that $\bigcup \mathcal{V} \neq X$ then $Y_\gamma \backslash (\bigcup \mathcal{V}) \neq \emptyset$.

Let $\mathcal{U}_\alpha = \bigcup\{\mathcal{O}_{F_\beta} : \beta < \alpha\}$ and consider the collection $W = \{\mathcal{V} \subset \mathcal{U}_\alpha : \mathcal{V}$ is finite and $\bigcup \mathcal{V} \neq X\}$. For each $\mathcal{V} \in W$, take any point $x(\mathcal{V}) \in X \backslash \bigcup \mathcal{V}$ and let $Y_\alpha = \bigcup\{Y_\beta : \beta < \alpha\} \cup \{x(\mathcal{V}) : \mathcal{V} \in W\}$. It is evident that the sets $\{Y_\beta : \beta \leq \alpha\}$ still satisfy (1)–(3), so our inductive construction can be continued to give us a collection $\{Y_\beta : \beta < \omega_1\}$ for which (1)–(3) are fulfilled for all $\beta < \omega_1$. Since all Y_β's are countable, for the set $Y = \bigcup\{Y_\beta : \beta < \omega_1\}$, we have $|Y| \leq \omega_1$, so it suffices to prove that Y is dense in X.

Observe that X is first countable (see Problem 327 of [TFS]) and therefore $\overline{Y} = \bigcup\{\overline{A} : A \subset Y$ and A is countable$\} = \bigcup\{\overline{Y}_\beta : \beta < \omega_1\} = \bigcup\{F_\beta : \beta < \omega_1\}$. If $\overline{Y} \neq X$, then fix any point $q \in X \backslash \overline{Y}$. For each $y \in \overline{Y}$ there is $\beta < \omega_1$ such that $y \in F_\beta$ and hence there is $O_y \in \mathcal{O}_{F_\beta}$ such that $y \in O_y$ and $q \notin O_y$. As a consequence, the family $\mathcal{O} = \{O_y : y \in \overline{Y}\}$ is an open cover of the compact space $\overline{Y}$; let $\mathcal{V} \subset \mathcal{O}$ be a finite subcover of $\overline{Y}$. There is $\alpha < \omega_1$ such that $\mathcal{V} \subset \mathcal{U}_\alpha$; besides, $V = \bigcup \mathcal{V} \neq X$ because $q \notin \bigcup \mathcal{V}$. The property (3) shows that $Y_{\alpha+1} \backslash V \neq \emptyset$ which is a contradiction with $Y_{\alpha+1} \subset Y \subset \overline{Y} \subset V$. Therefore Y is dense in X; since $|Y| \leq \omega_1$, we have $d(X) \leq \omega_1$ and our solution is complete.

T.096. *Let Y be a subspace of a perfectly normal compact space X. Prove that $nw(Y) = w(Y)$.*

Solution. Since $nw(Y) \leq w(Y)$ for any space Y, it suffices to establish the inequality $w(Y) \leq nw(Y)$; the case of $nw(Y) < \omega$ is trivial, so we assume that $\kappa = nw(Y) \geq \omega$. Take any network $\mathcal{N}$ in the space Y with $|\mathcal{N}| = \kappa$ and consider the family $\mathcal{F} = \{\overline{N} : N \in \mathcal{N}\}$. We have $\chi(F, X) = \psi(F, X) = \omega$ for every $F \in \mathcal{F}$ (see Problem 327 of [TFS]), so we can choose a countable outer base $\mathcal{O}_F$ for each $F \in \mathcal{F}$. If $\mathcal{B}' = \bigcup\{\mathcal{O}_F : F \in \mathcal{F}\}$, then $|\mathcal{B}'| \leq \kappa \cdot \omega = \kappa$, so it is sufficient to show that $\mathcal{B} = \{U \cap Y : U \in \mathcal{B}'\}$ is a base in Y.

Take any point $y \in Y$ and $W \in \tau(y, Y)$; there exists a set $V \in \tau(y, X)$ such that $V \cap Y = W$. By regularity of X there is $U \in \tau(y, X)$ with $\overline{U} \subset V$. Since $\mathcal{N}$ is a network in Y, we can find $N \in \mathcal{N}$ such that $y \in N \subset U \cap Y$. Consequently, $F = \overline{N} \in \mathcal{F}$ and $F \subset \overline{U} \subset V$; the family $\mathcal{O}_F$ being an outer base of F, there is $O \in \mathcal{O}_F$ such that $F \subset O \subset V$. It is clear that we have $y \in F \cap Y \subset O \cap Y \subset V \cap Y = W$. Since $W' = O \cap Y \in \mathcal{B}$, we proved that for any $y \in Y$ and any $W \in \tau(y, Y)$ there is $W' \in \mathcal{B}$ such that $y \in W' \subset W$. Thus $\mathcal{B}$ is a base in Y and hence $w(Y) \leq |\mathcal{B}| \leq \kappa = nw(Y)$.

T.097. *Prove that, under CH, there exists a strictly $<^*$-increasing ω_1-sequence $S = \{f_\alpha : \alpha < \omega_1\} \subset \mathbf{P}$ which is $<^*$-cofinal in ω^ω.*

Solution. It will be easy to construct the promised ω_1-sequence after we prove the following fact.

Fact 1. Let A be a countable subset of ω^ω. Then there exists $f \in \mathbf{P}$ such that $g <^* f$ for any $g \in A$.

Proof. We have $A = \{g_n : n \in \omega\}$ (this enumeration of A can have repetitions). Let $f(n) = \sum_{i,j=0}^{n} g_i(j) + n + 1$ for each $n \in \omega$. It is immediate that $f(n) < f(m)$ if $n < m$ so $f \in \mathbf{P}$. Given any $n \in \omega$, observe that $g_n(m) < f(m)$ for any $m \geq n$, so $g_n <^* f$ and Fact 1 is proved. $\qquad\square$

Returning to our solution apply CH to choose some enumeration $\{g_\alpha : \alpha < \omega_1\}$ of the set ω^ω. Apply Fact 1 to the set $\{g_0\}$ to find a function $f_0 \in \mathbf{P}$ such that $g_0 <^* f_0$. Assume that $\alpha < \omega_1$ and we have functions $\{f_\beta : \beta < \alpha\} \subset \mathbf{P}$ with the following properties:

(1) $f_\gamma <^* f_\beta$ if $\gamma < \beta < \alpha$.
(2) $g_\gamma <^* f_\beta$ whenever $\gamma \leq \beta < \alpha$.

For the countable set $A = \{g_\beta : \beta \leq \alpha\} \cup \{f_\beta : \beta < \alpha\}$ apply Fact 1 to find a function $f_\alpha \in \mathbf{P}$ such that $g <^* f_\alpha$ for each $g \in A$. It is clear that the set $\{f_\beta : \beta \leq \alpha\}$ still has properties (1) and (2), so our inductive construction can go on to give us an ω_1-sequence $S = \{f_\beta : \beta < \omega_1\} \subset \mathbf{P}$ for which the properties (1) and (2) hold for every $\beta < \omega_1$. It follows from (1) that S is strictly $<^*$-increasing. To prove cofinality of S take any $g \in \omega^\omega$; then $g = g_\alpha$ for some $\alpha < \omega_1$ and hence $g = g_\alpha <^* f_\alpha \in S$ by (2). Therefore S is cofinal in ω^ω, so our solution is complete.

T.098. *Assuming CH, prove that there exists a space Y with the following properties:*

(i) $|Y| = \omega_1$ *and Y is scattered;*
(ii) *Y is locally compact and locally countable, i.e., every point of Y has a countable neighborhood (and hence Y is not Lindelöf);*
(iii) *Y^k is hereditarily separable for every $k \in \mathbb{N}$.*

In particular, strong S-spaces exist under CH.

Solution. For an arbitrary $n \in \mathbb{N}$, let $M_n = \{1, \ldots, n\}$ and denote by S_n the set of all bijections from M_n onto itself. Given a space Z and a natural number $n \geq 2$, let $\Delta_{ij}^n(Z) = \{z = (z_1, \ldots, z_n) \in Z^n : z_i = z_j\}$ for any distinct $i, j \in M_n$. The set $\Delta_n(Z) = \bigcup\{\Delta_{ij}^n(Z) : 1 \leq i < j \leq n\}$ is called the n-*diagonal* of Z. It is convenient to consider that $\Delta_1(Z) = \emptyset$. Given any $\sigma \in S_n$ and any $z = (z_1, \ldots, z_n) \in Z^n$, let $T_\sigma(z) = (z_{\sigma(1)}, \ldots, z_{\sigma(n)})$. It is clear that $T_\sigma : Z^n \to Z^n$ is a homeomorphism and $T_\sigma(Z^n \backslash \Delta_n(Z)) = Z^n \backslash \Delta_n(Z)$.

Suppose that Z is a space which is faithfully indexed by the ordinals from ω_1, i.e., $Z = \{z_\alpha : \alpha < \omega_1\}$ (*faithfully indexed* means $z_\alpha \neq z_\beta$ whenever $\alpha \neq \beta$). Given any $n \in \mathbb{N}$ and $z = (z_{\beta_1}, \ldots, z_{\beta_n}) \in Z^n$, let $\max(z) = \max\{\beta_1, \ldots, \beta_n\}$ and $\min(z) = \min\{\beta_1, \ldots, \beta_n\}$. A point $z = (z_{\beta_1}, \ldots, z_{\beta_n}) \in Z^n \backslash \Delta_n(Z)$ is called *ordered* if $\beta_1 < \cdots < \beta_n$. Call two distinct points $p, q \in Z^n \backslash \Delta_n(Z)$ *comparable* if we have either $\max(p) < \min(q)$ or $\max(q) < \min(p)$. A set $P \subset Z^n \backslash \Delta_n(Z)$ is called *adequate* if each $p \in P$ is ordered and every two distinct elements of P are comparable.

Fact 1. Assume that we have a faithfully indexed space $Z = \{z_\alpha : \alpha < \omega_1\}$. Suppose that for any $n \in \mathbb{N}$, there is no uncountable adequate discrete subspace of $Z^n \backslash \Delta_n(Z)$. Then $s(Z^n) = \omega$ for all $n \in \mathbb{N}$.

Proof. We will prove this by induction on n. For each $z = z_\alpha \in Z$, let $\gamma(z) = \alpha$. If $n = 1$, then $\Delta_n(Z) = \emptyset$ and every uncountable $D \subset Z$ contains an ω_1-sequence $P = \{p_\alpha : \alpha < \omega_1\}$ such that $\gamma(p_\alpha) < \gamma(p_\beta)$ for any $\alpha < \beta < \omega_1$. It is evident that $\gamma(p) = \max(p) = \min(p)$ for any $p \in P$, so the set P is adequate and hence not discrete by our hypothesis. Consequently, the subspace D cannot be discrete either. Thus Z has no uncountable discrete subspaces, i.e., $s(Z) = \omega$.

Assume that our fact is proved for all $m < n$; we have $s(Z^n) = s(Z^n \backslash \Delta_n(Z))$ (see Fact 0 of T.019), so it suffices to establish that $s(Z^n \backslash \Delta_n(Z)) = \omega$. Given any $i \in M_n$, the map $q_i : Z^n \to Z$ is the natural projection of Z^n onto its ith factor and $Z_\alpha = \{z_\beta : \beta < \alpha\}$ for every $\alpha < \omega_1$.

Suppose that $n > 1$ and D is an uncountable discrete subspace of $Z^n \backslash \Delta_n(Z)$. It is clear that for any $z \in D$ there exists $\sigma(z) \in S_n$ such that $T_{\sigma(z)}(z)$ is an ordered point. Since S_n is finite, we can choose an uncountable $D' \subset D$ and $\sigma \in S_n$ such that $\sigma(z) = \sigma$ for all $z \in D'$ and therefore the point $T_\sigma(z)$ is ordered for every $z \in D'$. The mapping $T_\sigma : Z^n \backslash \Delta_n(Z) \to Z^n \backslash \Delta_n(Z)$ being a homeomorphism, the subspace $E = T_\sigma(D')$ is also discrete and consists of ordered points of $Z^n \backslash \Delta_n(Z)$. Since every subspace of E is discrete, it cannot contain an adequate uncountable set.

It is an easy exercise to see that a subset $A \subset Z^n \backslash \Delta_n(Z)$ of ordered points of $Z^n \backslash \Delta_n(Z)$ contains no adequate uncountable set if and only if the set $M(A) = \{\min(z) : z \in A\}$ is bounded in ω_1. Therefore $M(E)$ is bounded in ω_1 and hence there is $\alpha < \omega_1$ such that $\min(z) < \alpha$ for any $z \in E$. As a consequence, $E \subset Q = \bigcup\{q_i^{-1}(Z_\alpha) : i \in M_n\}$. It is evident that $q_i^{-1}(Z_\alpha)$ is a countable union of spaces homeomorphic to Z^{n-1} for each $i \in M_n$ and therefore Q is also a countable union of spaces homeomorphic to Z^{n-1}. Since $s(Z^{n-1}) = \omega$ by the induction hypothesis, we have $s(Q) = \omega$ (it is an easy exercise that a countable union of spaces of countable spread has countable spread) which is a contradiction with the fact that E is an uncountable discrete subspace of Q. Thus $s(Z^n) = \omega$; this completes our inductive step and shows that $s(Z^n) = \omega$ for all $n \in \mathbb{N}$. Fact 1 is proved.   □

Fact 2. Let Z be any space (no axioms of separation are assumed). If, for any $z \in Z$, there exists $U \in \tau(z, Z)$ such that $\overline{U}$ is compact and Hausdorff, then Z is a Tychonoff space.

Proof. Recall that any compact Hausdorff space is Tychonoff by Problem 124 of [TFS]; take an arbitrary point $z \in Z$ and any closed $F \subset Z$ with $z \notin F$. There exists $U \in \tau(z, Z)$ such that $\overline{U}$ is compact and Hausdorff. It is clear that $G = (\overline{U} \backslash U) \cup (F \cap \overline{U})$ is a closed subset of $\overline{U}$ such that $z \notin G$. Therefore there is a continuous function $g : \overline{U} \to [0, 1]$ such that $g(z) = 1$ and $g(G) = \{0\}$. Let $f(y) = 0$ for all $y \in Z \backslash \overline{U}$ and $f(y) = g(y)$ for any $y \in \overline{U}$. It is immediate that $f : Z \to [0, 1]$; besides, $f(z) = g(z) = 1$ and $f(F) \subset \{0\}$, so we must only check that f is continuous at each $y \in Z$. Fix an arbitrary $\varepsilon > 0$.

If $y \in Z \backslash \overline{U}$, then the set $V = Z \backslash \overline{U}$ is a neighborhood of the point y such that $f(V) = \{0\} \subset (-\varepsilon, \varepsilon)$ and hence f is continuous at y. If $y \in U$, then there is $W \in \tau(y, \overline{U})$ such that $g(W) \subset (g(y) - \varepsilon, g(y) + \varepsilon)$. The set $W' = W \cap U$ is open in U and hence in Z, so we have $W' \in \tau(y, Z)$ and $f(W) = g(W) \subset (g(y) - \varepsilon, g(y) + \varepsilon) = (f(y) - \varepsilon, f(y) + \varepsilon)$, so f is continuous at the point y. Finally, if $y \in \overline{U} \backslash U$, then $f(y) = 0$. Choose $O \in \tau(y, \overline{U})$ such that $f(O) = g(O) \subset (-\varepsilon, \varepsilon)$; there is $O' \in \tau(Z)$ such that $O' \cap \overline{U} = O$. It is clear that $O' \in \tau(y, Z)$ and it follows from $f(O' \backslash \overline{U}) = \{0\}$ that $f(O') \subset (-\varepsilon, \varepsilon)$, so f is continuous at y. Fact 2 is proved.   □

Returning to our solution, apply CH to fix a strictly $<^*$-increasing ω_1-sequence $F = \{f_\alpha : \alpha < \omega_1\} \subset \mathbf{P}$ which is cofinal in ω^ω (see Problem 097). If $f, g \in \omega^\omega$ are distinct, then $\Delta(f, g) = \min\{n \in \omega : f(n) \neq g(n)\}$. It is convenient to consider that $\Delta(f, f) = \omega$ for every $f \in \omega^\omega$. Observe that given any function $f \in \omega^\omega$ and any $n \in \omega$, the set $U_n(f) = \{g \in \omega^\omega : \Delta(f, g) \geq n\}$ is clopen in ω^ω and the family $\mathcal{U}_f = \{U_n(f) : n \in \omega\}$ is a local base at f in the space ω^ω. It is easy to check that the function Δ and the collection $\{\mathcal{U}_f : f \in \omega^\omega\}$ have the following properties:

(0) $\Delta(f, h) \geq \min\{\Delta(f, g), \Delta(g, h)\}$ for any functions $f, g, h \in \omega^\omega$; if, additionally, $\Delta(f, g) \neq \Delta(g, h)$, then $\Delta(f, h) = \min\{\Delta(f, g), \Delta(g, h)\}$.
(1) for any $n \in \omega$ and any $f \in \omega^\omega$ if $g \in U_n(f)$, then $U_n(g) \subset U_n(f)$;
(2) $U_{n+1}(f) \subset U_n(f)$ for any $f \in \omega^\omega$ and $n \in \omega$.

Choose an ω_1-sequence $\{s_\alpha : \alpha < \omega_1\}$ with the following properties:

(3) $s_\alpha : \alpha \to \omega$ is an injection for each $\alpha < \omega_1$;
(4) $s_\beta | \alpha \approx s_\alpha$ whenever $\alpha < \beta < \omega_1$.

The existence of such an ω_1-sequence (called an *Aronszajn coding*) was proved in Problem 068. Let $G_\alpha = \{\beta < \alpha : s_\alpha(\beta) \leq f_\alpha(\Delta(f_\alpha, f_\beta))\}$ for each $\alpha < \omega_1$. If we have some $f = f_\alpha \in F$, then let $\gamma(f) = \alpha$ and $H(f) = \{f_\beta : \beta \in G_\alpha\}$. Recall that each ordinal can be identified with the set of its predecessors and, in particular, $n = \{0, \ldots, n-1\}$ for any $n \in \mathbb{N}$. If $f \in \omega^n$ and $g \in \omega^\omega$, then $f \subset g$ says that $g|n = f$. If $A \subset \omega^\omega$, say that A *converges to a point* $f \in \omega^\omega$ if the set $A \backslash U$ is finite for any $U \in \tau(f, \omega^\omega)$. This terminology covers the usual convergence of a sequence to a point as well as the case when A is finite. Thus it is worth remembering that a finite subset of ω^ω "converges" to any point of ω^ω. We must first establish the following fundamental property of the function $H : F \to \exp(F)$.

(5) Given any $f \in F$ the set $H(f)$ converges to f in the topology of ω^ω or, equivalently, the set $\{g \in H(f) : \Delta(g, f) < n\}$ is finite for any $n \in \omega$.

   To see that (5) is true assume the contrary; then we can find $n \in \mathbb{N}$ and an infinite set $A \subset H(f) \backslash U_n(f)$. It is clear that $\Delta(g, f) < n$ for any $g \in A$, so for $r = f(0) + \cdots + f(n) + 1$ and $\alpha = \gamma(f)$, we have $s_\alpha(\gamma(g)) \leq f(\Delta(f, g)) < r$ for any $g \in A$ by definition of $H(f)$. However, $s_\alpha : \alpha \to \omega$ is injective as well as the function $\gamma : A \to \alpha$. It turns out that $s_\alpha \circ \gamma : A \to \omega$ is injective while $s_\alpha(\gamma(A))$ is finite, a contradiction. The property (5) is proved.

   In what follows we consider the space F with its indexation given above and fixed. This indexation is considered for applying the notions introduced before Fact 1. Our promised space Y will have F as the underlying set; the topology of Y will be constructed later using the function H. The following property of H will guarantee the hereditary separability of all finite powers of the space Y:

(6) For any $k \in \mathbb{N}$ and any uncountable adequate set $P \subset F^k$ there exists a point $p = (p_1, \ldots, p_k) \in P$ such that $W(p) = \{q = (q_1, \ldots, q_k) \in P : q_i \in H(p_i)$ for all $i \in M_k\}$ is an infinite set.

   The proof is not easy at all and will be done in several steps. Given any $p = (p_1, \ldots, p_k) \in (\omega^\omega)^k$, let $\pi_i(p) = p_i$ for all $i \in M_k$. Since every subset of an adequate set is an adequate set, we have

(7) for every uncountable $P' \subset P$, the set $\{\pi_i(p) : p \in P'\}$ is $<^*$-cofinal in ω^ω for every $i \in M_k$;

   Since P is an uncountable subspace of the second countable space $(\omega^\omega)^k$, there exists an uncountable $\tilde{P} \subset P$ which has no isolated points in the topology induced from $(\omega^\omega)^k$ (see Fact 1 of S.151). Thus there is no loss of generality to assume that $\tilde{P} = P$, i.e., P is dense-in-itself considered as a subspace of ω^ω. Choose a countable set $D \subset P$ which is dense in P; it is clear that D does not have isolated points as well. Fix any $\mu < \omega_1$ such that $\max(d) < \mu$ for each $d \in D$ and let $P_1 = \{p \in P : \min(p) > \mu\}$; it is clear that P_1 is uncountable. We are going to prove first that

(∗) for every number $l \in \mathbb{N}$ and any uncountable $P' \subset P_1$, there exist distinct points $d_1, \ldots, d_l \in D$ and $p \in P'$ such that $\pi_i(d_j) \in H(\pi_i(p))$ for all $i \in M_k$ and $j \in M_l$.

Observe that for any $p = (p_1, \ldots, p_k) \in P'$, we have $p_1 <^* \cdots <^* p_k$ and hence there exists $m(p) \in \mathbb{N}$ such that $p_i(n) < p_j(n)$ for any $n \geq m(p)$ and $i < j$. There is an uncountable $P_2 \subset P'$ and $m' \in \mathbb{N}$ such that $m(p) = m'$ for all $p \in P_2$.

Since for any $p = (p_1, \ldots, p_k) \in P_2$ we have $f_\mu <^* p_1$, there is $n(p) \in \mathbb{N}$ such that $f_\mu(n) < p_1(n)$ for all $n \geq n(p)$. Choose an uncountable $P_3 \subset P_2$ such that there is $m'' \in \mathbb{N}$ for which $n(p) = m''$ for all $p \in P_3$. Let $m = m' + m''$. Since the set $\{\pi_i(p) | m : p \in P_3\}$ is countable for each $i \in M_k$, we can find an uncountable $P_3^1 \subset P_3$ and $t_1 \in \omega^m$ such that $\pi_1(p) | m = t_1$ for all $p \in P_3^1$. If $h \geq 1$ and we have sets $P_3^1 \supset \cdots \supset P_3^h$, we can choose an uncountable $P_3^{h+1} \subset P_3^h$ such that there is $t_{h+1} \in \omega^m$ for which $\pi_{h+1}(p) | m = t_{h+1}$ for all $p \in P_3^{h+1}$. Making k successive choices as described above we obtain an uncountable set $P_4 = P_3^k \subset P_3$ such that there are functions $\{t_i : i \in M_k\} \subset \omega^m$ for which $\pi_i(p) | m = t_i$ for all $p \in P_4$ and $i \in M_k$.

Observe also that the set $\{s_{\gamma(\pi_i(p))} | \mu : p \in P_4\}$ is countable for every $i \in M_k$: this easily follows from (4). As in the previous paragraph we can refine k times the set P_4 to obtain an uncountable $P_5 \subset P_4$ and functions $u_i : \mu \to \omega$ such that $s_{\gamma(\pi_i(p))} | \mu = u_i$ for all $i \in M_k$ and $p \in P_5$. To sum up, we found an uncountable $P_5 \subset P'$, a number $m \in \mathbb{N}$, functions $t_1, \ldots, t_k \in \omega^m$ and functions $u_1, \ldots, u_k \in \omega^\mu$ such that

(8) for any $p = (p_1, \ldots, p_k) \in P_5$, we have $t_i \subset p_i$ for all $i \in M_k$;

(9) for any $p = (p_1, \ldots, p_k) \in P_5$ and any $i, j \in M_k$ such that $i < j$, we have $p_i(n) < p_j(n)$ for every $n \geq m$;

(10) for any $p = (p_1, \ldots, p_k) \in P_5$, we have $f_\mu(n) < p_1(n)$ for any $n \geq m$;

(11) for any $p = (p_1, \ldots, p_k) \in P_5$, we have $u_i \subset s_{\gamma(p_i)}$ for every $i \in M_k$.

Given an indexed set $A = \{n_u : u \in U\} \subset \omega$ say that A converges to infinity if U is infinite and the set $\{u \in U : n_u \leq h\}$ is finite for any $h \in \omega$. Observe that the set U has to be countable; besides, if $U' \subset U$ is infinite, then the set $\{n_u : u \in U'\}$ also converges to infinity.

The set $B(n) = \{\pi_1(p)(n) : p \in P_5\}$ cannot be finite for all $n \in \omega$; for otherwise, the function h defined by $h(n) = \sup B(n)$ for all $n \in \omega$ is a $<^*$-upper bound for the set $\{\pi_1(p) : p \in P_5\}$, a contradiction with the property (7). Let $l_1 = \min\{n \in \mathbb{N} :$ the set $B(n)$ is infinite$\}$. Observe that $l_1 \geq m$ and the set $\{\pi_1(p) | l_1 : p \in P_5\}$ is finite, so there is an infinite $R_1 \subset P_5$ and $w_1 \in \omega^{l_1}$ such that the set $\{\pi_1(p)(l_1) : p \in R_1\}$ converges to infinity and $\pi_1(p) | l_1 = w_1$ for all $p \in R_1$.

Assume that $1 \leq j < k$ and we have infinite sets $R_1, \ldots, R_j$, natural numbers $l_1, \ldots, l_j$ and functions $w_1, \ldots, w_j$ with the following properties:

(12) $P_5 \supset R_1 \supset \cdots \supset R_j$;

(13) $l_i \geq m$ and $w_i \in \omega^{l_i}$ for all $i \in M_j$;

(14) $\pi_i(p) | l_i = w_i$ for all $i \in M_j$;

(15) the set $\{\pi_i(p)(l_i) : p \in R_i\}$ converges to infinity for all $i \in M_j$.

The set $\{\pi_{j+1}(p)(l_j) : p \in R_j\}$ cannot be bounded in ω because, for each $p \in R_j$, we have $\pi_{j+1}(p)(l_j) > \pi_j(p)(l_j)$ by (9) and (13). Therefore we can define the number $l_{j+1} = \min\{n \in \mathbb{N} :$ the set $\{\pi_{j+1}(p)(n) : p \in R_j\}$ is unbounded$\}$. Observe that $l_{j+1} \geq m$ by (8). The set $\{\pi_{j+1}(p)|l_{j+1} : p \in R_j\}$ being finite, there is an infinite $R_{j+1} \subset R_j$ and $w_{j+1} \in \omega^{l_j+1}$ such that the set $\{\pi_{j+1}(p)(l_{j+1}) : p \in R_{j+1}\}$ converges to infinity and $\pi_{j+1}(p)|l_{j+1} = w_{j+1}$ for all $p \in R_{j+1}$. It is clear that the sets $R_1, \ldots, R_{j+1}$, the numbers $l_1, \ldots, l_{j+1}$ and functions $w_1, \ldots, w_{j+1}$ still satisfy (12)–(15).

Thus our inductive construction can go on to give us infinite sets $R_1, \ldots, R_k$, natural numbers $l_1, \ldots, l_k$ and functions $w_1, \ldots, w_k$ which satisfy (12)–(15) for all $i \in M_k$. The set $P_6 = R_k \subset P$ is infinite and we have

(16) for any $p = (p_1, \ldots, p_k) \in P_6$ we have $p_i|l_i = w_i$ for all $i \in M_k$;
(17) for any $n \in \omega$ there exists $p = (p_1, \ldots, p_k) \in P_6$ such that $p_i(l_i) > n$ for all $i \in M_k$.

The property (16) is evident and (17) can be easily deduced from the fact that the set $\{\pi_i(p)(l_i) : p \in P_6\}$ converges to infinity for all $i \in M_k$.

Since D is dense in P and has no isolated points (in the topology of $(\omega^\omega)^k$), there exist distinct points $d_1, \ldots, d_l \in D$ such that $w_i \subset \pi_i(d_j)$ for all $i \in M_k$ and $j \in M_l$. Since $\pi_i(d_j) <^* f_\mu$ for all $i \in M_k$, there is $r \in \mathbb{N}$ such that $r \geq m + \sum_{i=1}^{k} l_i + 1$ and $\pi_i(d_j)(n) < f_\mu(n)$ for all $n \geq r$ and $j \in M_l$. Apply (17) to conclude that there is $p = (p_1, \ldots, p_k) \in P_6$ such that

$$p_i(l_i) > \max\{u_i(\gamma(\pi_i(d_j))) : j \in M_l\} + f_\mu(r) \text{ for all } i \in M_k.$$

Observe that we have $p_i|l_i = w_i = \pi_i(d_j)|l_i$ for all $i \in M_k$ and $j \in M_l$ while $\pi_i(d_j)(l_i) < \pi_i(d_j)(r) < f_\mu(r) < p_i(l_i)$ which shows that $\Delta(\pi_i(d_j), p_i) = l_i$ and therefore

$$s_{\gamma(p_i)}(\gamma(\pi_i(d_j))) = u_{\gamma(p_i)}(\gamma(\pi_i(d_j))) < p_i(l_i) = p_i(\Delta(p_i, \pi_i(d_j)))$$

for all $i \in M_k$ and $j \in M_l$ which shows that $\pi_i(d_j) \in H(p_i)$ for all $i \in M_k$ and $j \in M_l$. This completes the proof of (∗).

Now assume that for any $p \in P_1$, the set $W(p)$ is finite. Then there is $l \in \mathbb{N}$ and an uncountable $P' \subset P_1$ such that $|W(p)| < l$ for all $p \in P'$. However, (∗) says that there exist distinct $d_1, \ldots, d_l \in D$ and $p \in P'$ such that $d_i \in W(p)$ for all $i \in M_l$; this contradiction shows that there exists $p \in P_1 \subset P$ such that $W(p)$ is infinite. The property (6) is proved.

Let us define recursively a set $C(f)$ such that $f \in C(f) \subset \{g \in F : \gamma(g) \leq \gamma(f)\}$ for any $f \in F$. If $f = f_0$, let $C(f) = \{f\}$. Assume that $f = f_\alpha$ and we have defined $C(g)$ for all $g \in F_\alpha = \{f_\beta : \beta < \alpha\}$. Let $C(f) = \{f\} \cup J(f)$ where $J(f) = \{p \in F_\alpha :$ there exists $q \in H(f)$ such that $p \in C(q)$ and $\Delta(p, q) > \Delta(p, r)$ for any $r \in (\{f\} \cup H(f))\setminus\{q\}\}$. Observe that

(18) $H(f) \subset C(f)$ for any $f \in F$

because for any $p \in H(f)$ we can let $q = p$; then $p \in C(q)$ and $\Delta(p,q) = \omega > \Delta(p,r)$ for any $r \in (H(f) \cup \{f\})\backslash\{q\}$.

Now let $C_n(f) = C(f) \cap U_n(f)$ for each $n \in \omega$ and $f \in F$. It turns out that

(19) for any $n \in \omega$ and $f \in F$, if $g \in C_n(f)$, then $C_m(g) \subset C_n(f)$ for some $m \in \omega$.

We will prove (19) by induction on $\gamma(f)$; if $f = f_0$, then (19) is evidently true. Assume that $\gamma(f) = \alpha < \omega_1$ and (19) is proved for all $g \in F_\alpha$. We must consider three cases.

Case 1: $g = f$. then $m = n$ shows that (19) is true.

Case 2: $g \in H(f)$. Then g is an isolated point of $H_f = \{f\} \cup H(f)$ by (5). Therefore there is $k \in \mathbb{N}$ such that $U_k(g) \cap H_f = \{g\}$ and hence $\Delta(p,g) < k$ for any $p \in H_f\backslash\{g\}$. If $m = k + n$, then $C_m(g) \subset C_n(f)$. To prove it, take any $h \in C_m(g)$; we have $\Delta(h,g) \geq m > \Delta(g,p)$ for any $p \in H_f\backslash\{g\}$. Applying (0) we conclude that $\Delta(h,p) = \min\{\Delta(h,g), \Delta(g,p)\} = \Delta(g,p) < m \leq \Delta(h,g)$ for every $p \in H_f\backslash\{g\}$ which shows that $h \in C(f)$. The property (0) implies that $\Delta(h,f) \geq \min\{\Delta(h,g), \Delta(g,f)\} = \Delta(g,f) \geq n$. Therefore $h \in C_n(f)$ and we proved (19) for this case.

Case 3: $g \notin H(f)$; then there exists a function $h \in H(f)$ such that $g \in C(h)$ and we have $\omega > k = \Delta(g,h) > \Delta(g,p)$ for any $p \in H_f\backslash\{h\}$. Consequently, $\Delta(h,p) = \min\{\Delta(h,g), \Delta(g,p)\} = \Delta(g,p) < k$ for all $p \in H_f\backslash\{h\}$. This shows that $U_k(h) \cap H_f = \{h\}$; since for any $g' \in U_k(h)$ we have $U_k(g') \subset U_k(h)$, we obtain the equality $(H_f\backslash\{h\}) \cap U_k(g') = \emptyset$. Thus, if $g' \in C_k(h)$, then $\Delta(g',h) \geq k > \Delta(g',p)$ for any $p \in H_f\backslash\{h\}$ and hence $g' \in C(f)$. As a consequence, $C_k(h) \subset C(f)$. Besides, $n \leq \Delta(g,f) < \Delta(g,h) = k$ whence $\Delta(h,f) \geq n$ which shows that $h \in U_n(f)$ and therefore $U_n(h) \subset U_n(f)$, so $C_k(h) \subset C(f) \cap U_n(h) \subset C(f) \cap U_n(f) = C_n(f)$. Since $g \in C_k(h)$ and $\gamma(h) < \gamma(f)$, we can apply the induction hypothesis to h to conclude that there exists $m \in \omega$ such that $C_m(g) \subset C_k(h) \subset C_n(f)$ which shows that the proof of (19) is complete.

The property (19) implies that the family $\{C_n(f) : f \in F \text{ and } n \in \omega\}$ generates a topology τ on F as a base (see Problem 006 of [TFS]); let $Y = (F, \tau)$. It is an easy exercise that the topology τ' inherited by F from ω^ω is contained in τ. An immediate consequence is that Y is Hausdorff. Let us prove that

(20) $C(f)$ is a compact subspace of Y for any $f \in F$.

We prove this again by induction on $\gamma(f)$. If $f = f_0$, then $C(f)$ is a singleton and hence compact. Assume that $\alpha < \omega_1$ and we proved (20) for all $g \in F$ with $\gamma(g) < \alpha$; let $f = f_\alpha$. Since $C(f)$ is countable, it suffices to show that it is countably compact, i.e., every infinite set $D \subset C(f)$ has an accumulation point in $C(f)$ (see Problem 132 of [TFS]).

Assume first that for some $n \in \omega$, the set $D_n = \{d \in D : \Delta(d, f) = n\}$ is infinite. By the definition of the set $C(f)$, for any $d \in D_n$, there is $f_d \in H(f)$ such that $d \in C(f_d)$ and $\Delta(d, f_d) > \Delta(d, h)$ for all $h \in H_f \setminus \{f_d\}$ and, in particular, $\Delta(d, f_d) > \Delta(d, f) = n$ for all $d \in D_n$. Applying the property (0) to the functions d, f_d, f, we obtain the equality $\Delta(f_d, f) = \min\{\Delta(f_d, d), \Delta(d, f)\} = n$ for all $d \in D_n$. The property (5) implies that the set $\{f_d : d \in D_n\}$ is finite, so there is $g \in H(f)$ such that $\Delta(g, f) = n$ and the set $N = \{d \in D_n : f_d = g\}$ is infinite.

Let $m = \min\{\Delta(d, g) : d \in N\}$; pick any $d \in N$ with $\Delta(d, g) = m$. It is evident that $N \subset C_m(g)$. Observe that $\Delta(g, h) = \min\{\Delta(g, d), \Delta(d, h)\} = \Delta(d, h)$ for every $h \in H_f \setminus \{g\}$ (we used (0) again). Since $\Delta(d, h) < \Delta(d, g) = m$, we have $\Delta(g, h) < m$ for any $h \in H_f \setminus \{g\}$. Now take any $g' \in C_m(g)$; we have $\Delta(g', g) \geq m$. On the other hand $\Delta(g, h) < m \leq \Delta(g', g)$ for all $h \in H_f \setminus \{g\}$, and hence we can apply (0) once more to conclude that $\Delta(g', h) = \min\{\Delta(g', g), \Delta(g, h)\} = \Delta(g, h) < m \leq \Delta(g', g)$. Thus $\Delta(g', h) < \Delta(g', g)$ for any $h \in H_f \setminus \{g\}$ so $g' \in C(f)$. The point $g' \in C_m(g)$ was taken arbitrarily so $C_m(g) \subset C(f)$. Therefore N is an infinite subset of $C_m(g)$; since $\gamma(g) < \gamma(f)$, we can apply the induction hypothesis and conclude that the set N has an accumulation point p in $C(g)$. But $N \subset C_m(g)$ and $C_m(g)$ is closed in $C(g)$, so $p \in C_m(g)$ and hence $p \in C(f)$, i.e., p is an accumulation point for D in $C(f)$. Finally, if the set D_n is finite for each $n \in \omega$, then every $C_n(f)$ contains infinitely many points of D; hence f is an accumulation point of D and our proof of (20) is complete.

Since $C(f)$ is a neighborhood of f in Y, the space Y is locally compact and hence Tychonoff by Fact 2. Let us show that $s^*(Y) = \omega$. Observe that we have an enumeration on Y because the enumerated set F is the underlying set of Y. Thus it suffices to show that any adequate discrete $E \subset Y^k$ is countable for any $k \in \mathbb{N}$ (see Fact 1). Assume the contrary and take any uncountable discrete adequate $E \subset Y^k$. For each $p = (p_1, \ldots, p_k) \in E$, there exists $n = n(p) \in \mathbb{N}$ such that $V_p \cap E = \{p\}$ where $V_p = C_n(p_1) \times \cdots \times C_n(p_k)$. There is an uncountable $E' \subset E$ and $n \in \mathbb{N}$ such that $n(p) = n$ for all $p \in E'$. The property (6) implies that there is $p = (p_1, \ldots, p_k) \in E'$ and an infinite $Q \subset E'$ such that $p \notin Q$ and $\pi_i(q) \in H(p_i)$ for any $i \in M_k$ and any $q \in Q$. Since the set E' is adequate, we have $\pi_i(q) \neq \pi_i(q')$ for any distinct $q, q' \in Q$ and any $i \in M_k$. The infinite set $\{\pi_i(q) : q \in Q\}$ converges to p_i (in the topology of ω^ω) for each $i \in M_k$, so there is $q = (p_1, \ldots, p_k) \in Q$ such that $q_i \in C_n(p_i)$ for every $i \in M_k$ and therefore $q \in (V_p \cap E) \setminus \{p\}$ which is a contradiction. Thus $s^*(Y) = \omega$.

Next observe that Y is scattered; to see it suffices to show that every closed $Z \subset Y$ has an isolated point. Take any $z \in Z$; the set $K = C(z) \cap Z$ is a compact open neighborhood of z in Z. Since K is countable, it has an isolated point w. It is clear that w is isolated in Z, so we established that Y is scattered and hence not hereditarily Lindelöf (see Problems 005 and 006). As a consequence, Y^n is not hereditarily Lindelöf for all $n \in \mathbb{N}$. It follows from $s(Y^n \times Y^n) = \omega$ that Y^n is either hereditarily separable or hereditarily Lindelöf by Problem 014. Since Y^n is not hereditarily Lindelöf, we have $hd(Y^n) = \omega$ for all $n \in \mathbb{N}$, i.e., $hd^*(Y) = \omega$. Given any $f \in Y$, the set $C(f)$ is a countable neighborhood of f, so Y is locally countable. Thus, we checked everything promised in (i)–(iii) for Y so our solution is complete.

T.099. *Prove that, under CH, there exists a scattered compact space X which is not first countable (and hence not metrizable) while X^ω is hereditarily separable and hence $(C_p(X))^\omega$ is hereditarily Lindelöf. This implies that under CH, there exist strong L-spaces and strong compact S-spaces. Observe that, under $MA+\neg CH$, any compact space X is metrizable whenever $(C_p(X))^\omega$ is hereditarily Lindelöf.*

Solution. Let X be the one-point compactification of the space Y constructed in Problem 098. The space X is well-defined because Y is locally compact. It is an easy exercise that adding one point to a scattered space gives a scattered space so X is scattered. Let a be the unique point of $X \backslash Y$. If $\psi(a, X) = \omega$, then the space $Y = X \backslash \{a\}$ is σ-compact and hence Lindelöf which contradicts Problem 098(ii). This proves that $\psi(a, X) > \omega$ and hence X is not first countable.

Fact 1. Let Z be any space; assume that $z \in Z$ and $hd^*(Z \backslash \{z\}) \leq \kappa$ for some infinite cardinal κ. Then $hd^*(Z) \leq \kappa$. In particular, if all finite powers of a space are hereditarily separable, then adding one point to the space does not destroy this property.

Proof. Let $M_n = \{1, \ldots, n\}$; we will prove by induction on n that $hd(Z^n) \leq \kappa$ for all $n \in \mathbb{N}$. Observe first that the space $Z = (Z \backslash \{z\}) \cup \{z\}$ is a union of two spaces of hereditary density $\leq \kappa$. It is evident that even a countable union of spaces of hereditary density $\leq \kappa$ has hereditary density $\leq \kappa$ so the case of $n = 1$ is clear.

Now assume that we proved that $hd(Z^k) \leq \kappa$ for all $k < n$. Given any $i \in M_n$, let $p_i : Z^n \to Z$ be the natural projection of Z^n onto its ith factor. Consider the set $Q = \bigcup \{p_i^{-1}(z) : i \in M_n\}$; it is easy to see that $Z^n = (Z \backslash \{z\})^n \cup Q$. The space Q is a finite union of spaces homeomorphic to Z^{n-1}, so $hd(Q) \leq \kappa$ by the induction hypothesis and the above observation about countable unions of spaces of whose hereditary density does not exceed κ. Since $hd((Z \backslash \{z\})^n) \leq \kappa$ by our hypothesis, we represented the space Z^n as a finite union of spaces of hereditary density $\leq \kappa$. Thus $hd(Z^n) \leq \kappa$ for all $n \in \mathbb{N}$ and hence Fact 1 is proved. $\square$

Returning to our solution apply Fact 1 to conclude that $hd^*(X) = \omega$ and hence $hl((C_p(X))^\omega) = hl^*(C_p(X)) = hd^*(X) = \omega$ (we applied Problems 011 and 027). Thus under CH we have a compact space X with all promised properties. If $MA+\neg CH$ holds and we have a compact space X such that $hl((C_p(X))^\omega) = \omega$, then $s(X \times X) \leq hd^*(X) = hl^*(C_p(X)) = \omega$ and hence we can apply Problem 062 to conclude that X is metrizable.

T.100. *Prove that, under CH, there exists a non-normal X such that the space $(C_p(X))^\omega$ is hereditarily Lindelöf.*

Solution. Let K be a non-metrizable compact space with $hd^*(K) = \omega$. The existence of such a space under CH is proved in Problem 099. Apply Problem 090 to conclude that K^3 is not hereditarily normal and hence there is $X \subset K^3$ such that X is not normal. It is evident that $hd^*(X) \leq hd^*(K^3) = hd^*(K) = \omega$ and hence $hl((C_p(X))^\omega) = hl^*(C_p(X)) = hd^*(X) = \omega$ (we applied Problems 011 and 027). Thus X is a non-normal space for which $(C_p(X))^\omega$ is hereditarily Lindelöf.

T.101. *Prove that any metrizable space is strongly monolithic.*

Solution. Let M be a metrizable space. If $A \subset M$ is an infinite set and $|A| = \kappa$, then $\overline{A}$ is also a metrizable space and therefore $w(\overline{A}) = d(\overline{A}) \leq \kappa$ by Problem 214 of [TFS]. Thus M is κ-monolithic for any infinite cardinal κ.

T.102. *Prove that if X is a metrizable space with $w(X) \leq \mathfrak{c}$, then X condenses onto a second countable space.*

Solution. We will need the following fact.

Fact 1. Let Z be an infinite space with $w(Z) = \kappa$. Then, for any base $\mathcal{B}$ for the space Z, there is $\mathcal{B}' \subset \mathcal{B}$ such that $|\mathcal{B}'| \leq \kappa$ and $\mathcal{B}'$ is a base in Z. In other words, any base of a space contains a base of minimal cardinality.

Proof. Take any base $\mathcal{C}$ in Z such that $|\mathcal{C}| = \kappa$. Call a pair $\mu = (U, V) \in \mathcal{C} \times \mathcal{C}$ *admissible* if there is $B \in \mathcal{B}$ such that $U \subset B \subset V$. If a pair $\mu = (U, V)$ is Admissible, then fix some $B = B(\mu) \in \mathcal{B}$ such that $U \subset B \subset V$. The family $\mathcal{B}' = \{B(\mu) : \mu$ is an admissible pair$\} \subset \mathcal{B}$ has cardinality at most $|\mathcal{C} \times \mathcal{C}| = |\mathcal{C}| = \kappa$. The family $\mathcal{B}'$ is a base in Z; indeed, if $x \in W \in \tau(Z)$, then there is $V \in \mathcal{C}$ such that $x \in V \subset W$ because $\mathcal{C}$ is a base in Z. By the same reason, there is $B' \in \mathcal{B}$ with $x \in B' \subset V$. Analogously, we can find $U \in \mathcal{C}$ such that $x \in U \subset B'$. Since $U \subset B' \subset V$, the pair $\mu = (U, V)$ is admissible and therefore $B = B(\mu) \in \mathcal{B}'$ and $x \in B \subset V \subset W$. Therefore $\mathcal{B}'$ is a base in Z and Fact 1 is proved.   □

Returning to our solution fix any σ-discrete base $\mathcal{B}$ in the space X (see Problem 221 of [TFS]). Fact 1 shows that we can assume, without loss of generality, that $|\mathcal{B}| \leq \mathfrak{c}$. We have $\mathcal{B} = \bigcup\{\mathcal{B}_n : n \in \omega\}$ where each family $\mathcal{B}_n$ is discrete. Since $|\mathcal{B}_n| \leq \mathfrak{c}$, for each $n \in \omega$, there exists an injection $\varphi_n : \mathcal{B}_n \to T$ where $T = \{(a, b) \in \mathbb{R} \times \mathbb{R} : a^2 + b^2 = 1\}$ is the unit circumference centered at the origin $z_0 = (0, 0)$ of $\mathbb{R} \times \mathbb{R}$. Fix any $n \in \omega$; for any $U \in \mathcal{B}_n$, the set $I_U = \{(ta, tb) \in \mathbb{R}^2 : t \in [0, 1]$ and $(a, b) = \varphi_n(U)\}$ is homeomorphic to the interval $[0, 1] \subset \mathbb{R}$ and hence there exists a continuous map $f_U : X \to I_U$ such that $X \backslash U = f_U^{-1}(z_0)$ (we used Fact 1 of S.358 and perfect normality of any metrizable space). Let $g_n : X \to \mathbb{R} \times \mathbb{R}$ be defined as follows: $g_n(x) = z_0$ if $x \notin \bigcup \mathcal{B}_n$; if $x \in U \in \mathcal{B}_n$, then $g_n(x) = f_U(x)$.

To see that g_n is continuous take any $x \in X$. Since the family $\mathcal{B}_n$ is discrete, there is $W \in \tau(x, X)$ such that W intersects at most one element of $\mathcal{B}_n$, say U. It is evident that $g_n|W = f_U|W$ is a continuous map, so we can apply Fact 1 of S.472 to see that g_n is continuous for all $n \in \omega$. The diagonal product $g = \Delta\{g_n : n \in \omega\}$ maps X onto a second countable space $Y \subset (\mathbb{R} \times \mathbb{R})^\omega$. Thus it suffices to show that g is an injection. Take any distinct $x, y \in X$; there is $n \in \omega$ and $U \in \mathcal{B}_n$ such that $x \in U$ and $y \notin U$. Therefore $g_n(y) = f_U(y) = z_0$ while $g_n(x) = f_U(x) \neq z_0$ because $x \notin X \backslash U$. As a consequence, $g(x)(n) = g_n(x) \neq g_n(y) = g(y)(n)$ and hence $g(x) \neq g(y)$, i.e., $g : X \to Y$ is a condensation.

T.103. *Suppose that A is a proper closed subset of a metric space (X,d) and let $d(x,A) = \inf\{d(x,a) : a \in A\}$ for any $x \in X$. Prove that there exists a family $\{U_s, a_s : s \in S\}$ (called a Dugundji system for $X\backslash A$) such that*

(i) $U_s \subset X\backslash A$ and $a_s \in A$ for any $s \in S$;
(ii) $\{U_s : s \in S\}$ is an open locally finite (in $X\backslash A$) cover of $X\backslash A$;
(iii) $d(x,a_s) \le 2d(x,A)$ for any $s \in S$ and any $x \in U_s$.

Solution. A family $\mathcal{A} \subset \exp(X)$ is called *inscribed* in a family $\mathcal{B} \subset \exp(X)$ if, for any $A \in \mathcal{A}$, there in $B \in \mathcal{B}$ such that $A \subset B$. As usual, if $x \in X$ and $r > 0$, then $B(x,r) = \{y \in X : d(x,y) < r\}$ is the open ball of radius r centered at x. It is evident that $r(x) = d(x,A) > 0$ for any $x \in X\backslash A$; it is also immediate that $B(x, r(x)) \subset X\backslash A$ for any $x \in X\backslash A$. Thus the family $\mathcal{V} = \{B(x, \frac{1}{4}r(x)) : x \in X\backslash A\}$ is an open cover of $X\backslash A$. Since any metrizable space is paracompact (Problem 218 of [TFS]), there is a locally finite (in $X\backslash A$) cover $\mathcal{U} = \{U_s : s \in S\}$ of $X\backslash A$ inscribed in $\mathcal{V}$. Therefore for each $s \in S$ there is $x_s \in X\backslash A$ such that $U_s \subset B(x_s, \frac{1}{4}r(x_s))$. By definition of $d(x_s, A)$, there is $a_s \in A$ such that $d(x_s, a_s) \le \frac{5}{4}d(x_s, A)$.

It is straightforward that the family $\mathcal{W} = \{U_s, a_s : s \in S\}$ satisfies the conditions (i)–(ii). To check that $\mathcal{W}$ also has (iii), take an arbitrary $s \in S$ and pick any $x \in U_s$. Let $r = d(x_s, A)$; it follows from $x \in B(x_s, \frac{1}{4}r)$ that $d(x, A) \ge \frac{3}{4}r$ (and hence $\frac{3}{2}r \le 2d(x, A)$) for otherwise there is $a \in A$ with $d(x, a) < \frac{3}{4}r$, so we have

$$r = d(x_s, A) \le d(x_s, a) \le d(x_s, x) + d(x, a) < \frac{1}{4}r + \frac{3}{4}r = r,$$

a contradiction. On the other hand, $d(x, a_s) \le d(x, x_s) + d(x_s, a_s) \le \frac{1}{4}r + \frac{5}{4}r = \frac{3}{2}r$, which shows that $d(x, a_s) \le \frac{3}{2}r \le 2d(x, A)$ so the family $\mathcal{W}$ is as promised.

T.104. *Let A be a closed subspace of a metrizable space X and suppose that $f : A \to L$ is a continuous map of A into a locally convex linear topological space L. Prove that there exists a continuous map $F : X \to L$ such that $F|A = f$ and $F(X) \subset conv(f(A))$.*

Solution. Let us first develop some technique of handling linear topological spaces and continuous maps into them. If M is a linear topological space, $x \in M$ and $P \subset M$, then $x + P = \{x + y : y \in P\}$. Given any $t \in \mathbb{R}$ and $P \subset M$, let $tP = \{ty : y \in P\}$; if $P_1, \ldots, P_n \subset M$, then

$$P_1 + \cdots + P_n = \{x_1 + \cdots + x_n : x_i \in P_i \text{ for all } i = 1, \ldots, n\}.$$

Assume that Z is a space and $\{b_s : s \in S\}$ is a family of real-valued functions on Z such that $S(z) = \{s \in S : b_s(z) \ne 0\}$ is finite for any $z \in Z$. Then we denote by $\sum_{s \in S} b_s$ the function $b : Z \to \mathbb{R}$ defined by $b(z) = \sum_{s \in S(z)} b_s(z)$ for each $z \in Z$.

Fact 1. Given a linear space M and $P \subset M$ we have :

(1) $P \subset \mathrm{conv}(P)$;
(2) $\mathrm{conv}(\mathrm{conv}(P)) = \mathrm{conv}(P)$;
(3) the set P is convex if and only if $\mathrm{conv}(P) \subset P$ and, in particular, $\mathrm{conv}(P)$ is a convex set for any $P \subset M$.

Proof. Given any $x \in P$ we have $x = 1 \cdot x \in \mathrm{conv}(P)$ which proves (1). Therefore $\mathrm{conv}(P) \subset \mathrm{conv}(\mathrm{conv}(P))$. Now assume that $u \in \mathrm{conv}(\mathrm{conv}(P))$ and hence there exist $s_1, \ldots, s_k \in [0, 1]$ with $s_1 + \cdots + s_k = 1$ such that $u = s_1 x_1 + \cdots + s_k x_k$ for some $x_1, \ldots, x_k \in \mathrm{conv}(P)$. For each $i \le k$ there are $t_{i1}, \ldots, t_{in_i} \in [0, 1]$ with $t_{i1} + \cdots + t_{in_i} = 1$ and $x_i = t_{i1} y_{i1} + \cdots + t_{in_i} y_{in_i}$ for some $y_{i1}, \ldots, y_{in_i} \in P$. Therefore $u = \sum_{i=1}^{k} (\sum_{j=1}^{n_i} s_i t_{ij} y_{ij})$; it is clear that $s_i t_{ij} \in [0, 1]$ for all $i \le k$ and $j \le n_i$. The equality $\sum_{i=1}^{k} (\sum_{j=1}^{n_i} s_i t_{ij}) = \sum_{i=1}^{k} s_i (\sum_{j=1}^{n_i} t_{ij}) = \sum_{i=1}^{k} s_i = 1$ implies $u \in \mathrm{conv}(P)$ so $\mathrm{conv}(\mathrm{conv}(P)) \subset \mathrm{conv}(P)$ and (2) is proved.

Assume that $\mathrm{conv}(P) \subset P$; then $\mathrm{conv}(P) = P$ by the property (1); observe that $tx + (1 - t)y \in \mathrm{conv}(P) = P$ for any $x, y \in P$ and $t \in [0, 1]$, so P has to be convex and we proved sufficiency in (3).

Now, if P is convex, then take any $u = t_1 x_1 + \cdots + t_n x_n \in \mathrm{conv}(P)$ such that $t_1, \ldots, t_n \in [0, 1]$, $x_1, \ldots, x_n \in P$ and $t_1 + \cdots + t_n = 1$; we will prove that $u \in P$ by induction on n. If $n = 1$, then $t_1 = 1$ and hence $u = x_1 \in P$.

Assume that $n > 1$, and, for any $k < n$ and any $s_1, \ldots, s_k \in [0, 1]$ such that $s_1 + \cdots + s_k = 1$, we have $\sum_{i=1}^{k} s_i y_i \in P$ for any $y_1, \ldots, y_k \in P$. Observe that $u = (1 - t_n)z + t_n x_n$ where $z = \frac{t_1}{1-t_n} x_1 + \cdots + \frac{t_{n-1}}{1-t_n} x_{n-1} \in P$ by the induction hypothesis because $\sum_{i=1}^{n-1} \frac{t_i}{1-t_n} = 1$. Therefore $u \in P$ because P is convex. Since the point $u \in \mathrm{conv}(P)$ was taken arbitrarily, we established that $\mathrm{conv}(P) \subset P$, so we settled (3) and hence Fact 1 is proved.   □

Fact 2. Assume that M is a linear topological space; denote by **0** the zero vector of M. Given a space Z, a map $f : Z \to M$ is continuous at a point $z \in Z$ if and only if for any $V \in \tau(\mathbf{0}, M)$ there is $U \in \tau(z, Z)$ such that $f(y) - f(z) \in V$ for any $y \in U$.

Proof. If $v \in M$, then the map $S_v : M \to M$ defined by $S_v(u) = u + v$ for all $u \in M$ is a homeomorphism (see Fact 1 of S.496) and S_{-v} is the inverse map for S_v.

Assume first that f is continuous at the point z; given any $V \in \tau(\mathbf{0}, M)$ let $W = f(z) + V = S_{f(z)}(V)$. Then $W \in \tau(f(z), M)$ being W a homeomorphic image of V under the map $S_{f(z)}$. By continuity of f at the point z there is $U \in \tau(z, Z)$ such that $f(U) \subset W$; now, for any $y \in U$ we have $f(y) \in W$ and therefore

$$f(y) - f(z) \in W - f(z) = S_{-f(z)}(W) = V,$$

which settles necessity.

Now assume that for any set $V \in \tau(\mathbf{0}, M)$ there is $U \in \tau(z, Z)$ such that $f(y) - f(z) \in V$ for any $y \in U$. Given any $W \in \tau(f(z), M)$ observe that we have

$V = W - f(z) \in \tau(\mathbf{0}, M)$, so there is $U \in \tau(z, Z)$ such that $f(y) - f(z) \in V$ for all $y \in U$. Consequently, $f(y) \in f(z) + V = W$ for any $y \in U$ and hence $f(U) \subset W$, i.e., f is continuous at the point z. Fact 2 is proved.   □

Fact 3. Take a space Z and a linear topological space M. Then

(4) if $f, g : Z \to M$ are continuous maps, then the map $h = f + g : Z \to M$ defined by $h(z) = f(z) + g(z)$ for each $z \in Z$ is continuous;
(5) if $f : Z \to M$ is a continuous map, then the map $m = b \cdot f : Z \to M$ defined by $m(z) = b(z) \cdot f(z)$ for each $z \in Z$ is continuous for any $b \in C(Z)$;
(6) if $f_1, \ldots, f_n : Z \to M$ are continuous maps and $b_1, \ldots, b_n \in C(Z)$, then the map $u = b_1 f_1 + \cdots + b_n f_n : Z \to M$ defined by $u(z) = b_1(z) f_1(z) + \cdots + b_n(z) f_n(z)$ for each $z \in Z$ is continuous.

Proof. Define a map $\alpha : M \times M \to M$ by $\alpha(p, q) = p + q$ for any $(p, q) \in M \times M$; let $\beta : \mathbb{R} \times M \to M$ be the map defined by $\beta(t, p) = t \cdot p$ for each $(t, p) \in \mathbb{R} \times M$. The maps α and β are continuous by the definition of the topology of a linear topological space. If $h'(x) = (f(x), g(x))$ and $m'(x) = (b(x), f(x))$ for every $x \in X$, then the maps $h' = \Delta\{f, g\} : X \to M \times M$ and $m' = \Delta\{b, f\} : X \to \mathbb{R} \times M$ are also continuous (see Introductory Part of V1.5). Therefore the maps $h = \alpha \circ h'$ and $m = \beta \circ m'$ are continuous so we proved (4) and (5). Finally, (6) is obtained from (4) and (5) by an evident induction. Fact 3 is proved.   □

Fact 4. Let Z be a paracompact space. Then, for any open cover $\mathcal{U} = \{U_s : s \in S\}$ of the space Z, there exists a closed locally finite cover $\{F_s : s \in S\}$ of the space Z such that $F_s \subset U_s$ for all $s \in S$.

Proof. By paracompactness of Z, there exists a closed locally finite refinement $\mathcal{F}$ of the cover $\mathcal{U}$. For each $F \in \mathcal{F}$ there is $s(F) \in S$ such that $F \subset U_{s(F)}$. Let $F_s = \bigcup\{F \in \mathcal{F} : s(F) = s\}$; since every locally finite family is closure-preserving (Fact 2 of S.221), the set F_s is closed for any $s \in S$. If $z \in Z$, then $z \in F$ for some $F \in \mathcal{F}$ and hence $z \in F_{s(F)}$ which shows that $\mathcal{F}'$ is a cover of Z. It is evident that $F_s \subset U_s$ for each $s \in S$; let us prove that the family $\mathcal{F}' = \{F_s : s \in S\}$ is locally finite.

Take any point $z \in Z$ and choose a set $W \in \tau(z, Z)$ such that the family $\mathcal{G} = \{F \in \mathcal{F} : W \cap F \neq \emptyset\}$ is finite. The set $S' = \{s(F) : F \in \mathcal{G}\}$ is also finite, and given any $s \in S \backslash S'$ and any $F \in \mathcal{F}$ with $s(F) = s$, we have $W \cap F = \emptyset$ and therefore $W \cap F_s = \emptyset$. This shows that W intersects only finitely many elements of $\mathcal{F}'$, so the family $\mathcal{F}'$ is locally finite and Fact 4 is proved.   □

Returning to our solution, choose a metric d on X with $\tau(d) = \tau(X)$ and a Dugundji system $\{U_s, a_s : s \in S\}$ for $X \backslash A$ (see Problem 103). Apply Fact 4 to find a locally finite (in $X \backslash A$) family $\{F_s : s \in S\}$ such that F_s is closed in $X \backslash A$ and $F_s \subset U_s$ for each $s \in S$. By normality of the space $X \backslash A$, there exists a continuous function $c_s : X \backslash A \to [0, 1]$ such that $c_s(F_s) \subset \{1\}$ and $c_s((X \backslash A) \backslash U_s) \subset \{0\}$ for every $s \in S$. Since the family $\{U_s : s \in S\}$ is locally finite in $X \backslash A$, the set $S(x) = \{s \in S : c_s(x) \neq 0\}$ is finite for every $x \in X \backslash A$ and hence the function

$c = \sum_{s \in S} c_s$ is well-defined. Given any $x \in X \backslash A$, there is $W \in \tau(x, X \backslash A)$ such that $S' = \{s \in S : W \cap U_s \neq \emptyset\}$ is finite; it is evident that $S(y) \subset S'$ for any $y \in W$ and hence $c|W = (\sum_{s \in S'} c_s)|W$ is continuous. Thus we can apply Fact 1 of S.472 to conclude that $c : X \backslash A \to \mathbb{R}$ is continuous.

Note that for any $x \in X \backslash A$, there is $s \in S$ such that $x \in F_s$ and hence $c_s(x) = 1$; this shows that $c(x) \geq 1 > 0$ for any $x \in X \backslash A$, so the function $b_s = \frac{c_s}{c}$ is well-defined. It is immediate that

(7) $(\sum_{s \in S} b_s)(x) = 1$ for any $x \in X \backslash A$.

Now, let $F(x) = f(x)$ for any $x \in A$ and $F(x) = \sum_{s \in S} b_s(x) f(a_s)$ for each $x \in X \backslash A$. Observe that $F(x)$ makes sense for any $x \in X \backslash A$ because $b_s(x) = 0$ for any $s \in S \backslash S(x)$. We have $F|A = f$ by our definition of F; besides, for any $x \in X \backslash A$, the point $F(x) = \sum_{s \in S} b_s(x) f(a_s) = \sum_{s \in S(x)} b_s(x) f(a_s)$ belongs to $\text{conv}(f(A))$ due to the fact that $\sum_{s \in S(x)} b_s(x) = 1$ and $b_s(x) \in [0, 1]$ for each $s \in S(x)$. This shows that $F(X) \subset \text{conv}(f(A))$, so we only must prove that F is continuous. Given any $p \in L$ let $\theta_p : X \backslash A \to L$ be the constant function defined by p, i.e., $\theta_p(x) = p$ for every $x \in X \backslash A$.

Observe first that

(8) for any finite set $S' \subset S$, the mapping $F_{S'} : X \backslash A \to L$ defined by the equality $F_{S'}(x) = \sum_{s \in S'} b_s(x) f(a_s)$ for each $x \in X \backslash A$ is continuous

because it is obtained from constant (and hence continuous) maps $\{\theta_{f(a_s)} : s \in S'\}$ by operations described in Fact 3. Given any $x \in X \backslash A$, there is $W \in \tau(x, X \backslash A)$ such that the set $T(x) = \{s \in S : W \cap U_s \neq \emptyset\}$ is finite. As a consequence, $S(y) \subset T(x)$ for every $y \in W$ which shows that $F|W = (\sum_{s \in T(x)} b_s \theta_{f(a_s)})|W = F_{T(x)}|W$ is continuous because so is the function $F_{T(x)}$ by the property (7). Therefore F is continuous at the point x.

Now take any point $a \in A$ and any $W \in \tau(0, L)$. Since L is locally convex, there is a convex $V \in \tau(0, L)$ such that $V \subset W$. Since f is continuous on A, we can apply Fact 2 to find $\delta > 0$ such that $a' \in A$ and $d(a, a') < \delta$ imply $f(a') - f(a) \in V$. Let $U = B(a, \frac{\delta}{3}) = \{a' \in X : d(a, a') < \frac{\delta}{3}\}$. If $x \in U \cap A$, then $F(x) - F(a) = f(x) - f(a) \in V$. Now, if $x \in U \backslash A$, then $d(x, A) \leq d(x, a) < \frac{\delta}{3}$. Besides, if $x \in U_s$, then $d(a_s, a) \leq d(a_s, x) + d(x, a) \leq 2d(x, A) + d(x, a) < \frac{2\delta}{3} + \frac{\delta}{3} = \delta$ (we used the fact that $\{U_s, a_s : s \in S\}$ is a Dugundji system and therefore $d(x, a_s) \leq 2d(x, A)$ because $x \in U_s$). As a consequence, $f(a_s) - f(a) \in V$ for any $s \in S$ such that $x \in U_s$. Let $S' = \{s \in S : x \in U_s\}$; then $S(x) \subset S'$ and we have

$$F(x) - F(a) = \sum_{s \in S'} b_s(x) f(a_s) - f(a) = \sum_{s \in S'} b_s(x)(f(a_s) - f(a)) \in \sum_{s \in S'} b_s(x) V \subset V;$$

we used the equality $\sum_{s \in S'} b_s(x) = 1$, the convexity of V and Fact 1. Thus we found $U \in \tau(a, X)$ such that $F(x) - F(a) \in V \subset W$ for any $x \in U$, so F is continuous at the point a by Fact 2. Therefore F has all required properties, so our solution is complete.

T.105. *Prove that every metrizable non-separable space can be mapped continuously onto a metrizable space of weight ω_1.*

Solution. Our solution will be based on the fact that there are sufficiently many metrizable locally convex spaces.

Fact 1. Given a space X, let $d(f, g) = \sup\{|f(x) - g(x)| : x \in X\}$ for any functions $f, g \in C^*(X)$. Then d is a metric on $C^*(X)$ and $M = (C^*(X), d)$ is a locally convex metrizable linear topological space.

Proof. It is evident that $C^*(X)$ is a linear space if we consider it with the usual operations of summing functions and multiplying them by a real number. The function d is indeed a metric on $C^*(X)$: this was proved in Problem 248 of [TFS]. To see that M is locally convex, it suffices to show that the ball $B_d(f, \varepsilon) = \{g \in C^*(X) : d(f, g) < \varepsilon\}$ is convex for any $f \in C^*(X)$ and $\varepsilon > 0$. Take any $g, h \in B_d(f, \varepsilon)$ and $t \in [0, 1]$; there is $\delta > 0$ such that $\max\{d(f, g), d(f, h)\} < \delta < \varepsilon$. Let $p = tg + (1 - t)h$; then, for any $x \in X$, we have $|f(x) - p(x)| = |t(f(x) - g(x)) + (1 - t)(f(x) - h(x)| \le td(f, g) + (1 - t)d(f, h) < t\delta + (1 - t)\delta = \delta$, so $d(f, p) \le \delta < \varepsilon$ and therefore $p \in B_d(f, \varepsilon)$. Thus $B_d(f, \varepsilon)$ is convex and hence L is locally convex.

Let $s : M \times M \to M$ be defined by $s(f, g) = f + g$ for any $f, g \in M$; define a function $m : \mathbb{R} \times M \to M$ by $m(t, f) = t \cdot f$ for any $t \in \mathbb{R}$ and $f \in M$. To finish the proof of our fact we must show that s and m are continuous.

Take arbitrary $f, g \in M$ and any $W \in \tau(f + g, M)$; there is $\varepsilon > 0$ such that $B_d(f + g, \varepsilon) \subset W$. The set $U = B_d(f, \frac{\varepsilon}{2}) \times B_d(g, \frac{\varepsilon}{2})$ is an open neighborhood of (f, g) in $M \times M$. If $(f_1, g_1) \in U$, then $d(f_1, f) < \frac{\varepsilon}{2}$ and $d(g_1, g) < \frac{\varepsilon}{2}$, so we can choose $\delta < \frac{\varepsilon}{2}$ such that $\max\{d(f, f_1), d(g, g_1)\} < \delta$. Given any $x \in X$ we have $|(f_1 + g_1)(x) - (f + g)(x)| \le |f(x) - f_1(x)| + |g(x) - g_1(x)| \le d(f, f_1) + d(g, g_1) < 2\delta$. Thus $d(f_1 + g_1, f + g) \le 2\delta < \varepsilon$ and hence $f_1 + g_1 = s(f_1, g_1) \in B_d(f + g, \varepsilon)$; the point $(f_1, g_1) \in U$ was taken arbitrarily, so $s(U) \subset B_d(f + g, \varepsilon) \subset W$ which proves continuity of s at the point (f, g). Therefore $s : M \times M \to M$ is continuous.

To prove continuity of m take any $(t, f) \in \mathbb{R} \times M$ and any $W \in \tau(tf, M)$; there is $\varepsilon > 0$ such that $B_d(tf, \varepsilon) \subset W$. Since f is a bounded function, there is $r > 0$ such that $|f(x)| < r$ for any $x \in X$. Take any $\delta > 0$ such that $\delta < \min\{1, \frac{\varepsilon}{3(r+1)}\}$ and $|\delta t| < \frac{\varepsilon}{3}$. The set $U = (t - \delta, t + \delta) \times B_d(f, \delta)$ is an open neighborhood of (t, f) in the space $\mathbb{R} \times M$. Take any $(t_1, f_1) \in U$; then $d(f, f_1) < \delta$, and hence, for any $x \in X$, we have $|f_1(x) - f(x)| < \delta$ whence $|f_1(x)| < |f(x)| + \delta < r + 1$. Thus $|f_1(x)| < r + 1$ for any $x \in X$. Furthermore, $|t_1 f_1(x) - tf(x)| = |(t_1 - t)f_1(x) + t(f_1(x) - f(x))| \le |t_1 - t||f_1(x)| + |t||f_1(x) - f(x)| \le \delta(r + 1) + |t\delta| < \frac{\varepsilon}{3} + \frac{\varepsilon}{3} = \frac{2\varepsilon}{3}$. As a consequence, $d(t_1 f_1, tf) \le \frac{2\varepsilon}{3} < \varepsilon$ so $m(t_1, f_1) = t_1 f_1 \in B_d(tf, \varepsilon) \subset W$. The point $(t_1, f_1) \in U$ was chosen arbitrarily, so we proved that $m(U) \subset W$ and hence m is continuous at the point (t, f). Fact 1 is proved. $\qquad\square$

Fact 2. For any infinite cardinal κ, there exists a metrizable locally convex space of weight κ.

Proof. Recall that $D(\kappa)$ is a discrete space of cardinality κ; let $\text{Fin}(D(\kappa))$ be the family of all finite subsets of $D(\kappa)$. Consider the space $L = \{f \in C(D(\kappa)) : D(\kappa)\backslash f^{-1}(0)$ is finite$\}$. In other words, L consists of functions on $D(\kappa)$ which take nonzero values at only finitely many points of $D(\kappa)$. It is evident that L is a linear subspace of a metrizable locally convex space $C^*(D(\kappa))$ (see Fact 1). Therefore L is a locally convex metrizable space. Let $P = \{f \in L : f(x) \in \mathbb{Q}$ for any $x \in D(\kappa)\}$. Given a finite $A \subset D(\kappa)$ the set $P_A = \{f \in P : f(D(\kappa)\backslash A) = \{0\}\}$ is countable because the map $f \to f|A$ maps P_A injectively onto $\mathbb{Q}^A$ which is countable. Thus $|P| = |\bigcup\{P_A : A \in \text{Fin}(D(\kappa))\}| \leq \omega \cdot |\text{Fin}(D(\kappa))| = \kappa \cdot \omega = \kappa$.

Furthermore, the set P is dense in L; to see this take any $f \in L$ and any $\varepsilon > 0$. The set $S = f^{-1}(\mathbb{R}\backslash\{0\})$ is finite, so we can choose a rational number q_s such that $|f(s) - q_s| < \frac{\varepsilon}{2}$ for each $s \in S$. Let $g(s) = q_s$ for each $s \in S$; if $t \in D(\kappa)\backslash S$, then let $g(t) = 0$. Then $g \in P$ and $|g(x) - f(x)| < \frac{\varepsilon}{2}$ for every $x \in D(\kappa)$. Therefore $d(f, g) \leq \frac{\varepsilon}{2} < \varepsilon$, so $B_d(f, \varepsilon) \cap P \neq \varnothing$ which proves that P is dense in L. As a consequence, $w(L) = d(L) \leq |P| \leq \kappa$.

On the other hand, for any $x \in D(\kappa)$, let $f_x(x) = 1$ and $f_x(y) = 0$ for all $y \in D(\kappa)\backslash\{x\}$. The subspace $D = \{f_x : x \in D(\kappa)\}$ of the space L is discrete because $B_d(f_x, 1) \cap D = \{f_x\}$ for each $x \in D(\kappa)$. Therefore $w(L) \geq s(L) \geq |D| = \kappa$ which shows that $w(L) = \kappa$ and Fact 2 is proved.   □

Returning to our solution, take any non-separable metric space X. Then $\text{ext}(X) = w(X) = d(X) \geq \omega_1$ (see Problem 214 of [TFS]) and hence there exists a closed discrete $D \subset X$ with $|D| = \omega_1$. Apply Fact 2 to take a metrizable locally convex space L with $w(L) = \omega_1$. Then $\text{ext}(L) = \omega_1$, so we can find a discrete $E \subset L$. Let $f : D \to E$ be any bijection. Then f is a continuous map from D to L because D is discrete. By Problem 104, there exists a continuous map $F : X \to L$ such that $F|D = f$. If $Y = F(X)$, then $E \subset Y$ and hence $w(Y) \geq s(Y) \geq |E| = \omega_1$ so $w(Y) = \omega_1$. Thus F maps X continuously onto a metric space Y with $w(Y) = \omega_1$, so our solution is complete.

T.106. *Prove that a metrizable space is ω-stable if and only if it is separable.*

Solution. Let X be a metrizable space with $w(X) > \omega$. There exists a continuous onto map $f : X \to Y$ of the space X onto a metrizable space Y with $w(Y) = \omega_1$ (see Problem 105). Since $w(Y) \leq \mathfrak{c}$, there exists a condensation $g : Y \to M$ such that $w(M) = \omega$ (see Problem 102). If X is stable, then $w(Y) = nw(Y) = \omega$ which is a contradiction. Thus no metrizable space of uncountable weight is stable, i.e., stability of a metrizable space implies its separability.

Now, if X is a metrizable separable space, then $w(X) = \omega$ and hence $nw(Y) \leq \omega$ whenever Y is a continuous image of X. Thus X is ω-stable.

T.107. *Prove that any Σ-product of spaces with a countable network is monolithic. In particular, any Σ-product of second countable spaces is monolithic.*

Solution. Let X_t be a space with a countable network for each $t \in T$. Fix any point $a \in X = \prod\{X_t : t \in T\}$; given any $x \in X$, let $\text{supp}(x) = \{t \in T : x(t) \neq a(t)\}$. For any $A \subset T$, let $p_A : X \to X_A = \prod\{X_t : t \in A\}$ be the natural projection onto the face X_A. We must prove that the space $\Sigma = \Sigma(X, a) = \{x \in X : |\mathrm{supp}(t)| \le \omega\}$ is κ-monolithic for any infinite cardinal κ.

Take any $Y \subset \Sigma(X, a)$ such that $|Y| \le \kappa$ and let $S = \bigcup\{\mathrm{supp}(y) : y \in Y\}$; it is clear that $|S| \le \kappa$. We have the equality $y(t) = a(t)$ for any $t \in T \backslash S$ and hence $Y \subset F = \{p_{T \backslash S}(a)\} \times X_S$. It is an easy exercise that F is a closed subspace of X and $nw(F) \le \kappa$ (observe that F is homeomorphic to X_S and use the fact that the product of $\le \kappa$-many spaces with a countable network has the network weight $\le \kappa$). Therefore $\mathrm{cl}_\Sigma(Y) \subset \mathrm{cl}_X(Y) \subset F$ and $nw(\mathrm{cl}_\Sigma(Y)) \le nw(F) \le \kappa$ which proves that $\Sigma(X, a)$ is κ-monolithic.

T.108. *Prove that any space X is κ-stable for any $\kappa \ge nw(X)$. In particular, any space with a countable network is stable.*

Solution. If $\lambda = nw(X)$ and Y is a continuous image of X, then $nw(Y) \le \lambda$. Thus we have $nw(Y) \le \lambda \le \kappa$ (the condensations of Y are irrelevant) which proves that X is κ-stable.

T.109. *Prove that any product of spaces with countable network is stable. In particular, any product of second countable spaces is stable.*

Solution. We will need the following general factorization theorem.

Fact 1. Suppose that X_t is a space such that $nw(X_t) = \omega$ for every $t \in T$ and let $X = \prod\{X_t : t \in T\}$; given any $A \subset T$, the map $p_A : X \to X_A = \prod\{X_t : t \in A\}$ is the natural projection of X onto the face X_A defined by $p_A(x) = x|A$ for every $x \in X$. Suppose that Y is a dense subspace of the space X and $f : Y \to M$ is a continuous map of Y onto a space M such that $w(M) = \kappa \ge \omega$. Then there is a set $S \subset T$ and a continuous map $g : p_S(Y) \to M$ such that $|S| \le \kappa$ and $f = g \circ (p_S|Y)$.

Proof. We will also need projections between faces of X, namely, if $A \subset B \subset T$, then we have a natural projection $p_A^B : X_B \to X_A$ defined by $p_A^B(x) = x|A$ for any $x \in X_B$. The map p_A^B is open, continuous and $p_A^B \circ P_B = p_A$ for all $A \subset B \subset T$.

Since $w(M) \le \kappa$, there is an embedding of M into $\mathbb{R}^\kappa$, so we can assume, without loss of generality, that $M \subset \mathbb{R}^\kappa$. For every $\alpha < \kappa$, the map $\pi_\alpha : \mathbb{R}^\kappa \to \mathbb{R}$ is the natural projection of $\mathbb{R}^\kappa$ onto its αth factor; let $q_\alpha = \pi_\alpha|M$. Given any $\alpha < \kappa$, we can apply Problem 299 of [TFS] to the map $f_\alpha = q_\alpha \circ f : Y \to \mathbb{R}$ to find a countable set $S_\alpha \subset T$ and a continuous map $g_\alpha : p_{S_\alpha}(Y) \to \mathbb{R}$ such that $f_\alpha = g_\alpha \circ (p_{S_\alpha}|Y)$. We claim that $S = \bigcup\{S_\alpha : \alpha < \kappa\}$ is as promised.

Indeed, given $z \in p_S(Y)$, let $g(z)(\alpha) = g_\alpha(z|S_\alpha)$ for each $\alpha < \kappa$; this defines a map $g : p_S(Y) \to \mathbb{R}^\kappa$. It is immediate that $g = \Delta\{g_\alpha \circ (p_{S_\alpha}^S|p_S(Y)) : \alpha < \kappa\}$ and hence the map g is continuous (see the introductory part of Sect. 1.4). Given any $y \in Y$, we have

$$f(y)(\alpha) = q_\alpha(f(y)) = f_\alpha(y) = g_\alpha(p_{S_\alpha}(y)) = g_\alpha(p_{S_\alpha}^S(p_S(y))) = g(p_S(y))(\alpha)$$

for every $\alpha < \kappa$ which shows that $f(y) = g(p_S(y))$ and hence $g(p_S(y)) \in M$ for each $y \in Y$. As a consequence, $g : p_S(Y) \to M$ and $f = g \circ (p_S|Y)$, so Fact 1 is proved. □

Returning to our solution, fix any infinite cardinal κ; suppose that $nw(X_t) = \omega$ for all $t \in T$ and take any continuous onto map $f : X = \prod_{t \in T} X_t \to Y$ for which there is a condensation $g : Y \to M$ such that $w(M) \leq \kappa$. If $h = g \circ f$, then $h : X \to M$, so we can apply Fact 1 to find $S \subset T$ such that $|S| \leq \kappa$, and there is a continuous map $b : X_S = \prod_{t \in S} X_t \to M$ with $b \circ p_S = h$. For the map $d = g^{-1} \circ b$, we have $d : X_S \to Y$ and $d \circ p_S = f$. Observe also that for any $U \in \tau(Y)$, the set $d^{-1}(U) = p_S(f^{-1}(U))$ is open in X_S because $f^{-1}(U)$ is open in X and the map p_S is open (see Problem 107 of [TFS]). Consequently, d is a continuous map and hence $nw(Y) \leq nw(X_S) \leq \kappa$ (see Problem 157 of [TFS] and observe that any product of $\leq \kappa$-many spaces with a countable network has network weight $\leq \kappa$). Therefore X is κ-stable and our solution is complete.

T.110. *Prove that any Σ-product of spaces with countable network is stable. In particular, any Σ-product of second countable spaces is stable.*

Solution. Given a product $X = \prod_{t \in T} X_t$, we know that natural projections of X onto its faces are open maps (see Problem 107 of [TFS]). It turns out that the same is true for certain subspaces of X.

Fact 1. Let X_t be a space for each $t \in T$; given a point $x \in X = \prod\{X_t : t \in T\}$, let $\sigma(X, x) = \{y \in X : |\{t \in T : x(t) \neq y(t)\}| < \omega\}$. The natural projection $p_S : X \to X_S = \prod_{t \in S} X_t$ is defined by $p_S(x) = x|S$ for any $x \in X$. Suppose that $Y \subset X$ and $\sigma(X, x) \subset Y$ for any $x \in Y$. Then the map $p_S|Y : Y \to p_S(Y)$ is open for any $S \subset T$.

Proof. The map p_S is continuous (see Problem 107 of [TFS]), so $p = p_S|Y$ is also continuous. It is clear that p is surjective, so we only must show that $p(U)$ is open in $Z = p_S(Y)$ for any $U \in \tau(Y)$. It suffices to find a base $\mathcal{B}$ in Y such that $p(V) \in \tau(Z)$ for any $V \in \mathcal{B}$. Let $\mathcal{C} = \{U \subset X : U = \prod_{t \in T} U_t : U_t \in \tau(X_t)$ for all $t \in T$ and the set $\mathrm{supp}(U) = \{t \in T : U_t \neq X_t\}$ is finite$\}$. The family $\mathcal{C}$ is a base of X (see Problem 101 of [TFS]), so $\mathcal{B} = \{U \cap Y : U \in \mathcal{C}\}$ is a base in Y. Take any $U = \prod_{t \in T} U_t \in \mathcal{C}$ and let $V = U \cap Y$. Observe that the set $U_S = \prod_{t \in S} U_t$ is open in X_S, so it suffices to show that $p_S(V) = U_S \cap Z$. Of course, there is no loss of generality to assume that $U \neq \emptyset$.

Take any $z \in U_S \cap Z$ and fix a point $y \in Y$ such that $p_S(y) = z$. We have $y(t) = z(t) \in U_t$ for any $t \in S$. Choose $a_t \in U_t$ for any $t \in \mathrm{supp}(U)\backslash S$ and define $u \in X$ as follows: $u|S = z$, $u(t) = a_t$ for any $t \in \mathrm{supp}(U)\backslash S$ and $u(t) = y(t)$ for any $t \in T\backslash(S \cup \mathrm{supp}(U))$. Then the set $A = \{t \in T : u(t) \neq y(t)\}$ is finite, so $u \in \Sigma(X, y) \subset Y$. Besides, $u(t) \in U_t$ for all $t \in T$ and therefore $u \in U \cap Y = V$. It is clear that $\pi_S(u) = z$ and hence $z \in p_S(V)$. This proves that $U_S \cap Z \subset p_S(V)$.

Now, if $y \in V$, then $y(t) \in U_t$ for all $t \in T$ and hence $p_S(y)(t) = y(t) \in U_t$ for all $t \in S$. Consequently, $p_S(y) \in U_S \cap Z$ so $p_S(V) \subset U_S \cap Z$. We proved that $p_S(V) = U_S \cap Z$ is an open subset of Z so Fact 1 is proved. □

Returning to our solution assume that $nw(X_t) = \omega$ for every $t \in T$ and take any $a \in X = \prod_{t \in T} X_t$; we must prove that the space

$$\Sigma = \Sigma(X, a) = \{x \in X : |\{t \in T : x(t) \neq a(t)\}| \leq \omega\}$$

is κ-stable for any infinite cardinal κ. It is an easy exercise that Σ is dense in X and $\sigma(X, x) \subset \Sigma$ for any $x \in \Sigma$. Take any continuous surjective map $f : \Sigma \to Y$ such that there is a condensation $g : Y \to M$ of Y onto a space M with $w(M) \leq \kappa$. We can apply Fact 1 of T.109 to the map $h = g \circ f$ to find $S \subset T$ such that $|S| \leq \kappa$ and there exists a continuous map $b : p_S(\Sigma) \to M$ such that $h = b \circ (p_S|\Sigma)$. For the map $d = g^{-1} \circ b$ we have $d : p_S(\Sigma) \to Y$ and $d \circ (p_S|\Sigma) = f$. Given any $U \in \tau(Y)$ the set $d^{-1}(U) = p_S(f^{-1}(U))$ is open in $\Sigma_S = p_S(\Sigma)$ because $f^{-1}(U)$ is open in Σ and $p_S|\Sigma$ is an open map by Fact 1. Consequently, d is a continuous map.

Since $\Sigma_S \subset X_S$, we have $nw(\Sigma_S) \leq nw(X_S) \leq \kappa$ (it is an easy exercise to see that a product of $\leq \kappa$-many spaces with countable network has network weight $\leq \kappa$). Since Y is a continuous image of Σ_S (under the mapping d), we have $nw(Y) \leq nw(\Sigma_S) \leq \kappa$ (see Problem 157 of [TFS]) which proves that Σ is κ-stable, so our solution is complete.

T.111. *Prove that any σ-product of spaces with countable network is stable. In particular, any σ-product of second countable spaces is stable.*

Solution. Assume that we have $nw(X_t) = \omega$ for every $t \in T$. Given any point $x \in X = \prod_{t \in T} X_t$, let $\sigma(X, x) = \{y \in X : |\{t \in T : y(t) \neq x(t)\}| < \omega\}$ and fix any point $a \in X$; we must prove that the space $\sigma = \sigma(X, a)$ is κ-stable for any infinite cardinal κ. Given any $S \subset T$ the map $p_S : X \to X_S = \prod_{t \in S} X_t$ is the natural projection defined by $p_S(x) = x|S$ for any $x \in X$. It is an easy exercise that σ is dense in X and $\sigma(X, x) \subset \sigma$ for any $x \in \sigma$.

Take any continuous surjective map $f : \sigma \to Y$ such that there is a condensation $g : Y \to M$ of Y onto a space M with $w(M) \leq \kappa$. We can apply Fact 1 of T.109 to the map $h = g \circ f$ to find $S \subset T$ such that $|S| \leq \kappa$ and there exists a continuous map $b : p_S(\sigma) \to M$ such that $h = b \circ (p_S|\sigma)$.

For the map $d = g^{-1} \circ b$ we have $d : p_S(\sigma) \to Y$ and $d \circ (p_S|\sigma) = f$. Given any $U \in \tau(Y)$ the set $d^{-1}(U) = p_S(f^{-1}(U))$ is open in $\sigma_S = p_S(\sigma)$ because $f^{-1}(U)$ is open in σ and $p_S|\sigma$ is an open map by Fact 1 of T.110. Consequently, d is a continuous map. Since $\sigma_S \subset X_S$, we have $nw(\sigma_S) \leq nw(X_S) \leq \kappa$ (it is an easy exercise to see that a product of $\leq \kappa$-many spaces with countable network has network weight $\leq \kappa$). Since Y is a continuous image of σ_S (under the mapping d), we have $nw(Y) \leq nw(\sigma_S) \leq \kappa$ (see Problem 157 of [TFS]) which proves that σ is κ-stable so our solution is complete.

T.112. *Prove that any σ-product of Lindelöf P-spaces is ω-stable.*

Solution. Our solution needs some insight into the products and σ-products of Lindelöf P-spaces.

Fact 1. If X and Y are Lindelöf P-spaces, then $X \times Y$ is a Lindelöf P-space. Therefore any finite product of Lindelöf P-spaces is a Lindelöf P-space.

Proof. Let P be a G_δ-set in $X \times Y$; fix a family $\mathcal{O} = \{O_n : n \in \omega\} \subset \tau(X \times Y)$ such that $P = \bigcap \mathcal{O}$. If $P = \emptyset$, then there is nothing to prove; if not, take an arbitrary $z = (x, y) \in P$. It is easy to construct families $\{U_n : n \in \omega\} \subset \tau(x, X)$ and $\{V_n : n \in \omega\} \subset \tau(y, Y)$ such that $U_n \times V_n \subset O_n$ for every $n \in \omega$. Since X and Y are P-spaces, we have $U = \bigcap\{U_n : n \in \omega\} \in \tau(X)$ and $V = \bigcap\{V_n : n \in \omega\} \in \tau(Y)$, so $W = U \times V \in \tau(z, X \times Y)$ and $W \subset P$. Therefore every $z \in P$ is in the interior of P, i.e., P is open in $X \times Y$. Consequently, $X \times Y$ is a P-space.

To see that $X \times Y$ is Lindelöf, take any open cover $\mathcal{W}$ of the space $X \times Y$; let $p_X : X \times Y \to X$ and $p_Y : X \times Y \to Y$ be the respective natural projections. Given any set $A \subset X \times Y$ let $A_X = p_X(A)$ and $A_Y = p_Y(A)$. We can assume that the elements of $\mathcal{W}$ belong to the standard base of the space $X \times Y$, i.e., $W = W_X \times W_Y$ for any $W \in \mathcal{W}$. For each $x \in X$ the space $Y_x = \{x\} \times Y$ is homeomorphic to Y, so there is a countable $\mathcal{W}_x \subset \mathcal{W}$ such that $Y_x \subset \bigcup \mathcal{W}_x$. Let $U_x = \bigcap\{W_X : W \in \mathcal{W}_x\}$ for every $x \in X$. The set U_x is open in X because X is a P-space; by the Lindelöf property of X, there is a countable $A \subset X$ such that $X = \bigcup\{U_x : x \in A\}$. The family $\mathcal{W}' = \bigcup\{\mathcal{W}_x : x \in A\} \subset \mathcal{W}$ is countable, so it suffices to show that $\bigcup \mathcal{W}' = X \times Y$.

Take any $z = (a, b) \in X \times Y$; there is $x \in X$ such that $a \in U_x$. Since $Y_x \subset \bigcup \mathcal{W}_x$, we can find $W \in \mathcal{W}_x$ such that $(x, b) \in W$ and hence $b \in W_Y$. By the definition of U_x, we have $W_X \supset U_x \ni a$ so $a \in W_X$ and therefore $(a, b) \in W \in \mathcal{W}'$. The point $z \in X \times Y$ was chosen arbitrarily, so $\mathcal{W}'$ is a countable subcover of $\mathcal{W}$ and hence $X \times Y$ is Lindelöf.

Finally, the same result for any finite number of Lindelöf P-spaces can be obtained from what we proved for two factors by a trivial induction so Fact 1 is proved. $\qquad\square$

Fact 2. Suppose that $X = \bigcup\{X_n : n \in \omega\}$ where, for each $n \in \omega$, the space X_n is κ-stable for some infinite cardinal κ. Then X is κ-stable.

Proof. Take any continuous onto map $f : X \to Y$ such that there is a condensation $g : Y \to M$ of the space Y onto some space M with $w(M) \le \kappa$. The space $Y_n = f(X_n)$ is a continuous image of X_n and $g|Y_n$ is a condensation of Y_n onto a space of weight $\le \kappa$ for every $n \in \omega$. Consequently, $nw(Y_n) \le \kappa$ for any $n \in \omega$; since $Y = \bigcup\{Y_n : n \in \omega\}$, we have $nw(Y) \le \kappa$, so X is κ-stable and Fact 2 is proved. $\qquad\square$

Fact 3. Every countable σ-product of Lindelöf P-spaces is ω-stable.

Proof. Let us check first that

(∗) any Lindelöf P-space is ω-stable.

To prove (∗) assume that Z is a Lindelöf P-space and take any continuous onto map $f : Z \to Y$ such that there is a condensation $g : Y \to M$ of Y onto some space M such that $w(M) = \omega$. Every point of Y is a G_δ-set in Y because so is every point of M. Thus $f^{-1}(y)$ is a G_δ-set in Z and hence an open subset of Z for any $y \in Y$. The family $\mathcal{U} = \{f^{-1}(y) : y \in Y\}$ is a disjoint open cover of Z; the space Z being Lindelöf there is a countable $\mathcal{U}' \subset \mathcal{U}$ with $\bigcup \mathcal{U}' = Z$. This shows that there is a countable $A \subset Y$ such that $f^{-1}(A) = Z$ which implies that $Y = f(Z) = A$ is countable. Therefore $nw(Y) \leq |Y| = \omega$ so Z is ω-stable and $(*)$ is proved.

Now assume that Z_n is a Lindelöf P-space for every $n \in \omega$ and choose any point $a \in Z = \prod\{Z_n : n \in \omega\}$. Given any $z \in Z$ let $\operatorname{supp}(z) = \{n \in \omega : z(n) \neq a(n)\}$. We must prove that the space $\sigma(Z, a) = \{z \in Z : |\operatorname{supp}(z)| < \omega\}$ is ω-stable. Consider the set $\sigma_n(Z, a) = \{z \in Z : z(m) = a(m) \text{ for all } m \geq n\}$ for every $n \in \omega$. It is evident that $\sigma(Z, a) = \bigcup\{\sigma_n(Z, a) : n \in \omega\}$. It is also straightforward that $\sigma_n(Z, a)$ is homeomorphic to $Z_0 \times \cdots \times Z_{n-1}$ so $\sigma_n(Z, a)$ is a Lindelöf P-space by Fact 1. The property $(*)$ shows that $\sigma_n(Z, a)$ is ω-stable for every $n \in \omega$ and hence $\sigma(Z, a)$ is ω-stable by Fact 2. Fact 3 is proved. $\qquad \square$

Fact 4. Every σ-product of Lindelöf P-spaces is Lindelöf.

Proof. Assume that X_t is a Lindelöf P-space for every $t \in T$; take any point $a \in X = \prod\{X_t : t \in T\}$ and let $\operatorname{supp}(x) = \{t \in T : x(t) \neq a(t)\}$ for every $x \in X$. We must prove that the space $\sigma = \sigma(X, a) = \{x \in X : |\operatorname{supp}(x)| < \omega\}$ is Lindelöf. Observe that $\sigma = \bigcup\{\sigma_n : n \in \omega\}$ where $\sigma_n = \{x \in \sigma : |\operatorname{supp}(x)| \leq n\}$ for every $n \in \omega$. Therefore it suffices to show that σ_n is Lindelöf for every $n \in \omega$. We will do this by induction on n. If $n = 0$, then $\sigma_n = \{a\}$ is a Lindelöf space; assume that $n > 0$ and we proved that σ_{n-1} is Lindelöf.

For any $A \subset T$ the map $p_A : X \to X_A = \prod_{t \in A} X_t$ is the natural projection of X onto its face X_A defined by $p_A(x) = x|A$ for each $x \in X$. Call a set $U \subset X$ *standard* if $U = \prod_{t \in T} U_t$ where $U_t \in \tau(X_t)$ for each $t \in T$ and the set $\operatorname{supp}(U) = \{t \in T : U_t \neq X_t\}$ is finite. The family $\mathcal{B}$ of all standard sets is a base in X (see Problem 101 of [TFS]), so to prove that σ_n is Lindelöf, it suffices to show that for every $\mathcal{U} \subset \mathcal{B}$ such that $\sigma_n \subset \bigcup \mathcal{U}$ there is a countable $\mathcal{U}' \subset \mathcal{U}$ such that $\sigma_n \subset \bigcup \mathcal{U}'$.

By the induction hypothesis, there is a countable $\mathcal{V} \subset \mathcal{U}$ such that $\sigma_{n-1} \subset \bigcup \mathcal{V}$. The set $A = \bigcup\{\operatorname{supp}(V) : V \in \mathcal{V}\}$ is countable; we claim that

$(**) \quad \operatorname{supp}(x) \subset A$ for any $x \in \sigma_n \backslash (\bigcup \mathcal{V})$.

Indeed, if $(**)$ is not true, then $t_0 \in \operatorname{supp}(x) \backslash A$ for some $x \in \sigma_n \backslash (\bigcup \mathcal{V})$. Let $y(t) = x(t)$ for all $t \in T \backslash \{t_0\}$ and $y(t_0) = a(t_0)$. Then $y \in \sigma_{n-1}$ and hence $y \in V$ for some $V \in \mathcal{V}$. We have $\operatorname{supp}(V) \subset A$ and therefore $p_A^{-1}((p_A(V)) = V$; since $p_A(x) = p_A(y)$ and $y \in V$, we have $x \in V$ which is a contradiction showing that $(**)$ is true.

It follows from $(**)$ that $P = \sigma_n \backslash (\bigcup \mathcal{V}) \subset Q = \sigma_n \cap (X_A \times \{p_{T \backslash A}(a)\})$. It is straightforward that the mapping $p_A|Q$ is a homeomorphism of Q onto the space $\sigma_n(X_A, p_A(a)) = \{y \in X_A : |\{t \in T : y(t) \neq a(t)\}| \leq n\}$. Furthermore $\sigma_n(X_A, p_A(a)) = \bigcup\{R_B : B \subset A \text{ and } |B| \leq n\}$ where $R_B = X_B \times \{p_{A \backslash B}(a)\}$ is homeomorphic to X_B for any $B \subset A$. The space R_B is Lindelöf being a finite product of Lindelöf P-spaces (see Fact 1). Any countable union of Lindelöf spaces is a Lindelöf space, so $\sigma_n(X_A, p_A(a))$ is Lindelöf and hence Q is also Lindelöf being homeomorphic to $\sigma_n(X_A, p_A(a))$. Since P is a closed subset of σ_n, it is also closed in Q and hence Lindelöf. Thus we can choose a countable $\mathcal{W} \subset \mathcal{U}$ such that $P \subset \bigcup \mathcal{W}$. It is clear that $\mathcal{U}' = \mathcal{V} \cup \mathcal{W}$ is a countable subfamily of $\mathcal{U}$ and $\sigma_n \subset \bigcup \mathcal{U}'$. The inductive step being carried out, we proved that every σ_n is Lindelöf and hence $\sigma = \bigcup \sigma_n$ is also Lindelöf so Fact 4 is proved. $\square$

We are finally ready to present our solution. Assume that X_t is a Lindelöf P-space for every $t \in T$ and take any point $a \in X = \prod\{X_t : t \in T\}$. Let $\operatorname{supp}(x) = \{t \in T : x(t) \neq a(t)\}$ for every $x \in X$. We must prove that the space $\sigma = \sigma(X, a) = \{x \in X : |\operatorname{supp}(x)| < \omega\}$ is ω-stable, so take any map $f : \sigma \to Y$ such that there is a condensation $g : Y \to M$ of Y onto some space M with $w(M) = \omega$. We know that σ is Lindelöf (see Fact 4), so apply Problem 298 of [TFS] to the map $h = g \circ f : \sigma \to M$ to obtain a countable $S \subset T$ and a continuous map $b : p_S(\sigma) \to M$ such that $b \circ p_S = h$. For the map $d = g^{-1} \circ b$ we have $d : p_S(\sigma) \to Y$ and $f = d \circ (p_S|\sigma)$. Given any $U \in \tau(Y)$ the set $d^{-1}(U) = p_S(f^{-1}(U))$ is open in $\sigma_S = p_S(\sigma)$ because $f^{-1}(U)$ is open in σ and $p_S|\sigma$ is an open map by Fact 1 of T.110. Therefore the map d is continuous. We must also note that $p_S(\sigma)$ is a countable σ-product of Lindelöf P-spaces and therefore $p_S(\sigma)$ is ω-stable by Fact 3. Since Y is a continuous image (under the map d) of an ω-stable space σ_S, we have $nw(Y) \leq \omega$ because Y condenses onto a second countable space M. This shows that σ is ω-stable and hence our solution is complete.

T.113. *Prove that for any cardinal κ, if X is κ-monolithic, then any $Y \subset X$ is κ-monolithic.*

Solution. Take any $A \subset Y$ with $|A| \leq \kappa$. Then $nw(\operatorname{cl}_Y(A)) \leq nw(\operatorname{cl}_X(A)) \leq \kappa$. Thus $nw(\operatorname{cl}_Y(A)) \leq \kappa$ for any $A \subset Y$ with $|A| \leq \kappa$ so Y is κ-monolithic.

T.114. *Let κ be an infinite cardinal. Suppose that X_α is a κ-monolithic space for each $\alpha < \kappa$. Prove that $\prod\{X_\alpha : \alpha < \kappa\}$ is also κ-monolithic. In particular, any finite product of κ-monolithic spaces is κ-monolithic.*

Solution. For the space $X = \prod\{X_\alpha : \alpha < \kappa\}$ the map $p_\alpha : X \to X_\alpha$ is the natural projection for every $\alpha < \kappa$. Take any $A \subset X$ with $|A| \leq \kappa$; then $A_\alpha = p_\alpha(A) \subset X_\alpha$ and $|A_\alpha| \leq \kappa$. It is straightforward that $\overline{A} \subset \prod_{\alpha < \kappa} \overline{A_\alpha}$.

The space X_α being κ-monolithic, we have $nw(\overline{A_\alpha}) \leq \kappa$ for each $\alpha < \kappa$. We leave it as an easy exercise for the reader that any product of $\leq \kappa$-many spaces of network weight $\leq \kappa$ has network weight $\leq \kappa$. Therefore $nw(\prod_{\alpha < \kappa} \overline{A_\alpha}) \leq \kappa$ whence $nw(\overline{A}) \leq nw(\prod_{\alpha < \kappa} \overline{A_\alpha}) \leq \kappa$ which shows that X is κ-monolithic.

T.115. *Prove that if a space X is covered with a locally finite family of closed monolithic subspaces, then X is monolithic.*

Solution. Let $\mathcal{F}$ be a locally finite closed cover of X such that every $F \in \mathcal{F}$ is monolithic. Fix an infinite cardinal κ and any set $A \subset X$ with $|A| \leq \kappa$. Each $x \in X$ belongs to finitely many elements of $\mathcal{F}$, so there is $\mathcal{F}' \subset \mathcal{F}$ such that $|\mathcal{F}'| \leq \kappa$

and $A \subset \bigcup \mathcal{F}'$. The family $\mathcal{A} = \{A \cap F : F \in \mathcal{F}'\}$ is locally finite and hence closure-preserving (see Fact 2 of S.221), so $\overline{A} = \overline{\bigcup \mathcal{A}} = \bigcup \{\overline{A \cap F} : F \in \mathcal{F}'\}$.

We have $nw(\overline{A \cap F}) \leq \kappa$ for any $F \in \mathcal{F}'$ because $\overline{A \cap F} \subset F$ and F is monolithic. Since $|\mathcal{F}'| \leq \kappa$, we have represented $\overline{A}$ as a union of $\leq \kappa$-many spaces of network weight $\leq \kappa$. It is evident that any union of $\leq \kappa$-many spaces of network weight $\leq \kappa$ has network weight $\leq \kappa$ so $nw(\overline{A}) \leq \kappa$. We proved that for any $A \subset X$ with $|A| \leq \kappa$, we have $nw(\overline{A}) \leq \kappa$, i.e., X is κ-monolithic. The cardinal κ was chosen arbitrarily so X is monolithic.

T.116. *Give an example of a space which is not ω-monolithic being a union of two monolithic subspaces.*

Solution. The Mrowka space M constructed in Problem 142 of [TFS] can be represented as $M = \omega \cup \mathcal{M}$ where ω is dense in M and all points of ω are isolated in M while the set $\mathcal{M}$ is uncountable and discrete in M. Thus M is a union of two discrete (and hence monolithic) subspaces; since $nw(\overline{\omega}) = nw(M) \geq s(M) \geq |\mathcal{M}| > \omega$, the space M is not ω-monolithic.

T.117. *Show that $\mathbb{R}^{\omega_1}$ is not ω-monolithic. Hence the product of uncountably many monolithic spaces can fail to be ω-monolithic.*

Solution. There exists a countable set $A \subset \mathbb{R}^{\omega_1}$ such that $\overline{A} = \mathbb{R}^{\omega_1}$ (see Problem 108 of [TFS]). The space $A(\omega_1)$ embeds in $\mathbb{I}^{\omega_1} \subset \mathbb{R}^{\omega_1}$ (see Problems 126 and 129 of [TFS]). As a consequence, $nw(\overline{A}) = nw(\mathbb{R}^{\omega_1}) \geq s(\mathbb{R}^{\omega_1}) \geq s(A(\omega_1)) = \omega_1$ and hence $\mathbb{R}^{\omega_1}$ is not ω-monolithic.

T.118. *Prove that any compact space is stable.*

Solution. Let X be a compact space. Given an infinite cardinal κ assume that we have a continuous onto map $f : X \to Y$ such that there is a condensation $g : Y \to M$ of Y onto some space M with $w(M) \leq \kappa$. The space Y is also compact; any condensation of a compact space is a homeomorphism (see Problem 123 of [TFS]), so Y is homeomorphic to M and therefore $nw(Y) \leq w(Y) = w(M) = \kappa$. This shows that X is κ-stable for any infinite cardinal κ.

T.119. *Prove that any pseudocompact space is ω-stable.*

Solution. Let X be a pseudocompact space and take any continuous onto map $f : X \to Y$ such that there is a condensation $g : Y \to M$ of the space Y onto a second countable space M. The space Y is also pseudocompact; since any condensation of a pseudocompact space onto a second countable space is a homeomorphism (see Problem 140 of [TFS]), the space Y is homeomorphic to M and therefore $nw(Y) \leq w(Y) = w(M) = \omega$. Thus X is ω-stable.

T.120. *Show that any compact ω-monolithic space of countable tightness must be Fréchet–Urysohn. Give an example of a monolithic compact space of uncountable tightness.*

Solution. Suppose that X is a compact ω-monolithic space with $t(X) = \omega$. Given any $A \subset X$ and any $x \in \overline{A}$, there is a countable $B \subset A$ such that $x \in \overline{B}$. By ω-monolithity of X, we have $nw(\overline{B}) = \omega$ and hence $w(\overline{B}) = \omega$ because $\overline{B}$ is compact (see Fact 4 of S.307). Thus $\overline{B}$ is a Fréchet–Urysohn space, so there is a sequence $\{x_n : n \in \omega\} \subset B$ such that $x_n \to x$. Since we also have $\{x_n : n \in \omega\} \subset A$, we found a sequence in A which converges to x. Therefore X is a Fréchet–Urysohn space.

Now let K be the set $\omega_1 + 1$ with the interval topology. Then K is compact (see Problem 314 of [TFS]); if $x = \omega_1$ and $A = \{\beta : \beta < \omega_1\}$, then $x \in \overline{A}$. If $B \subset A$ is countable, then there is $\beta < \omega_1$ such that $B \subset [0, \beta] = \{\alpha : \alpha \le \beta\}$. Since the set $[0, \beta]$ is closed and countable, we have $\overline{B} \subset [0, \beta]$ which shows that $x \notin \overline{B}$ for any countable $B \subset A$. Therefore $t(K) > \omega$.

Observe that $w(K) = \omega_1$, so if $P \subset K$ and $|P| = \omega_1$, then $nw(\overline{P}) \le w(K) = \omega_1$, so K is ω_1-monolithic. If $P \subset K$ is countable, then there exists $\beta < \omega_1$ such that $P \subset [0, \beta] \cup \{\omega_1\}$. Since $[0, \beta] \cup \{\omega_1\}$ is a countable closed set, the set $\overline{P}$ is also countable and therefore K is ω-monolithic. The space K has no subsets of infinite cardinalities other than ω and ω_1, so we established that K is monolithic.

T.121. *Let* $f : X \to Y$ *be a closed continuous onto map. Prove that if* X *is* κ-*monolithic, then so is* Y.

Solution. Take an arbitrary $A \subset Y$ with $|A| \le \kappa$ and choose any $B \subset X$ such that $|B| \le \kappa$ and $f(B) = A$. The map f is closed so $f(\overline{B}) = \overline{A}$. We have $nw(\overline{B}) \le \kappa$ because X is κ-monolithic. Therefore $nw(\overline{A}) \le nw(\overline{B}) \le \kappa$ (see Problem 157 of [TFS]) and hence Y is κ-monolithic.

T.122. *Give an example showing that a continuous image of a monolithic space is not necessarily* ω-*monolithic.*

Solution. Let X be the set $\mathbb{R}^{\omega_1}$ with the discrete topology. If $f : X \to \mathbb{R}^{\omega_1}$ is the identity map and $\mathbb{R}^{\omega_1}$ is considered with its natural product topology, then f is a continuous map because X is discrete. Any discrete space is metrizable and hence monolithic (see Problem 101). Thus the space $\mathbb{R}^{\omega_1}$ is a continuous image of a monolithic space X, so we only have to note that $\mathbb{R}^{\omega_1}$ is not ω-monolithic by Problem 117.

T.123. *Given an infinite cardinal* κ, *prove that any continuous image of a* κ-*stable space is* κ-*stable. In particular, any retract of a* κ-*stable space is* κ-*stable.*

Solution. Suppose that X is a κ-stable space and $f : X \to Y$ is a continuous onto map. Take any continuous map $g : Y \to Z$ such that there is a condensation $h : Z \to M$ of the space Z onto a space M with $w(M) \le \kappa$. The map $g \circ f$ maps X onto Z, so $nw(Z) \le \kappa$ because X is κ-stable. This shows that Y is also κ-stable.

T.124. *Suppose that* $X = \bigcup\{X_\alpha : \alpha < \kappa\}$ *and* X_α *is a* κ-*stable space for any* $\alpha < \kappa$. *Prove that* X *is* κ-*stable. In particular, any space which is a countable union of stable subspaces is stable.*

Solution. Take any continuous onto map $f : X \to Y$ such that there is a condensation $g : Y \to M$ of the space Y onto some space M with $w(M) \le \kappa$. For every $\alpha < \kappa$ the space $Y_\alpha = f(X_\alpha)$ is a continuous image of X_α and $g|Y_\alpha$ is a condensation of Y_α onto a space of weight $\le \kappa$. Consequently, $nw(Y_\alpha) \le \kappa$ for each $\alpha < \kappa$; since $Y = \bigcup\{Y_\alpha : \alpha < \kappa\}$, we have $nw(Y) \le \kappa$ so X is κ-stable.

T.125. *Give an example of an ω-stable space X such that some closed subspace of X is not ω-stable.*

Solution. Any pseudocompact space is ω-stable by Problem 119. The Mrowka space M constructed in Problem 142 of [TFS] is pseudocompact and hence ω-stable. On the other hand there is an uncountable closed discrete $D \subset M$. Since $|D| \le |M| \le \mathfrak{c}$, we have $w(D) \le \mathfrak{c}$. Since every discrete space is metrizable, the space D can be condensed onto a second countable space (see Problem 102); we have $nw(D) = |D| > \omega$ which shows that D is not ω-stable. Therefore D is a closed non-ω-stable subset of an ω-stable space M.

T.126. *Give an example of an ω-stable space whose square is not ω-stable.*

Solution. Given a set B denote by $P_\omega(B)$ the family of all countably infinite subsets of B.

Fact 1. Let D be a discrete space of cardinality ω_1. Then there exist countably compact subspaces X and Y of the space βD such that $D = X \cap Y$ and hence $X \times Y$ contains an uncountable clopen discrete subspace.

Proof. Given any set $A \subset \beta D$ denote by A^d the set of all accumulation points of A in βD. Let $X_0 = D$; assume that $\alpha < \omega_1$ and we have constructed sets $\{X_\beta : \beta < \alpha\}$ with the following properties:

(1) $X_\beta \subset \beta D$ for each $\beta < \alpha$;
(2) $X_0 = D$ and $X_\beta \subset X_\gamma$ whenever $\beta < \gamma < \alpha$';
(3) $|X_\beta| \le \mathfrak{c}$ for each $\beta < \alpha$;
(4) if $\beta < \gamma < \alpha$ and $A \in P_\omega(X_\beta)$, then $A^d \cap X_\gamma \ne \emptyset$.

Consider the set $Y_\alpha = \bigcup\{X_\beta : \beta < \alpha\}$; it is evident that $|Y_\alpha| \le \mathfrak{c}$ and therefore $|P_\omega(Y_\alpha)| \le \mathfrak{c}^\omega = \mathfrak{c}$. Given any $A \in P_\omega(Y_\alpha)$, the set A^d is non-empty because βD is a compact space. Choose a point $x_A \in A^d$ for each $A \in P_\omega(Y_\alpha)$ and let $X_\alpha = Y_\alpha \cup \{x_A : A \in P_\omega(Y_\alpha)\}$. It is immediate that the properties (1)–(4) hold for the family $\{X_\beta : \beta \le \alpha\}$, so our inductive construction can be continued to obtain a family $\{X_\beta : \beta < \omega_1\}$ for which the conditions (1)–(4) are satisfied for each $\alpha < \omega_1$.

The space $X = \bigcup\{X_\alpha : \alpha < \omega_1\}$ has cardinality $\le \mathfrak{c}$ because

$$|X| \le (\sup\{|X_\alpha| : \alpha < \omega_1\}) \cdot \omega_1 \le \mathfrak{c} \cdot \omega_1 = \mathfrak{c}$$

[we applied the property (3)]. Furthermore, X is countably compact. Indeed, if $A \subset P_\omega(X)$, then $A \subset X_\alpha$ for some $\alpha < \omega_1$ and hence there is $x \in A^d \cap X_{\alpha+1}$ by the property (4). It is clear that x is an accumulation point for A in X; this proves that every infinite subset of X has an accumulation point in X and hence X is countably compact (see Problem 132 of [TFS]).

We claim that the space $Y = D \cup (\beta D \setminus X)$ is also countably compact. To prove it take any countably infinite $A \subset Y$; there is an infinite discrete $B \subset A$ (see Fact 4 of S.382). Apply Fact 2 of S.382 and Fact 2 of S.286 to conclude that $\overline{B}$ is homeomorphic to $\beta\omega$ and hence $|\overline{B}| = 2^{\mathfrak{c}}$ by Problem 368 of [TFS]. Therefore $|B^d| = 2^{\mathfrak{c}}$, so it is impossible that $B^d \subset X$ because $|X| \leq \mathfrak{c}$. Thus there is a point $y \in B^d \cap Y$; it is clear that y is an accumulation point for A in Y. This proves that every infinite subset of Y has an accumulation point in Y and hence Y is countably compact.

Consider the space $X \times Y \subset \beta D \times \beta D$. The diagonal $\Delta = \{(x, x) : x \in \beta D\}$ is closed in βD and hence $E = \Delta \cap (X \times Y)$ is closed in $X \times Y$. Now, if $(x, x) \in X \times Y$, then $x \in X \cap Y = D$ and therefore $E = \{(x, x) : x \in D\}$. Every $x \in D$ is isolated in βD, so (x, x) is an isolated point of $\beta D \times \beta D$ which implies that (x, x) is isolated in $X \times Y$ for any $x \in D$. This shows that E is an uncountable clopen discrete subspace of $X \times Y$ so Fact 1 is proved.   □

Returning to our solution observe that

(5) for any space Z if $U \neq \emptyset$ is a clopen subset of Z, then U is a retract (and hence a continuous image) of Z.

Indeed, choose any $u \in U$; let $f(x) = x$ for every $x \in U$ and $f(x) = u$ for each $x \in Z \setminus U$. It is a trivial exercise that $f : Z \to U$ is a retraction.

Consider the space $K = X \oplus Y$ where X and Y are the countably compact spaces from Fact 1; it is evident that K is countably compact. Observe that $X \times Y$ is a clopen subspace of $K \times K$ and hence there is an uncountable clopen discrete $U \subset K \times K$ by Fact 1. The space K is ω-stable by Problem 119. If $K \times K$ is ω-stable, then the clopen set U is also ω-stable being a continuous image of $K \times K$ by (5) and Problem 123. The space U is discrete and hence metrizable, so ω-stability of U implies that U is separable (see Problem 106) and hence countable. This contradiction shows that $K \times K$ is not ω-stable, so our solution is complete.

T.127. *Prove that any Lindelöf P-space is ω-simple.*

Solution. Let X be a Lindelöf P-space. Assume that M is a second countable space and $f : X \to M$ is a continuous onto map. Since every point $x \in M$ is a G_δ-set in M, the set $f^{-1}(x)$ is a G_δ-set in X; since X is a P-space, the set $f^{-1}(x)$ is open in X for any $x \in M$. Therefore $\{f^{-1}(x) : x \in M\}$ is a disjoint open cover of X. The space X being Lindelöf, we can choose a countable $A \subset M$ such that $X = \bigcup\{f^{-1}(x) : x \in A\} = f^{-1}(A)$. Consequently, $M = f(X) = f(f^{-1}(A)) = A$ and therefore $M = A$ is a countable set. Thus X is ω-simple.

T.128. *Prove that for any scattered space X with $l(X) \leq \kappa$, the Lindelöf degree of the κ-modification of X does not exceed κ. In particular, if X is a Lindelöf (or compact!) scattered space, then the ω-modification of X is Lindelöf.*

Solution. Let $\tau = \tau(X)$ and denote by μ the topology of κ-modification of τ. Given any $B \subset X$ let $\tau|B = \{U \cap B : U \in \tau\}$, i.e., $\tau|B$ is the topology induced by τ on B. Fix any $\mathcal{U} \subset \mu$ with $\bigcup\mathcal{U} = X$; we must prove that X can be covered by a subfamily of $\mathcal{U}$ of cardinality $\leq \kappa$.

Consider the set $O = \{x \in X : \text{there is } \mathcal{U}' \subset \mathcal{U} \text{ and } O_x \in \tau \text{ such that } |\mathcal{U}'| \leq \kappa$ and $x \in O_x \subset \bigcup \mathcal{U}'\}$. If a set F is closed in (X, τ) and $F \subset O$, then the Lindelöf number of $(F, \tau | F)$ does not exceed κ, so we can choose a set $A \subset F$ such that $|A| \leq \kappa$ and $F \subset \bigcup \{O_x : x \in A\}$. For each $x \in A$ there is $\mathcal{U}_x \subset \mathcal{U}$ such that $|\mathcal{U}_x| \leq \kappa$ and $O_x \subset \bigcup \mathcal{U}_x$. The family $\mathcal{U}' = \bigcup \{\mathcal{U}_x : x \in A\} \subset \mathcal{U}$ has cardinality $\leq \kappa$ and covers F. Thus we have

(1) If F is closed in (X, τ) and $F \subset O$, then there is $\mathcal{U}' \subset \mathcal{U}$ such that $|\mathcal{U}'| \leq \kappa$ and $F \subset \bigcup \mathcal{U}'$. In particular, if $O = X$, then we can take $F = X$ to conclude that X can be covered by a subfamily of $\mathcal{U}$ of cardinality κ.

The property (1) shows that it suffices to establish that $O = X$. Assume the contrary; since X is scattered, there exists $a \in F = X \backslash O$ which is isolated in F. Consequently, there is $U \in \tau$ such that $a \in U$ and $\text{cl}_\tau(U) \cap F = \{a\}$. Pick any $P \in \mathcal{U}$ such that $a \in P$; since G_κ-subsets of X form a base in (X, μ) there is a family $\mathcal{V} \subset \tau$ such that $|\mathcal{V}| \leq \kappa$ and $a \in \bigcap \mathcal{V} \subset P$. Observe that $F_V = \text{cl}_\tau(U) \backslash V$ is a closed subset of (X, τ) and $F_V \subset O$ for each $V \in \mathcal{V}$. Therefore we can apply (1) to find a family $\mathcal{W}_V \subset \mathcal{U}$ such that $|\mathcal{W}_V| \leq \kappa$ and $F_V \subset \bigcup \mathcal{W}_V$ for every $V \in \mathcal{V}$. It is clear that $\bigcup \{F_V : V \in \mathcal{V}\}$ covers the set $\text{cl}_\tau(U) \backslash P$ and hence the family $\mathcal{U}' = \bigcup \{\mathcal{W}_V : V \in \mathcal{V}\} \cup \{P\} \subset \mathcal{U}$ covers the set $\text{cl}_\tau(U)$. This implies $U \subset \bigcup \mathcal{U}'$ which means that we have found a neighborhood of the point a in (X, τ) which is covered by a subfamily of $\mathcal{U}$ of cardinality $\leq \kappa$, i.e., $a \in O$ which is a contradiction. Thus $O = X$ and hence (1) can be applied to conclude that there is $\mathcal{U}_1 \subset \mathcal{U}$ such that $|\mathcal{U}_1| \leq \kappa$ and $X = \bigcup \mathcal{U}_1$. Since a μ-open cover of X was chosen arbitrarily, we proved that $l(X, \mu) \leq \kappa$.

T.129. *Prove that any Lindelöf scattered space (and, in particular, any compact scattered space) is a simple space.*

Solution. Let X be a Lindelöf scattered space and fix an infinite cardinal κ. Take any continuous onto map $f : X \to Y$ such that $w(Y) \leq \kappa$. Denote by μ the κ-modification of the topology $\tau(X)$; denote by X' the space (X, μ). We have $l(X) = \omega \leq \kappa$, so Problem 128 can be applied to conclude that $l(X') \leq \kappa$. Since $\tau(X) \subset \mu$, the identity map $i : X' \to X$ is continuous and hence $g = f \circ i$ is a continuous map of X' onto Y.

Observe that $\psi(Y) \leq w(Y) \leq \kappa$ and therefore every $y \in Y$ is a G_κ-set in Y. As a consequence, $g^{-1}(y)$ is open in X' being a G_κ-subset of X' (see Fact 2 of S.493). Thus the family $\{g^{-1}(y) : y \in Y\}$ is an open cover of the space X'; since $l(X') \leq \kappa$, there exists $A \subset Y$ such that $|A| \leq \kappa$ and $X' = \bigcup \{g^{-1}(y) : y \in A\} = g^{-1}(A)$. This shows that $Y = g(X') = g(g^{-1}(A)) = A$, so $|Y| = |A| \leq \kappa$ which proves that X is κ-simple. We proved that X is κ-simple for any infinite cardinal κ, i.e., X is simple.

T.130. *Give an example of a pseudocompact scattered non-simple space.*

Solution. Recall that a family $\mathcal{A}$ is called *almost disjoint* if every $A \in \mathcal{A}$ is infinite and $A \cap A'$ is finite for any distinct $A, A' \in \mathcal{A}$. A *Mrowka space* $M = M(\mathcal{M})$

is constructed as follows: $M = \omega \cup \mathcal{M}$ where $\mathcal{M}$ is a maximal almost disjoint subfamily of infinite subsets of ω. The points of ω are isolated in M and the local base at a point $x \in \mathcal{M}$ is the family $\{\{x\} \cup (x \backslash F) : F$ is a finite subset of $x\}$ (see Problem 142 of [TFS] for the details). Every Mrowka space is a locally countable, locally compact and first countable pseudocompact space such that ω is open, discrete and dense in M while $\mathcal{M}$ is a closed discrete subspace of M (see Problem 142 of [TFS]).

Fact 1. Every Mrowka space M is scattered.

Proof. Take any non-empty set $A \subset M$; if $A \cap \omega \neq \emptyset$, then any $x \in A \cap \omega$ is isolated in A. If $A \cap \omega = \emptyset$, then $A \subset \mathcal{M}$ and hence A is discrete, so all points of A are isolated in A. Fact 1 is proved. $\square$

It was proved in Fact 2 of S.154 that there exists an almost disjoint family $\mathcal{M}$ on ω such that the resulting Mrowka space $X = M(\mathcal{M})$ can be mapped continuously onto $\mathbb{I}$; since $w(\mathbb{I}) = \omega$ and $|\mathbb{I}| > \omega$, the space X is not ω-simple. Finally observe that the space X is pseudocompact by Problem 142 of [TFS] and scattered by Fact 1.

T.131. *Prove that the following are equivalent for any space X:*

  (i)   *the space X is weight(ω)-stable;*
 (ii)   *the space X is character(ω)-stable;*
(iii)   *the space X is Fréchet–Urysohn(ω)-stable;*
 (iv)   *the space X is sequential(ω)-stable;*
  (v)   *the space X is k-property(ω)-stable;*
 (vi)   *the space X is $\pi w(\omega)$-stable;*
(vii)   *the space X is $\pi\chi(\omega)$-stable;*
(viii) *the space X is pseudocompact.*

Solution. Our solution will be based on some simple facts.

Fact 1. Given any space Z, the space $C_p(Z)$ considered with its usual operations of summing functions and multiplying them by a real number is a locally convex linear topological space.

Proof. It is clear that $C_p(Z)$ is a linear space. Its operations are continuous by Problem 114 of [TFS] and Problem 115 of [TFS]; the local convexity of $C_p(Z)$ was established in Problem 069 of [TFS] so Fact 1 is proved. $\square$

Fact 2. Every infinite closed subspace of $\beta\omega$ has cardinality $2^{\mathfrak{c}}$ and hence $\beta\omega$ has no nontrivial convergent sequences.

Proof. Let F be an infinite closed subspace of $\beta\omega$. Choose an infinite discrete $B \subset F$ (see Fact 4 of S.382). Apply Fact 2 of S.382 and Fact 2 of S.286 to conclude that $\overline{B}$ is homeomorphic to $\beta\omega$ and hence $|\overline{B}| = 2^{\mathfrak{c}}$ by Problem 368 of [TFS]. Since $\overline{B} \subset F$, we have $|F| = 2^{\mathfrak{c}}$ so Fact 2 is proved. $\square$

Fact 3. Let M be a non-compact second countable space. Then there exists a continuous onto map $f : M \rightarrow Z$ such that Z is not a k-space. Consequently, M is not k-property(ω)-stable.

Proof. Since M is not compact, it is not countably compact and hence there exists a countably infinite closed discrete $D \subset M$. Take any $\xi \in \beta\omega \setminus \omega$; then $N = \omega \cup \{\xi\}$ is not a k-space. Indeed, ω is not closed in N but $K \cap \omega$ is closed for any compact $K \subset N$ because all compact subsets of N are finite by Fact 2.

If $g : D \rightarrow N$ is any surjection, then g is a continuous map because D is discrete. The space N embeds in $C_p(C_p(N))$ as a closed subspace by Problem 167 of [TFS], so we can consider that $N \subset C_p(C_p(N))$. Therefore we have a continuous map $g : D \rightarrow C_p(C_p(N))$; since $C_p(C_p(N))$ is locally convex by Fact 1 and M is metrizable, we can apply 104 to conclude that there exists a continuous map $f : M \rightarrow C_p(C_p(N))$ such that $f|D = g$. If $Z = f(M)$, then $f : M \rightarrow Z$ is an onto map. Observe that $N = g(D) = f(D) \subset Z$; since N is closed in $C_p(C_p(N))$, it is closed in Z, so Z is not a k-space (it is an easy exercise to prove that a closed subset of a k-space is a k-space, so if a space contains a closed subset which is not a k-space, then it cannot be a k-space either).

Since $w(M) = \omega$, we have $nw(Z) \leq \omega$ and hence the space Z condenses onto a second countable space (see Problems 156 and 157 of [TFS]). This proves that the space M is not k-property(ω)-stable so Fact 3 is proved.   □

Fact 4. Let M be a non-compact second countable space. Then there exists a continuous onto map $f : M \rightarrow Z$ such that $\pi\chi(Z) > \omega$. Consequently, M is not $\pi\chi(\omega)$-stable.

Proof. Since M is not compact, it is not countably compact and hence there exists a countably infinite closed discrete $D \subset M$. Now, the space $C_p(\mathbb{I})$ has uncountable π-character at zero and hence at all of its points by Problem 171 of [TFS]. It is an easy exercise that

(1) if Y is a space and T is dense in Y, then $\pi\chi(y, Y) = \pi\chi(y, T)$ for any $y \in T$.

So if we take a countable dense $A \subset C_p(\mathbb{I})$, then $\pi\chi(A) > \omega$. If $g : D \rightarrow A$ is any surjection, then g is a continuous map because D is discrete. Therefore we have a continuous map $g : D \rightarrow C_p(\mathbb{I})$; since $C_p(\mathbb{I})$ is locally convex by Fact 1 and M is metrizable, we can apply Problem 104 to conclude that there exists a continuous map $f : M \rightarrow C_p(\mathbb{I})$ such that $f|D = g$. If $Z = f(M)$, then $f : M \rightarrow Z$ is an onto map. Observe that $A = g(D) = f(D) \subset Z$; since A is dense in $C_p(\mathbb{I})$, it is dense in Z, so $\pi\chi(Z) = \pi\chi(A) > \omega$ by (1). Since $w(M) = \omega$, we have $nw(Z) \leq \omega$ and hence Z condenses onto a second countable space (see Problems 156 and 157 of [TFS]). This proves that M is not $\pi\chi(\omega)$-stable. Fact 4 is proved.   □

Returning to our solution, observe that

(weight $\leq \omega$) $\Longrightarrow$ (character $\leq \omega$) $\Longrightarrow$ (Fréchet–Urysohn) $\Longrightarrow$ (sequential) and (sequential) $\Longrightarrow$ (k-space) which shows that (i)$\Longrightarrow$(ii)$\Longrightarrow$(iii)$\Longrightarrow$(iv)$\Longrightarrow$(v). Now assume that X is not pseudocompact. Then there is a countably infinite closed discrete $D \subset X$ such that D is C-embedded in X (see Fact 1 of S.350). Let $D = \{d_n : n \in \omega\}$ and $f(n) = n$ for each $n \in \omega$; then $f : D \to \mathbb{R}$ is a continuous map because D is discrete. Since D is C-embedded in X, there is $g \in C(X)$ such that $g | D = f$. If $M = g(X)$, then M is not compact being unbounded in $\mathbb{R}$. Thus

(2) if X is not pseudocompact, then there is a continuous onto map $g : X \to M$ of
   X onto a non-compact second countable space M.

Apply Fact 3 to find a continuous onto map $h : M \to Z$ such that Z is not a k-space. Since $nw(Z) = \omega$, the space Z condenses onto a second countable space. Consequently, $h \circ g$ maps X onto the space Z with $iw(Z) = \omega$ while Z is not a k-space. Therefore X is not k-property(ω)-stable which shows that (v)$\Longrightarrow$(viii)

Now assume that X is pseudocompact; given any continuous map $f : X \to Y$, the space Y is also pseudocompact. If Y condenses onto a second countable space M, then this condensation is a homeomorphism by Problem 140 of [TFS] and hence $w(Y) = w(M) = \omega$. Therefore the space X is weight(ω)-stable, i.e., we proved that (viii)$\Longrightarrow$(i); as a consequence, (i) $\Longleftrightarrow$ (ii) $\Longleftrightarrow$ (iii) $\Longleftrightarrow$ (iv) $\Longleftrightarrow$ (v) $\Longleftrightarrow$ (viii).

It is immediate that (weight$\leq\omega$) $\Longrightarrow$ (π-weight$\leq\omega$) $\Longrightarrow$ (π-character$\leq\omega$), so (i)$\Longrightarrow$(vi)$\Longrightarrow$(vii). If X is not pseudocompact, then, by (2), there is a continuous onto map $g : X \to M$ of X onto a non-compact second countable space M. Apply Fact 4 to find a continuous onto map $h : M \to Z$ such that $\pi\chi(Z) > \omega$. Since $nw(Z) = \omega$, the space Z condenses onto a second countable space. Consequently, $h \circ g$ maps X onto the space Z with $iw(Z) = \omega$ while $\pi\chi(Z) > \omega$. Therefore X is not $\pi\chi(\omega)$-stable which shows that (vii)$\Longrightarrow$(viii). We saw already that (viii)$\Longrightarrow$(i), so (i) $\Longleftrightarrow$ (vi) $\Longleftrightarrow$ (vii) $\Longleftrightarrow$ (viii) and hence our solution is complete.

T.132. *Prove that the following conditions are equivalent for any space X:*

   (i) *for any countable $A \subset C_p(X)$, the space $\overline{A}$ is σ-compact;*
   (ii) *for any countable $A \subset C_p(X)$, the space $\overline{A}$ is σ-countably compact;*
   (iii) *for any countable $A \subset C_p(X)$, the space $\overline{A}$ is σ-pseudocompact;*
   (iv) *for any countable $A \subset C_p(X)$, the space $\overline{A}$ is locally compact;*
   (v) *$C_p(X)$ is a Hurewicz space;*
   (vi) *$C_p(X)$ is a σ-locally compact space;*
   (vii) *for any countable $A \subset C_p(X)$, the space $\overline{A}$ is Hurewicz;*
   (viii) *for any countable $A \subset C_p(X)$, the space $\overline{A}$ is σ-locally compact;*
   (ix) *X is finite.*

Solution. If Z is a space, call a family $\mathcal{A} = \{A_s : s \in S\} \subset \exp(Z)$ *indexed discrete* if, for any $z \in Z$, there is $W \in \tau(z, Z)$ such that the set $\{s \in S : W \cap A_s \neq \emptyset\}$ is finite. For the sake of brevity denote the space ω^ω by $\mathbb{P}$. A space is *zero-dimensional* if it has a base which consists of clopen sets. The expression $X \simeq Y$ says that the space X is homeomorphic to the space Y.

As usual, we identify any ordinal with the set of its predecessors and, in particular, $n = \{0, \ldots, n - 1\}$ for any $n \in \omega$. If $k, n \in \omega$ and $s \in \omega^k$, then $s^\frown n \in \omega^{k+1}$ is defined by $(s^\frown n)(k) = n$ and $(s^\frown n)|k = s$. If we have $s \in \omega^n$, $t \in \omega^k$ where $n \le k$ and $n, k \in \omega + 1$, then $s \subset t$ says that $t|n = s$. If d is a metric on a set M, then $\tau(d)$ is the topology generated by the metric d.

If we work with a metric space (M, d), then for any $x \in X$ and $r > 0$ the set $B_d(x, r) = \{y \in M : d(x, y) < r\}$ is the ball of radius r centered at x. If f is a function, then $\mathrm{dom}(f)$ is its domain. Suppose that we have a set of functions $\{f_i : i \in I\}$ such that $f_i|(\mathrm{dom}(f_i) \cap \mathrm{dom}(f_j)) = f_j|(\mathrm{dom}(f_i) \cap \mathrm{dom}(f_j))$ for any $i, j \in I$. Then we can define a function f with $\mathrm{dom}(f) = \bigcup_{i \in I} \mathrm{dom}(f_i)$ as follows: given any $x \in \mathrm{dom}(f)$, find any $i \in I$ with $x \in \mathrm{dom}(f_i)$ and let $f(x) = f_i(x)$. It is easy to check that the value of f at x does not depend on the choice of i, so we have consistently defined a function f which will be denoted by $\bigcup\{f_i : i \in I\}$

The concept of Hurewicz space is a new one, so let us get some insight into it:

Fact 1.

(1) any countable union of Hurewicz spaces is a Hurewicz space;
(2) any σ-compact space is Hurewicz;
(3) any closed subspace of a Hurewicz space is a Hurewicz space;
(4) any continuous image of a Hurewicz space is a Hurewicz space;
(5) the space $\mathbb{P}$ is not Hurewicz and hence not σ-compact.

Proof. (1) Assume that $Z = \bigcup\{Z_n : n \in \omega\}$ and Z_n is Hurewicz for each $n \in \omega$. Let $\{\mathcal{U}_n : n \in \omega\}$ be a sequence of open covers of the space Z. Represent ω as a disjoint infinite union of its infinite subsets, i.e., $\omega = \bigcup\{M_n : n \in \omega\}$ where $|M_n| = \omega$ for each $n \in \omega$ and $M_k \cap M_l = \emptyset$ whenever $k \ne l$. The collection $\mathcal{W}_k = \{\mathcal{U}_i : i \in M_k\}$ is a sequence of open covers of Z_k, so we can choose a finite $\mathcal{V}_i \subset \mathcal{U}_i$ for each $i \in M_k$ so that $Z_k \subset \bigcup(\bigcup\{\mathcal{V}_i : i \in M_k\})$ for each $k \in \omega$. After this choice is carried out for each $k \in \omega$, we obtain a finite $\mathcal{V}_i \subset \mathcal{U}_i$ for each $i \in \omega$ and

$$\bigcup\left(\bigcup\{\mathcal{V}_i : i \in \omega\}\right) = \bigcup\left\{\bigcup\left(\bigcup\{\mathcal{V}_i : i \in M_k\}\right) : k \in \omega\right\} \supset \bigcup\{Z_k : k \in \omega\} = Z,$$

which shows that Z is a Hurewicz space and settles (1). The statement (2) follows easily from (1) and the evident fact that every compact space is Hurewicz. The proof of (3) and (4) is an easy exercise for the reader so let us check (5).

Given $n \in \mathbb{N}$ and $s \in \omega^n$ the set $[s] = \{x \in \omega^\omega$ such that $s \subset x\}$ is clopen in ω^ω. The family $\mathcal{U}_n = \{[s] : s \in \omega^{n+1}\}$ is an open disjoint cover of $\mathbb{P}$ for every $n \in \omega$. Suppose that $\mathcal{V}_n$ is a finite subfamily of $\mathcal{U}_n$ and let $W_n = \bigcup \mathcal{V}_n$ for all $n \in \omega$. The family $\mathcal{U}_0$ is disjoint and infinite, so we can choose $s_0 \in \omega^1$ such that $[s_0] \notin \mathcal{V}_0$ and hence $[s_0] \cap W_0 = \emptyset$. Assume that $k > 0$ and we have chosen $s_0, \ldots s_{k-1}$ with the following properties:

(a) $s_i \in \omega^{i+1}$ for all $i < k$;
(b) $s_0 \subset \cdots \subset s_{k-1}$;
(c) $[s_i] \cap W_i = \emptyset$ for each $i < k$.

The family $\mathcal{U}'_k = \{[s] : s \in \omega^{k+1}$ and $s_{k-1} \subset s\} \subset \mathcal{U}_k$ is infinite and disjoint, so there is $s_k \in \omega^{k+1}$ such that $[s_k] \in \mathcal{U}'_k \backslash \mathcal{V}_k$ and therefore $[s_k] \cap W_k = \emptyset$. We have $s_{k-1} \subset s_k$ by our definition of $\mathcal{U}'_k$, so the conditions (a)–(c) are satisfied for the collection $\{s_i : i \leq k\}$. Thus our inductive construction can be continued to give us a sequence $\{s_i : i \in \omega\}$ with the properties (a)–(c). It follows from (b) that $s = \bigcup\{s_i : i \in \omega\} \in \mathbb{P}$; the condition (b) shows that $s \in \bigcap\{[s_i] : i \in \omega\}$ and hence $s \notin \bigcup\{W_i : i \in \omega\}$ by (c) which proves that the family $\bigcup\{\mathcal{V}_i : i \in \omega\}$ is not a cover of $\mathbb{P}$. Thus $\mathbb{P}$ is not Hurewicz and Fact 1 is proved. □

Fact 2. Let M be a zero-dimensional second countable space. Then every closed non-empty set $F \subset M$ is a retract of M.

Proof. Take any metric d on the set M which generates the topology of M; for each $x \in M\backslash F$, let $d(x, F) = \inf\{d(x, y) : y \in F\}$ and $r(x) = \frac{1}{4}d(x, F)$. Consider the family $\mathcal{U} = \{B_d(x, r(x)) : x \in M\backslash F\}$; since M is zero-dimensional, we can choose a clopen set $O_x \in \tau(x, M)$ such that $O_x \subset B_d(x, r(x))$ for every $x \in M\backslash F$. Observe that $O_x \subset B_d(x, r(x)) \subset M\backslash F$ for each $x \in M\backslash F$. Since $M\backslash F$ is Lindelöf, there is a countable $A \subset M\backslash F$ such that $\bigcup\{O_x : x \in A\} = M\backslash F$. Choose any enumeration $\{U_n : n \in \omega\}$ of the family $\{O_x : x \in A\}$ and let $V_n = U_n\backslash(\bigcup\{U_i : i < n\})$ for each $n \in \omega$. By our construction, for each $n \in \omega$, there is a point $y_n \in A$ such that $V_n \subset B_d(y_n, r_n)$ where $r_n = \frac{1}{4}d(y_n, F)$. Take any $a_n \in F$ such that $d(y_n, a_n) \leq \frac{5}{4}d(y_n, F) = 5r_n$.

Observe that

(6) given any $n \in \omega$, we have $d(x, a_n) \leq 2d(x, F)$ for every $x \in V_n$.

To prove (6) note that $d(x, F) \geq 3r_n$; for otherwise, there is a point $z \in F$ such that $d(x, z) < 3r_n$ and hence

$$d(y_n, F) \leq d(y_n, z) \leq d(y_n, x) + d(z, x) < r_n + 3r_n = 4r_n = d(y_n, F),$$

which is a contradiction proving that $d(x, F) \geq 3r_n$. Furthermore, it follows from $d(x, a_n) \leq d(x, y_n) + d(y_n, a_n) \leq r_n + 5r_n = 6r_n \leq 2d(x, F)$ that (6) is proved.

Now, let $f(x) = x$ for every $x \in F$. If $x \in M\backslash F$, then there is a unique $n \in \omega$ such that $x \in V_n$; let $f(x) = a_n$. It is evident that $f : M \to F$ and $f(x) = x$ for any $x \in F$, so we only must show that f is a continuous map.

If $x \in M\backslash F$, then $x \in V_n$ for some n so $f(x) = a_n$. Since V_n is a neighborhood of x such that $f(V_n) = \{a_n\}$, the function f is continuous at the point x. Now take any $x \in F$ and any $U \in \tau(x, F)$; there exists $\varepsilon > 0$ such that $B_d(x, \varepsilon) \cap F \subset U$. The set $W = B_d(x, \frac{\varepsilon}{3})$ is open in M and contains x. We claim that $f(W) \subset B_d(x, \varepsilon) \cap F$. Indeed, if $y \in W \cap F$, then $f(y) = y$ so $f(y) \in W \cap F \subset B_d(x, \frac{\varepsilon}{3}) \subset B_d(x, \varepsilon)$. Now, if $y \in W\backslash F$, then there is $n \in \omega$ such that $y \in V_n$ and hence $f(y) = a_n$. We have $d(x, a_n) \leq d(x, y) + d(y, a_n) \leq \frac{\varepsilon}{3} + 2d(y, F) \leq \frac{\varepsilon}{3} + 2d(x, y) < \frac{\varepsilon}{3} + \frac{2\varepsilon}{3} = \varepsilon$ and hence $f(W) \subset B_d(x, \varepsilon) \cap F \subset U$ which proves continuity of f at the point x. Thus $f : M \to F$ is a retraction and Fact 2 is proved. □

Fact 3. Call a space M *analytic* if M is a continuous image of the space $\mathbb{P}$. Then

(7) every closed subspace of an analytic space is an analytic space;

(8) every continuous image of an analytic space is an analytic space;

(9) any countable product of analytic spaces is an analytic space;

(10) any countable intersection of analytic spaces is an analytic space;

(11) any countable union of analytic spaces is an analytic space;

(12) every σ-compact space is analytic;

(13) every second countable completely metrizable space is analytic.

Proof. Take any analytic space M and any non-empty closed $A \subset M$. There is a continuous onto map $f : \mathbb{P} \to M$; let $F = f^{-1}(A)$ and $g = f|F$. Then g is a continuous map of F onto A. The space $\mathbb{P}$ is zero-dimensional, so there is a retraction $r : \mathbb{P} \to F$ by Fact 2. Then $h = g \circ r$ is a continuous map and $h(\mathbb{P}) = A$ so (7) is proved. Now, if $q : M \to N$ is a continuous onto map, then $p = q \circ f : \mathbb{P} \to N$ is a continuous map such that $N = p(\mathbb{P})$, so N is analytic and hence we settled (8).

To prove (9) note that $\mathbb{P}^\omega = (\omega^\omega)^\omega \simeq \omega^{\omega \times \omega}$ is homeomorphic to $\mathbb{P}$ (see Problem 103 of [TFS]). Given any family $\{M_n : n \in \omega\}$ of analytic spaces, let $M = \prod\{M_n : n \in \omega\}$ and fix a continuous onto map $f_n : \mathbb{P} \to M_n$ for all $n \in \omega$. Let $g(x)(n) = f_n(x(n))$ for any $x \in \mathbb{P}^\omega$ and $n \in \omega$. It is immediate that $g : \mathbb{P}^\omega \to M$ (called the product of the family $\{f_n : n \in \omega\}$) is an onto map. Continuity of any product of maps was proved in Fact 1 of S.271 so g is continuous. Since $\mathbb{P}^\omega \simeq \mathbb{P}$, the space M is analytic and we proved (9).

To prove (10) take an arbitrary family $\{M_n : n \in \omega\}$ of analytic spaces and let $M = \bigcap\{M_n : n \in \omega\}$. Observe that the space M embeds as a closed subspace in the space $\prod\{M_n : n \in \omega\}$ by Fact 7 of S.271, so we can apply (7) and (9) to conclude that M is analytic.

As to (11), assume that M_n is an analytic space for every $n \in \omega$ and let $M = \bigcup\{M_n : n \in \omega\}$. It is evident that the space $P_n = \{x \in \mathbb{P} : x(0) = n\}$ is homeomorphic to $\mathbb{P}$ for each $n \in \omega$ and $P \simeq \bigoplus\{P_n : n \in \omega\}$. Since each M_n is analytic, we can fix a continuous onto map $f_n : P_n \to M_n$ for every $n \in \omega$. Given any $x \in \mathbb{P}$ there is a unique $n \in \omega$ such that $x \in P_n$; let $f(x) = f_n(x)$. Then $f : \mathbb{P} \to M$ is an onto map and the family $\{P_n : n \in \omega\}$ is an open cover of $\mathbb{P}$ such that $f|P_n = f_n$ is continuous for each $n \in \omega$. Therefore f is continuous by Fact 1 of S.472. This shows that M is analytic so (11) is proved.

Now observe that $K = \{0,1\}^\omega$ is a closed subset of $\mathbb{P}$ and hence K is a continuous image of $\mathbb{P}$ by Fact 2. Since every metrizable compact space is a continuous image of K (see Problem 128 of [TFS]), we can apply (8) to conclude that every compact metrizable space is analytic. Applying (11) we conclude that every σ-compact second countable space is analytic so (12) is verified.

Finally, take any completely metrizable second countable space M. Then M is homeomorphic to a closed subspace in $\mathbb{R}^\omega$ by Problem 273 of [TFS]. The space $\mathbb{R}$ is analytic being σ-compact by (12). Applying (9) and (7) we conclude that $\mathbb{R}^\omega$ and M are analytic, so (13) is settled and Fact 3 is proved.   □

Fact 4. Let M be an analytic second countable non-σ-compact space. Then there is a closed subspace $F \subset M$ such that F is homeomorphic to $\mathbb{P}$.

Proof. The space $\mathbb{P}$ is completely metrizable (see Problem 204 and 207 of [TFS]), so we can fix a complete metric ρ on $\mathbb{P}$ such that $\tau(\rho) = \tau(\mathbb{P})$. Take any metric d on M with $\tau(d) = \tau(M)$. Since M is analytic, there is a continuous onto map $f : \mathbb{P} \to M$. We will need the family $\mathcal{S} = \{A \subset M :$ there is a σ-compact $P \subset M$ such that $A \subset P\}$. Let $A_\emptyset = \mathbb{P}$; assume that $n > 0$, and for each $k < n$, we have a family $\{A_s : s \in \omega^k\}$ of closed subsets of $\mathbb{P}$ with the following properties:

(14) if $0 < k < n$, then $\mathrm{diam}_\rho(A_s) < 2^{-k}$ for all $s \in \omega^k$;
(15) if $0 < k < n$, then $\mathrm{diam}_d(f(A_s)) < 2^{-k}$ for all $s \in \omega^k$;
(16) if $0 \le k < n$, then $\{A_s : s \in \omega^k\}$ is an indexed discrete family in $\mathbb{P}$;
(17) if $0 \le k < n$, then $\{f(A_s) : s \in \omega^k\}$ is an indexed discrete family in M;
(18) if $0 \le k < l < n$, $s \in \omega^k$, $t \in \omega^l$ and $s \subset t$, then $A_t \subset A_s$;
(19) if $0 \le k < n$, then $f(A_s) \notin \mathcal{S}$ for any $s \in \omega^k$.

Fix $s \in \omega^{n-1}$ and let $B_s = \{x \in A_s : f(U) \notin \mathcal{S}$ for any $U \in \tau(x, A_s)\}$. For each $x \in A_s \backslash B_s$ there is $W_x \in \tau(x, A_s)$ such that $f(W_x) \in \mathcal{S}$. The open cover $\mathcal{W} = \{W_x : x \in A_s \backslash B_s\}$ of the Lindelöf space $A_s \backslash B_s$ has a countable subcover $\mathcal{W}'$; it is evident that $f(\bigcup \mathcal{W}') = f(A_s \backslash B_s) \in \mathcal{S}$. Since $f(A_s) \notin \mathcal{S}$, we have $f(B_s) \notin \mathcal{S}$ and hence there exists an infinite set $D \subset f(B_s)$ which is discrete and closed in M. It is easy to find a set $V_y \in \tau(y, M)$ such that $\mathrm{diam}_d(V_y) < 2^{-n}$ for each $y \in D$ and the family $\{V_y : y \in D\}$ is indexed discrete in M. The family $\{f^{-1}(V_y) : y \in D\}$ is indexed discrete in $\mathbb{P}$; pick a point $x(y) \in f^{-1}(y) \cap B_s$ for each $y \in D$. It is clear that $x(y) \in f^{-1}(V_y)$, so we can find $U_y \in \tau(x(y), \mathbb{P})$ such that $U_y \subset f^{-1}(V_y)$ and $\mathrm{diam}_\rho(U_y) < 2^{-n}$ for every $y \in D$.

Take a faithful enumeration $\{y_n : n \in \omega\}$ of the set D and let $A_{s^\frown n} = \overline{U_{y_n} \cap B_s}$ for each $n \in \omega$. It follows from $U_{y_n} \cap B_s \neq \emptyset$ that $f(A_{s^\frown n}) \notin \mathcal{S}$ for each $n \in \omega$. It is clear that $\mathrm{diam}_\rho(A_{s^\frown n}) \le \mathrm{diam}_\rho(U_{y_n}) < 2^{-n}$. Besides, $\mathrm{diam}_d(f(A_{s^\frown n})) \le \mathrm{diam}_d(f(U_{y_n})) \le \mathrm{diam}_d(V_{y_n}) < 2^{-n}$ which shows that (14) and (15) are fulfilled for the sets $A_{s^\frown n}$ for all $n \in \omega$. The properties (16) and (17) hold because the families $\mathcal{U} = \{U_{y_n} : n \in \omega\}$ and $\mathcal{V} = \{V_{y_n} : n \in \omega\}$ are indexed discrete and the families $\{A_{s^\frown n} : n \in \omega\}$ and $\{f(A_{s^\frown n}) : n \in \omega\}$ are obtained by shrinking $\mathcal{U}$ and $\mathcal{V}$ respectively. After we define the family $\{A_{s^\frown n} : n \in \omega\}$ for each $s \in \omega^{n-1}$, we have the collection $\{A_t : t \in \omega^n\}$. This method of construction guarantees the property (18). We also assured that $f(A_{s^\frown n}) \notin \mathcal{S}$ for each $s \in \omega^{n-1}$ and $n \in \omega$ which shows that (19) is also true for the family $\{A_t : t \in \omega^n\}$.

Thus our inductive construction can be completed for all natural n to give us families $\{A_s : s \in \omega^n\}$ with (14)–(19) for all $n \in \omega$. Now we are ready to describe a closed subset of M which is homeomorphic to $\mathbb{P}$. Given any $x \in \mathbb{P}$ let $s_n = x|n$ for each $n \in \omega$; it follows from (14) and (18) that $\{A_{s_n} : n \in \omega\}$ is a decreasing family of closed subsets of $\mathbb{P}$ such that $\mathrm{diam}_\rho(A_{s_n}) \to 0$. As a consequence there is a unique point $g(x) \in \bigcap\{A_{s_n} : n \in \omega\}$ (see Problem 236 of [TFS]). Let $h(x) = f(g(x))$; this gives us a map $h : \mathbb{P} \to M$. We must check that $F = h(\mathbb{P})$ is closed in M and the map $h : \mathbb{P} \to F$ is a homeomorphism.

Take any point $x \in \mathbb{P}$ and an arbitrary $\varepsilon > 0$; there exists $m \in \omega$ such that $2^{-m} < \varepsilon$. The set $U = [x|m]$ is an open neighborhood of the point x in $\mathbb{P}$, and for any $y \in U$, we have $y|m = x|m$ and hence $h(y) = f(g(y)) \in f(A_{x|m})$. Since $\text{diam}_d(f(A_{x|m})) \leq 2^{-m} < \varepsilon$, we have $d(h(y), h(x)) \leq \text{diam}_d(f(A_{x|m})) < \varepsilon$, which shows that $h(U) \subset B_d(h(x), \varepsilon)$ and hence h is continuous at the point x.

If x and y are distinct points of $\mathbb{P}$, then there is $n \in \omega$ such that $x|n \neq y|n$. The property (17) implies that $f(A_{x|n}) \cap f(A_{y|n}) = \emptyset$; since $h(x) \in f(A_{x|n})$ and $h(y) \in f(A_{y|n})$, we have $h(y) \neq h(x)$ and hence h is a condensation.

To see that $h^{-1} : F \to \mathbb{P}$ is continuous, take any $y \in F$. There exists a sequence $\{s_i : i \in \omega\}$ such that $s_i \in \omega^i$ for all $i \in \omega$ and $y \in \bigcap\{f(A_{s_i}) : i \in \omega\}$. It follows from (17) that $s_i \subset s_{i+1}$ for all $i \in \omega$ and hence $x = \bigcup\{s_i : i \in \omega\}$ is well-defined. It is immediate that $x = h^{-1}(y)$; observe that the family $\{[x|n] : n \in \mathbb{N}\}$ is a local base at x in $\mathbb{P}$, so if we take any $W \in \tau(x, \mathbb{P})$, then there is $n \in \omega$ such that $[x|n] \subset W$. The set $U = f(A_{s_n}) \cap F \ni y$ is open in F by (17). If $z \in U$, then $h^{-1}(z)|n = s_n$ by (16); this implies $h^{-1}(z)|n = x|n$, i.e., $h^{-1}(z) \in [x|n] \subset W$. The point $z \in U$ was chosen arbitrarily so $h^{-1}(U) \subset [x|n]$. Thus we proved that for any $W \in \tau(x, \mathbb{P})$ there exists $U \in \tau(y, F)$ such that $h^{-1}(U) \subset W$. Therefore h^{-1} is continuous at every $y \in F$ and h is a homeomorphism.

To see that F is closed in M take any point $y \in \overline{F}$. If $F_n = \bigcup\{f(A_s) : s \in \omega^n\}$ for each $n \in \omega$, then $F = \bigcap\{F_n : n \in \omega\}$. Therefore $y \in \overline{F_n}$ for each $n \in \omega$; the family $\{f(A_s) : s \in \omega^n\}$ being discrete, there is $s_n \in \omega^n$ such that $y \in \overline{f(A_{s_n})}$ for all $n \in \omega$. The properties (17) and (18) imply that $s_n \subset s_{n+1}$ for any $n \in \omega$ and hence we can define $x = \bigcup\{s_n : n \in \omega\}$. Now observe that $h(x) \in \bigcap\{f(A_{s_n}) : n \in \omega\}$; the property (15) shows that $d(h(x), y) \leq \text{diam}_d\left(\overline{f(A_{s_n})}\right) = \text{diam}_d(f(A_{s_n})) < 2^{-n}$ for every $n \in \omega$. Thus $d(h(x), y) = 0$, i.e., $h(x) = y$ and hence $y \in F$. This proves that F is closed in M so Fact 4 is proved.   □

Fact 5. Given any space Z assume that $\mathcal{U} = \{U_t : t \in T\} \subset \tau^*(Z)$ is an indexed discrete family and $x_t \in U_t$ for every $t \in T$. Let $D = \{x_t : t \in T\}$ and take any function $b_t \in C(Z)$ such that $b_t(x_t) = 1$ and $b_t(Z \backslash U_t) = \{0\}$ for all $t \in T$. Define a map φ by $\varphi(h)(z) = \sum_{t \in T} h(x_t) b_t(z)$ for every $h : D \to \mathbb{R}$ and $z \in Z$. Then $\varphi(h) \in C(Z)$ and $\varphi(h)|D = h$ for each $h \in \mathbb{R}^D$; besides, the map $\varphi : \mathbb{R}^D \to C_p(Z)$ is linear and continuous.

Proof. For any $z \in Z$ there is $W \in \tau(z, Z)$ such that W intersects at most one element of $\mathcal{U}$, say U_t. Consequently, $\varphi(h)|W = h(x_t) b_t|W$ is a continuous function, so we can apply Fact 1 of S.472 to conclude that $\varphi(h)$ is continuous. By the definition of $\varphi(h)$ we have $\varphi(h)(x_t) = h(x_t) b_t(x_t) = h(x_t)$ for each $t \in T$ so $\varphi(h)|D = h$.

The linearity of φ is straightforward; to prove that it is continuous take any point $z \in Z$ and let $\pi_z(f) = f(z)$ for every $f \in C_p(Z)$. Observe that the mapping π_z is the restriction of the natural projection of $\mathbb{R}^Z$ onto its factor determined by z. Let $q_t : \mathbb{R}^D \to \mathbb{R}$ be the natural projection onto the factor determined by x_t, i.e., $q_t(h) = h(x_t)$ for every $t \in T$. Note that φ maps $\mathbb{R}^D$ to the product $\mathbb{R}^Z$, so it suffices to prove that $\pi_z \circ \varphi$ is continuous for each $z \in Z$ (see Problem 102 of [TFS]).

If $z \in Z \backslash (\bigcup \mathcal{U})$, then $\varphi(h)(z) = 0$ for all $h \in \mathbb{R}^D$, so $\pi_z \circ \varphi$ is continuous being a constant map. If $z \in \bigcup \mathcal{U}$, then $z \in U_t$ for some $t \in T$ and hence $b_u(z) = 0$ for all $u \in T \backslash \{t\}$. As a consequence, $\varphi(h)(z) = h(x_t) b_t(z)$ for any $h \in \mathbb{R}^D$ which shows that $\pi_z \circ \varphi$ is continuous because it coincides with $b_t(z) q_t$. Thus φ is continuous and hence Fact 5 is proved. $\qquad \square$

Fact 6. If a space Z is not pseudocompact, then $C_p(Z) \simeq C_p(Z) \times \mathbb{R}^\omega$. In particular, $\mathbb{R}^\omega$ embeds in $C_p(Z)$ as a closed subspace.

Proof. Since the space Z is not pseudocompact, we can fix an indexed discrete family $\mathcal{U} = \{U_n : n \in \omega\} \subset \tau^*(Z)$ and pick $x_n \in U_n$ for every $n \in \omega$; let $D = \{x_n : n \in \omega\}$. Choose a function $b_n \in C_p(Z, [0, 1])$ such that $b_n(x_n) = 1$ and $b_n(Z \backslash U_n) = \{0\}$ for all $n \in \omega$. Define a map φ by $\varphi(h)(z) = \sum_{n \in \omega} h(x_n) b_n(z)$ for every $h : D \to \mathbb{R}$ and $z \in Z$. Then $\varphi(h) \in C(Z)$ and $\varphi(h) | D = h$ for each $h \in \mathbb{R}^D$; besides, the map $\varphi : \mathbb{R}^D \to C_p(Z)$ is continuous (see Fact 5).

We will show that the space $I = \{f \in C_p(Z) : f(D) = \{0\}\}$ is a factor of $C_p(Z)$, namely, that $I \times \mathbb{R}^D \simeq C_p(Z)$. To do so, let $u(p) = g + \varphi(h)$ for any $p = (g, h) \in I \times \mathbb{R}^D$. It follows from Fact 5 that $u : I \times \mathbb{R}^D \to C_p(Z)$. Given any $f \in C_p(Z)$, let $h = f | D$ and $g = f - \varphi(h)$. It is immediate that $p = (g, h) \in I \times \mathbb{R}^D$ and $u(p) = f$, i.e., u is an onto map. Furthermore, let $w(f) = (f - \varphi(f | D), f | D)$ for each $f \in C_p(Z)$. It is straightforward that $w : C_p(Z) \to I \times \mathbb{R}^D$ and w is the inverse function of u. Thus u is a bijection, so it suffices to prove continuity of u and w.

Given any $p = (g, h) \in I \times \mathbb{R}^D$, let $u_0(p) = g$ and $u_1(p) = g$; it is clear that $u_0 : I \times \mathbb{R}^D \to I$ and $u_1 : I \times \mathbb{R}^D \to \mathbb{R}^D$ are continuous maps. For any pair $(f_1, f_2) \in C_p(Z) \times C_p(Z)$, let $m(f_1, f_2) = f_1 + f_2$; then $m : C_p(Z) \times C_p(Z) \to C_p(Z)$ is also continuous (see Problem 115 of [TFS]). Now observe that $u = m \circ (\Delta \{u_0, \varphi \circ u_1\})$ and hence u is continuous being obtained from continuous maps u_0, u_1, φ and m by applying compositions and Δ-product.

To prove continuity of w, observe that the map $\pi_D : C_p(Z) \to \mathbb{R}^D$ defined by $\pi_D(f) = f | D$ is continuous as well as the map $r : C_p(Z) \times C_p(Z) \to C_p(Z)$ defined by $r(f_1, f_2) = f_1 - f_2$ for any pair $(f_1, f_2) \in C_p(Z) \times C_p(Z)$. If $id : C_p(Z) \to C_p(Z)$ is the identity map, then $w = \Delta \{r \circ (\Delta \{id, \varphi \circ \pi_D\}), \pi_D\}$ is continuous because it is obtained from continuous maps π_D, φ, r and id by applying compositions and Δ-products. Thus u is a homeomorphism which shows that

$$C_p(Z) \simeq I \times \mathbb{R}^D \simeq I \times \mathbb{R}^\omega \simeq I \times \mathbb{R}^\omega \times \mathbb{R}^\omega \simeq C_p(Z) \times \mathbb{R}^\omega,$$

so Fact 6 is proved. $\qquad \square$

Fact 7. If Z is an infinite space, then there is $f \in C(Z)$ such that $f(Z)$ is infinite.

Proof. We can choose a faithfully indexed discrete $D = \{z_n : n \in \mathbb{N}\} \subset Z$ (see Fact 4 of S.382) and a disjoint family $\mathcal{U} = \{U_n : n \in \mathbb{N}\} \subset \tau(Z)$ such that $z_n \in U_n$ for each $n \in \mathbb{N}$ (see Fact 1 of S.369). Using the Tychonoff property of Z, for each $n \in \mathbb{N}$, we can find $f_n \in C(Z, [0, \frac{1}{n}])$ such that $f_n(z_n) = \frac{1}{n}$ and $f_n(Z \backslash U_n) = \{0\}$. We claim that the function $f = \sum_{n \in \mathbb{N}} f_n : Z \to \mathbb{R}$ is continuous.

To see it, take any $z \in Z$; if $z \in U_n$ for some $n \in \mathbb{N}$, then $f|U_n = f_n|U_n$ is a continuous map, so f is continuous at the point z. Now take any $z \in Z \backslash (\bigcup \mathcal{U})$ and $\varepsilon > 0$. The set $F_n = f_n^{-1}([\varepsilon, \frac{1}{n}])$ is closed in Z and $F_n \subset U_n$ for each $n \in \mathbb{N}$. Take any $m \in \mathbb{N}$ such that $\frac{1}{m} < \varepsilon$ and observe that $W = Z \backslash (F_1 \cup \cdots \cup F_m)$ is an open neighborhood of z. Given any $y \in W$ consider the following cases:

Case 1. $y \in U_n$ for some $n \leq m$. Then $y \notin F_n$ and hence $f(y) = f_n(y) < \varepsilon$;
Case 2. $y \in U_n$ for some $n > m$. Then $f(y) = f_n(y) \leq \frac{1}{n} < \frac{1}{m} < \varepsilon$;
Case 3. $y \in Z \backslash (\bigcup \mathcal{U})$. Then $f(y) = 0 < \varepsilon$.

Thus we established that $f(W) \subset [0, \varepsilon) \subset (f(x) - \varepsilon, f(x) + \varepsilon)$ and hence f is continuous at the point z. Since $z \in Z$ was chosen arbitrarily, the function f is continuous; we have $\{\frac{1}{n} : n \in \mathbb{N}\} \subset f(Z)$, so the set $f(Z)$ is infinite and Fact 7 is proved. $\square$

Returning to our solution observe that it is trivial that (i)$\Longrightarrow$(ii)$\Longrightarrow$(iii). Besides, (iii) implies that X is pseudocompact because otherwise $C_p(X)$ has a closed subspace F homeomorphic to $\mathbb{R}^\omega$ by Fact 6. The space $\mathbb{R}^\omega$ is separable, so we can take a dense countable $A \subset F$. Then $F = \overline{A} \simeq \mathbb{R}^\omega$ has to be σ-pseudocompact which contradicts Fact 2 of S.399.

Furthermore the condition (viii) also implies pseudocompactness of X because otherwise $\mathbb{R}^\omega$ embeds in $C_p(X)$ as a closed subspace. Since $\mathbb{R}^\omega$ is separable, it must be σ-locally compact, so $\mathbb{R}^\omega = \bigcup \{L_n : n \in \omega\}$ where L_n is locally compact for any $n \in \omega$. It is an easy exercise that every Lindelöf locally compact space is σ-compact; since $nw(L_n) \leq nw(\mathbb{R}^\omega) = \omega$, the space L_n is Lindelöf and hence σ-compact for each $n \in \omega$. This implies that $\mathbb{R}^\omega$ is σ-compact which is a contradiction (see Fact 2 of S.399). This proves that (viii)$\Longrightarrow$(ix); since (iv)$\Longrightarrow$(viii) and (vi)$\Longrightarrow$(viii), the conditions (iv) and (vi) also imply pseudocompactness of X.

The same reasoning shows that the condition (vii) also implies pseudocompactness of X because otherwise $\mathbb{R}^\omega$ embeds in $C_p(X)$ as a closed subspace; since $\mathbb{R}^\omega$ is separable, it has to Hurewicz by (vii). The space $\mathbb{P}$ is completely metrizable, so it embeds in $\mathbb{R}^\omega$ as a closed subspace by Problem 273 of [TFS]. Thus $\mathbb{P}$ is also Hurewicz which contradicts Fact 1. Since (v)$\Longrightarrow$(vii), the condition (v) also implies pseudocompactness of X. Thus, we established that each one of the conditions (i)–(viii) implies pseudocompactness of X. It is evident that (ix) implies each and every one of the conditions (i)–(viii).

Now assume (iii); if X is infinite, there exists $f : X \rightarrow \mathbb{R}$ such that $M = f(X)$ is infinite (see Fact 7). The map $f^* : C_p(M) \rightarrow C_p(X)$ embeds $C_p(M)$ in $C_p(X)$ as a closed subspace because X is pseudocompact and hence f is $\mathbb{R}$-quotient (see Problem 163 of [TFS] and Fact 3 of S.154). The space $C_p(M)$ (and hence $f^*(C_p(M))$) is separable, so if $A \subset f^*(C_p(M))$ is a dense countable subset of $f^*(C_p(M))$, then $\overline{A} = f^*(C_p(M))$ is σ-pseudocompact by (iii). Thus $C_p(M)$ is σ-pseudocompact and hence σ-compact because $nw(C_p(M)) = \omega$. This shows that M must be finite by Problem 186 of [TFS] which is a contradiction. Thus (iii)$\Longrightarrow$(ix) and therefore (i) $\Longleftrightarrow$ (ii) $\Longleftrightarrow$ (iii) $\Longleftrightarrow$ (ix).

Now assume (viii); if X is infinite, then there exists $f : X \to \mathbb{R}$ such that $M = f(X)$ is infinite (see Fact 7). The map $f^* : C_p(M) \to C_p(X)$ embeds $C_p(M)$ in $C_p(X)$ as a closed subspace because X is pseudocompact and hence f is $\mathbb{R}$-quotient (see Problem 163 of [TFS] and Fact 3 of S.154). The space $C_p(M)$ (and hence $f^*(C_p(M))$) is separable, so if $A \subset f^*(C_p(M))$ is a dense countable subset of $f^*(C_p(M))$, then $\overline{A} = f^*(C_p(M))$ is σ-locally compact by (viii). Thus $C_p(M)$ is σ-locally compact, so $C_p(M) = \bigcup\{L_n : n \in \omega\}$ where L_n is locally compact for any $n \in \omega$. Since $nw(L_n) \leq nw(C_p(M)) = \omega$, the space L_n is Lindelöf and hence σ-compact for each $n \in \omega$. This implies that $C_p(M)$ is σ-compact which is a contradiction because M is infinite (see Problem 186 of [TFS]). This proves that (viii)$\Longrightarrow$(ix). It is evident that (iv)$\Longrightarrow$(viii) and (vi)$\Longrightarrow$(viii) so (iv) $\Longleftrightarrow$ (vi) $\Longleftrightarrow$ (viii) $\Longleftrightarrow$ (ix).

Finally assume that (vii) holds. If X is infinite, then there exists $f : X \to \mathbb{R}$ such that $M = f(X)$ is infinite (see Fact 7). The map $f^* : C_p(M) \to C_p(X)$ embeds $C_p(M)$ in $C_p(X)$ as a closed subspace because X is pseudocompact and hence f is $\mathbb{R}$-quotient (see Problem 163 of [TFS] and Fact 3 of S.154). The space $C_p(M)$ (and hence $f^*(C_p(M))$) is separable, so if $A \subset f^*(C_p(M))$ is a dense countable subset of $f^*(C_p(M))$, then $\overline{A} = f^*(C_p(M))$ is Hurewicz by (vii). Thus $C_p(M)$ is Hurewicz. The space M being an infinite metrizable compact space, there is a nontrivial convergent sequence $S \subset M$ (we take it with its limit so S is infinite, compact and has a unique non-isolated point). The restriction map $\pi_S : C_p(M) \to C_p(S)$ is continuous and maps $C_p(M)$ onto $C_p(S)$ so $C_p(S)$ is also a Hurewicz space by Fact 1.

Given $f, g \in C(S) = C^*(S)$ let $d(f, g) = \sup\{|f(x) - g(x)| : x \in S\}$. Then d is a metric on $C(S)$ and $T = (C(S), d)$ is a complete metric space (see Problem 248 of [TFS]). We claim that T is separable; to show this, denote by a the unique non-isolated point of S and consider the set $D = \{f \in C(S) : f(S) \subset \mathbb{Q}$ and the set $\{x \in S : f(x) \neq f(a)\}$ is finite$\}$. It is easy to see that D is countable; to see that D is dense in T, take an arbitrary $f \in C(S)$ and any $\varepsilon > 0$. Since f is continuous at the point a, there is a finite $P \subset S \backslash \{a\}$ such that $|f(a) - f(x)| < \frac{\varepsilon}{2}$ for all $x \in S \backslash P$. Take any rational number $r \in (f(a), f(a) + \frac{\varepsilon}{2})$ and let $g(x) = r$ for all $x \in S \backslash P$. Given any $x \in P$ choose any $g(x) \in (f(x), f(x) + \varepsilon) \cap \mathbb{Q}$. It is clear that $g \in D$ and $|g(x) - f(x)| < \varepsilon$ for each $x \in P$. If $x \in S \backslash P$, then $|g(x) - f(x)| \leq |g(x) - f(a)| + |f(a) - f(x)| = |r - f(a)| + \frac{\varepsilon}{2} < \frac{\varepsilon}{2} + \frac{\varepsilon}{2} = \varepsilon$. This shows that $B_d(f, \varepsilon) \cap D \neq \emptyset$ for each $f \in C(S)$ and $\varepsilon > 0$. Therefore D is dense in T, so T is a completely metrizable second countable space. This implies that T is analytic by Fact 3. It is clear that the identity map from T to $C_p(S)$ is continuous, so $C_p(S)$ is analytic being a continuous image of the analytic space T (we used Fact 3 again). The space S is infinite, so $C_p(S)$ is not σ-compact by Problem 186 of [TFS]. Since $C_p(S)$ is second countable, we can apply Fact 4 to find a closed $F \subset C_p(S)$ such that F is homeomorphic to $\mathbb{P}$. Any closed subspace of a Hurewicz space is Hurewicz by Fact 1, so F is a Hurewicz space whence $\mathbb{P}$ is Hurewicz, a contradiction with another item of Fact 1. This proves that X is finite and hence (vii)$\Longrightarrow$(ix). It is evident that (v)$\Longrightarrow$(vii) so (v) $\Longleftrightarrow$ (vii) $\Longleftrightarrow$ (ix) and hence our solution is complete.

T.133. *Prove that any non-scattered countably compact space can be mapped continuously onto* $\mathbb{I}$.

Solution. We denote by $\mathbb{D}$ the set $\{0, 1\}$ with the discrete topology. As usual, we will identify any ordinal with the set of its predecessors and, in particular, $n = \{0, \ldots, n - 1\}$ for any $n \in \omega$. If $k \in \omega$, $i \in \mathbb{D}$ and $s \in \mathbb{D}^k$, then $s^\frown i \in \mathbb{D}^{k+1}$ is defined by $(s^\frown i)(k) = i$ and $(s^\frown i)|k = s$. If we have $s \in \mathbb{D}^n$, $t \in \mathbb{D}^k$ where $n \leq k$ and $n, k \in \omega + 1$, then $s \subset t$ says that $t|n = s$. For any $k \in \omega$ and $s \in \mathbb{D}^k$, let $[s] = \{x \in \mathbb{D}^\omega : s \subset x\}$. Given a point $x \in \mathbb{D}^\omega$, it is immediate that the family $\{[x|n] : n \in \omega\}$ is a local base at x in the space $\mathbb{D}^\omega$.

Let X be a non-scattered countably compact space. If X is not zero-dimensional, then it can be mapped continuously onto the space $[0, 1] \subset \mathbb{R}$ (see Fact 4 of T.063). Since $[0, 1]$ is homeomorphic to $\mathbb{I}$, the space X also maps continuously onto $\mathbb{I}$ so this case is clear.

Now assume that X is zero-dimensional and denote by $\mathcal{C}$ the family of all clopen subsets of X. Since X is not scattered, we can find a non-empty subspace $A \subset X$ which is dense-in-itself. Let $F_\emptyset = X$ and assume that, for some $n > 0$, we have constructed families $\{F_s : s \in \mathbb{D}^k\}$ for all $k < n$ with the following properties:

(1) given any $k < n$ we have $F_s \in \mathcal{C}$ for any $s \in \mathbb{D}^k$;
(2) for any $k < n$ and any distinct $s, t \in \mathbb{D}^k$ we have $F_s \cap F_t = \emptyset$;
(3) $\bigcup \{F_s : s \in \mathbb{D}^k\} = X$ for any $k < n$;
(4) for any $k < n$ we have $F_s \cap A \neq \emptyset$ for all $s \in \mathbb{D}^k$;
(5) if $k < m < n$, $s \in \mathbb{D}^k$, $t \in \mathbb{D}^m$ and $s \subset t$, then $F_t \subset F_s$.

Fix any $s \in \mathbb{D}^{n-1}$; since $F_s \cap A \neq \emptyset$, the set $F_s \cap A$ is a non-empty open subset of A, so it is dense-in-itself and hence infinite. Take distinct point $x, y \in A \cap F_s$. The space X being zero-dimensional, there exists $U \in \mathcal{C}$ such that $x \in U$ and $y \notin U$. The sets $F_{s^\frown 0} = U \cap F_s$ and $F_{s^\frown 1} = F_s \backslash U$ are clopen in X and disjoint. Since $x \in F_{s^\frown 0}$ and $y \in F_{s^\frown 1}$, we have $F_{s^\frown 0} \cap A \neq \emptyset$ and $F_{s^\frown 1} \cap A \neq \emptyset$; besides, $F_s = F_{s^\frown 0} \cup F_{s^\frown 1}$. After we construct the pair $\{F_{s^\frown 0}, F_{s^\frown 1}\}$ for all $s \in \mathbb{D}^{n-1}$ we obtain the family $\{F_s : s \in \mathbb{D}^n\}$ for which the property (5) holds. The condition (1) is satisfied because both sets $F_{s^\frown 0}, F_{s^\frown 1}$ are clopen in X for any $s \in \mathbb{D}^{n-1}$. The property (3) follows easily from the fact that $F_s = F_{s^\frown 0} \cup F_{s^\frown 1}$ for each $s \in \mathbb{D}^{n-1}$. The property (4) takes place because $F_{s^\frown 0} \cap A \neq \emptyset$ and $F_{s^\frown 1} \cap A \neq \emptyset$ for all $s \in \mathbb{D}^{n-1}$.

To show that the property (2) holds as well, take any distinct $s, t \in \mathbb{D}^n$; if $s' = s|(n - 1) \neq t' = t|(n - 1)$ then we have $F_{s'} \cap F_{t'} = \emptyset$ by the induction hypothesis, so $F_s \cap F_t \subset F_{s'} \cap F_{t'} = \emptyset$ (we used the property (5) here) which proves that $F_s \cap F_t = \emptyset$. Now, if $s|(n - 1) = t(n - 1) = u$, then $s = u^\frown i$ and $t = u^\frown(1 - i)$ for some $i \in \mathbb{D}$ and therefore $F_s \cap F_t = F_{u^\frown 0} \cap F_{u^\frown 1} = \emptyset$, so the property (2) holds in all possible cases.

Consequently, we can construct the family $\{F_s : s \in \mathbb{D}^k\}$ for every $k \in \omega$ in such a way that the conditions (1)–(5) are satisfied. Given any $x \in X$ and $k \in \omega$, there is a unique $s_k \in \mathbb{D}^k$ such that $x \in F_{s_k}$ by (2) and (3). Observe that $s_k \subset s_{k+1}$ for each $k \in \omega$ by (2) and (5), so the point $y = \bigcup\{s_i : i \in \omega\} \in \mathbb{D}^\omega$ is well-defined; let $f(x) = y$. Observe that it follows from our definition of $f(x)$ that

(6) if $k \in \omega$, $s \in \mathbb{D}^k$ and $x \in F_s$, then $s \subset f(x)$.

We claim that the map $f : X \to \mathbb{D}^\omega$ is continuous. To prove it take any $x \in X$ and $W \in \tau(f(x), \mathbb{D}^\omega)$. There is $n \in \omega$ such that $[f(x)|n] \subset W$. Let $s = f(x)|n$; then $U = F_s$ is an open neighborhood of the point x. Given any $y \in U$ we have $s \subset f(y)$ by (6) and hence $f(y)|n = s = f(x)|n$, i.e., $y \in [f(x)|n]$. This proves that $f(U) \subset [f(x)|n] \subset W$ and therefore f is continuous at the point x.

The map f is surjective; indeed, given any $y \in \mathbb{D}^\omega$, the family $\{F_{y|n} : n \in \omega\}$ is decreasing by (5). Since X is countably compact, there is $x \in \bigcap\{F_{y|n} : n \in \omega\}$. It is immediate that $f(x) = y$; the point y was taken arbitrarily, so we proved that f maps X onto $\mathbb{D}^\omega$. There exists a continuous onto map $g : \mathbb{D}^\omega \to \mathbb{I}$ by Problem 128 of [TFS], so $h = g \circ f$ maps X continuously onto $\mathbb{I}$.

T.134. *Prove that the following conditions are equivalent for any compact X:*

(i) $C_p(X)$ is a Fréchet–Urysohn space;
(ii) $\overline{A}$ is a Fréchet–Urysohn space for any countable $A \subset C_p(X)$;
(iii) X is scattered.

Solution. The implication (i)$\Longrightarrow$(ii) is obvious because the Fréchet–Urysohn property is hereditary. Now, if (ii) holds and X is not scattered, then there is a continuous onto map $f : X \to \mathbb{I}$ by Problem 133. Since f is closed, the set $f^*(C_p(\mathbb{I}))$ is a closed subspace of $C_p(X)$ homeomorphic to $C_p(\mathbb{I})$ (see Problem 163 of [TFS]). There is a dense countable $A \subset f^*(C_p(\mathbb{I}))$ and hence $f^*(C_p(\mathbb{I})) = \overline{A}$ is Fréchet–Urysohn. Therefore $C_p(\mathbb{I})$ is also Fréchet–Urysohn; this contradiction with Problem 147 of [TFS] proves that (ii)$\Longrightarrow$(iii).

For the implication (iii)$\Longrightarrow$(i) assume that X is scattered; we have to note first that X is ω-simple. Indeed, the ω-modification Y of the space X is a Lindelöf P-space by Problem 128 and the identity map of Y onto X is continuous. Thus every second countable continuous image of X is also a continuous image of Y, so it must be countable by Problem 127.

Now take an arbitrary set $A \subset C_p(X)$. Given any $g \in \overline{A}$, there is a countable $B \subset A$ such that $g \in \overline{B}$ (see Problem 149 of [TFS]). For each $x \in X$ let $\varphi(x)(f) = f(x)$ for any $f \in B$. Then $\varphi(x) \in C_p(B)$ for every $x \in X$ and $\varphi : X \to C_p(B)$ is a continuous map (see Problem 166 of [TFS]). Weight of $C_p(B)$ is countable, so $Z = \varphi(X)$ is countable because X is ω-simple. The map φ is closed so $\varphi^* : C_p(Z) \to C_p(X)$ embeds $C_p(Z)$ into $C_p(X)$ as a closed subspace by Problem 163 of [TFS]. It is easy to see that $B \subset \varphi^*(C_p(Z))$ so $\overline{B} \subset \varphi^*(C_p(Z))$. The space $\varphi^*(C_p(Z))$ is second countable being homeomorphic to $C_p(Z)$ so $\overline{B}$ is also second countable. Then $\{g\} \cup B$ is second countable as well and hence there is a sequence $S = \{g_n : n \in \omega\} \subset B$ such that $g_n \to g$. Thus we found a sequence $S \subset A$ which converges to g and hence $C_p(X)$ is Fréchet–Urysohn.

T.135. *Prove that for any Lindelöf P-space X, the space $C_p(X)$ is a strongly ω-monolithic Fréchet–Urysohn space.*

Solution. Take any countable $A \subset C_p(X)$ and let $\varphi(x)(f) = f(x)$ for any $x \in X$ and $f \in A$. Then $\varphi(x) \in C_p(A)$ for any $x \in X$ and the map $\varphi : X \to C_p(A)$ is continuous (see Problem 166 of [TFS]). If $Y = \varphi(X)$, then $|Y| = \omega$ because X is ω-simple by Problem 127. Every $y \in Y$ is a G_δ-set in Y so $\varphi^{-1}(y)$ is open in X being a G_δ-set in X.

Let Z be the set Y with the discrete topology; then the identity map $i : Z \to Y$ is continuous. Furthermore, the map $\delta = i^{-1} \circ \varphi : X \to Z$ is also continuous because $\delta^{-1}(z) = \varphi^{-1}(i(z))$ is open in X for any $z \in Z$. Every map onto a discrete space is open so δ is open and hence $\mathbb{R}$-quotient. The map $\delta^* : C_p(Z) \to C_p(X)$ embeds $C_p(Z)$ in $C_p(X)$ as a closed subspace (see Problem 163 of [TFS]). It is easy to verify that $A \subset \varphi^*(C_p(Y)) \subset \delta^*(C_p(Z))$ and therefore $\overline{A} \subset \delta^*(C_p(Z))$ which shows that $w(\overline{A}) \leq w(\delta^*(C_p(Z))) = w(C_p(Z)) = \omega$.

This proves that $C_p(X)$ is strongly ω-monolithic. Since X^n is Lindelöf for each $n \in \mathbb{N}$ (see Fact 1 of T.112), we have $t(C_p(X)) = \omega$ (see Problem 149 of [TFS]).

Now take an arbitrary set $A \subset C_p(X)$. Given any $f \in \overline{A}$, there is a countable $B \subset A$ such that $f \in \overline{B}$. The space $C_p(X)$ is strongly ω-monolithic so $\overline{B}$ is second countable. Then $\{f\} \cup B$ is second countable as well and hence there is a sequence $S = \{f_n : n \in \omega\} \subset B$ such that $f_n \to f$. Thus we found a sequence $S \subset A$ which converges to f and hence $C_p(X)$ is Fréchet–Urysohn.

T.136. *Prove that for any Lindelöf scattered space X, the space $C_p(X)$ is a strongly monolithic Fréchet–Urysohn space.*

Solution. Observe that the ω-modification $(X)_\omega$ of the space X is a Lindelöf P-space by Problem 128. Since the identity map $i : (X)_\omega \to X$ is continuous, the dual map $i^* : C_p(X) \to C_p((X)_\omega)$ embeds $C_p(X)$ in $C_p((X)_\omega)$ (see Problem 163 of [TFS]). Apply Problem 135 to conclude that $C_p((X)_\omega)$ is a Fréchet–Urysohn space so $C_p(X)$ is Fréchet–Urysohn as well.

To prove strong monolithity of $C_p(X)$ fix an infinite cardinal κ and let Y be the κ-modification of X; then Y is a P_κ-space (i.e., every G_κ-subset of Y is open in Y) and $l(Y) \leq \kappa$ by Problem 128. The topology of Y contains the topology of X, so the identity map $i : Y \to X$ is continuous. The map $i^* : C_p(X) \to C_p(Y)$ is an embedding by Problem 163 of [TFS] and therefore $C_p(X)$ is homeomorphic to a subspace of $C_p(Y)$. Since strong κ-monolithity is a hereditary property (we leave the easy proof as an exercise for the reader), it suffices to prove that Y is strongly κ-monolithic.

Take any $A \subset C_p(Y)$ with $|A| \leq \kappa$ and let $\varphi(y)(f) = f(y)$ for any $y \in Y$ and $f \in A$. Then $\varphi(y) \in C_p(A)$ for any $y \in Y$ and the map $\varphi : Y \to C_p(A)$ is continuous (see Problem 166 of [TFS]). If $T = \varphi(Y)$, then $w(T) \leq \kappa$ and therefore every $y \in T$ is a G_κ-set. As a consequence, $\varphi^{-1}(y)$ is open in Y being a G_κ-subset of Y. The family $\{\varphi^{-1}(y) : y \in T\}$ is an open cover of the space Y; since $l(Y) \leq \kappa$, there is $E \subset Y$ such that $|E| \leq \kappa$ and $Y = \bigcup\{\varphi^{-1}(y) : y \in E\} = \varphi^{-1}(E)$. Therefore $T = \varphi(Y) = \varphi(\varphi^{-1}(E)) = E$ which proves that $|T| = |E| \leq \kappa$.

Let Z be the set T with the discrete topology; then the identity map $j : Z \rightarrow T$ is continuous. Furthermore, the map $\delta = j^{-1} \circ \varphi : Y \rightarrow Z$ is also continuous because the set $\delta^{-1}(z) = \varphi^{-1}(j(z))$ is open in Y for any $z \in Z$. Every map onto a discrete space is open so δ is open and hence $\mathbb{R}$-quotient. Thus the dual mapping $\delta^* : C_p(Z) \rightarrow C_p(Y)$ embeds $C_p(Z)$ in $C_p(Y)$ as a closed subspace (see Problem 163 of [TFS]). It is easy to verify that $A \subset \varphi^*(C_p(T)) \subset \delta^*(C_p(Z))$ and therefore $\overline{A} \subset \delta^*(C_p(Z))$ which shows that $w(\overline{A}) \leq w(\delta^*(C_p(Z))) = w(C_p(Z)) = \kappa$. This proves that $C_p(Y)$ is strongly κ-monolithic. The infinite cardinal κ was chosen arbitrarily, so $C_p(X)$ is strongly κ-monolithic for any infinite cardinal κ, i.e., $C_p(X)$ is strongly monolithic.

T.137. *Let X be a Lindelöf P-space. Prove that $\overline{A}$ is Čech-complete for any countable $A \subset C_p(X)$.*

Solution. Take any countable $A \subset C_p(X)$ and let $\varphi(x)(f) = f(x)$ for any $x \in X$ and $f \in A$. Then $\varphi(x) \in C_p(A)$ for any $x \in X$ and the map $\varphi : X \rightarrow C_p(A)$ is continuous (see Problem 166 of [TFS]). If $Y = \varphi(X)$, then $|Y| = \omega$ because X is ω-simple by Problem 127. Every $y \in Y$ is a G_δ-set in Y, so $\varphi^{-1}(y)$ is open in X being a G_δ-set in X.

Let Z be the set Y with the discrete topology; then the identity map $i : Z \rightarrow Y$ is continuous. Furthermore, the map $\delta = i^{-1} \circ \varphi : X \rightarrow Z$ is also continuous because $\delta^{-1}(z) = \varphi^{-1}(i(z))$ is open in X for any $z \in Z$. Every map onto a discrete space is open so δ is open and hence $\mathbb{R}$-quotient. The map $\delta^* : C_p(Z) \rightarrow C_p(X)$ embeds $C_p(Z)$ in $C_p(X)$ as a closed subspace (see Problem 163 of [TFS]). It is easy to verify that $A \subset \varphi^*(C_p(Y)) \subset \delta^*(C_p(Z))$ and therefore $\overline{A} \subset \delta^*(C_p(Z))$. But Z is a countable discrete space, so $C_p(Z) = \mathbb{R}^Z$ is homeomorphic to $\mathbb{R}^\omega$ which is Čech-complete. Consequently, $\delta^*(C_p(Z))$ is also Čech-complete and hence so is $\overline{A}$ being a closed subspace of $\delta^*(C_p(Z))$.

T.138. *Let X be a pseudocompact space. Suppose that $\overline{A}$ is Čech-complete for any countable $A \subset C_p(X)$. Prove that X is finite.*

Solution. If X is infinite, then there is a continuous function $f : X \rightarrow \mathbb{R}$ such that $Y = f(X)$ is infinite (see Fact 7 of T.132). The map $f : X \rightarrow Y$ is $\mathbb{R}$-quotient (see Fact 3 of S.154) and hence the dual map $f^* : C_p(Y) \rightarrow C_p(X)$ embeds $C_p(Y)$ in $C_p(X)$ as a closed subspace (see Problem 163 of [TFS]). The space $C_p(Y)$ is separable and hence so is $f^*(C_p(Y))$; if $A \subset f^*(C_p(Y))$ is countable and dense in $f^*(C_p(Y))$, then the space $f^*(C_p(Y)) = \overline{A}$ must be Čech-complete. This implies that $C_p(Y)$ is also Čech-complete and hence Y is discrete by Problem 265 of [TFS]. However, Y is an infinite compact subspace of $\mathbb{R}$ which cannot be discrete; this contradiction shows that X has to be finite.

T.139. *Suppose that $\overline{A}$ is normal for any countable $A \subset C_p(X)$. Prove that $\overline{A}$ is collectionwise normal for any countable $A \subset C_p(X)$, i.e., if the space $C_p(X)$ is normal(ω)-monolithic, then it is collectionwise-normal(ω)-monolithic.*

Solution. We will need some general facts which involve non-Tychonoff spaces. Assume that Z is a set and $\mathcal{T}$ is a collection of topologies on Z. It is evident that the family $\bigcup \mathcal{T}$ generates a topology on Z as a subbase; this topology is denoted by sup $\mathcal{T}$ and called *the least upper bound of topologies from* $\mathcal{T}$. If τ is a topology on the set Z, then a set $H \subset Z$ is τ-*closed (τ-open)* if H is closed (or open respectively) in the space (Z, τ). Analogously, a function $f : Z \to \mathbb{R}$ is τ-*continuous* if f is continuous on the space (Z, τ).

Fact 1. The least upper bound of any family of completely regular (not necessarily Tychonoff) topologies on a set Z is a completely regular topology on Z.

Proof. Take any family $\mathcal{T}$ of completely regular topologies on Z and let $\tau = \sup \mathcal{T}$. Assume that we have $z \in Z$ and $F \subset Z$ such that F is τ-closed and $z \notin F$. Therefore $z \in U = Z \backslash F \in \tau$; since $\mathcal{T}$ is a subbase of τ, there are $\tau_1, \ldots, \tau_n \in \mathcal{T}$ and $U_1, \ldots, U_n$ such that $U_i \in \tau_i$ for each $i \leq n$ and $x \in \bigcap \{U_i : i \leq n\} \subset U$.

As τ_i is completely regular, there are τ_i-continuous functions $f_i : Z \to [0, 1]$ such that $f_i(z) = 1$ and $f_i(Z \backslash U_i) \subset \{0\}$ for all $i \leq n$. Each function f_i is also τ-continuous because $f_i^{-1}(H) \in \tau_i \subset \tau$ for every open set $H \subset [0, 1]$. Therefore the function $f = f_1 \cdot \ldots \cdot f_n : Z \to [0, 1]$ is τ-continuous while $f(z) = 1$ and $f(F) \subset \{0\}$ whence (Z, τ) is a completely regular space. Fact 1 is proved.   □

Fact 2. Assume that Z and T are Tychonoff spaces and $f : Z \to T$ is a continuous onto map. Then there exists a Tychonoff space T' such that for some $\mathbb{R}$-quotient continuous onto map $g : Z \to T'$ and a condensation $h : T' \to T$, we have $f = h \circ g$.

Proof. Let $\mathcal{F} = \{p \in \mathbb{R}^T : p \circ f \text{ is continuous}\}$; then $C(T) \subset \mathcal{F} \subset \mathbb{R}^T$. Given any $p \in \mathcal{F}$, let $\tau_p = \{p^{-1}(O) : O \in \tau(\mathbb{R})\}$. Any τ_p is a completely regular topology on T (see Problems 097 and 098 of [TFS]), so $\tau = \sup\{\tau_p : p \in \mathcal{F}\}$ is a completely regular topology on T by Fact 1. Observe that $\bigcup\{\tau_p : p \in C(T)\}$ is a base for $\tau(T)$ (see Fact 1 of S.437) and hence $\tau(T) = \sup\{\tau_p : p \in C(T)\} \subset \tau$. This implies that every $\tau(T)$-closed set is also τ-closed; since Z is Tychonoff, the set $\{z\}$ is $\tau(Z)$-closed and hence τ-closed for each $z \in Z$. Thus $T' = (T, \tau)$ is a Tychonoff space and the identity map $h : T' \to T$ is a condensation. If $g = h^{-1} \circ f$, then $f = h \circ g$, so we only must prove that g is continuous and $\mathbb{R}$-quotient.

Observe that f and g coincide if considered as mappings between sets, so it suffices to establish that f is continuous and $\mathbb{R}$-quotient considered as a map from Z to (T, τ). The family $\mathcal{B} = \bigcup\{\tau_p : p \in \mathcal{F}\}$ is a subbase for (T, τ), so it suffices to show that $f^{-1}(U) \in \tau(Z)$ for any $U \in \mathcal{B}$. Now, $U \in \tau_p$ for some $p \in \mathcal{F}$ and hence $U = p^{-1}(O)$ for some $O \in \tau(\mathbb{R})$. Therefore $f^{-1}(U) = f^{-1}(p^{-1}(O)) = (p \circ f)^{-1}(O)$ is open in Z because $p \circ f$ is continuous by the definition of $\mathcal{F}$. Thus the map f is continuous.

To see that f is $\mathbb{R}$-quotient, take any function $p : T \to \mathbb{R}$ such that $p \circ f$ is continuous. Then $p \in \mathcal{F}$ and hence $p^{-1}(O) \in \tau_p \subset \tau$ for every $O \in \tau(\mathbb{R})$. Thus p is continuous considered as a function from (T, τ) to $\mathbb{R}$ and hence the map $f : Z \to (T, \tau)$ is $\mathbb{R}$-quotient so Fact 2 is proved.   □

Returning to our solution fix a countable set $A \subset C_p(X)$ and let $\varphi(x)(f) = f(x)$ for any $x \in X$ and $f \in A$. Then $\varphi(x) \in C_p(A)$ for any $x \in X$ and $\varphi : X \to C_p(A)$ is a continuous map by Problem 166 of [TFS]. If $Y = \varphi(X)$, then $\varphi : X \to Y$ is a surjective map, so we can apply Fact 2 to find a space Z for which there exists an $\mathbb{R}$-quotient continuous onto map $p : X \to Z$ and a condensation $i : Z \to Y$ such that $i \circ p = \varphi$.

The dual map $p^* : C_p(Z) \to C_p(X)$ embeds $C_p(Z)$ in $C_p(X)$ as a closed subspace and the dual map $i^* : C_p(Y) \to C_p(Z)$ embeds $C_p(Y)$ in $C_p(Z)$ as a dense subspace (see Problem 163 of [TFS]). Since $w(Y) \leq w(C_p(A)) = \omega$, the space $C_p(Y)$ is separable and therefore so is $C_p(Z)$. If we choose a countable set $B \subset p^*(C_p(Z))$ which is dense in the subspace $p^*(C_p(Z))$, then $\overline{B} = p^*(C_p(Z))$, so the space $p^*(C_p(Z))$ is normal by normal(ω)-monolithity of $C_p(X)$. As a consequence, the space $C_p(Z)$ is also normal and hence collectionwise normal by Problem 295 of [TFS]. Recalling again that $p^*(C_p(Z))$ and $C_p(Z)$ are homeomorphic we conclude that $p^*(C_p(Z))$ is collectionwise normal. It is straightforward that $A \subset \varphi^*(C_p(Y)) \subset p^*(C_p(Z))$ and therefore $\overline{A} \subset p^*(C_p(Z))$. Since any closed subspace of a collectionwise normal space is collectionwise normal, the space $\overline{A}$ is collectionwise normal so our solution is complete.

T.140. *Prove that, under MA+¬CH, there exists a normal(ω)-monolithic space which is not collectionwise-normal(ω)-monolithic.*

Solution. If Z is a set and τ is a topology on Z, then a set is called τ-closed if it is closed in (Z, τ); a τ-interior of a set $A \subset Z$ is the interior of A in the space (Z, τ). Denote by D the discrete space of cardinality ω_1. The set D is dense in the space βD, so $w(\beta D) \leq 2^{\omega_1}$ (see Fact 2 of S.368) and therefore βD is homeomorphic to a subspace of $\mathbb{I}^{2^{\omega_1}}$ (see Problem 209 of [TFS]). Thus we can assume, without loss of generality, that $\beta D \subset \mathbb{I}^{2^{\omega_1}}$. Since MA+¬CH holds, we have $2^{\omega_1} = \mathfrak{c}$ (see Problem 056), so we can consider that $\beta D \subset \mathbb{I}^{\mathfrak{c}}$.

The set βD is nowhere dense in $\mathbb{I}^{\mathfrak{c}}$; for otherwise, the closure of the discrete subspace D contains some $W \in \tau^*(\mathbb{I}^{\mathfrak{c}})$ which implies that W has isolated points and hence $\mathbb{I}^{\mathfrak{c}}$ has isolated points which is false (we leave the proof of this easy fact to the reader). The space $\mathbb{I}^{\mathfrak{c}}$ is separable by Problem 108 of [TFS], so let us fix a dense countable $A \subset \mathbb{I}^{\mathfrak{c}} \backslash \beta D$; it is clear that A is also dense in $\mathbb{I}^{\mathfrak{c}}$.

Let $Z = A \cup D$ and consider the family $\mathcal{U} = \{\{a\} : a \in A\} \cup \{U \cap Z : U \in \tau(\mathbb{I}^{\mathfrak{c}})\}$. It is clear that $\bigcup \mathcal{U} = Z$, so there is a unique topology τ on Z for which $\mathcal{U}$ is a subbase (see Problem 008 of [TFS]). It follows from the definition of $\mathcal{U}$ that the topology τ_Z on Z induced from $\mathbb{I}^{\mathfrak{c}}$ is contained in τ. Since (Z, τ_Z) is Hausdorff and $\tau_Z \subset \tau$, the space (Z, τ) is also Hausdorff and hence T_1 (it is another trivial exercise to prove that if a topology is Hausdorff, then any stronger one is also Hausdorff). Let us show that $X = (Z, \tau)$ is a normal space. Observe first that

(∗) if $P, Q \subset D$ and $P \cap Q = \emptyset$, then there are $U, V \in \tau$ such that $P \subset U$, $Q \subset V$ and $U \cap V = \emptyset$.

To prove $(*)$ note that $\mathrm{cl}_{\beta D}(P) \cap \mathrm{cl}_{\beta D}(Q) = \emptyset$ (see Fact 1 of S.382) so $\mathrm{cl}_{\beta D}(P)$ and $\mathrm{cl}_{\beta D}(Q)$ are disjoint closed subsets of $\mathbb{I}^c$. The space $\mathbb{I}^c$ being normal, there are disjoint $U', V' \in \tau(\mathbb{I}^c)$ such that $\mathrm{cl}_{\beta D}(P) \subset U'$ and $\mathrm{cl}_{\beta D}(Q) \subset V'$. It is clear that $U = U' \cap Z$ and $V = V' \cap Z$ are as promised so $(*)$ is proved.

Now take any disjoint τ-closed $F, G \subset Z$ and let $P = F \cap D$, $Q = G \cap D$. Applying $(*)$ to the sets P and Q we can find disjoint $U_1, V_1 \in \tau$ such that $P \subset U_1$ and $Q \subset V_1$. Then $U = (U_1 \backslash G) \cup F \in \tau$ and $V = (V_1 \backslash F) \cup G \in \tau$. To see it observe that G is τ-closed so $U_1 \backslash G \in \tau$ and hence every $z \in U_1 \backslash G$ belongs to the τ-interior of U; besides, any point of $F \backslash (U_1 \backslash G)$ is isolated in (Z, τ) and hence it also belongs to the τ-interior of U. An identical argument shows that every point of V belongs to the τ-interior of V, i.e., $V \in \tau$. Finally, observe that $F \subset U$, $G \subset V$ and $U \cap V = \emptyset$ so the space $X = (Z, \tau)$ is a T_4-space.

Next observe that the countable set A is dense in X. Indeed, if $x \in D$ and $W \in \tau(x, X)$, then there are $U_1, \ldots, U_n \in \mathcal{U}$ such that $x \in \bigcap_{i \leq n} U_i \subset W$. Since $x \in U_i \cap D$, it is impossible that $U_i = \{a\}$ for some $a \in A$, so there is $V_i \in \tau(\mathbb{I}^c)$ such that $V_i \cap X = U_i$ for each $i \leq n$. If $V = \bigcap_{i \leq n} V_i$, then $V \in \tau(x, \mathbb{I}^c)$ and hence $V \cap A \neq \emptyset$ because A is dense in $\mathbb{I}^c$. If $a \in V \cap A$, then $a \in U \cap A \subset W \cap A$, so we proved that $W \cap A \neq \emptyset$ for any $W \in \tau(x, X)$, i.e., $x \in \overline{A}$. The point $x \in D$ was taken arbitrarily, so $D \subset \overline{A}$ and hence A is dense in X.

The family $\mathcal{D} = \{\{d\} : d \in D\}$ is discrete in X and consists of closed subsets of X while there is no disjoint family $\mathcal{W} = \{W_d : d \in D\} \subset \tau$ with $d \in W_d$ for each $d \in D$. Indeed, if such a family $\mathcal{W}$ exists, then $\mathcal{W} \subset \tau^*(X)$ which is impossible because X is separable and hence $c(X) = \omega$ which implies that no disjoint family of non-empty open sets is uncountable. This proves that X is not collectionwise normal; being separable, the space X is not collectionwise-normal(ω)-monolithic. Since normality implies being normal(ω)-monolithic, X is a normal(ω)-monolithic space which is not collectionwise-normal(ω)-monolithic.

T.141. *Suppose that $\overline{A}$ is normal for any countable $A \subset C_p(X)$. Prove that $\overline{A}$ is countably paracompact for any countable $A \subset C_p(X)$.*

Solution. Take any countable $A \subset C_p(X)$; let $\varphi(x)(f) = f(x)$ for any $x \in X$ and $f \in A$. Then $\varphi(x) \in C_p(A)$ and $\varphi : X \to C_p(A)$ is a continuous map by Problem 166 of [TFS]. If $Y = \varphi(X)$, then $\varphi : X \to Y$ is a surjective map, so we can apply Fact 2 of T.139 to find a space Z for which there exists an $\mathbb{R}$-quotient continuous onto map $p : X \to Z$ and a condensation $i : Z \to Y$ such that $i \circ p = \varphi$.

The dual map $p^* : C_p(Z) \to C_p(X)$ embeds $C_p(Z)$ in $C_p(X)$ as a closed subspace and the dual map $i^* : C_p(Y) \to C_p(Z)$ embeds $C_p(Y)$ in $C_p(Z)$ as a dense subspace (see Problem 163 of [TFS]). Since $w(Y) \leq w(C_p(A)) = \omega$, the space $C_p(Y)$ is separable and therefore so is $C_p(Z)$. If we choose a countable set $B \subset p^*(C_p(Z))$ which is dense in the subspace $p^*(C_p(Z))$, then $\overline{B} = p^*(C_p(Z))$, so the space $p^*(C_p(Z))$ is normal by normal(ω)-monolithity of $C_p(X)$. As a consequence, the space $C_p(Z)$ is also normal and hence countably paracompact by Problem 289 of [TFS]. Recalling again that $p^*(C_p(Z))$ and $C_p(Z)$

are homeomorphic, we conclude that $p^*(C_p(Z))$ is also countably paracompact. It is straightforward that $A \subset \varphi^*(C_p(Y)) \subset p^*(C_p(Z))$ and therefore $\overline{A} \subset p^*(C_p(Z))$. Since any closed subspace of a countably paracompact space is countably paracompact, the space $\overline{A}$ is countably paracompact.

T.142. *Suppose that* $\overline{A}$ *is hereditarily normal for any countable* $A \subset C_p(X)$. *Prove that* $\overline{A}$ *is perfectly normal for any countable* $A \subset C_p(X)$.

Solution. Take any countable $A \subset C_p(X)$; let $\varphi(x)(f) = f(x)$ for any $x \in X$ and $f \in A$. Then $\varphi(x) \in C_p(A)$ and $\varphi : X \rightarrow C_p(A)$ is a continuous map by Problem 166 of [TFS]. If $Y = \varphi(X)$, then $\varphi : X \rightarrow Y$ is a surjective map, so we can apply Fact 2 of T.139 to find a space Z for which there exists an $\mathbb{R}$-quotient continuous onto map $p : X \rightarrow Z$ and a condensation $i : Z \rightarrow Y$ such that $i \circ p = \varphi$.

The dual map $p^* : C_p(Z) \rightarrow C_p(X)$ embeds $C_p(Z)$ in $C_p(X)$ as a closed subspace and the dual map $i^* : C_p(Y) \rightarrow C_p(Z)$ embeds $C_p(Y)$ in $C_p(Z)$ as a dense subspace (see Problem 163 of [TFS]). Since $w(Y) \leq w(C_p(A)) = \omega$, the space $C_p(Y)$ is separable and therefore so is $C_p(Z)$. If we choose a countable set $B \subset p^*(C_p(Z))$ which is dense in the subspace $p^*(C_p(Z))$, then $\overline{B} = p^*(C_p(Z))$, so the space $p^*(C_p(Z))$ is hereditarily normal by hereditary-normality(ω)-monolithity of $C_p(X)$. As a consequence, the space $C_p(Z)$ is also hereditarily normal and hence perfectly normal by Problem 292 of [TFS]. Recalling again that $p^*(C_p(Z))$ and $C_p(Z)$ are homeomorphic, we conclude that $p^*(C_p(Z))$ is perfectly normal. It is straightforward that $A \subset \varphi^*(C_p(Y)) \subset p^*(C_p(Z))$ and therefore $\overline{A} \subset p^*(C_p(Z))$. Since any subspace of a perfectly normal space is perfectly normal (see Problem 003), the space $\overline{A}$ is perfectly normal.

T.143. *Given an infinite cardinal* κ, *prove that* $C_p(X, \mathbb{I})$ *is* κ-*stable if and only if* $C_p(X)$ *is* κ-*stable.*

Solution. If Z is a space and $E \subset C(Z)$ say that E *separates points from closed sets in* Z if, for any $z \in Z$ and any closed $F \subset Z$ with $z \notin F$, there is $f \in E$ such that $f(z) \notin \overline{f(F)}$. Given any set $A \subset X$ let $\pi_A : C_p(X) \rightarrow C_p(A)$ be the restriction map defined by $\pi_A(f) = f|A$ for any $f \in C_p(X)$. We will also need the space $C_p(A|X) = \pi_A(C_p(X)) \subset C_p(A)$. If $A \subset B \subset X$, then $\pi_A^B : C_p(B|X) \rightarrow C_p(A|X)$ is also the restriction map, i.e., $\pi_A^B(f) = f|B$ for any $f \in C_p(B|X)$.

The space $C_p(X, \mathbb{I})$ is a continuous image of $C_p(X)$ (see Problem 092 of [TFS]), so if $C_p(X)$ is κ-stable, then $C_p(X, \mathbb{I})$ is also κ-stable by Problem 123.

Now assume that the space $C_p(X, \mathbb{I})$ is κ-stable and take a continuous onto map $\varphi : C_p(X) \rightarrow Y$ such that there is a condensation $i : Y \rightarrow M$ for which $w(M) \leq \kappa$. We have a continuous map $\delta = i \circ \varphi : C_p(X) \rightarrow M$. The space $C_p(X)$ is a dense subspace of $\mathbb{R}^X$; given any $A \subset X$, the restriction map $\pi_A : C_p(X) \rightarrow C_p(A|X)$ coincides with the restriction to $C_p(X)$ of the natural projection of $\mathbb{R}^X$ onto its face $\mathbb{R}^A$, so we can apply Fact 1 of T.109 to find a set $A \subset X$ and a continuous map $\beta : C_p(A|X) \rightarrow M$ such that $|A| \leq \kappa$ and $\beta \circ \pi_A = \delta$. Let $B = \overline{A}$; then π_A^B is a condensation of $C_p(B|X)$ onto $C_p(A|X)$ because it is the restriction to $C_p(B|X)$

of the restriction map of $C_p(B)$ to $C_p(A)$ which is injective (see Problem 152 of [TFS]). Let $T = \pi_B(C_p(X, \mathbb{I}))$; then $\pi_A^B | T$ is also a condensation of T onto $\pi_A^B(T) \subset C_p(A)$ and hence $w(\pi_A^B(T)) \leq |A| \leq \kappa$. The space $C_p(X, \mathbb{I})$ being κ-stable, we have $nw(T) \leq \kappa$.

Given any $x \in B$ let $\varkappa(x)(f) = f(x)$ for any $f \in T$. Then $\varkappa(x) \in C_p(T)$ for each $x \in B$. The Tychonoff property of X implies that $C_p(X, \mathbb{I})$ separates points from closed sets in X; an immediate consequence is that $T = \pi^B(C_p(X, \mathbb{I}))$ separates points from closed sets in B, so we can apply Problem 166 of [TFS] to conclude that $\varkappa : B \to C_p(T)$ is a homeomorphism. Consequently, $nw(B) = nw(\varkappa(B)) \leq nw(C_p(T)) = nw(T) \leq \kappa$. Furthermore, $\gamma = \beta \circ \pi_A^B : C_p(B|A) \to M$ is a continuous onto map such that $\gamma \circ \pi_B = \delta$. We claim that the map $\varepsilon = i^{-1} \circ \gamma : C_p(B|A) \to Y$ is continuous.

To see it take an arbitrary set $W \in \tau(Y)$; then $W_1 = \varphi^{-1}(W) \in \tau(C_p(X))$ and hence $\pi_B(W_1)$ is open in $C_p(B|X)$ because the map $\pi_B : C_p(X) \to C_p(B|X)$ is open (see Problem 152 of [TFS]). It is easy to see that $\varepsilon \circ \pi_B = \varphi$ and therefore $\varepsilon^{-1}(W) = \pi_B(W_1)$ is open in $C_p(B|X)$, so the map ε is indeed continuous. Thus Y is a continuous image of $C_p(B|X)$ and therefore $nw(Y) \leq nw(C_p(B|X)) \leq nw(C_p(B)) \leq nw(B) \leq \kappa$ which shows that $nw(Y) \leq \kappa$ and hence $C_p(X)$ is κ-stable.

T.144. *Given an infinite cardinal κ, prove that $C_p(X, \mathbb{I})$ is κ-monolithic if and only if $C_p(X)$ is κ-monolithic.*

Solution. If $C_p(X)$ is κ-monolithic, then $C_p(X, \mathbb{I})$ has to be κ-monolithic being a subspace of $C_p(X)$ (see Problem 113). Now, if $C_p(X, \mathbb{I})$ is κ-monolithic, then the space $C_p(X, (-1, 1)) \subset C_p(X, \mathbb{I})$ is also κ-monolithic by Problem 113. Finally observe that the space $C_p(X, (-1, 1))$ is homeomorphic to $C_p(X)$ (see Fact 1 of S.295) so $C_p(X)$ is also κ-monolithic.

T.145. *Let θ be any cardinal function. Prove that the Hewitt realcompactification υX of a space X is $\theta(\omega)$-stable if and only if X is $\theta(\omega)$-stable. In particular, υX is ω-stable if and only if X is ω-stable.*

Solution. Suppose that the space X is $\theta(\omega)$-stable and take any continuous onto map $f : \upsilon X \to Y$ such that $iw(Y) \leq \omega$ and hence $\psi(Y) \leq \omega$. If there is any point $y \in Y \backslash f(X)$, then $\{y\}$ is a G_δ-set in Y and hence $P = f^{-1}(y) \subset \upsilon X \backslash X$ is a non-empty G_δ-set in υX. However, every non-empty G_δ-subset of υX intersects X (see Problem 417 of [TFS]); this contradiction shows that $f(X) = Y$. Since the space X is $\theta(\omega)$-stable, we have $\theta(Y) \leq \omega$ and therefore υX is $\theta(\omega)$-stable.

Now assume that υX is $\theta(\omega)$-stable and take any continuous onto mapping $f : X \to Y$ such that $iw(Y) \leq \omega$. Every space of countable i-weight is realcompact (see Problem 446 of [TFS]), so Y is realcompact and hence there is a continuous map $g : \upsilon X \to Y$ such that $g|X = f$ (see Problem 413 of [TFS]). Since $g(\upsilon X) \supset f(X) = Y$, we have $g(\upsilon X) = Y$ and hence we can apply $\theta(\omega)$-stability of υX to conclude that $\theta(Y) \leq \omega$. Thus X is $\theta(\omega)$-stable.

T.146. *Let η and θ be cardinal functions such that $\theta(Z) = \eta(C_p(Z))$ for any space Z. Suppose that η is hereditary, i.e., for any space Z and any $Y \subset Z$, we have $\eta(Y) \le \eta(Z)$. Prove that for any infinite cardinal κ, a space X is $\theta(\kappa)$-stable if and only if $C_p(X)$ is $\eta(\kappa)$-monolithic.*

Solution. Assume that the space X is $\theta(\kappa)$-stable and fix any set $A \subset C_p(X)$ with $|A| \le \kappa$. Given any point $x \in X$ let $\varphi(x)(f) = f(x)$ for all $f \in A$. Then $\varphi(x) \in C_p(A)$ for any $x \in X$ and $\varphi : X \to C_p(A)$ is a continuous map (see Problem 166 of [TFS]); let $Y = \varphi(X)$. Apply Fact 2 of T.139 to get a space Z such that there exists an $\mathbb{R}$-quotient map $\delta : X \to Z$ and a condensation $i : Z \to Y$ for which $i \circ \delta = \varphi$. We have $w(Y) \le w(C_p(A)) = |A| \le \kappa$ so we can apply $\theta(\kappa)$-stability of X to conclude that $\theta(Z) \le \kappa$. By our hypothesis, $\eta(C_p(Z)) = \theta(Z) \le \kappa$. Observe that the dual map $\varphi^* : C_p(Y) \to C_p(X)$ embeds $C_p(Y)$ in $C_p(X)$ and the dual map $\delta^* : C_p(Z) \to C_p(X)$ embeds $C_p(Z)$ in $C_p(X)$ as a closed subspace (see Problem 163 of [TFS]). It is easy to check that $A \subset \varphi^*(C_p(Y)) \subset \delta^*(C_p(Z))$, so $\overline{A} \subset \delta^*(C_p(Z))$ and hence $\eta(\overline{A}) \le \eta(\delta^*(C_p(Z))) = \eta(C_p(Z)) = \theta(Z) \le \kappa$ which shows that the space $C_p(X)$ is $\eta(\kappa)$-monolithic.

Now assume that the space $C_p(X)$ is $\eta(\kappa)$-monolithic and take a continuous onto map $\varphi : X \to Y$ of the space X onto a space Y such that $iw(Y) \le \kappa$. The dual mapping $\varphi^* : C_p(Y) \to C_p(X)$ embeds the space $C_p(Y)$ in $C_p(X)$ (see Problem 163 of [TFS]) and $d(\varphi^*(C_p(Y))) = d(C_p(Y)) = iw(Y) \le \kappa$; as an immediate consequence, there exists a set $A \subset \varphi^*(C_p(Y))$ such that $|A| \le \kappa$ and $\varphi^*(C_p(Y)) \subset \overline{A}$. Thus $\eta(\varphi^*(C_p(Y))) \le \eta(\overline{A}) \le \kappa$ by $\eta(\kappa)$-monolithity of $C_p(X)$ and the fact that η is hereditary. Consequently, $\theta(Y) = \eta(C_p(Y)) = \eta(\varphi^*(C_p(Y))) \le \kappa$ which shows that X is $\theta(\kappa)$-stable.

T.147. *Let η and θ be cardinal functions such that $\theta(Z) = \eta(C_p(Z))$ for any space Z. Suppose that η is closed-hereditary, i.e., for any space Z and any closed $Y \subset Z$, we have $\eta(Y) \le \eta(Z)$. Prove that for any infinite cardinal κ, a space X is $\theta(\kappa)$-$\mathbb{R}$-quotient-stable if and only if $C_p(X)$ is $\eta(\kappa)$-monolithic.*

Solution. Assume that the space X is $\theta(\kappa)$-$\mathbb{R}$-quotient-stable and fix an arbitrary set $A \subset C_p(X)$ with $|A| \le \kappa$. Given any point $x \in X$ let $\varphi(x)(f) = f(x)$ for all $f \in A$. Then $\varphi(x) \in C_p(A)$ for any $x \in X$ and $\varphi : X \to C_p(A)$ is a continuous map (see Problem 166 of [TFS]); let $Y = \varphi(X)$. Apply Fact 2 of T.139 to get a space Z such that there exists an $\mathbb{R}$-quotient map $\delta : X \to Z$ and a condensation $i : Z \to Y$ for which $i \circ \delta = \varphi$. The map δ is $\mathbb{R}$-quotient and $w(Y) \le w(C_p(A)) = |A| \le \kappa$, so we can apply the fact that X is $\theta(\kappa)$-$\mathbb{R}$-quotient-stable to conclude that $\theta(Z) \le \kappa$. By our hypothesis, $\eta(C_p(Z)) = \theta(Z) \le \kappa$. Observe that the dual map $\varphi^* : C_p(Y) \to C_p(X)$ embeds $C_p(Y)$ in $C_p(X)$ and the dual map $\delta^* : C_p(Z) \to C_p(X)$ embeds $C_p(Z)$ in $C_p(X)$ as a closed subspace (see Problem 163 of [TFS]). It is easy to check that we have the inclusions $A \subset \varphi^*(C_p(Y)) \subset \delta^*(C_p(Z))$, so $\overline{A} \subset \delta^*(C_p(Z))$ and we can use the fact that η is closed-hereditary to conclude that $\eta(\overline{A}) \le \eta(\delta^*(C_p(Z))) = \eta(C_p(Z)) = \theta(Z) \le \kappa$ which shows that the space $C_p(X)$ is $\eta(\kappa)$-monolithic.

Now assume that the space $C_p(X)$ is $\eta(\kappa)$-monolithic and take any $\mathbb{R}$-quotient continuous onto mapping $\varphi : X \to Y$ of the space X onto a space Y such that $iw(Y) \leq \kappa$. The dual map $\varphi^* : C_p(Y) \to C_p(X)$ embeds $C_p(Y)$ in $C_p(X)$ as a closed subspace (see Problem 163 of [TFS]) and $d(\varphi^*(C_p(Y))) = d(C_p(Y)) = iw(Y) \leq \kappa$ which shows that we can find a set $A \subset \varphi^*(C_p(Y))$ such that $|A| \leq \kappa$ and $\varphi^*(C_p(Y)) = \overline{A}$. Thus $\eta(\varphi^*(C_p(Y))) = \eta(\overline{A}) \leq \kappa$ by $\eta(\kappa)$-monolithity of $C_p(X)$. Consequently, $\theta(Y) = \eta(C_p(Y)) = \eta(\varphi^*(C_p(Y))) \leq \kappa$ and hence X is $\theta(\kappa)$-$\mathbb{R}$-quotient-stable.

T.148. *Suppose that θ and η are cardinal functions such that:*

(i) for any space Z, if a space Y is a continuous image of Z, then $\eta(Y) \leq \eta(Z)$;
(ii) for any space Z and any closed $A \subset Z$, we have $\theta(A) = \eta(C_p(A|Z))$.

Prove that for an arbitrary space X, the space $C_p(X)$ is $\eta(\kappa)$-stable if and only if X is $\theta(\kappa)$-monolithic.

Solution. Given any set $B \subset X$ let $\pi_B : C_p(X) \to C_p(B)$ be the restriction map defined by $\pi_B(f) = f|B$ for any $f \in C_p(X)$. We will also need the space $C_p(B|X) = \pi_B(C_p(X)) \subset C_p(B)$. If $B \subset A \subset X$, then $\pi_B^A : C_p(A|X) \to C_p(B|X)$ is also the restriction map, i.e., $\pi_B^A(f) = f|B$ for any $f \in C_p(A|X)$.

Assume that $C_p(X)$ is $\eta(\kappa)$-stable and take any set $B \subset X$ with $|B| \leq \kappa$; let $A = \overline{B}$. The map $\pi_A : C_p(X) \to C_p(A|X)$ is continuous and onto while the restriction π_B^A condenses $C_p(A|X)$ onto the space $C_p(B|X)$; it is clear that $w(C_p(B|X)) \leq w(C_p(B)) = |B| \leq \kappa$. The space $C_p(X)$ being $\eta(\kappa)$-stable, we have $\eta(C_p(A|X)) \leq \kappa$ and therefore $\theta(A) = \eta(C_p(A|X)) \leq \kappa$ which shows that X is $\theta(\kappa)$-monolithic.

Now, if the space X is $\theta(\kappa)$-monolithic, then consider any continuous onto map $\varphi : C_p(X) \to Y$ for which there is a condensation $i : Y \to M$ such that $w(M) \leq \kappa$. If $\delta = i \circ \varphi$, then $\delta : C_p(X) \to M$ and hence there exists a set $B \subset X$ such that $|B| \leq \kappa$ and there is a continuous map $\beta : C_p(B|X) \to M$ for which $\delta = \beta \circ \pi_B$ (see Fact 1 of T.109). If $A = \overline{B}$ then $\theta(A) \leq \kappa$ by $\theta(\kappa)$-monolithity of the space X. Therefore $\eta(C_p(A|X)) = \theta(A) \leq \kappa$.

Our next step is to prove that the map $\gamma = i^{-1} \circ \beta \circ \pi_B^A$ is continuous. Observe first that $\gamma : C_p(A|X) \to Y$ and $\gamma \circ \pi_A = \varphi$. Given any $W \in \tau(Y)$ the set $\varphi^{-1}(W)$ is open in $C_p(X)$, so the set $\gamma^{-1}(W) = \pi_A(\varphi^{-1}(W))$ is open in $C_p(A|X)$ because $\pi_A : C_p(X) \to C_p(A|X)$ is an open map (see Problem 152 of [TFS]). Thus γ is continuous and hence Y is a continuous image of $C_p(A|X)$. By the property (i) we have $\eta(Y) \leq \eta(C_p(A|X)) \leq \kappa$ which proves that $C_p(X)$ is $\eta(\kappa)$-stable.

T.149. *Suppose that θ and η are cardinal functions such that:*

(i) for any space Z, if a space Y is a quotient image of Z, then $\eta(Y) \leq \eta(Z)$;
(ii) for any space Z and any closed $A \subset Z$, we have $\theta(A) = \eta(C_p(A|Z))$.

Prove that for an arbitrary space X, the space $C_p(X)$ is $\eta(\kappa)$-quotient-stable if and only if X is $\theta(\kappa)$-monolithic.

Solution. Given any set $B \subset X$ let $\pi_B : C_p(X) \to C_p(B)$ be the restriction map defined by $\pi_B(f) = f|B$ for any $f \in C_p(X)$. We will also need the space $C_p(B|X) = \pi_B(C_p(X)) \subset C_p(B)$. If $B \subset A \subset X$, then $\pi_B^A : C_p(A|X) \to C_p(B|X)$ is also the restriction map, i.e., $\pi_B^A(f) = f|B$ for any $f \in C_p(A|X)$.

Assume that $C_p(X)$ is $\eta(\kappa)$-quotient-stable and take any set $B \subset X$ with $|B| \leq \kappa$; let $A = \overline{B}$. The map $\pi_A : C_p(X) \to C_p(A|X)$ is open, continuous and onto (and hence quotient, see Problem 152 of [TFS]) while the restriction π_B^A condenses $C_p(A|X)$ onto the space $C_p(B|X)$; it is clear that $w(C_p(B|X)) \leq w(C_p(B)) = |B| \leq \kappa$. The space $C_p(X)$ being $\eta(\kappa)$-quotient-stable, we have $\eta(C_p(A|X)) \leq \kappa$ and therefore $\theta(A) = \eta(C_p(A|X)) \leq \kappa$ which shows that X is $\theta(\kappa)$-monolithic.

Now, if the space X is $\theta(\kappa)$-monolithic, then consider any quotient mapping $\varphi : C_p(X) \to Y$ for which there is a condensation $i : Y \to M$ such that $w(M) \leq \kappa$. If $\delta = i \circ \varphi$, then $\delta : C_p(X) \to M$ and hence there exists a set $B \subset X$ such that $|B| \leq \kappa$ and there is a continuous map $\beta : C_p(B|X) \to M$ for which $\delta = \beta \circ \pi_B$ (see Fact 1 of T.109). If $A = \overline{B}$, then $\theta(A) \leq \kappa$ by $\theta(\kappa)$-monolithity of the space X. Therefore $\eta(C_p(A|X)) = \theta(A) \leq \kappa$.

Our next step is to prove that the map $\gamma = i^{-1} \circ \beta \circ \pi_B^A$ is quotient. Observe first that $\gamma : C_p(A|X) \to Y$ and $\gamma \circ \pi_A = \varphi$ which implies that γ is surjective. Given any $W \in \tau(Y)$ the set $\varphi^{-1}(W)$ is open in $C_p(X)$, so the set $\gamma^{-1}(W) = \pi_A(\varphi^{-1}(W))$ is open in $C_p(A|X)$ because $\pi_A : C_p(X) \to C_p(A|X)$ is an open map (see Problem 152 of [TFS]). Thus γ is continuous. Now, if $H \subset Y$ and $\gamma^{-1}(H)$ is open in $C_p(A|X)$, then $\varphi^{-1}(H) = \pi_A^{-1}(\gamma^{-1}(H))$ is open in $C_p(X)$ and hence H is open in Y because φ is a quotient map.

Therefore γ is a quotient map and hence Y is a quotient image of $C_p(A|X)$. By the property (i) we have $\eta(Y) \leq \eta(C_p(A|X)) \leq \kappa$ which proves that $C_p(X)$ is $\eta(\kappa)$-quotient-stable.

T.150. *Suppose that θ and η are cardinal functions such that:*

(i) for any space Z, if a space Y is an $\mathbb{R}$-quotient continuous image of Z, then $\eta(Y) \leq \eta(Z)$;

(ii) Ffor any space Z and any closed $A \subset Z$ we have $\theta(A) = \eta(C_p(A|Z))$.

Prove that for an arbitrary space X, the space $C_p(X)$ is $\eta(\kappa)$-$\mathbb{R}$-quotient-stable if and only if X is $\theta(\kappa)$-monolithic.

Solution. Given any set $B \subset X$ let $\pi_B : C_p(X) \to C_p(B)$ be the restriction map defined by $\pi_B(f) = f|B$ for any $f \in C_p(X)$. We will also need the space $C_p(B|X) = \pi_B(C_p(X)) \subset C_p(B)$. If $B \subset A \subset X$, then $\pi_B^A : C_p(A|X) \to C_p(B|X)$ is also the restriction map, i.e., $\pi_B^A(f) = f|B$ for any $f \in C_p(A|X)$.

Assume that $C_p(X)$ is $\eta(\kappa)$-$\mathbb{R}$-quotient-stable and take any set $B \subset X$ with $|B| \leq \kappa$; let $A = \overline{B}$. The map $\pi_A : C_p(X) \to C_p(A|X)$ is open, continuous and onto (and hence $\mathbb{R}$-quotient, see Problem 152 of [TFS]) while the restriction π_B^A condenses $C_p(A|X)$ onto the space $C_p(B|X)$; it is clear that $w(C_p(B|X)) \leq w(C_p(B)) = |B| \leq \kappa$. The space $C_p(X)$ being $\eta(\kappa)$-$\mathbb{R}$-quotient-stable, we have $\eta(C_p(A|X)) \leq \kappa$ and therefore $\theta(A) = \eta(C_p(A|X)) \leq \kappa$ which shows that X is $\theta(\kappa)$-monolithic.

Now, if the space X is $\theta(\kappa)$-monolithic, then consider any $\mathbb{R}$-quotient mapping $\varphi : C_p(X) \to Y$ for which there is a condensation $i : Y \to M$ such that $w(M) \leq \kappa$. If $\delta = i \circ \varphi$, then $\delta : C_p(X) \to M$ and hence there exists a set $B \subset X$ such that $|B| \leq \kappa$ and there is a continuous map $\beta : C_p(B|X) \to M$ for which $\delta = \beta \circ \pi_B$ (see Fact 1 of T.109). If $A = \overline{B}$, then $\theta(A) \leq \kappa$ by $\theta(\kappa)$-monolithity of the space X. Therefore $\eta(C_p(A|X)) = \theta(A) \leq \kappa$.

Our next step is to prove that the map $\gamma = i^{-1} \circ \beta \circ \pi_B^A$ is $\mathbb{R}$-quotient. Observe first that $\gamma : C_p(A|X) \to Y$ and $\gamma \circ \pi_A = \varphi$ which implies that γ is surjective. Given any $W \in \tau(Y)$, the set $\varphi^{-1}(W)$ is open in $C_p(X)$, so the set $\gamma^{-1}(W) = \pi_A(\varphi^{-1}(W))$ is open in $C_p(A|X)$ because $\pi_A : C_p(X) \to C_p(A|X)$ is an open map (see Problem 152 of [TFS]). Thus the mapping γ is continuous. Now, if $u : Y \to \mathbb{R}$ and $u \circ \gamma$ is continuous, then $u \circ \varphi = u \circ (\gamma \circ \pi_A) = (u \circ \gamma) \circ \pi_A$ is a continuous map and hence u is continuous because φ is an $\mathbb{R}$-quotient map.

Therefore γ is an $\mathbb{R}$-quotient map and hence Y is an $\mathbb{R}$-quotient image of $C_p(A|X)$. By the property (i) we have $\eta(Y) \leq \eta(C_p(A|X)) \leq \kappa$ which proves that $C_p(X)$ is $\eta(\kappa)$-$\mathbb{R}$-quotient-stable.

T.151. *Suppose that θ and η are cardinal functions such that:*

(i) for any space Z, if Y is an open continuous image of Z, then $\eta(Y) \leq \eta(Z)$;
(ii) for any space Z and any closed $A \subset Z$, we have $\theta(A) = \eta(C_p(A|Z))$.

Prove that for an arbitrary space X, the space $C_p(X)$ is $\eta(\kappa)$-open-stable if and only if X is $\theta(\kappa)$-monolithic.

Solution. Given any set $B \subset X$ let $\pi_B : C_p(X) \to C_p(B)$ be the restriction map defined by $\pi_B(f) = f|B$ for any $f \in C_p(X)$. We will also need the space $C_p(B|X) = \pi_B(C_p(X)) \subset C_p(B)$. If $B \subset A \subset X$, then $\pi_B^A : C_p(A|X) \to C_p(B|X)$ is also the restriction map, i.e., $\pi_B^A(f) = f|B$ for any $f \in C_p(A|X)$.

Assume that $C_p(X)$ is $\eta(\kappa)$-open-stable and take any set $B \subset X$ with $|B| \leq \kappa$; let $A = \overline{B}$. The map $\pi_A : C_p(X) \to C_p(A|X)$ is open, continuous and onto (see Problem 152 of [TFS]) while the restriction π_B^A condenses $C_p(A|X)$ onto $C_p(B|X)$; it is clear that $w(C_p(B|X)) \leq w(C_p(B)) = |B| \leq \kappa$. The space $C_p(X)$ being $\eta(\kappa)$-open-stable, we have $\eta(C_p(A|X)) \leq \kappa$ and therefore $\theta(A) = \eta(C_p(A|X)) \leq \kappa$ which shows that X is $\theta(\kappa)$-monolithic.

Now, assume that the space X is $\theta(\kappa)$-monolithic and consider any open map $\varphi : C_p(X) \to Y$ for which there is a condensation $i : Y \to M$ such that $w(M) \leq \kappa$. If $\delta = i \circ \varphi$, then $\delta : C_p(X) \to M$ and hence there exists a set $B \subset X$ such that $|B| \leq \kappa$ and there is a continuous map $\beta : C_p(B|X) \to M$ for which $\delta = \beta \circ \pi_B$ (see Fact 1 of T.109). If $A = \overline{B}$, then $\theta(A) \leq \kappa$ by $\theta(\kappa)$-monolithity of the space X. Therefore $\eta(C_p(A|X)) = \theta(A) \leq \kappa$.

Our next step is to prove that the map $\gamma = i^{-1} \circ \beta \circ \pi_B^A$ is open (and hence continuous and onto). Observe first that $\gamma : C_p(A|X) \to Y$ and $\gamma \circ \pi_A = \varphi$ which implies that γ is surjective. Given any $W \in \tau(Y)$ the set $\varphi^{-1}(W)$ is open in $C_p(X)$, so $\gamma^{-1}(W) = \pi_A(\varphi^{-1}(W))$ is open in $C_p(A|X)$ because $\pi_A : C_p(X) \to C_p(A|X)$

is an open map (see Problem 152 of [TFS]). Thus γ is continuous. Now, if U is open in $C_p(A|X)$, then $\gamma(U) = \varphi(\pi_A^{-1}(U))$ is open in Y because φ is an open map.

Therefore γ is an open map and hence Y is an open continuous image of $C_p(A|X)$. By the property (i) we have $\eta(Y) \leq \eta(C_p(A|X)) \leq \kappa$ which proves that $C_p(X)$ is $\eta(\kappa)$-open-stable.

T.152. *Let X be an arbitrary space and κ an infinite cardinal. Prove that X is κ-monolithic if and only if $C_p(X)$ is κ-stable. In particular, X is monolithic if and only if $C_p(X)$ is stable.*

Solution. If Z is a space and $E \subset C(Z)$, say that E *separates points from closed sets in Z* if, for any $z \in Z$ and any closed $F \subset Z$ with $z \notin F$, there is $f \in E$ such that $f(z) \notin \overline{f(F)}$. Given any set $A \subset X$ let $\pi_A : C_p(X) \to C_p(A)$ be the restriction map defined by $\pi_A(f) = f|A$ for any $f \in C_p(X)$. We will also need the space $C_p(A|X) = \pi_A(C_p(X)) \subset C_p(A)$.

To apply the result of Problem 148 let $\eta = \theta = nw$. Then $\eta(\kappa)$-stability coincides with κ-stability and $\theta(\kappa)$-monolithity coincides with κ-monolithity. Network weight is not increased by continuous images (see Problem 157 of [TFS]) which shows that η satisfies the condition (i) from Problem 148.

For any $A \subset X$ we have $nw(A) = nw(C_p(A)) \geq nw(C_p(A|X))$. Given an arbitrary point $x \in A$ let $\varphi(x)(f) = f(x)$ for each $f \in E = C_p(A|X)$. Then $\varphi(x) \in C_p(E)$ and $\varphi : A \to C_p(E)$ is a continuous map (see Problem 166 of [TFS]). It is easy to see that E separates points from closed sets in A, so the map φ is actually an embedding (this was also proved in Problem 166 of [TFS]). As a consequence, $nw(A) \leq nw(C_p(E)) = nw(E) = nw(C_p(A|X))$ and hence $nw(A) = nw(C_p(A|X))$ for any $A \subset X$. This shows that the condition (ii) from Problem 148 is also satisfied (in a stronger form) so we can apply Problem 148 to the cardinal functions $\eta = \theta = nw$. Thus $C_p(X)$ is $nw(\kappa)$-stable if and only if X is $nw(\kappa)$-monolithic, i.e., $C_p(X)$ is κ-stable if and only if X is κ-monolithic.

T.153. *Prove that if $C_p(X)$ is a stable space, then $(C_p(X))^\kappa$ is also a stable space for any cardinal κ.*

Solution. The expression $P \simeq Q$ says that the spaces P and Q are homeomorphic. The stability of the space $C_p(X)$ is equivalent to monolithity of X by Problem 152. Let X_α be a homeomorphic copy of the space X for each $\alpha < \kappa$ and consider the space $Y = \bigoplus\{X_\alpha : \alpha < \kappa\}$ in which we identify each X_α with the respective clopen subspace of Y (see Problem 113 of [TFS]). Then $(C_p(X))^\kappa \simeq C_p(Y)$ (see Problem 114 of [TFS]) and hence stability of $(C_p(X))^\kappa$ is equivalent to monolithity of Y by Problem 152. It is evident that $\{X_\alpha : \alpha < \kappa\}$ is a locally finite (in fact, discrete) closed cover of Y; since each X_α is monolithic, we can apply Problem 115 to conclude that Y is also monolithic and hence $C_p(Y) \simeq (C_p(X))^\kappa$ is stable.

T.154. *Suppose that X is an arbitrary space and κ is an infinite cardinal. Prove that $C_p(X)$ is κ-monolithic if and only if X is κ-stable. In particular, $C_p(X)$ is monolithic if and only if X is stable.*

Solution. To apply the result of Problem 146 let $\eta = \theta = nw$. Then $\eta(\kappa)$-stability coincides with κ-stability and $\theta(\kappa)$-monolithity coincides with κ-monolithity. Network weight is hereditary (see Problem 159 of [TFS]) and $nw(C_p(Z)) = nw(Z)$ for any space Z which shows that the pair (η, θ) satisfies the hypothesis of Problem 146. Therefore Problem 146 is applicable and hence $C_p(X)$ is $nw(\kappa)$-monolithic if and only if X is $nw(\kappa)$-stable, i.e., $C_p(X)$ is κ-monolithic if and only if X is κ-stable.

T.155. *Prove that X is a monolithic space if and only if so is $C_p C_p(X)$.*

Solution. The space X is monolithic if and only if $C_p(X)$ is stable by Problem 152. Now $C_p(X)$ is stable if and only if $C_p(C_p(X))$ is monolithic by Problem 154 applied to the spaces $Y = C_p(X)$ and $C_p(Y) = C_p(C_p(X))$. As a consequence, X is monolithic if and only if so is $C_p(C_p(X))$.

T.156. *Prove that X is a stable space if and only if so is $C_p C_p(X)$.*

Solution. The space X is stable if and only if $C_p(X)$ is monolithic by Problem 154. Now $C_p(X)$ is monolithic if and only if $C_p(C_p(X))$ is stable by Problem 152 applied to the spaces $Y = C_p(X)$ and $C_p(Y) = C_p(C_p(X))$. As a consequence, X is stable if and only if so is $C_p(C_p(X))$.

T.157. *Prove that X is κ-simple if and only if $C_p(X)$ is strongly κ-monolithic.*

Solution. It is an easy exercise to see that the space X is κ-simple if and only if X is cardinality(κ)-stable; besides, strong κ-monolithity of X coincides with its $w(\kappa)$-monolithity. Now observe that weight is hereditary and $|Z| = w(C_p(Z))$ for any infinite space Z (see Problem 169 of [TFS]). Therefore, if $\eta = w$ and $\theta = cardinality$, then the pair (η, θ) satisfies the hypothesis of Problem 146. Therefore Problem 146 is applicable and hence $C_p(X)$ is $w(\kappa)$-monolithic if and only if X is cardinality(κ)-stable, i.e., X is κ-simple if and only if $C_p(X)$ is strongly κ-monolithic.

T.158. *Prove that the following properties are equivalent for any space X:*

   (i) $C_p(X)$ is strongly κ-monolithic;
  (ii) $C_p(X)$ is $\pi w(\kappa)$-monolithic;
 (iii) $C_p(X)$ is $\pi \chi(\kappa)$-monolithic;
 (iv) $C_p(X)$ is $\chi(\kappa)$-monolithic.

Solution. As usual, given an arbitrary space Z, points $z_1, \ldots, z_n \in Z$ and sets $O_1, \ldots, O_n \in \tau(\mathbb{R})$, let $[z_1, \ldots, z_n; O_1, \ldots, O_n] = \{f \in C_p(Z) : f(z_i) \in O_i$ for all $i \leq n\}$. The sets $[z_1, \ldots, z_n; O_1, \ldots, O_n]$ are called *standard open subsets of* $C_p(Z)$. If $U = [z_1, \ldots, z_n; O_1, \ldots, O_n]$ is a standard open subset of $C_p(Z)$, then let $\operatorname{supp}(U) = \{z_1, \ldots, z_n\}$.

Fact 1. For any infinite space Z we have $w(C_p(Z)) = \pi w(C_p(Z)) = \pi \chi(C_p(Z))$.

Proof. We have $\pi \chi(C_p(Z)) \leq \pi w(C_p(Z)) \leq w(C_p(Z)) = |Z|$ (see Problem 169 of [TFS]) so it suffices to prove that $|Z| \leq \pi \chi(C_p(Z))$. Let $u \in C_p(Z)$ be the function which is identically zero on Z. Suppose that $\pi \chi(C_p(Z)) = \lambda$ and fix a

π-base $\mathcal{B}$ with $|\mathcal{B}| \leq \lambda$ at the point u in the space $C_p(Z)$. The family $\mathcal{C}$ of all standard open subsets of $C_p(Z)$ is a base of the space $C_p(Z)$, so each $U \in \mathcal{B}$ contains a non-empty $O_U \in \mathcal{C}$; it is evident that $\mathcal{B}' = \{O_U : U \in \mathcal{B}\}$ is also a π-base at the point u.

If $Y = \bigcup\{\text{supp}(W) : W \in \mathcal{B}'\}$, then $|Y| \leq |\mathcal{B}'| \leq |\mathcal{B}| \leq \lambda$. Furthermore, if $x \in Z\backslash Y$, then $G = [x, (-1, 1)]$ is an open neighborhood of the point u. For any $W = [z_1, \ldots, z_n; O_1, \ldots, O_n] \in \mathcal{B}'$ we have $x \notin \text{supp}(W)$ and hence there exists a function $f \in C(Z)$ such that $f(z_i) \in O_i$ and $f(x) = 1$. It is immediate that $f \in W\backslash G$, which proves that $W\backslash G \neq \emptyset$ for any $W \in \mathcal{B}'$, i.e., G witnesses that $\mathcal{B}'$ is not a π-base at u. This contradiction proves that $Y = Z$ and therefore $|Z| \leq \lambda$. We have established that $|Z| \leq \pi\chi(C_p(Z))$ and hence $\pi\chi(C_p(Z)) = \pi w(C_p(Z)) = w(C_p(Z)) = |Z|$ so Fact 1 is proved. □

Returning to our solution observe that for an arbitrary space Z, we have "$w(Z) \leq \kappa$" $\Longrightarrow$ "$\pi w(Z) \leq \kappa$" $\Longrightarrow$ "$\pi\chi(Z) \leq \kappa$," so (i)$\Longrightarrow$(ii)$\Longrightarrow$(iii). It is also evident that "$w(Z) \leq \kappa$" $\Longrightarrow$ "$\chi(Z) \leq \kappa$" $\Longrightarrow$ "$\pi\chi(Z) \leq \kappa$" for any space Z, so (i)$\Longrightarrow$(iv)$\Longrightarrow$(iii). Consequently, it suffices to show that (iii)$\Longrightarrow$(i).

Assume that $C_p(X)$ is $\pi\chi(\kappa)$-monolithic and take any set $A \subset C_p(X)$ with $|A| \leq \kappa$; let $\varphi(x)(f) = f(x)$ for any $x \in X$ and $f \in A$. Then $\varphi(x) \in C_p(A)$ and $\varphi : X \to C_p(A)$ is a continuous map by Problem 166 of [TFS]. If $Y = \varphi(X)$, then $\varphi : X \to Y$ is a surjective map, so we can apply Fact 2 of T.139 to find a space Z for which there exists an $\mathbb{R}$-quotient continuous onto map $p : X \to Z$ and a condensation $i : Z \to Y$ such that $i \circ p = \varphi$.

The dual map $p^* : C_p(Z) \to C_p(X)$ embeds $C_p(Z)$ in $C_p(X)$ as a closed subspace and the dual map $i^* : C_p(Y) \to C_p(Z)$ embeds $C_p(Y)$ in $C_p(Z)$ as a dense subspace (see Problem 163 of [TFS]). Since $w(Y) \leq w(C_p(A)) \leq \kappa$, the space $C_p(Y)$ is has density $\leq \kappa$ and therefore so does $C_p(Z)$. If we choose a set $B \subset p^*(C_p(Z))$ with $|B| \leq \kappa$ and dense in the subspace $p^*(C_p(Z))$, then $\overline{B} = p^*(C_p(Z))$ so $\pi\chi(p^*(C_p(Z))) \leq \kappa$ by $\pi\chi(\kappa)$-monolithity of $C_p(X)$. As a consequence, we have $\pi\chi(C_p(Z)) \leq \kappa$ and hence $w(C_p(Z)) \leq \kappa$ by Fact 1. Recalling again that the spaces $p^*(C_p(Z))$ and $C_p(Z)$ are homeomorphic we conclude that $w(p^*(C_p(Z))) \leq \kappa$; it is straightforward that $A \subset \varphi^*(C_p(Y)) \subset p^*(C_p(Z))$ and therefore $\overline{A} \subset p^*(C_p(Z))$ whence $w(\overline{A}) \leq \kappa$. This proves that $C_p(X)$ is strongly κ-monolithic, i.e., (iii)$\Longrightarrow$(i), so our solution is complete.

T.159. *Given a space X and an infinite cardinal κ, prove that X is $s^*(\kappa)$-monolithic if and only if $C_p(X)$ is $s^*(\kappa)$-stable.*

Solution. If Z is a space and $E \subset C(Z)$, say that E *separates points from closed sets in* Z if, for any $z \in Z$ and any closed $F \subset Z$ with $z \notin F$, there is $f \in E$ such that $f(z) \notin \overline{f(F)}$. Given any set $A \subset X$ let $\pi_A : C_p(X) \to C_p(A)$ be the restriction map defined by $\pi_A(f) = f|A$ for any $f \in C_p(X)$. We will also need the space $C_p(A|X) = \pi_A(C_p(X)) \subset C_p(A)$.

To apply the result of Problem 148 let $\eta = \theta = s^*$. The invariant s^* is not increased by continuous images because if Y is a continuous image of a space Z, then Y^n is also a continuous image of Z^n for every $n \in \mathbb{N}$ (see Problem 157 of [TFS]) which shows that η satisfies the condition (i) from Problem 148.

For any $A \subset X$ we have $s^*(A) = s^*(C_p(A)) \geq s^*(C_p(A|X))$ (see Problem 025). Given an arbitrary point $x \in A$ let $\varphi(x)(f) = f(x)$ for each $f \in E = C_p(A|X)$. Then $\varphi(x) \in C_p(E)$ and $\varphi : A \to C_p(E)$ is a continuous map (see Problem 166 of [TFS]). It is easy to see that E separates points from closed sets in A, so the map φ is actually an embedding (this was also proved in Problem 166 of [TFS]). As a consequence, $s^*(A) \leq s^*(C_p(E)) = s^*(E) = s^*(C_p(A|X))$ and hence $s^*(A) = s^*(C_p(A|X))$ for any $A \subset X$. This shows that the condition (ii) from Problem 148 of [TFS] is also satisfied (in a stronger form), so we can apply it to the cardinal functions $\eta = \theta = s^*$. Thus $C_p(X)$ is $s^*(\kappa)$-stable if and only if X is $s^*(\kappa)$-monolithic.

T.160. *Given an arbitrary space X and an infinite cardinal κ, prove that if $C_p(X)$ is spread(κ)-stable, then X is spread(κ)-monolithic.*

Solution. Given a space Z, points $z_1, \ldots, z_n \in Z$ and sets $O_1, \ldots, O_n \in \tau(\mathbb{R})$, let $[z_1, \ldots, z_n; O_1, \ldots, O_n]_Z = \{f \in C_p(Z) : f(z_i) \in O_i \text{ for all } i \leq n\}$. Recall that the sets $[z_1, \ldots, z_n; O_1, \ldots, O_n]_Z$ are called *standard open subsets of* $C_p(Z)$. If we have any set $A \subset Z$, let $\pi_A : C_p(Z) \to C_p(A)$ be the restriction map defined by $\pi_A(f) = f|A$ for any $f \in C_p(Z)$. We will also need the space $C_p(A|Z) = \pi_A(C_p(Z)) \subset C_p(A)$. If $A \subset B \subset Z$, then $\pi_A^B : C_p(B|X) \to C_p(A|X)$ is also the restriction map, i.e., $\pi_A^B(f) = f|A$ for any $f \in C_p(B|X)$.

Fact 1. For an arbitrary space Z and any $A \subset Z$ we have $s(A) \leq s(C_p(A|Z))$.

Proof. Assume that $s(C_p(A|Z)) = \kappa$ and take any discrete $D \subset A$. For any $d \in D$ there is $U_d \in \tau(d, Z)$ such that $U_d \cap D = \{d\}$. Fix any $f_d \in C_p(Z)$ such that $f_d(d) = 1$ and $f_d(Z \backslash U_d) \subset \{0\}$ for every $d \in D$. The set $W_d = [d, (0, 2)]_A$ is an open neighborhood of the function $g_d = \pi_A(f_d)$ in $C_p(A)$ for each $d \in D$. Observe also that $e \in D \backslash \{d\}$ implies $g_e(d) = f_e(d) = 0$ and hence $g_e \notin W_d$. In particular, the map $\pi_A|D$ is an injection and the set $E = \pi_A(D) = \{g_d : d \in D\} \subset C_p(A|Z)$ is discrete because $W_d \cap E = \{g_d\}$ for every point $d \in D$. As a consequence, $|D| = |E| \leq s(C_p(A|Z)) = \kappa$ which shows that $s(A) \leq \kappa$ so Fact 1 is proved. $\square$

Returning to our solution take any $A \subset X$ with $|A| \leq \kappa$ and let $B = \overline{A}$. The space $C_p(B|X)$ is a continuous image of $C_p(X)$ and the map π_A^B condenses the space $C_p(B|X)$ onto the space $C_p(A|X)$. Since $w(C_p(A|X)) \leq |A| \leq \kappa$, we can use spread($\kappa$)-stability of $C_p(X)$ to conclude that $s(C_p(B|X)) \leq \kappa$. Finally, apply Fact 1 to see that $s(\overline{A}) = s(B) \leq s(C_p(B|X)) \leq \kappa$ and therefore the space X is spread(κ)-monolithic.

T.161. *Give an example of a spread-monolithic space X such that $C_p(X)$ is not spread(ω)-stable.*

Solution. Let X be the Sorgenfrey line (see Problem 165 of [TFS]). Then $s(X) = \omega$ and hence X is spread-monolithic because for any infinite cardinal κ and any $A \subset X$ with $|A| \leq \kappa$, we have $s(\overline{A}) \leq s(X) = \omega \leq \kappa$. The space $C_p(X)$ has an uncountable closed discrete subspace and hence $s(C_p(X)) > \omega$; since X is separable, $C_p(X)$ condenses onto a second countable space. This, together with $s(C_p(X)) > \omega$, implies that $C_p(X)$ is not spread(ω)-stable.

T.162. *Given an arbitrary space X and an infinite cardinal κ, prove that X is $s^*(\kappa)$-stable if and only if $C_p(X)$ is $s^*(\kappa)$-monolithic.*

Solution. To apply the result of Problem 146 let $\eta = \theta = s^*$. The invariant s^* is hereditary (see Problem 159 of [TFS]) and $s^*(C_p(Z)) = s^*(Z)$ for any space Z (see Problem 025) which shows that the pair (η, θ) satisfies the hypothesis of Problem 146. Therefore Problem 146 is applicable and hence $C_p(X)$ is $s^*(\kappa)$-monolithic if and only if X is $s^*(\kappa)$-stable.

T.163. *Prove that if $C_p(X)$ is spread(κ)-monolithic, then X is spread(κ)-stable.*

Solution. Take a continuous onto map $f : X \to Y$ for which $iw(Y) \le \kappa$. The dual map $f^* : C_p(Y) \to C_p(X)$ embeds $C_p(Y)$ in $C_p(X)$ (see Problem 163 of [TFS]). Furthermore, $d(f^*(C_p(Y))) = d(C_p(Y)) = iw(Y) \le \kappa$ (we applied Problem 174 of [TFS] and the fact that $f^*(C_p(Y))$ is homeomorphic to $C_p(Y)$). Now apply spread(κ)-monolithity of the space $C_p(X)$ to conclude that we have $s(f^*(C_p(Y))) \le \kappa$. As a consequence, $s(Y) \le s(C_p(Y)) = s(f^*(C_p(Y))) \le \kappa$ (here we applied Problem 016 and the fact that $C_p(Y)$ is homeomorphic to $f^*(C_p(Y))$); this proves that Y is spread(κ)-stable.

T.164. *Give an example of a spread-stable space X such that the space $C_p(X)$ is not spread(ω)-monolithic.*

Solution. Let X be the Sorgenfrey line (see Problem 165 of [TFS]). Then $s(X) = \omega$ and hence X is spread-stable because for every continuous image Y of the space X, we have $s(Y) \le s(X) = \omega \le \kappa$ (see Problem 157 of [TFS]) for any infinite cardinal κ (the condensations of Y do not matter). On the other hand, the space $C_p(X)$ has an uncountable closed discrete subspace and hence $s(C_p(X)) > \omega$. The identity map condenses X onto $\mathbb{R}$ so $d(C_p(X)) = iw(X) = \omega$. This, together with $s(C_p(X)) > \omega$, implies that $C_p(X)$ is not spread(ω)-monolithic.

T.165. *Given an arbitrary space X and an infinite cardinal κ prove that X is $hd^*(\kappa)$-monolithic if and only if $C_p(X)$ is $hl^*(\kappa)$-stable.*

Solution. If Z is a space and $E \subset C(Z)$ say that E *separates points from closed sets in Z* if, for any $z \in Z$ and any closed $F \subset Z$ with $z \notin F$, there is $f \in E$ such that $f(z) \notin \overline{f(F)}$. Given any set $A \subset X$ let $\pi_A : C_p(X) \to C_p(A)$ be the restriction map defined by $\pi_A(f) = f|A$ for any $f \in C_p(X)$. We will also need the space $C_p(A|X) = \pi_A(C_p(X)) \subset C_p(A)$.

To apply the result of Problem 148 let $\eta = hl^*$ and $\theta = hd^*$. The invariant hl^* is not increased by continuous images because if Y is a continuous image of a space Z, then Y^n is a continuous image of Z^n for every $n \in \mathbb{N}$ (see Problem 157 of [TFS]). Thus η satisfies the condition (i) from Problem 148.

For any $A \subset X$ we have $hd^*(A) = hl^*(C_p(A)) \geq hl^*(C_p(A|X))$ (see Problem 027). Given an arbitrary point $x \in A$ let $\varphi(x)(f) = f(x)$ for each $f \in E = C_p(A|X)$. Then $\varphi(x) \in C_p(E)$ and $\varphi : A \to C_p(E)$ is a continuous map (see Problem 166 of [TFS]). It is easy to see that E separates points from closed sets in A, so the map φ is actually an embedding (this was also proved in Problem 166 of [TFS]). As a consequence, $hd^*(A) \leq hd^*(C_p(E)) = hl^*(E) = hl^*(C_p(A|X))$ and hence $hd^*(A) = hl^*(C_p(A|X))$ for any $A \subset X$. This shows that the condition (ii) from Problem 148 of [TFS] is also satisfied (in a stronger form), so we can apply it to the cardinal functions $\eta = hl^*$ and $\theta = hd^*$. Thus $C_p(X)$ is $hl^*(\kappa)$-stable if and only if X is $hd^*(\kappa)$-monolithic.

T.166. *Given an arbitrary space X and an infinite cardinal κ prove that if $C_p(X)$ is $hl(\kappa)$-stable, then X is $hd(\kappa)$-monolithic.*

Solution. Given a space Z, points $z_1, \ldots, z_n \in Z$ and sets $O_1, \ldots, O_n \in \tau(\mathbb{R})$, let $[z_1, \ldots, z_n; O_1, \ldots, O_n]_Z = \{f \in C_p(Z) : f(z_i) \in O_i \text{ for all } i \leq n\}$. Recall that the sets $[z_1, \ldots, z_n; O_1, \ldots, O_n]_Z$ are called *standard open subsets of* $C_p(Z)$. If we have any set $A \subset Z$, let $\pi_A : C_p(Z) \to C_p(A)$ be the restriction map defined by $\pi_A(f) = f|A$ for any $f \in C_p(Z)$. We will also need the space $C_p(A|Z) = \pi_A(C_p(Z)) \subset C_p(A)$. If $A \subset B \subset Z$, then $\pi_A^B : C_p(B|X) \to C_p(A|X)$ is also the restriction map, i.e., $\pi_A^B(f) = f|A$ for any $f \in C_p(B|X)$.

Fact 1. For an arbitrary space Z and any $A \subset Z$ we have $hd(A) \leq hl(C_p(A|Z))$.

Proof. Assume that $hl(C_p(A|Z)) = \kappa$ and we have a set $D = \{d_\alpha : \alpha < \kappa^+\} \subset A$ which is left-separated by its indexation. For any $\alpha < \kappa^+$ there is $U_\alpha \in \tau(d_\alpha, Z)$ such that $U_\alpha \cap D \subset \{d_\beta : \beta \geq \alpha\}$. Fix any $f_\alpha \in C_p(Z)$ such that $f_\alpha(d_\alpha) = 1$ and $f_\alpha(Z \backslash U_\alpha) \subset \{0\}$ for every $\alpha < \kappa^+$. The set $W_\alpha = [d_\alpha, (0, 2)]_A$ is an open neighborhood of $g_\alpha = \pi_A(f_\alpha)$ in $C_p(A)$ for each $\alpha < \kappa^+$. Observe also that $\alpha < \beta$ implies $g_\beta(d_\alpha) = f_\beta(d_\alpha) = 0$ and hence $g_\beta \notin W_\alpha$. This implies that, the map $\pi_A|D$ is an injection. Besides, the set $E = \pi_A(D) = \{g_\alpha : \alpha < \kappa^+\} \subset C_p(A|Z)$ is right-separated because $W_\alpha \cap E \subset \{g_\beta : \beta \leq \alpha\}$ for each $\alpha < \kappa^+$. As a consequence, $|E| = |D| = \kappa^+ > \kappa = hl(C_p(A|Z))$ which is a contradiction with Problem 005. Thus A has no left-separated subsets of cardinality κ^+ (see Fact 2 of T.004) which proves that $hd(A) \leq \kappa$ by Problem 004. Fact 1 is proved. $\square$

Returning to our solution take any $A \subset X$ with $|A| \leq \kappa$ and let $B = \overline{A}$. The space $C_p(B|X)$ is a continuous image of $C_p(X)$ and the map π_A^B condenses the space $C_p(B|X)$ onto the space $C_p(A|X)$. Since $w(C_p(A|X)) \leq |A| \leq \kappa$, we can use $hl(\kappa)$-stability of $C_p(X)$ to conclude that $hl(C_p(B|X)) \leq \kappa$. Finally, apply Fact 1 to see that $hd(\overline{A}) = hd(B) \leq hl(C_p(B|X)) \leq \kappa$ and therefore the space X is $hd(\kappa)$-monolithic.

T.167. *Give an example of an hd-monolithic space X such that $C_p(X)$ is not $hl(\omega)$-stable.*

Solution. Let X be the Sorgenfrey line (see Problem 165 of [TFS]). Then $hd(X) = \omega$ and hence X is hd-monolithic because for any infinite cardinal κ and any $A \subset X$ with $|A| \leq \kappa$, we have $hd(\overline{A}) \leq hd(X) = \omega \leq \kappa$. The space $C_p(X)$ has an uncountable closed discrete subspace and hence $hl(C_p(X)) > \omega$; since X is separable, $C_p(X)$ condenses onto a second countable space. This, together with $hl(C_p(X)) > \omega$, implies that $C_p(X)$ is not $hl(\omega)$-stable.

T.168. *Given an arbitrary space X and an infinite cardinal κ prove that X is $hd^*(\kappa)$-stable if and only if $C_p(X)$ is $hl^*(\kappa)$-monolithic.*

Solution. To apply the result of Problem 146 let $\eta = hl^*$ and $\theta = hd^*$. The invariant hl^* is hereditary (see Problem 159 of [TFS]) and $hl^*(C_p(Z)) = hd^*(Z)$ for any space Z (see Problem 027) which shows that the pair (η, θ) satisfies the hypothesis of Problem 146. Therefore Problem 146 is applicable and hence $C_p(X)$ is $hl^*(\kappa)$-monolithic if and only if X is $hd^*(\kappa)$-stable.

T.169. *Given an arbitrary space X and an infinite cardinal κ, prove that if $C_p(X)$ is $hl(\kappa)$-monolithic, then X is $hd(\kappa)$-stable.*

Solution. Take a continuous onto map $f : X \to Y$ for which $iw(Y) \leq \kappa$. The dual map $f^* : C_p(Y) \to C_p(X)$ embeds $C_p(Y)$ in $C_p(X)$ (see Problem 163 of [TFS]). Furthermore, $d(f^*(C_p(Y))) = d(C_p(Y)) = iw(Y) \leq \kappa$ (we applied Problem 174 of [TFS] and the fact that $f^*(C_p(Y))$ is homeomorphic to $C_p(Y)$). Now apply $hl(\kappa)$-monolithity of the space $C_p(X)$ to conclude that we have $hl(f^*(C_p(Y))) \leq \kappa$. As a consequence, $hd(Y) \leq hl(C_p(Y)) = hl(f^*(C_p(Y))) \leq \kappa$ (here we applied Problem 017 and the fact that $C_p(Y)$ is homeomorphic to $f^*(C_p(Y))$); this proves that Y is $hd(\kappa)$-stable.

T.170. *Give an example of an hd-stable space X such that the space $C_p(X)$ is not $hl(\omega)$-monolithic.*

Solution. Let X be the Sorgenfrey line (see Problem 165 of [TFS]). Then $hd(X) = \omega$ and hence X is hd-stable because for every continuous image Y of the space X, we have $hd(Y) \leq hd(X) = \omega \leq \kappa$ (see Problem 157 of [TFS]) for any infinite cardinal κ (the condensations of Y do not matter). On the other hand, the space $C_p(X)$ has an uncountable closed discrete subspace and hence $hl(C_p(X)) > \omega$. The identity map condenses X onto $\mathbb{R}$, so $d(C_p(X)) = iw(X) = \omega$. This, together with $hl(C_p(X)) > \omega$, implies that $C_p(X)$ is not $hl(\omega)$-monolithic.

T.171. *Given an arbitrary space X and an infinite cardinal κ, prove that X is $hl^*(\kappa)$-stable if and only if $C_p(X)$ is $hd(\kappa)$-monolithic.*

Solution. To apply the result of Problem 146 let $\eta = hd$ and $\theta = hl^*$. The invariant hd is hereditary and $hd(C_p(Z)) = hl^*(Z)$ for any space Z (see Problem 030) which shows that the pair (η, θ) satisfies the hypothesis of Problem 146. Therefore Problem 146 is applicable and hence $C_p(X)$ is $hd(\kappa)$-monolithic if and only if X is $hl^*(\kappa)$-stable.

T.172. *Give an example of an hl-stable space X such that the space $C_p(X)$ is not $hd(\omega)$-monolithic.*

Solution. Let X be the Sorgenfrey line (see Problem 165 of [TFS]). Then $hl(X) = \omega$ and hence X is hl-stable because for every continuous image Y of the space X, we have $hl(Y) \leq hl(X) = \omega \leq \kappa$ (see Problem 157 of [TFS]) for any infinite cardinal κ (the condensations of Y do not matter). On the other hand, the space $C_p(X)$ has an uncountable closed discrete subspace and hence $hd(C_p(X)) > \omega$. The identity map condenses X onto $\mathbb{R}$ so $d(C_p(X)) = iw(X) = \omega$. This, together with $hd(C_p(X)) > \omega$, implies that $C_p(X)$ is not $hd(\omega)$-monolithic.

T.173. *Given an arbitrary space X and an infinite cardinal v, prove that X is $hl^*(v)$-monolithic if and only if $C_p(X)$ is $hd(v)$-stable.*

Solution. For an arbitrary $n \in \mathbb{N}$, let $M_n = \{1, \ldots, n\}$. If Z is a space and $n \geq 2$, let $\Delta_{ij}^n(Z) = \{z = (z_1, \ldots, z_n) \in Z^n : z_i = z_j\}$ for any distinct $i, j \in M_n$. The set $\Delta_n(Z) = \bigcup\{\Delta_{ij}^n(Z) : 1 \leq i < j \leq n\}$ is called the *n-diagonal* of the space Z. Call a set $W \in \tau(Z^n)$ *marked* if $W = W_1 \times \cdots \times W_n$ where $\{W_i : i \in M_n\} \subset \tau(Z)$ and $W_i \cap W_j = \emptyset$ for any distinct $i, j \in M_n$. If Y is a space and $Z \subset Y$, let $\pi_Z : C_p(Y) \to C_p(Z)$ be the restriction map defined by $\pi_Z(f) = f|Z$ for any $f \in C_p(Y)$. We will also need the space $C_p(Z|Y) = \pi_Z(C_p(Y)) \subset C_p(Z)$.

Fact 1. Suppose that Y is a space and $Z \subset Y$; given $n \in \mathbb{N}$ and any marked set $W \subset Z^n$, we have $hl(W) \leq hd(C_p(Z|Y))$.

Proof. Assume that $hd(C_p(Z|Y)) = \lambda$; we have $W = W_1 \times \cdots \times W_n$ where the family $\{W_i : i \in M_n\}$ consists of open subsets of Z and $W_i \cap W_j = \emptyset$ for any distinct $i, j \in M_n$. If $hl(W) > \lambda$, then there exists a set $P = \{z_\alpha : \alpha < \lambda^+\} \subset W$ which is right-separated by its indexation (see Problem 005 and Fact 2 of T.005). For any $\alpha < \lambda^+$, we have $z_\alpha = (z_\alpha^1, \ldots, z_\alpha^n)$ where $z_\alpha^i \in W_i$ for any $i \in M_n$.

For each $\alpha < \lambda^+$, there exist disjoint $O_\alpha^1, \ldots, O_\alpha^n \in \tau(Y)$ with the following properties:

(1) $z_\alpha^i \in O_\alpha^i$ for all $i \in M_n$;
(2) $O_\alpha \cap P_\alpha = \emptyset$ where $O_\alpha = O_\alpha^1 \times \cdots \times O_\alpha^n$ and $P_\alpha = \{z_\beta : \alpha < \beta\}$;
(3) $O_\alpha^i \cap Z \subset W_i$ for all $i \in M_n$.

For any $\alpha < \lambda^+$ and $i \in M_n$, take a function $f_\alpha^i \in C(Y, [0, 1])$ such that $f_\alpha^i(z_\alpha^i) = 1$ and $f_\alpha^i(Y \setminus O_\alpha^i) \subset \{0\}$; let $f_\alpha = f_\alpha^1 + \cdots + f_\alpha^n$ and $g_\alpha = \pi_Z(f_\alpha)$. We will prove that the set $E = \{g_\alpha : \alpha < \lambda^+\} \subset C_p(Z|Y)$ is left-separated by its indexation.

Let $U_\alpha = \{f \in C_p(Z|Y) : f(z_\alpha^i) > 0 \text{ for all } i \in M_n\}$. It is clear that U_α is an open subset of $C_p(Z|Y)$ and $g_\alpha \in U_\alpha$ for all $\alpha < \lambda^+$. Assume that $\alpha < \beta$ and $g_\alpha \in U_\beta$. Then, for any $i \in M_n$, we have $g_\alpha(z_\beta^i) > 0$ and hence $z_\beta^i \in O_\alpha^j$ for some $j \in M_n$. However, $z_\beta^i \in W_i$ and $O_\alpha^j \cap W_i = \emptyset$ for any $j \neq i$. Therefore $z_\beta^i \in O_\alpha^i$ for all $i \in M_n$ whence $z_\beta \in O_\alpha$ which contradicts (2). This contradiction shows that we have

(*) $g_\alpha \notin U_\beta$ for any $\alpha < \beta < \lambda^+$ and, in particular, the map $z_\alpha \to g_\alpha$ is a bijection between P and E.

Take any $\beta < \lambda^+$; given any $\alpha < \beta$, we have $g_\alpha \notin U_\beta$ by $(*)$. Therefore $U_\beta \cap E_\beta = \emptyset$ where $E_\beta = \{g_\gamma : \gamma < \beta\}$. This shows that the set E_β is closed in E for each $\beta < \lambda^+$, and hence E is a left-separated (by its indexation) subspace of $C_p(Z|Y)$. Since $|E| = \lambda^+$ by $(*)$, we have a contradiction with $hd(C_p(Z|Y)) \leq \lambda$ (see Problem 004). Fact 1 is proved. □

Fact 2. Let Y be a space; then, for any $Z \subset Y$, we have $l^*(Z) = t(C_p(Z|Y))$.

Proof. We have $t(C_p(Z|Y)) \leq t(C_p(Z)) = l^*(Z)$ by Problem 149 of [TFS]. To prove the converse inequality assume that $t(C_p(Z|Y)) = \kappa$ and take any ω-cover $\mathcal{U}$ of the space Z (i.e., $\mathcal{U} \subset \tau(Z)$, and for any finite $K \subset Z$, there is $U \in \mathcal{U}$ such that $K \subset U$). Consider the set $A = \{f \in C_p(Z|Y) : \text{supp}(f) = f^{-1}(\mathbb{R}\backslash\{0\}) \subset U$ for some $U \in \mathcal{U}\}$.

If $K \subset Z$ is finite, then there exists $U \in \mathcal{U}$ with $K \subset U$; take any $V \in \tau(Y)$ such that $V \cap Z = U$ and a function $g \in C_p(Y)$ for which $g(K) \subset \{1\}$ and $g(Y\backslash V) \subset \{0\}$. For the function $f = \pi_Z(g)$ we have $\text{supp}(f) \subset U$ and $f|K \equiv 1$. The finite set K was chosen arbitrarily, so we proved that $w \in \overline{A}$ for the function $w \in C_p(Z|Y)$ defined by $w(z) = 1$ for every $z \in Z$. There is $B \subset A$ with $|B| \leq \kappa$ and $w \in \overline{B}$; for every $f \in B$, there is $U(f) \in \mathcal{U}$ such that $\text{supp}(f) \subset U(f)$. The family $\mathcal{U}' = \{U(f) : f \in B\} \subset \mathcal{U}$ has cardinality $\leq \kappa$. If $K \subset Z$ is finite, then the set $H = \{f \in C_p(Z|Y) : f(z) > 0$ for all $z \in K\}$ is an open neighborhood of w in $C_p(Z|Y)$ and hence $H \cap B \neq \emptyset$. If $f \in B \cap H$, then $f(z) > 0$ for every $z \in K$ and hence $K \subset \text{supp}(f) \subset U(f) \in \mathcal{U}'$. Thus $\mathcal{U}'$ is an ω-cover of Z, so we proved that every ω-cover of Z has an ω-subcover of cardinality $\leq \kappa$. Therefore $l^*(Z) \leq \kappa$ by Problem 148 of [TFS] and Fact 2 is proved. □

Fact 3. Suppose that Y is a space and $Z \subset Y$; given any natural number $n \geq 2$, we have $hl(Z^n\backslash U) \leq hd(C_p(Z|Y))$ for any $U \in \tau(\Delta_n(Z), Z^n)$.

Proof. Let $hd(C_p(Z|Y)) = \lambda$; observe that $t(C_p(Z|Y)) \leq hd(C_p(Z|Y)) = \lambda$ and therefore $l^*(Z) \leq \lambda$ by Fact 2. The set $Z^n\backslash U$ is closed in Z^n, so $l(Z^n\backslash U) \leq \lambda$. For every $z \in F = Z^n\backslash U$, there exists a marked set $V_z \in \tau(z, Z^n)$. The family $V = \{V_z : z \in F\}$ is an open cover of the space F, so there is $V' \subset V$ such that $|V'| \leq \lambda$ and $\bigcup V' \supset F$. We have $hl(V) \leq \lambda$ for each $V \in V'$ by Fact 1, so $hl(\bigcup V') \leq \lambda$ (it is an easy exercise that a union of $\leq \lambda$-many spaces with hereditary Lindelöf number $\leq \lambda$ has hereditary Lindelöf number $\leq \lambda$). Therefore $hl(Z^n\backslash U) \leq hl(\bigcup V') \leq \lambda$ and Fact 3 is proved. □

Fact 4. We have $\psi(\Delta_n(Z), Z^n) \leq \Delta(Z)$ for any space Z.

Proof. Fix $n \in \mathbb{N}$ with $n \geq 2$; take an arbitrary family $\mathcal{U} \subset \tau(Z \times Z)$ such that $|\mathcal{U}| \leq \kappa = \Delta(Z)$ and $\bigcap \mathcal{U} = \Delta = \Delta_2(Z)$. Given distinct $i, j \in M_n$, let $q_{ij} : Z^n \rightarrow Z \times Z$ be the natural projection onto the face defined by i and j, i.e., for any $z = (z_1, \ldots, z_n) \in Z^n$, we have $q_{ij}(z) = (z_i, z_j) \in Z \times Z$. It is clear that $\Delta_{ij}^n(Z) = q_{ij}^{-1}(\Delta)$ and therefore $\Delta_{ij}^n(Z) = \bigcap \mathcal{U}_{ij}$ where $\mathcal{U}_{ij} = \{q_{ij}^{-1}(U) : U \in \mathcal{U}\}$. If $B_n = \{(i, j) \in M_n \times M_n : i < j\}$, then the family $V = \{U = \bigcup\{U_{ij} : (i, j) \in B_n\} : U_{ij} \in \mathcal{U}_{ij}$ for all $(i, j) \in B_n\}$ consists of open subsets of Z^n and $\bigcap V = \Delta_n(Z)$. It is evident that $|V| \leq \kappa$, so Fact 4 is proved. □

Fact 5. If Y is a space and $Z \subset Y$, then $hl^*(Z) = hd(C_p(Z|Y))$.

Proof. We have $hd(C_p(Z|Y)) \leq hd(C_p(Z)) = hl^*(Z)$ by Problem 030. To prove the converse inequality assume that $hd(C_p(Z|Y)) = \kappa$; since $C_p(Z|Y)$ is dense in $C_p(Z)$ (see Problem 152 of [TFS]), we have $iw(Z) = d(C_p(Z)) \leq \kappa$ (see Problem 174 of [TFS]). It is an easy exercise that $\Delta(Z) \leq iw(Z)$, so we have $\Delta(Z) \leq \kappa$ and hence $\psi(\Delta_n(Z), Z^n) \leq \kappa$ for any $n \geq 2$ by Fact 4.

Fix a family $\mathcal{V} \subset \tau(Z^n)$ such that $\bigcap \mathcal{V} = \Delta_n(Z)$; it is clear that $Z^n \setminus \Delta_n(Z) = \bigcup\{Z^n \setminus V : V \in \mathcal{V}\}$. For any set $F \in \mathcal{F} = \{Z^n \setminus V : V \in \mathcal{V}\}$, we have $hl(F) \leq \kappa$ by Fact 3. It follows from $|\mathcal{F}| \leq \kappa$ and $\bigcup \mathcal{F} = Z^n \setminus \Delta_n(Z)$ that $hl(Z^n \setminus \Delta_n(Z)) \leq \kappa$. Now apply Fact 0 of T.021 to conclude that $hl(Z^n) = hl(Z^n \setminus \Delta_n(Z)) \leq \kappa$. Thus $hl(Z^n) \leq \kappa$ for every $n \geq 2$ and hence $hl^*(Z) \leq \kappa$. This implies $hl^*(Z) = \kappa$ so Fact 5 is proved. $\square$

Returning to our solution, let $\eta = hd$ and $\theta = hl^*$. It is evident that hereditary density is not raised by continuous maps, so the condition (i) of Problem 148 is satisfied. Furthermore, Fact 5 shows that the condition (ii) of Problem 148 is satisfied as well (even in a stronger form). Therefore it is applicable to the pair (η, θ), so $C_p(X)$ is $hd(\nu)$-stable if and only if X is $hl^*(\nu)$-monolithic.

T.174. *Give an example of an hl-monolithic space X such that $C_p(X)$ is not $hd(\omega)$-stable.*

Solution. Let X be the Sorgenfrey line (see Problem 165 of [TFS]). Then $hl(X) = \omega$, and hence X is hl-monolithic because for any infinite cardinal κ and any $A \subset X$ with $|A| \leq \kappa$, we have $hl(\overline{A}) \leq hl(X) = \omega \leq \kappa$. The space $C_p(X)$ has an uncountable closed discrete subspace and hence $hd(C_p(X)) > \omega$; since X is separable, $C_p(X)$ condenses onto a second countable space. This, together with $hd(C_p(X)) > \omega$, implies that $C_p(X)$ is not $hd(\omega)$-stable.

T.175. *Given an arbitrary space X and an infinite cardinal κ prove that X is $p(\kappa)$-stable if and only if $C_p(X)$ is $a(\kappa)$-monolithic.*

Solution. Let $\eta = a$ and $\theta = p$. It is evident that the Alexandroff number is hereditary; besides, $a(C_p(Z)) = p(Z)$ for any space Z by Problem 178 of [TFS]. Consequently, the pair (η, θ) satisfies the hypothesis of 146 which shows that X is $p(\kappa)$-stable if and only if $C_p(X)$ is $a(\kappa)$-monolithic.

T.176. *Given an arbitrary space X and an infinite cardinal κ, prove that X is $l^*(\kappa)$-stable if and only if $C_p(X)$ is $t(\kappa)$-monolithic.*

Solution. Let $\eta = t$ and $\theta = l^*$. It is evident that tightness is hereditary; besides, $t(C_p(Z)) = l^*(Z)$ for any space Z by Problem 149 of [TFS]. Consequently, the pair (η, θ) satisfies the hypothesis of Problem 146 which shows that X is $l^*(\kappa)$-stable if and only if $C_p(X)$ is $t(\kappa)$-monolithic.

T.177. *Given an arbitrary space X and an infinite cardinal κ, prove that X is $d(\kappa)$-stable if and only if $C_p(X)$ is $iw(\kappa)$-monolithic.*

Solution. Let $\eta = iw$ and $\theta = d$. It is evident that the i-weight is hereditary; besides, $iw(C_p(Z)) = d(Z)$ for any space Z by Problem 173 of [TFS]. Consequently, the pair (η, θ) satisfies the hypothesis of Problem 146 which shows that X is $d(\kappa)$-stable if and only if $C_p(X)$ is $iw(\kappa)$-monolithic.

T.178. *Given an arbitrary space X and an infinite cardinal κ, prove that the following conditions are equivalent:*

(i) $C_p(X)$ is $iw(\kappa)$-monolithic;
(ii) $C_p(X)$ is $\Delta(\kappa)$-monolithic;
(iii) $C_p(X)$ is $\psi(\kappa)$-monolithic.

Solution. It is evident that $(iw(Z) \le \kappa) \Longrightarrow (\Delta(Z) \le \kappa) \Longrightarrow (\psi(Z) \le \kappa)$ for any space Z which shows that (i)$\Longrightarrow$(ii)$\Longrightarrow$(iii).

Now assume that $C_p(X)$ is $\psi(\kappa)$-monolithic and take any $A \subset C_p(X)$ with $|A| \le \kappa$. Let $\varphi(x)(f) = f(x)$ for any $x \in X$ and $f \in A$. Then $\varphi(x) \in C_p(A)$ and $\varphi : X \to C_p(A)$ is a continuous map by Problem 166 of [TFS]. If $Y = \varphi(X)$, then $\varphi : X \to Y$ is a surjective map, so we can apply Fact 2 of T.139 to find a space Z for which there exists an $\mathbb{R}$-quotient continuous onto map $p : X \to Z$ and a condensation $i : Z \to Y$ such that $i \circ p = \varphi$.

The dual map $p^* : C_p(Z) \to C_p(X)$ embeds $C_p(Z)$ in $C_p(X)$ as a closed subspace and the dual map $i^* : C_p(Y) \to C_p(Z)$ embeds $C_p(Y)$ in $C_p(Z)$ as a dense subspace (see Problem 163 of [TFS]). Since $w(Y) \le w(C_p(A)) \le \kappa$, the space $C_p(Y)$ has density $\le \kappa$ and therefore $d(C_p(Z)) \le \kappa$.

If we choose a set $B \subset p^*(C_p(Z))$ with $|B| \le \kappa$ which is dense in the subspace $p^*(C_p(Z))$, then $\overline{B} = p^*(C_p(Z))$, so $\psi(p^*(C_p(Z))) \le \kappa$ by $\psi(\kappa)$-monolithity of $C_p(X)$. As a consequence, $\psi(C_p(Z)) \le \kappa$ and hence $iw(C_p(Z)) \le \kappa$ by Problem 173 of [TFS]. Recalling again that $p^*(C_p(Z))$ and $C_p(Z)$ are homeomorphic, we conclude that $iw(p^*(C_p(Z))) \le \kappa$. It is straightforward that $A \subset \varphi^*(C_p(Y)) \subset p^*(C_p(Z))$ and therefore $\overline{A} \subset p^*(C_p(Z))$ which implies $iw(\overline{A}) \le \kappa$. Thus $C_p(X)$ is $iw(\kappa)$-monolithic which proves (iii)$\Longrightarrow$(i) and hence (i) $\Longleftrightarrow$ (ii) $\Longleftrightarrow$ (iii).

T.179. *Give an example of a space X which is pseudocharacter-monolithic but not diagonal-number-monolithic.*

Solution. Let X be the double arrow space (see Problem 384 of [TFS]). Then X is a separable first countable non-metrizable space. The space X is even χ-monolithic because for any infinite cardinal κ and any $A \subset X$ with $|A| \le \kappa$, we have $\chi(\overline{A}) \le \chi(X) = \omega \le \kappa$. On the other hand, $\Delta(X) > \omega$ because X is not metrizable (see Problem 091). This, together with separability of X, implies that X is not $\Delta(\omega)$-monolithic.

T.180. *Give an example of a space X which is diagonal-number-monolithic but not i-weight-monolithic.*

Solution. Call a space Z *strongly σ-discrete* if $Z = \bigcup\{Z_n : n \in \omega\}$ where Z_n is closed and discrete in Z for each $n \in \omega$. It is easy to see that every subset of a strongly σ-discrete space Z is an F_σ-subset of Z. Another easy fact (which we leave to the reader as an exercise) is that the square of a strongly σ-discrete is also strongly σ-discrete. Thus, if a space Z is strongly σ-discrete, then every subset of Z^2 is an F_σ-set. In particular, if $\Delta = \{(z, z) : z \in Z\} \subset Z^2$ is the diagonal of Z, then $Z^2 \backslash \Delta$ is an F_σ-subset of Z^2 and hence Δ is a G_δ-set in Z^2, i.e., $\Delta(Z) = \omega$.

Now let X be the Mrowka space (see Problem 142 of [TFS]). Then X is a pseudocompact non-compact space such that $X = \omega \cup \mathcal{M}$ where all points of ω are isolated in X and $\mathcal{M}$ is closed and discrete in X. Consequently, the space X is strongly σ-discrete and therefore $\Delta(Z) = \omega$. This shows that Z is Δ-monolithic because for any infinite cardinal κ and any $A \subset X$, we have $\Delta(\overline{A}) \leq \Delta(X) = \omega \leq \kappa$.

Observe that $iw(X) > \omega$ because any condensation of a pseudocompact space onto a second countable one is a homeomorphism (see Problem 140 of [TFS]), so if $iw(X) = \omega$, then X is metrizable and hence compact which is a contradiction. The set ω is dense in X, so X is separable which, together with $\Delta(X) > \omega$, implies that X is not $\Delta(\omega)$-monolithic.

T.181. *Given an arbitrary space X and an infinite cardinal κ, prove that X is $t_m(\kappa)$-$\mathbb{R}$-quotient-stable if and only if $C_p(X)$ is $q(\kappa)$-monolithic.*

Solution. Let $\eta = q$ and $\theta = t_m$. The Hewitt–Nachbin number q is closed-hereditary by Problem 422 of [TFS]; besides, $q(C_p(Z)) = t_m(Z)$ for any space Z by Problem 429 of [TFS]. Consequently, the pair (η, θ) satisfies the hypothesis of Problem 147 which shows that X is $t_m(\kappa)$-$\mathbb{R}$-quotient-stable if and only if $C_p(X)$ is $q(\kappa)$-monolithic.

T.182. *Give an example of a space X for which $C_p(X)$ is q-monolithic and X is not $t_m(\omega)$-stable.*

Solution. Let X be a discrete space of cardinality $\mathfrak{c}$; then $q(C_p(X)) = \omega$ (see Problem 429 of [TFS]). Consequently, for any infinite cardinal κ and any set $A \subset C_p(X)$, we have $q(\overline{A}) \leq q(C_p(X)) = \omega \leq \kappa$ (see Problem 422 of [TFS]) so $C_p(X)$ is q-monolithic.

Now if M is the Mrowka space (see Problem 142 of [TFS]), then M is a pseudocompact non-compact separable space. This implies that $q(M) > \omega$ by Problem 407 of [TFS] and hence we have $t_m(C_p(M)) = q(M) > \omega$ (see Problem 434 of [TFS]). Since M is separable, the space $C_p(M)$ condenses onto a second countable space by Problem 173 of [TFS] and therefore $|C_p(M)| \leq \mathfrak{c}$. As a consequence, there is a surjective map $\varphi : X \rightarrow C_p(M)$ which is continuous because X is discrete. Thus $C_p(M)$ is a continuous image of the space X such that $iw(C_p(M)) = \omega < t_m(C_p(M))$ which proves that X is not $t_m(\omega)$-stable.

T.183. *Given an infinite cardinal κ, prove that X is $l^*(\kappa)$-monolithic if and only if $C_p(X)$ is $t(\kappa)$-quotient-stable.*

Solution. Let $\eta = t$ and $\theta = l^*$. Tightness t is an invariant of quotient maps by Problem 162 of [TFS]; this shows that the condition (i) of Problem 149 is satisfied for η. Besides, $t(C_p(A|X)) = l^*(A)$ for any $A \subset X$ by Fact 2 of T.173; hence the condition (ii) is also satisfied in a stronger form. Therefore Problem 149 is applicable to the pair (η, θ) and hence X is $l^*(\kappa)$-monolithic if and only if $C_p(X)$ is $t(\kappa)$-quotient-stable.

T.184. *Given an infinite cardinal* κ, *suppose that* $C_p(X)$ *is* $l(\kappa)$-*monolithic. Prove that the space* X *is tightness*$^*(\kappa)$-$\mathbb{R}$-*quotient stable.*

Solution. Take any $\mathbb{R}$-quotient continuous surjective map $\varphi : X \to Y$ such that $iw(Y) \leq \kappa$. The map $\varphi^* : C_p(Y) \to C_p(X)$ embeds $C_p(Y)$ in $C_p(X)$ as a closed subspace (see Problem 163 of [TFS]). We have $d(\varphi^*(C_p(Y))) = d(C_p(Y)) = iw(Y) \leq \kappa$, so there is a set $A \subset \varphi^*(C_p(Y))$ such that $|A| \leq \kappa$ and $\overline{A} = \varphi^*(C_p(Y))$. The space $C_p(X)$ being $l(\kappa)$-monolithic, we have $l(\varphi^*(C_p(Y))) \leq \kappa$ and hence $l(C_p(Y)) \leq \kappa$. Now apply Problem 189 of [TFS] to conclude that $t^*(Y) \leq \kappa$ which shows that X is $t^*(\kappa)$-$\mathbb{R}$-quotient-stable.

T.185. *Given an arbitrary space* X, *suppose that* $C_p(X)$ *is* $t(\kappa)$-*monolithic for some infinite cardinal* κ. *Prove that it is* $t^*(\kappa)$-*monolithic.*

Solution. Take any $A \subset C_p(X)$ with $|A| \leq \kappa$; for any $x \in X$ let $\varphi(x)(f) = f(x)$ for every $f \in A$. Then $\varphi(x) \in C_p(A)$ for each $x \in X$ and the map $\varphi : X \to C_p(A)$ is continuous (see Problem 166 of [TFS]). If $Y = \varphi(X)$, then we can apply Fact 2 of T.139 to find a space Z and a continuous $\mathbb{R}$-quotient onto map $p : X \to Z$ such that $i \circ p = \varphi$ for some condensation $i : Z \to Y$.

The dual map $p^* : C_p(Z) \to C_p(X)$ embeds $C_p(Z)$ in $C_p(X)$ as a closed subspace and the dual map $i^* : C_p(Y) \to C_p(Z)$ embeds $C_p(Y)$ in $C_p(Z)$ as a dense subspace (see Problem 163 of [TFS]). Since $w(Y) \leq w(C_p(A)) \leq \kappa$, the space $C_p(Y)$ has density $\leq \kappa$ and therefore $d(C_p(Z)) \leq \kappa$.

If we choose a set $B \subset p^*(C_p(Z))$ with $|B| \leq \kappa$ which is dense in the subspace $p^*(C_p(Z))$, then $\overline{B} = p^*(C_p(Z))$, so $t(p^*(C_p(Z))) \leq \kappa$ by $t(\kappa)$-monolithity of $C_p(X)$. As a consequence, $t(C_p(Z)) \leq \kappa$ and hence $t^*(C_p(Z)) \leq \kappa$ by Problem 150 of [TFS]. Recalling again that $p^*(C_p(Z))$ and $C_p(Z)$ are homeomorphic, we conclude that $t^*(p^*(C_p(Z))) \leq \kappa$. It is straightforward that $A \subset \varphi^*(C_p(Y)) \subset p^*(C_p(Z))$ and therefore $\overline{A} \subset p^*(C_p(Z))$ which implies $t^*(\overline{A}) \leq \kappa$. Thus $C_p(X)$ is $t^*(\kappa)$-monolithic

T.186. *Suppose that* $C_p(X)$ *is Fréchet–Urysohn*(κ)-*monolithic for some infinite cardinal* κ. *Prove that it is Fréchet–Urysohn*$^*(\kappa)$-*monolithic.*

Solution. Take any $A \subset C_p(X)$ with $|A| \leq \kappa$; for any $x \in X$, let $\varphi(x)(f) = f(x)$ for every $f \in A$. Then $\varphi(x) \in C_p(A)$ for each $x \in X$ and the map $\varphi : X \to C_p(A)$ is continuous (see Problem 166 of [TFS]). If $Y = \varphi(X)$, then we can apply Fact 2 of T.139 to find a space Z and a continuous $\mathbb{R}$-quotient onto map $p : X \to Z$ such that $i \circ p = \varphi$ for some condensation $i : Z \to Y$.

The dual map $p^* : C_p(Z) \to C_p(X)$ embeds $C_p(Z)$ in $C_p(X)$ as a closed subspace and the dual map $i^* : C_p(Y) \to C_p(Z)$ embeds $C_p(Y)$ in $C_p(Z)$ as a dense subspace (see Problem 163 of [TFS]). Since $w(Y) \leq w(C_p(A)) \leq \kappa$, the space $C_p(Y)$ has density $\leq \kappa$ and therefore $d(C_p(Z)) \leq \kappa$.

If we choose a set $B \subset p^*(C_p(Z))$ with $|B| \leq \kappa$ which is dense in the subspace $p^*(C_p(Z))$, then $\overline{B} = p^*(C_p(Z))$, so $p^*(C_p(Z))$ is a Fréchet–Urysohn space by Fréchet–Urysohn(κ)-monolithity of $C_p(X)$. As a consequence, $C_p(Z)$ is also a Fréchet–Urysohn space and hence $(C_p(Z))^\omega$ is also Fréchet–Urysohn by Problem 145 of [TFS]. Recalling again that $p^*(C_p(Z))$ and $C_p(Z)$ are homeomorphic, we conclude that $(p^*(C_p(Z)))^\omega$ is Fréchet–Urysohn; since $A \subset \varphi^*(C_p(Y)) \subset p^*(C_p(Z))$, we have $\overline{A} \subset p^*(C_p(Z))$ which implies that $(\overline{A})^\omega$ is a Fréchet–Urysohn space. Thus $C_p(X)$ is Fréchet–Urysohn*(κ)-monolithic.

T.187. *Given an infinite cardinal κ, prove that X is κ-scattered if and only if $C_p(X)$ is $w(\kappa)$-open-stable.*

Solution. The definition of κ-scattered space shows that being κ-scattered is the same as being cardinality(κ)-monolithic. Let $\eta = w$ and $\theta =$ cardinality. Since open continuous maps do not increase weight (see Problem 161 of [TFS]), the condition (i) of Problem 151 is satisfied for η.

Fact 1. Let Z be an arbitrary space. If Y is dense in Z, then $\pi w(Y) = \pi w(Z)$.

Proof. If $\mathcal{B}$ is a π-base in Z, then $\{U \cap Y : U \in \mathcal{B}\}$ is a π-base in Y, so $\pi w(Y) \leq \pi w(Z)$. Now take any π-base $\mathcal{C}$ in Y. For each $V \in \mathcal{C}$, fix $O_V \in \tau(Z)$ such that $O_V \cap Y = V$. We claim that the family $\mathcal{B} = \{O_V : V \in \mathcal{C}\}$ is a π-base in Z. To see this take any $W \in \tau^*(Z)$; there is $W_1 \in \tau^*(Z)$ such that $\overline{W}_1 \subset W$. Choose any $V \in \mathcal{C}$ with $V \subset W_1 \cap Y$. Then $U = O_V \in \mathcal{B}$ and $U \subset \overline{U} = \overline{O_V} \subset \overline{W}_1 \subset W$, and hence $\mathcal{B}$ is a π-base in Z. Since $|\mathcal{B}| \leq |\mathcal{C}|$, we showed that $\pi w(Z) \leq \pi w(Y)$ so Fact 1 is proved.   □

Fact 2. If Z is a space and $Y \subset Z$, then $w(C_p(Y|Z)) = \pi w(C_p(Y|Z)) = |Y|$.

Proof. It is evident that $\pi w(C_p(Y|Z)) \leq w(C_p(Y|Z)) \leq w(C_p(Y)) = |Y|$. On the other hand, $C_p(Y|Z)$ is dense in $C_p(Y)$ (see Problem 152 of [TFS]), so we can apply Fact 1 to conclude that $\pi w(C_p(Y)) = \pi w(C_p(Y|Z))$. Now we have $|Y| = w(C_p(Y)) = \pi w(C_p(Y))$ by Fact 1 of T.158. This implies $w(C_p(Y|Z)) = \pi w(C_p(Y|Z)) = |Y|$, so Fact 2 is proved.   □

Returning to our solution observe that Fact 2 shows that the condition (ii) of Problem 151 is also satisfied (in a stronger form) for the cardinal functions η and θ. Thus Problem 151 is applicable to the pair (η, θ) and therefore X is κ-scattered if and only if $C_p(X)$ is $w(\kappa)$-open-stable.

T.188. *Let X be a $d(\omega)$-stable space such that X^n is Hurewicz for all $n \in \mathbb{N}$. Prove that for any $A \subset C_p(X)$ and any $f \in \overline{A} \backslash A$, there is a discrete $D \subset A$ such that f is the only accumulation point of D.*

Solution. Given a space Z say that a family $\mathcal{U}$ is *an open ω-cover* of Z if $\mathcal{U} \subset \tau(Z)$ and for any finite $K \subset Z$, there is $U \in \mathcal{U}$ with $K \subset U$.

Fact 1. The following properties are equivalent for any space Z:

(1) Z^n is a Hurewicz space for any $n \in \mathbb{N}$.
(2) If $\mathcal{U}_k$ is an open ω-cover of Z for all $k \in \omega$. then we can choose a finite $\mathcal{U}'_k \subset \mathcal{U}_k$ for each $k \in \omega$ in such a way that $\bigcup\{\mathcal{U}'_k : k \in \omega\}$ is an ω-cover of Z.

Proof. $(1)\Longrightarrow(2)$. Find a disjoint family $\mathcal{A} = \{A_k : k \in \mathbb{N}\}$ of infinite subsets of ω such that $\bigcup \mathcal{A} = \omega$. Given any $n \in \mathbb{N}$, the collection $\mathcal{V}^n_k = \{U^n : U \in \mathcal{U}_k\}$ is an open cover of Z^n for any $k \in \omega$, so $\{\mathcal{V}^n_k : k \in A_n\}$ is a countably infinite family of open covers of Z^n. By the Hurewicz property of Z^n, for all $k \in A_n$, there exists a finite $\mathcal{W}^n_k \subset \mathcal{V}^n_k$ such that $\bigcup\{\mathcal{W}^n_k : k \in A_n\}$ is a cover of Z^n. Fix a finite $\mathcal{U}^n_k \subset \mathcal{U}_k$ such that $\mathcal{W}^n_k = \{U^n : U \in \mathcal{U}^n_k\}$ for all $n \in \mathbb{N}$ and $k \in A_n$.

Once we have the collections $\{\mathcal{U}^n_k : k \in A_n\}$ for each $n \in \mathbb{N}$, take any $k \in \omega$ and find the unique $n = n(k) \in \mathbb{N}$ such that $k \in A_n$. Letting $\mathcal{U}'_k = \mathcal{U}^n_k$, we obtain a finite $\mathcal{U}'_k \subset \mathcal{U}_k$ for every $k \in \omega$. Given any finite set $K = \{z_1, \ldots, z_n\} \subset Z$ we have $z = (z_1, \ldots, z_n) \in Z^n$ and hence there is $k \in A_n$ and $W \in \mathcal{W}^n_k$ such that $z \in W$. We have $W = U^n$ for some $U \in \mathcal{U}^n_k = \mathcal{U}'_k$, so $z \in U^n$ and hence $K \subset U$. This shows that $\bigcup\{\mathcal{U}'_k : k \in \omega\}$ is an ω-cover of Z.

$(2)\Longrightarrow(1)$. Fix an arbitrary $n \in \mathbb{N}$ and take any collection $\{\mathcal{V}_k : k \in \omega\}$ of open covers of Z^n. Call a finite family $\mathcal{U}$ of open subsets of Z k-*small* if for any $U_1, \ldots, U_n \in \mathcal{U}$ (which are not necessarily distinct), there is $V \in \mathcal{V}_k$ such that $U_1 \times \cdots \times U_n \subset V$. It is easy to see that for each k-small family $\mu \subset \tau(Z)$, there exists a finite $\mathcal{W}(\mu) \subset \mathcal{V}_k$ such that for any $U_1, \ldots, U_n \in \mu$ there is $V \in \mathcal{W}(\mu)$ for which $U_1 \times \cdots \times U_n \subset V$.

If $F \subset Z$ is a finite set, then for any $z_1, \ldots, z_n \in F$ we have $z = (z_1, \ldots, z_n) \in V$ for some $V \in \mathcal{V}_k$, so there are $U_1, \ldots, U_n \in \tau(Z)$ such that $z_i \in U_i$ for all $i \leq n$ and $U_1 \times \cdots \times U_n \subset V$. Taking all possible n-tuples of the elements of F and intersecting the respective neighborhoods of every $z \in F$, we obtain a k-small family $\mu = \{W_z : z \in F\}$ such that $F \subset \bigcup \mu$. As a consequence, the family $\mathcal{U}_k = \{\bigcup \mu : \mu \subset \tau(Z)$ is k-small$\}$ is an ω-cover of Z for every $k \in \omega$. The property (2) guarantees that we can choose a finite $\mathcal{U}'_k \subset \mathcal{U}_k$ for every $k \in \omega$ in such a way that $\bigcup\{\mathcal{U}'_k : k \in \omega\}$ is an ω-cover of Z. For each $U \in \mathcal{U}'_k$, take a k-small family μ_U such that $U = \bigcup \mu_U$ and let $\mathcal{V}'_k = \bigcup\{\mathcal{W}(\mu_U) : U \in \mathcal{U}'_k\}$. It is evident that $\mathcal{V}'_k \subset \mathcal{V}_k$ and $|\mathcal{V}'_k| < \omega$ for all $k \in \omega$.

To see that $\bigcup\{\mathcal{V}'_k : k \in \omega\}$ is a cover of Z^n, take any $z = (z_1, \ldots, z_n) \in Z^n$. There is $k \in \omega$ and $U \in \mathcal{U}'_k$ such that $F = \{z_1, \ldots, z_n\} \subset U = \bigcup \mu_U$. Therefore we can choose $U_1, \ldots, U_n \in \mu_U$ such that $z_i \in U_i$ for all $i \leq n$. There is $V \in \mathcal{W}(\mu_U)$ with $W = U_1 \times \cdots \times U_n \subset V$; since $\mathcal{W}(\mu_U) \subset \mathcal{V}'_k$, we have $z \in W \subset V \in \mathcal{V}'_k$. The point $z \in Z^n$ was chosen arbitrarily, so $\bigcup\{\mathcal{V}'_k : k \in \omega\}$ is a cover of Z^n which proves that Z^n is a Hurewicz space for each $n \in \mathbb{N}$. Fact 1 is proved. $\qquad\square$

Fact 2. The following properties are equivalent for any space Z:

(3) Z^n is a Hurewicz space for any $n \in \mathbb{N}$;
(4) if $A_k \subset C_p(Z)$ for every $k \in \omega$ and $u \in \bigcap\{\overline{A}_k : k \in \omega\}$, then for each $k \in \omega$, we can choose a finite $B_k \subset A_k$ in such a way that $u \in \overline{\bigcup\{B_k : k \in \omega\}}$.

Proof. (3)$\Longrightarrow$(4). Since $C_p(Z)$ is homogeneous (i.e., for any $f, g \in C_p(Z)$, there is a homeomorphism $\varphi : C_p(Z) \to C_p(Z)$ such that $\varphi(f) = g$ (see Problem 079 of [TFS])), it suffices to prove (4) only for the function u which is identically zero on Z. Let $\mathcal{U}_k^n = \{f^{-1}((-\frac{1}{n}, \frac{1}{n})) : f \in A_k\}$ for all $k \in \omega$ and $n \in \mathbb{N}$. Since $u \in \overline{A}_k$ for each $k \in \omega$, the family $\mathcal{U}_k^n$ is an ω-cover of Z for each $n \in \mathbb{N}$. Fact 1 shows that we can use the property (2) of Fact 1 instead of (3). For any $n \in \mathbb{N}$ apply (2) to the countable sequence $\{\mathcal{U}_k^n : k \geq n - 1\}$ of ω-covers of Z to find, for all $k \geq n - 1$, a finite family $\mathcal{V}_k^n \subset \mathcal{U}_k^n$ such that $\bigcup\{\mathcal{V}_k^n : k \geq n - 1\}$ is an ω-cover of Z.

For any $U \in \mathcal{V}_k^n$, there is a function $f_U \in A_k$ such that $U = f_U^{-1}(-(\frac{1}{n}, \frac{1}{n}))$. The set $B_k = \{f_U : U \in \mathcal{V}_k^n : n \leq k + 1\} \subset A_k$ is finite for each $k \in \omega$. We claim that $u \in \bigcup\{B_k : k \in \omega\}$. To see this, take any finite $P \subset Z$ and $\varepsilon > 0$; it suffices to show that there is a point $f \in B = \bigcup\{B_k : k \in \omega\}$ such that $|f(z)| < \varepsilon$ for all $z \in P$. Fix any $n \in \mathbb{N}$ with $\frac{1}{n} < \varepsilon$; since the family $\mathcal{W} = \bigcup\{\mathcal{V}_k^n : k \geq n - 1\}$ is an ω-cover of Z, there is $k \geq n - 1$ and $U \in \mathcal{V}_k^n$ such that $P \subset U$. Then $f_U \in B_k$ and $|f_U(z)| < \frac{1}{n} < \varepsilon$ for each $z \in P$. This proves that $u \in \overline{B}$.

(4)$\Longrightarrow$(3). Let $w \in C_p(Z)$ be defined by $w(z) = 1$ for all $z \in Z$. By Fact 1, it suffices to show that (4)$\Longrightarrow$(2), so take any sequence $\{\mathcal{U}_k : k \in \omega\}$ of open ω-covers of Z. If $A_k = \{f \in C_p(Z) : \text{supp}(f) = f^{-1}(\mathbb{R}\backslash\{0\}) \subset U$ for some $U \in \mathcal{U}_k\}$, then it is an easy exercise that $w \in \overline{A}_k$ for each $k \in \omega$. Apply (4) to choose, for each $k \in \omega$, a finite $B_k \subset A_k$ such that $w \in \overline{B}$ where $B = \bigcup\{B_k : k \in \omega\}$. For each $f \in B_k$ there is $U(f) \in \mathcal{U}_k$ such that $\text{supp}(f) \subset U(f)$. The family $\mathcal{U}'_k = \{U(f) : f \in B_k\} \subset \mathcal{U}_k$ is finite for each $k \in \omega$. To see that the family $\mathcal{U}' = \bigcup\{\mathcal{U}'_k : k \in \omega\}$ is an ω-cover of Z, take any finite $P \subset Z$. The set $O = \{f \in C_p(Z) : f(z) > 0$ for each $z \in P\}$ is an open neighborhood of w, so there are $k \in \omega$ and $f \in B_k$ such that $f \in O$. Therefore $P \subset \text{supp}(f) \subset U(f) \in \mathcal{U}'_k \subset \mathcal{U}'$ which shows that $\mathcal{U}'$ is an ω-cover of Z and Fact 2 is proved. $\square$

Returning to our solution, say that $vet(C_p(Z)) = \omega$ to abbreviate the fact that (4) is fulfilled for the space Z. We have $vet(C_p(X)) = \omega$ by Fact 2. Observe that any Hurewicz space is Lindelöf so $t(C_p(X)) = \omega$ by Problem 149 of [TFS]; this implies that there is a countable $B \subset A$ such that $f \in \overline{B}\backslash B$. Since X is $d(\omega)$-stable, the space $C_p(X)$ is $iw(\omega)$-monolithic by Problem 177 and hence the space $\overline{B}$ has countable pseudocharacter. Therefore we can find a family $\{O_n : n \in \omega\} \in \tau(f, C_p(X))$ such that $\overline{O}_{n+1} \subset O_n$ for each $n \in \omega$ and $(\bigcap\{O_n : n \in \omega\}) \cap \overline{B} = \{f\}$. If $C_n = O_n \cap B$, then $f \in \overline{C}_n$ for all $n \in \omega$. Since $vet(C_p(X)) = \omega$, we can find a finite $D_n \subset C_n$ such that $f \in \overline{D}$ where $D = \bigcup\{D_n : n \in \omega\}$. To see that D is discrete, observe that $D \subset B$ and hence all accumulation points of D belong to $\overline{B}$. Furthermore, the set $F_n = (C_p(X)\backslash\overline{O}_n) \cap D \subset D_0 \cup \cdots \cup D_{n-1}$ is finite and $F_n \in \tau(D)$ for each $n \in \omega$. Since $\bigcup\{F_n : n \in \omega\} = D$, each element of D has a finite neighborhood in D which shows that D is discrete. Finally, if $g \in \overline{B}\backslash\{f\}$, then $g \in C_p(X)\backslash\overline{O}_n$ for some $n \in \omega$ and hence g has a neighborhood whose intersection with D is finite. Therefore f is the unique accumulation point of D and our solution is complete.

T.189. *Let X be an $l(\omega)$-monolithic space of countable spread. Prove that X is Lindelöf.*

Solution. Let $\mathcal{U}$ be an open cover of the space X. Apply Fact 1 of T.007 to find a countable family $\mathcal{U}' \subset \mathcal{U}$ and a discrete $D \subset X$ such that $X = (\bigcup \mathcal{U}') \cup \overline{D}$. The set D is countable because $s(X) = \omega$; the space X being $l(\omega)$-monolithic, $\overline{D}$ is Lindelöf and therefore there is a countable $\mathcal{U}'' \subset \mathcal{U}$ such that $\overline{D} \subset \bigcup \mathcal{U}''$. The family $\mathcal{V} = \mathcal{U}' \cup \mathcal{U}''$ is, evidently, a countable subcover of $\mathcal{U}$ which proves that X is Lindelöf.

T.190. *Let X be an $hl(\omega)$-monolithic space of countable spread. Prove that X is hereditarily Lindelöf. In particular, if X is an ω-monolithic space of countable spread, then X is hereditarily Lindelöf.*

Solution. It is evident that $hl(\omega)$-monolithity is hereditary and, in particular, any $Y \subset X$ is $l(\omega)$-monolithic. We also have $s(Y) \leq s(X) = \omega$ which shows that Problem 189 can be applied to conclude that Y is Lindelöf. Therefore every $Y \subset X$ is Lindelöf, i.e., X is hereditarily Lindelöf.

T.191. *Let X be an ω-monolithic space such that $C_p(X)$ has countable spread. Prove that $hl^*(X) = \omega$ and hence $hd^*(C_p(X)) = \omega$.*

Solution. We have $s(X \times X) \leq s(C_p(X)) = \omega$ by Problem 016. The space $X \times X$ is ω-monolithic by Problem 114, so we can apply Problem 190 to conclude that $hl(X \times X) = \omega$ and hence $\Delta(X) = \omega$. Now Problems 025 and 028 show that $s^*(X) = s^*(C_p(X)) = \omega$. For any $n \in \mathbb{N}$, the space X^n is ω-monolithic by Problem 114; since $s(X^n) = \omega$, we can apply Problem 190 again to see that $hl(X^n) = \omega$. Thus $hl^*(X) = \omega$ and hence $hd^*(C_p(X)) = \omega$ by Problem 026.

T.192. *Suppose that a space X is ω-monolithic and ω-stable. Prove that the equality $s(C_p(X)) = \omega$ implies that $nw(X) = \omega$.*

Solution. Since X is ω-stable, the space $C_p(X)$ is ω-monolithic (see Problem 154). This, together with $s(C_p(X)) = \omega$, implies that $hl(C_p(X)) = \omega$ (see Problem 190) and hence $d(X) \leq hd(X) = \omega$ by Problem 017. Thus X is ω-monolithic and separable; an immediate consequence is that $nw(X) = \omega$.

T.193. *Assume SA. Prove that for any space X, if $C_p(X)$ is $s(\omega)$-monolithic, then it is $hl^*(\omega)$-monolithic. In particular, if $C_p(X)$ is $hl(\omega)$-monolithic, then it is $hl^*(\omega)$-monolithic.*

Solution. Take any countable $A \subset C_p(X)$; for any $x \in X$, let $\varphi(x)(f) = f(x)$ for every $f \in A$. Then $\varphi(x) \in C_p(A)$ for each $x \in X$ and the map $\varphi : X \to C_p(A)$ is continuous (see Problem 166 of [TFS]). If $Y = \varphi(X)$, then we can apply Fact 2 of T.139 to find a space Z and a continuous $\mathbb{R}$-quotient onto map $p : X \to Z$ such that $i \circ p = \varphi$ for some condensation $i : Z \to Y$.

The dual map $p^* : C_p(Z) \to C_p(X)$ embeds $C_p(Z)$ in $C_p(X)$ as a closed subspace and the dual map $i^* : C_p(Y) \to C_p(Z)$ embeds $C_p(Y)$ in $C_p(Z)$ as a dense subspace (see Problem 163 of [TFS]). Since $w(Y) \leq w(C_p(A)) = \omega$, the space $C_p(Y)$ is separable and therefore $d(C_p(Z)) = \omega$.

If we choose a countable set $B \subset p^*(C_p(Z))$ which is dense in the subspace $p^*(C_p(Z))$, then $\overline{B} = p^*(C_p(Z))$, so $s(p^*(C_p(Z))) \leq \kappa$ by $s(\omega)$-monolithity of $C_p(X)$. As a consequence, $s(C_p(Z)) = \omega$ and hence $hl^*(C_p(Z)) = \omega$ by SA and Problem 036. Recalling again that $p^*(C_p(Z))$ and $C_p(Z)$ are homeomorphic, we conclude that $hl^*(p^*(C_p(Z))) = \omega$. It is straightforward that $A \subset \varphi^*(C_p(Y)) \subset p^*(C_p(Z))$ and therefore $\overline{A} \subset p^*(C_p(Z))$ which implies $hl^*(\overline{A}) \leq \omega$. Thus the space $C_p(X)$ is $hl^*(\omega)$-monolithic.

T.194. *Give an example of an $hl(\omega)$-monolithic non-$hl^*(\omega)$-monolithic space.*

Solution. Let X be the Sorgenfrey line (see Problem 165 of [TFS]). Then $hl(X) = \omega$, and hence X is hl-monolithic because for any infinite cardinal κ and any $A \subset X$ with $|A| \leq \kappa$, we have $hl(\overline{A}) \leq hl(X) = \omega \leq \kappa$. The space $X \times X$ has an uncountable closed discrete subspace and hence $hl^*(X) \geq hl(X \times X) > \omega$. This, together with separability of X, implies that X is not $hl^*(\omega)$-monolithic.

T.195. *Assume SA. Let X be a ω-stable space such that $s(C_p(X)) = \omega$. Prove that X has a countable network.*

Solution. Since X is ω-stable, the space $C_p(X)$ is ω-monolithic by Problem 154. The axiom SA and $s(C_p(X)) = \omega$ imply that $hd^*(C_p(X)) = \omega$ (see Problem 036). In particular, the space $C_p(X)$ is separable, so $nw(C_p(X)) = \omega$ by ω-monolithity of $C_p(X)$. Finally, $nw(X) = nw(C_p(X)) = \omega$ by Problem 172 of [TFS].

T.196. *Assume SA. Prove that if $s(C_p(X)) = \omega$, then for any $Y \subset X$ we have $s(C_p(Y)) = \omega$.*

Solution. Let $\pi_Y : C_p(X) \to C_p(Y)$ be the restriction map defined by the equality $\pi_Y(f) = f|Y$ for each $f \in C_p(X)$. As usual, $C_p(Y|X) = \pi_Y(C_p(X)) \subset C_p(Y)$. Observe that SA and $s(C_p(X)) = \omega$ imply $hd(C_p(X)) \leq \omega$ by Problem 036. Consequently, $hd(C_p(Y|X)) \leq \omega$ and hence $hl^*(Y) = hd(C_p(Y|X)) \leq \omega$ (see Fact 5 of T.173). Thus we can apply Problem 026 to conclude that $s(C_p(Y)) \leq hd^*(C_p(Y)) = hl^*(Y) \leq \omega$.

T.197. *Assume MA+¬CH. Let X be an ω-monolithic space such that the spread of $C_p(X)$ is countable. Prove that X has a countable network.*

Solution. We have $s(X \times X) \leq s(C_p(X)) = \omega$ (see Problem 016). The space $X \times X$ being ω-monolithic (see Problem 114), we have $hl(X \times X) = \omega$ (see Problem 190) and therefore $\Delta(X) = \omega$. Thus $s^*(C_p(X)) = \omega$ (see Problem 028) and hence $hl^*(C_p(X)) = \omega$ by Problem 059. As a consequence, $hd^*(X) = hl^*(C_p(X)) = \omega$ (see Problem 027) and, in particular, X is separable. Since every separable ω-monolithic space has a countable network, we have $nw(X) = \omega$.

T.198. *Assume MA+¬CH and suppose that $C_p(X)$ contains no uncountable free sequences. Prove that for any $Y \subset X$, the space Y is hereditarily Lindelöf if and only if it is hereditarily separable.*

Solution. For an arbitrary set P we denote by $\text{Fin}(P)$ the family of all non-empty finite subsets of P. For the sake of brevity, a free sequence of length ω_1 is called *a free ω_1-sequence*. Given a space Z call a set $\{z_\alpha : \alpha < \omega_1\} \subset Z$ a *weak free ω_1-sequence* if $\{z_\beta : \beta \leq \alpha\} \cap \{z_\beta : \beta > \alpha\} = \emptyset$ for any $\alpha < \omega_1$. It is easy to see that every free ω_1-sequence is a weak free ω_1-sequence while the space ω_1 with its interval topology is an example of a weak free ω_1-sequence which is not a free ω_1-sequence. However, the following fact shows that for our purposes we can restrict ourselves to weak free ω_1-sequences.

Fact 1. A space Z does not contain an uncountable free sequence if and only if it does not contain a weak free ω_1-sequence.

Proof. Every free ω_1-sequence is a weak free ω_1-sequence so sufficiency is trivial. To prove necessity take any weak free ω_1-sequence $\{z_\alpha : \alpha < \omega_1\} \subset Z$. It is easy to find a function $\varphi : \omega_1 \to \omega_1$ such that $\varphi(\alpha)$ is a successor ordinal for each $\alpha < \omega_1$ and $\alpha < \beta$ implies $\varphi(\alpha) < \varphi(\beta)$. Now if $y_\alpha = z_{\varphi(\alpha)}$ for all $\alpha < \omega_1$, then $S = \{y_\alpha : \alpha < \omega_1\}$ is a free ω_1-sequence in Z. To see it, fix any $\alpha < \omega_1$; then $\varphi(\alpha) = \gamma + 1$ for some $\gamma < \omega_1$. Observe that $\{y_\beta : \beta < \alpha\} \subset A = \{z_\beta : \beta \leq \gamma\}$ and $\{y_\beta : \alpha \leq \beta\} \subset B = \{z_\beta : \beta > \gamma\}$ and hence $\overline{\{y_\beta : \beta < \alpha\}} \cap \overline{\{y_\beta : \beta \geq \alpha\}} \subset \overline{A} \cap \overline{B} = \emptyset$ which proves that S is a free ω_1-sequence. Fact 1 is proved. $\square$

Returning to our solution, suppose that some $Y \subset X$ is a hereditarily Lindelöf non-hereditarily separable space. Then there is a set $\{y_\alpha : \alpha < \omega_1\} \subset Y$ which is left-separated by its indexation (see Problem 004 and Fact 2 of T.004), so there is no loss of generality to assume that $Y = \{y_\alpha : \alpha < \omega_1\}$; let $Y_\alpha = \{y_\beta : \beta < \alpha\}$ for all $\alpha < \omega_1$. For every $y \in Y$ there is a unique $\alpha < \omega_1$ such that $y = y_\alpha$; let $\nu(y) = \alpha$ and fix sets $U_y, V_y \in \tau(y, X)$ and a function $f_y \in C(X, [0, 1])$ with the following properties:

(1) $\overline{U}_y \subset V_y$ and $f_y(X \setminus V_y) = \{0\}$;
(2) $f_y | \overline{U}_y \equiv 1$ and $z \notin V_y$ whenever $\nu(z) < \nu(y)$.

Let $\mathcal{P} = \{p \in \text{Fin}(Y) : \text{for any } y, z \in p, \text{ if } \nu(y) < \nu(z), \text{ then } z \notin U_y\}$. The order $\preceq$ on $\mathcal{P}$ is the reverse inclusion, i.e., given any $p, q \in \mathcal{P}$ we declare that $p \preceq q$ if and only if $q \subset p$. Observe first that

(3) the partially ordered set $(\mathcal{P}, \preceq)$ is not ccc.

Indeed, if $(\mathcal{P}, \preceq)$ is ccc, then MA+¬CH implies that for any uncountable $R \subset \mathcal{P}$ there is an uncountable centered subset of R (see Problem 049). In particular, there is an uncountable $Q \subset Y$ such that the family $\{\{y\} : y \in Q\}$ is centered. As a consequence,

(4) if $y, z \in Q$ and $\nu(y) < \nu(z)$, then $z \notin U_y$

because there is $p \in \mathcal{P}$ with $p \preceq \{y\}$ and $p \preceq \{z\}$ which implies $\{y, z\} \subset p$, so (4) holds by the definition of $\mathcal{P}$. Next observe that if $y \in Q$, then $U_y \cap Q = \{y\}$;

indeed, if $z \in Q$ and $v(z) < v(y)$, then $z \notin V_y$ by the definition of $V_y \supset U_y$. Now if $v(z) > v(y)$, then $z \notin U_y$ by (4). Therefore Q is an uncountable discrete subspace of a hereditarily Lindelöf space Y; this contradiction proves (3).

By (3), there is an uncountable antichain $\mathcal{A}$ in $(\mathcal{P}, \preceq)$. We can apply the Delta-lemma (see Problem 038) to find a finite set $w \subset Y$ (which may be empty) and an uncountable $\mathcal{A}' \subset \mathcal{A}$ such that $p \cap q = w$ for any distinct $p, q \in \mathcal{A}'$. Since the family $\mathcal{B} = \{p \backslash w : p \in \mathcal{A}'\}$ is disjoint, only countably many elements of $\mathcal{B}$ intersect Y_α for each $\alpha < \omega_1$. This makes it possible to choose, by an evident transfinite induction, a set $\mathcal{B}' = \{p_\alpha : \alpha < \omega_1\} \subset \mathcal{B}$ such that $\alpha < \beta < \omega_1$ implies $v(t) < v(y) < v(z)$ for any $y \in p_\alpha$, $z \in p_\beta$ and $t \in w$. For each ordinal $\alpha < \omega_1$ consider the sets $U(p_\alpha) = \bigcup\{U_y : y \in p_\alpha\}$ and $V(p_\alpha) = \bigcup\{V_y : y \in p_\alpha\}$; observe that if $\alpha < \beta$ and $U(p_\alpha) \cap p_\beta = \emptyset$, then $q = w \cup p_\alpha \cup p_\beta \in \mathcal{P}$. Moreover, we have $q \preceq w \cup p_\alpha$ and $q \preceq w \cup p_\beta$ which contradicts the incompatibility of $w \cup p_\alpha$ and $w \cup p_\beta$. Therefore

(5) if $\alpha < \beta < \omega_1$, then $V(p_\beta) \cap p_\alpha = \emptyset$ and $U(p_\alpha) \cap p_\beta \neq \emptyset$.

Now let $g_\alpha = \sum\{f_y : y \in p_\alpha\}$ for every $\alpha < \omega_1$. It is clear that $g_\alpha(x) \geq 1$ for every $x \in p_\alpha$ and $g_\alpha(X \backslash V(p_\alpha)) = \{0\}$. Let $G_\alpha = \{f \in C_p(X) : f(x) > \frac{1}{2}$ for some $x \in p_\alpha\}$ and $H_\alpha = \{f \in C_p(X) : f(x) > 0$ for some $x \in p_\alpha\}$. It is straightforward that G_α and H_α are open in $C_p(X)$ and $\overline{G}_\alpha \subset H_\alpha$ for each $\alpha < \omega_1$.

Now take any $\alpha < \omega_1$; if $\beta \leq \alpha$, then it follows from (5) that there exist $y \in p_\beta$ and $z \in p_\alpha$ such that $z \in U_y$ and therefore $g_\beta(z) \geq f_y(z) = 1 > \frac{1}{2}$ which shows that $g_\beta \in G_\alpha$. On the other hand, if $\beta > \alpha$, then $y \notin V_z$ for any $z \in p_\beta$, and hence $f_z(y) = 0$ for all $y \in p_\alpha$ and $z \in p_\beta$ which proves that $g_\beta \notin H_\alpha$. As a consequence,

$$\overline{\{g_\beta : \beta \leq \alpha\}} \cap \overline{\{g_\beta : \beta > \alpha\}} \subset \overline{G}_\alpha \cap \overline{C_p(X) \backslash H_\alpha} = \overline{G}_\alpha \cap (C_p(X) \backslash H_\alpha) = \emptyset$$

for every $\alpha < \omega_1$ which shows that $S = \{f_\alpha : \alpha < \omega_1\}$ is a weak free ω_1-sequence in $C_p(X)$. Thus there exists a free ω_1-sequence in $C_p(X)$ by Fact 1 which is a contradiction. Consequently, every hereditarily Lindelöf subspace of X is hereditarily separable.

Now suppose that some $Y \subset X$ is a hereditarily separable non-hereditarily Lindelöf space. There is a set $\{y_\alpha : \alpha < \omega_1\} \subset Y$ which is right-separated by its indexation (see Problem 005 and Fact 2 of T.005), so there is no loss of generality to assume that $Y = \{y_\alpha : \alpha < \omega_1\}$; let $Y_\alpha = \{y_\beta : \beta < \alpha\}$ for all $\alpha < \omega_1$.

For every $y \in Y$ there is a unique $\alpha < \omega_1$ such that $y = y_\alpha$; let $v(y) = \alpha$. Fix sets $U_y, V_y \in \tau(y, X)$ and a function $f_y \in C(X, [0, 1])$ with the following properties:

(6) $\overline{U}_y \subset V_y$ and $f_y(X \backslash V_y) = \{0\}$;
(7) $f_y | \overline{U}_y \equiv 1$ and $z \notin V_y$ whenever $v(z) > v(y)$.

Let $\mathcal{P} = \{p \in \text{Fin}(Y) :$ for any $y, z \in p$, if $v(y) > v(z)$, then $z \notin U_y\}$. The order $\preceq$ on $\mathcal{P}$ is the reverse inclusion, i.e., given any $p, q \in \mathcal{P}$ we declare that $p \preceq q$ if and only if $q \subset p$. Observe first that

(8) the partially ordered set $(\mathcal{P}, \preceq)$ is not ccc.

Indeed, if $(\mathcal{P}, \preceq)$ is ccc, then MA+¬CH implies that for any uncountable $R \subset \mathcal{P}$ there is an uncountable centered subset of R (see Problem 049). In particular, there is an uncountable $Q \subset Y$ such that the family $\{\{y\} : y \in Q\}$ is centered. As a consequence,

(9) if $y, z \in Q$ and $\nu(y) > \nu(z)$, then $z \notin U_y$

because there is $p \in \mathcal{P}$ with $p \preceq \{y\}$ and $p \preceq \{z\}$ which implies $\{y, z\} \subset p$ so (9) holds by the definition of $\mathcal{P}$. Next observe that if $y \in Q$, then $U_y \cap Q = \{y\}$; indeed, if $z \in Q$ and $\nu(z) > \nu(y)$, then $z \notin V_y$ by the definition of $V_y \supset U_y$. Now if $\nu(z) < \nu(y)$, then $z \notin U_y$ by (9). Therefore Q is an uncountable discrete subspace of a hereditarily separable space Y; this contradiction proves (8).

By (8), there exists an uncountable antichain $\mathcal{A}$ in $(\mathcal{P}, \preceq)$. We can apply the Delta-lemma (see Problem 038) to find a finite set $w \subset Y$ (which may be empty) and an uncountable $\mathcal{A}' \subset \mathcal{A}$ such that $p \cap q = w$ for any distinct $p, q \in \mathcal{A}'$. Since the family $\mathcal{B} = \{p \backslash w : p \in \mathcal{A}'\}$ is disjoint, only countably many elements of $\mathcal{B}$ intersect Y_α for each $\alpha < \omega_1$. This makes it possible to choose, by an evident transfinite induction, a set $\mathcal{B}' = \{p_\alpha : \alpha < \omega_1\} \subset \mathcal{B}$ such that $\alpha < \beta < \omega_1$ implies $\nu(t) < \nu(y) < \nu(z)$ for any $y \in p_\alpha$, $z \in p_\beta$ and $t \in w$. For each $\alpha < \omega_1$ consider the sets $U(p_\alpha) = \bigcup\{U_y : y \in p_\alpha\}$ and $V(p_\alpha) = \bigcup\{V_y : y \in p_\alpha\}$; observe that if $\alpha > \beta$ and $U(p_\alpha) \cap p_\beta = \emptyset$, then $q = w \cup p_\alpha \cup p_\beta \in \mathcal{P}$. Moreover, we have $q \preceq w \cup p_\alpha$ and $q \preceq w \cup p_\beta$ which contradicts the incompatibility of $w \cup p_\alpha$ and $w \cup p_\beta$. Therefore,

(10) if $\beta < \alpha < \omega_1$, then $V(p_\beta) \cap p_\alpha = \emptyset$ and $U(p_\alpha) \cap p_\beta \neq \emptyset$.

Now let $g_\alpha = \sum\{f_y : y \in p_\alpha\}$ for every $\alpha < \omega_1$. It is clear that $g_\alpha(x) \geq 1$ for any $x \in p_\alpha$ and $g_\alpha(X \backslash V(p_\alpha)) = \{0\}$. Let $G_\alpha = \{f \in C_p(X) : f(x) > \frac{1}{2}$ for some $x \in p_\alpha\}$ and $H_\alpha = \{f \in C_p(X) : f(x) > 0$ for some $x \in p_\alpha\}$. It is straightforward that G_α and H_α are open in $C_p(X)$ and $\overline{G}_\alpha \subset H_\alpha$ for each $\alpha < \omega_1$.

Now take any $\alpha < \omega_1$; if $\alpha \leq \beta$, then it follows from (10) that there exist $y \in p_\beta$ and $z \in p_\alpha$ such that $z \in U_y$ and therefore $g_\beta(z) \geq f_y(z) = 1 > \frac{1}{2}$ which shows that $g_\beta \in G_\alpha$. On the other hand, if $\beta < \alpha$, then $y \notin V_z$ for any $z \in p_\beta$ and hence $f_z(y) = 0$ for all $y \in p_\alpha$ and $z \in p_\beta$ which proves that $g_\beta \notin H_\alpha$. As a consequence,

$$\overline{\{g_\beta : \beta < \alpha\}} \cap \overline{\{g_\beta : \beta \geq \alpha\}} \subset \overline{G}_\alpha \cap \overline{C_p(X) \backslash H_\alpha} = \overline{G}_\alpha \cap (C_p(X) \backslash H_\alpha) = \emptyset$$

for every $\alpha < \omega_1$ which shows that $S = \{f_\alpha : \alpha < \omega_1\}$ is a free ω_1-sequence in $C_p(X)$ which is a contradiction. Consequently, every hereditarily separable subspace of X is hereditarily Lindelöf so our solution is complete.

T.199. *Assume MA+¬CH. Let X be an ω-monolithic space with $l^*(X) = \omega$. Suppose that $C_p(X)$ is Lindelöf and $Y \subset X$ has countable spread. Prove that Y has a countable network.*

Solution. We have $hl(Y) = \omega$ because the space Y is ω-monolithic and has countable spread (see Problem 190). It follows from $l^*(X) = \omega$ that $t(C_p(X)) = \omega$ (see Problem 149 of [TFS]). Now assume that $S = \{f_\alpha : \alpha < \omega_1\}$ is a free sequence in $C_p(X)$; let $S_\alpha = \{f_\beta : \beta < \alpha\}$ and $T_\alpha = S \backslash S_\alpha$ for every $\alpha < \omega_1$. The family $\mathcal{T} = \{\overline{T}_\alpha : \alpha < \omega_1\}$ is countably centered (i.e., the intersection of any countable subfamily of $\mathcal{T}$ is non-empty) and consists of non-empty closed subsets of $C_p(X)$. The space $C_p(X)$ being Lindelöf, we have $T = \bigcap \mathcal{T} \neq \emptyset$ (see Fact 2 of S.336). If $f \in T$, then $f \in \overline{S}$, and hence there is a countable $S' \subset S$ with $f \in \overline{S'}$. Choose $\alpha < \omega_1$ such that $S' \subset S_\alpha$ and observe that $f \in \overline{S}_\alpha \cap \overline{T}_\alpha$, a contradiction with the fact that S is a free sequence.

Therefore $C_p(X)$ has no uncountable free sequences and hence we can apply Problem 198 to the hereditarily Lindelöf space Y to conclude that Y is separable. Any separable ω-monolithic space has a countable network so $nw(Y) = \omega$.

T.200. *Assume* $MA+\neg CH$. *Prove that if* $C_p(X)$ *is hereditarily stable, then X has a countable network.*

Solution. A discrete space is ω-stable only if it is countable by Problem 106, so $C_p(X)$ contains no uncountable discrete subspaces, i.e., $s(C_p(X)) = \omega$. The space X is monolithic because $C_p(X)$ is stable (see Problem 152). Now apply Problem 197 to conclude that X has a countable network.

T.201. *Prove that any continuous image of a hereditarily normal compact space is hereditarily normal.*

Solution. Assume that Z is a space and $f : Z \to Y$ is a closed continuous onto map. For any open $U \subset Z$, let $f^\#(U) = Y \backslash f(Z \backslash U)$. It is straightforward that $f^\#(U)$ is an open subset of Y and $f^\#(U) = \{y \in Y : f^{-1}(y) \subset U\}$. An immediate consequence is that

(1) if $U, V \subset \tau(Z)$ and $U \cap V = \emptyset$ then $f^\#(U) \cap f^\#(V) = \emptyset$.

Now it is easy to prove the following statement.

Fact 1. A closed continuous image of a normal space is a normal space.

Proof. Let Z be a normal space; suppose that $f : Z \to Y$ is a closed continuous onto map. If we have disjoint closed sets $F, G \subset Y$, then the sets $F' = f^{-1}(F)$ and $G' = f^{-1}(G)$ are closed in Z and disjoint. By normality of Z there exist disjoint $U', V' \in \tau(Z)$ such that $F' \subset U'$ and $G' \subset V'$. Given any point $y \in F$, we have $f^{-1}(y) \subset f^{-1}(F) = F' \subset U'$ and therefore $y \in f^\#(U')$; this shows that $F \subset f^\#(U')$. Analogously, $G \subset f^\#(V')$, so we can apply (1) to conclude that the sets $U = f^\#(U')$ and $V = f^\#(V')$ are open disjoint neighborhoods of F and G respectively. Thus Y is normal and Fact 1 is proved. $\square$

Returning to our solution take any compact hereditarily normal space X and a continuous onto map $f : X \to Y$. The map f is closed (Problem 122 of [TFS]) and hence, for any $A \subset Y$, the map $f_A = f|f^{-1}(A) : f^{-1}(A) \to A$ is also closed by Fact 1 of S.261. By hereditary normality of X the space $f^{-1}(A)$ is normal and hence A is normal by Fact 1. Therefore every $A \subset Y$ is normal, i.e., Y is hereditarily normal.

T.202. *Let X be a compact space for which X^2 is hereditarily normal. Prove that X is perfectly normal and hence first countable.*

Solution. We can assume that X is infinite and hence the space X has a countable non-closed subset by Fact 2 of T.090. Therefore Fact 1 of T.090 is applicable to the spaces $Y = Z = X$ to conclude that there is no closed $F \subset X$ which is not a G_δ-set in X. In other words, every closed subset of X is a G_δ-set in X and hence X is perfectly normal.

T.203. *Let X be a compact space and denote by Δ the diagonal of X, i.e., $\Delta = \{(x, x) : x \in X\} \subset X^2$. Prove that if $X^2 \backslash \Delta$ is paracompact, then X is metrizable.*

Solution. Given a space Z the set $\Delta(Z) = \{(z, z) : z \in Z\} \subset Z^2$ is its diagonal; a family $\mathcal{U} \subset \exp(Z)$ is T_1-*separating* if $\bigcap \{U \in \mathcal{U} : z \in U\} = \{z\}$ for every $z \in Z$. The family $\mathcal{U}$ is called *point-countable* if every $z \in Z$ belongs to at most countably many elements of $\mathcal{U}$. Given a set $A \subset Z$ let $\mathcal{U}[A] = \{U \in \mathcal{U} : U \cap A \neq \emptyset\}$; for any $z \in Z$ we will write $\mathcal{U}[z]$ instead of $\mathcal{U}[\{z\}]$.

Fact 1. A countably compact space Z is metrizable if and only if there exists a point-countable T_1-separating family $\mathcal{U} \subset \tau(Z)$.

Proof. If the space Z is metrizable, then it has a countable base $\mathcal{U}$ which is, evidently, point-countable and T_1-separating so necessity is clear.

To prove sufficiency fix a point-countable T_1-separating family $\mathcal{U} \subset \tau(Z)$ and take any $z_0 \in Z$; let $A_0 = \{z_0\}$. Assume that $n > 0$ and we have countable sets $A_0, \ldots, A_{n-1}$ with the following properties:

(1) $A_0 \subset \cdots \subset A_{n-1} \subset Z$;
(2) if $k < n - 1$ and $\mathcal{V} \subset \mathcal{U}[A_k]$ is finite, then $\bigcup \mathcal{V} \neq Z$ implies $A_{k+1} \backslash (\bigcup \mathcal{V}) \neq \emptyset$.

The set A_{n-1} being countable, the family $\mathcal{U}[A_{n-1}]$ has to be countable as well because $\mathcal{U}$ is point-countable. Therefore the family

$$\mathcal{S} = \{\mathcal{V} : \mathcal{V} \text{ is a finite subfamily of } \mathcal{U}[A_{n-1}] \text{ such that } \bigcup \mathcal{V} \neq Z\}$$

is countable. For any family $\mathcal{V} \in \mathcal{S}$ we can choose a point $z(\mathcal{V}) \in Z \backslash (\bigcup \mathcal{V})$; then $A_n = A_{n-1} \cup \{z(\mathcal{V}) : \mathcal{V} \in \mathcal{S}\}$ is countable and the sets $A_0, \ldots, A_n$ still satisfy the conditions (1) and (2). Therefore our inductive construction can be continued to provide a sequence $\{A_n : n \in \omega\}$ with the properties (1) and (2) for each $n \in \omega$. We claim that the set $A = \bigcup \{A_n : n \in \omega\}$ is dense in Z. Observe first that an open set intersects A if and only if it intersects $\overline{A}$. Therefore $\mathcal{U}[\overline{A}] = \mathcal{U}[A]$ is countable. If $\overline{A} \neq Z$, then take any $u \in Z \backslash \overline{A}$; the family $\mathcal{U}$ being T_1-separating, for every $z \in \overline{A}$, there is $W_z \in \mathcal{U}[z]$ such that $W_z \not\ni u$. Since $W_z \in \mathcal{U}[A]$ for every $z \in \overline{A}$, the family $\mathcal{W} = \{W_z : z \in \overline{A}\}$ is a countable open cover of a countably compact space $\overline{A}$. Take any finite $\mathcal{V} \subset \mathcal{W}$ such that $\overline{A} \subset \bigcup \mathcal{V}$. Since $\mathcal{U}[A] = \bigcup \{\mathcal{U}[A_n] : n \in \omega\}$ and (1) holds, there is $k \in \omega$ such that $\mathcal{V} \subset \mathcal{U}[A_k]$. We have $u \in Z \backslash (\bigcup \mathcal{V})$ which implies $\bigcup \mathcal{V} \neq Z$ and hence $A_{k+1} \backslash (\bigcup \mathcal{V}) \neq \emptyset$ by (2) which is a contradiction with $A_{k+1} \subset A \subset \bigcup \mathcal{V}$.

Thus A is dense in Z and hence $\mathcal{U} = \mathcal{U}[A]$ is countable; therefore the family $\mathcal{F} = \{Z \backslash U : U \in \mathcal{U}\}$ is also countable and T_1-separating because $\mathcal{U}$ is T_1-separating. Thus $\bigcap \mathcal{F}[z] = \{z\}$ for every $z \in Z$. Observe also that $\mathcal{F}$ consists of closed subspaces of Z. Given any $z \in Z$ take an enumeration $\{F_n : n \in \omega\}$ of $\mathcal{F}[z]$; if $U \in \tau(z, Z)$, then let $G_n = F_0 \cap \cdots \cap F_n \cap (Z \backslash U)$ for all $n \in \omega$. It is clear that $G_{n+1} \subset G_n$ for all $n \in \omega$ and $\bigcap_{n \in \omega} G_n = \emptyset$. The space Z being countably compact, we have $G_n = \emptyset$ for some $n \in \omega$ and therefore $F_0 \cap \cdots \cap F_n \subset U$.

This proves that for any $z \in Z$ and $U \in \tau(z, Z)$ there is a finite $\mathcal{F}' \subset \mathcal{F}[z]$ such that $\bigcap \mathcal{F}' \subset U$. Consequently, the family of all finite intersections of the elements of $\mathcal{F}$ is a countable network in Z. Therefore Z condenses onto a second countable space (see Problem 156 of [TFS]); the respective condensation is a homeomorphism (see Problem 140 of [TFS]), so Z is second countable and hence metrizable. Fact 1 is proved.   □

Fact 2. If κ is an uncountable cardinal and $Z = A(\kappa)$, then $Z^2 \backslash \Delta(Z)$ is not normal.

Proof. Denote by a the unique non-isolated point of Z; let $w = (a, a) \in Z^2$. The sets $F = (Z \times \{a\}) \backslash \{w\}$ and $G = (\{a\} \times Z) \backslash \{w\}$ are closed in $Z^2 \backslash \Delta(Z)$ and disjoint. Assume that there are disjoint $U, V \in \tau(Z^2)$ such that $F \subset U$ and $G \subset V$. Choose any faithfully indexed sequence $\{a_n : n \in \omega\} \subset Z \backslash \{a\}$; since $(a_n, a) \in U$ there is a finite $B_n \subset Z$ such that $\{a_n\} \times (Z \backslash B_n) \subset U$ for all $n \in \omega$. The set $B = \bigcup \{B_n : n \in \omega\}$ is countable, so there is $z \in Z \backslash (B \cup \{a\})$. It is evident that $u_n = (a_n, z) \in U$ for each $n \in \omega$; since the sequence $S = \{u_n : n \in \omega\}$ converges to $(a, z) \in G$, we have $\overline{S} \cap G \neq \emptyset$ which contradicts the fact that $S \subset U \subset Z^2 \backslash V$ and hence $\overline{S} \subset Z^2 \backslash V \subset Z^2 \backslash G$. Therefore the disjoint closed sets F and G cannot be separated by disjoint open sets in $Z^2 \backslash \Delta(Z)$, i.e., $Z^2 \backslash \Delta(Z)$ is not normal and hence Fact 2 is proved.   □

Fact 3. If Z is a locally compact paracompact space, then there is a disjoint family $\mathcal{U}$ of clopen subsets of Z such that each $U \in \mathcal{U}$ is σ-compact and $\bigcup \mathcal{U} = Z$.

Proof. Each $z \in Z$ has an open neighborhood O_z such that $\overline{O_z}$ is compact. If $\mathcal{W}$ is a locally finite refinement of the open cover $\{O_z : z \in Z\}$ of the space Z, then $\overline{W}$ is compact for any $W \in \mathcal{W}$. Call a sequence $C = \{W_1 \ldots, W_n\}$ of elements of $\mathcal{W}$ a *chain* if $W_k \cap W_{k+1} \neq \emptyset$ for every $k = 1, \ldots, n - 1$. Given $y, z \in Z$, let $y \sim z$ if there is a chain $C = \{W_1, \ldots, W_n\}$ of elements of $\mathcal{W}$ such that $y \in W_1$ and $z \in W_n$ (we will say that the chain C *connects the points y and z*). It is evident that $\sim$ is an equivalence relationship on Z; let $\mathcal{U}$ be the set of the respective equivalence classes. Then $\mathcal{U}$ is disjoint and $\bigcup \mathcal{U} = Z$.

For any $z \in Z$ denote by U_z the equivalence class of z. Given any $t \in U_z$ there is a chain $C = \{W_1, \ldots, W_n\}$ of elements of $\mathcal{W}$ such that $z \in W_1$ and $t \in W_n$. It is clear that the same chain C connects z and any point $t' \in W_n$. Therefore $W_n \subset U_z$ and hence every $t \in U_z$ lies in U_z together with its neighborhood W_n which proves that U_z is open in Z for any $z \in Z$. The family $\mathcal{U}$ being disjoint, each $U \in \mathcal{U}$ is also closed in Z.

The last thing we must prove is that every $U \in \mathcal{U}$ is σ-compact. Observe that if a set $F \subset Z$ is compact, then $\mathcal{W}[F]$ is finite because the family $\{W \cap F : W \in \mathcal{W}\}$ is a locally finite family of open sets in a compact space F (see Problem 136 of [TFS]). Take any $z_0 \in U$, choose $W_0 \in \mathcal{W}$ with $z_0 \in W_0$ and let $\mathcal{A}_0 = \{W_0\}$. Assume that $n > 0$ and we have finite families $\mathcal{A}_0 \subset \cdots \subset \mathcal{A}_{n-1} \subset \mathcal{W}$ such that

(3) $\mathcal{W}[\bigcup \mathcal{A}_k] \subset \mathcal{A}_{k+1}$ for any $k < n - 1$.

The set $V_n = \bigcup \{\overline{V} : V \in \mathcal{A}_{n-1}\}$ is compact, so $\mathcal{A}_n = \mathcal{W}[V_n] = \mathcal{W}[\bigcup \mathcal{A}_{n-1}]$ is finite and the sequence $\{\mathcal{A}_0, \ldots, \mathcal{A}_n\}$ still satisfies (3). Thus we can construct a sequence $\{\mathcal{A}_n : n \in \omega\}$ with the property (3) for all $n \in \omega$. It is clear that the family $\mathcal{A} = \bigcup\{\mathcal{A}_n : n \in \omega\}$ is countable and hence $A = \bigcup\{\overline{W} : W \in \mathcal{A}\}$ is σ-compact. Given any $z \in U = U_{z_0}$, there is a chain $C = \{W_1, \ldots, W_n\} \subset \mathcal{W}$ which connects z_0 and z. It is immediate from (3) that $W_i \in \mathcal{A}_i$ for each $i = 1, \ldots, n$ and hence $z \in \bigcup \mathcal{A}_n$. As a consequence, the space $U \subset \bigcup \mathcal{A} \subset A$ is σ-compact being a closed subset of a σ-compact set A. Fact 3 is proved.   □

Returning to our solution assume that X is not metrizable and observe that $X^2 \backslash \Delta$ is a locally compact paracompact space, so we can apply Fact 3 to find a disjoint family $\mathcal{W}$ of open σ-compact sets such that $\bigcup \mathcal{W} = X^2 \backslash \Delta$. For any $W \in \mathcal{W}$ and any $z = (x, y) \in W$, we have $x \neq y$ and hence there exist disjoint $U_z, V_z \in \tau(X)$ such that $x \in U_z$, $y \in V_z$ and $U_z \times V_z \subset W$. The space W being σ-compact, there is a countable subcover of the cover $\{U_z \times V_z : z \in U\}$. Thus we can choose, for each $W \in \mathcal{W}$, countable families $\mathcal{U}_W = \{U(W, n) : n \in \omega\} \subset \tau(X)$ and $\mathcal{V}_W = \{V(W, n) : n \in \omega\} \subset \tau(X)$ such that $U(W, n) \cap V(W, n) = \emptyset$ for each $n \in \omega$ and $\bigcup\{U(W, n) \times V(W, n) : n \in \omega\} = W$.

We claim that the family $\mathcal{U} = \bigcup\{\mathcal{U}_W : W \in \mathcal{W}\}$ is T_1-separating in X. Indeed, if $x \in X$ and $y \neq x$, then $(x, y) \in X^2 \backslash \Delta$ and hence $(x, y) \in W$ for some $W \in \mathcal{W}$ which implies that there is $n \in \omega$ such that $(x, y) \in U(W, n) \times V(W, n)$ and therefore $x \in U(W, n)$ and $y \in V(W, n)$. The sets $U(W, n)$ and $V(W, n)$ being disjoint, we have $x \in U(W, n) \not\ni y$ and hence $\mathcal{U}$ is T_1-separating. Since X is not metrizable, the family $\mathcal{U}$ cannot be point-countable by Fact 1. Take any $x \in X$ such that $\mathcal{U}[x]$ is uncountable; there exists a faithfully indexed family $\{W_\alpha : \alpha < \omega_1\} \subset \mathcal{W}$ and $n \in \omega$ such that $x \in U(W_\alpha, n)$ for each $\alpha < \omega_1$. The family $\{W_\alpha : \alpha < \omega_1\}$ is discrete in $X^2 \backslash \Delta$ and $U(W_\alpha, n) \times V(W_\alpha, n) \subset W_\alpha$ for each $\alpha < \omega_1$. Therefore the family $\{U(W_\alpha, n) \times V(W_\alpha, n) : \alpha < \omega_1\}$ is also discrete in $X^2 \backslash \Delta$. Let $X_x = \{x\} \times X$; it is clear that $X_x \cap (U(W_\alpha, n) \times V(W_\alpha, n)) = \{x\} \times V(W_\alpha, n)$ for every $\alpha < \omega_1$ which implies that $\{\{x\} \times V(W_\alpha, n) : \alpha < \omega_1\}$ is discrete in $X_x \backslash \{(x, x)\}$ and therefore the family $\{V(W_\alpha, n) : \alpha < \omega_1\}$ is discrete in $X \backslash \{x\}$. Pick any $x_\alpha \in V(W_\alpha, n)$ for every $\alpha < \omega_1$. Then the set $D = \{x_\alpha : \alpha < \omega_1\}$ is closed and discrete in $X \backslash \{x\}$ and hence $K = D \cup \{x\}$ is homeomorphic to $A(\omega_1)$ being a compact space of cardinality ω_1 with a unique non-isolated point. It is easy to see that $K^2 \backslash \Delta(K)$ is a closed subset of $X^2 \backslash \Delta(X)$ and therefore $K^2 \backslash \Delta(K)$ is paracompact which contradicts Fact 2. This contradiction shows that X is metrizable and hence our solution is complete.

T.204. *Observe that any Fréchet–Urysohn space is Whyburn. Prove that any countably compact Whyburn space is Fréchet–Urysohn.*

Solution. If X is a Fréchet–Urysohn space, then for any $A \subset X$ and any $x \in \overline{A} \backslash A$, there exists a sequence $\{a_n : n \in \omega\} \subset A$ such that $a_n \to x$. It is immediate that $F = \{a_n : n \in \omega\}$ is an almost closed set with $x \in \overline{F}$ so X is Whyburn.

Now assume that X is a countably compact Whyburn space. Given a set $A \subset X$ and $x \in \overline{A} \backslash A$ choose an almost closed $F \subset A$ with $x \in \overline{F}$. Take a maximal family $\mathcal{U} \subset \tau^*(F)$ such that $\overline{U} \cap \overline{V} = \emptyset$ for any distinct $U, V \in \mathcal{U}$ and $x \notin \overline{U}$ for every $U \in \mathcal{U}$. It is evident that $G = \bigcup \mathcal{U}$ is dense in F and hence $x \in \overline{G}$.

Use the Whyburn property of X again to find an almost closed $H \subset G$ with $x \in \overline{H}$. Let $H_U = H \cap U$ for every $U \in \mathcal{U}$. Every $U \in \mathcal{U}$ is closed in G, so it follows from $x \notin \overline{H}_U$ that H_U is closed (maybe empty) in X for each $U \in \mathcal{U}$. The family $\mathcal{U}' = \{U \in \mathcal{U} : H_U \neq \emptyset\}$ has to be infinite; for otherwise, $x \notin \bigcup \{H_U : U \in \mathcal{U}'\} = \overline{H}$ which is a contradiction.

Now, take any $y \in X \backslash \{x\}$; if $y \notin H$, then $W = X \backslash \overline{H} \in \tau(y, X)$ and $W \cap H_U = \emptyset$ for any $U \in \mathcal{U}'$. If $y \in H$, then there is a unique $U \in \mathcal{U}'$ such that $y \in H_U$. Then $U \in \tau(y, X)$ and $U \cap H_V = \emptyset$ for any $V \in \mathcal{U}' \backslash \{U\}$. This shows that the family $\{H_U : U \in \mathcal{U}'\}$ is discrete in $X \backslash \{x\}$.

Thus for any set $W \in \tau(x, X)$ the family $\mathcal{H}_W = \{H_U \backslash W : (H_U \backslash W) \neq \emptyset$ and $U \in \mathcal{U}'\}$ is discrete in X. Since X is countably compact, $\mathcal{H}_W$ has to be finite and hence $H_U \subset W$ for all but finitely many $U \in \mathcal{U}'$. Pick a point $x_U \in H_U$ for any $U \in \mathcal{U}'$. Then the set $B = \{x_U : U \in \mathcal{U}'\}$ is infinite and $B \backslash W$ is finite for any $W \in \tau(X)$ with $x \in W$. This shows that for any faithfully indexed sequence $S = \{a_n : n \in \omega\} \subset B$ we have $S \subset A$ and $S \to x$, i.e., X is a Fréchet–Urysohn space.

T.205. *Give an example of a pseudocompact Whyburn space which is not Fréchet–Urysohn.*

Solution. Take an arbitrary countably infinite set D which will be identified with the discrete space whose underlying set is D. The letter K stands for the Cantor set (see Problem 128 of [TFS]); the mappings $\pi_D : D \times K \to D$ and $\pi_K : D \times K \to K$ are the respective natural projections. If $A \subset D \times K$, then $A(d)$ denotes the set $\pi_K(A \cap (\{d\} \times K))$ for any $d \in D$. A set $A \subset D \times K$ is called *admissible* if it has the following properties:

(1) the set $i(A) = \pi_D(A)$ is infinite;
(2) the set $A(d)$ is clopen in K for every $d \in i(A)$ and the family $\{A(d) : d \in i(A)\}$ is disjoint.

A family $\mathcal{A}$ of admissible subsets of $D \times K$ is called *essentially disjoint* if for any distinct $A, B \in \mathcal{A}$, the set $\{d \in D : A(d) \cap B(d) \neq \emptyset\}$ is finite. Denote by $\mathcal{A}(D)$ any maximal essentially disjoint family of admissible subsets of $D \times K$ (we leave to the reader a simple verification that such a maximal family exists). Assign some point $p_A \notin D \times K$ to every set $A \in \mathcal{A}(D)$ in such a way that $p_A \neq p_B$ for any distinct $A, B \in \mathcal{A}(D)$.

We can now define our auxiliary space $P(D) = (D \times K) \cup \{p_A : A \in \mathcal{A}(D)\}$. The local bases at the points of $D \times K$ are given by their usual local bases in $D \times K$

(remember that D is endowed with the discrete topology). Given $A \in \mathcal{A}(D)$, a local base at the point p_A is the family $\{\{p_A\} \cup (A\backslash(F \times K)) : F \subset D \text{ is finite}\}$.

Fact 1. The space $P(D)$ is Tychonoff, first countable, pseudocompact and zero-dimensional; $D \times K$ is an open subspace of $P(D)$ and the set $\{p_A : A \in \mathcal{A}(D)\}$ is closed and discrete.

Proof. We omit a simple proof that the local bases are well-defined. The elements of the local base of any $w \in D \times K$ are all contained in $D \times K$, which shows that every $w \in D \times K$ belongs to $D \times K$ together with its neighborhood; thus $D \times K$ is open in $P(D)$.

Let us check that the space $P(D)$ is Hausdorff. If x and y are distinct points from $D \times K$, then they have open disjoint neighborhoods in the space $D \times K$; since the same neighborhoods are open in $P(D)$, any pair of distinct points from $D \times K$ can be separated by disjoint open neighborhoods in $P(D)$.

Now, if $x \in D \times K$ and $y \in \{p_A : A \in \mathcal{A}(D)\}$, then $y = p_A$ for some $A \in \mathcal{A}(D)$ and hence the sets $\{p_A\} \cup (A\backslash(\{\pi_D(x)\} \times K))$ and $\{\pi_D(x)\} \times K$ are disjoint open neighborhoods of the points y and x respectively.

Finally, if $x = p_A$ and $y = p_B$ for distinct $A, B \in \mathcal{A}(D)$, then $F = i(A) \cap i(B)$ is finite because $\mathcal{A}(D)$ is essentially disjoint. Therefore $\{p_A\} \cup (A\backslash(F \times K))$ and $\{p_B\} \cup (B\backslash(F \times K))$ are open disjoint neighborhoods of x and y respectively. This proves that $P(D)$ is a Hausdorff space.

Next observe that K is zero-dimensional. Indeed, if $x \in K$ and $U \in \tau(x, K)$, then there is an interval $(a, b) \subset \mathbb{R}$ such that $x \in (a, b) \cap K \subset U$ (we consider that $K \subset \mathbb{R}$ (see Problem 128 of [TFS])). The interval (x, b) cannot be contained in K because the interior of K is empty. Thus there is $c \in (x, b)\backslash K$; analogously, there exists a point $d \in (a, x)\backslash K$. Consequently, $W = (c, d) \cap K = [c, d] \cap K$ is a clopen subset of K and $x \in W \subset (a, b) \cap K \subset U$ which shows that clopen subsets form a base in K, i.e., K is zero-dimensional. It is evident that $\{d\} \times U$ is a clopen subset of $D \times K$ for any $d \in D$ and clopen $U \subset K$. Since the family $\mathcal{O} = \{\{d\} \times U : d \in D \text{ and } U \text{ is clopen in } K\}$ is a base in $D \times K$, the space $D \times K$ is also zero-dimensional. It follows from the definition of local bases in $P(D)$ that every $O \in \mathcal{O}$ is clopen in $P(D)$ which proves that the family of all clopen subsets of $P(D)$ contains local bases at all points of $D \times K$.

Observe that given any $A \in \mathcal{A}(D)$, the set $U \cap (D \times K)$ is clopen in $D \times K$ for any basic neighborhood U of any point p_A and hence no point from $D \times K$ belongs to $\overline{U}\backslash U$. Furthermore, if $B \in \mathcal{A}(D)\backslash\{A\}$, then $F = i(A) \cap i(B)$ is finite because $\mathcal{A}(D)$ is essentially disjoint; therefore $\{p_B\} \cup (B\backslash(F \times K))$ is an open neighborhood of p_B which does not intersect U. As a consequence, p_B does not belong to $\overline{U}\backslash U$, i.e., U is a clopen subspace of $P(D)$. Therefore $P(D)$ is zero-dimensional and hence Tychonoff by Fact 1 of S.232.

It is evident that $P = \{p_A : A \in \mathcal{A}(D)\}$ is closed; to see that it is discrete, observe that $U \cap P = p_A$ for any basic neighborhood U of the point p_A.

Let us finally check that $P(D)$ is pseudocompact. The set $D \times K$ is open and dense in $P(D)$, so if $\{W_n : n \in \omega\} \subset \tau^*(P(D))$ is a discrete family, then we can choose a non-empty clopen $U_n \subset W_n \cap (D \times K)$ for each $n \in \omega$. It is evident that the family $\{U_n : n \in \omega\}$ is also Discrete, so it suffices to show that for any infinite family $\mathcal{U}$ of non-empty clopen subsets of $D \times K$, there exists an accumulation point for $\mathcal{U}$, i.e., a point $x \in P(D)$ such that every neighborhood of x meets infinitely many elements of $\mathcal{U}$.

If the family $\mathcal{U}_d = \{U \in \mathcal{U} : U(d) \neq \emptyset\}$ is infinite for some $d \in D$, then we have an infinite family $\mathcal{U}' = \{U \cap (\{d\} \times K) : U \in \mathcal{U}_d\}$ of non-empty open sets in the compact space $\{d\} \times K$. Thus there is an accumulation point x for $\mathcal{U}'$ in $\{d\} \times K$ which, evidently, is an accumulation point for $\mathcal{U}$ in $P(D)$.

If every $\mathcal{U}_d$ is finite, then the set $\{d \in D : U(d) \neq \emptyset$ for some $U \in \mathcal{U}\}$ is infinite and hence we can find an admissible set $B \subset D \times K$ and an injection $d \to U_d$ of the set $i(B)$ into $\mathcal{U}$ such that, $B(d) \subset U_d(d)$ for each $d \in i(B)$. The family $\mathcal{A}(D)$ being maximal essentially disjoint, we can find $A \in \mathcal{A}(D)$ such that $A(d) \cap B(d) \neq \emptyset$ for infinitely many d's. To finish our proof observe that each neighborhood of the point p_A intersects infinitely many elements of $\mathcal{U}$. Fact 1 is proved.   □

Returning to our solution denote by L the subset of $\mathfrak{c}^+$ which consists of limit ordinals of countable cofinality. Considering L with its order topology it is natural to say that a countably infinite set $S \subset L$ is *a sequence converging to* $\alpha \in L$ if $S \subset \alpha$, $\sup S = \alpha$ and $S \cap \beta$ is finite for any $\beta < \alpha$.

If $\alpha \in L$ and $\alpha = \beta + \omega$ for some $\beta < \mathfrak{c}^+$ (i.e., if α is an isolated point of L), let $\mathcal{D}_\alpha = \{[\beta, \alpha)\}$, where $[\beta, \alpha) = \{\gamma : \beta \leq \gamma < \alpha\}$. If α is not isolated in L, define $\mathcal{D}_\alpha$ to be any maximal almost disjoint family of sequences converging to α. We leave it to the reader as an exercise to prove that $\mathcal{D} = \bigcup\{\mathcal{D}_\alpha : \alpha \in L\}$ is a maximal almost disjoint family of countable subsets of $\mathfrak{c}^+$.

Apply Fact 1 to fix the family $\mathcal{A}(D)$, the points $\{p_A : A \in \mathcal{A}(D)\}$ and the space $P(D)$ for every set $D \in \mathcal{D}$. Choose a new point p and define

$$X = \{p\} \cup \{p_A : A \in \mathcal{A}(D), \ D \in \mathcal{D}\} \cup (\mathfrak{c}^+ \times K)$$

to be the underlying set of the space we want to construct. Given an arbitrary point $x = (\alpha, z) \in \mathfrak{c}^+ \times K$, take a clopen local base $\{W_n : n \in \omega\}$ of z in the space K and let $U_n = \{\alpha\} \times W_n$ for each $n \in \omega$. We declare the family $\mathcal{B}_x = \{U_n : n \in \omega\}$ to be a local base of the point x in X. Observe that the local bases thus defined generate the product topology on $\mathfrak{c}^+ \times K$ if $\mathfrak{c}^+$ is considered with the discrete topology.

If $A \in \mathcal{A}(D)$ for some $D \in \mathcal{D}$, then the local base $\mathcal{B}_x$ at the point $x = p_A$ is the same as in the space $P(D)$ (recall that $P(D) = D \times K \subset \mathfrak{c}^+ \times K$ is a subset of X). Finally, declare the local base of X at the point p to be the family $\mathcal{B}_p = \{V_\alpha : \alpha \in L\}$ where

$$V_\alpha = \{p\} \cup (\{\beta : \alpha < \beta < \mathfrak{c}^+\} \times K) \cup \{p_A : A \in \mathcal{A}(D), \ D \in \bigcup\{\mathcal{D}_\beta : \beta > \alpha\}\}$$

for each $\alpha \in L$.

We claim that X is a pseudocompact Whyburn space which is not Fréchet–Urysohn. It is immediate that X is a T_1-space, so to prove that X is Tychonoff, it suffices to show that it is zero-dimensional (see Fact 1 of S.232). We will establish a stronger fact, namely, that all elements of the local bases defined above are clopen in X.

If $x = (\alpha, z) \in \mathfrak{c}^+ \times K$ and $U \in \mathcal{B}_x$, then $W = (\mathfrak{c}^+ \times K) \backslash U$ is open in $\mathfrak{c}^+ \times K$ and therefore W is a neighborhood of any $y \in (\mathfrak{c}^+ \times K) \backslash U$. Given $A \in \mathcal{A}(D)$ for some $D \in \mathcal{D}$, it is immediate from the definition of the local base at $y = p_A$ that $\{p_A\} \cup ((\mathfrak{c}^+ \times K) \backslash U)$ is an open neighborhood of y which does not meet U. Now, if $y = p$, then $V_\beta \cap U = \emptyset$ for any $\beta \in L$, $\beta > \alpha$ and this finishes the proof that U is a clopen set.

We will show next that if $x = p_A$ for some $A \in \mathcal{A}(D)$, $D \in \mathcal{D}$, then every $U \in \mathcal{B}_x$ is a clopen set. Observe first that $W = (\mathfrak{c}^+ \times K) \backslash U$ is a clopen set in $\mathfrak{c}^+ \times K$, so every point from W is not in the closure of U. If $y = p_B$ is distinct from x, then either $B \in \mathcal{A}(D)$ or $i(A) \cap i(B)$ is finite because the family $\mathcal{D}$ is almost disjoint. Now observe that p_B belongs to the closure of a set $T \subset \mathfrak{c}^+ \times K$ if and only if T intersects infinitely many sets of the family $\{\{d\} \times B(d) : d \in i(B)\}$. If $B \in \mathcal{A}(D)$, this is impossible for $T = A$ because the family $\mathcal{A}(D)$ is essentially disjoint, and if $B \in \mathcal{A}(D')$ for some $D' \neq D$, then $i(B) \cap i(A)$ is finite being contained in the finite set $D \cap D'$. Finally, to see that p is not in the closure of U observe that $V_\alpha \cap U = \emptyset$ for any $\alpha \in L$ with $\alpha > \sup(D)$.

To establish that V_α is clopen for any $\alpha \in L$, observe that $V_\alpha \cap (\mathfrak{c}^+ \times K)$ is clopen in $\mathfrak{c}^+ \times K$, so no point of $(\mathfrak{c}^+ \times K) \backslash V_\alpha$ can be in the closure of V_α. If $p_A \notin V_\alpha$, then $A \in \mathcal{A}(D)$ for some $D \in \mathcal{D}_\beta$ with $\beta \leq \alpha$. By definition of $\mathcal{D}_\beta$, we have $\sup(D) \leq \alpha$ and therefore $\{p_A\} \cup (D \times K)$ is a neighborhood of p_A which misses V_α. Thus we have finally proved that X is a Tychonoff zero-dimensional space.

It is easy to see that $\mathcal{U} = \{\{\alpha\} \times K : \alpha < \mathfrak{c}^+\}$ is a family of clopen subsets of X such that $p \in \overline{\bigcup \mathcal{U}}$ but $p \notin \overline{\bigcup \mathcal{U}'}$ for any $\mathcal{U}' \subset \mathcal{U}$ with $|\mathcal{U}'| \leq \mathfrak{c}$. This shows that X has uncountable tightness and hence it is not Fréchet–Urysohn.

To establish that X is pseudocompact take any discrete family $\{U_n : n \in \omega\}$ of non-empty open subsets of X. Since $\mathfrak{c}^+ \times K$ is open and dense in X, we can assume (choosing a non-empty clopen subset in each $U_n \cap (\mathfrak{c}^+ \times K)$ if necessary) that U_n is a clopen subset of $\mathfrak{c}^+ \times K$ for each $n \in \omega$. Since each $K_\alpha = \{\alpha\} \times K$ is compact, only finitely many U_n's can intersect each K_α. Thus, choosing smaller clopen sets and passing to an appropriate infinite subfamily, we can construct a discrete family $\{V_n : n \in \omega\}$ of non-empty clopen subsets of the space X such that, for all $m, n \in \omega$ we have $|i(V_n)| = 1$, $i(V_n) \cap i(V_m) = \emptyset$ if $n \neq m$ and the family $\{\pi_K(V_n) : n \in \omega\}$ is disjoint. By maximality of $\mathcal{D}$ there exists a set $D \in \mathcal{D}$ such that the set $D' = D \cap (\bigcup\{i(V_n) : n \in \omega\})$ is infinite. It is easy to see that the set $B = \bigcup\{V_n : i(V_n) \subset D'\}$ is admissible. By maximality of the family $\mathcal{A}(D)$ there exists $A \in \mathcal{A}(D)$ such that $A(\alpha) \cap B(\alpha) \neq \emptyset$ for infinitely many α's. As a consequence, p_A is an accumulation point of the family $\{V_n : n \in \omega\}$ which is a contradiction. Thus X is pseudocompact.

Finally we show that X is a Whyburn space. It is evident that all points of X except for p have countable local bases, so we must only check the Whyburn property at p. Let T be the set $\{p_A : A \in \mathcal{A}(D), D \in \mathcal{D}\}$. Suppose that $p \in \overline{C} \backslash C$ for some $C \subset X$. If $p \in \overline{C \cap T}$, then $B = C \cap T$ is an almost closed subset of C

with $p \in \overline{B}$. If not, then $p \in \overline{C \cap (\mathfrak{c}^+ \times K)}$ and therefore the cardinality of the set $C' = C \cap (\mathfrak{c}^+ \times K)$ is equal to $\mathfrak{c}^+$. Since there are only $\mathfrak{c}$ points in K, there exists $t \in K$ such that the set $B = \{\alpha < \mathfrak{c}^+ : (\alpha, t) \in C'\}$ has cardinality $\mathfrak{c}^+$. It is easy to check that $F = B \times \{t\}$ is an almost closed subset of C with $p \in \overline{F}$. We finally established that X is a Whyburn pseudocompact space of uncountable tightness so our solution is complete.

T.206. *Observe that a continuous image of a Whyburn space need not be Whyburn. Prove that any image of a Whyburn space under a closed map is a Whyburn space. Prove that the same is true for weakly Whyburn spaces.*

Solution. The space $X = (\omega_1 + 1)$ with the interval topology is not Whyburn because $t(X) > \omega$ while any compact Whyburn space is Fréchet–Urysohn (see Problem 204). However, X is a continuous image of a discrete space D of cardinality ω_1 which is Whyburn vacuously because there exists no $x \in D$ with $x \in \overline{A} \backslash A$ for some $A \subset D$. This shows that the Whyburn property is not preserved by continuous maps.

Now assume that X is a Whyburn space and take any closed continuous onto map $f : X \to Y$. If $A \subset Y$ and $y \in \overline{A} \backslash A$, then let $B = f^{-1}(A)$. Observe that $P = f(\overline{B})$ is a closed set with $A \subset P$ and hence $\overline{A} \subset P$. Pick any $x \in \overline{B}$ with $f(x) = y$; it is clear that $x \in \overline{B} \backslash B$, so there is an almost closed $G \subset B$ such that $x \in \overline{G}$. The set $F = f(G)$ is contained in A; besides, $F \cup \{y\} = f(G \cup \{x\})$ is a closed set which together with $y \in \overline{F}$ implies that F is almost closed and hence Y is Whyburn.

To prove that the weak Whyburn property is also preserved by closed maps, take any weakly Whyburn space X and a closed continuous onto map $f : X \to Y$. If $A \subset Y$ is not closed in Y, then $B = f^{-1}(A)$ is not closed in X and hence there is $x \in \overline{B} \backslash B$ and an almost closed $G \subset B$ such that $x \in \overline{B}$. It is immediate that $y = f(x) \in \overline{A} \backslash A$; the set $F = f(G)$ is contained in A and $y \in \overline{F}$. Furthermore $F \cup \{y\} = f(G \cup \{x\})$ is a closed set which implies that F is almost closed and hence Y is weakly Whyburn.

T.207. *Prove that every space with a unique non-isolated point is Whyburn. In particular, there exist Whyburn spaces of uncountable tightness.*

Solution. Let X be a space with a unique non-isolated point a. Given a set $A \subset X$ and $x \in \overline{A} \backslash A$, we must have $x = a$ and hence $F = A$ is an almost closed subset of X with $x \in \overline{F}$, i.e., X is Whyburn.

Now, if $X = L(\omega_1)$ is the Lindelöfication of a discrete space of cardinality ω_1, then X is a Whyburn space of uncountable tightness. A much more complicated example of a Whyburn space of uncountable tightness is constructed in T.205.

T.208. *Prove that any submaximal space is Whyburn.*

Solution. Let X be a submaximal space; assume that $x \in \overline{A} \backslash A$ for some $A \subset X$. Observe first that if a set $P \subset X$ has empty interior, then $X \backslash P$ is dense in X and hence open. Therefore every $P \subset X$ with $\text{Int}(P) = \emptyset$ is closed and discrete in X. In particular, the set $\overline{A} \backslash A$ is closed and discrete in X, so there is $U \in \tau(x, X)$ such that $U \cap (\overline{A}\backslash A) = \{x\}$. By regularity of X there is $V \in \tau(x, X)$ with $\overline{V} \subset U$; let $F = \overline{V} \cap A$. It is immediate that $x \in \overline{F}\backslash F$; assume that $y \neq x$ and $y \in \overline{F}$. Then $y \in \overline{V} \cap \overline{A} \subset U \cap \overline{A}$; however, the unique point of U which can belong to $\overline{A}\backslash A$ is x which is distinct from y. Therefore $y \in A$ and hence $y \in \overline{V} \cap A = F$, i.e., x is the unique point of $\overline{F}\backslash F$. Thus F is an almost closed subset of A with $x \in \overline{F}$ which proves that X is a Whyburn space.

T.209. *Prove that any radial space is weakly Whyburn.*

Solution. Let X be a radial space and take any non-closed $A \subset X$. Denote by κ the minimal cardinality of a set $B \subset A$ such that $\overline{B}$ is not contained in A. Fix a set $B \subset A$ such that $|B| = \kappa$ and $\overline{B}\backslash A \neq \emptyset$; pick any $x \in \overline{B}\backslash A$. The space X being radial, there is a regular cardinal λ and a λ-sequence $S = \{x_\alpha : \alpha < \lambda\} \subset B$ with $S \to x$. Since $x \in \overline{S}$ and $S \subset B$, the choice of B shows that $\kappa \leq |S| \leq \kappa$ and hence $\lambda = \kappa$; let $S_\alpha = \{x_\beta : \beta < \alpha\}$ for each $\alpha < \kappa$. The set $F = \bigcup\{\overline{S_\alpha} : \alpha < \kappa\}$ contains S and hence $x \in \overline{F}$. Observe also that $|S_\alpha| < \kappa$ and hence $\overline{S_\alpha} \subset A$ for each $\alpha < \kappa$. Consequently, $F \subset A$ and the last thing we must prove is that F is almost closed.

Given any $y \in X\backslash\{x\}$ take disjoint $U, V \in \tau(X)$ such that $x \in U$ and $y \in V$. Since $S \to x$, there is $\alpha < \kappa$ such that $S\backslash S_\alpha \subset U$ and therefore $y \notin \overline{S\backslash S_\alpha}$. Thus $y \in \overline{F}$ implies $y \in \overline{S_\alpha} \subset F$ which proves that x is the unique point of $\overline{F}\backslash F$, i.e., F is almost closed.

T.210. *Prove that any k-space, which is a Whyburn, is Fréchet–Urysohn. In particular, any sequential Whyburn space as well as any Čech-complete Whyburn space is Fréchet–Urysohn.*

Solution. Assume that X is a Whyburn k-space; take any $A \subset X$ and $x \in \overline{A}\backslash A$. There is an almost closed $F \subset A$ such that $x \in \overline{F}$. Since F is not closed, there is a compact $K \subset X$ such that $P = K \cap F$ is not closed in K. The set $F \cup \{x\}$ is closed and hence so is $K \cap (F \cup \{x\}) = (K \cap F) \cup \{x\} = P \cup \{x\}$. Therefore $\overline{P} = P \cup \{x\}$, i.e., the unique point of $\overline{P}\backslash P$ is x. Furthermore, $\overline{P}$ is a compact Whyburn space (it is compact because it is a closed subspace of K and it is an easy exercise that a closed subspace of a Whyburn space is Whyburn). Thus $\overline{P}$ is a Fréchet–Urysohn space by Problem 204 and hence there is a sequence $\{a_n : n \in \omega\} \subset P \subset A$ with $a_n \to x$. This proves that X is a Fréchet–Urysohn space.

Fact 1. Every Čech-complete space is a k-space.

Proof. Take any Čech-complete space Z and assume that $A \subset Z$ and $A \cap K$ is closed in K for any compact $K \subset Z$. To obtain a contradiction, suppose that A is not closed in Z and fix a point $z \in \overline{A}\backslash A$. There exists a family $\{O_n : n \in \omega\} \subset \tau(\beta Z)$ such that $O_{n+1} \subset O_n$ for all $n \in \omega$ and $\bigcap\{O_n : n \in \omega\} = Z$. Choose a sequence $\{U_n : n \in \omega\} \subset \tau(z, \beta Z)$ such that $U_0 = O_0$ and $\overline{U}_{n+1} \subset U_n \cap O_n$ for every $n \in \omega$ (the bar denotes the closure in the space βZ). It is easy to see that the set $P = \bigcap\{U_n : n \in \omega\} = \bigcap\{\overline{U}_n : n \in \omega\}$ is a compact subspace of Z. The set $K = P \cap A$ is closed in P and hence compact, so there is $V \in \tau(z, \beta Z)$ such that $\overline{V} \cap K = \emptyset$.

Our next step is to construct a family $\{V_n : n \in \omega\} \subset \tau(z, \beta Z)$ such that $V_0 = V$ and $\overline{V}_{n+1} \subset V_n \cap U_n$ for every $n \in \omega$. The set $Q = \bigcap\{V_n : n \in \omega\} = \bigcap\{\overline{V}_n : n \in \omega\}$ is again a compact subspace of Z such that $Q \cap A = \emptyset$ and it is an easy consequence of Fact 1 of S.326 that the family $\mathcal{V} = \{V_n : n \in \omega\}$ is an outer base of Q in βZ.

Pick a point $y_n \in V_n \cap A$ for each $n \in \omega$; the subspace $L = \{y_n : n \in \omega\} \cup Q$ is compact. Indeed, if $\mathcal{U} \subset \tau(\beta Z)$ is an open cover of L, then there is a finite $\mathcal{U}' \subset \mathcal{U}$ such that $Q \subset W = \bigcup \mathcal{U}'$. Since $\mathcal{V}$ is an outer base of Q, we have $V_n \subset W$ for some $n \in \omega$ and therefore $L \backslash W \subset \{y_0, \ldots, y_{n-1}\}$ is a finite set which can be covered by a finite $\mathcal{U}'' \subset \mathcal{U}$. It is immediate that $\mathcal{U}' \cup \mathcal{U}''$ is a finite subcover of L and hence L is compact. The set $Y = L \cap A = \{y_n : n \in \omega\}$ is not closed in L; for otherwise, we have $Y \cap Q = \emptyset$ and we can apply again the fact that $\mathcal{V}$ is an outer base of Q to find $n \in \omega$ such that $V_n \cap Y = \emptyset$ which is a contradiction with $y_n \in Y \cap V_n$. Thus L is a compact subspace of Z such that $L \cap A$ is not closed in L; this contradiction shows that A is closed in Z and hence Z is a k-space. Fact 1 is proved.   □

Returning to our solution, observe that if X is sequential and $A \subset X$ is not closed, then there is a sequence $S = \{a_n : n \in \omega\} \subset A$ such that $S \to x$ for some $x \notin A$. It is clear that $K = S \cup \{x\}$ is a compact subset of X such that $K \cap A$ is not closed in K. Consequently, every sequential space is a k-space. Finally, every Čech-complete space is a k-space by Fact 1 so our solution is complete.

T.211. *Prove that any compact weakly Whyburn space is pseudoradial but not necessarily sequential.*

Solution. To prove that every compact weakly Whyburn space is pseudoradial, we will have to establish existence of sufficiently many convergent transfinite sequences.

Fact 1. Let Z be a compact space. If z is a non-isolated point of Z, then there exists a regular cardinal κ and a κ-sequence $S = \{z_\alpha : \alpha < \kappa\} \subset Z\backslash\{z\}$ such that $S \to z$.

Proof. Let $\chi(z, Z) = \lambda$ and fix a base $\mathcal{B} = \{B_\alpha : \alpha < \lambda\} \subset \tau(z, Z)$ at the point z in the space Z. There is nothing to prove if $\lambda = \omega$, so we assume that λ is uncountable. For each $\alpha < \lambda$, there is a closed G_δ-set F_α such that $z \in F_\alpha \subset B_\alpha$ (see Fact 2 of S.328).

The ordinal $\kappa = \mathrm{cf}(\lambda)$ is regular and we can choose an increasing κ-sequence $\{\nu_\alpha : \alpha < \kappa\} \subset \lambda$ which is cofinal in λ. Let $H_\alpha = \bigcap\{F_\beta : \beta < \nu_\alpha\}$ for all $\alpha < \kappa$. Since each H_α is the intersection of $< \lambda$-many open sets, it is impossible that $\{z\} = H_\alpha$ because $\psi(z, Z) = \chi(z, Z) = \lambda$ (see Problem 327 of [TFS]). Choose a point $z_\alpha \in H_\alpha\backslash\{z\}$ for each $\alpha < \kappa$. We have $\{z\} = \bigcap\{H_\alpha : \alpha < \kappa\}$, so Fact 1 of S.326 can be applied to conclude that $\mathcal{H} = \{H_\alpha : \alpha < \kappa\}$ is a net at the point z, i.e., for any $U \in \tau(z, Z)$, there is $\beta < \kappa$ such that $H_\beta \subset U$; an immediate consequence is that $z_\alpha \in H_\alpha \subset H_\beta \subset U$ for any $\alpha \geq \beta$ which shows that the κ-sequence $\{z_\alpha : \alpha < \kappa\} \subset Z\backslash\{z\}$ converges to the point z. Fact 1 is proved.   □

Fact 2. The space $Z = (\omega_1 + 1)$ with its interval topology is radial.

Proof. Take any $A \subset Z$ and $x \in \overline{A} \setminus A$. If $x \neq \omega_1$, then $\chi(x, Z) = \omega$, so there is a sequence $S = \{x_n : n \in \omega\} \subset A$ with $S \to x$. Now, if $x = \omega_1$, then A is cofinal in ω_1, and hence we can choose an increasing ω_1-sequence $T = \{x_\alpha : \alpha < \omega_1\} \subset A$ which is cofinal in ω_1. It is an easy exercise that $T \to \omega_1$ so Fact 2 is proved. $\square$

Given a weakly Whyburn space X let A be a non-closed subset of X. There is an almost closed $F \subset A$ such that $x \in \overline{F}$ for some $x \in X \setminus A$. We can apply Fact 1 to the compact space $Z = F \cup \{x\}$ to find a regular cardinal κ and a κ-sequence $S = \{x_\alpha : \alpha < \kappa\} \subset F \subset A$ such that $S \to x$. This shows that X is pseudoradial.

Finally, observe that the compact space $(\omega_1 + 1)$ is radial by Fact 2 and hence weakly Whyburn by Problem 209. However $(\omega_1 + 1)$ does not even have countable tightness. Therefore not every weakly Whyburn compact space is sequential.

T.212. *Give an example of a Whyburn space which is not pseudoradial.*

Solution. Take any $\xi \in \beta\omega \setminus \omega$ and let $X = \omega \cup \{\xi\}$. Assume that S is a nontrivial convergent sequence in X. Then S must converge to ξ and we can choose infinite disjoint sets $A, B \subset S \cap \omega$. The point ξ is still the limit of both sequences A and B and hence $\xi \in \overline{A} \cap \overline{B}$ while $\overline{A} \cap \overline{B} = \emptyset$ by Fact 1 of S.382. This contradiction shows that X has no nontrivial convergent sequences.

The space X is Whyburn because it has a unique non-isolated point (see Problem 207). If X is pseudoradial, then $\xi \in \text{cl}_X(\omega)$ implies that there is a regular cardinal κ and a κ-sequence $S = \{x_\alpha : \alpha < \kappa\} \subset \omega$ with $S \to \xi$. The space X being countable, if $\kappa > \omega$, then there is $x \in X$ such that $|\{\alpha < \kappa : x_\alpha = x\}| = \kappa$ and hence the set $X \setminus \{x\} \in \tau(\xi, X)$ witnesses the fact that S does not converge to ξ. Therefore $\kappa = \omega$ and hence S is a nontrivial convergent sequence in X, i.e., we obtained a contradiction again. Thus X is a Whyburn space which is not pseudoradial.

T.213. *Prove that any scattered space is weakly Whyburn.*

Solution. If Z is a space and $A \subset Z$, call the set A *Whyburn closed* if $\overline{F} \subset A$ for any almost closed $F \subset A$. It is evident that a space Z is weakly Whyburn if and only if every Whyburn closed subspace of Z is closed.

Let X be a scattered space; by the remark above, it suffices to show that every Whyburn closed subset of X is closed. For any $Y \subset X$ denote by $i(Y)$ the set of isolated points of the subspace Y. To obtain a contradiction assume that some non-closed set $A \subset X$ is Whyburn closed in X. Let $A_0 = \emptyset$ and $B_0 = \overline{A}$; suppose that $\beta > 0$ and we have families $\{A_\alpha : \alpha < \beta\}$ and $\{B_\alpha : \alpha < \beta\}$ with the following properties:

(1) $A_\alpha \cap B_\alpha = \emptyset$ and $A_\alpha \cup B_\alpha = \overline{A}$ for any $\alpha < \beta$;
(2) $\gamma < \alpha < \beta$ implies $A_\gamma \subset A_\alpha$;
(3) if $\alpha + 1 < \beta$, then $A_{\alpha+1} = A_\alpha \cup i(B_\alpha)$;
(4) if $\alpha < \beta$ is a limit ordinal, then $A_\alpha = \bigcup\{A_\gamma : \gamma < \alpha\}$.

If β is a successor ordinal, then take the ordinal γ such that $\beta = \gamma + 1$ and let $A_\beta = A_\gamma \cup i(B_\gamma)$, $B_\beta = \overline{A} \setminus A_\beta$. If β is a limit ordinal, let $A_\beta = \bigcup\{A_\alpha : \alpha < \beta\}$

and $B_\beta = \bigcap\{B_\alpha : \alpha < \beta\}$. It is evident that in both cases the properties (1)–(3) hold for all $\alpha \leq \beta$. Therefore this inductive construction can be accomplished for all $\alpha < \kappa = |\overline{A}|^+$.

The space X being scattered, the property (3) implies that for each $\alpha < \kappa$, the inclusion $A_\alpha \subset A_{\alpha+1}$ is strict, i.e., $A_\alpha \neq A_{\alpha+1}$ if $B_\alpha \neq \emptyset$. If $B_\alpha \neq \emptyset$ for all $\alpha < \kappa$, then $\{A_{\alpha+1}\backslash A_\alpha : \alpha < \kappa\}$ is a disjoint family of cardinality κ of non-empty subsets of $\overline{A}$ which contradicts $|\overline{A}| < \kappa$. Therefore $B_\alpha = \emptyset$ for some $\alpha < \kappa$ and hence $A_\alpha = \overline{A}$ by (1). Consequently, the ordinal $\beta = \min\{\alpha < \kappa : A_\alpha \cap (\overline{A}\backslash A) \neq \emptyset\}$ is well-defined. It follows from $A_0 = \emptyset$ and (4) that $\beta = \gamma + 1$ for some $\gamma < \kappa$.

Pick a point $x \in A_\beta \backslash A$. Since $A_\gamma \subset A$, we have $x \in i(B_\gamma)\backslash A$ by (3). The point x being isolated in B_γ, there is an open $U \subset X$ such that $U \cap B_\gamma = \{x\}$. Take any $V \in \tau(X)$ with $x \in V \subset \overline{V} \subset U$ and consider the set $F = \overline{V} \cap A$. Since $x \in \overline{A}$, we have $x \in \overline{F}$. We claim that F is an almost closed set. Indeed, $\overline{F} \subset \overline{A} = A_\gamma \cup B_\gamma$ and $\overline{F} \cap B_\gamma \subset \overline{V} \cap B_\gamma = \{x\}$. Therefore $\overline{F}\backslash\{x\} \subset A_\gamma \subset A$ and therefore $\overline{F}\backslash\{x\} \subset \overline{V} \cap A = F$ which proves that F is almost closed; since $x \in \overline{F}\backslash A$, the set A is not Whyburn closed which is a contradiction. Thus every Whyburn closed $A \subset X$ is closed in X and hence X is weakly Whyburn.

T.214. *Observe that any sequential space is a k-space. Prove that any hereditarily k-space (and hence any hereditarily sequential space) is Fréchet–Urysohn.*

Solution. If X is sequential and $A \subset X$ is not closed, then there is a sequence $S = \{a_n : n \in \omega\} \subset A$ such that $S \to x$ for some $x \notin A$. It is clear that $K = S \cup \{x\}$ is a compact subset of X such that $K \cap A$ is not closed in K. Consequently, every sequential space is a k-space.

Now assume that X is a hereditarily k-space. Given any $A \subset X$ and $x \in \overline{A}\backslash A$, the space $B = A \cup \{x\}$ is a k-space and hence there is a compact $K \subset B$ such that $F = K \cap A$ is not closed in B. It is evident that $\overline{F} \subset K = (K \cap A) \cup \{x\} = F \cup \{x\}$; besides, $x \in \overline{F}$ because F is not closed. Consequently, F is an almost closed subset of A such that $x \in \overline{F}$ which proves that X is a Whyburn space. Any Whyburn k-space is Fréchet–Urysohn by Problem 210, so X is a Fréchet–Urysohn space.

T.215. *Prove that there exist hereditarily weakly Whyburn spaces which are not Whyburn.*

Solution. The compact space $K = (\omega_1 + 1)$ is scattered because it cannot be continuously mapped on $\mathbb{I}$ (see Fact 1 of S.319 and Problem 133). Therefore K is hereditarily weakly Whyburn (see Problem 213) because every subspace of a scattered space is scattered. On the other hand, the space K is not Whyburn because it is not Fréchet–Urysohn (see Problem 204).

T.216. *Prove that if X is $d(\omega)$-stable and X^n is a Hurewicz space for each natural n, then $C_p(X)$ is a Whyburn space. In particular, if X is a σ-compact space, then $C_p(X)$ is a Whyburn space.*

Solution. Given a space Z say that $vet(C_p(Z)) = \omega$ if it has the following property:

$(*)$ if $A_k \subset C_p(Z)$ for every $k \in \omega$ and $f \in \bigcap\{\overline{A_k} : k \in \omega\}$, then for each $k \in \omega$, we can choose a finite $B_k \subset A_k$ in such a way that $f \in \overline{\bigcup\{B_k : k \in \omega\}}$.

It was proved in Fact 2 of T.188 that $vet(C_p(Z)) = \omega$ is equivalent to the fact that Z^n is a Hurewicz space for all $n \in \mathbb{N}$.

Thus it follows from our hypothesis $vet(C_p(X)) = \omega$; any Hurewicz space is Lindelöf so $t(C_p(X)) = \omega$ by Problem 149 of [TFS]. Take any $A \subset C_p(X)$ and $f \in \overline{A}\setminus A$. There is a countable $B \subset A$ such that $f \in \overline{B}\setminus B$. Since X is $d(\omega)$-stable, the space $C_p(X)$ is $iw(\omega)$-monolithic by Problem 177 and hence the space $\overline{B}$ has countable pseudocharacter. Therefore we can find a family $\{O_n : n \in \omega\} \in \tau(f, C_p(X))$ such that $\overline{O}_{n+1} \subset O_n$ for each $n \in \omega$ and $(\bigcap\{O_n : n \in \omega\}) \cap \overline{B} = \{f\}$. If $C_n = O_n \cap B$, then $f \in \overline{C}_n$ for all $n \in \omega$. Since $vet(C_p(X)) = \omega$, we can find a finite $D_n \subset C_n$ such that $f \in \overline{D}$ where $D = \bigcup\{D_n : n \in \omega\} \subset A$. Observe that $D \subset B$ and hence all accumulation points of D belong to $\overline{B}$. Furthermore, the set $(C_p(X)\setminus\overline{O}_n) \cap D \subset D_0 \cup \cdots \cup D_{n-1}$ is finite for any $n \in \omega$; since for any $g \in \overline{B}\setminus\{f\}$ we have $g \in W = C_p(X)\setminus\overline{O}_n$ for some $n \in \omega$, the function g has a neighborhood W whose intersection with D is finite. This proves that f is the unique accumulation point of D and hence D is almost closed.

We have proved that for any $A \subset C_p(X)$ and any $f \in \overline{A}\setminus A$, there is an almost closed $D \subset A$ such that $f \in \overline{D}$. Therefore $C_p(X)$ is Whyburn. To finish our solution, observe that if X is σ-compact, then X^n is also σ-compact for all $n \in \mathbb{N}$. Since every σ-compact space is Hurewicz by Fact 1 of T.132, the space $C_p(X)$ is Whyburn if X is σ-compact.

T.217. *Prove that for a paracompact space X, if $C_p(X)$ is Whyburn, then X is a Hurewicz space. In particular, if X is metrizable and $C_p(X)$ is Whyburn, then X is separable.*

Solution. If we have a space Z and $A \subset Z$, then $\pi_A : C_p(Z) \to C_p(A)$ is the restriction map defined by $\pi_A(f) = f|A$ for every $f \in C_p(Z)$. A *retraction* in Z is a map $r : Z \to Z$ such that $r \circ r = r$. A set $R \subset Z$ is *a retract of Z* if there is a retraction $r : Z \to Z$ such that $r(Z) = R$.

Fact 1. Given a space Z and a non-empty discrete family $\mathcal{U} \subset \tau^*(Z)$, choose a point $z_U \in U$ and a function $f_U \in C_p(Z)$ such that $f_U(z_U) = 1$ and $f_U(Z\setminus U) = \{0\}$ for every $U \in \mathcal{U}$. For the set $A = \{z_U : U \in \mathcal{U}\} \subset Z$ define a mapping φ by $\varphi(f)(z) = \Sigma\{f(z_U) \cdot f_U(z) : U \in \mathcal{U}\}$ for every $f \in \mathbb{R}^A$ and $z \in Z$. Then

(1) φ maps $\mathbb{R}^A$ into $C_p(Z)$ and $\pi_A(\varphi(f)) = f$ for any $f \in \mathbb{R}^A$;
(2) φ is a linear homeomorphism between $\mathbb{R}^A$ and $L = \varphi(\mathbb{R}^A)$;
(3) L is a closed linear subspace of $C_p(Z)$ and the map $r = \varphi \circ \pi_A$ is a linear retraction of $C_p(Z)$ onto L;
(4) for the set $I_A = \{f \in C_p(Z) : f|A \equiv 0\}$ define $\xi : L \times I_A \to C_p(Z)$ by $\xi(f, g) = f + g$ for any $(f, g) \in L \times I_A$; then ξ is a linear homeomorphism.

In particular, $C_p(Z)$ is linearly homeomorphic to $\mathbb{R}^A \times I_A$ and $\mathbb{R}^A$ is linearly homeomorphic to a closed linear subspace of $C_p(Z)$.

Proof. For any $z \in Z$ there is a set $V \in \tau(z, Z)$ such that V intersects at most one element of $\mathcal{U}$, so we can find $U \in \mathcal{U}$ such that $V \cap W = \emptyset$ for any $W \in \mathcal{U} \backslash \{U\}$. Then $\varphi(f)(t) = f_U(t) \cdot f(z_U)$ for every $t \in V$ and hence $\varphi(f) | V = (f(z_U) \cdot f_U) | V$ is a continuous function. Thus $\varphi(f)$ is continuous by Fact 1 of S.472 and hence $\varphi : \mathbb{R}^A \to C_p(Z)$. Furthermore, $\varphi(f)(z_U) = f(z_U) \cdot f_U(z_U) = f(z_U)$ for every $f \in \mathbb{R}^A$ and $U \in \mathcal{U}$ which shows that $\varphi(f) | A = f$ for any $f \in \mathbb{R}^A$, i.e., (1) is proved.

It is easy to see that the map φ is linear; since $\pi_A(\varphi(f)) = f$ for any $f \in \mathbb{R}^A$ by (1), for any $f, g \in \mathbb{R}^A$ if $\varphi(f) = \varphi(g)$, then $f = \pi_A(\varphi(f)) = \pi_A(\varphi(g)) = g$ and therefore $\varphi : \mathbb{R}^A \to L$ is a bijection while $\pi_A | L$ is the inverse map for φ.

Since φ maps $\mathbb{R}^A$ into the product $\mathbb{R}^Z$, to prove continuity of φ, it suffices to show that $\pi_z \circ \varphi$ is continuous for any $z \in Z$. Here $\pi_z : \mathbb{R}^Z \to \mathbb{R}$ is the natural projection to the factor determined by z; recall that $\pi_z(h) = h(z)$ for every $h \in \mathbb{R}^Z$. If $z \notin \bigcup \mathcal{U}$, then $f_U(z) = 0$ for every $U \in \mathcal{U}$ and hence $\pi_z(\varphi(f)) = \varphi(f)(z) = 0$ for every $f \in \mathbb{R}^A$, so the map $\pi_z \circ \varphi$ is constant and hence continuous. If $z \in U \in \mathcal{U}$, then $\pi_z(\varphi(f)) = \varphi(f)(z) = f_U(z) \cdot f(z_U)$ and hence $\pi_z \circ \varphi : \mathbb{R}^A \to \mathbb{R}$ is continuous because it coincides with the function $f_U(z) \cdot q_U$ where $q_U : \mathbb{R}^A \to \mathbb{R}$ is the natural projection to the factor determined by z_U. This shows that the map φ is continuous; since the inverse of φ is a continuous map $\pi_A | L$ (see Problem 152 of [TFS]), the map φ is a linear homeomorphism and we settled (2).

It follows from the linearity of φ that L is a linear subspace of $C_p(Z)$ and r is also a linear map. For any $f \in C_p(Z)$, we have $\pi_A(\varphi(\pi_A(f))) = \pi_A(f)$ by (1) and hence $r(r(f)) = \varphi(\pi_A(r(f))) = \varphi(\pi_A(\varphi(\pi_A(f)))) = \varphi(\pi_A(f)) = r(f)$ for every $f \in C_p(Z)$, i.e., r is indeed a retraction. It is straightforward that $r(C_p(Z)) = L$, so L is closed being a retract of $C_p(Z)$ (see Fact 1 of S.351); this completes the proof of (3).

It is evident that ξ is a linear continuous map. For any function $h \in C_p(Z)$, let $\zeta(h) = (\varphi(\pi_A(h)), h - \varphi(\pi_A(h)))$; we omit a trivial verification of the fact that $\zeta : C_p(Z) \to L \times I_A$ is a linear continuous map and $\zeta \circ \xi$ is the identity map on $L \times I_A$. Thus the maps ξ and ζ are linear homeomorphisms which settles (4) and completes the proof of Fact 1.   □

Fact 2. Suppose that $C_p(Z)$ is a Whyburn space. Then any discrete $\mathcal{U} \subset \tau^*(Z)$ is countable.

Proof. If there is an uncountable discrete family $\mathcal{U} \subset \tau^*(Z)$, then we can assume, without loss of generality, that $|\mathcal{U}| = \omega_1$ and hence $\mathbb{R}^{\omega_1}$ embeds in $C_p(Z)$ by Fact 1. The Whyburn property being hereditary, we conclude that $\mathbb{R}^{\omega_1}$ is Whyburn and hence every subspace of $\mathbb{R}^{\omega_1}$ is also Whyburn. The space $Y = (\omega_1 + 1)$ has weight ω_1 and hence it embeds in $\mathbb{R}^{\omega_1}$. Since Y is compact and has uncountable tightness, it cannot be Whyburn by Problem 204. The obtained contradiction shows that $\mathbb{R}^{\omega_1}$ is not Whyburn and therefore no discrete family of non-empty open subsets of Z is uncountable. Fact 2 is proved.   □

Fact 3. If Z is paracompact and $C_p(Z)$ is a Whyburn space, then Z is Lindelöf.

Proof. Every open cover $\mathcal{U}$ of Z has a σ-discrete open refinement $\mathcal{V}$ by Problem 230 of [TFS]. Since discrete families of non-empty open subsets of Z are countable by Fact 2, the family $\mathcal{V}$ is countable and hence Z is Lindelöf. Fact 3 is proved. □

Fact 4. Suppose that Z is normal and $C_p(Z)$ is a Whyburn space. Assume that we have a sequence $\{\mathcal{U}_n : n \in \mathbb{N}\}$ of open covers of Z with the following properties:

(5) $\mathcal{U}_n = \{U_m^n : m \in \omega\}$ and $U_m^n \subset U_{m+1}^n$ for each $m \in \omega$ and $n \in \mathbb{N}$;
(6) for each $n \in \mathbb{N}$ there exists a closed cover $\mathcal{F}_n = \{F_m^n : m \in \omega\}$ of the space Z such that $F_m^n \subset U_m^n$ and $F_m^n \subset F_{m+1}^n$ for all $m \in \omega$.

Then we can choose $W_n \in \mathcal{U}_n$ for each $n \in \mathbb{N}$ in such a way that $\{W_n : n \in \mathbb{N}\}$ is an ω-cover of Z, i.e., for any finite $K \subset Z$, there is $n \in \mathbb{N}$ such that $K \subset W_n$.

Proof. For each pair $(m, n) \in \omega \times \mathbb{N}$, choose $f_m^n \in C_p(X)$ such that $f_m^n|F_m^n \equiv \frac{1}{n}$ and $f_m^n|(X \backslash U_m^n) \equiv 1$. It is clear that the sequence $S_n = \{f_m^n : m \in \omega\}$ converges to the function $h_n \equiv \frac{1}{n}$. As a consequence, the function $h \equiv 0$ is in the closure of the set $S = \bigcup\{S_n : n \in \mathbb{N}\}$. Apply the Whyburn property of the space $C_p(X)$ to find an almost closed $F \subset S$ such that $h \in \overline{F}$. Observe that for any $n \in \mathbb{N}$, the set $F_n = F \cap S_n$ cannot be infinite because otherwise $h_n \in \overline{F} \backslash F$. Therefore, for each $n \in \mathbb{N}$, we have a natural $m(n)$ such that $F_n \subset \{f_m^n : m \leq m(n)\}$. For each $n \in \mathbb{N}$ let $W_n = U_{m(n)}^n$. We claim that the family $\{W_n : n \in \mathbb{N}\}$ is an ω-cover of X.

Indeed, let K be a finite subset of X. Since $h \in \overline{F}$, there exists $f_m^n \in F$ such that $f_m^n(x) < 1$ for every $x \in K$ and therefore $K \cap (X \backslash U_m^n) = \emptyset$. Consequently, $K \subset U_m^n \subset U_{m(n)}^n = W_n$ and Fact 4 is proved. □

Returning to our solution apply Fact 3 to conclude that X is a Lindelöf space. Let $\{\mathcal{L}_n : n \in \mathbb{N}\}$ be a sequence of open covers of the space X. Since X is Lindelöf, without loss of generality, we may assume that each $\mathcal{L}_n$ is countable and choose an enumeration $\{W_m^n : m \in \omega\}$ of $\mathcal{L}_n$ for each $n \in \mathbb{N}$. Let $U_m^n = \bigcup\{W_i^n : i \leq m\}$ for all $m \in \omega$ and $n \in \mathbb{N}$. It is clear that the family $\mathcal{U}_n = \{U_m^n : m \in \omega\}$ is a cover of the space X for every $n \in \mathbb{N}$ and $U_m^n \subset U_{m+1}^n$ for all $m \in \omega$.

For each $n \in \mathbb{N}$ there exists a precise closed shrinking $\{G_m^n : m \in \omega\}$ of the cover $\mathcal{L}_n$, i.e., $\{G_m^n : m \in \omega\}$ is a closed cover of X and $G_m^n \subset W_m^n$ for all $m \in \omega$ and $n \in \mathbb{N}$ (see Fact 1 of S.219 and Fact 2 of S.226). Now if $F_m^n = \bigcup\{G_i^n : i \leq m\}$, then the covers $\mathcal{U}_n = \{U_m^n : m \in \omega\}$ and $\mathcal{F}_n = \{F_m^n : m \in \omega\}$, constructed for all $n \in \mathbb{N}$, satisfy the hypothesis of Fact 4. Therefore we can choose $W_n \in \mathcal{U}_n$ so that $\{W_n : n \in \mathbb{N}\}$ is a(n) (ω-)cover of X. Since each W_n is covered by finitely many elements of $\mathcal{L}_n$, there exist finite families $\mathcal{N}_n \subset \mathcal{L}_n$ such that $\bigcup\{\mathcal{N}_n : n \in \mathbb{N}\} = X$. Thus X is Hurewicz and our solution is complete.

T.218. *Given a space X such that $C_p C_p(X)$ is Whyburn prove that X has to be finite if it is either countably compact or has a countable network.*

Solution. If we have a space Z and $A \subset Z$ then $\pi_A : C_p(Z) \to C_p(A)$ is the restriction map defined by $\pi_A(f) = f|A$ for every $f \in C_p(Z)$.

Fact 1. If Z is any space and $K \subset Z$ is compact, then K is C-embedded in Z, i.e., $\pi_K(C_p(Z)) = C_p(K)$.

Proof. Let $f \in C_p(K)$; since $Z \subset \beta Z$, the set K is a closed subspace of βZ. The space βZ being normal there is $h \in C_p(\beta Z)$ such that $h|K = f$. It is clear that $g = h|Z$ is continuous on Z and $g|K = f$ so Fact 1 is proved. $\square$

Returning to our solution, assume first that $nw(X) = \omega$. Then $nw(C_p(X)) = \omega$ and hence $C_p(X)$ is Lindelöf; apply Problem 217 to conclude that $C_p(X)$ is Hurewicz and hence X is finite by Problem 132. Thus,

(∗) if $nw(X) = \omega$ and $C_p(C_p(X))$ is Whyburn, then X is finite.

To prove the same for a countably compact space X note that X embeds in $C_p(C_p(X))$ by Problem 167 of [TFS] which implies that X is Whyburn and hence Fréchet–Urysohn by Problem 204. Assume towards a contradiction that X is infinite. Then it has a nontrivial convergent sequence $S = \{x_n : n \in \omega\}$. If x is the limit of S, then $K = S \cup \{x\}$ is a countably infinite compact subspace of X. The map $\pi_K : C_p(X) \to C_p(K)$ is surjective by Fact 1, and hence the dual map $(\pi_K)^*$ embeds $C_p(C_p(K))$ in $C_p(C_p(X))$ (see Problem 163 of [TFS]) which proves that $C_p(C_p(K))$ is also Whyburn. The set K is countable, so $nw(K) = \omega$ and hence K is finite by (∗). This contradiction shows that X also has to be finite.

T.219. *Prove that there exists a separable metrizable space X such that $C_p(X)$ is not weakly Whyburn.*

Solution. In this solution *no axioms of separation are assumed by default for the topological spaces under consideration.* A space Z is called *crowded* if every set $U \in \tau^*(Z)$ is infinite. Observe that a T_1-space is crowded if and only if it has no isolated points.

Call a space Z *maximal* if it is crowded while any strictly stronger topology on Z fails to be crowded. The space Z is *maximal Tychonoff* if Z is Tychonoff and dense-in-itself while any stronger *Tychonoff* topology on Z has isolated points. The space Z is *nodec* if every nowhere dense subset of Z is closed. Call Z *irresolvable* if it is crowded and no two dense subsets of Z are disjoint. A crowded space Z is *resolvable* if it is not irresolvable, i.e., there are disjoint $A, B \subset Z$ such that $\overline{A} = \overline{B} = Z$. A crowded space Z is *hereditarily irresolvable* if every crowded subspace of Z is irresolvable. A space Z is *extremally disconnected* if the closure of any open subset of Z is open. If Z is crowded and the closures of any two disjoint crowded subspaces of Z are disjoint, then Z is called *ultradisconnected*. The space Z is *perfectly disconnected* if no point of z is an accumulation point of two disjoint subsets of Z, i.e., if $A, B \subset Z$ and $A \cap B = \emptyset$, then $(\overline{A} \setminus A) \cap (\overline{B} \setminus B) = \emptyset$.

Given a set Z and a family $\mathcal{A} \subset \exp(Z)$ with $\bigcup \mathcal{A} = Z$, we denote by $\langle \mathcal{A} \rangle$ the topology generated by $\mathcal{A}$ as a subbase.

Fact 0. Any maximal space is T_1.

Proof. If Z is a maximal space, then the family $\mathcal{B} = \{U \setminus F : U \in \tau(Z)$ and $F \subset Z$ is finite$\}$ generates a topology μ on Z as a base; it is clear that $\tau(Z) \subset \mu$.

Since all elements of $\mathcal{B}$ are infinite, the topology μ is crowded and hence $\mu \subset \tau(Z)$ by maximality of Z. Thus $\mu = \tau(Z)$ and hence all finite sets are closed in Z, i.e., Z is a T_1-space. Fact 0 is proved. $\square$

Fact 1. Let Z be any space; if $\mathcal{A}$ is a family of resolvable subspaces of Z, then $\bigcup \mathcal{A}$ is also resolvable.

Proof. In the space $Y = \bigcup \mathcal{A}$ take a maximal disjoint family $\mathcal{D}$ of non-empty resolvable subspaces of Y. We claim that the set $\bigcup \mathcal{D}$ is dense in Y. Indeed, if $W = Y \backslash \text{cl}_Y(\bigcup \mathcal{D}) \neq \emptyset$, then there is $A \in \mathcal{A}$ with $W \cap A \neq \emptyset$. It is evident that any non-empty open subspace of a resolvable space is resolvable, so $A' = W \cap A$ is resolvable and hence $\mathcal{D} \subset \mathcal{D}' = \mathcal{D} \cup \{A'\} \neq \mathcal{D}$ while $\mathcal{D}'$ is still disjoint and consists of resolvable subspaces of Y; this gives a contradiction with the maximality of $\mathcal{D}$. Thus $T = \bigcup \mathcal{D}$ is dense in Y. For each $D \in \mathcal{D}$ choose disjoint $E_D, F_D \subset D$ such that $\text{cl}_D(E_D) = \text{cl}_D(F_D) = D$. It is straightforward that $E = \bigcup\{E_D : D \in \mathcal{D}\}$ and $F = \bigcup\{F_D : D \in \mathcal{D}\}$ are disjoint dense subsets of Y. Thus $Y = \bigcup \mathcal{A}$ is resolvable so Fact 1 is proved. $\square$

Fact 2. If Z is an irresolvable space, then there is a non-empty open hereditarily irresolvable $U \subset Z$.

Proof. Let $\mathcal{A}$ be the family of all non-empty resolvable subspaces of Z. The space $Y = \bigcup \mathcal{A}$ is resolvable by Fact 1. If a space has a dense resolvable subspace, then it is resolvable, so Y cannot be dense in Z. It is evident that $U = Z \backslash \overline{Y}$ is as promised so Fact 2 is proved. $\square$

Fact 3. A space Z is ultradisconnected if and only if for any crowded $A \subset Z$, if $Z \backslash A$ is also crowded, then A is clopen. In particular, every ultradisconnected T_1-space is irresolvable.

Proof. Necessity follows from the fact that $\overline{A} \cap \overline{Z \backslash A} = \emptyset$ implies that both A and $Z \backslash A$ have to be clopen in Z. To prove sufficiency assume that our hypothesis holds and take any disjoint crowded $A, B \subset Z$. If $A_1 = \overline{A} \backslash B$ and $B_1 = \overline{B} \backslash A_1$, then $A \subset A_1 \subset \overline{A}$ and $B \subset B_1 \subset \overline{B}$ which implies that both A_1 and B_1 are crowded. The spaces $C = \overline{A} \cup \overline{B}$ and $Z \backslash C$ are also crowded which shows that C is clopen in Z. Furthermore, $A_1 \cap B_1 = \emptyset$ and $A_1 \cup B_1 = \overline{A} \cup \overline{B}$ which implies that $Z \backslash A_1 = B_1 \cup (Z \backslash C)$ and therefore $Z \backslash A_1$ is also crowded being a union of two crowded subspaces. Thus A_1 is clopen and hence $Z \backslash A_1$ is clopen as well; therefore $\overline{A} = A_1$ and $\overline{B} \subset Z \backslash A_1$ so $\overline{A} \cap \overline{B} = \emptyset$.

To see that any ultradisconnected T_1-space Z is irresolvable, observe that given disjoint dense $A, B \subset Z$, the subspaces A and B have to be crowded because, in any T_1-space, a dense subspace of a crowded space is crowded. Since Z is ultradisconnected, we have $\overline{A} \cap \overline{B} = \emptyset$ which is a contradiction with $\overline{A} = \overline{B} = Z$. Thus Z is irresolvable and Fact 3 is proved. $\square$

Fact 4. Given a crowded space Z suppose that A and $Z \backslash A$ are crowded subspaces of Z for some $A \subset Z$. Then the topology $\mu = \langle \tau(Z) \cup \{A, Z \backslash A\} \rangle$ is also crowded and (Z, μ) is homeomorphic to the space $A \oplus (Z \backslash A)$.

Proof. Let $B = Z \backslash A$; if $\mathcal{C}_A = \{U \cap A : U \in \tau(Z)\}$ and $\mathcal{C}_B = \{U \cap B : U \in \tau(Z)\}$, then $\mathcal{C} = \mathcal{C}_A \cup \mathcal{C}_B$ is a base of μ; an immediate consequence is that the topology of subspace of (Z, μ) on each of the sets A and $Z \backslash A$ coincides with the topology on the respective set induced from Z. Since A and $Z \backslash A$ are open in (Z, μ), the space (Z, μ) is homeomorphic to $A \oplus (Z \backslash A)$.

Finally, observe that every non-empty element of $\mathcal{C}_A$ ($\mathcal{C}_B$) is infinite being an open subset of a crowded space A (or B respectively). Thus no element of $\mathcal{C}$ is finite which implies that μ is crowded and Fact 4 is proved.   □

Fact 5. The following conditions are equivalent for any crowded T_1-space Z:

(1) Z is maximal;
(2) Z is perfectly disconnected;
(3) every crowded subspace of Z is open;
(4) Z is ultradisconnected and nodec;
(5) Z is ultradisconnected and every discrete subspace of Z is closed;
(6) Z is submaximal and extremally disconnected;
(7) Z is extremally disconnected, hereditarily irresolvable and nodec.

Proof. Assume that (7) holds and consider any dense $A \subset Z$. If $\overline{Z \backslash A}$ has non-empty interior, then take any $V \in \tau^*(Z)$ with $V \subset \overline{Z \backslash A}$ and observe that the sets $(Z \backslash A) \cap V$ and $A \cap V$ are disjoint dense subsets of V which contradicts the fact that Z is hereditarily irresolvable. Thus $\text{Int}(\overline{Z \backslash A}) = \emptyset$; consequently, $Z \backslash A$ is nowhere dense and hence closed in Z because Z is nodec. This proves that A is open in Z, so Z is submaximal and we have established that $(7) \Longrightarrow (6)$.

Assume that Z is submaximal and extremally disconnected. If Y is a subspace of Z with $\text{Int}(Y) = \emptyset$, then $X \backslash Y$ is open being dense in Z. Consequently, every subspace of Z with empty interior is closed and discrete. In particular, if D is a discrete subspace of Z, then it has empty interior and hence D is closed. Now assume that A and $Z \backslash A$ are crowded subspaces of Z. Let $U = \text{Int}(A)$; suppose that there exists a point $z \in A \backslash U$. The set $E = A \backslash U$ has empty interior and hence E is closed and discrete. There is $W \in \tau(z, Z)$ such that $W \cap E = \{z\}$; it is clear that $V = W \backslash \overline{U} \in \tau(z, Z)$ and $V \cap A = \{z\}$ which contradicts the fact that A is crowded. Therefore $A \subset \overline{U}$; analogously, if $G = \text{Int}(Z \backslash A)$, then $Z \backslash A \subset \overline{G}$. The space Z being extremally disconnected, the set $\overline{U}$ is open and $\overline{U} \cap G = \emptyset$. Therefore $\overline{G} \cap \overline{U} = \emptyset$ which implies that $\overline{A} \cap \overline{Z \backslash A} = \emptyset$ and hence A and $Z \backslash A$ are clopen in Z which shows that Z is ultradisconnected by Fact 3. Thus $(6) \Longrightarrow (5)$.

Now assume that (5) holds for Z and take a nowhere dense $A \subset Z$. The set $B = Z \backslash A$ is dense in Z and hence crowded because Z is a T_1-space. As a consequence, every $B' \supset B$ is also crowded, so if the set I of isolated points of A is not dense in A, then $C = A \backslash \overline{I}$ is a non-empty crowded subspace of A, and by the above remark, the set $B' = Z \backslash C$ is also crowded which implies that C is clopen because Z is ultradisconnected. However, C is nowhere dense because $C \subset A$; this contradiction shows that the discrete subspace I is dense in A and hence $A = I$ because (5) holds for Z and hence I has to be closed. Thus every nowhere dense subspace of Z is closed, i.e., Z is nodec so we settled $(5) \Longrightarrow (4)$.

If Z is ultradisconnected and nodec, then take any crowded $A \subset Z$ and let E to be the set of all isolated points of $Z \setminus A$. The set E is nowhere dense and hence closed in Z because Z is nodec. Furthermore, the set $F = (Z \setminus A) \setminus E$ is crowded and hence $\overline{F} \cap A = \emptyset$ because Z is ultradisconnected. Consequently, $\overline{Z \setminus A} = \overline{F} \cup \overline{E} = \overline{F} \cup E$ is disjoint from A which shows that $\overline{Z \setminus A} = Z \setminus A$, i.e., A is open in Z and hence $(4) \Longrightarrow (3)$.

Now suppose that every crowded subspace of Z is open and take any disjoint sets $A, B \subset Z$ such that $z \in (\overline{A} \setminus A) \cap (\overline{B} \setminus B)$. If $U = \text{Int}(A)$, then $P = A \setminus U$ has empty interior and hence $Z \setminus P$ is crowded being dense in Z (we applied once more the fact that Z is T_1). By our hypothesis, the set $Z \setminus P$ is open and therefore P is closed in Z. Since $z \in \overline{A} = \overline{P} \cup \overline{U} = P \cup \overline{U}$, we have $z \in \overline{U}$ because $z \notin A$ while $P \subset A$. Thus the crowded space U is dense in $U \cup \{z\}$ whence $V = U \cup \{z\}$ is crowded and hence open. It turns out that V is a neighborhood of z which is disjoint from B. Therefore $z \notin \overline{B}$; this contradiction proves that $(3) \Longrightarrow (2)$.

Assume that Z is perfectly disconnected and take any crowded $A \subset Z$. If $z \in A$, then $z \in \overline{A \setminus \{z\}}$ because A is crowded. Now perfect disconnectedness of Z implies that $z \notin \overline{Z \setminus A}$; the point $z \in A$ was taken arbitrarily, so $\overline{Z \setminus A} \cap A = \emptyset$ and therefore $\overline{Z \setminus A} = Z \setminus A$, i.e., A is open in Z. This proves that $(2) \Longrightarrow (3)$ and hence $(2) \Longleftrightarrow (3)$.

Now, if (3) holds for Z and μ is a crowded topology on Z with $\tau(Z) \subset \mu$, assume that there is $A \in \mu \setminus \tau(Z)$. Then A is crowded in (Z, μ) and hence in Z because for every $U \in \tau(Z)$ we have $U \cap A \in \mu$ and hence $U \cap A$ is infinite whenever it is non-empty. Apply (3) to conclude that $A \in \tau(Z)$ which is a contradiction. Thus $\mu = \tau(Z)$ and hence Z is maximal, i.e., $(3) \Longrightarrow (1)$.

Finally, assume that Z is maximal. If $A \subset Z$ and both sets A and $Z \setminus A$ are crowded then the topology $\mu = \langle \tau(Z) \cup \{A, Z \setminus A\}\rangle$ is crowded by Fact 4 and hence $\mu = \tau(Z)$ which shows that $A \in \tau(Z)$ and $Z \setminus A \in \tau(Z)$, i.e., A is clopen in Z. Therefore Z is ultradisconnected by Fact 3. Given any open $U \subset Z$ both sets $\overline{U}$ and $Z \setminus \overline{U}$ are crowded, so $\overline{U}$ is also open which shows that Z is extremally disconnected.

Now if A is a crowded subspace of Z and $B, C \subset A$ are disjoint and dense in A, then they are both crowded because Z is a T_1-space. However, we have $A \subset \overline{B} \cap \overline{C} = \emptyset$ because Z is ultradisconnected; this contradiction shows that Z is hereditarily irresolvable. Finally, if A is nowhere dense in Z, then $D = Z \setminus A$ is dense in Z. Consider the topology $\delta = \langle \tau(Z) \cup \{D\}\rangle$. It is clear that $\tau(Z) \subset \delta$ and the base in (Z, δ) is given by the family $\mathcal{E} = \tau(Z) \cup \{U \cap D : U \in \tau(Z)\}$. The T_1-property of Z implies that $U \cap D$ is infinite for every $U \in \tau^*(Z)$ and hence all elements of $\mathcal{E}$ are infinite which shows that δ is a crowded topology on Z. By maximality of Z we have $\delta = \tau(Z)$ and therefore $D \in \tau(Z)$, i.e., A is closed in Z. Thus every nowhere dense subset of Z is closed in Z, i.e., Z is nodec and we established that $(1) \Longrightarrow (7)$ so Fact 5 is proved. $\qquad \square$

Fact 6. For any Tychonoff crowded space Z, there exists a maximal Tychonoff topology μ on the set Z such that $\tau(Z) \subset \mu$.

Proof. Let $\mathcal{T}$ be the family of all Tychonoff crowded topologies τ on Z such that $\tau(Z) \subset \tau$. If $\mathcal{C}$ is a chain in $\mathcal{T}$, then $\delta = \langle \bigcup \mathcal{C} \rangle$ contains all topologies from $\mathcal{C}$ and δ is Tychonoff by Problem 099 of [TFS]. The base of δ is given by all (non-empty) intersections $U = U_1 \cap \cdots \cap U_n$ where $U_i \in \tau_i$ and $\tau_i \in \mathcal{C}$ for all $i \leq n$. The collection $\mathcal{C}$ being a chain, there is $\tau \in \mathcal{C}$ such that $\tau_i \subset \tau$ and hence $U_i \in \tau$ for all $i \leq n$. Therefore $U \in \tau$ and hence U is infinite. Consequently, the topology δ is crowded because it has a base all elements of which are infinite. Thus Zorn's lemma is applicable to $\mathcal{T}$ to conclude that there is a maximal element $\mu \in \mathcal{T}$. It is immediate that μ is as promised so Fact 6 is proved. □

Fact 7. A space Z is maximal Tychonoff if and only if Z is Tychonoff and ultradisconnected.

Proof. Suppose that Z is a maximal Tychonoff space and take any $A \subset Z$ such that both A and $Z \backslash A$ are crowded. The topology $\mu = \langle \tau(Z) \cup \{A, Z \backslash A\} \rangle$ is stronger than $\tau(Z)$ and the space (Z, μ) is homeomorphic to $A \oplus (Z \backslash A)$ (see Fact 4) which shows that (Z, μ) is Tychonoff and crowded. By maximality of $\tau(Z)$ in the family of all Tychonoff crowded topologies on Z, we have $A \in \tau(Z)$ and $Z \backslash A \in \tau(Z)$, i.e., A is clopen in Z, so Z is ultradisconnected by Fact 3 which proves necessity.

To prove sufficiency, assume that Z is an ultradisconnected Tychonoff space and take any crowded Tychonoff topology μ on the set Z such that $\tau(Z) \subset \mu$. By regularity of (Z, μ) the topology μ has a base $\mathcal{B}$ such that $U = \text{Int}(\overline{U})$ for every $U \in \mathcal{B}$ (the bar denotes the closure in the space (Z, μ) and the interior is also taken in (Z, μ)).

Given any $U \in \mathcal{B}$, the set $X \backslash \overline{U}$ is dense in $X \backslash U$; for otherwise, there is a non-empty $W \in \mu$ with $W \subset \overline{U}$ such that W is not contained in U which contradicts the fact that $U = \text{Int}(\overline{U})$ is the largest set from μ contained in $\overline{U}$. Thus $Z \backslash U$ contains a dense crowded subspace $Z \backslash \overline{U}$ and hence $Z \backslash U$ is crowded in (Z, μ) for any $U \in \mathcal{B}$. It turns out that both sets U and $Z \backslash U$ are crowded in (Z, μ) and hence in Z for any $U \in \mathcal{B}$. Since Z is ultradisconnected, we have $U \in \tau(Z)$ for any $U \in \mathcal{B}$ and therefore $\mu = \tau(Z)$. This shows that Z is maximal Tychonoff and finishes the proof of Fact 7. □

Fact 8. Suppose that Z is a countable Tychonoff hereditarily irresolvable space. Then the set $T = \{z \in Z : z \in \overline{D} \backslash D$ for some discrete subspace $D \subset Z\}$ is nowhere dense in Z.

Proof. Observe that

$(*)$ if $\text{Int}(T) = \emptyset$, then T is nowhere dense.

Indeed if $\text{Int}(T) = \emptyset$, then $Z \backslash T$ is dense in Z, so if $T \cap U$ is dense in some subspace $U \in \tau^*(Z)$, then $T \cap U$ and $(Z \backslash T) \cap U$ are disjoint dense subsets of U which contradicts the fact that Z is hereditarily irresolvable.

Therefore it suffices to prove that T has empty interior; to obtain a contradiction, assume that $U \subset T$ for some $U \in \tau^*(Z)$. Then we can choose for any $z \in U$ a discrete $E(z) \subset U$ such that $z \in \overline{E(z)} \backslash E(z)$. Take any $z_0 \in U$ and let $D_0 = \{z_0\}$. Suppose that $n \geq 0$ and we have sets $D_0, \dots, D_n$ with the following properties:

(i) $D_i \subset U$ and D_i is discrete for any $i \le n$;

(ii) $\overline{D}_i \cap D_j = \emptyset$ if $i < j \le n$;

(iii) $D_i \subset \overline{D}_j$ for any $i < j \le n$.

To construct D_{n+1} observe that we can choose $U_d \in \tau(d, Z)$ for each $d \in D_n$ such that the family $\{U_d : d \in D_n\}$ is disjoint (see Fact 1 of S.369) and $U_d \cap \overline{D}_i = \emptyset$ for every $i < n$. Observe that $U_d \cap (\overline{D}_n \backslash D_n) = \emptyset$ for every $d \in D_n$ and let $A_d = E(d) \cap U_d$ for every $d \in D_n$. It is evident that A_d is discrete and $d \in \overline{A}_d \backslash A_d$. Let $D_{n+1} = \bigcup \{A_d : d \in D_n\}$. The conditions (ii) and (iii) are obviously satisfied. As to (i), we must only prove that D_{n+1} is discrete. Pick any $z \in D_{n+1}$; then $z \in A_d$ for some $d \in D_n$. Since A_d is discrete, there is $W \in \tau(z, Z)$ such that $W \cap A_d = \{z\}$; it is immediate that $(W \cap U_d) \cap D_{n+1} = \{z\}$, so D_{n+1} is also discrete and hence our inductive construction can be continued to provide a sequence $\{D_n : n \in \omega\}$ such that (i)–(iii) hold for all $n \in \omega$.

The set $D = \bigcup \{D_n : n \in \omega\}$ is dense-in-itself because given any $d \in D$, we have $d \in D_n$ for some $n \in \omega$ and hence $d \in \overline{D}_{n+1} \backslash D_{n+1} \subset \overline{D \backslash \{d\}}$ by (iii). Let $G = \bigcup \{D_{2n} : n \in \omega\}$ and $H = \bigcup \{D_{2n+1} : n \in \omega\}$; again, if $d \in D$, then $d \in D_n$ for some $n \in \omega$ and hence $d \in \overline{D}_{2n} \cap \overline{D}_{2n+1}$ by (iii). Therefore G and H are disjoint dense subsets of D which contradicts the fact that Z is hereditarily irresolvable. This contradiction proves that $\mathrm{Int}(T) = \emptyset$ and hence T is nowhere dense by $(*)$. Fact 8 is proved.   □

Fact 9. There exists a countable Tychonoff maximal space.

Proof. By Fact 6 there exists a maximal Tychonoff topology μ on the set $\mathbb{Q}$ such that $\tau(\mathbb{Q}) \subset \mu$. The space $Q = (\mathbb{Q}, \mu)$ is ultradisconnected by Fact 7 and hence irresolvable by Fact 3. Apply Fact 2 to find a hereditarily irresolvable open $R \subset Q$. The space $T = \{z \in R : z \in \overline{D} \backslash D$ for some discrete $D \subset R\}$ is nowhere dense in R by Fact 8, so there is $S \in \tau^*(R)$ such that $S \cap T = \emptyset$. It is obvious that no point of S is an accumulation point of a discrete subspace of S, so every discrete subspace of S is closed. The space Q being ultradisconnected, S is also ultradisconnected and hence maximal by Fact 5. Fact 9 is proved.   □

Fact 10. If Z is a Tychonoff submaximal countable space, then $Z \times A(\omega)$ is not weakly Whyburn.

Proof. Let $s = 0$ and $s_n = \frac{1}{n+1}$ for all $n \in \omega$; then the space $S = \{s_n : n \in \omega\} \cup \{s\}$, with the topology induced from $\mathbb{R}$, is homeomorphic to $A(\omega)$, so it suffices to show that $Z \times S$ is not weakly Whyburn. Let $S_n = \{s_0, \dots, s_n\}$ for any $n \in \omega$ and take a bijection $\varphi : Z \to S \backslash \{s\}$; then its graph $Y = \{(z, \varphi(z)) : z \in Z\} \subset Z \times S$ is not closed in $Z \times S$. Indeed, take any point $t = (z, s) \in Z \times \{s\}$. If $U \in \tau(t, Z \times S)$, then there is $n \in \omega$ and $U_1 \in \tau(z, Z)$ such that $U_1 \times (S \backslash S_n) \subset U$. The set $P = \varphi^{-1}(S_n)$ is finite and U_1 is infinite, so there is $z \in U_1 \backslash P$. It is obvious that $(z, \varphi(z)) \in U \cap Y$ which shows that $U \cap Y \ne \emptyset$ for any $U \in \tau(t, Z \times S)$ and hence $t \in \overline{Y} \backslash Y$.

If $Z \times S$ is weakly Whyburn, then there is an almost closed $F \subset Y$ such that $u \in \overline{F}$ for some $u \in (Z \times S) \backslash Y$. If $u = (z, s_n)$ for some $n \in \omega$ and $z \in Z$, then $W = Z \times \{s_n\}$ is an open neighborhood of u with $|W \cap Y| \le 1$, so u cannot be an accumulation point for F. Therefore $u = (z, s)$ for some $z \in Z$. It is clear that $u \in \overline{F}_1$ where $F_1 = F \backslash \{(z, \varphi(z))\}$; let $\pi : Z \times S \to Z$ be the natural projection. Then, for the set $H = \pi(F_1)$ we have $z \in \overline{H} \backslash H$. If $\text{Int}(H) = \emptyset$, then $Z \backslash H$ is dense in Z; since Z is submaximal, the set $Z \backslash H$ is open and hence H is closed which contradicts $z \in \overline{H} \backslash H$.

Therefore $O = \text{Int}(H) \neq \emptyset$ and hence we can choose $y \in O \backslash \{z\}$. Given any open neighborhood V of the point $u' = (y, s)$, there is $m \in \omega$ such that $V_1 \times (S \backslash S_m) \subset V$ for some $V_1 \in \tau(y, Z)$ with $V_1 \subset O$. The set $B = F_1 \cap (Z \times S_m)$ is finite, so there is $w = (a, s_k) \in F_1 \backslash B$ such that $a \in O$. It is immediate that $w \in V \cap F_1$ and therefore $V \cap F_1 \neq \emptyset$ for each $V \in \tau(u', Z \times S)$. As a consequence, $u' \in \overline{F}_1 \subset \overline{F}$, which is a contradiction because $u' \in (Z \times S) \backslash Y$ and $u' \neq u$. Thus $Z \times S$ is not weakly Whyburn and Fact 10 is proved.   □

Returning to our solution, observe that it follows from Facts 9 and 10 that there exists a countable space T which is not weakly Whyburn. The space $X = C_p(T)$ is separable and metrizable while T embeds in $C_p(X)$ as a closed subspace (see Problem 167 of [TFS]). It is an easy exercise that a closed subspace of a weakly Whyburn space is weakly Whyburn; therefore $C_p(X)$ is not weakly Whyburn and our solution is complete.

T.220. *Prove that there exists a compact weakly Whyburn space which is not hereditarily weakly Whyburn.*

Solution. If α and β are ordinals such that $\alpha \leq \beta$, then $[\alpha, \beta] = \{\gamma : \alpha \leq \gamma \leq \beta\}$ and $(\alpha, \beta) = \{\gamma : \alpha < \gamma < \beta\}$. Analogously, $[\alpha, \beta) = \{\gamma : \alpha \leq \gamma < \beta\}$ and $(\alpha, \beta] = \{\gamma : \alpha < \gamma \leq \beta\}$. Given a space Z and a set $A \subset Z$, say that A is *Whyburn closed* if $\overline{F} \subset A$ for any almost closed $F \subset A$. It is evident that a space Z is weakly Whyburn if and only if any Whyburn closed subset of Z is closed in Z. Since any nontrivial convergent sequence is almost closed, any Whyburn closed $A \subset Z$ is *sequentially closed* in Z, i.e., if $S = \{a_n : n \in \omega\} \subset A$ is a sequence with $S \to x$, then $x \in A$. It is also immediate that a space Z is sequential if and only if any sequentially closed subspace of Z is closed in Z.

Fact 1. If M is an uncountable second countable space, then there is an uncountable $M' \subset M$ such that every $U \in \tau^*(M')$ is uncountable.

Proof. Let $A = \{x \in M : \text{there is a countable } U_x \in \tau(x, M)\}$. Since A is Lindelöf, the open cover $\mathcal{U} = \{U_x : x \in A\}$ of the set A has a countable subcover $\mathcal{U}' \subset \mathcal{U}$. Since each element of $\mathcal{U}'$ is countable, we have $|A| \leq |\bigcup \mathcal{U}'| = \omega$ and therefore the set $M' = M \backslash A$ is uncountable.

Given any non-empty open subset U of the space M', find a set $V \in \tau(M)$ with $V \cap M' = U$. Then V is uncountable because there is a point $x \in V \cap M'$, and, by definition of M', every open neighborhood of x is uncountable. Thus $U = V \backslash A$ is also uncountable so Fact 1 is proved.   □

Fact 2. For any second countable uncountable space M the space $Z = L(\omega_1) \times M$ is not weakly Whyburn.

Proof. Recall that $L(\omega_1) = \omega_1 \cup \{a\}$ where a is the unique non-isolated point of $L(\omega_1)$ and the neighborhoods of a are the complements of countable subsets of ω_1. We will need the set $I = \{a\} \times M$; for any $t \in M$ let $x_t = (a, t) \in I$. Choose any injection $\varphi : \omega_1 \to M$; then its graph $A = \{(\alpha, \varphi(\alpha)) : \alpha \in \omega_1\}$ is not closed in Z. Indeed, the set $P = \varphi(\omega_1) \subset M$ is uncountable, so there is a point $t \in P$ such that $U \cap P$ is uncountable for any $U \in \tau(t, M)$ (see Fact 1). Given any $W \in \tau(x_t, Z)$ there is a countable $C \subset \omega_1$ and $U \in \tau(t, M)$ such that $(L(\omega_1) \backslash C) \times U \subset W$. Since $U \cap P$ is uncountable, there is $s \in (U \cap P) \backslash \varphi(C)$. If $\alpha = \varphi^{-1}(s)$, then $\alpha \notin C$ and therefore $(\alpha, s) \in ((L(\omega_1) \backslash C) \times U) \cap A \subset W \cap A$ which shows that $W \cap A \neq \emptyset$ for any $W \in \tau(x_t, Z)$ and hence $x_t \in \overline{A} \backslash A$.

Now assume that $F \subset A$ is almost closed non-closed subset of Z; if $x = (\alpha, t)$ for some $\alpha < \omega_1$ and $t \in M$, then $G = \{\alpha\} \times M$ is an open neighborhood of the point x such that $|G \cap F| \leq 1$, so $x \notin \overline{F} \backslash F$. Therefore F is uncountable and $\overline{F} \backslash F \subset I$. The set $H = \varphi(F) \subset M$ being uncountable, we can apply Fact 1 again to find two distinct points $t, s \in M$ such that $U \cap H$ is uncountable for any $U \in \tau(M)$ with $U \cap \{t, s\} \neq \emptyset$. Given any $W \in \tau(x_t, Z)$, there is a countable $C \subset \omega_1$ and $U \in \tau(t, M)$ such that $(L(\omega_1) \backslash C) \times U \subset W$. Since $U \cap \varphi(F)$ is uncountable, there is $t' \in (U \cap \varphi(F)) \backslash \varphi(C)$. If $\alpha = \varphi^{-1}(t')$, then $\alpha \notin C$ and hence $(\alpha, t') \in (L(\omega_1) \backslash C) \times U) \cap A \subset W \cap A$ which shows that $W \cap A \neq \emptyset$ for any $W \in \tau(x_t, Z)$ and hence $x_t \in \overline{F} \backslash F$. Analogously, $x_s \in \overline{F} \backslash F$ which contradicts the fact that F is an almost closed subset of Z. Thus Z is not weakly Whyburn and Fact 2 is proved. $\qquad\square$

Returning to our solution note that the space $X = (\omega_1 + 1) \times \mathbb{I}$ is compact and $Y = X \backslash (\{\omega_1\} \times \mathbb{I}) = \omega_1 \times \mathbb{I}$ is first countable and hence sequential. To see that X is weakly Whyburn take any Whyburn closed non-closed $A \subset X$. By the remark above, A is sequentially closed and hence $A \cap Y$ is sequentially closed in Y; by sequentiality of Y, the set $A' = A \cap Y$ is closed in Y. Analogously, if $I = \{\omega_1\} \times \mathbb{I}$, then I is second countable and hence sequential so $G = A \cap I$ is closed in I.

As a consequence, $\overline{A} \backslash A \subset I$; take any $x = (\omega_1, t) \in \overline{A} \backslash A$ and fix $\varepsilon > 0$ such that $(\{\omega_1\} \times (t - \varepsilon, t + \varepsilon)) \cap G = \emptyset$. It is easy to see that the set $P = (\omega_1 \times \{t\}) \cap A$ is almost closed in X. If P is uncountable, then $x \in \overline{P} \backslash P$ which contradicts the fact that A is Whyburn closed in X.

Now assume that P is countable and hence there exists $\beta < \omega_1$ such that $(\alpha, t) \notin A$ for any $\alpha \geq \beta$. Let $K_\alpha = \{(\gamma, t) : \beta \leq \gamma \leq \alpha\} = [\beta, \alpha] \times \{t\}$ for each $\alpha \in \omega_1 \backslash \beta$. It is clear that K_α is compact; since A' is closed in Y and $K_\alpha \cap A' = \emptyset$ for any $\alpha \geq \beta$, we can apply Fact 3 of S.271 to conclude that there is $n(\alpha) \in \mathbb{N}$ such that $[\beta, \alpha] \times (t - \frac{1}{n(\alpha)}, t + \frac{1}{n(\alpha)})$ does not intersect A'. There is an uncountable $R \subset [\beta, \omega_1)$ and $n \in \mathbb{N}$ such that $n(\alpha) = n$ for every $\alpha \in R$. It is immediate that the set $H = [\beta, \omega_1) \times (t - \frac{1}{n}, t + \frac{1}{n})$ does not intersect A' and hence $H \cap A = \emptyset$. Now, if $\delta = \min\{\frac{1}{n}, \varepsilon\}$, then $W = (\beta, \omega_1] \times (t - \delta, t + \delta)$ is an open neighborhood of x; since $W \subset H \cup (\{\omega_1\} \times (t - \varepsilon, t + \varepsilon))$, we have $W \cap A = \emptyset$ which is a contradiction with $x \in \overline{A}$. Therefore the set A has to be closed and hence X is weakly Whyburn.

To see that the space X is not hereditarily weakly Whyburn, consider its subspace $Z = (L \cup \{\omega_1\}) \times \mathbb{I}$ where $L = \{\alpha < \omega_1 : \alpha = \beta + 1 \text{ for some } \beta < \omega_1\}$.

Observe that $M = L \cup \{\omega_1\}$ is homeomorphic to $L(\omega_1)$ because ω_1 is the unique non-isolated point of M and a set $U \subset M$ is a neighborhood of ω_1 if and only if $\omega_1 \in U$ and $M \backslash U$ is countable. Consequently, the space Z is homeomorphic to $L(\omega_1) \times \mathbb{I}$ and therefore Z is not weakly Whyburn by Fact 2. Thus X is a compact weakly Whyburn space while $Z \subset X$ is not weakly Whyburn which shows that our solution is complete.

T.221. *Prove that any metrizable space is a p-space and a Σ-space at the same time.*

Solution. Let (X, d) be a metric space. Then $\mathcal{C} = \{\{x\} : x \in X\}$ is a closed cover of X with compact subsets; the space X has a σ-discrete base $\mathcal{B}$ (see Problem 221 of [TFS]) which is, evidently, a network with respect to $\mathcal{C}$. This proves that X is a Σ-space.

For each $x \in X$ and $n \in \mathbb{N}$, take $U_x^n \in \tau(\beta X)$ such that $U_x^n \cap X = B_d(x, \frac{1}{n})$. Then $\mathcal{U}_n = \{U_x^n : x \in X\} \subset \tau(\beta X)$ is a cover of X for each $n \in \omega$. Assume that $z \in \beta X \backslash X$ and $z \in \bigcap\{\text{St}(x, \mathcal{U}_n) : n \in \mathbb{N}\}$ for some $x \in X$. Take any $W \in \tau(x, \beta X)$ such that $z \notin \overline{W}$ (the bar denotes the closure in βX) and find $n \in \mathbb{N}$ for which $B_d(x, \frac{2}{n}) \subset W \cap X$. Since $z \in \text{St}(x, \mathcal{U}_n)$, there is $y \in X$ such that $x \in U_y^n$ and $z \in U_y^n$. Consequently, $d(x, y) < \frac{1}{n}$, and given any $t \in B_d(y, \frac{1}{n})$, we have $d(x, t) \leq d(x, y) + d(y, t) < \frac{2}{n}$ which shows that $t \in B_d(x, \frac{2}{n})$. Therefore $B_d(y, \frac{1}{n}) \subset B_d(x, \frac{2}{n}) \subset W$ which implies $z \in \overline{U_y^n} = \overline{B_d(y, \frac{1}{n})} \subset \overline{W}$, a contradiction with the choice of W. Therefore $\bigcap\{\text{St}(x, \mathcal{U}_n) : n \in \mathbb{N}\} \subset X$ for each $x \in X$ and hence X is a p-space.

T.222. *Prove that $C_p(X)$ is a p-space if and only if X is countable.*

Solution. If X is countable, then $C_p(X)$ is second countable and hence metrizable which shows that $C_p(X)$ is a p-space by Problem 221.

Fact 1. If Z is a p-space, then it is of pointwise countable type, i.e., for every $z \in Z$, there is a compact $K \subset Z$ such that $z \in K$ and $\chi(K, Z) \leq \omega$.

Proof. Let $\{\mathcal{U}_n : n \in \omega\}$ be the sequence of open (in βZ) covers of Z which witnesses that Z is a p-space. Choose $U_n \in \mathcal{U}_n$ such that $z \in U_n$ for all $n \in \omega$. It is easy to construct a sequence $\{V_n : n \in \omega\} \subset \tau(z, \beta Z)$ such that $V_n \subset U_n$ and $\overline{V}_{n+1} \subset V_n$ (the bar denotes the closure in βZ) for all $n \in \omega$. The set $K = \bigcap\{V_n : n \in \omega\} = \bigcap\{\overline{V}_n : n \in \omega\}$ is compact and

$$z \in K \subset \bigcap\{U_n : n \in \omega\} \subset \bigcap\{\text{St}(z, \mathcal{U}_n) : n \in \omega\} \subset Z.$$

Since K is a G_δ-set in βZ, we have $\chi(K, \beta Z) = \omega$ by Problem 327 of [TFS] and hence $\chi(K, Z) \leq \omega$. Fact 1 is proved. $\square$

Returning to our solution assume that $C_p(X)$ is a p-space. Then it is of pointwise countable type by Fact 1 an hence there is a compact $K \subset C_p(X)$ with $\chi(K, C_p(X)) = \omega$. Now apply Problem 170 of [TFS] to conclude that X is countable.

T.223. *Prove that every Lindelöf p-space is a Lindelöf Σ-space. Give an example of a p-space which is not a Σ-space.*

Solution. Suppose that we have a space Z and a family $\mathcal{A} \subset \exp(Z)$; then $\text{St}(z, \mathcal{A}) = \bigcup\{A : A \in \mathcal{A} \text{ and } z \in A\}$. Given a family $\mathcal{U}_n \subset \tau(\beta Z)$ for every $n \in \omega$, call the sequence $\{\mathcal{U}_n : n \in \omega\}$ a *feathering* of the space Z if $Z \subset \bigcup \mathcal{U}_n$ for each $n \in \omega$ and $\bigcap\{\text{St}(z, \mathcal{U}_n) : n \in \omega\} \subset Z$ for every $z \in Z$. It is clear that a sequence $\{\mathcal{U}_n : n \in \omega\}$ is a feathering of Z if and only if it witnesses the fact that Z is a p-space. Given families $\mathcal{U}$ and $\mathcal{V}$ of subsets of Z, say that $\mathcal{U}$ *is (strongly) inscribed in* $\mathcal{V}$ if for any $U \in \mathcal{U}$ there is $V \in \mathcal{V}$ such that $U \subset V$ (or $\overline{U} \subset V$ respectively). If $\mathcal{A} \subset \exp(Z)$ and $z \in Z$, then $\mathcal{A}(z) = \{A \in \mathcal{A} : z \in A\}$ and $\text{ord}(z, \mathcal{A}) = |\mathcal{A}(z)|$; given any $Y \subset Z$ we let $\mathcal{A}|Y = \{A \cap Y : A \in \mathcal{A}\}$.

Call Z *a strong Σ-space* if there exists a *compact* cover $\mathcal{C}$ of the space Z and a σ-discrete family $\mathcal{F} \subset \exp(Z)$ which is a network with respect to $\mathcal{C}$ (it is evident that strong Σ-spaces are Σ-spaces). A space Z is *subparacompact* if every open cover of Z has a (not necessarily open) σ-discrete refinement.

Fact 1. Suppose that L is a Lindelöf subspace of a space Z. Then, for any family $\mathcal{U} \subset \tau(Z)$ with $L \subset \bigcup \mathcal{U}$, there exists a countable $\mathcal{V} \subset \tau(Z)$ such that $L \subset \bigcup \mathcal{V}$, the family $\mathcal{V}$ is strongly inscribed in $\mathcal{U}$ and $\text{ord}(y, \mathcal{V}) < \omega$ for any $y \in L$.

Proof. For every $y \in L$ take $U_y \in \mathcal{U}$ such that $y \in U_y$ and use regularity of Z to find $V_y \in \tau(y, Z)$ such that $\text{cl}_Z(V_y) \subset U_y$. If $\mathcal{V}' = \{V_y : y \in L\}$, then $\mathcal{V}'|L$ is an open cover of L, so by paracompactness of L (see Fact 1 of S.219), there is locally finite open (in the space L) refinement $\mathcal{V}''$ of the cover $\mathcal{V}'|L$. By Lindelöfness of L we can assume that $\mathcal{V}''$ is countable. For any $W \in \mathcal{V}''$, there is $V_W \in \mathcal{V}'$ such that $W \subset V_W \cap L$; take any $O_W \in \tau(Z)$ such that $O_W \cap L = W$ and let $G_W = V_W \cap O_W$. It is straightforward that $\mathcal{V} = \{G_W : W \in \mathcal{V}''\}$ is as promised so Fact 1 is proved. $\square$

Fact 2. Every strong Σ-space is subparacompact.

Proof. Assume that Z is a strong Σ-space and take an open cover $\mathcal{U}$ of the space Z. Let $\mathcal{C}$ be a compact cover of Z such that there exists a σ-discrete family $\mathcal{F}$ which is a network with respect to $\mathcal{C}$. Thus $\mathcal{F} = \bigcup\{\mathcal{F}_n : n \in \omega\}$ where $\mathcal{F}_n$ is discrete for every $n \in \omega$. A set $F \in \mathcal{F}$ will be called $\mathcal{U}$-*small* if there is a finite $\mathcal{U}_F \subset \mathcal{U}$ such that $F \subset \bigcup \mathcal{U}_F$. In particular, there is $k_F \in \omega$ such that $\mathcal{U}_F = \{U(F, 0), \ldots, U(F, k_F)\}$ for each $\mathcal{U}$-small $F \in \mathcal{F}$; let $H(F, i) = F \cap U(F, i)$ for all $i \leq k_F$. The family $\mathcal{H}(i, n) = \{H(F, i) : F \in \mathcal{F}_n \text{ is } \mathcal{U}\text{-small and } i \leq k_F\}$ is discrete for any $i, n \in \omega$ because each $\mathcal{H}(i, n)$ is obtained from $\mathcal{F}_n$ by shrinking some of its elements. It is immediate from the definition that $\mathcal{H}(i, n)$ is inscribed in $\mathcal{U}$ for all $i, n \in \omega$. The family $\mathcal{H} = \bigcup\{\mathcal{H}(i, n) : i, n \in \omega\}$ being σ-discrete, it suffices to show that $\bigcup \mathcal{H} = Z$. Take any $z \in Z$; there is $K \in \mathcal{C}$ such that $z \in K$. The set K being compact there is a finite $\mathcal{U}' \subset \mathcal{U}$ such that $K \subset \bigcup \mathcal{U}'$. Since the family $\mathcal{F}$ is a network with respect to $\mathcal{C}$, there is $F \in \mathcal{F}$ with $K \subset F \subset \bigcup \mathcal{U}'$. Consequently, F is $\mathcal{U}$-small; since $F = \bigcup\{H(F, i) : i \leq k_F\}$, there is $i \leq k_F$ with $z \in H(F, i) \subset \bigcup \mathcal{H}$. Thus $\mathcal{H}$ is a σ-discrete refinement of $\mathcal{U}$ and Fact 2 is proved. $\square$

Fact 3. Every locally compact space is p-space.

Proof. If Z is locally compact, then is open in βZ, and we can take $\mathcal{U}_n = \{Z\}$ for any $n \in \omega$. It is evident that the sequence $\{\mathcal{U}_n : n \in \omega\}$ is a feathering of Z so Fact 3 is proved.   □

Fact 4. The space $(A(\omega_2) \times A(\omega_2)) \backslash \{(a, a)\}$ is not subparacompact.

Proof. Recall that $A(\omega_2) = \omega_2 \cup \{a\}$ where all points of ω_2 are isolated in $A(\omega_2)$ and the neighborhoods of the point $a \notin \omega_2$ are the complements of finite subsets of ω_2. Let $H_\alpha = A(\omega_2) \times \{\alpha\}$ and $V_\alpha = \{\alpha\} \times A(\omega_2)$ for every $\alpha < \omega_2$. If $h_\alpha = \{(a, \alpha)\}$ and $v_\alpha = \{(\alpha, a)\}$ for each $\alpha < \omega_2$, then H_α is a compact open neighborhood of h_α and V_α is a compact open neighborhood of v_α. Let $\mathcal{H} = \{H_\alpha : \alpha < \omega_2\}$ and $\mathcal{V} = \{V_\alpha : \alpha < \omega_2\}$; to obtain a contradiction assume that the open cover $\mathcal{H} \cup \mathcal{V}$ of the space $(A(\omega_2) \times A(\omega_2)) \backslash \{(a, a)\}$ has a refinement $\mathcal{F} = \bigcup \{\mathcal{F}_n : n \in \omega\}$ such that $\mathcal{F}_n$ is discrete for all $n \in \omega$. Every $F \in \mathcal{F}$ is contained in an element of $\mathcal{H} \cup \mathcal{V}$, so $\mathcal{F} = \mathcal{F}_H \cup \mathcal{F}_V$ where $\mathcal{F}_H = \{F \in \mathcal{F} : F \subset H_\alpha$ for some $\alpha < \omega_2\}$ and $\mathcal{F}_V = \{F \in \mathcal{F} : F \subset V_\alpha$ for some $\alpha < \omega_2\}$.

Observe that $|F \cap H_\alpha| \leq 1$ for every $F \in \mathcal{F}_V$; besides, $h_\alpha \in \overline{D} \backslash D$ for every infinite $D \subset H_\alpha \backslash \{h_\alpha\}$ and $h_\alpha \notin B_n = \bigcup (\mathcal{F}_n \cap \mathcal{F}_V)$ for all $\alpha < \omega_2$ and $n \in \omega$. The family $\mathcal{F}_n \cap \mathcal{F}_V$ being discrete for all $n \in \omega$, the set $B_n \cap H_\alpha$ is finite for all $\alpha < \omega_2$. Analogously, if $A_n = \bigcup (\mathcal{F}_n \cap \mathcal{F}_H)$, then $A_n \cap V_\alpha$ is finite for any $\alpha < \omega_2$. Let $A = \bigcup \{A_n : n \in \omega\}$ and $B = \bigcup \{B_n : n \in \omega\}$. Then $\bigcup \mathcal{F} = A \cup B$ while the sets $A \cap V_\alpha$ and $B \cap H_\alpha$ are countable for all $\alpha < \omega_2$.

For every $\alpha < \omega_1$, there is $\beta(\alpha) < \omega_2$ such that $(\alpha, \beta) \notin A \cap V_\alpha$ for every $\beta \geq \beta(\alpha)$. There is $\gamma < \omega_2$ such that $\beta(\alpha) < \gamma$ for all $\alpha < \omega_1$. Consequently, the set $P = \{(\alpha, \gamma) : \alpha < \omega_1\}$ does not intersect A and therefore $P \subset H_\gamma \cap B$ which is a contradiction because $H_\gamma \cap B$ has to be countable. Thus the open cover $\mathcal{H} \cup \mathcal{V}$ of the space $(A(\omega_2) \times A(\omega_2)) \backslash \{(a, a)\}$ has no σ-discrete refinement and Fact 4 is proved.   □

Returning to our solution, suppose that X be a Lindelöf p-space and fix its feathering $\{\mathcal{U}_n : n \in \omega\}$. Observe that if $\mathcal{U}_n' \subset \tau(\beta X)$ is inscribed in $\mathcal{U}_n$ and $X \subset \bigcup \mathcal{U}_n'$ for all $n \in \omega$, then the sequence $\{\mathcal{U}_n' : n \in \omega\}$ is still a feathering of X. The space X being a Lindelöf subspace of βX, we can construct inductively, using Fact 1, a sequence $\{\mathcal{V}_n : n \in \omega\}$ with the following properties:

(1) $\mathcal{V}_n \subset \tau(\beta X)$, $X \subset \bigcup \mathcal{V}_n$ and $\mathcal{V}_n$ is countable for each $n \in \omega$;
(2) $\mathcal{V}_{n+1}$ is strongly inscribed in the family $\mathcal{U}_n \wedge \mathcal{V}_n = \{U \cap V : U \in \mathcal{U}_n$ and $V \in \mathcal{V}_n\}$ for any $n \in \omega$;
(3) $\text{ord}(x, \mathcal{V}_n) < \omega$ for any $x \in X$ and $n \in \omega$.

Observe that it follows from (1) and (2) that the sequence $\{\mathcal{V}_n : n \in \omega\}$ is a feathering of X. Furthermore, (2) and (3) imply that

$$\overline{\text{St}(x, \mathcal{V}_{n+1})} = \bigcup \{\overline{V} : V \in \mathcal{V}_{n+1}(x)\} \subset \bigcup \{V : V \in \mathcal{V}_n(x)\} = \text{St}(x, \mathcal{V}_n)$$

for any point $x \in X$ and $n \in \omega$ (the bar denotes the closure in βX). Therefore $K_x = \bigcap \{\mathrm{St}(x, V_n) : n \in \omega\} = \bigcap \{\overline{\mathrm{St}(x, V_n)} : n \in \omega\}$ is a compact subset of X and it follows from Fact 1 of S.326 that $\mathcal{C}_x = \{\mathrm{St}(x, V_n) : n \in \omega\}$ is an outer base of K_x in βX. Consequently, $\mathcal{C}_x | X$ is an outer base of K_x in X. Since each $V_n(x)$ is finite, the family $\mathcal{C}'$ of all finite unions of the elements of the family $V = \bigcup \{V_n : n \in \omega\}$ is a countable network in βX with respect to the compact cover $\mathcal{F} = \{K_x : x \in X\}$ of the space X. Therefore $\mathcal{C} = \mathcal{C}' | X$ is a countable (and hence σ-discrete) network with respect to the family $\mathcal{F}$ in X which shows that X is a Σ-space and proves that any Lindelöf p-space is a Lindelöf Σ-space.

As to an example of a p-space which is not a Σ-space, observe that the space $Z = (A(\omega_2) \times A(\omega_2)) \backslash \{(a, a)\}$ is locally compact being open in the compact space $A(\omega_2) \times A(\omega_2)$. Therefore Z is a p-space by Fact 3. Let us show that

(∗) every closed countably compact subspace of Z is compact.

Let $p_1 : Z \to A(\omega_2)$ and $p_2 : Z \to A(\omega_2)$ be the restrictions of the relevant natural projections of the space $A(\omega_2) \times A(\omega_2)$ onto its first and second factors respectively. If $K \subset Z$ is countably compact, then there exists a finite $Q \subset \omega_2$ such that $K \subset p_1^{-1}(Q) \cup p_2^{-1}(Q)$. Indeed, if such Q does not exist, then we can choose inductively an infinite $W \subset \omega_2$ and an injective map $\varphi : W \to \omega_2$ such that $G = \{(\alpha, \varphi(\alpha)) : \alpha \in W\} \subset K$. It is immediate that G is a closed discrete subspace in Z and hence in K which contradicts countable compactness of K. Thus $K \subset R = p_1^{-1}(Q) \cup p_2^{-1}(Q)$ for some finite $Q \subset A(\omega_2)$. Since R is compact and K is closed in R, the space K is compact, i.e., (∗) is established.

An immediate consequence of (∗) is that if Z is a Σ-space, then it is a strong Σ-space and hence subparacompact by Fact 2. Since Z is not subparacompact by Fact 4, the space Z is an example of a p-space which is not a Σ-space, so our solution is complete.

T.224. *Prove that*

(i) *any closed subspace of a Σ-space is a Σ-space. In particular, any closed subspace of a Lindelöf Σ-space is a Lindelöf Σ-space;*

(ii) *any closed subspace of a p-space is a p-space. In particular, any closed subspace of a Lindelöf p-space is a Lindelöf p-space.*

Solution. Suppose that K is a compact space and Z is dense in K. If $\mathcal{U}_n \subset \tau(K)$ for every $n \in \omega$, call the sequence $\{\mathcal{U}_n : n \in \omega\}$ a *feathering* of the space Z in its compactification K if $Z \subset \bigcup \mathcal{U}_n$ for each $n \in \omega$ and $\bigcap \{\mathrm{St}(z, \mathcal{U}_n) : n \in \omega\} \subset Z$ for every $z \in Z$.

Fact 1. A space Z is a p-space if and only if there exists a compactification K of the space Z such that Z has a feathering in K.

Proof. If Z is a p-space, then it has a feathering in βZ so necessity is easy. Now, assume that Z has a feathering $\{\mathcal{U}_n : n \in \omega\}$ in some compactification K. There is a continuous onto map $f : \beta Z \to K$ such that $f(z) = z$ for each $z \in Z$ (see Problem 258 of [TFS]). The family $V_n = \{f^{-1}(U) : U \in \mathcal{U}_n\}$ consists of open subsets of βZ and $Z \subset \bigcup V_n$ for all $n \in \omega$. Given any $z \in Z$ observe that $\mathrm{St}(z, V_n) = f^{-1}(\mathrm{St}(z, \mathcal{U}_n))$ for each $n \in \omega$ and hence

$$\bigcap \{\mathrm{St}(z, V_n) : n \in \omega\} = \bigcap \{f^{-1}(\mathrm{St}(z, \mathcal{U}_n)) : n \in \omega\} \subset f^{-1}(Z) = Z$$

(see Fact 1 of S.259). Thus the sequence $\{V_n : n \in \omega\}$ is a feathering of Z in βZ and hence Z is a p-space. Fact 1 is proved. $\square$

(i) Assume that X is a Σ-space and take a closed cover $\mathcal{C}$ of the space X for which there exists a σ-discrete network $\mathcal{F}$ with respect to $\mathcal{C}$. If $P \subset X$ is closed, then the family $\mathcal{C}' = \{C \cap P : C \in \mathcal{C}\}$ is a closed cover of P with countably compact subsets of P and it is straightforward that the family $\mathcal{F}' = \{F \cap P : F \in \mathcal{F}\}$ is a σ-discrete network with respect to $\mathcal{C}'$. Thus P is a Σ-space.

(ii) Suppose that X is a p-space and F is a closed subset of X. Fix a feathering $\{\mathcal{U}_n : n \in \omega\}$ of the space X in βX and let $V_n = \{U \cap \mathrm{cl}_{\beta X}(F) : U \in \mathcal{U}_n\}$ for each $n \in \omega$. It is immediate that the sequence $\{V_n : n \in \omega\}$ is a feathering of F in its compactification $K = \mathrm{cl}_{\beta X}(F)$, so F is a p-space by Fact 1.

T.225. *Prove that X is a Lindelöf Σ-space if and only if X has a countable network with respect to a compact cover $\mathcal{C}$.*

Solution. Suppose that X is a Lindelöf Σ-space and fix a closed cover $\mathcal{C}$ of X with countably compact subspaces of X such that there exists a σ-discrete network $\mathcal{F}$ with respect to $\mathcal{C}$. Every closed subspace of a Lindelöf space is Lindelöf and every Lindelöf countably compact space is compact, so $\mathcal{C}$ consists of compact subsets of X. Furthermore, $\mathrm{ext}(X) \leq l(X) = \omega$ and hence every discrete family in X is countable. Consequently, $\mathcal{F}$ is a countable network with respect to the compact cover $\mathcal{C}$ of the space X. This proves necessity.

Now assume that X has a compact cover $\mathcal{C}$ for which there is a countable network $\mathcal{F}$ with respect to $\mathcal{C}$. It is clear that X is a Σ-space, so we only must establish that X is Lindelöf. Let $\mathcal{U}$ be an open cover of the space X. A set $F \in \mathcal{F}$ will be called $\mathcal{U}$-small if there is a finite family $\mathcal{U}_F \subset \mathcal{U}$ for which $F \subset \bigcup \mathcal{U}_F$. The family $\mathcal{U}' = \bigcup \{\mathcal{U}_F : F \in \mathcal{F}$ is $\mathcal{U}$-small$\}$ is countable and $\mathcal{U}' \subset \mathcal{U}$. Given any $x \in X$, there is $C \in \mathcal{C}$ such that $x \in C$. The set C being compact there is a finite $V \subset \mathcal{U}$ such that $C \subset \bigcup V$. Since the family $\mathcal{F}$ is a network with respect to $\mathcal{C}$, there is $F \in \mathcal{F}$ for which $C \subset F \subset \bigcup V$. It is evident that F is $\mathcal{U}$-small and hence $x \in C \subset F \subset \bigcup \mathcal{U}_F \subset \bigcup \mathcal{U}'$ which shows that we found a countable subcover $\mathcal{U}'$ of the cover $\mathcal{U}$ so X is Lindelöf and sufficiency is proved.

T.226. *Suppose that $X = \bigcup \{X_n : n \in \omega\}$ where X_n is countably compact and closed in X for any $n \in \omega$. Prove that X is a Σ-space. As a consequence, every σ-compact space is a Lindelöf Σ-space.*

Solution. Let $\mathcal{F} = \mathcal{C} = \{F_n : n \in \omega\}$; then $\mathcal{C}$ is a closed cover of X with countably compact subspaces of X and $\mathcal{F}$ is a network with respect to $\mathcal{C}$. This proves that X is a Σ-space. If $Y = \bigcup \{K_n : n \in \omega\}$ where each K_n is compact, then K_n is closed in Y, so Y is a Σ-space by the previous observation. Every σ-compact space is Lindelöf so Y is a Lindelöf Σ-space.

T.227. *Prove that the Sorgenfrey line and the space $L(\omega_1)$ are examples of Lindelöf spaces which are not Lindelöf Σ.*

Solution. A cover C of a space Z is called *compact* if all elements of C are compact; C is *a Σ-cover* of Z if there exists a σ-discrete $\mathcal{F} \subset \exp(Z)$ which is a network with respect to C.

Fact 1. If X and Y are Lindelöf Σ-spaces, then $X \times Y$ is also a Lindelöf Σ-space.

Proof. Apply Problem 225 to find compact Σ-covers C_X and C_Y of the spaces X and Y respectively and let $\mathcal{F}_X$ and $\mathcal{F}_Y$ be the respective σ-discrete networks. The spaces X and Y being Lindelöf, the families $\mathcal{F}_X$ and $\mathcal{F}_Y$ are countable. It is evident that $C = \{K \times L : K \in C_X$ and $L \in C_Y\}$ is a compact cover of the space $X \times Y$. To see that C is a Σ-cover, let $\mathcal{F} = \{F \times G : F \in \mathcal{F}_X$ and $G \in \mathcal{F}_Y\}$. If $C \in C$ and $O \in \tau(C, X \times Y)$, then $C = K \times L$ for some $K \in C_X$ and $L \in C_Y$, so we can apply Fact 3 of S.271 to find $U \in \tau(K, X)$ and $V \in \tau(L, Y)$ such that $U \times V \subset O$. Since $\mathcal{F}_X$ and $\mathcal{F}_Y$ are networks for C_X and C_Y, respectively, there exist $F \in \mathcal{F}_X$ and $G \in \mathcal{F}_Y$ such that $K \subset F \subset U$ and $L \subset G \subset V$. Then $H = F \times G \in \mathcal{F}$ and $C \subset H \subset O$ which proves that $\mathcal{F}$ is a network with respect to C. The family $\mathcal{F}$ is countable, so we can apply Problem 225 again to conclude that $X \times Y$ is a Lindelöf Σ-space. Fact 1 is proved.   □

Fact 2. Assume that X is a Lindelöf Σ-space such that every compact subspace of X is finite. Then X is countable.

Proof. Apply Problem 225 to take a compact cover C of the space X such that there exists a countable network $\mathcal{F}$ with respect to C. We do not lose generality assuming that $\mathcal{F}$ is closed under finite intersections. Let $C' = \{C \in C : C \in \mathcal{F}\}$; it is clear that C' is countable. Assume for a moment that there is an element $C \in C \backslash C'$; since $\mathcal{F}$ is a network for C, there is a sequence $S = \{F_n : n \in \omega\} \subset \mathcal{F}$ such that $F_{n+1} \subset F_n$ for all $n \in \omega$ and S is a network at C in the sense that for any $U \in \tau(C, X)$ there is $n \in \omega$ for which $C \subset F_n \subset U$. Since $F_n \neq C$ for all $n \in \omega$, we can pick $x_n \in F_n \backslash C$ for each $n \in \omega$.

The space $K = C \cup \{x_n : n \in \omega\}$ is compact. Indeed, if $\mathcal{U}$ is an open (in X) cover of K, then there is a finite $\mathcal{U}' \subset \mathcal{U}$ such that $C \subset U = \bigcup \mathcal{U}'$. There is $m \in \omega$ such that $F_m \subset U$ and hence $x_n \in F_n \subset F_m \subset U$ for all $n \geq m$ which shows that $K \backslash U$ is finite and hence there is a finite $\mathcal{U}'' \subset \mathcal{U}$ with $K \backslash U \subset \bigcup \mathcal{U}''$. It is immediate that $\mathcal{U}' \cup \mathcal{U}''$ is a finite subcover of $\mathcal{U}$ so K is, indeed, compact.

The set $P = \{x_n : n \in \omega\}$ cannot be finite; for otherwise, $X \backslash P \in \tau(C, X)$ and hence $x_n \in F_n \subset X \backslash P$ for some $n \in \omega$ which is a contradiction. Thus K is an infinite compact subset of X; this contradiction shows that $C = C'$ and hence C is countable. Since every element of C is finite, the set $X = \bigcup C$ is countable so Fact 2 is proved.   □

Returning to our solution, recall that it was proved in Problem 165 of [TFS] that Sorgenfrey line S is Lindelöf while $S \times S$ is not Lindelöf. Thus S cannot be a Lindelöf Σ-space by Fact 1. The space $L(\omega_1)$ is Lindelöf by Problem 354 of [TFS] and it is easy to see that every compact subspace of $L(\omega_1)$ is finite, so $L(\omega_1)$ is not Lindelöf Σ by Fact 2.

T.228. *Prove that any space with a σ-discrete network is a Σ-space. In particular, if $nw(X) \leq \omega$, then X is a Lindelöf Σ-space.*

Solution. If $\mathcal{F}$ is a σ-discrete network of a space Y, then it is a network with respect to the cover $\mathcal{C} = \{\{y\} : y \in Y\}$ of the space Y which consists of closed compact subsets of Y. Thus Y is a Σ-space. Every countable network is σ-discrete, so $nw(X) \leq \omega$ implies that X is a Σ-space. Since every space with a countable network has to be Lindelöf, the space X is Lindelöf Σ.

T.229. *Let X be a metrizable space. Prove that $C_p(X)$ is a Σ-space if and only if X is second countable.*

Solution. If we have a set P, then $\mathrm{Fin}(P)$ is the family of all finite subsets of P. Given a space Z say that a cover $\mathcal{C}$ of Z is *compact (closed)* if all elements of $\mathcal{C}$ are compact (or closed in Z, respectively). A cover $\mathcal{C}$ of a space Z is called *a Σ-cover* of Z if there exists a σ-discrete $\mathcal{F} \subset \exp(Z)$ which is a network with respect to $\mathcal{C}$. Thus Z is a Σ-space if and only if it has a closed Σ-cover whose elements are countably compact. Call a space Z *a strong Σ-space* if Z has a compact Σ-cover.

If $\mathcal{A} \subset \exp(Z)$, then $\mathcal{A}(z) = \{A : A \in \mathcal{A} \text{ and } z \in A\}$. Given a set $A \subset \omega_1$ and $s \in \omega^A$ let $[s, A] = \{t \in \omega^{\omega_1} : t|A = s\}$. It is evident that the family $\mathcal{B} = \{[s, A] : A \in \mathrm{Fin}(\omega_1) \text{ and } s \in \omega^A\}$ is a base of the space ω^{ω_1}.

If f is a function, then $\mathrm{dom}(f)$ is its domain. Suppose that we have a set of functions $\{f_i : i \in I\}$ such that $f_i|(\mathrm{dom}(f_i) \cap \mathrm{dom}(f_j)) = f_j|(\mathrm{dom}(f_i) \cap \mathrm{dom}(f_j))$ for any $i, j \in I$. Then we can define a function f with $\mathrm{dom}(f) = \bigcup_{i \in I} \mathrm{dom}(f_i)$ as follows: given any $x \in \mathrm{dom}(f)$, find any $i \in I$ with $x \in \mathrm{dom}(f_i)$ and let $f(x) = f_i(x)$. It is easy to check that the value of f at x does not depend on the choice of i, so we have consistently defined a function f which will be denoted by $\bigcup\{f_i : i \in I\}$

Fact 1. Suppose that $\mathcal{C}$ is a compact Σ-cover of a space Z. Then there is a σ-discrete network $\mathcal{F}$ with respect to $\mathcal{C}$ such that each $F \in \mathcal{F}$ is closed in Z.

Proof. There is some network $\mathcal{G}$ with respect to $\mathcal{C}$ such that $\mathcal{G} = \bigcup\{\mathcal{G}_n : n \in \omega\}$ and $\mathcal{G}_n$ is discrete for all $n \in \omega$. Let $\mathcal{F}_n = \{\overline{G} : G \in \mathcal{G}_n\}$; then the family $\mathcal{F}_n$ is discrete for all $n \in \omega$, so it suffices to show that $\mathcal{F} = \bigcup\{\mathcal{F}_n : n \in \omega\}$ is a network with respect to $\mathcal{C}$.

Take any $C \in \mathcal{C}$; if $O \in \tau(C, Z)$, then there is $V \in \tau(C, Z)$ such that $\overline{V} \subset O$ (this is an easy exercise on using compactness of C). The family $\mathcal{G}$ being a network with respect to $\mathcal{C}$ there is $G \in \mathcal{G}$ such that $C \subset G \subset V$. Then $F = \overline{G} \in \mathcal{F}$ and $C \subset F \subset \overline{V} \subset O$, so $\mathcal{F}$ is a σ-discrete network with respect to $\mathcal{C}$ and Fact 1 is proved. $\square$

Fact 2. Let Z be a strong Σ-space. Then there exists a σ-discrete family $\mathcal{F} \subset$ $\exp(Z)$ such that all elements of $\mathcal{F}$ are closed in Z and $C_z = \bigcap \mathcal{F}(z)$ is compact for any $z \in Z$.

Proof. Take a compact Σ-cover $\mathcal{C}$ of the space Z and a σ-discrete network $\mathcal{F}$ with respect to $\mathcal{C}$ such that all elements of $\mathcal{F}$ are closed in Z. This is possible by Fact 1. Given any $z \in Z$ there is $C \in \mathcal{C}$ with $z \in C$. The family $\mathcal{F}_C = \{F \in \mathcal{F} : C \subset F\}$ is contained in $\mathcal{F}(z)$; since $\mathcal{F}$ is a network with respect to $\mathcal{C}$, we have $\bigcap \mathcal{F}_C = C$ and hence $C_z = \bigcap\{F \in \mathcal{F} : z \in F\} \subset \bigcap \mathcal{F}_C = C$. Therefore C_z is compact being a closed subspace of a compact space C. Fact 2 is proved. $\qquad\square$

Returning to our solution observe that if $w(X) \leq \omega$, then $nw(C_p(X)) = \omega$ (see Problem 172 of [TFS]) and therefore $C_p(X)$ is a Lindelöf Σ-space by Problem 228.

Now, if $C_p(X)$ is a Σ-space and $w(X) > \omega$, then $\mathrm{ext}(X) > \omega$ and hence $\mathbb{R}^{\omega_1}$ embeds in $C_p(X)$ as a closed subspace by Fact 1 of S.215. This implies that $\mathbb{R}^{\omega_1}$ is a Σ-space by Problem 224 and hence ω^{ω_1} is also a Σ-space because it embeds in $\mathbb{R}^{\omega_1}$ as a closed subspace. To obtain a contradiction observe first that

(∗) every closed countably compact subspace of ω^{ω_1} is compact

because it is realcompact and pseudocompact (see Problems 401, 403 and 407 of [TFS]); consequently, if $\mathcal{C}$ is a closed Σ-cover of ω^{ω_1} by countably compact subspaces of ω^{ω_1}, then each $C \in \mathcal{C}$ has to be compact by (∗). Thus ω^{ω_1} has a compact Σ-cover, i.e., ω^{ω_1} is a strong Σ-space. Apply Fact 2 to find a σ-discrete family $\mathcal{F}$ of closed subspaces of ω^{ω_1} such that $C_s = \bigcap \mathcal{F}(s)$ is compact for every $s \in \omega^{\omega_1}$. We have $\mathcal{F} = \bigcup\{\mathcal{F}_n : n \in \omega\}$ where the family $\mathcal{F}_n$ is discrete for each $n \in \omega$. Let $\Phi_n = \mathcal{F}_0 \cup \cdots \cup \mathcal{F}_n$ for all $n \in \omega$. We have the following important property:

(∗∗) If $U \in \tau^*(\omega^{\omega_1})$ and $\mathcal{G}$ is a σ-discrete family of closed subspaces of ω^{ω_1} with $U \subset \bigcup \mathcal{G}$, then there is $G \in \mathcal{G}$ such that $\mathrm{Int}(G \cap U) \neq \emptyset$.

To prove (∗∗) recall that we have $\mathcal{G} = \bigcup\{\mathcal{G}_n : n \in \omega\}$ where the family $\mathcal{G}_n$ is discrete for every $n \in \omega$. The set $G_n = \bigcup \mathcal{G}_n$ is closed in ω^{ω_1} for each $n \in \omega$ and $U = \bigcup\{G_n \cap U : n \in \omega\}$. Since the space ω^{ω_1} has the Baire property (see Problems 464, 465 and 470 of [TFS]), some $G_n \cap U$ is not nowhere dense in U; being a closed subset of U, the set $G_n \cap U$ has a non-empty interior W in U; it is evident that $W \in \tau^*(\omega^{\omega_1})$. Each $G \in \mathcal{G}_n$ is open in G_n because $\mathcal{G}_n$ is discrete, so take any $G \in \mathcal{G}_n$ with $W' = W \cap G \neq \emptyset$; then W' is open in ω^{ω_1}, non-empty and $W' \subset G \cap U$. This shows that $\mathrm{Int}(G \cap U) \neq \emptyset$ so (∗∗) is proved.

Let $F_n = \bigcup \Phi_n$ and $P_n = \bigcup\{F_i : i < n\}$ for all $n \in \omega$; it follows from $\bigcup \mathcal{F} = \omega^{\omega_1}$ and the Baire property of ω^{ω_1} that the number $m_0 = \min\{n \in \omega : \mathrm{Int}(F_n) \neq \emptyset\}$ is well-defined. Apply (∗∗) to choose a set $H_0 \in \mathcal{F}_{m_0}$ such that $W = \mathrm{Int}(H_0) \neq \emptyset$. The set P_{m_0} is nowhere dense in ω^{ω_1}, so there is $V \in \tau^*(\omega^{\omega_1})$ such that $V \subset H_0 \backslash P_{m_0}$. Therefore there exists a finite set $A_0 \subset \omega_1$ and a function $s_0 \in \omega^{A_0}$ such that $[s_0, A_0] \subset H_0 \backslash P_{m_0}$.

Assume that $n \in \omega$ and we have constructed a sequence $m_0, \ldots, m_n$ of natural numbers, finite sets $A_0, \ldots, A_n$, functions $f_0, \ldots, f_n$ and sets $H_0, \ldots, H_n$ with the following properties:

(1) $A_0 \subset \cdots \subset A_n \subset \omega_1$ and $f_i \in \omega^{A_i}$ for all $i \leq n$;
(2) $m_0 < \cdots < m_n$ and $H_i \in \mathcal{F}_{m_i}$ for all $i \leq n$;
(3) $f_j | A_i = f_i$ whenever $0 \leq i < j \leq n$;
(4) $[f_i, A_i] \subset H_i$ for every $i \leq n$;
(5) $[f_i, A_i] \cap F_j = \emptyset$ for any $i \leq n$ and $j \in \{0, \ldots, m_n\} \setminus \{m_0, \ldots, m_n\}$.

Observe that it follows from (1) and (3) that $[f_j, A_j] \subset [f_i, A_i]$ if $0 \leq i < j \leq n$. Besides, $[f_n, A_n] \subset H_n$ which shows, together with (5), that $\bigcap \Phi_{m_n}(s) \supset [f_n, A_n]$ for any $s \in [f_n, A_n]$. If there is some $s \in [f_n, A_n]$ such that $s \notin F_k$ for all $k > m_n$, then the compact set $C_s = \bigcap \Phi_{m_n}(s)$ contains a non-compact closed subset $[f_n, A_n]$ which is a contradiction. Thus $[f_n, A_n] \subset \bigcup \{F_k : k > m_n\}$, so we can apply $(\ast\ast)$ to conclude that the number $m_{n+1} = \min\{k > m_n : \text{Int}(F_k \cap [f_n, A_n]) \neq \emptyset\}$ is well-defined. Take any $W \in \tau^*(\omega^{\omega_1})$ with $W \in F_{m_{n+1}} \cap [f_n, A_n]$; there is $G \in \mathcal{F}_{m_{n+1}}$ such that $G \cap W \neq \emptyset$. The set G is open in $\mathcal{F}_{m_{n+1}}$ because $\mathcal{F}_{m_{n+1}}$ is discrete. Therefore $W' = W \cap G$ is open in ω^{ω_1} and contained in $G \cap [f_n, A_n]$. Observe that the set $Q = \bigcup \{F_k \cap [f_n, A_n] : m_n < k < m_{n+1}\}$ is closed and nowhere dense, so we can find $V \in \mathcal{B}$ such that $V \subset W' \setminus Q$. By definition of $\mathcal{B}$, there is and finite $A_{n+1} \subset \omega_1$ and a function $f_{n+1} \in \omega^{A_{n+1}}$ such that $A_n \subset A_{n+1}$ and $f_{n+1} | A_n = f_n$. It is immediate that the conditions (1)–(5) are satisfied for the natural numbers $m_0, \ldots, m_{n+1}$, finite sets $A_0, \ldots, A_{n+1}$, functions $f_0, \ldots, f_{n+1}$ and sets $H_0, \ldots, H_{n+1}$. Therefore our inductive construction can be continued to obtain a sequence $\{m_i : i \in \omega\} \subset \omega$, a collection $\{A_i : i \in \omega\}$ of finite sets, a sequence $\{f_i : i \in \omega\}$ of functions and a family $\{H_i : i \in \omega\}$ such that the conditions (1)–(5) are satisfied for all $n \in \omega$.

Let $A = \bigcup \{A_i : i \in \omega\}$. It follows from the conditions (1) and (3) that the function $f = \bigcup \{f_i : i \in \omega\}$ is well-defined on A and $f | A_i = f_i$ for all $i \in \omega$. Therefore $[f, A] \subset [f_i, A_i]$ for all $i \in \omega$ and hence $[f, A] \subset \bigcap \{H_i : i \in \omega\}$ by the property (4). If $s \in [f, A]$, then $s \notin F_k$ for all $k \in \omega \setminus \{m_n : n \in \omega\}$ which shows that $C_s = \bigcap \{H_i : i \in \omega\}$ and hence a non-compact closed set $[f, A]$ is contained in a compact set C_s which is a contradiction. Therefore ω^{ω_1} is not a Σ-space; this final contradiction shows that $\mathbb{R}^{\omega_1}$ is not a Σ-space and hence $C_p(X)$ cannot be a Σ-space either. Thus X has to be separable and our solution is complete.

T.230. *Prove that any p-space is a k-space. Give an example of a Lindelöf Σ-space which is not a k-space.*

Solution. Suppose that K is a compact space and Z is dense in K. If $\mathcal{U}_n \subset \tau(K)$ for every $n \in \omega$, call the sequence $\{\mathcal{U}_n : n \in \omega\}$ a *feathering* of the space Z in its compactification K if $Z \subset \bigcup \mathcal{U}_n$ for each $n \in \omega$ and $\bigcap \{\text{St}(z, \mathcal{U}_n) : n \in \omega\} \subset Z$ for every $z \in Z$. It is clear that every p-space Z has a feathering in βZ.

Take any p-space X and assume that $A \subset X$ and $A \cap K$ is closed in K for any compact $K \subset X$. To obtain a contradiction, suppose that A is not closed in X and fix a point $z \in \overline{A} \setminus A$. Let $\mathcal{F} = \{\mathcal{U}_n : n \in \omega\}$ be a feathering of X in βX; take

$O_n \in \mathcal{U}_n$ such that $z \in O_n$ for every $n \in \omega$. Since $\mathcal{F}$ is a feathering, we have $\bigcap\{O_n : n \in \omega\} \subset X$. Choose a sequence $\{U_n : n \in \omega\} \subset \tau(z, \beta X)$ such that $U_0 = O_0$ and $\overline{U}_{n+1} \subset U_n \cap O_0 \cap \cdots \cap O_n$ for every $n \in \omega$ (the bar denotes the closure in the space βX). It is easy to see that the set $P = \bigcap\{U_n : n \in \omega\} = \bigcap\{\overline{U}_n : n \in \omega\}$ is a compact subspace of X. The set $K = P \cap A$ is closed in P and hence compact, so there is $V \in \tau(z, \beta X)$ such that $\overline{V} \cap K = \emptyset$.

Our next step is to construct a family $\{V_n : n \in \omega\} \subset \tau(z, \beta X)$ such that $V_0 = V$ and $\overline{V}_{n+1} \subset V_n \cap U_n$ for every $n \in \omega$. The set $Q = \bigcap\{V_n : n \in \omega\} = \bigcap\{\overline{V}_n : n \in \omega\}$ is again a compact subspace of X such that $Q \cap A = \emptyset$ and it is an easy consequence of Fact 1 of S.326 that the family $\mathcal{V} = \{V_n : n \in \omega\}$ is an outer base of Q in βX.

Pick a point $y_n \in V_n \cap A$ for each $n \in \omega$; the subspace $L = \{y_n : n \in \omega\} \cup Q$ is compact. Indeed, if $\mathcal{U} \subset \tau(\beta X)$ is an open cover of L, then there is a finite $\mathcal{U}' \subset \mathcal{U}$ such that $Q \subset W = \bigcup \mathcal{U}'$. Since $\mathcal{V}$ is an outer base of Q, we have $V_n \subset W$ for some $n \in \omega$ and therefore $L \backslash W \subset \{y_0, \ldots, y_{n-1}\}$ is a finite set which can be covered by a finite $\mathcal{U}'' \subset \mathcal{U}$. It is immediate that $\mathcal{U}' \cup \mathcal{U}''$ is a finite subcover of L and hence L is compact. The set $Y = L \cap A = \{y_n : n \in \omega\}$ is not closed in L; for otherwise, we have $Y \cap Q = \emptyset$ and we can apply again the fact that $\mathcal{V}$ is an outer base of Q to find $n \in \omega$ such that $V_n \cap Y = \emptyset$ which is a contradiction with $y_n \in Y \cap V_n$. Thus L is a compact subspace of X such that $L \cap A$ is not closed in L; this contradiction shows that A is closed in X and hence X is a k-space. This proves that every p-space is a k-space.

Any countable space is Lindelöf Σ because it has a countable network (see Problem 228), so to finish our solution it suffices to construct a countable space which is not a k-space. Take any $\xi \in \beta\omega \backslash \omega$ and consider the space $N = \omega \cup \{\xi\}$. All compact subspaces of N are closed in $\beta\omega$; since every infinite closed subspace of $\beta\omega$ is uncountable (see Fact 2 of T.131), no compact subset of N is infinite. It is clear that ω is dense in N while $K \cap \omega$ is closed in K for any compact $K \subset N$ because K has to be finite. This shows that N is a Lindelöf Σ-space which is not a k-space.

T.231. *Give an example of a countable space which is not a p-space. Note that this example shows that not every Lindelöf Σ-space is a p-space.*

Solution. Any countable space is Lindelöf Σ because it has a countable network (see Problem 228). Take any $\xi \in \beta\omega \backslash \omega$ and consider the space $N = \omega \cup \{\xi\}$. All compact subspaces of N are closed in $\beta\omega$; since every infinite closed subspace of $\beta\omega$ is uncountable (see Fact 2 of T.131), no compact subset of N is infinite. It is clear that ω is dense in N while $K \cap \omega$ is closed in K for any compact $K \subset N$ because K has to be finite. This shows that N is a countable space (and hence a Lindelöf Σ-space) which is not a k-space. Thus N is not a p-space either because every p-space is a k-space by Problem 230.

T.232. *Prove that any Čech-complete space is a p-space. Give an example of a p-space which is not Čech-complete.*

Solution. If X is Čech-complete, fix a family $\{U_n : n \in \omega\} \subset \tau(\beta X)$ such that $\bigcap\{U_n : n \in \omega\} = X$ and let $\mathcal{U}_n = \{U_n\}$ for all $n \in \omega$. Then $X \subset \bigcup \mathcal{U}_n$ for each $n \in \omega$ and $\bigcap\{St(x, \mathcal{U}_n) : n \in \omega\} = \bigcap\{U_n : n \in \omega\} = X$ so X is a p-space. This proves that every Čech-complete space is a p-space. Finally, the space $\mathbb{Q}$ is a p-space being second countable (see Problem 221) while it is not Čech-complete because it is of first category in itself (see Problem 274 of [TFS]).

T.233. *Prove that the following conditions are equivalent for any space X:*

(i) *for an arbitrary compactification bX of the space X, there exists a countable family of compact subspaces of bX which separates X from $bX \backslash X$;*

(ii) *there exists a compactification bX of the space X and a countable family of compact subspaces of bX which separates X from $bX \backslash X$;*

(iii) *there exists a compactification bX of the space X and a countable family of Lindelöf Σ-subspaces of bX which separates X from $bX \backslash X$;*

(iv) *there exists a space Z such that X is a subspace of Z and there is a countable family of compact subspaces of Z which separates X from $Z \backslash X$;*

(v) *X is a Lindelöf Σ-space.*

Solution. A family $\mathcal{C}$ of subsets of a space Z is *compact* if all elements of $\mathcal{C}$ are compact. A cover $\mathcal{C}$ of Z is called *a Σ-cover* if there is a σ-discrete $\mathcal{F} \subset \exp(Z)$ which is a network with respect to $\mathcal{C}$. Observe that in a Lindelöf space Z, any σ-discrete family is countable, so if Z is Lindelöf and $\mathcal{C}$ is a Σ-cover of Z, then there is actually a *countable* network with respect to $\mathcal{C}$.

The implication (i)$\Longrightarrow$(ii) is trivial; (ii)$\Longrightarrow$(iii) is an easy consequence of the fact that each compact space is Lindelöf Σ. To prove (iii)$\Longrightarrow$(iv) assume (iii) and let $\mathcal{F}$ be a countable family of Lindelöf Σ-subspaces of bX which separates X from $bX \backslash X$. For any element $F \in \mathcal{F}$ there is a compact Σ-cover $\mathcal{C}_F$ of the space F (see Problem 225); let $\mathcal{N}_F$ be the relevant countable network in F with respect to $\mathcal{C}_F$. The family $\mathcal{G} = \{\overline{P} : P \in \mathcal{N}_F$ for some $F \in \mathcal{F}\}$ consists of compact subsets of bX (the bar denotes the closure in bX). Observe that $F = \bigcup \mathcal{N}_F$ for any $F \in \mathcal{F}$ and hence $X \subset \bigcup \mathcal{F} \subset \bigcup \mathcal{G}$ which shows that $\mathcal{G}$ separates X from $bX \backslash X$ if $bX \backslash X = \emptyset$. Now, if $x \in X$ and $y \in bX \backslash X$, then there is $F \in \mathcal{F}$ such that $x \in F$ and $y \notin F$. There is $C \in \mathcal{C}_F$ such that $x \in C$; it is clear that $y \notin C$ and hence there is $U \in \tau(C, bX)$ such that $y \notin \overline{U}$. The family $\mathcal{N}_F$ being a network with respect to $\mathcal{C}_F$, there is $P \in \mathcal{N}_F$ such that $C \subset P \subset U \cap F$. Consequently, $x \in C \subset \overline{P} \subset \overline{U} \subset bX \backslash \{y\}$ which shows that $x \in \overline{P}$ while $y \notin \overline{P}$. Thus $\mathcal{G}$ is a countable family of compact subsets of $Z = bX$ which separates X from $Z \backslash X$, i.e., (iii)$\Longrightarrow$(iv) is proved.

To prove the implication (iv)$\Longrightarrow$(v) take a countable family $\mathcal{F}'$ of compact subsets of Z which separates X from $Z \backslash X$. We can assume, without loss of generality that the family $\mathcal{F}'$ is closed under finite intersections. Let $\mathcal{F} = \{F \cap X : F \in \mathcal{F}'\}$. Then $\mathcal{F}$ is a countable family of subsets of X. Given any point $x \in X$ let $C_x = \bigcap\{F : F \in \mathcal{F}'$ and $x \in F\}$. Since $\mathcal{F}'$ separates X from $Z \backslash X$, we have $C_x \subset X$ and hence $\mathcal{C} = \{C_x : x \in X\}$ is a compact cover of X. Take any $x \in X$

and $U \in \tau(C_x, X)$; there is $V \in \tau(Z)$ such that $V \cap X = U$. For the family $\mathcal{F}'_x = \{F : F \in \mathcal{F}' \text{ and } x \in F\}$, we have $C_x = \bigcap \mathcal{F}'_x \subset V$ and hence we can apply Fact 1 of S.326 (and the fact that $\mathcal{F}'$ is closed under finite intersections) to conclude that there is $F \in \mathcal{F}'_x \subset \mathcal{F}'$ such that $C_x \subset F \subset V$ and hence $C_x \subset F \cap X \subset U$ which, together with $F \cap X \in \mathcal{F}$, shows that $\mathcal{F}$ is a network with respect to C. Therefore X is a Lindelöf Σ-space by Problem 225 and hence we established that (iv)$\Longrightarrow$(v).

Finally, assume that X is a Lindelöf Σ-space and fix a compact cover C of the space X such that there is a countable network $\mathcal{F}$ with respect to C. The family $\mathcal{G} = \{\overline{F} : F \in \mathcal{F}\}$ is countable and consists of compact subsets of bX (the bar denotes the closure in bX). Since $\mathcal{F}$ is a cover of X, so is $\mathcal{G}$ which shows that we have the required separation of X from $bX \setminus X$ when $bX \setminus X = \emptyset$. If $bX \setminus X \neq \emptyset$, take any $x \in X$ and $y \in bX \setminus X$; there is $C \in \mathcal{C}$ with $x \in C$. Since $y \notin C$, there is $U \in \tau(C, bX)$ such that $y \notin \overline{U}$. Since $\mathcal{F}$ is a network with respect to C, there is $F \in \mathcal{F}$ such that $C \subset F \subset U \cap X$. Then $G = \overline{F} \in \mathcal{G}$ and we have $x \in C \subset G$ while $G \subset \overline{U}$ and hence $y \notin G$. Thus $\mathcal{G}$ is a countable family of compact subsets of bX which separates X from $bX \setminus X$, so (v)$\Longrightarrow$(i) is proved and hence our solution is complete.

T.234. *Let X be a space of countable tightness such that $C_p(X)$ is a Σ-space. Prove that if $C_p(X)$ is normal, then it is Lindelöf.*

Solution. Since $t(X) = \omega$, we have $t_m(X) \leq t(X) = \omega$ (see Problem 419 of [TFS]). Therefore $q(C_p(X)) = t_m(X) = \omega$, i.e., $C_p(X)$ is realcompact (see Problem 429 of [TFS]). The space $C_p(X)$ being a Σ-space, there is a cover $\mathcal{C}$ of $C_p(X)$ such that all elements of $\mathcal{C}$ are closed and countably compact while there exists a σ-discrete network $\mathcal{F}$ with respect to $\mathcal{C}$. Every closed subspace of a realcompact space is realcompact and every realcompact countably compact space is compact (see Problems 403 and 407 of [TFS]), so all elements of $\mathcal{C}$ are compact. The space $C_p(X)$ being normal, we have $\text{ext}(C_p(X)) = \omega$ (see Problem 295 of [TFS]) and hence the family $\mathcal{F}$ has to be countable. Thus we can apply Problem 225 to conclude that $C_p(X)$ is a Lindelöf Σ-space.

T.235. *Let X be a Σ-space with a G_δ-diagonal. Prove that X has a σ-discrete network.*

Solution. If Z is a space and $\mathcal{A} \subset \exp(Z)$, then $\text{St}(z, \mathcal{A}) = \bigcup\{A \in \mathcal{A} : z \in A\}$; let $\wedge\mathcal{A}$ be the family of all finite intersections of the elements of $\mathcal{A}$. If $\mathcal{A}_1, \ldots, \mathcal{A}_n$ are families of subsets of Z, then $\mathcal{A}_1 \wedge \cdots \wedge \mathcal{A}_n = \{A_1 \cap \cdots \cap A_n : A_i \in \mathcal{A}_i \text{ for all } i \leq n\}$. The space Z is *subparacompact* if every open cover of Z has a σ-discrete (not necessarily open) refinement.

Fact 1. A space Z has a G_δ-diagonal if and only if there is a sequence $\{\mathcal{D}_n : n \in \omega\}$ of open covers of Z such that $\bigcap\{\text{St}(z, \mathcal{D}_n) : n \in \omega\} = \{z\}$ for any $z \in Z$. The sequence $\{\mathcal{D}_n : n \in \omega\}$ is called *a G_δ-diagonal sequence for Z.*

Proof. Let $\Delta = \{(z,z) : z \in Z\}$ be the diagonal of Z. Given a G_δ-diagonal sequence $\{\mathcal{D}_n : n \in \omega\}$ for the space Z, consider the set $O_n = \bigcup\{U \times U : U \in \mathcal{D}_n\}$; it is evident that O_n is an open subset of $Z \times Z$ for every $n \in \omega$. For any $z \in Z$ and $n \in \omega$, there is $U \in \mathcal{D}_n$ such that $z \in U$ and hence $(z,z) \in U \times U \subset O_n$. Therefore $\Delta \subset O_n$ for every $n \in \omega$. If $z = (x,y) \in (Z \times Z) \backslash \Delta$, then $x \neq y$ and hence there exists $n \in \omega$ such that $y \notin \mathrm{St}(x, \mathcal{D}_n)$. Now, if $U \in \mathcal{D}_n$ and $z \in U \times U$, then $\{x,y\} \subset U$ which shows that $y \in \mathrm{St}(x, \mathcal{D}_n)$ which is a contradiction. Therefore $(x,y) \notin U \times U$ for any $U \in \mathcal{D}_n$, i.e., $z \notin O_n$ which proves that $\Delta = \bigcap\{O_n : n \in \omega\}$ and hence $\Delta(Z) = \omega$.

Now assume that $\Delta(Z) = \omega$ and fix a family $\{O_n : n \in \omega\} \subset \tau(Z \times Z)$ such that $O_{n+1} \subset O_n$ for all $n \in \omega$ and $\bigcap\{O_n : n \in \omega\} = \Delta$. Given any $n \in \omega$, consider the family $\mathcal{D}_n = \{U \in \tau(Z) : U \times U \subset O_n\}$. If $z \in Z$, then $(z,z) \in O_n$ and hence there is $U \in \tau(z, Z)$ such that $U \times U \subset O_n$ which shows that $\mathcal{D}_n$ is a cover of Z for all $n \in \omega$. Given distinct $x, y \in Z$, there is $n \in \omega$ such that $(x,y) \notin O_n$; if $U \in \mathcal{D}_n$, then $(x,y) \notin U \times U$ and therefore $\{x,y\}$ is not contained in U for any $U \in \mathcal{D}_n$. It is easy to see that this implies that $y \notin \mathrm{St}(x, \mathcal{D}_n)$ and hence $\{x\} = \bigcap\{\mathrm{St}(x, \mathcal{D}_n) : n \in \omega\}$ for any $x \in Z$, i.e., $\{\mathcal{D}_n : n \in \omega\}$ is a G_δ-diagonal sequence in Z, so Fact 1 is proved.   □

Fact 2. Every countably compact space Z with $\Delta(Z) = \omega$ is compact and hence metrizable.

Proof. Apply Fact 1 to fix a G_δ-diagonal sequence $\{\mathcal{D}_n : n \in \omega\}$ of open covers of the space Z. It is obvious that if $\mathcal{D}'_n$ is an open refinement of $\mathcal{D}_n$ for any $n \in \omega$, then the family $\{\mathcal{D}'_n : n \in \omega\}$ is also a G_δ-diagonal sequence in Z. Thus, letting $\mathcal{D}'_n = \mathcal{D}_0 \wedge \cdots \wedge \mathcal{D}_n$ for each $n \in \omega$, we obtain a G_δ-diagonal sequence $\{\mathcal{D}'_n : n \in \omega\}$ such that $\mathcal{D}'_i$ is a refinement of $\mathcal{D}'_j$ whenever $j < i$. To simplify the notation, we will assume that $\mathcal{D}_n = \mathcal{D}'_n$, i.e., $\mathcal{D}_i$ is a refinement of $\mathcal{D}_j$ whenever $j < i$.

To prove compactness of Z, it suffices to show that Z is Lindelöf. To obtain a contradiction, assume that there is an open cover $\mathcal{U}$ of the space Z which has no countable subcover. Call a set $Y \subset Z$ *small* if there is a countable $\mathcal{U}' \subset \mathcal{U}$ such that $Y \subset \bigcup \mathcal{U}'$. Otherwise the set Y will be called *large*. Observe that a countable union of small sets is a small set, and hence if a set $Y = \bigcup\{Y_n : n \in \omega\}$ is large, then some Y_n has to be large. Observe also that if a set Y is small and closed in Z, then there is a finite $\mathcal{U}' \subset \mathcal{U}$ such that $Y \subset \bigcup \mathcal{U}'$ because Y is countably compact. Choose any $z_0 \in Z$; since the set $Z \backslash \{z_0\} = \bigcup\{Z \backslash \mathrm{St}(z_0, \mathcal{D}_n) : n \in \omega\}$ is large, there exists $n_0 \in \omega$ such that $Z \backslash \mathrm{St}(z_0, \mathcal{D}_{n_0})$ is large. Suppose that $0 < \alpha < \omega_1$ and we have chosen sets $\{z_\beta : \beta < \alpha\} \subset Z$ and $\{n_\beta : \beta < \alpha\} \subset \omega$ with the following properties:

(1) $z_\gamma \notin \bigcup\{\mathrm{St}(z_\beta, \mathcal{D}_{n_\beta}) : \beta < \gamma\}$ for any $\gamma < \alpha$;
(2) the set $L_\gamma = Z \backslash (\bigcup\{\mathrm{St}(z_\beta, \mathcal{D}_{n_\beta}) : \beta \leq \gamma\})$ is large for any $\gamma < \alpha$.

If $\alpha = \gamma + 1$ for some $\gamma < \omega_1$, then choose any point $z_\alpha \in L_\gamma$ and observe that the set $L_\gamma \backslash \{z_\alpha\} = \bigcup\{L_\gamma \backslash \mathrm{St}(z_\alpha, \mathcal{D}_n) : n \in \omega\}$ is large and hence there is $n_\alpha \in \omega$ such that $L_\gamma \backslash \mathrm{St}(z_\alpha, \mathcal{D}_{n_\alpha})$ is large. It is immediate that the properties (1) and (2) are fulfilled for all $\gamma \leq \alpha$.

Now if α is a limit ordinal, then the set $M_\alpha = \bigcap\{L_\gamma : \gamma < \alpha\}$ is large. Indeed, suppose that there is a countable family $\mathcal{U}' \subset \mathcal{U}$ such that $M_\alpha \subset U = \bigcup \mathcal{U}'$; then $\{L_\gamma \backslash U : \gamma < \alpha\}$ is decreasing and consists of closed subsets of a countably compact space Z and $\bigcap\{L_\gamma \backslash U : \gamma < \alpha\} = \emptyset$. Take any increasing sequence $\{\gamma_n : n \in \omega\} \subset \alpha$ such that $\lim \gamma_n = \alpha$. If $F_n = L_{\gamma_n} \backslash U$, then $F_{n+1} \subset F_n$ for every $n \in \omega$ and $\bigcap\{F_n : n \in \omega\} = \emptyset$. By countable compactness of Z, there is $n \in \omega$ such that $F_n = \emptyset$ and hence $L_{\gamma_n} \subset U$ which shows that L_{γ_n} is small, a contradiction. Therefore M_α is large and reasoning as in the case of a successor α, we can choose $z_\alpha \in M_\alpha$ and $n_\alpha \in \omega$ such that the set $L_\alpha = M_\alpha \backslash \mathrm{St}(z_\alpha, \mathcal{D}_{n_\alpha})$ is large. It is clear that in this case we also have the properties (1) and (2) for all $\gamma \leq \alpha$ and therefore our construction can be continued to give us sets $\{z_\alpha : \alpha < \omega_1\} \subset Z$ and $\{n_\alpha : \alpha < \omega_1\} \subset \omega$ with the properties (1) and (2) for each $\alpha < \omega_1$.

There is $n \in \omega$ and an uncountable $A \subset \omega_1$ such that $n_\alpha = n$ for all $\alpha \in A$. The set $E = \{z_\alpha : \alpha \in A\}$ is closed and discrete in Z because, given $z \in Z$, there is $U \in \mathcal{D}_n$ with $z \in U$; it easily follows from (1) that U cannot contain more than one element of E. The space Z being countably compact, this is a contradiction which proves that Z is Lindelöf and hence compact. Finally, apply Problem 091 to see that Z is metrizable. Fact 2 is proved. $\square$

Fact 3. If Z is a space and $\mathcal{F} \subset \exp(X)$ is σ-discrete, then the family $\mathcal{G} = \wedge \mathcal{F}$ is also σ-discrete.

Proof. We have $\mathcal{F} = \bigcup\{\mathcal{F}_n : n \in \omega\}$ where $\mathcal{F}_n$ is discrete for all $n \in \omega$. If $G \in \mathcal{G}$, then there is $n \in \omega$ and distinct $j_0, \ldots, j_n \in \omega$ such that $G = F_0 \cap \cdots \cap F_n$ for some $F_0 \in \mathcal{F}_{j_0}, \ldots, F_n \in \mathcal{F}_{j_n}$.

Therefore $\wedge \mathcal{F} = \bigcup\{\mathcal{F}_{j_0} \wedge \cdots \wedge \mathcal{F}_{j_n} : j_0, \ldots, j_n \in \omega \text{ are distinct}\}$, so it suffices to show that the family $\mathcal{F}_{j_0} \wedge \cdots \wedge \mathcal{F}_{j_n}$ is discrete for any distinct $j_0, \ldots, j_n \in \omega$. Take any $z \in Z$ and $i \leq n$; since $\mathcal{F}_{j_i}$ is discrete, there is $U_i \in \tau(z, Z)$ and $F_i \in \mathcal{F}_{j_i}$ such that $U_i \cap F = \emptyset$ for any $F \in \mathcal{F}_{j_i} \backslash \{F_i\}$. Then $U = \bigcap_{i \leq n} U_i \in \tau(z, Z)$. If $G_i \in \mathcal{F}_{j_i}$ for all $i \leq n$ and $G = \bigcap_{i \leq n} G_i$, then $U \cap G \neq \emptyset$ implies $U \cap G_i \neq \emptyset$ and therefore $G_i = F_i$ for every $i \leq n$, i.e., U intersects at most one element of $\mathcal{F}_{j_0} \wedge \cdots \wedge \mathcal{F}_{j_n}$. Fact 3 is proved. $\square$

Fact 4. A space Z is subparacompact if and only if any open cover of Z has a closed σ-discrete refinement.

Proof. Sufficiency is obvious, so take any subparacompact space Z and $\mathcal{U} \subset \tau(Z)$ such that $\bigcup \mathcal{U} = Z$. For every $z \in Z$ take $U_z \in \mathcal{U}$ with $z \in U$ and $V_z \in \tau(z, Z)$ such that $\overline{V}_z \subset U_z$. The cover $\{V_z : z \in Z\}$ has a σ-discrete refinement $\mathcal{F}$; if $\mathcal{G} = \{\overline{F} : F \in \mathcal{F}\}$, then $\mathcal{G}$ is a closed σ-discrete refinement of $\mathcal{U}$ so Fact 4 is proved. $\square$

Returning to our solution fix a G_δ-diagonal sequence $\{\mathcal{D}_n : n \in \omega\}$ in the space X (which exists by Fact 1). Let $\mathcal{C}$ be a closed cover of X such that every $C \in \mathcal{C}$ is countably compact and there is a σ-discrete network $\mathcal{F}$ with respect to $\mathcal{C}$. The property $\Delta(X) = \omega$ implies that $\Delta(C) = \omega$ for any $C \in \mathcal{C}$ and therefore C is compact and metrizable by Fact 2. Therefore, for every $U \in \tau(C, X)$, there is

$V \in \tau(C, X)$ such that $\overline{V} \subset U$; as a consequence, $\mathcal{F}_1 = \{\overline{F} : F \in \mathcal{F}\}$ is also a σ-discrete network with respect to the family $\mathcal{C}$.

It follows from compactness of all elements of $\mathcal{C}$ that X is subparacompact (see Fact 2 of T.223). Apply Fact 4 to find a σ-discrete closed refinement $\mathcal{E}_n$ of the cover $\mathcal{D}_n$ for every $n \in \omega$. It is clear that the family $\mathcal{E} = \bigcup\{\mathcal{E}_n : n \in \omega\}$ is also σ-discrete and therefore $\mathcal{G} = \wedge\mathcal{E}$ is σ-discrete as well by Fact 3; besides, $\mathcal{G}$ is closed under finite intersections.

Let $\mathcal{N} = \mathcal{G} \wedge \mathcal{F}_1$; the family $\mathcal{N}$ is σ-discrete being contained in $\wedge(\mathcal{G} \cup \mathcal{F}_1)$ (see Fact 3), so it suffices to show that $\mathcal{N}$ is a network in the space X. Pick any $x \in X$ and $U \in \tau(x, X)$; there is $C \in \mathcal{C}$ such that $x \in C$. Furthermore, for each $n \in \omega$, there is $E_n \in \mathcal{E}_n$ such that $x \in E_n$; it follows from $\bigcap\{\mathrm{St}(x, \mathcal{D}_n) : n \in \omega\} = \{x\}$ that $\bigcap\{E_n : n \in \omega\} = \{x\}$. Therefore the family $\mathcal{S} = \{E_n \cap C : n \in \omega\}$ consists of compact subsets of X and $\bigcap \mathcal{S} = \{x\}$. Now apply Fact 1 of S.326 to conclude that there is a finite $\mathcal{S}' \subset \mathcal{S}$ such that $\bigcap \mathcal{S}' \subset U$ which shows that there is a finite $\mathcal{E}' \subset \mathcal{E}$ for which $x \in (\bigcap \mathcal{E}') \cap C \subset U$. The set $P = \bigcap \mathcal{E}'$ belongs to $\mathcal{G}$; since P is closed, the set $W = (X \backslash P) \cup U$ is an open neighborhood of C. The family $\mathcal{F}_1$ being a network with respect to $\mathcal{C}$, there is $F \in \mathcal{F}_1$ such that $C \subset F \subset W$. Now $N = P \cap F \in \mathcal{N}$ and $x \in N$. Besides, $N \subset F \subset W$ implies that $N \subset U \cup (X \backslash P)$ which together with $N \subset P$ implies $N \subset U$. Therefore, for any $x \in X$ and any $U \in \tau(x, X)$, there is $N \in \mathcal{N}$ such that $x \in N \subset U$. Consequently, $\mathcal{N}$ is a σ-discrete network of X, so our solution is complete.

T.236. *Let X be a Lindelöf Σ-space of countable pseudocharacter. Prove that $|X| \leq \mathfrak{c}$.*

Solution. Take a compact cover $\mathcal{C}$ of the space X such that there is a countable network $\mathcal{F}$ with respect to the family $\mathcal{C}$ (see Problem 225). It is easy to see that every $C \in \mathcal{C}$ is the intersection of a family $\mathcal{F}_C \subset \mathcal{F}$. Therefore $|\mathcal{C}| \leq |\exp \mathcal{F}| \leq |\exp \omega| = \mathfrak{c}$. We have $\chi(C) = \psi(C) \leq \psi(X) = \omega$ (see Problem 327 of [TFS]) and hence $|C| \leq 2^{\chi(C)} = 2^\omega = \mathfrak{c}$ for any $C \in \mathcal{C}$ (see Problem 329 of [TFS]). This implies $|X| \leq |\mathcal{C}| \cdot \mathfrak{c} = \mathfrak{c} \cdot \mathfrak{c} = \mathfrak{c}$.

T.237. *Prove that, under CH, there exists a hereditarily separable compact space X such that $C_p(X)$ does not have a dense Σ-subspace.*

Solution. It was proved in Problem 098 that, under CH, there exists a non-metrizable compact space X such that $hd^*(X) = \omega$. We have $hl^*(C_p(X)) = \omega$ by Problem 027, so if $S \subset C_p(X)$ is a dense Σ-subspace of $C_p(X)$, then S is Lindelöf Σ.

Given any $x \in X$ let $\varphi(x)(f) = f(x)$ for any $f \in S$. The function $\varphi(x)$ is continuous on S for any $x \in X$ and the map $\varphi : X \to C_p(S)$ is continuous by Problem 166 of [TFS]. It follows from the density of S is $C_p(X)$ that S separates the points of X and hence $\varphi : X \to Y = \varphi(X)$ is a condensation by Fact 2 of S.351. Since every condensation of a compact space is a homeomorphism, the space Y is homeomorphic to X, i.e., X embeds in $C_p(S)$.

Fact 1. Every continuous image of a Lindelöf Σ-space is a Lindelöf Σ-space.

Proof. Suppose that Z is a Lindelöf Σ-space and $f : Z \to T$ is a continuous onto map. By Problem 225 we can choose a compact cover $\mathcal{C}$ of the space Z such that there exists a countable network $\mathcal{F}$ with respect to $\mathcal{C}$. It is evident that $\mathcal{D} = \{f(C) : C \in \mathcal{C}\}$ is a compact cover of T; we claim that $\mathcal{G} = \{f(F) : F \in \mathcal{F}\}$ is a (countable) network with respect to $\mathcal{D}$. Indeed, if $D \in \mathcal{D}$ and $U \in \tau(D, T)$, then there is $C \in \mathcal{C}$ with $f(C) = D$ and therefore $V = f^{-1}(U) \in \tau(C, Z)$. The family $\mathcal{F}$ being a network with respect to $\mathcal{C}$, there is $F \in \mathcal{F}$ such that $C \subset F \subset V$; then $G = f(F) \in \mathcal{G}$ and $D \subset G \subset U$ which shows that $\mathcal{G}$ is a countable network with respect to the compact cover $\mathcal{D}$ of the space T. Thus T is a Lindelöf Σ-space and Fact 1 is proved. □

Fact 2. Every Lindelöf Σ-space is ω-stable.

Proof. Let Z be a Lindelöf Σ-space and take any continuous onto map $f : Z \to T$ such that there is a condensation of the space T onto a second countable space. It is immediate that $\Delta(T) \leq iw(T) = \omega$; the space T is Lindelöf Σ by Fact 1, so we can apply Problem 235 to conclude that T has a σ-discrete network $\mathcal{F}$. In a Lindelöf space, every σ-discrete family is Countable, so $nw(T) \leq |\mathcal{F}| = \omega$ which shows that Z is ω-stable and hence Fact 2 is proved. □

Returning to our solution, observe that the space S being ω-stable (see Fact 2) the space $C_p(S)$ is ω-monolithic (see Problem 154) and hence any separable subspace of $C_p(S)$ has a countable network. Since Y is separable, we have $w(X) = nw(X) = nw(Y) = \omega$ and hence X is metrizable which is a contradiction (see Fact 4 of S.307). Thus X is a hereditarily separable compact space such that there is no dense Lindelöf Σ-subspace in the space $C_p(X)$.

T.238. *Prove that, under CH, the space $C_p(\beta\omega)$ is not a Σ-space.*

Solution. It was proved in Problem 237 that, under CH, there exists a separable compact space X such that there is no dense Σ-subspace in $C_p(X)$. In particular, $C_p(X)$ is not a Σ-space. Take any countable dense $D \subset X$; if $p : \omega \to D$ is any surjection, then p is continuous and hence there is a continuous map $\varphi : \beta\omega \to X$ such that $\varphi|\omega = p$. Since $\varphi(\beta\omega)$ contains a dense subset D of the space X, it is dense in X and hence $\varphi(\beta\omega) = X$.

The dual map $\varphi^* : C_p(X) \to C_p(\beta\omega)$ embeds $C_p(X)$ in $C_p(\beta\omega)$ as a closed subspace (see Problem 163 of [TFS]) because the map φ is closed and hence $\mathbb{R}$-quotient. If $C_p(\beta\omega)$ is a Σ-space, then $\varphi^*(C_p(X))$ is also a Σ-space (see Problem 224) which is a contradiction because $\varphi^*(C_p(X))$ is homeomorphic to $C_p(X)$ while $C_p(X)$ is not a Σ-space.

T.239. *Prove that for any metrizable X, the space $C_p(X)$ has a dense Lindelöf Σ-subspace.*

Solution. Since X is metrizable, the space $C_p(X)$ has a dense σ-compact subspace by Problem 313 of [TFS]. Since every σ-compact subspace is Lindelöf Σ by Problem 226, the space $C_p(X)$ has a dense Lindelöf Σ-subspace.

T.240. *Let* $p : X \to Y$ *be compact-valued upper semicontinuous onto map. Prove that* $l(Y) \leq l(X)$.

Solution. Let $l(X) = \kappa$ and take any open cover $\mathcal{U}$ of the space Y. Given any point $x \in X$, the set $p(x)$ is compact and hence there is a finite family $\mathcal{U}_x \subset \mathcal{U}$ such that $p(x) \subset U_x = \bigcup \mathcal{U}_x$. The mapping p being upper semicontinuous, the set $V_x = \{x' \in X : p(x') \subset U_x\}$ is open in X and $x \in V_x$. The open cover $\{V_x : x \in X\}$ of the space X has a subcover of cardinality $\leq \kappa$, so there is $A \subset X$ such that $|A| \leq \kappa$ and $X = \bigcup \{V_x : x \in A\}$. If $\mathcal{U}' = \bigcup \{\mathcal{U}_x : x \in A\}$, then $\mathcal{U}' \subset \mathcal{U}$ and $|\mathcal{U}'| \leq \kappa$. Pick any $y \in Y$ and take $x \in X$ with $y \in p(x)$. There is $a \in A$ such that $x \in V_a$ and hence $p(x) \subset U_a \subset \bigcup \mathcal{U}'$. Therefore $y \in p(x) \subset \bigcup \mathcal{U}'$ which shows that $\mathcal{U}'$ is a subcover of $\mathcal{U}$ of cardinality κ. Thus $l(Y) \leq \kappa = l(X)$.

T.241. *Let* $p : X \to Y$ *be compact-valued upper semicontinuous onto map. Prove that if* X *is compact, then so is* Y.

Solution. Take an open cover $\mathcal{U}$ of the space Y. Given any $x \in X$, the set $p(x)$ is compact and hence there is a finite $\mathcal{U}_x \subset \mathcal{U}$ such that $p(x) \subset U_x = \bigcup \mathcal{U}_x$. The mapping p being upper semicontinuous, the set $V_x = \{x' \in X : p(x') \subset U_x\}$ is open in X and $x \in V_x$. The open cover $\{V_x : x \in X\}$ of the space X has a finite subcover, so there is a finite $A \subset X$ such that $X = \bigcup \{V_x : x \in A\}$. If $\mathcal{U}' = \bigcup \{\mathcal{U}_x : x \in A\}$, then $\mathcal{U}' \subset \mathcal{U}$ and $|\mathcal{U}'| < \omega$. Pick any $y \in Y$ and take $x \in X$ with $y \in p(x)$. There is $a \in A$ such that $x \in V_a$ and hence $p(x) \subset U_a \subset \bigcup \mathcal{U}'$. Therefore $y \in p(x) \subset \bigcup \mathcal{U}'$ which shows that $\mathcal{U}'$ is a finite subcover of $\mathcal{U}$. We proved that every open cover of Y contains a finite subcover of Y, i.e., Y is compact.

T.242. *Let* $p : X \to Y$ *be compact-valued upper semicontinuous onto map. Prove that if* X *is a Lindelöf* Σ*-space, then so is* Y.

Solution. Fix a compact cover $\mathcal{C}$ of the space X such that there is a countable network $\mathcal{F}$ with respect to $\mathcal{C}$. For any $A \subset X$, we let $p(A) = \bigcup \{p(x) : x \in A\}$; if $B \subset Y$, then $p^{-1}(B) = \{x \in X : p(x) \subset B\}$. Given $C \in \mathcal{C}$, let $C' = p(C)$; we claim that the map $q = p|C : C \to C'$ is upper semicontinuous. Indeed, if $U \in \tau(C')$, then take any $V \in \tau(Y)$ with $V \cap C' = U$ and observe that $q^{-1}(U) = p^{-1}(U) \cap C$ is an open subset of C. It is clear that q is compact-valued and onto, so C' is compact by Problem 241. Therefore $\mathcal{D} = \{p(C) : C \in \mathcal{C}\}$ is a compact cover of Y.

Let $\mathcal{G} = \{p(F) : F \in \mathcal{F}\}$; if $D \in \mathcal{D}$ and $U \in \tau(D, Y)$, then there is $C \in \mathcal{C}$ such that $D = p(C)$ and therefore $C \subset V = p^{-1}(U)$. The family $\mathcal{F}$ being a network with respect to $\mathcal{C}$ there is $F \in \mathcal{F}$ such that $C \subset F \subset V$. It is immediate that $G = p(F) \in \mathcal{G}$ and $D \subset G \subset U$ which shows that $\mathcal{G}$ is a countable network with respect to the compact cover $\mathcal{D}$ of the space Y. Thus Y is a Lindelöf Σ-space.

T.243. *Prove that*

(i) *any continuous image of a Lindelöf* Σ*-space is a Lindelöf* Σ*-space;*
(ii) *any perfect preimage of a Lindelöf* Σ*-space is a Lindelöf* Σ*-space.*

Solution. The statement of (i) was proved in Fact 1 of T.237. To prove (ii) assume that $f : X \to Y$ is a perfect map and Y is a Lindelöf Σ-space. Let $p(y) = f^{-1}(y)$ for every $y \in Y$. Then $p : Y \to X$ is a compact-valued map because f is perfect. Furthermore, $p(Y) = \bigcup \{p(y) : y \in Y\} = f^{-1}(Y) = X$ and hence p is an onto map. Given an open $U \subset X$ observe that

$$p^{-1}(U) = \{y \in Y : p(y) \subset U\} = \{y \in Y : f^{-1}(y) \subset U\} = Y \backslash f(X \backslash U)$$

is an open subset of Y because f is a closed map. Thus p is upper semicontinuous and therefore X is a Lindelöf Σ-space by Problem 242 so (ii) is proved.

T.244. *Prove that $w(X) = nw(X) = iw(X)$ for any Lindelöf p-space X. In particular, any Lindelöf p-space with a countable network has a countable base.*

Solution. Given a set Z, a family $\mathcal{A} \subset \exp(Z)$ and $P \subset Z$, recall that $\mathrm{St}(P, \mathcal{A}) = \bigcup \{A \in \mathcal{A} : A \cap P \neq \emptyset\}$. If $z \in Z$, then $\mathcal{A}(z) = \{A \in \mathcal{A} : z \in A\}$; we write $\mathrm{St}(z, \mathcal{A})$ instead of $\mathrm{St}(\{z\}, \mathcal{A})$. Given any $Z' \subset Z$, let $\mathcal{A}|Z' = \{A \cap Z' : A \in \mathcal{A}\}$; the family $\mathcal{A}$ is *locally finite in Z' at a point* $z \in Z'$ if there is $U \in \tau(z, Z')$ such that U intersects only finitely many elements of $\mathcal{A}$.

If Z is a space and we are given families $\mathcal{A}, \mathcal{B} \subset \exp(Z)$, say that $\mathcal{A}$ is *(strongly) inscribed* in $\mathcal{B}$ if for any $A \in \mathcal{A}$ there is $B \in \mathcal{B}$ such that $A \subset B$ (or $\overline{A} \subset B$ respectively). Say that $\mathcal{A}$ is *(strongly) barycentrically inscribed in $\mathcal{B}$* if the family $\{\mathrm{St}(z, \mathcal{A}) : z \in Z\}$ is (strongly) inscribed in $\mathcal{B}$.

Suppose that K is a compact space and Z is dense in K. If $\mathcal{U}_n \subset \tau(K)$ for every $n \in \omega$, call the sequence $\{\mathcal{U}_n : n \in \omega\}$ a *feathering* of the space Z in its compactification K if $Z \subset \bigcup \mathcal{U}_n$ for each $n \in \omega$ and $\bigcap \{\mathrm{St}(z, \mathcal{U}_n) : n \in \omega\} \subset Z$ for every $z \in Z$. A space Z is a *p-space* if and only if it has a feathering in some compactification of Z (see Fact 1 of T.224). It is easy to see that

(1) if $\{\mathcal{U}_n : n \in \omega\}$ a feathering of a space Z in βZ and we are given a sequence $\mathcal{S}' = \{\mathcal{U}'_n : n \in \omega\}$ such that $\mathcal{U}'_n \subset \tau(\beta Z)$, $Z \subset \bigcup \mathcal{U}'_n$ and $\mathcal{U}'_n$ is inscribed in $\mathcal{U}_n$ for each $n \in \omega$, then $\mathcal{S}'$ is also a feathering of Z in βZ.

Fact 1. Suppose that Z is a paracompact space. If $\mathcal{U} \subset \tau(\beta Z)$ and $Z \subset \bigcup \mathcal{U}$, then there exists a family $\mathcal{V} \subset \tau(\beta Z)$ such that $Z \subset \bigcup \mathcal{V}$, the family $\mathcal{V}$ is locally finite in βZ at each $z \in \bigcup \mathcal{V}$ and $\mathcal{V}$ is strongly barycentrically inscribed in $\mathcal{U}$.

Proof. For any $W \in \tau(Z)$, choose a set $O(W) \in \tau(\beta Z)$ such that $O(W) \cap Z = W$. Let $\mathcal{U}' = \{U' \in \tau(\beta Z) : \overline{U'} \subset U$ for some $U \in \mathcal{U}\}$ (the bar denotes the closure in βZ). It is clear that $Z \subset \bigcup \mathcal{U}'$ and hence there is a family $\mathcal{W} \subset \tau(Z)$ such that $\mathcal{W}$ is a barycentric refinement of $\mathcal{U}'|Z$ (see Problem 230 of [TFS]). Use paracompactness of Z again to find a locally finite (in Z) refinement $\mathcal{W}'$ of the cover $\mathcal{W}$. Observe that the family $\mathcal{V}' = \{O(W) : W \in \mathcal{W}'\}$ is locally finite in βZ at all points of Z. Indeed, given $z \in Z$, there is $U \in \tau(z, Z)$ such that the family $\mathcal{A} = \{W \in \mathcal{W}' : U \cap W \neq \emptyset\}$ is finite. It is an immediate consequence of density of Z in βZ that $\{W \in \mathcal{W}' : O(U) \cap O(W) \neq \emptyset\} = \mathcal{A}$, i.e., the set $O(U)$ witnesses that $\mathcal{V}'$ is locally finite at z in βZ. It is clear that the set $G = \{z \in \beta Z : \mathcal{V}'$ is locally finite at z in $\beta Z\}$ is open in βZ and $Z \subset G$. Let $\mathcal{V} = \{V \cap G : V \in \mathcal{V}'\}$; it is evident that $\mathcal{V} \subset \tau(\beta Z)$, $Z \subset \bigcup \mathcal{V}$ and $\mathcal{V}$ is locally finite in βZ at all points of G and hence at all points of $\bigcup \mathcal{V}$.

To prove that $\mathcal{V}$ is strongly barycentrically inscribed in $\mathcal{U}$, take any $z \in \beta Z$; the case when $\mathrm{St}(z, \mathcal{V}) = \emptyset$ is trivial, so assume that $\mathrm{St}(z, \mathcal{V}) \neq \emptyset$. Then $z \in \bigcup \mathcal{V}$ and therefore $\mathcal{V}$ is locally finite at z in βZ. In particular, the family $\mathcal{V}(z)$ is finite and hence the set $H = \bigcap \mathcal{V}(z)$ is open in βZ and non-empty; take any $t \in H \cap Z$. It turns out that every element of $\mathcal{V}$ which contains z also contains t, so $\mathrm{St}(z, \mathcal{V}) \subset \mathrm{St}(t, \mathcal{V})$. For every $V \in \mathcal{V}(t)$ we have $W_V = V \cap Z \in \mathcal{W}'$. By definition of $\mathcal{W}'$, there is $G_V \in \mathcal{W}$ such that $W_V \subset G_V$. The family $\mathcal{W}$ being a barycentric refinement of $\mathcal{U}'$, we have $\mathrm{St}(t, \mathcal{W}) \subset U'$ for some $U' \in \mathcal{U}'$; by definition of $\mathcal{U}'$ there is $U \in \mathcal{U}$ such that $\overline{U'} \subset U$. Consequently,

$$\overline{\mathrm{St}(z, \mathcal{V})} \subset \overline{\mathrm{St}(t, \mathcal{V})} = \bigcup \{\overline{V} : V \in \mathcal{V}(t)\} \subset \bigcup \{\overline{W}_V : V \in \mathcal{V}(t)\}$$

$$\subset \bigcup \{\overline{G}_V : V \in \mathcal{V}(t)\} \subset \overline{\mathrm{St}(t, \mathcal{W})} \subset \overline{U'} \subset U$$

and hence $\mathcal{V}$ is strongly barycentrically inscribed in $\mathcal{U}$, i.e., Fact 1 is proved. $\square$

Returning to our solution, observe that we have $iw(Z) \leq nw(Z) \leq w(Z)$ for any space Z, so it suffices to show that $w(X) \leq \kappa = iw(X)$. The space X is paracompact being Lindelöf, so we can apply (1) and Fact 1 to fix a feathering $\{\mathcal{U}_n : n \in \omega\}$ of X in βX such that each $\mathcal{U}_n$ is countable and locally finite in βX at all points of $\bigcup \mathcal{U}_n$ and $\mathcal{U}_{n+1}$ is strongly barycentrically inscribed in $\mathcal{U}_n$ for every $n \in \omega$.

There exists a condensation $f : X \to Y$ such that $w(Y) \leq \kappa$; let $\mathcal{B}$ be a base in Y such that any finite intersection of elements of $\mathcal{B}$ belongs to $\mathcal{B}$ and $|\mathcal{B}| \leq \kappa$. It follows from the fact that $\mathcal{B}$ is a base in Y that the family $\mathcal{C} = \{f^{-1}(U) : U \in \mathcal{B}\}$ has the following property:

(2) $\bigcap \{\overline{U} : U \in \mathcal{C}(x)\} = \{x\}$ for any $x \in X$ (the bar denotes the closure in X).

Since the family $\mathcal{U}$ of all finite unions of elements of $\bigcup \{\mathcal{U}_n : n \in \omega\}$ is countable, the family $\mathcal{E}$ of all finite intersections of elements of $\mathcal{C} \cup (\mathcal{U}|X)$ has cardinality $\leq \kappa$; we claim that $\mathcal{E}$ is a base in X. To prove it fix any $x \in X$ and $O \in \tau(x, X)$. For every $n \in \omega$, let $U_n = \mathrm{St}(x, \mathcal{U}_n)$; the family $\mathcal{U}_n(x)$ is finite, so it follows from the fact that $\mathcal{U}_{n+1}$ is strongly inscribed in $\mathcal{U}_n$ that $\mathrm{cl}_{\beta X}(U_{n+1}) \subset U_n$ for all $n \in \omega$. The set $K = \bigcap \{U_n : n \in \omega\} = \bigcap \{\mathrm{cl}_{\beta X}(U_n) : n \in \omega\}$ is contained in X and compact because $\{\mathcal{U}_n : n \in \omega\}$ is a feathering of X in βX. It is an easy consequence of Fact 1 of S.326 that the family $\{U_n : n \in \omega\}$ is an outer base of K in βX and therefore

(3) the family $\{U_n \cap X : n \in \omega\}$ is an outer base of K in X.

Apply Fact 1 of S.326 to the family $\mathcal{F} = \{\overline{U} \cap K : U \in \mathcal{C}(x)\}$; since $\bigcap \mathcal{F} = \{x\}$, there is $U \in \mathcal{C}(x)$ such that $\overline{U} \cap K \subset O$. Thus the set $\overline{U} \backslash O$ is closed in X and disjoint from K. The property (3) implies that there is $n \in \omega$ such that $(U_n \cap X) \cap (\overline{U} \backslash O) = \emptyset$. Therefore $V = (U_n \cap X) \cap U = U_n \cap U \in \mathcal{E}$ and

$x \in V \subset O$ which proves that $\mathcal{E}$ is a base in X. Thus $w(X) \leq \kappa \leq iw(X)$ whence $w(X) = nw(X) = iw(X)$ and therefore our solution is complete.

T.245. *Prove that any perfect image and any perfect preimage of a Lindelöf p-space is a Lindelöf p-space. Give an example of a closed continuous onto map $f : X \to Y$ such that X is a Lindelöf p-space and Y is not a p-space.*

Solution. Given a set Z, a family $\mathcal{A} \subset \exp(Z)$ and $P \subset Z$, recall that $\mathrm{St}(P, \mathcal{A}) = \bigcup\{A \in \mathcal{A} : A \cap P \neq \emptyset\}$. If $z \in Z$, then $\mathcal{A}(z) = \{A \in \mathcal{A} : z \in A\}$; we write $\mathrm{St}(z, \mathcal{A})$ instead of $\mathrm{St}(\{z\}, \mathcal{A})$. Given any $Z' \subset Z$, let $\mathcal{A}|Z' = \{A \cap Z' : A \in \mathcal{A}\}$; the family $\mathcal{A}$ is *locally finite in Z' at a point* $z \in Z'$ if there is $U \in \tau(z, Z')$ such that U intersects only finitely many elements of $\mathcal{A}$.

If Z is a space and we are given families $\mathcal{A}, \mathcal{B} \subset \exp(Z)$, say that $\mathcal{A}$ is *(strongly) inscribed* in $\mathcal{B}$ if for any $A \in \mathcal{A}$ there is $B \in \mathcal{B}$ such that $A \subset B$ (or $\overline{A} \subset B$ respectively). Say that $\mathcal{A}$ is *(strongly) barycentrically inscribed in $\mathcal{B}$* if the family $\{\mathrm{St}(z, \mathcal{A}) : z \in Z\}$ is (strongly) inscribed in $\mathcal{B}$.

Suppose that K is a compact space and Z is dense in K. If $\mathcal{U}_n \subset \tau(K)$ for every $n \in \omega$, call the sequence $\{\mathcal{U}_n : n \in \omega\}$ a *feathering* of the space Z in its compactification K if $Z \subset \bigcup \mathcal{U}_n$ for each $n \in \omega$ and $\bigcap\{\mathrm{St}(z, \mathcal{U}_n) : n \in \omega\} \subset Z$ for every $z \in Z$. A space Z is a p-space if and only if it has a feathering in some compactification of Z (see Fact 1 of T.224). If $f : Z \to T$ is a closed (and hence continuous and onto) map between the spaces Z and T, then $f^{\#}(U) = T \backslash f(Z \backslash U)$ for any $U \subset Z$. It is easy to see that $f^{\#}(U) = \{t \in T : f^{-1}(t) \subset U\}$ for any $U \subset Z$ and the set $f^{\#}(U)$ is open in T (maybe empty) for any open $U \subset Z$.

Fact 1. Let Z be a paracompact p-space; if $f : Z \to T$ is a perfect map (recall that our definition of a perfect map implies that f is continuous and onto), then T is also a paracompact p-space. In other words, a perfect image of a paracompact p-space is a paracompact p-space.

Proof. The space T is paracompact by Fact 4 of S.226. There exists a continuous map $g : \beta Z \to \beta T$ such that $g|Z = f$ (see Problem 258 of [TFS]). Furthermore, g is onto and $g(\beta Z \backslash Z) = \beta T \backslash T$ (see Fact 3 of S.261).

Using paracompactness of Z, we will improve its feathering in βZ which exists because Z is a p-space. It is easy to see that

(1) if $\{\mathcal{U}_n : n \in \omega\}$ a feathering of the space Z in βZ and we are given a sequence $\mathcal{S}' = \{\mathcal{U}'_n : n \in \omega\}$ such that $\mathcal{U}'_n \subset \tau(\beta Z)$, $Z \subset \bigcup \mathcal{U}'_n$ and $\mathcal{U}'_n$ is inscribed in $\mathcal{U}_n$ for each $n \in \omega$, then $\mathcal{S}'$ is also a feathering of Z in βZ.

We also have the following property:

(2) if $\mathcal{U} \subset \tau(\beta Z)$ and $Z \subset \bigcup \mathcal{U}$, then there exists a family $\mathcal{V} \subset \tau(\beta Z)$ such that $Z \subset \bigcup \mathcal{V}$, the family $\mathcal{V}$ is locally finite in βZ at each $z \in \bigcup \mathcal{V}$ and $\mathcal{V}$ is strongly barycentrically inscribed in $\mathcal{U}$

which is an immediate consequence of Fact 1 of T.244.

Now apply properties (1) and (2) to choose a feathering $\mathcal{U} = \{\mathcal{U}_n : n \in \omega\}$ of Z in βZ such that each $\mathcal{U}_n$ is locally finite in βZ at all points of $\bigcup \mathcal{U}_n$ and $\mathcal{U}_{n+1}$ is strongly barycentrically inscribed in $\mathcal{U}_n$ for all $n \in \omega$. For any $A \subset \beta Z$, let $\mathcal{U}[A] = \bigcap \{\mathrm{St}(A, \mathcal{U}_n) : n \in \omega\}$; if $A = \{z\}$ for some $z \in \beta Z$, then we write $\mathcal{U}[z]$ instead of $\mathcal{U}[\{z\}]$. Observe that

(3) $\mathcal{U}[z] \subset \beta Z \backslash Z$ for any $z \in \beta Z \backslash Z$,

because if $y \in \mathcal{U}[z] \cap Z$, then there is $U_n \in \mathcal{U}_n$ such that $\{y, z\} \subset U_n$ for all $n \in \omega$ which implies that $\{y, z\} \subset \mathcal{U}[y] \subset Z$ (the last inclusion holds because $\mathcal{U}$ is a feathering of Z in βZ). This contradiction proves (3).

Furthermore,

(4) if $K \subset Z$ is compact, then $\mathcal{U}[K] \subset Z$ and the family $\{\mathrm{St}(K, \mathcal{U}_n) : n \in \omega\}$ is an outer base of $\mathcal{U}[K]$ in βZ.

For any $n \in \omega$, the family $\mathcal{U}_n(K) = \{U \in \mathcal{U}_n : U \cap K \neq \emptyset\}$ is finite being $\mathcal{U}_n | K$ a locally finite family in a compact space K. If $n \in \omega$ and $U \in \mathcal{U}_{n+1}$, then there is $V_U \in \mathcal{U}_n$ such that $\overline{U} \subset V_U$ (the bar denotes the closure in βZ). Since only finitely many elements of $\mathcal{U}_{n+1}$ intersect K, we have

$$\overline{\mathrm{St}(K, \mathcal{U}_{n+1})} = \bigcup \{\overline{U} : U \in \mathcal{U}_{n+1}(K)\} \subset \bigcup \{V_U : U \in \mathcal{U}_{n+1}(K)\} \subset \mathrm{St}(K, \mathcal{U}_n)$$

for each $n \in \omega$ and therefore $\mathcal{U}[K] = \bigcap \{\overline{\mathrm{St}(K, \mathcal{U}_n)} : n \in \omega\}$. Now apply Fact 1 of S.326 to conclude that for any set $W \in \tau(\mathcal{U}[K], \beta Z)$, there is $n \in \omega$ such that $\mathrm{St}(K, \mathcal{U}_n) \subset \overline{\mathrm{St}(K, \mathcal{U}_n)} \subset W$, and hence $\{\mathrm{St}(K, \mathcal{U}_n) : n \in \omega\}$ is an outer base of $\mathcal{U}[K]$ in βZ. To show that $\mathcal{U}[K] \subset Z$, assume that there is $z \in \mathcal{U}[K] \backslash Z$. This implies that for any $n \in \omega$, there is $U_n \in \mathcal{U}_n$ such that $z \in U_n$ and $U_n \cap K \neq \emptyset$. As a consequence, $\mathrm{St}(z, \mathcal{U}_n) \cap K \neq \emptyset$ for every $n \in \omega$. Since the family $\mathcal{U}_{n+1}$ is strongly barycentrically inscribed in $\mathcal{U}_n$, we have $\overline{\mathrm{St}(z, \mathcal{U}_{n+1})} \subset \mathrm{St}(z, \mathcal{U}_n)$ for each $n \in \omega$ and therefore $\{\overline{\mathrm{St}(z, \mathcal{U}_n)} \cap K : n \in \omega\}$ is a decreasing sequence of non-empty compact subsets of K which shows that there exists a point $y \in K \cap (\bigcap \{\overline{\mathrm{St}(z, \mathcal{U}_n)} : n \in \omega\}) \subset Z$. However, we have $\bigcap \{\overline{\mathrm{St}(z, \mathcal{U}_n)} : n \in \omega\} = \bigcap \{\mathrm{St}(z, \mathcal{U}_n) : n \in \omega\} = \mathcal{U}[z] \subset \beta Z \backslash Z$ by (3); this contradiction finishes the proof of the property (4).

Let $\mathcal{V}_n = \{\mathrm{St}(f^{-1}(t), \mathcal{U}_n) : t \in T\}$ for all $n \in \omega$. It is clear that $\mathcal{V}_n \subset \tau(\beta Z)$ and $Z \subset \bigcup \mathcal{V}_n$ for all $n \in \omega$. We claim that

(5) $\mathcal{V}[K] = \bigcap \{\mathrm{St}(K, \mathcal{V}_n) : n \in \omega\} \subset Z$ for any compact $K \subset Z$.

Indeed, assume that $z \in \beta Z \backslash Z$ for some point $z \in \mathcal{V}[K]$; let $t = g(z)$. There exists a sequence $\{t_n : n \in \omega\} \subset Y$ such that $z \in \mathrm{St}(f^{-1}(t_n), \mathcal{U}_n)$ and $\mathrm{St}(f^{-1}(t_n), \mathcal{U}_n) \cap K \neq \emptyset$ for all $n \in \omega$. Therefore we can find $V_n, W_n \in \mathcal{U}_n$ such that $z \in W_n$, $W_n \cap f^{-1}(t_n) \neq \emptyset$, $V_n \cap f^{-1}(t_n) \neq \emptyset$ and $V_n \cap K \neq \emptyset$ for all $n \in \omega$. Choose any $a_n \in W_n \cap f^{-1}(t_n)$ and $b_n \in V_n \cap f^{-1}(t_n)$ for each $n \in \omega$. It follows from (3) and (4) that $F = \mathcal{U}[K]$ and $G = \mathcal{U}[z]$ are compact subsets of Z and $\beta Z \backslash Z$ respectively and hence $g(F) \cap g(G) = \emptyset$ because $g(F) = f(F) \subset T$

and $g(G) \subset \beta T \setminus T$. Take disjoint $U \in \tau(g(F), \beta T)$ and $V \in \tau(g(G), \beta T)$; then $U' = g^{-1}(U)$ and $V' = g^{-1}(V)$ are disjoint open (in βZ) neighborhoods of F and G respectively. Observe that $b_n \in V_n \subset \mathrm{St}(K, \mathcal{U}_n)$ for each $n \in \omega$ which, together with (4), implies that there is $m \in \omega$ such that $b_n \in \mathrm{St}(K, \mathcal{U}_n) \subset U'$ for all $n \geq m$.

The family $\mathcal{U}_{n+1}$ is strongly barycentrically inscribed in $\mathcal{U}_n$ for all $n \in \omega$, so it follows from Fact 1 of S.326 that the family $\{\mathrm{St}(z, \mathcal{U}_n) : n \in \omega\}$ is an outer base of $\mathcal{U}[z]$ in βZ. Thus there is $k \in \omega$ such that $a_n \in W_n \subset \mathrm{St}(z, \mathcal{U}_n) \subset V'$ for all $n \geq k$. For $n = m + k$ we have $a_n \in V'$ and $b_n \in U'$ and hence $t_n = g(a_n) = g(b_n) \in U \cap V$ which is a contradiction. The property (5) is proved.

Clearly, $\mathcal{W}_n = \{g^{\#}(V) : V \in \mathcal{V}_n\}$ is a family of open subsets of βT for every $n \in \omega$. We claim that the sequence $\mathcal{W} = \{\mathcal{W}_n : n \in \omega\}$ is a feathering of T in βT.

To prove it observe first that $g^{-1}(t) = f^{-1}(t) \subset \mathrm{St}(f^{-1}(t), \mathcal{U}_n)$ and therefore $t \in g^{\#}(\mathrm{St}(f^{-1}(t), \mathcal{U}_n)) \in \mathcal{W}_n$ for any $t \in T$ and $n \in \omega$ which shows that $T \subset \bigcup \mathcal{W}_n$ for all $n \in \omega$.

Now for an arbitrary $t \in T$, let $V_t^n = \mathrm{St}(f^{-1}(t), \mathcal{U}_n)$ and $W_t^n = g^{\#}(V_t^n)$. Then for any $t \in T$ we have $\mathrm{St}(t, \mathcal{W}_n) = \bigcup \{W_s^n : t \in W_s^n\}$. For every $s \in T$, if $t \in W_s^n$, then $f^{-1}(t) \subset g^{-1}(W_s^n) \subset V_s^n$. Thus $g^{-1}(\mathrm{St}(t, \mathcal{W}_n)) \subset \mathrm{St}(f^{-1}(t), \mathcal{V}_n)$ for any $n \in \omega$. Therefore $g^{-1}(\bigcap\{\mathrm{St}(t, \mathcal{W}_n) : n \in \omega\}) \subset V[f^{-1}(t)] \subset Z$ by (5). Consequently, $\bigcap\{\mathrm{St}(t, \mathcal{W}_n) : n \in \omega\} \subset T$, i.e., the sequence $\{\mathcal{W}_n : n \in \omega\}$ is indeed a feathering of T in βT, so T is a p-space and Fact 1 is proved. □

Fact 2. Let Z be a normal space and assume that F is a non-empty closed subset of Z; for any $A \subset Z$, let $A^* = (A \setminus F) \cup \{F\}$. Given $z \in Z$, let $p_F(z) = z$ if $z \in Z \setminus F$ and $p_F(z) = F$ if $z \in F$. It is clear that $p_F : Z \to Z_F = \{F\} \cup (Z \setminus F)$. Then

(i) the family $\tau_F = \{U \in \tau(Z) : U \subset Z \setminus F\} \cup \{U^* : U \in \tau(F, Z)\}$ is a topology on the set Z_F;

(ii) the space $Z/F = (Z_F, \tau_F)$ is T_1 and normal (and hence Tychonoff) and the map $p_F : Z \to Z/F$ is continuous, closed and onto.

The operation of obtaining the space Z/F from a space Z is called *collapsing the set F to a point*.

Proof. (i) Since $\emptyset \in \tau(Z)$ and $\emptyset \subset Z \setminus F$, we have $\emptyset \in \tau_F$. Since $Z_F = Z^*$, we have $Z_F \in \tau_F$, so the first axiom of topology is satisfied. Let $\mathcal{U} = \{U \in \tau(Z) : U \subset Z \setminus F\}$ and $\mathcal{V} = \{U^* : U \in \tau(F, Z)\}$. It is immediate that $U \cap U' \in \mathcal{U}$ for any $U, U' \in \mathcal{U}$ and $V \cap V' \in \mathcal{V}$ for any $V, V' \in \mathcal{V}$. Now, if $U \in \mathcal{U}$ and $V \in \mathcal{V}$, then $U \cap V \in \mathcal{U}$ which shows that the intersection of any two elements of τ_F is again in τ_F.

As to the third axiom of topology, observe first that $\bigcup \mathcal{U}' \in \mathcal{U}$ and $\bigcup \mathcal{V}' \in \mathcal{V}$ for any $\mathcal{U}' \subset \mathcal{U}$ and $\mathcal{V}' \subset \mathcal{V}$. Now, if $\mathcal{W} \subset \tau_F$, then let $\mathcal{U}' = \mathcal{W} \cap \mathcal{U}$ and $\mathcal{V}' = \mathcal{W} \cap \mathcal{V}$. We have $U = \bigcup \mathcal{U}' \in \mathcal{U}$ and $V = \bigcap \mathcal{V}' \in \mathcal{V}$. Besides, $\bigcup \mathcal{W} = U \cup V \in \mathcal{V}$ which shows that the union of any subfamily of τ_F belongs to τ_F, i.e., τ_F is a topology on Z_F and hence (i) is proved.

(ii) The surjectivity of p_F is obvious; if $U \in \mathcal{U}$, then $p_F^{-1}(U) = U$ is an open subset of Z; if $V \in \mathcal{V}$, then $V = U^*$ for some $U \in \tau(F, Z)$ and therefore $p_F^{-1}(V) = U$ is again an open subset of Z. Since $\tau_F = \mathcal{U} \cup \mathcal{V}$, this shows that the map p_F is continuous. To prove that p_F is closed, take any closed $G \subset Z$. If $G \cap F = \emptyset$, then $p_F(G) = G$ is a closed subset of Z/F because $(Z/F)\backslash G = (Z\backslash G)^*$ is an open subset of Z/F. If $G \cap F \neq \emptyset$, then $p_F(G) = G \cup \{F\}$ is closed in Z/F because $(Z/F)\backslash(G \cup \{F\}) = Z\backslash(F \cup G) \in \mathcal{U}$ is also an open subset of Z/F. Thus the map p_F is closed.

It is straightforward to check that Z/F is a T_1-space; let us prove that it is normal. Take any disjoint closed sets $H, G \subset Z/F$. Then $H' = p_F^{-1}(H)$ and $G' = p_F^{-1}(G)$ are disjoint closed subsets of Z. By normality of the space Z, there are $U', V' \in \tau(Z)$ such that $G' \subset U'$, $H' \subset V'$ and $U' \cap V' = \emptyset$. It is immediate that $U = p_F^{\#}(U') \in \tau_F$, $V = p_F^{\#}(V') \in \tau_F$ and $U \cap V = \emptyset$, so Z/F is a T_4-space and Fact 2 is proved.   □

Returning to our solution, assume that Z is a Lindelöf p-space and T is a perfect image of Z. Then T is a p-space by Fact 1 because any Lindelöf space is paracompact. Since any continuous image of a Lindelöf space is Lindelöf, the space T is a Lindelöf p-space and hence any perfect image of a Lindelöf p-space is a Lindelöf p-space.

Now assume that $f : Z \to T$ is a perfect map and T is a Lindelöf p-space. Then Z is a Lindelöf Σ-space by Problem 243. Let $\{\mathcal{V}_n : n \in \omega\}$ be a sequence of open (in βT) covers of T which witnesses that T is a p-space. There is a continuous map $g : \beta Z \to \beta T$ such that $g|Z = f$ (see Problem 258 of [TFS]); besides, $g(\beta Z\backslash Z) \subset \beta T\backslash T$ (see Fact 3 of S.261) and therefore

(6) for any $A \subset \beta T$, we have $A \subset T$ if and only if $g^{-1}(A) \subset Z$.

Let $\mathcal{U}_n = \{g^{-1}(U) : U \in \mathcal{V}_n\}$ for all $n \in \omega$. It is clear that $\{\mathcal{U}_n : n \in \omega\}$ is a sequence of open (in βZ) covers of Z. Furthermore, $\mathrm{St}(z, \mathcal{U}_n) = g^{-1}(\mathrm{St}(g(z), \mathcal{V}_n))$ for any $n \in \omega$ and $z \in Z$. This implies, together with the property (6), that $\bigcap\{\mathrm{St}(z, \mathcal{U}_n) : n \in \omega\} \subset Z$ for any $z \in Z$, i.e., Z is a p-space. Thus we have established that any perfect preimage of a Lindelöf p-space is a Lindelöf p-space.

To construct the promised example, let $X = \mathbb{R}$ and $F = \omega \subset \mathbb{R}$. The space X is Lindelöf p by Problem 221; let $Y = X/F$ and $f = p_F$. Then f is a closed map by Fact 2. It is clear that Y has a countable network, so if Y is a p-space, then $w(Y) = \omega$ and, in particular, $\chi(F, Y) = \omega$ (according to the context, the set F is considered either as a point of Y or a subset of X). Let $\mathcal{W}$ be a countable local base at the point F in Y. Given any $U \in \tau(F, X)$ the set $V = f^{\#}(U)$ is open in Y and $F \in V$. Therefore there is $W \in \mathcal{W}$ such that $F \in W \subset V$. As a consequence, $F \subset f^{-1}(W) \subset f^{-1}(V) \subset U$ which shows that the family $\mathcal{V} = \{f^{-1}(W) : W \in \mathcal{W}\}$ is a countable outer base of the set F in X. Let $\{V_n : n \in \omega\}$ be an enumeration of $\mathcal{V}$. For each $n \in \omega$, the set V_n is an open neighborhood of n in $\mathbb{R}$ and hence there is $\varepsilon_n \in (0, \frac{1}{3})$ such that $(n - \varepsilon_n, n + \varepsilon_n) \subset V_n$; choose any $r_n \in (n - \varepsilon_n, n + \varepsilon_n)$. The set $W = \mathbb{R}\backslash\{r_n : n \in \omega\}$ is open in $\mathbb{R}$ because $\{r_n : n \in \omega\}$ is closed and discrete. Since $F \subset W$, there is $n \in \omega$ such that $F \subset V_n \subset W$ and hence $r_n \notin V_n$ which is a contradiction. Thus $\chi(F, \mathbb{R}) > \omega$ and hence Y is not first countable. Therefore $nw(Y) = \omega < w(Y)$ which shows that Y is not a p-space (see Problem 244) and makes our solution complete.

T.246. *Suppose that $C_p(X)$ is a closed continuous image of a Lindelöf p-space. Prove that X is countable.*

Solution. Given spaces Y and Z call a continuous onto map $h : Y \to Z$ *irreducible* if, for any closed $F \subset Y$ with $F \neq Y$, we have $h(F) \neq Z$. For any $U \subset Y$, let $h^{\#}(U) = Z \backslash h(Y \backslash U)$. It is easy to see that $h^{\#}(U) = \{z \in Z : h^{-1}(z) \subset U\}$; if the map h is closed, then $h^{\#}(U)$ is open (maybe empty) for any $U \in \tau(Y)$. Another easy observation is that a closed map $h : Y \to Z$ is irreducible if and only if $h^{\#}(U) \neq \emptyset$ for any $U \in \tau^{*}(Y)$. A space Y is *of pointwise countable type* if for every $y \in Y$ there is a compact $P \subset Y$ such that $y \in P$ and $\chi(P, Y) = \omega$.

Let Y be a space; given any points $y_1, \ldots, y_n \in Y$ and sets $O_1, \ldots, O_n \in \tau^{*}(\mathbb{R})$, the set $[y_1, \ldots, y_n; O_1, \ldots, O_n] = \{f \in C_p(Y) : f(y_i) \in O_i$ for all $i \leq n\}$ is called *a standard open subset of $C_p(Y)$*. Standard open sets $[y_1, \ldots, y_n; O_1, \ldots, O_n]$ where $n \in \mathbb{N}$, $y_1, \ldots, y_n \in Y$ and $O_1, \ldots, O_n \in \tau(\mathbb{R})$ form a base in the space $C_p(Y)$ (see Problem 056 of [TFS]). If $U = [y_1, \ldots, y_n; O_1, \ldots, O_n]$ is a standard open subset of $C_p(Y)$, then $\text{supp}(U) = \{y_1, \ldots, y_n\}$.

Fact 1. Let Y be a paracompact space. Suppose that Z is a space in which any point is a limit of a nontrivial convergent sequence. Then any closed map $h : Y \to Z$ is irreducible on some closed subset of Y, i.e., there is a closed $F \subset Y$ such that $h(F) = Z$ and $h_F = h|F$ is irreducible.

Proof. For every $y \in Z$ fix a sequence $S_y = \{y_n : n \in \omega\} \subset Z \backslash \{y\}$ converging to y. We will prove first that the set $P_y = h^{-1}(y) \cap \overline{\bigcup \{h^{-1}(y_n) : n \in \omega\}}$ is compact for every $y \in Z$. Indeed, if for some $y \in Z$ the set P_y is not compact, then it is not countably compact being closed in Y and hence paracompact (it is an easy exercise that any countably compact paracompact space is compact). Therefore there is a countably infinite closed discrete set $D = \{x_n : n \in \omega\} \subset P_y$. Since Y is paracompact, it is collectionwise normal and hence we can find a discrete family $\gamma = \{U_n : n \in \omega\} \subset \tau^{*}(Y)$ with $x_n \in U_n \cap P_y$ for all $n \in \omega$.

If A is an arbitrary finite subset of ω, then for each natural number n, we have $U_n \cap (\bigcup \{h^{-1}(z_k) : k \in \omega \backslash A\}) \neq \emptyset$. This makes it possible to choose a point $z_n \in U_n \cap (\bigcup \{h^{-1}(z_k) : k \in \omega\})$ in such a way that $h(z_m) \neq h(z_n)$ if $n \neq m$.

The family γ being discrete the set $D = \{z_n : n \in \omega\}$ is closed and discrete in Y. The set $h(D)$ is also closed because h is a closed map. Note that $h(D)$ has also to be discrete because $h(C)$ is closed for any $C \subset D$. However $h(D)$ is a nontrivial sequence converging to y, a contradiction with the fact that $h(D)$ is closed and discrete. This proves P_y is compact for all $y \in Z$.

Claim. Suppose that H is a closed subset of Y such that $h(H) = Z$. Then $H \cap P_y \neq \emptyset$ for all $y \in Z$.

Proof of the claim. Fix any $y \in Z$; it follows from $h(H) = Z$ that it possible to choose a point $t_n \in H \cap h^{-1}(y_n)$ for all $n \in \omega$. The map h is closed and therefore $\overline{\{t_n : n \in \omega\}} \cap h^{-1}(y) \neq \emptyset$. But $H \supset \{t_n : n \in \omega\}$ and $\overline{\{t_n : n \in \omega\}} \cap h^{-1}(y) \subset P_y$. Thus $H \cap P_y \neq \emptyset$ and the claim is proved.   □

Suppose that we have a family $\mathcal{F}$ of closed subsets of Y such that $\mathcal{F}$ is totally ordered by inclusion and $h(H) = Z$ for every $H \in \mathcal{F}$. Then $h(\bigcap \mathcal{F}) = Z$. Indeed, $H \cap P_y \neq \emptyset$ for any $y \in Z$ and $H \in \mathcal{F}$. We proved that the set P_y is compact, so $(\bigcap \mathcal{F}) \cap h^{-1}(y) \supset (\bigcap \mathcal{F}) \cap P_y \neq \emptyset$ for all $y \in Z$; consequently, $(\bigcap \mathcal{F}) \cap h^{-1}(y) \neq \emptyset$ which implies $y \in h(\bigcap \mathcal{F})$ for every $y \in Z$, i.e., $h(\bigcap \mathcal{F}) = Z$. Finally, use Zorn's lemma to find a closed $F \subset Y$ which is maximal (with respect to the inverse inclusion) in the family of all closed sets $H \subset Y$ such that $h(H) = Z$. It is evident that h_F is irreducible so Fact 1 is proved.   □

Returning to our solution, suppose that Y is a Lindelöf p-space for which there is a closed continuous onto map $\varphi : Y \to C_p(X)$. Given $f \in C_p(X)$ observe that the sequence $\{f + \frac{1}{n}\}$ is nontrivial and converges to f. Therefore Fact 1 is applicable to the map $\varphi : Y \to C_p(X)$ to obtain a closed $F \subset Y$ such that $\varphi(F) = C_p(X)$ and $\varphi|F$ is irreducible. The space F is also Lindelöf p by Problem 224 which shows that $C_p(X)$ is a closed irreducible image of a Lindelöf p-space. Thus we can assume, without loss of generality, that the map φ is irreducible. The space Y is of pointwise countable type by Fact 1 of T.222, so there is a non-empty compact $P \subset Y$ such that $\chi(P, Y) = \omega$. Fix a decreasing outer base $\{U_n : n \in \omega\}$ of the set P in Y. The space $K = \varphi(P) \subset C_p(X)$ is compact and the family $\{\varphi^{\#}(U_n) : n \in \omega\}$ consists of non-empty open subsets of $C_p(X)$ with the following property:

(1) for any $W \in \tau(K, C_p(X))$ there is $m \in \omega$ such that $V_n = \varphi^{\#}(U_n) \subset W$ for every $n \geq m$.

To see that (1) holds observe that $U_m \subset \varphi^{-1}(W)$ for some $m \in \omega$ because the family $\{U_n : n \in \omega\}$ is an outer base of P in Y and $P \subset \varphi^{-1}(W)$. Therefore $U_n \subset U_m \subset \varphi^{-1}(W)$ and hence $V_n \subset W$ for all $n \geq m$. Observe that if $V'_n \subset V_n$ for all $n \in \omega$, then (1) still holds for the sequence $\{V'_n : n \in \omega\}$. Therefore there exists a family $\mathcal{O} = \{O_n : n \in \omega\}$ of standard non-empty open subsets of $C_p(X)$ such that (1) holds for $\mathcal{O}$.

The set $A = \bigcup \{\operatorname{supp}(O_n) : n \in \omega\}$ is countable; to prove that $X = A$ assume that $z \in X \backslash A$. The map $p_z : C_p(X) \to \mathbb{R}$ defined by $p_z(f) = f(z)$ for all $f \in C_p(X)$ is continuous (see Problem 166 of [TFS]) and therefore $Q = p_z(K)$ is a compact and hence bounded subspace of $\mathbb{R}$. Take any bounded $H \in \tau(\mathbb{R})$ such that $Q \subset H$. The set $W = [z, H]$ is open in $C_p(X)$ and contains K, so there is $n \in \omega$ for which $O_n \subset W$. Take any $f \in O_n$; we have $g \in O_n$ for any $g \in C_p(X)$ with $g|\operatorname{supp}(O_n) = f|\operatorname{supp}(O_n)$. Since $\operatorname{supp}(O_n) \subset A$, we have $z \notin B = \operatorname{supp}(O_n)$ and hence there is $g \in C_p(X)$ such that $g|B = f|B$ and $g(z) \notin H$ (see Problem 034 of [TFS]). As a consequence, $g \in O_n \backslash W$ which is a contradiction. Therefore $X = A$ is countable and our solution is complete.

T.247. *Show that an open continuous image of a p-space is not necessarily a p-space. Supposing that $C_p(X)$ is an open continuous image of a p-space, prove that X is countable (and hence $C_p(X)$ is a p-space).*

Solution. Let S be the Sorgenfrey line (see Problem 165 of [TFS]). Then S is a Lindelöf space with $iw(S) = \omega < nw(S)$. Thus S is not a p-space by Problem 244. The space S is first countable, so there is a metrizable space M and an open continuous onto map $\varphi : M \to S$ (see Problem 223 of [TFS]). The space M is a p-space (see Problem 221) which proves that S is an open image of a p-space which fails to be a p-space.

Now assume that $C_p(X)$ is an open continuous image of a p-space Y. The space Y is *of pointwise countable type*, i.e., for every $y \in Y$, there is a compact $P \subset Y$ such that $y \in P$ and $\chi(P, Y) = \omega$ (see Fact 1 of T.222). It is an easy exercise that any open image of a space of pointwise countable type is a space of pointwise countable type. Thus $C_p(X)$ is of pointwise countable type and therefore there is a non-empty compact $P \subset C_p(X)$ such that $\chi(P, C_p(X)) = \omega$. Finally, apply Problem 170 of [TFS] to conclude that X is countable.

T.248. *Prove that X is a Lindelöf Σ-space if and only if there exists a second countable space M and a compact K such that X is a continuous image of a closed subspace of $K \times M$.*

Solution. Suppose that there exists a compact space K and a second countable space M such that some closed $F \subset K \times M$ maps continuously onto X. Observe first that K and M are Lindelöf Σ-spaces by Problems 226 and 221. Therefore $K \times M$ is also a Lindelöf Σ-space by Fact 1 of T.227 and hence so is the space F by Problem 224. Since every continuous image of a Lindelöf Σ-space is Lindelöf Σ by Problem 243, the space X is Lindelöf Σ as well. This proves sufficiency.

Now assume that X is a Lindelöf Σ-space and fix a family $\mathcal{F} = \{F_n : n \in \omega\}$ of compact subsets of βX which separates X from $\beta X \backslash X$ (this is possible by Problem 233). Let $M = \{s \in \omega^\omega : P_s = \bigcap\{F_{s(n)} : n \in \omega\} \subset X\}$; it is clear that M is a second countable space. For $K = \beta X$ consider the set $F = \{(x,s) \in K \times M : x \in P_s\}$; it is clear that $F \subset K \times M$. If $z = (x,t) \in (K \times M) \backslash F$, then $x \notin P_t$ and hence there is $n \in \omega$ such that $x \notin F_{t(n)}$. Then $V = K \backslash F_{t(n)} \in \tau(x, K)$ and $W = \{s \in M : s(n) = t(n)\}$ is an open subset of M with $t \in W$. Thus $U = V \times W$ is an open neighborhood of z; if $(y,s) \in U$, then $y \in V = K \backslash F_{t(n)}$ while $F_{s(n)} = F_{t(n)}$ which shows that $y \notin F_{s(n)}$ and therefore $y \notin P_s$ which implies $(y,s) \notin F$. This proves that every $z \in (K \times M) \backslash F$ has a neighborhood U with $U \cap F = \emptyset$, i.e., F is a closed subspace of $K \times M$.

Let $p : K \times M \to K$ be the natural projection, i.e., $p(z) = x$ for any point $z = (x,s) \in K \times M$. Then $f = p|F : F \to \beta X$ is a continuous map; for any $z = (x,s) \in F$, we have $x \in P_s \subset X$ by the definition of M and hence $x = p(z) \in X$ whence $f(F) = p(F) \subset X$. Now, given any $x \in X$, let $A = \{n \in \omega : x \in F_n\}$; then $x \in P = \bigcap\{F_n : n \in A\} \subset X$ because the family $\mathcal{F}$ separates X from $\beta X \backslash X$. Choose any $s \in \omega^\omega$ such that $s(\omega) = A$; then $P_s = P$ and therefore $z = (x,s) \in F$

while $f(z) = p(z) = x$. This shows that $f(F) = X$ and hence X is a continuous image of F. We settled necessity so our solution is complete.

T.249. *Prove that the following properties are equivalent for any space X:*

(i) *there exists a second countable space M and a space Y such that Y maps perfectly onto M and continuously onto X;*

(ii) *there exists an upper semicontinuous compact-valued onto map $\varphi : M \to X$ for some second countable space M;*

(iii) *X is a Lindelöf Σ-space.*

Solution. Given spaces Z, T and a map $h : Z \to T$, let $h^{\#}(U) = T \backslash h(Z \backslash U)$ for any $U \subset Z$. It is easy to see that $h^{\#}(U) = \{t \in T : h^{-1}(t) \subset U\}$; if the map h is closed, then $h^{\#}(U)$ is open (maybe empty) for any $U \in \tau(Z)$.

(i)$\Longrightarrow$(ii). Assume that $w(M) = \omega$ while, for some space Y, there exists a perfect map $f : Y \to M$ and a continuous onto map $g : Y \to X$. Given any $s \in M$, let $\varphi(s) = g(f^{-1}(s))$. It is obvious that $\varphi : M \to X$ is a compact-valued map. If $x \in X$, then $g(y) = x$ for some $y \in Y$ and hence $x \in g(f^{-1}(f(y)))$, i.e., $x \in \varphi(s)$ for $s = f(y)$ which shows that the map φ is onto. To see that φ is upper semicontinuous, take any $U \in \tau(X)$. Then

$$\varphi^{-1}(U) = \{s \in M : \varphi(s) \subset U\} = \{s \in M : f^{-1}(s) \subset g^{-1}(U)\} = f^{\#}(g^{-1}(U))$$

is an open subset of M and therefore $\varphi : M \to X$ is an upper semicontinuous compact-valued onto map.

(ii)$\Longrightarrow$(iii). Assume that there is a second countable space M and an upper semicontinuous compact-valued onto map $\varphi : M \to X$. The space M is Lindelöf Σ by Problem 221, so we can apply Problem 242 to conclude that X is a Lindelöf Σ-space.

(iii)$\Longrightarrow$(i). If X is a Lindelöf Σ-space, then there exists a compact space K and a second countable space N such that X is a continuous image of some closed subspace Y of the space $K \times N$ (see Problem 248). Let $p : K \times N \to N$ be the natural projection; then p is a perfect map by Fact 3 of S.288. If $M = p(Y)$, then $f = p|Y : Y \to M$ is a perfect map because Y is closed in $K \times N$. Thus Y maps perfectly onto a second countable space M and continuously onto X. This completes the proof of (iii)$\Longrightarrow$(i) so our solution is complete.

T.250. *Give an example of a space X which embeds into $C_p(Y)$ for some Lindelöf p-space Y and is not embeddable into $C_p(Z)$ for any $K_{\sigma\delta}$-space Z.*

Solution. For any $n \in \mathbb{N}$, let $M_n = \{1, \ldots, n\}$. A space Z is *K-analytic* if it is a continuous image of a $K_{\sigma\delta}$-space. The Cantor set $\mathbb{K}$ is the space $\{0, 1\}^{\omega}$. Call a space Z *uniformly uncountable* if every non-empty open subset of Z is uncountable. A set $F \subset C_p(Z)$ is called *D-separating* if for any closed $P \subset Z$ and finite $K \subset Z$ with $K \cap P = \emptyset$, if $\varepsilon > 0$, then there exists a function $f \in F$ such that $f(K) \subset (-\varepsilon, \varepsilon)$ and $f(P) \subset [\frac{3}{4}, 1]$. The function $u_Z \in \mathbb{R}^Z$ is defined by $u_Z(z) = 0$ for all $z \in Z$.

Given a space Z and a point $p \in Z$, let $D(Z, p) = \{f \in \mathbb{I}^Z : f(p) = 0$ and $f(U) \subset [-\frac{1}{2}, \frac{1}{2}]$ for some $U \in \tau(p, Z)\}$. We consider $D(Z, p)$ to be a space with the topology induced from $\mathbb{I}^Z$.

If $\mathcal{C}$ is a class of spaces, then $Y \in \mathcal{C}$ says that a space Y belongs to $\mathcal{C}$; furthermore, $Y \in \mathrm{Im}[\mathcal{C}]$ if Y is a continuous image of a space from $\mathcal{C}$. The statement $Y \in \mathrm{Un}[\mathcal{C}]$ says that Y is a countable union of spaces from $\mathcal{C}$ and the expression $Y \in \mathrm{Cl}[\mathcal{C}]$ is a short way of saying that Y is a closed subspace of a space from $\mathcal{C}$. Besides, we write $Y \in \mathrm{Prod}[\mathcal{C}]$ if Y is a finite product of spaces from $\mathcal{C}$.

We will stick to the usual agreements of set theory in what concerns operations on classes of spaces. In particular, given classes $\mathcal{C}$ and $\mathcal{D}$, we write $\mathcal{C} \subset \mathcal{D}$ if every space from $\mathcal{C}$ belongs to $\mathcal{D}$; also, if $\mathcal{C}_t$ is a class of spaces for all $t \in T$, then the class $\bigcup\{\mathcal{C}_t : t \in T\}$ consists of spaces Y such that $Y \in \mathcal{C}_t$ for some $t \in T$.

Let Z be a space; then a space Y belongs to the class $\mathcal{K}_0(Z)$ if and only if $Y = Z$ or Y is compact. Assume that $\alpha < \omega_1$ and we have defined a class $\mathcal{K}_\beta(Z)$ for all $\beta < \alpha$. If α is a limit ordinal, then let $\mathcal{K}_\alpha(Z) = \bigcup\{\mathcal{K}_\beta(Z) : \beta < \alpha\}$; if $\alpha = \beta + 1$ for some $\beta < \omega_1$, then let $\mathcal{K}_\alpha(Z) = \mathrm{Im}[\mathcal{K}_\beta(Z)] \cup \mathrm{Cl}[\mathcal{K}_\beta(Z)] \cup \mathrm{Un}[\mathcal{K}_\beta(Z)] \cup \mathrm{Prod}[\mathcal{K}_\beta(Z)]$. Once we have defined $\mathcal{K}_\alpha(Z)$ for all $\alpha < \omega_1$, let $\mathcal{K}(Z) = \bigcup\{\mathcal{K}_\alpha(Z) : \alpha < \omega_1\}$.

Fact 1. Let $\mathcal{CS}$ be the class of compact spaces. For any space Z, the class $\mathcal{K}(Z)$ is the minimal class of spaces which contains $\{Z\} \cup \mathcal{CS}$ and is invariant under finite products, countable unions, closed subspaces and continuous images. To put it more rigorously, if $\mathcal{C}$ is a class of spaces such that

$$\{Z\} \cup \mathcal{CS} \subset \mathcal{C} \text{ and } \mathrm{Un}[\mathcal{C}] = \mathrm{Cl}[\mathcal{C}] = \mathrm{Prod}[\mathcal{C}] = \mathrm{Im}[\mathcal{C}] = \mathcal{C},$$

then $\mathcal{K}(Z) \subset \mathcal{C}$.

Proof. If $Y \in \mathcal{K}(Z)$ and T is a continuous image of Y, then $Y \in \mathcal{K}_\alpha(Z)$ for some $\alpha < \omega_1$ and therefore $T \in \mathcal{K}_{\alpha+1}(Z)$. If $Y \in \mathcal{K}(Z)$ and T is a closed subspace of Y, then $Y \in \mathcal{K}_\alpha(Z)$ for some $\alpha < \omega_1$ and therefore $T \in \mathcal{K}_{\alpha+1}(Z)$. If $Y_1, \ldots, Y_n \in \mathcal{K}(Z)$ and $T = Y_1 \times \cdots \times Y_n$, then for any $i \in M_n$, we have $Y_i \in \mathcal{K}_{\alpha_i}(Z)$ for some $\alpha_i < \omega_1$. If $\alpha = \max\{\alpha_i : i \in M_n\}$, then $Y_i \in \mathcal{K}_\alpha(Z)$ for all $i \in M_n$ and hence $T \in \mathcal{K}_{\alpha+1}(Z)$.   □

Proof. Now, if $\{Y_i : i \in \omega\} \subset \mathcal{K}(Z)$ and $T = \bigcup\{Y_i : i \in \omega\}$, then for any $i \in \omega$, we have $Y_i \in \mathcal{K}_{\alpha_i}(Z)$ for some $\alpha_i < \omega_1$. If $\alpha > \sup\{\alpha_i : i \in \omega\}$, then $Y_i \in \mathcal{K}_\alpha(Z)$ for all $i \in \omega$ and hence $T \in \mathcal{K}_{\alpha+1}(Z)$. This proves that $F[\mathcal{K}(Z)] = \mathcal{K}(Z)$ for all $F \in \{\mathrm{Im}, \mathrm{Cl}, \mathrm{Un}, \mathrm{Prod}\}$. Finally, if $\mathcal{C}$ is a class such that $\{Z\} \cup \mathcal{CS} \subset \mathcal{C}$ and $\mathrm{Un}[\mathcal{C}] = \mathrm{Cl}[\mathcal{C}] = \mathrm{Prod}[\mathcal{C}] = \mathrm{Im}[\mathcal{C}] = \mathcal{C}$, then by an evident induction, $\mathcal{K}_\alpha(Z) \subset \mathcal{C}$ for all $\alpha < \omega_1$ and therefore $\mathcal{K}(Z) \subset \mathcal{C}$. Fact 1 is proved.   □

Fact 2. If Z_i is a $K_{\sigma\delta}$-space for all $i \in \omega$, then $Z = \bigoplus\{Z_i : i \in \omega\}$ is also a $K_{\sigma\delta}$-space.

Proof. For each $i \in \omega$, there is a space Y_i such that $Z_i \subset Y_i$ and $Z_i = \bigcap\{Y_n^i : n \in \omega\}$ where Y_n^i is a σ-compact subspace of Y_i for all $n \in \omega$. It is evident that

Z is a subspace of $Y = \bigoplus\{Y_i : i \in \omega\}$. Furthermore, $Y_n = \bigcup\{Y_n^i : i \in \omega\}$ is a σ-compact subspace of Y for every $n \in \omega$ and $\bigcap\{Y_n : n \in \omega\} = Z$, so Z is a $K_{\sigma\delta}$-space and Fact 2 is proved.   □

Fact 3.

(i)   every closed subspace of a K-analytic space is K-analytic;
(ii) any countable product of K-analytic spaces is a K-analytic space;
(iii) any continuous image of a K-analytic spaces is a K-analytic space;
(iv) any countable union of K-analytic spaces is a K-analytic space.

Proof.   (i)   If Z is K-analytic and P is a closed subspace of Z, take any $K_{\sigma\delta}$-space Y such that there is a continuous onto map $f : Y \to Z$. The set $Y' = f^{-1}(P)$ is a $K_{\sigma\delta}$-space being closed in Y (see Problem 338 of [TFS]) and $f|Y'$ maps Y' continuously onto P. Thus P is K-analytic and (i) is proved.

(ii) If Z_i is a K-analytic space for all $i \in \omega$ and $Z = \prod\{Z_i : i \in \omega\}$, take a $K_{\sigma\delta}$-space Y_i and a continuous onto map $f_i : Y_i \to Z_i$ for all $i \in \omega$. Then $f = \prod\{f_i : i \in \omega\}$ maps the space $Y = \prod\{Y_i : i \in \omega\}$ continuously onto Z (see Fact 1 of S.271); since Y is a $K_{\sigma\delta}$-space by Problem 338 of [TFS], the space Z is K-analytic and (ii) is proved.

(iii) Assume that Z is K-analytic and fix a $K_{\sigma\delta}$-space Y and a continuous onto map $f : Y \to Z$. If $g : Z \to T$ is a continuous onto map and then $g \circ f$ maps Y continuously onto T, so T is also K-analytic and (iii) is proved.

(iv) Let Z be a space such that $Z = \bigcup\{Z_i : i \in \omega\}$ where Z_i is K-analytic for all $i \in \omega$. For every $i \in \omega$, take a $K_{\sigma\delta}$-space Y_i which maps continuously onto Z_i. Then $Y = \bigoplus\{Y_i : i \in \omega\}$ is a $K_{\sigma\delta}$-space (see Fact 2) which maps continuously onto Z, so Z is K-analytic and Fact 3 is proved.

                                                                                □

Fact 4. Let M be a second countable uncountable K-analytic space. Then the Cantor set $\mathbb{K}$ embeds in M.

Proof. Fix a metric d on M with $\tau(d) = \tau(M)$ and a continuous map $\varphi : Z \to M$ of some $K_{\sigma\delta}$-space Z onto M. Let $Y \supset Z$ be any space such that $Z = \bigcap\{Y_n : n \in \mathbb{N}\}$ and each Y_n is a σ-compact subspace of Y. A set $B \subset Z$ will be called n-*precompact* if $\mathrm{cl}_Y(B)$ is a compact subset of Y_n. Given a set $A \subset M$, the symbol $\overline{A}$ denotes the closure of A in M. For each $k \in \mathbb{N}$, denote by C_k the set of all functions from $k = \{0, \ldots, k-1\}$ to $\{0, 1\}$. For every $k \in \mathbb{N}$, we will construct by induction families $\{P_f : f \in C_k, k \in \mathbb{N}\} \subset \exp(M)$ and $\{Q_f : f \in C_k, k \in \mathbb{N}\} \subset \exp(Z)$ with the following properties:

(1) P_f is uniformly uncountable and $\mathrm{diam}(P_f) \leq \frac{1}{k}$ for any $k \in \mathbb{N}$ and $f \in C_k$;
(2) Q_f is k-precompact, closed in Z and $\varphi(Q_f) = P_f$ for any $f \in C_k$ and $k \in \mathbb{N}$;
(3) the family $\{P_f : f \in C_k\}$ is disjoint for any $k \in \mathbb{N}$;
(4) if $m, k \in \mathbb{N}$, $m < k$ and $f \in C_k$, then $Q_{f|m} \subset Q_f$.

Since the set $Y_1 \supset Z$ is a countable union of compact spaces, there is a compact $K_1 \subset Y_1$ such that the set $\varphi(K_1 \cap Z)$ is uncountable; it is clear that $Z_1 = K_1 \cap Z$ is closed in Z; let $\varphi_1 = \varphi|Z_1 : Z_1 \to M_1 = \varphi(Z_1)$. Take a uniformly uncountable $L_1 \subset M_1$ which is closed in M_1 (this is possible by Fact 1 of S.343). Pick distinct $x_0, x_1 \in L_1$ and choose $U \in \tau(x_0, M)$, $V \in \tau(x_1, M)$ such that $\overline{U} \cap \overline{V} = \varnothing$. There is $\varepsilon \in (0, 1/2)$ such that $B(x_0, \varepsilon) \subset U$ and $B(x_1, \varepsilon) \subset V$. We have $C_1 = \{f_0, f_1\}$ where $f_i(0) = i$ for $i \le 1$; let $P_{f_0} = \mathrm{cl}_{M_1}(B(x_0, \varepsilon) \cap L_1)$ and $P_{f_1} = \mathrm{cl}_{M_1}(B(x_1, \varepsilon) \cap L_1)$. Observe that $P_{f_i} \subset L_1$ because L_1 is closed in M_1; since P_{f_i} is closed in M_1, the set $Q_{f_i} = \varphi_1^{-1}(P_{f_i})$ is closed in Z_1 and hence in Z for all $i \in \{0, 1\}$.

Since the closure of any open set in a uniformly uncountable space is uniformly uncountable, the space P_{f_i} is uniformly uncountable for every $i \in \{0, 1\}$. Since $Q_{f_i} \subset K_1 \subset Y_1$, every Q_{f_i} is 1-precompact. Furthermore, $\mathrm{diam}(P_{f_i}) \le 1$, because $P_{f_i} \subset \overline{B(x_i, \varepsilon)}$ and $\mathrm{diam}(\overline{B(z, r)}) \le 2r$ for any $z \in M$ and $r > 0$. Therefore the properties (1)–(4) are satisfied for the families $\{P_f : f \in C_1\}$ and $\{Q_f : f \in C_1\}$.

Suppose that for each $k \le n$, we defined P_f for all $f \in C_k$ so that the properties (1)–(4) hold. Any function $f \in C_{n+1}$ is an extension of the function $f|n$ and there are exactly two such extensions. This shows that $C_{n+1} = \{f_0^g, f_1^g : g \in C_n\}$ where $f_i^g|n = g$ and $f_i^g(n) = i$ for $i = 0, 1$.

Now, take an arbitrary function $g \in C_n$; observe that Q_g is contained in Y_{n+1} which is σ-compact. Therefore there is a compact $K_{n+1} \subset Y_{n+1}$ such that the set $M_{n+1} = \varphi(K_{n+1} \cap Q_g) \subset P_g$ is uncountable. It is clear that $Z_{n+1} = K_{n+1} \cap Q_g$ is closed in Q_g and hence in Z. Let $\varphi_{n+1} = \varphi|Z_{n+1} : Z_{n+1} \to M_{n+1}$. Apply Fact 1 of S.343 to find a uniformly uncountable $L_{n+1} \subset M_{n+1}$ which is closed in M_{n+1}. The space L_{n+1} has no isolated points and hence we can take distinct points $x_0, x_1 \in L_{n+1}$. Fix any $U \in \tau(x_0, M)$, $V \in \tau(x_1, M)$ such that $\overline{U} \cap \overline{V} = \varnothing$. We can find a number $\varepsilon \in (0, 1/(2n + 2))$ such that $B(x_0, \varepsilon) \subset U$ and $B(x_1, \varepsilon) \subset V$. Let $P_{f_0^g} = \mathrm{cl}_{M_{n+1}}(B(x_0, \varepsilon) \cap L_{n+1})$ and $P_{f_1^g} = \mathrm{cl}_{M_{n+1}}(B(x_1, \varepsilon) \cap L_{n+1})$. Observe that $P_{f_i^g} \subset L_{n+1}$ because L_{n+1} is closed in M_{n+1}; since $P_{f_i^g}$ is closed in M_{n+1}, the set $Q_{f_i^g} = \varphi_{n+1}^{-1}(P_{f_i^g})$ is closed in Z_{n+1} and hence in Z for all $i \in \{0, 1\}$. Since the function $g \in C_n$ was taken arbitrarily, we indicated how to construct sets $P_{f_0^g}$, $P_{f_1^g}$ and $Q_{f_0^g}$, $Q_{f_1^g}$ for all $g \in C_n$. This gives the desired families $\{P_f : f \in C_{n+1}\}$ and $\{Q_f : f \in C_{n+1}\}$.

Since the closure of any open set in a uniformly uncountable space is uniformly uncountable, the space $P_{f_i^g}$ is uniformly uncountable for each $i \in \{0, 1\}$. Applying Fact 1 of S.236 to the sets $P_{f_i^g} \subset \overline{B(x_i, \varepsilon)} \cap M_{n+1}$, we conclude that

$$\mathrm{diam}(P_{f_i^g}) \le \mathrm{diam}(B(x_i, \varepsilon) \cap L_{n+1}) \le \mathrm{diam}(\overline{B(x_i, \varepsilon)}) = \mathrm{diam}(B(x_i, \varepsilon)) \le 2\varepsilon < \frac{1}{n+1}$$

so (1) is satisfied.

Since $Q_{f_i^g} \subset K_{n+1} \subset Y_{n+1}$, every set $Q_{f_i^g}$ has to be $(n + 1)$-precompact; the rest of the statements of property (2) are, evidently, true by our construction. Now, (3) has only to be checked for $k = n + 1$. Observe that if $f, g \in C_{n+1}$, $f \ne g$ and

$f|n = g|n$, then $P_f \cap P_g = \emptyset$ by our construction. If we have $f|n \neq g|n$, then $P_f \cap P_g \subset P_{f|n} \cap P_{g|n} = \emptyset$ by the induction hypothesis. Therefore (3) holds for $k = n+1$ and all $m \leq n$ by the induction hypothesis. The property (4) is guaranteed by our construction for $m = n$ and $k = n + 1$. Thus our inductive construction can be continued to give us families $\{P_f : f \in C_n, n \in \mathbb{N}\}$ and $\{Q_f : f \in C_n, n \in \mathbb{N}\}$ with the properties (1)–(4).

For each $f \in \mathbb{K}$, let $T_f = \bigcap \{\mathrm{cl}_Y(Q_{f|n}) : n \in \mathbb{N}\}$ and observe that the family $\{\mathrm{cl}_Y(Q_{f|n}) : n \in \mathbb{N}\}$ consists of non-empty decreasing compact sets by (2) and (4) so $T_f \neq \emptyset$. Besides, (2) implies that there is a compact set $K_n \subset Y_n$ such that $\mathrm{cl}_Y(Q_{f|n}) \subset K_n \subset Y_n$ and consequently $T_f \subset \bigcap \{Y_n : n \in \mathbb{N}\} = Z$. Since $T_f \subset \mathrm{cl}_Y(Q_{f|n})$ for every $n \in \mathbb{N}$, we have $T_f \subset \mathrm{cl}_Y(Q_{f|n}) \cap Z = \mathrm{cl}_Z(Q_{f|n}) = Q_{f|n}$. Thus $T_f \subset \bigcap \{Q_{f|n} : n \in \mathbb{N}\}$; take any $z_f \in T_f$ and let $\delta(f) = \varphi(z_f)$. This gives us a map $\delta : \mathbb{K} \to M$ such that $\delta(f) \in \varphi(Q_{f|n}) = P_{f|n}$ for all $n \in \mathbb{N}$. As a consequence, $\delta(f) \in \bigcap \{P_{f|n} : n \in \mathbb{N}\}$; since $\mathrm{diam}(P_{f|n}) \to 0$, we have $\{\delta(f)\} = \bigcap \{P_{f|n} : n \in \mathbb{N}\}$.

The map δ is injective because if $f \neq g$, then $f|n \neq g|n$ for some $n \in \mathbb{N}$; consequently, $\delta(f) \in P_{f|n}$ and $\delta(g) \in P_{g|n}$. Since $P_{f|n} \cap P_{g|n} = \emptyset$ by (3), we have $\delta(f) \neq \delta(g)$.

The map δ is continuous; to see this, take any $f \in \mathbb{K}$ and $\varepsilon > 0$. There exists $n \in \mathbb{N}$ such that $1/n < \varepsilon$. The set $W = \{g \in \mathbb{K} : g|n = f|n\}$ is open in $\mathbb{K}$ and $f \in W$. For any $g \in W$ we have $\delta(g) \in P_{g|n} = P_{f|n}$; since $\mathrm{diam}(P_{f|n}) \leq 1/n$, we have $d(\delta(g), \delta(f)) \leq 1/n < \varepsilon$ and hence $\delta(W) \subset B(\delta(f), \varepsilon)$ which proves continuity of δ at the point f. Thus $\delta : \mathbb{K} \to C = \delta(\mathbb{K})$ is a condensation and hence homeomorphism. This shows that $\mathbb{K}$ embeds in M so Fact 4 is proved. □

Fact 5. There is a subspace $Y \subset \mathbb{K}$ which is not K-analytic.

Proof. It follows from Fact 5 of S.151 that there exist disjoint sets $A, B \subset \mathbb{K}$ such that $A \cap P \neq \emptyset \neq B \cap P$ for any uncountable compact $P \subset \mathbb{K}$. Observe that A has to be uncountable. Indeed, if $|A| \leq \omega$, then $A' = \mathbb{K} \backslash A$ is a G_δ-subset of $\mathbb{K}$ and hence A' is $K_{\sigma\delta}$ because every open subset of $\mathbb{K}$ is σ-compact. Therefore A' is K-analytic and uncountable, so there is $K \subset A'$ homeomorphic to $\mathbb{K}$ by Fact 4. However, according to the definition of A, it must intersect every uncountable compact subset of $\mathbb{K}$; this contradiction proves that A is uncountable. Now, if the space $Y = A$ is K-analytic, then there is $P \subset A$ homeomorphic to $\mathbb{K}$ by Fact 4. However $B \cap P \neq \emptyset$ because P is uncountable, so $A \cap B \supset P \cap B \neq \emptyset$ which is a contradiction. Thus Y is not K-analytic and Fact 5 is proved. □

Fact 6. For any space Z with $nw(Z) \leq \kappa$, there exists a space M such that $w(M) \leq \kappa$ and M condenses onto Z.

Proof. Take any network $\mathcal{A}$ in Z with $|\mathcal{A}| \leq \kappa$. It is an easy consequence of the regularity of the space Z that the family $\mathcal{N} = \{\overline{N} : N \in \mathcal{A}\}$ is also a network in Z. Since $\bigcup \mathcal{N} = Z$, the family $\mathcal{N}$ generates a topology μ on Z as a subbase. Observe that $\tau(Z) \subset \mu$ because for every $U \in \tau(Z)$, there is $\mathcal{N}' \subset \mathcal{N} \subset \mu$ with $U = \bigcup \mathcal{N}'$ so $U \in \mu$. Therefore the identity map $i : (Z, \mu) \to Z$ is a condensation. Since $\tau(Z) \subset \mu$, every closed subspace of Z is also closed in (Z, μ) which shows that all elements of $\mathcal{N}$ are clopen in $M = (Z, \mu)$. The family $\mathcal{N}$ being a network for the space Z, for any distinct $x, y \in Z$, there are $P, Q \in \mathcal{N}$ such that $x \in P$, $y \in Q$ and $P \cap Q = \emptyset$. Thus the space M is Hausdorff. The family $\mathcal{B}$ of all finite intersections of the elements of $\mathcal{N}$ is a base in M and all elements of $\mathcal{B}$ are clopen, i.e., M is zero-dimensional. Clearly, $|\mathcal{B}| \leq \kappa$ and hence $w(M) \leq \kappa$. Every zero-dimensional Hausdorff space is Tychonoff (see Fact 1 of S.232), so M is a Tychonoff space with $w(Z) \leq \kappa$ which condenses onto Z. Fact 6 is proved. □

Fact 7. If N is a space with a countable network, then there is a second countable M such that N embeds in $C_p(M)$.

Proof. The space $C_p(N)$ has a countable network by Problem 172 of [TFS], so there is a second countable space M which can be mapped continuously onto $C_p(N)$ (see Fact 6). If $\varphi : M \to C_p(N)$ is a continuous onto map, then the dual map φ^* embeds $C_p(C_p(N))$ in $C_p(M)$ (see Problem 163 of [TFS]). Since N embeds in $C_p(C_p(N))$ by Problem 167 of [TFS], it also embeds in $C_p(M)$ so Fact 7 is proved. □

Fact 8. Given a space Z suppose that $T \subset Z$ and $p \in T$. Then the natural projection $\pi : \mathbb{I}^Z \to \mathbb{I}^T$ of $\mathbb{I}^Z$ onto its face $\mathbb{I}^T$ maps $D(Z, p)$ onto $D(T, p)$ and hence $D(T, p)$ is a continuous image of $D(Z, p)$.

Proof. Recall that $\pi(f) = f|T$ for any $f \in \mathbb{I}^Z$. If $f \in D(Z, p)$, then let $g = \pi(f)$; there is $U \in \tau(p, Z)$ such that $f(U) \subset [-\frac{1}{2}, \frac{1}{2}]$. It is obvious that $V = U \cap T \in \tau(p, T)$ and $g(V) \subset [-\frac{1}{2}, \frac{1}{2}]$. Therefore $\pi(D(Z, p)) \subset D(T, p)$. Now, if $g \in D(T, p)$, then there exists $V \in \tau(p, T)$ such that $g(V) \subset [-\frac{1}{2}, \frac{1}{2}]$. Choose any $U \in \tau(p, Z)$ with $U \cap T = V$ and let $f(z) = g(z)$ for all $z \in T$; if $z \in Z \backslash T$, then let $f(z) = 0$. It is immediate that $\pi(f) = g$ and $f(U) \subset [-\frac{1}{2}, \frac{1}{2}]$. Therefore $f \in D(Z, p)$ and Fact 8 is proved. □

Fact 9. Given a space Z suppose that $F \subset C_p(Z, \mathbb{I})$ is a D-separating set such that $u = u_Z \in F$. Then Z is homeomorphic to a closed subspace of $D(F, u)$.

Proof. Given any $z \in Z$, let $e(z)(f) = f(z)$ for any $f \in F$. Then $e(z) \in C_p(F, \mathbb{I})$ and the map $e : Z \to C_p(F, \mathbb{I})$ is continuous (see Problem 166 of [TFS]). Since $e(z)(u) = u(z) = 0$ for any $z \in Z$ and $e(z)$ is continuous at u, we have $e(z) \in D(F, u)$ for any $z \in Z$, i.e., $e(Z) \subset D(F, u)$. It is an evident consequence of the fact that F is D-separating that for any $z \in Z$ and any closed $P \subset Z$ with $z \notin P$, there is $f \in F$ such that $f(z) \notin \overline{f(P)}$. Therefore e is an embedding by Problem 166 of [TFS] and we only must prove that $e(Z)$ is closed in $D(F, u)$.

Take any element $\varphi \in D(F, u) \backslash e(Z)$. There exists $O \in \tau(u, F)$ such that $\varphi(O) \subset [-\frac{1}{2}, \frac{1}{2}]$. By definition of the pointwise convergence topology, there is a finite $K \subset Z$ and $\varepsilon > 0$ such that $u \in W = \{f \in F : f(K) \subset (-\varepsilon, \varepsilon)\} \subset O$ and hence $\varphi(W) \subset [-\frac{1}{2}, \frac{1}{2}]$. Since $\varphi \notin e(K)$, there is $U \in \tau(K, Z)$ such that $\varphi \notin \overline{e(U)}$. The family F being D-separating, there is $g \in F$ such that $g(K) \subset (-\varepsilon, \varepsilon)$ and $g(Z \backslash U) \subset [\frac{3}{4}, 1]$ and, in particular, $g \in W$. This implies $e(z)(g) = g(z) \in [\frac{3}{4}, 1]$ for all $z \in Z \backslash U$ while $\varphi(g) \in \varphi(W) \subset [-\frac{1}{2}, \frac{1}{2}]$. Consequently, $G = \{\delta \in D(F, u) :$

$\delta(g) < \frac{3}{4}\}$ is an open neighborhood of φ in $D(F, u)$ such that $G \cap e(Z \backslash U) = \emptyset$ and therefore $\varphi \notin \overline{e(Z \backslash U)}$. This implies $\varphi \notin \overline{e(U)} \cup \overline{e(Z \backslash U)} = \overline{e(Z)}$. The function $\varphi \in D(F, u) \backslash e(Z)$ was chosen arbitrarily, so $e(Z)$ is closed in $D(F, u)$ and Fact 9 is proved.   □

Fact 10. Given a space Z and any $F \subset C_p(Z, \mathbb{I})$ with $u = u_Z \in F$, the space $D(F, u)$ belongs to the class $\mathcal{K}(Z)$.

Proof. Given any number $n \in \mathbb{N}$ and a point $z = (z_1, \ldots, z_n) \in Z^n$, consider the set $B(z, n) = \{\varphi \in \mathbb{I}^F : \varphi(f) \in [-\frac{1}{2}, \frac{1}{2}]$ for any $f \in F$ such that $f(z_i) \in (-\frac{1}{n}, \frac{1}{n})$ for all $i \in M_n\}$. It is immediate that $B(z, n) \subset D(F, u)$ for any $n \in \mathbb{N}$ and $z = (z_1, \ldots, z_n) \in Z^n$. Let $B_n = \bigcup\{B(z, n) : z \in Z^n\}$ for every $n \in \mathbb{N}$; it is easy to see that $D(F, u) = \bigcup\{B_n : n \in \mathbb{N}\}$ and hence it suffices to establish that $B_n \in \mathcal{K}(Z)$ for all $n \in \mathbb{N}$.

Fix any $n \in \mathbb{N}$ and denote by $\pi_i : Z^n \to Z$ the natural projection of Z^n onto its ith factor, i.e., $\pi_i(z) = z_i$ for any $z = (z_1, \ldots, z_n) \in Z^n$. Consider the set $P_n = \{(z, \varphi) \in Z^n \times \mathbb{I}^F : z \in Z^n$ and $\varphi \in B(z, n)\}$. It turns out that

(∗) the set P_n is closed in $Z^n \times \mathbb{I}^F$.

To prove (∗) take any point $a = (a_1, \ldots, a_n) \in Z^n$ and $\varphi \in \mathbb{I}^F$ such that $(a, \varphi) \in (Z^n \times \mathbb{I}^F) \backslash P_n$. Then $\varphi \notin B(a, n)$ which implies that there is a function $f \in F$ such that $|f(a_i)| < \frac{1}{n}$ for all $i \in M_n$ while $|\varphi(f)| > \frac{1}{2}$. Recalling that f is continuous on Z and $\mathbb{I}^F$ carries the product topology, we conclude that the set $W = \{(b, \delta) \in Z^n \times \mathbb{I}^F : |f(\pi_i(b))| < \frac{1}{n}$ for all $i \in M_n$ and $|\delta(f)| > \frac{1}{2}\}$ is open in $Z^n \times \mathbb{I}^F$. It is evident that $(a, \varphi) \in W$ and $W \cap P_n = \emptyset$, so any point $(a, \varphi) \in (Z^n \times \mathbb{I}^F) \backslash P_n$ has a neighborhood W which does not meet P_n. Thus P_n is closed in $Z^n \times \mathbb{I}^F$ and (∗) is proved.

Observe that $Z^n \times \mathbb{I}^F \in \mathcal{K}(Z)$ because $\mathbb{I}^F$ is compact and hence $P_n \in \mathcal{K}(Z)$ by (∗). Now, if $\pi : Z^n \times \mathbb{I}^F \to \mathbb{I}^F$ is the natural projection, then $\pi(P_n) = B_n$ and hence B_n is a continuous image of P_n which implies that $B_n \in \mathcal{K}(Z)$. Therefore $D(F, u) = \bigcup\{B_n : n \in \mathbb{N}\}$ also belongs to $\mathcal{K}(Z)$ and Fact 10 is proved.   □

Fact 11. If Z is a K-analytic space, then every $Y \in \mathcal{K}(Z)$ is also K-analytic.

Proof. Let $\mathcal{CS}$ be the class of compact spaces; it is evident that every compact space is K-analytic, so the class $\mathcal{A}$ of K-analytic spaces contains $\{Z\} \cup \mathcal{CS}$. Besides, $\mathcal{A}$ is closed under finite (and even countable) products, closed subspaces, continuous images and countable unions by Fact 3, so $\mathcal{K}(Z) \subset \mathcal{A}$ by Fact 1. Therefore Fact 11 is proved.   □

Fact 12. If Z is a K-analytic space and $C_p(Y)$ embeds in $C_p(Z)$, then Y is also K-analytic.

Proof. It is easy to see that there exists an embedding $e : C_p(Z) \to C_p(Z, \mathbb{I})$ such that $e(u_Z) = u_Z$. The space $C_p(Z)$ being homogeneous (i.e., for any $f, g \in C_p(Z)$, there is a homeomorphism $\varphi : C_p(Z) \to C_p(Z)$ such that $\varphi(f) = g$ (see Problem 079 of [TFS])), there is an embedding $w : C_p(Y) \to C_p(Z)$ such that $w(u_Y) = u_Z$.

Therefore $i = e \circ w$ embeds $C_p(Y)$ in $C_p(Z, \mathbb{I})$ in such a way that $i(u_Y) = u_Z$. If $F = C_p(Z, \mathbb{I})$, $H = i(C_p(Y))$ and $p = u_Z$, then $D(H, p)$ is a continuous image of $D(F, p)$ by Fact 8. The space $D(F, p)$ belongs to the class $\mathcal{K}(Z)$ by Fact 10 and hence $D(F, p)$ is K-analytic by Fact 11. Being a continuous image of $D(F, p)$ (see Fact 8), the space $D(H, p)$ is also K-analytic by Fact 3. Now apply Fact 9 to conclude that Y embeds in $D(H, p)$ as a closed subspace and therefore Y is also K-analytic. Fact 12 is proved. $\square$

Now it is easy to finish our solution. There exists a space $M \subset \mathbb{K}$ which is not K-analytic (see Fact 5). The space $X = C_p(M)$ has countable network and hence there is a second countable space Y such that X embeds in $C_p(Y)$ (see Fact 7). It is clear that Y is a Lindelöf p-space (see Problem 221). If the space $X = C_p(M)$ embeds in $C_p(Z)$ for some K-analytic space Z, then M is K-analytic by Fact 12 which is a contradiction. Therefore X is not embeddable in $C_p(Z)$ if Z is K-analytic; hence X is a space which embeds in $C_p(Y)$ for a Lindelöf p-space Y while X is not embeddable in $C_p(Z)$ if Z is a $K_{\sigma\delta}$-space. Our solution is complete.

T.251. *Give an example of a p-space Y and a pseudo-open (and hence quotient) map $\varphi : Y \to C_p(X)$ of Y onto $C_p(X)$ for an uncountable space X.*

Solution. The space $X = A(\omega_1)$ is uncountable and it was proved in Problem 146 of [TFS] that $C_p(X)$ is a Fréchet–Urysohn space. Therefore we can find a metrizable space Y such that there is a pseudo-open map $f : Y \to C_p(X)$ (see Problem 225 of [TFS]). Since Y is a p-space (see Problem 221), our spaces X and Y have all required properties.

T.252. *Prove that X is a Lindelöf p-space if and only if it there is a perfect map of X onto a second countable space.*

Solution. If Z is a space and $A \subset Z$, then A is *a cozero (zero) set in Z* if there is $f \in C_p(Z)$ such that $A = f^{-1}(\mathbb{R} \backslash \{0\})$ (or $A = f^{-1}(0)$ respectively).

Suppose that K is a compact space and Z is dense in K. If $\mathcal{U}_n \subset \tau(K)$ for every $n \in \omega$, call the sequence $\{\mathcal{U}_n : n \in \omega\}$ *a feathering* of the space Z in its compactification K if $Z \subset \bigcup \mathcal{U}_n$ for each $n \in \omega$ and $\bigcap \{\mathrm{St}(z, \mathcal{U}_n) : n \in \omega\} \subset Z$ for every $z \in Z$. A space Z is a p-space if and only if it has a feathering in βZ.

Fact 1. For any space Z:

(1) a set $A \subset Z$ is cozero if and only if $Z \backslash A$ is a zero-set;
(2) if $\varphi : Z \to T$ is a continuous map and U is a cozero-set in T, then $\varphi^{-1}(U)$ is a cozero-set in Z;
(3) any finite intersection of cozero-sets in Z is a cozero-set in Z;
(4) any countable union of cozero-sets in Z is a cozero-set in Z;
(5) a set $A \subset Z$ is cozero if and only if there is $f \in C_p(Z)$ and $U \in \tau(\mathbb{R})$ such that $A = f^{-1}(U)$;
(6) the family $\mathrm{Coz}(Z)$ of all cozero subsets of Z is a base in Z.

Proof. It is evident that a function $f \in C_p(Z)$ witnesses that A is a cozero-set if and only if f witnesses that $Z \backslash A$ is zero-set. This proves (1). If $f \in C_p(T)$ and $U = f^{-1}(\mathbb{R} \backslash \{0\})$, then $g = f \circ \varphi \in C_p(Z)$ and $\varphi^{-1}(U) = g^{-1}(\mathbb{R} \backslash \{0\})$. This proves (2). The property (3) follows from (1) and the fact that any finite union of zero-sets is a zero-set by Fact 1 of S.499. Analogously, the property (4) holds because any countable intersection of zero-sets is a zero-set by Fact 1 of S.499.

To prove (5) observe that in a perfectly normal space, any open set is cozero (see Fact 2 of T.080). Therefore every $U \in \tau(\mathbb{R})$ is a cozero-set in $\mathbb{R}$ and hence we can apply (2) to conclude that $f^{-1}(U)$ is a cozero-set in Z for any $f \in C_p(Z)$. This proves sufficiency in (5); necessity in (5) is evident from the definition.

As to (6), take any $z \in Z$ and any $W \in \tau(z, Z)$. By the Tychonoff property of Z, there is $f \in C_p(Z, [0, 1])$ such that $f(z) = 1$ and $f(Z \backslash W) \subset \{0\}$. It is immediate that $U = f^{-1}(\mathbb{R} \backslash \{0\}) \in \text{Coz}(Z)$ and $z \in U \subset W$. This establishes (6) and completes the proof of Fact 1.   □

Returning to our solution, observe that any second countable space is Lindelöf p by Problem 221, so if there is perfect map of X onto a second countable space, then X is Lindelöf p by Problem 245. This proves sufficiency.

Now assume that X is a Lindelöf p-space and fix a feathering $\{\mathcal{U}_n : n \in \omega\}$ of the space X in βX. Given $n \in \omega$, for every $x \in X$, there is a cozero-set U_x in the space βX such that $x \in U_x \subset U$ for some $U \in \mathcal{U}_n$ (see Fact 1); let $\mathcal{V}_n$ be a countable subcover of the cover $\{U_x : x \in X\}$ of the space X. It is obvious that the sequence $\{\mathcal{V}_n : n \in \omega\}$ is also a feathering of X in βX. For each $V \in \mathcal{V} = \bigcup \{\mathcal{V}_n : n \in \omega\}$, take a function $f_V \in C_p(\beta X)$ with $V = f_V^{-1}(\mathbb{R} \backslash \{0\})$. The family $\mathcal{V}$ is countable, so the function $f = \Delta\{f_V : V \in \mathcal{V}\}$ maps βX into a second countable space $\mathbb{R}^{\mathcal{V}}$ and hence $Y = f(X)$ is also second countable. Thus $g = f|X$ maps X onto a second countable space Y, so it suffices to show that g is perfect.

Take any $x \in X$ and $y \in \beta X \backslash X$. Since $\{\mathcal{V}_n : n \in \omega\}$ is a feathering of X in βX, there is $n \in \omega$ such that $y \notin \text{St}(x, \mathcal{V}_n)$ and hence there is $V \in \mathcal{V}$ with $x \in V$ and $y \notin V$. As a consequence, $f_V(y) = 0 \neq f_V(x)$ and therefore $f(x) \neq f(y)$. This shows that $f(\beta X \backslash X) \cap Y = \emptyset$ and hence $X = f^{-1}(Y)$. Now apply Fact 2 of S.261 to conclude that g is a perfect map of X onto a second countable space Y. This settles necessity and makes our solution complete.

T.253. *Prove that X is a Lindelöf Σ-space if and only if it is a continuous image of a Lindelöf p-space.*

Solution. If X is a continuous image of a Lindelöf p-space, then X is a Lindelöf Σ-space by Problems 223 and 243. Now if X is a Lindelöf Σ-space, then there is a compact space K, a second countable space M and a closed subspace F of $K \times M$ such that X is a continuous image of F (see Problem 248). The natural projection $\pi : K \times M \to M$ is a perfect map by Fact 3 of S.288. The space M being a Lindelöf p-space by Problem 221, the product $K \times M$ is also a Lindelöf p-space because it is a perfect preimage of M (see Problem 245). Furthermore, F is a Lindelöf p-space by Problem 224, so X is a continuous image of a Lindelöf p-space F.

T.254. *Prove that the class of Lindelöf Σ-spaces is the smallest one which contains all compact spaces, all second countable spaces and is invariant with respect to finite products, continuous images and closed subspaces.*

Solution. Let $L(\Sigma)$ be the class of Lindelöf Σ-spaces. Apply Problem 226 to see that $L(\Sigma)$ contains the class CS of compact spaces; the class $\mathcal{M}$ of second countable spaces is contained in $L(\Sigma)$ by Problem 221. It was proved in Fact 1 of T.227 that $L(\Sigma)$ is invariant under finite products. The class $L(\Sigma)$ is invariant under continuous images by Problem 243 and under closed subspaces by Problem 224, so $L(\Sigma)$ contains $CS \cup \mathcal{M}$ and has the above mentioned invariance properties.

Now assume that a class $\mathcal{D}$ of spaces contains the class $CS \cup \mathcal{M}$ and is invariant under finite products, closed subspaces and continuous images. For any $X \in L(\Sigma)$ there is a compact space K, a second countable space M and a closed subspace F of $K \times M$ such that X is a continuous image of F (see Problem 248). Since $K \in CS \subset \mathcal{D}$ and $M \in \mathcal{M} \subset \mathcal{D}$, we have $K \times M \in \mathcal{D}$ because $\mathcal{D}$ is invariant under finite products. Besides, $F \in \mathcal{D}$ because $\mathcal{D}$ is invariant under closed subspaces and, finally, $X \in \mathcal{D}$ because $\mathcal{D}$ is invariant under continuous images. This shows that $L(\Sigma) \subset \mathcal{D}$ and hence $L(\Sigma)$ is the minimal class containing $CS \cup \mathcal{M}$ with the invariance properties in question.

T.255. *Suppose that X_n is a Lindelöf p-space for each $n \in \omega$. Prove that the product $\prod\{X_n : n \in \omega\}$ is a Lindelöf p-space.*

Solution. For any $n \in \omega$, there is a second countable space M_n and a perfect map $f_n : X_n \to M_n$ because X_n is Lindelöf p (see Problem 252). Then $f = \prod\{f_n : n \in \omega\}$ maps $X = \prod\{X_n : n \in \omega\}$ onto the second countable space $M = \prod\{M_n : n \in \omega\}$ and the map f is perfect by Fact 4 of S.271. Applying Problem 252 once more, we conclude that X is a Lindelöf p-space.

T.256. *Suppose that X_n is a Lindelöf Σ-space for each $n \in \omega$. Prove that $\prod\{X_n : n \in \omega\}$ is a Lindelöf Σ-space.*

Solution. For any $n \in \omega$, there is a Lindelöf p-space Y_n and a continuous onto map $f_n : Y_n \to X_n$ because the space X_n is Lindelöf Σ (see Problem 253). Then the function $f = \prod\{f_n : n \in \omega\}$ maps $Y = \prod\{Y_n : n \in \omega\}$ onto the space $X = \prod\{X_n : n \in \omega\}$; besides, the map f is continuous and onto by Fact 1 of S.271. The space Y is Lindelöf p by Problem 255, so we can apply Problem 253 once more to conclude that X is a Lindelöf Σ-space.

T.257. *Let X be a space such that $X = \bigcup\{X_n : n \in \omega\}$, where each X_n is a Lindelöf Σ-space. Prove that X is a Lindelöf Σ-space.*

Solution. It is straightforward that the space X is a continuous image of the space $Y = \bigoplus\{X_n : n \in \omega\} \subset Z = \bigoplus\{\beta X_n : n \in \omega\}$. We consider each βX_n to be the respective clopen subspace of Z (see Problem 113 of [TFS]). For each $n \in \omega$, there is a countable family $\mathcal{F}_n$ of compact subsets of βX_n which separates X_n from $\beta X_n \setminus X_n$ (see Problem 233). It is evident that the family $\mathcal{F} = \bigcup\{\mathcal{F}_n : n \in \omega\}$ is countable, consists of compact subsets of Z and separates Y from $Z \setminus Y$. Therefore Y is a Lindelöf Σ-space by Problem 233 and hence X is a Lindelöf Σ-space by Problem 243.

T.258. *Suppose that Z is a space and $X_n \subset Z$ is Lindelöf Σ for each $n \in \omega$. Prove that $X = \bigcap \{X_n : n \in \omega\}$ is a Lindelöf Σ-space.*

Solution. Observe that the space X is homeomorphic to a closed subspace of $Y = \prod \{X_n : n \in \omega\}$ by Fact 7 of S.271. Since Y is a Lindelöf Σ-space by Problem 256, the space X is also Lindelöf Σ by Problem 224.

T.259. *Let X be a Lindelöf Σ-space such that each compact subset of X is finite. Prove that X is countable.*

Solution. This is Fact 2 of T.227.

T.260. *Give an example of a Lindelöf p-space X such that $nw(X) > \omega$ and all compact subsets of X are countable.*

Solution. Given an arbitrary space Z recall that its *Alexandroff double* $AD(Z)$ is the space with the underlying set $Z \times \{0, 1\}$ in which all points of $Z \times \{1\}$ are isolated and the local base at any point $y = (z, 0)$ is given by the family $\mathcal{B}_y = \{(U \times \{0, 1\}) \setminus \{(z, 1)\} : U \in \tau(z, Z)\}$. Let $\pi : AD(Z) \to Z$ be the natural projection, i.e., $\pi(y) = z$ for any $y = (z, i) \in AD(X)$ with $i \in \{0, 1\}$. It is immediate that $\pi : AD(Z) \to Z$ is a continuous map for any space Z.

For the Cantor set $\mathbb{K} \subset \mathbb{R}$ apply Fact 5 of S.151 to find disjoint $A, B \subset \mathbb{K}$ such that $A \cap F \neq \emptyset \neq B \cap F$ for any uncountable closed $F \subset \mathbb{K}$. The projection $\pi : AD(\mathbb{K}) \to \mathbb{K}$ is perfect because $AD(\mathbb{K})$ is compact (see Problem 364 of [TFS]). If $X = \pi^{-1}(A)$, then $f = \pi|X : X \to A$ is also a perfect map by Fact 2 of S.261. Since A is second countable, the space X is Lindelöf p by Problem 245. Observe that the set $\{\{(a, 1)\} : a \in A\}$ is a disjoint uncountable family of non-empty open subsets of X and hence $nw(X) \geq c(X) > \omega$. Now, if G is an uncountable compact subspace of X, then $F = f(G)$ is an uncountable compact subset of A and hence $B \cap A \supset B \cap F \neq \emptyset$ which is a contradiction. Thus every compact subspace of X is countable, i.e., X is a Lindelöf p-space of uncountable network weight in which all compact subsets are countable.

T.261. *Prove that any $K_{\sigma\delta}$-space is Lindelöf Σ. Show, additionally, that there exists a $K_{\sigma\delta}$-space which is not Lindelöf p.*

Solution. If X is a $K_{\sigma\delta}$-space, then there exists a sequence $\{Y_n : n \in \omega\}$ such that Y_n is σ-compact for all $n \in \omega$ and X embeds in the product $Y = \prod \{Y_n : n \in \omega\}$ as a closed subspace (see Problem 338 of [TFS]). Each Y_n is Lindelöf Σ by Problem 226, so Y is a Lindelöf Σ-space by Problem 256. Therefore X is Lindelöf Σ by Problem 224.

It was proved in Problem 231 that there exists a countable space N which is not a p-space. Every countable space is $K_{\sigma\delta}$ being σ-compact. Thus N is an example of a $K_{\sigma\delta}$-space which is not a p-space.

T.262. *Let X be a $K_{\sigma\delta}$-space such that all compact subsets of X are countable and $\psi(X) \leq \omega$. Prove that X is countable.*

Solution. The space $\mathbb{D}$ is the set $\{0, 1\}$ with the discrete topology. We consider that $\mathbb{D}^0 = \{\emptyset\}$; for every $n \in \omega$, $f \in \mathbb{D}^n$ and $i \in \mathbb{D}$ let $(f^\frown i)|n = f$ and $(f^\frown i)(n) = i$. Then $f^\frown 0, f^\frown 1 \in \mathbb{D}^{n+1}$ and $D^{n+1} = \{f^\frown 0, f^\frown 1 : f \in \mathbb{D}^n\}$. As usual, we identify any ordinal with the set of its predecessors and, in particular, $n = \{0, \ldots, n-1\}$ for every $n \in \omega$.

Fact 1. If Z is an uncountable $K_{\sigma\delta}$-space with $\psi(Z) = \omega$, then there exist uncountable closed sets $F, G \subset Z$ such that $F \cap G = \emptyset$.

Proof. Assume first that every $z \in Z$ has a countable open neighborhood U_z. The space Z being Lindelöf, there is a countable subcover of the cover $\{U_z : z \in Z\}$, i.e., $Z = \bigcup\{U_z : z \in A\}$ for some countable $A \subset Z$ and hence Z is countable which is a contradiction.

Thus there is $z \in Z$ such that $|U| > \omega$ for any $U \in \tau(z, Z)$. It follows from $\psi(Z) = \omega$ that there is a sequence $\{U_n : n \in \omega\} \subset \tau(z, Z)$ such that $\overline{U}_{n+1} \subset U_n$ for all $n \in \omega$ and $\bigcap\{U_n : n \in \omega\} = \{z\}$. The set $Z \backslash \{z\} = \bigcup\{Z \backslash U_n : n \in \omega\}$ being uncountable, there is $n \in \omega$ such that $F = Z \backslash U_n$ is uncountable. The set $G = \overline{U}_{n+1}$ is also uncountable by our choice of z; it follows from $\overline{U}_{n+1} \subset U_n$ that $F \cap G = \emptyset$, so the sets F and G are like promised and Fact 1 is proved. □

Returning to our solution, assume that X is uncountable and fix a space Y such that $X \subset Y$ and $X = \bigcap\{Y_n : n \in \omega\}$ where each Y_n is a σ-compact subspace of Y. We have $Y_n = \bigcup\{K_n^m : m \in \omega\}$ where K_n^m is compact and $K_n^m \subset K_n^{m+1}$ for all $m, n \in \omega$. Since $X \subset Y_0$, there exists $m_0 \in \omega$ such that $H_\emptyset = K_0^{m_0} \cap X$ is uncountable. Given $n \in \omega$ assume that we constructed $\{m_i : i \leq n\} \subset \omega$ and families $\mathcal{H}_k = \{H_f : f \in \mathbb{D}^k\}$ for all $k \leq n$ with the following properties:

(1) $m_0 \leq \cdots \leq m_n$;
(2) $\mathcal{H}_k$ is disjoint for every $k \leq n$;
(3) H_f is closed in X and uncountable for any $f \in \mathbb{D}^k$ and $k \leq n$;
(4) if $m < k \leq n$, then $H_f \subset H_{f|m}$ for any $f \in \mathbb{D}^k$;
(5) $\bigcup \mathcal{H}_k \subset \bigcap\{K_i^{m_i} \cap X : i \leq k\}$ for any $k \leq n$.

To construct the family $\mathcal{H}_{n+1}$ fix any $f \in \mathbb{D}^n$; the set H_f being uncountable, there is $q_f \in \omega$ such that the set $Z = K_{n+1}^{q_f} \cap H_f$ is uncountable. It is clear that Z is a $K_{\sigma\delta}$-space, so there are uncountable closed disjoint sets $F, G \subset Z$ by Fact 1; let $H_{f^\frown 0} = F$ and $H_{f^\frown 1} = G$. After we have $H_{f^\frown 0}$ and $H_{f^\frown 1}$ for all $f \in \mathbb{D}^n$ we have the family $\{H_{f^\frown 0}, H_{f^\frown 1} : f \in \mathbb{D}^n\} = \{H_f : f \in \mathbb{D}^{n+1}\}$; let $m_{n+1} = m_n + \sum\{q_f : f \in \mathbb{D}^n\}$. We omit a straightforward verification of the fact that the set $\{m_i : i \leq n+1\}$ and the families $\{\mathcal{H}_k : k \leq n+1\}$ satisfy the conditions (1)–(5). Continuing our inductive construction, we obtain a set $\{m_i : i \in \omega\} \subset \omega$ and a sequence $\{\mathcal{H}_k : k \in \omega\}$ of families with the properties (1)–(5) fulfilled for any $n \in \omega$.

Observe that $Q_n = \bigcap \{K_i^{m_i} : i \leq n\}$ is a non-empty compact subspace of Y because the uncountable set $\bigcup \mathcal{H}_n$ is contained in Q_n by the property (5). Since the sequence $\{Q_n : n \in \omega\}$ is decreasing, we have $Q = \bigcap \{Q_n : n \in \omega\} \neq \emptyset$. Besides, $Q \subset \bigcap \{Y_n : n \in \omega\} = X$, so Q is a compact subspace of X. Given any $f \in \mathbb{D}^\omega$, observe that the sequence $\{H_{f|n} : n \in \omega\}$ is decreasing by (4); the set $\overline{H}_{f|n}$ is compact for each $n \in \omega$ because $H_{f|n} \subset K_n^{m_n}$ for each $n \in \omega$ by (5) (the bar denotes the closure in Y). As a consequence, $R_f = \bigcap \{\overline{H}_{f|n} : n \in \omega\} \neq \emptyset$ and $R_f \subset \bigcap \{Y_n : n \in \omega\} = X$ which shows that $R_f \subset \overline{H}_{f|n} \cap X = \text{cl}_X(H_{f|n}) = H_{f|n}$ [see (3)] for every $n \in \omega$, so $S_f = \bigcap \{H_{f|n} : n \in \omega\} \neq \emptyset$ for any $f \in \mathbb{D}^\omega$; take any $x_f \in S_f$. If $f, g \in \mathbb{D}^\omega$ and $f \neq g$, then $f|n \neq g|n$ for some $n \in \omega$ and hence $H_{f|n} \cap H_{g|n} = \emptyset$ by (2) which shows that $x_f \neq x_g$. Consequently, $f \to x_f$ is an injection of $\mathbb{D}^\omega$ into Q and hence Q is a compact subspace of X of cardinality at least continuum which contradicts the assumption that all compact subspaces of X are countable. Thus X is countable and our solution is complete.

T.263. *Suppose that X is a Lindelöf Σ-space and $C_p(X)$ has the Baire property. Prove that X is countable. In particular, if X is a space with a countable network and $C_p(X)$ is Baire, then X is countable.*

Solution. All compact subsets of X are finite by Problem 284 of [TFS]. Now apply Problem 259 to conclude that X is countable.

T.264. *Prove that there exists an uncountable Lindelöf space X for which $C_p(X)$ has the Baire property.*

Solution. The space $X = L(\omega_1)$ is the desired example. It is clear that X is Lindelöf and uncountable. Besides, all countable subsets of X are closed in X which, together with normality of X, implies that all countable subsets of X are closed and C-embedded in X. Thus $C_p(X)$ is pseudocomplete by Problem 485 of [TFS] and hence Baire by Problem 464 of [TFS].

T.265. *Suppose that $C_p(X)$ is a Lindelöf Σ-space and has the Baire property. Prove that X is countable.*

Solution. Given any points $x_1, \ldots, x_n \in X$ and sets $O_1, \ldots, O_n \in \tau^*(\mathbb{R})$, the set $[x_1, \ldots, x_n; O_1, \ldots, O_n] = \{f \in C_p(X) : f(x_i) \in O_i \text{ for all } i \leq n\}$ is called *a standard open subset of $C_p(X)$*. Standard open sets $[x_1, \ldots, x_n; O_1, \ldots, O_n]$ where $n \in \mathbb{N}$, $x_1, \ldots, x_n \in X$ and $O_1, \ldots, O_n \in \tau(\mathbb{R})$ form a base in the space $C_p(X)$ (see Problem 056 of [TFS]). If $U = [x_1, \ldots, x_n; O_1, \ldots, O_n]$ is a standard open subset of $C_p(X)$, then $\text{supp}(U) = \{x_1, \ldots, x_n\}$.

Choose any countable network $\mathcal{F}$ with respect to a compact cover $\mathcal{C}$ of the space $C_p(X)$ (see Problem 225). We can assume that $\mathcal{F}$ is closed with respect to finite intersections and all elements of $\mathcal{F}$ are closed in $C_p(X)$ (see Fact 1 of T.229). Let $\mathcal{F}' = \{F \in \mathcal{F} : \text{Int}(F) = \emptyset\}$. Then $\bigcup \mathcal{F}' \neq C_p(X)$ by the Baire property of $C_p(X)$; fix any $u \in C_p(X) \backslash (\bigcup \mathcal{F}')$. If $\mathcal{H} = \{F \in \mathcal{F} : u \in F\}$, then $\text{Int}(F) \neq \emptyset$ for any $F \in \mathcal{H}$, so we can choose a non-empty standard open $W_F \subset F$ for every $F \in \mathcal{H}$. The set $A = \bigcup \{\text{supp}(W_F) : F \in \mathcal{H}\}$ is countable; we claim that $A = X$.

Observe first that $\bigcap \mathcal{H}$ is a compact subspace of $C_p(X)$. Indeed, there is $C \in \mathcal{C}$ with $u \in C$; it follows from the fact that $\mathcal{F}$ is a network with respect to $\mathcal{C}$ that $\bigcap \{F \in \mathcal{F} : C \subset F\} = C$. Since $\{F \in \mathcal{F} : C \subset F\} \subset \mathcal{H}$, we have $\bigcap \mathcal{H} \subset \bigcap \{F \in \mathcal{F} : C \subset F\} = C$, i.e., the set $\bigcap \mathcal{H}$ is a closed subspace of the compact space C and therefore $K = \bigcap \mathcal{H}$ is compact. Thus we can apply Fact 1 of S.326 to conclude that $\mathcal{H}$ is a network at K, i.e., for any $U \in \tau(K, C_p(X))$, there is $H \in \mathcal{H}$ with $H \subset U$.

Now, if $A \neq X$, then take any $x \in X \backslash A$. For any $g \in C_p(X)$ let $\varphi(g) = g(x)$; then the map $\varphi : C_p(X) \to \mathbb{R}$ is continuous by Problem 166 of [TFS]. Therefore $\varphi(K)$ is a compact (and hence bounded) subset of $\mathbb{R}$; take any $m \in \mathbb{N}$ such that $\varphi(K) \subset (-m, m)$. The set $U = \{h \in C_p(X) : h(x) \in (-m, m)\}$ is open in $C_p(X)$ and $K \subset U$. Since $\mathcal{H}$ is network at K, there is $H \in \mathcal{H}$ with $H \subset U$ and hence $W_H \subset U$. We have $W_H = [x_1, \ldots, x_n; O_1, \ldots, O_n]$ where $x_i \in A$ for all $i \leq n$ and hence $x \notin \{x_1, \ldots, x_n\}$. Therefore there exists a function $h \in C_p(X)$ such that $h(x_i) \in O_i$ for all $i \leq n$ and $h(x) = m + 1$ (see Problem 034 of [TFS]). It is immediate that $h \in W_H \backslash U$ which contradicts $W_H \subset H \subset U$. This shows that $X = A$ is a countable space.

T.266. *Prove that every Lindelöf Σ-space is stable, and hence, for every Lindelöf Σ-space X, the space $C_p(X)$ is monolithic.*

Solution. If we establish that every Lindelöf Σ-space is stable, then $C_p(X)$ is monolithic for every Lindelöf Σ-space X by Problem 154. To show that every Lindelöf Σ-space is stable, we will need the following fact.

Fact 1. Let Z be a space and suppose that $f_a : Z \to Y_a$ is a continuous map for all $a \in A$. For the map $f = \Delta\{f_a : a \in A\} : Z \to Y = \prod\{Y_a : a \in A\}$, let $Z' = f(Z)$. If the map f_b is perfect for some $b \in A$, then $f : Z \to Z'$ is also perfect.

Proof. Clearly, there is nothing to prove if $A = \{b\}$. If $A \backslash \{b\} \neq \emptyset$, consider the map $g = \Delta\{f_a : a \in A \backslash \{b\}\} : Z \to T = \prod\{Y_a : a \in A \backslash \{b\}\}$. Since g is continuous, the graph $G(g) = \{(z, g(z)) : z \in Z\} \subset Z \times T$ is closed in $Z \times T$ and the restriction $\pi | G(g)$ of the natural projection $\pi : Z \times T \to Z$ is a homeomorphism of $G(g)$ onto Z (see Fact 4 of S.390). If $\varphi = (\pi | G(g))^{-1}$, then $\varphi : Z \to G(g)$ is also a homeomorphism. Define the identity map $\mathrm{id} : T \to T$ by $\mathrm{id}(z) = z$ for every $z \in T$. Then id is a homeomorphism and hence perfect which shows that the map $u = f_b \times \mathrm{id} : Z \times T \to Y_b \times T = Y$ is also perfect by Fact 4 of S.271. It is immediate that $Z' = u(G(g))$ and hence the map $u' = u | G(g) : G(g) \to Z'$ is also perfect being a restriction of a perfect map u to a closed subset $G(g)$ of the space $Z \times T$. Now observe that $f = u \circ \varphi$ and hence f is perfect as well. Fact 1 is proved. $\square$

Returning to our solution, let us show first that

(1) $nw(Z) \leq iw(Z)$ for any Lindelöf Σ-space Z.

To prove (1) take an infinite cardinal κ such that $iw(Z) \leq \kappa$. There is a Lindelöf p-space Y and a continuous onto map $f : Y \to Z$ (see Problem 253). There exists a second countable space M and a perfect map $g : Y \to M$ by Problem 252. Let $h : Z \to T$ be a condensation such that $w(T) \leq \kappa$.

The map $\varphi = (h \circ f) \Delta g : Y \to T \times M$ is perfect because so is g (see Fact 1); if $\pi : T \times M \to T$ is the natural projection, then it is immediate from the definition of the diagonal product of mappings that $\pi \circ \varphi = h \circ f$. Besides, for the space $H = \varphi(Y)$, we have $w(H) \leq w(T \times M) \leq w(T) + w(M) \leq \kappa + \omega = \kappa$. The map $p = \pi|H : H \to T$ is continuous and onto so $r = h^{-1} \circ p$ maps H onto Z. Given a closed subset F of the space Z, we have $r^{-1}(F) = \varphi(f^{-1}(F))$ because $r \circ \varphi = f$. The map φ being perfect, the set $r^{-1}(F)$ is closed in H for every closed $F \subset Z$, i.e., r is continuous. It turns out that Z is a continuous image of the space H; therefore $nw(Z) \leq w(H) \leq \kappa$ and (1) is proved.

Now it is easy to see that any Lindelöf Σ-space is stable. Indeed, if Z is Lindelöf Σ and $f : Z \to Y$ is a continuous onto map, assume that $iw(Y) \leq \kappa$ for some infinite cardinal κ. The space Y is also Lindelöf Σ by Problem 243, so we can apply (1) to conclude that $nw(Y) \leq iw(Y) \leq \kappa$. This shows that Z is stable and makes our solution complete.

T.267. *Prove that if υX is a Lindelöf Σ-space, then X is ω-stable.*

Solution. Take a continuous onto map $f : X \to Y$ such that $iw(Y) \leq \omega$. Apply Problem 446 of [TFS] to conclude that Y is realcompact and hence there is a continuous map $g : \upsilon X \to Y$ such that $g|X = f$ (see Problem 412 of [TFS]). Therefore Y is a Lindelöf Σ-space being a continuous image of a Lindelöf Σ-space υX (see Problem 243). Consequently, the space Y is stable by Problem 266 and hence $nw(Y) \leq iw(Y) \leq \omega$ which proves that X is ω-stable.

T.268. *Prove that any product and any σ-product of Lindelöf Σ-spaces is stable. Show that any Σ-product of Lindelöf Σ-spaces is ω-stable.*

Solution. Given a family $\{X_t : t \in T\}$ of spaces and a point $a \in X = \prod_{t \in T} X_t$, let $\sigma(X, a) = \{x \in X : |\{t \in T : x(t) \neq a(t)\}| < \omega\}$; we will also need the space $\Sigma(X, a) = \{x \in X : |\{t \in T : x(t) \neq a(t)\}| \leq \omega\}$. For any $S \subset T$, the map $p_S : X \to X_S = \prod_{t \in S} X_t$ is the natural projection defined by $p_S(x) = x|S$ for any $x \in X$.

Fact 1. Given spaces Y and Z, if $q : Y \to Z$ is an $\mathbb{R}$-quotient map, then for any space M, a map $p : Z \to M$ is continuous if and only if $p \circ q$ is continuous.

Proof. Necessity is evident so assume that $p \circ q$ is continuous for some mapping $p : Z \to M$. The family $\mathcal{B} = \{f^{-1}(O) : O \in \tau(\mathbb{R}), \ f \in C(M)\}$ is a base in M by Fact 1 of S.437, so it suffices to show that $p^{-1}(U)$ is open in Z for any $U \in \mathcal{B}$. There is $O \in \tau(\mathbb{R})$ and $f \in C(Z)$ such that $U = f^{-1}(O)$. Then $p^{-1}(U) = (f \circ p)^{-1}(O)$.

Observe that $f \circ (p \circ q) = (f \circ p) \circ q$ is continuous because $p \circ q$ is continuous. Therefore $f \circ p$ is also continuous because q is $\mathbb{R}$-quotient. Consequently, the set $p^{-1}(U) = (f \circ p)^{-1}(O)$ is open in Z and hence the map p is continuous. Fact 1 is proved.   □

Fact 2. Assume that we have a family $\{X_t : t \in T\}$ of spaces such that the product $X_A = \prod_{t \in A} X_t$ is Lindelöf for every finite $A \subset T$. Then for any $a \in X = \prod_{t \in T} X_t$, the space $\sigma = \sigma(X, a)$ is Lindelöf.

Proof. Let $\mathrm{supp}(x) = \{t \in T : x(t) \neq a(t)\}$ for every $x \in X$ and observe that $\sigma = \bigcup\{\sigma_n : n \in \omega\}$ where $\sigma_n = \{x \in \sigma : |\mathrm{supp}(x)| \leq n\}$ for every $n \in \omega$. Therefore it suffices to show that σ_n is Lindelöf for every $n \in \omega$. We will do this by induction on n. If $n = 0$, then $\sigma_n = \{a\}$ is a Lindelöf space; assume that $n > 0$ and we proved that σ_{n-1} is Lindelöf.

Call a set $U \subset X$ *standard* if $U = \prod_{t \in T} U_t$ where $U_t \in \tau(X_t)$ for each $t \in T$ and the set $\mathrm{supp}(U) = \{t \in T : U_t \neq X_t\}$ is finite. The family $\mathcal{B}$ of all standard sets is a base in X (see Problem 101 of [TFS]), so to prove that σ_n is Lindelöf, it suffices to show that for every $\mathcal{U} \subset \mathcal{B}$ such that $\sigma_n \subset \bigcup \mathcal{U}$, there is a countable $\mathcal{U}' \subset \mathcal{U}$ such that $\sigma_n \subset \bigcup \mathcal{U}'$.

By the induction hypothesis, there is a countable $\mathcal{V} \subset \mathcal{U}$ such that $\sigma_{n-1} \subset \bigcup \mathcal{V}$. The set $A = \bigcup\{\mathrm{supp}(V) : V \in \mathcal{V}\}$ is countable; we claim that

$(*)$ $\mathrm{supp}(x) \subset A$ for any $x \in \sigma_n \backslash (\bigcup \mathcal{V})$.

Indeed, if $(*)$ is not true, then $t_0 \in \mathrm{supp}(x) \backslash A$ for some $x \in \sigma_n \backslash (\bigcup \mathcal{V})$. Let $y(t) = x(t)$ for all $t \in T \backslash \{t_0\}$ and $y(t_0) = a(t_0)$. Then $y \in \sigma_{n-1}$ and hence $y \in V$ for some $V \in \mathcal{V}$. We have $\mathrm{supp}(V) \subset A$ and therefore $p_A^{-1}((p_A(V)) = V$; since $p_A(x) = p_A(y)$ and $y \in V$, we have $x \in V$ which is a contradiction showing that $(*)$ is true.

It follows from $(*)$ that $P = \sigma_n \backslash (\bigcup \mathcal{V}) \subset Q = \sigma_n \cap (X_A \times \{p_{T \backslash A}(a)\})$. It is straightforward that the mapping $p_A | Q$ is a homeomorphism of Q onto the space $\sigma(n, A) = \{y \in X_A : |\{t \in T : y(t) \neq a(t)\}| \leq n\}$. It is easy to see that we have the equality $\sigma(n, A) = \bigcup\{R_B : B \subset A \text{ and } |B| \leq n\}$ where $R_B = X_B \times \{p_{A \backslash B}(a)\}$ is Lindelöf being homeomorphic to X_B for any finite $B \subset A$. Any countable union of Lindelöf spaces is a Lindelöf space, so $\sigma(n, A)$ is Lindelöf and hence Q is also Lindelöf being homeomorphic to $\sigma(n, A)$. Since P is a closed subset of σ_n, it is also closed in Q which implies that P is Lindelöf. Thus we can choose a countable $\mathcal{W} \subset \mathcal{U}$ such that $P \subset \bigcup \mathcal{W}$. It is clear that $\mathcal{U}' = \mathcal{V} \cup \mathcal{W}$ is a countable subfamily of $\mathcal{U}$ and $\sigma_n \subset \bigcup \mathcal{U}'$. The inductive step being carried out, we proved that every σ_n is Lindelöf and hence $\sigma = \bigcup \sigma_n$ is also Lindelöf so Fact 2 is proved. $\square$

Fact 3. Given a space X_t for every $t \in T$, let $X = \prod\{X_t : t \in T\}$. Suppose that Y is a Lindelöf subspace of the space X and $f : Y \to M$ is a continuous map of Y onto a space M such that $w(M) = \kappa \geq \omega$. Then there is a set $S \subset T$ and a continuous map $g : p_S(Y) \to M$ such that $|S| \leq \kappa$ and $f = g \circ (p_S | Y)$.

Proof. We will also need projections between the faces of X, namely, if $A \subset B \subset T$, then we have a natural projection $p_A^B : X_B \to X_A$ defined by $p_A^B(x) = x | A$ for any $x \in X_B$. The map p_A^B is open, continuous and $p_A^B \circ P_B = p_A$ for all $A \subset B \subset T$.

Since $w(M) \leq \kappa$, there is an embedding of M into $\mathbb{R}^\kappa$, so we can assume, without loss of generality, that $M \subset \mathbb{R}^\kappa$. For every $\alpha < \kappa$, the map $\pi_\alpha : \mathbb{R}^\kappa \to \mathbb{R}$ is the natural projection of $\mathbb{R}^\kappa$ onto its αth factor; let $q_\alpha = \pi_\alpha | M$. Given any $\alpha < \kappa$, we can apply Problem 298 of [TFS] to the map $f_\alpha = q_\alpha \circ f : Y \to \mathbb{R}$ to find a countable set $S_\alpha \subset T$ and a continuous map $g_\alpha : p_{S_\alpha}(Y) \to \mathbb{R}$ such that $f_\alpha = g_\alpha \circ (p_{S_\alpha} | Y)$. We claim that $S = \bigcup \{ S_\alpha : \alpha < \kappa \}$ is as promised.

Indeed, given $z \in p_S(Y)$, let $g(z)(\alpha) = g_\alpha(z | S_\alpha)$ for each $\alpha < \kappa$; this defines a map $g : p_S(Y) \to \mathbb{R}^\kappa$. It is immediate that $g = \Delta \{ g_\alpha \circ (p_{S_\alpha}^S | p_S(Y)) : \alpha < \kappa \}$ and hence the map g is continuous. Given any $y \in Y$, we have

$$f(y)(\alpha) = q_\alpha(f(y)) = f_\alpha(y) = g_\alpha(p_{S_\alpha}(y)) = g_\alpha(p_{S_\alpha}^S(p_S(y))) = g(p_S(y))(\alpha)$$

for every $\alpha < \kappa$ which shows that $f(y) = g(p_S(y))$ for each $y \in Y$. As a consequence, $g : p_S(Y) \to M$ and $f = g \circ (p_S | Y)$ so Fact 3 is proved. □

Fact 4. Assume that we have a family $\{ X_t : t \in T \}$ of spaces such that the product $\prod_{t \in A} X_t$ is Lindelöf for every finite $A \subset T$. Let $X = \prod \{ X_t : t \in T \}$ and suppose that $Y \in \{ \sigma(X, a), \Sigma(X, a) \}$. Then, for any infinite cardinal κ and any continuous map $f : Y \to M$ of Y to a space M with $w(M) \leq \kappa$, there is a set $S \subset T$ and a continuous map $g : p_S(Y) \to M$ such that $|S| \leq \kappa$ and $f = g \circ (p_S | Y)$.

Proof. If $Y = \sigma(X, a)$, then Y is Lindelöf by Fact 1, so our statement is true for Y by Fact 3. Now, if $Y = \Sigma(X, a)$, then fix a set $S \subset T$ with $|S| \leq \kappa$ such that there is a continuous map $g_1 : p_S(\sigma(X, a)) \to M$ for which $g_1 \circ (p_S | \sigma(X, a)) = f | \sigma(X, a)$. Take any $x, y \in \Sigma(X, a)$ such that $p_S(x) = p_S(y)$; then $\text{supp}(x) \cap S = \text{supp}(y) \cap S$, so we can choose sets $P = \{ t_i : i \in \omega \} \subset T$ and $Q = \{ s_i : i \in \omega \} \subset T$ such that, for any $i \in \omega$, we have $t_i \in S$ if and only if $s_i \in S$ while $\text{supp}(x) \subset P$ and $\text{supp}(y) \subset Q$. Let $x_i(t) = a(t)$ for all $t \in T \setminus \{ t_k : k \leq i \}$ and $x_i(t) = x(t)$ otherwise. Analogously let $y_i(t) = a(t)$ for all $t \in T \setminus \{ s_k : k \leq i \}$ and $y_i(t) = y(t)$ otherwise. It is immediate that $x_i, y_i \in \sigma(X, a)$, $p_S(x_i) = p_S(y_i)$ for all $i \in \omega$ and $\lim x_i = x$, $\lim y_i = y$. By our choice of S, the equality $p_S(z) = p_S(z')$ implies $f(z) = f(z')$ for any $z, z' \in \sigma(X, a)$ so $f(x_i) = f(y_i)$ for all $i \in \omega$ and therefore $f(x) = f(y)$ by continuity of f.

We proved that $p_S(x) = p_S(y)$ implies $f(x) = f(y)$ for any $x, y \in \Sigma(X, a)$. As a consequence, there exists a map $g : p_S(\Sigma(X, a)) \to M$ with $g \circ (p_S | \Sigma(X, a)) = f$. Observe that $\sigma(X, y) \subset \Sigma(X, a)$ for any $y \in \Sigma(X, a)$ and hence the map $p_S | \Sigma(X, a)$ is open by Fact 1 of T.110. Since the map $f = g \circ (p_S | \Sigma(X, a))$ is continuous, we can apply Fact 1 to conclude that g is a continuous map (see Problems 153 and 154 of [TFS]). Fact 4 is proved. □

Fact 5. Assume that we have a family $\{ X_t : t \in T \}$ of spaces such that the product $X_A = \prod_{t \in A} X_t$ is Lindelöf for every finite $A \subset T$. Let $X = \prod \{ X_t : t \in T \}$ and suppose that $f : X \to M$ is a continuous map of X to a space M such that $w(M) \leq \kappa$. Then there is a set $S \subset T$ and a continuous map $g : X_S \to M$ such that $|S| \leq \kappa$ and $f = g \circ p_S$.

Proof. Assume first that $\kappa = \omega$. Choose any $a \in X$ and let $\Sigma = \Sigma(X, a)$. By Fact 4 there exists a countable set $S \subset T$ and a continuous mapping $g : p_S(\Sigma) = X_S \to M$ such that $g \circ (p_S|\Sigma) = f|\Sigma$. For any $x \in X$ let $h(x) = g(p_S(x))$; then $h : X \to M$ is a continuous map such that $h|\Sigma = f|\Sigma$. Since Σ is dense in X, we have $f = h$ by Fact 0 of S.351. Therefore $f = g \circ p_S$ and Fact 5 is proved for $\kappa = \omega$.

Now assume that κ is an arbitrary infinite cardinal. Since $w(M) \leq \kappa$, there is an embedding of M into $\mathbb{R}^\kappa$, so we can assume, without loss of generality, that $M \subset \mathbb{R}^\kappa$. For every $\alpha < \kappa$, the map $\pi_\alpha : \mathbb{R}^\kappa \to \mathbb{R}$ is the natural projection of $\mathbb{R}^\kappa$ onto its αth factor; let $q_\alpha = \pi_\alpha|M$. Given any $\alpha < \kappa$, we can apply the result proved for $\kappa = \omega$ to the map $f_\alpha = q_\alpha \circ f : X \to \mathbb{R}$ to find a countable set $S_\alpha \subset T$ and a continuous map $g_\alpha : X_{S_\alpha} \to \mathbb{R}$ such that $f_\alpha = g_\alpha \circ p_{S_\alpha}$. We claim that $S = \bigcup\{S_\alpha : \alpha < \kappa\}$ is as promised.

It is evident that $|S| \leq \kappa$. Given $z \in X_S$, let $g(z)(\alpha) = g_\alpha(z|S_\alpha)$ for each $\alpha < \kappa$; this defines a mapping $g : X_S \to \mathbb{R}^\kappa$. Obviously, $g = \Delta\{g_\alpha \circ p_{S_\alpha}^S : \alpha < \kappa\}$ and hence the map g is continuous. Given any $y \in X$, we have

$$f(y)(\alpha) = q_\alpha(f(y)) = f_\alpha(y) = g_\alpha(p_{S_\alpha}(y)) = g_\alpha(p_{S_\alpha}^S(p_S(y))) = g(p_S(y))(\alpha)$$

for every $\alpha < \kappa$ which shows that $f(y) = g(p_S(y))$ for each $y \in X$. As a consequence, $g : X_S \to M$ and $f = g \circ p_S$, so Fact 5 is proved. □

Returning to our solution assume that X_t is a Lindelöf Σ-space for any $t \in T$ and take any point $a \in X = \prod_{t \in T} X_t$. Let $\sigma = \sigma(X, a)$ and $\Sigma = \Sigma(X, a)$. Take any infinite cardinal κ and a continuous onto map $f : \sigma \to Z$ for which there is a condensation $h : Z \to M$ such that $w(M) \leq \kappa$.

We can apply Fact 4 to the space σ and the map $h \circ f$ to find a set $S \subset T$ and a continuous map $g : p_S(\sigma) \to M$ such that $|S| \leq \kappa$ and $g \circ (p_S|\sigma) = h \circ f$. Denote the map $p_S|\sigma : \sigma \to p_S(\sigma)$ by π. Then π is an open map because $\sigma(X, y) \subset \sigma$ for any $y \in \sigma$ (see Fact 1 of T.110). For the map $\varphi = h^{-1} \circ g$ the map $\varphi \circ \pi = f$ is continuous, so φ is continuous by Fact 1. Therefore Z is a continuous image of the space $p_S(\sigma)$.

For any $A \subset S$ let $Q_A = X_A \times \{p_{S \setminus A}(a)\}$. It is clear that Q_A is a subspace of X_S homeomorphic to X_A. Observe that $p_S(\sigma) = \bigcup\{Q_A : A$ is a finite subset of $S\}$. Each Q_A is a Lindelöf Σ-space by Problem 256 and hence $p_S(\sigma)$ is a union of $\leq \kappa$-many Lindelöf Σ-spaces. Since each element of this union is stable by Problem 266, the space $p_S(\sigma)$ is κ-stable by Problem 124. Thus Z is κ-stable by Problem 123 and hence $nw(Z) \leq \kappa$ which proves that σ is κ-stable. This proves stability of σ.

To prove stability of the space X observe first that for any $t \in T$, there exists a Lindelöf p-space Y_t and a continuous onto map $q_t : Y_t \to X_t$ (see Problem 253). Therefore $q = \prod_{t \in T} q_t$ maps $Y = \prod_{t \in T} Y_t$ continuously onto X (see Fact 1 of S.271). Now apply Problem 123 to conclude that it suffices to prove that Y is stable, and hence we can assume, without loss of generality, that X_t is a Lindelöf p-space for every $t \in T$.

Now take any continuous onto map $f : X \to Z$ such that there is a condensation $h : Z \to M$ where $w(M) \leq \kappa$. Apply Fact 5 to find a set $S \subset T$ and a continuous map $g : X_S \to M$ such that $|S| \leq \kappa$ and $g \circ p_S = h \circ f$. If $\varphi = h^{-1} \circ g$, then the map $\varphi \circ p_S = f$ is continuous and hence φ is continuous because p_S is an open map (see Fact 1). Thus Z is a continuous image of X_S.

We saw that each X_t can be assumed to be Lindelöf p, so there is a perfect map $u_t : X_t \to L_t$ of X_t onto a second countable space L_t for every $t \in T$ (see Problem 252) and hence $u = \prod_{t \in S} u_t$ maps X_S perfectly onto the space $L = \prod_{t \in S} L_t$ (see Fact 4 of S.271). It is immediate that $w(L) \leq \kappa$. Observe that $\xi = u \Delta g$ is perfect by Fact 1 of T.266 and maps X_S onto a subspace F of $L \times M$; it is clear that $w(F) \leq \kappa$. Let $\pi : F \to M$ be the restriction of the natural projection. Then $\pi(F) = M$ and $r = h^{-1} \circ \pi : F \to Z$. Observe that $r \circ \xi = \varphi$ is a continuous map, so r is continuous by Fact 1. Thus the space Z is a continuous image of F and therefore $nw(Z) \leq nw(F) \leq w(F) \leq \kappa$ which shows that $nw(Z) \leq \kappa$ and hence X is κ-stable. This shows that X is a stable space.

Finally take any continuous onto map $f : \Sigma \to Z$ for which there is a condensation $h : Z \to M$ such that $w(M) \leq \omega$. We can apply Fact 4 to the space Σ and the map $h \circ f$ to find a countable set $S \subset T$ and a continuous map $g : p_S(\Sigma) = X_S \to M$ such that $g \circ (p_S|\Sigma) = h \circ f$. Denote the map $p_S|\Sigma : \Sigma \to X_S$ by π. Then π is an open map because $\sigma(X, y) \subset \Sigma$ for any $y \in \Sigma$ (see Fact 1 of T.110). For the map $\varphi = h^{-1} \circ g$ the map $\varphi \circ \pi = f$ is continuous, so φ is continuous by Fact 1. Therefore Z is a continuous image of X_S which is a Lindelöf Σ-space by Problem 256. Thus Z is also Lindelöf Σ which shows that Z is stable by Problem 266. As a consequence, $nw(Z) \leq \omega$ which proves ω-stability of Σ and makes our solution complete.

T.269 (Baturov's theorem). *Let X be a Lindelöf Σ-space. Prove that for any set $Y \subset C_p(X)$, we have $\mathrm{ext}(Y) = l(Y)$.*

Solution. Suppose that we have spaces Z, T and a map $u : Z \to T$. Given any number $n \in \mathbb{N}$, let $u^n(z_1, \ldots, z_n) = (u(z_1), \ldots, u(z_n))$ for any point $z = (z_1, \ldots, z_n) \in Z^n$. Thus $u^n : Z^n \to T^n$. If P is a set, then $\mathrm{Fin}(P)$ is the family of all non-empty finite subsets of P. Consider the family $\mathcal{O} = \{(a, b) : a < b, \, a, b \in \mathbb{Q}\}$; given a number $n \in \mathbb{N}$ let $\mathcal{O}^n = \{O_1 \times \cdots \times O_n : O_i \in \mathcal{O} \text{ for any } i = 1, \ldots, n\}$. Let $\{O_k : k \in \omega\}$ be an enumeration of the family $\bigcup \{\mathcal{O}^n : n \in \mathbb{N}\}$. Thus, for each $k \in \omega$, there is $m_k \in \mathbb{N}$ and $O_1^k, \ldots, O_{m_k}^k \in \mathcal{O}$ such that $O_k = O_1^k \times \cdots \times O_{m_k}^k$. For any $x = (x_1, \ldots, x_{m_k}) \in X^{m_k}$, let $[x, O_k] = \{f \in C_p(X) : f^{m_k}(x) \in O^k\}$. If $\mathcal{B}_k = \{[x, O_k] : x \in X^{m_k}\}$, then the family $\mathcal{B} = \bigcup \{\mathcal{B}_k : k \in \omega\}$ is a base of the space $C_p(X)$.

There is a Lindelöf p-space X' such that X is a continuous image of X' (see Problem 253). The relevant dual map embeds $C_p(X)$ into $C_p(X')$ which shows that Y can be considered a subspace of $C_p(X')$. Therefore we can assume, without loss of generality, that $X = X'$, i.e., X is a Lindelöf p-space. Fix a perfect map $p : X \to M$ of X onto a second countable space M (see Problem 252). Then $p_k = p^{m_k}$ maps X^{m_k} perfectly onto the second countable space M^{m_k} for every $k \in \omega$ (see Fact 4 of S.271).

To prove that $\text{ext}(Y) = l(Y)$ it suffices to show that $l(Y) \leq \text{ext}(Y)$ so assume the contrary; then there exists an infinite cardinal $\kappa \geq \text{ext}(Y)$ and $\mathcal{U} \subset \tau(C_p(X))$ such that $Y \subset \bigcup \mathcal{U}$ and no subfamily of $\mathcal{U}$ of cardinality $\leq \kappa$ covers Y. It is easy to see that, without loss of generality, we can assume that $\mathcal{U} \subset \mathcal{B}$. We will need the set $A_k = \{x \in X^{m_k} : [x, O_k] \in \mathcal{U}\}$ for every $k \in \omega$; since $\mathcal{U}$ covers Y, we have

(1) for any $f \in Y$, there is $k \in \omega$ and $x \in A_k$ such that $f^{m_k}(x) \in O_k$.

On the other hand, no subfamily of $\mathcal{U}$ of cardinality $\leq \kappa$ covers Y, so we have

(2) if $B_k \subset A_k$ and $|B_k| \leq \kappa$ for any $k \in \omega$, then there is a function $f \in Y$ such that $f^{m_k}(B_k) \cap O_k = \emptyset$ for every $k \in \omega$.

Choose a function $f_0 \in Y$ arbitrarily and let $B(k, 0) = \emptyset$ for all $k \in \omega$. Proceeding inductively assume that $0 < \alpha < \kappa^+$ and we have chosen a set $\{f_\beta : \beta < \alpha\} \subset Y$ and a family $\{B(k, \beta) : \beta < \alpha, k \in \omega\}$ with the following properties:

(3) $B(k, \beta) \subset A_k$ and $|B(k, \beta)| \leq \kappa$ for all $k \in \omega$ and $\beta < \alpha$;
(4) if $\gamma < \beta < \alpha$, then $B(k, \gamma) \subset B(k, \beta)$ for every $k \in \omega$;
(5) for any $\beta < \alpha$, $k \in \omega$ and any $H \in \text{Fin}(\{f_\gamma : \gamma < \beta\})$, the set $u_H(B(k, \beta))$ is dense in $u_H(A_k)$ where $u_H = p_k \Delta(\Delta\{f^{m_k} : f \in H\}) : X^{m_k} \to M^{m_k} \times \mathbb{R}^{m_k \cdot |H|}$;
(6) $f_\beta^{m_k}(B(k, \beta)) \cap O_k = \emptyset$ for all $\beta < \alpha$ and $k \in \omega$.

To get f_α, let $F_\alpha = \{f_\beta : \beta < \alpha\}$ and fix any $k \in \omega$; for every $H \in \text{Fin}(F_\alpha)$, let $u_H = p_k \Delta(\Delta\{f^{m_k} : f \in H\}) : X^{m_k} \to M^{m_k} \times \mathbb{R}^{m_k \cdot |H|}$. The space $u_H(A_k)$ being second countable, there is a countable $B(H, k) \subset A_k$ such that $u_H(B(H, k))$ is dense in $u_H(A_k)$. The set $B(k, \alpha) = (\bigcup\{B(k, \beta) : \beta < \alpha\}) \cup (\bigcup\{B(H, k) : H \in \text{Fin}(F_\alpha)\})$ has cardinality $\leq \kappa$. Once we have a set $B(k, \alpha)$ for every $k \in \omega$, apply (2) to find a function $f_\alpha \in Y$ such that $f_\alpha^{m_k}(B(k, \alpha)) \cap O_k = \emptyset$ for all $k \in \omega$. It is immediate that the properties (3)–(6) still hold for the set $\{f_\beta : \beta \leq \alpha\}$ and the family $\{B(k, \beta) : \beta \leq \alpha, k \in \omega\}$, and therefore, our inductive construction can be continued to give us a set $D = \{f_\alpha : \alpha < \kappa^+\}$ and a family $\{B(k, \beta) : \beta < \kappa^+, k \in \omega\}$ such that the properties (3)–(6) hold for all $\alpha < \kappa^+$.

Assume that $\beta < \alpha < \kappa^+$; it follows from (5) that $(p_k \Delta f_\beta^{m_k})(B(k, \alpha))$ is dense in $(p_k \Delta f_\beta^{m_k})(A_k)$ and hence $f_\beta^{m_k}(B(k, \alpha))$ is dense in $f_\beta^{m_k}(A_k)$ for all $k \in \omega$. The property (1) shows that $f_\beta^{m_k}(A_k) \cap O_k \neq \emptyset$ and therefore $f_\beta^{m_k}(B(k, \alpha)) \cap O_k \neq \emptyset$ for some $k \in \omega$. On the other hand, $f_\alpha^{m_k}(B(k, \alpha)) \cap O_k = \emptyset$ for all $k \in \omega$ by the property (6). Consequently, $f_\alpha \neq f_\beta$ and therefore $|D| = \kappa^+$.

Our purpose is to prove that D is closed and discrete in Y, so assume, towards a contradiction, that g is an accumulation point in Y for the set D. Recall that $l(X^\omega) = \omega$ by Problem 256 and hence $t(Y) \leq t(C_p(X)) = \omega$ (see Problem 149 of [TFS]). Therefore g is also an accumulation point for some countable subset of D, and hence the ordinal $\alpha = \min\{\beta < \kappa^+ : g$ is an accumulation point for $F_\beta\}$

is well-defined. It is evident that α is a limit ordinal. There is $k \in \omega$ and $y \in A_k$ such that $g \in [y, O_k]$; it is evident that g is also an accumulation point for the set $G = F_\alpha \cap [y, O_k]$. The set $K = \bigcap\{(f^{m_k})^{-1}(f^{m_k}(y)) : f \in G\}$ is non-empty because $y \in K$.

Let $W = (g^{m_k})^{-1}(O_k)$ and assume that $K \backslash W \neq \emptyset$. Take any $x \in K \backslash W$ and observe that $g^{m_k}(x) \notin O_k$ while $g^{m_k}(y) \in O_k$ and therefore $g^{m_k}(x) \neq g^{m_k}(y)$. On the other hand, $f^{m_k}(x) = f^{m_k}(y)$ for all $f \in G$ which contradicts $g \in \overline{G}$. We proved that the case $K \backslash W \neq \emptyset$ is impossible, i.e., $K \subset W$.

Let $K_f = (f^{m_k})^{-1}(f^{m_k}(y))$ for all $f \in G$; the set $N = p_k^{-1}(p_k(y))$ is compact because the map p_k is perfect. Therefore $N \cap K_f$ is a non-empty compact set for all $f \in G$ and $y \in \bigcap\{N \cap K_f : f \in G\} \subset K \subset W$. Now we can apply Fact 1 of S.326 to conclude that there is a finite $H \subset G$ such that $Q = \bigcap\{N \cap K_f : f \in H\} \subset W$. Observe that for the map $u_H = p_k \Delta(\Delta\{f^{m_k} : f \in H\})$, we have $Q = u_H^{-1}(u_H(y))$. Now, if $Y = u_H(X)$, then the map $u_H : X \to Y$ is perfect because p_k is perfect (see Fact 1 of T.266).

Therefore Fact 1 of S.226 is applicable to conclude that there is $U \in \tau(Y)$ such that $u_H(y) \in U$ and $u_H^{-1}(U) \subset W$. Let $\gamma = \max\{\beta : f_\beta \in H\}$. Then $\gamma < \mu = \gamma + 1 < \alpha$ because α is a limit ordinal. We have $H \in \mathrm{Fin}(F_\mu)$ and therefore $u_H(B(k, \mu))$ is dense in $u_H(A_k)$ by (5). Furthermore, $u_H(y) \in U \cap u_H(A_k)$ which shows that $U \cap u_H(A_k)$ is a non-empty open subset of $u_H(A_k)$. The set $u_H(B(k, \mu))$ being dense in $u_H(A_k)$ by (5), we have $u_H(B(k, \mu)) \cap U \neq \emptyset$ and therefore there is $z \in B(k, \mu)$ for which $u_H^{-1}(u_H(z)) \subset W$ and, in particular, $z \in W$. This implies that $g^{m_k}(z) \in O_k$.

On the other hand, the condition (6) implies that $f_\mu^{m_k}(B(k, \mu)) \cap O_k = \emptyset$; the conditions (4) and (6) show that for any ordinal β with $\mu < \beta < \alpha$, we have $f_\beta^{m_k}(B(k, \mu)) \cap O_k \subset f_\beta^{m_k}(B(k, \beta)) \cap O_k = \emptyset$. Consequently, $f_\beta^{m_k}(z) \notin O_k$ whenever $\mu \leq \beta < \alpha$ which shows that $g \notin \overline{G \backslash F_\mu}$ and hence g is an accumulation point for the set F_μ which is a contradiction with $\mu < \alpha$ and the choice of α. This contradiction proves that D is a closed discrete subspace of Y. We already saw that $|D| = \kappa^+ > \mathrm{ext}(Y)$; this contradiction shows that $l(Y) \leq \mathrm{ext}(Y)$ and completes our solution.

T.270. *Prove that every subspace of X is a Lindelöf Σ-space if and only if X has a countable network.*

Solution. Given a space Z say that a family $\mathcal{F}$ of subsets of Z is T_0-*separating* if for any distinct $x, y \in Z$ there is $F \in \mathcal{F}$ such that $F \cap \{x, y\}$ consists of exactly one point, i.e., either $F \cap \{x, y\} = \{x\}$ or $F \cap \{x, y\} = \{y\}$. The family $\mathcal{F}$ is T_1-*separating* if $\bigcap\{F \in \mathcal{F} : z \in F\} = \{z\}$ for every $z \in Z$.

Fact 1. Let Z be a Lindelöf Σ-space. If there exists a countable T_1-separating family of closed subsets of Z, then $nw(Z) = \omega$.

Proof. Take a compact cover $\mathcal{C}$ be of the space Z such that some countable family $\mathcal{N}$ of closed subspaces of Z is a network with respect to $\mathcal{C}$ (see Problem 225 and Fact 1 of T.229). Let $\mathcal{F}$ be a countable T_1-separating family of closed subsets of Z.

The family $\mathcal{M}$ of all finite intersections of the elements of $\mathcal{N} \cup \mathcal{F}$ is countable, so it suffices to show that $\mathcal{M}$ is a network in Z.

Take any $z \in Z$ and $O \in \tau(z, Z)$; let $\mathcal{F}_z = \{F \in \mathcal{F} : z \in F\}$. There is $C \in \mathcal{C}$ with $z \in C$. It is evident that the family $\mathcal{K} = \{C \cap F : F \in \mathcal{F}_z\}$ consists of compact subsets of Z and $\{z\} = \bigcap \mathcal{K}$. Therefore we can apply Fact 1 of S.326 to conclude that there exists a finite $\mathcal{P} \subset \mathcal{F}_z$ such that $\bigcap \{C \cap F : F \in \mathcal{P}\} = (\bigcap \mathcal{P}) \cap C \subset O$. It is clear that $G = \bigcap \mathcal{P} \in \mathcal{M}$. Since $G \cap C \subset O$, the closed set $G \setminus O$ is disjoint from C, so there are $U, V \in \tau(Z)$ such that $C \subset U$, $G \setminus O \subset V$ and $U \cap V = \emptyset$. The family $\mathcal{N}$ being a network with respect to $\mathcal{C}$, there is $N \in \mathcal{N}$ such that $C \subset N \subset U$. We have $P = N \cap G \in \mathcal{M}$; it is clear that $z \in P$. Besides, given any $y \in P$, if $y \notin O$, then $y \in G \setminus O \subset V$ which, together with $y \in N \subset U$ gives us a contradiction. Therefore $y \in O$ for any $y \in P$, i.e., $z \in P \subset O$ which proves that $\mathcal{M}$ is a countable network in Z, so Fact 1 is proved.   □

Fact 2. Suppose that a hereditarily Lindelöf space Z has a countable T_0-separating family of closed subsets. Then Z has a countable T_1-separating family of closed subsets.

Proof. Take a countable T_0-separating family $\mathcal{F}$ of closed subsets of Z. Since Z is hereditarily Lindelöf, every open subset of Z is an F_σ-set in Z, so we can fix, for any $F \in \mathcal{F}$, a countable family $\mathcal{P}_F$ of closed subsets of Z such that $\bigcup \mathcal{P}_F = Z \setminus F$. It is clear that the family $\mathcal{G} = \mathcal{F} \cup (\bigcup \{\mathcal{P}_F : F \in \mathcal{F}\})$ is countable and consists of closed subsets of Z.

To see that $\mathcal{G}$ is T_1-separating, take any point $x \in Z$. If $y \neq x$, then there exists an element $F \in \mathcal{F}$ such that $F \cap \{x, y\}$ is a singleton. If $F \cap \{x, y\} = \{x\}$, then $x \in F$ and $y \notin F$; then $P = F \in \mathcal{G}$. Now, if $F \cap \{x, y\} = \{y\}$, then $x \in Z \setminus F$ and hence there is $Q \in \mathcal{P}_F$ such that $x \in Q \subset Z \setminus F$ which shows that $y \notin Q$; then $P = Q \in \mathcal{G}$. As a consequence, for any $x \in Z$ and any $y \neq x$, there is $P \in \mathcal{G}$ such that $x \in P$ and $y \notin P$. Thus $\bigcap \{G \in \mathcal{G} : x \in G\} = \{x\}$ for any $x \in Z$ and hence $\mathcal{G}$ is a T_1-separating countable family of closed subsets of Z. Fact 2 is proved.   □

Fact 3. Let Z be a hereditarily Lindelöf space. If $\mathcal{K}$ is the family of all compact subspaces of Z, then $|\mathcal{K}| \leq \mathfrak{c}$.

Proof. We have $|Z| \leq \mathfrak{c}$ by Problem 015. Since $\psi(Z) \leq hl(Z) = \omega$, we can fix, for every $z \in Z$, a family $\{U_n^z : n \in \omega\} \subset \tau(z, Z)$ such that $\mathrm{cl}_Z(U_{n+1}^z) \subset U_n^z$ for all $n \in \omega$ and $\bigcap \{U_n^z : n \in \omega\} = \{z\}$. The family $\mathcal{U} = \{U_n^z : n \in \omega, z \in Z\}$ has cardinality $\leq \mathfrak{c}$, so the cardinality of the family $\mathcal{V}$ of all countable unions of elements of $\mathcal{U}$ does not exceed $\mathfrak{c}$.

Given any $K \in \mathcal{K}$, take any $z \in Z \setminus K$; we have $\bigcap \{\mathrm{cl}_Z(U_n^z) : n \in \omega\} = \{z\}$ and hence the family of compact sets $\{\mathrm{cl}_Z(U_n^z) \cap K : n \in \omega\}$ has an empty intersection. This implies that there exists $m(z) \in \omega$ for which $\mathrm{cl}_Z(U_{m(z)}^z) \cap K = \emptyset$ and hence $U_{m(z)}^z \subset Z \setminus K$. Consequently, $Z \setminus K = \bigcup \{U_{m(z)}^z : z \in Z \setminus K\}$. The space $Z \setminus K$ being Lindelöf, there is a countable subfamily of the family $\{U_{m(z)}^z : z \in Z \setminus K\}$ which covers $Z \setminus K$. Thus $Z \setminus K \in \mathcal{V}$ which shows that $|\mathcal{K}| = |\{Z \setminus K : K \in \mathcal{K}\}| \leq |\mathcal{V}| \leq \mathfrak{c}$. Fact 3 is proved.   □

Fact 4. If Z is a hereditarily Lindelöf space, then there exist $A, B \subset Z$ with the following properties:

(1) $A \cup B = Z$;
(2) all compact subspaces of A are countable;
(3) all compact subspaces of B are also countable.

Proof. Let $\mathcal{K}$ be the family of all uncountable compact subsets of Z. By Fact 3, we can enumerate $\mathcal{K}$ (possibly with repetitions) as $\mathcal{K} = \{K_\alpha : \alpha < \mathfrak{c}\}$. Observe also that every $K \in \mathcal{K}$ has cardinality $\mathfrak{c}$. Indeed, $|K| \leq |Z| \leq \mathfrak{c}$ (see Problem 015). On the other hand, if $|K| < \mathfrak{c}$, then K cannot be mapped continuously (and even discontinuously) onto $\mathbb{I}$ because $|\mathbb{I}| = \mathfrak{c}$. Therefore K is scattered by Problem 133. Now apply Problem 128 to conclude that the ω-modification L of the space K is Lindelöf. However, L is discrete because $\psi(K) \leq \psi(Z) = \omega$. Since a discrete space is Lindelöf only if it is countable, we have $|K| = |L| \leq \omega$ and hence $K \notin \mathcal{K}$ which is a contradiction.

Now, choose arbitrarily distinct points $x_0, y_0 \in K_0$; assume that $\beta < \mathfrak{c}$ and we have chosen sets $A_\alpha = \{x_\beta : \beta < \alpha\}$ and $B_\alpha = \{y_\beta : \beta < \alpha\}$ such that

(a) $A_\alpha \cap B_\alpha = \emptyset$;
(b) $\{x_\beta, y_\beta\} \subset K_\beta$ for all $\beta < \alpha$.

Since $|K_\alpha| = \mathfrak{c}$, we can choose distinct points $x_\alpha, y_\alpha \in K_\alpha \backslash (A_\alpha \cup B_\alpha)$. It is immediate that the sets $\{x_\beta : \beta \leq \alpha\}$ and $\{y_\beta : \beta \leq \alpha\}$ still satisfy (a) and (b), so our inductive construction can be continued to furnish us sets $P = \{x_\beta : \beta < \mathfrak{c}\}$ and $Q = \{y_\beta : \beta < \mathfrak{c}\}$ with the properties (a) and (b) fulfilled for all $\alpha < \mathfrak{c}$. We claim that $A = P$ and $B = Z \backslash P$ are as promised.

Indeed, if K is an uncountable compact subspace of A, then $K = K_\alpha$ for some $\alpha < \mathfrak{c}$ and hence $y_\alpha \in Q \cap K \subset B \cap K$ which contradicts $K \subset A$ and $A \cap B = \emptyset$. Analogously, if $K \subset B$, then $x_\alpha \in P \cap K = A \cap K$ which contradicts $K \subset B$ and $A \cap B = \emptyset$. Therefore all properties (1)–(3) are fulfilled for A and B and hence Fact 4 is proved. $\square$

Returning to our solution, observe that if $nw(X) = \omega$, then $nw(Y) = \omega$ for every $Y \subset X$ and hence every $Y \subset X$ is a Lindelöf Σ-space by Problem 228. Now assume that every subspace of X is a Lindelöf Σ-space. By Fact 4, there are $A, B \subset X$ such that $X = A \cup B$, and if K is a compact subset contained either in A or in B, then K is countable. If we prove that $nw(A) = nw(B) = \omega$, then evidently, $nw(X) = \omega$, so we can carry out our proof for $X = A$ and $X = B$ which means that we can assume, without loss of generality, that all compact subspaces of X are countable.

Let $\mathcal{C}$ be a compact cover of X such that there is a countable family $\mathcal{F}$ of closed subsets of X which is a network with respect to $\mathcal{C}$. For any $x \in X$, let $\mathcal{F}_x = \{F \in \mathcal{F} : x \in F\}$; for any $x, y \in X$, let $x \sim y$ if $\mathcal{F}_x = \mathcal{F}_y$. It is evident that $\sim$ is an equivalence relationship on X; denote by $\mathcal{T}$ the set of its equivalence classes. Clearly, $T \in \mathcal{T}$ if and only if, there is $x \in X$ such that $T = T_x = \{y \in X : y \sim x\}$.

Given any point $x \in X$ take any $C \in \mathcal{C}$ with $x \in C$ and observe that the family $\mathcal{F}_C = \{F \in \mathcal{F} : C \subset F\}$ is contained in $\mathcal{F}_x$ and hence $\bigcap \mathcal{F}_x \subset \bigcap \mathcal{F}_C$. Besides, $\mathcal{F}_C$ is a network at C and therefore $\bigcap \mathcal{F}_C = C$ which shows that $F_x = \bigcap \mathcal{F}_x$ is compact being a closed subset of a compact set C. Besides, $T_x \subset F_x$ for any $x \in X$ and hence T_x is countable because all compact subsets of X are countable. Thus we can choose an enumeration (possibly with repetitions) $\{x_T^n : n \in \omega\}$ of every $T \in \mathcal{T}$.

For any $n \in \omega$ let $X_n = \{x_T^n : T \in \mathcal{T}\}$; it is clear that $X_n \cap T = \{x_T^n\}$ for every $T \in \mathcal{T}$ and $\bigcup\{X_n : n \in \omega\} = X$. The family $\mathcal{F}_n = \{F \cap X_n : F \in \mathcal{F}\}$ consists of closed subsets of X_n for all $n \in \omega$; we claim that $\mathcal{F}_n$ is T_0-separating.

Indeed, given distinct $x, y \in X_n$, there are distinct $T, S \in \mathcal{T}$ such that $x = x_T^n$ and $y = x_S^n$. Since $T = T_x \neq S = T_y$, we have $\mathcal{F}_x \neq \mathcal{F}_y$. Assume first that there is $F \in \mathcal{F}_x \backslash \mathcal{F}_y$ and hence $x \in F$ while $y \notin F$; then $F \cap \{x, y\} = \{x\}$. Analogously, if there is $F \in \mathcal{F}_y \backslash \mathcal{F}_x$, then $F \cap \{x, y\} = \{y\}$. This proves that $\mathcal{F}_n$ is a countable T_0-separating family in X_n for all $n \in \omega$. The space X is hereditarily Lindelöf, so we can apply Fact 2 to conclude that there is a countable T_1-separating family of closed subsets of X_n. Since X_n is also Lindelöf Σ, we have $nw(X_n) = \omega$ by Fact 1. It is an easy exercise that a countable union of spaces with a countable network also has a countable network so $nw(X) = \omega$ and hence our solution is complete.

T.271. *Prove that every subspace of X is a Lindelöf p-space if and only if X is second countable.*

Solution. If X is second countable, then every subspace of X is second countable and hence each $Y \subset X$ is a Lindelöf p-space by Problem 221. Now assume that every $Y \subset X$ is a Lindelöf p-space. Since every Lindelöf p-space is a Lindelöf Σ-space (see Problem 223), every subspace of X is a Lindelöf Σ-space. Therefore we can apply Problem 270 to conclude that $nw(X) \leq \omega$ and hence $w(X) = nw(X) = \omega$ by Problem 244.

T.272. *Observe that there exist hereditarily Čech-complete non-metrizable spaces. Therefore a hereditarily p-space need not be metrizable. Prove that any hereditarily Čech-complete space is scattered.*

Solution. We denote by $\mathbb{D}$ the set $\{0, 1\}$ with the discrete topology. As usual, we will identify any ordinal with the set of its predecessors and, in particular, $n = \{0, \ldots, n - 1\}$ for any $n \in \omega$. Recall that $\mathbb{D}^0 = \{\emptyset\}$; if $k \in \omega$, $i \in \mathbb{D}$ and $s \in \mathbb{D}^k$, then $s^\frown i \in \mathbb{D}^{k+1}$ is defined by $(s^\frown i)(k) = i$ and $(s^\frown i)|k = s$. If we have $s \in \mathbb{D}^n$, $t \in \mathbb{D}^k$ where $n \leq k$ and $n, k \in \omega + 1$, then $s \subset t$ says that $t|n = s$. For any $k \in \omega$ and $s \in \mathbb{D}^k$, let $[s] = \{x \in \mathbb{D}^\omega : s \subset x\}$. Given a point $x \in \mathbb{D}^\omega$, it is immediate that the family $\{[x|n] : n \in \omega\}$ is a local base at x in the space $\mathbb{D}^\omega$.

If f is a function, then $\text{dom}(f)$ is its domain. Suppose that we have a set of functions $\{f_i : i \in I\}$ such that $f_i|(\text{dom}(f_i) \cap \text{dom}(f_j)) = f_j|(\text{dom}(f_i) \cap \text{dom}(f_j))$ for any $i, j \in I$. Then we can define a function f with $\text{dom}(f) = \bigcup_{i \in I} \text{dom}(f_i)$ as follows: given any $x \in \text{dom}(f)$, find any $i \in I$ with $x \in \text{dom}(f_i)$ and let $f(x) = f_i(x)$. It is easy to check that the value of f at x does not depend on the choice of i, so we have consistently defined a function f which will be denoted by $\bigcup\{f_i : i \in I\}$ (this makes sense if we identify each function with its graph).

Fact 1. For any non-scattered Čech-complete space Z, there exists a dense-in-itself compact $K \subset Z$. In other words, if Z is Čech-complete and every compact subspace of Z is scattered, then Z is itself scattered.

Proof. The fact that Z is not scattered implies that there is a dense-in-itself subspace $Y \subset Z$. It is clear that $\overline{Y}$ is also dense-in-itself; besides, it is Čech-complete by Problem 260 of [TFS]. This shows that we can assume, without loss of generality, that $Z = \overline{Y}$, i.e., Z is a Čech-complete space without isolated points. There exists a family $\{O_n : n \in \omega\} \subset \tau(\beta Z)$ such that $O_0 = \beta Z$, $O_{n+1} \subset O_n$ for all $n \in \omega$ and $\bigcap\{O_n : n \in \omega\} = Z$. Denote by $\mathcal{C}$ the family of all non-empty closed subspaces of Z which have no isolated points.

Let $F_\emptyset = Z$ and assume that, for some $n > 0$, we have constructed families $\{F_s : s \in \mathbb{D}^k\} \subset \mathcal{C}$ for all $k < n$ with the following properties:

(1) for any $k < n$ and $s \in \mathbb{D}^k$, we have $\mathrm{cl}_{\beta Z}(F_s) \subset O_k$;
(2) for any $k < n$ and any distinct $s, t \in \mathbb{D}^k$, we have $F_s \cap F_t = \emptyset$;
(3) if $k < m < n$, $s \in \mathbb{D}^k$, $t \in \mathbb{D}^m$ and $s \subset t$ then $F_t \subset F_s$.

Fix any $s \in \mathbb{D}^{n-1}$; since F_s is dense-in-itself, it is infinite so we can take distinct points $x, y \in F_s$. There exist $U \in \tau(x, \beta Z)$, $V \in \tau(y, \beta Z)$ such that $\mathrm{cl}_{\beta Z}(U) \cup \mathrm{cl}_{\beta Z}(V) \subset O_n$ and $\mathrm{cl}_{\beta Z}(U) \cap \mathrm{cl}_{\beta Z}(V) = \emptyset$.

The sets $F_{s^\frown 0} = \overline{U \cap F_s}$ and $F_{s^\frown 1} = \overline{V \cap F_s}$ are closed in Z and disjoint (the bar denotes the closure in Z). Since F_s has no isolated points, neither $U \cap F_s$ nor $V \cap F_s$ have isolated points and hence $F_{s^\frown i}$ is dense-in-itself for each $i \in \mathbb{D}$. After we construct the pair $\{F_{s^\frown 0}, F_{s^\frown 1}\}$ for all $s \in \mathbb{D}^{n-1}$, we obtain the family $\{F_s : s \in \mathbb{D}^n\}$ for which the properties (1) and (3) hold. The condition (2) is satisfied because the sets $F_{s^\frown 0}$ and $F_{s^\frown 1}$ are disjoint for any $s \in \mathbb{D}^{n-1}$.

Consequently, we can construct the family $\mathcal{F}_k = \{F_s : s \in \mathbb{D}^k\}$ for every $k \in \omega$ in such a way that the conditions (1)–(3) are satisfied for every $n \in \omega$; let $P_k = \bigcup \mathcal{F}_k$ for every $k \in \omega$. Observe that the set $F = \bigcap\{P_k : k \in \omega\}$ is compact. Indeed, it follows from (1) that $\mathrm{cl}_{\beta Z}(P_k) \subset O_k$ for each $k \in \omega$ and therefore

$$\bigcap_{k \in \omega} P_k = \bigcap_{k \in \omega}(\mathrm{cl}_{\beta Z}(P_k) \cap Z) = \bigcap\{\mathrm{cl}_{\beta Z}(P_k) : k \in \omega\} \cap Z = \bigcap\{\mathrm{cl}_{\beta Z}(P_k) : k \in \omega\}$$

because $\bigcap\{\mathrm{cl}_{\beta Z}(P_k) : k \in \omega\} \subset \bigcap_{k \in \omega} O_k = Z$ [see (1)]. Thus F is compact being the intersection of compact subspaces of βZ. Besides, $F \neq \emptyset$ because the family $\{\mathrm{cl}_{\beta Z}(P_k) : k \in \omega\}$ is decreasing by (3). Given any $x \in F$ and $k \in \omega$, we have $x \in \bigcup \mathcal{F}_k$ and hence there is a unique $s_k \in \mathbb{D}^k$ such that $x \in F_{s_k}$ [see (2)]. Observe that $s_k \subset s_{k+1}$ for each $k \in \omega$ by (2) and (3) so the point $y = \bigcup\{s_i : i \in \omega\} \in \mathbb{D}^\omega$ is well-defined; let $f(x) = y$. Observe that it follows from our definition of $f(x)$ that

(4) if $k \in \omega$, $s \in \mathbb{D}^k$ and $x \in F_s \cap F$, then $s \subset f(x)$.

We claim that the map $f : F \to \mathbb{D}^\omega$ is continuous. To prove it take any $x \in F$ and $W \in \tau(f(x), \mathbb{D}^\omega)$. There is $n \in \omega$ such that $[f(x)|n] \subset W$. Let $s = f(x)|n$; then $U = F_s \cap F$ is an open neighborhood of the point x in the space F because it follows from (2) that $F_s \cap F$ is open in F. Given any $y \in U$, we have $s \subset f(y)$ by (4) and hence $f(y)|n = s = f(x)|n$, i.e., $y \in [f(x)|n]$. This proves that $f(U) \subset [f(x)|n] \subset W$ and therefore f is continuous at the point x.

The map f is surjective; indeed, given any $y \in \mathbb{D}^\omega$, the family $\{F_{y|n} : n \in \omega\}$ is decreasing by (3) and therefore

$$\bigcap_{n\in\omega} F_{y|n} = \bigcap_{n\in\omega}(\mathrm{cl}_{\beta Z}(F_{y|n}) \cap Z) = \left(\bigcap_{n\in\omega}\mathrm{cl}_{\beta Z}(F_{y|n})\right) \cap Z = \bigcap_{n\in\omega}\mathrm{cl}_{\beta Z}(F_{y|n})$$

because $\bigcap\{\mathrm{cl}_{\beta Z}(F_{y|n}) : n \in \omega\} \subset \bigcap_{n\in\omega} O_n = Z$ by the property (1). Since the family $\mathcal{A} = \{\mathrm{cl}_{\beta Z}(F_{y|n}) : n \in \omega\}$ is decreasing and consists of compact sets, we have $\bigcap\{F_{y|n} : n \in \omega\} = \bigcap \mathcal{A} \neq \emptyset$, and hence there is $x \in \bigcap\{F_{y|n} : n \in \omega\}$. It is immediate that $f(x) = y$; the point y was taken arbitrarily, so we proved that f maps F continuously onto $\mathbb{D}^\omega$. If F is scattered, then every second countable continuous image of F is countable by Problem 129 which is a contradiction. Thus F is a non-scattered compact subspace of Z and hence there exists a closed dense-in-itself subspace $K \subset F$. It is clear that K is the desired compact subspace of Z, so Fact 1 is proved.   □

Returning to our solution, observe that the space $X = A(\omega_1)$ is a non-metrizable compact space while every $Z \subset X$ is either compact (if it contains the unique non-isolated point of X) or discrete (if it does not contain the non-isolated point of X). Since both compact and discrete spaces are Čech-complete, every subspace of X is Čech-complete. Every Čech-complete space is a p-space, so X is an example of a non-metrizable hereditarily p-space.

Now assume that X is a non-scattered hereditarily Čech-complete space. By Fact 1 there is a non-scattered compact $K \subset X$. This implies that there is a continuous onto map $f : K \to \mathbb{I}$ (see Problem 133). The map f being perfect, there is a closed $F \subset K$ such that $f(F) = \mathbb{I}$ and $g = f|F$ is irreducible (see Problem 366 of [TFS]). Let A be a countable dense subset of $\mathbb{I}$; for any $a \in A$, choose a point $x_a \in g^{-1}(a)$ and let $Y = \{x_a : a \in A\}$.

Observe first that F has no isolated points. Indeed, if x is isolated in F, then $F_1 = F\backslash\{x\}$ is closed in F and hence $g(F_1)$ is closed in $\mathbb{I}$. The map g being irreducible, we have $g(F_1) \neq \mathbb{I}$ and therefore $\mathbb{I}\backslash g(F_1) = \{g(x)\}$. However, a complement of any point of $\mathbb{I}$ is dense in $\mathbb{I}$. Thus $g(F_1)$ is closed and dense in $\mathbb{I}$, so we have $g(F_1) = \mathbb{I}$ which is a contradiction showing that F is dense-in-itself.

Observe also that $A \subset g(Y)$ and hence $g(Y)$ is dense in $\mathbb{I}$. The map g being closed, we have $g(\overline{Y}) = \overline{g(Y)} = \mathbb{I}$. Since g is irreducible, we have $\overline{Y} = F$, i.e., Y is dense in F and therefore Y does not have isolated points. This implies that $\{y\}$ is nowhere dense in Y for any $y \in Y$. Since Y is countable, it turns out that Y is a countable union of its nowhere dense sets, i.e., Y does not have the Baire property.

This implies that Y is a subspace of X that is not Čech-complete (see Problem 274 of [TFS]) which is a contradiction, making our solution complete.

T.273. *Prove that $\omega_1 + 1$ is a scattered compact space which is not hereditarily Čech-complete.*

Solution. We proved in Problem 314 of [TFS] that $\omega_1 + 1$ is compact. Every second countable continuous image of $\omega_1 + 1$ is countable by Fact 1 of S.319, so $\omega_1 + 1$ is scattered by Problem 133. The set L of limit ordinals of ω_1 is stationary by Problem 065 and hence there exist stationary sets $A, B \subset L$ such that $L = A \cup B$ and $A \cap B = \emptyset$ (see Problem 066). Denote by D the set of successor ordinals of ω_1. We claim that the subspace $Y = (D \cup A) \cup \{\omega_1\}$ of the space $\omega_1 + 1$ is not Čech-complete.

To prove it assume towards a contradiction that Y is Čech-complete. Since Y is dense in $(\omega_1 + 1)$, the set $Z = (\omega_1 + 1)\backslash Y = B$ is an F_σ-set in $(\omega_1 + 1)$ and hence B is σ-compact. However, any compact $K \subset B$ is countable because the stationary set A is disjoint from B and intersects any uncountable compact subset of $\omega_1 + 1$. This shows that B is a countable union of countable compact sets and hence $|B| = \omega$ which is a contradiction (see Problem 065). Thus Y is a subspace of $\omega_1 + 1$ which is not Čech-complete.

T.274. *Prove that every subspace of X is σ-compact if and only if X is countable.*

Solution. If the space X is countable, then every subspace of X is σ-compact being countable. Now assume that every subspace of X is σ-compact. Every σ-compact space is Lindelöf Σ by Problem 226, so every subspace of X is a Lindelöf Σ-space and hence $nw(X) = \omega$ by Problem 270. This implies $hl(X) = \omega$ and therefore there exist $A, B \subset X$ such that $A \cup B = X$ and any compact $K \subset X$ is countable if it is contained either in A or in B (see Fact 4 of T.270). By our assumption about X, both sets A and B are σ-compact and hence countable. Therefore X is also countable.

T.275. *Prove that*

(i) *if an uncountable regular cardinal κ is a caliber of a space X, then κ is a precaliber of X;*

(ii) *if an infinite successor cardinal κ is a precaliber of a space X, then $c(X) < \kappa$. In particular, if ω_1 is a precaliber of X, then $c(X) = \omega$.*

Solution.

(i) Take a family $\mathcal{U} = \{U_\alpha : \alpha < \kappa\} \subset \tau^*(X)$. There exists a set $A \subset \kappa$ such that $|A| = \kappa$ and $\bigcap\{U_\alpha : \alpha \in A\} \neq \emptyset$. It is evident that the family $\{U_\alpha : \alpha \in A\}$ is centered and hence κ is a precaliber of X.

(ii) Assume that $\kappa = \lambda^+$ and $c(X) \geq \kappa$; if $|\mathcal{U}| < \kappa$ for any disjoint $\mathcal{U} \subset \tau^*(X)$, then $|\mathcal{U}| \leq \lambda$ for any such family $\mathcal{U}$ and hence $c(X) \leq \lambda$, a contradiction. Therefore there exists a disjoint $\mathcal{U} = \{U_\alpha : \alpha < \kappa\} \subset \tau^*(X)$. It is clear that no non-empty subfamily of $\mathcal{U}$ is centered and therefore $\mathcal{U}$ witnesses that κ is not a precaliber of X which is again a contradiction proving that $c(X) < \kappa$.

T.276. *Let κ be an uncountable regular cardinal. Prove that if κ is a precaliber (caliber) of X_n for every $n \in \omega$, then κ is a precaliber (caliber) of $\bigcup \{X_n : n \in \omega\}$.*

Solution. Assume that κ is a precaliber (caliber) of X_n for every $n \in \omega$. Let $X = \bigcup \{X_n : n \in \omega\}$ and take any family $\mathcal{U} = \{U_\alpha : \alpha < \kappa\} \subset \tau^*(X)$. If $A_n = \{\alpha < \kappa : U_\alpha \cap X_n \neq \emptyset\}$ for every $n \in \omega$, then $\bigcup \{A_n : n \in \omega\} = \kappa$ and therefore $|A_n| = \kappa$ for some $n \in \omega$; let $\varphi : \kappa \to A_n$ be a bijection.

By our choice of the set A_n, if $V_\alpha = U_{\varphi(\alpha)} \cap X_n$ for each $\alpha < \kappa$, then the family $\{V_\alpha : \alpha < \kappa\}$ consists of non-empty open subsets of X_n. Thus there exists $B \subset \kappa$ such that $|B| = \kappa$ and the family $\{V_\beta : \beta \in B\} = \{V_{\varphi^{-1}(\alpha)} : \alpha \in \varphi(B)\}$ is centered (has non-empty intersection). Since $V_{\varphi^{-1}(\alpha)} \subset U_\alpha$ for each $\alpha \in A = \varphi(B)$, the family $\{U_\alpha : \alpha \in A\}$ is also centered (has non-empty intersection); besides, $|A| = |B| = \kappa$ which shows that κ is a precaliber (caliber) of X.

T.277. *Let κ be an uncountable regular cardinal. Prove that if κ is a precaliber (caliber) of X, then κ is a precaliber (caliber) of every continuous image of X.*

Solution. Assume that κ is a precaliber (caliber) of the space X. Let $f : X \to Y$ be a continuous onto map. Given a family $\{U_\alpha : \alpha < \kappa\} \subset \tau^*(Y)$, the family $\{f^{-1}(U_\alpha) : \alpha < \kappa\}$ consists of non-empty open subsets of X and hence there is $A \subset \kappa$ such that $|A| = \kappa$ and the family $\{f^{-1}(U_\alpha) : \alpha \in A\}$ is centered (has non-empty intersection). It is immediate that the family $\{U_\alpha : \alpha \in A\}$ is also centered (or has non-empty intersection respectively) which proves that κ is a precaliber (caliber) of the space Y.

T.278. *Suppose that κ is an uncountable regular cardinal and Y is a dense subspace of X. Prove that*

(i) if κ is a caliber of Y, then it is a caliber of X;
(ii) κ is a precaliber of Y if and only if it is a precaliber of X.

Solution.

(i) If $\{U_\alpha : \alpha < \kappa\} \subset \tau^*(X)$, then $\{U_\alpha \cap Y : \alpha < \kappa\} \subset \tau^*(Y)$ because the subspace Y is dense in X. Therefore there is a set $A \subset \kappa$ such that $|A| = \kappa$ and $P = \bigcap \{U_\alpha \cap Y : \alpha \in A\} \neq \emptyset$. It is evident that $P \subset \bigcap \{U_\alpha : \alpha \in A\}$ and hence $\bigcap \{U_\alpha : \alpha \in A\} \neq \emptyset$ which proves that κ is a caliber of X.

(ii) Assume first that κ is a precaliber of Y. If $\{U_\alpha : \alpha < \kappa\} \subset \tau^*(X)$, then $\{U_\alpha \cap Y : \alpha < \kappa\} \subset \tau^*(Y)$ because Y is dense in X. Therefore there is a set $A \subset \kappa$ such that $|A| = \kappa$ and the family $\{U_\alpha \cap Y : \alpha \in A\}$ is centered. It is evident that $\{U_\alpha : \alpha \in A\}$ is also centered which proves that κ is a precaliber of X.

Now let κ be a precaliber of X and take any family $\{V_\alpha : \alpha < \kappa\} \subset \tau^*(Y)$. Take $U_\alpha \in \tau(X)$ such that $U_\alpha \cap Y = V_\alpha$ for each $\alpha < \kappa$. Then $\{U_\alpha : \alpha < \kappa\} \subset \tau^*(X)$ and therefore there exists $A \subset \kappa$ for which the family $\{U_\alpha : \alpha \in A\}$ is centered. Given

$\alpha_1, \ldots, \alpha_n \in A$, the set $U = U_{\alpha_1} \cap \cdots \cap U_{\alpha_n}$ is non-empty and hence $U \cap Y \neq \emptyset$ because Y is dense in X. Since $V_{\alpha_1} \cap \cdots \cap V_{\alpha_n} = U \cap Y \neq \emptyset$ and the ordinals $\alpha_1, \ldots \alpha_n \in A$ were chosen arbitrarily, we proved that the family $\{V_\alpha : \alpha \in A\}$ is centered and hence κ is a precaliber of Y.

T.279. *Show that an uncountable regular cardinal κ is a caliber of a compact space X if and only if it is a precaliber of X.*

Solution. Any caliber of X is also a precaliber of X by Problem 275, so assume that κ is a precaliber of X and take a family $\{U_\alpha : \alpha < \kappa\} \subset \tau^*(X)$. By regularity of X, there is a family $\{V_\alpha : \alpha < \kappa\} \subset \tau^*(X)$ such that $\overline{V}_\alpha \subset U_\alpha$ for all $\alpha < \kappa$. There is $A \subset \kappa$ such that $|A| = \kappa$ and the family $\{V_\alpha : \alpha \in A\}$ is centered. Therefore the family $\mathcal{F} = \{\overline{V}_\alpha : \alpha \in A\}$ is centered as well and hence there is $x \in \bigcap \mathcal{F}$ because X is compact. It is immediate that $x \in \bigcap\{U_\alpha : \alpha \in A\}$ and hence κ is a caliber of X.

T.280. *Let κ be an uncountable regular cardinal. Prove that if κ is a precaliber of X_t for every $t \in T$, then κ is a precaliber of $\prod\{X_t : t \in T\}$.*

Solution. If $X_s = \emptyset$ for some $s \in T$, then $X = \prod_{t \in T} X_t = \emptyset$ and there is nothing to prove, so we assume that $X_t \neq \emptyset$ for all $t \in T$. A set $U \subset X$ is called *a standard open subset of X* if $U = \prod_{t \in T} U_t$ where $U_t \in \tau^*(X_t)$ for all $t \in T$ and the set $\mathrm{supp}(U) = \{t \in T : U_t \neq X_t\}$ is finite. Standard open subsets of X form a base of X (see Problem 101 of [TFS]).

Fact 1. An uncountable regular cardinal λ is a precaliber of a space Z if and only if for any family $\{U_a : a \in A\} \subset \tau^*(Z)$ indexed by a set A with $|A| = \lambda$, there is a set $B \subset A$ such that $|B| = \lambda$ and the family $\{U_a : a \in B\}$ is centered.

Proof. Sufficiency is clear, so assume that λ is a precaliber of Z; given a set A with $|A| = \lambda$ and a family $\{U_a : a \in A\} \subset \tau^*(Z)$ fix any bijection $\varphi : \lambda \to A$. If $V_\alpha = U_{\varphi(\alpha)}$ for any $\alpha < \lambda$, then $\{V_\alpha : \alpha < \lambda\} \subset \tau^*(Z)$, and hence there is a set $C \subset \lambda$ such that $|C| = \lambda$ and the family $\{V_\alpha : \alpha \in C\}$ is centered. If $B = \varphi(C)$, then $B \subset A$, $|B| = \lambda$ and the family $\{U_a : a \in B\} = \{U_{\varphi(\alpha)} : \alpha \in C\} = \{V_\alpha : \alpha \in C\}$ is centered. Fact 1 is proved. $\square$

To prove that precalibers are preserved by any products, we have to show first that they are preserved by finite ones. To that end, let us prove that

(1) given spaces Y and Z if κ is a precaliber of both Y and Z, then κ is a precaliber of $Y \times Z$.

Take any set A such that $|A| = \kappa$ and a family $\mathcal{U} = \{U_a : a \in A\} \subset \tau^*(Y \times Z)$. For any $a \in A$ there exist $V_\alpha \in \tau^*(Y)$ and $W_\alpha \in \tau^*(Z)$ such that $U_\alpha \supset V_\alpha \times W_\alpha$. Since κ is a precaliber of Y, by Fact 1, there is a set $A_1 \subset A$ such that $|A_1| = \kappa$ and the family $\{V_\alpha : a \in A_1\}$ is centered; since κ is a precaliber of Z, we can choose a set $A_2 \subset A_1$ such that $|A_2| = \kappa$ and the family $\mathcal{W} = \{W_a : a \in A_2\}$ is also centered (here we used Fact 1 again).

We claim that the family $\mathcal{U}' = \{U_a : a \in A_2\}$ is centered as well. To see this, take any $U_{a_1}, \ldots, U_{a_n} \in \mathcal{U}'$; since $\{V_a : a \in A_2\}$ is centered, we can find a point $y \in V_{a_1} \cap \cdots \cap V_{a_n}$. The family $\mathcal{W}$ being centered, there is a point $z \in W_{a_1} \cap \cdots \cap W_{a_n}$. Hence $w = (y, z) \in (V_{a_1} \times W_{a_1}) \cap \cdots \cap (V_{a_n} \times W_{a_n}) \subset U_{a_1} \cap \cdots \cap U_{a_n}$, so the family $\mathcal{U}'$ is centered. This shows that κ is a precaliber of $Y \times Z$, i.e., the property (1) is proved. A trivial induction shows that

(2) given spaces $Y_1, \ldots, Y_n$, if κ is a precaliber of Y_i for all $i = 1, \ldots, n$, then κ is a precaliber of $Y_1 \times \cdots \times Y_n$.

Now take any set A with $|A| = \kappa$ and a family $\{U_a : a \in A\} \subset \tau^*(X)$. For any $a \in A$ there is a standard open set $V_a \subset U_a$; let $P_a = \mathrm{supp}(V_a)$ and fix a point $v_a \in V_a$. By the Δ-lemma (see Problem 038), there is a finite set $S = \{s_1, \ldots, s_n\} \subset T$ and $B \subset A$ such that $|B| = \kappa$ and $P_a \cap P_b = S$ for any distinct $a, b \in B$. The family $\{p_S(V_a) : a \in B\}$ consists of non-empty open subsets of $X_S = \prod_{s \in S} X_s$; the cardinal κ being a precaliber of X_S by (2), there exists $C \subset B$ such that $|C| = \kappa$ and the family $\{P_S(V_a) : a \in C\}$ is centered (we again applied Fact 1 here). To see that $\{U_a : a \in C\}$ is centered take any $a_1, \ldots, a_m \in C$. There is a point $y \in P_S(V_{a_1}) \cap \cdots p_S(V_{a_m})$. Since the family $\{P_{a_i} \setminus S : i = 1, \ldots, m\}$ is disjoint, we can choose a point $x \in X$ such that $x|S = y$ and $x(t) = v_{a_i}(t)$ for any $t \in P_{a_i} \setminus S$ and $i = 1, \ldots, m$. It is immediate that $x \in V_{a_1} \cap \cdots \cap V_{a_m} \subset U_{a_1} \cap \cdots \cap U_{a_m}$ and therefore $U_{a_1} \cap \cdots \cap U_{a_m} \neq \emptyset$. This proves that the family $\{U_a : a \in C\}$ is centered and hence κ is a precaliber of X.

T.281. *Let κ be an uncountable regular cardinal. Prove that if κ is a caliber of X_t for every $t \in T$, then κ is a caliber of $\prod\{X_t : t \in T\}$.*

Solution. If $X_s = \emptyset$ for some $s \in T$, then $X = \prod_{t \in T} X_t = \emptyset$ and there is nothing to prove so we assume that $X_t \neq \emptyset$ for all $t \in T$. A set $U \subset X$ is called *a standard open subset of X* if $U = \prod_{t \in T} U_t$ where $U_t \in \tau^*(X_t)$ for all $t \in T$ and the set $\mathrm{supp}(U) = \{t \in T : U_t \neq X_t\}$ is finite. Standard open subsets of X form a base of X (see Problem 101 of [TFS]).

Fact 1. An uncountable regular cardinal λ is a caliber of a space Z if and only if for any family $\{U_a : a \in A\} \subset \tau^*(Z)$ indexed by a set A with $|A| = \lambda$, there is a set $B \subset A$ such that $|B| = \lambda$ and $\bigcap\{U_a : a \in B\} \neq \emptyset$.

Proof. Sufficiency is clear so assume that λ is a caliber of Z; given a set A with $|A| = \lambda$ and a family $\{U_a : a \in A\} \subset \tau^*(Z)$, fix any bijection $\varphi : \lambda \to A$. If $V_\alpha = U_{\varphi(\alpha)}$ for any $\alpha < \lambda$, then $\{V_\alpha : \alpha < \lambda\} \subset \tau^*(Z)$, and hence there is a set $C \subset \lambda$ such that $|C| = \lambda$ and $\bigcap\{V_\alpha : \alpha \in C\} \neq \emptyset$. If $B = \varphi(C)$, then $B \subset A$, $|B| = \lambda$ and $\bigcap\{U_a : a \in B\} = \bigcap\{U_{\varphi(\alpha)} : \alpha \in C\} = \bigcap\{V_\alpha : \alpha \in C\} \neq \emptyset$, so Fact 1 is proved. $\square$

To prove that calibers are preserved by any products, we have to show first that they are preserved by finite ones. To that end let us prove that

(1) given spaces Y and Z if κ is a caliber of both Y and Z, then κ is a caliber of $Y \times Z$.

Take any set A such that $|A| = \kappa$ and a family $\mathcal{U} = \{U_a : a \in A\} \subset \tau^*(Y \times Z)$. For any $a \in A$ there exist $V_\alpha \in \tau^*(Y)$ and $W_\alpha \in \tau^*(Z)$ such that $U_\alpha \supset V_\alpha \times W_\alpha$. Since κ is a caliber of Y, by Fact 1, there is a set $A_1 \subset A$ such that $|A_1| = \kappa$ and $\bigcap\{V_a : a \in A_1\} \neq \emptyset$; since κ is a caliber of Z, we can choose a set $A_2 \subset A_1$ such that $|A_2| = \kappa$ and $\bigcap\{W_a : a \in A_2\} \neq \emptyset$ (here we used Fact 1 again).

Choose any points $y \in \bigcap\{V_a : a \in A_2\}$ and $z \in \bigcap\{W_a : a \in A_1\}$. Then $(y, z) \in \bigcap\{V_a \times W_a : a \in A_2\} \subset \bigcap\{U_a : a \in A_2\}$ and hence $\bigcap\{U_a : a \in A_2\} \neq \emptyset$. This shows that κ is a caliber of $Y \times Z$, i.e., the property (1) is proved. A trivial induction shows that

(2) given spaces $Y_1, \ldots, Y_n$, if κ is a caliber of Y_i for all $i = 1, \ldots, n$, then κ is a caliber of $Y_1 \times \cdots \times Y_n$.

Now take any set A with $|A| = \kappa$ and a family $\{U_a : a \in A\} \subset \tau^*(X)$. For any $a \in A$ there is a standard open set $V_a \subset U_a$; let $P_a = \mathrm{supp}(V_a)$ and fix a point $v_a \in V_a$. By the Δ-lemma (see Problem 038), there is a finite set $S = \{s_1, \ldots, s_n\} \subset T$ and $B \subset A$ such that $|B| = \kappa$ and $P_a \cap P_b = S$ for any distinct $a, b \in B$. The family $\{p_S(V_a) : a \in B\}$ consists of non-empty open subsets of $X_S = \prod_{s \in S} X_s$; the cardinal κ being a caliber of X_S by (2), there exists $C \subset B$ such that $|C| = \kappa$ and $\bigcap\{P_S(V_a) : a \in C\} \neq \emptyset$ (we again applied Fact 1 here). In fact, the set C witnesses that κ is a caliber of X, i.e., $\bigcap\{U_a : a \in C\} \neq \emptyset$. Indeed, take any point $y \in \bigcap\{P_S(V_a) : a \in C\}$. Since the family $\{P_a \backslash S : a \in C\}$ is disjoint, we can choose a point $x \in X$ such that $x|S = y$ and $x(t) = v_a(t)$ for any $t \in P_a \backslash S$ and $a \in C$. It is immediate that $x \in \bigcap\{V_a : a \in C\} \subset \bigcap\{U_a : a \in C\}$ and therefore $\bigcap\{U_a : a \in C\} \neq \emptyset$ which proves that κ is a caliber of X.

T.282. *Prove that any product X of separable spaces satisfies the Shanin condition, i.e., every uncountable regular cardinal is a caliber of X.*

Solution. Take any uncountable regular cardinal κ. By our hypothesis we have $X = \prod\{X_t : t \in T\}$ where each X_t is separable; fix a countable dense $D_t \subset X_t$ for every $t \in T$. Take any $t \in T$ and a family $\{U_\alpha : \alpha < \kappa\} \subset \tau^*(X_t)$. Let $P_d = \{\alpha < \kappa : d \in U_\alpha\}$ for every $d \in D_t$. Since $U_\alpha \cap D_t \neq \emptyset$ for each $\alpha < \kappa$, we have $\bigcup\{P_d : d \in D_t\} = \kappa$. The set D_t being countable, there is $d \in D_t$ such that $A = P_d$ has cardinality κ. As a consequence $d \in \bigcap\{U_\alpha : \alpha \in A\}$, which, together with $|A| = \kappa$, proves that κ is a caliber of X_t for all $t \in T$. Finally apply Problem 281 to conclude that κ is a caliber of X.

T.283. *Prove that any uncountable regular cardinal is a precaliber of $C_p(X)$ for any space X.*

Solution. Any uncountable regular cardinal κ is a caliber of $\mathbb{R}^X$ by Problem 282. Since $C_p(X)$ is dense in $\mathbb{R}^X$, we can apply Problems 275 and 278 to conclude that κ is a precaliber of $C_p(X)$.

T.284. *Prove that there exists a space X such that ω_1 is a precaliber of X while the point-finite cellularity of X is uncountable. Observe that if ω_1 is a caliber of X, then $p(X) = \omega$.*

Solution. It is evident that ω_1 is a caliber of X if and only if any point-countable family of non-empty open subsets of X is countable. This clearly implies that any point-finite family of non-empty open subsets of X is also countable, i.e., $p(X) \le \omega$.

Now, if $X = C_p(A(\omega_1))$, then ω_1 is a precaliber of X by Problem 283. Furthermore, $A(\omega_1)$ embeds in $C_p(C_p(A(\omega_1))) = C_p(X)$ which shows that $p(X) \ge \omega_1$ by Problem 178 of [TFS].

T.285. *Let X be a metrizable space. Prove that any regular uncountable cardinal is a caliber of $C_p(X)$.*

Solution. For any number $n \in \mathbb{N}$ consider the set $M_n = \{1, \dots, n\}$; we will also need the family $\mathcal{O} = \{(a,b) : a, b \in \mathbb{Q} \text{ and } a < b\}$. Given any points $x_1, \dots, x_n \in X$ and sets $O_1, \dots, O_n \in \mathcal{O}$, let $[x_1, \dots, x_n; O_1, \dots, O_n] = \{f \in C_p(X) : f(x_i) \in O_i \text{ for all } i \in M_n\}$. The family $\mathcal{B} = \{[x_1, \dots, x_n; O_1, \dots, O_n] : n \in \mathbb{N}, \ x_i \in X \text{ and } O_i \in \mathcal{O} \text{ for all } i \in M_n\}$ is a base in the space $C_p(X)$. It is straightforward to see that the family $\mathcal{C} = \{[x_1, \dots, x_n; O_1, \dots, O_n] \in \mathcal{B} : x_i \ne x_j \text{ and } O_i \cap O_j = \emptyset$ for any distinct $i, j \in M_n\}$ is a π-base in $C_p(X)$, i.e., any non-empty open subset of $C_p(X)$ contains an element of $\mathcal{C}$.

Fact 1. Let Z be a metrizable space. If λ is not a countably cofinal cardinal and $w(Z) \ge \lambda$, then there is a closed discrete $D \subset Z$ such that $|D| = \lambda$.

Proof. Let $\mathcal{U} = \bigcup\{\mathcal{U}_n : n \in \omega\} \subset \tau^*(Z)$ be a base in Z such that $\mathcal{U}_n$ is discrete for all $n \in \omega$; it is clear that $|\mathcal{U}| \ge \lambda$. Since λ is not ω-cofinal, we have $|\mathcal{U}_n| \ge \lambda$ for some $n \in \omega$. Now take a point $z(U) \in U$ for every $U \in \mathcal{U}_n$; then the set $D' = \{z(U) : U \in \mathcal{U}_n\}$ is closed and discrete. Since $|D'| \ge \lambda$, we can choose a set $D \subset D'$ with $|D| = \lambda$. It is evident that D is as promised so Fact 1 is proved. □

Fact 2. For any space Z if λ is a regular uncountable cardinal such that $d(Z) < \lambda$, then λ is a caliber of Z.

Proof. Fix a set $D \subset Z$ such that $\overline{D} = Z$ and $|D| = \mu = d(Z)$. If we are given a family $\{U_\alpha : \alpha < \lambda\} \subset \tau^*(Z)$, then let $E_d = \{\alpha < \lambda : d \in U_\alpha\}$ for every $d \in D$. Since D is dense in Z, we have $\bigcup\{E_d : d \in D\} = \lambda$. The cardinal $\lambda > \mu$ being regular there is $d \in D$ such that $|E_d| = \lambda$. Since $d \in \bigcap\{U_\alpha : \alpha \in E_d\}$, the set E_d witnesses that λ is a caliber of Z. Fact 2 is proved. □

Returning to our solution, fix a metric d on X which generates $\tau(X)$ and take a regular uncountable cardinal κ. It is an easy exercise that

$(*)$ if D is a discrete subset of X with $|D| = \kappa$, then there is $E \subset D$ such that E is closed (and, obviously, discrete) in X and $|E| = \kappa$.

Given a family $\{U_a : a \in A\} \subset \tau^*(C_p(X))$ such that $|A| = \kappa$, we can choose a set $V_a = [x_1^a, \dots, x_{n(a)}^a; O_1^a, \dots, O_{n(a)}^a] \in \mathcal{C}$ such that $V_a \subset U_a$; let $\text{supp}(V_a) = \{x_1^a, \dots, x_{n(a)}^a\}$ for all $a \in A$. Since $\mathcal{O}$ is countable, there exists a set $B \subset A$, a number $n \in \mathbb{N}$ and $O_1, \dots, O_n \in \mathcal{O}$ such that $|B| = \kappa$, $n(a) = n$ and $O_i^a = O_i$ for any $a \in B$ and $i \in M_n$.

Now we can apply the Delta-lemma (see Problem 038) to find a finite set $P = \{p_1, \ldots, p_m\} \subset X$ and $B_1 \subset B$ such that $|B_1| = \kappa$ and $\text{supp}(V_a) \cap \text{supp}(V_b) = P$ for any distinct $a, b \in B_1$ (it possible that $m = 0$, i.e., $P = \emptyset$). Since the order of points and open sets in the definition of V_a does not influence the definition of V_a and there are only countably many possibilities of assigning an element of $\mathcal{O}$ to a point of P, we can consider that $V_a = [p_1, \ldots, p_m, x_1^a, \ldots, x_{n-m}^a; G_1, \ldots, G_m, O_1, \ldots, O_{n-m}]$ for each $a \in B_1$.

Given any $E \subset B_1$ we will need the set $R_i(E) = \{x_i^a : a \in E\}$ for all $i \in M_{n-m}$; then the family $\{R_i(E) : i \in M_{n-m}\}$ is disjoint for any $E \subset B_1$. Consider the set $H = \{i \in M_{n-m} : w(R_i(B_1)) < \kappa\}$; it follows from Fact 1 and $(*)$ that for each $i \in M_{n-m} \backslash H$, we can find a set $D_i \subset R_i(B_1)$ of cardinality κ which is discrete and closed in X. Observe also that $w(Y) = w(\overline{Y})$ for any $Y \subset X$ and hence $|\overline{R_k(B_1)} \cap D_i| < \kappa$ for any $k \in H$ and $i \in M_{n-m} \backslash H$. Therefore we can choose a set $B_2 \subset B_1$ and $\varepsilon > 0$ with the following properties:

(1) $d(x, y) > \varepsilon$ for any $x \in P \cup \overline{R_k(B_2)}$ and $y \in R_i(B_2)$ whenever $k \in H$ and $i \in M_{n-m} \backslash H$;
(2) $|B_2| = \kappa$ and $R_i(B_2)$ is closed and discrete in X for any $i \in M_{n-m} \backslash H$.

Let $Y = P \cup (\bigcup\{\overline{R_i(B_2)} : i \in H\})$; then $w(Y) < \kappa$. Choose $i_1, \ldots, i_l \in M_{n-m}$ such that $H = \{i_1, \ldots, i_l\}$ and consider the set $W_a = \{f \in C_p(Y) : f(p_i) \in G_i$ for all $i \in M_m$ and $f(x_{i_j}^a) \in O_{i_j}$ for all $j \in M_l\}$ for all $a \in B_2$. It is evident that $\{W_a : a \in B_2\} \subset \tau^*(C_p(Y))$. Since $d(C_p(Y)) \leq nw(C_p(Y)) = nw(Y) = w(Y) < \kappa$, the cardinal κ is a caliber of $C_p(Y)$ by Fact 2. Consequently, there is $B_3 \subset B_2$ and $f \in C_p(Y)$ such that $|B_3| = \kappa$ and $f \in \bigcap\{W_a : a \in B_3\}$.

The set Y is closed in X and hence so is $Y \cup (\bigcup\{R_i(B_2) : i \in M_{n-m} \backslash H\})$. Since the set $F = \bigcup\{R_i(B_2) : i \in M_{n-m} \backslash H\}$ is closed, discrete and disjoint from Y, there exists $g \in C(Y \cup F)$ such that $g|Y = f$ and $g(x) \in O_i$ for every $i \in M_{n-m} \backslash H$ and $x \in R_i(B_2)$. The set $Y \cup F$ closed and hence C-embedded in the normal space X, so there is a function $h \in C_p(X)$ with $h|(Y \cup F) = g$. It is immediate that $h \in \bigcap\{V_a : a \in B_3\} \subset \bigcap\{U_a : a \in B_3\}$ which proves that κ is a caliber of $C_p(X)$. Since an uncountable regular cardinal κ was taken arbitrarily, we proved that every uncountable regular cardinal is a caliber of $C_p(X)$, i.e., our solution is complete.

T.286. *Prove that an uncountable regular cardinal κ is a precaliber of X if and only if it is a caliber of βX.*

Solution. If κ is a caliber of βX, then it is a precaliber of X by Problem 278. Now, if κ is a precaliber of X, then it is also a precaliber of βX by Problem 278. Therefore κ is a caliber of βX by Problem 279.

T.287. *Suppose that X is a compact space of countable tightness. Prove that if ω_1 is a caliber of X, then X is separable. Give an example of a non-separable compact space X such that ω_1 is a caliber of X.*

Solution. It follows from $t(X) = \omega$ that the space X has a point-countable π-base $\mathcal{B}$ (see Problem 332 of [TFS]). Since ω_1 is a caliber of X, the family $\mathcal{B}$ must be countable and therefore $d(X) \le \pi w(X) = \omega$.

Now if $\kappa = 2^{\mathfrak{c}}$, then $Y = \mathbb{I}^{\kappa}$ is a compact space such that ω_1 is a caliber of Y by Problem 282. However, Y is not separable because otherwise $w(Y) \le 2^{d(Y)} = 2^{\omega} = \mathfrak{c}$ while $w(Y) = 2^{\mathfrak{c}}$ (see Fact 2 and Fact 3 of S.368).

T.288. *Assuming MA+¬CH, prove that ω_1 is a precaliber of any space which has the Souslin property.*

Solution. Let Z be an arbitrary space such that $c(Z) = \omega$ and take any family $\{U_{\alpha} : \alpha < \omega_1\} \subset \tau^*(Z)$. Consider the set $\mathcal{P} = \tau^*(Z)$ where $U \le V$ iff $U \subset V$. It is clear that $\le$ is a partial order on $\tau^*(Z)$; it is immediate that $U, V \in \mathcal{P}$ are compatible if and only if $U \cap V \ne \emptyset$. An immediate consequence is that $\mathcal{P}$ is ccc because $c(Z) = \omega$.

If there is $\alpha < \omega_1$ such that the set $A(\alpha) = \{\beta : U_{\beta} = U_{\alpha}\}$ is uncountable, then for any $x \in U_{\alpha}$ we have $x \in \bigcap\{U_{\beta} : \beta \in A(\alpha)\}$ and hence $A(\alpha)$ witnesses that ω_1 is a caliber of Z. Now if $A(\alpha)$ is countable for any $\alpha \in \omega_1$, then we can choose an uncountable $B \subset \omega_1$ such that $U_{\alpha} \ne U_{\beta}$ for any distinct $\alpha, \beta \in B$. Observe also that a set $\mathcal{E} \subset \mathcal{P}$ is centered in the sense of Problem 049 if and only if the family $\mathcal{E}$ is centered as a family of subsets of Z. Therefore we can apply Problem 049 to conclude that the uncountable family $\mathcal{U} = \{U_{\alpha} : \alpha \in B\}$ contains an uncountable centered subfamily, i.e., there is an uncountable $C \subset B$ for which $\{U_{\alpha} : \alpha \in C\}$ is centered. This proves that ω_1 is a precaliber of Z.

T.289. *Assume the axiom of Jensen ($\diamondsuit$). Prove that there exists a space X with $c(X) = \omega$ while ω_1 is not a precaliber of X.*

Solution. Under the axiom of Jensen, there exists a hereditarily Lindelöf non-separable compact space X (see Problem 073). Therefore $c(X) \le hl(X) = \omega$; furthermore, $t(X) \le \chi(X) = \psi(X) = \omega$, so if ω_1 is a precaliber of X, then it is a caliber of X by Problem 279 and hence X is separable by Problem 287 which is a contradiction. Thus $c(X) = \omega$ while ω_1 is not a precaliber of X.

T.290. *Prove that for any uncountable regular cardinal κ, the diagonal of $C_p(X)$ is κ-small if and only if κ is a caliber of X. In particular, ω_1 is a caliber of X if and only if $C_p(X)$ has a small diagonal.*

Solution. For any $n \in \mathbb{N}$, denote by M_n the set $\{1, \ldots, n\}$; given $x_1, \ldots, x_n \in X$ and $\varepsilon > 0$, let $O(x_1, \ldots, x_n, \varepsilon) = \{f \in C_p(X) : |f(x_i)| < \varepsilon \text{ for all } i \in M_n\}$. The function $u \in C_p(X)$ is defined by $u(x) = 0$ for all $x \in X$; it is evident that the family $\{O(x_1, \ldots, x_n, \varepsilon) : n \in \mathbb{N}, \ x_1, \ldots, x_n \in X \text{ and } \varepsilon > 0\}$ is a local base of $C_p(X)$ at the point u. Let $\Delta = \{(f, f) : f \in C_p(X)\} \subset C_p(X) \times C_p(X)$ be the diagonal of the space $C_p(X)$.

To prove necessity assume, towards a contradiction, that the diagonal Δ of $C_p(X)$ is κ-small while κ is not a caliber of the space X, i.e., there exists a family $\mathcal{U} = \{U_{\alpha} : \alpha < \kappa\} \subset \tau^*(X)$ such that

(∗) $\bigcap\{U_\alpha : \alpha \in A\} = \emptyset$ for any $A \subset \kappa$ with $|A| = \kappa$.

Take any point $x_\alpha \in U_\alpha$ and a function $f_\alpha \in C_p(X, [0,1])$ such that $f_\alpha(x_\alpha) = 1$ and $f_\alpha(X \backslash U_\alpha) \subset \{0\}$ for all $\alpha < \kappa$. Let $P_\alpha = \{\beta < \kappa : f_\beta(x_\alpha) = 1\}$; it is immediate from (∗) that $|P_\alpha| < \kappa$ for every $\alpha < \kappa$. Using regularity of κ, it is easy to construct by transfinite induction a set $A \subset \kappa$ such that $|A| = \kappa$ and $f_\alpha(x_\beta) \neq 1$ (which, evidently, implies $f_\alpha \neq f_\beta$) for any distinct $\alpha, \beta \in A$. For the set $F = \{f_\alpha : \alpha \in A\}$ we claim that

(1) $|\{\alpha \in A : f_\alpha \notin W\}| < \kappa$ for any $W \in \tau(u, C_p(X))$.

Indeed, $W_1 = O(x_1, \ldots, x_n, \varepsilon) \subset W$ for some $x_1, \ldots, x_n \in X$ and $\varepsilon > 0$. Each x_i belongs to strictly less than κ elements of $\mathcal{U}$ by (∗) and hence the cardinality of the set $B = \{\alpha \in A : f_\alpha(x_i) \neq 0$ for some $i \in M_n\}$ is strictly less than κ. Since $\{\alpha \in A : f_\alpha \notin W\} \subset \{\alpha \in A : f_\alpha \notin W_1\} \subset B$, the property (1) is proved.

It follows from the choice of the set A that F is faithfully indexed and hence $|F| = \kappa$. Therefore the set $H = \{(f_\alpha, u) : \alpha \in A\} \subset (C_p(X) \times C_p(X)) \backslash \Delta$ also has cardinality κ. Given any set $O \in \tau(\Delta, C_p(X) \times C_p(X))$, we have $(u, u) \in O$ and hence there is $W \in \tau(u, C_p(X))$ such that $W \times W \subset O$. It follows from (1) that there exists a set $B \subset A$ with $|B| < \kappa$ such that $f_\alpha \in W$ and hence $(f_\alpha, u) \in W \times W \subset O$ for all $\alpha \in A \backslash B$. This proves that $|H \backslash O| \leq |B| < \kappa$; since the set $O \in \tau(\Delta, C_p(X) \times C_p(X))$ was chosen arbitrarily, the set H witnesses that Δ is not κ-small which is a contradiction. Thus the family $\mathcal{U} = \{U_\alpha : \alpha < \kappa\} \subset \tau^*(X)$ with the property (∗) cannot exist and hence we established necessity.

Now assume that the cardinal κ is a caliber of X and the diagonal Δ of $C_p(X)$ is not κ-small, i.e., there exists a set $P = \{(g_\alpha, h_\alpha) : \alpha < \kappa\} \subset (C_p(X))^2 \backslash \Delta$ such that $(g_\alpha, h_\alpha) \neq (g_\beta, h_\beta)$ for distinct $\alpha, \beta < \kappa$ and $|P \backslash O| < \kappa$ for any set $O \in \tau(\Delta, C_p(X) \times C_p(X))$. Let $f_\alpha = g_\alpha - h_\alpha$ for every $\alpha < \kappa$; it turns out that

(2) $|\{\alpha \in \kappa : f_\alpha \notin W\}| < \kappa$ for any $W \in \tau(u, C_p(X))$.

Indeed, $O = \{(f, g) \in (C_p(X))^2 : f - g \in W\} \in \tau(\Delta, C_p(X) \times C_p(X))$ and hence $|P \backslash O| < \kappa$ which shows that $|\{\alpha < \kappa : g_\alpha - h_\alpha = f_\alpha \notin W\}| < \kappa$ and therefore (2) holds for the set $F = \{f_\alpha : \alpha < \kappa\}$.

Given any $n \in \mathbb{N}$, let $U_\alpha^n = f_\alpha^{-1}(\mathbb{R} \backslash [-\frac{1}{n}, \frac{1}{n}])$ for all $\alpha < \kappa$. For each $\alpha < \kappa$, we have $f_\alpha \neq u$, and hence there is $n(\alpha) \in \mathbb{N}$ such that $U_\alpha^{n(\alpha)} \neq \emptyset$. The cardinal κ being regular, there is $n \in \mathbb{N}$ and $A \subset \kappa$ such that $|A| = \kappa$ and $n(\alpha) = n$ for any $\alpha \in A$. As a consequence, $\mathcal{U} = \{U_\alpha^n : \alpha \in A\} \subset \tau^*(X)$ and $|A| = \kappa$.

Furthermore, given $x \in X$, we have $W = O(x, \frac{1}{n}) \in \tau(u, C_p(X))$, so it follows from (2) that the set $\{\alpha \in A : x \in U_\alpha^n\} \subset \{\alpha \in A : |f_\alpha(x)| \geq \frac{1}{n}\} \subset \{\alpha : f_\alpha \notin W\}$ has cardinality $< \kappa$. The point $x \in X$ was chosen arbitrarily, so any point of X belongs to strictly less than κ-many elements of $\mathcal{U}$. Thus for any $B \subset A$ with $|B| = \kappa$, we have $\bigcap\{U_\alpha^n : \alpha \in B\} = \emptyset$ and hence the family $\mathcal{U}$ witnesses that κ is not a caliber of X which is a contradiction. This settles sufficiency and makes our solution complete.

T.291. *Prove that an uncountable regular cardinal κ is a caliber of X if and only if it is a caliber of $C_p(C_p(X))$.*

Solution. Say that a space Z has a property $\mathcal{P}$ if κ is a caliber of Z. It follows from Problems 277, 281 and 282 that $\mathcal{P}$ is a *complete* property, i.e.,

(1) any metrizable compact space has $\mathcal{P}$;
(2) if $n \in \mathbb{N}$ and Z_i has $\mathcal{P}$ for all $i = 1, \ldots, n$, then $Z_1 \times \cdots \times Z_n$ has $\mathcal{P}$;
(3) if Z has $\mathcal{P}$, then every continuous image of Z has $\mathcal{P}$.

Assume that κ is a caliber of X, i.e., X has $\mathcal{P}$. For any $x \in X$, let $e_x(f) = f(x)$ for every $f \in C_p(X)$. Then $e_x \in C_p(C_p(X))$ for all $x \in X$ and the map $x \to e_x$ is a homeomorphism of X onto the subspace $E = \{e_x : x \in X\}$ of the space $C_p(C_p(X))$ (see Problem 167 of [TFS]).

Since E is homeomorphic to X, the cardinal κ is a caliber of E. There exists an algebra $R(E)$ in the space $C_p(C_p(X))$ such that $E \subset R(E)$ (see Fact 1 of S.312); since $\mathcal{P}$ is a complete property, the space $R(E)$ has σ-$\mathcal{P}$ by Fact 2 of S.312 and hence $R(E)$ has $\mathcal{P}$ because $\mathcal{P}$ is preserved by countable unions (see Problem 276). It is easy to see that E (and hence $R(E)$) separates the points of $C_p(X)$ and therefore $R(E)$ is dense in $C_p(C_p(X))$ by Problem 192 of [TFS]. Finally apply Problem 278 to conclude that $C_p(C_p(X))$ has $\mathcal{P}$, i.e., κ is a caliber of $C_p(C_p(X))$. This proves necessity.

Now, if the cardinal κ is a caliber of $C_p(C_p(X))$, then the diagonal of the space $C_p(C_p(C_p(X)))$ is κ-small by Problem 290. It is immediate that having a κ-small diagonal is a hereditary property, so $C_p(X)$ has a κ-small diagonal being homeomorphic to a subspace of $C_p(C_p(C_p(X)))$ (see Problem 167 of [TFS]). Finally, apply Problem 290 again to conclude that κ is a caliber of X; this settles sufficiency and makes our solution complete.

T.292. *Suppose that an uncountable regular cardinal κ is a caliber of $C_p(X)$. Prove that for any $Y \subset X$ the cardinal κ is a caliber of $C_p(Y)$.*

Solution. Let $\pi : C_p(X) \to C_p(Y)$ be the restriction map, i.e., $\pi(f) = f|Y$ for all $f \in C_p(X)$. Then π is continuous and $Z = \pi(C_p(X))$ is dense in $C_p(Y)$ (see Problem 152 of [TFS]). Since Z is a continuous image of $C_p(X)$, the cardinal κ is a caliber of Z (see Problem 277). Finally, κ is a caliber of $C_p(Y)$ because Z is dense in $C_p(Y)$ (see Problem 278).

T.293. *Let κ be an uncountable regular cardinal. Prove that if κ is a caliber of $C_p(X)$, then the diagonal of X is κ-small. In particular, if ω_1 is a caliber of $C_p(X)$, then the diagonal of X is small.*

Solution. Since κ is a caliber of $C_p(X)$, the diagonal of the space $C_p(C_p(X))$ is κ-small by Problem 290. It is an easy exercise to see that having a κ-small diagonal is a hereditary property, so X has a κ-small diagonal because it is homeomorphic to a subspace of $C_p(C_p(X))$ by Problem 167 of [TFS].

T.294. *Let κ be an uncountable regular cardinal. Prove that if all finite powers of X are Lindelöf and X has a κ-small diagonal, then κ is a caliber of $C_p(X)$. As a consequence, if $l^*(X) = \omega$, then X has a κ-small diagonal if and only if κ is a caliber of $C_p(X)$. In particular, if X is compact, then the diagonal of X is small if and only if ω_1 is a caliber of $C_p(X)$.*

Solution. For any number $n \in \mathbb{N}$ consider the set $M_n = \{1, \ldots, n\}$; we will also need the family $\mathcal{O} = \{(a, b) : a, b \in \mathbb{Q} \text{ and } a < b\}$. Given points $x_1, \ldots, x_n \in X$ and sets $O_1, \ldots, O_n \in \mathcal{O}$, let $[x_1, \ldots, x_n; O_1, \ldots, O_n] = \{f \in C_p(X) : f(x_i) \in O_i \text{ for all } i \in M_n\}$. The family $\mathcal{B} = \{[x_1, \ldots, x_n; O_1, \ldots, O_n] : n \in \mathbb{N}, \ x_i \in X$ and $O_i \in \mathcal{O}$ for all $i \in M_n\}$ is a base in the space $C_p(X)$. It is straightforward to see that the family $\mathcal{C} = \{[x_1, \ldots, x_n; O_1, \ldots, O_n] \in \mathcal{B} : x_i \neq x_j$ and $O_i \cap O_j = \emptyset$ for any distinct $i, j \in M_n\}$ is a π-base in $C_p(X)$, i.e., any non-empty open subset of $C_p(X)$ contains an element of $\mathcal{C}$.

If Z is a space and $n \geq 2$, let $\Delta_{ij}^n(Z) = \{z = (z_1, \ldots, z_n) \in Z^n : z_i = z_j\}$ for any distinct $i, j \in M_n$. The set $\Delta_n(Z) = \bigcup\{\Delta_{ij}^n(Z) : 1 \leq i < j \leq n\}$ is called the *n-diagonal* of the space Z.

Given a space Z and a set $F \subset Z$, say that F is *κ-small* if for any set $A \subset Z \backslash F$ with $|A| = \kappa$, there is $U \in \tau(F, Z)$ such that $|A \backslash U| = \kappa$. Evidently, F is κ-small if and only if for any $A \subset Z \backslash F$ with $|A| = \kappa$, there is $B \subset A$ such that $|B| = \kappa$ and $\overline{B} \cap F = \emptyset$.

Fact 1.

(i) given a space Z any finite union of κ-small subsets of Z is a κ-small subset of Z;

(ii) if $f : Z \to Y$ is a continuous mapping and F is a κ-small subset of Y, then $G = f^{-1}(F)$ is a κ-small subset of Z.

Proof. (i) Suppose that F_i is a κ-small subset of Z for any $i \in M_n$; we must prove that the set $F = \bigcup\{F_i : i \in M_n\}$ is κ-small. Take any $A \subset Z \backslash F$ with $|A| = \kappa$; since F_1 is κ-small, there is $B_1 \subset A$ such that $|B_1| = \kappa$ and $\overline{B}_1 \cap F_1 = \emptyset$. Assume that $1 \leq j < n$ and we have a set $B_j \subset A$ such that

(1) $|B_j| = \kappa$ and $\overline{B}_j \cap F_i = \emptyset$ for all $i \leq j$.

Since F_{j+1} is κ-small, there is $B_{j+1} \subset B_j$ such that $|B_{j+1}| = \kappa$ and $\overline{B}_{j+1} \cap F_{j+1} = \emptyset$. It is evident that (1) still holds for the set B_{j+1}, so we can continue this inductive construction to obtain a set $B = B_n \subset A$ such that $|B_n| = \kappa$ and $\overline{B}_n \cap F_i = \emptyset$ for all $i \in M_n$. This implies $\overline{B} \cap F = \emptyset$, i.e., B witnesses that F is κ-small and hence (i) is proved.

(ii) If $A \subset Z \backslash G$, then $f(A) \subset Y \backslash F$. If $|f(A)| < \kappa$, then by regularity of κ, there is some $y \in f(A)$ such that $B = f^{-1}(y) \cap A$ has cardinality κ. Since $\overline{B} \subset f^{-1}(y) \subset Z \backslash G$, the set B witnesses that G is κ-small. If, on the other hand, we have $|f(A)| = \kappa$, then there is $C \subset f(A)$ such that $|C| = \kappa$ and $\overline{C} \cap F = \emptyset$. It is immediate that for the set $B = f^{-1}(C) \cap A$, we have $|B| = \kappa$ and $\overline{B} \cap G = \emptyset$, so (ii) is settled and Fact 1 is proved. $\qquad \square$

Fact 2. If the diagonal of a space Z is κ-small, then the set $\Delta_n(Z)$ is κ-small in Z^n for all $n \in \mathbb{N}$, $n \geq 2$.

Proof. Let $\Delta = \{(z, z) : z \in Z\} \subset Z \times Z$ be the diagonal of the space Z. Let $q_{ij} : Z^n \to Z \times Z$ be the natural projection onto the face defined by i and j, i.e., for any $z = (z_1, \ldots, z_n) \in Z^n$ we have $q_{ij}(z) = (z_i, z_j) \in Z \times Z$. It is clear that $\Delta_{ij}^n(Z) = q_{ij}^{-1}(\Delta)$ and therefore $\Delta_{ij}^n(Z)$ is κ-small in Z^n for all distinct $i, j \in M_n$ (see Fact 1). Consequently, $\Delta_n(Z) = \bigcup\{\Delta_{ij}^n(Z) : 1 \leq i < j \leq n\}$ is a finite union of κ-small sets, so it is κ-small by Fact 1. Fact 2 is proved. $\square$

Returning to our solution, take any family $\{U_a : a \in A\} \subset \tau^*(C_p(X))$ such that $|A| = \kappa$. We can choose a set $V_a = [x_1^a, \ldots, x_{n(a)}^a; O_1^a, \ldots, O_{n(a)}^a] \in \mathcal{C}$ such that $V_a \subset U_a$ for each $a \in A$. Since $\mathcal{O}$ is countable, there exists a set $B \subset A$, a number $n \in \mathbb{N}$ and $O_1, \ldots, O_n \in \mathcal{O}$ such that $|B| = \kappa$, $n(a) = n$ and $O_i^a = O_i$ for any $a \in B$ and $i \in M_n$. Let $x_a = (x_1^a, \ldots, x_n^a)$ for all $a \in B$; it is evident that $P = \{x_a : a \in B\} \subset X^n \backslash \Delta_n(X)$. The set $\Delta_n(X)$ is κ-small by Fact 2 and therefore there is $D \subset B$ such that $|D| = \kappa$ and the closure of the set $Q = \{x_a : a \in D\}$ does not intersect $\Delta_n(X)$.

For any $y = (y_1, \ldots, y_n) \in \overline{Q}$ fix a function $h_y \in C_p(X)$ such that $h_y(y_i) \in O_i$ for all $i \in M_n$ (such a function exists because $y_i \neq y_j$ whenever $i, j \in M_n$, $i \neq j$ and hence we can apply Problem 034 of [TFS]). Let $W_y = h_y^{-1}(O_1) \times \cdots \times h_y^{-1}(O_n)$; it is clear that $W_y \in \tau(y, X^n)$ for all $y \in \overline{Q}$. The space $\overline{Q}$ being Lindelöf, there is a countable $R \subset \overline{Q}$ such that $\overline{Q} \subset \bigcup\{W_y : y \in R\}$. The cardinal κ being regular there exists a set $E \subset D$ and $y \in R$ such that $|E| = \kappa$ and $x_a \in W_y$ for all $a \in E$. As a consequence, $h_y(x_i^a) \in O_i$ for all $a \in E$ and $i \in M_n$ which shows that $h_y \in V_a \subset U_a$ for all $a \in E$. Thus we have found a set $E \subset A$ such that $|E| = \kappa$ and $\bigcap\{U_a : a \in E\} \neq \emptyset$. This proves that κ is a caliber of $C_p(X)$.

Finally, apply 293 together with our proved result to see that for a space X with $l^*(X) = \omega$ the diagonal of X is κ-small if and only if κ is a caliber of $C_p(X)$.

T.295. *Prove that any compact space of weight $\leq \omega_1$ with a small diagonal is metrizable.*

Solution. Let X be a compact space such that the diagonal of X is small and $w(X) \leq \omega_1$. If X is not metrizable, then the diagonal $\Delta = \{(x, x) : x \in X\} \subset X \times X$ is a not a G_δ-set of $X \times X$ (see Problem 091). It is easy to see that $\chi(\Delta, X \times X) = \omega_1$ and hence there exists an outer base $\mathcal{U} = \{U_\alpha : \alpha < \omega_1\} \subset \tau(\Delta, X \times X)$ of Δ in $X \times X$.

For each $\alpha < \omega_1$ it is easy to construct a sequence $\{U_\alpha^n : n \in \omega\} \subset \tau(\Delta, X \times X)$ such that $U_\alpha^0 = U_\alpha$ and $\overline{U_\alpha^{n+1}} \subset U_\alpha^n$ for all $n \in \omega$. It is immediate that the set $F_\alpha = \bigcap\{U_\alpha^n : n \in \omega\} = \bigcap\{\overline{U_\alpha^n} : n \in \omega\}$ is a closed G_δ-subset of $X \times X$ and $\Delta \subset F_\alpha \subset U_\alpha$ for all $\alpha < \omega_1$.

If $G_\alpha = \bigcap\{F_\beta : \beta \leq \alpha\}$ for all $\alpha < \omega_1$, then each G_α is also a closed G_δ-set of $X \times X$; besides, $\Delta \subset G_\alpha \subset U_\alpha$ and $G_\beta \subset G_\alpha$ whenever $\alpha < \beta < \omega_1$. Observe also that

(*) for any countable $P \subset (X \times X) \backslash \Delta$, there is $\alpha < \omega_1$ such that $G_\alpha \cap P = \emptyset$.

Indeed, since $\mathcal{U}$ is a base of Δ in $X \times X$, for every $p \in P$, there is $\alpha_p < \omega_1$ such that $p \notin U_{\alpha_p}$; if $\alpha > \sup\{\alpha_p : p \in P\}$, then $G_\alpha \cap P = \emptyset$ which proves $(*)$.

Choose any $x_0 \in G_0 \backslash \Delta$; this is possible because $G_0 \neq \Delta$; for otherwise, Δ is a G_δ-set, a contradiction. Now, assume that $\alpha < \omega_1$ and we have chosen points $\{x_\beta : \beta < \alpha\}$ so that

(1) $x_\beta \in G_\beta \backslash \Delta$ for every $\beta < \alpha$;
(2) $x_\beta \neq x_\gamma$ if $\beta, \gamma < \alpha$ and $\beta \neq \gamma$.

The property $(*)$ shows that there is $\gamma \geq \alpha$ such that $G_\gamma \cap \{x_\beta : \beta < \alpha\} = \emptyset$; if we take any $x_\alpha \in G_\gamma \backslash \Delta$ (this is possible again because Δ is not a G_δ-subset of $X \times X$) then $x_\alpha \in G_\alpha$ and the conditions (1) and (2) are satisfied for the set $\{x_\beta : \beta \leq \alpha\}$. Thus our inductive construction can be continued to provide us a set $Y = \{x_\beta : \beta < \omega_1\} \subset (X \times X) \backslash \Delta$ such that (1) and (2) hold for all $\alpha < \omega_1$.

It follows from (2) that $|Y| = \omega_1$; given any $U \in \tau(\Delta, X \times X)$, there is $\alpha < \omega_1$ such that $U_\alpha \subset U$. Consequently, $x_\beta \in G_\beta \subset G_\alpha \subset U_\alpha \subset U$ for all $\beta \geq \alpha$ which shows that $Y \backslash U \subset \{x_\beta : \beta < \alpha\}$ and therefore the set $Y \backslash U$ is countable for every $U \in \tau(\Delta, X \times X)$. This contradiction with the fact that the diagonal of X is small shows that X is metrizable and completes our solution.

T.296. *Let X be a compact space with a small diagonal. Prove that if X is ω-monolithic and has countable tightness, then it is metrizable.*

Solution. If X is not metrizable, then there is $Y = \{x_\alpha : \alpha < \omega_1\} \subset X$ which is not metrizable (see Problem 092). Let $F_\alpha = \overline{\{x_\beta : \beta < \alpha\}}$ for each $\alpha < \omega_1$; observe that the set $F = \bigcup\{F_\alpha : \alpha < \omega_1\}$ is closed. Indeed, if $x \in \overline{F}$, then there is a countable $A \subset F$ such that $x \in \overline{A}$. For any $z \in A$, there is $\alpha_z \in \omega_1$ such that $z \in F_{\alpha_z}$; if $\alpha > \sup\{\alpha_z : z \in A\}$, then $\overline{A} \subset \overline{F_\alpha} = F_\alpha \subset F$ and hence $x \in F$.

Since $Y \subset F$, the subspace F is not metrizable; by ω-monolithity of the space X, we have $nw(F_\alpha) = \omega$ for every $\alpha < \omega_1$. It is an easy exercise that a union of $\leq \omega_1$-many spaces with a countable network has a network weight $\leq \omega_1$ which proves that $w(F) = nw(F) \leq \omega_1$ (see Fact 4 of S.307). It is easy to see that having a small diagonal is a hereditary property and therefore F is a space with a small diagonal. Finally, apply Problem 295 to conclude that F is metrizable; this contradiction shows that X is metrizable.

T.297. *Let X be a compact space with a small diagonal. Prove that if X is monolithic, then it is metrizable.*

Solution. If X is not metrizable, then there is $Y = \{x_\alpha : \alpha < \omega_1\} \subset X$ which is not metrizable (see Problem 092). Since X is ω_1-monolithic, we have $w(\overline{Y}) = nw(\overline{Y}) \leq \omega_1$ (see Fact 4 of S.307). The property of having a small diagonal is hereditary, so $\overline{Y}$ has a small diagonal which, together with Problem 095, implies that $\overline{Y} \supset Y$ is metrizable. This contradiction shows that X is metrizable.

T.298. *Prove that, under CH, any compact space with a small diagonal is metrizable.*

Solution. For each $n \in \mathbb{N}$, let $M_n = \{1, \dots, n\}$ and denote by $\mathbb{D}$ the set $\{0, 1\}$ with the discrete topology; if f is a function, then $\mathrm{dom}(f)$ is its domain. If f and g are functions, then $f \subset g$ says that $\mathrm{dom}(f) \subset \mathrm{dom}(g)$ and $g|\mathrm{dom}(f) = f$. Given a set A let $\mathrm{Fn}(A)$ be the set of functions f such that $\mathrm{dom}(f)$ is a non-empty finite subset of A and $f : \mathrm{dom}(f) \to \mathbb{D}$. For any $h \in \mathrm{Fn}(A)$, let $[h] = \{s \in \mathbb{D}^A : h \subset s\}$. It is easy to see that $\mathcal{H} = \{[h] : h \in \mathrm{Fn}(A)\}$ is a base in the space $\mathbb{D}^A$. A space Z is called *a convergent ω_1-sequence* if $|Z| = \omega_1$ and there is a point $z \in Z$ such that $|Z \backslash U| \leq \omega$ for any $U \in \tau(z, Z)$. Such a point z will be called ω_1-*small* and hence a convergent ω_1-sequence is a space of cardinality ω_1 which has an ω_1-small point.

Let $f : X \to Y$ be a closed continuous onto map; then f is called *irreducible* if for any closed $F \subset X$ with $F \neq X$, we have $f(F) \neq Y$. For any open non-empty $U \subset X$, we will need the sets $f^{\#}(U) = Y \backslash f(X \backslash U)$ and $U^* = f^{-1}(f^{\#}(U))$. It is straightforward that $f^{\#}(U) \subset f(U)$, $U^* \subset U$ and $f^{\#}(U) \in \tau(Y)$, $U^* \in \tau(X)$. Besides, if f is irreducible, then the set U^* is dense in U and hence $f^{\#}(U)$ is dense in $f(U)$ (see Fact 1 of S.383).

Given an infinite cardinal κ, say that a family $\mathcal{F} = \{F_t^0, F_t^1 : t \in T\}$ of closed subsets of a space X is κ-*dyadic* if we have $|T| = \kappa$ and $F_t^0 \cap F_t^1 = \emptyset$ for every $t \in T$ while $I(\mathcal{F}, h) = \bigcap\{F_t^{h(t)} : t \in \mathrm{dom}(h)\} \neq \emptyset$ for any $h \in \mathrm{Fn}(T)$. In particular, $F_t^i \neq \emptyset$ for any $t \in T$ and $i \in \mathbb{D}$. If $\mathcal{A}$ is a family of sets, then $\bigwedge \mathcal{A}$ is the family of all non-empty finite intersections of the elements of $\mathcal{A}$.

Fact 1. If a space X has a small diagonal, then no convergent ω_1-sequence is embeddable in X.

Proof. Let $\Delta = \{(x, x) : x \in X\}$ be the diagonal of X. Assume the contrary and take any $Z \subset X$ for which $|Z| = \omega_1$ and there is $z \in Z$ such that $|Z \backslash U| \leq \omega$ for any $U \in \tau(z, Z)$ and hence for any $U \in \tau(z, X)$. Then $A = \{(t, z) : t \in Z \backslash \{z\}\}$ is contained in $(X \times X) \backslash \Delta$ and $|A| = \omega_1$; given any $O \in \tau(\Delta, X \times X)$, we have $(z, z) \in O$ and therefore there exists $U \in \tau(z, X)$ for which $U \times U \subset O$. The point z being ω_1-small, the set $Z \backslash U$ is countable and therefore $A \backslash O \subset A \backslash (U \times U) \subset (Z \backslash U) \times \{z\}$ which shows that $A \backslash O$ is also countable and hence the set A witnesses that the diagonal of X is not small, a contradiction. Fact 1 is proved. □

Fact 2. The space $\mathbb{I}^{\kappa}$ is a continuous image of $\mathbb{D}^{\kappa}$ for any infinite cardinal κ.

Proof. Fix a continuous onto map $p : \mathbb{D}^{\omega} \to \mathbb{I}$ (see Problem 128 of [TFS]) and let $\varphi_{\alpha} = p$ for each $\alpha < \kappa$; then the product $\varphi = \prod_{\alpha < \kappa} \varphi_{\alpha}$ maps $(\mathbb{D}^{\omega})^{\kappa}$ continuously onto $\mathbb{I}^{\kappa}$ (see Fact 1 of S.271). The space $(\mathbb{D}^{\omega})^{\kappa}$ is homeomorphic to $\mathbb{D}^{\kappa \times \omega}$; the cardinal κ being infinite, we have $|\kappa| = |\kappa \times \omega|$ and hence $\mathbb{D}^{\kappa \times \omega}$ is homeomorphic to $\mathbb{D}^{\kappa}$. Thus φ maps $\mathbb{D}^{\kappa}$ continuously onto $\mathbb{I}^{\kappa}$ so Fact 2 is proved. □

Fact 3. Given a cardinal κ, any subspace of $\mathbb{D}^{\kappa}$ is zero-dimensional, i.e., has a base which consists of clopen subsets.

Proof. Given any $Y \subset \mathbb{D}^\kappa$ if $\mathcal{B}$ is a base in $\mathbb{D}^\kappa$, then $\mathcal{B}_Y = \{U \cap Y : U \in \mathcal{B}\}$ is a base in Y; it is straightforward that if all elements of $\mathcal{B}$ are clopen in $\mathbb{D}^\kappa$, then all elements of $\mathcal{B}_Y$ are clopen in Y, so it suffices to show that $\mathbb{D}^\kappa$ has a base whose elements are clopen.

Given $\alpha < \kappa$ and $i \in \mathbb{D}$, let $O_\alpha^i = \{x \in \mathbb{D}^\kappa : x(\alpha) = i\}$. It is evident that $\mathcal{S} = \{O_\alpha^i : \alpha < \kappa, \ i \in \mathbb{D}\}$ is a subbase of $\mathbb{D}^\kappa$. Each $O_\alpha^i \in \mathcal{S}$ is clopen in $\mathbb{D}^\kappa$ being an inverse image of the clopen set $\{i\}$ under the natural projection of $\mathbb{D}^\kappa$ onto the αth factor. Since each finite intersection of clopen sets is a clopen set, the family $\bigwedge \mathcal{S}$ is a base of $\mathbb{D}^\kappa$ which consists of clopen sets so Fact 3 is proved.   □

Fact 4. Given an infinite cardinal κ the following conditions are equivalent for any compact space X:

(a) the space X can be mapped continuously onto $\mathbb{I}^\kappa$;
(b) there exists a continuous onto map $f : F \to \mathbb{D}^\kappa$ for some closed $F \subset X$;
(c) there exists a κ-dyadic family of closed subsets of X.

Proof. (a) $\Longrightarrow$(b). Let $g : X \to \mathbb{I}^\kappa$ be a continuous onto map. It is clear that $\mathbb{D}^\kappa \subset \mathbb{I}^\kappa$, so the set $F = g^{-1}(\mathbb{D}^\kappa)$ is closed in X and the map $f = g|F$ is as promised.

(b) $\Longrightarrow$(a). There is a continuous onto map $\varphi : \mathbb{D}^\kappa \to \mathbb{I}^\kappa$ by Fact 2; clearly $q = \varphi \circ f$ maps F continuously onto $\mathbb{I}^\kappa$. For any $\alpha < \kappa$, let $\pi_\alpha : \mathbb{I}^\kappa \to \mathbb{I}$ be the natural projection of $\mathbb{I}^\kappa$ onto its αth factor, i.e., $\pi_\alpha(x) = x(\alpha)$ for every $x \in \mathbb{I}^\kappa$. The map $q_\alpha = \pi_\alpha \circ q$ is continuous and hence, by normality of X, there is a continuous map $r_\alpha : X \to \mathbb{I}$ such that $r_\alpha|F = q_\alpha$ for all $\alpha < \kappa$. It is clear that $r = \Delta\{r_\alpha : \alpha < \kappa\}$ maps X to $\mathbb{I}^\kappa$; since $r|F = q$, we have $r(X) \supset q(F) = \mathbb{I}^\kappa$ and hence X can be continuously mapped onto $\mathbb{I}^\kappa$.

(b) $\Longrightarrow$(c). Let $G_\alpha^i = \{x \in \mathbb{D}^\kappa : x(\alpha) = i\}$ for each $\alpha < \kappa$ and $i \in \mathbb{D}$. It is immediate that $\{G_\alpha^0, G_\alpha^1 : \alpha < \kappa\}$ is a κ-dyadic family in $\mathbb{D}^\kappa$, and it is straightforward that if $F_\alpha^i = f^{-1}(G_\alpha^i)$ for all $\alpha < \kappa$ and $i \in \mathbb{D}$, then $\{F_\alpha^0, F_\alpha^1 : \alpha < \kappa\}$ is a κ-dyadic family of closed subsets of X.

(c) $\Longrightarrow$(b). Let $\mathcal{F}$ be a κ-dyadic family of closed subsets of X. After an evident reindexation, we can assume that $\mathcal{F} = \{F_\alpha^0, F_\alpha^1 : \alpha < \kappa\}$. If $s \in \mathbb{D}^\kappa$, then consider the family $\mathcal{F}_s = \{F_\alpha^{s(\alpha)} : \alpha < \kappa\}$. Since $\mathcal{F}$ is dyadic, the family $\mathcal{F}_s$ is centered and hence $F_s = \bigcap \mathcal{F}_s \neq \emptyset$; let $f(x) = s$ for any $x \in F_s$. If $s, t \in \mathbb{D}^\kappa$ and $s \neq t$, then $s(\alpha) \neq t(\alpha)$ for some $\alpha < \kappa$ and hence $F_s \cap F_t \subset F_\alpha^{s(\alpha)} \cap F_\alpha^{t(\alpha)} = \emptyset$. This shows that the map $f : F = \bigcup\{F_s : s \in \mathbb{D}^\kappa\} \to \mathbb{D}^\kappa$ is well-defined. It is straightforward that $F = \bigcap\{F_\alpha^0 \cup F_\alpha^1 : \alpha < \kappa\}$, so the subspace F is compact and hence closed in X.

It follows from the definition of f that f is surjective, so we must only prove continuity of f. Take any $x \in F$ and let $s = f(x)$; given any $O \in \tau(s, \mathbb{D}^\kappa)$, there is $h \in \mathrm{Fn}(\kappa)$ such that $s \in [h] \subset O$. Let $W_\alpha^i = F_\alpha^i \cap F$ for all $\alpha < \kappa$ and $i \in \mathbb{D}$. It is evident that each W_α^i is closed in F; besides, $F = W_\alpha^0 \cup W_\alpha^1$ while $W_\alpha^0 \cap W_\alpha^1 = \emptyset$ which shows that each W_α^i is a complement of a closed subset of F and hence W_α^i

is also open in F. As a consequence, the set $W = \bigcap\{W_\alpha^{h(\alpha)} : \alpha \in \mathrm{dom}(h)\}$ is clopen in F. Observe also that for any $y \in W$, if $t = f(y)$, then $h \subset t$ and hence $f(y) \subset [h] \subset O$ which proves that $f(W) \subset O$, i.e., W witnesses continuity of the mapping f at the point x. Thus f is a continuous onto map and Fact 4 is proved.

□

Fact 5. If X is a compact space such that $\pi\chi(x, X) > \omega$ for each $x \in X$, then there is a closed subset $P \subset X$ for which there is a continuous onto map $f : P \to \mathbb{D}^{\omega_1}$.

Proof. Denote by $\mathcal{C}$ the family of all closed non-empty G_δ-subsets of X. Each $G \in \mathcal{C}$ has a countable outer base $\mathcal{B}_G$ in X by Problem 327 of [TFS]. This shows that $G \subset U \in \tau(X)$ implies that there is $V \in \mathcal{B}_G$ with $V \subset U$.

Suppose that $\mathcal{C}' \subset \mathcal{C}$ and $|\mathcal{C}'| \leq \omega$. Given a point $x \in X$, if every $U \in \tau(x, X)$ contains some $G \in \mathcal{C}'$, then it also contains some $V \in \mathcal{B}_G$ by the previous remark. This shows that $\bigcup\{\mathcal{B}_G : G \in \mathcal{C}'\}$ is a countable local π-base at x which is a contradiction. This proves that

(∗)   for any $x \in X$ and any countable $\mathcal{C}' \subset \mathcal{C}$, there is $W \in \tau(x, X)$ such that $G \backslash W \neq \emptyset$ for all $G \in \mathcal{C}'$.

To construct an ω_1-dyadic family in X, we will also need the following property:

(∗∗)   for any countable family $\mathcal{C}' \subset \mathcal{C}$, there are $F, G \in \mathcal{C}$ such that $F \cap C \neq \emptyset$ and $G \cap C \neq \emptyset$ for any $C \in \mathcal{C}'$ but there exists $B \in \mathcal{C}'$ such that $F \cap G \cap B = \emptyset$.

To prove (∗∗), apply (∗) for every $x \in X$ to obtain a set $W_x \in \tau(x, X)$ such that $C \backslash W_x \neq \emptyset$ for every $C \in \mathcal{C}'$. Taking a smaller W_x if necessary, we can assume that W_x is an F_σ-set and therefore $P_x = X \backslash W_x$ is a G_δ-set for all $x \in X$. Take a finite subcover $\{W_{x_1}, \ldots, W_{x_n}\}$ of the cover $\{W_x : x \in X\}$; then $P_{x_i} \cap C \neq \emptyset$ for any $C \in \mathcal{C}'$ and $i \in M_n$ while $P_{x_1} \cap \cdots \cap P_{x_n} = \emptyset$. Let $k \in M_n$ be the minimal number for which there exist $Q_1, \ldots, Q_k \in \{P_{x_1}, \ldots, P_{x_n}\}$ such that $Q_1 \cap \cdots \cap Q_k \cap B = \emptyset$ for some $B \in \mathcal{C}'$. Then $k \geq 2$ and the sets $F = Q_1$ and $G = Q_2 \cap \cdots \cap Q_k$ are as promised so (∗∗) is proved.

Take any disjoint $K_0^0, K_0^1 \in \mathcal{C}$ and let $C_0^0 = K_0^0$, $C_0^1 = K_0^1$. Assume that for some $\alpha < \omega_1$, we have chosen the sets $K_\beta^0, K_\beta^1, C_\beta^0, C_\beta^1 \in \mathcal{C}$ for all $\beta < \alpha$ and a function $h_\beta \in \mathrm{Fn}(\beta)$ for each $\beta \in [1, \alpha) = \{\gamma \in \omega_1 : 1 \leq \gamma < \alpha\}$ so that

(1) if $\mathcal{K}_\beta = \{C_\gamma^0, C_\gamma^1 : \gamma < \beta\}$, then $I_\beta = I(\mathcal{K}_\beta, h_\beta) \neq \emptyset$ for all $\beta \in [1, \alpha)$;
(2) $C_\beta^0 = I_\beta \cap K_\beta^0$, $C_\beta^1 = I_\beta \cap K_\beta^1$ for all $\beta \in [1, \alpha)$ and $C_\beta^0 \cap C_\beta^1 = \emptyset$ for all $\beta < \alpha$;
(3) $K_\beta^i \cap H \neq \emptyset$ for any $H \in \bigwedge \mathcal{K}_\beta$ whenever $0 < \beta < \alpha$ and $i \in \mathbb{D}$.

The family $\mathcal{K}_\alpha = \{C_\beta^0, C_\beta^1 : \beta < \alpha\}$ is countable and hence so is $\mathcal{C}' = \bigwedge \mathcal{K}_\alpha$. Thus (∗∗) is applicable to find $K_\alpha^0, K_\alpha^1 \in \mathcal{C}$ such that $K_\alpha^i \cap H \neq \emptyset$ for any $H \in \bigwedge \mathcal{C}'$ and $i \in \mathbb{D}$ while there is $F \in \bigwedge \mathcal{C}'$ such that $K_\alpha^0 \cap K_\alpha^1 \cap F = \emptyset$. It follows from (2) that there is $h_\alpha \in \mathrm{Fn}(\alpha)$ such that $F = I_\alpha = I(\mathcal{K}_\alpha, h_\alpha)$. Finally, if we let $C_\alpha^i = K_\alpha^i \cap F$ for every $i \in \mathbb{D}$, then the conditions (1)–(3) are satisfied for all $\beta \leq \alpha$. Thus our inductive construction can be continued to obtain a family $\{K_\beta^0, K_\beta^1 : \beta < \omega_1\}$ and a set $\{h_\beta : 1 \leq \beta < \omega_1\}$ such that (1)–(3) hold for all $\alpha < \omega_1$.

Given any $\alpha \in [1, \omega_1)$, let $\varphi(\alpha) = \max\{\text{dom}(h_\alpha)\}$; it is clear that $\varphi(\alpha) < \alpha$ for all $\alpha \in [1, \omega_1)$ and hence we can apply the pressing-down lemma (see Problem 067) to find $\beta < \omega_1$ such that the set $A' = \{\alpha < \omega_1 : \varphi(\alpha) = \beta\}$ is uncountable. Since $\text{dom}(h_\alpha) \subset \beta + 1$ for every $\alpha \in A'$ and the set $\text{Fn}(\beta + 1)$ is countable, there is an uncountable $A'' \subset A'$ and $h \in \text{Fn}(\beta + 1)$ such that $h_\alpha = h$ for any $\alpha \in A''$; let $A = A'' \backslash (\beta + 1)$ and observe that

$(***)$ $I_\alpha = I_\gamma = I = I(\mathcal{K}_{\beta+1}, h)$ for any $\alpha, \gamma \in A$.

We claim that the family $\mathcal{D} = \{C_\alpha^0, C_\alpha^1 : \alpha \in A\}$ is ω_1-dyadic. The property (2) shows that, to prove it, we only have to establish that $I(\mathcal{D}, g) \neq \emptyset$ for any $g \in \text{Fn}(A)$. Let $\text{dom}(g) = \{\alpha_1, \ldots, \alpha_n\}$ where $\alpha_1 < \cdots < \alpha_n$ and $k_i = g(\alpha_i)$ for all $i \in M_n$.

Observe (1)–(3) imply that $C_{\alpha_1}^{k_1} \neq \emptyset$; assume that $1 \leq m \leq n$ and we proved that $H = C_{\alpha_1}^{k_1} \cap \cdots \cap C_{\alpha_{m-1}}^{k_{m-1}} \neq \emptyset$. Then $H \in \bigwedge\{C_\beta^0, C_\beta^1 : \beta < \alpha_m\}$ and hence $K_{\alpha_m}^{k_m} \cap H \neq \emptyset$ which implies that

$$C_{\alpha_1}^{k_1} \cap \cdots \cap C_{\alpha_{m-1}}^{k_{m-1}} \cap C_{\alpha_m}^{k_m} = C_{\alpha_1}^{k_1} \cap \cdots \cap C_{\alpha_{m-1}}^{k_{m-1}} \cap K_{\alpha_m}^{k_m} \cap I_{\alpha_m} = C_{\alpha_1}^{k_1} \cap \cdots \cap C_{\alpha_{m-1}}^{k_{m-1}} \cap K_{\alpha_m}^{k_m}$$

because $C_{\alpha_1}^{k_1} \cap I_{\alpha_m} = K_{\alpha_1}^{k_1} \cap I \cap I = C_{\alpha_1}^{k_1}$ (we used $(***)$). Now it follows from (3) that $C_{\alpha_1}^{k_1} \cap \cdots \cap C_{\alpha_{m-1}}^{k_{m-1}} \cap C_{\alpha_m}^{k_m} = K_{\alpha_m}^{k_m} \cap H \neq \emptyset$ and hence we can go on inductively to finally establish that $C_{\alpha_1}^{k_1} \cap \cdots \cap C_{\alpha_n}^{k_n} \neq \emptyset$ and hence the family $\mathcal{D}$ is, indeed, ω_1-dyadic. Now apply Fact 4 to conclude that there is a closed $P \subset X$ which maps continuously onto $\mathbb{D}^{\omega_1}$. Fact 5 is proved. $\square$

Fact 6. Let M_t be a second countable space for every $t \in T$. If $M = \prod_{t \in T} M_t$, then let $p_S : M \to M_S = \prod_{t \in S} M_t$ be the natural projection of M onto its face M_S (recall that p_S is defined by $p_S(x) = x|S$ for any $x \in M$). Then, for any open $U \subset M$, the set $\overline{U}$ depends on countably many coordinates, i.e., there exists a countable $S \subset T$ such that $p_S^{-1}(p_S(\overline{U})) = \overline{U}$.

Proof. The sets U and $V = M \backslash \overline{U}$ are open and disjoint and hence separated in M, i.e., $\overline{U} \cap V = \overline{V} \cap U = \emptyset$. Thus we can apply Fact 3 of S.291 to find a countable $S \subset T$ such that $\pi_S(U)$ and $\pi_S(V)$ are separated in M_S. Therefore $\pi_S(V) \cap \text{cl}_{M_S}(\pi_S(U)) = \emptyset$; by continuity of p_S, we have $\pi_S(\overline{U}) \subset \text{cl}_{M_S}(p_S(U))$ and hence $\pi_S(\overline{U}) \cap \pi_S(M \backslash \overline{U}) = \emptyset$ which implies that $p_S^{-1}(p_S(\overline{U})) = \overline{U}$ so Fact 6 is proved. $\square$

Fact 7. Suppose that X is a compact space such that there is a continuous onto map $f : X \to \mathbb{D}^{\omega_1}$. Then X contains a convergent ω_1-sequence.

Proof. The map f is perfect and hence there is a closed $Y \subset X$ such that $f(Y) = \mathbb{D}^{\omega_1}$ and $f|Y$ is irreducible (see Problem 366 of [TFS]). Since any convergent ω_1-sequence in Y is also a convergent ω_1-sequence in X, we can assume, without loss of generality, that $Y = X$, i.e., the map f is irreducible. Let us assume first that X is zero-dimensional and denote by $\mathcal{C}$ the family of all non-empty clopen subsets of X.

For any open non-empty $U \subset X$ we will need the sets $f^{\#}(U) = \mathbb{D}^{\omega_1} \backslash f(X \backslash U)$ and $U^{*} = f^{-1}(f^{\#}(U))$. It is straightforward that $f^{\#}(U) \subset f(U)$, $U^{*} \subset U$ and $f^{\#}(U) \in \tau(\mathbb{D}^{\omega_1})$, $U^{*} \in \tau(X)$. Besides, the set U^{*} is dense in U and hence $f^{\#}(U)$ is dense in $f(U)$ (see Fact 1 of S.383). An immediate consequence is

(4) $\mathrm{Int}(f(U))$ is dense in $f(U)$ for any $U \in \mathcal{C}$

because $f^{\#}(U) \subset \mathrm{Int}(f(U))$. Another important property is

(5) $\mathrm{Int}(f(U \cap V)) = \mathrm{Int}(f(U)) \cap \mathrm{Int}(f(V))$ for any $U, V \in \mathcal{C}$.

To prove (5) observe that the inclusion $\mathrm{Int}(f(U \cap V)) \subset \mathrm{Int}(f(U)) \cap \mathrm{Int}(f(V))$ is straightforward, so assume that $(\mathrm{Int}(f(U)) \cap \mathrm{Int}(f(V))) \backslash \mathrm{Int}(f(U \cap V)) \neq \emptyset$ and therefore $W = (\mathrm{Int}(f(U)) \cap \mathrm{Int}(f(V))) \backslash f(U \cap V) \neq \emptyset$. Since W is a non-empty open subset of $f(U)$ and $f^{\#}(U)$ is dense in $f(U)$, the set $W_1 = f^{\#}(U) \cap W$ is non-empty, open and contained in $f(V)$. The set $f^{\#}(V)$ being dense in $f(V)$, we have $W_2 = f^{\#}(V) \cap W_1 \in \tau^{*}(\mathbb{D}^{\omega_1})$ and hence $f^{-1}(W_2) \subset f^{-1}(f^{\#}(U) \cap f^{\#}(V)) = U^{*} \cap V^{*} \subset U \cap V$. Consequently, $W_2 \subset f(U \cap V)$ which contradicts $W_2 \subset W$ and $W \cap f(U \cap V) = \emptyset$. This contradiction shows that (5) is true.

Let $u \in \mathbb{D}^{\omega_1}$ be defined by $u(\alpha) = 0$ for all $\alpha < \omega_1$; furthermore, for all $\alpha, \beta < \omega_1$, let $u_{\alpha}(\beta) = 0$ if $\alpha \neq \beta$ and $u_{\alpha}(\alpha) = 1$. The space $K = \{u\} \cup \{u_{\alpha} : \alpha < \omega_1\} \subset \mathbb{D}^{\omega_1}$ has the unique non-isolated point u and $K \backslash O$ is finite for any $O \in \tau(u, \mathbb{D}^{\omega_1})$. Pick any $x \in f^{-1}(u)$ and let $F_{\alpha} = f^{-1}(u_{\alpha})$ for all $\alpha < \omega_1$.

For every $\alpha < \omega_1$ denote by $p_{\alpha} : \mathbb{D}^{\omega_1} \to \mathbb{D}^{\alpha}$, the natural projection of $\mathbb{D}^{\omega_1}$ onto its face $\mathbb{D}^{\alpha}$, and in general, if $S \subset \omega_1$, then $p_S : \mathbb{D}^{\omega_1} \to \mathbb{D}^{S}$ is the natural projection of $\mathbb{D}^{\omega_1}$ onto its face $\mathbb{D}^{S}$. Observe that if a set $E \subset \mathbb{D}^{\omega_1}$ depends on a set $S \subset \omega_1$ (recall that this means that $E = p_S^{-1}(p_S(E))$), then E depends on S' for any $S' \supset S$. Now it follows from (4) and Fact 6 that, for any $U \in \mathcal{C}$, the set $f(U)$ depends on some countable $S \subset \omega_1$ and therefore

(6) for any $U \in \mathcal{C}$ there is $\alpha < \omega_1$ such that $f(U)$ depends on first α coordinates.

Denote by $\mathcal{C}(x)$ the family $\{U \in \mathcal{C} : x \in U\}$ and let $\mathcal{C}_{\alpha} = \{U \in \mathcal{C} : x \in U$ and $f(U)$ depends on first α coordinates$\}$. We already saw that $\mathcal{C}(x) = \bigcup\{\mathcal{C}_{\alpha} : \alpha < \omega_1\}$. An easy consequence of (6) is that if $t \in \mathbb{D}^{\omega_1}$, $U \in \mathcal{C}_{\alpha}$ and $p_{\alpha}(t) \in p_{\alpha}(f(U))$, then $t \in f(U)$. In particular,

(7) if $U \in \mathcal{C}_{\alpha}$, then $u_{\beta} \in f(U)$ for any $\beta \geq \alpha$

because $u = f(x) \in f(U)$ and $p_{\alpha}(u) = p_{\alpha}(u_{\beta})$. Next we claim that

(8) if $U, V \in \mathcal{C}_{\alpha}$, then $W = U \cap V \in \mathcal{C}_{\alpha}$.

To prove (8), let $U' = \text{Int}(f(U))$ and observe that $f(U) = p_\alpha^{-1}(p_\alpha(f(U)))$ implies that $p_\alpha^{-1}(p_\alpha(U')) \subset f(U)$ and hence $p_\alpha^{-1}(p_\alpha(U')) \subset \text{Int}(f(U)) = U'$. Thus $p_\alpha^{-1}(p_\alpha(U')) = U'$, and analogously, if $V' = \text{Int}(f(V))$, then $p_\alpha^{-1}(p_\alpha(V')) = V'$. Now let $W' = \text{Int}(f(W))$ and apply (5) to conclude that $W' = U' \cap V'$ and therefore $p_\alpha^{-1}(p_\alpha(W')) = W'$. Finally, W' is dense in $f(W)$ which, together with the openness of the map p_α implies that

$$p_\alpha^{-1}(p_\alpha(f(W))) \subset p_\alpha^{-1}(\overline{p_\alpha(W')}) = \overline{p_\alpha^{-1}(p_\alpha(W'))} = \overline{W'} = f(W)$$

which proves that $p_\alpha^{-1}(p_\alpha(f(W))) = f(W)$ and settles (8). Observe also that we have $\mathcal{C}_\alpha \subset \mathcal{C}_\beta$ if $\alpha \le \beta$ and therefore

(9) $\bigcap \mathcal{C}_\beta \subset \bigcap \mathcal{C}_\alpha$ whenever $\alpha \le \beta < \omega_1$.

Given any $\alpha < \omega_1$ it follows from (7) that $U \cap F_\alpha \neq \emptyset$ for any $U \in \mathcal{C}_\alpha$; besides, the family $\{U \cap F_\alpha : U \in \mathcal{C}_\alpha\}$ is centered by (8) which shows that there exists $x_\alpha \in F_\alpha \cap (\bigcap \mathcal{C}_\alpha)$ because F_α is compact.

We claim that $Z = \{x_\alpha : \alpha < \omega_1\} \cup \{x\}$ is a convergent ω_1-sequence. In first place, $|Z| = \omega_1$ because the map $f|Z$ condenses Z onto K. Furthermore, for any $U \in \tau(x, X)$, there is $\alpha < \omega_1$ and $V \in \mathcal{C}_\alpha$ such that $x \in V \subset U$. Given any $\beta \ge \alpha$, we have $x_\beta \in \bigcap \mathcal{C}_\beta \subset \bigcap \mathcal{C}_\alpha \subset V$ and hence $|Z \backslash U| \le |Z \backslash V| \le |\{x_\beta : \beta < \alpha\}| \le \omega$. Moreover, we proved that there is a convergent ω_1-sequence Z in X such that $f|Z$ condenses Z onto the compact space $K \subset \mathbb{D}^{\omega_1}$.

This finishes the proof of our fact for compact zero-dimensional spaces. Now assume that X is an arbitrary compact space; if $w(X) = \kappa$, then we can assume that $X \subset \mathbb{I}^\kappa$. Apply Fact 2 to take a continuous onto map $\varphi : \mathbb{D}^\kappa \to \mathbb{I}^\kappa$. Then $Y = \varphi^{-1}(X)$ is a zero-dimensional compact space by Fact 3 and $\varphi_1 = \varphi|Y$ maps Y continuously onto X. Now, $f \circ \varphi_1$ maps Y continuously onto $\mathbb{D}^{\omega_1}$, so we can apply what we have proved for zero-dimensional compact spaces to conclude that there is a convergent ω_1-sequence $Z_1 \subset Y$ such that $f \circ \varphi$ condenses Z_1 onto K. An immediate consequence is that φ condenses Z_1 onto $Z = \varphi(Z_1) \subset X$. It is immediate that a continuous image of a convergent ω_1-sequence is also a convergent ω_1-sequence provided that it is uncountable. Since $\varphi|Z_1 : Z_1 \to Z$ is a bijection, we have $|Z| = |Z_1| = \omega_1$ and hence the space Z is also a convergent ω_1-sequence in X, i.e., Fact 7 is proved. $\qquad\square$

Fact 8. Given an infinite cardinal κ suppose that X is a compact space and $S \subset X$ is a free κ^+-sequence. Then the set $\overline{S}$ can be continuously mapped onto $\kappa^+ + 1$. In particular, if X is a compact space with $t(X) > \kappa$, then there is a closed $Y \subset X$ which maps continuously onto $\kappa^+ + 1$.

Proof. If $t(X) > \kappa$, then there is a free sequence $S = \{x_\alpha : \alpha < \kappa^+\} \subset X$ (see Problem 328 of [TFS]); let $S_\alpha = \{x_\beta : \beta < \alpha\}$ for every $\alpha \le \kappa^+$. To prove the second part of our fact, it suffices to take $Y = \overline{S}$, i.e., it is sufficient to show that $\overline{S}$ maps continuously onto $\kappa^+ + 1$.

To do that, let $f(x_n) = n$ for any $n \in \omega$; if $\alpha \in \kappa^+ \backslash \omega$, then let $f(x_\alpha) = \alpha + 1$. If $x \in \overline{S} \backslash S$, then the ordinal $f(x) = \min\{\alpha \leq \kappa^+ : x \in \overline{S_\alpha}\}$ is well-defined; besides, $f(x)$ is a limit ordinal for any $x \in \overline{S} \backslash S$ because if $\alpha = \gamma + 1$, then $x \in \overline{S_\alpha}$ implies $x \in \overline{S_\alpha} \backslash S_\alpha$ and therefore $x \in \overline{S_\alpha \backslash \{x_\gamma\}} = \overline{S_\gamma}$. This defines a map $f : \overline{S} \to \kappa^+ + 1$.

Observe that x_α is an isolated point of Y for any $\alpha < \kappa^+$, so it suffices to prove continuity of f at any $x \in Y \backslash S$. If $f(x) = \alpha$ and $U \in \tau(\alpha, \kappa^+ + 1)$, then there is $\beta < \alpha$ such that $(\beta, \alpha] = \{\gamma \in \kappa^+ : \beta < \gamma \leq \alpha\} \subset U$. The set S being a free sequence, it follows from the definition of $f(x)$ that the set $W = X \backslash \overline{S_{\beta+1}} \cup (S \backslash S_\alpha)$ is an open neighborhood of x. Observe that $x_\gamma \in W$ implies $\gamma \in (\beta, \alpha)$ and therefore $f(x_\gamma) \in (\beta, \alpha) \subset U$. Furthermore, if $y \in (Y \backslash S) \cap W$, then $y \notin \overline{S \backslash S_\alpha}$ and therefore $y \in \overline{S_\alpha}$ whence $f(y) \leq \alpha$. Now, it follows from $y \notin \overline{S_{\beta+1}}$ that $f(y) \geq \beta + 1 > \beta$, i.e., $f(y) \in (\beta, \alpha]$. This proves that $f(W \cap Y) \subset (\beta, \alpha] \subset U$ and hence f is continuous at the point x. The point $x \in Y \backslash S$ was chosen arbitrarily, so f is a continuous map. The set $f(S)$ is dense in $\kappa^+ + 1$, so $f(\overline{S}) = \kappa^+ + 1$ and hence Fact 8 is proved. □

Fact 9. If X is a compact space with $t(X) > \omega$, then X contains a convergent ω_1-sequence.

Proof. It follows from Fact 8 that there is a closed $H \subset X$ which maps continuously onto $\omega_1 + 1$. By Problem 366 of [TFS], there is closed $H' \subset H$ and a continuous irreducible onto map $f : H' \to (\omega_1 + 1)$. Since it suffices to find a convergent ω_1-sequence in H', we can assume, without loss of generality, that $X = H'$, i.e., that there is a continuous irreducible onto map $f : X \to (\omega_1 + 1)$.

Let $X_\alpha = \{x \in X : f(x) = \alpha\}$ for each $\alpha \leq \omega_1$. Observe first that X_{ω_1} is nowhere dense in X because otherwise there is an open non-empty $U \subset X_{\omega_1}$ and hence $f^\#(U)$ is a non-empty open subset of $\{\omega_1\}$ which is a contradiction. Let $Y_\alpha = \bigcup \{X_\beta : \alpha \leq \beta \leq \omega_1\}$ for every $\alpha < \omega_1$. If $\pi\chi(x, X_{\omega_1}) > \omega$ for any $x \in X_{\omega_1}$, then there is a closed $P \subset X_{\omega_1}$ that maps continuously onto $\mathbb{D}^{\omega_1}$ by Fact 5 and hence P contains a convergent ω_1-sequence by Fact 7. Therefore we can assume, without loss of generality, that there is $x \in X_{\omega_1}$ such that $\pi\chi(x, X_{\omega_1}) = \omega$.

Fix a family $\{O_n : n \in \omega\} \subset \tau(X)$ such that $\{O_n \cap X_{\omega_1} : n \in \omega\}$ is a π-base of X_{ω_1} at the point x. Choose a set $U_n \in \tau(X)$ such that $U_n \cap X_{\omega_1} \neq \emptyset$ and $\overline{U}_n \subset O_n$ for all $n \in \omega$.

We saw already that $Y = X \backslash X_{\omega_1}$ is dense in X and hence $Y \cap U_n$ is dense in U_n for all $n \in \omega$. As a consequence, the set $A_n = \{\alpha < \omega_1 : U_n \cap X_\alpha \neq \emptyset\}$ is uncountable for each $n \in \omega$. Furthermore, $A_n = f(U_n \cap Y)$; since the map $f|Y$ is closed (see Fact 1 of S.261), the set $B_n = f(\overline{U}_n \cap Y)$ is closed and unbounded in ω_1 for all $n \in \omega$. Therefore the set $B = \bigcap \{B_n : n \in \omega\}$ is also closed and unbounded by Problem 064.

Now let $V_\alpha = \{V \in \tau(x, X) : $ for any $n \in \omega$; if $\overline{U}_n \cap X_{\omega_1} \subset V$, then $\overline{U}_n \cap Y_\alpha \subset V\}$ for every $\alpha < \omega_1$; observe that $V_1, V_2 \in V_\alpha$ implies $V_1 \cap V_2 \in V_\alpha$. Besides $V_\alpha \subset V_\beta$ whenever $\alpha < \beta < \omega_1$.

Now let $F_\alpha = \bigcap \{\overline{V} : V \in V_\alpha\}$ for all $\alpha < \omega_1$. Observe that for any set $V \in \tau(x, X)$, there is $\alpha < \omega_1$ such that $V \in V_\alpha$. Indeed, $X_{\omega_1} = \bigcap \{Y_\alpha : \alpha < \omega_1\}$, so

$\overline{U}_n \cap X_{\omega_1} \subset V$ implies that $\bigcap \{\overline{U}_n \cap Y_\alpha : \alpha < \omega_1\} \subset V$ and hence we can apply Fact 1 of S.326 to conclude that there is $\alpha_n < \omega_1$ for which $\overline{U}_n \cap Y_{\alpha_n} \subset V$. Now, if $\alpha > \sup\{\alpha_n : \overline{U}_n \cap X_{\omega_1} \subset V\}$, then $V \in \mathcal{V}_\alpha$.

Consequently, we have a family $\{F_\alpha : \alpha < \omega_1\}$ of closed subsets of X such that $F_\beta \subset F_\alpha$ whenever $\alpha < \beta < \omega_1$ and $\bigcap \{F_\alpha : \alpha < \omega_1\} = \bigcap \{\overline{V} : V \in \tau(x, X)\} = \{x\}$.

Besides, for every ordinal $\alpha < \omega_1$ and any $V \in \mathcal{V}_\alpha$, there is $n \in \omega$ such that $\overline{U}_n \cap X_{\omega_1} \subset O_n \cap X_{\omega_1} \subset V$ and hence $\overline{U}_n \cap Y_\alpha \subset V$. This implies, however, that $B \backslash \alpha \subset B_n \backslash \alpha \subset f(\overline{U}_n \cap Y_\alpha) \subset f(\overline{V})$ which shows that $f(\overline{V}) \supset B \backslash \alpha$ for each $V \in \mathcal{V}_\alpha$ and hence $B \backslash \alpha \subset f(F_\alpha)$. As a consequence, $F_\alpha \neq \{x\}$ for all $\alpha < \omega_1$.

Observe also that if $Q \subset X \backslash \{x\}$ is a countable set, then for any $y \in Q$ there is $\alpha_y < \omega_1$ such that $y \notin F_{\alpha_y}$; if $\alpha > \sup\{\alpha_y : y \in Q\}$, then $F_\alpha \cap Q = \emptyset$. This makes it possible to construct, by an evident induction, a set $Z' = \{x_\alpha : \alpha < \omega_1\}$ such that $x_\alpha \in F_\alpha \backslash \{x\}$ and $x_\alpha \neq x_\beta$ whenever $\alpha, \beta < \omega_1$ and $\alpha \neq \beta$.

To see that $Z = \{x_\alpha : \alpha < \omega_1\} \cup \{x\}$ is a convergent ω_1-sequence observe first that $|Z| = \omega_1$ because the enumeration of the set Z' is faithful. Furthermore, if $W \in \tau(x, X)$, then we can apply Fact 1 of S.326 to conclude that there is $\alpha < \omega_1$ for which $F_\alpha \subset W$. Therefore $Z \backslash W \subset \{x_\beta : \beta < \alpha\}$ and hence Z is a convergent ω_1-sequence. Fact 9 is proved.   □

Now it is easy to finish our solution. Assume that CH holds and X is a compact space with a small diagonal. It follows from Fact 1 and Fact 9 that $t(X) \leq \omega$. Given a countable $A \subset X$, we have $w(\overline{A}) \leq 2^{|A|} \leq \mathfrak{c}$ by Fact 2 of S.368 and hence $w(\overline{A}) \leq \omega_1$ by CH. Recall that having a small diagonal is hereditary, so $\overline{A}$ has a small diagonal; thus we can apply Problem 295 to conclude that $\overline{A}$ is metrizable and therefore X is ω-monolithic. Finally, X is metrizable by Problem 296 so our solution is complete.

T.299. *Assume that* $2^{\omega_1} = \omega_2$. *Prove that any compact* X, *with* ω_1 *and* ω_2 *calibers of* $C_p(X)$, *is metrizable.*

Solution. If the space X is not metrizable, then there is $Y \subset X$ such that $|Y| \leq \omega_1$ and Y is not metrizable (see Problem 092). Thus $Z = \overline{Y}$ is not metrizable either and $w(Z) \leq 2^{|Y|} \leq 2^{\omega_1} = \omega_2$ (see Fact 2 of S.368). Furthermore, both cardinals ω_1 and ω_2 are calibers of $C_p(Z)$ by Problem 292. If $C_p(Z)$ is separable, then Z is metrizable (see Problems 174 and 140 of [TFS]) which is a contradiction with the fact that $Y \subset Z$ is not metrizable.

As a consequence, $\omega < d(C_p(Z)) \leq nw(C_p(Z)) = nw(Z) \leq w(Z) \leq \omega_2$ and hence $d(C_p(Z)) = \kappa \in \{\omega_1, \omega_2\}$. Let $D = \{f_\alpha : \alpha < \kappa\}$ be a dense subset of $C_p(Z)$ and denote by D_α the set $\{f_\beta : \beta < \alpha\}$ for every $\alpha < \kappa$. Since $|D_\alpha| < \kappa$ for each $\alpha < \kappa$, the set $U_\alpha = C_p(Z) \backslash \overline{D_\alpha}$ is open and non-empty for all $\alpha < \kappa$.

Given any $f \in C_p(Z)$, there is a countable $A \subset D$ with $f \in \overline{A}$ (recall that $t(C_p(Z)) = \omega$ because Z is compact). Since $\kappa = \text{cf}(\kappa) > \omega$, there is $\alpha < \kappa$ such that $A \subset D_\alpha$ and hence $f \in \overline{D_\alpha}$. This proves that $\bigcup \{\overline{D_\alpha} : \alpha < \kappa\} = Z$ and hence $\bigcap \{U_\alpha : \alpha < \kappa\} = \emptyset$. This, together with the fact that $U_\beta \subset U_\alpha$ whenever $\alpha < \beta$, implies that every $f \in C_p(Z)$ belongs to strictly less than κ-many elements of the family $\mathcal{U} = \{U_\alpha : \alpha < \kappa\}$. Therefore $\mathcal{U}$ witnesses that κ is not a caliber of $C_p(Z)$. This contradiction shows that a non-metrizable $Y \subset X$ with $|Y| \leq \omega_1$ cannot exist and hence X is metrizable.

T.300. *Observe that any Lindelöf Σ-space with a diagonal G_δ has a countable network. Prove that, under CH, any Lindelöf Σ-space with a small diagonal has a countable network.*

Solution. For every $n \in \mathbb{N}$ let $M_n = \{1, \ldots, n\}$; if $\mathcal{A}$ is a family of sets, then $\bigwedge \mathcal{A}$ is the family of all non-empty finite intersections of the elements of $\mathcal{A}$. If Z is a space, then a family $\mathcal{A} \subset \exp(Z)$ is called *closed (compact)* if all elements of $\mathcal{A}$ are closed in Z (or compact, respectively). Given a space Z, a family $\mathcal{F} \subset \exp(Z)$ is a network with respect to a family $\mathcal{C} \subset \exp(Z)$ if for any $C \in \mathcal{C}$ and $U \in \tau(C, Z)$, there is $F \in \mathcal{F}$ such that $C \subset F \subset U$. The family $\mathcal{F}$ is a called *a network at any point of a set* $A \subset Z$ if it is a network with respect to $\{\{a\} : a \in A\}$. Say that distinct points $x, y \in Z$ are T_1-*separated* by a family $\mathcal{A} \subset \exp(Z)$ if there are $A, B \in \mathcal{A}$ such that $x \in A \not\ni y$ and $y \in B \not\ni x$. Say that $\mathcal{A}$ is a T_1-separating family for the points of $Y \subset Z$ if any distinct $x, y \in Y$ are T_1-separated by the family $\mathcal{A}$.

A space Z is Lindelöf Σ if and only if there is a countable family $\mathcal{F}$ of closed subsets of Z which is a network with respect to a compact cover $\mathcal{C}$ of the space Z (see Problem 225 and Fact 1 of T.229).

Fact 1. Suppose that Z is a space in which there exists a countable closed network $\mathcal{F}$ with respect to a compact cover $\mathcal{C}$ of the space Z such that every $C \in \mathcal{C}$ is metrizable. If, additionally, the space Z has a small diagonal, then $nw(Z) = \omega$.

Proof. There is no loss of generality to consider that $\bigwedge \mathcal{F} = \mathcal{F}$, i.e., $\mathcal{F}$ is closed under finite intersections. For every $C \in \mathcal{C}$, let $\mathcal{F}_C = \{F \in \mathcal{F} : C \subset F\}$. It turns out that

(1) given any $C \in \mathcal{C}$, there exists a countable family $\mathcal{U}_C$ of cozero-sets of Z such that $\bigwedge \mathcal{U}_C = \mathcal{U}_C$ and $\bigwedge(\mathcal{F}_C \cup \mathcal{U}_C)$ is a network at all points of C.

To prove (1) take a countable base $\mathcal{B}$ of the space C and call a pair $(U, V) \in \mathcal{B}$ *strongly disjoint* if $\overline{U} \cap \overline{V} = \emptyset$ (the bar denotes the closure in Z). The space Z is Lindelöf Σ and hence normal, so if $U, V \in \mathcal{B}$ are strongly disjoint, then we can fix disjoint cozero-sets G_U, G_V of the space Z such that $\overline{U} \subset G_U$ and $\overline{V} \subset G_V$. The family $\mathcal{U}_C = \bigwedge\{G_U, G_V :$ the pair $(U, V) \in \mathcal{B} \times \mathcal{B}$ is strongly disjoint$\}$ is, evidently, countable and $\bigwedge \mathcal{U}_C = \mathcal{U}_C$. To see that $\mathcal{U}_C$ is as promised take any $x \in C$ and $W \in \tau(x, Z)$. Let $\{P_n : n \in \omega\}$ be an enumeration of all elements of the family $\bigwedge(\mathcal{F}_C \cup \mathcal{U}_C)$ which contain x. Since the family $\bigwedge(\mathcal{F}_C \cup \mathcal{U}_C)$ is closed under finite intersections, if $P_n \backslash W \neq \emptyset$ for every $n \in \omega$, then we can choose a point $x_n \in (\bigcap\{P_i : i \leq n\}) \backslash W$ for every $n \in \omega$; let $S = \{x_n : n \in \omega\}$. For any $A \subset Z$ let $S(A) = \{n \in \omega : x_n \in A\}$.

Assume first that, for any point $y \in C$, there exists $U_y \in \tau(y, Z)$ such that the set $S(U_y)$ is finite. The subspace C being compact we can choose a finite subcover $\{U_{y_1}, \ldots, U_{y_k}\}$ of the open cover $\{U_y : y \in C\}$ of the space C. Since $\mathcal{F}_C$ is a network with respect to C, there is $n \in \omega$ such that $P_n \subset U_{y_1} \cup \cdots \cup U_{y_k}$ and hence $S(P_n) \subset S(U_{y_1}) \cup \cdots \cup S(U_{y_k})$ which shows that $S(P_n)$ is a finite set, a contradiction with $S(P_n) \supset \{m \in \omega : m \geq n\}$.

As a consequence, there is a point $y \in C$ such that $S(G)$ is infinite for any $G \in \tau(y, Z)$. It is evident that $y \notin W$ and hence $y \neq x$. The family $\mathcal{B}$ being a base in C, there are $U, V \in \mathcal{B}$ such that $x \in U$, $y \in V$ and $\overline{U} \cap \overline{V} = \emptyset$. This implies $x \in G_U$, $y \in G_V$ and $G_U \cap G_V = \emptyset$. Observe that $G_U = P_n$ for some $n \in \omega$ and therefore $S(P_n) \supset \{m \in \omega : m \geq n\}$ whence $|\omega \backslash S(P_n)| \leq n$. However, the set $S(G_V)$ is infinite by the choice of y and $S(G_V) \cap S(G_U) = \emptyset$, i.e., the infinite set $S(G_V)$ is contained in the finite set $\omega \backslash S(G_U)$. This contradiction shows that $x \in P_n \subset W$ for some $n \in \omega$ and hence (1) is proved.

Denote by $\mathcal{U}$ the family of all cozero-sets of Z. Observe that each $U \in \mathcal{U}$ is an F_σ-set of Z and the family $\mathcal{U}$ is a base in Z (see Fact 1 of T.252). We can assume without loss of generality that

(2) if $\mathcal{U}' \subset \mathcal{U}$ is countable, then $\mathcal{F} \cup \mathcal{U}'$ does not T_1-separate the points of Z.

Indeed, if it does, then let $\mathcal{A}_U$ be a countable family of closed subsets of Z such that $\bigcup \mathcal{A}_U = U$ for every $U \in \mathcal{U}'$. The family $\mathcal{F} \cup (\bigcup \{\mathcal{A}_U : U \in \mathcal{U}'\})$ is also countable, consists of closed subsets of Z and T_1-separates the points of Z, so Fact 1 of T.270 is applicable to conclude that Z has a countable network.

Given a family $\mathcal{A} \subset \exp(Z)$, let $\mathcal{A}[z] = \bigcap \{A \in \mathcal{A} : z \in A\}$ for all $z \in Z$. Observe also that

(3) if $x, y \in Z$ and $y \in \mathcal{F}[x]$, then $y \in \mathcal{C}[x]$

because if $x \in C \not\ni y$ for some $C \in \mathcal{C}$, then the family $\mathcal{F}$ being a network with respect to $\mathcal{C}$, there is $F \in \mathcal{F}$ for which $C \subset F \subset Z \backslash \{y\}$ and hence $x \in F \not\ni y$ which is a contradiction.

Since the family $\mathcal{F}$ does not T_1-separates the points of Z, there exist distinct $a_0, b_0 \in Z$ such that $b_0 \in \mathcal{F}[a_0]$. There is $C_0 \in \mathcal{C}$ with $a_0 \in C_0$; it follows from (3) that $b_0 \in C_0$.

Assume that $\alpha < \omega_1$ and we have chosen a family $\{C_\beta : \beta < \alpha\} \subset \mathcal{C}$ and points $\{a_\beta, b_\beta : \beta < \alpha\} \subset Z$ such that

(4) $a_\beta \neq b_\beta$ and $\{a_\beta, b_\beta\} \subset C_\beta$ for all $\beta < \alpha$;
(5) For any $\beta < \alpha$ if $\mathcal{V}_\beta = \mathcal{F} \cup (\bigcup \{\mathcal{U}_{C_\gamma} : \gamma < \beta\})$, then $b_\beta \in \mathcal{V}_\beta[a_\beta]$.

Applying (2) to the family $\mathcal{V}_\alpha = \mathcal{F} \cup (\bigcup \{\mathcal{U}_{C_\beta} : \beta < \alpha\})$, we can find distinct points $a_\alpha, b_\alpha \in Z$ such that $b_\alpha \in \mathcal{V}_\alpha[a_\alpha]$. There is $C_\alpha \in \mathcal{C}$ such that $a_\alpha \in C_\alpha$; it follows from (3) that $b_\alpha \in C_\alpha$. It is immediate that the conditions (4) and (5) are satisfied for the family $\{C_\beta : \beta \leq \alpha\}$ and the set $\{a_\beta, b_\beta : \beta \leq \alpha\}$ which shows that our inductive construction can be continued to give us a family $\{C_\beta : \beta < \omega_1\} \subset \mathcal{C}$ and a set $\{a_\beta, b_\beta : \beta < \omega_1\} \subset Z$ such that (4) and (5) are fulfilled for any $\alpha < \omega_1$.

Let $\Delta = \{(z, z) : z \in Z\}$ be the diagonal of the space Z; since Δ is small and $D = \{(a_\alpha, b_\alpha) : \alpha < \omega_1\} \subset (Z \times Z) \backslash \Delta$, there is an uncountable $T \subset \omega_1$ such that the closure R of the set $\{(a_\alpha, b_\alpha) : \alpha \in T\}$ does not meet Δ. For each $z = (x, y) \in R$, take $U_z \in \tau(x, Z)$ and $V_z \in \tau(y, Z)$ such that $\overline{U}_z \cap \overline{V}_z = \emptyset$.

The space Z is Lindelöf Σ so $Z \times Z$ is Lindelöf (see Problem 256) and hence there is a countable $Q \subset R$ such that $R \subset \bigcup\{U_z \times V_z : z \in Q\}$. Consequently, there is an uncountable $E \subset \omega_1$ and $z \in R$ for which $\{(a_\alpha, b_\alpha) : \alpha \in E\} \subset U_z \times V_z$, i.e., $\{a_\alpha : \alpha \in E\} \subset U_z$ and $\{b_\alpha : \alpha \in E\} \subset V_z$.

Given an $\alpha \in E$, the family $\bigwedge(\mathcal{F} \cup \mathcal{U}_{C_\alpha})$ is a network at all points of C_α by (1); it follows from $\bigwedge \mathcal{F} = \mathcal{F}$ and $\bigwedge \mathcal{U}_{C_\alpha} = \mathcal{U}_{C_\alpha}$ that for every point $x \in C_\alpha$, there exists $O_x \in \mathcal{U}_{C_\alpha}$ and $F_x \in \mathcal{F}_{C_\alpha}$ such that $O_x \cap F_x$ misses either $\overline{U}_z$ or $\overline{V}_z$. The set C_α is compact, so there are finite subcollections $\{U_1^\alpha, \ldots, U_{n_\alpha}^\alpha\} \subset \mathcal{U}_{C_\alpha}$ and $\{F_1^\alpha, \ldots, F_{n_\alpha}^\alpha\} \subset \mathcal{F}_{C_\alpha}$ such that $C_\alpha \subset U^\alpha = U_1^\alpha \cup \cdots \cup U_{n_\alpha}^\alpha$ and $U_i^\alpha \cap F_i^\alpha$ misses either $\overline{U}_z$ or $\overline{V}_z$ for every $i \in M_{n_\alpha}$. Since $\mathcal{F}$ is a network with respect to $\mathcal{C}$, we can choose $F^\alpha \in \mathcal{F}$ for which $C_\alpha \subset F^\alpha \subset F_1^\alpha \cap \cdots \cap F_{n_\alpha}^\alpha \cap U^\alpha$ for all $\alpha \in E$.

The family $\mathcal{F}$ being countable, there exist $\alpha, \beta \in \omega_1$ such that $\alpha < \beta$ and $F^\alpha = F^\beta = F$. Therefore $C_\alpha \cup C_\beta \subset F \subset U^\alpha$ and hence $a_\beta \in U_i^\alpha$ for some $i \in M_{n_\alpha}$. The set $U_i^\alpha \cap F_i^\alpha$ misses either $\overline{U}_z$ or $\overline{V}_z$ while $a_\beta \in U_i^\alpha \cap F_i^\alpha$. Consequently, $U_i^\alpha \cap F_i^\alpha$ misses $\overline{V}_z$ which implies that $b_\beta \notin U_i^\alpha \cap F_i^\alpha$. However, $U_i^\alpha \cap F_i^\alpha \in \mathcal{V}_\beta$ which shows that $b_\beta \notin \mathcal{V}_\beta[a_\beta]$ which gives a contradiction with the property (5). Therefore Z has a T_1-separating family $\mathcal{F} \cup \mathcal{U}'$ for some countable $\mathcal{U}' \subset \mathcal{U}$ which shows that $nw(Z) = \omega$ and hence Fact 1 is proved. $\qquad \square$

Returning to our solution, observe that an evident consequence of Problem 235 is that every Lindelöf Σ-space with a G_δ-diagonal has a countable network. Now, if CH holds and X is a Lindelöf Σ-space with a small diagonal, then every compact $K \subset X$ also has a small diagonal and hence K is metrizable by Problem 298. Take a countable family $\mathcal{F}$ of closed subsets of X such that $\mathcal{F}$ is a network with respect to a compact cover $\mathcal{C}$ of the space X (see Problem 225 and Fact 1 of T.229). Every element of $\mathcal{C}$ is metrizable, so we can apply Fact 1 to conclude that $nw(X) = \omega$ and finish our solution.

T.301. *Let X be a zero-dimensional space. Prove that any subspace of X is also zero-dimensional.*

Solution. Fix a base $\mathcal{B}$ in the space X such that every $B \in \mathcal{B}$ is clopen in X. Given any $Y \subset X$, the family $\mathcal{B}_Y = \{B \cap Y : B \in \mathcal{B}\}$ is a base in Y and it is immediate that all elements of $\mathcal{B}_Y$ are clopen in Y. Therefore Y is zero-dimensional.

T.302. *Prove that an arbitrary product of zero-dimensional spaces must be a zero-dimensional space.*

Solution. Denote by M_n the set $\{1, \ldots, n\}$ for all $n \in \mathbb{N}$. Given a family $\mathcal{S}$, the family $\bigwedge \mathcal{S}$ is the collection of all finite intersections of the elements of $\mathcal{S}$. Assume that X_t is a zero-dimensional space for every $t \in T$; we must prove that the space $X = \prod\{X_t : t \in T\}$ is also zero-dimensional. Let $\mathcal{B}_t$ be a base in X_t such that every $B \in \mathcal{B}_t$ is clopen in X_t. The map $p_t : X \to X_t$ is the natural projection for all $t \in T$.

The family $\mathcal{S} = \{p_t^{-1}(B) : B \in \mathcal{B}_t, t \in T\}$ consists of clopen subsets of X because each p_t is a continuous map. Take any $x \in X$ and $U \in \tau(x, X)$. By the definition of the product topology there exists a set $T' = \{t_1, \ldots, t_n\} \subset T$ and

$O_i \in \tau(X_{t_i})$ for all $i \in M_n$ such that $x \in O = \prod \{O_i : i \in M_n\} \times \prod_{t \in T \setminus T'} X_t \subset U$. Since each X_t is zero-dimensional, for every $i \in M_n$, there is $V_i \in \mathcal{B}_{t_i}$ such that $x(t_i) \in V_i \subset O_i$.

Observe also that if $W_i = p_{t_i}^{-1}(V_i)$ for all $i \in M_n$, then $V = W_1 \cap \cdots \cap W_n \in \bigwedge \mathcal{S}$ and $x \in V \subset O \subset U$ which shows that $\bigwedge \mathcal{S}$ is a base in X, i.e., $\mathcal{S}$ is a subbase of X. Since every finite intersection of clopen sets is a clopen set, the base $\bigwedge \mathcal{S}$ of the space X consists of clopen subsets of X and hence X is zero-dimensional.

T.303. *Given a cardinal κ and an infinite space X with $w(X) \leq \kappa$, prove that X is zero-dimensional if and only if it can be embedded in $\mathbb{D}^\kappa$.*

Solution. If X is a subspace of $\mathbb{D}^\kappa$, then X is zero-dimensional by Fact 3 of T.298. Now assume that X is zero-dimensional and fix a base $\mathcal{B}'$ of the space X such that every $B \in \mathcal{B}'$ is a clopen subset of X. It follows from Fact 1 of T.102 that there is $\mathcal{B} \subset \mathcal{B}'$ such that $\mathcal{B}$ is still a base of X and $|\mathcal{B}| \leq \kappa$.

For every $B \in \mathcal{B}$ let $\chi_B(x) = 1$ if $x \in B$ and $\chi_B(x) = 0$ for all $x \in X \setminus B$. Then $\chi_B : X \to \mathbb{D}$ is a continuous map because $\{B, X \setminus B\}$ is an open cover of X on the elements of which χ_B is constant (see Fact 1 of S.472).

Consequently, $\varphi = \Delta\{\chi_B : B \in \mathcal{B}\} : X \to \mathbb{D}^\mathcal{B}$ is a continuous map; let $Y = \varphi(X)$. We claim that the map $\varphi : X \to Y$ is a homeomorphism. In the first place observe that for any distinct $x, y \in X$, there is $B \in \mathcal{B}$ such that $x \in B$ and $y \notin B$. Thus $\chi_B(x) = 1$ while $\chi_B(y) = 0$ which shows that $\varphi(x) \neq \varphi(y)$ and therefore φ is a bijection. For any $B \in \mathcal{B}$ let $p_B : \mathbb{D}^\mathcal{B} \to \mathbb{D}$ be the natural projection of $\mathbb{D}^\mathcal{B}$ onto its Bth factor.

To see that φ^{-1} is continuous take any $y \in Y$ and $U \in \tau(\varphi^{-1}(y), X)$; let $x = \varphi^{-1}(y)$. Since $\mathcal{B}$ is a base in X, there is $B \in \mathcal{B}$ such that $x \in B \subset U$. The set $W = \{z \in Y : p_B(z) = 1\} = p_B^{-1}(1) \cap Y$ is an open subset of Y because $p_B^{-1}(1)$ is open in $\mathbb{D}^\mathcal{B}$. Now, if $t \in W$ and $s = \varphi^{-1}(t)$, then, by definition of the diagonal product, $1 = p_B(t) = \chi_B(s)$ and hence $s \in B$ which implies that $s = \varphi^{-1}(t) \in U$ for any $t \in W$. Thus $\varphi^{-1}(W) \subset U$, i.e., W witnesses continuity at the point y. The point y has been chosen arbitrarily, so the map φ^{-1} is continuous and hence φ is an embedding of X in $\mathbb{D}^\mathcal{B}$. Since $|\mathcal{B}| \leq \kappa$, the space $\mathbb{D}^\mathcal{B}$ embeds in $\mathbb{D}^\kappa$ (this is an easy exercise that we leave to the reader) so X embeds in $\mathbb{D}^\kappa$ as well.

T.304. *Prove that any space X is a perfect image of a zero-dimensional space Y such that $w(Y) \leq w(X)$.*

Solution. If $\kappa = w(X)$ is finite, then X is zero-dimensional, so there is nothing to prove. We will assume, therefore, that κ is an infinite cardinal; there is no loss of generality to consider that X is a subspace of $\mathbb{I}^\kappa$ (see Problem 209 of [TFS]).

There exists a continuous onto map $\varphi : \mathbb{D}^\kappa \to \mathbb{I}^\kappa$ by Fact 2 of T.298; let $Y = \varphi^{-1}(X)$. The map $f = \varphi|Y : Y \to X$ is perfect because so is φ (see Problem 122 of [TFS] and Fact 2 of S.261). The space Y is zero-dimensional by Problem 303 and $w(Y) \leq w(\mathbb{D}^\kappa) \leq \kappa$ which shows that X is a perfect image of a zero-dimensional space Y such that $w(Y) \leq \kappa = w(X)$.

T.305. *Prove that any non-zero-dimensional space can be continuously mapped onto* $\mathbb{I}$.

Solution. Let X be a non-zero-dimensional space. It was proved in Fact 4 of T.063 that there exists a continuous onto map $f : X \to I = [0, 1]$ where I carries the topology induced from $\mathbb{R}$. If $g(x) = 2x - 1$ for all $x \in I$, then g maps I continuously onto $\mathbb{I}$ and hence $g \circ f$ maps X continuously onto $\mathbb{I}$.

T.306. *Prove that any Lindelöf space is zero-dimensional if and only if it is strongly zero-dimensional. In particular, compact zero-dimensional spaces and second countable zero-dimensional spaces are strongly zero-dimensional.*

Solution. Given a space Z let $\mathcal{C}(Z)$ be the family of all clopen subsets of Z.

Fact 1. If a space Z is strongly zero-dimensional, then it is zero-dimensional.

Proof. To prove that $\mathcal{C}(Z)$ is a base in Z, take any $z \in Z$ and $U \in \tau(z, Z)$. There is $V \in \tau(z, Z)$ such that $\overline{V} \subset U$ and hence the family $\mathcal{U} = \{U, X \backslash \overline{V}\}$ is a finite open cover of Z. The space Z being strongly zero-dimensional, there is a disjoint refinement $\mathcal{V} \subset \tau(X)$ of the cover $\mathcal{U}$. Observe first that every $W \in \mathcal{V}$ is clopen in Z because $Z \backslash W = \bigcup(\mathcal{V} \backslash \{W\})$ is an open set. Take any $W \in \mathcal{V}$ with $z \in W$. Since $\mathcal{V}$ is a refinement of $\mathcal{U}$, we have either $W \subset U$ or $W \subset Z \backslash \overline{V}$. The second inclusion is impossible because $z \in W \backslash (Z \backslash \overline{V})$. Therefore $x \in W \subset U$ which implies, together with $W \in \mathcal{C}(Z)$, that $\mathcal{C}(Z)$ is a base of Z and hence Z is zero-dimensional. Fact 1 is proved.   □

Returning to our solution observe that if X is strongly zero-dimensional, then it is zero-dimensional by Fact 1; for this implication, there is no need to assume that X is Lindelöf.

Now let X be a Lindelöf zero-dimensional space. To prove that X is strongly zero-dimensional, take any open cover $\mathcal{U}$ of the space X (we should formally assume that $\mathcal{U}$ is finite, but proving it without this assumption gives a stronger statement). Since the space X is zero-dimensional, the family $\mathcal{V} = \{V \in \mathcal{C}(X) :$ there exists $U(V) \in \mathcal{U}$ with $V \subset U(V)\}$ is a cover of X. The space X being Lindelöf there is a family $\mathcal{V}' = \{V_n : n \in \omega\} \subset \mathcal{V}$ such that $\bigcup \mathcal{V}' = X$. It is easy to see that the set $W_n = V_n \backslash (\bigcup_{i < n} V_i)$ is clopen for all $n \in \mathbb{N}$; let $W_0 = V_0$ and observe that $\mathcal{W} = \{W_n : n \in \omega\}$ is a disjoint open refinement of $\mathcal{U}$. This proves that every Lindelöf zero-dimensional space is strongly zero-dimensional and finishes our solution.

T.307. *Let X be a space with $|X| < \mathfrak{c}$. Prove that X is zero-dimensional. In particular, any countable space is strongly zero-dimensional.*

Solution. If X is not zero-dimensional, then there exists a continuous onto map $f : X \to \mathbb{I}$ by Problem 305 and hence $|X| \geq |\mathbb{I}| = \mathfrak{c}$ which is a contradiction. Therefore X is zero-dimensional. Furthermore, if X is countable, then X is Lindelöf and $|X| < \mathfrak{c}$. Therefore X is strongly zero-dimensional by Problem 306.

T.308. *For an arbitrary space X, prove that X is strongly zero-dimensional if and only if $\mathrm{Ind}\,X=0$. Observe that, as a consequence, any strongly zero-dimensional space is normal.*

Solution. Given a space Z let $\mathcal{C}(Z)$ be the family of all clopen subsets of Z. Suppose that Z is a space such that $\mathrm{Ind}(Z) = 0$. Given disjoint closed sets $F, G \subset Z$ we have $U = Z \backslash G \in \tau(F, Z)$ and hence there is $O \in \mathcal{C}(Z)$ such that $F \subset O \subset Z \backslash G$. As a consequence, O and $Z \backslash O$ are open disjoint neighborhoods of F and G respectively which proves that

(1) for any space Z if $\mathrm{Ind}(Z) = 0$, then Z is normal.

To prove sufficiency assume that $\mathrm{Ind}(X) = 0$ and take any open cover $\mathcal{U} = \{U_1, \ldots, U_n\}$ of the space X. The set $F = X \backslash (U_2 \cup \cdots \cup U_n)$ is closed in X and $F \subset U_1$; besides $F \cup U_2 \cup \cdots \cup U_n = X$. Since $\mathrm{Ind}(X) = 0$, there is $V_1 \in \mathcal{C}(X)$ such that $F \subset V_1 \subset U_1$; it is clear that $V_1 \cup U_2 \cup \cdots \cup U_n = X$.

Assume that $1 \le k < n$ and we have clopen sets $V_1, \ldots, V_k$ such that $V_i \subset U_i$ for all $i \le k$ and $V_1 \cup \cdots \cup V_i \cup U_{i+1} \cup \cdots \cup U_n = X$ for any $i \le k$. The set $F = X \backslash (V_1 \cup \cdots \cup V_k \cup U_{k+2} \cup \cdots \cup U_n)$ is closed in X and $F \subset U_{k+1}$. Since $\mathrm{Ind}(X) = 0$, there is $V_{k+1} \in \mathcal{C}(X)$ such that $F \subset V_{k+1} \subset U_{k+1}$; it is clear that $V_1 \cup \cdots \cup V_{k+1} \cup U_{k+2} \cup \cdots \cup U_n = X$ and hence our inductive construction can be continued to obtain a family $\{V_1, \ldots, V_n\} \subset \mathcal{C}(X)$ such that $V_i \subset U_i$ for every $i \le n$ and $V_1 \cup \cdots \cup V_n = X$. Finally, if $W_1 = V_1$ and $W_k = V_k \backslash (V_1 \cup \cdots \cup V_{k-1})$ for all $k \in \{2, \ldots, n\}$, then $\{W_i : i \le n\}$ is a disjoint open refinement of $\mathcal{U}$ which proves that X is strongly zero-dimensional and settles sufficiency.

As to necessity, assume that X is strongly zero-dimensional and take any closed $F \subset X$ and $U \in \tau(F, X)$. The family $\mathcal{U} = \{U, X \backslash F\}$ is a finite open cover of X, so there exists a disjoint open refinement $\mathcal{V}$ of the cover $\mathcal{U}$. Observe first that for every $\mathcal{V}' \subset \mathcal{V}$, the set $W = \bigcup \mathcal{V}'$ is clopen in X because $X \backslash W = \bigcup (\mathcal{V} \backslash \mathcal{V}')$ is an open set.

Take any $W \in \mathcal{V}' = \{V \in \mathcal{V} : V \cap F \ne \emptyset\}$. Since $\mathcal{V}$ is a refinement of $\mathcal{U}$, we have either $W \subset U$ or $W \subset X \backslash F$. The second inclusion is impossible because there is $x \in W \cap F$ and hence $x \in W \backslash (X \backslash F)$. Thus $W \subset U$ for any $W \in \mathcal{V}'$ and hence $V = \bigcup \mathcal{V}'$ is a clopen set such that $F \subset V \subset U$ which implies, together with $V \in \mathcal{C}(X)$, that $\mathrm{Ind}(X) = 0$ and hence necessity is established. To finish our solution observe that $\mathrm{Ind}(X) = 0$ for any strongly zero-dimensional space X and hence X is normal by (1).

T.309. *Prove that any strongly zero-dimensional space is zero-dimensional. Give an example of a normal zero-dimensional space which fails to be strongly zero-dimensional.*

Solution. Any strongly zero-dimensional space is zero-dimensional: this was proved in Fact 1 of T.306. Denote by I the set $[0, 1]$ with the topology inherited from $\mathbb{R}$. For any space Z let $\mathcal{C}(Z)$ be the family of all clopen subsets of Z. When we work with ordinals, then each ordinal α considered as a set is identified with the set of its predecessors, i.e., $\alpha = \{\beta : \beta < \alpha\}$. For any $\alpha, \beta \in \omega_1 + 1$, let $[\alpha, \beta) = \{\gamma : \alpha \leq \gamma < \beta\}$.

Fact 1. Given any points $a, b \in \mathbb{R}$ such that $a < b$, the interval $[a, b]$ is connected, i.e., $\mathcal{C}([a, b]) = \{\emptyset, [a, b]\}$. As a consequence, the space $[a, b]$ is not zero-dimensional.

Proof. Assume, towards a contradiction, that there is a clopen subset U of the space $[a, b]$ such that $U \neq \emptyset$ and $U \neq [a, b]$. Then the sets U and $V = [a, b]\backslash U$ are non-empty, compact, disjoint and $U \cup V = [a, b]$. Take any $x \in U$ and $y \in V$; then $x \neq y$ because the sets U and V are disjoint. Interchanging U and V if necessary, we can assume that $x < y$.

The sets $P = U \cap [x, y]$ and $Q = V \cap [x, y]$ are compact, disjoint, their union is $[x, y]$ and $x \in P$, $y \in Q$. Let $\xi = \inf Q$. The set $Q \subset [x, y]$ is bounded and non-empty so $\xi \in [x, y]$. The point ξ has to belong to P or to Q. Observe that P and Q are complementary closed sets in $[x, y]$ and hence they are both open in $[x, y]$. Now, if $\xi \in P$, then $\xi < y$, and by openness of P, there is $\varepsilon > 0$ such that $[\xi, \xi + \varepsilon) \subset P$, and hence there are no points of Q in $[\xi, \xi + \varepsilon)$ which shows that $\inf Q \geq \xi + \varepsilon$, a contradiction. If $\xi \in Q$, then $x < \xi$, and, by openness of Q, we have $(\xi - \varepsilon, \xi] \in Q$ for some $\varepsilon > 0$. Therefore all points of $(\xi - \varepsilon, \xi]$ belong to Q and hence $\inf Q \leq \xi - \varepsilon$ and we again obtained a contradiction which shows that $[a, b]$ is connected.

Finally, observe that if $[a, b]$ is zero-dimensional, then there is $U \in \mathcal{C}([a, b])$ such that $a \in U$ and $b \notin U$. Therefore U is a clopen set in $[a, b]$ such that $U \neq \emptyset$ and $U \neq [a, b]$; this contradiction completes the proof of Fact 1. □

Fact 2. If $Z \subset \mathbb{R}$, then Z is zero-dimensional if and only if $\text{Int}(Z) = \emptyset$.

Proof. Suppose that Z is zero-dimensional and $\text{Int}(Z) \neq \emptyset$. Then there are $a, b \in \mathbb{R}$ such that $a < b$ and $[a, b] \subset Z$. Now, it follows from Problem 301 that $[a, b]$ is zero-dimensional which contradicts Fact 1. This proves necessity.

Now assume that $\text{Int}(Z) = \emptyset$; given any $z \in Z$ and $U \in \tau(z, Z)$ there are $a, b \in \mathbb{R}$ such that $a < b$ and $z \in (a, b) \cap Z \subset U$. It follows from $\text{Int}(Z) = \emptyset$ that $(a, z)\backslash Z \neq \emptyset$ and $(z, b)\backslash Z \neq \emptyset$. Pick any $c \in (a, z)\backslash Z$ and $d \in (z, b)\backslash Z$. Then $z \in W = (c, d) \cap Z$; it is evident that $W \in \tau(Z)$. Furthermore, $W = [c, d] \cap Z$ and hence W is also closed in Z. Since $z \in W \subset U$ and $W \in \mathcal{C}(Z)$, we showed that $\mathcal{C}(Z)$ is a base in Z, i.e., that Z is zero-dimensional. Fact 2 is proved. □

Fact 3. Let $Q = \mathbb{Q} \cap I$; then there exists a disjoint family $\{Q_\alpha : \alpha < \omega_1\}$ of countable dense subsets of I such that $Q_\alpha \cap Q = \emptyset$ for all $\alpha < \omega_1$.

Proof. Let $Q_{-1} = Q$ and suppose that for some ordinal $\alpha < \omega_1$, we have a disjoint family $\{Q_\beta : -1 \leq \beta < \alpha\}$ of countable dense subsets of I. Observe that the set $R = \bigcup\{Q_\beta : -1 \leq \beta < \alpha\}$ is countable and therefore $I\backslash R$ is dense in I. Since the space $I\backslash R$ is second countable, there is a countable dense $Q_\alpha \subset I\backslash R$. It is easy to see that Q_α is also dense in I and hence our inductive construction can be continued to provide a disjoint family $\{Q_\beta : -1 \le \beta < \omega_1\}$ of countable dense subsets of I such that $Q_{-1} = Q$. Clearly, the family $\{Q_\alpha : 0 \le \alpha < \omega_1\}$ is as promised, so Fact 3 is proved. $\square$

Fact 4. For any space Z, if $K \subset Z$ is compact, $F \subset Z$ is closed and $K \cap F = \emptyset$, then K and F are open-separated, i.e., there are $U \in \tau(K, Z)$ and $V \in \tau(F, Z)$ such that $U \cap V = \emptyset$.

Proof. For every $x \in K$, there is $U_x \in \tau(x, Z)$ such that $\overline{U}_x \cap F = \emptyset$. By compactness of the space K, there are $x_1, \ldots, x_n \in K$ such that $K \subset U_{x_1} \cup \cdots \cup U_{x_n}$. For the set $U = U_{x_1} \cup \cdots \cup U_{x_n}$, we have $\overline{U} = \overline{U}_{x_1} \cup \cdots \cup \overline{U}_{x_n} \subset Z \setminus F$ because $\overline{U}_{x_i} \subset Z \setminus F$ for all $i \le n$. As a consequence, $U \in \tau(K, Z)$ and $V = Z \setminus \overline{U} \in \tau(F, Z)$; since the sets U and V are disjoint, the proof of Fact 4 is complete. $\square$

Fact 5. If Z is a space which can be covered by its clopen zero-dimensional subspaces, then Z is zero-dimensional.

Proof. Take any $\mathcal{U} \subset \mathcal{C}(Z)$ such that each $U \in \mathcal{U}$ is zero-dimensional and $\bigcup \mathcal{U} = Z$. Given any $z \in Z$ and $W \in \tau(z, Z)$, there is $U \in \mathcal{U}$ such that $z \in U$. Evidently, $W \cap U \in \tau(z, U)$; since U is zero-dimensional, there exists $O \in \mathcal{C}(U)$ such that $z \in O \subset U \cap W$. It is immediate that $O \in \mathcal{C}(Z)$ and $z \in O \subset U$ which shows that Z is zero-dimensional. Fact 5 is proved. $\square$

Returning to our solutions let $\{Q_\alpha : \alpha < \omega_1\}$ be a disjoint family of countable dense subsets of I such that $Q_\alpha \cap Q = \emptyset$ for all $\alpha < \omega_1$, where $Q = \mathbb{Q} \cap I$ (see Fact 3). Observe that the set $S_\alpha = I \setminus (\bigcup \{Q_\beta : \beta \ge \alpha\})$ is dense in I for every $\alpha < \omega_1$ because $Q \subset S_\alpha$. Besides, $S_\alpha \subset S_\beta$ whenever $\alpha < \beta < \omega_1$ and $\bigcup \{S_\alpha : \alpha < \omega_1\} = I$.

Let $X = \bigcup \{\{\alpha\} \times S_\alpha : \alpha < \omega_1\} \subset (\omega_1 + 1) \times I$. We consider X with the topology inherited from the compact space $(\omega_1 + 1) \times I$ so X is a Tychonoff space; let $X_\alpha = (\alpha \times I) \cap X$ for all $\alpha < \omega_1$. To see that X is normal, consider the space $Y = X \cup (\{\omega_1\} \times I)$. We show first that

(1) the space Y is normal.

Let F, G be closed disjoint subsets of Y. Then the sets $F' = F \cap (\{\omega_1\} \times I)$ and $G' = G \cap (\{\omega_1\} \times I)$ are compact disjoint subsets of $J = \{\omega_1\} \times I$ which is homeomorphic to I. By Fact 4, there exist $U \in \tau(F', Y)$ and $V \in \tau(G', Y)$ such that $U \cap V = \emptyset$. We claim that $F \setminus U \subset X_\alpha$ for some $\alpha < \omega_1$. Indeed, suppose for a moment that for any $\alpha < \omega_1$, there is $z_\alpha = (\beta_\alpha, t_\alpha) \in F \setminus U$ for some $\beta_\alpha \ge \alpha$. The space I being compact, there is a point $t \in I$ such that the set $A_\varepsilon = \{\alpha < \omega_1 : |t - t_\alpha| < \varepsilon\}$ is uncountable for any $\varepsilon > 0$. Let $z = (\omega_1, t)$; if $W \in \tau(z, Y)$, then there is $\varepsilon > 0$ and $\gamma < \omega_1$ such that $z \in O = \{(\alpha, u) : \gamma < \alpha$ and $|u - t| < \varepsilon\} \subset W$. Since A_ε is uncountable, there is $\alpha \in A_\varepsilon$ with $\alpha > \gamma$ and hence $\beta_\alpha \ge \alpha > \gamma$ which shows that $z_\alpha \in O \cap F \subset W \cap F$ and hence z is in the closure of $F \setminus U$ in Y. The set $F \setminus U$ being closed, we have $z \in (F \setminus U) \cap J$ which is a contradiction with $F' = F \cap J \subset U$. The same reasoning applied to $G \setminus V$ shows that there exists $\alpha < \omega_1$ such that $(F \backslash U) \cup (G \backslash V) \subset X_\alpha$. Since $X_{\alpha+1}$ is clopen in Y, the sets $U' = U \backslash X_{\alpha+1}$ and $V' = V \backslash X_{\alpha+1}$ are open in Y.

The space $X_{\alpha+1} \subset (\alpha + 1) \times I$ is second countable and hence normal; since $F \cap X_{\alpha+1}$ and $G \cap X_{\alpha+1}$ are disjoint closed subsets of $X_{\alpha+1}$, there are disjoint $U'', V'' \in \tau(X_{\alpha+1})$ such that $F \cap X_{\alpha+1} \subset U''$ and $F \cap X_{\alpha+1} \subset V''$. It is evident that $U'', V'' \in \tau(Y)$; since $U' \cup U''$ and $V' \cup V''$ are disjoint open neighborhoods of F and G respectively, we proved that Y is normal.

To establish that X is also normal we need to prove first that

(2) if F, G are disjoint closed subsets of X, then $\overline{F} \cap \overline{G} = \emptyset$ (the bar denotes the closure in Y).

It is clear that $\overline{F} \cap \overline{G} \cap X = \emptyset$ so assume, towards a contradiction, that $z \in \overline{F} \cap \overline{G}$ for some $z = (\omega_1, t) \in J$. Since $\bigcup \{ S_\alpha : \alpha < \omega_1 \} = I$, there is $\alpha < \omega_1$ such that $t \in S_\beta$ for all $\beta \geq \alpha$ and hence $P = [\alpha, \omega_1) \times \{t\} \subset Y$. It is clear that P is homeomorphic to $[\alpha, \omega_1)$; any two closed unbounded subsets of $[\alpha, \omega_1)$ have nonempty intersection (see Problem 064) and therefore both sets $F \cap P$ and $G \cap P$ cannot be uncountable. Assume, for example, that $F \cap P$ is countable and hence there is $\gamma < \omega_1$ for which $F \cap P \subset \gamma \times \{t\}$.

It is easy to see that $z \in \overline{F}$ implies that there exists a strictly increasing sequence $\{\alpha_n : n \in \omega\} \subset \omega_1 \backslash \gamma$ and $\{t_n : n \in \omega\} \subset I$ such that $z_n = (\alpha_n, t_n) \in F$ and $|t_n - t| < \frac{1}{n+1}$ for all $n \in \omega$. Letting $\alpha = \sup\{\alpha_n : n \in \omega\}$ it is straightforward to show that $z' = (\alpha, t) \in \overline{F} = F$ which contradicts $F \cap P \subset \gamma \times \{t\}$.

Analogously, if $G \cap P$ is countable, then $z \notin \overline{G}$; this final contradiction shows that (2) is true.

Now it is easy to prove that X is normal. Indeed, take any disjoint closed $F, G \subset X$. The sets $\overline{F}$ and $\overline{G}$ are closed in Y and disjoint by (2). The property (1) guarantees that there exist disjoint $U' \in \tau(\overline{F}, Y)$ and $V' \in \tau(\overline{G}, Y)$; it is clear that $U = U' \cap X$ and $V = V' \cap X$ are disjoint open (in X) neighborhoods of F and G, respectively, so X is normal.

Given any $\alpha < \omega_1$, the set S_α has empty interior in I (and hence in $\mathbb{R}$) because $Q_\alpha \subset I \backslash S_\alpha$ is dense in I. Consequently, S_α is zero-dimensional by Fact 2. The space α is zero-dimensional being countable by Problem 307 and hence $\alpha \times S_\alpha$ is zero-dimensional by Problem 302. Therefore $X_\alpha \subset \alpha \times S_\alpha$ is zero-dimensional by Problem 301 for every $\alpha < \omega_1$. Thus $\{X_{\alpha+1} : \alpha < \omega_1\}$ is a clopen cover of X by its zero-dimensional subspaces which shows that X is zero-dimensional by Fact 5.

To finally see that X is not strongly zero-dimensional, assume the contrary; then $\mathrm{Ind}(X) = 0$ by Problem 308. Let $F = \omega_1 \times \{0\}$ and $G = \omega_1 \times \{1\}$; it is evident that F and G are closed disjoint subsets of X. Since $\mathrm{Ind}(X) = 0$, there exists $U \in \mathcal{C}(X)$ such that $F \subset U \subset X \backslash G$, i.e., U and $V = X \backslash U$ are disjoint clopen subsets of X. Observe also that $p = (\omega_1, 0) \in \overline{F} \subset \overline{U}$ and $q = (\omega_1, 1) \in \overline{G} \subset \overline{V}$ which shows that $U' = \overline{U} \cap J$ and $V' = \overline{V} \cap J$ are non-empty disjoint closed subsets of J. Since $U \cup V = X$ and X is dense in Y, we have $\overline{U} \cup \overline{V} = Y$ and therefore $U' \cup V' = J$. The set J being homeomorphic to I, there exist non-empty complementary closed subsets $A, B \subset I$. Therefore A is clopen in I and $A \notin \{\emptyset, I\}$ which contradicts

Fact 1. Consequently, there is no $U \in C(X)$ such that $F \subset U \subset X \backslash G$ which shows that X is not strongly zero-dimensional and finishes our solution.

T.310. *Prove that a closed subspace of a strongly zero-dimensional space is strongly zero-dimensional. Give an example of a strongly zero-dimensional space X such that some $Y \subset X$ is not strongly zero-dimensional.*

Solution. Given a space Z, let $C(Z)$ be the family of all clopen subsets of Z. Suppose that Z is a strongly zero-dimensional space and Y is a closed subspace of Z. It suffices to show that $\text{Ind} Y = 0$ by Problem 308, so take any closed subset F of the space Y and let $U \in \tau(F, Y)$. There is $V \in \tau(Z)$ such that $V \cap Y = U$; it is evident that F is also closed in Z and $V \in \tau(F, Z)$. Since $\text{Ind} Z = 0$, there is $W' \in C(Z)$ such that $F \subset W' \subset V$. It is immediate that $W = W' \cap Y \in C(Y)$; since $F \subset W \subset U$, we established that $\text{Ind} Y = 0$ and hence Y is strongly zero-dimensional.

It was proved in Problem 309 that there exists a zero-dimensional space Y such that Y is not strongly zero-dimensional. There exists a cardinal κ such that Y embeds in $\mathbb{D}^{\kappa}$ by Problem 303, so we can consider that Y is a subspace of $X = \mathbb{D}^{\kappa}$. The space X is compact and zero-dimensional by Problem 303, so it is strongly zero-dimensional by Problem 306. Thus the strongly zero-dimensional space X contains a subspace Y which is not strongly zero-dimensional.

T.311. *Let X be a normal space such that $X = \bigcup \{X_n : n \in \omega\}$, where each X_n is strongly zero-dimensional and closed in X. Prove that X is strongly zero-dimensional.*

Solution. For every $n \in \mathbb{N}$, denote by M_n the set $\{1, \ldots, n\}$. Given a space Z denote by $C(Z)$ the family of all clopen subsets of Z; if $\mathcal{U} = \{U_1, \ldots, U_n\}$ is an open cover of Z, say that an open cover $\mathcal{V}$ of the space Z is *a strong precise refinement* of $\mathcal{U}$ if $\mathcal{V} = \{V_1, \ldots, V_n\}$ and $\overline{V}_i \subset U_i$ for all $i \in M_n$. If $\mathcal{A} \subset \exp(Z)$, then $\overline{\mathcal{A}} = \{\overline{A} : A \in \mathcal{A}\}$; if $Y \subset Z$, then $\mathcal{A}|Y = \{A \cap Y : A \in \mathcal{A}\}$.

Fact 1. Every open cover $\mathcal{U} = \{U_1, \ldots, U_n\}$ of a normal space Z has a strong precise refinement.

Proof. It is evident that the set $F = Z \backslash (U_2 \cup \cdots \cup U_n)$ is closed in Z and $F \subset U_1$; besides $F \cup U_2 \cup \cdots \cup U_n = Z$. Since Z is normal, there is $V_1 \in \tau(F, Z)$ such that $F \subset V_1 \subset \overline{V}_1 \subset U_1$; it is clear that $V_1 \cup U_2 \cup \cdots \cup U_n = Z$.

Assume that $1 \leq k < n$ and we have open sets $V_1, \ldots, V_k$ such that $\overline{V}_i \subset U_i$ and $V_1 \cup \cdots \cup V_i \cup U_{i+1} \cup \cdots \cup U_n = Z$ for any $i \leq k$. It is clear that the set $F = Z \backslash (V_1 \cup \cdots \cup V_k \cup U_{k+2} \cup \cdots \cup U_n)$ is closed in Z and $F \subset U_{k+1}$. Since Z is normal, there is $V_{k+1} \in \tau(Z)$ such that $F \subset V_{k+1} \subset \overline{V}_{k+1} \subset U_{k+1}$; it is evident that $V_1 \cup \cdots \cup V_{k+1} \cup U_{k+2} \cup \cdots \cup U_n = Z$ and hence our inductive construction can be continued to obtain a family $\{V_1, \ldots, V_n\} \subset \tau(Z)$ such that $\overline{V}_i \subset U_i$ for every $i \leq n$ and $V_1 \cup \cdots \cup V_n = Z$. Thus $\mathcal{V} = \{V_1, \ldots, V_n\}$ is a strong precise refinement of $\mathcal{U}$ and Fact 1 is proved.   □

Fact 2. For every open cover $\mathcal{U} = \{U_1, \ldots, U_n\}$ of a strongly zero-dimensional space Z, there is a disjoint family $\mathcal{V} = \{V_1, \ldots, V_n\} \subset \mathcal{C}(Z)$ such that $\bigcup \mathcal{V} = Z$ and $V_i \subset U_i$ for all $i \in M_n$.

Proof. There is a disjoint refinement $\mathcal{W}$ of the cover $\mathcal{U}$. Observe first that for any $\mathcal{W}' \subset \mathcal{W}$, the set $\bigcup \mathcal{W}'$ is clopen in Z because its complement $\bigcup(\mathcal{W} \backslash \mathcal{W}')$ is open in Z. For every $W \in \mathcal{W}$, there is $O(W) \in \mathcal{U}$ such that $W \subset O(W)$. For each $i \in M_n$, let $V_i = \bigcup \{W \in \mathcal{W} : O(W) = U_i\}$. By our first observation, each V_i is clopen in Z; since $V_1 \cup \cdots \cup V_n = \bigcup \mathcal{W} = Z$ and $V_i \subset U_i$ for every $i \in M_n$, the family $\mathcal{V} = \{V_1, \ldots, V_n\}$ is as promised and Fact 2 is proved. ☐

Returning to our solution, take any open cover $\mathcal{U} = \{U_1, \ldots, U_n\}$ of the space X. Since X_0 is strongly zero-dimensional, we can apply Fact 2 to find a family $\mathcal{V} = \{V_1, \ldots, V_n\} \subset \mathcal{C}(X_0)$ such that $V_1 \cup \cdots \cup V_n = X_0$ and $V_i \subset U_i \cap X_0$ for all $i \in M_n$. It is easy to see that the set $U_i' = (U_i \backslash X_0) \cup V_i$ is open in X for all $i \in M_n$; besides, the family $\{U_i' \cap X_0 : i \in M_n\}$ is disjoint because so is $\mathcal{V}$ and $U_i' \cap X_0 = V_i$ for all $i \in M_n$.

Another easy observation is that the family $\mathcal{U}' = \{U_i' : i \in M_n\}$ is a cover of X and therefore Fact 1 can be applied to conclude that there exists a strong precise refinement $\mathcal{U}_0 = \{U_1^0, \ldots, U_n^0\}$ of the cover $\mathcal{U}'$. Furthermore, $\overline{U_i^0} \cap X_0 \subset U_i' \cap X_0 = V_i$ for all $i \in M_n$ and hence $\mathcal{U}_0 | X_0$ is disjoint.

Now assume that $k \in \omega$ and we have a collection $\{\mathcal{U}_i : i \in \{0, \ldots, k\}\}$ of open covers of X such that

(1) $\mathcal{U}_i = \{U_1^i, \ldots, U_n^i\}$ for all $i \leq k$;
(2) $\mathcal{U}_0$ is a strong precise refinement of $\mathcal{U}$ and $\mathcal{U}_{i+1}$ is a strong precise refinement of $\mathcal{U}_i$ for all $i < k$;
(3) $\overline{\mathcal{U}}_i | X_i$ is disjoint for any $i \leq k$.

Since the space X_{k+1} is strongly zero-dimensional and $\mathcal{U}_k | X_{k+1}$ is an open cover of X_{k+1}, there is a disjoint family $\mathcal{V} = \{V_1 \ldots, V_n\} \subset \mathcal{C}(X_{k+1})$ such that $\bigcup \mathcal{V} = X_{k+1}$ and $V_i \subset U_i^k \cap X_{k+1}$ for all $i \leq n$ (see Fact 2). It is straightforward that the set $U_i' = (U_i^k \backslash X_{k+1}) \cup V_i$ is open in X for all $i \in M_n$; besides, the family $\{U_i' \cap X_{k+1} : i \in M_n\}$ is disjoint because so is $\mathcal{V}$ and $U_i' \cap X_{k+1} = V_i$ for all $i \in M_n$.

Another easy observation is that the family $\mathcal{U}' = \{U_i' : i \in M_n\}$ is a cover of X and therefore Fact 1 can be applied to conclude that there exists a strong precise refinement $\mathcal{U}_{k+1} = \{U_1^{k+1}, \ldots, U_n^{k+1}\}$ of the cover $\mathcal{U}'$. It is immediate that the properties (1) and (2) hold for the collection $\{\mathcal{U}_0, \ldots, \mathcal{U}_{k+1}\}$; the property (3) holds because $\overline{U_i^{k+1}} \cap X_{k+1} \subset U_i' \cap X_{k+1} = V_i$ for all $i \in M_n$ and hence $\overline{\mathcal{U}}_{k+1} | X_{k+1}$ is disjoint.

Therefore we can construct inductively a collection $\{\mathcal{U}_i : i \in \omega\}$ of open covers of X such that (1)–(3) are satisfied for all $k \in \omega$. Given $i \in M_n$ let $O_i = \bigcap \{U_i^j : j \in \omega\}$; it is clear that O_i is a closed subset of X and $O_i \subset U_i^j$ for all $j \in \omega$ which, together with (2), implies that $O_i \subset U_i$. If $x \in X$, then $x \in X_n$ for some $n \in \omega$; given distinct $i, j \in M_n$, we have $O_i \cap O_j \subset \overline{U_i^n} \cap \overline{U_j^n}$ and hence $x \notin O_i \cap O_j$

because $\overline{U_i^n} \cap \overline{U_j^n} \cap X_n = \emptyset$ by (3). This shows that the family $\mathcal{O} = \{O_1, \ldots, O_n\}$ is disjoint. On the other hand, the family $\mathcal{U}_i$ is a cover of X and hence there is $m(i) \in M_n$ such that $x \in U_{m(i)}^i$ for every $i \in \omega$. There is an infinite $A \subset \omega$ and $m \in M_n$ such that $m(i) = m$ for all $i \in A$. It follows from (2) that $x \in U_m^i$ for all $i \in \omega$ and hence $x \in \bigcap \{U_m^i : i \in \omega\} = \bigcap \{\overline{U_m^i} : i \in \omega\} = O_m$.

Therefore $\mathcal{O}$ is a disjoint cover of X; observe that every $O \in \mathcal{O}$ is open in X because its complement $\bigcup(\mathcal{O}\setminus\{O\})$ is closed in X. Thus $\mathcal{O}$ is a disjoint open refinement of $\mathcal{U}$ which proves that X is strongly zero-dimensional and makes our solution complete.

T.312. *Prove that there exists a space X which is not zero-dimensional while $X = \bigcup\{X_n : n \in \omega\}$, where each X_n is strongly zero-dimensional and closed in X.*

Solution. For every $n \in \mathbb{N}$ denote by M_n the set $\{1, \ldots, n\}$. Given a space Z let $\mathcal{C}(Z)$ be the family of all clopen subsets of Z. The space Z is called *connected* if $\mathcal{C}(Z) = \{\emptyset, Z\}$. If A is a set, then $\text{Fin}(A)$ is the family of all non-empty finite subsets of A.

Fact 1. (a) If Z is a connected space and $|Z| > 1$, then Z is not zero-dimensional.
(b) If Z is a space such that some connected $Y \subset Z$ is dense in Z, then Z is connected.

Proof. Assume that Z is connected and zero-dimensional; if a, b are distinct points of Z, then there exists a set $U \in \mathcal{C}(Z)$ such that $a \in U$ and $b \notin U$. It is evident that $U \notin \{\emptyset, Z\}$; this contradiction shows that (a) is true.

Now assume that Y is a connected dense subspace of Z. If $U \in \mathcal{C}(Z)\setminus\{\emptyset, Z\}$, then $U' = U \cap Y \neq \emptyset$. Moreover, $Z \setminus U \in \tau^*(Z)$ and hence $Y \setminus U' = (Z \setminus U) \cap Y \neq \emptyset$ which shows that $U' \in \mathcal{C}(Y)\setminus\{\emptyset, Y\}$; the obtained contradiction settles (b) and completes the proof of Fact 1. $\square$

Fact 2. The space $\sigma(A) = \{x \in \mathbb{R}^A : |\{a \in A : x(a) \neq 0\}| < \omega\}$ is connected for any infinite set A.

Proof. Suppose that $U \in \mathcal{C}(\sigma(A))\setminus\{\emptyset, \sigma(A)\}$. Then $V = \sigma(A)\setminus U$ is a non-empty clopen subset of $\sigma(A)$. Take any points $x \in U$ and $y \in V$ and consider the set $J = \{tx + (1-t)y : t \in [0,1]\} \subset \mathbb{R}^A$. It is evident that $J \subset \sigma(A)$ and hence $U \cap J$ and $V \cap J$ are non-empty disjoint clopen subsets of J which shows that $U \cap J \neq \emptyset$ and $U \cap J \neq J$, i.e., J is not connected.

Since $\mathbb{R}^A$ can be identified with $C_p(A)$ where A carries the discrete topology, we can apply Fact 1 of S.301 to conclude that J is homeomorphic to $[0,1] \subset \mathbb{R}$. It was proved in Fact 1 of T.309 that any interval $[p,q] \subset \mathbb{R}$ is connected, so the connected space $[0,1]$ is homeomorphic to the space J which is not connected. This contradiction shows that $\sigma(A)$ is connected and proves Fact 2. $\square$

Returning to our solution, note first that $|\mathbb{R}^B| = \mathfrak{c}$ for any finite non-empty set $B \subset \mathfrak{c}$ and hence $|\bigcup\{\mathbb{R}^B : B \in \text{Fin}(\mathfrak{c})\}| = \mathfrak{c}$. Therefore we can choose an enumeration $\{s_\alpha : \alpha < \mathfrak{c}\}$ of the set $E = \bigcup\{\mathbb{R}^B : B \in \text{Fin}(\mathfrak{c})\}$ such that each $s \in E$

occurs $\mathfrak{c}$-many times in this enumeration, i.e., $|\{\alpha < \mathfrak{c} : s_\alpha = s\}| = \mathfrak{c}$ for any $s \in E$ (see Fact 3 of S.286). For each $\alpha < \mathfrak{c}$, let $B_\alpha \in \text{Fin}(\mathfrak{c})$ be set for which $s_\alpha \in \mathbb{R}^{B_\alpha}$.

It is easy to construct a disjoint family $\mathcal{T} = \{T_\alpha : \alpha < \mathfrak{c}\} \subset \exp \mathfrak{c}$ such that $\bigcup \mathcal{T} = \mathfrak{c}$ and $|T_\alpha| = \mathfrak{c}$ for every $\alpha \in \mathfrak{c}$. Given $\alpha < \mathfrak{c}$, define a point $x_\alpha \in \mathbb{R}^{\mathfrak{c}}$ as follows: $x_\alpha|B_\alpha = s_\alpha$; if $\beta \in T_\alpha \backslash B_\alpha$, then $x_\alpha(\beta) = 1$, and if $\beta \in T_\gamma \backslash B_\alpha$ for some $\gamma \neq \alpha$, then $x_\alpha(\beta) = 0$. We claim that the space $X = \{x_\alpha : \alpha < \mathfrak{c}\}$ (considered with the topology induced from $\mathbb{R}^{\mathfrak{c}}$) is as promised. The proof is not easy and will be done in several steps.

Let $X_n = \{x_\alpha \in X : |B_\alpha| = n\}$ for any $n \in \mathbb{N}$; since the set B_α is finite for every $\alpha < \mathfrak{c}$, we have $X = \bigcup\{X_n : n \in \mathbb{N}\}$. Given any $x = x_\alpha \in X$, choose any distinct $v_1, \ldots, v_{n+1} \in T_\alpha \backslash B_\alpha$ and let $U_x = \{x_\beta \in X : x_\beta(v_i) > 0 \text{ for all } i \in M_{n+1}\}$. It is evident that $U_x \in \tau(x, X)$. If $\beta \neq \alpha$ and $x_\beta \in X_n$, then $x_\beta(\gamma) = 0$ for all $\gamma \in T_\alpha \backslash B_\beta$. Since $|B_\beta| = n$, there is $i \in M_{n+1}$ such that $v_i \notin B_\beta$ and hence $x_\beta(v_i) = 0$.

This shows that $x_\beta \notin U_x$ if $x_\beta \in X_n$ and $\beta \neq \alpha$, i.e., $U_x \cap X_n \subset \{x_\alpha\}$. Thus every point of X has a neighborhood which contains at most one point of X_n and hence X_n is closed and discrete in X for every $n \in \mathbb{N}$. It is evident that every discrete space is strongly zero-dimensional, so X_n is strongly zero-dimensional for all $n \in \mathbb{N}$. Letting $X_0 = X_1$ we have $X = \bigcup\{X_n : n \in \omega\}$ where X_n is closed in X and strongly zero-dimensional for all $n \in \omega$.

For any $A \subset \mathfrak{c}$ let $\pi_A : X \to \mathbb{R}^A$ be the projection of X to the face $\mathbb{R}^A$ defined by $\pi_A(x) = x|A$ for any $x \in X$. Observe first that X is dense in $\mathbb{R}^{\mathfrak{c}}$. Indeed, given a non-empty finite $B \subset \mathfrak{c}$ and $s \in \mathbb{R}^B$, there is $\alpha < \mathfrak{c}$ with $B = B_\alpha$ and $s = s_\alpha$. Recalling the definition of x_α, we can see that $\pi_B(x_\alpha) = s$. This proves that $\pi_B(X) = \mathbb{R}^B$ for any $B \in \text{Fin}(\mathfrak{c})$; it is an easy exercise that this implies density of X in $\mathbb{R}^{\mathfrak{c}}$.

To prove that X is connected, assume that $U \in \mathcal{C}(X) \backslash \{\emptyset, X\}$. Then $V = X \backslash U$ is a non-empty clopen subset of X; let $f(x) = 0$ for all $x \in U$ and $f(x) = 1$ for every $x \in V$. It is immediate that $f : X \to \mathbb{D}$ is a continuous onto map, so we can apply Problem 299 of [TFS] to find a countable $A \subset \mathfrak{c}$ and a continuous map $g : \pi_A(X) \to \mathbb{D}$ such that $f = g \circ \pi_A$. Since $f(X) = g(\pi_A(X)) = \mathbb{D}$, the sets $U' = g^{-1}(0)$ and $V' = g^{-1}(1)$ are non-empty disjoint and clopen in $\pi_A(X)$. An immediate consequence is that $\pi_A(X)$ is not connected.

Next we prove that $\pi_A(X) \supset \sigma(A)$. Take any $z \in \sigma(A)$ and a set $B \in \text{Fin}(\mathfrak{c})$ such that $\{\alpha < \mathfrak{c} : z(\alpha) \neq 0\} \subset B$. The set $Q = \{\alpha < \mathfrak{c} : T_\alpha \cap A \neq \emptyset\}$ is countable because the family $\mathcal{T}$ is disjoint. By the choice of our enumeration, there is $\alpha < \mathfrak{c}$ such that $T_\alpha \cap A = \emptyset$, $B = B_\alpha$ and $s_\alpha = z|B$. Then $x_\alpha|B = s_\alpha|B = z|B$ and $x_\alpha(\beta) = 0$ for any $\beta \in A \backslash B$ because $\beta \notin T_\alpha$. This proves that $\pi_A(x_\alpha) = z$ and hence $z \in \pi_A(X)$. The point $z \in \sigma(A)$ was chosen arbitrarily, so we proved that $\sigma(A) \subset \pi_A(X)$. The space $\sigma(A)$ is connected by Fact 2 and dense in $\mathbb{R}^A$ (this is an easy exercise) and hence $\sigma(A)$ is dense in $\pi_A(X)$. A space which contains a dense connected subspace is itself connected by Fact 1, so $\pi_A(X)$ is connected which is a contradiction. Therefore X not zero-dimensional because it is connected (see Fact 1). We finally proved that X is a connected (and hence nonzero-dimensional) space which is a countable union of its closed discrete (and hence strongly zero-dimensional) subspaces, so our solution is complete.

T.313. *Prove that the space $\mathbb{P}$ of the irrationals is homeomorphic to ω^ω and hence $\mathbb{P}$ is zero-dimensional.*

Solution. Given a function f the set $\mathrm{dom}(f)$ is its domain; if f and g are functions, then $f \subset g$ says that $\mathrm{dom}(f) \subset \mathrm{dom}(g)$ and $g|\mathrm{dom}(f) = f$. Note that if f is an empty function (in this case we write $f = \emptyset$), then $f \subset g$ for any function g.

Now suppose that we have a collection of functions $\{f_i : i \in I\}$ such that $f_i|(\mathrm{dom}(f_i) \cap \mathrm{dom}(f_j)) = f_j|(\mathrm{dom}(f_i) \cap \mathrm{dom}(f_j))$ for any $i, j \in I$. Then we can define a function f with $\mathrm{dom}(f) = \bigcup_{i\in I} \mathrm{dom}(f_i)$ as follows: given any $x \in \mathrm{dom}(f)$, find any $i \in I$ with $x \in \mathrm{dom}(f_i)$ and let $f(x) = f_i(x)$. It is easy to check that the value of f at x does not depend on the choice of i, so we have consistently defined a function f which will be denoted by $\bigcup\{f_i : i \in I\}$ (this makes sense if we identify each function with its graph).

Let $\omega^0 = \{\emptyset\}$ and denote by $\omega^{<\omega}$ the set $\bigcup\{\omega^n : n \in \omega\}$. We will also need the set $\omega^{\leq n} = \bigcup\{\omega^k : 0 \leq k \leq n\}$ for every $n \in \omega$. Given any $s \in \omega^{<\omega}$, the set $[s] = \{f \in \omega^\omega : f|n = s\}$ is open in ω^ω and the family $\{[f|n] : n \in \omega\}$ is a local base at f in ω^ω. Suppose that $k \in \omega$ and $s \in \omega^k$; for any $n \in \omega$, define the function $t = s^\frown n$ by $t|k = s$ and $t(k) = n$.

If (X, ρ) is a metric space, call $\mathcal{U} \subset \tau^*(X)$ a $\omega^{<\omega}$-*directed family* if

(1) $\mathcal{U} = \{U_s : s \in \omega^{<\omega}\}$;
(2) the family $\mathcal{U}(n) = \{U_s : s \in \omega^n\}$ is disjoint for any $n \in \omega$;
(3) for any $s, t \in \omega^{<\omega}$ if $s \subset t$ and $s \neq t$, then $\overline{U}_t \subset U_s$;
(4) $\mathrm{diam}(U_s) \leq \frac{1}{n}$ for any $s \in \omega^n$ and $n \in \mathbb{N}$.

Fact 1. If (X, ρ) is a complete metric space and $\mathcal{U}$ is a $\omega^{<\omega}$-directed family in (X, ρ), then $L(\mathcal{U}) = \bigcap\{\bigcup\mathcal{U}(n) : n \in \omega\}$ is homeomorphic to ω^ω.

Proof. For any $f \in \omega^\omega$ consider the sequence $\{U_{f|n} : n \in \omega\}$. It follows from (4) that $\mathrm{diam}(U_{f|n}) \to 0$ and hence $Q_f = \bigcap\{U_{f|n} : n \in \omega\} = \bigcap\{\overline{U}_{f|n} : n \in \omega\} \neq \emptyset$ (we applied (3) and Problem 236 of [TFS]). It follows from (4) that there is a point $x_f \in X$ such that $Q_f = \{x_f\}$; let $\varphi(f) = x_f$ for any $f \in \omega^\omega$.

It is immediate from the definition of φ that $\varphi(\omega^\omega) \subset L(\mathcal{U})$. If $x \in L(\mathcal{U})$, then for any $n \in \omega$ we have $x \in \bigcup\mathcal{U}(n)$ and hence there is $s_n \in \omega^n$ such that $x \in U_{s_n}$; it follows from (2) that this s_n is unique for every $n \in \omega$. Furthermore, if $n < m$, then $x \in U_{s_m|n} \cap U_{s_n}$ [see (3)] and hence $s_m|n = s_n$ by (2). Thus the function $f = \bigcup\{s_n : n \in \omega\} \in \omega^\omega$ is well-defined and $s_n = f|n$ for all $n \in \omega$ which shows that $x = x_f$. As a consequence, $L(\mathcal{U}) \subset \varphi(\omega^\omega)$ and hence $L(\mathcal{U}) = \varphi(\omega^\omega)$.

To see that the mapping φ is continuous, take any $f \in \omega^\omega$ and $\varepsilon > 0$. Choose $n \in \mathbb{N}$ such that $\frac{1}{n} < \varepsilon$ and observe that $\varphi([f|n]) \subset U_{f|n} \subset B_\rho(\varphi(f), \varepsilon)$ because $\mathrm{diam}(U_{f|n}) \leq \frac{1}{n} < \varepsilon$ by (4). Thus the open neighborhood $[f|n]$ of the point f in ω^ω witnesses continuity of φ at the point f.

Now, if f, g are distinct points of ω^ω, then there is $n \in \omega$ such that $f \mid n \neq g \mid n$; by the definition of φ, we have $\varphi(f) \in U_{f\mid n}$ and $\varphi(g) \in U_{g\mid n}$. Since $U_{f\mid n} \cap U_{g\mid n} = \emptyset$ by (2), we have $\varphi(f) \neq \varphi(g)$, i.e., φ is a bijection. To finally see that φ^{-1} is continuous take any $x \in L(\mathcal{U})$ and let $f = \varphi^{-1}(x)$. Given any $O \in \tau(f, \omega^\omega)$ there is $n \in \omega$ such that $[f \mid n] \subset O$. Furthermore, $V = U_{f\mid n} \cap L(\mathcal{U}) \in \tau(x, L(\mathcal{U}))$; if $y \in V$ and $g = \varphi^{-1}(y)$, then $y = \varphi(g)$ and hence $y \in U_{g\mid n} \cap U_{f\mid n}$. Applying property (2) again we conclude that $f \mid n = g \mid n$ and hence $g \in [f \mid n] \subset O$. The point $y \in V$ was chosen arbitrarily, so we proved that $\varphi^{-1}(V) \subset [f \mid n] \subset O$, i.e., V witnesses continuity of φ^{-1} at the point x. Thus φ is a homeomorphism and Fact 1 is proved.

□

Returning to our solution, take a faithful enumeration $\{O_n : n \in \omega\}$ of the family $\{(n, n+1) : n \in \omega\} \cup \{(-n-1, -n) : n \in \omega\}$; it is easy to choose an enumeration $\{q_n : n \in \omega\}$ of the set $\mathbb{Q}$ such that $q_0 = 0$. Our first step is to define $U_\emptyset = \mathbb{R}$; for any $f \in \omega^1$, let $U_f = O_{f(n)}$. Assume that $n \in \mathbb{N}$ and we have defined a family $\{U_s : s \in \omega^{\leq n}\}$ such that

(5) the set U_s is a non-empty open interval with rational endpoints for any $s \neq \emptyset$;
(6) the family $\mathcal{U}(k) = \{U_s : s \in \omega^k\}$ is disjoint and $\{q_0, \ldots, q_k\} \subset \mathbb{R} \backslash (\bigcup \mathcal{U}(k)) \subset \mathbb{Q}$ whenever $1 \leq k \leq n$;
(7) for any $s, t \in \omega^{\leq n}$ if $s \subset t$ and $s \neq t$ then $\overline{U}_t \subset U_s$;
(8) if $1 \leq k \leq n$, then $\mathrm{diam}(U_s) \leq \frac{1}{k}$ for any $s \in \omega^k$.

To construct the family $\mathcal{U}(n+1)$, it suffices to define $U_{s \frown k}$ for all $s \in \omega^n$ and $k \in \omega$. Fix any $s \in \omega^n$; by the induction hypothesis, $U_s = (a, b)$ where $a, b \in \mathbb{Q}$. Thus $b_0 = \frac{a+b}{2} \in \mathbb{Q}$ and the diameters of both intervals (a, b_0) and (b_0, b) do not exceed $\frac{1}{2} \cdot \frac{1}{n} \leq \frac{1}{n+1}$. Choose a strictly decreasing sequence $\{a_n : n \in \mathbb{N}\} \subset \mathbb{Q} \cap (a, b_0)$ and a strictly increasing sequence $\{b_n : n \in \mathbb{N}\} \subset \mathbb{Q} \cap (b_0, b)$ such that $a_n \to a$ and $b_n \to b$. Let $a_0 = b_0$ and observe that if $q_{n+1} \in U_s$, then we can choose our sequences in such a way that q_{n+1} be listed either in $A_s = \{a_n : n \in \omega\}$ or in $B_s = \{b_n : n \in \omega\}$. Choose a faithful enumeration $\{W_n : n \in \omega\}$ of the family $\{(a_{n+1}, a_n) : n \in \omega\} \cup \{(b_n, b_{n+1}) : n \in \omega\}$ and let $U_{s \frown n} = W_n$ for each $n \in \omega$.

After defining $U_{s \frown k}$ for all $k \in \omega$ and $s \in \omega^n$, we obtain a family $\{U_s : s \in \omega^{n+1}\}$. Observe that $U_s \backslash (\bigcup \{W_n : n \in \omega\}) = A_s \cup B_s \subset \mathbb{Q}$ for all $s \in \omega^n$; besides, if $q_{n+1} \in U_s$, then $q_{n+1} \in A_s \cup B_s$.

This guarantees that (6) is fulfilled for the family $\{U_s : s \in \omega^{\leq n+1}\}$; it is evident that (5), (7) and (8) are also satisfied, so we can continue our inductive construction to obtain a family $\mathcal{U} = \{U_s : s \in \omega^{<\omega}\}$ with the properties (5)–(8). It is evident that $\mathcal{U}$ is $\omega^{<\omega}$-directed and hence $L(\mathcal{U})$ is homeomorphic to ω^ω by Fact 1. It easily follows from (6) that $L(\mathcal{U}) = \mathbb{P}$ which proves that $\mathbb{P}$ is homeomorphic to ω^ω. Finally, observe that the discrete space ω is zero-dimensional and hence so is ω^ω by Problem 302. This shows that $\mathbb{P}$ is also zero-dimensional being homeomorphic to a zero-dimensional space ω^ω and completes our solution.

T.314. *Let X be a paracompact space. Prove that X is strongly zero-dimensional if and only if every open cover of X has a disjoint open refinement.*

Solution. For sufficiency observe that it follows from the definition of a strongly zero-dimensional space that if any cover of X has a disjoint open refinement, then X is strongly zero-dimensional (paracompactness of X is not needed here). Now assume that X is paracompact and strongly zero-dimensional and let $C(X)$ be the family of all clopen subsets of X.

If $\mathcal{U}$ is an open cover of X, then there exists a locally finite refinement $\mathcal{V}$ of the cover $\mathcal{U}$. Let us index $\mathcal{V}$ by some well-ordered set $(S, <)$, i.e., consider that $\mathcal{V} = \{V_s : s \in S\}$. There exists a closed cover $\{F_s : s \in S\}$ of the space X such that $F_s \subset V_s$ for every $s \in S$ (see Fact 2 of S.226). Since $\text{Ind}X = 0$ by 308, there exists $O_s \in C(X)$ such that $F_s \subset O_s \subset V_s$ for all $s \in S$. It is evident that the family $\mathcal{O} = \{O_s : s \in S\}$ is a locally finite refinement of $\mathcal{U}$. Now let $W_s = O_s \backslash (\bigcup \{O_t : t < s\})$ for every $s \in S$. The family $\mathcal{O}$ is closure-preserving, so $\bigcup \mathcal{O}'$ is closed (and, evidently, open) for any $\mathcal{O}' \subset \mathcal{O}$. This proves that W_s is a clopen set for any $s \in S$. It is immediate that the family $\mathcal{W} = \{W_s : s \in S\}$ is a disjoint open refinement of $\mathcal{U}$ so we settled necessity.

T.315. *Let P be a strongly zero-dimensional paracompact space and suppose that M is a completely metrizable space. Denote by $CL(M)$ the set of all closed non-empty subsets of M and let $\varphi : P \to CL(M)$ be a lower semicontinuous map. Prove that φ has a continuous selection, i.e., there exists a continuous map $f : P \to M$ such that $f(x) \in \varphi(x)$ for any $x \in P$.*

Solution. Choose a complete metric d on the space M such that $d(x, y) \leq 1$ for any $x, y \in M$ (such a choice is possible by Problem 206 of [TFS]). For any set $A \subset M$, let $\varphi^{-1}(A) = \varphi_l^{-1}(A) = \{x \in P : \varphi(x) \cap A \neq \emptyset\}$. Then $\varphi^{-1}(U)$ is open in P for any $U \in \tau(M)$. For each fixed non-empty $A \subset M$, define the map $d_A : M \to \mathbb{R}$ by $d_A(x) = \inf\{d(x, y) : y \in A\}$; then, for any closed set $A \subset M$, we have $x \in A$ if and only if $d_A(x) = 0$. If we are given functions $f, g \in C(P, M) = C^*(P, M)$, let $\sigma(f, g) = \sup\{d(f(x), g(x)) : x \in P\}$. Then σ is a complete metric on $C(P, M)$ (see Problem 248 of [TFS]).

Take a point $u_0 \in M$ and let $f_0(x) = u_0$ for all $x \in P$. The function $f_0 : P \to M$ is constant and hence continuous. Now assume that $k \in \omega$ and we have defined continuous functions $f_0, \ldots, f_k$ from P to M with the following properties:

(1) $f_n^{-1}(u)$ is a clopen (maybe empty) subset of P for any $n \leq k$ and $u \in M$;
(2) $\sigma(f_n, f_{n+1}) \leq \frac{1}{2^n}$ for any $n < k$;
(3) for all $n \leq k$ and $x \in P$ there is point $y \in \varphi(x)$ such that $d(f_n(x), y)) \leq \frac{1}{2^n}$.

Apply the property (3) to choose, for any $x \in P$, a point $y_x \in \varphi(x)$ such that $d(f_k(x), y_x) \leq \frac{1}{2^k}$. Then $O_x = \varphi^{-1}(B_d(y_x, \frac{1}{2^{k+1}})) \in \tau(x, P)$ for any $x \in P$.

Fix any $w \in M$; the family $\mathcal{V}_w = \{O_x \cap f_k^{-1}(w) : x \in f_k^{-1}(w)\}$ is an open cover of the set $f_k^{-1}(w)$ which is clopen in P and hence paracompact and strongly zero-dimensional (see Problem 310). Apply Problem 314 to find a disjoint open refinement $\mathcal{U}_w$ of the cover $\mathcal{V}_w$. For any $U \in \mathcal{U}_w$, there is $x(U) \in f_k^{-1}(w)$ such that $U \subset O_{x(U)}$; let $f_{k+1}^w(x) = y_{x(U)}$ for any $x \in f_k^{-1}(w)$. Observe that

$(*)$ $d(f_{k+1}^w(x), f_k(x)) = d(f_{k+1}^w(x), w) = d(y_{x(U)}, w) = d(y_{x(U)}, f_k(x(U))) \leq$
$\frac{1}{2^k}$

for any $x \in f_k^{-1}(w)$. After we have f_{k+1}^w for all $w \in M$ (observe that nothing has to be done if $f_k^{-1}(w) = \emptyset$), we can define a map $f_{k+1} : P \to M$ as follows: for any $x \in P$ let $w = f_k(x)$ and $f_{k+1}(x) = f_{k+1}^w(x)$. It is immediate that the function $f_{k+1}|U$ is constant for any $U \in \bigcup\{\mathcal{U}_w : w \in M\}$; therefore f_{k+1} is continuous by Fact 1 of S.472 and it is easy to see that $f_{k+1}^{-1}(u)$ is open for any $u \in M$. Consequently, (1) holds for $n = k + 1$. Furthermore, it follows from $(*)$ that $d(f_{k+1}(x), f_k(x)) \leq \frac{1}{2^k}$ for all $x \in P$ and hence (2) is fulfilled for $k = n$. Given any $x \in P$ let $w = f_k(x)$; then there is $U \in \mathcal{U}_w$ such that $x \in U \subset O_{x(U)}$ and hence $E = \varphi(x) \cap B_d(y_{x(U)}, \frac{1}{2^{k+1}}) \neq \emptyset$. If $y \in E$, then $d(f_{k+1}(x), y) = d(y_{x(U)}, y) < \frac{1}{2^{k+1}}$, so (3) is fulfilled as well for $n = k + 1$.

As a consequence, our inductive construction can be continued to provide us a sequence $S = \{f_n : n \in \omega\}$ for which (1)–(3) are satisfied for all $k \in \omega$. It is an easy exercise to check that (2) implies that S is a Cauchy sequence in the space $(C(P, M), \sigma)$ and hence there is a function $f \in C(P, M)$ such that, for all $x \in P$, we have $f(x) = \lim f_n(x)$ when $n \to \infty$.

Fix any $x \in P$ and let $A = \varphi(x)$. It follows from (3) that $d_A(f_n(x)) \leq \frac{1}{2^n}$ for all $n \in \omega$; the function d_A being continuous by Fact 1 of S.212, we have $d_A(f(x)) = 0$ and therefore $f(x) \in A = \varphi(x)$. Thus f is as promised and our solution is complete.

T.316. *Let M be a strongly zero-dimensional completely metrizable space. Prove that any closed non-empty $F \subset M$ is a closed retract of M, i.e., there exists a closed continuous map $f : M \to F$ such that $f(x) = x$ for any $x \in F$.*

Solution. It is easy to prove that there exists *some* retraction of M onto F. To see it denote by $CL(F)$ the family of all non-empty closed subsets of F; define a map $\varphi : M \to CL(F)$ as follows: $\varphi(x) = \{x\}$ if $x \in F$ and $\varphi(x) = F$ for any $x \in M \backslash F$. The map φ is lower semicontinuous for if $U \in \tau^*(F)$, then $\varphi_l^{-1}(U) = U \cup (M \backslash F)$ is an open subset of M. The space M is paracompact by Problem 218 of [TFS] and F is completely metrizable, so we can apply Problem 315 to conclude that there is a continuous $g : M \to F$ such that $g(x) \in \varphi(x)$ for all $x \in X$. In particular, $g(x) \in \varphi(x) = \{x\}$ and hence $g(x) = x$ for any $x \in F$, i.e., g is a retraction of M onto F. However, it takes a much harder proof to actually find a *closed retraction* of M onto F.

Fact 1. If Y is a sequential space, then a continuous onto map $f : Z \to Y$ is closed if and only if $f(D)$ is not a nontrivial convergent sequence for any closed discrete $D \subset Z$.

Proof. If f is closed and $D \subset Z$ is closed and discrete, then $f(D)$ is also closed and discrete, so it cannot be a nontrivial convergent sequence.

To prove sufficiency, assume that for any closed discrete $D \subset Z$ the set $f(D)$ is not a nontrivial convergent sequence. If f is not closed, then there is a closed $A \subset Z$ such that the set $f(A)$ is not closed in the space Y and hence there is a sequence $S = \{y_n : n \in \omega\} \subset f(A)$ such that $y_n \to y$ for some $y \in Y \setminus f(A)$. Take $z_n \in f^{-1}(y_n) \cap A$ for every $n \in \omega$; then the set $D = \{z_n : n \in \omega\}$ is closed and discrete in Z while $f(D) = S$ is a nontrivial convergent sequence. This contradiction shows that f is a closed map and hence Fact 1 is proved.   □

Returning to our solution, fix a complete metric d on the space M such that $\tau(d) = \tau(M)$ and $d(x, y) \leq 1$ for all $x, y \in M$. Given functions $p, q \in C(M, F)$, let $\sigma(p, q) = \sup\{d(p(x), q(x)) : x \in M\}$. The function σ is a complete metric on $C(M, F)$ by Problem 248 of [TFS] (observe that $C(M, F) = C^*(M, F)$ because $(F, d|(F \times F))$ is a bounded metric space).

Choose $x_0 \in F$; let $\mathcal{U}_0 = \{M\}$ and $f_0(x) = x_0$ for all $x \in M$. Assume that $k \in \omega$ and we have chosen maps $f_0, \ldots, f_k$ and disjoint open covers $\mathcal{U}_0, \ldots, \mathcal{U}_k$ of the space M with the following properties:

(1) $f_i : M \to F$ for all $i \leq k$;
(2) for any $i \leq k$ and $U \in \mathcal{U}_i$, there is $y_U^i \in F$ such that $f_i(x) = y_U^i$ for all $x \in U$;
(3) $\mathcal{U}_{i+1}$ is a refinement of $\mathcal{U}_i$ for any $i < k$;
(4) For any $i \leq k$, if $U \in \mathcal{U}_i$ and $U \cap F \neq \emptyset$, then $y_U^i \in U \cap F$.
(5) for any $i < k$, if $U \in \mathcal{U}_{i+1}$ and $U \cap F = \emptyset$, then $y_U^{i+1} = y_V^i$ where V is the unique element of $\mathcal{U}_i$ for which $U \subset V$;
(6) $\mathrm{diam}(U) \leq \frac{1}{2^i}$ for all $U \in \mathcal{U}_i$ and $i \leq k$.

Observe that the property (2) implies that each f_i is continuous (see Fact 1 of S.472) and fix any $U \in \mathcal{U}_k$; since M is strongly zero-dimensional, there is a disjoint open cover $\mathcal{V}_U$ of the space U such that $\mathrm{diam}(V) \leq \frac{1}{2^{k+1}}$ for every $V \in \mathcal{V}_U$. If $V \cap F = \emptyset$, then let $f_{k+1}(x) = y_U^k$ for all $x \in V$; if $V \cap F \neq \emptyset$, then choose any $y_V^{k+1} \in V \cap F$ and let $f_{k+1}(x) = y_V^{k+1}$ for all $x \in V$. This defines $f_{k+1}(x)$ for all $x \in U$; the set U was chosen arbitrarily, so we can define $f_{k+1}(x)$ for all $x \in M$. Letting $\mathcal{U}_{k+1} = \bigcup\{\mathcal{V}_U : U \in \mathcal{U}_k\}$, we obtain the desired disjoint open cover of the space M. It is evident that the properties (1)–(6) still hold for the maps $f_0, \ldots, f_{k+1}$ and covers $\mathcal{U}_0, \ldots, \mathcal{U}_{k+1}$. Thus our inductive construction can be continued to obtain sequences $\{f_i : i \in \omega\}$ and $\{\mathcal{U}_i : i \in \omega\}$ for which (1)–(6) are fulfilled for all $k \in \omega$.

Let us prove by induction on i that

(7) the set $f_i(M)$ is closed and discrete (in F and hence in M) for all $i \in \omega$.

This is true for $i = 0$ because $f(M) = \{x_0\}$. Now if $n < \omega$ and we proved that $f_n(M)$ is closed and discrete, then consider the family $\mathcal{V} = \{V \in \mathcal{U}_{n+1} : V \cap F \neq \emptyset\}$ and let $O = \bigcup \mathcal{V}$. It is clear that the set $f_{n+1}(O) = \{y_V^{n+1} : V \in \mathcal{V}\}$ is closed and discrete because the family $\mathcal{V}$ is discrete and $y_V^{n+1} \in V$ for all $V \in \mathcal{V}$ by (4). The property (5) implies that $f_{n+1}(M \setminus O) \subset f_n(M \setminus O)$; since $f_n(M \setminus O)$ is closed and discrete by the induction hypothesis, the set $f_{n+1}(M \setminus O)$ is closed and discrete as well. Thus $f_{n+1}(M) = f_{n+1}(O) \cup f_{n+1}(M \setminus O)$ is also closed and discrete so (7) is proved. Next, let us show that

(8) $\sigma(f_i, f_{i+1}) \leq \frac{1}{2^i}$ for all $i \in \omega$.

Take any $x \in M$; there are $U \in \mathcal{U}_i$ and $V \in \mathcal{U}_{i+1}$ such that $x \in U \cap V$. It follows from (3) that $V \subset U$. If $V \cap F \neq \emptyset$, then $f_{i+1}(x) = y_V^{i+1} \in V \cap F$ by (4) and hence $f_{i+1}(x) \in U$. If $V \cap F = \emptyset$, then $f_{i+1}(x) = y_U^i = f_i(x)$. As a consequence, in the second case, we have $d(f_i(x), f_{i+1}(x)) = 0$ while in the first one $y_U^i \in U \cap F$ because $U \cap F \neq \emptyset$ and hence $d(f_i(x), f_{i+1}(x)) = d(y_V^{i+1}, y_U^i) \leq \mathrm{diam}(U) \leq \frac{1}{2^i}$. Thus $d(f_i(x), f_{i+1}(x)) \leq \frac{1}{2^i}$ for all $x \in M$ and hence (8) is proved.

It is an easy consequence of (8) that the sequence $T = \{f_n : n \in \omega\} \subset C(M, F)$ is fundamental in the complete metric space $(C(M, F), \sigma)$. Therefore there exists a function $f \in C(M, F)$ such that f_n converges to f in $(C(M, F), \sigma)$ which is equivalent to uniform convergence of T to f. It is an easy exercise to check, using the property (8) that

(9) $d(f_n(x), f(x)) \leq \frac{1}{2^{n-1}}$ for all $x \in M$ and $n \in \mathbb{N}$.

If $x \in F$, then let U_i be the unique element of $\mathcal{U}_i$ which contains x; then $U_i \cap F \neq \emptyset$ for all $i \in \omega$ and hence $f_i(x) = y_{U_i}^i \in U_i$ by the property (4). This implies $d(x, f_i(x)) \leq \mathrm{diam}(U_i) \leq \frac{1}{2^i}$ and hence the sequence $\{f_i(x) : i \in \omega\}$ converges to x. Therefore $f(x) = \lim_{i \to \infty} f_i(x) = x$ for any $x \in F$ and hence f is a retraction of the space M onto F.

To see that f is a closed map assume that $D = \{d_n : n \in \omega\}$ is a closed discrete subset of M such that $f(D)$ is a nontrivial convergent sequence. We can assume that $d_i \neq d_j$ if $i \neq j$. If $D \cap F$ is infinite, then it is closed and discrete and hence $f(D)$ contains a closed discrete subspace $f(D \cap F) = D \cap F$ of the space F which is a contradiction with the fact that $f(D)$ is a nontrivial convergent sequence. Thus $D \cap F$ is finite, so if necessary, we can throw away finitely many point of D to guarantee that $D \cap F = \emptyset$. Let $z \in F$ be the limit of $f(D)$; since $K = f(D) \cup \{z\}$ is a compact subspace disjoint from D,

(10) there is $\varepsilon > 0$ such that $d(x, y) \geq \varepsilon$ for any $x \in K$ and $y \in D$.

Pick $n \in \omega$ with $\frac{1}{2^n} < \frac{\varepsilon}{4}$ and take any $x \in D$; there is $U \in \mathcal{U}_n$ with $x \in U$. If $U \cap F \neq \emptyset$, then $f_n(x) = y_U^n \in U$ and hence $d(x, f_n(x)) \leq \mathrm{diam}(U) \leq \frac{1}{2^n}$. It follows from (9) that $d(x, f(x)) \leq d(x, f_n(x)) + d(f_n(x), f(x)) \leq \frac{1}{2^{n-1}} + \frac{1}{2^n} < \frac{3\varepsilon}{4} < \varepsilon$ which is a contradiction with (10). As a consequence, $U \cap F = \emptyset$ whenever $U \in \mathcal{U}_n$ and $U \cap D \neq \emptyset$. If $i \geq n$ and $V \cap D \neq \emptyset$ for some $V \in \mathcal{U}_i$, then there is $U \in \mathcal{U}_n$ with $V \subset U$ and hence $V \cap F \subset U \cap F = \emptyset$. This proves that

(11) for any $i \geq n$ if $U \in \mathcal{U}_i$ and $U \cap D \neq \emptyset$, then $U \cap F = \emptyset$.

It is easy to prove by an evident induction applying (11) and the property (5) that $f_i(x) = f_n(x)$ for all $x \in D$ and $i \geq n$. Thus $f(x) = \lim_{k \to \infty} f_k(x) = f_n(x)$ for all $x \in D$. As a consequence, $f(D) = f_n(D)$ is a closed discrete subset of F by (7) which is a contradiction. Thus $f(D)$ cannot be a nontrivial convergent sequence and hence the map f is closed by Fact 1. Our solution is complete.

T.317. *Let F be a closed subspace of $\mathbb{P}$ and suppose that a space X is a continuous image of F. Prove that X is also a continuous image of $\mathbb{P}$.*

Solution. Let $g : F \to X$ be a continuous onto map. The space $\mathbb{P}$ is zero-dimensional by Problem 313; being second countable, it is also strongly zero-dimensional by Problem 306, so we can apply Problem 316 to conclude that there is a continuous onto map $f : \mathbb{P} \to F$ (which is a retraction but we don't need that). Then $g \circ f$ maps $\mathbb{P}$ continuously onto X.

T.318. *Prove that for any second countable space X and every countable ordinal ξ, there exists a set $U \in \Sigma_\xi^0(X \times \mathbb{K})$ such that $\Sigma_\xi^0(X) = \{U[y] : y \in \mathbb{K}\}$, where $U[y] = \{x \in X : (x, y) \in U\}$ for any $y \in \mathbb{K}$. Observe that, as an easy consequence, for any second countable space X and every countable ordinal ξ, there exists a set $V \in \Pi_\xi^0(X \times \mathbb{K})$ such that $\Pi_\xi^0(X) = \{V[y] : y \in \mathbb{K}\}$.*

Solution. We consider that $\mathbb{K} = \mathbb{D}^\omega$ (see Problem 128 of [TFS]). Given an ordinal $\alpha < \omega_1$ and a space Z, a set $U \subset Z \times \mathbb{K}$ will be called $\Sigma_\alpha^0(Z)$-*universal* if $U \in \Sigma_\alpha^0(Z \times \mathbb{K})$ and $\Sigma_\alpha^0(Z) = \{U[y] : y \in \mathbb{K}\}$. Analogously, a set $V \subset Z \times \mathbb{K}$ is called $\Pi_\alpha^0(Z)$-*universal* if $V \in \Pi_\alpha^0(Z \times \mathbb{K})$ and $\Pi_\alpha^0(Z) = \{U[y] : y \in \mathbb{K}\}$.

Fact 1. Given spaces Z and Y let $f : Z \to Y$ be a continuous map. Then, for any $\alpha < \omega_1$ and $A \in \Sigma_\alpha^0(Y)$ (or $A \in \Pi_\alpha^0(Y)$), we have $f^{-1}(A) \in \Sigma_\alpha^0(Z)$ (or $f^{-1}(A) \in \Pi_\alpha^0(Z)$ respectively).

Proof. Since f is continuous, $f^{-1}(U)$ is open (closed) in Z for any open (closed) $U \subset Y$. In terms of Borel subsets this means that $A \in \Sigma_0^0(Y)$ (or $A \in \Pi_0^0(Y)$) implies $f^{-1}(A) \in \Sigma_0^0(Z)$ (or $f^{-1}(A) \in \Pi_0^0(Z)$ respectively). Suppose that $\alpha < \omega_1$ and, for any $\beta < \alpha$, we proved that $f^{-1}(A) \in \Sigma_\beta^0(Z)$ ($f^{-1}(A) \in \Pi_\beta^0(Z)$) for any $A \in \Sigma_\beta^0(Y)$ (or $A \in \Pi_\beta^0(Y)$ respectively).

Now, if $A \in \Sigma_\alpha^0(Y)$, then there is a sequence $\{\beta_n : n \in \omega\} \subset \alpha$ and $A_n \in \Pi_{\beta_n}^0(Y)$ for every $n \in \omega$ such that $A = \bigcup\{A_n : n \in \omega\}$. Thus $f^{-1}(A_n) \in \Pi_{\beta_n}^0(Z)$ for all $n \in \omega$ by the induction hypothesis so $f^{-1}(A) = \bigcup\{f^{-1}(A_n) : n \in \omega\} \in \Sigma_\alpha^0(Z)$. Thus $f^{-1}(A) \in \Sigma_\alpha^0(Z)$ for any set $A \in \Sigma_\alpha^0(Y)$. Observe finally that if $A \in \Pi_\alpha^0(Y)$, then $B = Y \backslash A \in \Sigma_\alpha^0(Y)$ which implies $f^{-1}(B) \in \Sigma_\alpha^0(Z)$. As a consequence, $f^{-1}(A) = Z \backslash f^{-1}(B) \in \Pi_\alpha^0(Z)$ and hence our inductive proof can be carried out for all $\alpha < \omega_1$. Fact 1 is proved. $\square$

Fact 2. Suppose that Z is a space and $Y \subset Z$. If α is a countable ordinal and $A \in \Sigma_\alpha^0(Z)$ ($A \in \Pi_\alpha^0(Z)$), then $A \cap Y \in \Sigma_\alpha^0(Y)$ (or $A \cap Y \in \Pi_\alpha^0(Y)$ respectively).

Proof. If A is open (closed) in Z, then $A \cap Y$ is open (closed) in Y, so our fact is evident for $\alpha = 0$. Assume that $\alpha < \omega_1$ and we proved our fact for all $\beta < \alpha$. If $A \in \Sigma_\alpha^0(Z)$, then there is a sequence $\{\beta_n : n \in \omega\} \subset \alpha$ and $A_n \in \Pi_{\beta_n}^0(Z)$ for all $n \in \omega$ such that $A = \bigcup\{A_n : n \in \omega\}$. By the induction hypothesis, we have $B_n = A_n \cap Y \in \Pi_{\beta_n}^0(Y)$ for all $n \in \omega$ and therefore $A \cap Y = \bigcup\{B_n : n \in \omega\} \in \Sigma_\alpha^0(Y)$.

Now, if $B \in \Pi_\alpha^0(Z)$, then $A = Z \backslash B \in \Sigma_\alpha^0(Z)$ and hence $C = A \cap Y \in \Sigma_\alpha^0(Y)$. Consequently, $B \cap Y = Y \backslash C \in \Pi_\alpha^0(Y)$ and hence our fact holds for all $\beta \leq \alpha$. Thus our inductive procedure shows that our fact is true for all $\alpha < \omega_1$ and hence Fact 2 is proved. $\square$

Returning to our solution, suppose that for some ordinal $\alpha < \omega_1$, there exists a $\Sigma^0_\alpha(X)$-universal set $U \subset X \times \mathbb{K}$. Then, for the set $V = (X \times \mathbb{K})\backslash U$, we have $V \in \Pi^0_\alpha(X \times \mathbb{K})$ and $V[y] = X\backslash U[y]$ for any $y \in \mathbb{K}$. Therefore

$$\{V[y] : y \in \mathbb{K}\} = \{X\backslash U[y] : y \in \mathbb{K}\} = \{X\backslash W : W \in \Sigma^0_\alpha(X)\} = \Pi^0_\alpha(X),$$

i.e., the set V is $\Pi^0_\alpha(X)$-universal. Consequently,

(1) if $\alpha < \omega_1$ and there is a $\Sigma^0_\alpha(X)$-universal set, then there exists a $\Pi^0_\alpha(X)$-universal set as well.

Thus it suffices to construct a $\Sigma^0_\xi(X)$-universal set for all $\xi < \omega_1$. We will do this by induction on ξ.

Fix a base $\mathcal{B} = \{B_n : n \in \omega\}$ in the space X; let $O_n = \{y \in \mathbb{K} : y(n) = 1\}$ for all $n \in \omega$. Then $O_n \in \tau(\mathbb{K})$ and hence the set $U_n = B_n \times O_n$ is open in the space $X \times \mathbb{K}$ for every $n \in \omega$. As a consequence, $U = \bigcup\{U_n : n \in \omega\} \in \Sigma^0_0(X \times \mathbb{K})$. For any point $y \in \mathbb{K}$ the natural projection $\pi : X \times \mathbb{K} \rightarrow X$ maps $X \times \{y\}$ homeomorphically onto X and hence $U[y] = \pi(U \cap (X \times \{y\}))$ is an open subset of X for any $y \in Y$. Thus $\{U[y] : y \in \mathbb{K}\} \subset \Sigma^0_0(X)$.

Now, if $W \in \tau(X)$, then let $A = \{n \in \omega : B_n \subset W\}$. Since $\mathcal{B}$ is a base of X, we have $W = \bigcup\{B_n : n \in A\}$; let $y \in \mathbb{D}^\omega$ be defined as follows: $y(n) = 1$ for all $n \in A$ and $y(n) = 0$ whenever $n \in \omega\backslash A$. Observe that if $x \in X$, then $(x, y) \in U$ if and only if $(x, y) \in U_n$ for some $n \in \omega$ which takes place if and only if $n \in A$ and $x \in B_n$. Therefore $(x, y) \in U$ if and only if $x \in \bigcup\{B_n : n \in A\} = W$ which shows that $W = U[y]$ and hence $\{U[y] : y \in \mathbb{K}\} = \Sigma^0_0(X)$, i.e., the set U is $\Sigma^0_0(X)$-universal.

Now assume that $\xi < \omega_1$ and there exists a $\Sigma^0_\beta(X)$-universal set for each $\beta < \xi$. Thus there also exists a $\Pi^0_\beta(X)$-universal set for each $\beta < \xi$ by (1). Let K_i be a homeomorphic copy of the space $\mathbb{K}$ for all $i \in \omega$ and choose an enumeration $\{\alpha_n : n \in \omega\}$ of the set $\xi = \{\alpha : \alpha < \xi\}$ such that each $\alpha < \xi$ occurs infinitely many times in this enumeration.

There is a set $U_n \subset X \times K_n$ which is $\Pi^0_{\alpha_n}(X)$-universal for all $n \in \omega$. In the space $T = X \times \prod\{K_n : n \in \omega\}$ let $\pi_n : T \rightarrow X \times K_n$ be the natural projection for every $n \in \omega$. Then $\pi_n^{-1}(U_n) \in \Pi^0_{\alpha_n}(T)$ for all $n \in \omega$ by Fact 1 and therefore $U = \bigcup\{\pi_n^{-1}(U_n) : n \in \omega\} \in \Sigma^0_\xi(T) = \Sigma^0_\xi(X \times K)$ where $K = \prod\{K_n : n \in \omega\}$; it is evident that K is homeomorphic to $\mathbb{K}$. Let $\pi : X \times K \rightarrow X$ be the natural projection. Given any $y \in K$ we have $U[y] = \pi(U \cap (X \times \{y\}))$ and $\pi|(X \times \{y\})$ is a homeomorphism. Since $U \cap (X \times \{y\}) \in \Sigma^0_\xi(X \times \{y\})$ by Fact 2, we have $U[y] \in \Sigma^0_\xi(X)$ for any $y \in K$. Thus $\{U[y] : y \in K\} \subset \Sigma^0_\xi(X)$.

To prove the inverse inclusion take any $W \in \Sigma^0_\xi(X)$. There exists a sequence $\{\beta_n : n \in \omega\} \subset \xi$ and $W_n \in \Pi^0_{\beta_n}(X)$ for all $n \in \omega$ such that $W = \bigcup\{W_n : n \in \omega\}$. It is easy to choose a sequence $S = \{m(n) : n \in \omega\} \subset \omega$ such that $\alpha_{m(n)} = \beta_n$ for all $n \in \omega$. Since $U_{m(n)}$ is $\Pi^0_{\beta_n}(X)$-universal, there is $y_{m(n)} \in K_{m(n)}$ such that $U_{m(n)}[y_{m(n)}] = W_n$ for all $n \in \omega$. Observe that the empty set belongs to $\Pi^0_\beta(X)$ for any $\beta < \omega_1$ and hence we can take, for every $n \in \omega\backslash S$, a point $y_n \in K_n$ such that

$U_n[y_n] = \emptyset$. Now, if we let $y(n) = y_n$ for every $n \in \omega$, then $y \in K$. Observe that $x \in U[y]$ if and only if $(x, y) \in \pi_n^{-1}(U_n)$ for some $n \in \omega$ which takes place if and only if $\pi_n(x, y) \in U_n$, or, in other words, $(x, y_n) \in U_n$, i.e., $x \in U_n[y_n]$. Since $U_n[y_n] \subset W$ for all $n \in \omega$, we have $U[y] \subset W$.

On the other hand, if $x \in W$, then $x \in W_n = U_{m(n)}[y_{m(n)}]$ for some $n \in \omega$ and therefore $(x, y_{m(n)}) \in U_{m(n)}$ which shows that $(x, y) \in \pi_{m(n)}^{-1}(U_{m(n)}) \subset U$. Thus $x \in U[y]$ for any $x \in W$ that $x \in U[y]$ and hence $W \subset U[y]$, i.e., $W = U[y]$. Thus $\{U[y] : y \in K\} = \Sigma_\xi^0(X)$, so identifying the spaces K and $\mathbb{K}$, we can conclude that U is $\Sigma_\xi^0(X)$-universal. Our inductive procedure shows that we proved the existence of a $\Sigma_\xi^0(X)$-universal set for any ordinal $\xi < \omega_1$ which demonstrates, together with (1), that our solution is complete.

T.319. *Prove that for any uncountable Polish space X and every countable ordinal ξ, the classes $\Sigma_\xi^0(X)$ and $\Pi_\xi^0(X)$ do not coincide.*

Solution. If Z is a space, $\mathcal{A} \subset \exp(Z)$ and $Y \subset Z$, then $\mathcal{A}|Y = \{A \cap Y : A \in \mathcal{A}\}$. Given an ordinal $\alpha < \omega_1$, a set $U \subset Z \times \mathbb{K}$ will be called $\Sigma_\alpha^0(Z)$-*universal* if $U \in \Sigma_\alpha^0(Z \times \mathbb{K})$ and $\Sigma_\alpha^0(Z) = \{U[y] : y \in \mathbb{K}\}$. Here $U[y] = \{z \in Z : (z, y) \in U\}$ for any $y \in \mathbb{K}$.

Fact 1. Given a countable ordinal ξ, for an arbitrary space Z and any $Y \subset Z$, we have $\Sigma_\xi^0(Z)|Y = \Sigma_\xi^0(Y)$ and $\Pi_\xi^0(Z)|Y = \Pi_\xi^0(Y)$.

Proof. The inclusions $\Sigma_\xi^0(Z)|Y \subset \Sigma_\xi^0(Y)$ and $\Pi_\xi^0(Z)|Y \subset \Pi_\xi^0(Y)$ were proved in Fact 2 of T.318. It is an immediate consequence of the definition of subspace topology that $\tau(Z)|Y = \tau(Y)$, i.e., $\Sigma_0^0(Z)|Y = \Sigma_0^0(Y)$. Furthermore, if F is a closed subset of Y, then $\text{cl}_Z(F) \cap Y = F$ which shows that $\Pi_0^0(Z)|Y = \Pi_0^0(Y)$.

Now assume that $\xi < \omega_1$ and we proved the equalities $\Sigma_\alpha^0(Z)|Y = \Sigma_\alpha^0(Y)$ and $\Pi_\alpha^0(Z)|Y = \Pi_\alpha^0(Y)$ for all $\alpha < \xi$. Given any $U \in \Sigma_\xi^0(Y)$, there is a sequence $\{\alpha_n : n \in \omega\} \subset \xi$ and $U_n \in \Pi_{\alpha_n}^0(Y)$ for all $n \in \omega$ such that $U = \bigcup\{U_n : n \in \omega\}$. By the induction hypothesis, there is $V_n \in \Pi_{\alpha_n}^0(Z)$ such that $V_n \cap Y = U_n$ for all $n \in \omega$. Then $V = \bigcup\{V_n : n \in \omega\} \in \Sigma_\xi^0(Z)$ and $V \cap Y = U$ which shows that $\Sigma_\xi^0(Z)|Y = \Sigma_\xi^0(Y)$. Passing to complements, we obtain $\Pi_\xi^0(Z)|Y = \Pi_\xi^0(Y)$, so our inductive proof is valid for all $\xi < \omega_1$. Fact 1 is proved.   □

Returning to our solution observe that the space X is a continuous image of $\mathbb{P}$ by Fact 3 of T.132; since $\mathbb{P} = \bigcap\{\mathbb{R}\backslash\{q\} : q \in \mathbb{Q}\}$, the set $\mathbb{P}$ is G_δ in $\mathbb{R}$. Every open subset of $\mathbb{R}$ is a countable union of compact sets, so $\mathbb{P}$ is a $K_{\sigma\delta}$-space, and hence X is K-analytic. Therefore we can apply Fact 4 of T.250 to conclude that $\mathbb{K}$ embeds in X. If $\Sigma_\xi^0(X) = \Pi_\xi^0(X)$ for some $\xi < \omega_1$, then we can apply Fact 1 to conclude that

(1) $\Sigma_\xi^0(\mathbb{K}) = \Sigma_\xi^0(X)|\mathbb{K} = \Pi_\xi^0(X)|\mathbb{K} = \Pi_\xi^0(\mathbb{K})$.

Let $\Delta = \{(y, y) : y \in \mathbb{K}\}$ be the diagonal of the space $\mathbb{K}$. Now apply Problem 318 to find a set $U \subset \mathbb{K} \times \mathbb{K}$ which is $\Sigma_\xi^0(\mathbb{K})$-universal. Then $V = (\mathbb{K} \times \mathbb{K})\backslash U \in \Pi_\xi^0(\mathbb{K} \times \mathbb{K})$ and hence $V \cap \Delta \in \Pi_\xi^0(\Delta)$ by Fact 1. Let $\pi : \mathbb{K} \times \mathbb{K} \to \mathbb{K}$ be the natural projection onto the first factor. It is evident that $\pi|\Delta$ is a homeomorphism of Δ onto $\mathbb{K}$ and therefore $W = \pi(V \cap \Delta) \in \Pi^0_\xi(\mathbb{K})$. The property (1) shows that $W \in \Sigma^0_\xi(\mathbb{K})$ and, by universality of U, there is $y \in \mathbb{K}$ such that $U[y] = W$.

If $y \in W$, then by the definition of W, we have $(y, y) \notin U$. On the other hand, $y \in W = U[y]$ which shows that $(y, y) \in U$ which is a contradiction. Now, if $y \notin W$, then $(y, y) \notin V$ and hence $(y, y) \in U$. Therefore $y \in U[y] = W$; this final contradiction shows that the equality $\Sigma^0_\xi(\mathbb{K}) = \Pi^0_\xi(\mathbb{K})$ is impossible and hence the equality $\Sigma^0_\xi(X) = \Pi^0_\xi(X)$ is impossible as well.

T.320. *Let X be a second countable space. Given countable ordinals ξ and $\eta > \xi$, prove that $\Sigma^0_\xi(X) \cup \Pi^0_\xi(X) \subset \Delta^0_\eta(X)$. Show that if X is an uncountable Polish space, then $\Sigma^0_\xi(X) \cup \Pi^0_\xi(X) \neq \Delta^0_\eta(X)$.*

Solution. If Z is a space and $\mathcal{A} \subset \exp(Z)$, then $\mathcal{A}|Y = \{A \cap Y : A \in \mathcal{A}\}$ for every $Y \subset Z$. The expression $Z \simeq T$ says that the spaces Z and T are homeomorphic.

Fact 1. Let Z be a second countable space. Then, for any closed $Y \subset Z$, we have $\Sigma^0_\beta(Y) \subset \Sigma^0_\beta(Z)$ for all countable ordinals $\beta > 0$ and $\Pi^0_\beta(Y) \subset \Pi^0_\beta(Z)$ for all $\beta < \omega_1$. As a consequence, $\Delta^0_\beta(Z)|Y = \Delta^0_\beta(Y)$ for any countable ordinal $\beta > 0$.

Proof. Any closed subset of Y is closed in Z, so any F_σ-subset of Y is an F_σ-subset of Z as well. This proves that $\Pi^0_0(Y) \subset \Pi^0_0(Z)$ and $\Sigma^0_1(Y) \subset \Sigma^0_1(Z)$. Furthermore, if $A \in \Pi^0_1(Y)$, then $Y \backslash A$ is an F_σ-set in Y and hence in X. Any open set in Z is also an F_σ-set because Z is second countable. Thus $Z \backslash A = (Y \backslash A) \cup (Z \backslash Y)$ is an F_σ-set, i.e., $Z \backslash A \in \Sigma^0_1(Z)$. Consequently, $A \in \Pi^0_1(Z)$ and therefore $\Pi^0_1(Y) \subset \Pi^0_1(Z)$.

Assume that $1 < \beta < \omega_1$ and we proved that $\Pi^0_\alpha(Y) \subset \Pi^0_\alpha(Z)$ for all $\alpha < \beta$ and $\Sigma^0_\alpha(Y) \subset \Sigma^0_\alpha(Z)$ whenever $0 < \alpha < \beta$. Given any $A \in \Sigma^0_\beta(Y)$, there is a sequence $\{\alpha_n : n \in \omega\} \subset \beta$ and $A_n \in \Pi^0_{\alpha_n}(Y)$ for all $n \in \omega$ such that $A = \bigcup\{A_n : n \in \omega\}$. By the induction hypothesis, we have $A_n \in \Pi^0_{\alpha_n}(Z)$ for all $n \in \omega$ and therefore $A \in \Sigma^0_\beta(Z)$. This proves that $\Sigma^0_\beta(Y) \subset \Sigma^0_\beta(Z)$.

Now if $B \in \Pi^0_\beta(Y)$, then $A = Y \backslash B \in \Sigma^0_\beta(Y) \subset \Sigma^0_\beta(Z)$ and hence $A \in \Sigma^0_\beta(Z)$. There is a sequence $\{\alpha_n : n \in \omega\} \subset \beta$ and $A_n \in \Pi^0_{\alpha_n}(Z)$ for all $n \in \omega$ such that $A = \bigcup\{A_n : n \in \omega\}$. Furthermore, we have $Z \backslash Y = \bigcup\{P_n : n \in \omega\}$ where each P_n is closed in Z and hence $P_n \in \Pi^0_0(Z)$. Choose an enumeration $\{Q_n : n \in \omega\}$ of the family $\{A_n : n \in \omega\} \cup \{P_n : n \in \omega\}$; it is evident that $Q_n \in \Pi^0_{\beta_n}$ where $\beta_n < \beta$ for all $n \in \omega$. Thus $Z \backslash B = A \cup (Z \backslash Y) = \bigcup\{Q_n : n \in \omega\}$ and hence $Z \backslash B \in \Sigma^0_\beta(Z)$. This shows that $B \in \Pi^0_\beta(Z)$, i.e., $\Pi^0_\beta(Y) \subset \Pi^0_\beta(Z)$, so our inductive procedure completes the proof the first part of our fact.

It follows from Fact 1 of T.319 that $\Delta^0_\beta(Z)|Y \subset \Delta^0_\beta(Y)$ for all $\beta < \omega_1$. Now, if $\beta > 0$ and $A \in \Delta^0_\beta(Y)$, then $A \in \Sigma^0_\beta(Z)$ and $A \in \Pi^0_\beta(Z)$ by the first part of our fact. Thus $A \in \Delta^0_\beta(Z)$ and $A = A \cap Y \in \Delta^0_\beta(Z)|Y$. Therefore $\Delta^0_\beta(Z)|Y = \Delta^0_\beta(Y)$ and Fact 1 is proved. $\square$

Returning to our solution, observe that it follows from the inductive definition of $\Sigma_\beta^0(X)$ that

(1) $\Pi_\alpha^0(X) \subset \Sigma_\beta^0(X)$ whenever $\alpha < \beta < \omega_1$.

Given a family $\mathcal{A} \subset \exp(X)$, let $\mathcal{A}^c = \{X \backslash A : A \in \mathcal{A}\}$. It is evident that $\mathcal{A} \subset \mathcal{B} \subset \exp(X)$ implies $\mathcal{A}^c \subset \mathcal{B}^c$, so it follows from (2) that

(2) $\Sigma_\alpha^0(X) \subset \Pi_\beta^0(X)$ whenever $\alpha < \beta < \omega_1$,

because $\Sigma_\alpha^0(X) = (\Pi_\alpha^0(X))^c$ and $\Pi_\beta^0(X) = (\Sigma_\beta^0(X))^c$.

Since X is second countable, every $U \in \tau(X)$ is an F_σ-set which shows that $\Sigma_0^0(X) \subset \Sigma_1^0(X)$. Assume that $\beta < \omega_1$ and we established, for every ordinal $\alpha < \beta$, that $\Sigma_\alpha^0(X) \subset \Sigma_{\alpha+1}^0(X)$. If $U \in \Sigma_\beta^0$, then we can find a sequence $\{\alpha_n : n \in \omega\} \subset \beta$ and $F_n \in \Pi_{\alpha_n}^0(X)$ for all $n \in \omega$ such that $U = \bigcup\{F_n : n \in \omega\}$. It is evident that $\{\alpha_n : n \in \omega\} \subset (\beta + 1)$, so $U \in \Sigma_{\beta+1}^0(X)$ by the definition of $\Sigma_{\beta+1}^0(X)$. Therefore $\Sigma_\beta^0(X) \subset \Sigma_{\beta+1}^0(X)$ and our inductive procedure shows that

(3) $\Sigma_\beta^0(X) \subset \Sigma_{\beta+1}^0(X)$ for any ordinal $\beta < \omega_1$.

Therefore $(\Sigma_\beta^0(X))^c \subset (\Sigma_{\beta+1}^0(X))^c$ so we obtain

(4) $\Pi_\beta^0(X) \subset \Pi_{\beta+1}^0(X)$ for any ordinal $\beta < \omega_1$.

Assume first that $\eta = \xi + 1$; since $\Sigma_\xi^0(X) \subset \Sigma_\eta^0(X)$ by (3) and $\Sigma_\xi^0(X) \subset \Pi_\eta^0(X)$ by (2), we have $\Sigma_\xi^0(X) \subset \Delta_\eta^0(X)$. Furthermore, it follows from the property (4) that $\Pi_\xi^0(X) \subset \Pi_\eta^0(X)$; besides, $\Pi_\xi^0(X) \subset \Sigma_\eta^0(X)$ by (1) which implies $\Pi_\xi^0(X) \subset \Delta_\eta^0(X)$. Thus $\Sigma_\xi^0(X) \cup \Pi_\xi^0(X) \subset \Delta_\eta^0(X)$.

Now, if $\xi + 1 < \eta$, then $\Sigma_\xi^0(X) \subset \Pi_{\xi+1}^0(X) \subset \Sigma_\eta^0(X)$ by (2) and (1). The property (2) also implies that $\Sigma_\xi^0(X) \subset \Pi_\eta^0(X)$, so $\Sigma_\xi^0(X) \subset \Delta_\eta^0(X)$. Analogously, $\Pi_\xi^0(X) \subset \Sigma_{\xi+1}^0(X) \subset \Pi_\eta^0(X)$ by (1) and (2). The property (1) also implies that $\Pi_\xi^0(X) \subset \Sigma_\eta^0(X)$ so $\Pi_\xi^0(X) \subset \Delta_\eta^0(X)$. Thus $\Sigma_\xi^0(X) \cup \Pi_\xi^0(X) \subset \Delta_\eta^0(X)$ whenever $\xi < \eta < \omega_1$.

To finish our solution, assume towards a contradiction that X is an uncountable Polish space such that $\Sigma_\xi^0(X) \cup \Pi_\xi^0(X) = \Delta_\eta^0(X)$ for some ordinals ξ and η with $\xi < \eta$. The space X is a continuous image of $\mathbb{P}$ by Fact 3 of T.132; it follows from $\mathbb{P} = \bigcap\{\mathbb{R}\backslash\{q\} : q \in \mathbb{Q}\}$ that $\mathbb{P}$ is a G_δ-set in $\mathbb{R}$. Every open subset of $\mathbb{R}$ is a countable union of compact sets, so $\mathbb{P}$ is a $K_{\sigma\delta}$-space, and hence X is K-analytic. Therefore we can apply Fact 4 of T.250 to conclude that $\mathbb{K}$ embeds in X.

Restricting the respective families to $\mathbb{K}$ and applying Fact 1 of this solution and Fact 1 of T.319, we conclude that $\Sigma_\xi^0(\mathbb{K}) \cup \Pi_\xi^0(\mathbb{K}) = \Delta_\eta^0(\mathbb{K})$. It is an easy exercise that there are disjoint clopen $U, V \subset \mathbb{K}$ such that $U \simeq V \simeq \mathbb{K}$.

We have $\Pi_\xi^0(U) \neq \Sigma_\xi^0(U)$ by Problem 319; besides, the inclusion $\Pi_\xi^0(U) \subset \Sigma_\xi^0(U)$ implies $\Sigma_\xi^0(U) \subset \Pi_\xi^0(U)$ (and vice versa) because each of the families $\Pi_\xi^0(U)$ and $\Sigma_\xi^0(U)$ is obtained from the other by passing to the relevant complements. Thus it follows from $\Pi_\xi^0(U) \subset \Sigma_\xi^0(U)$ that $\Pi_\xi^0(U) = \Sigma_\xi^0(U)$ which is a contradiction. Therefore neither of the families $\Pi_\xi^0(U)$ and $\Sigma_\xi^0(U)$ is contained in the other. Of course, the same is true for the families $\Pi_\xi^0(V)$ and $\Sigma_\xi^0(V)$. This makes it possible to choose $F \in \Pi_\xi^0(U)\backslash\Sigma_\xi^0(U)$ and $G \in \Sigma_\xi^0(V)\backslash\Pi_\xi^0(V)$. Now Fact 1 of this solution and Fact 1 of T.319 imply that $F \in \Pi_\xi^0(\mathbb{K})$ and $G \in \Sigma_\xi^0(\mathbb{K})$ (we must use the fact that U and V are open in $\mathbb{K}$ if $\xi = 0$). Now, the first part of our solution implies that $F, G \in \Delta_\eta^0(\mathbb{K})$ and hence $H = F \cup G \in \Delta_\eta^0(\mathbb{K})$ (we leave it as an easy exercise for the reader to prove that the union of any two elements of $\Delta_\eta^0(\mathbb{K})$ belongs to $\Delta_\eta^0(\mathbb{K})$). However, $H \notin \Sigma_\xi^0(\mathbb{K}) \cup \Pi_\xi^0(\mathbb{K})$; indeed, if $H \in \Sigma_\xi^0(\mathbb{K})$, then $F = H \cap U \in \Sigma_\xi^0(U)$ which is a contradiction. Analogously, if $H \in \Pi_\xi^0(\mathbb{K})$, then $G = H \cap V \in \Pi_\xi^0(V)$ which contradicts the choice of G.

This contradiction shows that the equality $\Sigma_\xi^0(\mathbb{K}) \cup \Pi_\xi^0(\mathbb{K}) = \Delta_\eta^0(\mathbb{K})$ is impossible and hence $\Sigma_\xi^0(X) \cup \Pi_\xi^0(X) \neq \Delta_\eta^0(X)$, i.e., our solution is complete.

T.321. *Suppose that X is a second countable space. Prove that for every countable limit ordinal η, we have $\bigcup\{\Sigma_\xi^0(X) : \xi < \eta\} \subset \Delta_\eta^0(X)$. Show that if X is uncountable and Polish, then the inclusion is strict, i.e., $\bigcup\{\Sigma_\xi^0(X) : \xi < \eta\} \neq \Delta_\eta^0(X)$.*

Solution. We have $\Sigma_\xi^0(X) \subset \Delta_\eta^0(X)$ for all $\xi < \eta$ by Problem 320; as a consequence, $\bigcup\{\Sigma_\xi^0(X) : \xi < \eta\} \subset \Delta_\eta^0(X)$.

Now assume, towards a contradiction, that X is uncountable and Polish, and we have $\bigcup\{\Sigma_\xi^0(X) : \xi < \eta\} = \Delta_\eta^0(X)$ for some countable limit ordinal η. The space X is a continuous image of $\mathbb{P}$ by Fact 3 of T.132; it follows from $\mathbb{P} = \bigcap\{\mathbb{R}\backslash\{q\} : q \in \mathbb{Q}\}$ that $\mathbb{P}$ is a G_δ-set in $\mathbb{R}$. Every open subset of $\mathbb{R}$ is a countable union of compact sets, so $\mathbb{P}$ is a $K_{\sigma\delta}$-space, and hence X is K-analytic. Therefore we can apply Fact 4 of T.250 to conclude that $\mathbb{K}$ embeds in X.

Restricting the relevant families to $\mathbb{K}$ and applying Fact 1 of T.320 and Fact 1 of T.319, we conclude that $\bigcup\{\Sigma_\xi^0(\mathbb{K}) : \xi < \eta\} = \Delta_\eta^0(\mathbb{K})$. We leave it to the reader to verify that we can choose a disjoint family $\{U_n : n \in \omega\}$ of clopen subsets of $\mathbb{K}$ such that U_n is homeomorphic to $\mathbb{K}$ for every $n \in \omega$. Let $U = \bigcup\{U_n : n \in \omega\}$ and $F = \overline{U}\backslash U$. Since $\mathbb{K}\backslash\overline{U}$ is open in $\mathbb{K}$, there is a sequence $\{F_n : n :\in \omega\} \subset \Pi_0^0(\mathbb{K})$ with $\bigcup\{F_n : n \in \omega\} = \mathbb{K}\backslash\overline{U}$. Take a strictly increasing sequence $\{\xi_n : n \in \omega\} \subset \eta$ such that $\eta = \sup\{\xi_n : n \in \omega\}$. It follows from Problem 320 that there exists a set $A_n \in \Delta_{\xi_n+1}^0(U_n)\backslash\Sigma_{\xi_n}^0(U_n)$ for all $n \in \omega$. Apply Fact 1 of T.320 to conclude that $A_n \in \Delta_{\xi_n+1}^0(\mathbb{K})$ for all $n \in \omega$. Consequently, $A = \bigcup\{A_n : n \in \omega\} \in \Sigma_\eta^0(\mathbb{K})$.

Observe also that $B_n = U_n\backslash A_n \in \Pi_{\xi_n+1}^0(\mathbb{K})$ for all $n \in \omega$. Furthermore, the set $\mathbb{K}\backslash A = (\bigcup\{B_n : n \in \omega\}) \cup F \cup (\bigcup\{F_n : n \in \omega\})$ is a union of the family $\{B_n : n \in \omega\} \cup \{F\} \cup \{F_n : n \in \omega\}$ which can be enumerated as $\{Q_n : n \in \omega\}$ where $Q_n \in \Pi_{\beta_n}^0(\mathbb{K})$ and $\beta_n < \eta$ for all $n \in \omega$. Thus $\mathbb{K}\backslash A \in \Sigma_\eta^0(\mathbb{K})$ which shows that $A \in \Pi_\eta^0(\mathbb{K})$ and hence $A \in \Delta_\eta^0(\mathbb{K})$. Now, if $\xi < \eta$ and $A \in \Sigma_\xi^0(\mathbb{K})$, then $A_n = A \cap U_n \in \Sigma_\xi^0(U_n)$ for all $n \in \omega$ (see Fact 1 of T.319). There is $n \in \omega$ such that $\xi < \xi_n$ and therefore $A_n \in \Sigma^0_{\xi_n}(U_n)$ by Problem 320 which is a contradiction with the choice of A_n. Therefore the equality $\bigcup\{\Sigma^0_\xi(\mathbb{K}) : \xi < \eta\} = \Delta^0_\eta(\mathbb{K})$ is false and hence $\bigcup\{\Sigma^0_\xi(X) : \xi < \eta\} \neq \Delta^0_\eta(X)$ so our solution is complete.

T.322. *Prove that there exists a countable space which cannot be embedded into* $C_p(B)$ *for any Borel set* B.

Solution. Given a space Z let $\mathcal{F}(Z)$ be the family of all closed subsets of Z; if $\mathcal{A} \subset \exp(Z)$, then $\mathcal{A}|Y = \{A \cap Y : A \in \mathcal{A}\}$ for every $Y \subset Z$. The expression $Z \simeq T$ says that the spaces Z and T are homeomorphic.

Fact 1. If Z is a space and $w(Z) = \kappa$, then $|\mathcal{F}(Z)| = |\tau(Z)| \leq 2^\kappa$. In particular, $|\mathcal{F}(Z)| \leq \mathfrak{c}$ for any second countable space Z.

Proof. It is evident that $|\mathcal{F}(Z)| = |\tau(Z)|$; furthermore, if $\mathcal{B}$ is a base in Z with $|\mathcal{B}| \leq \kappa$, then for every $U \in \tau(Z)$ there is $\mathcal{B}' \subset \mathcal{B}$ such that $U = \bigcup \mathcal{B}'$. As a consequence, $|\tau(Z)| \leq |\exp(\mathcal{B})| \leq 2^\kappa$ so Fact 1 is proved.   □

Fact 2. There are at most $\mathfrak{c}$-many pairwise non-homeomorphic Polish spaces.

Proof. Every Polish space is homeomorphic to a closed subspace of $\mathbb{R}^\omega$ by Problem 273 of [TFS]. Thus the cardinality of any family of pairwise non-homeomorphic Polish spaces cannot be greater than $|\mathcal{F}(\mathbb{R}^\omega)| \leq \mathfrak{c}$ (see Fact 1). Fact 2 is proved.   □

Fact 3. For any second countable space X, we have $|\mathbb{B}(X)| \leq \mathfrak{c}$. An easy consequence is that there are at most $\mathfrak{c}$-many pairwise non-homeomorphic Borel sets.

Proof. If X is a second countable space, then $|\Sigma^0_0(X)| = |\tau(X)| \leq \mathfrak{c}$ by Fact 1. Suppose that $0 < \beta < \omega_1$ and we proved that $|\Sigma^0_\alpha(X)| \leq \mathfrak{c}$ for all $\alpha < \beta$. Then $|\Pi^0_\alpha(X)| \leq \mathfrak{c}$ for all ordinals $\alpha < \beta$ and therefore the cardinality of the family $\mathcal{U} = \bigcup\{\Pi^0_\alpha(X) : \alpha < \beta\}$ does not exceed $\mathfrak{c}$ as well. Since $\Sigma^0_\beta(X)$ is the family of unions of countable subfamilies of $\mathcal{U}$, we have $|\Sigma^0_\beta(X)| \leq |\mathcal{U}^\omega| \leq \mathfrak{c}^\omega = \mathfrak{c}$. Thus our inductive procedure shows that $|\Sigma^0_\beta(X)| \leq \mathfrak{c}$ for all $\beta < \omega_1$. Consequently, $|\mathbb{B}(X)| \leq \mathfrak{c} \cdot \omega_1 = \mathfrak{c}$.

By Fact 2, there is an enumeration (with possible repetitions) $\{X_\alpha : \alpha < \mathfrak{c}\}$ of all Polish spaces. By the first part of our fact, the family $\mathcal{V} = \bigcup\{\mathbb{B}(X_\alpha) : \alpha < \mathfrak{c}\}$ has cardinality at most $\mathfrak{c} \cdot \mathfrak{c} = \mathfrak{c}$; since any Borel set is an element of $\mathcal{V}$, Fact 3 is proved.   □

Fact 4. In the family $\{\omega \cup \{\xi\} : \xi \in \beta\omega\backslash\omega\}$ there exists a subfamily of $2^\mathfrak{c}$-many pairwise non-homeomorphic spaces. Here $\omega \cup \{\xi\}$ is considered with the subspace topology inherited from $\beta\omega$ for all $\xi \in \beta\omega\backslash\omega$.

Proof. Denote by N_ξ the space $\omega \cup \{\xi\}$ and let $E(\xi) = \{\eta \in \beta\omega\backslash\omega : N_\xi \simeq N_\eta\}$ for all $\xi \in \beta\omega\backslash\omega$. For any $\eta \in E(\xi)$ fix a homeomorphism $f_\eta : N_\xi \to N_\eta$. Then $f_\eta(\omega) = \omega$ and $f_\eta(\xi) = \eta$; besides, for a set $A \subset \omega$, we have $A \cup \{\xi\} \in \tau(\xi, N_\xi)$ if and only if $f_\eta(A) \cup \{\eta\} \in \tau(\eta, N_\eta)$. If $\mathcal{U}_\xi = \tau(\xi, N_\xi)|\omega$ for every $\xi \in \beta\omega\backslash\omega$, then $f_\eta(\mathcal{U}_\xi) = \mathcal{U}_\eta$ for any $\eta \in E(\xi)$.

Observe that $P_\xi = \bigcap \{\overline{A} : A \in \mathcal{U}_\xi\} = \{\xi\}$ for any $\xi \in \beta\omega \backslash \omega$ (the bar denotes the closure in $\beta\omega$). Indeed, it is clear that $\xi \in P_\xi$; if $\eta \neq \xi$, then there are disjoint $U, V \in \tau(\beta\omega)$ such that $\xi \in U$ and $\eta \in V$. Then $U \cap \omega \in \mathcal{U}_\xi$; it is straightforward that $\eta \notin \overline{U \cap \omega}$ and hence $\eta \notin P_\xi$.

It follows from $\{\eta\} = P_\eta$ that the correspondence $\eta \to \mathcal{U}_\eta = f_\eta(\mathcal{U}_\xi)$ is an injection on the set $E(\xi)$ and therefore the cardinality of $E(\xi)$ does not exceed the cardinality of the set of all bijections from ω to ω. Thus $|E(\xi)| \leq |\omega^\omega| = \mathfrak{c}$ and we established that

(1) $|E(\xi)| \leq \mathfrak{c}$ for any $\xi \in \beta\omega \backslash \omega$.

Furthermore, if we let $\xi \sim \eta$ if and only if $N_\xi \simeq N_\eta$, then $\sim$ is an equivalence relationship on $\beta\omega \backslash \omega$ and $E(\xi)$ is the equivalence class of ξ. Since $|\beta\omega \backslash \omega| = 2^\mathfrak{c}$ and the cofinality of $2^\mathfrak{c}$ is greater than $\mathfrak{c}$, the property (1) implies that there are $2^\mathfrak{c}$-many distinct equivalence classes. If we enumerate them as $\{E_\alpha : \alpha < 2^\mathfrak{c}\}$, then choosing $\xi_\alpha \in E_\alpha$ for each $\alpha < 2^\mathfrak{c}$, we obtain a family $\mathcal{N} = \{N_{\xi_\alpha} : \alpha < 2^\mathfrak{c}\}$ of pairwise non-homeomorphic spaces with $|\mathcal{N}| = 2^\mathfrak{c}$. Fact 4 is proved.   □

Returning to our solution, take a family $\mathcal{S} = \{S_\alpha : \alpha < 2^\mathfrak{c}\}$ of pairwise non-homeomorphic countable spaces whose existence is guaranteed by Fact 4. Assume that, for every $\alpha < 2^\mathfrak{c}$, the space S_α embeds in $C_p(B_\alpha)$ for some Borel set B_α. Since there are at most $\mathfrak{c}$-many Borel sets by Fact 3, there is a Borel set B such that the set $Q = \{\alpha < 2^\mathfrak{c} : B_\alpha = B\}$ has cardinality $2^\mathfrak{c}$. The space B being second countable, we have $|C_p(B)| = \mathfrak{c}$ because $nw(C_p(B)) = nw(B) \leq \omega$. Therefore there are at most $\mathfrak{c}$-many countable subspaces of $C_p(B)$ (because $\mathfrak{c}^\omega = \mathfrak{c}$). As a consequence, there are distinct $\alpha, \beta \in Q$ such that S_α and S_β are homeomorphic to the same countable subspace of $C_p(B)$. Therefore $S_\alpha \simeq S_\beta$ which is a contradiction. Consequently, some of the spaces from the family $\mathcal{S}$ does not embed in $C_p(B)$ for any Borel set B, so our solution is complete.

T.323. *Prove that a second countable space X is an absolute F_σ if and only if X is σ-compact.*

Solution. Suppose that $X = \bigcup\{K_n : n \in \omega\}$ where K_n is compact for all $n \in \omega$. If $X \subset Y$, then K_n is closed in Y (recall that all our spaces are Tychonoff and a compact space is closed in any larger Hausdorff space). Thus X is an F_σ-set in Y, i.e., X is an absolute F_σ. This proves sufficiency.

Since X is second countable, we can consider that $X \subset \mathbb{I}^\omega$; if X is absolute F_σ, then X is an F_σ-set in $\mathbb{I}^\omega$. Since all closed subsets of $\mathbb{I}^\omega$ are compact, the space X has to be σ-compact; this settles necessity.

T.324. *Prove that a second countable space X is an absolute G_δ if and only if X is Čech-complete.*

Solution. We can consider that $X \subset \mathbb{I}^\omega$; if X is absolute G_δ, then X is Čech-complete being a G_δ-set in the compact space $\mathbb{I}^\omega$ (see Problem 260 of [TFS]). This proves necessity.

Now, assume that X is Čech-complete and $X \subset M$ where M is a metrizable space. Let $Y = \overline{X}$ (the bar denotes the closure in M). Then Y is an extension of X and hence X is a G_δ-set in Y (see Problem 259 of [TFS]). Since M is metrizable, the set Y is G_δ in M. Finally, apply Fact 2 of S.358 to see that X is also a G_δ-set in M which shows that X is absolute G_δ.

T.325. *Suppose that X is a Polish space and $f : X \to Y$ is a perfect map. Prove that Y is Polish (remember that any perfect map is onto).*

Solution. Being completely metrizable is equivalent to being metrizable and Čech-complete by Problem 269 of [TFS], so to prove that Y is Polish, we must establish that Y is Čech-complete and $w(Y) = \omega$.

A perfect image of a Čech-complete space is a Čech-complete space by Problem 261 of [TFS] so Y is Čech-complete. Furthermore, $nw(Y) \leq \omega$ (see Problem 157 of [TFS]) and hence we can apply Problem 270 of [TFS] to conclude that $w(Y) = \omega$. Thus Y is a second countable Čech-complete space, i.e., Y is Polish.

T.326. *Let X be a Polish space. Suppose that $f : X \to Y$ is a continuous surjective open map. Prove that there is a closed $F \subset X$ such that $f(F) = Y$ and $f|F$ is a perfect map.*

Solution. The space X is Čech-complete by Problem 269 of [TFS] and $nw(Y) \leq nw(X) = \omega$ by Problem 157 of [TFS], so the space Y is Lindelöf and hence paracompact. Therefore Fact 1 of S.491 is applicable to conclude that there exists a closed $F \subset X$ such that $f(F) = Y$ and $f|F$ is a perfect map.

T.327. *Prove that any open continuous image of a Polish space is a Polish space.*

Solution. Suppose that X is Polish and $f : X \to Y$ is an open continuous onto map. We can apply Problem 326 to find a closed $F \subset X$ such that $f(F) = Y$ and $f|F$ is a perfect map. The space F being Polish, we can apply Problem 325 to conclude that Y is also Polish.

T.328. *Prove that X is a Polish space if and only if it is an open continuous image of $\mathbb{P}$.*

Solution. The space $\mathbb{P}$ is Čech-complete being a G_δ-subspace of $\mathbb{R}$ (see Problem 260 of [TFS]). Thus $\mathbb{P}$ is Polish by Problem 269 of [TFS]. If X is an open continuous image of $\mathbb{P}$, then X is Polish by Problem 327.

Now assume that X is a Polish space and fix a complete bounded metric d on the space X for which $\tau(d) = \tau(X)$. Given $s \in \omega^{<\omega}$, let $l(s) = n$ if $s \in \omega^n$ for some $n \in \mathbb{N}$; if $s = \emptyset$, let $l(s) = 0$. It was proved in Fact 3 of S.491 that there exists *an A-system in X*, i.e., a family $\mathcal{U} = \{U_s : s \in \omega^{<\omega}\} \subset \tau(X)$ with the following properties:

(A1) $U_\emptyset = X$ and $U_s = \bigcup\{U_t : t \in \omega^{n+1}$ and $t|n = s\}$ for any $n \in \omega$ and $s \in \omega^n$;
(A2) if $s \in \omega^{<\omega}$ and $l(s) = n \in \mathbb{N}$, then $\mathrm{diam}(U_s) < \frac{1}{n}$;
(A3) for any $f \in \omega^\omega$, we have $\bigcap\{U_{f|n} : n \in \omega\} \neq \emptyset$.

We will identify $\mathbb{P}$ with ω^ω (see Problem 313). For any $f \in \omega^\omega$, observe that the set $P_f = \bigcap \{U_{f|n} : n \in \omega\}$ consists of precisely one point because $P_f \neq \emptyset$ by (A3) and the diameters of the sets $U_{f|n}$ approach zero. Let $\varphi(f) = x$ where $x \in X$ is the point for which $P_f = \{x\}$. We claim that $\varphi : \omega^\omega \to X$ is a continuous open onto map.

Fix $m \in \omega$ and $s \in \omega^m$ and let $W_s = \{f \in \omega^\omega : f|m = s\}$. It is straightforward from the definition of φ that $\varphi(W_s) \subset U_s$. Take any $x \in U_s$ and let $s_m = s$. Assume that $m \leq n \in \omega$ and we have constructed functions $\{s_k : m \leq k \leq n\}$ with the following properties:

(i) $s_k \in \omega^k$ for each $k \in \{m, \ldots, n\}$;
(ii) $x \in U_{s_k}$ for each $k \in \{m, \ldots, n\}$;
(iii) $s_{k+1}|k = s_k$ for all $k \in \{m, \ldots, n-1\}$.

Since $x \in U_{s_n} = \bigcup \{U_t : t \in \omega^{n+1}$ and $t|n = s_n\}$, there is $s_{n+1} \in \omega^{n+1}$ such that $s_{n+1}|n = s_n$ and $x \in U_{s_{n+1}}$. It is evident that (i)–(iii) hold for the sequence $\{s_k : k \leq n+1\}$, so our inductive construction can be continued, giving us a sequence $\{s_n : m \leq n < \omega\}$ with the properties (i)–(iii). It follows from (iii) that there exists $f \in \omega^\omega$ such that $f|n = s_n$ for each $n \geq m$. Condition (ii) implies that $x \in \bigcap \{U_{f|n} : n \in \omega\}$ and therefore $\varphi(f) = x$, so $\varphi(W_s) = U_s$. Since $s \in \omega^{<\omega}$ has been chosen arbitrarily, we proved that

(∗) $\varphi(W_s) = U_s$ for any $s \in \omega^{<\omega}$.

If we take $s = \emptyset$, then we obtain $\varphi(\omega^\omega) = X$, i.e., the map φ is onto. For any point $x \in X$ take any $f \in \omega^\omega$ such that $\varphi(f) = x$. Observe that the family $C_x = \{U_{f|n} : n \in \omega\}$ is a local base of X at x. Indeed, all elements of C_x are open in X and contain x. Given any $\varepsilon > 0$, choose $n \in \mathbb{N}$ with $\frac{1}{n} < \varepsilon$ and note that, for any $y \in U_{f|n}$, we have $d(x, y) \leq \operatorname{diam}(U_{f|n}) < \frac{1}{n}$ so $x \in U_{f|n} \subset B(x, \varepsilon)$ which proves that C_x is a local base at x. Besides, the family $\mathcal{B}_f = \{\{g \in \omega^\omega : g|n = f|n\} : n \in \omega\}$ is a local base of ω^ω at f and $\{\varphi(U) : U \in \mathcal{B}_f\} = C_x$ by (∗), so we can apply Fact 2 of S.491 to conclude that the map φ is continuous and open. We already saw that φ is surjective, so X is an open continuous image of $\mathbb{P}$ and hence our solution is complete.

T.329. *Prove that a second countable space is Polish if and only if it is a closed continuous image of* $\mathbb{P}$. *Show that a closed continuous image of* $\mathbb{P}$ *is not necessarily first countable.*

Solution. Assume that X is a second countable space and $f : \mathbb{P} \to X$ is a closed continuous onto map. For every $x \in X$ let $K_x = f^{-1}(x) \backslash \operatorname{Int}(f^{-1}(x))$. It turns out that

(1) the set K_x is compact (maybe empty) for every $x \in X$.

Indeed, if the set K_x is not compact for some point $x \in X$, then there is a set $A = \{a_n : n \in \omega\} \subset K_x$ which is closed and discrete in $\mathbb{P}$. Fix a local base $\{U_n : n \in \omega\}$ of X at x such that $U_{n+1} \subset U_n$ for all $n \in \omega$. There exists a discrete family $\mathcal{V} = \{V_n : n \in \omega\} \subset \tau(\mathbb{P})$ such that $a_n \in V_n$ for all $n \in \omega$. By continuity of f, making the sets V_n smaller if necessary, we can assume that $f(V_n) \subset U_n$ for all $n \in \omega$. The set A is contained in the boundary of $f^{-1}(x)$ so we can take $y_n \in V_n \setminus f^{-1}(x)$ for each $n \in \omega$.

The family $\mathcal{V}$ being discrete, the set $D = \{y_n : n \in \omega\}$ is closed and discrete in $\mathbb{P}$; besides, $f(D) \not\ni x$ by our choice of D. Furthermore, $f(y_n) \in U_n$ for all $n \in \omega$ which shows that the sequence $\{f(y_n) : n \in \omega\}$ converges to x which is a contradiction because f is a closed map and hence $f(D)$ must be a closed set. This proves (1).

Now, take any $x \in X$; if $f^{-1}(x)$ is open, then choose $a(x) \in f^{-1}(x)$ and let $P_x = \{a(x)\}$. If not, then let $P_x = K_x$. It is immediate from (1) that P_x is a compact set for all $x \in X$, so if $Q = \bigcup \{P_x : x \in X\}$ and $g = f|Q$, then $g(Q) = X$ and $g^{-1}(x) = P_x$ is a compact set for all points $x \in X$. Observe that $\mathbb{P} \setminus Q = \bigcup \{f^{-1}(x) \setminus P_x : x \in X\}$ is an open subset of $\mathbb{P}$ because $f^{-1}(x) \setminus P_x$ is open in $\mathbb{P}$ for each $x \in X$. Thus Q is closed in $\mathbb{P}$ and therefore g is a perfect map. The space Q is Polish being closed in $\mathbb{P}$ and hence we can apply Problem 325 to conclude that X is Polish finishing the proof of sufficiency.

Now assume that X is a Polish space. Then there is a open map $g : \mathbb{P} \to X$ by Problem 328. Apply Problem 326 to find a closed $F \subset \mathbb{P}$ such that $g(F) = X$ and $g|F$ is a perfect (and hence closed) map. The space $\mathbb{P}$ is Polish and strongly zero-dimensional, so there is a closed retraction $h : \mathbb{P} \to F$ by Problem 316. Consequently, $f = (g|F) \circ h : \mathbb{P} \to X$ is a continuous closed onto map (it is evident that a compositions of two closed maps is a closed map).

To finally construct the desired example, let $F = \{\sqrt{2} + n : n \in \omega\} \subset \mathbb{P}$. Then F is a closed discrete subset of $\mathbb{P}$. Let us construct a space $\mathbb{P}/F$ as follows: for any $A \subset \mathbb{P}$, let $A^* = (A \setminus F) \cup \{F\}$. Given $z \in \mathbb{P}$, let $p_F(z) = z$ if $z \in \mathbb{P} \setminus F$ and $p_F(z) = F$ if $z \in F$. It is clear that $p_F : \mathbb{P} \to \mathbb{P}_F = \{F\} \cup (\mathbb{P} \setminus F)$. Then the family $\tau_F = \{U \in \tau(\mathbb{P}) : U \subset \mathbb{P} \setminus F\} \cup \{U^* : U \in \tau(F, \mathbb{P})\}$ is a topology on the set $\mathbb{P}_F$; the space $\mathbb{P}/F = (\mathbb{P}_F, \tau_F)$ is T_1 and normal (and hence Tychonoff) and the map $p_F : \mathbb{P} \to \mathbb{P}/F$ is continuous, closed and onto (see Fact 2 of T.245).

To distinguish between the point F of the space $\mathbb{P}/F$ and the set F, let us denote by z_F the set F when it is considered as a point of the space $Z = \mathbb{P}/F$. We claim that Z is not first countable at the point z_F. To prove it assume, towards a contradiction, that there is a family $\{O_n : n \in \omega\} \subset \tau(z_F, Z)$ which is a base of Z at z_F. Let $x_n = \sqrt{2} + n$ for all $n \in \omega$; it is easy to find $U_n \in \tau(x_n, \mathbb{P})$ for every $n \in \omega$ such that the family $\{U_n : n \in \omega\}$ is discrete.

Take a point $y_n \in (O_n \cap U_n) \setminus \{x_n\}$ for each $n \in \omega$. It is immediate that the set $E = \{y_n : n \in \omega\}$ is closed, discrete and $E \cap F = \emptyset$. Therefore $\mathbb{P} \setminus E \in \tau(z_F, Z)$ and hence there is $n \in \omega$ for which $O_n \cap E = \emptyset$. However, $y_n \in O_n \cap E$ which is a contradiction. Thus Z is a closed continuous image of $\mathbb{P}$ such that $\chi(Z) > \omega$ and hence our solution is complete.

T.330. *Prove that X is homeomorphic to a Borel subset of some Polish space if and only if it is homeomorphic to a Borel subset of $\mathbb{R}^\omega$.*

Solution. Suppose that X is a Borel subset of a Polish space M. We can assume that M is a closed subspace of $\mathbb{R}^\omega$ by Problem 273 of [TFS]. There is a ordinal $\xi < \omega_1$ such that $X \in \Sigma^0_{\xi+1}(M)$; since $\Sigma^0_{\xi+1}(M) \subset \Sigma^0_{\xi+1}(\mathbb{R}^\omega)$ by Fact 1 of T.320, we have $X \in \Sigma^0_{\xi+1}(\mathbb{R}^\omega)$ and therefore X is a Borel subset of $\mathbb{R}^\omega$.

T.331. *Let X be a Borel set. Prove that every $Y \in \mathbb{B}(X)$ is also a Borel set. In particular, any closed and any open subspace of a Borel set is a Borel set.*

Solution. We can assume that X is a Borel subspace of a Polish space M and hence there is $\xi < \omega_1$ such that $X \in \Sigma^0_\xi(M)$ and, consequently, $X \in \Pi^0_{\xi+1}(M)$ by Problem 320. Recall that a space Z is called *perfect* if every $U \in \tau(Z)$ is an F_σ-set.

Fact 1. For any perfect space Z we have

(a) $\Sigma^0_\xi(Z) \cup \Pi^0_\xi(Z) \subset \Delta^0_\eta(Z)$ whenever $\xi < \eta < \omega_1$;
(b) $\mathbb{B}(Z) = \bigcup\{\Sigma^0_\xi(Z) : \xi < \omega_1\} = \bigcup\{\Pi^0_\xi(Z) : \xi < \omega_1\} = \bigcup\{\Delta^0_\xi(Z) : \xi < \omega_1\}$;
(c) if $A \in \mathbb{B}(Z)$, then $Z \backslash A \in \mathbb{B}(Z)$;
(d) if $B_i \in \mathbb{B}(Z)$ for any $i \in \omega$, then $B = \bigcup\{B_i : i \in \omega\} \in \mathbb{B}(Z)$;
(e) if $C_i \in \mathbb{B}(Z)$ for any $i \in \omega$, then $C = \bigcap\{C_i : i \in \omega\} \in \mathbb{B}(Z)$.

Proof. Given an ordinal β with $0 < \beta < \omega_1$, it follows from the inductive definition of $\Sigma^0_\beta(Z)$ that

(1) $\Pi^0_\alpha(Z) \subset \Sigma^0_\beta(Z)$ whenever $\alpha < \beta < \omega_1$.

Given a family $\mathcal{A} \subset \exp(Z)$, let $\mathcal{A}^c = \{Z \backslash A : A \in \mathcal{A}\}$. It is evident that $\mathcal{A} \subset \mathcal{B} \subset \exp(Z)$ implies $\mathcal{A}^c \subset \mathcal{B}^c$, so it follows from (2) that

(2) $\Sigma^0_\alpha(Z) \subset \Pi^0_\beta(Z)$ whenever $\alpha < \beta < \omega_1$,

because $\Sigma^0_\alpha(Z) = (\Pi^0_\alpha(Z))^c$ and $\Pi^0_\beta(Z) = (\Sigma^0_\beta(Z))^c$.

Since the space Z is perfect, every $U \in \tau(Z)$ is an F_σ-set which shows that $\Sigma^0_0(Z) \subset \Sigma^0_1(Z)$. Assume that $\beta < \omega_1$ and we established, for every ordinal $\alpha < \beta$, that $\Sigma^0_\alpha(Z) \subset \Sigma^0_{\alpha+1}(Z)$. If $U \in \Sigma^0_\beta$, then we can find a sequence $\{\alpha_n : n \in \omega\} \subset \beta$ and $F_n \in \Pi^0_{\alpha_n}(Z)$ for all $n \in \omega$ such that $U = \bigcup\{F_n : n \in \omega\}$. It is evident that $\{\alpha_n : n \in \omega\} \subset (\beta + 1)$, so $U \in \Sigma^0_{\beta+1}(Z)$ by the definition of $\Sigma^0_{\beta+1}(Z)$. Therefore $\Sigma^0_\beta(Z) \subset \Sigma^0_{\beta+1}(Z)$ and our inductive procedure shows that

(3) $\Sigma^0_\beta(Z) \subset \Sigma^0_{\beta+1}(Z)$ for any ordinal $\beta < \omega_1$.

Therefore $(\Sigma^0_\beta(Z))^c \subset (\Sigma^0_{\beta+1}(Z))^c$ so we obtain

(4) $\Pi^0_\beta(Z) \subset \Pi^0_{\beta+1}(Z)$ for any ordinal $\beta < \omega_1$.

Assume first that $\eta = \xi + 1$; since $\Sigma^0_\xi(Z) \subset \Sigma^0_\eta(Z)$ by (3) and $\Sigma^0_\xi(Z) \subset \Pi^0_\eta(Z)$ by (2), we have $\Sigma^0_\xi(Z) \subset \Delta^0_\eta(Z)$. Furthermore, it follows from the property (4) that $\Pi^0_\xi(Z) \subset \Pi^0_\eta(Z)$; besides, $\Pi^0_\xi(Z) \subset \Sigma^0_\eta(Z)$ by (1) which implies $\Pi^0_\xi(Z) \subset \Delta^0_\eta(Z)$. Thus $\Sigma^0_\xi(Z) \cup \Pi^0_\xi(Z) \subset \Delta^0_\eta(Z)$.

Now, if $\xi + 1 < \eta$, then $\Sigma_\xi^0(Z) \subset \Pi_{\xi+1}^0(Z) \subset \Sigma_\eta^0(Z)$ by (2) and (1). The property (2) also implies that $\Sigma_\xi^0(Z) \subset \Pi_\eta^0(Z)$ so $\Sigma_\xi^0(Z) \subset \Delta_\eta^0(Z)$. Analogously, $\Pi_\xi^0(Z) \subset \Sigma_{\xi+1}^0(Z) \subset \Pi_\eta^0(Z)$ by (1) and (2). The property (1) also implies that $\Pi_\xi^0(Z) \subset \Sigma_\eta^0(Z)$ so $\Pi_\xi^0(Z) \subset \Delta_\eta^0(Z)$. Thus $\Sigma_\xi^0(Z) \cup \Pi_\xi^0(Z) \subset \Delta_\eta^0(Z)$ whenever $\xi < \eta < \omega_1$ and (a) is proved. The property (b) is an evident consequence of (a).

It follows from (b) that if $A \in \mathbb{B}(Z)$, then $A \in \Sigma_\xi^0(Z)$ for some $\xi < \omega_1$ and therefore $Z \setminus A \in \Pi_\xi^0(Z) \in \mathbb{B}(Z)$ which shows that $Z \setminus A \in \mathbb{B}(Z)$, i.e., (c) is proved.

As to (d), apply (b) to conclude that for every $i \in \omega$, there is an ordinal $\alpha_i < \omega_1$ such that $B_i \in \Pi_{\alpha_i}^0(Z)$. If $\alpha = \sup\{\alpha_i : i \in \omega\} + 1$, then by definition of $\Sigma_\alpha^0(Z)$ we have $B \in \Sigma_\alpha^0(Z) \subset \mathbb{B}(Z)$ and hence (d) is settled. Finally, the property (e) is an immediate consequence of (c) and (d), so Fact 1 is proved. $\square$

Returning to our solution, suppose first that F is a closed (open) subset of X; then there is a closed (open) subset G of the space M such that $G \cap X = F$. Since $G \in \Sigma_0^0(M) \cup \Pi_0^0(M) \subset \Delta_1^0(M)$, by Problem 320, we have $G \in \Pi_{\xi+1}^0(M)$ and therefore $F = G \cap X \in \Pi_{\xi+1}^0(M)$ (it is easy to see that the intersection of any two elements of $\Pi_\alpha^0(M)$ belongs to $\Pi_\alpha^0(M)$ for every $\alpha < \omega_1$) and hence F is a Borel subset of M.

Now assume that $\alpha < \omega_1$ and we established that for all ordinals $\beta < \alpha$, if $F \in \Sigma_\beta^0(X) \cup \Pi_\beta^0(X)$, then F is a Borel subset of the space M. If $F \in \Sigma_\alpha^0(X)$, then there is a sequence $\{\beta_n : n \in \omega\} \subset \alpha$ and $F_n \in \Pi_{\beta_n}^0(X)$ for all $n \in \omega$ such that $F = \bigcup\{F_n : n \in \omega\}$. Since $F_n \in \mathbb{B}(M)$ for all $n \in \omega$ by the induction hypothesis, we have $F \in \mathbb{B}(M)$ by Fact 1.

Now if $F \in \Pi_\alpha^0(X)$, then $H = X \setminus F \in \Sigma_\alpha^0(X)$ and hence there is a sequence $\{\gamma_n : n \in \omega\} \subset \alpha$ and $H_n \in \Pi_{\gamma_n}^0(X)$ for all $n \in \omega$ such that $H = \bigcup\{H_n : n \in \omega\}$. Since $H_n \in \mathbb{B}(M)$ for all $n \in \omega$ by the induction hypothesis, the sets H and $M \setminus H$ are Borel subsets of M (see Fact 1). Therefore we can apply Fact 1 once more to conclude that $F = X \setminus H = X \cap (M \setminus H)$ is a Borel subset of M. Thus our inductive procedure shows that for any $\alpha < \omega_1$, if $Y \in \Sigma_\alpha^0(X)$, then Y is a Borel subset of M. In other words, every $Y \in \mathbb{B}(X)$ is a Borel subset of M so our solution is complete.

T.332. *Given second countable spaces X and Y and a continuous map $f : X \to Y$, prove that for every Borel subset A of the space Y, the set $f^{-1}(A)$ is Borel in X.*

Solution. If A is open (closed) in Y, then $f^{-1}(A)$ is also open (closed) in Y because the map f is continuous. This proves that if $A \in \Sigma_0^0(Y) \cup \Pi_0^0(Y)$, then $f^{-1}(A)$ is a Borel subset of X. Now assume that $\alpha < \omega_1$ and we proved that for any $\beta < \alpha$, if $A \in \Sigma_\beta^0(Y) \cup \Pi_\beta^0(Y)$, then $f^{-1}(A) \in \mathbb{B}(X)$.

If $A \in \Sigma_\alpha^0(Y)$, then there is a sequence $\{\beta_n : n \in \omega\} \subset \alpha$ and $A_n \in \Pi_{\beta_n}^0(Y)$ for all $n \in \omega$ such that $A = \bigcup\{A_n : n \in \omega\}$. Since $f^{-1}(A_n) \in \mathbb{B}(X)$ for all $n \in \omega$ by the induction hypothesis, we have $f^{-1}(A) = \bigcup\{f^{-1}(A_n) : n \in \omega\}$ and hence $f^{-1}(A)$ is a Borel subset of X by Fact 1 of T.331.

If, on the other hand, we have a set $A \in \Pi^0_\alpha(Y)$, then $Y \setminus A \in \Sigma^0_\alpha(Y)$ and hence $X \setminus f^{-1}(A) = f^{-1}(Y \setminus A)$ is a Borel subset of X. Thus $f^{-1}(A)$ is also a Borel subset of X by Fact 1 of T.331, so our inductive procedure shows that, for any $A \in \mathbb{B}(Y)$, we have $f^{-1}(A) \in \mathbb{B}(X)$.

T.333. *Prove that any countable product of Borel sets is a Borel set. Show that for any second countable space X, if $X = \bigcup \{X_i : i \in \omega\}$ and each X_i is a Borel set, then X is also a Borel set.*

Solution. Given a space Z and $Y \subset Z$, let $\mathrm{id}^Y_Z : Y \to Z$ be the identity embedding, i.e., $\mathrm{id}^Y_Z(z) = z$ for all $z \in Y$.

Fact 1. Suppose that M is a complete metric space. Given any space Z and a continuous map $f : A \to M$ for some dense $A \subset Z$, there is a G_δ-set G of the space Z such that $A \subset G$ and there is a continuous map $g : G \to M$ with $g|A = f$.

Proof. Take a complete metric d on the space M which generates the topology of M. Given any $n \in \mathbb{N}$ let $U_n = \{z \in Z :$ there exists $W \in \tau(z, Z)$ such that $\mathrm{diam}(f(W \cap A)) \le \frac{1}{n}\}$. It is evident that U_n is open in Z; the map f being continuous, for any $z \in A$, there is $V \in \tau(z, A)$ such that $f(V) \subset B(f(z), \frac{1}{2n})$. If W is an open subset of Z such that $W \cap A = V$, then $\mathrm{diam}(f(W \cap A)) = \mathrm{diam}(f(V)) \le \mathrm{diam}(B(f(z), \frac{1}{2n})) \le \frac{1}{n}$ which shows that $z \in U_n$. Consequently, $A \subset U_n$ for any $n \in \mathbb{N}$.

Thus $G = \bigcap \{U_n : n \in \mathbb{N}\}$ is a G_δ-subset of Z and $A \subset G$. To prove that the map f can be continuously extended over the set G, take any $z \in G$ and consider the family $\mathcal{F}_z = \{\overline{f(W \cap A)} : W \in \tau(z, Z)\}$. It is clear that $\mathcal{F}_z$ consists of closed subsets of the space M; if $W_1, \ldots, W_n \in \tau(z, Z)$ and $W = W_1 \cap \cdots \cap W_n$, then $\overline{f(W \cap A)} \subset \overline{f(W_1 \cap A)} \cap \cdots \cap \overline{f(W_n \cap A)}$ which shows that the family $\mathcal{F}_z$ is centered.

Given any $\varepsilon > 0$ take $n \in \mathbb{N}$ with $\frac{1}{n} < \varepsilon$. Since $z \in U_n$, there is $W \in \tau(z, Z)$ such that $\mathrm{diam}(f(W \cap A)) \le \frac{1}{n} < \varepsilon$. Therefore $\mathcal{F}_z$ contains elements of arbitrarily small diameter and hence $\bigcap \mathcal{F}_z \ne \emptyset$ by Problem 236 of [TFS]. It is easy to see that $\bigcap \mathcal{F}_z$ cannot contain more than one point, so there is a point $g(z)$ such that $\bigcap \mathcal{F}_z = \{g(z)\}$.

Once we have the function $g : G \to M$, observe first that $g|A = f$ because if $z \in A$, then $f(z) \in f(W \cap A) \subset \overline{f(W \cap A)}$ for any $W \in \tau(z, Z)$ and hence $f(z) \in \bigcap \mathcal{F}_z$, i.e., $g(z) = f(z)$.

To see that g is continuous, take any $z \in G$ and $\varepsilon > 0$. Choose $n \in \mathbb{N}$ and $W \in \tau(z, Z)$ such that $\frac{1}{n} < \varepsilon$ and $\mathrm{diam}(\overline{f(W \cap A)}) = \mathrm{diam}(f(W \cap A)) \le \frac{1}{n}$. If $W' = W \cap G$, then $W' \in \tau(z, G)$; given any $t \in W'$, we have $t \in W \in \tau(t, Z)$ and hence $\overline{f(W \cap A)} \in \mathcal{F}_t \cap \mathcal{F}_z$. Consequently, $\{g(z), g(t)\} \subset \overline{f(W \cap A)}$ and therefore $d(g(z), g(t)) \le \mathrm{diam}(\overline{f(W \cap A)}) \le \frac{1}{n} < \varepsilon$ which proves that $g(W') \subset B(g(z), \varepsilon)$, i.e., g is continuous at the point z. Since $z \in G$ was taken arbitrarily, the function g is continuous on G and Fact 1 is proved.   □

Fact 2 (Lavrentieff theorem). Let M and N be complete metric spaces. If we are given sets $A \subset M$, $B \subset N$ and a homeomorphism $f : A \to B$, then there exist $A' \subset M$, $B' \subset N$ such that $A \subset A'$, $B \subset B'$, the set A' is G_δ in M, the set B' is G_δ in N and there is a homeomorphism $h : A' \to B'$ such that $h|A = f$.

Proof. If we find the promised sets for the spaces $M_1 = \overline{A}$ and $N_1 = \overline{B}$, then they will be G_δ-sets in M and N respectively by Fact 2 of S.358 (recall that every closed subset in a metric space is G_δ). Thus we can assume that $M_1 = M$ and $N_1 = N$, i.e., A is dense in M and B is dense in N.

Apply Fact 1 to find a G_δ-set A_1 in the space M and a G_δ-set B_1 in the space N such that $A \subset A_1$, $B \subset B_1$ and there are continuous maps $f_1 : A_1 \to N$ and $g_1 : B_1 \to M$ for which $f_1|A = f$ and $g_1|B = g = f^{-1}$. Apply Fact 2 of S.358 once more to conclude that $A' = f_1^{-1}(B_1)$ is a G_δ-set in M and $B' = g_1^{-1}(A_1)$ is a G_δ-set in N. Since $f_1(A) = f(A) = B \subset B_1$, we have $A \subset A'$; it follows from $g_1(B) = g(B) = A \subset A_1$ that $B \subset B'$.

Let $h = f_1|A'$; then $h(A') = f_1(A') \subset B_1$ and therefore the function $g_1 \circ h$ is well-defined. Observe also that $g_1 \circ h : A' \to g_1(B_1) = A_1$. Furthermore, $g_1(h(x)) = g(f(x)) = x$ for any $x \in A$ and hence the identity map $i = \text{id}_{A_1}^{A'} : A' \to A_1$ coincides with $g_1 \circ h$ on a dense set A of the space A' which shows that $g_1 \circ h = i$ (see Fact 0 of S.351), i.e.,

(1) $g_1(h(x)) = x$ for all $x \in A'$.

In particular, $g_1(h(x)) \in A' \subset A_1$ and hence $g_1(h(x)) \in A_1$, i.e., $h(x) \in B'$ for all $x \in A'$; thus $h(A') \subset B'$.

Analogously, for the map $u = g_1|B'$, we have $u(B') = g_1(B') \subset A_1$ and hence the function $f_1 \circ u$ is well-defined. Furthermore, $f_1(u(y)) = f(g(y)) = y$ for every $y \in B$ and therefore $f_1 \circ u$ coincides with the identity map $j = \text{id}_{B_1}^{B'}$ on a dense set B of the space B' which shows that $f_1 \circ u = j$ (see Fact 0 of S.351), i.e.,

(2) $f_1(u(y)) = y$ for all $y \in B'$.

An immediate consequence is that $f_1(u(y)) \in B_1$ and hence $u(y) \in A'$ for each $y \in B'$; thus $u(B') \subset A'$.

Given any $y \in B'$, we have $y = f_1(u(y)) = h(x)$ where $x = u(y) \in A'$; consequently, $h(A') = B'$. Analogously, $x = g_1(h(x)) = u(y)$ for $y = h(x) \in B'$ and therefore $u(B') = A'$. Finally, the properties (1) and (2) show that we have $u(h(x)) = g_1(h(x)) = x$ for all $x \in A'$ and $h(u(y)) = f_1(u(y)) = y$ for all $y \in B'$, i.e., the map h is the promised homeomorphism and $u = h^{-1}$. Fact 2 is proved. $\square$

Fact 3. Suppose that M is a Polish space and $Z \in \Sigma_\xi^0(M)$ (or $Z \in \Pi_\xi^0(M)$) for some countable ordinal $\xi \geq 2$. Then, for any Polish space N, if Z embeds in N, then $Z \in \Sigma_\xi^0(N)$ (or $Z \in \Pi_\xi^0(N)$ respectively). In particular, if Z embeds as a Borel subset in some Polish space M, then for any Polish space N, if Z embeds in N, then $Z \in \mathbb{B}(N)$.

Proof. Denote by Y the subspace of N homeomorphic to Z; let $f : Z \to Y$ be the respective homeomorphism. Apply Fact 2 to find a G_δ-set Z' in the space M and a G_δ-set Y' in the space N such that there is a homeomorphism $h : Z' \to Y'$ for which $h|Z = f$.

Since $Z \in \Sigma^0_\xi(M)$ (or $Z \in \Pi^0_\xi(M)$ respectively), we have $Z = Z \cap Z' \in \Sigma^0_\xi(Z')$ (or $Z = Z \cap Z' \in \Pi^0_\xi(Z')$ respectively) by Fact 1 of T.319. Therefore $Y \in \Sigma^0_\xi(Y')$ (or $Y \in \Pi^0_\xi(Y')$ respectively) (it is evident that homeomorphisms preserve both additive and multiplicative classes). Now apply Fact 1 of T.319 once more to see that there is $A \in \Sigma^0_\xi(N)$ (or $A \in \Pi^0_\xi(N)$ respectively) such that $Y = A \cap Y'$. Since Y' is a G_δ-subset of N, we have $Y' \in \Delta^0_\xi(N)$ and hence $Y \in \Sigma^0_\xi(N)$ (or $Y \in \Pi^0_\xi(N)$ respectively) by Fact 1 of T.341 and hence Fact 3 is proved. $\square$

Returning to our solution, assume that Y_i is a Borel set for every $i \in \omega$ and fix a Polish space M_i such that $Y_i \in \mathbb{B}(M_i)$ for all $i \in \omega$. Let $\pi_i : M = \prod_{n \in \omega} M_n \to M_i$ be the natural projection for all $i \in \omega$. To prove that $Y = \prod_{i \in \omega} Y_i \subset M$ is a Borel subset of M observe that $Y'_i = \pi_i^{-1}(Y_i)$ is a Borel subset of M for every $i \in \omega$ by Problem 332. Since $Y = \bigcap\{Y'_i : i \in \omega\}$, we can apply Fact 1 of T.331 to conclude that $Y \in \mathbb{B}(M)$, i.e., Y is a Borel set. This proves that any countable product of Borel sets is a Borel set.

Now, consider any embedding of X in a Polish space N. There is an embedding of X_i into a Polish space N_i such that $X_i \in \mathbb{B}(N_i)$ for all $i \in \omega$. Since X_i is also embedded in N, we have $X_i \in \mathbb{B}(N)$ by Fact 3. Thus $X \in \mathbb{B}(N)$ by Fact 1 of T.331 which shows that X is a Borel set and hence our solution is complete.

T.334. *Prove that every Borel set is an analytic space.*

Solution. Let X be a Borel set; we can assume that $X \subset M$ for some Polish space M and hence there is $\xi < \omega_1$ such that $X \in \Sigma^0_\xi(M)$. Let us prove, by induction on ξ, that X is analytic.

If X is closed or open in M, then X is Polish and hence it is a continuous image of $\mathbb{P}$ by Problem 328, so X is analytic. Now assume that $0 < \xi < \omega_1$ and we have proved, for all $\alpha < \xi$, that if $X \in \Sigma^0_\alpha(M) \cup \Pi^0_\alpha(M)$, then X is analytic.

Now, if $X \in \Sigma^0_\xi(M)$, then there exists a sequence $\{\alpha_n : n \in \omega\} \subset \xi$ and $X_n \in \Pi^0_{\alpha_n}(M)$ for all $n \in \omega$ such that $X = \bigcup\{X_n : n \in \omega\}$. It is an easy exercise to see that $\mathbb{P} = \bigoplus\{P_n : n \in \omega\}$ where P_n is homeomorphic to $\mathbb{P}$ for all $n \in \omega$. By the induction hypothesis, there is a continuous onto map $f_n : P_n \to X_n$ for all $n \in \omega$. Define $f : \mathbb{P} \to X$ as follows; if $x \in \mathbb{P}$, then there is a unique $n \in \omega$ with $x \in P_n$; let $f(x) = f_n(x)$. It is easy to prove, using Fact 1 of S.472, that f is a continuous onto map and hence X is analytic.

Finally, if $X \in \Pi^0_\xi(M)$, then $M \setminus X \in \Sigma^0_\xi(M)$ and therefore there exist a sequence of ordinals $\{\beta_n : n \in \omega\} \subset \xi$ and a set $Y_n \in \Pi^0_{\alpha_n}(M)$ for all $n \in \omega$ such that $M \setminus X = \bigcup\{Y_n : n \in \omega\}$. Therefore $X = \bigcap\{M \setminus Y_n : n \in \omega\}$. The space $M \setminus Y_n \in \Sigma^0_{\beta_n}(M)$ is analytic for all $n \in \omega$ by the induction hypothesis and hence X is analytic by Fact 3 of T.132. Thus our inductive procedure shows that all elements of $\mathbb{B}(M)$ are analytic and hence X is analytic.

T.335. *Prove that*

(i) *any closed subspace of an analytic space is an analytic space;*
(ii) *any open subspace of an analytic space is an analytic space;*
(iii) *any countable product of analytic spaces is an analytic space.*

Solution. The items (i) and (iii) were proved in Fact 3 of T.132. To prove that
(ii) also holds, assume that X is an analytic space and $U \in \tau(X)$. Since X has a
countable network, there is a family $\{F_n : n \in \omega\}$ of closed subsets of X such that
$U = \bigcup_{n \in \omega} F_n$. Each F_n is analytic by (i) and a countable union of analytic spaces
is analytic: this was also proved in Fact 3 of T.132. Thus U is analytic and the proof
of (ii) is complete.

T.336. *Assume that Y is a space and $X_i \subset Y$ is an analytic space for all $i \in \omega$.*
Prove that $X = \bigcap\{X_i : i \in \omega\}$ is also an analytic space.

Solution. This was proved in Fact 3 of T.132.

T.337. *Assume that $X = \bigcup\{X_i : i \in \omega\}$ and X_i is an analytic space for every*
$i \in \omega$. *Prove that X is an analytic space.*

Solution. This was proved in Fact 3 of T.132.

T.338. *Let X and Y be Polish spaces. Suppose that $f : X \to Y$ is a continuous*
map. Prove that for any analytic set $B \subset Y$, the set $f^{-1}(B)$ is also analytic.

Solution. Every Polish space is analytic by Problem 334 and hence the space $X \times Y$
is analytic by Problem 335. The graph $G(f) = \{(x, f(x)) : x \in X\}$ of the map f
is a closed subspace of $X \times Y$ (see Fact 4 of S.390) and hence $G(f)$ is also analytic
by Problem 335. The space $X \times B$ is analytic as well being a product of two analytic
spaces (see Problem 335) and therefore the space $F = (X \times B) \cap G(f)$ is analytic
by Problem 336. Now denote by $\pi : X \times Y \to X$ the relevant natural projection and
observe that $\pi(F) = f^{-1}(B)$. It is evident that a continuous image of an analytic
space is an analytic space so $f^{-1}(B)$ is analytic.

T.339. *Let A and B be two disjoint analytic subsets of a Polish space M. Prove that*
there exist Borel subsets A' and B' of the space M such that $A \subset A'$, $B \subset B'$ and
$A' \cap B' = \emptyset$.

Solution. Given a Polish space N, call sets $C, D \subset N$ *Borel-separated in N* if
there exist disjoint $C', D' \in \mathbb{B}(N)$ such that $C \subset C'$ and $D \subset D'$. We identify
$\mathbb{P}$ with the space ω^ω (see Problem 313). If f is a function, then $\mathrm{dom}(f)$ is its
domain; given a function g, the expression $f \subset g$ says that $\mathrm{dom}(f) \subset \mathrm{dom}(g)$ and
$g|\mathrm{dom}(f) = f$. Given any $s \in \omega^{<\omega}$, let $O_s = \{x \in \mathbb{P} : s \subset x\}$.

If $k, n \in \omega$ and $s \in \omega^k$, then $s{}^\frown n \in \omega^{k+1}$ is defined by $(s{}^\frown n)(k) = n$ and
$(s{}^\frown n)|k = s$. Suppose that we have a set of functions $\{f_i : i \in I\}$ such that
$f_i|(\mathrm{dom}(f_i) \cap \mathrm{dom}(f_j)) = f_j|(\mathrm{dom}(f_i) \cap \mathrm{dom}(f_j))$ for any $i, j \in I$. Then we
can define a function f with $\mathrm{dom}(f) = \bigcup_{i \in I} \mathrm{dom}(f_i)$ as follows: given any $x \in$
$\mathrm{dom}(f)$, find any $i \in I$ with $x \in \mathrm{dom}(f_i)$ and let $f(x) = f_i(x)$. It is easy to check

that the value of f at x does not depend on the choice of i, so we have consistently defined a function f which will be denoted by $\bigcup\{f_i : i \in I\}$.

Fact 1. Suppose that N is a Polish space and $C_n, D_n \subset N$ for all $n \in \omega$. If the sets C_n and D_m are Borel-separated in N for all $m, n \in \omega$, then the sets $C = \bigcup_{n \in \omega} C_n$ and $D = \bigcup_{n \in \omega} D_n$ are also Borel-separated in N.

Proof. Take disjoint Borel sets $U(m, n)$ and $V(m, n)$ in the space N such that $C_n \subset U(m, n)$ and $D_m \subset V(m, n)$ for all $m, n \in \omega$ and observe that the sets $U = \bigcup_{n \in \omega}(\bigcap_{m \in \omega} U(m, n))$ and $V = \bigcup_{m \in \omega}(\bigcap_{n \in \omega} V(m, n))$ are Borel in N (see Fact 1 of T.331) and disjoint while $C \subset U$ and $D \subset V$. Fact 1 is proved. □

Returning to our solution, fix continuous onto maps $f : \mathbb{P} \to A$ and $g : \mathbb{P} \to B$. Assume, towards a contradiction, that A and B are not Borel-separated in M. Since $\mathbb{P} = \bigcup\{O_s : s \in \omega^1\}$, we can apply Fact 1 to conclude that there exist $s_1, t_1 \in \omega^1$ such that $f(O_{s_1})$ is not Borel-separated in M from $g(O_{t_1})$.

Suppose that $n \in \mathbb{N}$ and we have constructed $s_1, \ldots, s_n, t_1, \ldots, t_n \in \omega^{<\omega}$ with the following properties:

(1) $s_i, t_i \in \omega^i$ for every $i \leq n$;
(2) $s_{i-1} \subset s_i$ and $t_{i-1} \subset t_i$ if $2 \leq i \leq n$;
(3) the sets $f(O_{s_i})$ and $g(O_{t_i})$ are not Borel-separated in M for all $i \leq n$.

Since $O_{s_n} = \bigcup\{O_{s_n \frown k} : k \in \omega\}$ and $O_{t_n} = \bigcup\{O_{t_n \frown k} : k \in \omega\}$, we can apply Fact 1 to conclude that there are $k, l \in \omega$ such that $f(O_{s_n \frown k})$ is not Borel-separated in M from $g(O_{t_n \frown l})$. Letting $s_{n+1} = s_n \frown k$ and $t_{n+1} = t_n \frown l$, we obtain a sequence $s_1, \ldots, s_n, s_{n+1}, t_1, \ldots, t_n, t_{n+1} \in \omega^{<\omega}$ for which the conditions (1)–(3) hold for all $i \leq n + 1$.

Thus our inductive construction can go on to provide sequences $\{s_i : i \in \mathbb{N}\}$ and $\{t_i : i \in \mathbb{N}\}$ such that (1)–(3) are fulfilled for all $i \in \mathbb{N}$. It follows from (2) that the points $s = \bigcup\{s_i : i \in \mathbb{N}\} \in \mathbb{P}$ and $t = \bigcup\{t_i : i \in \mathbb{N}\} \in \mathbb{P}$ are well-defined. The sets A and B being disjoint, we have $f(s) \neq g(t)$ and hence there are disjoint $U, V \in \tau(M)$ such that $f(s) \in U$ and $g(t) \in V$. The families $\{O_{s_n} : n \in \mathbb{N}\}$ and $\{O_{t_n} : n \in \mathbb{N}\}$ are local bases at the points s and t respectively, so, by continuity of f and g, there is $n \in \mathbb{N}$ such that $f(O_{s_n}) \subset U$ and $g(O_{t_n}) \subset V$. The sets U and V being Borel in M, we conclude that $f(O_{s_n})$ and $g(O_{t_n})$ are Borel-separated in M which contradicts (3). Therefore the sets A and B are Borel-separated in M and hence our solution is complete.

T.340. *Let X be a subspace of a Polish space M. Prove that X is Borel if and only if X and $M \setminus X$ are analytic.*

Solution. If X and $M \setminus X$ are analytic, then there are disjoint $A, B \in \mathbb{B}(M)$ such that $X \subset A$ and $M \setminus X \subset B$ (see Problem 339). It is immediate that $X = A$ and $M \setminus X = B$, so X is a Borel subset of M.

Now, if $X \in \mathbb{B}(M)$, then $M \setminus X \in \mathbb{B}(M)$ by Fact 1 of T.331, so both X and $M \setminus X$ are analytic by Problem 334.

T.341. *Prove that X is a Borel set if and only if there exists a closed subspace of* $\mathbb{P}$ *which condenses onto X. As a consequence, if a Borel set X can be condensed onto a second countable space Y, then Y is a also Borel set.*

Solution. The expression $Z \simeq T$ says that the spaces Z and T are homeomorphic. A family of sets $\mathcal{A}$ is *closed under finite (countable) intersections* if for any finite (countable) $\mathcal{A}' \subset A$, we have $\bigcap \mathcal{A}' \in \mathcal{A}$. Analogously, $\mathcal{A}$ is *closed under finite (countable) unions* if for any finite (countable) $\mathcal{A}' \subset A$, we have $\bigcup \mathcal{A}' \in \mathcal{A}$. Say that a second countable space Z is $\mathbb{P}$-*representable* if there is a closed $F \subset \mathbb{P}$ which condenses onto Z. We must prove that Z is a Borel set if and only if it is $\mathbb{P}$-representable.

We identify $\mathbb{P}$ with the space ω^ω (see Problem 313). If f is a function, then $\mathrm{dom}(f)$ is its domain; given a function g, the expression $f \subset g$ says that $\mathrm{dom}(f) \subset \mathrm{dom}(g)$ and $g|\mathrm{dom}(f) = f$. Given any $s \in \omega^{<\omega}$ let $O_s = \{x \in \mathbb{P} : s \subset x\}$. If $k, n \in \omega$ and $s \in \omega^k$, then $s^\frown n \in \omega^{k+1}$ is defined by $(s^\frown n)(k) = n$ and $(s^\frown n)|k = s$.

Fact 1. If Z is an arbitrary space and ξ is a countable ordinal, then

(a) the class $\Sigma_\xi^0(Z)$ is closed under countable unions and finite intersections;

(b) the class $\Pi_\xi^0(Z)$ is closed under finite unions and countable intersections.

Proof. It is evident that a finite intersection of open subsets of Z is open in Z. Besides, any union of open subsets of Z is open in Z which shows that (a) is proved for $\xi = 0$. Analogously, (b) is true for $\xi = 0$ because any finite union of closed subsets of Z is closed in Z and any intersection of closed subsets of Z is again closed in Z.

Assume that $\alpha < \omega_1$ and we proved the properties (a) and (b) for all $\xi < \alpha$. Let $\Pi_\alpha = \bigcup \{\Pi_\beta^0(Z) : \beta < \alpha\}$; if $A_n \in \Sigma_\alpha^0(Z)$ for all $n \in \omega$, then $A_n = \bigcup_{m \in \omega} A_n^m$ where $A_n^m \in \Pi_\alpha$ for all $m, n \in \omega$ and hence $A = \bigcup_{n \in \omega} A_n = \bigcup \{A_n^m : m, n \in \omega\} \in \Sigma_\alpha^0(Z)$ which shows that $\Sigma_\alpha^0(Z)$ is closed under countable unions.

To prove that $\Sigma_\alpha^0(Z)$ is closed under finite intersections, it suffices to establish it for intersections of any two elements of $\Sigma_\alpha^0(Z)$, so take any $A, B \in \Sigma_\alpha^0(Z)$. Choose families $\{A_n : n \in \omega\} \subset \Pi_\alpha$ and $\{B_n : n \in \omega\} \subset \Pi_\alpha$ such that $A = \bigcup_{n \in \omega} A_n$ and $B = \bigcup_{n \in \omega} B_n$. By the induction hypothesis, we have $A_n \cap B_m \in \Pi_\alpha$ for all $m, n \in \omega$ and hence $A \cap B = \bigcup \{A_n \cap B_m : n, m \in \omega\} \in \Sigma_\alpha^0(Z)$ which proves (a) for $\xi = \alpha$.

Now assume that $B_n \in \Pi_\alpha^0(Z)$ and hence $A_n = Z \backslash B_n \in \Sigma_\alpha^0(Z)$ for all $n \in \omega$. For every $n \in \omega$, there is a collection $\{A_n^m : m \in \omega\} \subset \Pi_\alpha$ such that $A_n = \bigcup_{m \in \omega} A_n^m$ and hence $A = \bigcup_{n \in \omega} A_n = \bigcup \{A_n^m : n, m \in \omega\} \in \Sigma_\alpha^0(Z)$. Thus $B = \bigcap_{n \in \omega} B_n = Z \backslash A \in \Pi_\alpha^0(Z)$ which shows that $\Pi_\alpha^0(Z)$ is closed under countable intersections.

Finally, if $A, B \in \Pi_\alpha^0(Z)$, then $A' = Z \backslash A \in \Sigma_\alpha^0(Z)$ and $B' = Z \backslash B \in \Sigma_\alpha^0(Z)$ which implies that $A' \cap B' \in \Sigma_\alpha^0(Z)$ because we proved that $\Sigma_\alpha^0(Z)$ is closed under finite intersections. Therefore $A \cup B = Z \backslash (A' \cap B') \in \Pi_\alpha^0(Z)$ and hence $\Pi_\alpha^0(Z)$

is closed under finite unions. This settles (a) and (b) for all $\xi \leq \alpha$, so our inductive procedure shows that (a) and (b) hold for all $\xi < \omega_1$. Fact 1 is proved.   □

Fact 2. Suppose that Z is a second countable space. If $\{A_n : n \in \omega\}$ is a disjoint family of analytic subspaces of Z, then there is a disjoint family $\{B_n : n \in \omega\} \subset \mathbb{B}(Z)$ such that $A_n \subset B_n$ for all $n \in \omega$.

Proof. We can consider that there is a Polish space M such that $Z \subset M$. For every $n \in \omega$, the sets A_n and $A'_n = \bigcup\{A_k : k \in \omega\backslash\{n\}\}$ are analytic and disjoint, so there are disjoint $B'_n, C'_n \in \mathbb{B}(M)$ such that $A_n \subset B'_n$ and $A'_n \subset C'_n$ by Problem 339. Then $D_n = B'_n\backslash(\bigcup_{i<n} B'_i) \in \mathbb{B}(M)$ (see Fact 1 of T.331) for every $n \in \omega$. It follows from the choice of the sets B'_n that $A_n \subset D_n$ for all $n \in \omega$; it is immediate that the family $\{D_n : n \in \omega\}$ is disjoint. If $B_n = D_n \cap Z$, then $B_n \in \mathbb{B}(Z)$ for all $n \in \omega$ and the family $\{B_n : n \in \omega\}$ is as promised. Fact 2 is proved.   □

Our proof of necessity will be accomplished in several steps:

Step 1.   The space $\mathbb{R}$ is $\mathbb{P}$-representable.
It is easy to see that we can represent $\mathbb{P}$ as $\bigoplus_{i\in\omega} P_i$ where P_i is homeomorphic to $\mathbb{P}$ for all $i \in \omega$. Let $\{q_i : i \in \mathbb{N}\}$ be a faithful enumeration of $\mathbb{Q}$; choose a point $z_i \in P_i$ for all $i \in \mathbb{N}$ and fix a homeomorphism $f_0 : P_0 \to \mathbb{P}$.
It is evident that $F = \{z_i : i \in \mathbb{N}\} \cup P_0$ is a closed subset of $\mathbb{P}$. Let $f(z_i) = q_i$ for all $i \in \mathbb{N}$; if $z \in P_0$, then let $f(z) = f_0(z)$. It is immediate that $f : F \to \mathbb{R}$ is a condensation.

Step 2.   Any countable product of $\mathbb{P}$-representable spaces is $\mathbb{P}$-representable and, in particular, the space $\mathbb{R}^\omega$ is $\mathbb{P}$-representable.
If Z_i is $\mathbb{P}$-representable for all $i \in \omega$, then there is a closed $F_i \subset \mathbb{P}$ and a condensation $f_i : F \to Z_i$. It is easy to see that the map $g = \prod_{i\in\omega} f_i$ condenses $F = \prod_{i\in\omega} F_i$ onto $\prod_{i\in\omega} Z_i$. The set F is closed in $\mathbb{P}^\omega$ and the latter space is homeomorphic to $\mathbb{P}$. Thus there is a closed $F \subset \mathbb{P}$ which condenses onto $\prod_{i\in\omega} Z_i$. Finally, recall that $\mathbb{R}$ is $\mathbb{P}$-representable (see Step 1) and hence $\mathbb{R}^\omega$ is also $\mathbb{P}$-representable.

Step 3.   If a second countable space Z is $\mathbb{P}$-representable and A is closed in Z, then A is also $\mathbb{P}$-representable.
Indeed, if there is a closed $G \subset \mathbb{P}$ and a condensation $f : G \to Z$, then $F = f^{-1}(A)$ is a closed subset of G and hence of $\mathbb{P}$ which condenses onto A.

Step 4.   Any Polish space is $\mathbb{P}$-representable.
Indeed, if M is a Polish space, then we can consider that M is a closed subspace of $\mathbb{R}^\omega$ (see Problem 273 of [TFS]). Now apply the results of Steps 2 and 3.

Step 5.   Suppose that Z is a second countable space such that $Z = \bigcup\{Z_i : i \in \omega\}$ and $Z_i \cap Z_j = \emptyset$ whenever $i \neq j$. If Z_i is $\mathbb{P}$-representable for all $i \in \omega$, then Z is also $\mathbb{P}$-representable.
For each $i \in \omega$, let F_i be a closed subspace of a space $P_i \simeq \mathbb{P}$ for which there exists a condensation $f_i : F_i \to Z_i$. It is immediate that the space $F = \bigoplus_{i\in\omega} F_i$ condenses onto Z; since F is a closed subspace of $P = \bigoplus_{i\in\omega} P_i$ and $P \simeq \mathbb{P}$, the space Z is $\mathbb{P}$-representable.

Step 6. Suppose that Z is a second countable space and a subspace $Z_i \subset Z$ is ℙ-representable for all $i \in \omega$. Then $A = \bigcap \{Z_i : i \in \omega\}$ is also ℙ-representable. Indeed, the space A is homeomorphic to a closed subspace of $\prod_{i \in \omega} Z_i$ (see Fact 7 of S.271), so A is ℙ-representable (see Steps 2 and 3).

Now let X be a Borel subset of a Polish space M. To start an inductive proof assume that $X \in \Sigma_0^0(M) \cup \Pi_0^0(M)$; then X is either closed or open in M and hence Polish, so it is ℙ-representable by Step 4.

Suppose that we have an ordinal $\xi < \omega_1$ such that any $A \in \Sigma_\beta^0(M) \cup \Pi_\beta^0(M)$ is ℙ-representable for each $\beta < \xi$. Let $\Delta_\beta = \{A \cap B : A \in \Sigma_\beta^0(Z) \text{ and } B \in \Pi_\beta^0(Z)\}$ for all $\beta < \xi$. Then every element of Δ_β is ℙ-representable (see Step 6) for all $\beta < \xi$.

If $X \in \Sigma_\xi^0(M)$, then there is a sequence $\{\alpha_n : n \in \omega\} \subset \xi$ and $A_n \in \Pi_{\alpha_n}(M)$ for all $n \in \omega$ such that $X = \bigcup_{n \in \omega} A_n$. We can assume that $\alpha_n \leq \alpha_{n+1}$ and hence $A_0, \ldots, A_n \in \Pi_{\alpha_n}^0(M)$ for all $n \in \omega$ (see Problem 320). Thus $A_n' = A_0 \cup \cdots \cup A_n \in \Pi_{\alpha_n}^0(M)$ for all $n \in \omega$ (see Fact 1). Let $X_0 = A_0$ and $X_{n+1} = A_{n+1} \setminus A_n'$ for all $n \in \omega$. It is easy to see that $X_n \in \Delta_{\alpha_n}$ for all $n \in \omega$; besides, each X_n is ℙ-representable, the family $\{X_n : n \in \omega\}$ is disjoint and $X = \bigcup_{n \in \omega} X_n$. Therefore X is ℙ-representable by Step 5.

Finally, assume that $X \in \Pi_\xi^0(M)$; then $M \setminus X \in \Sigma_\xi^0(M)$ and hence there is a sequence of ordinals $\{\beta_n : n \in \omega\} \subset \xi$ and $B_n \in \Pi_{\alpha_n}(M)$ for every $n \in \omega$ such that $M \setminus X = \bigcup_{n \in \omega} B_n$. Therefore $C_n = M \setminus B_n \in \Sigma_{\beta_n}^0(M)$ is ℙ-representable for all $n \in \omega$ by the induction hypothesis and $X = \bigcap_{n \in \omega} C_n$ which shows that X is ℙ-representable by Step 6. Thus our inductive procedure can be continued to prove that every $X \in \mathbb{B}(M)$ is ℙ-representable. Since every Borel set X belongs to $\mathbb{B}(M)$ for some Polish space M, we proved that every Borel set X is ℙ-representable, i.e., there is a closed subspace $F \subset \mathbb{P}$ which condenses onto X. This settles necessity.

Now suppose that $F \subset \mathbb{P}$ is closed and $f : F \to X$ is a condensation; we can assume that $X \subset M$ for some Polish space M. Consider the set $B_s = f(O_s \cap F)$ for any $s \in \omega^{<\omega}$. The set O_s is analytic being closed in ℙ; therefore $O_s \cap F$ is also analytic and hence B_s is analytic as well (maybe empty) for all $s \in \omega^{<\omega}$.

The family $\{B_s : s \in \omega^1\}$ is countable and disjoint (because f is a condensation), so we can apply Fact 2 to find a disjoint family $\{B_s' : s \in \omega^1\} \subset \mathbb{B}(M)$ such that $B_s \subset B_s'$ for all $s \in \omega^1$; let $C_s = B_s' \cap \overline{B}_s$ for all $s \in \omega^1$ (the bar denotes the closure in M); then the family $\{C_s : s \in \omega^1\} \subset \mathbb{B}(M)$ is still disjoint and $B_s \subset C_s$ for every $s \in \omega^1$.

Assume that $1 \leq n < \omega$ and we have a family $\{C_s : s \in \omega^i\} \subset \mathbb{B}(M)$ for every $i \leq n$ with the following properties:

(1) $\{C_s : s \in \omega^i\}$ is disjoint for all $i \leq n$;
(2) $B_s \subset C_s \subset \overline{B}_s$ for every $s \in \omega^i$ and $i \leq n$;
(3) $C_s \subset C_{s|j}$ whenever $s \in \omega^i$ and $1 \leq j < i \leq n$.

Fix any $s \in \omega^n$; since the family $\{B_{s \frown k} : k \in \omega\}$ is disjoint and consists of analytic sets, we can find a disjoint family $\{A_k : k \in \omega\} \subset \mathbb{B}(M)$ such that $B_{s \frown k} \subset A_k$ for all $k \in \omega$ (see Fact 2). If $C_{s \frown k} = A_k \cap C_s \cap \overline{B}_s$, then $C_{s \frown k} \in \mathbb{B}(M)$ and

$B_{s\frown k} \subset C_{s\frown k}$ for all $k \in \omega$. After we construct $C_{s\frown k}$ for all $s \in \omega^n$ and $k \in \omega$, we will have a family $\{C_s : s \in \omega^{n+1}\}$ and it is immediate that (1)–(3) still hold for all $i \leq n+1$. Thus our inductive construction can be continued to provide families $\{C_s : s \in \omega^i\}$ for all $i \in \mathbb{N}$ such that the conditions (1)–(3) are satisfied for all $n \in \omega$.

We claim that $X = Q = \bigcap\{\bigcup\{C_s : s \in \omega^i\} : i \in \mathbb{N}\}$; of course, if we prove this, then X is a Borel subset of M by Fact 1 of T.331. Since $X \subset \bigcup\{B_s : s \in \omega^i\}$ for al $i \in \mathbb{N}$, the condition (2) implies that $X \subset Q$.

Now take any $z \in Q$; there is $s \in \mathbb{P}$ such that $z \in \bigcap\{C_{s|n} : n \in \mathbb{N}\}$. As a consequence, $z \in \bigcap\{\overline{B}_{s|n} : n \in \mathbb{N}\}$ by (2) and hence $O_{s|n} \cap F \neq \emptyset$ for all $n \in \mathbb{N}$. The family $\{O_{s|n} : n \in \mathbb{N}\}$ is a local base of $\mathbb{P}$ at s, so $s \in \overline{F} = F$ (recall that F is closed in $\mathbb{P}$). If $f(s) \neq z$, then, by continuity of f, there is $n \in \mathbb{N}$ such that $z \notin \overline{f(O_{s|n} \cap F)} = \overline{B}_{s|n}$ which is a contradiction. Therefore $z = f(s) \in X$ and hence $Q \subset X$ which shows that $Q = X$ and hence the space X is a Borel set. This finishes the proof of sufficiency.

Finally, assume that X is a Borel set and $g : X \to Y$ is a condensation of X onto a second countable space Y. We proved that there exists a closed $F \subset \mathbb{P}$ for which there is a condensation $f : F \to X$. Therefore $g \circ f$ is a condensation of F onto Y and hence Y is a Borel set. Our solution is complete.

T.342. *Show that there exists a subspace $X \subset \mathbb{R}$ which is not analytic (and hence not Borel).*

Solution. In Fact 5 of T.250, we constructed a space $Y \subset \mathbb{K} \subset \mathbb{R}$ which is not K-analytic. Observe that $\mathbb{P}$ is a G_δ-subset of $\mathbb{R}$; since every open subset of $\mathbb{R}$ is K_σ, the space $\mathbb{P}$ is $K_{\sigma\delta}$ and hence any analytic space is K-analytic. This shows that Y is not analytic. Since every Borel set in analytic by Problem 334, the space Y is not Borel either.

T.343. *Prove that*

(i) any closed subset of a K-analytic space is a K-analytic space;
(ii) any countable product of K-analytic spaces is a K-analytic space.

Solution. This was proved in Fact 3 of T.250.

T.344. *Assume that Y is a space and $X_i \subset Y$ is a K-analytic space for all $i \in \omega$. Prove that $X = \bigcap\{X_i : i \in \omega\}$ is also a K-analytic space.*

Solution. The space X embeds in $Z = \prod\{X_i : i \in \omega\}$ as a closed subspace by Fact 7 of S.271. The space Z is K-analytic by Fact 3 of T.250 and hence X is also K-analytic by Problem 343 (or by Fact 3 of T.250).

T.345. *Assume that $X = \bigcup\{X_i : i \in \omega\}$ and X_i is a K-analytic space for all $i \in \omega$. Prove that X is a K-analytic space.*

Solution. This was proved in Fact 3 of T.250.

T.346. *Observe that there exist K-analytic non-analytic spaces. Show that any analytic space is a K-analytic space. Prove that for any space X with a countable network, X is analytic if and only if it is K-analytic.*

Solution. Given a space Z denote by $\mathcal{K}(Z)$ the family of all compact subsets of Z. We identify $\mathbb{P}$ with the space ω^ω (see Problem 313). If f is a function, then $\text{dom}(f)$ is its domain; given a function g the expression $f \subset g$ says that $\text{dom}(f) \subset \text{dom}(g)$ and $g|\text{dom}(f) = f$. Given any $s \in \omega^{<\omega}$, let $O_s = \{x \in \mathbb{P} : s \subset x\}$. If Z and Y are spaces and $\varphi : Z \to \exp(Y)$, then $\varphi(A) = \bigcup\{\varphi(z) : z \in A\}$ for any $A \subset Z$ and $\varphi^{-1}(B) = \{z \in Z : \varphi(z) \subset B\}$ for every $B \subset Y$. Note that φ is called onto map if $\varphi(Z) = \bigcup\{\varphi(z) : z \in Z\} = Y$ which need not imply $\{\varphi(z) : z \in Z\} = \exp(Y)$.

Fact 1. Suppose that Z and T are spaces and we have a map $\varphi : Z \to \exp(T)$. Then φ is upper semicontinuous if and only if it is *upper semicontinuous at any* $z \in Z$, i.e., for any $z \in Z$ and $U \in \tau(\varphi(z), T)$ there is $V \in \tau(z, Z)$ such that $\varphi(V) \subset U$.

Proof. If the mapping φ is upper semicontinuous and $U \in \tau(\varphi(z), T)$, then the set $V = \varphi^{-1}(U)$ is open in Z and $z \in V$. It is evident that $\varphi(V) \subset U$, so necessity is proved.

Now assume that φ is upper semicontinuous at any $z \in Z$ and take any set $U \in \tau(T)$. If $z \in \varphi^{-1}(U)$, then $\varphi(z) \subset U$ and hence there is $V_z \in \tau(z, Z)$ such that $\varphi(V_z) \subset U$ which implies $V_z \subset \varphi^{-1}(U)$. Thus $\varphi^{-1}(U) = \bigcup\{V_z : z \in \varphi^{-1}(U)\}$ is an open set, so φ is upper semicontinuous. Fact 1 is proved. □

Fact 2. For any K-analytic space Z there is an upper semicontinuous onto map $\varphi : \mathbb{P} \to \mathcal{K}(Z)$.

Proof. Fix a $K_{\sigma\delta}$-space Y and a continuous onto map $f : Y \to Z$. There is a space T such that $Y \subset T$ and a family $\{Y_n : n \in \omega\}$ of σ-compact subspaces of T for which $Y = \bigcap_{n \in \omega} Y_n$. We can assume that $Y_n = \bigcup_{m \in \omega} K_n^m$ where $K_n^m \subset K_n^{m+1}$ and K_n^m is compact for all $m, n \in \omega$. Given any $s \in \mathbb{P}$, let $L(s) = \bigcap\{K_n^{s(n)} : n \in \omega\}$. Then $L(s) \subset Y$ is compact and therefore $K(s) = f(L(s))$ is a compact subspace of Z for every $s \in \mathbb{P}$. Let us prove that the map $\varphi : \mathbb{P} \to \mathcal{K}(X)$ defined by $\varphi(s) = K(s)$, is upper semicontinuous.

Take any $s \in \mathbb{P}$ and $U \in \tau(K(s), Z)$; the set $W = f^{-1}(U)$ is open in Y, so there is $W' \in \tau(T)$ such that $W' \cap Y = W$. Since $K(s) = f(L(s)) \subset U$, we have $L(s) \subset W'$ and hence there is $k \in \omega$ such that $K' = \bigcap\{K_n^{s(n)} : n < k\} \subset W'$ (see Fact 1 of S.326). Then $V = O_{s|k} \in \tau(s, \mathbb{P})$; given any point $t \in V$, we have $t|k = s|k$ and hence $L(t) \subset \bigcap\{K_n^{t(n)} : n < k\} = \bigcap\{K_n^{s(n)} : n < k\} = K'$ which shows that $\varphi(t) = f(L(t)) \subset f(K' \cap Y) \subset f(W) = U$. Thus φ is upper semicontinuous at any point of $\mathbb{P}$ so it is upper semicontinuous by Fact 1. To finally see that $\varphi(\mathbb{P}) = Z$, take any $z \in Z$ and $y \in Y$ with $f(y) = z$. For every $n \in \omega$, we have $y \in Y_n$ and hence there is $s(n) \in \omega$ such that $y \in K_n^{s(n)}$. This defines a function $s \in \mathbb{P}$ such that $y \in L(s)$ and therefore $x \in K(s) = f(L(s)) = \varphi(s)$ so Fact 2 is proved. □

Fact 3. Suppose that Z is a space and $K_0, \ldots, K_n$ are compact subspaces of Z. If $K = \bigcap_{i \leq n} K_i$ and $U \in \tau(K, Z)$, then there exists $U_i \in \tau(K_i, Z)$ for every $i \leq n$ such that $\bigcap_{i \leq n} U_i \subset U$. In particular, if $K = \emptyset$, then $\bigcap_{i \leq n} U_i = \emptyset$.

Proof. Fix a set $V \in \tau(\beta Z)$ such that $V \cap Z = U$; then $K \subset V$. Consider the family $\mathcal{F} = \{\overline{V}_0 \cap \cdots \cap \overline{V}_n : V_i \in \tau(K_i, \beta Z) \text{ for all } i \leq n\}$ (the bar denotes the closure in βZ). If $z \in \beta Z \backslash K$, then $z \notin K_i$ for some $i \leq n$, and hence, by regularity of βZ, there is $V_i \in \tau(K_i, \beta Z)$ such that $z \notin \overline{V}_i$. Letting $V_j = \beta Z$ for all $j \in (n + 1) \backslash \{i\}$, we obtain a set $F = \overline{V}_0 \cap \cdots \cap \overline{V}_n = \overline{V}_i \in \mathcal{F}$ with $z \notin F$. This proves that $\mathcal{F}$ is a family of compacts sets with $\bigcap \mathcal{F} = K \subset V$.

Now we can apply Fact 1 of S.326 to find a finite $\mathcal{U}_i \subset \tau(K_i, \beta Z)$ for each $i \leq n$ such that $\bigcap \{\overline{W} : W \in \mathcal{U}_0 \cup \cdots \cup \mathcal{U}_n\} \subset V$. If $V_i = \bigcap \mathcal{U}_i$ for all $i \leq n$, then $V_0 \cap \cdots \cap V_n \subset V$ and hence, if $U_i = V_i \cap Z$, then $U_i \in \tau(K_i, Z)$ for all $i \leq n$ and $\bigcap_{i \leq n} U_i \subset U$. Fact 3 is proved.   □

Fact 4. Given spaces Z and Y let $\varphi : Z \to \exp(Y)$ be an upper semicontinuous map. If F is a closed subset of Y and $G = \{z \in Z : \varphi(z) \cap F \neq \emptyset\}$, then G is a closed subset of Z and the map $\varphi_1 : G \to \exp(Y)$ defined by $\varphi_1(z) = \varphi(z) \cap F$ is also upper semicontinuous.

Proof. The set G is closed because $G = Z \backslash \varphi^{-1}(Y \backslash F)$ and $\varphi^{-1}(Y \backslash F)$ is open by upper semicontinuity of φ. Now take any $z \in G$ and $U \in \tau(\varphi_1(z), Y)$. Then $U_1 = U \cup (Y \backslash F) \in \tau(\varphi(z), Y)$ and hence there is $V_1 \in \tau(z, Z)$ such that $\varphi(V_1) \subset U_1$. Evidently, $V = V_1 \cap G \in \tau(z, G)$ and $\varphi_1(V) \subset \varphi(V) \cap F \subset U_1 \cap F \subset U$ and we proved that φ_1 is upper semicontinuous at the point z. This, together with Fact 1, shows that Fact 4 is proved.   □

Returning to our solution observe that any non-metrizable compact space Z is K-analytic but not analytic because $nw(Z) = w(Z) > \omega$ while $nw(T) \leq \omega$ for any analytic space T. Thus $\mathbb{I}^{\omega_1}$ is K-analytic but not analytic.

Furthermore, $\mathbb{P}$ is a G_δ-subset of $\mathbb{R}$ because $\mathbb{P} = \bigcap \{\mathbb{R} \backslash \{q\} : q \in \mathbb{Q}\}$; since every open subset of $\mathbb{R}$ is σ-compact, the space $\mathbb{P}$ is $K_{\sigma\delta}$ and hence any analytic space is K-analytic.

Now assume that X is a K-analytic space with a countable network. Then there is a weaker second countable (Tychonoff!) topology μ on the space X (see Problem 156 of [TFS]); take a base $\mathcal{B} = \{B_n : n \in \omega\}$ of the space (X, μ) and let $F_n^0 = X \backslash B_n$, $F_n^1 = \overline{B}_n$ for all $n \in \omega$ (the bar denotes the closure in X).

By Fact 2, there is an upper semicontinuous onto map $\varphi : \mathbb{P} \to \mathcal{K}(X)$. The sets $P_n^0 = \{s \in \mathbb{P} : \varphi(s) \cap F_n^0 \neq \emptyset\}$ and $P_n^1 = \{s \in \mathbb{P} : \varphi(s) \cap F_n^1 \neq \emptyset\}$ are closed in $\mathbb{P}$ by Fact 4. Define a map $\varphi_n : P_n^0 \oplus P_n^1 \to X$ as follows: if $s \in P_n^0$, then $\varphi_n(s) = \varphi(s) \cap F_n^0$; if $s \in P_n^1$, then $\varphi_n(s) = \varphi(s) \cap F_n^1$. Observe that in the space $\mathbb{P}$, the sets P_n^0 and P_n^1 may intersect while they are considered disjoint in the space $P_n^0 \oplus P_n^1$. It is an easy consequence of Fact 4 that φ_n is upper semicontinuous for all $n \in \omega$. Evidently, φ_n is compact-valued, and it follows from $\varphi(\mathbb{P}) = X$ that $\varphi_n(P_n^0 \oplus P_n^1) = X$ for every $n \in \omega$.

The space $Q_n = P_n^0 \oplus P_n^1$ is homeomorphic to a closed subset of $\mathbb{P}$ for every $n \in \omega$, and hence $Q = \prod_{n \in \omega} Q_n$ is also homeomorphic to a closed subset of $\mathbb{P}$. For any $s \in Q$, let $\theta(s) = \bigcap_{n \in \omega} \varphi_n(s(n))$; we claim that

(1) $\theta(s)$ contains at most one element of X for every $s \in Q$.

To prove (1) suppose, towards a contradiction, that there are distinct points $x, y \in \theta(s)$. Since $\mathcal{B}$ is a base in (X, μ), there is $n \in \omega$ such that $x \in B_n$ and $y \notin \overline{B}_n$; thus $x \notin F_n^0$ and $y \notin F_n^1$. If $s(n) \in P_n^0$, then $\theta(s) \subset \varphi(s(n)) \cap F_n^0$ and hence $x \notin \theta(s)$, a contradiction. If $s(n) \in P_n^1$, then $\theta(s) \subset \varphi(s(n)) \cap F_n^1$ and hence $y \notin \theta(s)$; this final contradiction proves (1).

Let $F = \{s \in Q : \theta(s) \neq \emptyset\}$; it follows from (1) that for any $s \in F$ there is $h(s) \in X$ such that $\theta(s) = \{h(s)\}$ and therefore we have a map $h : F \to X$. Let us prove first that

(2) the set F is closed in Q.

Take any $s \in Q \backslash F$; then $\bigcap_{n \in \omega} \varphi_n(s(n)) = \emptyset$. Since all sets $\varphi_n(s(n))$ are compact, there is $k \in \omega$ such that $\bigcap_{n < k} \varphi_n(s(n)) = \emptyset$. By Fact 3 there exists a set $W_n \in \tau(\varphi_n(s(n)), X)$ for each $n < k$ such that $\bigcap_{n < k} W_n = \emptyset$. Since the map φ_n is upper semicontinuous, there is $V_n \in \tau(s(n), Q_n)$ such that $\varphi_n(V_n) \subset W_n$ for all $n < k$. The set $O = \{t \in Q : t(n) \in V_n \text{ for all } n < k\}$ is open in Q and $s \in O$; if $t \in O$, then $\theta(t) \subset \bigcap_{n < k} \varphi_n(t(n)) \subset \bigcap_{n < k} \varphi_n(V_n) \subset \bigcap_{n < k} W_n = \emptyset$ and hence $\theta(t) = \emptyset$. Thus every $s \in Q \backslash F$ has a neighborhood $O \subset Q \backslash F$ which proves that $Q \backslash F$ is open in Q, so F is closed in Q and (2) is proved. Our next step is to show that

(3) the map h is continuous.

Take any $s \in F$ and $U \in \tau(h(s), X)$. Since $\theta(s) \subset U$, there is $k \in \omega$ such that $\bigcap_{n < k} \varphi_n(s(n)) \subset U$ (see Fact 1 of S.326). Apply Fact 3 again to find, for each $n < k$, a set $U_n \in \tau(\varphi_n(s(n)), X)$ such that $\bigcap_{n < k} U_n \subset U$. The map φ_n being upper semicontinuous, there is $V_n \in \tau(s(n), Q_n)$ such that $\varphi_n(V_n) \subset U_n$ for all $n < k$.

The set $W' = \{t \in Q : t(n) \in V_n \text{ for all } n < k\}$ is open in Q and $s \in W'$; thus $W = W' \cap F \in \tau(s, F)$. If $t \in W$, then $t(n) \in V_n$ for all $n < k$ and hence $\theta(t) \subset \bigcap_{n < k} \varphi_n(t(n)) \subset \bigcap_{n < k} \varphi_n(V_n) \subset \bigcap_{n < k} U_n \subset U$, i.e., $h(W) \subset U$ which proves that h is continuous at the point s.

Furthermore, h is surjective; to see it take any $x \in X$ and $t \in \mathbb{P}$ with $x \in \varphi(t)$. For every $n \in \omega$, there is $i(n) \in \{0, 1\}$ such that $x \in F_n^{i(n)}$ and hence $\varphi(t) \cap F_n^{i(n)} \neq \emptyset$ which implies $t_n = t \in P_n^{i(n)}$ for all $n \in \omega$. Thus, for a point $s \in Q$ defined by $s(n) = t_n$ for all $n \in \omega$, we have $x \in \bigcap_{n \in \omega} \varphi_n(s(n))$ and therefore $h(s) = x$.

This proves that $h(F) = X$, i.e., X is a continuous image of F. We already saw that Q is analytic being homeomorphic to a closed subspace of $\mathbb{P}$. Therefore F is analytic by Problem 335 and hence X is also analytic which shows that our solution is complete.

T.347. *Prove that a non-empty Polish space X is homeomorphic to $\mathbb{P}$ if and only if X is zero-dimensional and any compact subspace of X has empty interior.*

Solution. If $k, n \in \omega$ and $s \in \omega^k$, then $s^\frown n \in \omega^{k+1}$ is defined by $(s^\frown n)(k) = n$ and $(s^\frown n)|k = s$.

We proved in Problem 313 that $\mathbb{P}$ is zero-dimensional. Considering $\mathbb{P}$ as a subspace of $\mathbb{R}$ assume that $K \subset \mathbb{P}$ is compact. Since $K \cap \mathbb{Q} = \emptyset$, the interior of $\mathbb{K}$ in $\mathbb{R}$ is empty and hence $U = \mathbb{R} \backslash K$ is a dense open subspace of $\mathbb{R}$. Since $\mathbb{P}$ is dense in $\mathbb{R}$, the set $\mathbb{P} \cap U$ is also dense in $\mathbb{R}$ and hence $\mathbb{P} \cap U = \mathbb{P} \backslash K$ is dense in $\mathbb{P}$. Consequently, the interior of K in $\mathbb{P}$ has to be empty. This settles necessity.

Now assume that X is a zero-dimensional Polish space such that $\text{Int}(K) = \emptyset$ for any compact $K \subset X$. Fix a complete metric d on X such that $\tau(d) = \tau(X)$ and $d(x, y) \le 1$ for any $x, y \in X$. Denote by $\mathcal{C}(X)$ the family of all non-empty clopen subsets of X. It is evident that

(0) any $U \in \mathcal{C}(X)$ is a complete metric zero-dimensional space such that every compact $K \subset U$ has empty interior.

Let $U_\emptyset = X$ and $\mathcal{U}(0) = \{U_\emptyset\}$; assume that $n \in \omega$ and we have a family $\mathcal{U}(k) = \{U_s : s \in \omega^k\} \subset \mathcal{C}(X)$ for all $k \le n$ with the following properties:

(1) $\mathcal{U}(k)$ is disjoint and $\bigcup \mathcal{U}(k) = X$ for all $k \le n$;
(2) if $0 \le k < n$ and $s \in \omega^k$, then $U_s = \bigcup\{U_{s^\frown i} : i \in \omega\}$;
(3) for any $k \le n$ and $s \in \omega^k$, we have $\text{diam}(U_s) \le \frac{1}{2^k}$.

Fix any $s \in \omega^n$; since U_s is not compact by (0), there is an infinite closed discrete $D \subset U_s$. Using paracompactness of U_s it is easy to construct a disjoint open cover $\mathcal{W}_s \subset \mathcal{C}(X)$ of U_s such that $\text{diam}(W) \le \frac{1}{2^{n+1}}$ and $|W \cap D| \le 1$ for each $W \in \mathcal{W}_s$. As a consequence, $\mathcal{W}_s$ is infinite and hence we can choose a faithful enumeration $\{W_n : n \in \omega\}$ of the family $\mathcal{W}_s$. Let $U_{s^\frown k} = W_k$ for all $k \in \omega$. After we construct $\mathcal{W}_s$ for all $s \in \omega^n$, we obtain a family $\mathcal{U}(n + 1) = \bigcup\{\mathcal{W}_s : s \in \omega^n\} = \{U_s : s \in \omega^{n+1}\}$ and it is evident that the conditions (1)–(3) are satisfied for all $k \le n + 1$. Therefore our inductive construction can be continued to give us families $\mathcal{U}(k)$ such that (1)–(3) hold for all $k \in \omega$.

Observe that our family $\mathcal{U} = \bigcup_{k \in \omega} \mathcal{U}(k) = \{U_s : s \in \omega^{<\omega}\}$ is $\omega^{<\omega}$-directed, i.e., has the properties (1)–(4) from T.313, so Fact 1 of T.313 is applicable to conclude that the space $L(\mathcal{U}) = \bigcap\{\bigcup \mathcal{U}(k) : k \in \omega\}$ is homeomorphic to $\mathbb{P}$. However, $\bigcup \mathcal{U}(k) = X$ for all $k \in \omega$ by the property (1). Therefore $L(\mathcal{U}) = X$ and hence X is homeomorphic to $\mathbb{P}$ which proves sufficiency and makes our solution complete.

T.348. *Prove that a metrizable compact X is homeomorphic to the Cantor set if and only if X is zero-dimensional and has no isolated points.*

Solution. We identify the Cantor set $\mathbb{K}$ with the space $\mathbb{D}^\omega$ (see Problem 128 of [TFS]). As usual, $\mathbb{D}^0 = \{\emptyset\}$ and $\mathbb{D}^{<\omega} = \bigcup\{\mathbb{D}^n : n \in \omega\}$. If f is a function, then $\text{dom}(f)$ is its domain; given a function g the expression $f \subset g$ says that $\text{dom}(f) \subset \text{dom}(g)$ and $g|\text{dom}(f) = f$. For any $s \in \mathbb{D}^{<\omega}$, let $O_s = \{x \in \mathbb{D}^\omega : s \subset x\}$. Given $k \in \omega$, $i \in \mathbb{D}$ and $s \in \mathbb{D}^k$, let $s^\frown i \in \mathbb{D}^{k+1}$ be defined by $(s^\frown i)(k) = i$ and $(s^\frown i)|k = s$. If $\mathcal{U}$ is a family of sets in a metric space M, then $\text{diam}(\mathcal{U}) = \sup\{\text{diam}(U) : U \in \mathcal{U}\}$.

If we have a set of functions $\{f_i : i \in I\}$ such that $f_i|(\text{dom}(f_i) \cap \text{dom}(f_j)) = f_j|(\text{dom}(f_i) \cap \text{dom}(f_j))$ for any $i, j \in I$, then we can define a function f with $\text{dom}(f) = \bigcup_{i \in I} \text{dom}(f_i)$ as follows: given any $x \in \text{dom}(f)$, find any $i \in I$ with $x \in \text{dom}(f_i)$ and let $f(x) = f_i(x)$. It is easy to check that the value of f at x does not depend on the choice of i, so we have consistently defined a function f which will be denoted by $\bigcup\{f_i : i \in I\}$.

Fact 1. Let (M, ρ) be a complete metric space. Suppose that we have a family $\mathcal{U} = \{U_s : s \in \mathbb{D}^{<\omega}\} \subset \tau^*(M)$ with the following properties:

(a) the family $\mathcal{U}(n) = \{U_s : s \in \mathbb{D}^n\}$ is disjoint for all $n \in \omega$;
(b) if $0 \le k < n < \omega$, $s \in \mathbb{D}^k$, $t \in \mathbb{D}^n$ and $s \subset t$, then $\overline{U}_t \subset U_s$;
(c) $\text{diam}(\mathcal{U}(n)) \to 0$ when $n \to \infty$.

Then the space $L(\mathcal{U}) = \bigcap\{\bigcup \mathcal{U}(n) : n \in \omega\}$ is homeomorphic to $\mathbb{K}$.

Proof. Given any $s \in \mathbb{D}^\omega$, the family $\mathcal{V}_s = \{U_{s|n} : n \in \omega\}$ is decreasing by (b) and the diameters of its elements approach zero by (c). Therefore we can apply Problem 236 of [TFS] to conclude that $\bigcap \mathcal{V}_s = \bigcap\{\overline{U}_{s|n} : n \in \omega\} \ne \emptyset$. An evident consequence of (c) is that $\bigcap \mathcal{V}_s$ can have at most one element and therefore there is $x(s) \in M$ such that $\bigcap \mathcal{V}_s = \{x(s)\}$; let $\varphi(s) = x(s)$ for any $s \in \mathbb{D}^\omega$.

If $s, t \in \mathbb{D}^\omega$ and $s \ne t$, then there is $n \in \omega$ for which $s|n \ne t|n$; then $\varphi(s) \in U_{s|n}$ and $\varphi(t) \in U_{t|n}$ while $U_{s|n} \cap U_{t|n} = \emptyset$ by (a) and therefore $\varphi(s) \ne \varphi(t)$, i.e., the map φ is injective. If $x \in L(\mathcal{U})$, then there is $s_n \in \mathbb{D}^n$ such that $x \in U_{s_n}$ for all $n \in \omega$. It follows from (a) and (b) that $s_n \subset s_{n+1}$ for every $n \in \omega$ and hence $s = \bigcup_{n \in \omega} s_n \in \mathbb{D}^\omega$ is well-defined; it is immediate that $\varphi(s) = x$ and hence $\varphi : \mathbb{D}^\omega \to L(\mathcal{U})$ is a bijection.

Finally, if we have an arbitrary $s \in \mathbb{D}^\omega$ and $U \in \tau(\varphi(s), M)$, then there is $\varepsilon > 0$ such that $B_\rho(\varphi(s), \varepsilon) \subset U$. Take $n \in \omega$ with $\text{diam}(\mathcal{U}(n)) < \varepsilon$. The set $O_{s|n}$ is an open neighborhood of the point s in $\mathbb{D}^\omega$ and, for any $t \in O_{s|n}$, we have $t|n = s|n$ and therefore $\varphi(t) \in U_{t|n} = U_{s|n}$. Furthermore, if $y \in U_{s|n}$, then $\rho(\varphi(s), y) \le \text{diam}(U_{s|n}) < \varepsilon$ which shows that $U_{s|n} \subset B_\rho(\varphi(s), \varepsilon) \subset U$. Thus $\varphi(O_{s|n}) \subset U$ which proves continuity of φ at the point s. We established that φ is a continuous bijection and hence homeomorphism because $\mathbb{D}^\omega$ is compact. Therefore $\mathbb{K} = \mathbb{D}^\omega$ is homeomorphic to $L(\mathcal{U})$ and Fact 1 is proved.   □

Returning to our solution observe that we proved in Problem 128 of [TFS] that the Cantor set $\mathbb{K}$ is homeomorphic to $\mathbb{D}^\omega$. This, together with Problem 302, implies that the space $\mathbb{K}$ is zero-dimensional. If $s \in \mathbb{D}^{<\omega}$, then O_s is an infinite open subset of $\mathbb{D}^\omega$; since the family $\{O_s : s \in \mathbb{D}^{<\omega}\}$ is a base in $\mathbb{D}^\omega$, the space $\mathbb{D}^\omega$ has no isolated points. Thus $\mathbb{K}$ is a zero-dimensional second countable compact space without isolated points.

Now assume that X is a metrizable compact zero-dimensional space without isolated points and denote by $\mathcal{C}(X)$ the family of all non-empty clopen subsets of X. Let d be a metric on X such that $\tau(d) = \tau(X)$ and $d(x, y) \le 1$ for all $x, y \in X$. It is easy to construct a sequence $\{\mathcal{W}_n : n \in \omega\}$ of finite disjoint open covers of X such that $\mathcal{W}_{n+1}$ is a refinement of $\mathcal{W}_n$ and $\text{diam}(\mathcal{W}_n) \le \frac{1}{2^n}$ for all $n \in \omega$.

If $A \subset X$, let $p(n, A)$ be the minimal number of elements of $\mathcal{W}_n$ needed to cover A. Since $\mathcal{W}_n$ is a finite cover of the space X, the number $p(n, A)$ is well-defined and $p(n, A) \leq |\mathcal{W}_n|$ for every $A \subset X$ and $n \in \omega$. For any $\mathcal{A} \subset \exp(X)$, let $P(n, \mathcal{A}) = \max\{p(n, A) : A \in \mathcal{A}\}$. It is evident that $P(n, \mathcal{A}) \leq 1$ says that $\mathcal{A}$ is *inscribed in* $\mathcal{W}_n$, i.e., for any $A \in \mathcal{A}$ there is $W \in \mathcal{W}_n$ such that $A \subset W$. Consequently, if $P(n, A) = 1$, then $\text{diam}(\mathcal{A}) \leq \text{diam}(\mathcal{W}_n) \leq \frac{1}{2^n}$.

Now, if we have a non-empty family $\mathcal{A} \subset \tau^*(X)$, then every $A \in \mathcal{A}$ is infinite because X has no isolated points. Fix $A \in \mathcal{A}$ and pick distinct $x, y \in A$; then $r = d(x, y) > 0$. There is $m \in \omega$ such that $\frac{1}{2^m} < r$ and hence $\text{diam}(W) < r$ for any $W \in \mathcal{W}_m$. This shows that no element of $\mathcal{W}_m$ can contain A and hence the set $N(\mathcal{A}) = \{n \in \omega : P(n, \mathcal{A}) > 1\}$ is non-empty. As a consequence, the number $i(\mathcal{A}) = \min N(\mathcal{A})$ is well-defined for any non-empty $\mathcal{A} \subset \tau^*(X)$.

Let $U_\emptyset = X$ and $\mathcal{U}(0) = \{U_\emptyset\}$. Suppose that $n \in \omega$ and have families $\mathcal{U}(k) = \{U_s : s \in \mathbb{D}^k\} \subset \mathcal{C}(X)$ for all $k \leq n$ with the following properties:

(1) $\mathcal{U}(k)$ is disjoint and $\bigcup \mathcal{U}(k) = X$ for all $k \leq n$;
(2) if $0 \leq m < k \leq n$, $s \in \mathbb{D}^m, t \in \mathbb{D}^k$ and $s \subset t$, then $U_t \subset U_s$;
(3) for each $k < n$ if $m = i(\mathcal{U}(k))$, then $P(m, \mathcal{U}(k+1)) < P(m, \mathcal{U}(k))$.

To construct the family $\mathcal{U}(n+1)$ take any $s \in \mathbb{D}^n$ and let $m = i(\mathcal{U}(n))$. If there is $W \in \mathcal{W}_m$ such that $U_s \subset W$, then split U_s into two disjoint clopen sets arbitrarily, e.g., take distinct $x, y \in U_s$ (which is possible because X has no isolated points), find $V \in \mathcal{C}(X)$ with $x \in V \subset X \setminus \{y\}$ and let $V_0 = V \cap U_s$, $V_1 = U_s \setminus V_0$. If we let $U_{s \frown j} = V_j$ for all $j \in \mathbb{D}$, then we get a family $\mathcal{V}_s = \{U_{s \frown 0}, U_{s \frown 1}\}$ such that $P(m, \mathcal{V}_s) = 1 < P(m, \mathcal{U}(n))$.

If $k = p(m, U_s) > 1$, then take $W_1, \ldots, W_k \in \mathcal{W}_m$ such that $U_s \subset W_1 \cup \cdots \cup W_k$ and let $V_0 = W_1 \cap U_s$ and $V_1 = U_s \setminus V_0$ (observe that $V_0 \neq \emptyset$ and $V_1 \neq \emptyset$ for otherwise U_s can be covered by less than k-many elements of $\mathcal{W}_m$). If we let $U_{s \frown j} = V_j$ for all $j \in \mathbb{D}$, then $p(m, V_0) = 1$ and $p(m, V_1) = k - 1$, i.e., for the family $\mathcal{V}_s = \{U_{s \frown 0}, U_{s \frown 1}\}$ the inequality $P(m, \mathcal{V}_s) < P(m, \mathcal{U}(n))$ still holds.

After we construct the sets $U_{s \frown j}$ for all $s \in \mathbb{D}^n$ and $j \in \mathbb{D}$, we obtain the family $\mathcal{U}(n+1) = \bigcup \{\mathcal{V}_s : s \in \mathbb{D}^n\} = \{U_s : s \in \mathbb{D}^{n+1}\}$; it is evident that the conditions (1) and (2) are now satisfied for all $k \leq n + 1$. The property (3) also holds for $k = n$ because $P(m, \mathcal{U}(n+1)) = \max\{P(m, \mathcal{V}_s) : s \in \mathbb{D}^n\} < P(m, \mathcal{U}(n))$ (recall that our construction guarantees that $P(m, \mathcal{V}_s) < P(m, \mathcal{U}(n))$ for every $s \in \mathbb{D}^n$). Thus our inductive procedure can be continued to obtain a sequence $\{\mathcal{U}(k) : k \in \omega\}$ such that (1)–(3) are fulfilled for all $n \in \omega$.

Let $|\mathcal{W}_j| = n_j$ and $m_j = n_0 + \cdots + n_j$ for all $j \in \omega$; observe that we have the following property:

(4) for any $j \in \omega$ and $n \geq m_j$, we have the equality $P(j, \mathcal{U}(n)) = 1$. As a consequence, $\text{diam}(\mathcal{U}(n)) \leq \frac{1}{2^j}$ for all $n \geq m_j$ and hence $\text{diam}(\mathcal{U}(n)) \to 0$ when $n \to \infty$.

Indeed, it follows from (2) that if $P(j, \mathcal{U}(n)) = 1$, then $P(j, \mathcal{U}(k)) = 1$ for all $k \geq n$. This, together with (3), implies that if $j = i(\mathcal{U}(n))$, then $P(j, \mathcal{U}(n+n_j)) = 1$ and hence (4) is true.

Thus our family $\mathcal{U} = \{U_s : s \in \mathbb{D}^{<\omega}\}$ satisfies (1)–(4) and hence the conditions (a)–(c) of Fact 1 are fulfilled for $\mathcal{U}$. Therefore Fact 1 can be applied to conclude that $L(\mathcal{U})$ is homeomorphic to $\mathbb{K} = \mathbb{D}^\omega$. The property (1) implies that $L(\mathcal{U}) = X$, so X is homeomorphic to $\mathbb{K}$ and hence our solution is complete.

T.349. *Prove that a countable metrizable space X is homeomorphic to $\mathbb{Q}$ if and only if X has no isolated points.*

Solution. We identify $\mathbb{P}$ with the space ω^ω (see Problem 313). As usual, $\omega^0 = \{\emptyset\}$ and $\omega^{<\omega} = \bigcup\{\omega^n : n \in \omega\}$. If f is a function, then $\text{dom}(f)$ is its domain; given a function g the expression $f \subset g$ says that $\text{dom}(f) \subset \text{dom}(g)$ and $g|\text{dom}(f) = f$. If we have a set of functions $\{f_i : i \in I\}$ such that $f_i|(\text{dom}(f_i) \cap \text{dom}(f_j)) = f_j|(\text{dom}(f_i) \cap \text{dom}(f_j))$ for any $i, j \in I$, then we can define a function f with $\text{dom}(f) = \bigcup_{i \in I} \text{dom}(f_i)$ as follows: given any $x \in \text{dom}(f)$, find any $i \in I$ with $x \in \text{dom}(f_i)$ and let $f(x) = f_i(x)$. It is easy to check that the value of f at x does not depend on the choice of i, so we have consistently defined a function f which will be denoted by $\bigcup\{f_i : i \in I\}$.

Given a space Z denote by $\mathcal{C}(Z)$ the family of all non-empty clopen subsets of Z. If Z is a metric space and we have a family $\mathcal{A} \subset \exp(Z)$, then $\text{diam}(\mathcal{A}) = \sup\{\text{diam}(A) : A \in \mathcal{A}\}$. The expression $Y \simeq Z$ says that the spaces Y and Z are homeomorphic. If $k, n \in \omega$ and $s \in \omega^k$, then $s^\frown n \in \omega^{k+1}$ is defined by $(s^\frown n)(k) = n$ and $(s^\frown n)|k = s$. For any $s \in \omega^{<\omega}$, let $O_s = \{x \in \omega^\omega : s \subset x\}$.

Fact 1. If P is a countable dense subspace of $\mathbb{R}$, then there exists a homeomorphism $f : \mathbb{Q} \to P$ such that $x, y \in \mathbb{Q}$ and $x < y$ implies $f(x) < f(y)$. In particular, if P and Q are countable dense subspaces of $\mathbb{R}$, then $P \simeq Q$.

Proof. Choose some faithful enumerations $\{q_n : n \in \omega\}$ and $\{p_n : n \in \omega\}$ of the sets $\mathbb{Q}$ and P respectively. If $x, y \in \mathbb{R}$ and $x \neq y$, then the expression $\text{ord}(x, y) = R$ says that xRy where R is one of the symbols " $<$ " or " $>$ ".

Letting $r_0 = p_0$ assume that $n \in \omega$ and we have defined a point $r_i \in P$ for all $i \leq n$ in such a way that

(1) for any $i < n$ if $R_k = \text{ord}(q_{i+1}, q_k)$ for each $k \leq i$, then $r_{i+1} = p_j$ where $j = \min\{l \in \omega : \text{ord}(p_l, r_k) = R_k \text{ for all } k \leq i\}$.

Let $R_k = \text{ord}(q_{n+1}, q_k)$ for all $k \leq n$. Then $U = \{p \in P : \text{ord}(p, p_k) = R_k$ for all $k \leq n\}$ is a non-empty open subset of P (it is evident that we have three possibilities for U: either $U = (p_k, +\infty) \cap P$ for some $k \leq n$ or $U = (-\infty, p_k) \cap P$ for some $k \leq n$ or else $U = (p_k, p_m) \cap P$ for some $k, m \leq n$). Therefore the number $j = \min\{l \in \omega : p_l \in U\}$ is well-defined; let $r_{n+1} = p_j$. It is straightforward that (1) is fulfilled for all $i \leq n$ which shows that we can continue our inductive construction to obtain a set $\{r_i : i \in \omega\}$ with the property (1) fulfilled for all $i \in \omega$. Observe that it follows from (1) that

(2) if $i \neq j$, then $\text{ord}(q_i, q_j) = \text{ord}(r_i, r_j)$ and, in particular, $q_i < q_j$ implies $r_i < r_j$.

Let $f(q_i) = r_i$ for all $i \in \omega$; then $f : \mathbb{Q} \to P$. An immediate consequence of (2) is that $p, q \in \mathbb{Q}$ and $p < q$ implies $f(p) < f(q)$. To see that $P' = \{r_i : i \in \omega\} = P$ suppose that it is false and consider the numbers $n = \min\{i \in \omega : p_i \in P \backslash P'\}$ and $k = \min\{i \in \omega : \{p_0, \dots, p_{n-1}\} \subset \{r_0, \dots, r_i\}\}$. For the set $A = \{r_0, \dots, r_k\}$ we have $\{p_0, \dots, p_{n-1}\} \subset A$; it is evident that the sets $L = \{r \in A : r < p_n\}$ and $R = \{r \in A : r > p_n\}$ are disjoint and $x < y$ for any $x \in L$ and $y \in R$ (note that it is possible that some of the sets L, R is empty). For the sets $A' = \{q_0, \dots, q_k\}$, $L' = \{q_i \in A' : r_i \in L\}$ and $R' = \{q_i \in A' : r_i \in R\}$ we have $f(A') = A$, $f(L') = L$ and $f(R') = R$. The property (2) shows that $a < b$ for any $a \in L'$ and $b \in R'$ and therefore $J = \{x \in \mathbb{R} : a < x < b$ for any $a \in L'$ and $b \in R'\}$ is a non-empty open subset of $\mathbb{R}$. We have $J \cap A' = \emptyset$; the set $\mathbb{Q}$ being dense in $\mathbb{R}$, the number $m = \min\{i \in \omega : q_i \in J\}$ is well-defined and $k < m$. For any $i \leq k$, if $q_m < q_i$, then $q_i \in R'$ and hence $r_i \in R$ which implies $p_n < r_i$. Analogously, if $i \leq k$ and $q_i < q_m$, then $r_i \in L$ and hence $r_i < p_n$.

Now, if $k < i < m$ and $q_i < q_m$ (or $q_i > q_m$), then it follows from $q_i \notin J$ that we have two possibilities: either there is $a \in L'$ with $q_i \leq a$ or there is $b \in R'$ such that $b \leq q_i$ (in fact, the inequalities have to be strict because our enumeration of $\mathbb{Q}$ is faithful, but that does not matter). Observe that the second (first) possibility cannot take place because then $q_m < b \leq q_i$ ($q_i \leq a < q_m$) which is a contradiction. Thus $q_i \leq a$ ($q_i \geq b$) for some $a \in L'$ (for some $b \in R'$ respectively) and hence $r_i \leq f(a) < p_n$ (or $p_n < f(b) \leq r_i$ respectively). This proves that

(3) $\operatorname{ord}(q_m, q_i) = \operatorname{ord}(p_n, r_i)$ for all $i < m$,

which implies, together with (2), that $r_m = p_n$ because the numbers $p_0, \dots, p_{n-1}$ belong to the set $\{r_i : i < m\}$ and hence n is a minimal natural number for which (3) holds. As a consequence, $p_n \in P'$; this contradiction shows that $f(\mathbb{Q}) = P$ and hence f is a bijection.

Now let $\mathcal{B} = \{(p, q) \cap \mathbb{Q} : p, q \in \mathbb{Q}$ and $p < q\}$. It follows from density of $\mathbb{Q}$ in $\mathbb{R}$ that $\mathcal{B}$ is a base in $\mathbb{Q}$. An immediate consequence of (2) is that $\{f(U) : U \in \mathcal{B}\} = \mathcal{C} = \{(a, b) \cap P : a, b \in P$ and $a < b\}$. The set P is also dense in $\mathbb{R}$ and therefore $\mathcal{C}$ is a base in P. Thus we can apply Fact 2 of S.491 to see that f is continuous. The map f being a bijection, we have $\{f^{-1}(V) : V \in \mathcal{C}\} = \mathcal{B}$, so Fact 2 of S.491 can be applied again to conclude that f^{-1} is continuous and hence f is a homeomorphism. Fact 1 is proved.   □

Returning to our solution, observe that it is evident that if X is homeomorphic to $\mathbb{Q}$, then X is a countable metrizable space without isolated points. Now assume that X is a countable metrizable space without isolated points and fix a metric d on X such that $\tau(d) = \tau(X)$ and $d(x, y) \leq 1$ for all $x, y \in X$.

Note first that

(4) if $U \in \tau^*(X)$, then $\overline{U}$ is not compact and hence there is an infinite $D \subset U$ which is closed and discrete in X.

Indeed, if $V \in \tau^*(X)$ and $\overline{V}$ is compact, then it must have an isolated point x (see Problem 133); the set V being dense in $\overline{V}$, we have $\{x\} \cap V \neq \emptyset$, i.e., $x \in V$.

Thus $\{x\} \in \tau(V)$ which implies $\{x\} \in \tau(X)$, i.e., the point x is isolated in X; this contradiction shows that $\overline{V}$ is not compact. Since X is zero-dimensional by Problem 307, there is a clopen $V \subset U$. We proved above that $V = \overline{V}$ is not compact and hence there is an infinite $D \subset V$ which is closed and discrete in V. The set V is closed, so $D \subset V \subset U$ is closed and discrete in X and hence (4) is proved. As an easy consequence we obtain the following property:

(5) For any $W \in C(X)$ and $\varepsilon > 0$, there is a disjoint $\mathcal{W} = \{W_n : n \in \omega\} \subset C(X)$ such that diam($\mathcal{W}$) $< \varepsilon$ and $W = \bigcup \mathcal{W}$.

Indeed, let $D \subset W$ be an infinite closed and discrete subset of X which exists by (4). Using paracompactness and zero-dimensionality of W, we can find a disjoint cover $\mathcal{W}$ of the space W such that diam($\mathcal{W}$) $< \varepsilon$ and $|V \cap D| \le 1$ for any $V \in \mathcal{W}$ (see Problem 314). It is evident that $\mathcal{W}$ cannot be finite, so we can choose a faithful enumeration $\{W_n : n \in \omega\}$ of the family $\mathcal{W}$ finishing the proof of (5).

Letting $U_\emptyset = X$ assume that $n \in \omega$ and we have, for all $k \le n$, a family $\mathcal{U}(k) = \{U_s : s \in \omega^k\} \subset C(X)$ such that

(6) $\mathcal{U}(k)$ is disjoint and $\bigcup \mathcal{U}(k) = X$ for all $k \le n$;
(7) if $k < n$ and $s \in \omega^k$, then $U_s = \bigcup\{U_{s^\frown i} : i \in \omega\}$;
(8) diam($\mathcal{U}(k)$) $\le \frac{1}{2^k}$ for all $k \le n$.

To construct a family $\mathcal{U}(n + 1)$ take any $s \in \omega^n$; the property (5) guarantees existence of a disjoint family $\mathcal{W} = \{W_i : i \in \omega\} \subset C(X)$ such that $U_s = \bigcup \mathcal{W}$ and diam($\mathcal{W}$) $\le \frac{1}{2^{n+1}}$. Let $U_{s^\frown i} = W_i$ for all $i \in \omega$. After the family $\{U_{s^\frown i} : i \in \omega\}$ is constructed for all $s \in \omega^n$, we obtain $\mathcal{U}(n + 1) = \{U_s : s \in \omega^{n+1}\}$ and it is straightforward that the conditions (6)–(8) are fulfilled for all $k \le n + 1$. Thus our inductive construction can be continued to obtain a sequence $\{\mathcal{U}(k) : k \in \omega\}$ for which (6)–(8) hold for all $n \in \omega$.

Consider the set $S = \{s \in \mathbb{P} : F_s = \bigcap\{U_{s|n} : n \in \omega\} \ne \emptyset\}$. Since for every $s \in \mathbb{P}$ we have diam($U_{s|n}$) $\to 0$ when $n \to \infty$, the set F_s can have at most one point. Therefore, for any $s \in S$, there is $x(s) \in X$ such that $F_s = \{x(s)\}$; let $\varphi(s) = x(s)$. This gives us a map $\varphi : S \to X$. Given any $x \in X$, there is $s_n \in \omega^n$ such that $x \in U_{s_n}$ for all $n \in \omega$. It follows from (6) and (7) that $s_n \subset s_{n+1}$ for all $n \in \omega$ and hence the point $s \in \bigcup_{n \in \omega} s_n$ is well-defined. It is immediate that $x = \varphi(s)$ which proves that φ is an onto map.

If s and t are distinct points of S, then $s|n \ne t|n$ for some $n \in \omega$ and therefore $\varphi(s) \in U_{s|n}$ while $\varphi(t) \in U_{t|n}$. The sets $U_{s|n}$ and $U_{t|n}$ being disjoint by (6), we have $\varphi(s) \ne \varphi(t)$ and hence φ is a bijection.

Now take a point $s \in S$ and $\varepsilon > 0$; there is $n \in \omega$ such that $\frac{1}{2^n} < \varepsilon$. The set $V = O_{s|n} \cap S$ is an open neighborhood of s in S. If $t \in V$, then $\varphi(t) \in U_{t|n} = U_{s|n}$; besides, for any $y \in U_{s|n}$, we have $d(\varphi(s), y) \le$ diam($U_{s|n}$) $< \frac{1}{2^n} < \varepsilon$ and therefore $\varphi(V) \subset U_{s|n} \subset B_d(\varphi(s), \varepsilon)$ which proves continuity of φ at the point s.

Now assume that $x \in X$ and $\varphi^{-1}(x) = s$; given any $V \in \tau(s, S)$, there is $n \in \omega$ such that $O_{s|n} \cap S \subset V$. Observe that $x \in U_{s|n}$ and $\varphi^{-1}(U_{s|n}) \subset O_{s|n}$ which shows that φ^{-1} is continuous at the point x. Thus φ is a homeomorphism. Observe also that $\{O_s : s \in \omega^{<\omega}\}$ is a base in $\mathbb{P}$; for any $s \in \omega^{<\omega}$, the set U_s is non-empty, and if $x \in U_s$, then $\varphi^{-1}(x) \in O_s \cap S$ which proves that S is dense in $\mathbb{P}$.

Thus we established that X is homeomorphic to a dense subspace of $\mathbb{P}$; recalling that $\mathbb{P}$ is dense in $\mathbb{R}$, we conclude that X is homeomorphic to a dense subspace of $\mathbb{R}$ and hence $X \simeq \mathbb{Q}$ by Fact 1. Our solution is complete.

T.350. *Prove that every countable metrizable space is homeomorphic to a closed subspace of* $\mathbb{Q}$.

Solution. Let $J = (-1, 1) \subset \mathbb{I}$ and take any countable metrizable space X. Pick a countable dense subspace D of the space $J^\omega \subset \mathbb{I}^\omega$. It is evident that D is also dense in $\mathbb{I}^\omega$ and hence $D \times D$ is a dense countable subspace of $\mathbb{I}^\omega \times \mathbb{I}^\omega$. Since X is homeomorphic to a subspace of $\mathbb{I}^\omega$ (see Problem 209 of [TFS]), we can consider that $X \subset I_0 = \{-1\} \times \mathbb{I}^\omega \subset \mathbb{I}^\omega \times \mathbb{I}^\omega$. Since $E = D \times D \subset J^\omega \times J^\omega$, we have $E \cap I_0 = \emptyset$ and hence $E \cap X = \emptyset$.

Furthermore, E is dense-in-itself because $\mathbb{I}^\omega \times \mathbb{I}^\omega$ does not have isolated points and therefore the space $F = E \cup X$ is also dense-in-itself. The set I_0 is closed in $\mathbb{I}^\omega \times \mathbb{I}^\omega$ and hence $X = I_0 \cap F$ is closed in F. Finally, observe that F is homeomorphic to $\mathbb{Q}$ because it is a countable metrizable space without isolated points (see Problem 349). Thus X can be embedded in $\mathbb{Q}$ as a closed subspace.

T.351. *Let X be a second countable σ-compact space. Prove that X is not Polish if and only if it contains a closed subspace homeomorphic to* $\mathbb{Q}$.

Solution. The expression $Y \simeq Z$ says that the spaces Y and Z are homeomorphic. If $\mathcal{U}$ is a family of non-empty subsets of Z and $z \in Z$, say that $\mathcal{U}$ *converges to z* (denoting this by $\mathcal{U} \to z$) if, for any $V \in \tau(z, Z)$, the family $\{U \in \mathcal{U} : U \backslash V \neq \emptyset\}$ is finite. Note that any finite $\mathcal{U} \subset \exp(Z) \backslash \{\emptyset\}$ converges to any point of Z.

Fact 1. Suppose that Z is a space, $\mathcal{U} \subset \exp(Z) \backslash \{\emptyset\}$ and $\mathcal{U} \to z$ for some $z \in Z$. Then $\overline{\bigcup \mathcal{U}} \subset \bigcup \{\overline{U} : U \in \mathcal{U}\} \cup \{z\}$.

Proof. If $y \in Z \backslash \{z\}$ and $y \in \overline{\bigcup \mathcal{U}}$, then take $W \in \tau(z, Z)$ such that $y \notin \overline{W}$. Then $V = Z \backslash \overline{W} \in \tau(y, Z)$ and V intersects only finitely many elements of $\mathcal{U}$ say, $U_0, \ldots, U_n$. As a consequence, $y \in \overline{U}_0 \cup \cdots \cup \overline{U}_n$ which shows that $y \in \overline{U}_i$ for some $i \leq n$ and hence $y \in \bigcup \{\overline{U} : U \in \mathcal{U}\}$ so Fact 1 is proved. $\square$

Fact 2. Suppose that Z is a first countable space; if $z \in Z$, $O \in \tau(Z)$ and $z \in \overline{O} \backslash O$, then there is family $\mathcal{U} = \{U_n : n \in \omega\} \subset \tau^*(O)$ such that $\overline{U}_n \cap \overline{U}_m = \emptyset$ if $n \neq m$ and $\mathcal{U} \to z$.

Proof. Fix a local base $\mathcal{B} = \{B_n : n \in \omega\}$ of Z at the point z such that $\overline{B}_{n+1} \subset B_n$ for every $n \in \omega$. We can choose a sequence $\{z_n : n \in \omega\} \subset O$ such that $z_n \to z$ when $n \to \infty$. Let $j(0) = 0$; there is $i(0) \in \omega$ such that $z_{i(0)} \in B_0$.

Suppose that $n \in \omega$ and we have $i(0) < \cdots < i(n)$ and $j(0) < \cdots < j(n)$ such that $z_{i(k)} \in B_{j(k)}$ for all $k \leq n$ while $z_{i(k)} \notin \overline{B}_{j(k+1)}$ for all $k < n$. Since $z_{i(n)} \neq z$, there exists $j(n+1) > j(n)$ such that $z_{i(n)} \notin \overline{B}_{j(n+1)}$; since $z_n \to z$, there is $i(n + 1) > i(n)$ for which $z_{i(n+1)} \in B_{j(n+1)}$. Thus we can construct sequences $\{i(n) : n \in \omega\} \subset \omega$ and $\{j(n) : n \in \omega\} \subset \omega$ such that

(1) $i(n) < i(n + 1)$ and $j(n) < j(n + 1)$ for all $n \in \omega$;
(2) $z_{i(n)} \in B_{j(n)} \setminus \overline{B}_{j(n+1)}$ for all $n \in \omega$.

Let $V_n = (B_{j(n)} \setminus \overline{B}_{j(n+1)}) \cap O$ for all $n \in \omega$. It is easy to check that the family $\{V_n : n \in \omega\} \subset \tau^*(O)$ is disjoint and converges to z. If we take a set $U_n \in \tau(z_{i(n)}, Z)$ such that $\overline{U}_n \subset V_n$ for all $n \in \omega$, then the family $\mathcal{U} = \{U_n : n \in \omega\}$ is as promised so Fact 2 is proved. □

Fact 3. Let Z be a first countable space. Suppose that P is a nowhere dense subset of Z and $A \subset Z$ is finite. Then, for any $O \in \tau(Z)$ with $A \subset \overline{O} \setminus O$, there is $H \in \tau(O)$ such that $A \subset \overline{H} \setminus H$ and $\overline{H} \cap P \subset A$.

Proof. We can find $W_a \in \tau(a, Z)$ for every $a \in A$ such that the family $\{\overline{W}_a : a \in A\}$ is disjoint; let $V_a = (W_a \cap O) \setminus \overline{P}$ for all $a \in A$. It is clear that $a \in \overline{V}_a \setminus V_a$ and therefore there is a family $\mathcal{G}_a = \{G_a^n : n \in \omega\} \subset \tau^*(V_a)$ such that $\overline{G_a^n} \cap \overline{G_a^m} = \emptyset$ if $n \neq m$ and $\mathcal{G}_a \to a$ (see Fact 2).

Choose $H_a^n \in \tau^*(Z)$ such that $\overline{H_a^n} \subset G_a^n$ for all $a \in A$ and $n \in \omega$. It is clear that $\mathcal{H}_a = \{H_a^n : n \in \omega\}$ still converges to a and $\overline{H_a^n} \cap P = \emptyset$ for all $a \in A$ and $n \in \omega$. Therefore, for the set $H_a = \bigcup \mathcal{H}_a$, we have $\overline{H}_a \subset \bigcup \{\overline{H_a^n} : n \in \omega\} \cup \{a\} \subset \bigcup \{G_a^n : n \in \omega\} \cup \{a\}$ (see Fact 1) whence $\overline{H}_a \cap P \subset \{a\}$. Consequently, the set $H = \bigcup_{a \in A} H_a$ is as promised and Fact 3 is proved. □

Fact 4. Let Z be a second countable space without isolated points. If Z is of first category in itself, then some closed $Y \subset Z$ is homeomorphic to $\mathbb{Q}$.

Proof. We have $Z = \bigcup \{Z_n : n \in \omega\}$ where Z_n is closed, nowhere dense in Z and $Z_n \subset Z_{n+1}$ for all $n \in \omega$. Let d be a metric on Z such that $\tau(d) = \tau(Z)$ and $d(x, y) \leq 1$ for all $x, y \in Z$.

Take $z_0 \in Z$ arbitrarily and let $A_0 = \{z_0\}$. By Fact 3 applied to the sets $P = Z_0$, $A = A_0$ and $O = Z \setminus \{z_0\}$, there is $V_0 \in \tau^*(O)$ such that $A_0 \subset \overline{V}_0 \setminus V_0$ and $\overline{V}_0 \cap Z_0 \subset A_0$. If $U_0 = Z \setminus \overline{V}_0$ then $Z_0 \setminus A_0 \subset U_0$ and $U_0 \cap V_0 = \emptyset$.

Assume that $n \in \omega$ and we have chosen non-empty finite sets $A_k \subset Z$ and $U_k, V_k \in \tau(Z)$ for all $k \leq n$ with the following properties:

(3) the family $\{A_0, \ldots, A_n\}$ is disjoint;
(4) if $k < n$, then for any $z \in B_k = A_0 \cup \cdots \cup A_k$, there is $y \in A_{k+1}$ such that $d(z, y) \leq \frac{1}{2^k}$;
(5) $Z_k \setminus B_k \subset U_k$, $B_k \subset \overline{V}_k \setminus V_k$ and $U_k \cap V_k = \emptyset$ for all $k \leq n$;
(6) $V_{k+1} \subset V_k$ for all $k < n$.

Since $B_n \subset \overline{V}_n \setminus V_n$, for any $z \in B_n$, we can choose $y(z) \in V_n$ such that $d(z, y(z)) \leq \frac{1}{2^n}$; let $A_{n+1} = \{y(z) : z \in B_n\}$. It is immediate that the properties (3) and (4) hold for $k = n$. It is easy to see that for the set $O = V_n \setminus A_{n+1}$, we have $B_{n+1} = B_n \cup A_{n+1} \subset \overline{O} \setminus O$, so we can apply Fact 3 to find $V_{n+1} \in \tau^*(O)$ such that $B_{n+1} \subset \overline{V}_{n+1} \setminus V_{n+1}$ and $\overline{V}_{n+1} \cap Z_{n+1} \subset B_{n+1}$. If $U_{n+1} = Z \setminus \overline{V}_{n+1}$ then

$Z_{n+1} \backslash B_{n+1} \subset U_{n+1}$, $V_{n+1} \subset V_n$ and $U_{n+1} \cap V_{n+1} = \emptyset$. It is clear that (3)–(6) are fulfilled for all $k \leq n + 1$ and therefore our inductive construction can be continued to give us families $\{A_i : i \in \omega\}$, $\{U_i : i \in \omega\}$ and $\{V_i : i \in \omega\}$ with the conditions (3)–(6) satisfied for all $n \in \omega$.

Let $Y = \bigcup_{i \in \omega} A_i$; given any $x \in Y$ and $m \in \omega$, we have $x \in A_n$ for some $n \in \omega$ and hence there is $x_m \in A_{n+m+1}$ such that $d(x, x_m) \leq \frac{1}{2^{n+m}}$. It follows from (3) that $\{x_m : m \in \omega\} \subset Y \backslash \{x\}$; since $x_m \to x$, the point x is not isolated in Y and hence Y has no isolated points.

Observe also that $Y \cap U_i = \emptyset$ for all $i \in \omega$. Indeed, if $x \in Y \cap U_i$ for some $i \in \omega$, then there is $n \in \omega$ such that $x \in A_n$. The property (5) implies that $x \in \overline{V}_{i+n+1} \subset \overline{V}_i$ [the last inclusion holds by (6)]. However, $V_i \cap U_i = \emptyset$ by (5) and hence $\overline{V}_i \cap U_i = \emptyset$ which contradicts $x \in \overline{V}_i \cap U_i$. Thus $Y \cap (\bigcup_{i \in \omega} U_i) = \emptyset$. Now, if $z \in Z \backslash Y$, then $z \in Z_n$ for some $n \in \omega$ and therefore $z \in Z_n \backslash B_n$. It follows from (5) that $z \in U_n$; the point $z \in Z \backslash Y$ was chosen arbitrarily, so $Z \backslash Y \subset \bigcup_{i \in \omega} U_i$. This proves that $Z \backslash Y = \bigcup_{i \in \omega} U_i$ and hence Y is a closed subset of X. Since Y is a countable metrizable space without isolated points, we have $Y \simeq \mathbb{Q}$ by Problem 349 and Fact 4 is proved. $\qquad\square$

Returning to our solution suppose that a closed $F \subset X$ is homeomorphic to $\mathbb{Q}$. If X is Polish, then so is F (see Problem 260 of [TFS]) and hence $\mathbb{Q}$ is Polish which is false because $\mathbb{Q}$ is of first category in itself (see Problem 274 of [TFS]). This contradiction shows that X is not Polish and proves sufficiency.

Now assume that X is not Polish; we can consider that X is a dense subspace of a metrizable compact space K. Since X is not a G_δ-set in K (see Problem 259 of [TFS]), the space $K \backslash X$ is not σ-compact. However, $K \backslash X$ is a G_δ-set in K because X is σ-compact. Consequently, $K \backslash X$ is a Borel set and hence it is analytic by Problem 334.

Therefore we can apply Fact 4 of T.132 to conclude that there is a subspace $P \subset K \backslash X$ which is closed in $K \backslash X$ and homeomorphic to $\mathbb{P}$. The space $Z = \overline{P}$ (the bar denotes the closure in K) is compact and hence $F = Z \cap X$ is closed in X. Besides, $Z \cap (K \backslash X) = P$ because P is closed in $K \backslash X$. The space P is not σ-compact (see Fact 1 of T.132) and hence not locally compact (it is an easy exercise that any second countable locally compact space is σ-compact). Besides, P is homeomorphic to ω^ω and hence any non-empty open $U \subset P$ contains a closed subspace homeomorphic to $\mathbb{P}$ (this also follows from Fact 4 of T.132). This proves that $\mathrm{cl}_P(U)$ is not compact for any $U \in \tau^*(P)$; an easy consequence is that F is dense in Z. The space Z does not have isolated points because it is a closure of P which is dense-in-itself. Since F is also dense in Z, it does not have isolated points either.

Furthermore, F is σ-compact being closed in X. Let $\{F_n : n \in \omega\}$ be a family of compact subsets of F such that $F = \bigcup_{n \in \omega} F_n$. If $W = \mathrm{Int}_F(F_n) \neq \emptyset$ for some $n \in \omega$, then there is $W' \in \tau(Z)$ such that $W' \cap F = W$. The density of F in Z implies that $\mathrm{cl}_Z(W') = \mathrm{cl}_Z(W) \subset \mathrm{cl}_Z(F_n) = F_n$ (the last equality holds because F_n is compact). Thus $W' \subset F_n$ and hence $W' \cap P = \emptyset$ which contradicts density of P in Z.

This contradiction shows that F_n is nowhere dense in F for all $n \in \omega$, i.e., F is a dense-in-itself second countable space of first category in itself. Thus we can apply Fact 4 to find a set $Y \subset F$ which is closed in F and $Y \simeq \mathbb{Q}$. It is immediate that Y is also closed in X and hence X contains a closed set homeomorphic to $\mathbb{Q}$. This settles necessity and makes our solution complete.

T.352. *Let X be an analytic non-σ-compact space. Prove that some closed subspace of X is homeomorphic to $\mathbb{P}$.*

Solution. If Z is a space, call a family $\mathcal{A} = \{A_s : s \in S\} \subset \exp(Z)$ *indexed discrete* if for any $z \in Z$ there is $W \in \tau(z, Z)$ such that the set $\{s \in S : W \cap A_s \neq \emptyset\}$ is finite. As usual, we identify the space $\mathbb{P}$ with ω^ω (see Problem 313). The expression $X \simeq Y$ says that the space X is homeomorphic to the space Y. As usual, we identify any ordinal with the set of its predecessors and, in particular, $n = \{0, \ldots, n - 1\}$ for any $n \in \omega$. If $k, n \in \omega$ and $s \in \omega^k$, then $s^\frown n \in \omega^{k+1}$ is defined by $(s^\frown n)(k) = n$ and $(s^\frown n)|k = s$. If we have $s \in \omega^n$, $t \in \omega^k$ where $n \leq k$ and $n, k \in \omega + 1$, then $s \subset t$ says that $t|n = s$. For any $s \in \omega^{<\omega}$, let $O_s = \{x \in \omega^\omega : s \subset x\}$.

If d is a metric on a set M, then $\tau(d)$ is the topology generated by the metric d. If we work with a metric space (M, d), then for any $x \in X$ and $r > 0$ the set $B_d(x, r) = \{y \in M : d(x, y) < r\}$ is the ball of radius r centered at x. If f is a function, then $\mathrm{dom}(f)$ is its domain. Suppose that we have a set of functions $\{f_i : i \in I\}$ such that $f_i|(\mathrm{dom}(f_i) \cap \mathrm{dom}(f_j)) = f_j|(\mathrm{dom}(f_i) \cap \mathrm{dom}(f_j))$ for any $i, j \in I$. Then we can define a function f with $\mathrm{dom}(f) = \bigcup_{i \in I} \mathrm{dom}(f_i)$ as follows: given any $x \in \mathrm{dom}(f)$, find any $i \in I$ with $x \in \mathrm{dom}(f_i)$ and let $f(x) = f_i(x)$. It is easy to check that the value of f at x does not depend on the choice of i so we have consistently defined a function f which will be denoted by $\bigcup\{f_i : i \in I\}$.

Since X is analytic, there is a continuous onto map $f : \mathbb{P} \to X$. Fix a complete metric ρ on $\mathbb{P}$ such that $\tau(\rho) = \tau(\mathbb{P})$. We will need the family $\mathcal{S} = \{A \subset X :$ there is a σ-compact $P \subset X$ such that $A \subset P\}$. Let $A_\emptyset = \mathbb{P}$; assume that $n > 0$, and for each $k < n$, we have a family $\{A_s : s \in \omega^k\}$ of closed subsets of $\mathbb{P}$ with the following properties:

(1) if $0 < k < n$, then $\mathrm{diam}_\rho(A_s) < 2^{-k}$ for all $s \in \omega^k$;
(2) if $0 \leq k < n$, then $\{A_s : s \in \omega^k\}$ is an indexed discrete family in $\mathbb{P}$;
(3) if $0 \leq k < n$, then $\{f(A_s) : s \in \omega^k\}$ is an indexed discrete family in X;
(4) if $0 \leq k < l < n$, $s \in \omega^k$, $t \in \omega^l$ and $s \subset t$, then $A_t \subset A_s$;
(5) if $0 \leq k < n$, then $f(A_s) \notin \mathcal{S}$ for any $s \in \omega^k$.

Fix $s \in \omega^{n-1}$ and consider the set $B_s = \{x \in A_s : f(U) \notin \mathcal{S}$ for any $U \in \tau(x, A_s)\}$. For each $x \in A_s \setminus B_s$ there is $W_x \in \tau(x, A_s)$ such that $f(W_x) \in \mathcal{S}$. The open cover $\mathcal{W} = \{W_x : x \in A_s \setminus B_s\}$ of the Lindelöf space $A_s \setminus B_s$ has a countable subcover $\mathcal{W}'$; it is evident that $f(\bigcup \mathcal{W}') = f(A_s \setminus B_s) \in \mathcal{S}$. Since $f(A_s) \notin \mathcal{S}$, we have $f(B_s) \notin \mathcal{S}$ and hence there exists an infinite set $D \subset f(B_s)$ which is discrete and closed in X. It is easy to find, for every $y \in D$, a set $V_y \in \tau(y, X)$ such that the family $\{V_y : y \in D\}$ is indexed discrete in X. The family $\{f^{-1}(V_y) : y \in D\}$ is indexed discrete in $\mathbb{P}$; pick a point $x(y) \in f^{-1}(y) \cap B_s$ for each $y \in D$. It is clear that $x(y) \in f^{-1}(V_y)$, so we can find $U_y \in \tau(x(y), \mathbb{P})$ such that $U_y \subset f^{-1}(V_y)$ and $\text{diam}_\rho(U_y) < 2^{-n}$ for every $y \in D$.

Take a faithful enumeration $\{y_n : n \in \omega\}$ of the set D and let $A_{s \frown n} = \overline{U_{y_n} \cap B_s}$ for each $n \in \omega$. It follows from $U_{y_n} \cap B_s \neq \emptyset$ that $f(A_{s \frown n}) \notin S$ for each $n \in \omega$. It is clear that $\text{diam}_\rho(A_{s \frown n}) \leq \text{diam}_\rho(U_{y_n}) < 2^{-n}$ which shows that (1) is fulfilled for the sets $A_{s \frown n}$ for all $n \in \omega$. The properties (2) and (3) hold because the families $\mathcal{U} = \{U_{y_n} : n \in \omega\}$ and $\mathcal{V} = \{V_{y_n} : n \in \omega\}$ are indexed discrete and the families $\{A_{s \frown n} : n \in \omega\}$ and $\{f(A_{s \frown n}) : n \in \omega\}$ are obtained by shrinking $\mathcal{U}$ and $\mathcal{V}$ respectively. After we define the family $\{A_{s \frown n} : n \in \omega\}$ for each $s \in \omega^{n-1}$, we have the collection $\{A_t : t \in \omega^n\}$. This method of construction guarantees the property (4). We also assured that $f(A_{s \frown n}) \notin S$ for each $s \in \omega^{n-1}$ and $n \in \omega$ which shows that (5) is also true for the family $\{A_t : t \in \omega^n\}$.

Thus our inductive construction can be completed for all natural n to give us families $\{A_s : s \in \omega^n\}$ with (1)–(5) for all $n \in \omega$. Now we are ready to describe a closed subset of X which is homeomorphic to $\mathbb{P}$. Given any $x \in \mathbb{P}$ let $s_n = x|n$ for each $n \in \omega$; it follows from (1) and (4) that $\{A_{s_n} : n \in \omega\}$ is a decreasing family of closed subsets of $\mathbb{P}$ such that $\text{diam}_\rho(A_{s_n}) \to 0$. As a consequence there is a unique point $g(x) \in \bigcap\{A_{s_n} : n \in \omega\}$ (see Problem 236 of [TFS]). Let $h(x) = f(g(x))$; this gives us a map $h : \mathbb{P} \to X$. We must check that $F = h(\mathbb{P})$ is closed in X and the map $h : \mathbb{P} \to F$ is a homeomorphism.

Take any point $x \in \mathbb{P}$ and an arbitrary $W \in \tau(h(x), X)$; since $h(x) = f(g(x))$ and f is a continuous map, there is $\varepsilon > 0$ such that $f(B_\rho(g(x), \varepsilon)) \subset W$. We saw that $\{g(x)\} = \bigcap\{A_{x|n} : n \in \omega\}$; since $\text{diam}_\rho(A_{x|n}) \to 0$ when $n \to \infty$, there exists $m \in \omega$ such that $A_{x|m} \subset B_\rho(g(x), \varepsilon)$. The set $U = O_{x|m}$ is an open neighborhood of x in $\mathbb{P}$ and, for any $y \in U$, we have $y|m = x|m$ which implies $g(y) \in A_{y|m} = A_{x|m}$ and therefore $h(y) = f(g(y)) \in f(A_{x|m}) \subset f(B_\rho(g(x), \varepsilon)) \subset W$ which shows that $h(U) \subset W$ and hence h is continuous at the point x.

If x and y are distinct points of $\mathbb{P}$, then there is $n \in \omega$ such that $x|n \neq y|n$. The property (3) implies that $f(A_{x|n}) \cap f(A_{y|n}) = \emptyset$; since $h(x) \in f(A_{x|n})$ and $h(y) \in f(A_{y|n})$, we have $h(y) \neq h(x)$ and hence h is a condensation.

To show that $h^{-1} : F \to \mathbb{P}$ is continuous, take any $y \in F$ and $x \in \mathbb{P}$ with $h(x) = y$; then $h^{-1}(y) = x$. If $s_i = x|i$ for all $i \in \omega$, then $y \in \bigcap\{f(A_{s_i}) : i \in \omega\}$. Observe that the family $\{O_{x|n} : n \in \omega\}$ is a local base at x in $\mathbb{P}$, so if we take any $W \in \tau(x, \mathbb{P})$, then there is $n \in \omega$ such that $O_{x|n} \subset W$. The set $U = f(A_{s_n}) \cap F \ni y$ is open in F by (3). If $z \in U$, then $h^{-1}(z)|n = s_n$ by (2) and (3); this implies $h^{-1}(z)|n = x|n$, i.e., $h^{-1}(z) \in O_{x|n} \subset W$. The point $z \in U$ was chosen arbitrarily, so $h^{-1}(U) \subset O_{x|n} \subset W$. Thus we proved that for any $W \in \tau(x, \mathbb{P})$, there exists $U \in \tau(y, F)$ such that $h^{-1}(U) \subset W$. Therefore h^{-1} is continuous at every $y \in F$ and h is a homeomorphism.

To see that F is closed in X take any point $y \in \overline{F}$. If $F_n = \bigcup\{f(A_s) : s \in \omega^n\}$ for each $n \in \omega$, then $F \subset \bigcap\{F_n : n \in \omega\}$. Therefore $y \in \overline{F_n}$ for each $n \in \omega$; the family $\{f(A_s) : s \in \omega^n\}$ being discrete, there is $s_n \in \omega^n$ such that $y \in \overline{f(A_{s_n})}$

for all $n \in \omega$. The properties (3) and (4) imply that $s_n \subset s_{n+1}$ for any $n \in \omega$ and hence we can define $x = \bigcup\{s_n : n \in \omega\} \in \mathbb{P}$. Now observe that $h(x) \in \bigcap\{f(A_{s_n}) : n \in \omega\}$; if $y \neq h(x)$, then there is $W \in \tau(h(x), X)$ such that $y \notin \overline{W}$. We have $h(x) = f(g(x))$, so continuity of f implies that there is $\varepsilon > 0$ such that $f(B_\rho(g(x), \varepsilon)) \subset W$; take $m \in \omega$ such that $2^{-m} < \varepsilon$. Now, $\{g(x)\} = \bigcap_{n \in \omega} A_{x|n}$ and hence $g(x) \in A_{x|m}$ which shows that $\rho(g(x), z) \leq \mathrm{diam}_\rho(A_{x|m}) < 2^{-m} < \varepsilon$ for every $z \in A_{x|m}$ and hence $A_{s_m} = A_{x|m} \subset B_\rho(g(x), \varepsilon)$.

As a consequence, $f(A_{s_m}) \subset f(B_\rho(g(x), \varepsilon)) \subset W$ and therefore $\overline{f(A_{s_m})} \subset \overline{W}$ which contradicts $y \notin \overline{W}$ and $y \in \overline{f(A_{s_m})}$. This contradiction shows that $y = h(x)$ and hence $y \in F$. The point $y \in \overline{F}$ was chosen arbitrarily, so $\overline{F} \subset F$ which shows that F is closed in X. Thus there exists a closed $F \subset X$ with $F \simeq \mathbb{P}$ and our solution is complete.

T.353. *Prove that any uncountable analytic space contains a closed subspace which is homeomorphic to the Cantor set.*

Solution. The expression $Y \simeq Z$ says that the spaces Y and Z are homeomorphic. Let X be an uncountable analytic space. Suppose first that X is σ-compact, i.e., $X = \bigcup\{K_n : n \in \omega\}$ where K_n is compact for all $n \in \omega$; it follows from $|X| > \omega$ that there is $n \in \omega$ such that K_n is uncountable. Every analytic space has a countable network, so $w(K_n) = nw(K_n) = \omega$ (see Fact 4 of S.307) which shows that K_n is an uncountable second countable K-analytic space. It was proved in Fact 4 of T.250 that every second countable uncountable K-analytic space contains a subspace homeomorphic to the Cantor set, so $\mathbb{K}$ embeds in K_n and hence in X.

Now, if X is not σ-compact, then $\mathbb{P}$ embeds in X by Problem 352. Since $\mathbb{P} \simeq \omega^\omega$ (see Problem 313) and $\mathbb{K} \simeq \mathbb{D}^\omega \subset \omega^\omega$, the Cantor set $\mathbb{K}$ embeds in $\mathbb{P}$ and hence in X.

T.354. *Prove that any non-σ-compact Borel set can be condensed onto $\mathbb{I}^\omega$ as well as onto $\mathbb{R}^\omega$.*

Solution. The expression $Y \simeq Z$ says that the spaces Y and Z are homeomorphic. We denote by J any of the spaces $\mathbb{I}$ or $\mathbb{R}$, so if a statement is proved for the symbol J, then we actually give two identical proofs: one for $\mathbb{I}$ and another one for $\mathbb{R}$. Suppose that we have a set of functions $\{f_i : i \in I\}$ such that $f_i|(\mathrm{dom}(f_i) \cap \mathrm{dom}(f_j)) = f_j|(\mathrm{dom}(f_i) \cap \mathrm{dom}(f_j))$ for any $i, j \in I$. Then we can define a function f with $\mathrm{dom}(f) = \bigcup_{i \in I} \mathrm{dom}(f_i)$ as follows: given any $x \in \mathrm{dom}(f)$, find $i \in I$ with $x \in \mathrm{dom}(f_i)$ and let $f(x) = f_i(x)$. It is easy to check that the value of f at x does not depend on the choice of i, so we have consistently defined a function f which will be denoted by $\bigcup\{f_i : i \in I\}$.

Fact 1. Let M be a second countable space. Suppose that $M_i \subset M$, $M_i \subset M_{i+1}$ for all $i \in \omega$ and $M = \bigcup_{i \in \omega} M_i$. Then, for every $i \in \omega$, we can choose a continuous function $h_i : M_i \to J$ so that the family $\{h_i : i \in \omega\}$ separates the points of M, i.e., for any distinct $x, y \in M$ there is $i \in \omega$ such that $x, y \in M_i$ and $h_i(x) \neq h_i(y)$.

Proof. The space M embeds in J^ω, so let $e : M \to J^\omega$ be the respective embedding. Define $e_i : M \to J$ by $e_i(x) = e(x)(i)$ for all $x \in M$. It is clear that the set $E = \{e_i : i \in \omega\}$ separates the points of M. Choose an enumeration $\{h'_i : i \in \omega\}$ of the set E in which every $e \in E$ occurs ω-many times and let $h_i = h'_i | M_i$ for all $i \in \omega$.

To see that the sequence $\{h_i : i \in \omega\}$ is as promised, take two distinct $x, y \in M$. There are $i \in \omega$ and $e \in E$ such that $\{x, y\} \subset M_i$ and $e(x) \neq e(y)$. There is $j > i$ such that $h'_j = e$ and hence $h_j(x) = e(x) \neq e(y) = h_j(y)$ so Fact 1 is proved. $\square$

Fact 2. Given spaces Z and Y, assume that $\mathcal{F} = \{F_t : t \in T\}$ is a closed locally finite cover of Z and we have a family $\{f_t : t \in T\}$ of functions such that $f_t : F_t \to Y$ is continuous for all $t \in T$ and $f_t | (F_t \cap F_s) = f_s | (F_t \cap F_s)$ for any $s, t \in T$. Then the function $f = \bigcup_{t \in T} f_t : Z \to Y$ is also continuous.

Proof. If H is a closed subset of Y, then $f^{-1}(H) = \bigcup_{t \in T} f_t^{-1}(H)$ is closed because the family $\{f_t^{-1}(H) : t \in T\}$ is locally finite (and hence closure-preserving by Fact 2 of S.221) being a shrinking of the family $\mathcal{F}$. Therefore f is continuous and Fact 2 is proved. $\square$

Returning to our solution, assume that we are given a non-σ-compact Borel set X. We will construct a condensation of X onto J^ω. Given a point $a \in J^\omega$, let $\sigma(a) = \{x \in J^\omega : |\{n \in \omega : x(n) \neq a(n)\}| < \omega\}$. It is evident that if $a, b \in J^\omega$ and the set $\{n \in \omega : a(n) \neq b(n)\}$ is infinite, then $\sigma(a) \cap \sigma(b) = \emptyset$. This observation makes it possible to find a set $\{a_i : i \in \omega\} \subset J^\omega$ such that $\sigma(a_i) \cap \sigma(a_j) = \emptyset$ for any distinct $i, j \in \omega$. It is an easy exercise that $\sigma(a_i)$ is a countable union of nowhere dense compact subspaces of J^ω; thus $\sigma(a_i)$ is of first category in J^ω for all $i \in \omega$, and hence $\bigcup_{i \in \omega} \sigma(a_i) \neq J^\omega$ because J^ω is Čech-complete and hence has the Baire property.

Let $S_n = J^\omega \backslash (\bigcup \{\sigma(a_i) : i \geq n\})$ for all $n \in \omega$. Since the set $\sigma(a_i)$ is σ-compact for every $i \in \omega$, the set S_n Borel in J^ω for all $n \in \omega$. Observe that $\bigcup_{n \in \omega} S_n = J^\omega$; besides, $S_n \subset S_{n+1}$ and $S_{n+1} \backslash S_n = \sigma(a_n)$ for all $n \in \omega$.

Any Borel set is analytic by Problem 334, so we can apply Problem 352 to find a closed $P \subset X$ with $P \simeq \mathbb{P}$. It is easy to find a disjoint family $\mathcal{P} = \{P_i : i \in \omega\}$ of clopen subsets of P such that $P = \bigcup_{i \in \omega} P_i$ and $P_i \simeq \mathbb{P}$ for each $i \in \omega$. It is immediate that $\mathcal{P}$ is a discrete family of closed subsets of X. Define a function $\alpha : P \to \mathbb{R}$ by letting $\alpha(x) = i$ for all $x \in P_i$ and $i \in \omega$; it is evident that α is continuous.

Since X is normal, there is a continuous function $\beta : X \to \mathbb{R}$ such that $\beta | P = \alpha$. Let $F_0 = \beta^{-1}((-\infty, \frac{1}{2}])$ and $F_n = \beta^{-1}([n - \frac{1}{2}, n + \frac{1}{2}])$ for every $n \in \mathbb{N}$; we will also need the sets $O_0 = \beta^{-1}((-\infty, \frac{1}{2}))$ and $O_n = \beta^{-1}((n - \frac{1}{2}, n + \frac{1}{2}))$ for every $n \in \mathbb{N}$. It is clear that $\mathcal{F} = \{F_n : n \in \omega\}$ is a locally finite family of closed subsets of X such that $\bigcup \mathcal{F} = X$ while the family $\mathcal{O} = \{O_n : n \in \omega\} \subset \tau(X)$ is disjoint and $P_n \subset O_n \subset F_n$ for all $n \in \omega$. Let $G_n = F_0 \cup \cdots \cup F_n$ for all $n \in \omega$.

Next we will inductively construct, for every $n \in \omega$, a continuous mapping $\varphi_n : G_n \to J^\omega \times J^\omega$ such that

(1) φ_n is injective for all $n \in \omega$;

(2) $\varphi_{n+1}|G_n = \varphi_n$ for all $n \in \omega$;

(3) $J^\omega \times S_n \subset \varphi_n(G_n) \subset J^\omega \times S_{n+1}$ for all $n \in \omega$.

The set $J^\omega \times S_0$ is Borel (see Problem 333), so there is closed $E_0 \subset P_0$ and a condensation $f_0 : E_0 \to J^\omega \times S_0$ (see Problem 341). There is a family $\{H_i : i \in \omega\}$ of closed subsets of X such that $H_i \subset H_{i+1}$ for all $i \in \omega$ and $\bigcup_{i \in \omega} H_i = F_0 \backslash E_0$.

In the product $J^\omega \times J^\omega$ let $p : J^\omega \times J^\omega \to J^\omega$ and $q : J^\omega \times J^\omega \to J^\omega$ be the natural projections onto the first and the second factor respectively. Apply Fact 1 to find a sequence $\mathcal{H} = \{h_i : i \in \omega\}$ of functions such that $h_i \in C(H_i, J)$ for all $i \in \omega$ and $\mathcal{H}$ separates the points of $F_0 \backslash E_0$. For each $i \in \omega$, define a function $u_i : E_0 \cup H_i \to J$ as follows: $u_i(x) = p(f_0(x))(i)$ if $x \in E_0$ and $u_i(x) = h_i(x)$ if $x \in H_i$; let $v_i \in C(F_0, J)$ be any continuous extension of u_i for every $i \in \omega$ (which exists because the space F_0 is normal, the set $E_0 \cup H_i$ is closed in F_0 and $u_i \in C(E_0 \cup H_i, J)$).

Furthermore, for each $i \in \omega$, define $s_i : E_0 \cup H_i \to J$ as follows: $s_i(x) = q(f_0(x))(i)$ if $x \in E_0$ and $s_i(x) = a_0(i)$ for all $x \in H_i$. Let $t_i \in C(F_0, J)$ be a continuous extension of s_i for all $i \in \omega$. Now let $\varphi_0 = (\Delta_{i \in \omega} v_i) \Delta (\Delta_{i \in \omega} t_i)$; it is immediate that $\varphi_0 : F_0 \to J^\omega \times J^\omega$ and $\varphi_0(x) = (v(x), t(x))$ where $v(x)(i) = v_i(x)$ and $t(x)(i) = t_i(x)$ for all $x \in F_0$ and $i \in \omega$. Since any diagonal product of continuous maps is continuous, the map φ_0 is continuous. It is immediate from the definition that $\varphi_0|E_0 = f_0$ and hence $\varphi_0(G_0) = \varphi_0(F_0) \supset f_0(E_0) = J^\omega \times S_0$.

Now, if $x \in F_0 \backslash E_0$, then there is $m \in \omega$ such that $x \in H_i$ for all $i \geq m$ and hence $t(x)(i) = a_0(i)$ for all $i \geq m$. Therefore $t(x) \in \sigma(a_0)$ for all $x \in F_0 \backslash E_0$. Since $t(x) \in S_0$ for all $x \in E_0$, we have $\varphi_0(x) = (v(x), t(x)) \in J^\omega \times (S_0 \cup \sigma(a_0)) = J^\omega \times S_1$, so the condition (3) is satisfied for $n = 0$. Since (2) is vacuous for $n = 0$, let us check that (1) is fulfilled, i.e., φ_0 is injective.

Given two distinct $x, y \in F_0$ assume first that $x, y \in E_0$. The map f_0 being injective, we have $\varphi_0(x) = f_0(x) \neq f_0(y) = \varphi_0(y)$. If $x \in E_0$ and $y \in F_0 \backslash E_0$, then $\varphi_0(x) \in J^\omega \times S_0$ while $\varphi_0(y) \in J^\omega \times \sigma(a_0)$ which shows that $\varphi_0(x) \neq \varphi_0(y)$. Finally, if $x, y \in F_0 \backslash E_0$, then by our choice of the sequence $\{h_i : i \in \omega\}$ there is $i \in \omega$ such that $x, y \in H_i$ and $h_i(x) \neq h_i(y)$. Consequently, $v_i(x) \neq v_i(y)$ which shows that $v(x) \neq v(y)$ and hence $\varphi_0(x) \neq \varphi_0(y)$. Thus the conditions (1)–(3) are satisfied for $n = 0$.

Now assume that we have $\varphi_0, \ldots, \varphi_{k-1}$ for some $k \in \mathbb{N}$ with the properties (1)–(3) for all $n \leq k - 1$. To construct φ_k, observe that $\varphi_{k-1}(G_{k-1})$ is a Borel subset of $J^\omega \times J^\omega$ because φ_{k-1} is a condensation (see Problem 314 and Fact 3 of T.333). Therefore $A = (J^\omega \times S_k) \backslash \varphi_{k-1}(G_{k-1})$ is a Borel subset of $J^\omega \times J^\omega$ and hence there exists a closed $E_k \subset P_k$ and a condensation $f_k : E_k \to A$ (see Problem 341; note that it is possible that $A = E_k = \emptyset$).

There is a family $\{H_i : i \in \omega\}$ of closed subsets of X such that $H_i \subset H_{i+1}$ for all $i \in \omega$ and $\bigcup_{i \in \omega} H_i = F_k \backslash (G_{k-1} \cup E_k)$.

Apply Fact 1 to find a sequence $\mathcal{H} = \{h_i : i \in \omega\}$ of functions such that $h_i \in C(H_i, J)$ for all $i \in \omega$ and $\mathcal{H}$ separates the points of $F_k \backslash (G_{k-1} \cup E_k)$. For each $i \in \omega$ define a function $u_i : G_{k-1} \cup E_k \cup H_i \to J$ as follows: $u_i(x) = p(f_k(x))(i)$

if $x \in E_k$; if $x \in G_{k-1}$, then $u_i(x) = p(\varphi_{k-1}(x))(i)$; and $u_i(x) = h_i(x)$ if $x \in H_i$; let $v_i \in C(G_k, J)$ be any continuous extension of u_i for every $i \in \omega$ (which exists because the space G_k is normal, the set $G_{k-1} \cup E_k \cup H_i$ is closed in F_k and, evidently, $u_i \in C(G_{k-1} \cup E_k \cup H_i, J)$).

Furthermore, for each $i \in \omega$, define $s_i : G_{k-1} \cup E_k \cup H_i \to J$ as follows: $s_i(x) = q(f_k(x))(i)$ if $x \in E_k$; if $x \in G_{k-1}$, then $s_i(x) = q(\varphi_{k-1}(x))(i)$ and $s_i(x) = a_k(i)$ for all $x \in H_i$. Let $t_i \in C(G_k, J)$ be a continuous extension of s_i for all $i \in \omega$. Now let $\varphi_k = (\Delta_{i \in \omega} v_i) \Delta (\Delta_{i \in \omega} t_i)$; it is immediate that $\varphi_k : G_k \to J^\omega \times J^\omega$ and $\varphi_k(x) = (v(x), t(x))$ where $v(x)(i) = v_i(x)$ and $t(x)(i) = t_i(x)$ for all $x \in G_k$ and $i \in \omega$. Since any diagonal product of continuous maps is continuous, the map φ_k is continuous. It is immediate from the definition that $\varphi_k | G_{k-1} = \varphi_{k-1}$, $\varphi_k | E_k = f_k$ and hence $\varphi_k(G_k) \supset \varphi_{k-1}(G_{k-1}) \cup f_k(E_k) = J^\omega \times S_k$.

Now, if $x \in F_k \backslash (G_{k-1} \cup E_k)$, then there exists $m \in \omega$ such that $x \in H_i$ and hence $t(x)(i) = a_k(i)$ for all $i \geq m$. Therefore $t(x) \in \sigma(a_k)$ for all $x \in F_k \backslash (G_{k-1} \cup E_k)$. Since $t(x) \in S_k$ for all $x \in E_k$, we have $\varphi_k(x) = (v(x), t(x)) \in J^\omega \times (S_k \cup \sigma(a_k)) = J^\omega \times S_{k+1}$, so the condition (3) is satisfied for $n = k$. The condition (2) is satisfied for $n = k - 1$ by our construction; let us check that (1) is fulfilled, i.e., φ_k is injective.

Given two distinct points $x, y \in G_k$, assume first that $x, y \in G_{k-1}$; then $\varphi_k(x) = \varphi_{k-1}(x) \neq \varphi_{k-1}(y) = \varphi_k(y)$. If $x \in G_{k-1}$ and $y \in F_k \backslash G_{k-1}$, then $\varphi_k(x) = \varphi_{k-1}(x) \in \varphi_{k-1}(G_{k-1})$ while $\varphi_k(y) \notin \varphi_{k-1}(G_{k-1})$ because $\varphi_k(y) = f_k(y) \notin \varphi_{k-1}(G_{k-1})$ if $y \in E_k$ and $\varphi_k(y) \in J^\omega \times \sigma(a_k) \subset (J^\omega \times J^\omega) \backslash \varphi_{k-1}(G_{k-1})$ for all $y \in F_k \backslash (G_{k-1} \cup E_k)$.

Next observe that if $x, y \in E_k$, then $\varphi_k(x) = f_k(x) \neq f_k(y) = \varphi_k(y)$ because f_k is a condensation. If $x \in E_k$ and $y \in F_k \backslash (G_{k-1} \cup E_k)$, then $\varphi_k(x) \in J^\omega \times S_k$ while we have $\varphi_k(y) \in J^\omega \times \sigma(a_k)$ which shows that $\varphi_k(x) \neq \varphi_k(y)$. Finally, if $x, y \in F_k \backslash (G_{k-1} \cup E_k)$, then by our choice of the sequence $\{h_i : i \in \omega\}$ there is $i \in \omega$ such that $x, y \in H_i$ and $h_i(x) \neq h_i(y)$. Consequently, $v_i(x) \neq v_i(y)$ which shows that $v(x) \neq v(y)$ and hence $\varphi_k(x) \neq \varphi_k(y)$.

Thus the conditions (1)–(3) are satisfied for $n = k$ and hence our inductive construction gives us a sequence of mappings $\{\varphi_n : n \in \omega\}$ with the properties (1)–(3). It follows from (2) that a map $\varphi = \bigcup_{i \in \omega} \varphi_n : X \to J^\omega \times J^\omega$ is well-defined. Since $\bigcup_{n \in \omega} S_n = J^\omega$, the condition (3) guarantees that $\varphi(X) = J^\omega \times J^\omega \simeq J^\omega$. The conditions (1) and (2) imply that φ is injective. Finally observe that it follows from (2) that the maps $\{\psi_n = \varphi_n | F_n : n \in \omega\}$ satisfy the assumptions of Fact 2 and hence $\varphi = \bigcup_{n \in \omega} \psi_n$ is continuous. Thus φ condenses X onto J^ω and hence our solution is complete.

T.355. *Give an example of a non-σ-compact subspace of $\mathbb{R}$ which cannot be condensed onto a compact space.*

Solution. If CH holds, then there exists a Luzin space $X \subset \mathbb{R}$ (see Fact 1 of T.046). If some compact $K \subset X$ is uncountable, then K is a Luzin space which is a contradiction because a Luzin space cannot even condense onto a compact space by Problem 045. Thus every compact subspace of X is countable;

since X is uncountable, it is not σ-compact. Consequently, X is an example of a non-σ-compact subspace of $\mathbb{R}$ which cannot be condensed onto a compact space by Problem 045.

Now, if CH is false, then take any $X \subset \mathbb{R}$ with $|X| = \omega_1$. Since CH does not hold, we have $|X| < \mathfrak{c}$. If a compact $K \subset X$ is uncountable, then $\mathbb{K}$ embeds in K by Problem 353 and hence $|K| \geq |\mathbb{K}| = \mathfrak{c} > \omega_1 = |X|$, a contradiction. Thus every compact $K \subset X$ is countable and hence X is not σ-compact. If X condenses onto a compact space Y, then $w(Y) = nw(Y) \leq nw(X) = \omega$ (see Fact 4 of S.307), i.e., Y is a second countable uncountable compact space. Therefore we can apply Problem 353 again to find a subspace $Z \subset Y$ which is homeomorphic to the Cantor set $\mathbb{K}$. As a consequence, $|X| = |Y| \geq |Z| = |\mathbb{K}| = \mathfrak{c} > \omega_1$; this final contradiction shows that X is a non-σ-compact subspace of $\mathbb{R}$ which does not condense onto a compact space.

T.356. *Prove that $\mathbb{Q}$ cannot be condensed onto a compact space.*

Solution. If there is a condensation $f : \mathbb{Q} \to K$ of $\mathbb{Q}$ onto a compact space K, then K is countable and hence scattered because otherwise it can be continuously mapped onto the space $\mathbb{I}$ which is uncountable (see Problem 133). Thus there is a point $x \in K$ which is isolated in K. Since f is a condensation, there is $q \in \mathbb{Q}$ such that $\{q\} = f^{-1}(x)$; since $\{x\} \in \tau(K)$, the set $\{q\}$ is open in $\mathbb{Q}$, i.e., q is an isolated point of $\mathbb{Q}$ which is a contradiction.

T.357. *Prove that for any metrizable compact X, the space $C_p(X)$ condenses onto a compact space.*

Solution. Given a space Z let ρ_Z be a metric on the set $C^*(Z)$ defined by the formula $\rho_Z(f, g) = \sup\{|f(z) - g(z)| : z \in Z\}$ for all $f, g \in C^*(Z)$. It was proved in Problem 248 of [TFS] that $(C^*(Z), \rho_Z)$ is a complete metric space. Given any set $A \subset C(Z)$ let $\mathrm{cl}_u(A) = \{f \in C(Z) :$ there is a sequence $\{f_n : n \in \omega\} \subset A$ such that $f_n \Rightarrow f\}$. It was proved in Problem 084 of [TFS] that there exists a unique topology τ_Z^u on $C(Z)$ such that $\mathrm{cl}_u(A)$ is the closure of A in $(C(Z), \tau_Z^u)$ for any $A \subset C(Z)$. The space $(C(Z), \tau_Z^u)$ is denoted by $C_u(Z)$; the space $C_u^*(Z)$ is the set $C^*(Z)$ with the topology induced from $C_u(Z)$.

Fact 1. For any space Z, the identity map $i : C_u^*(Z) \to C^*(Z)$ is a homeomorphism between $C_u^*(Z)$ and $(C^*(Z), \rho_Z)$ or, in other words, $\tau(C_u^*(Z)) = \tau(\rho_Z)$.

Proof. Let $B(f, \varepsilon) = \{g \in C^*(Z) : \rho_Z(f, g) < \varepsilon\}$ for all $f \in C^*(Z)$ and $\varepsilon > 0$, i.e., $B(f, \varepsilon)$ is the ε-ball (with respect to ρ_Z) centered at f. Suppose that F is a closed subset of $C_u^*(Z)$. If F is not closed in $(C^*(Z), \rho_Z)$, then take any $f \notin F$ which is in the $\tau(\rho_Z)$-closure of F. For every $n \in \mathbb{N}$ there exists $f_n \in B(f, \frac{1}{n}) \cap F$; it is immediate that $f_n \Rightarrow f$ and hence f is in the closure of F in the space $C_u^*(Z)$ which is a contradiction. Thus every closed subset of $C_u^*(Z)$ is also closed in $(C^*(Z), \rho_Z)$ and therefore $\tau(C_u^*(Z)) \subset \tau(\rho_Z)$.

Now assume that a set F is closed in $(C^*(Z), \rho_Z)$ and there is a function $f \in \mathrm{cl}_u(F) \backslash F$, i.e., there exists a sequence $S = \{f_n : n \in \omega\} \subset F$ with $f_n \Rightarrow f$. For any $\varepsilon > 0$, there is $m \in \omega$ such that $|f_m(z) - f(z)| < \frac{\varepsilon}{2}$ for all $z \in Z$ and hence $\rho_Z(f_m, f) \leq \frac{\varepsilon}{2} < \varepsilon$. Therefore $f_m \in B(f, \varepsilon) \cap S$ which shows that f is in the closure of S in the space $(C^*(Z), \rho_Z)$ and hence f is in the closure of F in $(C^*(Z), \rho_Z)$; this contradiction proves that any closed subset of $(C^*(Z), \rho_Z)$ is closed in $C_u^*(Z)$, i.e., $\tau(\rho_Z) \subset \tau(C_u^*(Z))$. Thus $\tau(C_u^*(Z)) = \tau(\rho_Z)$ and Fact 1 is proved. □

Fact 2. If K is a compact space, then $w(C_u(K)) = w(K)$. In particular, if K is metrizable, then $C_u(K)$ is a Polish space.

Proof. Observe first that $C_u(K) = C_u^*(K)$ and hence $C_u(K)$ is a completely metrizable space by Fact 1. As a consequence, $w(C_u(K)) = d(C_u(K))$; choose a dense $A \subset C_u(K)$ with $|A| = d(C_u(K))$. Since the identity map $i : C_u(K) \to C_p(K)$ is continuous (see Problem 086 of [TFS]), the set A has to be dense in $C_p(K)$ as well and hence $w(K) = iw(K) = d(C_p(K)) \leq |A| = w(C_u(K))$. Therefore $w(K) \leq w(C_u(K))$.

Now if $w(K) = \kappa$, then there is a set $B \subset C_p(K)$ with $|B| \leq \kappa$ which is dense in $C_p(K)$. It is evident that B separates the points of K. Let $P(B) = \{f_1 \cdots f_k : k \in \mathbb{N} \text{ and } f_i \in B \text{ for all } i \leq k\}$. We will need the sets $R(B) = \{\lambda_0 + \lambda_1 g_1 + \cdots + \lambda_n g_n : n \in \mathbb{N}, g_i \in P(B) \text{ and } \lambda_i \in \mathbb{R} \text{ for all } i \leq n\}$ and $Q(B) = \{q_0 + q_1 g_1 + \cdots + q_n g_n : n \in \mathbb{N}, g_i \in P(B) \text{ and } q_i \in \mathbb{Q} \text{ for all } i \leq n\}$. It is easy to check that $R(B)$ is an algebra in $C(K)$; it follows from $B \subset R(B)$ that $R(B)$ separates the points of K and hence $R(B)$ is dense in $C_u(K)$ (see Problem 191 of [TFS]).

Observe that $|Q(B)| = |B| \leq \kappa$, so it suffices to show that $Q(B)$ is dense in $C_u(K)$. We will do it in several steps.

(1) If $\{r_n : n \in \omega\} \subset \mathbb{R}$ and $r_n \to r$ then $r_n f \rightrightarrows rf$ for any $f \in C(K)$.

Indeed, there is $A > 0$ such that $|f(x)| < A$ for all $x \in K$. Given $\varepsilon > 0$, there is $m \in \omega$ such that $|r_n - r| < \frac{\varepsilon}{A}$ for all $n \geq m$. Then $|r_n f(x) - rf(x)| = |f(x)||r_n - r| < A \cdot \frac{\varepsilon}{A} = \varepsilon$ for all $n \geq m$ and $x \in K$ which shows that (1) is true.

(2) $f \in \mathrm{cl}_u(Q(B))$ for any $f \in R(B)$.

There are $g_1, \ldots, g_n \in P(B)$ such that $f = \lambda_0 + \lambda_1 g_1 + \cdots + \lambda_n g_n$. For each $i = 0, \ldots, n$, choose a sequence $\{r_k^i : k \in \omega\} \subset \mathbb{Q}$ such that $r_k^i \to \lambda_i$. Now, $r_k^i g_i \in Q(B)$ for all $k \in \omega$ and $i \leq n$, so we can apply (1) to conclude that $r_k^i g_i \rightrightarrows \lambda_i g_i$ for all $i \leq n$; therefore $f_k = r_k^0 + r_k^1 g_1 + \cdots + r_k^n g_n$ belongs to $Q(B)$ for all $k \in \omega$ and $f_k \rightrightarrows f$ by Problem 035 of [TFS]. Thus $f \in \mathrm{cl}_u(Q(B))$, i.e., (2) is proved.

It is an immediate consequence of (2) that we have $R(B) \subset \mathrm{cl}_u(Q(B))$ and therefore $C_u(K) = \mathrm{cl}_u(R(B)) \subset \mathrm{cl}_u(Q(B))$, i.e., the set $Q(B)$ is dense in the space $C_u(K)$. Thus $d(C_u(K)) \leq |Q(B)| \leq |B| = \kappa = w(K)$ so Fact 2 is proved. □

Fact 3. Every locally compact space condenses onto a compact space.

Proof. Let Z be a locally compact space; if Z is compact, then there is nothing to prove, so we assume that Z is not compact. Then the one-point compactification K

of the space Z is a compact space such that $|K \backslash Z| = 1$; let $a \in K$ be the point for which $Z \cup \{a\} = K$. Fix any point $b \in Z$. In the space K collapse the closed set $F = \{a, b\}$ to a point to obtain the space K/F. Recall that $K/F = (K \backslash F) \cup \{F\}$ (i.e., the set F is considered to be a point in the space K/F; denote this point by z_F). Define a map $p_F : K \to K/F$ as follows: $p(x) = x$ for all $x \in K \backslash F$ and $p(x) = z_F$ for all $x \in F$. It was proved in Fact 2 of T.245 that K/F is a Tychonoff space and the map p_F is continuous. Thus K/F is compact; it is immediate that $p_F | Z$ is a condensation of Z onto the compact space K/F, so Fact 3 is proved. □

Returning to our solution observe first that if X is finite, then $C_p(X) = \mathbb{R}^X$ is a locally compact space; therefore $C_p(X)$ condenses onto a compact space by Fact 3.

Now, if X is infinite, then fix a nontrivial convergent sequence $S \subset X$ (we consider that S contains its limit and hence S is a countably infinite compact subspace of X). If $Y \subset Z \subset X$, let $\pi_Y^Z : C_p(Z) \to C_p(Y)$ be the restriction map, i.e., $\pi_Y^Z(f) = f | Y$ for all $f \in C_p(Z)$; if $Z = X$, then π_Y^Z will be denoted by π_Y. Let $C_p(Y|X) = \pi_Y(C_p(X))$ for any $Y \subset X$.

Let A be a countable dense subset of X such that $S \subset A$. Observe that $C_p(S|X) = C_p(S)$ because S is compact (see Fact 1 of T.218). It follows from $\pi_S = \pi_S^A \circ \pi_A$ that π_S^A maps $C_p(A|X)$ onto $C_p(S)$; the space $C_p(S)$ is not σ-compact by Problem 186 of [TFS], so the space $C_p(A|X)$ is not σ-compact either.

Furthermore, the Polish space $C_u(X)$ condenses onto $C_p(X)$ (see Fact 2 and Problem 086 of [TFS]). Since the map π_A condenses $C_p(X)$ onto $C_p(A|X)$, there exists a condensation of $C_u(X)$ onto $C_p(A|X)$. Therefore $C_p(A|X)$ is a non-σ-compact Borel set (see Problem 341) and we can apply Problem 354 to conclude that there is a condensation $\varphi : C_p(A|X) \to K$ of $C_p(A|X)$ onto a compact space K. Then $\psi = \varphi \circ \pi_A$ condenses $C_p(X)$ onto the compact space K and hence our solution is complete.

T.358. *Prove that a Polish space X is dense-in-itself if and only if $\mathbb{P}$ condenses onto X.*

Solution. For the sake of brevity, we say that a space is *crowded* if it is non-empty and dense-in-itself, i.e., has no isolated points. We identify $\mathbb{P}$ with the space ω^ω (see Problem 313). Given any $s \in \omega^{<\omega}$ let $O_s = \{t \in \mathbb{P} : s \subset t\}$. If $\mathcal{A}$ is a family of sets in a metric space (Z, ρ), then $\mathrm{diam}_\rho(\mathcal{A}) = \sup\{\mathrm{diam}_\rho(A) : A \in \mathcal{A}\}$.

Fact 1. If F and G are closed subsets of a crowded Polish space (Z, d) and $F \backslash G$ is crowded, then for any $\varepsilon > 0$, there exists a family $\{F_n : n \in \omega\}$ of closed subsets of Z such that $\mathrm{diam}_d(F_n) < \varepsilon$, the set $F_n \backslash (\bigcup \{F_i : i < n\})$ is crowded for every $n \in \omega$ and $\bigcup \{F_n : n \in \omega\} = F \backslash G$.

Proof. For each $x \in F \backslash G$ fix a set $U_x \in \tau(x, Z)$ such that $\mathrm{diam}_d(U_x) < \varepsilon$ and $\overline{U}_x \cap G = \emptyset$ (the bar denotes the closure in Z). Choose a countable subcover $\{V_i : i \in \omega\}$ of the cover $\{U_x : x \in F \backslash G\}$ of the space $F \backslash G$. The set $W_i = V_i \cap (F \backslash G)$ is crowded being a non-empty open subset of a crowded set $F \backslash G$ for every $i \in \omega$. Therefore $H_i = \overline{W}_i$ is also crowded; it is clear that H_i is a closed subset of Z and $H_i \subset F \backslash G$ for all $i \in \omega$. Besides, $\bigcup_{i \in \omega} H_i = F \backslash G$.

If no finite subfamily of the family $\mathcal{H} = \{H_i : i \in \omega\}$ covers the space $F \setminus G$, then start off with the set $F_0 = H_0$, and for any $n \in \omega$, let $F_{n+1} = H_k$ where $k = \min\{i \in \omega : H_i$ is not contained in $F_0 \cup \cdots \cup F_n\}$. It is easy to see that $\bigcup_{n \in \omega} F_n = F \setminus G$; every F_n is crowded and $F_n' = F_n \setminus (\bigcup_{i<n} F_i)$ is a non-empty open subset of F_n by our definition of the family $\mathcal{F} = \{F_n : n \in \omega\}$. Thus F_n' is crowded for every $n \in \omega$ and hence the family $\mathcal{F}$ is as promised.

Now assume that $F \setminus G = H_0 \cup \cdots \cup H_n$ for some $n \in \omega$. Throwing away the empty elements of this union, we can consider that $H_k' = H_k \setminus (\bigcup_{i<k} H_i) \neq \emptyset$ for all $k \leq n$. The set H_n' is crowded and has diameter $< \varepsilon$, so it suffices to split it into countably many pieces as required in our fact without caring about the diameter.

To see that it is possible, observe that the space H_n' being crowded, we can choose a faithfully enumerated convergent sequence $A = \{a_n : n \in \omega\} \subset H_n'$ such that $a_n \to a \in H_n' \setminus A$. It is an easy exercise that it is possible to construct a family $\mathcal{O} = \{O_i : i \in \omega\} \subset \tau(H_n')$ such that $\overline{O}_i \subset H_n' \setminus \{a\}$ and $a_i \in O_i$ for all $i \in \omega$ while the family $\{\overline{O}_i : i \in \omega\}$ is disjoint and $\mathcal{O} \to a$ in the sense that for any $O \in \tau(a, Z)$, there are only finitely many elements of $\mathcal{O}$ which are not contained in O. It is easy to see that $\mathcal{O}' \to a$ for any $\mathcal{O}' \subset \mathcal{O}$.

Observe that if $E = \overline{\bigcup\{O_{2k} : k \in \omega\}}$, then $\overline{E} = E \cup \{a\}$ by Fact 1 of T.351. Since $a \in \overline{\bigcup\{O_{2k+1} : k \in \omega\}} \subset H_n' \setminus E$, we have $P_0 = \overline{H_n' \setminus (E \cup \{a\})} = \overline{H_n' \setminus E}$ and hence P_0 is crowded being a closure of a non-empty open subset of a crowded space H_n'. The set $P_i = \overline{O}_{2i-2}$ is also crowded for the same reason for all $i \in \mathbb{N}$. Finally, let $F_i = H_i$ for all $i < n$ and $F_{n+i} = P_i$ for all $i \in \omega$. It is immediate that the sequence $\{F_i : i \in \omega\}$ witnesses that Fact 1 is proved.   □

Returning to our solution, assume that $f : \mathbb{P} \to X$ is a condensation and some $x \in X$ is isolated in X. Then there is $s \in \mathbb{P}$ such that $\{s\} = f^{-1}(x)$ is open in $\mathbb{P}$, i.e., the point s is isolated in $\mathbb{P}$. However $\mathbb{P}$ has no isolated points by Problem 347; this contradiction proves that X has no isolated points either.

Now suppose that X is a dense-in-itself Polish space and fix a complete metric d on X such that $\tau(d) = \tau(X)$ and $d(x, y) \leq 1$ for all $x, y \in X$. Let $Q_\emptyset = X$; it is evident that we can apply Fact 1 to the sets $F = X$ and $G = \emptyset$ to find a sequence $\{P_n : n \in \omega\}$ of closed subsets of X such that $\text{diam}_d(P_n) \leq \frac{1}{2}$ and $P_n' = P_n \setminus (\bigcup_{i<n} P_i)$ is crowded for all $n \in \omega$. Now let $Q_s = P_{s(0)}$ for each $s \in \omega^1$.

Assume that $k > 1$ and we have, for every $i < k$, a family $\mathcal{Q}_i = \{Q_s : s \in \omega^i\}$ with the following properties:

(1) $Q_\emptyset = X$ and $Q_s = \bigcup\{Q_{s \frown i} : i \in \omega\}$ for all $s \in \omega^m$ whenever $m + 1 < k$;
(2) the set Q_s is crowded for every $s \in \omega^i$ whenever $i < k$;
(3) the family $\mathcal{Q}_i$ is disjoint and $\text{diam}(\mathcal{Q}_i) \leq \frac{1}{2^i}$ for all $i < k$;
(4) $Q_{s \frown 0} \cup \cdots \cup Q_{s \frown j}$ is closed in X for any $j \in \omega$ and $s \in \omega^m$ with $m + 1 < k$.

To construct $\mathcal{Q}_k$ fix any $t \in \omega^{k-1}$; let $s = t|(k-2)$ and $m = t(k-2)$. It follows from (4) that the sets $G = Q_{s \frown 0} \cup \cdots \cup Q_{s \frown (m-1)}$ and $F = Q_{s \frown 0} \cup \cdots \cup Q_{s \frown (m-1)} \cup Q_t$ are closed in X and $F \setminus G = Q_t$ is a crowded set. Apply Fact 1 to find a sequence $\mathcal{P} = \{P_n : n \in \omega\}$ of closed subsets of X such that $\bigcup_{i \in \omega} P_i = Q_t$, $\text{diam}(\mathcal{P}) \leq \frac{1}{2^k}$ and $P_n' = P_n \setminus (\bigcup_{i<n} P_i)$ is crowded for all $n \in \omega$. Let $Q_{t \frown i} = P_i'$ for every

$i \in \omega$; after we have the set $Q_{t \frown i}$ for all $t \in \omega^{k-1}$ and $i \in \omega$, we obtain a family $\mathcal{Q}_k = \{Q_s : s \in \omega^k\}$.

It is straightforward that the conditions (1)–(4) are still satisfied for the families $\mathcal{Q}_0, \ldots, \mathcal{Q}_k$ and hence we can construct inductively a sequence $\{\mathcal{Q}_i : i \in \omega\}$ with the properties (1)–(4). Given $s \in \mathbb{P}$, let $F = \bigcap \{\overline{Q}_{s|n} : n \in \omega\}$. It follows from (1) that the family $\{\overline{Q}_{s|n} : n \in \omega\}$ is decreasing. Since (X, d) is complete and $\mathrm{diam}_d (\overline{Q}_{s|n}) \to 0$ by (3), the set F consists of a single point $x(s)$; let $f(s) = x(s)$. This defines a function $f : \mathbb{P} \to X$.

Take any $s \in \mathbb{P}$ and $\varepsilon > 0$; there is $n \in \omega$ such that $\mathrm{diam}_d (\overline{Q}_{s|n}) < \varepsilon$. If $y \in \overline{Q}_{s|n}$, then $d(f(s), y) \leq \mathrm{diam}_d (\overline{Q}_{s|n}) < \varepsilon$ and hence $y \in B_d(f(s), \varepsilon)$ which shows that $\overline{Q}_{s|n} \subset B_d(f(s), \varepsilon)$. The set $W = O_{s|n}$ is an open neighborhood of s in $\mathbb{P}$. If $t \in W$, then $t|n = s|n$ and hence $f(t) \in \overline{Q}_{t|n} = \overline{Q}_{s|n} \subset B_d(f(s), \varepsilon)$, i.e., $f(W) \subset B_d(f(s), \varepsilon)$ which proves that f is continuous at the point s.

If $s, t \in \mathbb{P}$ and $s \neq t$, then there is $n \in \omega$ such that $s|n \neq t|n$; we can assume without loss of generality that $s(n-1) < t(n-1)$. It follows from the properties (3) and (4) that $\overline{Q}_{s|n} \cap Q_{t|n} = \emptyset$; besides, $\overline{Q}_{t|(n+1)} \subset Q_{t|n}$ and hence $\overline{Q}_{s|n} \cap \overline{Q}_{t|(n+1)} = \emptyset$. Since $f(s) \in \overline{Q}_{s|n}$ and $f(t) \in \overline{Q}_{t|(n+1)}$, we have $f(s) \neq f(t)$, i.e., f is an injection.

Finally, if $x \in X$, then it follows from the property (1) that there is a sequence $\{s_n : n \in \omega\} \subset \omega^{<\omega}$ such that $s_n \in \omega^n$ and $x \in Q_{s_n}$ for all $n \in \omega$. The property (3) shows that $s_n \subset s_{n+1}$ for all $n \in \omega$ and hence $s = \bigcup_{n \in \omega} s_n$ is well-defined. Thus $x \in \bigcap_{n \in \omega} Q_{s_n} = \bigcap_{n \in \omega} Q_{s|n} \subset \bigcap_{n \in \omega} \overline{Q}_{s|n}$ and therefore $f(s) = x$, i.e., f is a condensation of $\mathbb{P}$ onto X. Our solution is complete.

T.359. *Prove that for any metrizable compact X, the space $\mathbb{P}$ condenses onto $C_p(X)$.*

Solution. Since X is compact and metrizable, the space $C_u(X)$ is Polish (see Fact 2 of T.357). It is evident that $C_u(X)$ has no isolated points, so there is a condensation $f : \mathbb{P} \to C_u(X)$ by Problem 358. If $i : C_u(X) \to C_p(X)$ is the identity map, then i condenses $C_u(X)$ onto $C_p(X)$ and hence $i \circ f$ is a condensation of $\mathbb{P}$ onto the space $C_p(X)$.

T.360. *Prove that $C_p(X)$ is analytic if and only if $\mathbb{R}^\omega$ maps continuously onto $C_p(X)$. Observe that not every analytic space is a continuous image of $\mathbb{R}^\omega$.*

Solution. The space $\mathbb{R}^\omega$ is Polish and hence analytic (see Problem 334). If $\mathbb{R}^\omega$ maps continuously onto $C_p(X)$, then $C_p(X)$ is analytic because, evidently, any continuous image of an analytic space is an analytic space.

Now assume that the space $C_p(X)$ is analytic and fix a continuous onto map $f : \mathbb{P} \to C_p(X)$. We can consider that $\mathbb{P}$ is a closed subset of $\mathbb{R}^\omega$ (see Problem 273 of [TFS]). Since $C_p(X)$ is a locally convex space, we can apply Problem 104 to find a continuous map $g : \mathbb{R}^\omega \to C_p(X)$ such that $g|\mathbb{P} = f$. It is clear that $g(\mathbb{R}^\omega) \supset g(\mathbb{P}) = f(\mathbb{P}) = C_p(X)$, so $C_p(X)$ is a continuous image of $\mathbb{R}^\omega$.

Recall that a space Z is called *connected* if every clopen subset of Z belongs to the family $\{\emptyset, Z\}$. Call a clopen $U \subset Z$ *nontrivial* if $U \notin \{\emptyset, Z\}$. In this terminology, a space Z is connected if and only if it has no nontrivial clopen sets. Let $\sigma(\omega) = \{x \in \mathbb{R}^\omega : |\{n \in \omega : x(n) \neq 0\}| < \omega\}$. The set $\sigma(\omega)$ is connected by Fact 2 of T.312; besides, $\sigma(\omega)$ is dense in $\mathbb{R}^\omega$, so we can apply Fact 1 of T.312 to conclude that $\mathbb{R}^\omega$ is connected as well. The two-point space $\mathbb{D}$ is analytic being compact and metrizable; however it is impossible to map $\mathbb{R}^\omega$ continuously onto $\mathbb{D}$. Indeed, if $h : \mathbb{R}^\omega \to \mathbb{D}$ is a continuous onto map, then $h^{-1}(0)$ is a nontrivial clopen subset of $\mathbb{R}^\omega$ which is a contradiction because $\mathbb{R}^\omega$ is connected. Thus, an analytic space $\mathbb{D}$ is not a continuous image of $\mathbb{R}^\omega$.

T.361. *Suppose that X is an infinite space such that $C_p(X)$ is analytic. Prove that $C_p(X)$ can be continuously mapped onto $\mathbb{R}^\omega$. Deduce from this fact that if X and Y are infinite spaces such that $C_p(X)$ and $C_p(Y)$ are analytic, then each of the spaces $C_p(X)$ and $C_p(Y)$ maps continuously onto the other one.*

Solution. For every $i \in \omega$ let $\pi_i : \mathbb{R}^\omega \to \mathbb{R}$ be the natural projection of $\mathbb{R}^\omega$ onto its ith factor. The space $C_p(X)$ is not σ-compact by Problem 186 of [TFS], so we can apply 352 to conclude that there is a closed $F \subset C_p(X)$ such that $F \simeq \mathbb{P}$. Since $\mathbb{R}^\omega$ is analytic, there is a continuous onto map $f : F \to \mathbb{R}^\omega$; let $f_i = \pi_i \circ f$ for all $i \in \omega$. The space $C_p(X)$ is normal because it has a countable network; therefore there is a continuous function $g_i : C_p(X) \to \mathbb{R}$ such that $g_i | F = f_i$ for all $i \in \omega$. Then $g = \Delta_{i \in \omega} g_i : C_p(X) \to \mathbb{R}^\omega$. It is easy to see that g extends f and therefore $g(C_p(X)) \supset g(F) = f(F) = \mathbb{R}^\omega$ and hence g maps $C_p(X)$ continuously onto $\mathbb{R}^\omega$.

Now if $C_p(Y)$ is analytic, then $\mathbb{R}^\omega$ maps continuously onto $C_p(Y)$ by Problem 360. We proved that $C_p(X)$ maps continuously onto $\mathbb{R}^\omega$, so the composition of the relevant functions maps $C_p(X)$ continuously onto $C_p(Y)$. Analogously, if Y is infinite and $C_p(Y)$ is analytic, then $C_p(Y)$ maps continuously onto $C_p(X)$.

T.362. *Prove that for any second countable σ-compact space X, the space $C_p(X)$ is $K_{\sigma\delta}$.*

Solution. We have $X = \bigcup\{X_n : n \in \mathbb{N}\}$ where X_n is compact for all $n \in \mathbb{N}$. Let d be a metric on X such that $\tau(d) = \tau(X)$ and $d(x, y) \leq 1$ for all $x, y \in X$. For any $k, l, n \in \mathbb{N}$, let $M_{kln} = \{f \in \mathbb{I}^X : |f(x) - f(y)| \leq \frac{1}{k}$ whenever x and y are points of X such that $d(x, z) < \frac{1}{n}$ and $d(y, z) < \frac{1}{n}$ for some $z \in X_l\}$. Our first observation is that

(1) the set M_{kln} is compact for any $k, l, n \in \mathbb{N}$.

Indeed, if $f \in \mathbb{I}^X \setminus M_{kln}$, then there exist points $x, y \in X$ such that $d(x, z) < \frac{1}{n}$ and $d(y, z) < \frac{1}{n}$ for some $z \in X_l$ while $|f(x) - f(y)| > \frac{1}{k}$; then $r = |f(x) - f(y)| - \frac{1}{k} > 0$. The set $O_f = \{g \in \mathbb{I}^X : |g(x) - f(x)| < \frac{r}{2}$ and $|g(y) - f(y)| < \frac{r}{2}\}$ is open in the space $\mathbb{I}^X$ and $f \in O_f$. For any function $g \in O_f$, we have $|f(x) - f(y)| \leq |f(x) - g(x)| + |g(x) - g(y)| + |g(y) - f(y)|$ and therefore

$$|g(x)-g(y)| \geq |f(x)-f(y)|-|f(x)-g(x)|-|f(y)-g(y)| > r+\frac{1}{k}-\frac{r}{2}-\frac{r}{2} = \frac{1}{k}$$

which shows that $g \in \mathbb{I}^X \setminus M_{kln}$. Since a function $g \in O_f$ was chosen arbitrarily, we have $O_f \subset \mathbb{I}^X \setminus M_{kln}$. Thus, for every $f \in \mathbb{I}^X \setminus M_{kln}$, there is $O_f \in \tau(\mathbb{I}^X)$ such that $O_f \subset \mathbb{I}^X \setminus M_{kln}$. As a consequence, $\mathbb{I}^X \setminus M_{kln} = \bigcup\{O_f : f \in \mathbb{I}^X \setminus M_{kln}\}$ is open in $\mathbb{I}^X$ and hence M_{kln} is compact being closed in $\mathbb{I}^X$.

The property (1) implies that the set $P_{kl} = \bigcup\{M_{kln} : n \in \mathbb{N}\}$ is σ-compact; we claim that $C_p(X, \mathbb{I}) = \bigcap\{P_{kl} : k, l \in \mathbb{N}\}$ and hence $C_p(X, \mathbb{I})$ is $K_{\sigma\delta}$.

Take any $f \in \bigcap\{P_{kl} : k, l \in \mathbb{N}\}$; given a point $z \in Z$ and $\varepsilon > 0$, there is $l \in \mathbb{N}$ with $z \in X_l$. Pick any $k \in \mathbb{N}$ such that $\frac{1}{k} < \varepsilon$. Since $f \in P_{kl}$, there is $n \in \mathbb{N}$ such that $|f(x) - f(y)| \leq \frac{1}{k} < \varepsilon$ whenever $d(x, t) < \frac{1}{n}$ and $d(y, t) < \frac{1}{n}$ for some $t \in X_l$. In particular, $|f(x) - f(z)| < \varepsilon$ if $d(x, z) < \frac{1}{n}$ which proves continuity of f at the point z. Thus f is continuous and hence $\bigcap\{P_{kl} : k, l \in \mathbb{N}\} \subset C_p(X, \mathbb{I})$.

Now if $f \in C_p(X, \mathbb{I})$, fix any $k, l \in \mathbb{N}$. If $f \notin P_{kl}$, then for any $n \in \mathbb{N}$, there exist $x_n, y_n \in X$ and $z_n \in X_l$ such that $|f(x_n) - f(y_n)| > \frac{1}{k}$ while $d(x_n, z_n) < \frac{1}{n}$ and $d(y_n, z_n) < \frac{1}{n}$. Since X_l is compact, the sequence $\{z_n : n \in \mathbb{N}\}$ has a subsequence which converges to a point $z \in X_l$. Passing to that subsequence and reindexing the sequences $\{x_n : n \in \mathbb{N}\}$ and $\{y_n : n \in \mathbb{N}\}$ if necessary, we can assume, without loss of generality, that $z_n \to z$. We also have $x_n \to z$ and $y_n \to z$ because $d(x_n, z_n) < \frac{1}{n}$ and $d(y_n, z_n) < \frac{1}{n}$ for all $n \in \mathbb{N}$. Since f is continuous at the point z, there is $W \in \tau(z, X)$ such that $|f(w) - f(z)| < \frac{1}{2k}$ for all $w \in W$. There is $m \in \mathbb{N}$ such that $x_m, y_m \in W$ and hence $|f(x_m) - f(y_m)| \leq |f(x_m) - f(z)| + |f(z) - f(y_m)| < \frac{1}{2k} + \frac{1}{2k} = \frac{1}{k}$ which is a contradiction. Thus $f \in P_{kl}$ for all $k, l \in \mathbb{N}$ and therefore $f \in \bigcap\{P_{kl} : k, l \in \mathbb{N}\}$. This proves that $\bigcap\{P_{kl} : k, l \in \mathbb{N}\} = C_p(X, \mathbb{I})$.

Let $I_n = [-1 + \frac{1}{n+1}, 1 - \frac{1}{n+1}]$; it is evident that $I_n \subset \mathbb{I}$, $I_n \subset \text{Int}(I_{n+1})$ for all $n \in \mathbb{N}$ and $\bigcup\{I_n : n \in \mathbb{N}\} = J = (-1, 1)$. It is easy to see that the set $Q_{ln} = \{f \in \mathbb{I}^X : f(X_l) \subset I_n\}$ is closed in $\mathbb{I}^X$ and hence compact for all $l, n \in \mathbb{N}$. Therefore $E_l = \bigcup\{Q_{ln} : n \in \mathbb{N}\}$ is σ-compact for all $l \in \mathbb{N}$ which shows that the set $E = \bigcap\{E_l : l \in \mathbb{N}\}$ is $K_{\sigma\delta}$.

If $f \in C_p(X, J)$ and $l \in \mathbb{N}$, then $f(X_l)$ is a compact subset of J which implies that there is $n \in \mathbb{N}$ such that $f(X_l) \subset I_n$; thus $f \in E_l$ for all $l \in \mathbb{N}$ and therefore $f \in E$. Consequently, $C_p(X, J) \subset E$ and hence $C_p(X, J) = C_p(X, J) \cap C_p(X, \mathbb{I}) \subset E \cap P$ where $P = \bigcap\{P_{kl} : k, l \in \mathbb{N}\}$.

On the other hand, if $f \in E$, then $f(X) = \bigcup\{f(X_l) : l \in \mathbb{N}\} \subset J$ because $f(X_l) \subset J$ for every $l \in \mathbb{N}$. Therefore $f \in J^X$. If, additionally, $f \in P$, then f has to be continuous because so are all functions from P. As a consequence, $E \cap P \subset C_p(X, J)$ which shows that $E \cap P = C_p(X, J)$. We proved that the space $C_p(X, J)$ is the intersection of two $K_{\sigma\delta}$-subsets of $\mathbb{I}^X$ and hence $C_p(X, J)$ is $K_{\sigma\delta}$. Since $C_p(X)$ is homeomorphic to $C_p(X, J)$ (see Problem 025 of [TFS]), we proved that $C_p(X)$ is also a $K_{\sigma\delta}$-space.

T.363. *Let X be a space with a countable network. Prove that X is analytic if and only if every second countable continuous image of X is analytic.*

Solution. In this solution we will need some (possibly) discontinuous maps. Given spaces Y and Z call a map $f : Y \to Z$ *measurable* if $f^{-1}(B) \in \mathbb{B}(Y)$ for any $B \in \mathbb{B}(Z)$. A space is called *cosmic* if it has a countable network. A space Z is *perfect* if every $U \in \tau(Z)$ is an F_σ-set in Z.

Fact 1. Given perfect spaces Y and Z a map $f : Y \to Z$ is measurable if and only if $f^{-1}(U) \in \mathbb{B}(Y)$ for any $U \in \tau(Z)$.

Proof. We have $\mathbb{B}(Z) = \bigcup\{\Sigma_\xi^0(Z) : \xi < \omega_1\}$ by Fact 1 of T.331. By our assumption, $f^{-1}(U) \in \mathbb{B}(Y)$ for any $U \in \Sigma_0^0(Z)$; if $F \in \Pi_0^0(Z)$, then $U = Z \backslash F \in \Sigma_0^0(Z)$ and hence $U' = f^{-1}(U) \in \mathbb{B}(Y)$. Since $F' = f^{-1}(F) = Y \backslash U'$, we have $F' \in \mathbb{B}(Y)$ by Fact 1 of T.331.

Now assume that $\alpha < \omega_1$ and, for any $\beta < \alpha$ we have $f^{-1}(U) \in \mathbb{B}(Y)$ for any $U \in \Sigma_\beta^0(Z) \cup \Pi_\beta^0(Z)$. If $U \in \Sigma_\alpha^0(Z)$, then there is a sequence $\{\beta_n : n \in \omega\} \subset \alpha$ and $U_n \in \Pi_{\beta_n}^0(Z)$ for every $n \in \omega$ such that $U = \bigcup_{n \in \omega} U_n$. Therefore $f^{-1}(U) = \bigcup\{f^{-1}(U_n) : n \in \omega\}$; since $f^{-1}(U_n) \in \mathbb{B}(Y)$ for every $n \in \omega$ by the induction hypothesis, we have $f^{-1}(U) \in \mathbb{B}(Y)$ by Fact 1 of T.331.

Finally, if $F \in \Pi_\alpha^0(Z)$, then $U = Z \backslash F \in \Sigma_\alpha^0(Z)$ and hence $U' = f^{-1}(U) \in \mathbb{B}(Y)$. Since $F' = f^{-1}(F) = Y \backslash U'$, we have $F' \in \mathbb{B}(Y)$ by Fact 1 of T.331. Thus our inductive procedure can be carried out to establish that $f^{-1}(U) \in \mathbb{B}(Y)$ for any $U \in \Sigma_\alpha^0(Z) \cup \Pi_\alpha^0(Z)$ for all $\alpha < \omega_1$. Therefore f is a measurable map and Fact 1 is proved. □

Fact 2. Given a space Z for any $A \subset Z$, denote by Z_A the space with the underlying set Z and the topology generated by $\tau(Z) \cup \{A, Z \backslash A\}$. Suppose that $\mathcal{A} \subset \exp(Z)$ is a non-empty family. In the product $P = \prod\{Z_A : A \in \mathcal{A}\}$ consider the set $T(\mathcal{A}) = \{f \in P : f(A) = f(B)$ for any $A, B \in \mathcal{A}\}$. Then

(a) the space Z_A is Tychonoff for any $A \in \mathcal{A}$ and hence both spaces P and $T(\mathcal{A})$ are Tychonoff as well.
(b) $T(\mathcal{A})$ is a closed subspace of P homeomorphic to the space $Z(\mathcal{A})$ whose underlying set is Z and whose topology μ is generated (as a subbase) by the family $\mathcal{S} = \tau(Z) \cup \mathcal{A} \cup \{Z \backslash A : A \in \mathcal{A}\}$.

Proof. The Tychonoff property of every Z_A follows from $Z_A \simeq A \oplus (Z \backslash A)$ (we leave it to the reader to verify that any free union of Tychonoff spaces is a Tychonoff space). Since products and subspaces of Tychonoff spaces are Tychonoff, both spaces P and $T(\mathcal{A})$ are Tychonoff.

To see that $T(\mathcal{A})$ is closed in P take any $f \in P \backslash T(\mathcal{A})$. Then for some distinct $A, B \in \mathcal{A}$, we have $x = f(A) \neq y = f(B)$. Pick disjoint $U, V \in \tau(Z)$ such that $x \in U$ and $y \in V$. The set $W = \{f \in P : f(A) \in U$ and $f(B) \in V\}$ is an open neighborhood of f in P such that $W \cap T(\mathcal{A}) = \emptyset$. This shows that $P \backslash T(\mathcal{A})$ is open in P and hence $T(\mathcal{A})$ is a closed subset of P.

Let $q_A : P \to Z_A$ be the natural projection for every $A \in \mathcal{A}$. For each $z \in Z$, let $f_z(A) = z$ for all $A \in \mathcal{A}$ and observe that $T(\mathcal{A}) = \{f_z : z \in Z\}$. Furthermore, $q_A(f_z) = z$ for every $z \in Z$ and $A \in \mathcal{A}$. If $i(z) = f_z$ for every $z \in Z$, then $i : Z \to T(\mathcal{A})$ is a bijection. Let $\mathrm{id} : Z \to Z$ be the identity map.

To check that the map $i : (Z, \mu) \to T(\mathcal{A})$ is continuous, it suffices to show that $q_A \circ i$ is continuous for every $A \in \mathcal{A}$. It is clear that, $q_A \circ i = \mathrm{id}$; besides, the map $\mathrm{id} : Z(\mathcal{A}) \to Z_A$ is continuous. Indeed, $\mathcal{B} = \tau(Z) \cup \{A\} \cup \{Z \backslash A\}$ is a subbase of Z_A, so it suffices to show that $\mathrm{id}^{-1}(U) \in \mu$ for all $U \in \mathcal{B}$. We have $\mathcal{B} \subset \mathcal{S}$, so $\mathrm{id}^{-1}(U) = U \in \mathcal{S} \subset \mu$ which shows that $q_A \circ i$ is continuous for each $A \in \mathcal{A}$ and hence the map $i : (Z, \mu) \to T(\mathcal{A})$ is continuous.

Furthermore, observe that $j = i^{-1} = r_A = q_A | T(\mathcal{A})$ for every $A \in \mathcal{A}$. It is sufficient to prove that $j^{-1}(U)$ is open in $T(\mathcal{A})$ for every $U \in \mathcal{S}$. Now, if $U \in \mathcal{S}$, then there is $A \in \mathcal{A}$ such that $U \in \tau(Z_A)$ and hence $j^{-1}(U) = r_A^{-1}(U)$ is open in $T(\mathcal{A})$ because the map r_A is continuous. Therefore $j = i^{-1}$ is also a continuous map and hence $i : (Z, \mu) \to T(\mathcal{A})$ is a homeomorphism. Fact 2 is proved. $\square$

Fact 3. Let Y and Z be second countable spaces. If $f : Y \to Z$ is a measurable onto map and Y is analytic, then the space Z is also analytic.

Proof. Fix a continuous onto map $h : \mathbb{P} \to Y$; if $B \in \mathbb{B}(Z)$, then $B' = f^{-1}(B) \in \mathbb{B}(Y)$ and hence $B'' = h^{-1}(B') \in \mathbb{B}(\mathbb{P})$ by Problem 332. Therefore $(f \circ h)^{-1}(B)$ is a Borel subset of $\mathbb{P}$ for any Borel set $B \subset Z$, i.e., the map $g = f \circ h$ is measurable. Let $\mathcal{B} = \{B_n : n \in \omega\}$ be a base in Z; then, by our observation, the set $C_n = g^{-1}(B_n)$ is Borel in $\mathbb{P}$ for all $n \in \omega$. Therefore C_n is a Borel set (as a space) for every $n \in \omega$. Thus $\mathbb{P} \backslash C_n$ is also a Borel set (see Fact 1 of T.331) and therefore the space $T_n = C_n \oplus (\mathbb{P} \backslash C_n)$ is a Borel set as well for all $n \in \omega$ (see Problem 333).

The family $\tau(\mathbb{P}) \cup \{C_n : n \in \omega\} \cup \{\mathbb{P} \backslash C_n : n \in \omega\}$ generates a topology μ on $\mathbb{P}$ as a subbase. The space $P = (\mathbb{P}, \mu)$ is homeomorphic to a closed subspace F of the space $T = \prod_{n \in \omega} T_n$ by Fact 2. Each T_n is analytic by Problem 334, so F (and hence P) is an analytic space by 335. By our construction of μ, we have $g^{-1}(B_n) = C_n \in \mu$ for all $n \in \omega$ and therefore the map $g : P \to Z$ is continuous. Every continuous image of an analytic space is an analytic space, so Z is analytic and Fact 3 is proved. $\square$

Fact 4. If Z is a cosmic space, then there exist second countable spaces M and N and condensations $f : M \to Z$ and $g : Z \to N$ such that the mappings f^{-1} and g^{-1} are measurable.

Proof. Take a countable network $\mathcal{F}$ in the space Z such that all elements of $\mathcal{F}$ are closed in Z and any finite intersection of the elements of $\mathcal{F}$ belongs to $\mathcal{F}$. Let μ be a topology generated on Z by the family $\mathcal{S} = \tau(Z) \cup \mathcal{F}$. Observe that $Z \backslash F \in \tau(Z) \subset \mu$ for every $F \in \mathcal{F}$, and hence we can apply Fact 2 to conclude that (Z, μ) is a Tychonoff space.

The family $\mathcal{F}$ is a base in the space $M = (Z, \mu)$. To see it take any $U \in \mu$ and $x \in U$. The family $\mathcal{S}$ is a subbase of μ, so there are $V_1, \ldots, V_n \in \tau(Z)$ and $F_1, \ldots, F_k \in \mathcal{F}$ such that $x \in V \cap F \subset U$ where $V = V_1 \cap \cdots \cap V_n \in \tau(Z)$ and $F = F_1 \cap \cdots \cap F_n \in \mathcal{F}$. The family $\mathcal{F}$ being a network in Z, there is $G \in \mathcal{F}$ for which $x \in G \subset V$. Then $H = G \cap F \in \mathcal{F}$ and $x \in H \subset U$ which shows that $\mathcal{F}$ is a base in (Z, μ).

Let $f : M \to Z$ be the identity map. Since the topology of M is stronger than the topology of Z, the map f is a condensation. Given any $U \in \tau(M)$ there is

$\mathcal{F}' \subset \mathcal{F}$ such that $\bigcup \mathcal{F}' = U$. All elements of $\mathcal{F}'$ are closed in Z, so U is an F_σ-set of Z. Thus $U = (f^{-1})^{-1}(U)$ is an F_σ-set in Z for every $U \in \mu = \tau(M)$. Since every F_σ-set of Z is a Borel subset of Z, the map f^{-1} is measurable by Fact 1.

For any $F \in \mathcal{F}$ fix a continuous function $p_F : Z \to I = [0, 1]$ such that $p_F^{-1}(0) = F$. This is possible because Z is cosmic and hence perfectly normal. The diagonal product $g = \Delta\{p_F : F \in \mathcal{F}\} : Z \to I^{\mathcal{F}}$ is a continuous map; let $N = g(Z)$. For every $F \in \mathcal{F}$, let $q_F : I^{\mathcal{F}} \to I$ be the natural projection.

If $x, y \in Z$ and $x \neq y$, then there is $F \in \mathcal{F}$ such that $x \in F$ and $y \notin F$. Then $p_F(x) = 0 \neq p_F(y)$ and hence $g(x) \neq g(y)$. Thus $g : Z \to N$ is a condensation. For every $F \in \mathcal{F}$, let $F' = g(F)$. We have $F' = q_F^{-1}(0) \cap N$ which shows that F' is closed in N for every $F \in \mathcal{F}$.

For any set $U \in \tau(Z)$ there is a family $\mathcal{F}' \subset \mathcal{F}$ such that $U = \bigcup \mathcal{F}'$. Then $g(U) = \bigcup\{g(F) : F \in \mathcal{F}'\}$ is an F_σ-set in N because $g(F) = F'$ is closed in N for each $F \in \mathcal{F}'$. This proves that $(g^{-1})^{-1}(U) = g(U)$ is an F_σ-set in N for every $U \in \tau(Z)$, so we can apply Fact 1 again to conclude that g^{-1} is also a measurable map. Fact 4 is proved.   □

Returning to our solution observe that any continuous image of an analytic space is analytic, so if X is analytic, then every second countable continuous image of X is analytic. Now assume that every second countable continuous image of X is analytic and choose condensations $f : M \to X$ and $g : X \to N$ such that $w(M) = w(N) = \omega$ and the maps $h = g^{-1}$ and $u = f^{-1}$ are measurable (see Fact 4). Since every second countable continuous image of X is analytic, the space N is analytic. It is an easy exercise that a composition of measurable maps is a measurable map, so $u \circ h : N \to M$ is a measurable map. Therefore M is analytic by Fact 3 and hence X is analytic being a continuous image of an analytic space M. Our solution is complete.

T.364. *Let X be a space with a countable network. Prove that X is σ-compact if and only if every second countable continuous image of X is σ-compact.*

Solution. Necessity is evident so assume that every second countable continuous image of X is σ-compact. Then X is analytic by Problem 363 because every σ-compact second countable space is analytic. If X is not σ-compact, then there is a closed $F \subset X$ with $F \simeq \mathbb{P}$ by Problem 352. Since $\mathbb{P}$ is a subset of $\mathbb{R}$, there exists a mapping $h : F \to \mathbb{R}$ which is a homeomorphism of F onto $h(F)$. The space X is normal, so there is a continuous function $h_1 : X \to \mathbb{R}$ such that $h_1|F = h$.

By perfect normality of X there is a continuous function $f : X \to \mathbb{R}$ such that $F = f^{-1}(0)$. If $g = h_1 \Delta f$ and $M = g(X)$, then $M \subset \mathbb{R} \times \mathbb{R}$ is second countable. For any $z \in F$, we have $g(z) = (h_1(z), f(z)) = (h(z), 0)$; besides, if $g(y) = (t, 0) \in M$ for some $y \in X$, then $f(y) = 0$ and hence $y \in F$. This shows that $g(F) = M \cap (\mathbb{R} \times \{0\})$ is a closed subspace of M; since $g(F) = h(F) \times \{0\}$, we have $g(F) \simeq h(F) \simeq \mathbb{P}$. It turns out that $g(F)$ is a closed non-σ-compact subspace of M which is a second countable continuous image of X. Therefore M is not σ-compact; this contradiction settles sufficiency.

T.365. *Prove that a second countable space M is Polish if and only if there exists a map $f : \mathbb{P} \to \exp(M)$ with the following properties:*

(a) $f(s)$ is compact for any $s \in \mathbb{P}$;
(b) for any $s, t \in \mathbb{P}$, if $s \leq t$, then $f(s) \subset f(t)$;
(c) for any compact $K \subset M$, there is $s \in \mathbb{P}$ such that $K \subset f(s)$.

Solution. The space $\mathbb{P}$ is identified with ω^ω (see Problem 313). For any $n \in \mathbb{N}$ we let $\omega^{<n} = \bigcup\{\omega^k : k < n\}$. For any $s \in \omega^{<\omega}$ there is $n \in \omega$ such that $s \in \omega^n$; let $l(s) = n$. If $n, i \in \omega$ and $s \in \omega^n$, then $s^\frown i \in \omega^{n+1}$ is defined by $(s^\frown i)|n = s$ and $(s^\frown i)(n) = i$.

If M is Polish, then there exists an open continuous onto map $h : \mathbb{P} \to M$ by Problem 328. By Fact 1 of S.491, there is a closed $F \subset \mathbb{P}$ such that $h(F) = M$ and $h_1 = h|F$ is a perfect map. For any $s \in \mathbb{P}$ let $K(s) = \prod_{n \in \omega}(s(n) + 1)$ (here we consider the integer $(s(n)+1)$ to be the set $\{0, \ldots, s(n)\}$). Then $K(s)$ is compact for any $s \in \mathbb{P}$ and it is immediate that $K(s) \subset K(t)$ is $s \leq t$. The set $f(s) = h(K(s))$ is compact and it is clear that $f(s) \subset f(t)$ if $s \leq t$. Now, if K is a compact subset of M, then $L = h_1^{-1}(K)$ is a compact subset of F by Fact 2 of S.259. If $\pi_i : \mathbb{P} \to \omega$ is the natural projection onto the ith factor, then the set $\pi_i(L)$ is bounded for every $i \in \omega$ because any compact subspace of ω is finite.

As a consequence there is $s(i) \in \omega$ such that $\pi_i(L) \subset s(i)$ for all $i \in \omega$. This defines $s \in \omega^\omega = \mathbb{P}$ for which $L \subset K(s)$ and hence $f(s) = h(K(s)) \supset h(L) = K$ which implies $K \subset f(s)$, so we proved that the function f has the properties (a)–(c). This settles necessity.

Now assume that $f : \mathbb{P} \to \exp(M)$ is a function with the properties (a)–(c) and fix a metric d on M such that $\tau(d) = \tau(M)$. As usual, for any $x \in M$ and $\varepsilon > 0$, the set $B(x, \varepsilon) = \{y \in M : d(x, y) < \varepsilon\}$ is the ball of radius ε centered at x. We are going to construct a map $\delta : \omega^{<\omega} \to \omega^{<\omega}$ and a family $\{U_s : s \in \omega^{<\omega}\} \subset \tau^*(M)$ with the following properties:

(1) $U_\emptyset = M$ and $U_s = \bigcup\{\overline{U}_{s^\frown i} : i \in \omega\}$ for every $s \in \omega^{<\omega}$;
(2) if $s \in \omega^{<\omega}$ and $l(s) = i \geq 1$, then $\text{diam}(U_s) \leq \frac{1}{i}$;
(3) $\delta(\omega^n) \subset \omega^n$ for all $n \in \omega$ and if $s, t \in \omega^{<\omega}$ and $s \subset t$, then $\delta(s) \subset \delta(t)$;
(4) if $s \in \omega^{<\omega}$ and $K \subset U_s$ is compact, then there is $t \in \mathbb{P}$ such that $\delta(s) \subset t$ and $K \subset f(t)$.

To start our inductive construction let $\delta(\emptyset) = \emptyset$ and $U_\emptyset = M$. Assume that $n \in \mathbb{N}$ and we have a family $\{U_s : s \in \omega^{<n}\}$ and a map $\delta : \omega^{<n} \to \omega^{<n}$ such that (1) is fulfilled for all $s \in \omega^{<(n-1)}$ and the statements (2)–(4) are true for all $s \in \omega^{<n}$.

Fix any $s \in \omega^{n-1}$; we claim that for any $x \in U_s$ we can find $V_x \in \tau(x, M)$ and $k \in \omega$ such that $\overline{V}_x \subset U_s$, $\text{diam}(V_x) \leq \frac{1}{n}$ and for any compact $K \subset V_x$ there is $t \in \mathbb{P}$ such that $\delta(s)^\frown k \subset t$ and $K \subset f(t)$. Indeed, if this were not true, then we can choose a sequence $\{V_i : i \in \omega\} \subset \tau(x, M)$ such that for all $i \in \omega$ we have $\overline{V}_i \subset U_s \cap B(x, \frac{1}{2n+i})$ and there is a compact $K_i \subset V_i$ for which $K_i \not\subset f(t)$ for any $t \in \mathbb{P}$ with $\delta(s)^\frown i \subset t$. It is easy to see that the set $K = (\bigcup_{i \in \omega} K_i) \cup \{x\} \subset U_s$ is compact and hence it follows from the induction hypothesis that there is $t \in \mathbb{P}$ such that $\delta(s) \subset t$ and $K \subset f(t)$. If $i = t(n - 1)$, then $\delta(s)^\frown i \subset t$ and $K_i \subset K \subset f(t)$ which is a contradiction.

Thus we can construct sequences $\{V_i : i \in \omega\} \subset \tau^*(M)$ and $\{k_i : i \in \omega\} \subset \omega$ such that $U_s = \bigcup\{\overline{V}_i : i \in \omega\}$, $\mathrm{diam}(V_i) \leq \frac{1}{n}$ for every $i \in \omega$ and, for any compact $K \subset V_i$, there is $t \in \mathbb{P}$ such that $\delta(s)^\frown k_i \subset t$ and $K \subset f(t)$. Now let $U_{s^\frown i} = V_i$ and $\delta(s^\frown i) = \delta(s)^\frown k_i$ for all $i \in \omega$. After we accomplish this construction for all $s \in \omega^{n-1}$, we obtain a family $\{U_s : s \in \omega^n\}$ and a map $\delta : \omega^{<(n+1)} \to \omega^{<(n+1)}$ such that (1) holds for all $s \in \omega^{<n}$ and the assertions (2)–(4) are true for all $s \in \omega^{<(n+1)}$.

Therefore our inductive construction can be carried out for all $n \in \omega$ to give us a map $\delta : \omega^{<\omega} \to \omega^{<\omega}$ and a family $\{U_s : s \in \omega^{<\omega}\}$ with the properties (1)–(4).

Take any $t \in \mathbb{P}$ and assume that $\bigcap\{U_{t|n} : n \in \omega\} = \emptyset$. Let $s_n = t|n$ and pick a point $x_n \in U_{s_n}$ for every $n \in \omega$. The property (4) implies that there exists $t_n \in \mathbb{P}$ such that $\delta(s_n) \subset t_n$ and $x_n \in f(t_n)$ for all $n \in \omega$. Observe that $\delta(s_n) \subset \delta(s_{n+1})$ by (3) and therefore $t_m|n = \delta(s_n) = t_n|n$ for all $m \geq n$. Thus if we define a function $u \in \omega^\omega$ by $u(i) = t_0(i) + \cdots + t_i(i)$ for all $i \in \omega$, then $t_n \leq u$ for all $n \in \omega$. Therefore $x_n \in f(t_n) \subset f(u)$ for all $n \in \omega$, i.e., $S = \{x_n : n \in \omega\} \subset f(u)$. Observe that $x_n \in \overline{U}_{s_n}$ for all $n \in \omega$ which implies that $F_n = \overline{U}_{s_n} \cap f(u) \neq \emptyset$ for all $n \in \omega$. Since the sequence of compact sets $\{F_n : n \in \omega\}$ is decreasing by (1), there is $x \in \bigcap_{n\in\omega} F_n \subset \bigcap_{n\in\omega} \overline{U}_{t|n} = \bigcap_{n\in\omega} U_{t|n}$. This contradiction proves that

(2′) for any $t \in \mathbb{P}$, we have $\bigcap_{n\in\omega} U_{t|n} \neq \emptyset$,

and hence the family $\mathcal{U} = \{U_s : s \in \omega^{<\omega}\}$ is an A-system in M (which means precisely that the properties (1),(2) and (2′) hold for $\mathcal{U}$). It was proved in Fact 3 of S.491 that any second countable space with an A-system is Čech-complete, so M is Polish and hence our solution is complete.

T.366. *Prove that if $C_p(X)$ is analytic, then X is σ-compact.*

Solution. The space $\mathbb{P}$ is identified with ω^ω (see Problem 313). If Z is a set and $f, g \in \mathbb{R}^Z$, then $f \leq g$ if $f(z) \leq g(z)$ for all $z \in Z$.

Assume, towards a contradiction, that $C_p(X)$ is analytic while the space X is not σ-compact. We have $nw(X) = nw(C_p(X)) = \omega$ and hence there exists a continuous onto map $\varphi : X \to M$ such that $w(M) = \omega$ and M is not σ-compact (see Problem 364). Define the dual map $\varphi^* : \mathbb{R}^M \to \mathbb{R}^X$ by $\varphi^*(f) = f \circ \varphi$ for all $f \in \mathbb{R}^M$. Let X' and M' be the discrete spaces with the underlying sets X and M respectively. Then $C_p(X') = \mathbb{R}^X$ and $C_p(M') = \mathbb{R}^M$; since the map $\varphi : X' \to M'$ is open, the set $\varphi^*(\mathbb{R}^M)$ is closed in $\mathbb{R}^X$ (see Problem 163 of [TFS]). Observe that $\varphi^*(C_p(M)) \subset P = \varphi^*(\mathbb{R}^M) \cap C_p(X)$. The set P is analytic being closed in the analytic space $C_p(X)$. Since φ^* is an embedding, we proved that

(∗) there is an analytic space Q such that $C_p(M) \subset Q \subset \mathbb{R}^M$

because it suffices to take $Q = (\varphi^*)^{-1}(P)$. Let $R_+ = (0, +\infty) \subset \mathbb{R}$ and fix a homeomorphism $\alpha : \mathbb{R} \to R_+$ (we can take, e.g., $\alpha(t) = e^t$ for all $t \in \mathbb{R}$). The map $\gamma : \mathbb{R}^M \to R_+^M$ defined by $\gamma(f) = \alpha \circ f$ for all $f \in \mathbb{R}^M$ is a homeomorphism (see Problem 091 of [TFS]) and $\gamma(C_p(M)) = C_p(M, R_+)$. An immediate consequence of (∗) is that

(∗∗) the space $Q' = \gamma(Q)$ is analytic and $C_p(M, R_+) \subset Q' \subset R_+^M$.

We can consider that M is a metric subspace of a metric compact space (C, d). Fix a continuous onto map $h : \mathbb{P} \to Q'$. For any $s \in \mathbb{P}$ let $K(s) = \prod_{n \in \omega}(s(n) + 1)$ (here we consider the integer $(s(n) + 1)$ to be the set $\{0, \ldots, s(n)\}$). Then $K(s)$ is compact for any $s \in \mathbb{P}$ and it is immediate that $K(s) \subset K(t)$ if $s \leq t$. The set $w(s) = h(K(s))$ is a compact subspace of Q' and it is clear that $w(s) \subset w(t)$ if $s \leq t$.

For any $x \in M$ let $\pi_x : R_+^M \to R_+$ be the natural projection onto the factor determined by x. Given any $s \in \mathbb{P}$, the set $\pi_x(w(s))$ is a compact subspace of R_+ and therefore $r_s^x = \inf(\pi_x(w(s))) \in R_+$. We have $r_s \leq f(x)$ for any $f \in w(s)$; let $b_s(x) = r_s^x$ for every $x \in M$. This gives us a set $\{b_s : s \in \mathbb{P}\} \subset R_+^M$ such that

(1) if $s, t \in \mathbb{P}$ and $s \leq t$, then $b_t \leq b_s$;
(2) for any $f \in C_p(M, R_+)$, there is $s \in \mathbb{P}$ such that $b_s(x) \leq f(x)$ for all $x \in M$.

The property (1) follows from the inclusion $w(s) \subset w(t)$ and the definition of b_s and b_t. If $f = h(s)$ for some $s \in \mathbb{P}$, then $f \in w(s)$ and hence $b_s(x) \leq \pi_x(f) = f(x)$ for all $x \in M$, i.e., (2) is proved. As usual, given a point $x \in C$ and $\varepsilon > 0$ the set $B(x, \varepsilon) = \{y \in C : d(x, y) < \varepsilon\}$ is the ε-ball in C centered at x. For any $s \in \mathbb{P}$, let $W_s = \bigcup\{B(x, b_s(x)) : x \in M\}$ and $K(s) = C \backslash W_s$. Then $K(s)$ is a compact subset of $C \backslash M$; besides, $s, t \in \mathbb{P}$ and $s \leq t$ implies $K(s) \subset K(t)$ because it follows from (1) that $W_s \supset W_t$.

Now, if $K \subset C \backslash M$ is compact, let $f(x) = d(x, K) = \inf\{d(x, y) : y \in K\}$ for every $x \in M$. Then $f \in C_p(M, R_+)$ (see Fact 1 of S.212), so there exists $s \in \mathbb{P}$ such that $b_s \leq f$ [see (2)]. As a consequence, $W_s \cap K = \emptyset$ and hence $K \subset K(s)$. We proved that the map $K : \mathbb{P} \to \exp(C \backslash M)$ satisfies the conditions (a)–(c) of Problem 365. Thus $C \backslash M$ is a Polish space which implies that $C \backslash M$ is a G_δ-set of $E = \text{cl}_C(C \backslash M)$ (see Problem 259 of [TFS]). Any closed subset of a metric space is a G_δ-set, so E is a G_δ-set in C and therefore $C \backslash M$ is a G_δ-set in C (see Fact 2 of S.358). Consequently, M is an F_σ-subset of C, i.e., M is σ-compact. This contradiction shows that X is also σ-compact and finishes our solution.

T.367. *Prove that the following are equivalent for an arbitrary second countable space X:*

  (i) *$C_p(X)$ is analytic;*
 (ii) *$C_p(X)$ is a $K_{\sigma\delta}$-space;*
(iii) *X is σ-compact.*

Observe that, as a consequence, the spaces $C_p(\mathbb{P})$ and $C_p(\mathbb{R}^\omega)$ are not analytic.

Solution. The implication (i)$\Longrightarrow$(iii) was proved in Problem 366 while (iii)$\Longrightarrow$(ii) was established in Problem 362. Since every second countable $K_{\sigma\delta}$-space is analytic by Problem 346, we also have the implication (ii)$\Longrightarrow$(i).

Finally observe that $\mathbb{P}$ is not σ-compact by Fact 1 of T.132 and the space $\mathbb{R}^\omega$ is not σ-compact because $\mathbb{P}$ embeds in $\mathbb{R}^\omega$ as a closed subspace by Problem 273 of [TFS]. Therefore neither of the spaces $C_p(\mathbb{R}^\omega)$ and $C_p(\mathbb{P})$ is analytic.

T.368. *For a second countable space X let A be a countable dense subspace of X. Prove that the following conditions are equivalent:*

(i) $C_p(A|X)$ *is a Borel set;*
(ii) $C_p(A|X)$ *is analytic;*
(iii) X *is σ-compact.*

Solution. Given spaces Y and Z, a map $f : Y \to Z$ is called *measurable* if $f^{-1}(U) \in \mathbb{B}(Y)$ for any $U \in \mathbb{B}(Z)$. If $\mathcal{A} \subset \exp(Z)$, then $\bigwedge \mathcal{A}$ is the family of all finite intersections of elements of $\mathcal{A}$. A space Z is *perfect* if every $U \in \tau(Z)$ is an F_σ-subset of Z. The spaces with a countable network are called *cosmic*.

Fact 1. If Y is a perfect space and $hl(Z) = \omega$, then a map $f : Y \to Z$ is measurable if and only if there is a subbase $\mathcal{S}$ in the space Z such that $f^{-1}(U) \in \mathbb{B}(Y)$ for any $U \in \mathcal{S}$.

Proof. If f is measurable, then we can let $\mathcal{S} = \tau(Z)$ so necessity is clear. Now, suppose that there is a subbase $\mathcal{S}$ in the space Z such that $f^{-1}(U) \in \mathbb{B}(Y)$ for any $U \in \mathcal{S}$. The family $\mathcal{B} = \bigwedge \mathcal{S}$ is a base of Z and $U \in \mathcal{B}$ implies that there are $U_1, \dots, U_n \in \mathcal{S}$ with $U = \bigcap_{i \leq n} U_i$. Thus $f^{-1}(U) = \bigcap_{i \leq n} f^{-1}(U_i) \in \mathbb{B}(X)$ (see Fact 1 of T.331) which proves that $f^{-1}(U) \in \mathbb{B}(Y)$ for any $U \in \mathcal{B}$.

Now if $U \in \tau(Z)$, then it follows from $l(U) \leq hl(Z) = \omega$ that there is a countable $\mathcal{B}' \subset \mathcal{B}$ such that $U = \bigcup \mathcal{B}'$. Therefore $f^{-1}(U) = \bigcup \{f^{-1}(V) : V \in \mathcal{B}'\}$, i.e., $f^{-1}(U)$ is a Borel subset of Y being a countable union of Borel subsets of Y (here we applied again Fact 1 of T.331).

We proved that $f^{-1}(U) \in \mathbb{B}(Y)$ for any $U \in \tau(Z)$, so we can apply Fact 1 of T.363 to conclude that f is a measurable map. This settles sufficiency and finishes the proof of Fact 1. $\qquad\square$

Fact 2. Given a cosmic space Z, suppose that B is a dense subset of Z and, for any point $z \in Z$, there is a sequence $\{b_n : n \in \omega\} \subset B$ such that $b_n \to z$. If $\pi : C_p(Z) \to C_p(B|Z)$ is the restriction map, then $e = \pi^{-1} : C_p(B|Z) \to C_p(Z)$ is measurable. In particular, if Z is second countable and B is dense in Z, then the map e is measurable.

Proof. For any $O \in \tau(\mathbb{R})$ and $z \in Z$ let $[z, O] = \{f \in C_p(Z) : f(z) \in O\}$; then the family $\mathcal{S} = \{[z, O] : z \in Z \text{ and } O \in \tau(\mathbb{R})\}$ is a subbase in $C_p(Z)$. Take any $U = [z, O] \in \mathcal{S}$.

It is easy to find a family $\{O_n : n \in \omega\} \subset \tau(\mathbb{R})$ such that $\overline{O}_n \subset O_{n+1}$ for every $n \in \omega$ and $\bigcup_{n \in \omega} O_n = O$. Fix a sequence $\{b_n : n \in \omega\} \subset B$ such that $b_n \to z$. We claim that $\pi(U) = H_U = \bigcup_{n \in \omega} \bigcup_{k \in \omega} \bigcap_{i \geq k} \{f \in C_p(B|Z) : f(b_i) \in O_n\}$.

Indeed, if $f = \pi(g)$ for some $g \in U$, then $g(z) \in O$ and hence $g(z) \in O_n$ for some $n \in \omega$. Furthermore, $f(b_i) = g(b_i)$ for all $i \in \omega$; by continuity of g, we have $g(b_i) \to g(z)$ and therefore $f(b_i) \to g(z)$ which shows that there is $k \in \omega$ such that $f(b_i) \in O_n$ for all $i \geq k$. Thus $f \in H_U$ and we established that $\pi(U) \subset H_U$.

Conversely, if $f \in H_U$ and $f = \pi(g)$ for some $g \in C_p(Z)$, then there are $n, k \in \omega$ such that for all $i \geq k$ we have $g(b_i) = f(b_i) \in O_n$. Since g is a continuous function, $g(z) = \lim g(b_i)$ when $i \to \infty$ and therefore $g(z) \in \overline{O}_n \subset O_{n+1} \subset O$ which shows that $g(z) \in O$, i.e., $g \in U$. Thus $f \in \pi(U)$ and we proved that $H_U \subset \pi(U)$, i.e., $\pi(U) = H_U$.

Finally observe that the set $S_{ni} = \{f \in C_p(B|Z) : f(b_i) \in O_n\}$ is open in $C_p(B|Z)$ for all $n, i \in \omega$, so the set $H_U = \bigcup_{n\in\omega} \bigcup_{k\in\omega} \bigcap_{i\geq k} S_{ni}$ is $G_{\delta\sigma}$ in $C_p(B|Z)$. Consequently, $e^{-1}(U) = \pi(U) = H_U$ is a Borel subset of $C_p(B|Z)$ for any $U \in S$; since both spaces $C_p(Z)$ and $C_p(B|Z)$ are cosmic, we can apply Fact 1 to conclude that the map e is measurable. Fact 2 is proved. $\qquad\square$

Fact 3. Given any space Z, suppose that (M, d) is a complete metric space and $f : A \to M$ is a continuous map for some dense $A \subset Z$. For a point $z \in Z$ say that $\mathrm{osc}(f, z) = 0$ if for any $\varepsilon > 0$, there is $W \in \tau(z, Z)$ such that $\mathrm{diam}(f(W \cap A)) < \varepsilon$. Then $G = \{z \in Z : \mathrm{osc}(f, z) = 0\}$ is a G_δ-subset of Z such that $A \subset G$ and there is a continuous map $g : G \to M$ with $g|A = f$.

Proof. As usual, for any $x \in M$ and $\varepsilon > 0$, the set $B(x, \varepsilon) = \{y \in M : d(x, y) < \varepsilon\}$ is the ε-ball centered at x. Let $U_n = \{z \in Z :$ there exists $W \in \tau(z, Z)$ such that $\mathrm{diam}(f(W \cap A)) \leq \frac{1}{n}\}$; it is evident that the set U_n is open in Z for any $n \in \mathbb{N}$. The map f being continuous, for any $z \in A$, there is $V \in \tau(z, A)$ such that $f(V) \subset B(f(z), \frac{1}{2n})$. If W is an open subset of Z such that $W \cap A = V$, then $\mathrm{diam}(f(W \cap A)) = \mathrm{diam}(f(V)) \leq \mathrm{diam}(B(f(z), \frac{1}{2n})) \leq \frac{1}{n}$ which shows that $z \in U_n$. Consequently, $A \subset U_n$ for any $n \in \mathbb{N}$.

It is immediate that $G = \bigcap\{U_n : n \in \mathbb{N}\}$, so G is a G_δ-subset of Z and $A \subset G$. To prove that the map f can be continuously extended over the set G take any $z \in G$ and consider the family $\mathcal{F}_z = \{\overline{f(W \cap A)} : W \in \tau(z, Z)\}$. It is clear that $\mathcal{F}_z$ consists of closed subsets of the space M; if $W_1, \ldots, W_n \in \tau(z, Z)$ and $W = W_1 \cap \cdots \cap W_n$, then $\overline{f(W \cap A)} \subset \overline{f(W_1 \cap A)} \cap \ldots \cap \overline{f(W_n \cap A)}$ which shows that the family $\mathcal{F}_z$ is centered.

Given any $\varepsilon > 0$, take $n \in \mathbb{N}$ with $\frac{1}{n} < \varepsilon$. Since $z \in U_n$, there is $W \in \tau(z, Z)$ such that $\mathrm{diam}(f(W \cap A)) \leq \frac{1}{n} < \varepsilon$. Therefore $\mathcal{F}_z$ contains elements of arbitrarily small diameter and hence $\bigcap \mathcal{F}_z \neq \emptyset$ by Problem 236 of [TFS]. It is easy to see that $\bigcap \mathcal{F}_z$ cannot contain more than one point, so there is a point $g(z)$ such that $\bigcap \mathcal{F}_z = \{g(z)\}$. This defines a map $g : G \to M$.

Observe first that $g|A = f$ because if $z \in A$, then $f(z) \in \overline{f(W \cap A)} \subset \overline{f(W \cap A)}$ for any $W \in \tau(z, Z)$ and hence $f(z) \in \bigcap \mathcal{F}_z$, i.e., $g(z) = f(z)$.

To see that g is continuous, take any $z \in G$ and $\varepsilon > 0$. Choose $n \in \mathbb{N}$ and $W \in \tau(z, Z)$ such that $\frac{1}{n} < \varepsilon$ and $\mathrm{diam}(\overline{f(W \cap A)}) = \mathrm{diam}(f(W \cap A)) \leq \frac{1}{n}$. If $W' = W \cap G$, then $W' \in \tau(z, G)$; given any $t \in W'$ we have $t \in W \in \tau(t, Z)$ and hence $\overline{f(W \cap A)} \in \mathcal{F}_t \cap \mathcal{F}_z$. Consequently, $\{g(z), g(t)\} \subset \overline{f(W \cap A)}$ and therefore $d(g(z), g(t)) \leq \mathrm{diam}(\overline{f(W \cap A)}) \leq \frac{1}{n} < \varepsilon$ which proves that $g(W') \subset B(g(z), \varepsilon)$, i.e., g is continuous at the point z. Since $z \in G$ was taken arbitrarily, the function g is continuous on G and Fact 3 is proved. $\qquad\square$

Returning to our solution, observe that the implication (i)$\Longrightarrow$(ii) is obvious. Now, assume that $C_p(A|X)$ is analytic and let $\pi : C_p(X) \to C_p(A|X)$ be the restriction map. It follows from Fact 2 that the map $e = \pi^{-1}$ is measurable, so

$C_p(X)$ is an image of an analytic space $C_p(A|X)$ under a measurable map. Thus $C_p(X)$ is analytic by Fact 3 of T.363, and hence X is σ-compact by Problem 366, so we settled (ii)$\Longrightarrow$(iii).

Now, if (iii) holds, then $X = \bigcup\{X_n : n \in \mathbb{N}\}$ where X_n is compact for all $n \in \mathbb{N}$. Let d be a metric on X such that $\tau(d) = \tau(X)$ and $d(x,y) \leq 1$ for all $x, y \in X$. For any $k, l, n \in \mathbb{N}$, let $M_{kln} = \{f \in \mathbb{R}^A : |f(x) - f(y)| \leq \frac{1}{k}$ whenever x and y are points of A such that $d(x,z) < \frac{1}{n}$ and $d(y,z) < \frac{1}{n}$ for some $z \in X_l\}$. Our first observation is that

(1) the set M_{kln} is closed in $\mathbb{R}^A$ for any $k, l, n \in \mathbb{N}$.

Indeed, if $f \in \mathbb{R}^A \backslash M_{kln}$, then there exist points $x, y \in A$ such that $d(x,z) < \frac{1}{n}$ and $d(y,z) < \frac{1}{n}$ for some $z \in X_l$ while $|f(x) - f(y)| > \frac{1}{k}$; then $r = |f(x) - f(y)| - \frac{1}{k} > 0$. The set $O_f = \{g \in \mathbb{R}^A : |g(x) - f(x)| < \frac{r}{2}$ and $|g(y) - f(y)| < \frac{r}{2}\}$ is open in the space $\mathbb{R}^A$ and $f \in O_f$. For any function $g \in O_f$ we have $|f(x) - f(y)| \leq |f(x) - g(x)| + |g(x) - g(y)| + |g(y) - f(y)|$ and therefore

$$|g(x) - g(y)| \geq |f(x) - f(y)| - |f(x) - g(x)| - |f(y) - g(y)| > r + \frac{1}{k} - \frac{r}{2} - \frac{r}{2} = \frac{1}{k}$$

which shows that $g \in \mathbb{R}^A \backslash M_{kln}$. Since a function $g \in O_f$ was chosen arbitrarily, we have $O_f \subset \mathbb{R}^A \backslash M_{kln}$. Thus, for every $f \in \mathbb{R}^A \backslash M_{kln}$, there $O_f \in \tau(\mathbb{R}^A)$ such that $O_f \subset \mathbb{R}^A \backslash M_{kln}$. As a consequence, $\mathbb{R}^A \backslash M_{kln} = \bigcup\{O_f : f \in \mathbb{R}^A \backslash M_{kln}\}$ is open in $\mathbb{R}^A$ and hence M_{kln} is closed in $\mathbb{R}^A$.

The property (1) implies that $P_{kl} = \bigcup\{M_{kln} : n \in \mathbb{N}\}$ is an F_σ-subset of $\mathbb{R}^A$. We claim that $C_p(A|X) = \bigcap\{P_{kl} : k, l \in \mathbb{N}\}$ and hence $C_p(A|X)$ is an $F_{\sigma\delta}$-subset of $\mathbb{R}^A$.

Take any $f \in \bigcap\{P_{kl} : k, l \in \mathbb{N}\}$; given a point $z \in X$ and $\varepsilon > 0$, there is $l \in \mathbb{N}$ with $z \in X_l$. Pick any $k \in \mathbb{N}$ such that $\frac{1}{k} < \varepsilon$. Since $f \in P_{kl}$, there is $n \in \mathbb{N}$ such that $|f(x) - f(y)| \leq \frac{1}{k} < \varepsilon$ whenever $d(x,t) < \frac{1}{n}$ and $d(y,t) < \frac{1}{n}$ for some $t \in X_l$. As a consequence, for the set $W = \{x \in X : d(z,x) < \frac{1}{n}\}$ we have $d(x,z) < \frac{1}{n}$ and $d(y,z) < \frac{1}{n}$ for all $x, y \in W$ and hence $|f(x) - f(y)| \leq \frac{1}{k}$ whenever $x, y \in W \cap A$. This shows that $\mathrm{diam}(f(W \cap A)) \leq \frac{1}{k} < \varepsilon$ and therefore $\mathrm{osc}(f,z) = 0$. The point $z \in X$ was chosen arbitrarily so $\mathrm{osc}(f,z) = 0$ for all $z \in X$ and hence we can apply Fact 3 to conclude that there is a continuous function $g : X \to \mathbb{R}$ such that $g|A = f$. Therefore $f \in C_p(A|X)$ and hence $\bigcap\{P_{kl} : k, l \in \mathbb{N}\} \subset C_p(A|X)$.

Now if $f \in C_p(A|X)$, fix $g \in C_p(X)$ such that $g|A = f$ and take any $k, l \in \mathbb{N}$. If $f \notin P_{kl}$, then, for any $n \in \mathbb{N}$, there exist $x_n, y_n \in A$ and $z_n \in X_l$ such that $|f(x_n) - f(y_n)| > \frac{1}{k}$ while $d(x_n, z_n) < \frac{1}{n}$ and $d(y_n, z_n) < \frac{1}{n}$. Since X_l is compact, the sequence $\{z_n : n \in \mathbb{N}\}$ has a subsequence which converges to a point $z \in X_l$. Passing to that subsequence and reindexing the sequences $\{x_n : n \in \mathbb{N}\}$ and $\{y_n : n \in \mathbb{N}\}$ if necessary, we can assume, without loss of generality, that $z_n \to z$. We also have $x_n \to z$ and $y_n \to z$ because $d(x_n, z_n) < \frac{1}{n}$ and $d(y_n, z_n) < \frac{1}{n}$ for all $n \in \mathbb{N}$. Since g is continuous at the point z, there is $W \in \tau(z, X)$ such that $|g(w) - g(z)| < \frac{1}{2k}$ for all $w \in W$. There is $m \in \mathbb{N}$ such that $x_m, y_m \in W$ and

hence $|f(x_m) - f(y_m)| = |g(x_m) - g(y_m)| \leq |g(x_m) - g(z)| + |g(z) - g(y_m)| <$
$\frac{1}{2k} + \frac{1}{2k} = \frac{1}{k}$ which is a contradiction. Thus $f \in P_{kl}$ for all $k, l \in \mathbb{N}$ and therefore
$f \in \bigcap \{P_{kl} : k, l \in \mathbb{N}\}$. This proves that $\bigcap \{P_{kl} : k, l \in \mathbb{N}\} = C_p(A|X)$ and hence
$C_p(A|X)$ is an $F_{\sigma\delta}$-subset of $\mathbb{R}^A$. Consequently, $C_p(A|X)$ is a Borel set (see Fact 1
of T.331), which finishes the proof of (iii)$\Longrightarrow$(i) and makes our solution complete.

T.369. *Given a countable space X prove that $C_p(X)$ is analytic if and only if
$C_p(X, \mathbb{I})$ is analytic.*

Solution. The subspace $C_p(X, \mathbb{I})$ is closed in $C_p(X)$, so if $C_p(X)$ is analytic, then
$C_p(X, \mathbb{I})$ is also analytic by Problem 335. This proves necessity and shows that
countability of X is not needed for this part of the proof.

Now assume that the space $C_p(X, \mathbb{I})$ is analytic. For any point $x \in X$ the set
$P_x = \{f \in C_p(X, \mathbb{I}) : f(x) \in (-1, 1)\}$ is open in $C_p(X, \mathbb{I})$ and it is immediate
that $C_p(X, (-1, 1)) = \bigcap \{P_x : x \in X\}$. Thus $C_p(X, (-1, 1))$ is a G_δ-subset of
$C_p(X, \mathbb{I})$. Any open subspace of an analytic space is analytic by Problem 335, so
$C_p(X, (-1, 1))$ is analytic being a G_δ-subset of an analytic space $C_p(X, \mathbb{I})$ (see
336). Finally, $C_p(X)$ is analytic because it is homeomorphic to an analytic space
$C_p(X, (-1, 1))$.

T.370. *Prove that a countable space X embeds into $C_p(\mathbb{P})$ if and only if $C_p(X)$ is
analytic.*

Solution. If the space $C_p(X)$ is analytic, then there is a continuous onto map φ :
$\mathbb{P} \to C_p(X)$. The dual map $\varphi^* : C_p(C_p(X)) \to C_p(\mathbb{P})$ embeds $C_p(C_p(X))$ in
$C_p(\mathbb{P})$ (see Problem 163 of [TFS]). Since X embeds in $C_p(C_p(X))$ by Problem 167
of [TFS], it also embeds in $C_p(\mathbb{P})$ so sufficiency is proved.

Now assume that X is a countable subspace of $C_p(\mathbb{P})$. For any $x \in X$ and
$m, n \in \mathbb{N}$, let $Q(x, m, n) = \{\varphi \in \mathbb{I}^X : \text{there is a point } s = (s_1, \ldots, s_n) \in \mathbb{P}^n \text{ for}$
which $|\varphi(y) - \varphi(x)| \leq \frac{1}{m}$ whenever $|y(s_i) - x(s_i)| < \frac{1}{n}$ for all $i \leq n\}$. We will
also need the set $R(x, m, n) = \{(s, \varphi) \in \mathbb{P}^n \times \mathbb{I}^X : s = (s_1, \ldots, s_n) \in \mathbb{P}^n$ and
$|\varphi(y) - \varphi(x)| \leq \frac{1}{m}$ whenever $|y(s_i) - x(s_i)| < \frac{1}{n}$ for all $i \leq n\}$. It is immediate
that

(1) if $m, n \in \mathbb{N}$ and the mapping $p : \mathbb{P}^n \times \mathbb{I}^X \to \mathbb{I}^X$ is the natural projection, then
    $p(R(x, m, n)) = Q(x, m, n)$ for any $x \in X$.

Our next step is to prove that

(2) the set $R(x, m, n)$ is closed in $\mathbb{P}^n \times \mathbb{I}^X$ for all $x \in X$ and $m, n \in \mathbb{N}$.

Indeed, if $w = (s, \varphi) \in (\mathbb{P}^n \times \mathbb{I}^X) \backslash R(x, m, n)$, then there exists $y \in X$ such
that $|y(s_i) - x(s_i)| < \frac{1}{n}$ for all $i \leq n$ while $|\varphi(y) - \varphi(x)| > \frac{1}{m}$. Since y and x are
continuous functions on $\mathbb{P}$, the set $O_s = \{t = (t_1, \ldots, t_n) \in \mathbb{P}^n : |y(t_i) - x(t_i)| < \frac{1}{n}$
for all $i \leq n\}$ is open in $\mathbb{P}^n$; it is clear that $s \in O_s$.

Furthermore, the set $O_\varphi = \{\delta \in \mathbb{I}^X : |\delta(y) - \delta(x)| > \frac{1}{m}\}$ is open in $\mathbb{I}^X$ (the proof
is an easy exercise left to the reader). Therefore the set $O = O_s \times O_\varphi$ is an open
neighborhood of w in $\mathbb{P}^n \times \mathbb{I}^X$ such that $O \cap R(x, m, n) = \emptyset$. Thus every point of

$A = (\mathbb{P}^n \times \mathbb{I}^X) \backslash R(x, m, n)$ has a neighborhood which is contained in A. Therefore A is open in $\mathbb{P}^n \times \mathbb{I}^X$ which shows that $R(x, m, n)$ is closed in $\mathbb{P}^n \times \mathbb{I}^X$, i.e., (2) is proved.

For any $x \in X$ let $C_x = \{\varphi \in \mathbb{I}^X : \varphi$ is continuous at the point $x\}$; for any $m \in \mathbb{N}$, let $Q(x, m) = \bigcup\{Q(x, m, n) : n \in \mathbb{N}\}$. The following property is a key fact for our proof.

(3) $C_x = \bigcap\{Q(x, m) : m \in \mathbb{N}\}$ for any $x \in X$.

To prove (3) assume that $\varphi \in C_x$ and take any $m \in \mathbb{N}$. By the definition of the topology of pointwise convergence on X and continuity of the function φ at x, there exist $k \in \mathbb{N}, \varepsilon > 0$ and $(s_1, \ldots, s_k) \in \mathbb{P}^k$ for which $|\varphi(y) - \varphi(x)| < \frac{1}{m}$ for any $y \in X$ such that $|y(s_i) - x(s_i)| < \varepsilon$ for all $i \le k$. Pick $n \in \mathbb{N}$ such that $k \le n$ and $\frac{1}{n} < \varepsilon$ and let $s_{k+1} = \ldots = s_n = s_k$; it is clear that, for the point $s = (s_1, \ldots, s_n) \in \mathbb{P}^n$, we have $|\varphi(y) - \varphi(x)| < \frac{1}{m}$ for any $y \in X$ such that $|y(s_i) - x(s_i)| < \frac{1}{n}$ for all $i \le n$. This proves that $\varphi \in Q(x, m, n)$; since $m \in \mathbb{N}$ was chosen arbitrarily, we have $\varphi \in \bigcap\{Q(x, m) : m \in \mathbb{N}\}$, i.e., $C_x \subset \bigcap\{Q(x, m) : m \in \mathbb{N}\}$.

Now, if $\varphi \in \bigcap\{Q(x, m) : m \in \mathbb{N}\}$, then take any $\varepsilon > 0$ and choose $m \in \mathbb{N}$ such that $\frac{1}{m} < \varepsilon$. Since $\varphi \in Q(x, m)$, there exists $n \in \mathbb{N}$ for which $\varphi \in Q(x, m, n)$ and hence there is $s = (s_1, \ldots, s_n) \in \mathbb{P}^n$ such that $|\varphi(y) - \varphi(x)| \le \frac{1}{m} < \varepsilon$ for any $y \in X$ with $|y(s_i) - x(s_i)| < \frac{1}{n}$ for all $i \le n$. The set $W = \{y \in X : |y(s_i) - x(s_i)| < \frac{1}{n}$ for all $i \le n\}$ is an open neighborhood of x in X such that $\varphi(W) \subset (\varphi(x) - \varepsilon, \varphi(x) + \varepsilon)$ and hence φ is continuous at the point x. Thus $\bigcap\{Q(x, m) : m \in \mathbb{N}\} \subset C_x$ and (3) is proved.

Observe that the space $\mathbb{P}^n \times \mathbb{I}^X$ is analytic and hence (2) implies that so is $R(x, m, n)$ for any $x \in X$ and $m, n \in \mathbb{N}$ (see Problem 335). The property (1) shows that for any $x \in X$ and $m, n \in \mathbb{N}$, the space $Q(x, m, n)$ is analytic being a continuous image of $R(x, m, n)$. This implies that $Q(x, m)$ is analytic for any $x \in X$ and $m \in \mathbb{N}$ (see Problem 337). Consequently, for any $x \in X$, the space C_x is analytic by (3) and Problem 336. Furthermore, the space X is countable, so $C_p(X, \mathbb{I}) = \bigcap\{C_x : x \in X\}$ is also analytic by Problem 336; apply Problem 369 to conclude that $C_p(X)$ is analytic as well. This settles necessity and completes our solution.

T.371. *Take any $\xi \in \beta\omega\backslash\omega$ and consider the space $X = \omega \cup \{\xi\}$ with the topology inherited from $\beta\omega$. Prove that neither $C_p(X)$ nor $\mathbb{R}^X\backslash C_p(X)$ is analytic. As a consequence, X cannot be embedded into $C_p(\mathbb{P})$.*

Solution. Given a space Z and $A, B \subset Z$ let $A \sim B$ if $A \triangle B = (A\backslash B) \cup (B\backslash A)$ is a set of first category in Z. Say that $A \subset Z$ is a B-*set* if there is an open $U \subset Z$ such that $A \sim U$.

Fact 1. If Z is a space and $A \in \mathbb{B}(Z)$, then A is a B-set.

Proof. If A is open, then $U = A$ witnesses that A is a B-set. If A is closed, then, for $U = \text{Int}(A)$, the set $A \triangle U = A\backslash U$ is nowhere dense and hence $A \sim U$.

Now assume that $\alpha < \omega_1$ and we proved for any $\beta < \alpha$ and $A \in \Sigma_\beta^0(Z) \cup \Pi_\beta^0(Z)$ that A is a B-set. If $A \in \Sigma_\alpha^0(Z)$, then there is a sequence $\{\beta_n : n \in \omega\} \subset \alpha$ and $A_n \in \Pi_{\beta_n}^0(Z)$ for all $n \in \omega$ such that $A = \bigcup_{n \in \omega} A_n$. By the induction hypothesis, there is $U_n \in \tau(Z)$ such that $A_n \sim U_n$ for all $n \in \omega$. For the set $U = \bigcup_{n \in \omega} U_n$, we have $A \Delta U \subset \bigcup_{n \in \omega}(A_n \Delta U_n)$ which shows that $A \Delta U$ is of first category and hence $A \sim U$, i.e., A is a B-set.

Now if $A \in \Pi_\alpha^0(Z)$, then $A' = Z \backslash A \in \Sigma_\alpha^0(Z)$ and hence there is $U' \in \tau(Z)$ such that $A' \sim U'$. Observe that for the set $F = Z \backslash U'$ we have $A \Delta F = A' \Delta U'$ and therefore $A \sim F$. If $U = \mathrm{Int}(F)$, then the set $A \Delta U \subset (A \Delta F) \cup (F \Delta U)$ is of first category because so is $A \Delta F$ and $F \Delta U = F \backslash U$ is nowhere dense. Thus $A \sim U$ and hence A is a B-set. Therefore our inductive procedure can be carried out for all $\alpha < \omega_1$ which shows that every $A \in \mathbb{B}(Z)$ is a B-set. Fact 1 is proved.   □

Fact 2. Let D be an infinite discrete space. Then

(a) for any $A \subset D$, the set $\overline{A}$ is clopen in βD (the bar denotes the closure in the space βD);
(b) if $z \in \beta D \backslash D$ and $\mathcal{D}_z = \{U \cap D : U \in \tau(z, \beta Z)\}$, then a set $A \subset D$ belongs to $\mathcal{D}_z$ if and only if $z \in \overline{A}$;
(c) the family $\mathcal{D}_z$ is an ultrafilter on D for any $z \in \beta D \backslash D$.

Proof. For an arbitrary set $A \subset D$, we have $\overline{A} \cap \overline{D \backslash A} = \emptyset$ by Fact 1 of S.382. It is evident that $\overline{A} \cup \overline{D \backslash A} = \beta D$, so both sets $U = \overline{A}$ and $V = \overline{D \backslash A}$ are clopen in βD; this proves (a).

It is clear that $z \in \overline{A}$ for any $A \in \mathcal{D}_z$. Now if $A \subset D$ and $z \in \overline{A}$, then $U = \overline{A}$ is a clopen subset of βD by (a); since $z \in U$ and $U \cap D = A$, the item (b) is also settled.

To prove (c) observe that $\mathcal{D}_z$ is a centered family and take any $B \subset D$. The sets $U = \overline{B}$ and $V = \overline{D \backslash B}$ are clopen in βD by (a) and only one of them contains z. If $z \in U$, then $U \cap D = B \in \mathcal{D}_z$ and $z \in V$ implies that $V \cap D = (D \backslash B) \in \mathcal{D}_z$. Thus $B \in \mathcal{D}_z$ or $D \backslash B \in \mathcal{D}_z$ for any $B \subset D$ and hence $\mathcal{D}_z$ is an ultrafilter by Problem 117 of [TFS]. Fact 2 is proved.   □

Fact 3. Let Z be a homogeneous space, i.e., for every $x, y \in Z$, there is a homeomorphism $h : Z \to Z$ such that $h(x) = y$. If Z is of second category in itself, then it is a Baire space.

Proof. Assume that some $U \in \tau^*(Z)$ is of first category in Z. Consider a family $\mathcal{U} = \{V \in \tau^*(Z) : V$ there is a homeomorphism $h : Z \to Z$ such that $h(V) \subset U\}$. It is clear that every $V \in \mathcal{U}$ is of first category in Z. Let $\mathcal{V}$ be a maximal disjoint subfamily of $\mathcal{U}$. We claim that $W = \bigcup \mathcal{V}$ is dense in Z.

Indeed, if this is not the case, then there is a point $z \in Z \backslash \overline{W}$. Take a point $x \in U$ and fix a homeomorphism $h : Z \to Z$ such that $h(z) = x$. By continuity of h, there is $V \in \tau(z, Z)$ such that $V \subset Z \backslash \overline{W}$ and $h(V) \subset U$. Then the family $\mathcal{V}' = \mathcal{V} \cup \{V\}$ is still disjoint and $\mathcal{V}' \subset \mathcal{U}$ which contradicts maximality of $\mathcal{V}$.

This proves that W is dense in Z and hence $F = Z \backslash W$ is nowhere dense in Z. Since every $V \in \mathcal{V}$ is of first category in Z, there is a family $\mathcal{U}_V = \{P_V^n : n \in \omega\}$ of nowhere dense subsets of Z such that $V = \bigcup \mathcal{U}_V$. The set $P_n = \bigcup\{P_V^n : V \in \mathcal{V}\}$ is nowhere dense in Z for every $n \in \omega$ (we leave an easy verification of this fact to the reader) and $\bigcup_{n \in \omega} P_n = W$ whence $Z = (\bigcup_{n \in \omega} P_n) \cup F$ is of first category in itself which is a contradiction.

This contradiction shows that every $U \in \tau^*(Z)$ is of second category in Z and hence Z is a Baire space. Fact 3 is proved.   □

Returning to our solution, let $\pi : \mathbb{D}^X \to \mathbb{D}^\omega$ be the restriction map and consider the set $D = \{f \in \mathbb{D}^X : f(\xi) = 0\}$. It is evident that $\pi_1 = \pi|D : D \to \mathbb{D}^\omega$ is a homeomorphism. The family $\mathcal{D}_\xi = \{U \cap \omega : U \in \tau(\xi, X)\}$ is an ultrafilter on ω by Fact 2.

Now assume that $C_p(X)$ is analytic; since $C_p(X, \mathbb{D})$ is closed in $C_p(X)$, the space $C_p(X, \mathbb{D})$ is analytic too. Therefore $F = \{f \in C_p(X, \mathbb{D}) : f(\xi) = 0\}$ is also analytic. For any $f \in F$ the set $f^{-1}(0)$ is open in X and contains ξ, so $f^{-1}(0) \cap \omega \in \mathcal{D}_\xi$ for any $f \in F$.

On the other hand, if we have $f \in D$ and $W = f^{-1}(0) \cap \omega \in \mathcal{D}_\xi$, then $f^{-1}(0) = W \cup \{\xi\} \in \tau(\xi, X)$ because $\xi \in V = \mathrm{cl}_{\beta\omega}(W)$ by Fact 2 while the set V is open in $\beta\omega$ (here we used Fact 2 again) and $V \cap X = W \cup \{\xi\}$. Therefore $f \in F$ if and only if $f \in D$ and $f^{-1}(0) \cap \omega \in \mathcal{D}_\xi$. This shows that $F_1 = \{f \in \mathbb{D}^\omega : f^{-1}(0) \in \mathcal{D}_\xi\}$ is an analytic space because $F_1 = \pi_1(F)$.

Given any $f \in \mathbb{D}^\omega$, let $f'(n) = 1 - f(n)$ for all $n \in \omega$. If $e(f) = f'$ for each $f \in \mathbb{D}^\omega$, then $e : \mathbb{D}^\omega \to \mathbb{D}^\omega$ is a homeomorphism; it is easy to see that $F_2 = e(F_1)$ is disjoint from F_1 and $F_1 \cup F_2 = \mathbb{D}^\omega$.

Observe that $G = (\mathbb{R}^X \backslash C_p(X)) \cap D$ coincides with the set $\{f \in \mathbb{D}^X : f(\xi) = 0$ and f is discontinuous at $\xi\}$, so $G = \{f \in D : f^{-1}(0) \cap \omega \notin \mathcal{D}_\xi\}$ (here we used Fact 2 which guarantees that $\mathcal{D}_\xi$ is an ultrafilter). It is evident that G is a closed subset of $\mathbb{R}^X \backslash C_p(X)$ and $\pi(G) = F_2$. As a consequence, we proved that

(∗) if $C_p(X)$ is analytic, then F_1 is analytic; if $\mathbb{R}^X \backslash C_p(X)$ is analytic, then F_2 is analytic. Since F_1 and F_2 are homeomorphic, it follows from analyticity of either of spaces $C_p(X)$ or $\mathbb{R}^X \backslash C_p(X)$ that F_1 is analytic.

It is easy to see that F_1 and F_2 are both dense in $\mathbb{D}^\omega$, so if $i \in \{1, 2\}$, then $A \subset F_i$ is nowhere dense in F_i if and only if it is nowhere dense in $\mathbb{D}^\omega$. Consequently, if $F_1 = \bigcup_{n \in \omega} P_n$ where P_n is nowhere dense in F_1 for each $n \in \omega$, then $Q_n = e(P_n)$ is nowhere dense in F_2 for each $n \in \omega$ and therefore the space $\mathbb{D}^\omega = F_1 \cup F_2 = (\bigcup_{n \in \omega} P_n) \cup (\bigcup_{n \in \omega} Q_n)$ is of first category in itself which is false by Problem 274 of [TFS]. This contradiction shows that

(1) both spaces F_1 and F_2 are of second category in themselves and in $\mathbb{D}^\omega$.

We next prove that F_1 is homogeneous. Given any $f, g \in F_1$ define $h = f \oplus g$ by the equalities $h(n) = 1$ if $f(n) \neq g(n)$ and $h(n) = 0$ if $f(n) = g(n)$. For any $f \in \mathbb{D}^\omega$, let $L_f(g) = f \oplus g$ for every $g \in F_1$. It is easy to see that $f \oplus g \in F_1$ for any $f, g \in F_1$. Furthermore, the map $L_f : F_1 \to F_1$ is a homeomorphism for any $f \in F_1$; given any $f, g \in F_1$, if $h = f \oplus g$, then $L_h(f) = g$ which shows that F_1 is a homogeneous space. Now, Fact 3 and (1) imply that

(2) both spaces F_1 and F_2 have the Baire property.

Observe that if F_1 is analytic, then $F_2 = \mathbb{D}^\omega \backslash F_1$ is also analytic, so we can apply Problem 340 to conclude that F_1 is a Borel subset of $\mathbb{D}^\omega$. Therefore F_1 is a B-set by Fact 1; pick $O \in \tau(\mathbb{D}^\omega)$ such that $O \Delta F_1$ is of first category in $\mathbb{D}^\omega$. If $O = \emptyset$, then $F_1 = F_1 \Delta O$ is of first category in $\mathbb{D}^\omega$ which is impossible by (1). Thus $O \neq \emptyset$ and hence $O \cap F_2 = O \backslash F_1 \subset O \Delta F_1$ is a non-empty open subset of F_2 which is of first category in $\mathbb{D}^\omega$ and hence in itself. However, F_2 has the Baire property by (2), so all non-empty open subsets of F_2 are of second category; this contradiction shows that F_1 is not analytic. Finally, apply $(*)$ to conclude that neither of the spaces $C_p(X)$ and $\mathbb{R}^X \backslash C_p(X)$ is analytic. Since X is countable, it cannot be embedded in $C_p(\mathbb{P})$ by Problem 370 and hence our solution is complete.

T.372. *Prove that if $\alpha < \omega_1$, then there exists a countable space X with a unique non-isolated point such that $C_p(X) \in \mathbb{B}(\mathbb{R}^X) \backslash (\bigcup_{\beta < \alpha} \Sigma_\beta^0(\mathbb{R}^X))$, i.e., the space $C_p(X)$ can have an arbitrarily high Borel complexity for a countable space X with a unique non-isolated point.*

Solution. If Y and Z are spaces, the expression $Y \simeq Z$ says that Y is homeomorphic to Z; a map $f : Y \to Z$ is called *measurable* if $f^{-1}(A) \in \mathbb{B}(Y)$ for any $A \in \mathbb{B}(Z)$. Given a set Z and a family $\mathcal{A} \subset \exp(Z)$, we denote by $\bigwedge \mathcal{A}$ ($\bigvee \mathcal{A}$) the family of all finite intersections (unions) of the elements of $\mathcal{A}$.

For any space Z, let $\mathcal{F}(Z)$ be the family of all closed subsets of Z; if $n \in \omega$, then $L_Z^n = \{Y \subset Z : |Y| \leq n\}$ and $\mathrm{Fin}(Z) = \bigcup \{L_Z^n : n \in \omega\}$. For each $U \in \tau(Z)$, let $I(U) = \{F \in \mathcal{F}(Z) : F \subset U\}$ and $M(U) = \{F \in \mathcal{F}(Z) : F \cap U \neq \emptyset\}$. Denote by $\mathcal{V}(Z)$ *the Vietoris topology* on the set $\mathcal{F}(Z)$, i.e., the topology generated by the family $\mathcal{S}(Z) = \{I(U) : U \in \tau(Z)\} \cup \{M(U) : U \in \tau(Z)\}$ as a subbase (it is immediate that $\bigcup \mathcal{S}(Z) = \mathcal{F}(Z)$, so the family $\mathcal{S}(Z)$ can, indeed, generate a topology on $\mathcal{F}(Z)$ as a subbase).

Fact 1. If Z is a compact Hausdorff space, then $T_Z = (\mathcal{F}(Z), \mathcal{V}(Z))$ is also a compact Hausdorff (and hence Tychonoff) space.

Proof. It follows from Problem 118 of [TFS] that to prove compactness of the space $(\mathcal{F}(Z), \mathcal{V}(Z))$, it suffices to show that any cover of $\mathcal{F}(Z)$ with the elements of $\mathcal{S}(Z)$ has a finite subcover. Let $\mathcal{S}_I = \{I(U) : U \in \tau(Z)\}$ and $\mathcal{S}_M = \{M(U) : U \in \tau(Z)\}$; if $\mathcal{U} \subset \mathcal{S}(Z)$ and $\bigcup \mathcal{U} = \mathcal{F}(Z)$, then let $\mathcal{U}_I = \mathcal{U} \cap \mathcal{S}_I$ and $\mathcal{U}_M = \mathcal{U} \cap \mathcal{S}_M$. The set $O = \bigcup \{U : M(U) \in \mathcal{U}\}$ is open in Z and hence $H = Z \backslash O \in \mathcal{F}(Z)$. It is immediate from the definition of O that $H \notin \bigcup \mathcal{U}_M$, so $H \in \bigcup \mathcal{U}_I$, i.e., there is $V \in \tau(Z)$ such that $I(V) \in \mathcal{U}$ and $H \subset V$. The set $G = Z \backslash V$ is compact and $G \subset \bigcup \{U : M(U) \in \mathcal{U}\}$ which shows that there are $U_1, \ldots, U_n \in \tau(Z)$ such that $M(U_i) \in \mathcal{U}$ for all $i \leq n$ and $G \subset U_1 \cup \cdots \cup U_n$.

We claim that $\mathcal{U}' = \{I(V), M(U_1), \ldots, M(U_n)\}$ is a (finite) subcover of $\mathcal{U}$. Indeed, take any $F \in \mathcal{F}(Z)$; if $F \subset V$, then $F \in I(V) \subset \bigcup \mathcal{U}'$. If not, then $F \cap G \neq \emptyset$ and hence there is $i \leq n$ for which $F \cap U_i \neq \emptyset$, i.e., $F \in M(U_i)$ and therefore $F \in \bigcup \mathcal{U}'$. Thus $\mathcal{U}'$ is a finite subcover of $\mathcal{U}$, so we proved that the space T_Z is compact.

To see that T_Z is Hausdorff take distinct $F, G \in \mathcal{F}(Z)$. We can assume, without loss of generality, that $F \backslash G \neq \emptyset$; pick any $x \in F \backslash G$. The space Z is regular (see Problem 124 of [TFS]), so there are disjoint $U, V \in \tau(Z)$ such that $x \in U$ and $G \subset V$. It is straightforward that $F \in M(U)$, $G \in I(V)$ and $M(U) \cap I(V) = \emptyset$ which shows that T_Z is Hausdorff (and hence Tychonoff by Problem 124 of [TFS]). Fact 1 is proved. $\qquad\square$

Fact 2. If Z is a compact space, $w(Z) = \kappa \geq \omega$ and $T_Z = (\mathcal{F}(Z), \mathcal{V}(Z))$, then $w(T_Z) \leq \kappa$.

Proof. Take a base $\mathcal{B}$ in Z such that $|\mathcal{B}| = \kappa$ and let $\mathcal{C} = \bigvee \mathcal{B}$; then $|\mathcal{C}| = \kappa$. We claim that $\mathcal{U} = \{I(U) : U \in \mathcal{C}\} \cup \{M(U) : U \in \mathcal{B}\}$ is a subbase in T_Z. Indeed, take any $F \in \mathcal{F}(Z)$ and $O \in \tau(F, T_Z)$. Since $\mathcal{S}(Z)$ is a subbase of T_Z, there exist $U_1, \ldots, U_n, V_1, \ldots, V_k \in \tau(Z)$ such that

$$F \in O' = I(U_1) \cap \cdots \cap I(U_n) \cap M(V_1) \cap \cdots \cap M(V_k) \subset O.$$

This implies that $F \subset U = U_1 \cap \cdots \cap U_n$ and we can choose $x_i \in F \cap V_i$ for each $i \leq k$. There are $W_1 \ldots, W_k \in \mathcal{B}$ such that $x_i \in W_i \subset V_i$ for all $i \leq k$; besides, the family $\{W \in \mathcal{B} : W \subset U\}$ is, evidently, an open cover of the compact space F. Consequently, there are $H_1, \ldots, H_m \in \mathcal{B}$ such that $F \subset H = \bigcup_{i \leq m} H_i \subset U$. It is clear that $H \in \mathcal{C}$ and $F \in I(H) \cap M(W_1) \cap \cdots \cap M(W_k) \subset O' \subset O$ which proves that $\bigwedge \mathcal{U}$ is a base in T_Z, i.e., $\mathcal{U}$ is a subbase of T_Z. Since $|\mathcal{U}| = |\bigwedge \mathcal{U}| \leq |\mathcal{C}| = \kappa$, we have $w(T_Z) \leq \kappa$ so Fact 2 is proved. $\qquad\square$

Fact 3. For any space Z the set $L_Z^n = \{F \in \mathrm{Fin}(Z) : |F| \leq n\}$ is closed in the space $T_Z = (\mathcal{F}(Z), \mathcal{V}(Z))$. As a consequence, if Z is compact and we consider $\mathrm{Fin}(Z)$ as a subspace of T_Z, then $\mathrm{Fin}(Z)$ is σ-compact.

Proof. If $F \in \mathcal{F}(Z) \backslash L_Z^n$, then there exists distinct points $z_1, \ldots, z_{n+1} \in F$. Pick $U_i \in \tau(z_i, Z)$ for all $i \leq n + 1$ so that the family $\{U_i : i \leq n + 1\}$ is disjoint. It is immediate that $F \subset W = \bigcap_{i \leq n+1} M(U_i)$ while $W \cap L_Z^n = \emptyset$. This proves that L_Z^n is closed in T_Z. If Z is compact, then T_Z is also compact by Fact 1 so L_Z^n is compact for all $n \in \omega$ and hence $\mathrm{Fin}(Z) = \bigcup_{n \in \omega} L_Z^n$ is σ-compact. Fact 3 is proved. $\qquad\square$

Fact 4. Let T be an infinite set and $z \notin T$. For an arbitrary filter $\mathcal{C}$ on T with $\bigcap \mathcal{C} = \emptyset$, consider the family $\tau(\mathcal{C}) = \exp(T) \cup \{\{z\} \cup A : A \in \mathcal{C}\}$. Then $\tau(\mathcal{C})$ is a topology on $T^+ = T \cup \{z\}$, the space $E[T, \mathcal{F}, z] = (T^+, \tau(\mathcal{C}))$ is normal and T_1 and z is the unique non-isolated point of $E[T, \mathcal{F}, z]$.

Proof. Since $\emptyset \in \exp(T)$, we have $\emptyset \in \tau(\mathcal{C})$ and it follows from $T \in \mathcal{C}$ that $T^+ \in \tau(\mathcal{C})$. If $A, B \in \tau(\mathcal{C})$ and $z \notin A \cap B$, then $A \cap B \in \tau(\mathcal{C})$; if $z \in A \cap B$, then $A \cap T \in \mathcal{C}$ and $B \cap T \in \mathcal{C}$ and hence $C = A \cap B \cap T \in \mathcal{C}$ which shows that $A \cap B = C \cup \{z\} \in \tau(\mathcal{C})$.

Finally, if $\mathcal{U} \subset \tau(\mathcal{C})$ and $\bigcup \mathcal{U} \subset T$, then $\bigcup \mathcal{U} \in \tau(\mathcal{C})$; if, on the other hand, $z \in \bigcup \mathcal{U}$, then there is $A \in \mathcal{C}$ such that $A \cup \{z\} \in \mathcal{U}$ and hence $A \subset A' = (\bigcup \mathcal{U}) \cap T$

which implies $A' \in C$ proving that $\bigcup \mathcal{U} = A' \cup \{z\} \in \tau(C)$. Thus $\tau(C)$ is indeed a topology on T^+.

Since $\{t\} \in \tau(C)$ for each $t \in T$, all point of T are isolated in $E[T, \mathcal{F}, z]$; every neighborhood of z is an element of $\mathcal{F}$, so the point z is not isolated in $E[T, \mathcal{F}, z]$. It is straightforward to see that all points of $E[T, \mathcal{F}, z]$ are closed in $E[T, \mathcal{F}, z]$. Furthermore, every open subset which contains z is also closed in $E[T, \mathcal{F}, z]$ and $\{\{t\}\}$ is a clopen base at any $t \in T$. This proves that $E[T, \mathcal{F}, z]$ is zero-dimensional and hence Tychonoff by Fact 1 of S.232. Consequently, $E[T, \mathcal{F}, z]$ is normal by Claim 2 of S.018, so Fact 4 is proved.   □

The notions of the following construction will be used all through the rest of this solution.

Fact 5 (Basic construction and its properties). For each $n \in \omega$ let $\Omega_n = \mathbb{D}^n$ and $\Omega = \bigcup_{n \in \omega} \Omega_n$. We fix a point $w \notin \Omega$ and let $\Omega_* = \Omega \cup \{w\}$. For any $s \in \mathbb{D}^\omega$, let $A(s) = \{s|n : n \in \omega\} \subset \Omega$ and define $\delta(s) \in \mathbb{D}^{\Omega_*}$ by $\delta(s)(p) = 1$ for all $p \in A(s)$ and $\delta(s)(p) = 0$ if $p \in \Omega_* \setminus A(s)$. In other words, $\delta(s)$ is the characteristic function of the set $A(s)$ in Ω_*. The correspondence $s \to \delta(s)$ defines a map $\delta : \mathbb{D}^\omega \to \mathbb{D}^{\Omega_*}$. Then

(1) the map δ is continuous and injective and hence $\delta : \mathbb{D}^\omega \to \delta(\mathbb{D}^\omega)$ is a homeomorphism;
(2) for any distinct $s, t \in \mathbb{D}^\omega$ the set $A(s) \cap A(t)$ is finite;
(3) for any $S \subset \mathbb{D}^\omega$ the family $\mathcal{F}_S = \{\Omega \setminus A : A \subset \bigcup\{A(s) : s \in S'\}\} \cup P$ for some $S' \in \mathrm{Fin}(S)$ and $P \in \mathrm{Fin}(\Omega)\}$ is a filter on Ω;
(4) if $X[S] = E[\Omega, \mathcal{F}_S, w]$ (see Fact 4), then S embeds in $C_p(X[S])$ as a closed subspace.

Proof. (1) Fix any $s_0 \in \mathbb{D}^\omega$, let $t_0 = \delta(s_0)$ and take a set $U \in \tau(t_0, \mathbb{D}^{\Omega_*})$. There exists a finite $B \subset \Omega_*$ such that $U' = \{t \in \mathbb{D}^{\Omega_*} : t|B = t_0|B\} \subset U$. As a consequence, there is $n \in \omega$ for which $B \cap \Omega \subset \bigcup_{i \leq n} \Omega_i$. It is clear that the set $V = \{s \in \mathbb{D}^\omega : s|n = s_0|n\}$ is an open neighborhood of s_0 in $\mathbb{D}^\omega$; it is evident that also $s|i = s_0|i$ for all $i \leq n$, so, by definition of the map δ, we have $A(s) \cap (\bigcup_{i \leq n} \Omega_i) = A(s_0) \cap (\bigcup_{i \leq n} \Omega_i)$ which implies $A(s_0) \cap B = A(s) \cap B$ and therefore $\delta(s)|B = \delta(s_0)|B$ for all $s \in V$. This proves that $\delta(V) \subset U' \subset U$ and hence δ is continuous at the point s_0.

To see that δ is injective take distinct $s, t \in \mathbb{D}^\omega$ and fix $n \in \omega$ such that $a = s|n \neq t|n = b$. Then $a, b \in \Omega_n$ and $\delta(s)(a) = 1 \neq 0 = \delta(t)(a)$ (because from all points of Ω_n, the function $\delta(t)$ only equals 1 at the point b) which shows that $\delta(t) \neq \delta(s)$, so δ is injective and (1) is settled.

To see that (2) is true take $n \in \omega$ such that $s|n \neq t|n$ and observe that $A(s) \cap A(t) \subset \bigcup_{i \leq n} \Omega_i$ and $\bigcup_{i \leq n} \Omega_i$ is a finite set.

As to (3), letting $S' = \emptyset$ and $P = \emptyset$, we can see that $\Omega \in \mathcal{F}_S$, i.e., $\mathcal{F}_S \neq \emptyset$. Now, observe that $|A(s) \cap \Omega_i| = 1$ for any $s \in \mathbb{D}^\omega$ and $i \in \omega$; thus, if $S' \in \mathrm{Fin}(S)$ and $|S'| = n$, then $\Omega_i \setminus (\bigcup\{A(s) : s \in S'\})$ has at least $2^i - n$ elements for all $i > n$ which proves that $\Omega \setminus (\bigcup\{A(s) : s \in S'\})$ is an infinite set and hence $F \neq \emptyset$ for any $F \in \mathcal{F}_S$. It is straightforward that $F \in \mathcal{F}_S$ and $F \subset F'$ implies $F' \in \mathcal{F}$ and $F, G \in \mathcal{F}_S$ implies $F \cap G \in \mathcal{F}_S$, so $\mathcal{F}_S$ is, indeed, a filter and (3) is proved.

Given any $s \in S$, the set $U = \Omega_* \backslash A(s)$ is a clopen neighborhood of w; since $\delta(s)(U) = \{0\}$, the function $\delta(s)$ is continuous at the point w and therefore $\delta(s) \in C_p(X[S])$ for any $s \in S$. On the other hand, if $f \in C_p(X[S]) \cap \delta(\mathbb{D}^\omega)$, then $f \in \mathbb{D}^{\Omega_*}$, $f(w) = 0$ and f is continuous at the point w. Consequently, there exists $U \in \tau(w, X[S])$ such that $f(U) \subset (-\frac{1}{2}, \frac{1}{2})$, i.e., $f(U) = \{0\}$.

By the definition of the topology of $X[S]$, there is a finite $P \subset \Omega$ and a set $S' \in \text{Fin}(S)$ such that $\Omega \backslash U \subset \bigcup \{A(s) : s \in S'\} \cup P$. We have $f = \delta(t)$ for some $t \in \mathbb{D}^\omega$ and hence $f^{-1}(1) = A(t) \subset \bigcup \{A(s) : s \in S'\} \cup P$ which shows that $A(t) \cap A(s)$ is infinite for some $s \in S'$ and therefore $t = s$ [see (2)], i.e., $t \in S$. Consequently, $f \in \delta(S)$ and we established that $C_p(X[S]) \cap \delta(\mathbb{D}^\omega) = \delta(S)$; since $\delta(\mathbb{D}^\omega)$ is compact by (1), the set $\delta(S)$ is a closed subspace of $C_p(X[S])$ which is a homeomorphic copy of S by (1). This settles (4) so Fact 5 is proved. □

Fact 6. If S is a Borel subspace of $\mathbb{D}^\omega$ and $X[S] = E[\Omega, \mathcal{F}_S, w]$ (see Fact 4 and Fact 5), then $C_p(X[S])$ is a Borel subset of the space $\mathbb{R}^{\Omega_*}$.

Proof. By Fact 3 of T.333 being a Borel set as a space is equivalent to being a Borel subset of every larger Polish space, so we will use both terms interchangeably when proving that, in some spaces, certain subsets are Borel.

If $A \subset \Omega$, then $\chi_A \in \mathbb{D}^\Omega$ is the characteristic function of A, i.e., $\chi_A(p) = 1$ for all $p \in A$ and $\chi_A(p) = 0$ if $p \in \Omega \backslash A$. Consider the set $C' = \{f \in C_p(X[S]) : f(w) = 0\}$; it follows from Fact 1 of S.409 that $C_p(X[S]) \simeq C' \times \mathbb{R}$, so it suffices to show that C' is a Borel set (see Problem 333 and Fact 3 of T.333). Let $\pi : \mathbb{R}^{\Omega_*} \to \mathbb{R}^\Omega$ be the restriction map; if $D = \{f \in \mathbb{R}^{\Omega_*} : f(w) = 0\}$, then $\pi_1 = \pi | D : D \to \mathbb{R}^\Omega$ is a homeomorphism; it is evident that $\pi_1(D) = \mathbb{R}^\Omega$, so it suffices to show that $C = \pi_1(C')$ is a Borel subset of $\mathbb{R}^\Omega$.

For any $f \in \mathbb{R}^\Omega$ and $m \in \mathbb{N}$, let $H(f, m) = f^{-1}(\mathbb{R} \backslash (-\frac{1}{m}, \frac{1}{m}))$; we will also need the family $\mathcal{H}_S = \{A \subset \Omega : A \subset \bigcup \{A(s) : s \in S'\} \cup P$ for some $S' \in \text{Fin}(S)$ and $P \in \text{Fin}(\Omega)\}$ which, evidently, consists of the complements of the elements of $\mathcal{F}_S$; let $H[S] = \{\chi_A : A \in \mathcal{H}_S\}$.

For every $m \in \mathbb{N}$ define a map $\psi_m : \mathbb{R}^\Omega \to \mathbb{D}^\Omega$ by letting $\psi_m(f)$ to be the characteristic function of the set $H(f, m)$, i.e., $\psi_m(f)(p) = 1$ if $p \in H(f, m)$ and $\psi_m(f)(p) = 0$ for all $p \in \Omega \backslash H(f, m)$. It is evident that $\psi_m(f) = f$ for any $f \in \mathbb{D}^\Omega$, so $\psi_m(\mathbb{R}^\Omega) = \mathbb{D}^\Omega$ for each $m \in \mathbb{N}$.

Observe that $f \in D$ belongs to C' if and only if $f^{-1}((-\frac{1}{m}, \frac{1}{m})) \backslash \{w\} \in \mathcal{F}_S$ for all $m \in \mathbb{N}$, or, equivalently, if and only if $f^{-1}(\mathbb{R} \backslash (-\frac{1}{m}, \frac{1}{m})) \backslash \{w\} \in \mathcal{H}_S$ for all $m \in \mathbb{N}$. Therefore a function $f \in \mathbb{R}^\Omega$ belongs to the set C if and only if $H(f, m) \in \mathcal{H}_S$ or, equivalently, $\psi_m(f) \in H[S]$ for all $m \in \mathbb{N}$. Thus, we established that

(5) $C = \bigcap \{\psi_m^{-1}(H[S]) : m \in \mathbb{N}\}$.

We will show next that

(6) ψ_m is a measurable map for all $m \in \mathbb{N}$.

To do this, for any $p \in \Omega$ and $i \in \mathbb{D}$, let $O(p,i) = \{f \in \mathbb{D}^\Omega : f(p) = i\}$; it is evident that the family $S = \{O(p,i) : p \in \Omega \text{ and } i \in \mathbb{D}\}$ is a subbase of $\mathbb{D}^\Omega$. Therefore it suffices to show that $\psi_m^{-1}(O(p,i))$ is a Borel subset of $\mathbb{R}^\Omega$ for all $m \in \mathbb{N}$, $p \in \Omega$ and $i \in \mathbb{D}$ (see Fact 1 of T.368).

Now, $\psi_m^{-1}(O(p,0)) = \{f \in \mathbb{R}^\Omega : \psi_m(f)(p) = 0\} = \{f \in \mathbb{R}^\Omega : p \notin H(f,m)\}$, i.e., $\psi_m^{-1}(O(p,0)) = \{f \in \mathbb{R}^\Omega : |f(p)| < \frac{1}{m}\}$ is an open (and hence Borel) subset of $\mathbb{R}^\Omega$ for every $p \in \Omega$.

Analogously, the equality $\psi_m^{-1}(O(p,1)) = \{f \in \mathbb{R}^\Omega : p \in H(f,m)\}$ shows that $\psi_m^{-1}(O(p,1)) = \{f \in \mathbb{R}^\Omega : |f(p)| \geq \frac{1}{m}\}$ is a closed (and hence Borel) subset of $\mathbb{R}^\Omega$ for each $p \in \Omega$ so (6) is proved.

Our plan is to establish that $H[S]$ is a Borel subset of $\mathbb{D}^\Omega$. Since this is not easy, we will need several intermediate steps. The first one is to prove that

(7) the space $H = H[\mathbb{D}^\omega] = \{\chi_A : A \subset \bigcup\{A(s) : s \in S'\} \cup P \text{ for some } S' \in \mathrm{Fin}(\mathbb{D}^\omega) \text{ and } P \in \mathrm{Fin}(\Omega)\}$ is σ-compact.

Fix an arbitrary number $n \in \mathbb{N}$ and a finite set $P \subset \Omega$ and consider a subset $H_n^P = \{(f, g_1, \ldots, g_n) \in \mathbb{D}^\Omega \times (\delta(\mathbb{D}^\omega))^n : f^{-1}(1) \subset g_1^{-1}(1) \cup \cdots \cup g_n^{-1}(1) \cup P\}$ of the compact space $\mathbb{D}^\Omega \times (\delta(\mathbb{D}^\omega))^n$. The set H_n^P is closed in $\mathbb{D}^\Omega \times (\delta(\mathbb{D}^\omega))^n$; indeed, given $a = (f, g_1, \ldots, g_n) \in (\mathbb{D}^\Omega \times (\delta(\mathbb{D}^\omega))^n) \backslash H_n^P$, pick $p \in f^{-1}(1) \backslash ((\bigcup_{i \leq n} g_i^{-1}(1)) \cup P)$ and observe that $W = \{(h, u_1, \ldots, u_n) : h(p) = 1 \text{ and } u_i(p) = 0 \text{ for all } i \leq n\}$ is an open neighborhood of a in $\mathbb{D}^\Omega \times (\delta(\mathbb{D}^\omega))^n$ with $W \cap H_n^P = \emptyset$. Thus the space H_n^P is compact being closed in $\mathbb{D}^\Omega \times (\delta(\mathbb{D}^\omega))^n$.

Let $\pi_n : \mathbb{D}^\Omega \times (\delta(\mathbb{D}^\omega))^n \to \mathbb{D}^\Omega$ be the natural projection for all $n \in \mathbb{N}$. It is straightforward that $H = \bigcup\{\pi_n(H_n^P) : n \in \mathbb{N} \text{ and } P \in \mathrm{Fin}(\Omega)\}$ so H is σ-compact and (7) is proved.

Denote by V_C the set of all closed subsets of $\delta(\mathbb{D}^\omega)$ endowed with the Vietoris topology; then V_C is a metrizable compact space (see Fact 1 and Fact 2). Now let Φ be the set $\mathrm{Fin}(\delta(\mathbb{D}^\omega))$ with the topology induced from V_C; then Φ is a metrizable σ-compact space by Fact 3. Let $\Phi[S] = \mathrm{Fin}(\delta(S))$ considered to be a subspace of Φ. Then

(8) $\Phi[S]$ is a Borel subset of Φ.

Let $\Delta = \delta(\mathbb{D}^\omega)$, $\Phi_n = \{F \in \Phi : |F| \leq n\}$ and $\Phi_n[S] = \Phi_n \cap \Phi[S]$ for every $n \in \omega$. Given any $n \in \mathbb{N}$, we have "an order forgetting map" $\varphi_n : \Delta^n \to \Phi_n$ defined by $\varphi_n(x) = \{x(0), \ldots, x(n-1)\}$ for any $x \in \Delta^n$. Recall that we identify any $n \in \mathbb{N}$ with the set $\{0, \ldots, n-1\}$ and consider the product Δ^n to be the set of functions from $\{0, \ldots, n-1\}$ to Δ.

The sets $I(U) = \{F \in \Phi : F \subset U\}$ and $M(U) = \{F \in \Phi : F \cap U \neq \emptyset\}$ form a subbase S of Φ when U runs through all open subsets of Δ. Fix $U \in \tau(\Delta)$ and $x \in \varphi_n^{-1}(I(U))$. Then $F = \{x(0), \ldots, x(n-1)\} \subset U$ and therefore $V = U^n$ is an open neighborhood of x in Δ^n such that $\varphi_n(V) \subset I(U)$. This proves that $\varphi_n^{-1}(I(U))$ is open in Δ^n.

Furthermore, if $x \in \varphi_n^{-1}(M(U))$, then $F \cap U \neq \emptyset$ where $F = \{x(0), \ldots, x(n - 1)\}$. Pick $j < n$ such that $x(j) \in U$ and let $V = \{y \in \Delta^n : y(j) \in U\}$. Then $V \in \tau(x, \Delta^n)$ and $y(j) \in \varphi_n(y) \cap U$, i.e., $\varphi_n(y) \in M(U)$ for any $y \in V$. Therefore $\varphi_n(V) \subset M(U)$ and hence $\varphi_n^{-1}(M(U))$ is also open in Δ^n for every $U \in \tau(\Delta)$. Now that we proved that $\varphi_n^{-1}(O)$ is open in Δ^n for any $O \in \mathcal{S}$, we can conclude that the map φ_n is continuous for every $n \in \mathbb{N}$.

Consider the set $G_n = \{x \in \Delta^n : x(i) \neq x(j) \text{ if } i \neq j\} \in \tau(\Delta^n)$. It follows from Fact 3 that the set $G_n' = \Phi_n \backslash \Phi_{n-1}$ is open in Φ_n; given any element $F = \{x_0, \ldots, x_{n-1}\} \in G_n'$, the point $x = (x_0, \ldots, x_{n-1})$ belongs to G_n and $\varphi_n(x) = F$. Choose $\{U_0, \ldots, U_{n-1}\} \subset \tau(\Delta)$ such that the family $\{\overline{U}_0, \ldots, \overline{U}_{n-1}\}$ is disjoint and $x_i \in U_i$ for all $i < n$. The set $K = \overline{U}_0 \times \cdots \times \overline{U}_{n-1} \subset G_n$ is compact and for any distinct $y, z \in K$ we have $\varphi_n(y) \neq \varphi_n(z)$. Therefore the map $\varphi_n' = \varphi_n | K$ is an injection and hence $\varphi_n' : K \to \varphi_n'(K)$ is a homeomorphism. Besides, for the sets $U = U_0 \times \cdots \times U_{n-1}$ and $V = \bigcup_{i<n} U_i$, we have $F \in \varphi_n(U) = W \cap G_n'$ where $W = I(V) \cap (\bigcap_{i<n} M(U_i))$. Thus, every $F \in \Phi_n \backslash \Phi_{n-1}$ has an open neighborhood (in $\Phi_n \backslash \Phi_{n-1}$) which is homeomorphic to an open subset of Δ^n.

Since $S_1 = \delta(S)$ is a Borel set, the space S_1^n a Borel set as well by Problem 333. Therefore $S_1^n \cap G_n$ is a Borel set too and it is immediate that $\Phi_n[S] \backslash \Phi_{n-1}[S] = \varphi_n(S_1^n \cap G_n)$. We proved in the previous paragraph that any point $F \in \Phi_n \backslash \Phi_{n-1}$ has a neighborhood $W_F \in \tau(F, \Phi_n \backslash \Phi_{n-1})$ such that for some $U_F \in \tau(G_n)$, the map $\varphi_n | U_F : U_F \to W_F$ is a homeomorphism. As a consequence, if $F \in \Phi_n[S] \backslash \Phi_{n-1}[S]$, then $\varphi_n | (U_F \cap S_1^n) : (U_F \cap S_1^n) \to W_F' = W_F \cap (\Phi_n[S] \backslash \Phi_{n-1}[S])$ is a homeomorphism. This proves that any point of $\Phi_n^*[S] = \Phi_n[S] \backslash \Phi_{n-1}[S]$ has a neighborhood which is a Borel set being homeomorphic to a Borel set $U_F \cap S_1^n$. This gives us an open cover $\mathcal{C}$ of $\Phi_n^*[S]$ which consists of Borel sets; choosing a countable subcover of $\mathcal{C}$, we represent $\Phi_n^*[S]$ as a countable union of Borel sets. Therefore $\Phi_n^*[S]$ is a Borel set by Problem 333 and hence $\Phi[S] = \{\emptyset\} \cup (\bigcup \{\Phi_n^*[S] : n \in \mathbb{N}\})$ is also a Borel set (here we used Problem 333 again) so (8) is proved.

Recall that $H = \{\chi_A : A \subset \bigcup \{A(s) : s \in S'\} \cup P$ for some $S' \in \text{Fin}(\mathbb{D}^\omega)$ and $P \in \text{Fin}(\Omega)\}$; for any $h \in H$, let $\mu(h) = \{f \in \Delta : h^{-1}(1) \cap f^{-1}(1) \text{ is infinite}\}$. Then

(9) $\mu(h) \in \Phi$ for any $h \in H$ and the map $\mu : H \to \Phi$ is measurable.

Take any $h \in H$; by definition there are $s_1, \ldots, s_n \in \mathbb{D}^\omega$ and a finite $P \subset \Omega$ such that $h^{-1}(1) \subset A(s_1) \cup \cdots \cup A(s_n) \cup P$. If $h^{-1}(1)$ if finite, then $\mu(h) = \emptyset \in \Phi$. If $h^{-1}(1)$ is infinite and $f \in \mu(h)$, then $f^{-1}(1) \cap h^{-1}(1)$ is infinite and therefore $f^{-1}(1) \cap A(s_i)$ is infinite for some $i \leq n$. This implies $f = \delta(s_i)$ by (2) and hence we have the inclusion $\mu(h) \subset \{\delta(s_1), \ldots, \delta(s_n)\} \in \Phi$ which proves that $\mu(h) \in \Phi$ for any $h \in H$.

To prove that the function μ is measurable, we will use again the subbase $\mathcal{S}$ constructed in the proof of (8). Assume that $U \in \tau(\Delta)$ and consider the set $Q = \mu^{-1}(I(U)) = \{h \in H : \mu(h) \subset U\}$. Since Δ is compact and metrizable and U is open in Δ, it is σ-compact. Fix any $n \in \mathbb{N}$ and a finite set $P \subset \Omega$; the space H being also σ-compact by (7), the product $H \times U^n$ is σ-compact as well. The set

$Q(n, P) = \{(h, f_0, \ldots, f_{n-1}) \in H \times U^n : h^{-1}(1) \subset (\bigcup_{i<n} f_i^{-1}(1)) \cup P\}$ is closed in $H \times U^n$ because for any $a = (h, f_0, \ldots, f_{n-1}) \in (H \times U^n) \backslash Q(n, P)$, we can choose $p \in h^{-1}(1) \backslash ((\bigcup_{i<n} f_i^{-1}(1)) \cup P)$ and observe that $G = \{(k, g_0, \ldots, g_n) \in H \times U^n : k(p) = 1$ and $g_i(p) = 0$ for all $i < n\}$ is an open neighborhood of a in $H \times U^n$ and $G \cap Q(n, P) = \emptyset$.

Consequently, the set $Q(n, P)$ is σ-compact for all $n \in \mathbb{N}$ and $P \in \text{Fin}(\Omega)$; if $\pi_n : H \times U^n \to H$ is the natural projection, then $Q = \bigcup\{\pi_n(Q(n, P)) : n \in \mathbb{N}$ and $P \in \text{Fin}(\Omega)\}$. Thus

(9a) $Q = \mu^{-1}(I(U))$ is σ-compact and hence Borel for any $U \in \tau(\Delta)$.

Now assume that U is a clopen subset of Δ; then $G = \mu^{-1}(M(U)) = \{h \in H : \mu(h) \cap U \neq \emptyset\}$ and therefore $G' = H \backslash G = \{h \in H : \mu(h) \subset \Delta \backslash U\} = \mu^{-1}(I(\Delta \backslash U))$ is a Borel set by (9a). Consequently, $G = H \backslash G'$ is also a Borel set, so we proved that

(9b)   for any compact $U \in \tau(\Delta)$, the set $\mu^{-1}(M(U))$ is a Borel subset of H.

Finally, take an arbitrary set $U \in \tau(\Delta)$; it is evident that we can represent it as $U = \{U_n : n \in \omega\}$ where U_n is a clopen subset of Δ for all $n \in \omega$. Then $\mu^{-1}(M(U)) = \bigcup\{\mu^{-1}(M(U_n)) : n \in \omega\}$ is a Borel set because $\mu^{-1}(M(U_n))$ is Borel for all $n \in \omega$ by (9b). Thus we proved that $\mu^{-1}(U)$ is a Borel subset of H for any $U \in S$ whence μ is measurable so (9) is proved.

Our last step is to show that

(10) $\mu^{-1}(\Phi[S]) = H[S]$.

To do this take any $h \in H[S]$; by definition there are $s_1, \ldots, s_n \in S$ and a finite set $P \subset \Omega$ such that $h^{-1}(1) \subset A(s_1) \cup \cdots \cup A(s_n) \cup P$. If $h^{-1}(1)$ if finite, then $\mu(h) = \emptyset \in \Phi[S]$. If $h^{-1}(1)$ is infinite and $f \in \mu(h)$, then $f^{-1}(1) \cap h^{-1}(1)$ is infinite and therefore $f^{-1}(1) \cap A(s_i)$ is infinite for some $i \leq n$. This implies $f = \delta(s_i)$ by (2) and hence we have the inclusion $\mu(h) \subset \{\delta(s_1), \ldots, \delta(s_n)\} \in \Phi[S]$ which proves that $\mu(h) \in \Phi[S]$ for any $h \in H[S]$.

To prove the remaining inclusion suppose that $h \in H$ and $\mu(h) \in \Phi[S]$. By definition of H there is a finite $E \subset \mathbb{D}^\omega$ such that $h^{-1}(1) \subset \bigcup\{A(t) : t \in E\} \cup P$ for some finite $P \subset \Omega$. Observe that if $h^{-1}(1)$ is Infinite, then $h^{-1}(1) \cap A(t)$ is infinite for some $t \in E$ and hence $\delta(t) \in \mu(h) \subset \delta(S)$. This shows that $\mu(h) = \emptyset$ implies that $h^{-1}(1)$ is finite and hence $h \in H[S]$.

Now if we have $\{s_1, \ldots, s_n\} \subset S$ and $\mu(h) = \{\delta(s_1), \ldots, \delta(s_n)\}$ while the set $P = h^{-1}(1) \backslash (A(s_1) \cup \cdots \cup A(s_n))$ is infinite, then $P \cap A(t)$ is infinite for some $t \in E$ which shows that $\delta(t) \in \mu(h)$ and hence $t \in \{s_1, \ldots, s_n\}$ which is a contradiction. Thus the set P is finite and therefore $h^{-1}(1) \subset A(s_1) \cup \cdots \cup A(s_n) \cup P$ which shows that $h \in H[S]$ and (10) is proved.

Observe finally that $H[S]$ is a Borel subset of H because the map $\mu : H \to \Phi$ is measurable by (9), the set $\Phi[S]$ is Borel in Φ by (8) and $H[S] = \mu^{-1}(\Phi[S])$ by (10). Since H is σ-compact by (7), it is Borel in $\mathbb{D}^\Omega$, so the set $H[S]$ is also Borel in $\mathbb{D}^\Omega$ (see Fact 1 of T.319 and Fact 1 of T.331).

The map $\psi_m : \mathbb{R}^\Omega \to \mathbb{D}^\Omega$ is measurable by (6), so $\psi_m^{-1}(H[S])$ is Borel in $\mathbb{R}^\Omega$ for all $m \in \mathbb{N}$ which implies that $C = \bigcap\{\psi_m^{-1}(H[S]) : m \in \mathbb{N}\}$ is also a Borel subset of $\mathbb{R}^\Omega$. We already noted in the first paragraph of the proof of this fact that this implies that $C_p(X[S])$ is a Borel subset of $\mathbb{R}^{\Omega*}$ so Fact 6 is proved. □

Returning to our solution, take a set $S \in \mathbb{B}(\mathbb{D}^\omega)\backslash\Sigma_{\alpha+2}^0(\mathbb{D}^\omega)$ which exists by Problem 321 and let $X = X[S]$; then X is a space with a unique non-isolated point. By Fact 5 the space S is homeomorphic to a closed subspace T of the space $C_p(X)$.

The set $C_p(X)$ is Borel in $\mathbb{R}^X$ by Fact 6. Now, if $C_p(X) \in \Sigma_{\alpha+1}^0(\mathbb{R}^X)$, then the set $T = \overline{T} \cap C_p(X)$ (the bar denotes the closure in $\mathbb{R}^X$) also belongs to $\Sigma_{\alpha+1}^0(\mathbb{R}^X)$ because $\overline{T} \in \Pi_0^0(\mathbb{R}^X) \subset \Sigma_{\alpha+1}^0(\mathbb{R}^X)$ and finite intersections preserve Borel classes (see Problem 320 and Fact 1 of T.341).

Apply Fact 2 of T.333 to find a G_δ-subset S' of the space $\mathbb{D}^\omega$ and a G_δ-subset T' of $\mathbb{R}^X$ such that $S \subset S'$, $T \subset T'$ and there is a homeomorphism $h : S' \to T'$. We have $T \in \Sigma_{\alpha+1}^0(T')$ by Fact 1 of T.319 and therefore $S \in \Sigma_{\alpha+1}^0(S')$ (it is evident that homeomorphisms preserve Borel classes). Since S' is a G_δ-set in $\mathbb{D}^\omega$, we have $S' \in \Pi_1^0(\mathbb{D}^\omega) \subset \Sigma_{\alpha+2}^0(\mathbb{D}^\omega)$ by Problem 320. Apply Fact 1 of T.319 again to find a set $E \in \Sigma_{\alpha+1}^0(\mathbb{D}^\omega)$ such that $E \cap S' = S$. Since Borel classes are preserved by finite intersections (see Fact 1 of T.341), we have $S \in \Sigma_{\alpha+2}^0(\mathbb{D}^\omega)$ which contradicts the choice of S. Thus $C_p(X) \in \mathbb{B}(\mathbb{R}^X)\backslash\Sigma_{\alpha+1}^0(\mathbb{R}^X) \subset \mathbb{B}(\mathbb{R}^X)\backslash(\bigcup\{\Sigma_\beta^0(\mathbb{R}^X) : \beta < \alpha\})$ (see Problem 321) and hence our solution is complete.

T.373. *Prove that the following are equivalent for any metrizable space X:*

(i) X is an absolute $F_{\sigma\delta}$;
(ii) there is a completely metrizable space M such that X is an $F_{\sigma\delta}$-subset of M;
(iii) X has a complete sequence of σ-discrete closed covers.

Solution. If Z is a space and $\mathcal{A}_1, \ldots, \mathcal{A}_n$ are families of subsets of Z, then let $\mathcal{A}_1 \wedge \cdots \wedge \mathcal{A}_n = \{A \subset Z : A \neq \emptyset$ and $A = A_1 \cap \cdots \cap A_n : A_i \in \mathcal{A}_i$ for all $i \leq n\}$. If (Z, d) is a metric space, call a family $\mathcal{A} \subset \exp(Z)$ *uniformly discrete* if there is $\delta > 0$ such that the ball $B_d(z, \delta)$ meets at most one element of $\mathcal{A}$ for any $z \in Z$. We will say that δ *witnesses uniform discreteness* of $\mathcal{A}$. It is evident that any uniformly discrete family is discrete. A family $\mathcal{A}'$ is *a refinement of the family* $\mathcal{A}$ if $\bigcup \mathcal{A}' = \bigcup \mathcal{A}$, and for any $A' \in \mathcal{A}'$, there is $A \in \mathcal{A}$ such that $A' \subset A$. If $\mathcal{A} \subset \exp(Z)$ and $Y \subset Z$, then $\mathcal{A}|Y = \{A \cap Y : A \in \mathcal{A}$ and $A \cap Y \neq \emptyset\}$.

Fact 1. Let (Z, d) be a metric space. Then every open cover $\mathcal{U} = \{U_s : s \in S\}$ of the space Z has a refinement $\mathcal{B} = \bigcup\{\mathcal{B}_i : i \in \omega\}$ such that every $\mathcal{B}_i$ is uniformly discrete for all $i \in \omega$.

Proof. Fix a well-order $<$ on the set S and let $H_s = U_s\backslash(\bigcup\{U_t : t < s\})$ for each $s \in S$. Given $i \in \omega$ and $s \in S$, let $B_{s,i} = \{c \in H_s : B(c, 3/2^i) \subset U_s\}$. Next we define by induction on $i \in \omega$ the sets $V_{s,i}$ for all $s \in S$. The first step is to define $V_{s,0} = \bigcup\{B(c, 1) : c \in B_{s,0}\}$ for all $s \in S$. If we have constructed $V_{s,j}$ for each $j < i$ and $s \in S$, consider the sets $V_{s,i} = \bigcup\{B(c, 1/2^i) : c \in B_{s,i}\backslash(\bigcup\{V_{s,j} : s \in S, j < i\})\}$ for all $s \in S$. Observe that $V_{s,i} \subset U_s$ is an open set for all $s \in S$ and

$i \in \omega$. Let $\mathcal{B}_i = \{V_{s,i} : s \in S\}$ and $\mathcal{B} = \bigcup\{\mathcal{B}_i : i \in \omega\}$. For any $z \in Z$ there is a minimal $s \in S$ with $z \in U_s$. This implies $z \in H_s$. Pick $i \in \omega$ such that $B(z, 3/2^i) \subset U_s$; for this i we have $z \in B_{s,i}$. Now, if $z \in \bigcup\{V_{s,j} : s \in S, \ j < i\}$, then $z \in \bigcup \mathcal{B}$. If not, then $B(z, 1/2^i) \subset V_{s,i}$ and again $z \in \bigcup \mathcal{B}$. This yields $Z = \bigcup \mathcal{B}$ and hence $\mathcal{B}$ is a refinement of $\mathcal{U}$.

To see that every $\mathcal{B}_i$ is uniformly discrete, observe that

(1) if $x \in V_{s,i}, \ y \in V_{t,i}$ where $s < t$, then $d(x, y) > 1/2^i$.

Indeed, there is a point $c \in B_{s,i}$ such that $x \in B(c, 1/2^i)$ and $c' \in B_{t,i}$ with $y \in B(c', 1/2^i)$. We have $B(c, 3/2^i) \subset U_s$ while $c' \notin U_s$. Thus $d(c, c') \geq 3/2^i$, and if $d(x, y) \leq 1/2^i$, then $d(c, c') \leq d(c, x) + d(x, y) + d(y, c') < 1/2^i + 1/2^i + 1/2^i = 3/2^i$; this is a contradiction which proves (1).

Now take any $z \in Z$ and $U_z = B(z, 1/2^{i+1}) \in \tau(z, Z)$. If there exist $s, t \in S$ such that $s < t$ and $U_z \cap V_{s,i} \neq \emptyset \neq U_z \cap V_{t,i}$, then pick $x \in U_z \cap V_{s,i}$ and $y \in U_z \cap V_{t,i}$; our claim implies $d(x, y) > 1/2^i$. However

$$d(x, y) \leq d(x, z) + d(z, y) \leq 1/2^{i+1} + 1/2^{i+1} = 1/2^i$$

which is a contradiction. Thus, for $\delta = 2^{-i-1}$ the ball $B_d(z, \delta)$ meets at most one element of $\mathcal{B}_i$ for all $z \in Z$. Therefore each $\mathcal{B}_i$ is uniformly discrete and Fact 1 is proved.   □

Fact 2. Let (Z, d) be a metric space. Given $Y \subset Z$ and family $\mathcal{A} \subset \exp(Y)$ which is uniformly discrete in $(Y, d \,|\, (Y \times Y))$, the family $\mathcal{A}$ is also uniformly discrete (and hence discrete) in (Z, d).

Proof. There is $\delta_0 > 0$ which witnesses uniform discreteness of $\mathcal{A}$ in $(Y, d \,|\, (Y \times Y))$. Then $\delta = \frac{\delta_0}{2}$ witnesses uniform discreteness of $\mathcal{A}$ in (Z, d). Indeed, assume that $z \in Z$ and there are distinct $A, B \in \mathcal{A}$ such that $B_d(z, \delta)$ meets both A and B. Then there are $a \in A$ and $b \in B$ such that $d(a, z) < \delta$ and $d(b, z) < \delta$ which implies $d(a, b) \leq d(a, z) + d(z, b) < 2\delta = \delta_0$ which is a contradiction because the δ_0-ball of the point a in Y cannot intersect both sets A and B. Thus the family $\mathcal{A}$ is uniformly discrete (and hence discrete) in (Z, d), so Fact 2 is proved.   □

Fact 3. If (Z, d) is a metric space and a family $\mathcal{A}_i$ is uniformly discrete in (Z, d) for all $i < n \in \omega$, then the family $\mathcal{A} = \mathcal{A}_0 \wedge \cdots \wedge \mathcal{A}_{n-1}$ is also uniformly discrete.

Proof. Let $\delta_i > 0$ be the number which witnesses uniform discreteness of $\mathcal{A}_i$ for all $i < n$. If $\delta = \min\{\delta_0, \ldots, \delta_{n-1}\}$, then take any $z \in Z$ and assume that there are distinct $A, B \in \mathcal{A}$ such that $W = B_d(z, \delta)$ meets both A and B. For all $i < n$ choose $A_i, B_i \in \mathcal{A}_i$ such that $A = \bigcap_{i<n} A_i$ and $B = \bigcap_{i<n} B_i$. Since $A \neq B$, there is $i < n$ such that $A_i \neq B_i$. It follows from $W \cap A \neq \emptyset$ and $W \cap B \neq \emptyset$ that $W \cap A_i \neq \emptyset$ and $W \cap B_i \neq \emptyset$ which contradicts the choice of $\delta_i \geq \delta$. Thus δ witnesses uniform discreteness of $\mathcal{A}$ and Fact 3 is proved.   □

Returning to our solution, assume that $\{\mathcal{F}_n : n \in \omega\}$ is a complete sequence of σ-discrete closed covers of X and X is a subspace of a metric space (Y, d); let

$d_1 = d|(X \times X)$. Every F_σ-subset of $\text{cl}_Y(X)$ is an F_σ-subset of Y, so in proving that X is an $F_{\sigma\delta}$-subset of Y, there is no loss of generality to assume that X is dense in Y.

It is easy to see that

(2) if $\mathcal{S}' = \{\mathcal{F}'_n : n \in \omega\}$ is a sequence of covers of X and $\mathcal{F}'_n$ is a refinement of $\mathcal{F}_n$ for each $n \in \omega$, then $\mathcal{S}'$ is also complete.

Another evident property is

(3) if $\{P_t : t \in T\}$ is a uniformly discrete family in (Y, d) and $Q_t \subset P_t$ for all $t \in T$, then the family $\{Q_t : t \in T\}$ is uniformly discrete.

By our assumption $\mathcal{F}_n = \bigcup\{\mathcal{F}^m_n : m \in \omega\}$ where $\mathcal{F}^m_n$ is a discrete family of X, so we can choose an open cover $\mathcal{U}^m_n$ of the space X such that every $U \in \mathcal{U}^m_n$ has diameter $\leq 2^{-n}$ and intersects at most one element of $\mathcal{F}^m_n$. There is a refinement $\mathcal{V}^m_n = \bigcup_{k \in \omega} \mathcal{V}^m_n(k)$ of the cover $\mathcal{U}^m_n$ such that each $\mathcal{V}^m_n(k)$ is uniformly discrete in the space (X, d_1) (see Fact 1).

For all $n, m, k \in \omega$ the family $\mathcal{G}(n, m, k) = \mathcal{V}^m_n(k) \wedge \mathcal{F}^m_n$ is uniformly discrete by (3) because every set $V \in \mathcal{V}^m_n(k)$ contains at most one element of $\mathcal{G}(n, m, k)$. It is clear that $\mathcal{G}^m_n = \bigcup_{k \in \omega} \mathcal{G}(n, m, k)$ is a refinement of the family $\mathcal{F}^m_n$ and hence $\mathcal{G}'_n = \bigcup_{m \in \omega} \mathcal{G}^m_n$ is a refinement of the cover $\mathcal{F}_n$ for every $n \in \omega$. Since $\mathcal{F}_n$ consists of closed subsets of X, the family $\mathcal{G}_n = \{\text{cl}_X(G) : G \in \mathcal{G}'_n\}$ is also a refinement of $\mathcal{F}_n$ for all $n \in \omega$.

The property (2) shows that the sequence $\{\mathcal{G}_n : n \in \omega\}$ is complete; observe that we also have

(4) $\text{diam}(V) \leq 2^{-n}$ for every $n \in \omega$ and $V \in \mathcal{G}_n$;
(5) $\mathcal{G}_n = \bigcup\{\mathcal{G}_n(k) : k \in \omega\}$ where $\mathcal{G}_n(k)$ is uniformly discrete for every $k \in \omega$.

Furthermore, the family $\mathcal{H}(n, k) = \{\overline{P} : P \in \mathcal{G}_n(k)\}$ is uniformly discrete in (Y, d) by Fact 2 and an evident observation that the closures of the elements of a uniformly discrete family form a uniformly discrete family (the bar denotes the closure in Y).

Now if $m \in \mathbb{N}$ and we are given distinct numbers $n_1, \ldots, n_m \in \omega$ and arbitrary $k_1 \ldots, k_m \in \omega$, then let

$$\mathcal{Q}(n_1, \ldots, n_m, k_1, \ldots, k_m) = \{Q \in \mathcal{H}(n_1, k_1) \wedge \cdots \wedge \mathcal{H}(n_m, k_m) : Q \cap X = \emptyset\}.$$

It follows from Fact 3 that $\mathcal{Q}(n_1, \ldots, n_m, k_1, \ldots, k_m)$ is uniformly discrete and hence the set $Q(n_1, \ldots, n_m, k_1, \ldots, k_m) = \bigcup \mathcal{Q}(n_1, \ldots, n_m, k_1, \ldots, k_m)$ is closed in Y for any distinct $n_1, \ldots, n_m \in \omega$ and arbitrary $k_1 \ldots, k_m \in \omega$. Consequently, $D = \bigcup\{Q(n_1, \ldots, n_m, k_1, \ldots, k_m) : n_1, \ldots, n_m$ are distinct elements of ω and $k_1, \ldots, k_m \in \omega\}$ is an F_σ-subset of Y whence $R = Y \backslash D$ is a G_δ-subset of Y which contains X.

Every open subset of Y is an F_σ-set, so R is an $F_{\sigma\delta}$-subset of Y. Since $\mathcal{H}(n, k)$ is uniformly discrete, the set $H^k_n = \bigcup \mathcal{H}(n, k)$ is closed in Y for any $n, k \in \omega$.

Consequently, $H_n = \bigcup_{k \in \omega} H_n^k$ is an F_σ-subset of Y. Therefore $H = (\bigcap_{n \in \omega} H_n) \cap R$ is an $F_{\sigma\delta}$-subset of Y so it suffices to show that $X = H$.

The inclusion $X \subset H$ is evident, so assume that there is a point $y \in H \backslash X$. There exists a sequence $\{k_n : n \in \omega\} \subset \omega$ such that $y \in H_n^{k_n}$ for all $n \in \omega$. The family $\mathcal{H}(n, k_n)$ being uniformly discrete, we can choose $P_n \in \mathcal{G}_n(k_n)$, so that $y \in \overline{P_n}$ for every $n \in \omega$. Now, for any distinct $n_1, \dots, n_m \in \omega$, we have $y \in Q = \overline{P_{n_1}} \cap \dots \cap \overline{P_{n_m}}$ and hence $Q \cap R \neq \emptyset$, i.e., $Q \not\subset D$ which shows that $P_{n_1} \cap \dots \cap P_{n_m} = Q \cap X \neq \emptyset$ because D contains all elements of $\mathcal{G}_{n_1}(k_{n_1}) \wedge \dots \wedge \mathcal{G}_{n_m}(k_{n_m})$ which do not meet X.

This proves that the family $\mathcal{P} = \{P_n : n \in \omega\}$ is centered and hence we can choose a filter $\mathcal{F} \supset \mathcal{P}$ (see Problem 117 of [TFS]). Since $P_n \in \mathcal{F} \cap \mathcal{G}_n$ and the sequence $\{\mathcal{G}_n : n \in \omega\}$ is complete, we have $P = \bigcap_{n \in \omega} \text{cl}_X(P_n) \neq \emptyset$ because $P \supset \bigcap \{\text{cl}_X(F) : F \in \mathcal{F}\} \neq \emptyset$. It follows from (4) that there is $x \in X$ such that $\{x\} = P$ and therefore $x \in \overline{P_n}$ for all $n \in \omega$. However, $y \in \overline{P_n}$ and hence $d(x, y) \leq \text{diam}(\overline{P_n}) = \text{diam}(P_n) = 2^{-n}$ for all $n \in \omega$ which implies $x = y$, a contradiction with $x \in X$ and $y \in Y \backslash X$. This contradiction shows that $H = X$, i.e., X is an $F_{\sigma\delta}$-subset of Y and proves that (iii)$\Longrightarrow$(i).

The implication (i)$\Longrightarrow$(ii) follows from the fact that any metrizable space X can be embedded into a completely metrizable space (see Problem 237 of [TFS]).

Now assume that there is a complete metric space (M, d) such that $X \subset M$ and $X = \bigcap_{n \in \omega} M_n$ where each M_n is an F_σ-subset of M.

Fix $n \in \omega$; we have $M_n = \bigcup_{k \in \omega} M_n^k$ where M_n^k is closed in M for all $k \in \omega$. It is easy to find, for every $k \in \omega$, a family $\mathcal{G}_n^k \subset \tau(M_n^k)$ such that $\bigcup \mathcal{G}_n^k = M_n^k$, the family $\mathcal{G}_n^k$ is σ-discrete in M_n^k and $\text{diam}(U) \leq 2^{-n}$ for any $U \in \mathcal{G}_n^k$. Now, if $\mathcal{G}_n = \bigcup_{n \in \omega} \mathcal{G}_n^k$ and $\mathcal{G}_n' = \{\overline{G} : G \in \mathcal{G}_n\}$ (the bar denotes the closure in M), then

(6) $\bigcup \mathcal{G}_n' \supset X$, the family $\mathcal{F}_n = \mathcal{G}_n' | X$ is σ-discrete in X and consists of closed subsets of X and for any $F \in \mathcal{F}_n$ we have $\text{diam}(F) \leq 2^{-n}$ and $\overline{F} \subset M_n$.

We claim that $\{\mathcal{F}_n : n \in \omega\}$ is a complete sequence in X. Indeed, if $\mathcal{F}$ is a filter on X and $\mathcal{F} \cap \mathcal{F}_n \neq \emptyset$, then fix $P_n \in \mathcal{F} \cap \mathcal{F}_n$ for every $n \in \omega$. It is clear that $\mathcal{G} = \{\overline{F} : F \in \mathcal{F}\}$ is a centered family of closed subsets of M; besides, if $\varepsilon > 0$ and $2^{-n} < \varepsilon$, then $G = \overline{P_n} \in \mathcal{G}$ while $\text{diam}(G) = \text{diam}(P_n) \leq 2^{-n} < \varepsilon$. We proved that $\mathcal{G}$ has elements of arbitrarily small diameter and hence $\bigcap \mathcal{G} \neq \emptyset$ (see Problem 236 of [TFS]). It is easy to see that there is $p \in M$ such that $\bigcap \mathcal{G} = \{p\}$. Since $p \in \overline{P_n} \subset M_n$ for all $n \in \omega$ [see (6)], we have $p \in \bigcap_{n \in \omega} M_n = X$ and therefore $p \in X$. Thus $\bigcap \{\text{cl}_X(F) : F \in \mathcal{F}\} = \bigcap \{\overline{F} \cap X : F \in \mathcal{F}\} = (\bigcap \mathcal{G}) \cap X = \{p\} \cap X = \{p\}$ which shows that $\bigcap \{\text{cl}_X(F) : F \in \mathcal{F}\} \neq \emptyset$ and hence $\{\mathcal{F}_n : n \in \omega\}$ is a complete sequence of closed σ-discrete covers of X. We settled the implication (ii)$\Longrightarrow$(iii) so our solution is complete.

T.374. *Prove that $C_p(X)$ is an absolute $F_{\sigma\delta}$ for any countable metrizable X.*

Solution. Let d be a metric on X such that $\tau(d) = \tau(X)$. We first prove that

(1) given $n, k \in \mathbb{N}$ and $x \in X$ the set $P_{nk}(x) = \{f \in \mathbb{R}^X : |f(x) - f(y)| \leq \frac{1}{n}$ whenever $d(y, x) < \frac{1}{k}\}$ is closed in $\mathbb{R}^X$.

To see that (1) holds, take any $f \in \mathbb{R}^X \setminus P_{nk}(x)$; there exists $y \in X$ with $d(y, x) < \frac{1}{k}$ while $|f(x) - f(y)| > \frac{1}{n}$. If e_x and e_y are the natural projections of $\mathbb{R}^X$ onto the factors determined by x and y respectively, then they are continuous and hence so is the map $\varphi : \mathbb{R}^X \to \mathbb{R}$ defined by $\varphi(g) = |e_x(g) - e_y(g)| = |g(x) - g(y)|$ for every $g \in \mathbb{R}^X$ (it is immediate that $\varphi = |e_x - e_y|$). Now it follows from continuity of φ and the equality $W_f = \{g \in \mathbb{R}^X : |g(x) - g(y)| > \frac{1}{n}\} = \varphi^{-1}(\mathbb{R} \setminus [-\frac{1}{n}, \frac{1}{n}])$ that W_f is an open neighborhood of f which does not intersect $P_{nk}(x)$. Thus $P_{nk}(x)$ is closed in $\mathbb{R}^X$ and (1) is proved.

It follows from (1) that $P_n^x = \bigcup_{k \in \omega} P_{nk}(x)$ is an F_σ-subset of $\mathbb{R}^X$ for every $n \in \mathbb{N}$ and $x \in X$. Let us show next that

(2) $C_p(X) = P = \bigcap \{P_n^x : n \in \mathbb{N}$ and $x \in X\}$.

Pick any $f \in C_p(X)$; if $x \in X$, then f is continuous at the point x, so, for any $n \in \mathbb{N}$, there exists $\delta > 0$ such that $d(x, y) < \delta$ implies $|f(x) - f(y)| < \frac{1}{n}$. Now, if $k \in \mathbb{N}$ and $\frac{1}{k} < \delta$, then $f \in P_{nk}(x) \subset P_n^x$; thus, for any $n \in \mathbb{N}$ and $x \in X$, we have $f \in P_n^x$, i.e., $f \in \bigcap \{P_n^x : n \in \mathbb{N}$ and $x \in X\} = P$. The function $f \in C_p(X)$ was chosen arbitrarily, so we proved that $C_p(X) \subset P$.

Now if $f \in P$, then take any $x \in X$ and $\varepsilon > 0$. There exists a number $n \in \mathbb{N}$ such that $\frac{1}{n} < \varepsilon$; since $f \in P_n^x$, there is $k \in \mathbb{N}$ such that $d(y, x) < \frac{1}{k}$ implies $|f(x) - f(y)| \leq \frac{1}{n} < \varepsilon$ which shows that f is continuous at x. The point x was chosen arbitrarily, so $f \in C_p(X)$ which establishes that $P \subset C_p(X)$ and therefore $P = C_p(X)$, i.e., (2) is proved.

It immediate from (1) and (2) that $C_p(X)$ is an $F_{\sigma\delta}$-set in the completely metrizable space $\mathbb{R}^X$. Therefore we can apply Problem 373 to conclude that $C_p(X)$ is an absolute $F_{\sigma\delta}$.

T.375. *Let K be a compact space. Given a countable $X \subset C_p(K)$, prove that $C_p(X)$ is an absolute $F_{\sigma\delta}$.*

Solution. For any $m, n, \in \mathbb{N}$ and $f \in X$ consider the set $P_{mn}(f) = \{\varphi \in \mathbb{R}^X : $ there exist $x_1, \ldots, x_n \in K$ for which $|\varphi(f) - \varphi(g)| \leq \frac{1}{m}$ for any $g \in X$ such that $|f(x_i) - g(x_i)| < \frac{1}{n}$ for all $i \leq n\}$. Let us first prove that

(1) the set $P_{mn}(f)$ is closed in $\mathbb{R}^X$ for any $f \in X$ and $m, n \in \mathbb{N}$.

We will need the set $Q_{mn}(f) = \{(\varphi, x) \in \mathbb{R}^X \times K^n : |\varphi(f) - \varphi(g)| \leq \frac{1}{m}$ whenever $x = (x_1, \ldots, x_n)$ and $|g(x_i) - f(x_i)| < \frac{1}{n}$ for all $i \leq n\}$. If $\pi : \mathbb{R}^X \times K^n \to \mathbb{R}^X$ is the natural projection, then $\pi(Q_{mn}(f)) = P_{mn}(f)$.

To see that the set $Q_{mn}(f)$ is closed in the space $\mathbb{R}^X \times K^n$, take any point $w = (\varphi, x) \in (\mathbb{R}^X \times K^n) \setminus Q_{mn}(f)$ where $x = (x_1, \ldots x_n)$; there exists $g \in X$ such that $|g(x_i) - f(x_i)| < \frac{1}{n}$ for all $i \leq n$ while $|\varphi(f) - \varphi(g)| > \frac{1}{m}$.

Since the functions f and g are continuous on the space K, the set $W = \{(y_1, \ldots, y_n) \in K^n : |g(y_i) - f(y_i)| < \frac{1}{n}$ for all $i \leq n\}$ is open in K^n; clearly, $x \in W$. It is an easy exercise to see that the set $V = \{\xi \in \mathbb{R}^X : |\xi(f) - \xi(g)| > \frac{1}{m}\}$ is open in $\mathbb{R}^X$ and contains φ. Therefore $U = V \times W$ is an open neighborhood of w in $\mathbb{R}^X \times K^n$ such that $U \cap Q_{mn}(f) = \emptyset$. This shows that $Q_{mn}(f)$ is closed in $\mathbb{R}^X \times K^n$; since K^n is compact, the projection π is a closed map (see Fact 3 of S.288) so $P_{mn}(f) = \pi(Q_{mn}(f))$ is a closed subset of $\mathbb{R}^X$ and (1) is proved.

Let $R_m(f) = \bigcup_{n \in \mathbb{N}} P_{mn}(f)$ for all $m \in \mathbb{N}$ and $f \in X$. Our last step is to show that

(2) $C_p(X) = P = \bigcap \{R_m(f) : m \in \mathbb{N}$ and $f \in X\}$.

Take any $\varphi \in C_p(X)$; for any $f \in X$, the map φ is continuous at the point f, so for any $m \in \mathbb{N}$ there exists $y_1, \ldots, y_k \in K$ and $\varepsilon > 0$ such that $|\varphi(g) - \varphi(f)| < \frac{1}{m}$ for any $g \in X$ with $|g(y_i) - f(y_i)| < \varepsilon$ for all $i \leq k$. It is easy to find $n \in \mathbb{N}$ such that $k \leq n$ and $\frac{1}{n} < \varepsilon$; if $x = (x_1, \ldots, x_n) \in K$ and $\{y_1, \ldots, y_k\} \subset \{x_1, \ldots, x_n\}$, then $|\varphi(f) - \varphi(g)| < \frac{1}{m}$ for any $g \in X$ such that $|g(x_i) - f(x_i)| < \frac{1}{n}$ for all $i \leq n$. Consequently, $f \in P_{mn}(f) \subset R_m(f)$; we have chosen $f \in X$ and $m \in \mathbb{N}$ arbitrarily so $f \in P$ and hence $C_p(X) \subset P$.

Now, if $\varphi \in P$, take any $f \in X$ and $\varepsilon > 0$; there is $m \in \mathbb{N}$ such that $\frac{1}{m} < \varepsilon$. Since $\varphi \in R_m(f)$, there is $n \in \mathbb{N}$ such that $\varphi \in P_{mn}(f)$ and hence there exists $(x_1, \ldots, x_n) \in K^n$ such that $|\varphi(f) - \varphi(g)| \leq \frac{1}{m} < \varepsilon$ for any point $g \in X$ such that $|g(x_i) - f(x_i)| < \frac{1}{n}$ for all $i \leq n$. If $O = \{g \in X : |g(x_i) - f(x_i)| < \frac{1}{n}$ for all $i \leq n\}$, then $O \in \tau(f, X)$ and $\varphi(O) \subset (\varphi(f) - \varepsilon, \varphi(f) + \varepsilon)$, i.e., O witnesses continuity of φ at the point f. Thus φ is continuous at every $f \in X$ and therefore $\varphi \in C_p(X)$. This shows that $P \subset C_p(X)$ and hence $P = C_p(X)$.

Finally, observe that each $R_m(f)$ is an F_σ-subset of $\mathbb{R}^X$ by (1) so $C_p(X)$ is an $F_{\sigma\delta}$-subset of a completely metrizable space $\mathbb{R}^X$ by (2). Thus $C_p(X)$ is an absolute $F_{\sigma\delta}$ by Problem 373.

T.376. *Prove that any second countable space embeds into $C_p(\mathbb{K})$, where $\mathbb{K}$ is the Cantor set.*

Solution. It suffices to embed $\mathbb{I}^\omega$ in $C_p(\mathbb{K})$ by Problem 209 of [TFS]. Let $a = 0$ and $a_n = \frac{1}{n}$ for all $n \in \mathbb{N}$. Then $S = \{a\} \cup \{a_n : n \in \mathbb{N}\}$ is a convergent sequence and hence a compact metrizable space. There exists a continuous onto map $r : \mathbb{K} \to S$ by Problem 128 of [TFS]; the dual map $r^* : C_p(S) \to C_p(\mathbb{K})$ embeds $C_p(S)$ in $C_p(\mathbb{K})$ (see Problem 163 of [TFS]), so it suffices to embed $\mathbb{I}^\omega$ in $C_p(S)$.

Let $F = \{f \in \mathbb{I}^S : f(a) = 0$ and $f(a_n) \in [0, \frac{1}{n}]$ for all $n \in \mathbb{N}\}$. It is immediate that F is homeomorphic to the product $\{0\} \times \prod\{[0, \frac{1}{n}] : n \in \mathbb{N}\}$ which is a compact subspace of $\mathbb{I}^S$. Besides, $F \subset C_p(S)$ because $f(a_n) \to 0$ when $n \to \infty$ for every $f \in F$. Another easy observation is that F is homeomorphic to $\prod\{[0, \frac{1}{n}] : n \in \mathbb{N}\}$ which in its turn is homeomorphic to $\mathbb{I}^\omega$ because $[0, \frac{1}{n}]$ is homeomorphic to $\mathbb{I}$ for all $n \in \mathbb{N}$. Thus F is a subspace of $C_p(S)$ homeomorphic to $\mathbb{I}^\omega$, so $\mathbb{I}^\omega$ also embeds in $C_p(\mathbb{K})$ and hence any second countable space embeds in $C_p(\mathbb{K})$.

T.377. *Give an example of a second countable X such that for any compact K, the space X cannot be embedded in $C_p(K)$ as a closed subspace.*

Solution. The existence of our X can be easily deduced from the following fact.

Fact 1. If K is a compact space and M is a second countable closed subspace of $C_p(K)$, then M is a $K_{\sigma\delta}$-space.

Proof. Let $\varphi(x)(f) = f(x)$ for any $x \in K$ and $f \in M$. Then $\varphi(x) \in C_p(M)$ for any $x \in X$ and the map $\varphi : K \to C_p(M)$ is continuous (see Problem 166 of [TFS]). If $L = \varphi(K)$, then L is compact and $w(L) = nw(L) \leq nw(C_p(M)) = nw(M) \leq w(M) = \omega$ which shows that L is a metrizable compact space.

The dual mapping $\varphi^* : C_p(L) \to C_p(K)$ defined by $\varphi^*(g) = g \circ \varphi$ for every $g \in C_p(L)$ is an embedding by Problem 163 of [TFS] and it is immediate that $M \subset \varphi^*(C_p(L))$. Since M is closed in a larger space $C_p(K)$, it is also closed in $\varphi^*(C_p(L))$. This proves that M embeds as a closed subspace in $C_p(L)$. Since $C_p(L)$ is a $K_{\sigma\delta}$-space by Problem 362, so is M by V1.338 and Fact 1 is proved. $\square$

Returning to our solution, take a second countable space X which is not $K_{\sigma\delta}$ (such a space exists by Problems 334, 342 and 346) and observe that X cannot be homeomorphic to a closed subspace of $C_p(K)$ for any compact K by Fact 1.

T.378. *Prove that any countable second countable space embeds into $C_p(\mathbb{K})$ as a closed subspace.*

Solution. Let $a = 0$ and $a_n = \frac{1}{n}$ for all $n \in \mathbb{N}$. Then $S = \{a\} \cup \{a_n : n \in \mathbb{N}\}$ is a convergent sequence and hence it is a compact countable (and hence metrizable) space. There exists a continuous onto map $r : \mathbb{K} \to S$ by Problem 128 of [TFS]; the dual map $r^* : C_p(S) \to C_p(\mathbb{K})$ embeds $C_p(S)$ in $C_p(\mathbb{K})$ as a closed subspace (see Problem 163 of [TFS]). The space $C_p(S)$ is dense-in-itself, second countable and of first category in itself (see Problem 278 and 284 of [TFS]), so we can apply Fact 4 of T.351 to conclude that there is a closed $F \subset C_p(S)$ which is homeomorphic to $\mathbb{Q}$. Since $C_p(S)$ embeds in $C_p(\mathbb{K})$ as a closed subspace, there is a closed $G \subset C_p(\mathbb{K})$ which is homeomorphic to $\mathbb{Q}$.

Now, if M is a countable metrizable space, then there is a closed $N \subset G$ which is homeomorphic to M (see Problem 350). It is evident that N is also a closed subspace of $C_p(\mathbb{K})$, so M embeds in $C_p(\mathbb{K})$ as a closed subspace.

T.379. *Given a space X and a function $f : X \to \mathbb{R}$, consider the following conditions:*

(i) for any open $U \subset \mathbb{R}$, the set $f^{-1}(U)$ is an F_σ-set in X;
(ii) there exists a sequence $\{f_n : n \in \omega\} \subset C_p(X)$ which converges to f.

Prove that (ii)$\Longrightarrow$(i) for any space X. Show that if X is second countable, then also (i)$\Longrightarrow$(ii) and hence (i) $\Longleftrightarrow$ (ii).

Solution. Given a space Z let $B^*(Z) = \{f \in \mathbb{R}^Z : f \text{ is bounded on } Z\}$. It is easy to see that $B^*(Z) = C^*(Z')$ where Z' is the set Z with the discrete topology. For a second countable space Z and $A \subset Z$ the statements "A is an F_σ-subset of Z" and "$A \in \Sigma_1^0(Z)$" are identical, so we will use them alternatively to avoid repetitions and shorten our expressions.

Fact 1. Let Z be a second countable space; given an $n \in \omega$ assume that $A_i \in \Sigma_1^0(Z)$ for all $i < n$. Then there exists a disjoint family $\{B_i : i < n\} \subset \Sigma_1^0(Z)$ such that $B_i \subset A_i$ for all $i < n$ and $\bigcup_{i<n} A_i = \bigcup_{i<n} B_i$.

Proof. It is an easy exercise that

($*$)   the intersection of two F_σ-sets of Z is an F_σ-set of Z and if F and G are closed in Z then $F \backslash G$ is an F_σ-subset of Z.

In this proof we will use ($*$) many times without explicitly referring to it. Our fact is evident for $n \in \{0, 1\}$; to prove it for $n = 2$, take any sets $P, Q \in \Sigma_1^0(Z)$. There exist families $\mathcal{P} = \{P(i) : i \in \omega\}$, $\mathcal{Q} = \{Q(i) : i \in \omega\}$ of closed subsets of Z such that $\bigcup \mathcal{P} = P$, $\bigcup \mathcal{Q} = Q$ and $P(i) \subset P(i+1)$, $Q(i) \subset Q(i+1)$ for all $i \in \omega$. Let $P_0 = P(0)$, $Q_0 = Q(0)$ and $P_{i+1} = P(i+1)\backslash P(i)$, $Q_{i+1} = Q(i+1)\backslash Q(i)$ for all $i \in \omega$. Then $\{P_i : i \in \omega\} \cup \{Q_i : i \in \omega\} \subset \Sigma_1^0(Z)$.

Furthermore, the sets $P_i' = P(i)\backslash Q(i)$ and $Q_i' = Q(i)\backslash P(i)$ are F_σ in Z for all $i \in \omega$; it is straightforward that $P' = \bigcup_{i \in \omega} P_i'$ and $Q' = \bigcup_{i \in \omega} Q_i'$ are disjoint and F_σ while $P\backslash Q \subset P'$ and $Q\backslash P \subset Q'$. Now, the set $R = (P \cup Q)\backslash(P' \cup Q')$ is also F_σ because $R = \bigcup_{i \in \omega} R_i$ where $R_i = P_i \cap Q_i$ for all $i \in \omega$. As a consequence, $P'' = P' \cup R$ and Q' are disjoint F_σ-subsets of Z for which $P \subset P''$, $Q \subset Q'$ and $P'' \cup Q' = P \cup Q$. This, evidently, settles our fact for $n = 2$.

Now assume that $n = k + 1 > 2$ and we have proved our fact for all numbers $n \leq k$. Given a family $\{A_i : i < k + 1\}$ of F_σ-sets of Z, we can apply the induction hypothesis to find disjoint sets $B_0', \ldots, B_{k-1}' \in \Sigma_1^0(Z)$ such that $B_i' \subset A_i$ for all $i < k$ and $\bigcup_{i<k} B_i' = \bigcup_{i<k} A_i$. Since we proved our fact for $n = 2$, we can apply it to the F_σ-sets $P = \bigcup_{i<k} A_i$ and $Q = A_k$ to obtain disjoint F_σ-sets $P' \subset P$, $Q' \subset Q$ such that $P \cup Q = P' \cup Q'$. If we let $B_k = Q'$ and $B_i = B_i' \cap P'$ for all $i < k$, then it is immediate that we obtain the desired sets $B_0, \ldots, B_k$; this verifies our fact for $n = k + 1$ and shows that our inductive procedure guarantees that it is fulfilled for all $n \in \omega$. Thus Fact 1 is proved. $\square$

Fact 2. For any space Z the set $B_1(Z)$ is closed in the uniform topology on $\mathbb{R}^Z$. In other words, if $f_n \in B_1(Z)$ for all $n \in \omega$ and $f_n \rightrightarrows f$, then $f \in B_1(Z)$.

Proof. For any $g \in B^*(Z)$ let $||g|| = \sup\{g(z) : z \in Z\}$. If Z' is the set Z with the discrete topology, then $B^*(Z) = C^*(Z')$ and the function $\rho(g, h) = ||g - h||$ is a metric on $B^*(Z)$ (see Problem 248 of [TFS]). If $\mathbf{0} \in \mathbb{R}^Z$ is identically zero on Z, then $||g|| = \rho(g, \mathbf{0})$ for any $g \in B^*(Z)$. As a consequence,

(1) $||u + v|| \leq ||u|| + ||v||$ for any $u, v \in B^*(Z)$.

It follows from $f_n \rightrightarrows f$ that $||f_n - f|| \to 0$, so choosing an appropriate subsequence of the sequence $\{f_n\}_{n\in\omega}$, we can assume, without loss of generality, that $||f_n - f|| \leq 2^{-n-2}$ for all $n \in \omega$, i.e.,

(2) $|f_n(x) - f(x)| \leq 2^{-n-2}$ for any $n \in \omega$ and $x \in Z$.

Let $p_0 = f_0$ and $p_{n+1} = f_{n+1} - f_n$ for all $n \in \omega$. Then $p_n \in B_1(Z)$ and $||p_{n+1}|| \leq ||f_{n+1} - f|| + ||f - f_n|| \leq 2^{-n-1}$ for all $n \in \omega$ (we applied the property (1) here). Let $u_n(x) = 2^{-n}$ and $v_n(x) = -2^{-n}$ for all $n \in \mathbb{N}$ and $x \in Z$; then

(3) $v_n(x) \leq p_n(x) \leq u_n(x)$ for all $n \in \mathbb{N}$ and $x \in Z$.

For every $n \in \omega$ there is a sequence $\{r_n^m : m \in \mathbb{N}\} \subset C(Z)$ such that $r_n^m \to p_n$ as $m \to \infty$. It follows from (2) that if $q_n^m = \max(r_n^m, v_n)$ and $p_n^m = \min(q_n^m, u_n)$ for all $m \in \mathbb{N}$, then

(4) for all $n \in \omega$ and $m \in \mathbb{N}$, we have $p_n^m \in C(Z)$ and $||p_n^m|| \leq 2^{-n}$; besides, $p_n^m \to p_n$ when $m \to \infty$ for all $n \in \omega$.

Now let $s_n^m = \sum_{i=0}^n p_i^m$ for all $n, m \in \mathbb{N}$. Apply (3) and Problem 030 of [TFS] to conclude that for any $m \in \mathbb{N}$, there is a function s_m such that $s_n^m \rightrightarrows s_m$ (when $n \to \infty$). Therefore $s_m \in C(Z)$ for every $m \in \mathbb{N}$ by Problem 029 of [TFS]; let us prove that $s_m \to f$.

Observe first that for any $x \in Z$ and $n \in \mathbb{N}$, we have the equality $|s_m(x) - s_n^m(x)| = \lim_{k \to \infty} |s_k^m(x) - s_n^m(x)|$. Now,

$$||s_k^m(x) - s_n^m(x)|| = || \sum_{i=n+1}^k p_i^m|| \leq \sum_{i=n+1}^k ||p_i^m|| \leq \sum_{i=n+1}^\infty 2^{-i} = 2^{-n}$$

for any $x \in Z$ and $k \geq n$ which shows that

(5) $|s_m(x) - s_n^m(x)| \leq 2^{-n}$ for any $n, m \in \mathbb{N}$ and $x \in Z$.

Fix a point $z \in Z$ and an arbitrary $\varepsilon > 0$; there is $n \in \mathbb{N}$ such that $2^{-n} < \frac{\varepsilon}{3}$ and hence $|s_m(z) - s_n^m(z)| \leq 2^{-n} < \frac{\varepsilon}{3}$ for all $m \in \mathbb{N}$. Observe that $\sum_{i=0}^n p_i = f_n$, so it follows from $p_i^m \to p_i$ (for each $i \leq n$ as $m \to \infty$) that there is $k \in \mathbb{N}$ such that $|s_n^m(z) - f_n(z)| = |\sum_{i=0}^n p_i^m(z) - f_n(z)| < \frac{\varepsilon}{3}$ for all $m \geq k$. Therefore

$$|s_m(z) - f(z)| \leq |s_m(z) - s_n^m(z)| + |s_n^m(z) - f_n(z)| + |f_n(z) - f(z)| \leq 2^{-n} + \frac{\varepsilon}{3} + 2^{-n} = \varepsilon$$

for any $m \geq k$ [we used the properties (2) and (5)] which proves that the sequence $\{s_m(z) : m \in \mathbb{N}\} \subset C_p(Z)$ converges to $f(z)$ for every $z \in Z$. Now apply Problem 143 of [TFS] to conclude that $s_m \to f$ and hence $f \in B_1(Z)$. Fact 2 is proved. $\square$

Returning to our solution assume that $w(X) = \omega$ and $f^{-1}(U)$ is an F_σ-set in X for any open $U \subset \mathbb{R}$. Let $\varphi : \mathbb{R} \to (0, 1)$ be a homeomorphism; we will also need the maps $u_n, v_n \in C_p(X)$ defined by $u_n(x) = \frac{1}{n+1}$ and $v_n(x) = 1 - \frac{1}{n+1}$ for all $x \in X$ and $n \in \mathbb{N}$. If we prove that $g = \varphi \circ f \in B_1(X)$, then there is $\{p_n : n \in \omega\} \subset C_p(X)$ with $p_n \to g$. Letting $q_n = \max(p_n, u_n)$ and $r_n = \min(q_n, v_n)$ for all $n \in \mathbb{N}$, we obtain a sequence $\{r_n : n \in \mathbb{N}\} \subset C_p(X, (0, 1))$ with $r_n \to g$. It is straightforward that if $f_n = \varphi^{-1} \circ r_n$ for each $n \in \mathbb{N}$, then $\{f_n : n \in \mathbb{N}\} \subset C_p(X)$ and $f_n \to f$.

Therefore we can assume, without loss of generality, that $f(X) \subset (0, 1)$. Call a finite family $\mathcal{A} \subset \exp(X)$ *a partition* of X if $\mathcal{A}$ is disjoint and $\bigcup \mathcal{A} = X$. Our next step is to prove the following property.

(6) If $\mathcal{A}$ is a partition of X and each $A \in \mathcal{A}$ is an F_σ-subset of X, then every $g \in \mathbb{R}^X$ which is constant on all elements of $\mathcal{A}$, belongs to $B_1(X)$.

To see that (6) is true let $\mathcal{A} = \{A_1, \ldots, A_n\}$ and assume that $r_1, \ldots, r_n \in \mathbb{R}$ are chosen so that for every $i \leq n$ we have $g(x) = r_i$ for any $x \in A_i$. There is a family $\mathcal{A}_i = \{A_i^j : j \in \omega\}$ of closed subsets of X such that $\bigcup \mathcal{A}_i = A_i$ and $A_i^j \subset A_i^{j+1}$ for all $j \in \omega$ and $i \leq n$.

For any $j \in \omega$, the family $\mathcal{B}_j = \{A_i^j : i \leq n\}$ is finite, disjoint and consists of closed subsets of X, so by normality of X, there is a continuous function $g_j : X \to \mathbb{R}$ such that $g_j | (\bigcup \mathcal{B}_j) = g$. Given $x \in X$ it is immediate that $g_j(x) = g(x)$ for all but finitely many $j \in \omega$ and hence the sequence $\{g_j : j \in \omega\} \subset C_p(X)$ converges to g. This settles (6).

For any $n \in \mathbb{N}$ it is easy to choose an open finite cover $\{U_1, \ldots, U_k\}$ of the space $(0, 1)$ in such a way that $\mathrm{diam}(U_i) \leq \frac{1}{n}$ for all $i \leq k$. The set $V_i = f^{-1}(U_i)$ is an F_σ-set in X for every $i \leq k$ and $V_1 \cup \cdots \cup V_k = X$. By Fact 1 there is a partition $\mathcal{P} = \{P_1, \ldots, P_k\}$ of the space X such that P_i is an F_σ-subset of X and $P_i \subset U_i$ for all $i \leq n$. Choose $r_i \in U_i$ for all $i \leq k$ and given $x \in X$ let $f_n(x) = r_i$ if $x \in P_i$. Since $\mathcal{P}$ is a disjoint family, this consistently defines a function $f_n : X \to \mathbb{R}$ which is constant on all elements of $\mathcal{P}$. Therefore $f_n \in B_1(X)$ by (6); furthermore, if $x \in X$, then $x \in P_i$ for some $i \leq n$ and hence $f(x) \in U_i$ which shows that $|f(x) - f_n(x)| = |f(x) - r_i| \leq \mathrm{diam}(U_i) \leq \frac{1}{n}$. Thus $|f_n(x) - f(x)| \leq \frac{1}{n}$ for every $x \in X$ and $n \in \mathbb{N}$ which implies that the sequence $\{f_n : n \in \mathbb{N}\}$ converges uniformly to f. Now apply Fact 2 to conclude that $f \in B_1(X)$, i.e., we established that (i)$\Longrightarrow$(ii) for any second countable space X.

To prove the remaining implication assume that $f_n \in C_p(X)$ for all $n \in \omega$ and the sequence $\{f_n : n \in \omega\}$ converges to f. Given an open set $U \subset \mathbb{R}$, choose a family $\mathcal{V} = \{V_n : n \in \omega\} \subset \tau(\mathbb{R})$ such that $\overline{V}_n \subset V_{n+1}$ for all $n \in \omega$ and $\bigcup \mathcal{V} = U$. It turns out that

(7) $f^{-1}(U) = H = \bigcup_{n \in \omega} \bigcap_{k \geq n} f_k^{-1}(\overline{V}_n)$.

Indeed, if $x \in H$, then there is $n \in \omega$ such that $f_k(x) \in \overline{V}_n$ for all $k \geq n$. Since the sequence $\{f_k(x) : k \geq n\}$ converges to $f(x)$, we have $f(x) \in \overline{V}_n \subset U$, i.e., $f(x) \in U$ which shows that $x \in f^{-1}(U)$ and therefore $H \subset f^{-1}(U)$.

On the other hand, if $x \in f^{-1}(U)$, then $f(x) \in U$ and hence $f(x) \in V_m$ for some $m \in \omega$. The sequence $\{f_k(x) : k \in \omega\}$ converges to $f(x)$, so there is $n \in \omega$ such that $n \geq m$ and $f_k(x) \in V_m$ for all $k \geq n$. We have $V_m \subset V_n$, so $f_k(x) \in V_n \subset \overline{V}_n$ for all $k \geq n$ and therefore $x \in H$. Thus $f^{-1}(U) \subset H$, i.e., $f^{-1}(U) = H$ so (7) is proved.

Finally, apply (7) to conclude that $f^{-1}(U)$ is an F_σ-subset of X for any open $U \subset \mathbb{R}$; this proves the implication (ii)$\Longrightarrow$(i) for an arbitrary space X and makes our solution complete.

T.380. *Prove that if* $X = \mathbb{R}$, *then* $B_1(X) \neq \mathbb{R}^X$.

Solution. Let $f(x) = 0$ if $x \in \mathbb{Q}$ and $f(x) = 1$ for all $x \in \mathbb{P}$. Then $f : \mathbb{R} \to \mathbb{R}$; the sets $U = (-\frac{1}{2}, \frac{1}{2})$ and $V = (\frac{1}{2}, \frac{3}{2})$ are open in $\mathbb{R}$ while $\mathbb{Q} = f^{-1}(U)$ and $\mathbb{P} = f^{-1}(V)$. Since $\mathbb{Q}$ is not Čech-complete (see Problem 274 of [TFS]), it is not a G_δ-subset of $\mathbb{R}$ (see Problem 260 of [TFS]). As a consequence $\mathbb{P} = f^{-1}(V)$ is not an F_σ-subset of $\mathbb{R}$ which shows that $f \notin B_1(\mathbb{R})$ (see Problem 379), i.e., $f \in \mathbb{R}^X \backslash B_1(X)$.

T.381. *Prove that a compact space* X *is countable if and only if* $B_1(X) = \mathbb{R}^X$.

Solution. If X is a compact countable space, then X is second countable by Fact 4 of S.307. If $f \in \mathbb{R}^X$, then $f^{-1}(U)$ is an F_σ-subset of X (because all subsets of X are F_σ in X) for any $U \subset \mathbb{R}$ and hence $f \in B_1(X)$ by Problem 379. This proves necessity.

Now assume that X is a compact space such that $B_1(X) = \mathbb{R}^X$. Given any $A \subset X$, let f be the characteristic function of A in X, i.e., $f(x) = 1$ if $x \in A$ and $f(x) = 0$ for all $x \in X \backslash A$. Then $A = f^{-1}((-\frac{1}{2}, \frac{1}{2}))$; since $f \in B_1(X)$ by our assumption, the set A is F_σ in X by Problem 379. The space X being compact we proved that every $A \subset X$ is σ-compact, so we can apply Problem 274 to conclude that X is countable. This proves sufficiency and completes our solution.

T.382. *Prove that, under* MA+¬CH, *there exists an uncountable* $X \subset \mathbb{R}$ *such that* $B_1(X) = \mathbb{R}^X$.

Solution. Assume MA+¬CH and let X be a subset of $\mathbb{R}$ of cardinality $\omega_1 < \mathfrak{c}$. It follows from MA+¬CH that every $A \subset X$ is a G_δ-subset of X (see Problem 055). Thus every subset of X is F_σ being a complement of a G_δ-subset of X. As a consequence, if $f \in \mathbb{R}^X$, then $f^{-1}(U)$ is an F_σ-subset of X for any $U \subset \mathbb{R}$. This shows that $f \in B_1(X)$ (see Problem 379) and therefore $B_1(X) = \mathbb{R}^X$.

T.383. *Prove that the two arrows space is Rosenthal compact.*

Solution. For any $t \in (0, 1]$ let $a_t^0(x) = 0$ if $0 \leq x < t$ and $a_t^0(x) = 1$ for all $x \in [t, 1]$; this defines a function $a_t^0 : [0, 1] \to \mathbb{R}$. Now, if $t \in [0, 1)$, then consider the function $a_t^1 : [0, 1] \to \mathbb{R}$ defined by $a_t^1(x) = 0$ for all $x \in [0, t]$ and $a_t^1(x) = 1$ if $x \in (t, 1]$. If $A_0 = \{a_t^0 : t \in (0, 1]\}$ and $A_1 = \{a_t^1 : t \in [0, 1)\}$, then $A_0 \cup A_1 \subset B_1([0, 1])$ because $f^{-1}(U)$ is an F_σ-subset of $[0, 1]$ for any (not necessarily open) $U \subset \mathbb{R}$ and $f \in D = A_0 \cup A_1$ (see Problem 379).

Recall that the underlying set T of the two arrows space (T, τ) is defined by $T = ((0, 1] \times \{0\}) \cup ([0, 1) \times \{1\}) \subset \mathbb{R}^2$ while the topology τ is generated on T by families $\{\mathcal{B}_z : z \in T\}$ as local bases where the collection $\mathcal{B} = \{\mathcal{B}_z : z \in T\}$ is defined as follows. Given $z = (t, 0) \in T$, let $\mathcal{B}_z = \{((\varepsilon, t] \times \{0\}) \cup ((\varepsilon, t) \times \{1\}) : 0 < \varepsilon < t\}$; if $z = (t, 1) \in T$, then $\mathcal{B}_z = \{([t, \varepsilon) \times \{1\}) \cup ((t, \varepsilon) \times \{0\}) : t < \varepsilon < 1\}$. Observe that for any $x \in (0, 1)$, the sets $O_x^0 = \{(t, i) \in T : i \in \{0, 1\}$ and $t < x\} \cup \{(x, 0)\}$ and $O_x^1 = \{(t, i) \in T : i \in \{0, 1\}$ and $t > x\} \cup \{(x, 1)\}$ are disjoint, clopen in T and $T = O_x^0 \cup O_x^1$. For technical reasons it is convenient to consider also the sets $O_0^0 = \emptyset$, $O_0^1 = T$ and $O_1^0 = T$, $O_1^1 = \emptyset$.

Let $\varphi((t, 0)) = a_t^0$ for all $t \in (0, 1]$ and $\varphi((t, 1)) = a_t^1$ for all $t \in [0, 1)$. It is evident that the map $\varphi : T \to D$ is a bijection. Besides, φ maps T into the product space $\mathbb{R}^{[0,1]}$; for any $x \in [0, 1]$, let $\pi_x : \mathbb{R}^{[0,1]} \to \mathbb{R}$ be the natural projection onto the factor determined by x. If $z = (t, i)$, then $(\pi_x \circ \varphi)(z) = a_t^i(x) = 1$ if $t < x$ and $(\pi_x \circ \varphi)(z) = a_t^i(x) = 0$ if $t > x$.

Since also $(\pi_x \circ \varphi)((x, 0)) = 1$ and $(\pi_x \circ \varphi)((x, 1)) = 0$ for any $x \in (0, 1)$, we have the equalities $(\pi_x \circ \varphi)^{-1}(0) = O_x^1$ and $(\pi_x \circ \varphi)^{-1}(1) = O_x^0$ for all $x \in [0, 1]$ which show that the sets $(\pi_x \circ \varphi)^{-1}(0)$ and $(\pi_x \circ \varphi)^{-1}(1)$ are clopen in T. This implies that $\pi_x \circ \varphi$ is continuous for every $x \in [0, 1]$ and therefore the map φ is continuous by Problem 102 of [TFS]; being a bijection, it is a homeomorphism because T is compact by Problem 384 of [TFS]. Thus T is homeomorphic to the subspace D of $B_1([0, 1])$, i.e., the two arrows space T is Rosenthal compact.

T.384. *Prove that every Rosenthal compact space is Fréchet–Urysohn.*

Solution. If Z is a space and $A \subset Z$, then the expressions "A is an F_σ-subset of Z" and "$A \in \Sigma_1^0(Z)$" say the same thing, so we will use them interchangeably to avoid repetitions and save the space. Given $f : Z \to \mathbb{R}$ let $\mathrm{osc}(f, U) = \mathrm{diam}(f(U))$ for any $U \subset Z$. If $z \in Z$, then $\mathrm{osc}(f, z) = \inf\{\mathrm{osc}(f, U) : U \in \tau(z, Z)\}$. It is easy to see that f is continuous at a point z if and only if $\mathrm{osc}(f, z) = 0$. If we have infinite sets A and B, say that $A \subset^* B$ if $A \backslash B$ is finite. If $S = \{z_n : n \in \omega\} \subset Z$, then $z \in Z$ is *accumulation point of the sequence* S if the set $\{n \in \omega : z_n \in U\}$ is infinite for any $U \in \tau(z, Z)$. Observe that if $z \notin S$, then z is an accumulation point for the set S if and only if it is an accumulation point for S considered as a sequence (the difference is that S can be finite as a set, so it has no accumulation points while it has accumulation points as a sequence).

Fact 1. If Z is a set and we have a sequence $\{A_i : i \in \omega\} \subset \exp(Z)$ of infinite subsets of Z such that $A_{i+1} \subset^* A_i$ for all $i \in \omega$, then there exists an infinite $A \subset Z$ such that $A \subset^* A_i$ for all $i \in \omega$.

Proof. Let $B_0 = A_0$ and suppose that we have infinite sets $B_0 \supset \cdots \supset B_k$ such that $B_i \subset A_i$ and $A_i \backslash B_i$ is finite for all $i \leq k$. Since $C = A_{k+1} \backslash B_k \subset (A_{k+1} \backslash A_k) \cup (A_k \backslash B_k)$ is finite, for the set $B_{k+1} = A_{k+1} \backslash C$ our assumption still holds. Thus we can construct a sequence $\{B_i : i \in \omega\}$ of infinite subsets of Z such that $B_{i+1} \subset B_i$ and $B_i \subset A_i$ for all $i \in \omega$. Take $a_0 \in B_0$ arbitrarily and, if we have distinct points $a_0, \ldots, a_k$ such that $a_i \in B_i$ for each $i \leq k$, then we can choose a point $a_{k+1} \in B_{k+1} \backslash \{a_0, \ldots, a_k\}$ which shows that we can construct a set $A = \{a_i : i \in \omega\}$ such that $a_i \neq a_j$ if $i \neq j$ and $a_i \in B_i$ for all $i \in \omega$. It is straightforward that the set A is as promised so Fact 1 is proved. $\square$

Fact 2. Suppose that we have sequences $\{A_i : i \in \omega\}$ and $\{B_i : i \in \omega\}$ of infinite subsets of ω such that $A_{i+1} \subset^* A_i$ for all $i \in \omega$ and $B_j \subset^* A_i$ for all $i, j \in \omega$. Then there is an infinite $A \subset \omega$ such that $B_j \subset^* A \subset^* A_i$ for all $i, j \in \omega$.

Proof. It is easy to see that for any infinite sets $P_1, \ldots, P_n, Q \subset \omega$, if $P_i \subset^* Q$ for all $i \leq n$, then $P_1 \cup \cdots \cup P_n \subset^* Q$. Furthermore, if $Q \subset^* P_i$ for all $i \leq n$, then $Q \subset^* \bigcap \{P_i : i \leq n\}$. These observations prove that if $D_i = \bigcap_{k \leq i} A_k$ and $E_i = \bigcup_{k \leq i} B_i$ for all $i \in \omega$, then $E_i \subset^* D_j$ for all $i, j \in \omega$. It is easy to choose a sequence $\{m_i : i \in \omega\} \subset \omega$ such that $m_i < m_{i+1}$ and $E_i \subset D_i \cap (m_i, +\infty)$ for all $i \in \omega$. Finally, observe that the set $A = \bigcup \{E_i \cap (m_i, m_{i+1}] : i \in \omega\}$ is as promised so Fact 2 is proved. $\qquad \square$

Fact 3. Given a Polish space M, a function $f : M \to \mathbb{R}$ belongs to $B_1(M)$ if and only if for any closed $F \subset M$, the function $f|F$ has a point of continuity in the space F.

Proof. Assume first that $f \in B_1(M)$ and fix a sequence $\{f_n : n \in \omega\} \subset C_p(M)$ with $f_n \to f$. Let $g = f|F$ and $g_n = f_n|F$ for every $n \in \omega$. Since the restriction map is continuous, we have $g_n \to g$ and hence $g \in B_1(F)$. This shows that, to prove necessity, we can assume, without loss of generality, that $F = M$, i.e., it suffices to show that every $f \in B_1(M)$ is continuous at some point of M.

Let $\mathcal{B} = \{O_n : n \in \omega\}$ be a some base in $\mathbb{R}$. The set $f^{-1}(O_n)$ is F_σ in M for each $n \in \omega$ (see Problem 379) so fix a family $\{P_{mn} : m \in \omega\}$ of closed subsets of M such that $f^{-1}(O_n) = \bigcup \{P_{mn} : m \in \omega\}$. The set $E_{mn} = P_{mn} \backslash \text{Int}(P_{mn})$ is nowhere dense in M for all $m, n \in \omega$; since M is complete, we can choose a point $x \in M \backslash (\bigcup \{E_{mn} : m, n \in \omega\})$.

To see that f is continuous at x take any $\varepsilon > 0$; there is $n \in \omega$ such that $f(x) \in O_n \subset (f(x) - \varepsilon, f(x) + \varepsilon)$. Since $x \in f^{-1}(O_n)$, there is $m \in \omega$ such that $x \in P_{mn}$ and hence $x \in \text{Int}(P_{mn})$. Now, the open set $V = \text{Int}(P_{mn})$ witnesses continuity of f at the point x because $x \in V$ and $f(V) \subset O_n \subset (f(x) - \varepsilon, f(x) + \varepsilon)$. This proves necessity.

To prove sufficiency, assume that $f \in \mathbb{R}^M$ and, for every closed non-empty $F \subset M$, the function $f|F$ has a point of continuity in the space F. It suffices to show that $f^{-1}(O)$ is an F_σ-subset of M for any $O \in \tau(\mathbb{R})$ (see Problem 379). To do this we will prove an auxiliary statement. Say that sets $A, B \subset M$ are F_σ-separated if there are disjoint $A', B' \subset M$ such that $A \subset A'$, $B \subset B'$ and $A', B' \in \Sigma_1^0(M)$. Then

(1) if $A, B \subset M$ are not F_σ-separated, then there is a non-empty closed $F \subset M$ such that $\overline{A \cap F} = \overline{B \cap F} = F$.

Let $F_0 = \overline{A} \cap \overline{B}$; then $A \backslash F_0 \subset A_0' = \overline{A} \backslash F_0$, $B \backslash F_0 \subset B_0' = \overline{B} \backslash F_0$, the sets A_0', B_0' are disjoint and we have $A_0' \cup B_0' \subset M \backslash F_0$ while $A_0', B_0' \in \Sigma_1^0(M)$. Now assume that $0 < \alpha < \omega_1$ and we have constructed a family $\{F_\beta : \beta < \alpha\}$ of closed subsets of M and a family $\{A_\beta', B_\beta' : \beta < \alpha\} \subset \Sigma_1^0(M)$ with the following properties:

(2) if $0 \leq \gamma < \beta < \alpha$, then $F_\beta \subset F_\gamma$, $A_\gamma' \subset A_\beta'$ and $B_\gamma' \subset B_\beta'$;
(3) for every ordinal $\beta < \alpha$, we have $A_\beta' \cap B_\beta' = \emptyset$ and $A_\beta', B_\beta' \in \Sigma_1^0(M)$ while $A \backslash F_\beta \subset A_\beta'$, $B \backslash F_\beta \subset B_\beta'$ and $A_\beta' \cup B_\beta' \subset M \backslash F_\beta$;
(4) if $\beta = \gamma + 1 < \alpha$, then $F_\beta = \overline{A \cap F_\gamma} \cap \overline{B \cap F_\gamma}$.

If α is a limit ordinal, then the sets $F_\alpha = \bigcap\{F_\beta : \beta < \alpha\}$, $A'_\alpha = \bigcup\{A'_\beta : \beta < \alpha\}$ and $B'_\alpha = \bigcup\{B'_\beta : \beta < \alpha\}$ witness that (2)–(4) hold for all $\beta \leq \alpha$. If $\alpha = \gamma + 1$ then the sets $F_\alpha = \overline{A \cap F_\gamma} \cap \overline{B \cap F_\gamma}$, $A'_\alpha = (\overline{A \cap F_\gamma}\setminus F_\alpha) \cup A'_\gamma$ and $B'_\alpha = (\overline{B \cap F_\gamma}\setminus F_\alpha) \cup B'_\gamma$ show that we also have (2)–(4) for all $\beta \leq \alpha$.

Consequently, we can continue our inductive construction to obtain families $\{F_\beta : \beta < \omega_1\}$ and $\{A'_\beta, B'_\beta : \beta < \omega_1\}$ such that (2)–(4) hold for all $\beta < \omega_1$. Since M is a second countable space, there is $\alpha < \omega_1$ such that $F_\beta = F_\alpha$ for all $\beta \geq \alpha$. If $F = F_\alpha = \emptyset$, then (3) applied for $\beta = \alpha$ implies that A and B are F_σ-separated which is a contradiction. Thus the set F is non-empty and the property (4) applied for $\beta = \alpha + 1$, implies that $F = \overline{A \cap F} = \overline{B \cap F}$, i.e., (1) is proved.

To finish the proof of our fact take any open $O \subset \mathbb{R}$ and represent it as $O = \bigcup\{R_n : n \in \omega\}$ where R_n is closed in $\mathbb{R}$ and $R_n \subset R_{n+1}$ for all $n \in \omega$. Let $U = f^{-1}(O)$ and $Q_n = f^{-1}(R_n)$ for every $n \in \omega$. Assume first that the sets $M\setminus U$ and Q_n are F_σ-separated for all $n \in \omega$. Then for each $n \in \omega$, there is $S_n \in \Sigma^0_1(M)$ such that $Q_n \subset S_n$ and $S_n \cap (M\setminus U) = \emptyset$, i.e., $S_n \subset U$. Therefore $U = \bigcup\{S_n : n \in \omega\}$ is an F_σ-subset of M, i.e., we proved that $f \in B_1(M)$. Now if Q_n and $M\setminus U$ are not F_σ-separated for some $n \in \omega$, then we can apply (1) to find a non-empty closed $F \subset M$ such that $\overline{Q_n \cap F} = \overline{(M\setminus U)\cap F} = F$. By our assumption, there is a point $x \in F$ such that $f|F$ is continuous at x. Choose sequences, $\{q_i : i \in \omega\} \subset Q_n \cap F$ and $\{r_i : i \in \omega\} \subset (M\setminus U) \cap F$ such that $q_i \to x$ and $r_i \to x$. By continuity of $f|F$, we have $f(q_i) \to f(x)$ and hence $f(x) \in R_n$ because $\{f(q_i) : i \in \omega\} \subset R_n \subset O$ and R_n is closed in $\mathbb{R}$.

On the other hand, $\{f(r_i) : i \in \omega\} \subset \mathbb{R}\setminus O$ and $\mathbb{R}\setminus O$ is closed in $\mathbb{R}$ which implies that $f(x) = \lim_{i\to\infty} f(r_i) \in \mathbb{R}\setminus O$; this contradiction shows that $f^{-1}(O)$ is an F_σ-subset of M for every open $O \subset \mathbb{R}$, i.e., $f \in B_1(M)$ so Fact 3 is proved.

□

Fact 4. Given a Polish space M, a function $f : M \to \mathbb{R}$ belongs to $\mathbb{R}^M \setminus B_1(M)$ if and only if there is a countable $A \subset M$ and $\delta > 0$ such that $\mathrm{osc}(f, U) > \delta$ for any $U \in \tau^*(A)$.

Proof. If such a set A exists, then let $F = \overline{A}$ and $g = f|F$. It is straightforward that $\mathrm{osc}(g, x) \geq \delta$ for any $x \in F$ and hence g is discontinuous at x. Therefore $f \notin B_1(M)$ by Fact 3 and hence we proved sufficiency.

Now assume that $f \in \mathbb{R}^M \setminus B_1(M)$. By Fact 3 there exists a closed non-empty $F \subset M$ such that $g = f|F$ is discontinuous at all points of F. Thus $\mathrm{osc}(g, x) > 0$ for any $x \in F$. If $F_n = \{x \in F : \mathrm{osc}(g, x) \geq \frac{1}{n+1}\}$, then F_n is closed in F for each $n \in \omega$ and $\bigcup_{n\in\omega} F_n = F$. The space F being Polish, there is $U \in \tau^*(F)$ with $U \subset F_n$ for some $n \in \omega$. For $\delta = \frac{1}{n+2}$ we have $\mathrm{osc}(g, W) \geq \frac{1}{n+1} > \delta$ for any $W \in \tau^*(U)$. Take a base $\{V_i : i \in \omega\} \subset \tau^*(U)$ for the space U and choose $y_i, z_i \in V_i$ so that $|f(y_i) - f(z_i)| \geq \frac{1}{n+1}$ for all $i \in \omega$. It is immediate that $A = \{y_i, z_i : i \in \omega\}$ and $\delta = \frac{1}{n+2}$ are as promised so we settled necessity. Fact 4 is proved.

□

Fact 5. Given a set Z and $S = \{f_j : j \in \omega\} \subset \mathbb{R}^Z$, assume that the sequence S is pointwise bounded (i.e., the set $S(z) = \{f_j(z) : j \in \omega\}$ is bounded in $\mathbb{R}$ for every $z \in Z$) and has no convergent subsequences (in $\mathbb{R}^Z$). Then there exists an infinite set $A \subset \omega$ and $p, q \in \mathbb{R}$ such that $p < q$ and for every infinite $B \subset A$, there is $z \in Z$ such that the sets $P(B, p, z) = \{m \in B : f_m(z) < p\}$ and $Q(B, q, z) = \{m \in B : f_m(z) > q\}$ are both infinite.

Proof. For any infinite set $A \subset \omega$ and $z \in Z$ let $L_A(z) = \inf\{t \in \mathbb{R} : \text{the set} \{m \in A : f_m(z) > t\} \text{ is finite}\}$ and $l_A(z) = \sup\{t \in \mathbb{R} : \text{the set} \{m \in A : f_m(z) < t\}$ is finite$\}$. Observe that $L_A(z)$ and $l_A(z)$ make sense because the set $\{f_j(z) : j \in \omega\}$ is bounded in $\mathbb{R}$. It is evident that $l_A(z) \leq L_A(z)$ and $l_A(z) = L_A(z)$ if and only if the sequence $\{f_m(z) : m \in A\}$ is convergent.

Let $\{(p_n, q_n) : n \in \omega\}$ be an enumeration of all pairs (p, q) of rational numbers such that $p < q$. Given an infinite $A \subset \omega$ and $p, q \in \mathbb{R}$ with $p < q$ say that a triad (A, p, q) is *adequate* if the conclusion of our fact holds for (A, p, q), i.e., for any infinite $B \subset A$ there is $z \in Z$ for which the sets $P(B, p, z)$ and $Q(B, q, z)$ are infinite.

To obtain a contradiction suppose that no triad (A, p, q) is adequate. Then (ω, p_0, q_0) is not adequate and therefore we can find an infinite $A_0 \subset \omega$ such that $P(A_0, p_0, z)$ and $Q(A_0, q_0, z)$ are not both infinite for any $z \in Z$. Assume that we have constructed infinite sets $A_0 \supset \cdots \supset A_k$ such that $P(A_j, p_j, z)$ and $Q(A_j, q_j, z)$ are not both infinite for all $z \in Z$ and $j \leq k$. Since the triad (A_k, p_{k+1}, q_{k+1}) is not adequate, there exists an infinite $A_{k+1} \subset A_k$ such that the sets $P(A_{k+1}, p_{k+1}, z)$ and $Q(A_{k+1}, q_{k+1}, z)$ are not both infinite for any $z \in Z$.

After we have the sequence $\{A_i : i \in \omega\}$ apply Fact 1 to find an infinite set $A \subset \omega$ with $A \subset^* A_i$ for all $i \in \omega$. The sequence $\{f_n : n \in A\}$ is not convergent by our hypothesis, so there exists $z \in Z$ such that $l_A(z) < L_A(z)$ which implies that there are $p, q \in \mathbb{Q}$ for which $l_A(z) < p < q < L_A(z)$. We have $p = p_n$ and $q = q_n$ for some $n \in \omega$ and hence one of the sets $P(A_n, p_n, z)$ and $Q(A_n, q_n, z)$ is finite. It follows from $A \subset^* A_n$ that one of the sets $P(A, p_n, z)$ and $Q(A, q_n, z)$ is finite which is impossible by the definition of $l_A(z)$ and $L_A(z)$. This contradiction shows that there exists an infinite $A \subset \omega$ and $p, q \in \mathbb{R}$ such that $p < q$ and the triad (A, p, q) is adequate. Fact 5 is proved. □

Fact 6. Given a Polish space M and $F \subset B_1(M)$ suppose that every infinite $A \subset F$ has an accumulation point in $B_1(M)$. Then $\overline{F}$ is a compact subset of $B_1(M)$ (the bar denotes the closure in $\mathbb{R}^M$).

Proof. It is evident that $\overline{F}$ is pseudocompact, so it is compact by Problems 401 and 407 of [TFS]). To see that $\overline{F} \subset B_1(M)$ assume that there is $f \in \overline{F} \backslash B_1(M)$. Then there is a countable $A \subset M$ and $\delta > 0$ such that $\mathrm{osc}(f, U) > \delta$ for any $U \in \tau^*(A)$ (see Fact 4). Since $\mathbb{R}^A$ is second countable, there is a sequence $S = \{f_n : n \in \omega\} \subset F$ such that $f_n|A \to f|A$. By our hypothesis, there is $g \in B_1(M)$ such that g is an accumulation point of the sequence S. It is straightforward that $g|A = f|A$ and hence $g \notin B_1(M)$ by Fact 4; this contradiction proves that $f \in B_1(M)$, so the compact set $\overline{F}$ is contained in $B_1(M)$ and Fact 6 is proved. □

Fact 7. If M is a Polish space and F is a compact subspace of $B_1(M)$, then F is sequentially compact, i.e., every sequence in F contains a convergent subsequence.

Proof. Suppose that, on the contrary, there is a sequence $S = \{f_i : i \in \omega\} \subset F$ which has no convergent subsequence. It follows from compactness of F that the set $S(x) = \{f_i(x) : i \in \omega\}$ is bounded in $\mathbb{R}$ for all $x \in M$, so we can apply Fact 5 to find an infinite $A \subset \omega$ and $p, q \in \mathbb{R}$ such that the triad (A, p, q) is adequate, i.e., for every infinite $B \subset A$ there is $x \in M$ for which the sets $P(B, p, x) = \{m \in M : f_m(x) < p\}$ and $Q(B, q, x) = \{m \in M : f_m(x) > q\}$ are both infinite.

For every $B \subset A$ let $K(B) = \{x \in M : P(B, p, x)$ and $Q(B, q, x)$ are infinite$\}$. Then $K(B) \neq \emptyset$ for every infinite $B \subset A$ and it is evident that $B \subset^* B' \subset A$ implies $K(B) \subset K(B')$. As a consequence,

(5) there is an infinite $B \subset A$ such that $K(B) = K(B')$ for any infinite $B' \subset^* B$.

Indeed, if (5) is not true, then let $B_0 = A$ and assume that for some $\alpha < \omega_1$ with $\alpha > 0$, we have constructed a family $\{B_\beta : \beta < \alpha\}$ of infinite subsets of A such that

(6) $\beta < \gamma < \alpha$ implies $B_\gamma \subset^* B_\beta$ and $K(B_\gamma) \neq K(B_\beta)$.

If α is a limit ordinal, choose a sequence $\{\alpha_n : n \in \omega\}$ for which $\alpha_n < \alpha_{n+1}$ for all $n \in \omega$ and $\alpha_n \to \alpha$; by Fact 1 there exists an infinite $B_\alpha \subset A$ such that $B_\alpha \subset^* B_{\alpha_n}$ for every $n \in \omega$. Now, if $\beta < \alpha$, then there is $n \in \omega$ such that $\beta < \alpha_n$ and hence $B_\alpha \subset^* B_{\alpha_n} \subset^* B_\beta$; besides, $K(B_\alpha) \subset K(B_{\alpha_n}) \subset K(B_\beta)$ and $K(B_\beta) \neq K(B_{\alpha_n})$ by (6). This proves that $B_\alpha \subset^* B_\beta$ and $K(B_\alpha) \neq K(B_\beta)$ for all $\beta < \alpha$. Thus (6) holds for all $\beta \leq \alpha$.

If $\alpha = \gamma + 1$, then (5) does not hold for B_γ and therefore there exists an infinite $B_\alpha \subset B_\gamma$ such that $K(B_\alpha) \neq K(B_\gamma)$. It is immediate that (6) holds for all $\beta \leq \alpha$ in this case as well. Consequently, our inductive procedure can be carried out to obtain a family $\{B_\beta : \beta < \omega_1\}$ of infinite subsets of A such that (6) holds for all $\beta < \omega_1$. Then $\{K(B_\beta) : \beta < \omega_1\}$ is a strictly decreasing family of non-empty closed subsets of M. Since such a family cannot exist in a second countable space, we have a contradiction which proves (5).

Now fix an infinite $B \subset A$ such that (5) holds for B and let $K = K(B)$. If $U \in \tau^*(K)$ and C is an infinite subset of B, then $\overline{K(C)} = K$ and hence there is $x \in U \cap K(C)$. Since $P(C, p, x)$ is infinite, there is an infinite $C' \subset C$ such that the sequence $S[C', x] = \{f_m(x) : m \in C'\}$ converges to some $r_x \leq p$. We also have $\overline{K(C')} = K$, so there is $y \in U \cap K(C')$; the set $Q(C', q, y)$ being infinite, there is an infinite $C'' \subset C'$ such that the sequence $S[C'', y]$ converges to some $r_y \geq q$. Thus

(7) for any infinite $C \subset B$ and $U \in \tau^*(K)$, there is an infinite $C'' \subset C$ such that for some points $x, y \in U$, the sequences $S[C'', x]$ and $S[C'', y]$ converge to r_x and r_y respectively while $r_x \leq p$ and $r_y \geq q$.

Pick a base $\{U_i : i \in \omega\} \subset \tau^*(K)$ of the space K. Let $C_0 = B$; suppose that we have infinite sets $C_0 \supset \cdots \supset C_k$ such that for any $i \leq k$ there are $r_i \leq p$ and

$s_i \geq q$ for which $S[C_i, x_i] \to r_i$ and $S[C_i, y_i] \to s_i$ for some $x_i, y_i \in U_i$. Now apply (7) to $C = C_k$ to choose an infinite set $C_{k+1} \subset C_k$ such that there are points $x_{k+1}, y_{k+1} \in U_{k+1}$ for which $S[C_{k+1}, x_{k+1}] \to r_{k+1}$ and $S[C_{k+1}, y_{k+1}] \to s_{k+1}$ where $r_{k+1} \leq p$ and $s_{k+1} \geq q$. This shows that we can inductively construct a decreasing sequence $\{C_i : i \in \omega\}$ of infinite subsets of B as well as sequences $\{r_i, s_i : i \in \omega\} \subset \mathbb{R}$ and $\{x_i, y_i : i \in \omega\} \subset M$ such that $r_i \leq p$, $s_i \geq q$, $\{x_i, y_i\} \subset U_i$ while $S[C_i, x_i] \to r_i$ and $S[C_i, y_i] \to s_i$ for every $i \in \omega$.

Now apply Fact 1 again to find an infinite $C \subset B$ such that $C \subset^* C_i$ for all $i \in \omega$. If $L = \{x_i, y_i : i \in \omega\}$, then it is straightforward that $S[C, x_i] \to r_i \leq p$ and $S[C, y_i] \to s_i \geq q$ for every $i \in \omega$. Define a function $f \in \mathbb{R}^M$ as follows: $f(x_i) = r_i$ and $f(y_i) = s_i$ for all $i \in \omega$; if $x \in M \setminus L$, then let $f(x) = 0$. Then, for every $W \in \tau^*(L)$, we have $U_i \cap L \subset W$ for some $i \in \omega$ and hence $\{x_i, y_i\} \subset W$ which implies $\mathrm{osc}(f, W) \geq |f(x_i) - f(y_i)| \geq \delta = q - p > 0$. This shows that $f \in \mathbb{R}^M \setminus B_1(M)$ by Fact 4. By compactness of F the sequence $\{f_m : m \in C\}$ has an accumulation point $g \in F \subset B_1(M)$. However, $g|L = f|L$ and hence we can apply Fact 4 to the function g to conclude that $g \notin B_1(M)$. This contradiction shows that S must contain a convergent subsequence and hence Fact 7 is proved. $\qquad\square$

Fact 8. Let Z be a space; if $m \in \mathbb{N}$ and $f \in \mathbb{R}^Z$, let $\varphi_m(f)(z) = |f(z_1)| + \cdots + |f(z_m)|$ for any $z = (z_1, \ldots, z_m) \in Z^m$. Then $\varphi_m : \mathbb{R}^Z \to \mathbb{R}^{Z^m}$ is a continuous map such that $\varphi_m(C_p(Z)) \subset C_p(Z^m)$ and $\varphi_m(B_1(Z)) \subset B_1(Z^m)$.

Proof. For every $i \in \{1, \ldots, m\}$ let $\pi_i : Z^m \to Z$ be the natural projection of Z^m onto its ith factor. Then $\varphi_m(f) = \Sigma_{i=1}^m |f \circ \pi_i|$ which shows that $\varphi_m(f)$ is a continuous function on Z^m if f is continuous on Z. For each $z \in Z$ let $p_z : \mathbb{R}^Z \to \mathbb{R}$ be the natural projection onto the factor determined by z. Analogously, if $w \in Z^m$, let $q_w : \mathbb{R}^{Z^m} \to \mathbb{R}$ be the natural projection onto the factor determined by w. To prove continuity of φ_m it suffices to show that $q_w \circ \varphi_m$ is continuous for any $w = (w_1, \ldots, w_m) \in Z^m$ (see Problem 102 of [TFS]). Now, $q_w(\varphi_m(f)) = \varphi_m(f)(w) = \sum_{i=1}^m |f(w_i)|$ for any $f \in \mathbb{R}^Z$ which shows that $q_w \circ \varphi_m = \sum_{i=1}^m |p_{w_i}|$ is a continuous map because each p_{w_i} is continuous. This proves continuity of φ_m.

Finally, if $f \in B_1(Z)$, then there is a sequence $\{f_n : n \in \omega\} \subset C_p(Z)$ with $f_n \to f$. Then $\{\varphi_m(f_n) : n \in \omega\} \subset C_p(Z^m)$ and $\varphi_m(f_n) \to \varphi_m(f)$ by continuity of φ_m. Thus $\varphi_m(f) \in B_1(Z^m)$ and therefore $\varphi_m(B_1(Z)) \subset B_1(Z^m)$. Fact 8 is proved. $\qquad\square$

Fact 9. Suppose that M is a Polish space and F is a compact subspace of $B_1(M)$. Define $u \in \mathbb{R}^M$ by $u(x) = 0$ for all $x \in M$ and assume that $u \in \overline{H}$ where $H \subset F$ and $h(x) \geq 0$ for any $x \in M$ and $h \in H$. Then, for any $\varepsilon > 0$, there is a countable $G \subset H$ such that $\inf\{g(x) : g \in G\} < \varepsilon$ for all $x \in M$.

Proof. If this is false, then the set $M(G) = \{x \in M : \inf\{g(x) : g \in G\} \geq \varepsilon\}$ is non-empty for every countable non-empty $G \subset H$. Observe that $G \subset G'$ implies $M(G) \supset M(G')$. As a consequence, there exists a countable non-empty $G \subset H$ such that

(8) $\overline{M(G)} = \overline{M(G')}$ for any countable $G' \supset G$.

Indeed, if such a G does not exist, then it is easy to construct by transfinite induction a family $\{G_\alpha : \alpha < \omega_1\}$ of countable subsets of H in such a way that $\alpha < \beta$ implies $G_\alpha \subset G_\beta$ and $\overline{M(G_\alpha)} \neq \overline{M(G_\beta)}$. Therefore $\{\overline{M(G_\alpha)} : \alpha < \omega_1\}$ is a strictly decreasing ω_1-sequence of non-empty closed subsets of M which is a contradiction.

Now choose a non-empty countable $G \subset H$ such that (8) holds and take a countable dense subset D of the space $K = \overline{M(G)}$. Since $u \in \overline{H}$, there exists a sequence $S = \{f_i : i \in \omega\} \subset H$ such that $f_i(d) \to 0$ for every $d \in D$. If g is an accumulation point of the sequence S, then $g(d) = 0$ for every $d \in D$. On the other hand, if $G' = G \cup S$, then $\overline{M(S)} \supset \overline{M(G')}$ and hence $g(x) \geq \varepsilon$ for all $x \in \overline{M(G')}$. However, $\overline{M(G')}$ is also dense in K by (8) which shows that $g|K$ has no points of continuity and therefore $g \notin B_1(M)$ by Fact 3. This contradiction shows that Fact 9 is proved.   □

Fact 10. Let M be a Polish space. Then $t(F) \leq \omega$ for any compact $F \subset B_1(M)$.

Proof. Define $u \in \mathbb{R}^M$ by $u(x) = 0$ for all $x \in M$. Suppose that $A \subset F$ and $f \in \overline{A}$. Define a map $L_f : \mathbb{R}^M \to \mathbb{R}^M$ by $L_f(g) = g - f$ for all $g \in \mathbb{R}^M$. Then L_f is a homeomorphism because $\mathbb{R}^M$ is a linear topological space (see Problems 115 and 116 of [TFS] together with Fact 1 of S.491). Then $L_f(F)$ is homeomorphic to F, and for any $B \subset A$, we have $u \in \overline{L_f(B)}$ if and only if $f \in \overline{B}$. Thus we can assume, without loss of generality, that $f = u$.

For any $m \in \mathbb{N}$ and $f \in \mathbb{R}^M$, let $\varphi_m(f)(z) = |f(z_1)| + \cdots + |f(z_m)|$ for any point $z = (z_1, \ldots, z_m) \in M^m$. Then $\varphi_m : \mathbb{R}^M \to \mathbb{R}^{M^m}$ is a continuous map by Fact 8. If u_m is the function which is identically zero on M^m, then $u_m = \varphi_m(u)$ and $u_m \in \overline{\varphi_m(A)}$. The space M^m is Polish and $\varphi_m(A)$ is contained in a compact subspace $\varphi_m(F)$ of $B_1(M^m)$. Since also $\varphi_m(f)$ is non-negative for any $f \in A$, we can apply Fact 9 to find a countable $B_m \subset A$ such that $\inf\{\varphi_m(f)(z) : f \in B_m\} < \frac{1}{m}$ for any $z \in M^m$ and therefore

(9) for any $z_1, \ldots, z_m \in M$, there is $f \in B_m$ such that $|f(z_1)| + \cdots + |f(z_m)| < \frac{1}{m}$
    and hence $|f(z_i)| < \frac{1}{m}$ for all $i \leq m$.

Consider the set $B = \bigcup\{A_m : m \in \omega\}$; then B is a countable subset of A. To see that $u \in \overline{B}$, take any $U \in \tau(u, F)$. There exist a set $P = \{x_1, \ldots, x_k\} \subset M$ and $\varepsilon > 0$ such that $u \in V \subset U$ where $V = \{f \in F : |f(x_i)| < \varepsilon$ for all $i \leq k\}$. Choose $m \in \mathbb{N}$ with $k \leq m$ and $\frac{1}{m} < \varepsilon$; let $z = (z_1, \ldots, z_m) \in M^m$ be a point with $\{x_i : i \leq k\} \subset \{z_j : j \leq m\}$. By (9), there is $f \in B_m$ such that $|f(z_j)| < \frac{1}{m} < \varepsilon$ for all $j \leq m$ and hence $|f(x_i)| < \varepsilon$ for all $i \leq k$, i.e., $f \in V \cap B \subset U \cap B$. This proves that $U \cap B \neq \emptyset$ for any $U \in \tau(u, F)$ and therefore $u \in \overline{B}$. Thus $t(F) \leq \omega$ and Fact 10 is proved.   □

Fact 11. Given a Polish space M let $u(x) = 0$ for all $x \in M$. Assume that $S = \{f_n : n \in \omega\} \subset C_p(M)\backslash\{u\}$ and u is an accumulation point of the sequence S. Assume also that for every infinite $A \subset \omega$, there is an infinite $B \subset A$ such that

$S[B] = \{f_n : n \in B\}$ is a convergent sequence (in $\mathbb{R}^M$). Then there is an infinite $E \subset \omega$ such that $S[E]$ is a convergent sequence and $S[E] \to u$.

Proof. Observe first that every infinite subset of S has an accumulation point in $B_1(M)$ and therefore $T = \overline{S}$ is a compact subset of $B_1(M)$ by Fact 6 (the bar denotes the closure in $\mathbb{R}^M$). Say that a set $A \subset \omega$ is *big* if u is an accumulation point of $S[A]$. It is clear that every big set is infinite. If $A \subset \omega$, then the expression $S[A] \to f$ says that $S[A]$ is a convergent sequence and the limit of $S[A]$ is f. We claim that

(10) if $\{B_n : n \in \omega\} \subset \exp(\omega)$ where B_n is big and $B_{n+1} \subset^* B_n$ for all $n \in \omega$, then there is a big $B \subset \omega$ such that $B \subset^* B_n$ for all $n \in \omega$.

Let $A_0 = B_0$ and suppose that we have infinite sets $A_0 \supset \cdots \supset A_k$ such that $A_i \subset B_i$ and $B_i \backslash A_i$ are finite for all $i \leq k$. Since $C = B_{k+1} \backslash A_k \subset (B_{k+1} \backslash B_k) \cup (B_k \backslash A_k)$ is finite, for the set $A_{k+1} = B_{k+1} \backslash C$ our assumption still holds. Thus we can construct a sequence $\{A_i : i \in \omega\} \subset \omega$ such that $A_{i+1} \subset A_i$, $A_i \subset B_i$ and $|B_i \backslash A_i| < \omega$ for all $i \in \omega$. It is evident that taking away finitely many points from a big set, we still obtain a big set, so A_i is big for every $i \in \omega$.

Let $F = \{f \in T : \text{there is an infinite } A \subset \omega \text{ such that } A \subset^* A_n \text{ for all } n \in \omega$ and $S[A] \to f\}$. Then $u \in \overline{F}$. Indeed, it suffices to show that for any finite $P \subset M$ and $\varepsilon > 0$, there is $f \in F$ such that $|f(x)| < \varepsilon$ for all $x \in P$. So, fix a finite $P \subset M$ and $\varepsilon > 0$. Since every A_n is big, the set $A_n' = \{m \in A_n : |f_m(x)| < \frac{\varepsilon}{2}$ for all $x \in P\}$ is also big for any $n \in \omega$, so we can choose a sequence $\{m_i : i \in \omega\} \subset \omega$ such that $m_i \in A_i'$ and $m_i < m_{i+1}$ for every $i \in \omega$. By our hypothesis, there is an infinite $A \subset \{m_i : i \in \omega\}$ and $f \in T$ such that $S[A] \to f$. It is immediate that $A \subset^* A_n$ (and hence $A \subset^* B_n$) for all $n \in \omega$ so $f \in F$. Besides, $|f(x)| \leq \frac{\varepsilon}{2} < \varepsilon$ for all $x \in P$ which proves that $u \in \overline{F}$.

Since F is a subspace of a compact $T \subset B_1(M)$, we can apply Fact 10 to find a countable $G \subset F$ such that $u \in \overline{G}$. For every $g \in G$, there is an infinite $B_g \subset \omega$ such that $B_g \subset^* A_n$ (and hence $B_g \subset^* B_n$) for all $n \in \omega$ and $S[B_g] \to g$. By Fact 2, there is an infinite $B \subset \omega$ such that $B_g \subset^* B$ for all $g \in G$ and $B \subset^* B_n$ for all $n \in \omega$. Thus $g \in \overline{S[B]}$ for every $g \in G$ and therefore $u \in \overline{G} \subset \overline{S[B]}$ which shows that B is the promised big set and hence (10) is proved.

For any infinite set $A \subset \omega$ and any point $z \in M$, let $L_A(z) = \inf\{t \in \mathbb{R} : \text{the set } \{m \in A : |f_m(z)| > t\}$ is finite$\}$ and $l_A(z) = \sup\{t \in \mathbb{R} : \text{the set } \{m \in A : |f_m(z)| < t\}$ is finite$\}$. Observe that $L_A(z)$ and $l_A(z)$ make sense because it follows from compactness of T that the set $\{f_j(z) : j \in \omega\}$ is bounded in $\mathbb{R}$.

Now, fix $\varepsilon > 0$ and, for any infinite $A \subset \omega$, let $Q_A = \overline{\{x \in M : L_A(x) > \varepsilon\}}$. It is clear that for any infinite $A, B \subset \omega$ such that $A \subset^* B$, we have $Q_A \subset Q_B$. As a consequence,

(11) for any big set $A \subset \omega$, there is a *minimal big set* $B \subset A$ in the sense that $Q_C = Q_B$ for any big set $C \subset^* B$.

Indeed, if (11) is not true, then let $A_0 \subset A$ be a big set which witnesses that it is false. Suppose that $\alpha < \omega_1$ and we have a family $\{A_\beta : \beta < \alpha\}$ of big sets such that $\beta < \gamma < \alpha$ implies $A_\gamma \subset^* A_\beta$ and $Q_{A_\gamma} \neq Q_{A_\beta}$. If α is a limit ordinal, then choose a sequence $\{\alpha_n : n \in \omega\}$ such that $\alpha_n < \alpha_{n+1}$ for all $n \in \omega$ and $\alpha_n \to \alpha$. It follows from (10) that there is a big set $A_\alpha \subset \omega$ such that $A_\alpha \subset^* A_{\alpha_n}$ for all $n \in \omega$. If $\beta < \alpha$, then there is $n \in \omega$ such that $\beta < \alpha_n$ and hence $A_\alpha \subset^* A_{\alpha_n} \subset^* A_\beta$. Furthermore, $Q_{A_{\alpha_n}} \subset Q_{A_\beta}$ and $Q_{A_{\alpha_n}} \neq Q_{A_\beta}$ which implies $Q_{A_\alpha} \neq Q_{A_\beta}$, so our inductive assumption holds for all $\beta \leq \alpha$.

Now, if $\alpha = \gamma + 1$, then it follows from $A_\gamma \subset A$ that there is a big set $A_\alpha \subset A_\gamma$ such that $Q_{A_\alpha} \neq Q_{A_\gamma}$. It is clear that our inductive assumption is also fulfilled for all $\beta \leq \alpha$ and hence we can construct an ω_1-sequence $\{A_\alpha : \alpha < \omega_1\}$ such that $\{Q_{A_\alpha} : \alpha < \omega_1\}$ is a strictly decreasing ω_1-sequence of closed subsets of M which is a contradiction. Thus (11) is true. Our next step is to show that

(12) if A is a minimal big set, then $Q = Q_A = \emptyset$, i.e., $L_A(x) \leq \varepsilon$ for every $x \in X$.

To prove the property (12), assume that $Q \neq \emptyset$ and take a countable dense set $Y = \{y_i : i \in \omega\} \subset Q$ (repetitions are possible in the enumeration of Y to cover the case when Y is finite). Let us formulate explicitly an evident property, we already used in the proof of (10).

(13) If B is a big set, then, for any finite $K \subset M$ and $\delta > 0$, the set $B' = \{m \in B : |f_m(x)| < \delta$ for all $x \in K\}$ is also big.

Using (13) it is easy to choose big sets $\{B_n : n \in \omega\} \subset \exp(A)$ in such a way that $B_{n+1} \subset B_n$ for all $n \in \omega$ and $|f_m(y_i)| < \frac{1}{n}$ for all $m \in B_n$ and $i \leq n$.

Now apply (10) to find a big set $B \subset \omega$ such that $B \subset^* B_n$ for all $n \in \omega$. It is straightforward that

(∗)   if f is an accumulation point of $S[B]$, then $f(y) = 0$ for every $y \in Y$.

Choose a complete metric d_0 on the space M with $\tau(d_0) = \tau(M)$. Then the space Q with the metric $d = d_0|(Q \times Q)$ is also complete. Observe that given $U \in \tau^*(Q)$ and a big set $C \subset B$, the set $C' = \{m \in C : |f_m(u)| > \varepsilon$ for some $u \in U\}$ is also big because otherwise $C_0 = C \backslash C'$ is big and $|f_m(u)| \leq \varepsilon$ for all $u \in U$ and $m \in C_0$; since $C_0 \subset B$, we have $Q_{C_0} = Q$, i.e., the set $\{x \in Q : L_{C_0}(x)| > \varepsilon\}$ is dense in Q and therefore there is $m \in C_0$ with $|f_m(x)| > \varepsilon$ for some $x \in U$ which is a contradiction. An evident induction shows that if $k \in \mathbb{N}$ then

(14) for any $U_1, \ldots, U_k \in \tau^*(Q)$ the set $C = \{m \in B :$ for any $i \leq k$ we have $|f_m(x)| > \varepsilon$ for some $x \in U_i\}$ is big.

Fix a base $\{O_i : i \in \omega\} \subset \tau^*(Q)$ of the space Q. We will construct a family $\mathcal{U} = \{U(i, j) : i \in \omega, i \leq j < \omega\} \subset \tau^*(Q)$ and a sequence $\{m_i : i \in \omega\} \subset B$ with the following properties:

(15) $U(i, i) = O_i$ for all $i \in \omega$;
(16) $\overline{U(i, j + 1)} \subset U(i, j)$ for any $i \in \omega$ and $j \geq i$;
(17) $\mathrm{diam}(U(i, j)) \leq \frac{1}{j}$ for every $i \in \omega$ and $j \geq i + 1$;

(18) $m_{j+1} > m_i$ for all $j \in \omega$;

(19) if $i \in \omega$ and $j > i$, then $|f_{m_j}(u)| > \varepsilon$ for any $u \in U(i, j)$.

To start the construction pick an arbitrary $m_0 \in B$ and let $U(0, 0) = O_0$. Assume that $n \in \omega$ and we have the numbers $m_0, \ldots, m_n \in B$ and sets $U(i, j)$ for all $i \leq n$ and $i \leq j \leq n$ such that (15)–(19) are fulfilled where applicable. To satisfy (15) let $U(n + 1, n + 1) = O_{n+1}$; the property (14) implies that the set $C = \{m \in B :$ for any $i \in \{0, \ldots, n\}$ we have $|f_m(x)| > \varepsilon$ for some $x \in U(i, n)\}$ is big and hence there exists $m_{n+1} > m_n$ such that $m_{n+1} \in C$. Fix $i \in \{0, \ldots, n\}$; the choice of C implies that the set $\{x \in U(i, n) : |f_{m_{n+1}}(x)| > \varepsilon\}$ is non-empty. Since the function $f_{m_{n+1}}$ is continuous, there is a set $U(i, n + 1) \in \tau^*(Q)$ such that $\overline{U(i, n + 1)} \subset U(i, n)$, $\operatorname{diam}(U(i, n + 1)) \leq \frac{1}{n+1}$ and $|f_{m_{n+1}}(x)| > \varepsilon$ for any $x \in U(i, n + 1)$. This completes the inductive construction of the family $\mathcal{U}$. Let $D = \{m_i : i \in \omega\}$.

It follows from the completeness of (Q, d) that for each $i \in \omega$, we can choose a point $z_i \in \bigcap\{\overline{U(i, j)} : j \in \omega \backslash i\} = \bigcap\{U(i, j) : j \in \omega \backslash i\} \subset O_i$. Since the set $Z = \{z_i : i \in \omega\}$ intersects O_i for every $i \in \omega$, it is dense in Q. It follows from (19) that $f_{m_j}(z_i) > \varepsilon$ for all $j > i$ and hence $l_D(z_i) \geq \varepsilon$ for all $i \in \omega$. Since $S[D] \subset T$ and T is compact, we can take an accumulation point $f \in B_1(M)$ of the set $S[D]$. It is immediate that $|f(z)| \geq \varepsilon$ for all $z \in Z$. Clearly f is also an accumulation point of $S[B]$ so $f(y) = 0$ for all $y \in Y$ by $(*)$. Thus we have found two dense sets Y and Z in the space Q such that $f|Y$ is identically zero and $|f(z)| \geq \varepsilon$ for all $z \in Z$. Therefore f has no point of continuity on Q which contradicts Fact 3. This contradiction shows that (12) is true.

Finally, it follows from the properties (11) and (12) that we can construct a sequence $\{A_n : n \in \omega\}$ of big sets such that $A_{n+1} \subset A_n$ and $L_{A_n}(x) \leq \frac{1}{n}$ for all $x \in X$. Choose a sequence $J = \{k_i : i \in \omega\} \subset \omega$ such that $k_i < k_{i+1}$ and $k_i \in A_i$ for all $i \in \omega$. It is easy to see that $S[J] \to u$ and hence Fact 11 is proved. $\qquad \square$

Fact 12. Suppose that M is a Polish space and N is a second countable space. Let $f : M \to N$ be a measurable map, i.e., $f^{-1}(A) \in \mathbb{B}(M)$ for any $A \in \mathbb{B}(N)$. Then there is a Polish topology μ on the set M such that $\tau(M) \subset \mu$ and the map $f : (M, \mu) \to N$ is continuous.

Proof. Let $\mathcal{B}$ be a countable base in N. For every $B \in \mathcal{B}$, consider the space M_B whose underlying set is M and the topology $\tau(M_B)$ is generated by the family $\tau_B = \tau(M) \cup \{f^{-1}(B), M \backslash f^{-1}(B)\}$. It is evident that M_B is homeomorphic to the space $f^{-1}(B) \oplus (M \backslash f^{-1}(B))$ and therefore M_B is a Borel set. Now consider the space M' whose underlying set is M and the topology $\tau(M')$ is generated by the family $\bigcup\{\tau_B : B \in \mathcal{B}\}$. Then M' is homeomorphic to a closed subspace of the space $\prod\{M_B : B \in \mathcal{B}\}$ by Fact 2 of T.363 and hence M' is a Borel set by Problems 333 and 331. There exists a Polish topology μ on the set M such that $\tau(M') \subset \mu$ by Problem 341. We have $\tau(M) \subset \tau(M') \subset \mu$; besides, $f^{-1}(B) \in \tau(M_B) \subset \tau(M')$ for every $B \in \mathcal{B}$ which shows that $f^{-1}(B)$ is open in (M, μ) for every $B \in \mathcal{B}$ and hence the map $f : (M, \mu) \to N$ is continuous. Fact 12 is proved. $\qquad \square$

Now it is easy to finish our solution. Suppose that M is a Polish space and $X \subset B_1(M)$ is compact; denote by u the function which is identically zero on M. Take any $A \subset X$ and $f \in \overline{A}\backslash A$. Define a map $L_f : \mathbb{R}^M \to \mathbb{R}^M$ by $L_f(g) = g - f$ for all $g \in \mathbb{R}^M$. Then L_f is a homeomorphism because $\mathbb{R}^M$ is a linear topological space (see Problems 115 and 116 of [TFS] together with Fact 1 of S.491). Then $L_f(X)$ is homeomorphic to X, and for any sequence $\{f_n : n \in \omega\} \subset A$, we have $f_n \to f$ if and only if $L_f(f_n) \to u$. Thus we can assume, without loss of generality, that $f = u$.

Since $t(X) \leq \omega$ by Fact 10, there is a sequence $S = \{f_n : n \in \omega\} \subset A$ such that u is an accumulation point of S. There is a Polish topology μ on M such that f_n is continuous on $M' = (M, \mu)$ for all $n \in \omega$ (see Fact 12). Thus $C_p(M) \subset C_p(M') \subset \mathbb{R}^M$ and $S \subset C_p(M')$. It is clear that $B_1(M) \subset B_1(M')$ and hence S is contained in a compact space X which in turn is a subspace of $B_1(M')$. Thus X is sequentially compact by Fact 7 and therefore every infinite subset of S contains a convergent subsequence. We proved that $S \subset C_p(M')$ satisfies all premises of Fact 11, so we can apply it to conclude that some subsequence S' of the sequence S converges to u. Since $S' \subset A$, we proved that there is a sequence in A which converges to $u = f$ and hence X is Fréchet–Urysohn, i.e., our solution is complete.

T.385. *Let X be a separable compact space. Prove that X is Rosenthal compact if and only if, for any dense countable $A \subset X$, the space $C_p(A|X)$ is analytic.*

Solution. Given a space Z, a map $f : Z \to \mathbb{R}$ is *measurable* if $f^{-1}(P) \in \mathbb{B}(Z)$ for any $P \in \mathbb{B}(\mathbb{R})$; denote by $B(Z)$ the set of all measurable real-valued functions on Z. If $A \subset Z$, then χ_A is *a characteristic function of A* defined by $\chi_A(A) \subset \{1\}$ and $\chi_A(Z\backslash A) \subset \{0\}$; besides, for any $f \in \mathbb{R}^Z$, the function $g = 1 - f \in \mathbb{R}^Z$ is defined by $g(x) = 1 - f(x)$ for all $x \in Z$. If $A \subset \mathbb{R}^Z$, then let $L_0(A) = A$; if $\alpha < \omega_1$ and we have $\{L_\beta(A) : \beta < \alpha\}$, let $L_\alpha(A) = \{f \in \mathbb{R}^Z$: there are sequences $\{\alpha_n : n \in \omega\} \subset \alpha$ and $\{f_n : n \in \omega\} \subset \mathbb{R}^Z$ such that $f_n \in L_{\alpha_n}(A)$ for all $n \in \omega$ and $f_n \to f\}$. This gives us a family of sets $\{L_\alpha(A) : \alpha < \omega_1\}$; let $L(A) = \bigcup\{L_\alpha(A) : \alpha < \omega_1\}$.

Fact 1. If M is a metrizable space, then $B(M) = L(C_p(M))$.

Proof. It follows from Fact 1 of T.363 that $L_0 = L_0(C_p(M)) = C_p(M) \subset B(M)$. Suppose that $0 < \alpha < \omega_1$ and we proved that $\{L_\beta : \beta < \alpha\} \subset B(M)$ where $L_\beta = L_\beta(C_p(M))$ for all $\beta < \alpha$. If $f \in L_\alpha = L_\alpha(C_p(M))$, then there are sequences $\{\alpha_n : n \in \omega\} \subset \alpha$ and $\{f_n : n \in \omega\} \subset \mathbb{R}^M$ such that $f_n \in L_{\alpha_n}$ for all $n \in \omega$ and $f_n \to f$. Given $U \in \tau^*(\mathbb{R})$ it is easy to find a sequence $\{U_n : n \in \omega\} \subset \tau^*(\mathbb{R})$ such that $\overline{U}_n \subset U_{n+1}$ for all $n \in \omega$ and $\bigcup_{n\in\omega} U_n = U$. It is straightforward that $f^{-1}(U) = \bigcup_{n\in\omega} \bigcap_{k \geq n} f_k^{-1}(U_n)$ and therefore $f^{-1}(U)$ is a Borel subset of M because so is $f_k^{-1}(U_n)$ for all $k, n \in \omega$ by the induction hypothesis.

Thus we proved that $f^{-1}(U) \in \mathbb{B}(M)$ for every $U \in \tau(\mathbb{R})$, so f is measurable by Fact 1 of T.363. Therefore $L_\alpha(C_p(M)) \subset B(M)$ for all $\alpha < \omega_1$ and hence $L(C_p(M)) \subset B(M)$.

To prove that $B(M) \subset L = L(C_p(M))$, let $L_\alpha = L_\alpha(C_p(M))$ for all $\alpha < \omega_1$. Our first step is to check that for any $A \subset \mathbb{B}(M)$, the function χ_A belongs to L. We will need the following properties of L:

(1) if $f, g \in L$, then $f \cdot g \in L$ and $af + bg \in L$ for any $a, b \in \mathbb{R}$;
(2) if $\{f_n : n \in \omega\} \subset L$ and $f_n \to f$, then $f \in L$;
(3) if $\mathcal{A} \subset \exp(M)$ and $\chi_A \in L$ for any $A \in \mathcal{A}$, then $\{\chi_{M \setminus A}, \chi_{A \cup B}, \chi_{A \cap B}, \chi_{A \setminus B}\} \subset L$ for any $A, B \in \mathcal{A}$.

The property (1) can be proved by an evident transfinite induction; we leave it as an exercise to the reader. To see that (2) is true, choose $\alpha_n < \omega_1$ such that $f_n \in L_{\alpha_n}$ for all $n \in \omega$ and observe that if $\alpha = \sup\{\alpha_n : n \in \omega\} + 1$, then $f \in L_\alpha$.

Now, for any $A, B \in \mathcal{A}$, we have $\chi_{M \setminus A} = 1 - \chi_A$, $\chi_{A \cap B} = \chi_A \cdot \chi_B$ and $\chi_{A \setminus B} = (1 - \chi_B) \cdot \chi_A$. Furthermore, $\chi_{A \cup B} = 1 - (1 - \chi_A) \cdot (1 - \chi_B)$ so (1) implies that (3) holds.

If A is open, then there is a sequence $\{A_n : n \in \omega\}$ of closed subsets of M such that $A_n \subset A_{n+1}$ for every $n \in \omega$ and $\bigcup_{n \in \omega} A_n = A$. By normality of M there is a continuous $f_n : M \to [0, 1]$ such that $f_n(A_n) \subset \{1\}$ and $f_n(M \setminus A) \subset \{0\}$ for all $n \in \omega$. After a minute's reflection, one can see that $f_n \to \chi_A$. Now, if A is closed in M, then $M \setminus A$ is open in M and hence $\chi_{M \setminus A} \in L$; now, (3) implies $\chi_A \in L$. Thus $\chi_A \in L$ for every $A \in \Sigma_0^0(M) \cup \Pi_0^0(M)$.

Assume that $0 < \alpha < \omega_1$ and we have established that $\chi_A \in L$ whenever $A \in \Sigma_\beta^0(M) \cup \Pi_\beta^0(M)$ for some $\beta < \alpha$. If $A \in \Sigma_\alpha^0(M)$, then there is a sequence $\{\alpha_n : n \in \omega\} \subset \alpha$ and $A_n \in \Pi_{\alpha_n}^0(M)$ for each $n \in \omega$ such that $\bigcup_{n \in \omega} A_n = A$. Let $B_0 = A_0$ and $B_n = A_n \setminus (\bigcup_{i < n} A_i)$ for all $n \geq 1$. The family $\{B_n : n \in \omega\}$ is disjoint and $A = \bigcup_{n \in \omega} B_n$. It follows from (3) and the induction hypothesis that $\chi_{B_n} \in L$ and hence $g_n = \sum_{i=0}^n \chi_{B_i} \in L$ for any $n \in \omega$. Besides, $g_n \to \chi_A$ so (2) implies that $\chi_A \in L$ for any $A \in \Sigma_\alpha^0(M)$.

Finally, observe that if $A \in \Pi_\alpha^0(M)$, then $M \setminus A \in \Sigma_\alpha^0(M)$ and hence $\chi_{M \setminus A} \in L$, so it follows from (3) that $\chi_A \in L$. Thus our inductive procedure can be continued to show that $\chi_A \in L$ for every $A \in \mathbb{B}(M)$; an evident consequence is

(4) if $r_1, \ldots, r_k \in \mathbb{R}$ and $A_1, \ldots, A_k \in \mathbb{B}(M)$, then $r_1 \chi_{A_1} + \ldots + r_k \chi_{A_k} \in L$.

Now take an arbitrary $f \in B(M)$. Letting $f_+ = \frac{|f|+f}{2}$ and $f_- = \frac{|f|-f}{2}$ we have $f = f_+ - f_-$ where $f_+(x) \geq 0$ and $f_-(x) \geq 0$ for any $x \in M$. The property (1) shows that it is sufficient to prove that $f_+, f_- \in L$, i.e., we can assume, without loss of generality, that $f(x) \geq 0$ for any $x \in M$.

For every $n \in \mathbb{N}$, let $k_n = n \cdot 2^n$ and for each $i \leq k_n$ consider the number $l_i = \frac{i}{2^n}$. Then $\{l_0, \ldots, l_{k_n}\}$ divide $[0, n]$ into intervals of length 2^{-n}. Let $A_i = [l_i, l_{i+1})$ for all $i < k_n$; then the family $\{A_i : i < k_n\}$ is disjoint and $[0, n) = \bigcup\{A_i : i < k_n\}$. If $A_{k_n} = [n, \infty)$ and $B_i = f^{-1}(A_i)$ for all $i \leq k_n$, then $\{B_i : i \leq k_n\}$ is a partition of the space M; note that this partition consists of Borel subsets of M because every interval is a Borel subset of $\mathbb{R}$. If $i < k_n$, then $f(B_i) \subset [l_i, l_{i+1})$, so for the function $g_i = l_i \cdot \chi_{B_i}$, we have $|f(x) - g_i(x)| < 2^{-n}$ for any $x \in B_i$. As a consequence, if $f_n = g_0 + \cdots + g_{k_n - 1}$, then $|f(x) - f_n(x)| < 2^{-n}$ for any $x \in M_n = B_0 \cup \cdots \cup B_{k_n - 1}$.

It is clear that $M = \bigcup\{M_n : n \in \mathbb{N}\}$ and $M_n \subset M_{n+1}$ for all $n \in \mathbb{N}$, so $f_n \to f$. The property (4) implies that $f_n \in L$ for every $n \in \mathbb{N}$, so $f \in L$ by (2) and hence Fact 1 is proved. □

Fact 2. Given a Čech-complete space Z suppose that $\mathcal{A}$ is a family of pairs of open subsets of Z, i.e., $\mathcal{A} \subset \tau(Z) \times \tau(Z)$. Assume that there is a non-empty $Y \subset Z$ such that $\mathcal{A}$ is *weakly dense* on Y in the sense that for any $G_0, \ldots, G_n \in \tau^*(Y)$, there is $(U, V) \in \mathcal{A}$ such that $U \cap G_i \neq \emptyset \neq V \cap G_i$ for all $i \leq n$. Then there is a family $\mathcal{A}' = \{(H_i^0, H_i^1) : i \in \omega\} \subset \mathcal{A}$ and a compact $K \subset Z$ such that $\mathcal{A}'$ is *independent* on K, i.e., for every $s \in \mathbb{D}^\omega$ we have $\bigcap\{H_i^{s(i)} : i \in \omega\} \cap K \neq \emptyset$.

Proof. Choose a family $\{O_n : n \in \omega\} \subset \tau(\beta Z)$ such that $O_{n+1} \subset O_n$ for all $n \in \omega$ and $Z = \bigcap\{O_n : n \in \omega\}$. If $\mathcal{B} = \{(U, V) \in \tau(\beta Z) \times \tau(\beta Z) : (U \cap Z, V \cap Z) \in \mathcal{A}\}$, then $\mathcal{B}$ is also weakly dense on Y.

Let $\mathbb{D}^0 = \{\emptyset\}$ and $\mathbb{D}^{\leq n} = \bigcup\{\mathbb{D}^i : i \leq n\}$ for all $n \in \mathbb{N}$; we will also need the set $\mathbb{D}^{<\omega} = \bigcup\{\mathbb{D}^n : n \in \omega\}$. If $n \in \omega$, $j \in \mathbb{D}$ and $s \in \mathbb{D}^n$, then $s^\frown j \in \mathbb{D}^{n+1}$ is defined by $(s^\frown j)|n = s$ and $s(n) = j$.

Our purpose is to construct inductively families $\{(G_i^0, G_i^1) : i \in \omega\} \subset \mathcal{B}$ and $\{W_s : s \in \mathbb{D}^{<\omega}\} \subset \tau^*(\beta Z)$ with the following properties (the bar denotes the closure in βZ):

(5) $W_s \cap Y \neq \emptyset$ for any $s \in \mathbb{D}^{<\omega}$;
(6) if $k \in \omega$, then $\overline{W}_s \subset (\bigcap\{G_i^{s(i)} : i \leq k\}) \cap O_k$ for every $s \in \mathbb{D}^{k+1}$;
(7) if $s, t \in \mathbb{D}^{<\omega}$ and $t \subset s$, then $\overline{W}_s \subset W_t$.

To start off, let $W_\emptyset = \beta Z$ and observe that the family $\mathcal{B}$ being dense on Y, there is $(G_0^0, G_0^1) \in \mathcal{B}$ such that $G_0^0 \cap Y \neq \emptyset \neq G_0^1 \cap Y$. We have $\mathbb{D}^1 = \{s_0, s_1\}$ where $s_i(0) = i$ for each $i \in \mathbb{D}$. It is easy to find $W_{s_i} \in \tau^*(\beta Z)$ such that $W_{s_i} \cap Y \neq \emptyset$ and $\overline{W}_{s_i} \subset O_0 \cap G_0^i$ for every $i \in \mathbb{D}$. Now assume that $n \in \mathbb{N}$ and we have families $\{(G_i^0, G_i^1) : i < n\}$ and $\{W_s : s \in \mathbb{D}^{\leq n}\}$ such that (5) is true for all $s \in \mathbb{D}^{\leq n}$, the property (6) holds for all $k < n$ and (7) is fulfilled for all $s, t \in \mathbb{D}^{\leq n}$ such that $t \subset s$.

It follows from the weak density of $\mathcal{B}$ on Y that there is $(G_n^0, G_n^1) \in \mathcal{B}$ such that $G_n^j \cap W_s \cap Y \neq \emptyset$ for any $j \in \mathbb{D}$ and $s \in \mathbb{D}^n$. For any $j \in \mathbb{D}$ we can find a set $W_{s^\frown j} \in \tau^*(\beta Z)$ such that $W_{s^\frown j} \cap Y \neq \emptyset$ and $\overline{W}_{s^\frown j} \subset W_s \cap G_n^j \cap O_n$. Once we have $W_{s^\frown j}$ for all $s \in \mathbb{D}^n$ and $j \in \mathbb{D}$, we obtain a family $\{W_s : s \in \mathbb{D}^{n+1}\}$ such that (5) is true for all $s \in D^{\leq n+1}$, the property (6) holds for all $k \leq n$ and (7) is fulfilled for all $s, t \in \mathbb{D}^{\leq n+1}$ such that $t \subset s$.

Consequently, our inductive procedure can be continued to provide families $\{(G_i^0, G_i^1) : i \in \omega\} \subset \mathcal{B}$ and $\{W_s : s \in \mathbb{D}^{<\omega}\} \subset \tau^*(\beta Z)$ for which the properties (5)–(7) hold. The property (6) implies that the set $K_n = \bigcup\{\overline{W}_s : s \in \mathbb{D}^n\}$ is compact and $K_{n+1} \subset O_n$ for all $n \in \omega$. Therefore $K = \bigcap\{K_n : n \in \omega\}$ is a non-empty compact subset of the space $\bigcap_{n \in \omega} O_n = Z$. Let $H_k^i = G_k^i \cap Z$ for all $k \in \omega$ and $i \in \mathbb{D}$. To see that the family $\mathcal{A}' = \{(H_i^0, H_i^1) : i \in \omega\} \subset \mathcal{A}$ is independent of K, take any $s \in \mathbb{D}^\omega$. For every $n \in \omega$ the set $F_n = \overline{W}_{s|n}$ is closed in βZ and non-empty because it contains $W_{s|n} \cap Y \neq \emptyset$ [see (5)]. The property (7) implies that

$F_n \supset F_{n+1}$ for all $n \in \omega$ and therefore $F = \bigcap_{n \in \omega} F_n \neq \emptyset$. Finally observe that $F \subset (\bigcap \{H_i^{s(i)} : i \in \omega\}) \cap K$ by (6), so $(\bigcap \{H_i^{s(i)} : i \in \omega\}) \cap K \neq \emptyset$ and Fact 2 is proved. $\qquad\square$

Fact 3. If M is a Polish space and $f \in \mathbb{R}^M \backslash B_1(M)$, then there is a closed $F \subset M$ and $p, q \in \mathbb{R}$ such that $p < q$ and there are $A, B \subset F$ for which $\overline{A} = \overline{B} = F$ while $f(x) \leq p$ for all $x \in A$ and $f(x) \geq q$ for every $x \in B$.

Proof. There exists a closed $P \subset M$ such that $f | P$ is discontinuous at every $x \in P$ (see Fact 3 of T.384). Let $\{(p_n, q_n) : n \in \omega\}$ be an enumeration of all pairs (p, q) of rational numbers such that $p < q$. It is an easy exercise to see that for any $x \in P$ there are $a_x, b_x \in \mathbb{R}$ such that $a_x < b_x$ and for every $U \in \tau(x, P)$ there are $y, z \in U$ with $f(y) \leq a_x$ and $f(z) \geq b_x$. Therefore $P = \bigcup_{n \in \omega} P_n$ where $P_n = \{x \in P :$ for every $U \in \tau(x, P)$ there are $y, z \in U$ with $f(y) \leq p_n$ and $f(z) \geq q_n\}$. It is straightforward that P_n is closed in P for every $n \in \omega$; since P is Polish, there is $m \in \omega$ such that $U = \text{Int}_P(P_m) \neq \emptyset$. Fix a base $\{W_n : n \in \omega\} \subset \tau^*(P)$ in the space U and let $p = p_m, q = q_m$. By definition of U, there are $x_k, y_k \in W_k$ such that $f(x_k) \leq p$ and $f(y_k) \geq q$ for all $k \in \omega$. It is immediate that the sets $A = \{x_k : k \in \omega\}$, $B = \{y_k : k \in \omega\}$ and $F = \overline{A} = \overline{B}$ are as promised so Fact 3 is proved. $\qquad\square$

Fact 4. Given a Polish space M suppose that $E \subset C_p(M)$ is a countable set and $K = \overline{E}$ is compact (the bar denotes the closure in $\mathbb{R}^M$). If the set K is not contained in $B_1(M)$, then $\beta\omega$ embeds in K.

Proof. If every infinite $E' \subset E$ has an accumulation point in the space $B_1(M)$, then $\overline{E} \subset B_1(M)$ by Fact 6 of T.384, so there is an infinite $D \subset E$ such that D has no accumulation points in $B_1(M)$; in particular, D is a discrete subspace of $C_p(M)$. On the other hand, the set $\overline{D}$ is compact, so we can fix a function $f \in \overline{D} \backslash B_1(M)$. By Fact 3, we can find a non-empty closed $F \subset M$ and $p, q \in \mathbb{R}$ such that $p < q$ and there are $A, B \subset F$ with $\overline{A} = \overline{B} = F$ for which $f(x) \leq p$ for every $x \in A$ and $f(x) \geq q$ for all $x \in B$. Take $r, t \in \mathbb{R}$ with $p < r < t < q$ and consider the family $\mathcal{A} = \{(U_h^0, U_h^1) : h \in D\}$ where $U_h^0 = \{x \in F : h(x) < r\}$ and $U_h^1 = \{x \in F : h(x) > t\}$ are open subsets of F for any $h \in D$. Assume that $W_1, \ldots, W_n \in \tau^*(F)$ and take points $x_i \in A \cap W_i$ and $y_i \in B \cap W_i$ for all $i \leq n$. The set $O = \{g \in \mathbb{R}^M : g(x_i) < r$ and $g(y_i) > t$ for all $i \leq n\}$ is open in $\mathbb{R}^M$ and $f \in O$. Since $f \in \overline{D}$, there is $h \in D$ such that $h \in O$. Consequently, $x_i \in U_h^0 \cap W_i$ and $y_i \in U_h^1 \cap W_i$ for all $i \leq n$ which shows that the family $\mathcal{A}$ is weakly dense on F (see Fact 2 for the definition of weakly dense families). Therefore we can apply Fact 2 to find a set $H = \{h_n : n \in \omega\} \subset D$ such that the family $\mathcal{A}' = \{(U_{h_n}^0, U_{h_n}^1) : n \in \omega\}$ is independent on some compact $Q \subset F$. We claim that $\overline{H}$ is homeomorphic to $\beta\omega$.

To prove it take any disjoint sets $R, S \subset H$ and choose a point $s \in \mathbb{D}^\omega$ such that $R \subset \{h_n : n \in s^{-1}(0)\}$ and $S \subset \{h_n : n \in s^{-1}(1)\}$. By independence of the family $\mathcal{A}'$ on F, there is a point $z \in \bigcap \{U_{h_i}^{s(i)} : i \in \omega\}$. If $h = h_n \in R$ then $s(n) = 0$ and hence $z \in U_{h_n}^0$, i.e., $h(z) = h_n(z) < r$. This implies $g(z) \leq r$ for any $g \in \overline{R}$.

Now, if $h = h_n \in S$, then $s(n) = 1$ and hence $z \in U^1_{h_n}$, i.e., $h(z) = h_n(z) > t$, so $g(z) \geq t$ for any $g \in \overline{S}$. As a consequence, $\overline{R} \cap \overline{S} = \emptyset$, i.e., we proved that $\overline{H}$ is a compact extension of a countable discrete space H for which $\overline{R} \cap \overline{S} = \emptyset$ for any disjoint $R, S \subset H$. Now apply Fact 2 of S.286 to conclude that $\overline{H} \subset K$ is homeomorphic to $\beta\omega$. Fact 4 is proved. $\square$

Fact 5. Suppose that M is a Polish space and $E \subset C_p(M)$ is a countable set such that $K = \overline{E}$ is compact (the bar denotes the closure in $\mathbb{R}^M$) and $K \subset B(M)$. Then $K \subset B_1(M)$ and hence K is Rosenthal compact.

Proof. Given a set A let $P_\omega(A) = \{B \subset A : |B| \leq \omega\}$. It is evident that if $|A| \leq \mathfrak{c}$, then $|P_\omega(A)| \leq \mathfrak{c}$. It follows from $nw(C_p(M)) = \omega$ that $|C_p(M)| \leq \mathfrak{c}$. Now, if $\alpha < \omega_1$ and we proved that $|L_\beta(C_p(M))| \leq \mathfrak{c}$, then for the set $L'_\alpha = \bigcup\{L_\beta(C_p(M)) : \beta < \alpha\}$, we have $|L'_\alpha| \leq \mathfrak{c}$. Since the set $L_\alpha(C_p(M))$ consists of limits of sequences from L'_α, we have $|L_\alpha(C_p(M))| \leq |P_\omega(L'_\alpha)| \leq \mathfrak{c}$. This proves that $|L_\alpha(C_p(M))| \leq \mathfrak{c}$ for each $\alpha < \omega_1$ and hence $|B(M)| = |\bigcup\{L_\alpha(C_p(M)) : \alpha < \omega_1\}| \leq \mathfrak{c} \cdot \omega_1 = \mathfrak{c}$ (see Fact 1).

Now, if K is not contained in $B_1(M)$, then $\beta\omega$ embeds in K by Fact 4 and hence $|K| \geq |\beta\omega| = 2^\mathfrak{c}$ (see Problem 368 of [TFS]). On the other hand, $|K| \leq |B(M)| \leq \mathfrak{c}$ which is a contradiction showing that $K \subset B_1(M)$ and hence Fact 5 is proved. $\square$

Returning to our solution assume that X is a separable Rosenthal compact and hence there is a Polish space N such that $X \subset B_1(N)$. Fix an arbitrary dense countable $A \subset X$. By Fact 12 of T.384, there is a Polish topology μ on the set N such that $\tau(N) \subset \mu$ and every $f \in A$ is continuous on the space $M = (N, \mu)$. It is evident that we still have $X \subset B_1(M)$ while $A \subset C_p(M)$. Let $C = C_p(X)$ and $C_B = C_p(B|X)$ for any $B \subset X$. We claim that

(8) a function $\varphi : X \to \mathbb{R}$ belongs to C if and only if for any $\varepsilon > 0$ there is $n \in \mathbb{N}$ and $p = (p_1, \ldots, p_n) \in M^n$ such that for any $x, y \in X$ with $|x(p_i) - y(p_i)| < \frac{1}{n}$ for all $i \leq n$, we have $|\varphi(x) - \varphi(y)| < \varepsilon$.

If $x \in X$ is fixed, then the set $U = \{y \in X : |x(p_i) - y(p_i)| < \frac{1}{n}$ for all $i \leq n\}$ is an open neighborhood of x in X such that $\varphi(U) \subset (\varphi(x) - \varepsilon, \varphi(x) + \varepsilon)$, i.e., U witnesses continuity of φ at the point x. Therefore any $\varphi \in \mathbb{R}^X$ which satisfies (8) is continuous on X.

Let us prove that any function $\varphi \in C_p(X)$ satisfies (8). For any $n \in \mathbb{N}, \delta > 0$ and $p = (p_1, \ldots, p_n) \in M^n$, let $O(x, p, \delta) = \{y \in X : |y(p_i) - x(p_i)| < \delta$ for all $i \leq n\}$. For any $x \in X$ the family $\{O(x, p, \delta) : p \in M^n$ for some $n \in \mathbb{N}$ and $\delta > 0\}$ is a local base of X at the point x. Thus there exist $p_x = (p^x_1, \ldots, p^x_{n(x)}) \in M^{n(x)}$ and $\delta_x > 0$ such that $|\varphi(y) - \varphi(x)| < \varepsilon/2$ for every $y \in O(x, p_x, \delta_x)$. Since X is compact, there is a finite $E \subset X$ such that $\bigcup\{O(x, p_x, \delta_x/3) : x \in E\} = X$. It is easy to find a finite set $D = \{p_1, \ldots, p_n\} \subset M$ such that $\frac{1}{n} < \min\{\delta_x/3 : x \in E\}$ and $p^x_i \in D$ for all $x \in E$ and $i \leq n(x)$. Now assume that $u, v \in X$ and $|u(p_i) - v(p_i)| < \frac{1}{n}$ for all $i \leq n$. There is $x \in E$ such that $u \in O(x, p_x, \delta_x/3)$. Then $|v(p^x_i) - x(p^x_i)| \leq |v(p^x_i) - u(p^x_i)| + |u(p^x_i) - x(p^x_i)| \leq \frac{1}{n} + \frac{\delta_x}{3} < \delta_x$ and therefore

$v \in O(x, p_x, \delta_x)$ which shows that $|\varphi(u) - \varphi(v)| \leq |\varphi(u) - \varphi(x)| + |\varphi(x) - \varphi(v)| < \varepsilon/2 + \varepsilon/2 = \varepsilon$. Thus the point $p = (p_1, \ldots, p_n)$ witnesses that (8) holds for the function φ and hence (8) is proved.

Now we are ready to characterize the functions from C_A. It turns out that

(9) a function $\varphi : A \to \mathbb{R}$ belongs to C_A if and only if for any $\varepsilon > 0$ there is $n \in \mathbb{N}$ and $p = (p_1, \ldots, p_n) \in M^n$ such that for any $x, y \in A$ with $|x(p_i) - y(p_i)| < \frac{1}{n}$ for all $i \leq n$, we have $|\varphi(x) - \varphi(y)| < \varepsilon$.

If $\varphi \in C$, then φ satisfies (8); it is immediate that $\varphi|A$ satisfies (9) (note that (8) and (9) are identical except that A is substituted everywhere in place of X).

Now assume that $\varphi : A \to \mathbb{R}$ and φ has the property introduced in (9). For every $x \in X$ let $\operatorname{osc}(\varphi, x) = \inf\{\operatorname{diam}(\varphi(U \cap A) : U \in \tau(x, X)\}$. If $\varepsilon > 0$, then (9) implies that we can choose $n \in \mathbb{N}$ and $p = (p_1, \ldots, p_n) \in M^n$ such that for any $u, v \in A$ with $|u(p_i) - v(p_i)| < \frac{1}{n}$ for all $i \leq n$, we have $|\varphi(u) - \varphi(v)| < \varepsilon/2$. The set $U = \{y \in X : |y(p_i) - x(p_i)| < \frac{1}{2n}$ for all $i \leq n\}$ is an open neighborhood of x in X; if $u, v \in U \cap A$, then $|u(p_i) - v(p_i)| \leq |u(p_i) - x(p_i)| + |x(p_i) - v(p_i)| < \frac{1}{2n} + \frac{1}{2n} = \frac{1}{n}$ which shows that $|\varphi(u) - \varphi(v)| < \frac{\varepsilon}{2}$ and therefore $\operatorname{diam}(\varphi(U \cap A)) \leq \frac{\varepsilon}{2} < \varepsilon$. Since $x \in X$ and $\varepsilon > 0$ were chosen arbitrarily, we established that $\operatorname{osc}(\varphi, x) = 0$ for all $x \in X$ and hence there is a continuous $\varphi_1 : X \to \mathbb{R}$ such that $\varphi_1|A = \varphi$ (see Fact 3 of T.368). Thus $\varphi \in C_A$ and (9) is proved.

For all $k, n \in \mathbb{N}$ consider the set $R(k, n) = \{(\varphi, p) \in \mathbb{R}^A \times M^n : p = (p_1, \ldots, p_n)$ and $|\varphi(u) - \varphi(v)| \leq \frac{1}{k}$ whenever $u, v \in A$ and $|u(p_i) - v(p_i)| < \frac{1}{n}$ for all $i \leq n\}$. The set $R(k, n)$ is closed in $\mathbb{R}^A \times M^n$; to prove it take $p = (p_1, \ldots, p_n) \in M^n$ and $\varphi \in \mathbb{R}^A$ such that $(\varphi, p) \in (\mathbb{R}^A \times M^n) \backslash R(k, n)$. Then there exist $u, v \in A$ such that $|u(p_i) - v(p_i)| < \frac{1}{n}$ for all $i \leq n$ and $|\varphi(u) - \varphi(v)| > \frac{1}{k}$. Observe that the set $W = \{\theta \in \mathbb{R}^A : |\theta(u) - \theta(v)| > \frac{1}{k}\}$ is open in $\mathbb{R}^A$ and $\varphi \in W$. The functions u and v being continuous on M, the set $G = \{q = (q_1, \ldots, q_n) \in M^n : |u(q_i) - v(q_i)| < \frac{1}{n}$ for all $i \leq n\}$ is open in M^n and $p \in G$. It is easy to check that $H = W \times G$ is an open neighborhood of (φ, p) in $\mathbb{R}^A \times M^n$ such that $H \cap R(n, k) = \emptyset$. This proves that $R(n, k)$ is closed in the Polish space $\mathbb{R}^A \times M^n$ for all $n, k \in \mathbb{N}$. Let $\pi_n : \mathbb{R}^A \times M^n \to \mathbb{R}^A$ be the natural projection.

It turns out that

(10) if $S(k) = \bigcup\{\pi_n(R(n, k)) : n \in \mathbb{N}\}$ for all $k \in \mathbb{N}$, then $C_A = \bigcap\{S(k) : k \in \mathbb{N}\}$.

Checking the property (10) is a simple application of (9). If $\varphi \in C_A$, then, for any $k \in \mathbb{N}$, there exists $n \in \mathbb{N}$ and $p = (p_1, \ldots, p_n) \in M^n$ such that $|\varphi(u) - \varphi(v)| < \frac{1}{k}$ whenever $|u(p_i) - v(p_i)| < \frac{1}{n}$ for all $i \leq n$. Therefore $(\varphi, p) \in R(n, k)$ and hence $\varphi \in \pi_n(R(n, k))$ which shows that $\varphi \in \bigcap\{S(k) : k \in \mathbb{N}\}$. This proves the inclusion $C_A \subset \bigcap\{S(k) : k \in \mathbb{N}\}$.

Now if $\varphi \in \bigcap\{S(k) : k \in \mathbb{N}\}$, then take $\varepsilon > 0$; there exists $k \in \mathbb{N}$ for which $\frac{1}{k} < \varepsilon$. Since $\varphi \in \pi_n(R(n, k))$ for some $n \in \mathbb{N}$, there is $p = (p_1, \ldots, p_n) \in M^n$ such that $(\varphi, p) \in R(n, k)$ and therefore $|\varphi(u) - \varphi(v)| \leq \frac{1}{k} < \varepsilon$ for any $u, v \in A$

such that $|u(p_i) - v(p_i)| < \frac{1}{n}$ for all $i \leq n$. As a consequence, (9) holds for φ so $\varphi \in C_A$. Thus $C_A = \bigcap \{S(k) : k \in \mathbb{N}\}$ and (10) is proved.

Finally, observe that $R(n, k)$ is Polish and hence $\pi_n(R(n, k))$ is analytic for all $n, k \in \mathbb{N}$. Therefore $S(k)$ is analytic by Problem 337 and hence C_A is analytic by Problem 336. We established that for any separable Rosenthal compact X and any dense countable $A \subset X$, the space $C_A = C_p(A|X)$ is analytic, i.e., we settled necessity.

Now assume that the space C_B is analytic for any countable dense $B \subset X$; fix some countable dense set $A \subset X$ and a continuous onto map $r : M \to C_A$ of some Polish space M onto C_A. Let $\pi : C \to C_A$ is the relevant restriction map. Since π is a condensation, the map $\xi = \pi^{-1} : C_A \to C$ is well-defined. If $s = \xi \circ r$, then $s : M \to C$; the map s might be discontinuous, but it is continuous if M and C are considered with their respective discrete topologies. Therefore the map $s^* : \mathbb{R}^C \to \mathbb{R}^M$ defined by $s(f) = f \circ s$ for each $f \in \mathbb{R}^C$ is an embedding by Problem 163 of [TFS].

Another easy observation is that if $x \in X$ and $e_x(f) = f(x)$ for any $f \in C$, then the correspondence $x \to e_x$ is an embedding of X into $C_p(C) \subset \mathbb{R}^C$ (see Problem 167 of [TFS]). If $X' = \{e_x : x \in X\} \subset \mathbb{R}^C$, then X' is a homeomorphic copy of X and hence $Y = s^*(X') \subset \mathbb{R}^M$ is also a homeomorphic copy of X.

Given any $a \in A$ the map $e_a \circ \xi : C_A \to \mathbb{R}$ is continuous on C_A because for any $O \in \tau(\mathbb{R})$ we have $(e_a \circ \xi)^{-1}(O) = \{f \in C_A : f(a) \in O\}$ is an open set of C_A. Therefore $s^*(e_a) = e_a \circ s = e_a \circ (\xi \circ r) = (e_a \circ \xi) \circ r$ is a continuous map on M which shows that $A' = s^*(A) \subset C_p(M)$.

Now if $x \in X \setminus A$, then $e_x \circ \xi$ is a measurable map on C_A. Indeed, for any $O \in \tau(\mathbb{R})$, we have $U = (e_x \circ \xi)^{-1}(O) = \{f \in C_A : \xi(f)(x) \in O\}$. We know that for $B = A \cup \{x\}$, the set C_B is analytic. Let $\pi' : C \to C_B$ be the relevant restriction map and $\xi' = (\pi')^{-1}$. As before, we can establish that the map $e_x \circ \xi'$ is a continuous map on C_B because $U' = (e_a \circ \xi')^{-1}(O) = \{f \in C_B : f(x) \in O\}$ is an open subset of C_B for any $O \in \tau(\mathbb{R})$. We also have the natural restriction map $\pi_0 : C_B \to C_A$ defined by the formula $\pi_0(f) = f|A$ for all $f \in C_B$. Let $F' = C_B \setminus U'$ and $F = C_A \setminus U$; it is immediate that $U = \pi_0(U')$ and $F = \pi_0(F')$. The space C_B being analytic, the sets F' and U' are also analytic and hence F and U are analytic as well.

Since π_0 is a condensation, we have $F \cap U = \emptyset$, so we can apply Problem 339 to the Polish space $\mathbb{R}^A$ and disjoint analytic sets U and F. As a consequence, there are disjoint $U_0, F_0 \in \mathbb{B}(\mathbb{R}^A)$ such that $U \subset U_0$, $F \subset F_0$. But $F \cup U = C_A$ and therefore $F_0 \cap C_A = F$ and $U_0 \cap C_A = U$ which shows that U is a Borel subset of C_A (see Fact 1 of T.319). Thus $(e_x \circ \xi)^{-1}(O)$ is a Borel subset of C_A for any open $O \subset \mathbb{R}$ which implies that $e_x \circ \xi$ is a measurable map (see Fact 1 of T.363). Consequently, $(e_x \circ \xi) \circ r$ is also measurable because r is continuous (in fact, it is a trivial exercise that a composition of two measurable maps is a measurable map).

Therefore $s^*(e_x) = e_x \circ s = e_x \circ (\xi \circ r) = (e_x \circ \xi) \circ r$ is a measurable map on M which shows $Y \subset B(M)$. Thus Y is a closure in $\mathbb{R}^M$ of the set $A' \subset C_p(M)$ and we can apply Fact 5 to conclude that $Y \subset B_1(M)$ and hence Y is Rosenthal compact. Since Y is homeomorphic to X, the space X is also Rosenthal compact and hence we proved sufficiency and completed our solution.

T.386. *Let X be a compact space. Assume that A and B are dense countable subsets of X such that $C_p(A|X)$ is analytic and $C_p(B|X)$ is not. Prove that X contains a subspace homeomorphic to $\beta\omega$.*

Solution. Let $C = C_p(X)$ and $C_A = C_p(A|X)$; fix a continuous surjective map $r : M \to C_A$ of some Polish space M onto C_A. Let $\pi : C \to C_A$ is the relevant restriction map. Since π is a condensation, the map $\xi = \pi^{-1} : C_A \to C$ is well-defined. If $s = \xi \circ r$, then $s : M \to C$; the map s might be discontinuous, but it is continuous if M and C are considered with their respective discrete topologies. Therefore the map $s^* : \mathbb{R}^C \to \mathbb{R}^M$ defined by $s(f) = f \circ s$ for each $f \in \mathbb{R}^C$ is an embedding by Problem 163 of [TFS].

Another easy observation is that if $x \in X$ and $e_x(f) = f(x)$ for any $f \in C$, then the correspondence $x \to e_x$ is an embedding of X into $C_p(C) \subset \mathbb{R}^C$ (see Problem 167 of [TFS]). If $X' = \{e_x : x \in X\} \subset \mathbb{R}^C$, then X' is a homeomorphic copy of X and hence $Y = s^*(X') \subset \mathbb{R}^M$ is also a homeomorphic copy of X.

Given any $a \in A$ the map $e_a \circ \xi : C_A \to \mathbb{R}$ is continuous on C_A because for any $O \in \tau(\mathbb{R})$ we have $(e_a \circ \xi)^{-1}(O) = \{f \in C_A : f(a) \in O\}$ is an open set of C_A. Therefore

$$s^*(e_a) = e_a \circ s = e_a \circ (\xi \circ r) = (e_a \circ \xi) \circ r$$

is a continuous map on M which shows that $A' = s^*(A) \subset C_p(M)$.

Since A is dense in X, the set A' is dense in Y; if $Y \subset B_1(M)$, then Y Rosenthal compact and hence so is X which implies that $C_p(B|X)$ is analytic (see Problem 385). This contradiction shows that Y is not contained in $B_1(M)$ and therefore $\beta\omega$ embeds in Y by Fact 4 of T.385. Since Y is homeomorphic to X, the space $\beta\omega$ embeds in X as well.

T.387. *Suppose that X is a compact space and A is a dense countable subset of X such that $C_p(A|X)$ is analytic. Prove that X is Rosenthal compact or else $\beta\omega$ embeds in X.*

Solution. If the space X is not Rosenthal compact, then there is a countable dense $B \subset X$ such that $C_p(B|X)$ is not analytic (see Problem 385). Now we can apply Problem 386 to conclude that $\beta\omega$ embeds in X.

T.388. *Prove that a space X is K-analytic if and only if X is an image of $\mathbb{P}$ under a compact-valued upper semicontinuous map, i.e., when there exists a compact-valued upper semicontinuous $\varphi : \mathbb{P} \to X$ such that $\bigcup\{\varphi(p) : p \in \mathbb{P}\} = X$.*

Solution. Necessity was proved in Fact 2 of T.346. To prove sufficiency, assume that $\varphi : \mathbb{P} \to \exp(X)$ is a compact-valued upper semicontinuous map such that $\bigcup\{\varphi(p) : p \in \mathbb{P}\} = X$. For any $A \subset \mathbb{P}$ let $\varphi(A) = \bigcup\{\varphi(p) : p \in A\}$.

The set $F = \{(p, x) \in \mathbb{P} \times \beta Z : x \in \varphi(p)\}$ is closed in the space $\mathbb{P} \times \beta X$. Indeed, if we are given a point $(q, y) \in (\mathbb{P} \times \beta Z) \setminus F$, then $y \notin \varphi(q)$; since $\varphi(q)$ is compact, there is $U \in \tau(y, \beta X)$ such that $\overline{U} \cap F = \emptyset$ (the bar denotes the closure in βX). We have $F \subset V = (\beta X \backslash \overline{U}) \cap X \in \tau(F, X)$ and therefore there is a set $G \in \tau(q, \mathbb{P})$ for which $\varphi(G) \subset V$. The set $W = G \times U$ is an open neighborhood of the point (q, y) in the space $\mathbb{P} \times \beta X$.

If $(r, z) \in W$, then $\varphi(r) \subset V$ while $z \in U \subset \beta X \backslash V$ which shows that $z \notin \varphi(r)$. Thus $W \cap F = \emptyset$ and therefore every point of $(\mathbb{P} \times \beta X) \backslash F$ has a neighborhood contained in $(\mathbb{P} \times \beta X) \backslash F$. This proves that the set $(\mathbb{P} \times \beta X) \backslash F$ is open and hence F is closed in $\mathbb{P} \times \beta X$.

Let $\pi : \mathbb{P} \times \beta X \to \beta X$ be the natural projection. The spaces $\mathbb{P}$ and βX are K-analytic, so F is also K-analytic (see Problem 343). It is immediate that $\pi(F) = X$, so X is K-analytic being a continuous image of a K-analytic space.

T.389. *Prove that an arbitrary space X is K-analytic if and only if there exists a family $\mathcal{K} = \{K_f : f \in \omega^\omega\}$ of compact subsets of X with the following properties:*

(i)   the family $\mathcal{K}$ is a cover of X, i.e., $\bigcup \mathcal{K} = X$;
(ii) if a sequence $f_n \in \omega^\omega$ converges to $f \in \omega^\omega$ and $x_n \in K_{f_n}$ for all $n \in \omega$, then the sequence $\{x_n\}$ has an accumulation point which belongs to K_f.

Solution. We identify $\mathbb{P}$ with ω^ω. To prove sufficiency, assume that there exists a family $\mathcal{K} = \{K_f : f \in \mathbb{P}\}$ with the properties (i) and (ii). For any $f \in \mathbb{P}$, let $\varphi(f) = K_f$. Then $\varphi : \mathbb{P} \to X$ is a compact-valued onto map; for any $A \subset \mathbb{P}$, let $\varphi(A) = \bigcup \{\varphi(f) : f \in A\}$. To see that φ is upper semicontinuous, fix a point $f \in \mathbb{P}$ and $U \in \tau(\varphi(f), X)$. For every $n \in \omega$, let $O(f, n) = \{g \in \mathbb{P} : g | n = f | n\}$. Then the family $\{O(f, n) : n \in \omega\}$ is a local base of $\mathbb{P}$ at f.

Assume that $\varphi(O(f, n))$ is not contained in U and fix a point $x_n \in \varphi(O(f, n)) \backslash U$ for every $n \in \omega$. For each $n \in \omega$, there is $f_n \in O(f, n)$ such that $x_n \in \varphi(f_n)$. It is evident that $f_n \to f$ and hence we can apply (2) to conclude that there is there is an accumulation point $x \in \varphi(f)$ of the sequence $S = \{x_n : n \in \omega\}$. However, $S \subset X \backslash U$ and hence all accumulation points of S lie in $X \backslash U \subset X \backslash \varphi(f)$; this contradiction shows that $\varphi(O(f, n)) \subset U$ for some $n \in \omega$ and hence φ is upper semicontinuous (see Fact 1 of T.346). Thus X is K-analytic by Problem 388 and we proved sufficiency.

Now, if X is K-analytic, then there is an upper semicontinuous compact-valued onto map $\varphi : \mathbb{P} \to X$ (see Problem 388); let $K_f = \varphi(f)$ for any $f \in \mathbb{P}$. We claim that the family $\mathcal{K} = \{K_f : f \in \mathbb{P}\}$ satisfies (i) and (ii). Indeed, the property (i) is an immediate consequence of the fact that φ is onto. Suppose that a sequence $S = \{f_n : n \in \omega\} \subset \mathbb{P}$ converges to some $f \in \omega$ and take $x_n \in \varphi(f_n)$ for all $n \in \omega$. If S has no accumulation points in set $\varphi(f) = K_f$, then for any $y \in K_f$ there is $U_y \in \tau(y, X)$ such that the set $P(U_y) = \{i \in \omega : x_i \in U_y\}$ is finite. The set K_f being compact there are $y_1, \ldots, y_k \in K_f$ such that $K_f \subset U = U_{y_1} \cup \cdots \cup U_{y_k}$. The map φ is upper semicontinuous, so we can find $V \in \tau(f, \mathbb{P})$ for which $\varphi(V) \subset U$.

Since the sequence S converges to f, there is a number $m \in \omega$ such that the set $S' = \{f_i : i \geq m\}$ is contained in V. As a consequence, $\varphi(f_i) \subset U$ and hence there is $s(i) \in N_k = \{1, \ldots, k\}$ such that $x_i \in U_{y_{s(i)}}$ for all $i \geq m$. The set N_k

being finite, there is $l \in N_k$ such that $s(i) = l$ for infinitely many i. Consequently, $P(U_{y_l})$ is infinite which is a contradiction. Thus the sequence $\{x_n : n \in \omega\}$ has an accumulation point in K_f and hence the family $\mathcal{K}$ has the property (2). This proves necessity and makes our solution complete.

T.390. *Prove that an arbitrary space X is K-analytic if and only if there exists a space Y which maps perfectly onto $\mathbb{P}$ and continuously onto X.*

Solution. Suppose that some Y maps perfectly onto $\mathbb{P}$ and continuously onto X. Since every continuous image of a K-analytic space is K-analytic, it suffices to show that Y is K-analytic. Take a perfect map $f : Y \to \mathbb{P}$ and let $\varphi(p) = f^{-1}(p)$ for each $p \in \mathbb{P}$. Then $\varphi : \mathbb{P} \to Y$ is a compact-valued onto map. If $p \in \mathbb{P}$ and $U \in \tau(\varphi(p), Y)$, then there is $V \in \tau(p, \mathbb{P})$ for which $f^{-1}(V) \subset U$ (see Fact 1 of S.226). In terms of φ this says that $\varphi(V) = \bigcup\{\varphi(q) : q \in V\} \subset U$ and hence φ is upper semicontinuous (see Fact 1 of T.346). Thus we can apply Problem 388 to conclude that the space Y is K-analytic.

To prove necessity assume that X is K-analytic and fix a compact-valued upper semicontinuous onto map $\varphi : \mathbb{P} \to X$. The set $Y = \{(p, x) \in \mathbb{P} \times \beta Z : x \in \varphi(p)\}$ is closed in $\mathbb{P} \times \beta X$. Indeed, if we are given a point $(q, y) \in (\mathbb{P} \times \beta Z)\backslash Y$, then $y \notin \varphi(q)$; since $\varphi(q)$ is compact, there is $U \in \tau(y, \beta X)$ such that $\overline{U} \cap Y = \emptyset$ (the bar denotes the closure in βX). We have $Y \subset V = (\beta X\backslash\overline{U}) \cap X \in \tau(Y, X)$ and therefore there is $G \in \tau(q, \mathbb{P})$ for which $\varphi(G) \subset V$. The set $W = G \times U$ is an open neighborhood of (q, y) in $\mathbb{P} \times \beta X$. If $(r, z) \in W$, then $\varphi(r) \subset V$ while $z \in U \subset \beta X\backslash V$ which shows that $z \notin \varphi(r)$. Thus $W \cap Y = \emptyset$ and therefore every point of $(\mathbb{P} \times \beta X)\backslash Y$ has a neighborhood contained in $(\mathbb{P} \times \beta X)\backslash Y$. This proves that the set $(\mathbb{P} \times \beta X)\backslash Y$ is open and hence Y is closed in $\mathbb{P} \times \beta X$.

If $\pi_1 : \mathbb{P} \times \beta X \to \beta X$ be the natural projection, then $\pi(Y) = X$, so X is a continuous image of the space Y. On the other hand, if $\pi_0 : \mathbb{P} \times \beta X \to \mathbb{P}$ is the other natural projection, then π_0 is a perfect map (see Fact 3 of S.288). Therefore $f = \pi_0|Y$ is a perfect map as well because $\pi_0(Y) = \mathbb{P}$ and Y is closed in $\mathbb{P} \times \beta X$. Thus X is a continuous image of the space Y which maps perfectly onto $\mathbb{P}$.

T.391. *Prove that an arbitrary space X is K-analytic if and only if X is realcompact and has a $\mathbb{P}$-directed compact cover.*

Solution. We identify $\mathbb{P}$ with the space ω^ω, letting, as usual, $\omega^0 = \{\emptyset\}$ and $\omega^{<\omega} = \bigcup\{\omega^n : n \in \omega\}$; if $p, q \in \mathbb{P}$, then $p \leq q$ stands for "$p(i) \leq q(i)$ for all $i \in \omega$". If f is a function, then $\mathrm{dom}(f)$ is its domain; given functions f and g the expression $f \subset g$ says that $\mathrm{dom}(f) \subset \mathrm{dom}(g)$ and $g|\mathrm{dom}(f) = f$. If Z is a space and we have a sequence $S = \{z_n : n \in \omega\} \subset Z$, then $z \in Z$ is *an accumulation point for the sequence S* if for any $U \in \tau(z, Z)$ the set $\{n \in \omega : z_n \in U\}$ is infinite. For the sake of brevity, we will call a space $\mathbb{P}$-*dominated* if it has a $\mathbb{P}$-directed compact cover. A cover $\{K_p : p \in \mathbb{P}\}$ of a space Z is *sequentially subcontinuous* if for any sequence $\{p_n : n \in \omega\} \subset \mathbb{P}$ which converges to a point $p \in \mathbb{P}$ and any sequence $S = \{z_n : n \in \omega\} \subset Z$ such that $z_n \in K_{p_n}$ for all $n \in \omega$, there is $z \in K_p$ which is an accumulation point for the sequence S.

Given a $\mathbb{P}$-directed compact cover $\mathcal{K} = \{K_p : p \in \mathbb{P}\}$ of a space Z and $s \in \omega^{<\omega}$, let $H_s = \bigcup\{K_p : s \subset p\}$. Once we have the family $\mathcal{H} = \{H_s : s \in \omega^{<\omega}\}$ (which we will call *the envelope of the family* $\mathcal{K}$), define the set $F_p = \bigcap\{H_{p|n} : n \in \omega\}$ for any $p \in \mathbb{P}$ and call the family $\mathcal{K}' = \{F_p : p \in \mathbb{P}\}$ *the regularization* of the family $\mathcal{K}$.

Fact 1. Given a $\mathbb{P}$-directed compact cover $\mathcal{K} = \{K_p : p \in \mathbb{P}\}$ of a space Z, let $\mathcal{H} = \{H_s : s \in \omega^{<\omega}\}$ be it envelope and $\mathcal{K}' = \{F_p : p \in \mathbb{P}\}$ its regularization. Then, for any $p \in \mathbb{P}$ and any sequence $S = \{z_n : n \in \omega\} \subset Z$ such that $z_n \in H_{p|n}$ for all $n \in \omega$, there is $z \in F_p$ which is an accumulation point for the sequence S. In particular, F_p is countably compact for all $p \in \mathbb{P}$.

Proof. It follows from $z_n \in H_{p|n}$ that there is $q_n \in \mathbb{P}$ such that $p|n \subset q_n$ and $z_n \in F_{q_n}$ for all $n \in \omega$. For any $k \in \omega$ we have $q_n(k) = p(k)$ for all $n \geq k$ which shows that the set $\{q_n(k) : n \in \omega\}$ is finite for any $k \in \omega$. Therefore the number $q(k) = \sum\{q_n(k) : n \in \omega\}$ is well-defined for every $k \in \omega$ and hence we have a function $q \in \mathbb{P}$ such that $q_n \leq q$ for all $n \in \omega$. The cover $\mathcal{K}$ being $\mathbb{P}$-directed, we have $K_{q_n} \subset K_q$ for all $n \in \omega$ and hence $\{z_n : n \in \omega\} \subset K_q$. Since the set K_q is compact, we proved that

(1) the set L of accumulation points of the sequence S is non-empty.

Our next step is to establish that $L \subset F_p$, which will, evidently, finish the proof of this fact. So take any $z \in L$ and fix $n \in \omega$. It is clear that z is also an accumulation point of the sequence $S' = \{z_k : k \geq n\}$. For any $y = z_k \in S'$ we have $q_k|n = p|n$ and the set $\{q_k(i) : k \geq n\}$ is finite for any $i \geq n$. Thus we can define a point $r \in \mathbb{P}$ by $r(i) = p(i)$ for all $i < n$ and $r(i) = \sum\{q_k(i) : k \geq n\}$ for all $i \geq n$. Then $q_k \leq r$ and hence $F_{q_k} \subset F_r$ for all $k \geq n$, i.e., the sequence $\{z_k : k \geq n\}$ is contained in the set F_r. Since $p|n \subset r$, we have $F_r \subset H_{p|n}$, so all accumulation points of S' belong to $H_{p|n}$ and, in particular, $z \in H_{p|n}$. The number $n \in \omega$ was chosen arbitrarily, so we proved that $z \in H_{p|n}$ for all $n \in \omega$ and therefore $z \in F_p$, i.e., $L \subset F_p$. In particular, if $S \subset F_p$, then $S \subset H_{p|n}$ and hence $z_n \in H_{p|n}$ for all $n \in \omega$. By what we proved above, S has accumulation points in F_p, so F_p is countably compact and Fact 1 is proved. $\qquad\square$

Fact 2. Given a $\mathbb{P}$-directed compact cover $\mathcal{K} = \{K_p : p \in \mathbb{P}\}$ of a space Z its regularization $\mathcal{K}' = \{F_p : p \in \mathbb{P}\}$ is sequentially subcontinuous.

Proof. Let $\{H_s : s \in \omega^{<\omega}\}$ be the envelope of $\mathcal{K}$. Take an arbitrary sequence $T = \{p_n : n \in \omega\} \subset \mathbb{P}$ which converges to a point $p \in \mathbb{P}$ and assume that $z_n \in F_{p_n}$ for all $n \in \omega$. It is evident that it suffices to prove our statement for a subsequence of the sequence T. It follows from $p_n \to p$ that for any $n \in \omega$, there is $k_n \in \omega$ such that $p_i|n = p|n$ for all $i \geq k_n$. Thus, passing to an appropriate subsequence of the sequence T, we can assume, without loss of generality, that

(2) $p_i|n = p|n$ for all $i \geq n$.

Now it follows from $z_n \in F_{p_n}$ that $z_n \in H_{p_n|n} = H_{p|n}$ for all $n \in \omega$ and hence we can apply Fact 1 to conclude that the sequence $\{z_n : n \in \omega\}$ has an accumulation point $z \in F_p$. Fact 2 is proved.   □

Fact 3. If Z is $\mathbb{P}$-dominated and every pseudocompact closed $E \subset Z$ is compact, then Z is K-analytic.

Proof. Fix a $\mathbb{P}$-directed compact cover $\mathcal{K} = \{L_p : p \in \mathbb{P}\}$ of the space Z. The regularization $\mathcal{K}' = \{F_p : p \in \mathbb{P}\}$ of the cover $\mathcal{K}$ consists of countably compact subsets of Z by Fact 1. For any $p \in \mathbb{P}$ let $\xi(p) = F_p$; then $\xi : \mathbb{P} \to Z$ is a multivalued map. To see that it is onto take any $z \in Z$; then there is $p \in \mathbb{P}$ such that $z \in L_p$. It is evident that $L_p \subset F_p$ so $z \in F_p$ and hence ξ is an onto map. It turns out that

(3)   the map ξ is upper semicontinuous.

Indeed, take any $p \in \mathbb{P}$ and $U \in \tau(\xi(p), Z)$. For any $n \in \omega$, consider the set $O_n = \{q \in \mathbb{P} : q|n = p|n\}$. Then $\{O_n : n \in \omega\}$ is a local base of $\mathbb{P}$ at p. If $\xi(O_n)$ is not contained in U for any $n \in \omega$, then pick $p_n \in O_n$ and $z_n \in \xi(p_n)\backslash U = F_{p_n}\backslash U$ for all $n \in \omega$. Since $p_n \to p$, there is a point $z \in F_p$ which is an accumulation point of the sequence $S = \{z_n : n \in \omega\}$ (see Fact 2). However, $S \subset X\backslash U$ and hence all accumulation points of S belong to $X\backslash U \subset X\backslash F_p$; this contradiction shows that $\xi(O_n) \subset U$ for some $n \in \omega$ and hence ξ is upper semicontinuous (see Fact 1 of T.346).

By our assumption about the space Z, the closure of every countably compact subset of Z is pseudocompact and hence compact (see Fact 18 of S.351). Thus the set $K_p = \overline{F}_p$ is compact for any $p \in \mathbb{P}$ and therefore $\{K_p : p \in \mathbb{P}\}$ is a compact cover of the space Z. Let $\varphi(p) = K_p$ for every $p \in \mathbb{P}$. Then $\varphi : \mathbb{P} \to Z$ is a compact-valued onto map. To see that φ is upper semicontinuous take any $p \in \mathbb{P}$ and $U \in \tau(\varphi(p), Z)$. It follows from compactness of $K_p = \varphi(p)$ that there is $V \in \tau(K_p, Z)$ such that $\overline{V} \subset U$. Since $\xi(p) = F_p \subset K_p$, we have $\xi(p) \subset V$ and hence we can apply (3) to find $O \in \tau(p, \mathbb{P})$ such that $\xi(O) \subset V$. Now, for any $q \in O$ we have $F_q = \xi(q) \subset V$ and hence $\varphi(q) = K_q = \overline{F}_q \subset \overline{V} \subset U$. This shows that $\varphi(O) \subset U$ and hence the map φ is also upper semicontinuous. Therefore Problem 388 is applicable to conclude that Z is K-analytic. Fact 3 is proved.   □

Returning to our solution let us show first that if the space X is K-analytic, then it is $\mathbb{P}$-dominated. To do this, observe that there exists a compact-valued upper semicontinuous onto map $\varphi : \mathbb{P} \to X$ (see Problem 388). For any $p \in \mathbb{P}$ and $i \in \omega$ let $N_i = \{0, \ldots, p(i)\}$; then $M_p = \{q \in \mathbb{P} : q \le p\} = \prod\{N_i : i \in \omega\}$ is a compact subset of $\mathbb{P}$. It is evident that $p \le q$ implies $M_p \subset M_q$ so $\{M_p : p \in \mathbb{P}\}$ is a $\mathbb{P}$-directed compact cover of $\mathbb{P}$. The set $K_p = \varphi(M_p) = \bigcup\{\varphi(q) : q \in M_p\}$ is compact for every $p \in \mathbb{P}$ (to see this observe that $\varphi_p = \varphi|M_p : M_p \to \varphi(M_p)$ is a compact-valued upper semicontinuous onto map and apply Problem 241). It is immediate that $p \le q$ implies $K_p \subset K_q$, so $\{K_p : p \in \mathbb{P}\}$ is a $\mathbb{P}$-directed compact cover of X. Thus, if X is K-analytic, then it is $\mathbb{P}$-dominated; besides it is Lindelöf and hence realcompact (see Problems 338 and 406 of [TFS]) so we settled necessity.

Finally, if X is realcompact and $\mathbb{P}$-dominated, then every pseudocompact closed subspace of X is compact (see Problems 403 and 407 of [TFS]), so X is K-analytic by Fact 3. This proves sufficiency and makes our solution complete.

T.392. *Suppose that X can be condensed onto a metrizable space. Prove that X is analytic if and only if it has a $\mathbb{P}$-directed compact cover.*

Solution. If the space X is analytic, then it is K-analytic by Problem 346 and hence it has a $\mathbb{P}$-directed compact cover by Problem 391.

Now, assume that X has a $\mathbb{P}$-directed compact cover and there exists a condensation $f : X \to M$ of X onto some metrizable space M. Then, for any pseudocompact $P \subset X$, the set $P' = f(P)$ is compact and hence second countable (because it is pseudocompact and metrizable, see Problem 212 of [TFS]) which implies that the map $g = f|P : P \to P'$ is a homeomorphism (see Problem 140 of [TFS]). Therefore any pseudocompact $P \subset X$ is Compact, so we can apply Fact 3 of T.391 to conclude that X is K-analytic.

Thus X is Lindelöf and hence so is M; any Lindelöf metrizable space is second countable, so X condenses onto a second countable space. Even a Lindelöf Σ-space of countable i-weight has a countable network (see Problem 266) so $nw(X) = \omega$ (because every K-analytic space is Lindelöf Σ, see Problem 261) and hence X is analytic by Problem 346.

T.393. *Let X be a compact space. Prove that $C_p(X)$ is K-analytic if and only if it has a $\mathbb{P}$-directed compact cover.*

Solution. If $C_p(X)$ is K-analytic, then it has a $\mathbb{P}$-directed compact cover by Problem 391. On the other hand, if $P \subset C_p(X)$ is pseudocompact, then P is compact by Fact 19 of S.351, so if $C_p(X)$ has a $\mathbb{P}$-directed compact cover, then we can apply Fact 3 of T.391 to conclude that $C_p(X)$ is K-analytic.

T.394. *Give an example of a non-K-analytic space which has a $\mathbb{P}$-directed compact cover.*

Solution. We identify $\mathbb{P}$ with ω^ω; if $f, g \in \mathbb{P}$, then $f \leq g$ if $f(n) \leq g(n)$ for all $n \in \omega$. The expression $f \leq^* g$ says that the set $\{n \in \omega : f(n) > g(n)\}$ is finite. Denote by D the set $\mathbb{P}$ with the discrete topology and let $A_p = \{q \in D : q \leq^* p\}$ for any $p \in \mathbb{P}$. It is evident that $p, q \in \mathbb{P}$ and $p \leq q$ implies $A_p \subset A_q$ so if $K_p = \overline{A}_p$ for all $p \in \mathbb{P}$ (the bar denotes the closure in βD) then $\mathcal{K} = \{K_p : p \in \mathbb{P}\}$ is a $\mathbb{P}$-directed family of compact subsets of βD.

Letting $X = \bigcup \mathcal{K}$, we obtain a space $X \subset \beta D$ which has a $\mathbb{P}$-directed compact cover $\mathcal{K}$. Since $p \in A_p$ for any $p \in \mathbb{P}$, we have $D \subset X$ and hence X is dense in βD. It follows from Fact 1 of T.097 that

(1) for any countable $A \subset D$ there is $p \in \mathbb{P}$ such that $q \leq^* p$ for any $q \in A$.

Take any countable set $B \subset X$; then there is a countable $A \subset \mathbb{P}$ such that $B \subset \bigcup\{K_q : q \in A\}$. The property (1) implies that there is $p \in \mathbb{P}$ for which $q \leq^* p$ for any $q \in A$ and therefore $K_q \subset K_p$ for any $q \in A$. This shows that

$\overline{B} \subset \bigcup\{K_q : q \in A\} \subset K_p \subset X$, i.e., the closure of every countable subset of X is compact and therefore X is countably compact.

For any $p \in \mathbb{P}$ let $D_p = D\backslash A_p$ and consider the family $\mathcal{D} = \{D_p : p \in \mathbb{P}\}$. Given any finite $B \subset \mathbb{P}$ let $p(n) = \sum\{q(n) : q \in B\} + 1$ for any $n \in \omega$; this defines a point $p \in \mathbb{P}$ and it is immediate that $p \notin A_q$ for any $q \in B$. Therefore $p \in D\backslash(\bigcup_{q\in B} A_q) = \bigcap_{q\in B} D_q$ which shows that $\mathcal{D}$ is a centered family. The space βD being compact the set $F = \bigcap\{\overline{D}_p : p \in \mathbb{P}\}$ is non-empty.

Now observe that $\overline{D}_p \cap K_p = \emptyset$ for any $p \in \mathbb{P}$ (see Fact 1 of S.382). As a consequence, $\emptyset \neq F \subset \beta D\backslash X$ and hence $X \neq \beta D$. Therefore X is a countably compact non-compact space (because $\overline{X} = \beta D$ and $X \neq \beta D$). This shows that X cannot be Lindelöf and therefore X is a non-K-analytic space with a $\mathbb{P}$-directed compact cover.

T.395. *Assuming MA+¬CH, prove that if X is a K-analytic space such that every compact subspace of X is metrizable, then X has a countable network. Observe that if every compact subspace of an analytic space X is countable, then X is countable.*

Solution. We identify $\mathbb{P}$ with ω^ω; given $p, q \in \mathbb{P}$, let $p \leq^* q$ if there is $k \in \omega$ such that $p(n) \leq q(n)$ for all $n \geq k$. A set $E \subset \mathbb{P}$ is *bounded* if there is $p \in \mathbb{P}$ such that $q \leq^* p$ for any $q \in E$.

Fact 1. Under MA, if $E \subset \mathbb{P}$ and $|E| < \mathfrak{c}$, then the set E is bounded.

Proof. For each $q \in E$ let $A_q = \{(i, j) \in \omega \times \omega : j \leq q(i)\} \subset \omega \times \omega$ and consider the set $B_n = \{(n, j) : j \in \omega\}$ for any $n \in \omega$. For the families $\mathcal{A} = \{A_q : q \in E\}$ and $\mathcal{B} = \{B_n : n \in \omega\}$, we have $|\mathcal{A}| < \mathfrak{c}$, $|\mathcal{B}| < \mathfrak{c}$ and $B\backslash(\bigcup\mathcal{A}')$ is infinite for any $B \in \mathcal{B}$ and finite $\mathcal{A}' \subset \mathcal{A}$. Therefore we can apply Problem 051 to find an infinite set $L \subset \omega \times \omega$ such that $A\backslash L$ is finite for any $A \in \mathcal{A}$ and $B\backslash L$ is infinite for any $B \in \mathcal{B}$. Observe that Problem 051 is formulated for subsets of ω, but, of course, it is valid for any countably infinite set N; we must take a bijection $\varphi : N \to \omega$ and find the respective set L' for the families $\mathcal{A}' = \{\varphi(A) : A \in \mathcal{A}\}$ and $\mathcal{B}' = \{\varphi(B) : B \in \mathcal{B}\}$. Then $L = \varphi^{-1}(L')$ is as promised.

If $M = (\omega \times \omega)\backslash L$, then $M \cap A$ is finite for any $A \in \mathcal{A}$ and $M \cap B$ is infinite for every $B \in \mathcal{B}$. Consequently, there is a function $p \in \mathbb{P}$ such that $(n, p(n)) \in M$ for any $n \in \omega$. It is immediate that $q \leq^* p$ for any $q \in E$, so Fact 1 is proved.   □

Fact 2. If set $E \subset \mathbb{P}$ is bounded, then there is a σ-compact $Z \subset \mathbb{P}$ such that $E \subset Z$.

Proof. Choose a point $p \in \mathbb{P}$ for which $q \leq^* p$ for all $q \in E$. For any $q \in \mathbb{P}$ the set $H_q = \{r \in \mathbb{P} : r \leq^* q\}$ is compact because $H_q = \prod\{G_n : n \in \omega\}$ where $G_n = \{0, \dots, q(n)\}$ for all $n \in \omega$. The set $W_p = \{q \in \mathbb{P} : \text{there exists } n \in \omega \text{ such that } q(i) = p(i) \text{ for all } i \geq n\}$ is countable so $Z = \bigcup\{H_q : q \in W_p\}$ is σ-compact. It is easy to see that $E \subset Z$, so Fact 2 is proved.   □

Fact 3. Under MA+¬CH, if $E \subset \mathbb{P}$ is an uncountable set, then there is a compact $K \subset \mathbb{P}$ such that $E \cap K$ is uncountable.

Proof. We can consider that $|E| = \omega_1 < \mathfrak{c}$, so Fact 1 is applicable to conclude that E is bounded. By Fact 2, there exists a σ-compact set $Z \subset \mathbb{P}$ with $E \subset Z$. Now, $Z = \bigcup_{n \in \omega} K_n$ where K_n is compact for all $n \in \omega$. Consequently, $E = \bigcup_{n \in \omega} (E \cap K_n)$ and hence $E \cap K_n$ is uncountable for some $n \in \omega$. Letting $K = K_n$, we obtain the promised compact $K \subset \mathbb{P}$ so Fact 3 is proved.   □

Returning to our solution assume that MA+¬CH holds and all compact subspaces of a K-analytic space Z are metrizable. There is an upper semicontinuous compact-valued onto map $\varphi : \mathbb{P} \to Z$ by Problem 388. Assume first that $hl(Z) > \omega$ and therefore there is a right-separated $Y \subset Z$ with $|Y| = \omega_1 < \mathfrak{c}$ (see Problem 005). For any $p \in \mathbb{P}$ the space $\varphi(p)$ is metrizable and hence $Y \cap \varphi(p)$ is countable. Thus there are sets $Q = \{q_\alpha : \alpha < \omega_1\} \subset \mathbb{P}$ and $R = \{z_\alpha : \alpha < \omega_1\} \subset Z$ such that $z_\alpha \in \varphi(q_\alpha)$ for all $\alpha < \omega_1$ and $z_\alpha \neq z_\beta$ for any distinct $\alpha, \beta < \omega_1$.

Fact 3 guarantees that there is a compact subspace $K \subset \mathbb{P}$ such that the set $U = \{\alpha < \omega_1 : q_\alpha \in K\}$ is uncountable. Therefore $R' = \{z_\alpha : \alpha \in U\}$ is also uncountable and $R' \subset \varphi(K) = \bigcup\{\varphi(q) : q \in K\}$. The set $K' = \varphi(K)$ is compact because the map $\varphi|K : K \to K'$ is upper semicontinuous, compact-valued and onto (see Problem 241). By our assumption about Z the space K' is metrizable and hence second countable. However, R' is an uncountable right-separated subspace of K' which implies $hl(K') > \omega = w(K')$ (see Problem 005); this contradiction shows that

(1) under MA+¬CH any K-analytic space whose compact subsets are metrizable is hereditarily Lindelöf.

Finally, assume again that MA+¬CH holds and all compact subspaces of a K-analytic space X are metrizable. The space X^n is K-analytic (see Problem 343) and it is clear that all compact subspaces of X^n are metrizable for every $n \in \mathbb{N}$, so we can apply (1) to conclude that $hl(X^n) = \omega$ for all $n \in \mathbb{N}$. Thus $hd(C_p(X)^n) = \omega$ for all $n \in \omega$ by Problem 026; in particular, $C_p(X)$ is separable and hence X condenses onto a second countable space. The space X is Lindelöf Σ by Problem 261, so it is stable by Problem 266 which implies that $nw(X) = \omega$ and finishes our solution.

T.396. *Suppose that X is a compact space such that some outer base $\mathcal{B}$ of its diagonal $\Delta = \{(x, x) : x \in X\}$ in $X \times X$ can be indexed as $\mathcal{B} = \{O_p : p \in \omega^\omega\}$ in such a way that $O_p \subset O_q$ whenever $p, q \in \omega^\omega$ and $q \leq p$. Prove that X is metrizable.*

Solution. We identify $\mathbb{P}$ with ω^ω. For any $p \in \mathbb{P}$ and $n \in \omega$ define $s[p, n] \in \mathbb{P}$ by $s[p, n](m) = p(n + m)$ for all $m \in \omega$. A space Z is called $\mathbb{P}$-*dominated* if Z has a $\mathbb{P}$-directed compact cover.

Fact 1. For any sequence $\{p_k : k \in \omega\} \subset \mathbb{P}$ there exists $p \in \mathbb{P}$ such that $p_k \leq s[p, k]$ for all $k \in \omega$.

Proof. Let $p(n) = \sum_{i=0}^{n} p_i(n-i)$ for every $n \in \omega$ and observe that if $k \in \omega$, then

$$s[p,k](n) = p(k+n) = \sum_{i=0}^{k+n} p_i(k+n-i) \geq p_k(n)$$

for any $n \in \omega$; it is clear that this implies $p_k \leq s[p,k]$ for all $k \in \omega$, so Fact 1 is proved.   □

Returning to our solution, consider the set $K_p = \{f \in \mathbb{R}^X : |f(x)| \leq p(0)$ for all $x \in X$ and $|f(x) - f(y)| \leq \frac{1}{n+1}$ for any $n \in \omega$ and $x, y \in X$ with $(x, y) \in O_{s[p,n]}\}$ for any $p \in \mathbb{P}$. We claim that

(1) K_p is a compact subset of $C_p(X)$ for any $p \in \mathbb{P}$.

To prove the property (1) fix $p \in \mathbb{P}$ and take any $f \in \mathbb{R}^X \backslash K_p$. If $|f(x)| > p(0)$, then $V = \{g \in \mathbb{R}^X : |g(x)| > p(0)\}$ is an open subset of $\mathbb{R}^X$ such that $f \in V$ and $V \cap K_p = \emptyset$. If $|f(x)| \leq p(0)$ for all $x \in X$, then there is $n \in \omega$ such that $|f(x) - f(y)| > \frac{1}{n+1}$ for some $(x, y) \in O_{s[p,n]}$. Again, the set $V = \{g \in \mathbb{R}^X : |g(x) - g(y)| > \frac{1}{n+1}\}$ is open in $\mathbb{R}^X$ while $f \in V$ and $V \cap K_p = \emptyset$. It turns out that any $f \in \mathbb{R}^X \backslash K_p$ has a neighborhood contained in $\mathbb{R}^X \backslash K_p$, so $\mathbb{R}^X \backslash K_p$ is open and hence K_p is closed in $\mathbb{R}^X$. Furthermore, for $r = p(0)$ we have $K_p \subset [-r, r]^X$ and hence K_p is closed in a compact space $[-r, r]^X$ so K_p is compact.

To see that $K_p \subset C_p(X)$ take any $f \in K_p$; given $x \in X$ and $\varepsilon > 0$ choose $n \in \omega$ such that $\frac{1}{n+1} < \varepsilon$. Then $W = \{y \in X : (x, y) \in O_{s[p,n]}\}$ is an open subset of X and $x \in W$. For any $y \in W$ we have $(x, y) \in O_{s[p,n]}$ and therefore $|f(x) - f(y)| \leq \frac{1}{n+1} < \varepsilon$ which shows that f is continuous at the point x. Since the point $x \in X$ was chosen arbitrarily, we established that every $f \in K_p$ is continuous, i.e., $K_p \subset C_p(X)$ and (1) is proved.

Now take any function $f \in C_p(X)$; since X is compact, f is bounded on X, i.e., there is a real number $N > 0$ such that $|f(x)| \leq N$ for all $x \in X$. The set $W_k = \{(x, y) \in X \times X : |f(x) - f(y)| < \frac{1}{k+1}\}$ is an open neighborhood of Δ in $X \times X$ for every $k \in \omega$. Since $\mathcal{B}$ is a base of Δ in $X \times X$, for any $k \in \omega$, there is $p_k \in \mathbb{P}$ such that $O_{p_k} \subset W_k$. Apply Fact 1 to find $q \in \mathbb{P}$ for which $p_k \leq s[q, k]$ and therefore $O_{s[q,k]} \subset O_{p_k} \subset W_k$ for all $k \in \omega$. Now define a point $p \in \mathbb{P}$ by $p(n) = q(n) + N$ for every $n \in \omega$. It is immediate that $|f(x)| \leq N \leq p(0)$ for all $x \in X$ and $O_{s[p,k]} \subset O_{s[q,k]} \subset W_k$ for all $k \in \omega$; this implies $|f(x) - f(y)| \leq \frac{1}{k+1}$ whenever $k \in \omega$ and $x, y \in O_{s[p,k]}$ which shows that $f \in K_p$. Thus the family $\mathcal{K} = \{K_p : p \in \mathbb{P}\}$ is a compact cover of $C_p(X)$. It follows from the definition of $\mathcal{K}$ that $p \leq q$ implies $K_p \subset K_q$, i.e., $\mathcal{K}$ is $\mathbb{P}$-directed.

Thus $C_p(X)$ is K-analytic and hence Lindelöf Σ (see Problems 393 and 261); since X embeds in $C_p(C_p(X))$ (see Problem 167 of [TFS]), we can apply Baturov's theorem (see Problem 269 of [TFS]) to conclude that

(2) $\text{ext}(Y) = l(Y)$ for any $Y \subset X$.

The set $F_p = (X \times X) \backslash O_p$ is compact for all $p \in \mathbb{P}$ and the family $\{F_p : p \in \mathbb{P}\}$ is a $\mathbb{P}$-directed compact cover of $X' = (X \times X) \backslash \Delta$. It is evident that the property of being $\mathbb{P}$-dominated is closed-hereditary and therefore every closed

$H \subset X'$ is $\mathbb{P}$-dominated. In particular, if D is a closed discrete subspace of X', then it is metrizable and $\mathbb{P}$-dominated. Applying Problem 392 we can see that D is analytic and hence Lindelöf; any Lindelöf discrete space is countable, so we proved that every closed discrete $D \subset X'$ is countable, i.e., $\text{ext}(X') = \omega$.

The proof of the following statement is standard and easy; we leave it to the reader as an exercise.

(3) If a space Z is compact and F is closed in Z, then $\chi(F, Z) = \omega$ if and only if $Z \backslash F$ is Lindelöf.

Fix any $x \in X$ and observe that the subspace $Y_x = (\{x\} \times X) \backslash \Delta$ is closed in X' and homeomorphic to $X \backslash \{x\}$. Therefore $\text{ext}(Y_x) \leq \text{ext}(X') = \omega$ which, together with (2), implies that $X \backslash \{x\}$ is Lindelöf and hence $\chi(x, X) = \omega$ for any $x \in X$.

As a consequence, $\chi((x, x), X \times X) \leq \omega$ for any $x \in X$ and hence the diagonal $\Delta = \bigcup \{\{(x, x)\} : x \in X\}$ is a union of G_δ-subsets of a compact (and hence realcompact) space $X \times X$. Thus X' is realcompact by Problem 408 of [TFS]; since it is $\mathbb{P}$-dominated, it is K-analytic by Problem 391. As a consequence, $X' = (X \times X) \backslash \Delta$ is Lindelöf and hence the diagonal of X is a G_δ-set in $X \times X$ by (3). Finally apply Problem 091 to conclude that X is metrizable and finish our solution.

T.397. *Suppose that $C_p(X)$ is K-analytic and X is separable. Prove that $C_p(X)$ is analytic.*

Solution. Since X is separable, the space $C_p(X)$ condenses onto a second countable space (see Problem 173 of [TFS]). Now, $C_p(X)$ is Lindelöf Σ by Problem 261; therefore it is stable and hence $nw(C_p(X)) = \omega$ (see Problem 266). Finally, apply Problem 346 to see that $C_p(X)$ is analytic.

T.398. *Let X be a compact space such that $C_p(X)$ is K-analytic. Prove that X is a Fréchet–Urysohn space.*

Solution. The space $C_p(X)$ has the Lindelöf Σ-property and hence it is stable (see Problems 261 and 266). Thus the space $C_p(C_p(X))$ monolithic by Problem 154; since X embeds in $C_p(C_p(X))$ (see Problem 167 of [TFS]), it is also monolithic. Furthermore, $l^*(C_p(X)) = \omega$ (see Problem 256) and hence we have the inequality $t(X) \leq t(C_p(C_p(X))) = \omega$ (see Problem 149 of [TFS]). Therefore X is a monolithic compact space of countable tightness; now Problem 120 implies that the space X is Fréchet–Urysohn.

T.399. *Prove that the following conditions are equivalent for any space X:*

(i) $C_p(C_p(X))$ is K-analytic;
(ii) $C_p(C_p(X))$ is analytic;
(iii) X is finite.

Solution. If Z is a space, $A \subset Z$ and $\mathcal{B} \subset \exp(Z)$, then the family $\mathcal{B}$ *separates A from $Z \backslash A$* if for any $y, z \in Z$ such that $y \in A$ and $z \in Z \backslash A$ there is $B \in \mathcal{B}$ for which $y \in B$ and $z \notin B$.

Fact 1. If Z is a Lindelöf Σ-space, then there exists a Lindelöf Σ-space S such that $C_p(Z) \subset S \subset \mathbb{R}^Z$.

Proof. Fix a family $\mathcal{P} = \{P_n : n \in \omega\}$ which is a network with respect to a compact cover $\mathcal{C}$ of the space Z (see Problem 225). A function $f : Z \to \mathbb{R}$ will be called $\mathcal{P}$-bounded if for any $z \in Z$ there is $P \in \mathcal{P}$ such that $z \in P$ and $f(P)$ is bounded in $\mathbb{R}$. Let S be the set of all $\mathcal{P}$-bounded functions.

To see that $C_p(Z) \subset S$ take any $f \in C_p(Z)$ and $z \in Z$. There is $C \in \mathcal{C}$ with $z \in C$. By continuity of f and compactness of C, the set $C' = f(C)$ is bounded in $\mathbb{R}$, i.e., there is $a > 0$ such that $|f(y)| < a$ for any $y \in C$. The set $U = \{y \in Z : |f(y)| < a\}$ is an open neighborhood of C, so there is $P \in \mathcal{P}$ such that $C \subset P \subset U$ and, in particular, $z \in P$. It follows from $P \subset U$ that $f(P) \subset f(U) \subset (-a, a)$ and hence $f(P)$ is bounded in $\mathbb{R}$. This proves that $f \in S$ and therefore $C_p(Z) \subset S$.

To check that S is a Lindelöf Σ-space consider the map $\xi : \mathbb{R} \to J = (-1, 1)$ defined by $\xi(t) = \frac{2}{\pi} \arctan(t)$ for all $t \in \mathbb{R}$. It is clear that ξ is a homeomorphism and hence the map $q : C_p(Z) \to C_p(Z, J)$ defined by $q(f) = \xi \circ f$ for any $f \in C_p(Z)$ is a homeomorphism as well (see Problem 091 of [TFS]). Let $S' = q(S)$; it is easy to see that

(1) $S' = \{f \in J^Z :$ for any $z \in Z$ there is $P \in \mathcal{P}$ such that $z \in P$ and $f(P) \subset [-a, a]$ for some $a \in (0, 1)\}$.

Observe also that we have the inclusion $J^Z \subset \mathbb{I}^Z$ and therefore S' is a subspace of a compact space $\mathbb{I}^Z$. Let $a_k = 1 - 2^{-k-1} \in (0, 1)$ for all $k \in \omega$ and consider the set $B_{n,k} = \{f \in \mathbb{I}^Z : f(P_n) \subset [-a_k, a_k]\}$ for all $n, k \in \omega$. It is straightforward that $B_{n,k}$ is closed in $\mathbb{I}^Z$ and hence compact for any $n, k \in \omega$; we claim that the family $\mathcal{B} = \{B_{n,k} : n, k \in \omega\}$ separates $\mathbb{I}^Z \setminus S'$ from S'.

Indeed, let $f \in S'$ and $g \in \mathbb{I}^Z \setminus S'$. If $g(z) \in \{-1, 1\}$ for some $z \in Z$, then, by (1) there is $n \in \omega$ such that $z \in P_n$ and $f(P_n) \subset [-a, a]$ for some $a \in (0, 1)$. There is $k \in \omega$ with $a < a_k$ and hence $f(P_n) \subset [-a_k, a_k]$ which proves that $f \in B_{n,k}$. On the other hand, it follows from $g(z) \notin J$ that $g \notin B_{n,k}$.

Now, if $g(z) \notin \{-1, 1\}$ for all $z \in Z$, then $g \in J^Z$ and hence we can apply (1) to find $z \in Z$ such that for any $n \in \omega$, if $z \in P_n$, then the set $g(P_n)$ is not contained in $[-a, a]$ for any $a \in (0, 1)$. Again, it follows from $f \in S'$ that there is $n \in \omega$ for which $z \in P_n$ and $f(P_n) \subset [-a, a]$ for some $a \in (0, 1)$. There is $k \in \omega$ with $a < a_k$ and hence $f(P_n) \subset [-a_k, a_k]$ which proves that $f \in B_{n,k}$. On the other hand, it follows from $g(P_n) \not\subset [-a_k, a_k]$ that $g \notin B_{n,k}$.

This shows that $\mathcal{B}$ separates S' from $\mathbb{I}^Z \setminus S'$ and hence S' is a Lindelöf Σ-space by Problem 233. Finally, S is also Lindelöf Σ because it is homeomorphic to S'. Fact 1 is proved. □

Returning to our solution observe that if X is finite then $C_p(X) = \mathbb{R}^X$ is second countable and σ-compact, so the space $C_p(C_p(X))$ is analytic by Problem 367; this proves (iii)$\Longrightarrow$(ii). The implication (ii)$\Longrightarrow$(i) follows from Problem 346 so let us prove that (i)$\Longrightarrow$(iii).

Assume, towards a contradiction, that $C_p(C_p(X))$ is K-analytic while X is infinite and fix a countably infinite $A \subset X$. Since X is homeomorphic to a closed subspace of $C_p(C_p(X))$ (see Problem 167 of [TFS]), it is also K-analytic and hence normal. The set $Y = \overline{A}$ is closed in X and hence the restriction map $\pi : C_p(X) \to C_p(Y)$ is open and onto (see Problem 152 of [TFS]). This shows that the dual map $\pi^* : C_p(C_p(Y)) \to C_p(C_p(X))$ is a homeomorphic embedding such that $\pi^*(C_p(C_p(Y)))$ is closed in $C_p(C_p(X))$ (see Problem 163 of [TFS]).

Thus $C_p(C_p(Y))$ is K-analytic and hence Lindelöf Σ. Since Y is separable, we have $iw(C_p(Y)) = \omega$ and hence $C_p(Y)$ is a realcompact space (see Problem 446 of [TFS]). For every $f \in C_p(Y)$, let $e_f(\varphi) = \varphi(f)$ for any $\varphi \in C_p(C_p(Y))$. Then the correspondence $f \to e_f$ is an embedding of $C_p(Y)$ in $C_p(C_p(C_p(Y))) \subset \mathbb{R}^{C_p(C_p(Y))}$; let C be the respective image of $C_p(Y)$. By Fact 1, there exists a Lindelöf Σ-space S such that $C_p(C_p(C_p(Y))) \subset S \subset \mathbb{R}^{C_p(C_p(Y))}$.

By definition of realcompactness, the set C is closed in $\mathbb{R}^{C_p(C_p(Y))}$; since also $C \subset S$, it closed in S as well and therefore C is a Lindelöf Σ-space. The space $C_p(Y)$ is also a Lindelöf Σ because it is homeomorphic to C. An immediate consequence is that $C_p(Y)$ is stable (see Problem 266) and hence Y is monolithic by Problem 152. Now it follows from separability of the space Y that $nw(Y) = \omega$ and hence $nw(C_p(Y)) = nw(C_p(C_p(Y))) = \omega$. Thus $C_p(C_p(Y))$ is analytic by Problem 346; this implies that $C_p(Y)$ is σ-compact by 366 and hence Y is finite by Problem 186 of [TFS]. However, Y contains an infinite subset A; this contradiction shows that X cannot be infinite bringing to an end the proof of (i)$\Longrightarrow$(iii) and making our solution complete.

T.400. *Prove that the following properties are equivalent for any space X:*

  (i) X is hereditarily K-analytic;
  (ii) X is hereditarily analytic;
  (iii) X is countable.

Solution. The implications (iii)$\Longrightarrow$(ii) and (ii)$\Longrightarrow$(i) are evident, so let us prove that (i)$\Longrightarrow$(iii). Assume that X is hereditarily K-analytic and uncountable; then it is hereditarily Lindelöf Σ because every K-analytic space is Lindelöf Σ (see Problem 261). However, every hereditarily Lindelöf Σ-space has a countable network (see Problem 270), so $nw(X) = \omega$ and hence X is analytic by Problem 346.

Now apply Problem 353 to conclude that the Cantor set $\mathbb{K}$ embeds in X. By Fact 5 of T.250, there exists $Y \subset \mathbb{K}$ which is not K-analytic. Since Y also embeds in X, we have proved that there exists a non-K-analytic subspace in the space X; this contradiction shows that X is countable and settles the remaining implication (i)$\Longrightarrow$(iii).

T.401. *Observe that $A(\omega_1)$ is a union of two discrete (and hence metrizable) subspaces of countable i-weight. Deduce from this fact that metrizability, first axiom of countability, i-weight, P-property and pseudocharacter are not finitely additive.*

Solution. We have $A(\omega_1) = \omega_1 \cup \{a\}$ where all points of $Y = \omega_1$ are isolated and therefore Y is a discrete subspace of $A(\omega_1)$. Since the singleton $Z = \{a\}$ is also discrete, the space $A(\omega_1)$ is a union of two discrete and hence metrizable subspaces. Since $w(Y) = \omega_1 \leq \mathfrak{c}$, the space Y condenses onto a second countable space by Problem 102 and hence $iw(Y) = \omega$. It is clear that $w(Z) = \omega$.

Consequently, $A(\omega_1)$ is a union of two metrizable spaces; it is not first countable at a because if $\{U_n : n \in \omega\} \subset \tau(a, A(\omega_1))$, then $\omega_1 \backslash U_n$ is finite for every $n \in \omega$ and hence for $G = \bigcap_{n \in \omega} U_n$ we have $|\omega_1 \backslash G| \leq \omega$ so $G \neq \{a\}$ and therefore $\psi(a, A(\omega_1)) = \chi(a, A(\omega_1)) > \omega$.

This shows that if $\mathcal{P} \in \{$metrizability, countable character, countable pseudocharacter$\}$, then $\mathcal{P}$ is not finitely additive. We also saw that $A(\omega_1) = Y \cup Z$ where $iw(Y) \leq \omega$ and $iw(Z) \leq \omega$. However, $iw(A(\omega_1)) > \omega$ because any condensation of the compact space is a homeomorphism (see Problem 123 of [TFS]) so $A(\omega_1)$ cannot be condensed onto a space of countable weight (not even onto a space of countable character because $\chi(A(\omega_1)) > \omega$). Thus i-weight is not finitely additive.

Since Y and Z are both discrete, they are P-spaces. However, $A(\omega_1)$ is not a P-space because $A(\omega_1) \backslash \omega$ is a G_δ-subset of $A(\omega_1)$ which is not open in $A(\omega_1)$. This proves that P-property is not finitely additive either.

T.402. *Representing $L(\omega_1)$ as a union of two metrizable subspaces, observe that sequentiality, π-character, the Fréchet–Urysohn property, Čech-completeness and k-property are not finitely additive.*

Solution. We have $L(\omega_1) = \omega_1 \cup \{a\}$ where all points of $Y = \omega_1$ are isolated and therefore Y is a discrete subspace of $A(\omega_1)$. Since the singleton $Z = \{a\}$ is also discrete, we have $L(\omega_1) = Y \cup Z$ where Y and Z discrete and hence completely metrizable subspaces. This shows that Y and Z are Čech-complete (see Problems 204 and 269 of [TFS]) and have countable character and hence countable π-character; they are also k-spaces because they have no non-closed subsets to disprove this. It is also immediate that Y and Z are Fréchet–Urysohn and hence sequential.

Now, if $B = \omega_1$, then B is not closed in $L(\omega_1)$ while there is no sequence in B which converges outside of B because every countable subset of B is closed in $L(\omega_1)$. Therefore $L(\omega_1)$ is not sequential and hence not Fréchet–Urysohn. Thus both sequentiality and Fréchet–Urysohn property fail to be finitely additive.

Observe also that all compact subsets of $L(\omega_1)$ are finite because if $K \subset L(\omega_1)$ is infinite, then there is a countably infinite $D \subset K \backslash \{a\}$; it is evident that D is a closed discrete subset of K, so K is not compact. Therefore B is a non-closed subset of $L(\omega_1)$ such that $K \cap B$ is finite and hence closed in K for any compact $K \subset L(\omega_1)$. This proves that $L(\omega_1)$ is not a k-space. Since every Čech-complete space is a k-space (see Fact 1 of T.210), the space $L(\omega_1)$ is not Čech-complete either. This shows that Čech-completeness and k-property are not finitely additive.

To finally see that $\pi\chi(a, L(\omega_1)) > \omega$ assume that there exists a family $\mathcal{U} = \{U_n : n \in \omega\} \subset \tau^*(L(\omega_1))$ which is a π-base at a in $L(\omega_1)$. Then $U_n \cap B \neq \emptyset$ and hence we can choose a point $z_n \in U_n \cap B$ for all $n \in \omega$. Let $V_n = \{z_n\}$ for all $n \in \omega$; then $\{V_n : n \in \omega\}$ is also a π-base at a in $L(\omega_1)$. However, if $A = \{z_n : n \in \omega\}$, then $W = L(\omega_1) \backslash A$ is a neighborhood of a such that $V_n \not\subset W$ for any $n \in \omega$. This contradiction shows that $\pi\chi(a, L(\omega_1)) > \omega$ and hence π-character is not finitely additive.

T.403. *Let $\xi \in \beta\omega \setminus \omega$ and observe that the space $\omega \cup \{\xi\}$ is a union of two second countable spaces while $w(\omega \cup \{\xi\}) > \omega$. Therefore weight is not finitely additive.*

Solution. If $X = \omega \cup \{\xi\}$, then $X = Y \cup Z$ where $Y = \omega$ and $Z = \{\xi\}$. Since Y and Z are countable discrete spaces, we have $w(Y) \leq \omega$ and $w(Z) \leq \omega$. To see that $w(X) > \omega$ observe that $\xi \in \overline{Y}$ and assume that we have a sequence $S = \{y_n : n \in \omega\} \subset Y$ with $x_n \to \xi$. For any $y \in Y$ the set $U = X \setminus \{y\}$ is an open neighborhood of ξ, so there is $m_y \in \omega$ such that $x_n \in U$ for all $n \geq m_y$. Therefore the set $A_y = \{n \in \omega : x_n = y\}$ is finite for any $y \in Y$ which makes it possible to choose a sequence $\{k_i : i \in \omega\}$ such that $k_i < k_{i+1}$ for all $i \in \omega$ and $x_{k_i} \neq x_{k_j}$ if $i \neq j$. If $y_i = x_{k_i}$ for all $i \in \omega$ then the sequence $\{y_i : i \in \omega\}$ still converges to ξ.

Now let $A = \{i \in \omega : i \text{ is even}\}$ and $B = \{i \in \omega : i \text{ is odd}\}$. Then $P = \{y_i : i \in A\}$ and $Q = \{y_i : i \in B\}$ are infinite disjoint subsets of ω. But they are also sequences which both converge to ξ and hence $\xi \in \overline{P} \cap \overline{Q}$ (the bar denotes the closure in $\beta\omega$). However, $\overline{P} \cap \overline{Q} = \emptyset$ by Fact 1 of S.382; this contradiction shows that there is no sequence in Y which converges to ξ and hence $\chi(\xi, X) > \omega$. Therefore $w(X) \geq \chi(X) > \omega$, i.e., X is not second countable. This proves that the property of having countable weight is not finitely additive.

T.404. *Give an example of a non-realcompact space which is a union of two hereditarily realcompact subspaces.*

Solution. Recall that a Mrowka space M can be represented as $M = E \cup D$ where E is a countable dense set of isolated points of M while D is closed and discrete in M and $|M| \leq \mathfrak{c}$ (see Problem 142 of [TFS]). Since $w(E) = \omega$ and $iw(D) = \omega$ (see Problem 102), both spaces E and D are hereditarily realcompact by Problem 446 of [TFS]. However, the Mrowka space M is not realcompact because it is pseudocompact and non-compact (see Problems 142 and 407 of [TFS]). Thus the Mrowka space M is not realcompact while being a union of two hereditarily realcompact subspaces.

T.405. *Prove that if φ is a cardinal function and $\varphi \in \{$network weight, spread, Lindelöf number, hereditary Lindelöf number, density, hereditary density, extent, Souslin number, point-finite cellularity$\}$, then φ is completely additive and hence countably additive.*

Solution. Fix an infinite cardinal κ and a space $X = \bigcup \{X_\alpha : \alpha < \kappa\}$. Assume first that $nw(X_\alpha) \leq \kappa$ for any $\alpha < \kappa$. There exists a network $\mathcal{N}_\alpha$ in the space X_α such that $|\mathcal{N}_\alpha| \leq \kappa$ for all $\alpha < \kappa$. If $\mathcal{N} = \bigcup \{\mathcal{N}_\alpha : \alpha < \kappa\}$, then $\mathcal{N}$ is a network in X and $|\mathcal{N}| \leq \kappa$ whence $nw(X) \leq \kappa$. This proves that network weight is completely additive.

Now let $s(X_\alpha) \leq \kappa$ for all $\alpha < \kappa$. If $D \subset X$ is discrete and $|D| = \kappa^+$, then there is $\alpha < \kappa$ such that $|D \cap X_\alpha| = \kappa^+$; since $D \cap X_\alpha$ is a discrete subspace of X_α, this contradicts $s(X_\alpha) \leq \kappa$. Therefore $s(X) \leq \kappa$, i.e., we proved that spread is completely additive.

In case when $l(X_\alpha) \leq \kappa$ for any $\alpha < \kappa$ take an open cover $\mathcal{U}$ of the space X. Since $\{U \cap X_\alpha : U \in \mathcal{U}\}$ is an open cover of X_α, there is $\mathcal{U}_\alpha \subset \mathcal{U}$ such that $X_\alpha \subset \bigcup \mathcal{U}_\alpha$ and $|\mathcal{U}_\alpha| \leq \kappa$ for any $\alpha < \kappa$. It is clear that $\mathcal{U}' = \bigcup \{\mathcal{U}_\alpha : \alpha < \kappa\} \subset \mathcal{U}$ while $\bigcup \mathcal{U}' = X$ and $|\mathcal{U}'| \leq \kappa$. Therefore $l(X) \leq \kappa$ which proves that the Lindelöf number is completely additive.

To deal with hereditary Lindelöf number, assume that $hl(X_\alpha) \leq \kappa$ for every $\alpha < \kappa$ and take any $Y \subset X$. Then $Y = \bigcup \{Y_\alpha : \alpha < \kappa\}$ where $Y_\alpha = Y \cap X_\alpha$ for all $\alpha < \kappa$. Since $hl(X_\alpha) \leq \kappa$, we have $l(Y_\alpha) \leq \kappa$ for all $\alpha < \kappa$. We have already proved that the Lindelöf number is completely additive, so $l(Y) \leq \kappa$; since $Y \subset X$ was taken arbitrarily, we established that $hl(X) \leq \kappa$, so hereditary Lindelöf number is also completely additive.

If $d(X_\alpha) \leq \kappa$ for every $\alpha < \kappa$, then take $Y_\alpha \subset X_\alpha$ such that $|Y_\alpha| \leq \kappa$ and Y_α is dense in X_α. Then $Y = \bigcup \{Y_\alpha : \alpha < \kappa\}$ is dense in X and $|Y| \leq \kappa$. This shows that $d(X) \leq \kappa$ and proves that density is completely additive.

As to hereditary density, let $hd(X_\alpha) \leq \kappa$ for all $\alpha < \kappa$ and take any $Y \subset X$. Then $Y = \bigcup \{Y_\alpha : \alpha < \kappa\}$ where $Y_\alpha = Y \cap X_\alpha$ for all $\alpha < \kappa$. Since $hd(X_\alpha) \leq \kappa$, we have $d(Y_\alpha) \leq \kappa$ for all $\alpha < \kappa$. We have already proved that density is completely additive, so $d(Y) \leq \kappa$; since $Y \subset X$ was taken arbitrarily, we established that $hd(X) \leq \kappa$ so hereditary density is also completely additive.

To settle the case of extent, suppose that $\text{ext}(X_\alpha) \leq \kappa$ for all $\alpha < \kappa$. If $D \subset X$ is closed, discrete and $|D| = \kappa^+$, then there is $\alpha < \kappa$ such that $|D \cap X_\alpha| = \kappa^+$; since $D \cap X_\alpha$ is a closed discrete subspace of X_α, this contradicts $\text{ext}(X_\alpha) \leq \kappa$. Therefore $\text{ext}(X) \leq \kappa$, i.e., we proved that extent is completely additive.

Let $c(X_\alpha) \leq \kappa$ for all $\alpha < \kappa$. If $c(X) > \kappa$, then there is a disjoint family $\mathcal{U} \subset \tau^*(X)$ such that $|\mathcal{U}| = \kappa^+$. Observe that for each $U \in \mathcal{U}$, there is $\alpha_U < \kappa$ such that $U \cap X_{\alpha_U} \neq \emptyset$. Consequently, there is $\alpha < \kappa$ such that the family $\mathcal{U}' = \{U \in \mathcal{U} : \alpha_U = \alpha\}$ has cardinality κ^+. As a consequence, $\mathcal{V} = \{U \cap X_\alpha : U \in \mathcal{U}'\} \subset \tau^*(X_\alpha)$ is a disjoint family with $|\mathcal{V}| = \kappa^+$; this contradiction with $c(X_\alpha) \leq \kappa$ shows that $c(X) \leq \kappa$ and hence the Souslin number is completely additive.

Finally, assume that $p(X_\alpha) \leq \kappa$ for all $\alpha < \kappa$. If $p(X) > \kappa$, then there is a point-finite family $\mathcal{U} \subset \tau^*(X)$ such that $|\mathcal{U}| = \kappa^+$. Observe that for each $U \in \mathcal{U}$, there is $\alpha_U < \kappa$ such that $U \cap X_{\alpha_U} \neq \emptyset$. Consequently, there is $\alpha < \kappa$ such that the family $\mathcal{U}' = \{U \in \mathcal{U} : \alpha_U = \alpha\}$ has cardinality κ^+. As a consequence, $\mathcal{V} = \{U \cap X_\alpha : U \in \mathcal{U}'\} \subset \tau^*(X_\alpha)$ is a point-finite family with $|\mathcal{V}| = \kappa^+$; this contradiction with $p(X_\alpha) \leq \kappa$ shows that $p(X) \leq \kappa$ and hence the point-finite cellularity is completely additive.

T.406. *Prove that pseudocompleteness, π-weight and the Baire property are finitely additive but not countably additive.*

Solution. Take a countable dense subspace D of the space $C_p(\mathbb{I})$. It follows from $D = \bigcup \{\{d\} : d \in D\}$ that the space D is a countable union of compact spaces of countable π-weight. Since D has no isolated points, it is of first category in Itself and hence a countable union can destroy both pseudocompleteness and the Baire property, i.e., neither pseudocompleteness nor the Baire property is countably additive.

Now, $\pi w(C_p(\mathbb{I})) > \omega$ by Problem 171 of [TFS] and $\pi w(D) = \pi w(C_p(\mathbb{I}))$ by Fact 1 of T.187. Therefore $\pi w(D) > \omega$ and hence π-weight is not countably additive.

Fact 1. Suppose that X is a non-empty space with $X = A_1 \cup \cdots \cup A_n$. Then there exists a disjoint family $\mathcal{U} = \{U_1, \ldots, U_n\} \subset \tau(X)$ such that $\bigcup \mathcal{U}$ is dense in X and $U_i \subset \overline{A_i \cap U_i}$ for any $i \leq n$.

Proof. Consider the family $\mathcal{V} = \{V \in \tau^*(X) : \text{there is } i \leq n \text{ such that } V \subset \overline{V \cap A_i}\}$. We claim that $G = \bigcup \mathcal{V}$ is dense in X. Indeed, if $\overline{G} \neq X$, then $H = X \backslash \overline{G} \in \tau^*(X)$ and $A_i \cap H$ is nowhere dense for any $i \leq n$. Let $W_0 = W$ and construct, for any $i = 0, \ldots, n$, a set $W_i \in \tau^*(X)$ such that $W_{i+1} \subset W_i$ for all $i < n$ and $W_i \cap A_i = \emptyset$ for all $i = 1, \ldots, n$. Then $W_n \in \tau^*(X)$ and $W_n \cap A_i = \emptyset$ for all $i \leq n$ which is a contradiction with $X = A_1 \cup \cdots \cup A_n$; consequently, the set G is dense in X.

Now consider a maximal disjoint subfamily $\mathcal{W}$ of the family $\mathcal{V}$. Then the set $W = \bigcup \mathcal{W} \subset G$ is dense in G. Indeed, otherwise $W' = G \backslash \overline{W} \neq \emptyset$ and there is $x \in W'$ such that $V \cap A_i$ is dense in V for some $i \leq n$ and $V \in \tau(x, X)$. As a consequence, $V' = V \cap W' \in \mathcal{V}$ and the family $\mathcal{W} \cup \{V'\} \subset \mathcal{V}$ is disjoint and strictly larger than $\mathcal{W}$ which is a contradiction. Thus W is dense in G and hence in X. For every $W \in \mathcal{W}$ fix $i = i_W \leq n$ such that $W \subset \overline{A_i \cap W}$ and let $U_k = \bigcup \{W \in \mathcal{W} : i_W = k\}$ for every $k \leq n$. It is immediate that the family $\mathcal{U} = \{U_1, \ldots, U_n\}$ is as promised so Fact 1 is proved.   □

Returning to our solution fix a space X such that $X = X_1 \cup \cdots \cup X_n$ for some $n \in \mathbb{N}$; by Fact 1 there exists a disjoint family $\mathcal{U} = \{U_1, \ldots, U_n\} \subset \tau(X)$ such that $G = \bigcup \mathcal{U}$ is dense in X and $U_i \subset \overline{U_i \cap X_i}$ for all $i \leq n$. We can assume, without loss of generality that $U_i \neq \emptyset$ for all $i \leq n$.

Assume first that X_i is pseudocomplete for all $i \leq n$. Then the space $U_i \cap X_i$ is pseudocomplete being open in X_i (see Problem 466 of [TFS]); since $X_i \cap U_i$ is dense in U_i, the space U_i is pseudocomplete for every $i \leq n$ by Problem 467 of [TFS]. Since $G = U_1 \cup \cdots \cup U_n$, every point of G has a pseudocomplete neighborhood, so G is pseudocomplete by Fact 2 of S.488. Since G is dense in X, the space X is also pseudocomplete by Problem 467 of [TFS]. Therefore pseudocompleteness is finitely additive.

Now suppose that each X_i has the Baire property. Then, for any $i \leq n$, the space $U_i \cap X_i$ is Baire being open in a Baire space X_i (see Problem 275 of [TFS]). Thus we can apply Problem 275 of [TFS] again to conclude that U_i is Baire for every $i \leq n$. If G is not Baire, then some $H \in \tau^*(G)$ is of first category; then $H' = H \cap U_i \neq \emptyset$ for some $i \leq n$ and hence H' is also of first category which is a contradiction because H' is a non-empty open subset of a Baire space U_i. This contradiction shows that G is Baire and hence X is also Baire (here we used Problem 275 of [TFS] once more). This completes the proof of finite additivity of the Baire property.

In case when $\pi w(X_i) \leq \kappa$ for every $i \leq n$ we have $\pi w(X_i \cap U_i) \leq \kappa$ for any $i \leq n$ (it is an easy exercise that π-weight of an open subspace of a space does not exceed the π-weight of the space). Therefore $\pi w(U_i) \leq \kappa$ for all $i \leq k$ by Fact 1 of T.187. Take a π-base $\mathcal{B}_i$ in the space U_i for which $|\mathcal{B}_i| \leq \kappa$ for every $i \leq n$. If $\mathcal{B} = \mathcal{B}_1 \cup \cdots \cup \mathcal{B}_n$, then $\mathcal{B}$ is a π-base in G with $|\mathcal{B}| \leq \kappa$. Therefore $\pi w(X) = \pi w(G) \leq \kappa$ (we used Fact 1 of T.187 again) which proves finite additivity of π-weight and finishes our solution.

T.407. *Considering any Mrowka space, prove that normality is not finitely additive.*

Solution. The Mrowka space M can be represented as $E \cup D$ where E is the set of isolated points of M and D is a closed discrete subspace of M (see Problem 142 of [TFS]). The subspaces E and D are discrete and hence normal. However, the Mrowka space is not normal because it is pseudocompact but not countably compact (see Problem 137 and 142 of [TFS]). Thus normality is not finitely additive.

T.408. *Suppose that $X \times X = Y \cup Z$, where Y and Z are first countable. Prove that X is also first countable.*

Solution. Fix an arbitrary $x \in X$; we can assume, without loss of generality, that $a = (x, x) \in Y$. Let $V_x = \{x\} \times X$ and assume that $(x, y) \notin \overline{Y}$ for some $y \in X$. Then we can find $U \in \tau(x, X)$ such that $(U \times \{y\}) \cap Y = \emptyset$ and therefore $U \times \{y\} \subset Z$ which implies $\chi(U \times \{y\}) = \omega$. Since $U \times \{y\}$ is homeomorphic to U, we have $\chi(U) = \omega$ and hence $\chi(x, X) = \chi(x, U) = \omega$.

Now, if $V_x \subset \overline{Y}$, then $\omega = \chi(a, Y) = \chi(a, \overline{Y})$ (see Fact 1 of S.158) and therefore $\chi(a, V_x) \leq \chi(a, \overline{Y}) = \omega$. It is evident that the projection π onto the second factor maps V_x homeomorphically onto X in such a way that $\pi(a) = x$. Thus $\chi(x, X) = \chi(a, V_x) = \omega$. The point $x \in X$ was chosen arbitrarily, so we proved that $\chi(x, X) = \omega$ for any $x \in X$ and hence X is first countable.

T.409. *Suppose that $X \times X = Y \cup Z$, where Y and Z have countable pseudocharacter. Prove that $\psi(X) \leq \omega$.*

Solution. We must prove that every $x \in X$ is a G_δ-set so take an arbitrary point $x \in X$. We can assume, without loss of generality, that $a = (x, x) \in Y$ and hence there is a G_δ-set H in the space $X \times X$ such that $H \cap Y = \{a\}$. If $\mathcal{O} = \{O_n : n \in \omega\} \subset \tau(X \times X)$ and $\bigcap \mathcal{O} = H$, then we can choose $U_n \in \tau(x, X)$ such that $U_n \times U_n \subset O_n$ for all $n \in \omega$. Then $P = \bigcap\{U_n : n \in \omega\}$ is a G_δ-set in X such that $a \in P \times P \subset H$ and therefore $(P \times P) \cap Y = \{a\}$.

If $P = \{x\}$, then $\psi(x, X) = \omega$ and we are done; if there is $y \in P \setminus \{x\}$ then $(P \times \{y\}) \cap Y = \emptyset$ and hence $P \times \{y\} \subset Z$ which implies $\psi(P \times \{y\}) = \omega$. Since $P \times \{y\}$ is homeomorphic to P, we have $\psi(P) = \omega$; this shows that $\psi(x, X) \leq \psi(x, P) = \omega$ by Fact 2 of S.358. We proved that $\psi(x, X) = \omega$ for any $x \in X$ and hence $\psi(X) \leq \omega$.

T.410. *Suppose that $X \times X = Y \cup Z$, where Y and Z have countable tightness. Prove that $t(X) \leq \omega$.*

Solution. Take a set $A \subset X$ and $x \in \overline{A} \setminus A$; we can assume, without loss of generality, that $a = (x, x) \in Y$. For every $y \in X$ let $H_y = X \times \{y\}$; if there is a point $y \in X$ such that $(x, y) \notin \overline{Y \cap H_y}$, then there is $U \in \tau(x, X)$ for which

$(U \times \{y\}) \cap Y = \emptyset$ and hence $U \times \{y\} \subset Z$ which shows that $t(U \times \{y\}) = \omega$. Since U is homeomorphic to $U \times \{y\}$, we have $t(U) = \omega$; it is evident that $x \in \overline{A \cap U}$ and hence there is a countable $B \subset A \cap U \subset A$ such that $x \in \overline{B}$.

Now assume that $(x, y) \in \overline{Y \cap H_y}$ for any $y \in X$ and let $\pi : X \times X \to X$ be the natural projection onto the second factor. Since $a \in \{x\} \times A$ and $(x, y) \in \overline{Y \cap H_y}$ for any $y \in A$, we have $a \in \overline{Y \cap \pi^{-1}(A)}$. It follows from $t(Y) = \omega$ that there is a countable $C \subset Y \cap \pi^{-1}(A)$ such that $a \in \overline{C}$. By continuity of π, we have $x = \pi(a) \in \overline{\pi(C)}$. Thus $B = \pi(C)$ is a countable subset of A such that $x \in \overline{B}$. We established that for any $A \subset X$ and $x \in \overline{A}$, there is a countable $B \subset A$ with $x \in \overline{B}$. This proves that $t(X) \leq \omega$.

T.411. *Suppose that $X \times X = Y \cup Z$, where Y and Z have countable weight. Prove that $w(X) \leq \omega$.*

Solution. We have $nw(Y) = nw(Z) = \omega$; since the network weight is finitely additive by Problem 405, we have $nw(X) \leq nw(X \times X) = \omega$. We will often use the following property (whose proof is left to the reader as an exercise):

(1) If T is a space and D is dense in T, then $\overline{U \cap D} = \overline{U}$ for any $U \in \tau(T)$.

We have to establish first that

(2) for the set $F = \overline{Y} \cap \overline{Z}$, we have $w(F) = \omega$.

Let $\mathcal{B}_Y$ and $\mathcal{B}_Z$ be countable bases in Y and Z respectively. For any $U \in \mathcal{B}_Y$ take $G_U \in \tau(\overline{Y})$ such that $G_U \cap Y = U$; for any $V \in \mathcal{B}_Z$, take $H_V \in \tau(\overline{Z})$ such that $H_V \cap Z = V$. To see that $\mathcal{B} = \{G_U \cap F : U \in \mathcal{B}_Y\} \cup \{H_V \cap F : V \in \mathcal{B}_Z\}$ is a base in F take any $a \in F$ and $W \in \tau(a, F)$. Since $a \in Y \cup Z$, we have $a \in Y$ or $a \in Z$. We will consider only the case when $a \in Y$ because the proof for $a \in Z$ is identical.

Since $F \subset \overline{Y}$, we can pick $W' \in \tau(\overline{Y})$ with $W' \cap F = W$; there exists $O \in \tau(a, \overline{Y})$ such that $\overline{O} \subset W'$ (observe that the closures of subsets of $\overline{Y}$ in $X \times X$ and in $\overline{Y}$ coincide, so we use the same bar for both). Since $\mathcal{B}_Y$ is a base in Y, there is $U \in \mathcal{B}_Y$ such that $a \in U \subset O \cap Y$. Since $\overline{G_U} = \overline{U}$ by (1), we have $a \in U \subset \overline{G_U} = \overline{U} \subset \overline{O \cap Y} \subset \overline{O} \subset W'$. As a consequence, $a \in G_U \cap F \subset \overline{G_U} \cap F \subset W' \cap F = W$ which shows that for any $a \in F$ and $W \in \tau(a, F)$, there is $B \in \mathcal{B}$ such that $a \in B \subset W$, i.e., $\mathcal{B}$ is a countable base in F so (2) is proved.

Now let $P_x = \{x\} \times X$ for any $x \in X$; if $P_x \subset \overline{Y} \cap \overline{Z}$, then $w(P_x) \leq w(F) \leq \omega$ by (2); since P_x is homeomorphic to X, we have $w(X) = \omega$ and our proof is over. Thus we can assume that

(3) for any $x \in X$, there is $y \in X$ such that $a = (x, y) \notin F$ and hence $a \notin \overline{Y}$ or $a \notin \overline{Z}$.

It follows from (3) that for every $x \in X$ there is $y \in X$ and $U_x \in \tau(x, X)$ such that $(U_x \times \{y\}) \cap Y = \emptyset$ or $(U_x \times \{y\}) \cap Z = \emptyset$ and hence $U_x \times \{y\}$ is contained in one of the sets Y, Z, so $w(U_x \times \{y\}) = \omega$ and therefore $w(U_x) = \omega$ because U_x is homeomorphic to $U_x \times \{y\}$. Since $l(X) \leq nw(X) = \omega$, there is a countable $A \subset X$

such that $X = \bigcup\{U_x : x \in A\}$. Now if C_x is a countable base in U_x for all $x \in A$, then $C = \bigcup\{C_x : x \in A\}$ is a countable base in X, so $w(X) = \omega$ and our solution is complete.

T.412. *Suppose that X is a separable space such that $X \times X = Y \cup Z$, where Y and Z are metrizable. Prove that X is metrizable.*

Solution. A family $\mathcal{A}$ of subsets of a set T is σ-*disjoint* if $\mathcal{A} = \bigcup\{\mathcal{A}_n : n \in \omega\}$ and $\mathcal{A}_n$ is disjoint for every $n \in \omega$. If T is a space, $\mathcal{A} \subset \exp(T)$ and $S \subset T$, then $\mathcal{A}|S = \{A \cap S : A \in \mathcal{A}\}$. We will often use the following properties (their easy proof is left to the reader as an exercise):

(1) If T is a space and D is dense in T, then $\overline{U} = \overline{U \cap D}$ for any $U \in \tau(T)$.
(2) If T is a space and D is dense in T, then a family $\mathcal{U} \subset \tau(T)$ is disjoint if and only if $\mathcal{U}|D$ is disjoint.

Fact 1. Suppose that T is a space such that $T = T_0 \cup \cdots \cup T_n$ and T_i has a σ-disjoint base for every $i \leq n$. Then $F = \overline{T}_0 \cap \cdots \cap \overline{T}_n$ has a σ-disjoint base.

Proof. Let $\mathcal{B}_i$ be a σ-disjoint base of T_i for each $i \leq n$. For every $i \leq n$ and $B \in \mathcal{B}_i$, fix a set $O_B \in \tau(\overline{T}_i)$ such that $O_B \cap T_i = B$; it follows from (2) that the family $C_i = \{O_B : B \in \mathcal{B}_i\}$ is σ-disjoint. Therefore the family $C = \bigcup\{C_i|F : i \leq n\} \subset \tau(F)$ is σ-disjoint; let us prove that C is a base in F.

To that end, take any $x \in F$ and $U \in \tau(x, F)$; there is $j \leq n$ such that $x \in T_j$. Since $F \subset \overline{T}_j$, there is $V \in \tau(\overline{T}_j)$ such that $V \cap F = U$; apply regularity of $\overline{T}_j$ to find $W \in \tau(x, \overline{T}_j)$ such that $\overline{W} \subset V$. The family $\mathcal{B}_j$ being a base in T_j, there is $B \in \mathcal{B}_j$ such that $x \in B \subset W \cap T_j$. We have $\overline{B} = \overline{O}_B$ by (1) and therefore $x \in O_B \subset \overline{O}_B = \overline{B} \subset \overline{W \cap T_j} \subset \overline{W} \subset V$ which implies $x \in O_B \cap F \subset V \cap F = U$; thus, for arbitrary $x \in F$ and $U \in \tau(x, F)$, we found a set $C = O_B \cap F \in C$ such that $x \in C \subset U$; this implies that C is a base in F so Fact 1 is proved. □

Fact 2. Suppose that T is a space and $\mathcal{U}$ is an open cover of T. If $|\mathcal{U}| \leq \kappa$ and $w(U) \leq \kappa$ for every $U \in \mathcal{U}$, then $w(T) \leq \kappa$.

Proof. Choose a base $\mathcal{B}_U$ of the space U such that $|\mathcal{B}_U| \leq \kappa$ for any $U \in \mathcal{U}$. It is immediate that $\mathcal{B} = \bigcup\{\mathcal{B}_U : U \in \mathcal{U}\}$ is a base in T and $|\mathcal{B}| \leq \kappa$ so Fact 2 is proved. □

Returning to our solution observe that it suffices to show that the space X is second countable; to start off, let $B = (X \times X)\backslash\overline{Y}$ and $C = (X \times X)\backslash\overline{Z}$. Then $B, C \in \tau(X \times X)$ while $B \subset Z$ and $C \subset Y$. This shows that both B and C are metrizable; since $X \times X$ is separable, they are both separable and hence second countable. Let $\pi : X \times X \to X$ be the natural projection onto the first factor. Then π is open so both $\pi|B$ and $\pi|C$ are open maps. It is evident that open maps do not raise weight so $w(\pi(B)) = w(\pi(C)) = \omega$. The sets $B' = \pi(B)$ and $C' = \pi(C)$ are open in X, so we can apply Fact 2 to conclude that $w(B' \cup C') \leq \omega$. If $B' \cup C' = X$, then we are done; if not, then there is $x \in X\backslash(B' \cup C')$. It is evident that $\pi^{-1}(x) = \{x\} \times X \subset F = \overline{Y} \cap \overline{Z}$. The space F has a σ-disjoint base by Fact 1 and hence $\{x\} \times X$ also has a σ-disjoint base. The space X being homeomorphic to $\{x\} \times X$, we have a σ-disjoint base in X as well. But any σ-disjoint family of open sets in a separable space is countable (this is an easy exercise), so X has a countable base and our solution is complete.

T.413. *Suppose that X is a compact space such that $X \times X = Y \cup Z$, where Y and Z are metrizable. Prove that X is metrizable.*

Solution. It suffices to show that X is second countable. Any metrizable space has a σ-discrete base by Problem 221 of [TFS]; since any σ-discrete base is σ-disjoint, both spaces Y and Z have a σ-disjoint base. Thus the space $F = \overline{Y} \cap \overline{Z}$ also has a σ-disjoint base by Fact 1 of T.412. It is evident that a σ-disjoint base is point-countable; besides, any base $\mathcal{B}$ of the space F is a T_1-separating family in the sense that $\bigcap \{U \in \mathcal{B} : x \in U\} = \{x\}$ for any $x \in F$. Therefore F has a point-countable T_1-separating family of open sets, so it is metrizable and hence second countable by Fact 1 of T.203.

Let $\pi : X \times X \to X$ be the natural projection onto the first factor. If $p(F) = X$, then X is metrizable by Fact 4 of S.368. If there is $x \in X \backslash p(F)$, then, for any $y \in X$, the point (x, y) does not belong either to $\overline{Y}$ or to $\overline{Z}$ which shows that there is $U_y \in \tau(y, X)$ for which $\{x\} \times U_y$ is contained either in Y or in Z and hence $\{x\} \times U_y$ is metrizable. Since $\{x\} \times U_y$ is homeomorphic to U_y, the space U_y is metrizable. The space X is compact, so there is a σ-compact $V_y \in \tau(X)$ such that $y \in V_y \subset U_y$. The space V_y is also metrizable; being σ-compact, it is second countable. Finally, apply compactness of X to choose a finite subcover $\{V_{y_1}, \ldots, V_{y_n}\}$ of the open cover $\{V_y : y \in X\}$ of the space X. Now, Fact 2 of T.412 implies that $w(X) \leq \omega$ which completes our solution.

T.414. *Give an example of a non-metrizable space X such that $X \times X$ is a union of two metrizable subspaces.*

Solution. Denote by L the set of limit ordinals of ω_1, and choose for any $\alpha \in L$, a sequence $S_\alpha = \{\xi_i(\alpha) : i \in \omega\} \subset (\omega_1 \backslash L) \cap \alpha$ which converges to α. If $\alpha \in I = \omega_1 \backslash L$, then let $\mathcal{B}_\alpha = \{\{\alpha\}\}$; if $\alpha \in L$, then $\mathcal{B}_\alpha = \{O_n(\alpha) \cup \{\alpha\} : n \in \omega\}$ where $O_n(\alpha) = \{\xi_i(\alpha) : i \geq n\}$ for every $n \in \omega$.

If τ is the topology generated by the family $\mathcal{B} = \bigcup \{\mathcal{B}_\alpha : \alpha < \omega_1\}$ as a base, then $X = (\omega_1, \tau)$ is a Tychonoff zero-dimensional space for which $X = L \cup I$ where all points of I are isolated and L is closed and discrete in X. It is also evident that X is first countable, locally compact and the family $\mathcal{B}_\alpha$ is a local base at α for any $\alpha < \omega_1$.

Assume that the space X is collectionwise normal. Then there is a disjoint family $\{U_\alpha : \alpha \in L\} \subset \tau$ such that $\alpha \in U_\alpha$ for all $\alpha \in L$. Making every U_α smaller if necessary we can assume, without loss of generality, that $U_\alpha = O_{n(\alpha)}(\alpha) \cup \{\alpha\}$ for each $\alpha \in L$. We have $\xi_{n(\alpha)}(\alpha) \in O_{n(\alpha)}(\alpha)$ for any $\alpha \in I$ and hence the map $\varphi : L \to \omega_1$ defined by $\varphi(\alpha) = \xi_{n(\alpha)}$ for every $\alpha \in L$, is an injection.

However, L is a stationary subset of ω_1 by Problem 064 and $\varphi(\alpha) < \alpha$ for any $\alpha \in L$, so there exists $\beta < \omega_1$ such that the set $\{\alpha \in L : \varphi(\alpha) = \beta\}$ is stationary (see Problem 067). Therefore φ is not injective; this contradiction proves that X is not collectionwise normal which implies that X is not paracompact and hence not metrizable (see Problem 231 of [TFS]).

Now let us consider the sets $Y = \{(\alpha, \beta) \in X \times X : \max\{\alpha, \beta\} \in L\}$ and $Z = \{(\alpha, \beta) \in X \times X : \max\{\alpha, \beta\} \in I\}$. It is immediate that $Y \cup Z = X \times X$ and $Y \cap Z = \emptyset$. For any $\alpha < \omega_1$ let $M_\alpha = \{(\beta, \gamma) \in X \times X : \max\{\beta, \gamma\} = \alpha\}$. It is clear that $M_\alpha = ((\alpha + 1) \times \{\alpha\}) \cup (\{\alpha\} \times (\alpha + 1))$ for any $\alpha \in \omega_1$. The set $\alpha + 1 = \{\beta : \beta \le \alpha\}$ is open in X for any $\alpha \in \omega_1$; if $\alpha \in I$, then $\{\alpha\}$ in open in X. As a consequence, M_α is open in $X \times X$ for any $\alpha \in I$. We have $Z = \bigcup\{M_\alpha : \alpha \in I\}$ and the family $\{M_\alpha : \alpha \in I\}$ is disjoint, so Z is a union of its clopen countable subspaces. Since X is first countable, every M_α is second countable and hence metrizable. The space Z is homeomorphic to $\bigoplus\{M_\alpha : \alpha \in I\}$ (see Problem 113 of [TFS]) so it is metrizable by Fact 1 of S.234.

It turns out that M_α is also open in Y for any $\alpha \in L$. To see it, assume first that $\beta < \alpha$ and $z = (\alpha, \beta) \in M_\alpha$. There is $n \in \omega$ such that $\xi_i(\alpha) > \beta$ for all $i \ge n$. As a consequence, $U = (O_n(\alpha) \cup \{\alpha\}) \times (\beta + 1)$ is an open neighborhood of z; given $y = (\gamma, \beta') \in U$, if $\gamma = \alpha$, then $y \in M_\alpha$. If $\gamma \in O_n(\alpha)$ then $\gamma = \max\{\gamma, \beta'\} \in I$ and hence $y \notin Y$ which shows that $U \cap Y \subset M_\alpha$.

Analogously, if $z = (\beta, \alpha) \in M_\alpha$, then $V = (\beta + 1) \times (O_n(\alpha) \cup \{\alpha\})$ is an open neighborhood of z in $X \times X$ such that $V \cap Y \subset M_\alpha$. The last case is when $z = (\alpha, \alpha) \in M_\alpha$; here the set $W = (O_0(\alpha) \cup \{\alpha\}) \times (O_0(\alpha) \cup \{\alpha\})$ is an open neighborhood of z in $X \times X$ and $W \backslash M_\alpha \subset I \times I \subset Z$ which shows that $W \cap Y \subset M_\alpha$. We proved that every point of M_α has a neighborhood in Y which is contained in M_α. Therefore $\{M_\alpha : \alpha \in L\}$ is a disjoint open cover of Y.

Since X is first countable, every M_α is second countable and hence metrizable. The space Y is homeomorphic to $\bigoplus\{M_\alpha : \alpha \in L\}$ (see Problem 113 of [TFS]), so it is metrizable by Fact 1 of S.234. This proves that for a non-metrizable space X, we have $X \times X = Y \cup Z$ where both Y and Z are metrizable.

T.415. *Suppose that $X^\omega = \bigcup\{X_n : n \in \omega\}$. Prove that for some $n \in \omega$, there is $Y \subset X_n$ such that there exists an open continuous map of Y onto X^ω and hence there exists an open continuous map of Y onto X. As a consequence, if X^ω is a countable union of first countable subspaces, then X is first countable.*

Solution. If we have a product $Y = \prod\{Y_t : t \in T\}$ and a set $S \subset T$, then $p_S : Y \to Y_S = \prod\{Y_t : t \in S\}$ is the natural projection defined by $p_S(f) = f|S$ for any $f \in Y$. Say that a set $H \subset Y$ *covers all finite faces* of Y if $p_A(H) = Y_A$ for any finite $A \subset T$.

Fact 1. Suppose that we have a product $Y = \prod\{Y_t : t \in T\}$ and a set $H \subset Y$ which covers all finite faces of Y. Then the map $p = p_A|H : H \to Y_A$ is open for any finite $A \subset T$.

Proof. Call a non-empty set $U \subset Y$ *standard open* if $U = \prod\{U_t : t \in T\}$ where $U_t \in \tau(X_t)$ for every $t \in T$ and the set $\mathrm{supp}(U) = \{t \in T : U_t \ne X_t\}$ is finite. The family $\mathcal{B}$ of all standard open subsets of Y is a base in Y, so it suffices to prove that $p(U \cap H)$ is open in Y_A for any $U \in \mathcal{B}$.

We have $U = \prod_{t \in T} U_t$ and hence $p_A(x)(t) = x(t) \in U_t$ for any $x \in U$ and $t \in A$. Therefore $p_A(x) \in V = \prod_{t \in A} U_t$ for any $x \in U$ which shows that $p(U \cap H) \subset p_A(U) \subset V$. Now, if $z \in V$, then apply the fact that H covers all finite faces of Y to find $y \in H$ such that $y(t) = z(t)$ for any $t \in A$ and $y(t) \in U_t$ for any $t \in \mathrm{supp}(U) \backslash A$. It is evident that $y \in U \cap H$ and $p(y) = z$; the point $z \in V$ was chosen arbitrarily, so $V \subset p(U \cap H)$, i.e., $V = p(U \cap H)$ is an open subset of Y_A. Thus the map p is open and Fact 1 is proved.   □

Fact 2. For an arbitrary space Z assume that $Z^\omega = \bigcup\{Z_n : n \in \omega\}$. Then, for some $n \in \omega$, there is $H \subset Z_n$ such that H is closed in Z_n and homeomorphic to some $G \subset Z^\omega$ which covers all finite faces of Z^ω.

Proof. Let $p_A : Z^\omega \to Z^A$ be the natural projection for any $A \subset \omega$. Observe that for any $y \in Z_A = \prod\{Z_n : n \in A\}$, the set $p_A^{-1}(y)$ is a product because $p_A^{-1}(y) = \prod\{Z'_n : n \in \omega\}$ where $Z'_n = \{y(n)\}$ for all $n \in A$ and $Z'_n = Z$ for $n \in \omega \backslash A$. Assume first that

(1) there is a finite $A \subset \omega$ and a point $y \in Z_A$ such that some $Z_n \cap p_A^{-1}(y)$ covers all finite faces of $P = p_A^{-1}(y)$.

Let $H = Z_n \cap p_A^{-1}(y)$; then H is closed in Z_n and it is easy to see that the map $p = p_{\omega \backslash A} | p_A^{-1}(y) : p_A^{-1}(y) \to Z^{\omega \backslash A}$ is a homeomorphism such that $G' = p(H)$ covers all finite faces of $Z^{\omega \backslash A}$. For any bijection $\varphi : \omega \to \omega \backslash A$ the map $\Phi : Z^{\omega \backslash A} \to Z^\omega$ defined by $\Phi(f) = f \circ \varphi$ for every $f \in Z^{\omega \backslash A}$ is a homeomorphism such that $G = \Phi(G')$ covers all finite faces of Z^ω. This shows that the set $H \subset Z_n$ is as promised, so if (1) holds, then our fact is true.

Now assume that (1) is false and hence Z_0 does not cover all finite faces of Z^ω; choose a finite $A_0 \subset \omega$ and $y_0 \in Z^{A_0}$ such that $p_{A_0}^{-1}(y_0) \cap Z_0 = \emptyset$. Suppose that for some $n > 0$, we have finite sets $A_0, \ldots, A_n$ and points $y_0, \ldots, y_n$ with the following properties:

(2) $A_i \subset \omega$ is finite and $y_i \in Z^{A_i}$ for all $i \leq n$;
(3) $A_i \subset A_{i+1}$ and $y_{i+1} | A_i = y_i$ for all $i < n$;
(4) $p_{A_i}^{-1}(y_i) \cap Z_i = \emptyset$ for all $i \leq n$.

Since (1) fails, the set $Z_{n+1} \cap p_{A_n}^{-1}(y_n)$ does not cover all finite faces of $p_{A_n}^{-1}(y_n)$, so we can choose a finite set $B \subset \omega \backslash A_n$ and a point $z \in Z^B$ for which $p_B^{-1}(z) \cap p_{A_n}^{-1}(y_n)$ does not meet Z_{n+1}. Let $A_{n+1} = A_n \cup B$ and define $y_{n+1} \in Z^{A_{n+1}}$ by $y_{n+1} | A_n = y_n$ and $y_{n+1} | B = z$. It is immediate that (2)–(4) are still fulfilled for all $i \leq n + 1$, so we can continue our inductive construction to obtain a family $\{A_i : i \in \omega\}$ of finite subsets of ω and a set $\{y_i : i \in \omega\}$ for which (2)–(4) hold for all $i \in \omega$. It follows from (3) that there is $y \in Z^\omega$ such that $p_{A_n}(y) = y_n$ for all $n \in \omega$. As a consequence, for all $n \in \omega$, we have $y \notin Z_n$ by (4) which implies $y \notin \bigcup_{n \in \omega} Z_n$; this contradiction shows that (1) is true and hence Fact 2 is proved.   □

Now it is easy to finish our solution. Let $Z = X^\omega$; then $Z^\omega = (X^\omega)^\omega = X^\omega = \bigcup\{X_n : n \in \omega\}$, so by Fact 2, there is $n \in \omega$ and $Y \subset X_n$ such that

Y is homeomorphic, so some $G \subset Z^\omega$ which covers all finite faces of Z^ω. By Fact 1, the restriction to G of the natural projection onto the first factor is open (and, evidently, continuous and onto). Thus G maps openly onto $Z = X^\omega$ and hence so does Y.

Finally, if X_n is first countable for all $n \in \omega$, then choose $n \in \omega$ such that some $Y \subset X_n$ maps openly onto X. It is clear that Y is first countable and it is an easy exercise that an open continuous image of a first countable space is first countable. Thus X is first countable and our solution is complete.

T.416. *Given an arbitrary space X, suppose that X^ω is a finite union of metrizable subspaces. Prove that X is metrizable.*

Solution. Given a space Z and a family $\mathcal{A}$ of subsets of Z, let $\mathcal{A}|Y = \{A \cap Y : A \in \mathcal{A}\}$ for any $Y \subset Z$.

Fact 1. If Z is a space and Y is dense in Z, then for any $\mathcal{B} \subset \tau(Z)$ and $y \in Y$ the family $\mathcal{B}$ contains a local base at y in Z if and only if $\mathcal{B}|Y$ contains a local base at y in Y.

Proof. If $\mathcal{C} \subset \mathcal{B}$ is a local base at y in Z, then $\mathcal{C}|Y \subset \mathcal{B}|Y$ and it is evident that $\mathcal{C}|Y$ is a local base at y in Y. Now assume that $\mathcal{C} \subset \mathcal{B}$ and $\mathcal{C}|Y$ is a local base at y in Y. Given any $U \in \tau(y, Z)$, pick $V \in \tau(y, Z)$ such that $\overline{V} \subset U$ and $W \in \mathcal{C}$ for which $y \in W \cap Y \subset V \cap Y$. Then $y \in W \subset \overline{W} = \overline{W \cap Y} \subset \overline{V \cap Y} \subset \overline{V} \subset U$ and hence $y \in W \subset U$ which proves that $\mathcal{C}$ is a local base at y in Z, so Fact 1 is proved.   □

For any $A \subset \omega$ the map $p_A : X^\omega \to X^A$ is the natural projection of X^ω onto its face X^A. Call a set $Y \subset X^\omega$ *strongly dense in X^ω* if Y covers all finite faces of X^ω, i.e., $p_A(Y) = X_A$ for any finite $A \subset \omega$.

Suppose that $X^\omega = Y_1 \cup \cdots \cup Y_n$ and Y_i is metrizable for each $i \leq n$. Then

(1) there is $k \leq n$ and metrizable subspaces $Y_1', \ldots, Y_k'$ of the space X^ω such that $X^\omega = Y_1' \cup \cdots Y_k'$ and Y_i' is strongly dense in X^ω for every $i \leq k$.

We will prove (1) by induction on n; the case of $n = 1$ being evident assume that $n > 1$ and we proved (1) for all $k < n$. If there is $i \leq n$ such that Y_i is not strongly dense in X^ω, then there is a finite $A \subset \omega$ and $z \in X^A$ such that $p_A^{-1}(z) \cap Y_i = \emptyset$. The space $p_A^{-1}(z)$ is homeomorphic to X^ω and $p_A^{-1}(z) = Z_1 \cup \cdots \cup Z_{i-1} \cup Z_{i+1} \cup \cdots \cup Z_n$ where $Z_j = Y_j \cap p_A^{-1}(z)$ for any $j \leq n$, $j \neq i$.

Thus X^ω is represented as a union of $\leq n - 1$ metrizable subspaces, so the induction hypothesis is applicable to guarantee existence of metrizable $Y_1', \ldots, Y_k'$ as in (1). Now, if every Y_i is strongly dense in X^ω, then we can take $Y_i' = Y_i$ for all $i \leq n$ so (1) is proved.

It follows from (1) that we can assume, without loss of generality, that Y_i is strongly dense in X^ω for every $i \leq n$. Fix a family $\mathcal{B}_i \subset \tau(X^\omega)$ such that $\mathcal{B}_i|Y_i$ is a σ-discrete base of Y_i for all $i \leq n$. Every Y_i is dense in X^ω so $\mathcal{B}_i$ contains a local base in X^ω at every point of Y_i by Fact 1; therefore $\mathcal{B} = \mathcal{B}_1 \cup \cdots \cup \mathcal{B}_n$ is a base in X^ω.

For any $k \in \omega$, we identify k with the set $\{0, \ldots, k-1\}$ so $p_k : X^\omega \to X^k$ is the respective natural projection p_A for $A = \{0, \ldots, k-1\}$. For any $k \in \omega$, let $\mathcal{C}_k = \{p_k^{-1}(U) : U \in \tau(X^k)\}$; then $\mathcal{C} = \bigcup \{\mathcal{C}_k : k \in \omega\}$ is a base in X^ω. For any $U \in \tau(X^\omega)$ let $U_k = \bigcup \{V \in \mathcal{C}_k : V \subset U\}$ for every $k \in \omega$. It is clear that $U = \bigcup_{k \in \omega} U_k$ and $p_k^{-1}(U_k) = U_k$ for all $k \in \omega$; besides, if $U \subset W$, then $U_k \subset W_k$ for all $k \in \omega$.

We have $\mathcal{B}_i = \bigcup \{\mathcal{B}_i^m : m \in \omega\}$ where $\mathcal{B}_i^m | Y_i$ is discrete in Y_i for any $m \in \omega$ and $i \le n$. Let $\mathcal{B}_i^m(k) = \{U_k : U \in \mathcal{B}_i^m\}$ for every $k, m \in \omega$ and $i \le n$. It follows from discreteness of $\mathcal{B}_i^m | Y_i$ that the family $\mathcal{E}_i^m(k) = \mathcal{B}_i^m(k) | Y_i$ is discrete in Y_i for all $m, k \in \omega$ and $i \le n$.

Let $\mathcal{H}_i^m(k) = \{p_k(O) : O \in \mathcal{B}_i^m(k)\}$ and take any $y \in X^k$; since Y_i is strongly dense in X^ω, there is $z \in Y_i$ with $p_k(z) = y$. The family $\mathcal{E}_i^m(k)$ is discrete in Y_i, so there is $H \in \tau(z, Y_i)$ which intersects at most one element of $\mathcal{E}_i^m(k)$. The map $p_k | Y_i$ is open by Fact 1 of T.415, so $G = p_k(H)$ is an open neighborhood of y in X^k. Now, if $O \in \mathcal{B}_i^m(k)$ and $G \cap p_k(O) \neq \emptyset$, then $O \cap H \neq \emptyset$ because $O = p_k^{-1}(p_k(O))$. Furthermore, $H \subset Y_i$ implies that $H \cap (O \cap Y_i) \neq \emptyset$; since $O \cap Y_i \in \mathcal{E}_i^m(k)$, discreteness of $\mathcal{E}_i^m(k)$ implies that there is at most one $O \in \mathcal{B}_i^m(k)$ such that $O \cap H \neq \emptyset$. This proves that G intersects at most one element of $\mathcal{H}_i^m(k)$ and therefore $\mathcal{H}_i^m(k)$ is discrete in X^k for any $m, k \in \omega$ and $i \le n$.

It turns out that the images of the elements of $\mathcal{B}_i^m(k)$ under p_k form a discrete family in X^k; an immediate consequence is that $\mathcal{B}_i^m(k)$ is discrete in X^ω for all $m, k \in \omega$ and $i \le n$. Thus $\mathcal{B}' = \bigcup \{\mathcal{B}_i^m(k) : m, k \in \omega, \ i \le n\}$ is a σ-discrete family of open subsets of X^ω. It is easy to see that every element of $\mathcal{B}$ is a union of a subfamily of $\mathcal{B}'$, so $\mathcal{B}' | Y_i$ is a base in Y_i for all $i \le n$. Finally, apply Fact 1 to conclude that $\mathcal{B}'$ is a σ-discrete base of X^ω which shows that X^ω is metrizable (see Problem 221 of [TFS]). This implies that X is metrizable so our solution is complete.

T.417. *Given a countably compact space X, suppose that X^ω is a countable union of metrizable subspaces. Prove that X is metrizable.*

Solution. Given a space Z and a family $\mathcal{A}$ of subsets of Z, let $\mathcal{A} | Y = \{A \cap Y : A \in \mathcal{A}\}$ for any $Y \subset Z$. If $A \subset \omega$, then $p_A : X^\omega \to X^A$ is the natural projection of X^ω onto its face X^A. Call a set $Y \subset X^\omega$ *strongly dense in X^ω* if Y covers all finite faces of X^ω, i.e., $p_A(Y) = X^A$ for any finite $A \subset \omega$.

We have $X^\omega = \bigcup \{X_n : n \in \omega\}$ where X_n is metrizable for all $n \in \omega$. This implies $\chi(X) = \omega$ (see Problem 415) and hence it follows from Fact 1 of S.322 that X^n is countably compact for every $n \in \mathbb{N}$. Now apply Fact 2 of T.415 to find $n \in \omega$ and $X'_n \subset X_n$ which is homeomorphic to a strongly dense $M \subset X^\omega$.

Since the space M is metrizable, we can choose a family $\mathcal{B} \subset \tau(X^\omega)$ such that $\mathcal{B} = \bigcup_{n \in \omega} \mathcal{B}_n$ where $\mathcal{B}_n | M$ is discrete in M and $\mathcal{B} | M$ is a base of M.

For any $k \in \omega$ we identify k with the set $\{0, \ldots, k-1\}$, so $p_k : X^\omega \to X^k$ is the respective natural projection p_A for $A = \{0, \ldots, k-1\}$. For any $k \in \omega$, let $\mathcal{C}_k = \{p_k^{-1}(U) : U \in \tau(X^k)\}$; then $\mathcal{C} = \bigcup \{\mathcal{C}_k : k \in \omega\}$ is a base in X^ω. For any $U \in \tau(X^\omega)$, let $U_k = \bigcup \{V \in \mathcal{C}_k : V \subset U\}$ for every $k \in \omega$. It is clear that

$U = \bigcup_{k \in \omega} U_k$ and $p_k^{-1}(U_k) = U_k$ for all $k \in \omega$; besides, if $U \subset W$, then $U_k \subset W_k$ for all $k \in \omega$.

Let $\mathcal{B}_n(k) = \{U_k : U \in \mathcal{B}_n\}$ and $\mathcal{C}_n(k) = \{p_k(U) : U \in \mathcal{B}_n(k)\}$ for every $n, k \in \omega$. It follows from discreteness of $\mathcal{B}_n|M$ that the family $\mathcal{B}_n(k)|M$ is discrete in M for all $n, k \in \omega$.

The set M being strongly dense in X^ω, for any $z \in X^k$, there is $y \in M$ for which $p_k(y) = z$. Since the family $\mathcal{B}_n(k)|M$ is discrete, there is $H \in \tau(y, M)$ which intersects at most one element of $\mathcal{B}_n(k)|M$; the set $G = p_k(H)$ is open in X^k because the map $p_k|M : M \to X^k$ is open by Fact 1 of T.415. If $G \cap p_k(O) \neq \emptyset$ for some $O \in \mathcal{B}_n(k)$, then $H \cap O \neq \emptyset$ because $p_k^{-1}(p_k(O)) = O$. Since $H \subset M$, we have $H \cap (O \cap M) \neq \emptyset$; furthermore, $O \cap M \in \mathcal{B}_n(k)|M$, so $O \cap M$ is the unique element of $\mathcal{B}_n(k)|M$ which meets H. The family $\mathcal{B}_n(k)$ is disjoint (because so is $\mathcal{B}_n(k)|M$), so $p_k(O)$ is the unique element of $\mathcal{C}_n(k)$ which meets G. This proves that $\mathcal{C}_n(k)$ is a discrete family of open sets of X^k; since X^k is countably compact, $\mathcal{C}_n(k)$ has to be finite and hence so is $\mathcal{B}_n(k)$ for all $n, k \in \omega$.

As a consequence, $\mathcal{B}' = \bigcup\{\mathcal{B}_n(k) : n, k \in \omega\}$ is countable; observe that every element of $\mathcal{B}$ is a union of a subfamily of $\mathcal{B}'$, so $\mathcal{B}'|M$ is a countable base in M. The space X is an open continuous image of a second countable space M (see Fact 1 of T.415) so $w(X) = \omega$ and hence X is metrizable.

T.418. *Give an example of a non-metrizable space X such that X^ω is a countable union of its metrizable subspaces.*

Solution. If $\mathcal{A}$ is a family of subsets of a space Z and we have a subspace $Y \subset Z$, then $\mathcal{A}|Y = \{A \cap Y : A \in \mathcal{A}\}$. Any $n \in \omega$ is identified with the set $\{0, \ldots, n-1\}$. For the sake of brevity we call a space σ-*metrizable* if it is a countable union of its metrizable subspaces. Denote by L the set of all limit ordinals of ω_1. For any $\alpha \in I = \omega_1 \backslash L$ there exist unique $n \in \omega$ and $\beta \in L \cup \{0\}$ such that $\alpha = \beta + n$. To see it, let $\alpha_0 = \alpha$; if $\alpha_0 = 0$ then we are done. If not, then there is $\alpha_1 \in \omega_1$ with $\alpha_0 = \alpha_1 + 1$. Suppose that we have a sequence $\alpha_0, \ldots, \alpha_n \subset \omega_1$ such that $\alpha_{k+1} + 1 = \alpha_k$ for any $k < n$. If α_n is a limit ordinal or zero, then $\alpha = \alpha_0 = \alpha_n + n$ and we are done. If not, then there is $\alpha_{n+1} \in \omega_1$ for which $\alpha_{n+1} + 1 = \alpha_n$, so we can continue to construct our sequence inductively. However, this construction cannot have infinitely many steps because ω_1 has no infinite decreasing sequences. Thus α_n has to be limit or zero for some n and hence $\alpha = \alpha_n + n$, so we can take $\beta = \alpha_n$. The uniqueness of β and n is an easy exercise.

Let $I_0 = \{0\}$ and $I_n = \{\alpha \in I : \alpha = \beta + n$ for some $\beta \in L \cup \{0\}\}$ for every $n \in \mathbb{N}$. It is clear that $I_n \cap I_m = \emptyset$ if $n \neq m$ and $\bigcup\{I_n : n \in \omega\} = I$. Observe that

(1) for every $\alpha \in L$, there is a sequence $S_\alpha = \{\xi_n(\alpha) : n \in \omega\} \subset I \cap \alpha$ such that $\xi_n(\alpha) \to \alpha$ and $\xi_n(\alpha) \in I_n$ for every $n \in \omega$.

Indeed, if α is an isolated point of L (with the interval topology induced from ω_1), then for the ordinal $\beta = \sup\{\gamma \in L \cup \{0\} : \gamma < \alpha\}$ we have $\beta \in L \cup \{0\}$ and $\beta + n \to \alpha$, so we can take $\xi_0(\alpha) = 0$ and $\xi_n(\alpha) = \beta + n$ for all $n \in \mathbb{N}$.

If $\sup\{\gamma \in L : \gamma < \alpha\} = \alpha$, then choose a sequence $\{\beta_n : n \in \omega\} \subset L$ such that $\beta_n \to \alpha$ and $\beta_n < \beta_{n+1}$ for every $n \in \omega$. It is immediate that $\beta_n + n \in I_n \cap \alpha$ and $\beta_n + n \to \alpha$, so letting $\xi_0(\alpha) = 0$ and $\xi_n(\alpha) = \beta_n + n$ for all $n \in \mathbb{N}$ we conclude the proof of (1).

For any $\alpha \in I$, let $\mathcal{B}_\alpha = \{\{a\}\}$; if $\alpha \in L$, then $\mathcal{B}_\alpha = \{O_n(\alpha) \cup \{\alpha\} : n \in \omega\}$ where $O_n(\alpha) = \{\xi_i(\alpha) : i \geq n\}$ for every $n \in \omega$.

If τ is the topology generated by the family $\mathcal{B} = \bigcup\{\mathcal{B}_\alpha : \alpha < \omega_1\}$ as a base, then $X = (\omega_1, \tau)$ is a Tychonoff zero-dimensional space for which $X = L \cup I$ where all points of I are isolated and L is closed and discrete in X. It is also evident that X is first countable, locally compact and the family $\mathcal{B}_\alpha$ is a local base at α for any $\alpha < \omega_1$.

Assume that the space X is collectionwise normal. Then there is a disjoint family $\{U_\alpha : \alpha \in L\} \subset \tau$ such that $\alpha \in U_\alpha$ for all $\alpha \in L$. Making every U_α smaller if necessary we can assume, without loss of generality, that $U_\alpha = O_{n(\alpha)}(\alpha) \cup \{\alpha\}$ for each $\alpha \in L$. We have $\xi_{n(\alpha)}(\alpha) \in O_{n(\alpha)}(\alpha)$ for any $\alpha \in L$ and hence the map $\varphi : L \to \omega_1$ defined by $\varphi(\alpha) = \xi_{n(\alpha)}(\alpha)$ for every $\alpha \in L$ is an injection.

However, L is a stationary subset of ω_1 by Problem 064 and $\varphi(\alpha) < \alpha$ for any $\alpha \in L$, so there exists $\beta < \omega_1$ such that the set $\{\alpha \in L : \varphi(\alpha) = \beta\}$ is stationary (see Problem 067). Thus φ is not injective; this contradiction proves that X is not collectionwise normal which implies that X is not paracompact and hence not metrizable (see Problem 231 of [TFS]).

We will prove that X^ω is σ-metrizable. As a first step note that $X = L \cup I$ is a union of two discrete and hence metrizable subspaces. Therefore X^n is a union of finitely many metrizable subspaces which shows that $X^n \times L^\omega$ is also a finite union of metrizable spaces for any $n \in \omega$. Thus the set $Y = \bigcup\{X^n \times L^{\omega \setminus n} : n \in \mathbb{N}\} \subset X^\omega$ is σ-metrizable.

For any $f \in X^\omega$ and $n \in \mathbb{N}$ let $M_n(f) = \max\{f(i) : i < n\}$ and consider the set $Z = \{f \in X^\omega : \text{for any } k \in \omega, \text{ there is } n > k \text{ such that } f(n) \in I \text{ and } M_n(f) \in I\}$. Our purpose is to show that Z is metrizable.

Call a set $U \subset X^\omega$ standard if $U = \prod_{i \in \omega} U_i$ and there exists $n = l(U) \in \mathbb{N}$ such that $U_i = X$ for all $i \geq n$ and $U_i \in \mathcal{B}$ for all $i < n$. Thus U_i is a basic neighborhood of its "top point" $\max(U_i)$ for all $i < n$. Denote by $\mathcal{E}$ the family of all standard subsets of X^ω; it is evident that $\mathcal{E}$ is a base in X^ω.

We will need the map $\pi_n : X^\omega \to X$ which is the natural projection of the space X^ω onto its nth factor. For fixed ordinal $\alpha \in I$ and $m, n \in \omega$ such that $m < n$ the family $\mathcal{O}(\alpha, n, m) = \{U = \prod_{i \in \omega} U_i \in \mathcal{E} : l(U) = n, U_m = \{\alpha\} \text{ and } \max(U_j) \leq \alpha \text{ for all } j < n\}$ is countable, so we can enumerate it as $\{U(\alpha, n, m, i) : i \in \omega\}$. Now, consider the family $\mathcal{C}(n, m, i, j, k) = \{U(\alpha, n, m, i) \cap \pi_n^{-1}(\beta) : \alpha \in I_j, \beta \in I_k\}$. We claim that

(2) $\mathcal{C} = \bigcup\{\mathcal{C}(n, m, i, j, k) : n, m, i, j, k \in \omega \text{ and } m < n\}$ contains a local base in X^ω at every point of Z.

To prove (2) take any $f \in Z$ and $W = \prod_{i \in \omega} W_i \in \mathcal{E}$ such that $f \in W$. It is clear that we can consider that $W_i \in \mathcal{B}_{f(i)}$ for every $i < l(W)$. It follows from the definition of Z that there is $n \in \omega$ such that $n > l(W)$, $f(n) \in I$ and $M_n(f) = \alpha \in I$. There are $j, k \in \omega$ such that $\alpha \in I_j$ and $\beta = f(n) \in I_k$. Furthermore, $f(m) = \alpha$ for some $m < n$; let $V_i = W_i$ for all $i < l(W)$. For every $i \in \{l(W), \ldots, n\}$ choose $V_i \in \mathcal{B}_{f(i)}$ arbitrarily and let $V_i = X$ for all $i > n$. It is immediate that $V = \prod_{i \in \omega} V_i \in \mathcal{O}(\alpha, n, m)$ and hence there is $i \in \omega$ for which $V = U(\alpha, n, m, i)$ so $G = V \cap \pi_n^{-1}(\beta) \in \mathcal{C}(n, m, i, j, k)$ and $f \in G \subset W$ so (2) is proved.

To show that $\mathcal{C}(n, m, i, j, k)$ is discrete in X^ω for any $n, m, i, j, k \in \omega$ with $m < n$ take any $f \in X^\omega$ and assume that $f \in \overline{\bigcup \mathcal{C}(n, m, i, j, k)}$. Let $\alpha = f(m)$ and $\beta = f(n)$; if $\alpha \notin I_j$, then $\pi_m^{-1}(\omega_1 \backslash I_j)$ is an open neighborhood of f in X^ω which does not meet $\bigcup \mathcal{C}(n, m, i, j, k)$. Analogously, if $\beta \notin I_k$, then $\pi_n^{-1}(\omega_1 \backslash I_k)$ is an open neighborhood of f in X^ω which does not meet $\bigcup \mathcal{C}(n, m, i, j, k)$.

Thus $\alpha \in I_j$ and $\beta \in I_k$ and it is easy to see that $H = \pi_m^{-1}(\alpha) \cap \pi_n^{-1}(\beta)$ is an open neighborhood of f in X^ω which meets at most one element of $\mathcal{C}(n, m, i, j, k)$ because no $H \in \mathcal{C}(n, m, i, j, k)$ distinct from $U(\alpha, n, m, i) \cap \pi_n^{-1}(\beta)$ can meet H. This proves that $\mathcal{C}|Z$ is a σ-discrete base of Z, so the space Z is metrizable.

Now consider the set $T = \{f \in X^\omega \backslash Y : $ there is $k \in \omega$ such that for any $n > k$ if $f(n) \in I$, then $M_n(f) \in L\}$. It is easy to see that $X^\omega = Y \cup Z \cup T$, so it remains to show that T is σ-metrizable. Let $T_k = \{f \in X^\omega \backslash Y : $ for any $n > k$ if $f(n) \in I$, then $M_n(f) \in L\}$ for any $k \in \omega$. It is immediate that $T = \bigcup_{k \in \omega} T_k$, so it suffices to show that every T_k is metrizable.

Given a number $n \in \omega$, an ordinal $\alpha \in L$ and a non-empty set $F \subset n$ let $\mathcal{G}(\alpha, F, n) = \{U = \prod_{i \in \omega} U_i \in \mathcal{E} : l(U) = n, \max(U_j) = \alpha$ for any $j \in F$ and $\max(U_i) < \min(U_j)$ for any $j \in F$ and $i \in n \backslash F\}$; since this family is countable, we can enumerate its elements as $\{V(\alpha, F, n, i) : i \in \omega\}$. Now let $\mathcal{D}(F, n, i, m) = \{V(\alpha, F, n, i) \cap \pi_n^{-1}(\beta) : \alpha \in L$ and $\beta \in I_m\}$ for any $n, m, i \in \omega$ and a finite non-empty $F \subset n$. Next we show that

(3) the family $\mathcal{D} = \bigcup \{\mathcal{D}(F, n, i, m) : n, i, m \in \omega, n > k$ and $\emptyset \neq F \subset n\}$
    contains a local base in X^ω at any point $f \in T_k$.

To prove (3) take any $f \in T_k$ and $W = \prod_{i \in \omega} W_i \in \mathcal{E}$ such that $f \in W$; as before, we can assume that $W_i \in \mathcal{B}_{f(i)}$ for any $i < l(W)$. Since $f \notin Y$, there is $n > k + l(W)$ such that $\beta = f(n) \in I$. By definition of T_k the ordinal $\alpha = M_n(f)$ belongs to L; clearly, $F = \{i < n : f(i) = \alpha\}$ is a non-empty subset of n. Observe that $f(i) = \alpha$ is a limit ordinal for any $i \in F$ while $f(j) < \alpha$ for any $j \in n \backslash F$. Thus we can find a family $\{V_0, \ldots, V_{n-1}\} \subset \mathcal{B}$ such that $f(i) \in V_i \subset W_i$ for all $i < n$ and $\max(V_i) < \min(V_j)$ whenever $j \in F$ and $i \in n \backslash F$. Now, if $V_i = X$ for all $i \geq n$, then $V = \prod_{i \in \omega} V_i \in \mathcal{G}(\alpha, F, n)$. Therefore $V = V(\alpha, F, n, i)$ for some $i \in \omega$; besides, there is $m \in \omega$ for which $\beta = f(n) \in I_m$, so $W' = V \cap \pi_n^{-1}(\beta) \in \mathcal{D}(F, n, i, m)$ and $f \in W' \subset W$ so (3) is proved.

To finish the proof of metrizability of T_k it suffices to show that the family $\mathcal{D}(F, n, i, m)|T_k$ is discrete in T_k for any $n, m, i \in \omega$, $n > k$ and finite non-empty $F \subset n$. So take $f \in T_k$ such that every neighborhood of f in T_k meets an element of $\mathcal{D}(F, n, i, m)|T_k$. If $\beta = f(n) \notin I_m$, then $\pi_n^{-1}(\omega_1 \backslash I_m)$ is an open neighborhood of f in X^ω which does not meet any element of $\mathcal{D}(F, n, i, m)$. Thus $\beta \in I_m$; by definition of T_k, we have $\alpha = M_n(f) \in L$; let $F' = \{j \in n : f(j) = \alpha\}$. For any $j < n$, choose $U_j \in \mathcal{B}_{f(j)}$ so that $\max(U_i) < \min(U_j)$ whenever $j \in F'$ and $i \in n \backslash F'$ and let $U_j = X$ for all $j \geq n$. If $U = \prod_{i \in \omega} U_i$ and $W = U \cap \pi_n^{-1}(\beta)$, then W is a neighborhood of f in X^{ω}. Assume that $g \in T_k \cap W \cap V(\alpha', F, n, i) \cap \pi_n^{-1}(\gamma)$ for some $\alpha' \in L$ and $\gamma \in I_m$. Then $\gamma = g(n) = \beta$. Furthermore, $M_n(g) \in L$ because $g \in T_k$. Now $g \in U$ implies $g(j) \in U_j$ for every $j < n$ and hence $M_n(g) \in U_j$ for some $j \in F'$ because if $j \in n \backslash F'$, then $g(j) \leq \max(U_j) < \min U_i \leq g(i)$ for any $i \in F'$.

It follows from the definition of $V(\alpha', F, n, i)$ that $g \in V(\alpha', F, n, i)$ implies $M_n(g) = \alpha'$; since $\alpha' \in U_j$ for some $j \in F'$, we have $\alpha' = \alpha$ because the unique limit ordinal of U_j is α. Therefore the only element of $\mathcal{D}(F, n, i, m)$ which meets $W \cap T_k$ is $V(\alpha, F, n, i) \cap \pi_n^{-1}(\beta)$. This shows that $\mathcal{D}(F, n, i, m)|T_k$ is discrete in T_k and hence T_k is metrizable. Thus $X^{\omega} = Y \cup Z \cup T$ is σ-metrizable and our solution is complete.

T.419. *For an arbitrary space X and any $f, g \in C^*(X)$, let*

$$d(f, g) = \sup\{|f(x) - g(x)| : x \in X\}.$$

Prove that d is a complete metric on $C^*(X)$ and the topology, generated by d, coincides with the topology of $C_u^*(X)$.

Solution. It follows from what was proved in Problem 248 of [TFS] that d is a complete metric on $C^*(X)$. The coincidence of τ_u and the topology $\tau(d)$ generated by the metric d was proved in Fact 1 of T.357. It is worth mentioning that the correspondence $A \to \overline{A}^u$ satisfies all conditions of Problem 004 of [TFS], so it is indeed a closure operator which generates the topology τ_u on $C^*(X)$; this was proved in Problem 084 of [TFS].

T.420. *Let $\mathcal{P}$ be an F_{σ}-hereditary property, i.e., $X \vdash \mathcal{P}$ implies $Y \vdash \mathcal{P}$ whenever Y is an F_{σ}-subspace of X. Suppose that $C_p(X)$ is a finite union of subspaces which have the property $\mathcal{P}$. Prove that $C_p(X)$ is a finite union of dense subspaces each one of which has the property $\mathcal{P}$.*

Solution. It is evident that it suffices to prove by induction on $n \in \mathbb{N}$ the following statement:

$\mathcal{S}(n)$: if $C_p(X) = Y_1 \cup \cdots \cup Y_n$ and every Y_i has $\mathcal{P}$, then there is $m \leq n$ such that $C_p(X) = Y_1' \cup \cdots \cup Y_m'$ where every Y_i' is dense in $C_p(X)$ and has $\mathcal{P}$.

Since $\mathcal{S}(1)$ is evident, assume that we have proved $\mathcal{S}(n)$ for all $n < k$ and $C_p(X) = Y_1 \cup \cdots \cup Y_k$ where $Y_i \vdash \mathcal{P}$ for all $i \leq k$. If every Y_i is dense in $C_p(X)$ there is nothing to prove, so assume that $\overline{Y}_j \neq C_p(X)$ for some $j \leq k$. There exist $x_1, \ldots, x_p \in X$ and rational open intervals $O_1, \ldots, O_p \subset \mathbb{R}$ such that the set $W = \{f \in C_p(X) : f(x_i) \in O_i$ for all $i \leq p\}$ does not intersect Y_j and hence $W \subset Y_1 \cup \cdots \cup Y_{j-1} \cup Y_{j+1} \cup \cdots \cup Y_k$.

Let $K = \{x_1, \ldots, x_p\}$ and $C_0 = \{f \in C_p(X) : f|K \equiv 0\}$. It was proved in Fact 1 of S.409 that $C_p(X) \simeq C_0 \times \mathbb{R}^K$. On the other hand, $W \simeq C_0 \times (\prod_{i \leq k} O_i)$

(see Fact 1 of S.494). Since $O_i \simeq \mathbb{R}$ for every $i \leq k$, we have $W \simeq C_p(X)$. Furthermore, every O_i is an F_σ-set in $\mathbb{R}$, so W is an F_σ-subset of $C_p(X)$ (this is a consequence of the following easy fact which we leave to the reader as an exercise: if $A_1, \ldots, A_m$ are spaces and $B_i \subset A_i$ is an F_σ-subset of A_i for all $i \leq m$, then $\prod_{i \leq m} B_i$ is an F_σ-subset of $\prod_{i \leq m} A_i$).

As a consequence, the set $Z_i = Y_i \cap W$ is an F_σ-subset of Y_i and hence $Z_i \vdash \mathcal{P}$ for every $i \leq k$. Thus $W = Z_1 \cup \cdots \cup Z_{j-1} \cup Z_{j+1} \cup \cdots \cup Z_k$; since W is homeomorphic to $C_p(X)$, we proved that $C_p(X)$ can be represented as a union $\leq k - 1$ spaces with the property $\mathcal{P}$. Applying the induction hypothesis, we can find $m \leq k - 1$ and dense sets $Y_1', \ldots, Y_m' \subset C_p(X)$ such that $C_p(X) = Y_1' \cup \cdots \cup Y_m'$ and $Y_i' \vdash \mathcal{P}$ for every $i \leq m$. This proves $\mathcal{S}(k)$ and shows that $\mathcal{S}(n)$ is true for every $n \in \mathbb{N}$.

T.421. *Let $\mathcal{P}$ be a hereditary property. Suppose that $C_p(X)$ is a finite union of subspaces which have the property $\mathcal{P}$. Prove that there is $n \in \mathbb{N}$ and $\varepsilon > 0$ such that $C_p(X, (-\varepsilon, \varepsilon)) = Y_1 \cup \cdots \cup Y_n$, where $Y_i \vdash \mathcal{P}$ and $\overline{Y_i}^u \supset C(X, (-\varepsilon, \varepsilon))$ for each $i \in \{1, \ldots, n\}$.*

Solution. Given a space Z let $d_Z(f, g) = \sup\{|f(z) - g(z)| : z \in Z\}$ for any functions $f, g \in C^*(Z)$; then d_Z is a metric on $C^*(Z)$ which generates the topology τ_u^Z of the space $C_u^*(Z)$ (see Problem 419). For any function $f \in C^*(Z)$ and $r > 0$, let $B_Z(f, r) = \{g \in C^*(Z) : d_Z(f, g) < r\}$ and denote by $\mathbf{0}_Z$ the function which is identically zero on Z. If $P \subset Q \subset C^*(Z)$, we will say that P is τ_u^Z-dense in Q if the closure of P in $C_u^*(Z)$ contains Q.

Fact 1. If Z is a space, then for any $\varepsilon > 0$ we have $B_Z(\mathbf{0}_Z, \varepsilon) \subset C(Z, (-\varepsilon, \varepsilon))$ and the set $B_Z(\mathbf{0}_Z, \varepsilon)$ is τ_u^Z-dense in $C(Z, [-\varepsilon, \varepsilon])$ for any $\varepsilon > 0$.

Proof. If $f \in B_Z(\mathbf{0}_Z, \varepsilon)$, then $|f(z)| = |f(z) - \mathbf{0}_Z(z)| \leq d_Z(f, \mathbf{0}_Z) < \varepsilon$ for any $z \in Z$ which proves that $f \in C(Z, (-\varepsilon, \varepsilon))$ and hence $B_Z(\mathbf{0}_Z, \varepsilon) \subset C(Z, (-\varepsilon, \varepsilon))$.

Choose a sequence $\{\varepsilon_n : n \in \omega\}$ such that $0 < \varepsilon_n < \varepsilon$ for all $n \in \omega$ and $\varepsilon_n \to \varepsilon$. For any $f \in C(Z, [-\varepsilon, \varepsilon])$ define $f_n : Z \to \mathbb{R}$ as follows: $f_n(z) = f(z)$ if $|f(z)| < \varepsilon_n$; if $f(z) \geq \varepsilon_n$, then $f_n(z) = \varepsilon_n$, and if $f(z) \leq -\varepsilon_n$, then $f_n(z) = -\varepsilon_n$. It is an easy consequence of Problem 028 of [TFS] that $f_n \in C(Z)$ and hence $f_n \in C^*(Z)$ for any $n \in \omega$; besides, $d(f_n, \mathbf{0}_Z) \leq \varepsilon_n < \varepsilon$ for all $n \in \omega$ and hence $\{f_n : n \in \omega\} \subset B_Z(\mathbf{0}_Z, \varepsilon)$. We omit an easy verification of the fact that $f_n \rightrightarrows f$ which shows that every $f \in C(Z, [-\varepsilon, \varepsilon])$ is a uniform limit of some sequence from $B_Z(\mathbf{0}_Z, \varepsilon)$ and therefore $B_Z(\mathbf{0}_Z, \varepsilon)$ is τ_u^Z-dense in $C(Z, [-\varepsilon, \varepsilon])$. Fact 1 is proved.
□

Fact 2. Given a space Z and $f \in C^*(Z)$ let $L(g) = g + f$ for any $g \in C^*(Z)$. Then $d_Z(g, h) = d_Z(L(g), L(h))$ for any $g, h \in C^*(Z)$ and hence $L : C_u^*(Z) \to C_u^*(Z)$ is a homeomorphism.

Proof. We have $|L(g)(z) - L(h)(z)| = |g(z) + f(z) - (h(z) + f(z))| = |g(z) - h(z)|$ for any $z \in Z$ and $g, h \in C^*(Z)$; taking the respective suprema, we obtain the equality $d_Z(g, h) = d_Z(L(g), L(h))$. An immediate consequence is that $g_n \rightrightarrows g$

if and only if $L(g_n) \rightrightarrows L(g)$ for any sequence $\{g_n : n \in \omega\} \subset C^*(Z)$ and $g \in C^*(Z)$. This shows that $L : C_u^*(Z) \to C_u^*(Z)$ is a homeomorphism so Fact 2 is proved. $\square$

Fact 3. If Z is a non-empty space and $\mathcal{A}$ is a finite cover of Z, then there is $\mathcal{A}' \subset \mathcal{A}$ and $U \in \tau^*(Z)$ such that $U \subset \bigcup \mathcal{A}'$ and $A \cap U$ is dense in U for any $A \in \mathcal{A}'$.

Proof. If this is not true, then

$(*)$ for any $\mathcal{A}' \subset \mathcal{A}$ and $U \in \tau^*(Z)$ with $U \subset \bigcup \mathcal{A}'$ there is $A \in \mathcal{A}'$ and $V \subset \tau^*(Z)$ such that $V \subset U$ and $V \cap A = \emptyset$.

We have $\mathcal{A} = \{A_1, \ldots, A_n\}$ for some $n \in \mathbb{N}$; by $(*)$, there exists $V_1 \in \tau^*(Z)$ with $V_1 \cap A_1 = \emptyset$. Suppose that $k < n$ and we have $V_1, \ldots, V_k \in \tau^*(Z)$ such that $V_1 \supset \cdots \supset V_k$ and $V_i \cap A_i = \emptyset$ for all $i \le k$. Then $V_k \subset A_{k+1} \cup \cdots \cup A_n$, so we can apply $(*)$ to the family $\mathcal{A}' = \{A_{k+1}, \ldots, A_n\}$ and the set V_k to find $V_{k+1} \in \tau^*(V_k)$ such that $V_{k+1} \cap A_{k+1} = \emptyset$. This shows that we can inductively construct $V_1, \ldots, V_n \in \tau^*(Z)$ such that $V_1 \supset \cdots \supset V_n$ and $V_i \cap A_i = \emptyset$ for all $i \le n$. It is evident that $V_n \subset Z \backslash (\bigcup \mathcal{A})$ which contradicts $\bigcup \mathcal{A} = Z$. Fact 3 is proved. $\square$

Returning to our solution suppose that $C_p(X) = B_1 \cup \cdots \cup B_m$ and $B_i \vdash \mathcal{P}$ for every $i \le m$. Letting $A_i = B_i \cap C^*(X)$ for every $i \le m$, we obtain the equality $C_u^*(X) = \bigcup_{i \le m} A_i$; since $\mathcal{P}$ is hereditary, every A_i has the property $\mathcal{P}$ (as a subspace of $C_p(X)$). Applying Fact 3 to the space $C_u^*(X)$ and the family $\mathcal{A} = \{A_1, \ldots, A_m\}$ we can find a non-empty $W \in \tau_u^X$ and $\mathcal{A}' \subset \mathcal{A}$ such that $W \subset \bigcup \mathcal{A}'$ and $A \cap W$ is τ_u^X-dense in W for any $A \in \mathcal{A}'$.

Thus there is $f \in W$ and $\delta > 0$ for which $V = B_X(f, \delta) \subset W$. Since V is open in W, we have $V \subset \bigcup \mathcal{A}'$ and $A \cap V$ is τ_u^X-dense in V for any $A \in \mathcal{A}'$. Let $L(g) = g - f$ for any $g \in C^*(X)$. Then $L : C^*(X) \to C^*(X)$ is a homeomorphism both with respect to τ_u^X and $\tau(C_p^*(X))$. For $\tau(C_p^*(X))$ this follows from Problem 079 of [TFS] and for τ_u^X this was proved in Fact 2.

Observe that $G = L(V) \subset C(X, (-\delta, \delta))$; it follows from Fact 2 that $G = B_X(\mathbf{0}_X, \delta)$. If $\mathcal{B}' = \{L(A) : A \in \mathcal{A}'\}$, then $G \subset \bigcup \mathcal{B}'$ and $B \cap G$ is τ_u^X-dense in G for any $B \in \mathcal{B}'$. Therefore, for any $O \in \tau_u^X$ with $O \subset G$ the set $B \cap O$ is τ_u^X-dense in O for any $B \in \mathcal{B}'$. For $\varepsilon = \frac{\delta}{2}$, we have $C(X, (-\varepsilon, \varepsilon)) \subset B(\mathbf{0}_X, \delta) \subset \bigcup \mathcal{B}'$. Besides, $B \cap B_X(\mathbf{0}_X, \varepsilon)$ is τ_u^X-dense in $B_X(\mathbf{0}_X, \varepsilon)$ for any $B \in \mathcal{B}'$; applying Fact 2 we conclude that $B \cap B_X(\mathbf{0}_X, \varepsilon)$ is τ_u^X-dense in $C(X, (-\varepsilon, \varepsilon))$ and hence $B \cap C(X, (-\varepsilon, \varepsilon))$ has to be τ_u^X-dense in $C(X, (-\varepsilon, \varepsilon))$ for any $B \in \mathcal{B}'$.

If $\mathcal{B}' = \{B_1, \ldots, B_n\}$ and $Y_i = B_i \cap C(X, (-\varepsilon, \varepsilon))$ for all $i \le n$, then it follows from the inclusion $C(X, (-\varepsilon, \varepsilon)) \subset \bigcup \mathcal{B}'$ that $C(X, (-\varepsilon, \varepsilon)) = Y_1 \cup \cdots \cup Y_n$. Every B_i, considered as a subspace of $C_p(X)$, is homeomorphic to $L^{-1}(B_i) \in \mathcal{A}$. Thus $B_i \vdash \mathcal{P}$ for all $i \le n$. Since $\mathcal{P}$ is a hereditary property, the set Y_i has $\mathcal{P}$ for all $i \le n$. We already saw that Y_i is τ_u^X-dense in $C(X, (-\varepsilon, \varepsilon))$, so our solution is complete.

T.422. *Suppose that $C_p(X)$ is a finite union of its paracompact (not necessarily closed) subspaces. Prove that $C_p(X)$ is Lindelöf and hence paracompact.*

Solution. If Z is a space, $\mathcal{A} \subset \exp(Z)$ and $Y \subset Z$, then $\mathcal{A}|Y = \{A \cap Y : A \in \mathcal{A}\}$. If $\mathcal{A}, \mathcal{B} \subset \exp(Z)$, then $\mathcal{A}$ is *inscribed in* $\mathcal{B}$ if for any $A \in \mathcal{A}$ there is $B \in \mathcal{B}$ such that $A \subset B$.

Fact 1. If Z is a paracompact space and $Y \subset Z$ is an F_σ-set in Z, then Y is also paracompact.

Proof. Let $\mathcal{U}$ be an open cover of Y; for any $U \in \mathcal{U}$, choose $O_U \in \tau(Z)$ with $O_U \cap Y = U$. We have $Y = \bigcup_{n \in \omega} Y_n$ where Y_n is closed in Z for any $n \in \omega$. The family $\mathcal{V}_n = \{O_U : U \in \mathcal{U}\} \cup \{Z \backslash Y_n\}$ is an open cover of Z for any $n \in \omega$, so we can find an open locally finite refinement $\mathcal{W}_n$ of the cover $\mathcal{V}_n$.

If $\mathcal{W}_n' = \{W \in \mathcal{W}_n : W \cap Y_n \neq \emptyset\}$, then $\mathcal{W}_n'$ is inscribed in $\mathcal{V} = \{O_U : U \in \mathcal{U}\}$ and $Y_n \subset \bigcup \mathcal{W}_n'$ for every $n \in \omega$. Thus $\bigcup\{\mathcal{W}_n'|Y : n \in \omega\}$ is a σ-locally finite open refinement of $\mathcal{U}$ and hence Y is paracompact by Problem 230 of [TFS]. Fact 1 is proved. □

Returning to our solution assume that $C_p(X)$ is a finite union of it paracompact subspaces. It follows from Fact 1 and Problem 420 that there is a finite family $\mathcal{A}$ such that $C_p(X) = \bigcup \mathcal{A}$ while every $A \in \mathcal{A}$ is paracompact and dense in $C_p(X)$. Therefore $c(A) = c(C_p(X)) = \omega$ and hence A is Lindelöf for any $A \in \mathcal{A}$ (see Fact 2 of S.219). Consequently, $C_p(X)$ is Lindelöf being a finite union of Lindelöf spaces (see Problem 405). Since any Lindelöf space is paracompact (see Problem 230 of [TFS]), this proves that $C_p(X)$ is paracompact.

T.423. *Suppose that $C_p(X) = Y_1 \cup \cdots \cup Y_n$, where Y_i is realcompact for each $i \leq n$. Prove that $C_p(X)$ is realcompact.*

Solution. Any F_σ-subspace of a realcompact space is realcompact by Problem 408 of [TFS]. Therefore we can apply Problem 420 to convince ourselves that we can assume, without loss of generality, that every Y_i is dense in $C_p(X)$ and hence in $\mathbb{R}^X$. Since $\mathbb{R}^X$ is a Moscow space by Problem 424 of [TFS] (applied to X with the discrete topology), every Y_i is ω-placed in $\mathbb{R}^X$, i.e., for any $f \in \mathbb{R}^X \backslash Y_i$, there is a G_δ-set H in $\mathbb{R}^X$ such that $f \in H \subset \mathbb{R}^X \backslash Y_i$ (see Problem 425 of [TFS]).

Therefore we can find for any $f \in \mathbb{R}^X \backslash C_p(X)$ and $i \leq n$ a G_δ-subset H_i of the space $\mathbb{R}^X$ such that $f \in H_i \subset \mathbb{R}^X \backslash Y_i$. If $H_f = H_1 \cap \cdots \cap H_n$, then H_f is a G_δ-subset of $\mathbb{R}^X$ such that $f \in H_f \subset \mathbb{R}^X \backslash C_p(X)$. This shows that in a realcompact space $\mathbb{R}^X$, the set $\mathbb{R}^X \backslash C_p(X) = \bigcup\{H_f : f \in \mathbb{R}^X \backslash C_p(X)\}$ is a union of G_δ-subsets of $\mathbb{R}^X$. Therefore $C_p(X)$ is realcompact by Problem 408 of [TFS].

T.424. *Suppose that $C_p(X) = Y_1 \cup \cdots \cup Y_n$, where Y_i is homeomorphic to $\mathbb{R}^{\kappa_i}$ for each $i \leq n$. Prove that X is discrete.*

Solution. It follows from Problem 401, 465 and 470 of [TFS] that $\mathbb{R}^\kappa$ is pseudocomplete and realcompact for any cardinal κ. Applying Problem 423 we can see that $C_p(X)$ is realcompact. It is an immediate consequence of Problem 406 that $C_p(X)$ is pseudocomplete. Finally apply Problem 486 of [TFS] to conclude that X is discrete.

T.425. *Suppose that* $C_p(X) = Y_1 \cup \cdots \cup Y_n$, *where* Y_i *is hereditarily realcompact for each* $i \leq n$. *Prove that* $iw(C_p(X)) = \psi(C_p(X)) = \omega$ *and hence* $C_p(X)$ *is hereditarily realcompact.*

Solution. It follows from Problem 420 that we can assume, without loss of generality, that every Y_i is dense in $C_p(X)$. Fix any $f \in C_p(X)$; it is clear that $Y_i' = Y_i \setminus \{f\}$ is realcompact and dense in $C_p(X)$ for any $i \leq n$.

Since $C_p(X)$ is a Moscow space by Problem 424 of [TFS], we can apply Problem 425 of [TFS] to see that Y_i' is ω-placed in $C_p(X)$ and hence there is a G_δ-subset H_i of the space $C_p(X)$ such that $f \in H_i \subset C_p(X) \setminus Y_i'$ for every $i \leq n$. Consequently, $H = H_1 \cap \cdots \cap H_n$ is a G_δ-subset of $C_p(X)$ such that $f \in H \subset C_p(X) \setminus (\bigcup_{i \leq n} Y_i') = \{f\}$. Thus $\{f\}$ is a G_δ-set in $C_p(X)$ for any $f \in C_p(X)$ and hence $iw(C_p(X)) = \psi(C_p(X)) = \omega$. Any space of countable i-weight is hereditarily realcompact by Problem 446 of [TFS] so $C_p(X)$ is hereditarily realcompact.

T.426. *Given an infinite cardinal* κ *suppose that* $C_p(X)$ *is a finite union of its* κ-*monolithic (not necessarily closed) subspaces. Prove that* $C_p(X)$ *is* κ-*monolithic.*

Solution. Since κ-monolithity is a hereditary property, there exist $n \in \mathbb{N}$, $\varepsilon > 0$ and κ-monolithic spaces $Y_1, \ldots, Y_n$ such that $C_p(X, (-\varepsilon, \varepsilon)) = Y_1 \cup \cdots \cup Y_n$ and $C = C_p(X, (-\varepsilon, \varepsilon)) \subset \overline{Y_i}^u$ for every $i \leq n$ (see Problem 421). If $A \subset C$ and $|A| \leq \kappa$, then let $A_i = \overline{A} \cap Y_i$ for any $i \leq n$ (the bar denotes the closure in C). Since Y_i is uniformly dense in C, for every $f \in C$, there is a sequence $S_i^f \subset Y_i$ which converges uniformly to f. However, it is sufficient for us to know that S_i^f is countable and $f \in \mathrm{cl}_C(S_i^f)$ for any $f \in C$ and $i \leq n$.

Now, if $B_i = \bigcup \{S_i^f : f \in A\}$, then $B_i \subset Y_i$ and $|B_i| \leq \kappa$; besides, we have $A_i \subset \overline{A} \subset \overline{B_i}$ and therefore $A_i \subset \mathrm{cl}_{Y_i}(B_i)$ which shows that $nw(A_i) \leq \kappa$ because Y_i is κ-monolithic for all $i \leq n$. Since network weight is finitely additive (see Problem 405), we have $nw(\overline{A}) = nw(A_1 \cup \cdots \cup A_n) \leq \kappa$ which proves that C is κ-monolithic. Thus the space $C_p(X)$ is also κ-monolithic being homeomorphic to C.

T.427. *Given an infinite cardinal* κ *suppose that* $C_p(X)$ *is a finite union of its* spread(κ)-*monolithic (not necessarily closed) subspaces. Prove that* $C_p(X)$ *itself is* spread(κ)-*monolithic.*

Solution. Since spread(κ)-monolithity is hereditary, there exist $n \in \mathbb{N}$, $\varepsilon > 0$ and spread(κ)-monolithic spaces $Y_1, \ldots, Y_n$ such that $C_p(X, (-\varepsilon, \varepsilon)) = Y_1 \cup \cdots \cup Y_n$ and $C = C_p(X, (-\varepsilon, \varepsilon)) \subset \overline{Y_i}^u$ for every $i \leq n$ (see Problem 421). If $A \subset C$ and $|A| \leq \kappa$, then let $A_i = \overline{A} \cap Y_i$ for any $i \leq n$ (the bar denotes the closure in C). Since Y_i is uniformly dense in C, for every $f \in C$, there is a sequence $S_i^f \subset Y_i$ which converges uniformly to f. However, it is sufficient for us to know that S_i^f is countable and $f \in \mathrm{cl}_C(S_i^f)$ for any $f \in C$ and $i \leq n$.

Now, if $B_i = \bigcup\{S_i^f : f \in A\}$, then $B_i \subset Y_i$ and $|B_i| \leq \kappa$; besides, we have $A_i \subset \overline{A} \subset \overline{B_i}$ and therefore $A_i \subset \mathrm{cl}_{Y_i}(B_i)$ which shows that $s(A_i) \leq \kappa$ because Y_i is spread(κ)-monolithic for all $i \leq n$. Since spread is finitely additive (see Problem 405), we have $s(\overline{A}) = s(A_1 \cup \cdots \cup A_n) \leq \kappa$ which proves that C is spread(κ)-monolithic. Thus the space $C_p(X)$ is also spread(κ)-monolithic being homeomorphic to C.

T.428. *Given an infinite cardinal κ suppose that $C_p(X)$ is a finite union of its $hd(\kappa)$-monolithic (not necessarily closed) subspaces. Prove that the space $C_p(X)$ is $hd(\kappa)$-monolithic.*

Solution. Since $hd(\kappa)$-monolithity is hereditary, there exist $n \in \mathbb{N}$, $\varepsilon > 0$ and $hd(\kappa)$-monolithic spaces $Y_1, \ldots, Y_n$ such that $C_p(X, (-\varepsilon, \varepsilon)) = Y_1 \cup \cdots \cup Y_n$ and $C = C_p(X, (-\varepsilon, \varepsilon)) \subset \overline{Y}_i^u$ for every $i \leq n$ (see Problem 421). If $A \subset C$ and $|A| \leq \kappa$, then let $A_i = \overline{A} \cap Y_i$ for any $i \leq n$ (the bar denotes the closure in C). Since Y_i is uniformly dense in C, for every $f \in C$, there is a sequence $S_i^f \subset Y_i$ which converges uniformly to f. However, it is sufficient for us to know that S_i^f is countable and $f \in \mathrm{cl}_C(S_i^f)$ for any $f \in C$ and $i \leq n$.

Now, if $B_i = \bigcup\{S_i^f : f \in A\}$, then $B_i \subset Y_i$ and $|B_i| \leq \kappa$; besides, we have $A_i \subset \overline{A} \subset \overline{B_i}$ and therefore $A_i \subset \mathrm{cl}_{Y_i}(B_i)$ which shows that $hd(A_i) \leq \kappa$ because Y_i is $hd(\kappa)$-monolithic for all $i \leq n$. Since hereditary density is finitely additive (see Problem 405), we have $hd(\overline{A}) = hd(A_1 \cup \cdots \cup A_n) \leq \kappa$ which proves that the space C is $hd(\kappa)$-monolithic. Thus the space $C_p(X)$ is also $hd(\kappa)$-monolithic being homeomorphic to C.

T.429. *Given an infinite cardinal κ suppose that $C_p(X)$ is a finite union of its $hl(\kappa)$-monolithic (not necessarily closed) subspaces. Prove that the space $C_p(X)$ is $hl(\kappa)$-monolithic.*

Solution. Since $hl(\kappa)$-monolithity is hereditary, there exist $n \in \mathbb{N}$, $\varepsilon > 0$ and $hl(\kappa)$-monolithic spaces $Y_1, \ldots, Y_n$ such that $C_p(X, (-\varepsilon, \varepsilon)) = Y_1 \cup \cdots \cup Y_n$ and $C = C_p(X, (-\varepsilon, \varepsilon)) \subset \overline{Y}_i^u$ for every $i \leq n$ (see Problem 421). If $A \subset C$ and $|A| \leq \kappa$, then let $A_i = \overline{A} \cap Y_i$ for any $i \leq n$ (the bar denotes the closure in C). Since Y_i is uniformly dense in C, for every $f \in C$ there is a sequence $S_i^f \subset Y_i$ which converges uniformly to f. However, it is sufficient for us to know that S_i^f is countable and $f \in \mathrm{cl}_C(S_i^f)$ for any $f \in C$ and $i \leq n$.

Now, if $B_i = \bigcup\{S_i^f : f \in A\}$, then $B_i \subset Y_i$ and $|B_i| \leq \kappa$; besides, we have $A_i \subset \overline{A} \subset \overline{B_i}$ and therefore $A_i \subset \mathrm{cl}_{Y_i}(B_i)$ which shows that $hl(A_i) \leq \kappa$ because Y_i is $hl(\kappa)$-monolithic for all $i \leq n$. Since hereditary Lindelöf number is finitely additive (see Problem 405), we have $hl(\overline{A}) = hl(A_1 \cup \cdots \cup A_n) \leq \kappa$ which proves that the space C is $hl(\kappa)$-monolithic. Thus the space $C_p(X)$ is also $hl(\kappa)$-monolithic being homeomorphic to C.

T.430. *Suppose that $C_p(X)$ is a finite union of its Dieudonné complete subspaces. Prove that $C_p(X)$ is realcompact and hence Dieudonné complete.*

Solution. If Z is a Dieudonné complete space and $Y \subset Z$ is an F_σ-subspace of Z, then Y is also Dieudonné complete (see Problem 460 of [TFS]). Thus we can apply Problem 420 to conclude that there exist $n \in \mathbb{N}$, $\varepsilon > 0$ and Dieudonné complete spaces $Y_1, \ldots, Y_n$ such that $C_p(X) = Y_1 \cup \cdots \cup Y_n$ and Y_i is dense in $C_p(X)$ for each $i \leq n$.

As a consequence, $c(Y_i) = c(C_p(X)) = \omega$ and therefore Y_i is realcompact for all $i \leq k$ (see Problem 458 of [TFS]). Thus we can apply Problem 423 to conclude that $C_p(X)$ is realcompact and hence Dieudonné complete (see Problem 454 of [TFS]).

T.431. *Let X be an arbitrary space. Suppose that $C_p(X) = \bigcup \{Z_n : n \in \omega\}$. Prove that there exists $f \in C_p(X)$ and $\varepsilon > 0$ such that for some $n \in \omega$, the set $(Z_n + f) \cap C(X, (-\varepsilon, \varepsilon))$ is dense in $C_u(X, (-\varepsilon, \varepsilon))$ and hence also in $C_p(X, (-\varepsilon, \varepsilon))$.*

Solution. Given $g, h \in C^*(X)$ let $d(g, h) = \sup\{|g(x) - h(x)| : x \in X\}$. Then d is a complete metric on $C^*(X)$ which generates the topology τ_u of uniform convergence on X (see Problem 419). We denote by $\mathbf{0}$ the function which is identically zero on X. Let $B(g, r) = \{h \in C^*(X) : d(g, h) < r\}$ for any $f \in C^*(X)$ and $r > 0$. The set $B(g, r)$ is a ball with respect to d, so it is open in $C_u^*(X)$ for any $f \in C^*(X)$ and $r > 0$. If $P \subset Q \subset C^*(X)$, say that P is τ_u-dense in Q if $Q \subset \overline{P}^u$.

Let $Y_n = Z_n \cap C^*(X)$ for every $n \in \omega$; then $C_u^*(X) = \bigcup_{n \in \omega} Y_n$. Since $C_u^*(X)$ is Čech-complete, it has the Baire property, so it is impossible that all elements of the sequence $\{Y_n : n \in \omega\}$ be nowhere dense in $C_u^*(X)$.

As a consequence, there is a non-empty set $V \in \tau_u$ with $V \subset \overline{Y_n \cap V}^u$. Since the metric d generates the topology τ_u, we can find $v \in V$ and $\varepsilon > 0$ such that $W = B(v, \varepsilon) \subset V$ and hence $W \subset \overline{Y_n \cap W}^u$. The map $L : C^*(X) \to C^*(X)$ defined by $L(g) = g - v$ for all $g \in C^*(X)$ is an isometry with respect to the metric d, i.e., $d(g, h) = d(L(g), L(h))$ for any functions $g, h \in C^*(X)$ (see Fact 2 of T.421). An immediate consequence is that $L(W) = B(\mathbf{0}, \varepsilon)$.

Furthermore, the map L is a homeomorphism, so the set $L(Y_n) \cap B(\mathbf{0}, \varepsilon) = L(Y_n \cap W)$ is τ_u-dense in $B(\mathbf{0}, \varepsilon)$. It is evident that $B(\mathbf{0}, \varepsilon) \subset C(X, (-\varepsilon, \varepsilon))$; since it was proved in Fact 1 of T.421 that $B(\mathbf{0}, \varepsilon)$ is τ_u-dense in $C = C(X, (-\varepsilon, \varepsilon))$, the set $L(Y_n) \cap B(\mathbf{0}, \varepsilon)$ is τ_u-dense in C as well and therefore a larger set $L(Y_n) \cap C$ is τ_u-dense in C too. Finally, if $f = -v$, then $L(Y_n) \cap C = (Y_n + f) \cap C = (Z_n + f) \cap C$ which shows that $(Z_n + f) \cap C$ is τ_u-dense in C. Since the identity map $i : C_u(X, (-\varepsilon, \varepsilon)) \to C_p(X, (-\varepsilon, \varepsilon))$ is continuous (see Problem 086 of [TFS]), the set $(Z_n + f) \cap C$ is dense in $C_p(X, (-\varepsilon, \varepsilon))$, so we proved all the promised properties of Z_n and f.

T.432. *Let $C_p(X) = \bigcup \{Z_n : n \in \omega\}$, where each Z_n is closed in $C_p(X)$. Prove that some Z_n contains a homeomorphic copy of $C_p(X)$.*

Solution. Apply Problem 431 to find $n \in \omega$, $f \in C_p(X)$ and $\varepsilon > 0$ such that the set $(Z_n + f) \cap C(X, (-\varepsilon, \varepsilon))$ is dense in $C = C(X, (-\varepsilon, \varepsilon))$. It was proved in Problem 079 of [TFS] that the map $L : C_p(X) \to C_p(X)$ defined by $L(g) = f + g$

for all $g \in C_p(X)$ is a homeomorphism, so $Z_n + f = L(Z_n)$ is closed in $C_p(X)$. Thus $C \subset L(Z_n)$ and hence $C' = L^{-1}(C) \subset Z_n$. Since $C \simeq C_p(X)$ and $C' \simeq C$, the space C' is a homeomorphic copy of $C_p(X)$ which is contained in Z_n.

T.433. *Let $\mathcal{P}$ be a hereditary property. Suppose that $C_p(X) = \bigcup \{Z_n : n \in \omega\}$, where each Z_n is closed in $C_p(X)$ and has $\mathcal{P}$. Prove that $C_p(X)$ also has $\mathcal{P}$.*

Solution. By Problem 432, there is $n \in \omega$ such that $C_p(X)$ embeds in Z_n. Since Z_n has $\mathcal{P}$ and $\mathcal{P}$ is hereditary, the space $C_p(X)$ has $\mathcal{P}$ as well.

T.434. *Suppose that $C_p(X) = \bigcup \{Z_n : n \in \omega\}$, where each Z_n is locally compact. Prove that X is finite.*

Solution. If we are given a function $f \in \mathbb{R}^X$, a set $K \subset X$ is finite and $\delta > 0$, then $O(f, K, \delta) = \{g \in \mathbb{R}^X : |g(x) - f(x)| < \delta$ for all $x \in K\}$. It is clear that the family $\{O(f, K, \delta) : K$ is a finite subset of X and $\delta > 0\}$ is a local base in $\mathbb{R}^X$ at the point f for any $f \in \mathbb{R}^X$. If $A \subset C_p(X)$, then $\overline{A}$ is its closure in $C_p(X)$ while $\mathrm{cl}(A)$ denotes the closure of A in $\mathbb{R}^X$.

Fact 1. The space $\mathbb{R}^\omega$ cannot be represented as a countable union of its locally compact subspaces.

Proof. Every locally compact second countable spaces is σ-compact, so if $\mathbb{R}^\omega$ is a countable union of its locally compact subspaces, then it has to be σ-compact which it is not (see Fact 2 of S.186). Fact 1 is proved.   □

By the result of Problem 431, there exist $n \in \omega$, $\varepsilon > 0$ and $v \in C_p(X)$ such that the set $(Z_n + v) \cap C(X, (-\varepsilon, \varepsilon))$ is dense in $C = C(X, (-\varepsilon, \varepsilon))$. Since $Y = (Z_n + v)$ is locally compact, we can choose a point $f \in Y \cap C$ and $H \in \tau(f, Y)$ such that $\mathrm{cl}_Y(H)$ is compact. Consequently, $\overline{H} = \mathrm{cl}_Y(H)$ is compact as well.

The set $Z = Y \cap C$ is dense in C and $G = H \cap Z = H \cap C \in \tau^*(Z)$, so we can choose $O \in \tau(C)$ such that $O \cap Z = G$. It follows from density of Z in C that $\mathrm{cl}_C(G) = \mathrm{cl}_C(O)$. Therefore $\overline{O} = \mathrm{cl}_C(O) = \mathrm{cl}_C(G) = \overline{G} \subset \overline{H}$ is a compact subset of $C_p(X)$. Thus $\overline{A} \subset \overline{O}$ is compact for any $A \subset O$.

Since the set O is open in the space C and $f \in O$, we can choose $\delta > 0$ and a finite set $K \subset X$ such that $(f(x) - \delta, f(x) + \delta) \subset (-\varepsilon, \varepsilon)$ for all $x \in K$ while $U = O(f, K, \delta) \cap C \subset O$. It turns out that

(1) $E = (-\varepsilon, \varepsilon)^{X \setminus K} \times \prod_{x \in K} (f(x) - \delta, f(x) + \delta)) \subset \mathrm{cl}(U)$.

Indeed, for any $g \in E$, $r > 0$ and a finite $L \subset X$, there exists $h \in C$ such that $h|L = g|L$; then $h \in U \cap O(g, L, r)$, i.e., $U \cap O(g, L, r) \neq \emptyset$ for any basic neighborhood $O(g, L, r)$ of the point g in $\mathbb{R}^X$. Therefore $g \in \mathrm{cl}(U)$; since $g \in E$ was chosen arbitrarily, the property (1) is proved.

Observe also that $\overline{U}$ is a compact dense subset of $\mathrm{cl}(U)$ so $\overline{U} = \mathrm{cl}(U)$ which shows that $E \subset \overline{U} \subset C_p(X)$. Now let $g(x) = 0$ for any $x \in X \setminus K$ and $g(x) = f(x)$ for every $x \in K$. Then $g \in E$ and hence $g \in C$. It is easy to see that for the set $I = \{h \in C_p(X) : |g(x) - h(x)| \leq \frac{\delta}{2}$ for all $x \in X\}$, we have $I \subset C$; besides, I is closed in $C_p(X)$ and $I \subset E \subset \overline{U}$, so I is closed in the compact space $\overline{U}$ which implies that I is compact.

We have $I \simeq C_p(X, \mathbb{I})$ by Fact 3 of S.398, so $C_p(X, \mathbb{I})$ is compact and hence X is discrete (see Problem 396 of [TFS]). If X is infinite, then $\mathbb{R}^\omega$ embeds in $C_p(X) = \mathbb{R}^X$ as a closed subspace so $\mathbb{R}^\omega$ is σ-locally compact which contradicts Fact 1 and makes our solution complete.

T.435. *Suppose that* $C_p(X) = \bigcup\{Z_n : n \in \omega\}$, *where each* Z_n *is locally pseudocompact. Prove that* $C_p(X)$ *is* σ-*pseudocompact.*

Solution. If we are given a function $f \in C_p(X)$, a set $K \subset X$ is finite and $\delta > 0$, then $O(f, K, \delta) = \{g \in C_p(X) : |g(x) - f(x)| < \delta$ for all $x \in K\}$. It is clear that the family $\{O(f, K, \delta) : K$ is a finite subset of X and $\delta > 0\}$ is a local base in $C_p(X)$ at the point f for any $f \in C_p(X)$.

By the result of Problem 431, there exist $n \in \omega$, $\varepsilon > 0$ and $v \in C_p(X)$ such that the set $(Z_n + v) \cap C(X, (-\varepsilon, \varepsilon))$ is dense in $C = C(X, (-\varepsilon, \varepsilon))$. Since $Y = (Z_n + v)$ is locally pseudocompact, we can choose a point $f \in Y \cap C$ and $H \in \tau(f, Y)$ such that $\mathrm{cl}_Y(H)$ is pseudocompact. Since $\mathrm{cl}_Y(H)$ is a dense subspace of $\overline{H}$ (the bar denotes the closure in $C_p(X)$), the set $\overline{H}$ is also pseudocompact (see Fact 18 of S.351).

The set $Z = Y \cap C$ is dense in C and $G = H \cap Z = H \cap C \in \tau^*(Z)$, so we can choose $O \in \tau(C)$ such that $O \cap Z = G$. It follows from density of Z in C that $\mathrm{cl}_C(G) = \mathrm{cl}_C(O)$. Therefore $\overline{O} = \overline{\mathrm{cl}_C(O)} = \overline{\mathrm{cl}_C(G)} = \overline{G} \subset \overline{H}$.

Since the set O is open in the space C and $f \in O$, we can choose $\delta > 0$ and a finite set $K \subset X$ such that $[f(x) - 2\delta, f(x) + 2\delta] \subset (-\varepsilon, \varepsilon)$ for all points $x \in K$ while $U = O(f, K, 2\delta) \cap C \subset O$. If $r = \max\{|f(x)| : x \in K\} + \delta$ then $[-r - \delta, r + \delta] \subset (-\varepsilon, \varepsilon)$ and $f(x) \in [-r, r]$ for all $x \in K$. Therefore there exists a function $g \in C(X, [-r, r])$ such that $g|K = f|K$.

It is easy to see that for the set $I = \{h \in C_p(X) : |g(x) - h(x)| \leq \delta$ for all $x \in X\}$, we have $I \subset C$.

Furthermore, I is a retract of the space $C_p(X)$ by Fact 3 of S.398; an easy consequence of $I \subset \overline{O} \subset \overline{H}$ is that I is also a retract (and hence a continuous image) of a pseudocompact space $\overline{H}$. This implies that I is pseudocompact. We have $I \simeq C_p(X, \mathbb{I})$ by Fact 3 of S.398, so $C_p(X, \mathbb{I})$ is also pseudocompact.

If X is not pseudocompact, then $C_p(X) \simeq C_p(X) \times \mathbb{R}^\omega$ by Fact 6 of T.132. Therefore there is $R \subset C_p(X)$ such that R is a retract of $C_p(X)$ and $R \simeq \mathbb{R}^\omega$. If $T_i = Z_i \cap R$ for all $i \in \omega$, then $\bigcup_{i \in \omega} T_i = R$. Since R has the Baire property, there is $W \in \tau^*(R)$ such that $T_k \cap W$ is dense in W for some $k \in \omega$.

Choose a point $w \in T_k \cap W$ and $V \in \tau(w, T_k)$ such that $\mathrm{cl}_{T_k}(V)$ is pseudocompact. Since $\mathrm{cl}_{T_k}(V)$ is a dense subset of $\overline{V}$, the set $\overline{V}$ is also pseudocompact (see Fact 18 of S.351). The set $N = V \cap W = V \cap (W \cap T_k)$ is open in $T_k \cap W \subset R$, so there is $M \in \tau(R)$ such that $M \subset W$ and $M \cap (T_k \cap W) = M \cap T_k = N$. Since $T_k \cap W$ is dense in W, the set $N = T_k \cap M$ is dense in M, so $\mathrm{cl}_R(M) = \mathrm{cl}_R(N)$.

Let $r : C_p(X) \to R$ be a retraction. Then $r(\overline{V}) \supset \overline{V} \cap R \supset V \cap W = N$. The set $r(\overline{V})$ is pseudocompact in a second countable space R, so it is compact; since $N \subset r(\overline{V})$, the set $\mathrm{cl}_R(N) = \mathrm{cl}_R(M)$ is compact. It turns out that the closure of some non-empty open subset of R is compact and hence the same is true for some $E \in \tau^*(\mathbb{R}^\omega)$ (because R is homeomorphic to $\mathbb{R}^\omega$). Therefore every $h \in E$ is a point of local compactness of $\mathbb{R}^\omega$; since $\mathbb{R}^\omega$ is homogeneous, it is locally compact and hence σ-compact which is a contradiction (see Fact 2 of S.186).

Thus X is pseudocompact, so applying Problems 398 and 399 of [TFS], we can conclude that $C_p(X)$ is σ-pseudocompact and finish our solution.

T.436. *Suppose that $C_p(X) = \bigcup \{Z_n : n \in \omega\}$, where each Z_n is realcompact and closed in $C_p(X)$. Prove that $C_p(X)$ is realcompact.*

Solution. Our solution can be easily derived from the following fact.

Fact 1. For any space Z, if $C_p(Z, \mathbb{I})$ is realcompact, then $C_p(Z)$ is also realcompact.

Proof. Let $O_n = [-1, -1 + 2^{-n-1}) \cup (1 - 2^{-n-1}, 1]$; it is clear that $O_n \in \tau(\mathbb{I})$, so the set $W_n^z = \{f \in C_p(Z) : f(z) \in O_n\}$ is an open subset of $C_p(Z)$ for every $n \in \omega$ and $z \in Z$. The equality $F_z = \{f \in C_p(Z) : f(z) \in \{-1, 1\}\} = \bigcap_{n \in \omega} W_n^z$ implies that F_z is a G_δ-subset of $C_p(Z)$ for any $z \in Z$.

Now observe that $C_p(Z, (-1, 1)) = C_p(Z, \mathbb{I}) \setminus (\bigcup \{F_z : z \in Z\})$ and hence the space $C_p(Z, (-1, 1))$ is realcompact by Problem 408 of [TFS]. Since $C_p(Z)$ is homeomorphic to $C_p(Z, (-1, 1))$, the space $C_p(Z)$ is also realcompact so Fact 1 is proved. □

By the result of Problem 431, there exist $n \in \omega$, $\varepsilon > 0$ and $v \in C_p(X)$ such that the set $(Z_n + v) \cap C(X, (-\varepsilon, \varepsilon))$ is dense in $C = C(X, (-\varepsilon, \varepsilon))$. Since $Y = (Z_n + v)$ is closed in $C_p(X)$ we have $C_p(X, [-\varepsilon, \varepsilon]) = \overline{C} \subset Y$; the space $E = C_p(X, [-\varepsilon, \varepsilon])$ is closed in $C_p(X)$ and hence in Y, so E is realcompact by Problem 403 of [TFS].

Furthermore E is homeomorphic to $C_p(X, \mathbb{I})$ by Fact 3 of S.398, so $C_p(X, \mathbb{I})$ is realcompact which shows that $C_p(X)$ is also realcompact (see Fact 1).

T.437. *Prove that any metacompact collectionwise normal space is paracompact.*

Solution. If Z is a space, $\mathcal{A} \subset \exp(Z)$ and $z \in Z$, then $\mathrm{ord}(z, \mathcal{A})$ is the cardinality of the family $\mathcal{A}(z) = \{A \in \mathcal{A} : z \in A\}$. A family $\mathcal{B} \subset \exp(Z)$ is *inscribed* in $\mathcal{A}$ if for any $B \in \mathcal{B}$, there is $A \in \mathcal{A}$ such that $B \subset A$.

Suppose that X is a collectionwise normal metacompact space. Given an open cover $\mathcal{U}$ of the space X, it has a point-finite open refinement $\mathcal{V}$. For any $n \in \mathbb{N}$, let $X_n = \{x \in X : \mathrm{ord}(x, \mathcal{V}) = n\}$. It is clear that $X = \bigcup_{n \in \omega} X_n$. Observe first that

(1)   the set $Y_n = X_1 \cup \cdots \cup X_n$ is closed in X for any $n \in \mathbb{N}$.

Indeed, if $x \in X \setminus Y_n$, then there are distinct sets $V_1, \ldots, V_{n+1} \in \mathcal{V}$ such that $x \in V = \bigcap_{i \leq n+1} V_i$ which shows that $V \subset X \setminus Y_n$ and hence $X \setminus Y_n$ is open in X so (1) is proved.

For any point $x \in X$ there is a unique number $n \in \mathbb{N}$ such that $x \in X_n$; let $W_x = \bigcap \{V \in \mathcal{V} : x \in V\}$ and $V_x = W_x \cap X_n$. We claim that

(2) if $n \in \mathbb{N}$ and $x, y \in X_n$, then either $V_x = V_y$ or $V_x \cap V_y = \emptyset$.

To see that (2) holds, assume that $z \in V_x \cap V_y$; then $z \in (\bigcap \mathcal{V}(x)) \cap (\bigcap \mathcal{V}(y))$ However, $|\mathcal{V}(x)| = n$ and $|\mathcal{V}(y)| = n$, so if $\mathcal{V}(x) \backslash \mathcal{V}(y) \neq \emptyset$, then z belongs to more than n elements of $\mathcal{V}$ which contradicts $z \in X_n$. Analogously, it is impossible that $\mathcal{V}(y) \backslash \mathcal{V}(x) \neq \emptyset$, so $\mathcal{V}(x) = \mathcal{V}(y)$ and hence $V_x = V_y$ so (2) is proved.

It follows from (1) and (2) that $\mathcal{E}_n = \{V_x : x \in X_n\}$ is a disjoint open cover of X_n. Consequently, V_x is closed in X_n for every $x \in X_n$. For every $n \in \mathbb{N}$, choose an indexation $\{V_s^n : s \in S_n\}$ of the family $\mathcal{E}_n$ such that $V_s^n \neq V_t^n$ for any distinct $s, t \in S_n$. Given $n \in \mathbb{N}$ and $s \in S_n$, there is $x \in X_n$ with $V_s^n = V_x$; let $W_s^n = W_x$. Then $V_s^n \subset W_s^n$ for any $s \in S_n$ and the family $\mathcal{W}_n = \{W_s^n : s \in S_n\} \subset \tau(X)$ is inscribed in $\mathcal{V}$ and hence in $\mathcal{U}$.

Observe that all elements of $\mathcal{E}_1$ are closed in X_1 and hence in X by (1). Any disjoint open cover of a space is a discrete family in that space, so $\mathcal{E}_1$ is discrete in X_1 and hence in X. The space X being collectionwise normal, we can find a discrete family $\mathcal{H}_1 = \{U_s^1 : s \in S_1\} \subset \tau(X)$ such that $V_s^1 \subset U_s^1 \subset W_s^1$ for any $s \in S_1$.

Assume that $n \in \mathbb{N}$ and for any $i < n$ we have constructed a discrete family $\mathcal{H}_i = \{U_s^i : s \in S_n\} \subset \tau(X)$ which is inscribed in $\mathcal{U}$ and $Y_i \subset H_i = \bigcup_{j \leq i}(\bigcup \mathcal{H}_j)$ for all $i < n$. The set $H = H_1 \cup \cdots \cup H_{n-1}$ is open in X, so $P = X_n \backslash H = Y_n \backslash H$ is closed in X by (1). Consequently, $\mathcal{G} = \{V_s^n \cap P : s \in S_n\}$ is a disjoint open cover of P, so all elements of $\mathcal{G}$ are closed in P and hence in X.

Any disjoint open cover of a space is a discrete family in that space, so $\mathcal{G}$ is discrete in P and hence in X. The space X being collectionwise normal, we can find a discrete family $\mathcal{H}_n = \{U_s^n : s \in S_n\} \subset \tau(X)$ such that $V_s^n \cap P \subset U_s^n \subset W_s^n$ for any $s \in S_n$.

Thus $\mathcal{H}_n$ covers P and therefore $Y_n = H \cup P$ is covered by $H \cup Q_n$ where $Q_n = \bigcup \mathcal{H}_n$; since $\mathcal{W}_n$ is inscribed in $\mathcal{U}$, the family $\mathcal{H}_n$ is also inscribed in $\mathcal{U}$. This shows that we can construct inductively a sequence $\{\mathcal{H}_n : n \in \omega\}$ of discrete families of open subsets of X such that $\mathcal{H}_n$ is inscribed in $\mathcal{U}$ for all $n \in \omega$ and $Y_n \subset \bigcup_{i \leq n}(\bigcup \mathcal{H}_i)$ for all $n \in \mathbb{N}$.

Therefore $\mathcal{H} = \bigcup_{n \in \omega} \mathcal{H}_n$ is a σ-discrete open cover of X inscribed in $\mathcal{U}$, i.e., $\mathcal{H}$ is a σ-discrete open refinement of $\mathcal{U}$. The open cover $\mathcal{U}$ of the space X was chosen arbitrarily, so every open cover of X has a σ-discrete open refinement; applying Problem 230 of [TFS] we conclude that X is paracompact.

T.438. *Prove that if $C_p(X)$ is normal and metacompact, then it is Lindelöf.*

Solution. Normality of $C_p(X)$ implies that $C_p(X)$ is collectionwise normal (see Problem 295 of [TFS]). Therefore $C_p(X)$ is paracompact (see Problem 437) and hence Lindelöf by Problem 219 of [TFS].

T.439. *Prove that $C_p(\beta\omega)$ is not metacompact.*

Solution. An immediate consequence of Problem 382 of [TFS] is that $p(C_p(\beta\omega)) = \omega$, i.e., every point-finite family of non-empty open subsets of $C_p(\beta\omega)$ is countable. If $C_p(\beta\omega)$ is metacompact, then every open cover of $C_p(\beta\omega)$

has a point-finite open refinement which has to be countable by our observation. Thus $C_p(\beta\omega)$ is Lindelöf.

If T is the "two arrows" space (see Problem 384 of [TFS]), then T is separable, compact and $\text{ext}(C_p(T)) = \mathfrak{c}$. However, $C_p(T)$ embeds in $C_p(\beta\omega)$ as a closed subspace by Problem 375 of [TFS]. Consequently, $\text{ext}(C_p(\beta\omega)) \geq \text{ext}(C_p(T)) = \mathfrak{c}$ which contradicts the Lindelöf property of $C_p(\beta\omega)$. This contradiction shows that $C_p(\beta\omega)$ is not metacompact.

T.440. *Prove that $C_p(L(\kappa))$ is not metacompact for any uncountable κ.*

Solution. Recall that $L(\kappa) = \kappa \cup \{a\}$ where all point of the set κ are isolated and $U \in \tau(a, L(\kappa))$ if and only if $a \in U \subset L(\kappa)$ and $L(\kappa)\backslash U$ is countable. For any $A \subset L(\kappa)$, let $\pi_A : C_p(L(\kappa)) \to C_p(A)$ be the restriction map defined by $\pi_A(f) = f|A$ for any $f \in C_p(L(\kappa))$. Since $L(\kappa)$ is a P-space, the space $C_p(L(\kappa), \mathbb{I})$ is countably compact (see Problem 397 of [TFS]), so it follows from Problem 479 of [TFS] that $C_p(L(\kappa))$ is pseudocomplete and hence Baire by Problem 464 of [TFS].

Thus $p(C_p(L(\kappa))) = c(C_p(L(\kappa))) = \omega$ by Problem 282 of [TFS]. Now, if $C_p(L(\kappa))$ is metacompact, then every open cover of $C_p(L(\kappa))$ has a point-finite open refinement which has to be countable by $p(C_p(L(\kappa))) = \omega$. Therefore $C_p(L(\kappa))$ is Lindelöf and hence $t(L(\kappa)) = \omega$ by Problem 189 of [TFS] which is a contradiction because $a \in \text{cl}(\kappa)$ while no countable subset of κ contains a in its closure.

T.441. *Prove that neither the Baire property nor pseudocompleteness is countably additive in spaces $C_p(X)$.*

Solution. Recall that there exists an infinite pseudocompact space X such that $C_p(X)$ is σ-pseudocompact, i.e., $C_p(X) = \bigcup_{n\in\omega} Z_n$ where Z_n is pseudocompact for all $n \in \omega$ (see Problem 400 of [TFS]). The space $C_p(X)$ does not have the Baire property by Problem 284 of [TFS]. Since every pseudocompact space is pseudocomplete and hence Baire (see Problem 472 and 464 of [TFS]), every Z_n is pseudocomplete, so $C_p(X)$ witnesses that neither pseudocompleteness nor the Baire property is countably additive in $C_p(X)$.

T.442. *Prove that π-weight and π-character are not countably additive in spaces $C_p(X)$.*

Solution. Let X be the Cantor set $\mathbb{K}$ and denote by $\mathcal{C}$ the family of all non-empty clopen subsets of X. Since X is second countable and zero-dimensional, we can choose a countable base $\mathcal{B} \subset \mathcal{C}$ of the space X. Call a non-empty family $\mathcal{B}' \subset \mathcal{B}$ *adequate* if $\mathcal{B}'$ is finite, disjoint and $\bigcup \mathcal{B}' \neq X$. It is easy to see that the set $Q = \{(B_1, \ldots, B_n, q_1, \ldots, q_n) : n \in \mathbb{N}, \{B_1, \ldots, B_n\}$ is an adequate family and $q_i \in \mathbb{Q} \cap \mathbb{I}$ for all $i \leq n\}$ is countable, so we can enumerate it as $Q = \{w_i : i \in \omega\}$. For each $i \in \omega$ define a function $f_i : X \to \mathbb{R}$ as follows: if $w_i = (B_1, \ldots, B_n, q_1, \ldots, q_n)$, then let $f_i(x) = q_k$ if $x \in B_k$ for some $k \leq n$ and $f_i(x) = i + 2$ for any $x \in X\backslash(\bigcup_{i \leq n} B_i)$.

Observe that the set $D = \{f_i : i \in \omega\}$ is a discrete subspace of $C_p(X)$. To see it take any $i \in \omega$; if $w_i = (B_1, \ldots, B_n, q_1, \ldots, q_n)$ and $B = \bigcup_{i \le n} B_i$, then pick any $x \in X \backslash B$. The set $O_i = \{f \in C_p(X) : |f(x) - f_i(x)| < \frac{1}{2}\}$ is open in $C_p(X)$ and $O_i \cap D = \{f_i\}$ for all $i \in \omega$, so D is discrete and hence $Z = \overline{D}$ has countable π-weight because $\{\{f_i\} : i \in \omega\}$ is a π-base in Z.

Given any function $f \in C_p(X, \mathbb{I})$ consider any $U \in \tau(f, C_p(X))$. There are $x_1, \ldots, x_n \in X$ and $\varepsilon > 0$ such that $V = \{g \in C_p(X) : |f(x_i) - g(x_i)| < \varepsilon$ for all $i \le n\} \subset U$. It is easy to find an adequate family $\{B_1, \ldots, B_n\} \subset \mathcal{B}$ such that $x_k \in B_k$ for all $k \le n$. Choose $q_1, \ldots, q_n \in \mathbb{Q} \cap \mathbb{I}$ such that $|f(x_k) - q_k| < \varepsilon$ for all $k \le n$. There is $i \in \omega$ such that $(B_1, \ldots, B_n, q_1, \ldots, q_n) = w_i$ and hence $|f_i(x_k) - f(x_k)| < \varepsilon$ for all $k \le n$ which proves that $f_i \in V \cap D \subset U \cap D$. Thus $U \cap D \ne \emptyset$ for every $U \in \tau(f, C_p(X))$ and therefore $f \in \overline{D}$. Since the function $f \in C_p(X, \mathbb{I})$ was chosen arbitrarily, we proved that $C_p(X, \mathbb{I}) \subset Z$.

For any $n \in \mathbb{N}$, the map $\varphi_n : C_p(X) \to C_p(X)$ defined by $\varphi_n(f) = n \cdot f$ for every $f \in C_p(X)$ is a homeomorphism, so $Z_n = \varphi_n(Z)$ has a countable π-weight for all $n \in \mathbb{N}$. Since $C_p(X, [-n, n]) = \varphi_n(C_p(X, \mathbb{I})) \subset Z_n$ for all $n \in \mathbb{N}$, we have $\bigcup \{Z_n : n \in \mathbb{N}\} = C_p(X)$ because $C_p(X) = \bigcup \{C_p(X, [-n, n]) : n \in \mathbb{N}\}$; this last equality is due to the fact that X is compact.

It turns out that $C_p(X) = \bigcup \{Z_n : n \in \mathbb{N}\}$ and $\pi\chi(Z_n) \le \pi w(Z_n) = \omega$ for all $n \in \mathbb{N}$. However, $\pi\chi(C_p(X)) = w(C_p(X)) = |X| = \mathfrak{c} > \omega$ (see Fact 1 of T.158) and hence $C_p(X)$ witnesses that both π-weight and π-character are not countably additive in $C_p(X)$.

T.443. *Suppose that $C_p(X)$ is a countable union of its Čech-complete (not necessarily closed) subspaces. Prove that X is countable and discrete (and hence $C_p(X)$ is Čech-complete).*

Solution. Assume that $C_p(X) = \bigcup \{P_n : n \in \omega\}$ and every P_n is Čech-complete. It follows from Problem 431 that there is $n \in \omega$, $\varepsilon > 0$ and $f \in C_p(X)$ such that the set $(P_n + f) \cap C(X, (-\varepsilon, \varepsilon))$ is dense in $C = C(X, (-\varepsilon, \varepsilon))$. Since $Y = P_n + f$ is Čech-complete and C is dense in $E = C(X, [-\varepsilon, \varepsilon])$, the set $Y \cap E$ is dense in E. Furthermore, E is closed in $C_p(X)$ and therefore $Z = E \cap Y$ is a Čech-complete dense subspace of E.

The space E is homeomorphic to $C_p(X, \mathbb{I})$ (see Problem 287 of [TFS]), so $C_p(X, \mathbb{I})$ also has a dense Čech-complete subspace and hence X is discrete by Problem 287 of [TFS]. This implies that $C_p(X) = \mathbb{R}^X$ has the Baire property and therefore all sets P_m cannot be nowhere dense; fix $U \in \tau^*(C_p(X))$ such that $P_m \cap U$ is dense in U for some $m \in \omega$. There exists a set $K = \{x_1, \ldots, x_k\} \subset X$ and nonempty rational intervals $O_1, \ldots, O_k$ such that $V = \{f \in C_p(X) : f(x_i) \in O_i$ for all $i \le k\} \subset U$. It is immediate that $V = \mathbb{R}^{X \backslash K} \times O_1 \times \cdots \times O_k$; since $O_i \simeq \mathbb{R}$ for all $i \le k$, the space V is homeomorphic to $\mathbb{R}^X$. Besides, the set $P_m \cap V$ is Čech-complete and dense in V which shows that $C_p(X) = \mathbb{R}^X$ has a dense Čech-complete subspace. Finally apply Problem 265 of [TFS] to conclude that X is discrete and countable, so $C_p(X) \simeq \mathbb{R}^\omega$ is Čech-complete.

T.444. *Given an infinite cardinal κ suppose that $C_p(X)$ is a union of countably many (not necessarily closed) subspaces of character $\leq \kappa$. Prove that $\chi(C_p(X)) \leq \kappa$ and hence $|X| \leq \kappa$.*

Solution. Assume that $C_p(X) = \bigcup_{n \in \omega} Z_n$ and $\chi(Z_n) \leq \kappa$ for all $n \in \omega$. By Problem 431 there exist $n \in \omega$, $\varepsilon > 0$ and $f \in C_p(X)$ such that $(Z_n + f) \cap C(X, (-\varepsilon, \varepsilon))$ is dense in $C = C(X, (-\varepsilon, \varepsilon))$. If $Y = (Z_n + f) \cap C$, then $\chi(Y) \leq \kappa$ and Y is dense in C. Consequently, $\chi(g, C) = \chi(g, Y) \leq \kappa$ for any $g \in Y$ (see Fact 2 of S.265).

Since C is homeomorphic to the space $C_p(X)$ (see Fact 1 of S.295), there is some $h \in C_p(X)$ such that $\chi(h, C_p(X)) \leq \kappa$ and hence $\chi(C_p(X)) \leq \kappa$ because $C_p(X)$ is homogeneous (see Problem 079 of [TFS]). Therefore $|X| = \chi(C_p(X)) \leq \kappa$ by Problem 169 of [TFS].

T.445. *Prove that weight is countably additive in spaces $C_p(X)$.*

Solution. Given an infinite cardinal κ assume that $C_p(X) = \bigcup_{n \in \omega} Z_n$ and $w(Z_n) \leq \kappa$ for all $n \in \omega$. Since $\chi(Z_n) \leq w(Z_n) \leq \kappa$ for all $n \in \omega$, we can apply Problem 444 to conclude that $w(C_p(X)) = |X| = \chi(C_p(X)) \leq \kappa$ (see Problem 169 of [TFS]) and therefore $w(C_p(X)) \leq \kappa$. This proves that weight is countably additive in spaces $C_p(X)$.

T.446. *Prove that metrizability is countably additive in spaces $C_p(X)$.*

Solution. Assume that $C_p(X) = \bigcup_{n \in \omega} Z_n$ and Z_n is metrizable for all $n \in \omega$. Then $\chi(Z_n) \leq \omega$ for all $n \in \omega$, so we can apply Problem 444 to conclude that $w(C_p(X)) = |X| = \chi(C_p(X)) \leq \omega$ (see Problem 169 of [TFS]) and hence the space $C_p(X)$ is metrizable.

T.447. *Prove that tightness is countably additive in spaces $C_p(X)$.*

Solution. Given a function $f \in C_p(X)$, a finite set $K \subset X$ and a number $\delta > 0$, let $O(f, K, \delta) = \{g \in C_p(X) : |g(x) - f(x)| < \delta \text{ for all } x \in K\}$. It is clear that the family $\{O(f, K, \delta) : K \subset X \text{ is finite and } \delta > 0\}$ is a local base of $C_p(X)$ at the point f. Given any sets $A \subset B \subset C_p(X)$ we will say that A *is uniformly dense in* B if $B \subset \overline{A}^u$.

Let κ be an infinite cardinal; assume that $C_p(X) = \bigcup_{n \in \omega} Z_n$ and $t(Z_n) \leq \kappa$ for all $n \in \omega$. By Problem 431 there exist $n \in \omega$, $\varepsilon > 0$ and $v \in C_p(X)$ such that $Y = (Z_n + v) \cap C(X, (-\varepsilon, \varepsilon))$ is uniformly dense in $C = C(X, (-\varepsilon, \varepsilon))$.

Recall that a family $\mathcal{U} \subset \exp(X)$ is *an ω-cover of X* if for any finite $K \subset X$ there is $U \in \mathcal{U}$ such that $K \subset U$. If all elements of $\mathcal{U}$ are open in X, then $\mathcal{U}$ is called *an open ω-cover of X*. Fix an open ω-cover $\mathcal{U}$ of the space X and let $D = \{f \in Y : f^{-1}((\frac{\varepsilon}{9}, \varepsilon)) \subset U \text{ for some } U \in \mathcal{U}\}$. Let $w(x) = \frac{8}{9}\varepsilon$ for every $x \in X$; then $w \in C$ and hence there is $f \in Y$ such that $|f(x) - w(x)| < \frac{1}{9}\varepsilon$ for all $x \in X$. It is evident that $f(x) \geq \frac{7}{9}\varepsilon$ for all $x \in X$.

To show that $f \in \overline{D}$, take any $W \in \tau(f, C_p(X))$; there is a finite $K \subset X$ and $\delta > 0$ such that $O(f, K, \delta) \subset W$. The family $\mathcal{U}$ being an ω-cover of X, there is $U \in \mathcal{U}$ such that $K \subset U$. It is easy to construct a function $g \in C$ such that

$g|(X \setminus U) \equiv 0$ and $g(x) = f(x)$ for every $x \in K$. Let $\mu = \min\{\delta, \frac{1}{9}\varepsilon\}$; since Y is uniformly dense in C, there is $h \in Y$ such that $|h(x) - g(x)| < \mu$ for all $x \in X$. Thus $|h(x) - f(x)| < \delta$ for all $x \in K$ and therefore $h \in O(f, K, \delta)$. Besides, $h(x) < \frac{1}{9}\varepsilon$ for all $x \in X \setminus U$ which shows that $h^{-1}((\frac{1}{9}\varepsilon, \varepsilon)) \subset U$, i.e., $h \in D \cap O(f, K, \delta) \subset D \cap W$. Consequently, $D \cap W \neq \emptyset$ for any $W \in \tau(f, C_p(X))$ so $f \in \overline{D}$.

Since $t(Y) \leq \kappa$, there is $E \subset D$ such that $|E| \leq \kappa$ and $f \in \overline{E}$. For every $h \in E$ there is $U_h \in \mathcal{U}$ for which $S_h = h^{-1}((\frac{1}{9}\varepsilon, \varepsilon)) \subset U_h$. Then $\mathcal{U}' = \{U_h : h \in E\} \subset \mathcal{U}$ and $|\mathcal{U}'| \leq \kappa$. If $K \subset X$ is finite, then $G = \{g \in C_p(X) : g(x) > \frac{1}{9}\varepsilon$ for any $x \in K\}$ is an open neighborhood of f in $C_p(X)$, so there is $e \in E$ for which $e \in G$. Since $e(x) > \frac{1}{9}\varepsilon$ for all $x \in K$, we have $K \subset S_e \subset U_e \in \mathcal{U}'$ and therefore $\mathcal{U}'$ is an ω-cover of X. We proved that for any open ω-cover $\mathcal{U}$ of the space X, there is $\mathcal{U}' \subset \mathcal{U}$ such that $|\mathcal{U}'| \leq \kappa$ and $\mathcal{U}'$ is an ω-cover of X. Now apply Problems 148 and 149 of [TFS] to conclude that $t(C_p(X)) \leq \kappa$ and finish our solution.

T.448. *Prove that pseudocharacter is countably additive in spaces* $C_p(X)$.

Solution. If Z is a space, $f \in C^*(Z)$ and $\varepsilon > 0$, let $I_Z(f, \varepsilon) = \{g \in C^*(Z) : |f(z) - g(z)| \leq \varepsilon$ for all $z \in Z\}$. For any $f, g \in C^*(Z)$ we write $f \leq g$ if $f(z) \leq g(z)$ for all $z \in Z$.

Fact 1. Suppose that Z is a space, $f_0, \ldots, f_n \in C^*(Z)$ and $\varepsilon_0, \ldots, \varepsilon_n$ are positive numbers. Let $Y_i = \bigcap_{k \leq i} I_Z(f_k, \varepsilon_k)$ for every $i \leq n$ and assume additionally that $f_{i+1} \in Y_i$ for all $i < n$. Then $\psi(f, Y_n) = \psi(C_p(Z))$ for any $f \in Y_n$.

Proof. The inequality $\psi(f, Y_n) \leq \psi(C_p(Z))$ is true for any $f \in Y_n$ because Y_n is a subspace of $C_p(Z)$. Now, assume that $\psi(f, Y_n) \leq \kappa$ for some $f \in Y_n$ and an infinite cardinal κ. It follows from Fact 1 of S.426 that there exists $A \subset Z$ such that $|A| \leq \kappa$ and $G(A, f) \cap Y_n = \{f\}$ where $G(A, f) = \{g \in C^*(Z) : g|A = f|A\}$.

If A is not dense in Z, then fix any $z \in Z \setminus \overline{A}$ and a function $p \in C(Z, [0, 1])$ such that $p(z) = 1$ and $p(A) = \{0\}$. Let $m(y) = \max\{f_i(y) - \varepsilon_i : i \leq n\}$ and $M(y) = \min\{f_i(y) + \varepsilon_i : i \leq n\}$ for any point $y \in Z$. Then $m, M \in C^*(Z)$ and

(1) $m(y) < M(y)$ for any $y \in Z$.

Indeed, it follows from $f_n \in Y_n$ that $f_i(y) - \varepsilon_i \leq f_n(y) \leq f_i(y) + \varepsilon_i$ for all $i \leq n$ and therefore $m(y) \leq M(y)$ for all $y \in Z$. Now if $m(y) = M(y)$ for some $y \in Z$, then there exist distinct $l, k \leq n$ such that $f_l(y) - \varepsilon_l = f_k(y) + \varepsilon_k$ and hence $|f_l(z) - f_k(z)| = \varepsilon_l + \varepsilon_k > \max\{\varepsilon_l, \varepsilon_k\}$. If $l < k$, then we obtain a contradiction with $f_k \in I_Z(f_l, \varepsilon_l)$; if $k < l$, then our inequality contradicts $f_l \in I_Z(f_k, \varepsilon_k)$ so (1) is proved.

Observe that for any $g \in C^*(Z)$ we have $g \in Y_n$ if and only if $m \leq g \leq M$. In particular, $m(z) \leq f(z) \leq M(z)$, so either $m(z) < f(z)$ or $f(z) < M(z)$.

In the first case choose a positive number δ with $\delta < f(z) - m(z)$ and consider the functions $g = f - \delta p$ and $h = \max(g, m)$. It is immediate that $m \leq h \leq f$, so

$h \in Y_n$. If $y \in A$, then $p(y) = 0$ and hence $g(y) = f(y) = h(y)$, i.e., $h|A = f|A$. However, $h(z) = f(z) - \delta \neq f(z)$ which shows that $h \neq f$ while $h \in G(A, f) \cap Y_n$ which is a contradiction.

Finally, if $f(z) < M(z)$, then choose a positive number δ with $\delta < M(z) - f(z)$ and consider the functions $g = f + \delta p$ and $h = \min(g, M)$. It is immediate that $f \leq h \leq M$, so $h \in Y_n$. If $y \in A$, then $p(y) = 0$ and hence $g(y) = f(y) = h(y)$, i.e., $h|A = f|A$. However, $h(z) = f(z) + \delta \neq f(z)$ which shows that $h \neq f$ while $h \in G(A, f) \cap Y_n$ which is again a contradiction. Thus A is dense in Z and therefore $\psi(C_p(Z)) = d(Z) \leq \kappa$ so Fact 1 is proved.   □

Returning to our solution, assume that X is a space and κ is an infinite cardinal such that $C_p(X) = \bigcup_{n \in \omega} C'_n$ and $\psi(C'_n) \leq \kappa$ for every $n \in \omega$. Let $C_n = C'_n \cap C^*(X)$ for all $n \in \omega$; then $C^*_p(X) = \bigcup_{n \in \omega} C_n$. Since "pseudocharacter$\leq \kappa$" is a hereditary property, we can assume that $C_n \cap C_m = \emptyset$ whenever $m \neq n$.

If $f, g \in C^*(X)$, let $d(f, g) = \sup\{|f(x) - g(x)| : x \in X\}$. Then d is a complete metric on $C^*(X)$ and the topology $\tau(d)$ generated by d coincides with the uniform convergence topology τ_u on $C^*(X)$ (see Problem 419). For any $A \subset C^*(X)$ we denote by $\text{diam}(A)$ the diameter of A with respect to the metric d, i.e., $\text{diam}(A) = \sup\{d(f, g) : f, g \in A\}$. Assume, towards a contradiction, that $\psi(C_p(X)) > \kappa$ and denote by $\mathbf{0}_X$ the function which is identically zero on X. An immediate consequence of Fact 1 is that $\psi(C_p(X, [-n, n])) = \psi(I_X(\mathbf{0}_X, n)) = \psi(C_p(X))$ for all $n \in \mathbb{N}$ and hence $\psi(C^*_p(X)) = \psi(C_p(X)) > \kappa$. Besides, the space $C^*_p(X)$ is homogeneous by Problem 079 of [TFS], so $\psi(f, C^*_p(X)) = \psi(C^*_p(X)) > \kappa$ for any $f \in C^*_p(X)$.

Denote by $\mathcal{C}$ the family of all closed non-empty G_κ-subsets of $C^*_p(X)$. We will often implicitly use the fact that $\mathcal{C}$ is a base in the family of all G_κ-subsets of $C^*_p(X)$ in the sense that if A is a G_κ-subset of $C^*_p(X)$ and $f \in A$, then $f \in H \subset A$ for some $H \in \mathcal{C}$ (see Fact 2 of S.328).

Suppose that $Y \subset C^*_p(X)$ and $\psi(f, Y) > \kappa$ for any $f \in Y$. If A is a G_κ-set in $C^*_p(X)$ and $\psi(f, A \cap Y) \leq \kappa$ for some $f \in A \cap Y$, then $\psi(f, Y) \leq \kappa$ which is a contradiction. Therefore

(2) if $Y \subset C^*_p(X)$ and $\psi(f, Y) > \kappa$ for any $f \in Y$, then for any G_κ-set $A \subset C^*_p(X)$ we have $\psi(f, Y \cap A) > \kappa$ for any $f \in Y \cap A$.

For the sake of brevity, call a set $Y \subset C^*_p(X)$ deep if $\psi(f, Y) > \kappa$ for any $f \in Y$. Choose any $g \in C_0$ and let P be a closed G_κ-set in the space $C^*_p(X)$ such that $P \cap C_0 = \{g\}$. The set P is deep by (2) so choose $f_0 \in P_0 \setminus \{g\}$ and $H \in \mathcal{C}$ such that $f_0 \in H \subset C^*_p(X) \setminus \{g\}$. It is clear that $f_0 \in H_0 = H \cap P \in \mathcal{C}$ and $H_0 \cap C_0 = \emptyset$. Suppose that $n \in \mathbb{N}$ and we have $f_0, \ldots, f_{n-1} \in C^*_p(X)$ and $H_0, \ldots, H_{n-1} \in \mathcal{C}$ with the following properties:

(3) $H_0 \supset \cdots \supset H_{n-1}$;
(4) $f_i \in H_i \cap Y_i$ where $Y_i = \bigcap_{k \leq i} I_X(f_k, 2^{-k})$ for all $i \leq n - 1$;
(5) $H_i \cap Y_i \cap C_i = \emptyset$ for all $i < n$.

If $H_{n-1} \cap Y_{n-1} \cap C_n = \emptyset$, then for the sets $H_n = H_{n-1}$ and $f_n = f_{n-1}$ the conditions (3)–(5) still hold so the inductive step is fulfilled.

Otherwise, take any function $g \in C_n \cap H_{n-1} \cap Y_{n-1}$ and pick $P \in C$ such that $P \cap C_n = \{g\}$. The set $Q = H_{n-1} \cap P \cap Y_{n-1}$ is deep by Fact 1 and (2), so we can choose $f_n \in Q \backslash \{g\}$ and $H \in C$ such that $f_n \in H$ and $g \notin H$. Then $H_n = H_{n-1} \cap P \cap H \in C$ and it is immediate that (3)–(5) still hold for the functions $f_0, \ldots, f_n$ and the sets $H_0, \ldots, H_n$. Therefore our inductive procedure provides sequences $\{f_i : i \in \omega\} \subset C_p^*(X)$ and $\{H_i : i \in \omega\} \subset C$ such that (3)–(5) hold for all $i \in \omega$. Let $F_i = H_i \cap Y_i$ for all $i \in \omega$. Then F_i is closed in $C_p^*(X)$ and hence in $C_u^*(X)$; besides, $F_i \neq \emptyset$ because $f_i \in F_i$ by (4). Furthermore, $\mathrm{diam}(I_X(f, \varepsilon)) \leq 2\varepsilon$ for any $\varepsilon > 0$ and $f \in C_p^*(X)$ which shows that $\mathrm{diam}(F_i) \leq 2^{-i+1}$ for all $i \in \omega$ by (4). Thus $\{F_i : i \in \omega\}$ is a decreasing sequence of non-empty closed subsets of a complete metric space $C_u^*(X)$. Since $\mathrm{diam}(F_i) \to 0$, we have $F = \bigcap_{i \in \omega} F_i \neq \emptyset$ (see Problem 236 of [TFS]). Now, if $f \in F$, then $f \notin \bigcup_{i \in \omega} C_i$ by (5); this contradiction shows that $\psi(C_p(X)) \leq \kappa$ and makes our solution complete.

T.449. *Prove that i-weight and diagonal number are countably additive in spaces $C_p(X)$.*

Solution. Given an infinite cardinal κ, suppose that $C_p(X) = \bigcup\{C_n : n \in \omega\}$ and $\Delta(C_n) \leq \kappa$ (or $iw(C_n) \leq \kappa$) for all $n \in \omega$. It is easy to see that pseudocharacter does not exceed diagonal number (or i-weight respectively) for any space, so $\psi(C_n) \leq \Delta(C_n) \leq \kappa$ (or $\psi(C_n) \leq iw(C_n) \leq \kappa$ respectively) for all $n \in \omega$. Therefore we can apply Problem 448 to conclude that $\psi(C_p(X)) \leq \kappa$. Consequently, $\Delta(C_p(X)) = \psi(C_p(X)) \leq \kappa$ (or $iw(C_p(X)) = \psi(C_p(X)) \leq \kappa$ respectively) by Problem 173 of [TFS].

T.450. *Prove that the Fréchet–Urysohn property is countably additive in spaces $C_p(X)$.*

Solution. Suppose that $C_p(X) = \bigcup_{n \in \omega} C_n$ where C_n is a Fréchet–Urysohn space for each $n \in \omega$. By Problem 431, we can find $\varepsilon > 0$, $n \in \omega$ and $v \in C_p(X)$ such that $D = (C_n + v) \cap C(X, (-\varepsilon, \varepsilon))$ is uniformly dense in $C = C(X, (-\varepsilon, \varepsilon))$ (recall that this means that every function from C is a uniform limit of a sequence from the set D). If we are given a function $f \in C_p(X)$, a finite set $K \subset X$ and $\delta > 0$, then $O(f, K, \delta) = \{g \in C_p(X) : |g(x) - f(x)| < \delta$ for all $x \in K\}$. It is evident that the family $\{O(f, K, \delta) : K$ is a finite subset of X and $\delta > 0\}$ is a local base of $C_p(X)$ at the point f.

Since D is uniformly dense in C, there is $w \in D$ such that $w(x) \geq \frac{8}{9}\varepsilon$ for every $x \in X$. Recall that a family $\mathcal{A}$ is an ω-cover of X if for any finite $K \subset X$ there is $A \in \mathcal{A}$ such that $K \subset A$. Suppose that $\mathcal{U} \subset \tau(X)$ is an ω-cover of X and consider the set $P = \{f \in D : f^{-1}((\frac{1}{9}\varepsilon, \varepsilon)) \subset U$ for some $U \in \mathcal{U}\}$.

To see that $w \in \overline{P}$, take any $W \in \tau(w, C_p(X))$. There exists a finite $K \subset X$ and $\delta \in (0, \frac{1}{9}\varepsilon)$ such that $O(w, K, \delta) \subset W$. Since $\mathcal{U}$ is an ω-cover of X, we can take $U \in \mathcal{U}$ with $K \subset U$. It is easy to construct a function $f \in C$ such that $f|K = w|K$ and $f(X \backslash U) = \{0\}$. Since D is uniformly dense in C, there is $g \in D$ for which $|g(x) - f(x)| < \delta$ for all $x \in X$. It is immediate that $|g(x)| < \frac{1}{9}\varepsilon$ for all $x \in X \backslash U$ and hence $f^{-1}((\frac{1}{9}\varepsilon, \varepsilon)) \subset U$ which shows that $g \in P$. Besides, $|g(x) - w(x)| =$

$|g(x) - f(x)| < \delta$ for all $x \in K$ and hence $g \in O(w, K, \delta) \cap P \subset W \cap P$. Thus any $W \in \tau(w, C_p(X))$ intersects the set P, i.e., $w \in \overline{P}$.

The space D being Fréchet–Urysohn there is a sequence $\{f_k : k \in \omega\} \subset P$ such that $f_k \to w$. Pick $U_k \in \mathcal{U}$ such that $f_k^{-1}((\frac{1}{9}\varepsilon, \varepsilon)) \subset U_k$ for all $k \in \omega$. Given any $x \in X$ the set $G = \{h \in C_p(X) : h(x) > \frac{1}{9}\varepsilon\}$ is an open neighborhood of the point w, so there is $m \in \omega$ such that $f_k \in G$ for all $k \geq m$. Thus $x \in f_k^{-1}((\frac{1}{9}\varepsilon, \varepsilon)) \subset U_k$ for all $k \geq m$. Therefore every point of X belongs to all sets U_k except finitely many (this is usually denoted by $\lim\{U_k : k \in \omega\} = X$). It turns out that for every open ω-cover $\mathcal{U}$ of the space X, there is a sequence $\{U_k : k \in \omega\} \subset \mathcal{U}$ such that $\lim\{U_k : k \in \omega\} = X$. Therefore we can apply Problem 144 of [TFS] to conclude that $C_p(X)$ is a Fréchet–Urysohn space.

T.451. *Suppose that X is a metrizable space and $C_p(X) = \bigcup\{Y_i : i \in \omega\}$, where Y_i is hereditarily realcompact (not necessarily closed) for every $i \in \omega$. Prove that $nw(C_p(X)) = iw(C_p(X)) = \omega$ and hence $C_p(X)$ is hereditarily realcompact.*

Solution. If $\text{ext}(X) > \omega$, then $\mathbb{R}^{\omega_1}$ embeds in $C_p(X)$ by Fact 1 of S.215 and therefore $\mathbb{R}^{\omega_1} = \bigcup\{Z_n : n \in \omega\}$ where every Z_n is hereditarily realcompact. Observe first that

(1) if $\mathbb{R}^{\omega_1} = \bigcup_{n \in \omega} R_n$ and every R_n is dense in $\mathbb{R}^{\omega_1}$, then all R_n cannot be hereditarily realcompact.

Indeed, every R_n is a Moscow space (see Problems 423 and 424 of [TFS]). If R_n is hereditarily realcompact for every $n \in \omega$, then given a point $x \in R_n$, the space $R_n \backslash \{x\}$ is realcompact and dense in a Moscow space R_n, so $R_n \backslash \{x\}$ must be ω-placed in R_n (see Problem 425 of [TFS]) which implies that $\{x\}$ is a G_δ-set in R_n. This proves that $\psi(R_n) \leq \omega$ for all $n \in \omega$. Since $\mathbb{R}^{\omega_1} = C_p(D(\omega_1))$ (where $D(\omega_1)$ is the set ω_1 with the discrete topology), we can apply Problem 448 to conclude that $\psi(\mathbb{R}^{\omega_1}) = \omega$; this contradiction shows that (1) is true.

For every $A \subset \omega_1$ the map $\pi_A : \mathbb{R}^{\omega_1} \to \mathbb{R}^A$ is the natural projection. By (1), there is $n_0 \in \omega$ such that Z_{n_0} is not dense in $\mathbb{R}^{\omega_1}$, so there is a finite set $A_0 \subset \omega_1$ and $x_0 \in \mathbb{R}^{A_0}$ for which $\pi_{A_0}^{-1}(x) \cap Z_{n_0} = \emptyset$. Suppose that $\alpha < \omega_1$, and for every $\beta < \alpha$, we have $n_\beta \in \omega$, a finite set $A_\beta \subset \omega_1$ and $x_\beta \in \mathbb{R}^{A_\beta}$ with the following properties:

(2) $\beta < \gamma < \alpha$ implies $n_\beta \neq n_\gamma$ and $A_\beta \cap A_\gamma = \emptyset$;
(3) for any $\beta < \alpha$, let $B_\beta = \bigcup\{A_\gamma : \gamma \leq \beta\}$ and $y_\beta = \bigcup\{x_\gamma : \gamma \leq \beta\} \in \mathbb{R}^{B_\beta}$; then $\pi_{B_\beta}^{-1}(y_\beta) \cap Z_{n_\beta} = \emptyset$.

Let $y = \bigcup\{x_\beta : \beta < \alpha\}$; the set $B = \bigcup\{A_\beta : \beta < \alpha\}$ is countable and $y \in \mathbb{R}^B$, so $|\omega_1 \backslash B| = \omega_1$ and therefore $\pi_B^{-1}(y)$ is homeomorphic to $\mathbb{R}^{\omega_1}$. Furthermore, $\pi_B^{-1}(y) \subset \bigcap\{\pi_{B_\beta}^{-1}(y_\beta) : \beta < \alpha\}$ which shows that $\pi_B^{-1}(y) \cap Z_{n_\beta} = \emptyset$ for any $\beta < \alpha$.

If $N = \{n \in \omega : Z_n \cap \pi_B^{-1}(y) \neq \emptyset\}$, then $N \cap \{n_\beta : \beta < \alpha\} = \emptyset$ and (1) implies that it is impossible that $Z_n \cap \pi_B^{-1}(y)$ be dense in $\pi_B^{-1}(y)$ for all $n \in N$. Therefore there exists a finite set $A_\alpha \subset \omega_1 \backslash B$ and $x_\alpha \in \mathbb{R}^{A_\alpha}$ such that for $y_\alpha = y \cup x_\alpha$ and

$B_\alpha = B \cup A_\alpha$, we have $\pi_{B_\alpha}^{-1}(y_\alpha) \cap Z_{n_\alpha} = \emptyset$ for some $n_\alpha \in N$. It is evident that (2) and (3) still hold for all $\beta \le \alpha$, so our inductive construction can be continued to obtain, among other things, a set $\{n_\beta : \beta < \omega_1\} \subset \omega$ such that $n_\beta \ne n_\gamma$ for distinct $\beta, \gamma \in \omega_1$ [see (2)].

This contradiction shows that $\mathbb{R}^{\omega_1}$ cannot be represented as a countable union of its hereditarily realcompact subspaces and hence $\text{ext}(X) = \omega$. Since X is metrizable, we have $nw(C_p(X)) = nw(X) = w(X) = \omega$ (see Problem 172 and 214 of [TFS]) and hence $iw(C_p(X)) = d(X) = \omega$ which, together with Problem 446 of [TFS], implies that $C_p(X)$ is hereditarily realcompact.

T.452. *Let X be a pseudocompact space. Suppose that $C_p(X) = \bigcup\{Z_n : n \in \omega\}$, where each Z_n is paracompact and closed in $C_p(X)$. Prove that $C_p(X)$ is Lindelöf.*

Solution. By Problem 431, there exists $n \in \omega$, $\varepsilon > 0$ and $v \in C_p(X)$ such that the set $(Z_n + v) \cap C(X, (-\varepsilon, \varepsilon))$ is dense in $C = C(X, (-\varepsilon, \varepsilon))$. Since $Z = Z_n + v$ is closed in $C_p(X)$, we have $C_p(X, [-\varepsilon, \varepsilon]) = \overline{C} \subset Z$ (see Fact 1 of T.421) and hence $C_p(X, [-\varepsilon, \varepsilon])$ is paracompact being a closed subspace of a paracompact space Z. The space $C_p(X, \mathbb{I})$ is homeomorphic to $C_p(X, [-\varepsilon, \varepsilon])$ (see Fact 3 of S.398), so $C_p(X, \mathbb{I})$ is paracompact and hence Lindelöf because $c(C_p(X, \mathbb{I})) = \omega$ (see Problems 111 and 092 of [TFS] together with Fact 2 of S.219).

Therefore $C_p(X, [-n, n])$ is also Lindelöf for any $n \in \mathbb{N}$ (here we use Fact 3 of S.398 again) and hence $C_p(X) = \bigcup\{C_p(X, [-n, n]) : n \in \mathbb{N}\}$ is a Lindelöf space as well (the last equality holds because X is pseudocompact).

T.453. *Give an example of a non-normal space which is a countable union of its closed normal subspaces.*

Solution. The Mrowka space M can be represented as $M = D \cup E$ where E is countable and the set D is discrete and closed in M (see Problem 142 of [TFS]). Thus the family $\{D\} \cup \{\{x\} : x \in E\}$ is countable and consists of discrete (and hence normal) closed subspaces of M. Therefore M is a countable union of its closed normal subspaces. However, M is not normal because it is pseudocompact but not countably compact (see Problems 142 and 137 of [TFS]).

T.454. *Let X be a compact space. Suppose that $C_p(X) = \bigcup\{Z_n : n \in \omega\}$, where each Z_n is normal and closed in $C_p(X)$. Prove that $C_p(X)$ is Lindelöf.*

Solution. By Problem 431, there exists $n \in \omega$, $\varepsilon > 0$ and $v \in C_p(X)$ such that the set $(Z_n + v) \cap C(X, (-\varepsilon, \varepsilon))$ is dense in $C = C(X, (-\varepsilon, \varepsilon))$. Since $Z = Z_n + v$ is closed in $C_p(X)$, we have $C_p(X, [-\varepsilon, \varepsilon]) = \overline{C} \subset Z$ (see Fact 1 of T.421), and hence $C_p(X, [-\varepsilon, \varepsilon])$ is normal being a closed subspace of a normal space Z. The space $C_p(X, \mathbb{I})$ is homeomorphic to $C_p(X, [-\varepsilon, \varepsilon])$ (see Fact 3 of S.398), so $C_p(X, \mathbb{I})$ is also normal and hence $\text{ext}(C_p(X, \mathbb{I})) = \omega$ by Problem 296 of [TFS].

Since X is compact, we can apply Baturov's theorem (see 269) to conclude that $C_p(X, \mathbb{I})$ is Lindelöf. Thus $C_p(X, [-n, n])$ is also Lindelöf for any $n \in \mathbb{N}$ (here we used Fact 3 of S.398 again) and therefore $C_p(X) = \bigcup\{C_p(X, [-n, n]) : n \in \mathbb{N}\}$ is Lindelöf as well.

T.455. *Let X be a metrizable space. Suppose that $C_p(X)$ is a countable union of its (not necessarily closed) normal subspaces. Prove that X is second countable and hence $C_p(X)$ is normal.*

Solution. Let M_t be a space for any $t \in T$; in the product $M = \prod_{t \in T} M_t$ the map $p_A : M \to M_A = \prod_{t \in A} M_t$ is the natural projection for any $A \subset T$. A set $E \subset M$ *covers all countable faces* of M if $p_A(E) = M_A$ for any countable $A \subset T$.

Fact 1. If M_t is a second countable space for any $t \in T$ and $E \subset M = \prod_{t \in T} M_t$ covers all countable faces of M, then E is C-embedded in M.

Proof. It is easy to see that any set which covers all countable faces of a product is dense in that product, so E is dense in M. Given a continuous $f : E \to \mathbb{R}$, we can apply Problem 299 of [TFS] to find a countable $A \subset T$ and a continuous map $g : p_A(E) = M_A \to \mathbb{R}$ such that $f = g \circ (p_A|E)$. It is immediate that $h = g \circ p_A \in C(M)$ and $h|E = f$, so Fact 1 is proved. □

Fact 2. Suppose that a property $\mathcal{P}$ is closed-hereditary and $\mathbb{R}^{\omega_1}$ is a countable union of its subspaces with the property $\mathcal{P}$. Then there exists a sequence $\mathcal{S} = \{Y_n : n \in \omega\}$ of subspaces of $\mathbb{R}^{\omega_1}$ such that $\mathbb{R}^{\omega_1} = \bigcup \mathcal{S}$ and the set Y_i covers all countable faces of $\mathbb{R}^{\omega_1}$ and has $\mathcal{P}$ for any $i \in \omega$. The sequence $\mathcal{S}$ will be called $\mathcal{P}$-*representation of the space* $\mathbb{R}^{\omega_1}$.

Proof. For every $A \subset \omega_1$ the map $\pi_A : \mathbb{R}^{\omega_1} \to \mathbb{R}^A$ is the natural projection. Assume that no $\mathcal{P}$-representation of $\mathbb{R}^{\omega_1}$ exists while $\mathbb{R}^{\omega_1} = \bigcup\{Z_n : n \in \omega\}$ and $Z_n \vdash \mathcal{P}$ ($\equiv Z_n$ has $\mathcal{P}$) for all $n \in \omega$. For any countable $A \subset \omega_1$ and $x \in \mathbb{R}^A$, the space $\pi_A^{-1}(x)$ is closed in $\mathbb{R}^{\omega_1}$ and homeomorphic to $\mathbb{R}^{\omega_1 \setminus A}$; it is easy to construct a homeomorphism $h : \mathbb{R}^{\omega_1 \setminus A} \to \mathbb{R}^{\omega_1}$ such that $h(P)$ covers all countable faces of $\mathbb{R}^{\omega_1}$ whenever it covers all countable faces of $\mathbb{R}^{\omega_1 \setminus A}$. Thus our assumption implies that

(1) for any countable set $A \subset \omega_1$ and $x \in \mathbb{R}^A$, there exists $m \in \omega$ such that $Z_m \cap \pi_A^{-1}(x) \neq \emptyset$ while $\pi_{A \cup B}^{-1}(y) \cap Z_m = \emptyset$ for some countable $B \subset \omega_1 \setminus A$ and $y \in \mathbb{R}^{A \cup B}$ such that $y|A = x$.

It follows from (1) that there is $n_0 \in \omega$ such that Z_{n_0} does not cover all countable faces of $\mathbb{R}^{\omega_1}$, so there is a countable set $A_0 \subset \omega_1$ and $x_0 \in \mathbb{R}^{A_0}$ for which $\pi_{A_0}^{-1}(x_0) \cap Z_{n_0} = \emptyset$. Suppose that $\alpha < \omega_1$, and for every $\beta < \alpha$ we have $n_\beta \in \omega$, a countable set $A_\beta \subset \omega_1$ and $x_\beta \in \mathbb{R}^{A_\beta}$ with the following properties:

(2) $\beta < \gamma < \alpha$ implies $n_\beta \neq n_\gamma$ and $A_\beta \cap A_\gamma = \emptyset$;
(3) for any $\beta < \alpha$, let $B_\beta = \bigcup\{A_\gamma : \gamma \leq \beta\}$ and $y_\beta = \bigcup\{x_\gamma : \gamma \leq \beta\} \in \mathbb{R}^{B_\beta}$. then $\pi_{B_\beta}^{-1}(y_\beta) \cap Z_{n_\beta} = \emptyset$.

Let $y = \bigcup\{x_\beta : \beta < \alpha\}$; the set $B = \bigcup\{A_\beta : \beta < \alpha\}$ is countable and $y \in \mathbb{R}^B$. Furthermore, $\pi_B^{-1}(y) \subset \bigcap\{\pi_{B_\beta}^{-1}(y_\beta) : \beta < \alpha\}$ which shows that $\pi_B^{-1}(y) \cap Z_{n_\beta} = \emptyset$ for any $\beta < \alpha$.

If $N = \{n \in \omega : Z_n \cap \pi_B^{-1}(y) \neq \emptyset\}$, then $N \cap \{n_\beta : \beta < \alpha\} = \emptyset$ and (1) implies that it is impossible for $Z_n \cap \pi_B^{-1}(y)$ to cover all countable faces of $\pi_B^{-1}(y)$ for all $n \in N$. Therefore there exists a countable set $A_\alpha \subset \omega_1 \setminus B$ and $x_\alpha \in \mathbb{R}^{A_\alpha}$ such that, for $y_\alpha = y \cup x_\alpha$ and $B_\alpha = B \cup A_\alpha$ we have $\pi_{B_\alpha}^{-1}(y_\alpha) \cap Z_{n_\alpha} = \emptyset$ for some $n_\alpha \in N$. It is evident that (2) and (3) still hold for all $\beta \leq \alpha$ so our inductive construction can be continued to obtain, among other things, a set $\{n_\beta : \beta < \omega_1\} \subset \omega$ such that $n_\beta \neq n_\gamma$ for distinct $\beta, \gamma \in \omega_1$ [see (2)].

This contradiction shows that $\mathbb{R}^{\omega_1}$ has a $\mathcal{P}$-representation and hence Fact 2 is proved.   □

Fact 3. There exists a closed discrete $D \subset \mathbb{R}^{\omega_1}$ such that $|D| = \omega_1$ while $\pi_A(D)$ is countable for any countable $A \subset \omega_1$. Here $\pi_B : \mathbb{R}^{\omega_1} \to \mathbb{R}^B$ is the natural projection for any $B \subset \omega_1$.

Proof. Since ω is a closed subset of $\mathbb{R}$, the space ω^{ω_1} is a closed subset of $\mathbb{R}^{\omega_1}$, so it suffices to construct the promised set in ω^{ω_1}. It was proved in Problem 068 that there exists an ω_1-sequence $\{s_\alpha : \alpha < \omega_1\}$ such that $s_\alpha : \alpha \to \omega$ is an injection for each $\alpha < \omega_1$ and $s_\beta|\alpha \approx s_\alpha$ whenever $\alpha < \beta < \omega_1$ (recall that if we have sets T, S and maps $p, q : R \to S$, then $p \approx q$ says that the set $\{r \in R : p(r) \neq q(r)\}$ is finite). For every countable ordinal $\alpha \geq \omega$ fix a bijection $b_\alpha : \omega \to \alpha$ and define an element $f_\alpha \in \omega^{\omega_1}$ as follows: $f_\alpha(\beta) = s_{\alpha+1}(\beta)$ for all $\beta \leq \alpha$; if $\beta > \alpha$, then let $f_\alpha(\beta) = b_\beta^{-1}(\alpha)$.

We claim that the set $D = \{f_\alpha : \omega \leq \alpha < \omega_1\}$ is as promised. First assume that $\alpha < \beta < \omega_1$ and take any $\gamma > \beta$. Then $f_\alpha(\gamma) = b_\gamma^{-1}(\alpha) \neq b_\gamma^{-1}(\beta) = f_\beta(\gamma)$ (the inequality in the middle is true because b_γ^{-1} is a bijection). This proves that $f_\alpha \neq f_\beta$ if $\alpha \neq \beta$ so $|D| = \omega_1$.

To prove that D is closed and discrete in ω^{ω_1}, it suffices to show that every $f \in \omega^{\omega_1}$ has a neighborhood which intersects at most one element of D. So, take any $f \in \omega^{\omega_1}$ and let $\xi(\alpha) = b_\alpha(f(\alpha))$ for any $\alpha \geq \omega$. By our definition of b_α, we have $\xi(\alpha) < \alpha$ for all $\alpha \geq \omega$ and hence there is a stationary $S \subset \omega_1$ and $\delta < \omega_1$ such that $\xi(\alpha) = \delta$ for all $\alpha \in S$.

Therefore there exist $\mu, \eta \in S$ such that $\delta < \mu < \eta$ and $f(\mu) = f(\eta) = n$. The set $W = \{g \in \omega^{\omega_1} : g(\mu) = g(\eta) = n\}$ is an open neighborhood of f in ω^{ω_1}. If $\alpha \geq \eta$, then $f_\alpha \notin W$ because $f_\alpha|(\eta + 1)$ is an injection so $f_\alpha(\mu) \neq f_\alpha(\eta)$.

If $\mu \leq \alpha < \eta$ then $f_\alpha(\eta) = b_\eta^{-1}(\alpha)$ while $n = b_\eta^{-1}(\delta)$; since b_η^{-1} is a bijection and $\delta < \alpha$, we have $f_\alpha(\eta) \neq n$, i.e., again $f_\alpha \notin W$.

Finally, if $\alpha < \mu$, then $f_\alpha(\mu) = b_\mu^{-1}(\alpha)$, so if $f_\alpha \in W$, then $b_\mu^{-1}(\alpha) = b_\mu^{-1}(\delta)$ and hence $\alpha = \delta$ because b_μ is a bijection. Thus the only element of D which can belong to W is f_δ, so the set D is, indeed, closed and discrete in ω^{ω_1}.

To see that $\pi_B(D)$ is countable for any countable $B \subset \omega_1$, it suffices to show that the set $Q = \{f_\alpha|\beta : \omega \leq \alpha < \omega_1\}$ is countable for any $\beta < \omega_1$. Now, $Q \subset \{f_\alpha|\beta : \alpha \leq \beta\} \cup \{f_\alpha|\beta : \beta < \alpha < \omega_1\}$. The first set in this union is evidently, countable. The second one is equal to the set $P = \{s_{\alpha+1}|\beta : \beta < \alpha < \omega_1\}$. However, $p \approx s_\beta$ for any $p \in P$ and therefore $P \subset P' = \{s \in \omega^\beta : s \approx s_\beta\}$. Since it is evident that P' is countable, the set D has all required properties, so Fact 3 is proved.   □

Returning to our solution, suppose first that $\text{ext}(X) > \omega$; then $\mathbb{R}^{\omega_1}$ embeds in $C_p(X)$ as a closed subspace (see Fact 1 of S.215). An immediate consequence is that $\mathbb{R}^{\omega_1}$ is a countable union of its normal subspaces. This makes it possible to apply Fact 2 to conclude that $\mathbb{R}^{\omega_1} = \bigcup\{Z_n : n \in \omega\}$ where Z_n is normal and covers all countable faces of $\mathbb{R}^{\omega_1}$ for all $n \in \omega$. By Fact 1 every Z_n is C-embedded in $\mathbb{R}^{\omega_1}$.

Now apply Fact 3 to find a closed discrete $D \subset \mathbb{R}^{\omega_1}$ such that $|D| = \omega_1$, but $\pi_B(D)$ is countable for any countable $B \subset \omega_1$. There is $n \in \omega$ such that $D \cap Z_n$ is uncountable, so we can assume, without loss of generality, that $D \subset Z_n$. Take any injection $u : D \to \mathbb{R}$; since u is continuous on D, by normality of Z_n there is a continuous $v : Z_n \to \mathbb{R}$ such that $v|D = u$. Since Z_n is C-embedded in $\mathbb{R}^{\omega_1}$, there is a continuous $w : \mathbb{R}^{\omega_1} \to \mathbb{R}$ such that $w|Z_n = v$ and hence $w|D = u$. By Problem 299 of [TFS] there is a countable $B \subset \omega_1$ and a continuous map $s : \mathbb{R}^B \to \mathbb{R}$ such that $s \circ \pi_B = w$. Thus $w(D) = u(D) = s(\pi_B(D))$ is a countable subset of $\mathbb{R}$ because $\pi_B(D)$ is countable. However $u(D)$ is uncountable because u is an injection. This contradiction shows that our assumption is false, i.e., $\mathrm{ext}(X) = \omega$ and hence $w(X) = \omega$ (see Problem 214 of [TFS]). As a consequence, $nw(C_p(X)) = nw(X) = \omega$; since any space with a countable network is Lindelöf, $C_p(X)$ is Lindelöf and hence normal so our solution is complete.

T.456. *Let X be an arbitrary space. Given a uniformly dense $Y \subset C_p(X)$, prove that $t(X) \leq l(Y)$.*

Solution. Take any $A \subset X$ and $x \in \overline{A}$. Let $\kappa = l(Y)$; it is evident that the set $F = \{f \in Y : f(x) \geq \frac{1}{2}\}$ is closed in Y so $l(F) \leq \kappa$. For every $a \in A$ the set $V_a = \{f \in Y : f(a) > \frac{1}{3}\}$ is open in Y. Besides, $F \subset \bigcup\{V_a : a \in A\}$; indeed, if $g \in F$, then $g(x) \geq \frac{1}{2}$ and hence the set $W = g^{-1}((\frac{1}{3}, +\infty))$ is an open neighborhood of x. Since $x \in \overline{A}$, there is $a \in A \cap W$ and hence $g \in V_a$.

Now, $l(F) \leq \kappa$ implies that there is a set $B \subset A$ such that $|B| \leq \kappa$ and $F \subset \bigcup\{V_a : a \in B\}$. If $x \notin \overline{B}$, then there is $f \in C_p(X)$ such that $f(x) = 1$ and $f(a) = 0$ for all $a \in B$. The set Y is uniformly dense in $C_p(X)$, so there is $g \in Y$ for which $|g(y) - f(y)| < \frac{1}{3}$ for all $y \in X$. Consequently, $g(x) > \frac{2}{3} \geq \frac{1}{2}$ and $g(a) < \frac{1}{3}$ for any $a \in B$ which shows that $g \in F\backslash(\bigcup\{V_a : a \in B\})$; this contradiction proves that $x \in \overline{B}$. Thus, for any $A \subset X$ and $x \in \overline{A}$, we found a set $B \subset A$ such that $|B| \leq \kappa$ and $x \in \overline{B}$. Therefore $t(X) \leq \kappa = l(Y)$.

T.457. *For an arbitrary space X and a uniformly dense $Y \subset C_p(X)$, prove that $nw(Y) = nw(C_p(X))$ and $d(Y) = d(C_p(X))$.*

Solution. Our solution will be derived from the following statement.

Fact 1. If Z is a space and A is uniformly dense in $C_p(Z)$, then Z is homeomorphic to a closed subspace of $C_p(A)$.

Proof. It is easy to see that $A + f$ is uniformly dense in $C_p(Z)$ for any $f \in C_p(Z)$; since $A + f$ is homeomorphic to A, we can assume that $u \in A$ where u is identically zero on Z, i.e., $u(z) = 0$ for all $z \in Z$.

If $z \in Z$ and F is a closed subspace of Z with $z \notin F$, then there is $f \in C(Z)$ such that $f(z) = 1$ and $f(F) \subset \{0\}$. Since A is uniformly dense in Z, we can find

$g \in A$ such that $|g(y) - f(y)| < \frac{1}{3}$ for any $y \in Z$. It is evident that $g(z) \in [\frac{2}{3}, \frac{4}{3}]$ while $g(F) \in [-\frac{1}{3}, \frac{1}{3}]$ and therefore $g(z) \notin \overline{g(F)}$. Thus A separates the points and closed sets in Z and hence the map $\varphi : Z \to C_p(A)$ defined by $\varphi(z)(f) = f(z)$ for any $z \in Z$ and $f \in A$ is a homeomorphic embedding (see Problem 166 of [TFS]).

To see that $\varphi(Z)$ is closed in $C_p(A)$ take any $\xi \in C_p(A) \backslash \varphi(Z)$. The function ξ is continuous at u, so there is $U \in \tau(u, A)$ such that $\xi(U) \subset (-\frac{1}{2}, \frac{1}{2})$; recalling that $A \subset C_p(Z)$, we can see that there exists a finite $K \subset Z$ and $\varepsilon > 0$ such that $O = \{f \in A : f(K) \subset (-\varepsilon, \varepsilon)\} \subset U$. Since $\xi \neq \varphi(z)$ for any $z \in K$, there is $V \in \tau(K, Z)$ for which $\xi \notin \overline{\varphi(V)}$. It is easy to construct a function $f \in C_p(Z)$ such that $f(K) \subset \{0\}$ and $f(Z \backslash V) \subset \{1\}$. For $\delta = \min\{\frac{1}{4}, \varepsilon\}$ take $g \in A$ such that $|g(z) - f(z)| < \delta$ for any $z \in Z$. Then $g(K) \subset (-\varepsilon, \varepsilon)$ and $g(Z \backslash V) \subset [\frac{3}{4}, \frac{5}{4}]$. Consequently, $g \in O \subset U$ which implies $\xi(g) \in [-\frac{1}{2}, \frac{1}{2}]$ while $\varphi(z)(g) \in [\frac{3}{4}, \frac{5}{4}]$ for each $z \in Z \backslash V$ which shows that $\xi \notin \overline{\varphi(Z \backslash V)}$.

Consequently, $\xi \notin \overline{\varphi(Z)} = \overline{\varphi(V)} \cup \overline{\varphi(Z \backslash V)}$, i.e., we established that $\xi \notin \overline{\varphi(Z)}$ for any point $\xi \in C_p(A) \backslash \varphi(Z)$, i.e., $\varphi(Z)$ is closed in $C_p(A)$ and Fact 1 is proved. □

Returning to our solution observe that $nw(Y) \leq nw(C_p(X))$ because network weight is hereditary (see Problem 159 of [TFS]). On the other hand, the space X embeds in $C_p(Y)$ by Fact 1 and therefore $nw(C_p(X)) = nw(X) \leq nw(C_p(Y)) = nw(Y)$ which proves that $nw(Y) = nw(C_p(X))$ (it is worth mentioning that we did not need a *closed* embedding in $C_p(Y)$: the existence of *any* embedding is sufficient).

As to densities, the set Y is dense in $C_p(X)$ (see Problem 344 of [TFS]) which, evidently, implies $d(C_p(X)) \leq d(Y)$. Now assume that $d(C_p(X)) = \kappa$ and take a dense $A \subset C_p(X)$ with $|A| \leq \kappa$. It follows from uniform density of Y in the space $C_p(X)$ that for any $f \in A$ there exists a sequence $S_f \subset Y$ such that $S_f \to f$ and hence $f \in \overline{S}_f$. If $B = \bigcup\{S_f : f \in A\}$, then $B \subset Y$ and $|B| \leq \kappa$.

Thus $Y \subset C_p(X) = \overline{A} \subset \overline{B}$ which shows that the set B is dense in Y; since $|B| \leq \kappa = d(C_p(X))$, we proved that $d(Y) \leq d(C_p(X))$, i.e., $d(Y) = d(C_p(X))$ and hence our solution is complete.

T.458. *For an arbitrary space X and a uniformly dense $Y \subset C_p(X)$, prove that $hd(Y) = hd(C_p(X))$, $hl(Y) = hl(C_p(X))$ and $s(Y) = s(C_p(X))$.*

Solution. It is evident that $\varphi(Y) \leq \varphi(C_p(X))$ for any $\varphi \in \{hd, hl, s\}$. Now assume that $s(Y) = \kappa$ and there is a discrete $D \subset C_p(X)$ with $|D| = \kappa^+$. For any $d \in D$ there is a finite set $K_d \subset X$ and $n_d \in \mathbb{N}$ such that $O(d, K_d, \frac{1}{n_d}) \cap D = \{d\}$ where $O(d, K_d, \frac{1}{n_d}) = \{f \in C_p(X) : |f(x) - d(x)| < \frac{1}{n_d}$ for any $x \in K_d\}$ for every $d \in D$. There are $E \subset D$ and $n \in \mathbb{N}$ such that $|E| = \kappa^+$ and $n_d = n$ for any $d \in E$.

It follows from uniform density of Y in $C_p(X)$ that for any $d \in E$ there is $w_d \in Y$ such that $|w_d(x) - d(x)| < \frac{1}{3n}$ for any $x \in X$. Since $s(Y) \leq \kappa$, the set $\{w_d : d \in E\} \subset Y$ cannot be discrete, so there are distinct $d, e \in E$ for which $w_e \in O(w_d, K_d, \frac{1}{3n})$. Consequently, $|e(x) - d(x)| = |e(x) - w_e(x)| + $

$|w_e(x) - w_d(x)| + |w_d(x) - d(x)| < \frac{1}{3n} + \frac{1}{3n} + \frac{1}{3n} = \frac{1}{n}$ for any $x \in K_d$, i.e., $e \in O(d, K_d, \frac{1}{n})$ which is a contradiction with the choice of the set $O(d, K_d, \frac{1}{n_d})$. This contradiction shows that there are no discrete subsets of $C_p(X)$ of cardinality κ^+, i.e., $s(C_p(X)) \leq \kappa = s(Y)$. Therefore $s(Y) = s(C_p(X))$.

Now assume that $hd(Y) = \kappa$ while $hd(C_p(X)) > \kappa$ and hence there exists a left-separated $L \subset C_p(X)$ such that $|L| = \kappa^+$ (see Problem 004). Let $<$ be a well-order on L which witnesses that L is left-separated. For any $h \in L$ there is a finite $P_h \subset X$ and $m_h \in \mathbb{N}$ such that $O(h, P_h, \frac{1}{m_h}) \cap L \subset L_h = \{g \in L : h \leq g\}$. There are $M \subset L$ and $m \in \mathbb{N}$ for which $|M| = \kappa^+$ and $m_h = m$ for any $h \in M$. It follows from uniform density of Y in $C_p(X)$ that for any $h \in M$ there is $u_h \in Y$ such that $|u_h(x) - h(x)| < \frac{1}{3m}$ for any $x \in X$. Given distinct $g, h \in M$ let $u_g < u_h$ if and only if $g < h$. It is clear that $<$ is a well-order on the set $N = \{u_h : h \in M\}$. Since $hd(Y) \leq \kappa$, the set N cannot be left-separated, so there are $g, h \in M$ such that $g < h$ and $u_g \in O(u_h, P_h, \frac{1}{3m})$.

Consequently, $|g(x) - h(x)| \leq |g(x) - u_g(x)| + |u_g(x) - u_h(x)| + |u_h(x) - h(x)| < \frac{1}{3m} + \frac{1}{3m} + \frac{1}{3m} = \frac{1}{m}$ for any $x \in P_h$ and therefore $g \in O(h, P_h, \frac{1}{m})$ which contradicts the choice of $O(h, P_h, \frac{1}{m})$. This contradiction shows that there are no left-separated subsets of $C_p(X)$ of cardinality κ^+, so $hd(C_p(X)) \leq \kappa = hd(Y)$ (see Problem 004) and therefore $hd(Y) = hd(C_p(X))$.

Finally, suppose that $hl(Y) = \kappa$ while $hl(C_p(X)) > \kappa$ and hence there is a right-separated $R \subset C_p(X)$ such that $|R| = \kappa^+$ (see Problem 005). Let $<$ be a well-order on R which witnesses that R is right-separated. For any $h \in R$ there is a finite $F_h \subset X$ and $k_h \in \mathbb{N}$ such that $O(h, F_h, \frac{1}{k_h}) \cap R \subset R_h = \{g \in L : g \leq h\}$. There are $S \subset R$ and $k \in \mathbb{N}$ for which $|S| = \kappa^+$ and $k_h = k$ for any $h \in S$. It follows from uniform density of Y in $C_p(X)$ that for any $h \in S$ there is $v_h \in Y$ such that $|v_h(x) - h(x)| < \frac{1}{3m}$ for any $x \in X$. Given distinct $g, h \in M$ let $v_g < v_h$ if and only if $g < h$. It is clear that $<$ is a well-order on the set $T = \{v_h : h \in S\}$. Since $hl(Y) \leq \kappa$, the set T cannot be right-separated, so there are $g, h \in M$ such that $h < g$ and $v_g \in O(v_h, F_h, \frac{1}{3k})$.

Consequently, $|g(x) - h(x)| \leq |g(x) - v_g(x)| + |v_g(x) - v_h(x)| + |v_h(x) - h(x)| < \frac{1}{3k} + \frac{1}{3k} + \frac{1}{3k} = \frac{1}{k}$ for any $x \in F_h$, and therefore $g \in O(h, F_h, \frac{1}{k})$ which contradicts the choice of $O(h, F_h, \frac{1}{k})$. This contradiction shows that there are no right-separated subsets of $C_p(X)$ of cardinality κ^+, so $hl(C_p(X)) \leq \kappa = hl(Y)$ (see Problem 005) and therefore $hl(Y) = hl(C_p(X))$.

T.459. *Suppose that X is a space and $Y \subset C_p(X)$ is uniformly dense in $C_p(X)$. Prove that if Y is a Lindelöf Σ-space, then $C_p(X)$ is also Lindelöf Σ.*

Solution. The statement of this problem is an easy consequence of the following fact.

Fact 1. Suppose that Z is a space and A is a uniformly dense subset of $C_p(Z)$. Then $C_p(Z) = \bigcap\{C_n : n \in \mathbb{N}\}$ where $C_n \subset \mathbb{R}^Z$ is a continuous image of $A \times [-\frac{1}{n}, \frac{1}{n}]^Z$ for every $n \in \mathbb{N}$.

Proof. Define a map $\varphi_n : A \times [-\frac{1}{n}, \frac{1}{n}]^Z \to \mathbb{R}^Z$ by $\varphi_n(f, g) = f + g$ for each $(f, g) \in A \times [-\frac{1}{n}, \frac{1}{n}]^Z$. It is clear that φ_n is a continuous map for all $n \in \mathbb{N}$; let $C_n = \varphi_n\left(A \times [-\frac{1}{n}, \frac{1}{n}]^Z\right)$. Now, if $n \in \mathbb{N}$ and $h \in C_p(Z)$, then there exists $f \in A$ such that $|h(z) - f(z)| < \frac{1}{n}$ for all $z \in Z$. Consequently, $g = h - f \in [-\frac{1}{n}, \frac{1}{n}]^Z$ and $h = f + g$ which shows that $C_p(Z) \subset C_n$ for all $n \in \mathbb{N}$, i.e., $C_p(Z) \subset \bigcap\{C_n : n \in \mathbb{N}\}$.

Now, if $h \in \bigcap\{C_n : n \in \mathbb{N}\}$, then there is a sequence $S = \{f_n : n \in \mathbb{N}\} \subset C_p(Z)$ such that $|f_n(z) - h(z)| \leq \frac{1}{n}$ for all $z \in Z$ and $n \in \mathbb{N}$. Thus the sequence S converges uniformly to h and hence $h \in C_p(Z)$ (see Problem 029 of [TFS]). Fact 1 is proved.   □

Returning to our solution apply Fact 1 to find a sequence $\{C_n : n \in \mathbb{N}\}$ of subsets of $\mathbb{R}^X$ such that $C_p(X) = \bigcap\{C_n : n \in \mathbb{N}\}$ and C_n is a continuous image of $Z_n = Y \times [-\frac{1}{n}, \frac{1}{n}]^X$ for every $n \in \mathbb{N}$. Since Y is a Lindelöf Σ-space, the space Z_n is also Lindelöf Σ (see Problem 256) for any $n \in \mathbb{N}$. Therefore C_n is a Lindelöf Σ-space for all $n \in \mathbb{N}$ by Problem 243 and hence $C_p(X)$ is Lindelöf Σ by Problem 258.

T.460. *Suppose that X is a space and $Y \subset C_p(X)$ is uniformly dense in $C_p(X)$. Prove that*

(i) if Y is K-analytic, then $C_p(X)$ is K-analytic;
(ii) if Y is analytic, then $C_p(X)$ is analytic.

Solution. Apply Fact 1 of T.459 to find a sequence $\{C_n : n \in \mathbb{N}\}$ of subsets of $\mathbb{R}^X$ such that $C_p(X) = \bigcap\{C_n : n \in \mathbb{N}\}$ and C_n is a continuous image of the space $Z_n = Y \times [-\frac{1}{n}, \frac{1}{n}]^X$ for every $n \in \mathbb{N}$.

If Y is a K-analytic space, then the space Z_n is also K-analytic (see Problem 343) for any $n \in \mathbb{N}$. Therefore C_n is a K-analytic space for every $n \in \mathbb{N}$ (because a continuous image of a K-analytic space is a K-analytic space) and hence $C_p(X)$ is K-analytic by Problem 344. This proves (i).

As to (ii), if Y is analytic, then $nw(C_p(X)) = \omega$ by Problem 457; besides, $C_p(X)$ is K-analytic by (i), so it is analytic by Problem 346.

T.461. *For an arbitrary space X and a uniformly dense $Y \subset C_p(X)$, prove that $t(Y) = t(C_p(X))$.*

Solution. It follows from Problem 159 of [TFS] that $t(Y) \leq t(C_p(X))$; to prove the inverse inequality, assume that $t(Y) = \kappa$. Recall that a family $\mathcal{A} \subset \exp(X)$ is called *an ω-cover of X* if for every finite $K \subset X$, there is $A \in \mathcal{A}$ such that $K \subset A$. An ω-cover $\mathcal{A}$ of a space is *open* if all its elements are open.

Now take an arbitrary open ω-cover $\mathcal{U}$ of the space X; for any $f \in Y$, let $S(f) = \{x \in X : f(x) \geq \frac{1}{3}\}$. Consider the set $P = \{f \in Y : S(f) \subset U \text{ for some } U \in \mathcal{U}\}$. Since Y is uniformly dense in $C_p(X)$, there is $w \in Y$ such that $w(x) \geq \frac{2}{3}$ for all $x \in X$. We claim that $w \in \overline{P}$. Indeed, if $O \in \tau(w, C_p(X))$, then there is a finite $K \subset X$ and $\varepsilon > 0$ such that $W = \{f \in C_p(X) : |f(x) - w(x)| < \varepsilon$ for all $x \in K\} \subset O$. Since $\mathcal{U}$ is an ω-cover of X, there is $U \in \mathcal{U}$ such that

$K \subset U$; it is easy to construct a function $f \in C_p(X)$ for which $f|K = w|K$ and $f(X \backslash U) \subset \{0\}$. Let $\delta = \min\{\frac{1}{3}, \varepsilon\}$; there is $g \in Y$ such that $|g(x) - f(x)| < \delta$ for all $x \in X$. Therefore $|g(x) - w(x)| < \varepsilon$ for all $x \in K$ and hence $g \in W$. Furthermore, $|g(x) - f(x)| < \frac{1}{3}$ for all $x \in X \backslash U$ and hence $g(x) < \frac{1}{3}$ for all $x \in X \backslash U$ which shows that $S(g) \subset U$. Thus $g \in W \cap P \subset O \cap P$, so $O \cap P \neq \emptyset$ for any $O \in \tau(w, C_p(X))$ and hence $w \in \overline{P}$.

We have $t(Y) \le \kappa$, so there exists $Q \subset P$ such that $|Q| \le \kappa$ and $w \in \overline{Q}$. For every $f \in Q$ there is $U_f \in \mathcal{U}$ for which $S(f) \subset U_f$; let $\mathcal{U}' = \{U_f : f \in Q\}$. Then $\mathcal{U}' \subset \mathcal{U}$ and $|\mathcal{U}'| \le \kappa$. Given a finite $K \subset X$, the set $V = \{f \in Y : |f(x) - w(x)| < \frac{1}{3}\}$ is an open neighborhood of w in $C_p(X)$. Since $w \in \overline{Q}$, there is $f \in Q$ such that $f \in V$ and therefore $|f(x) - w(x)| < \frac{1}{3}$ for all $x \in K$ which implies $f(x) > \frac{1}{3}$ for all $x \in K$ and hence $K \subset S(f) \subset U_f$. This proves that $\mathcal{U}'$ is also an ω-cover of X. Thus every open ω-cover of X has an ω-subcover of cardinality $\le \kappa$ and therefore $t(C_p(X)) \le \kappa = t(Y)$ (see Problems 148 and 149 of [TFS]). As a consequence, $t(Y) = t(C_p(X))$.

T.462. *Let X be an arbitrary space with $\mathrm{ext}^*(X) \le \kappa$. Prove that $t(Y) \le \kappa$ for any compact $Y \subset C_p(X)$.*

Solution. For any $x \in X$ let $e_x(f) = f(x)$ for any $f \in Y$. Then $e_x \in C_p(Y)$ for any $x \in X$ and the map $e : X \to C_p(Y)$ defined by $e(x) = e_x$ for all $x \in X$ is continuous (see Problem 166 of [TFS]). If $Z = e(X)$, then $\mathrm{ext}^*(Z) \le \kappa$ (it is an easy exercise that if a space P is a continuous image of a space Q, then $\mathrm{ext}^*(P) \le \mathrm{ext}^*(Q)$).

Since $Z \subset C_p(Y)$, we have $Z^n \subset (C_p(Y))^n$ for any $n \in \mathbb{N}$; if Y_i is a homeomorphic copy of Y for each $i = 1, \ldots, n$, then $(C_p(Y))^n$ is homeomorphic to the space $C_p(Y_1 \oplus \cdots \oplus Y_n)$ (see Problem 114 of [TFS]). Since $K_n = Y_1 \oplus \cdots \oplus Y_n$ is a compact space, we proved that Z^n embeds in $C_p(K_n)$ for some compact space K_n for any $n \in \mathbb{N}$. Thus we can apply Baturov's theorem (see Problem 269) to conclude that $l(Z^n) = \mathrm{ext}(Z^n) \le \kappa$ for all $n \in \mathbb{N}$. As a consequence $l^*(Z) \le \kappa$ and hence $t(C_p(Z)) \le \kappa$ (see Problem 149 of [TFS]).

Given $f, g \in Y$ with $f \neq g$ there is a point $x \in X$ such that $f(x) \neq g(x)$. Therefore $e_x \in Z$ and $e_x(f) \neq e_x(g)$ which proves that Z separates the points of Y and hence Y embeds in $C_p(Z)$ (see Fact 2 of S.351). As a consequence, $t(Y) \le t(C_p(Z)) = \kappa$.

T.463. *Suppose that X has the Gerlits property φ. Prove that all continuous images and all closed subspaces of X have φ.*

Solution. Let $f : X \to Y$ be a continuous onto map; suppose that a family $\mathcal{U} \subset \tau(Y)$ is an ω-cover of Y such that $\mathcal{U} = \bigcup_{n \in \omega} \mathcal{U}_n$ and $\mathcal{U}_n \subset \mathcal{U}_{n+1}$ for all $n \in \omega$. If $\mathcal{V}_n = \{f^{-1}(U) : U \in \mathcal{U}_n\}$ for all $n \in \omega$, then $\mathcal{V} = \bigcup_{n \in \omega} \mathcal{V}_n$ is an open ω-cover of X such that $\mathcal{V}_n \subset \mathcal{V}_{n+1}$ for any $n \in \omega$. The space X has the Gerlits property φ, so there is a sequence $\{X_n : n \in \omega\} \subset \exp(X)$ such that $X_n \to X$ and X_n is ω-covered by $\mathcal{V}_n$ for every $n \in \omega$. Let $Y_n = f(X_n)$; it is immediate that $Y_n \to Y$ and Y_n is ω-covered by $\mathcal{U}_n$ for each $n \in \omega$. Therefore $Y \vdash \varphi$.

Now assume that F is a closed subset of X and a family $\mathcal{U} \subset \tau(F)$ is an ω-cover of F such that $\mathcal{U} = \bigcup_{n \in \omega} \mathcal{U}_n$ and $\mathcal{U}_n \subset \mathcal{U}_{n+1}$ for all $n \in \omega$. Let $O(U) = X \backslash (F \backslash U)$ for any $U \in \tau(F)$; it is evident that $O(U) \in \tau(X)$ and $O(U) \cap F = U$. The family $\mathcal{V}_n = \{O(U) : U \in \mathcal{U}_n\}$ consists of open subsets of X for any $n \in \omega$; if $\mathcal{V} = \bigcup_{n \in \omega} \mathcal{V}_n$, then $\mathcal{V}$ is an open ω-cover of X such that $\mathcal{V}_n \subset \mathcal{V}_{n+1}$ for all $n \in \omega$.

It follows from $X \vdash \varphi$ that there is a sequence $\{X_n : n \in \omega\} \subset \exp(X)$ such that $X_n \to X$ and X_n is ω-covered by $\mathcal{V}_n$ for any $n \in \omega$. Let $Y_n = X_n \cap F$ for all $n \in \omega$; it is evident that $Y_n \to F$ and Y_n is ω-covered by $\mathcal{U}_n$ for every $n \in \omega$. This proves that $F \vdash \varphi$.

T.464. *Prove that $C_p(X)$ is a Fréchet–Urysohn space if and only if X has the Gerlits property φ and $t(C_p(X)) = \omega$.*

Solution. We will first establish that φ is equivalent to some formally weaker property φ'.

Fact 1. If Z is an arbitrary space, then $Z \vdash \varphi$ if and only if $Z \vdash \varphi'$ where φ' is the following property: for any open ω-cover $\mathcal{U}$ of the space Z such that $\mathcal{U} = \bigcup_{n \in \omega} \mathcal{U}_n$ and $\mathcal{U}_n \subset \mathcal{U}_{n+1}$ for any $n \in \omega$, there are sequences $\{Z_n : n \in \omega\} \subset \exp(Z)$ and $\{k_n : n \in \omega\} \subset \omega$ such that $Z_n \to Z$, $k_n < k_{n+1}$ and Z_n is ω-covered by $\mathcal{U}_{k_n}$ for any $n \in \omega$.

Proof. It is evident that φ implies φ', so assume that $Z \vdash \varphi'$ and take any open ω-cover $\mathcal{U}$ of the space Z such that $\mathcal{U} = \bigcup_{n \in \omega} \mathcal{U}_n$ and $\mathcal{U}_n \subset \mathcal{U}_{n+1}$ for every $n \in \omega$. It follows from the property φ' that there are sequences $\{Z'_n : n \in \omega\} \subset \exp(Z)$ and $\{k_n : n \in \omega\} \subset \omega$ such that $Z'_n \to Z$, $k_n < k_{n+1}$ and Z'_n is ω-covered by $\mathcal{U}_{k_n}$ for all $n \in \omega$.

Let $Z_i = \emptyset$ for all $i < k_0$; assume that $m \in \omega$ and we have defined Z_i for all $i < k_m$ in such a way that Z_i is ω-covered by $\mathcal{U}_i$ for all $i < k_m$. Letting $Z_i = Z'_{k_m}$ for all $i = k_m, \ldots, k_{m+1} - 1$ we define the set Z_i for all $i < k_{m+1}$ and it follows from $\mathcal{U}_{k_m} \subset \mathcal{U}_i$ that Z_i is ω-covered by $\mathcal{U}_i$ for all $i = k_m, \ldots, k_{m+1} - 1$. Consequently, Z_i is ω-covered by U_i for all $i < k_{m+1}$, and hence our inductive construction can be continued giving us a sequence $\{Z_i : i \in \omega\} \subset \exp(Z)$ such that $Z_i \to Z$ and Z_i is ω-covered by $\mathcal{U}_i$ for all $i \in \omega$. Therefore $Z \vdash \varphi$ and Fact 1 is proved. □

Returning to our solution assume that $C_p(X)$ is a Fréchet–Urysohn space. Then $t(C_p(X)) \leq \omega$ because every Fréchet–Urysohn space has countable tightness. Now suppose that $\mathcal{U}$ is an open ω-cover of the space X such that $\mathcal{U} = \bigcup_{n \in \omega} \mathcal{U}_n$ and $\mathcal{U}_n \subset \mathcal{U}_{n+1}$ for any $n \in \omega$. By Problem 144 of [TFS] there is a sequence $\{U_n : n \in \omega\} \subset \mathcal{U}$ such that $U_n \to X$. It is easy to choose a strictly increasing sequence $\{k_n : n \in \omega\} \subset \omega$ such that $U_n \in \mathcal{U}_{k_n}$ for all $n \in \omega$. It is clear that U_n is ω-covered by $\mathcal{U}_{k_n}$ for each $n \in \omega$, so $X \vdash \varphi$ by Fact 1. This proves necessity.

Now assume that $t(C_p(X)) = \omega$ and $X \vdash \varphi$. Given an open ω-cover $\mathcal{U}$ of the space X, there is a countable $\mathcal{V} \subset \mathcal{U}$ such that $\mathcal{V}$ is also an ω-cover of X (see Problem 148 and 149 of [TFS]); let $\mathcal{V} = \{V_n : n \in \omega\}$ and $\mathcal{V}_n = \{V_i : i \leq n\}$ for every $n \in \omega$. Then $\bigcup_{n \in \omega} \mathcal{V}_n = \mathcal{V}$ and $\mathcal{V}_n \subset \mathcal{V}_{n+1}$ for every $n \in \omega$. Since X has φ, there is a sequence $\{X_n : n \in \omega\} \subset \exp(X)$ such that $X_n \to X$ and X_n is ω-covered by $\mathcal{V}_n$. However, a set is ω-covered by a finite family if and only if it is contained in an element of that family. Thus there is $U_n \in \mathcal{V}_n$ for which $X_n \subset U_n$ for all $n \in \omega$. An immediate consequence is that $U_n \to X$, so for any open ω-cover $\mathcal{U}$ of the space X, we have a sequence $\{U_n : n \in \omega\} \subset \mathcal{U}$ such that $U_n \to X$. Therefore $C_p(X)$ is a Fréchet–Urysohn space (see Problem 144 of [TFS]) so sufficiency is settled and hence our solution is complete.

T.465 (Gerlits–Pytkeev theorem). *Prove that the following conditions are equivalent for any space X:*

  (i) $C_p(X)$ *is a Fréchet–Urysohn space;*
 (ii) $C_p(X)$ *is a sequential space;*
(iii) $C_p(X)$ *is a k-space.*

Solution. The implications (i)$\Longrightarrow$(ii)$\Longrightarrow$(iii) are clear.

Fact 1. Let Z be any space; then, for any pseudocompact $P \subset C_p(Z)$ and $z \in Z$ the set $P(z) = \{f(z) : f \in P\}$ is compact and hence bounded in $\mathbb{R}$.

Proof. The map $e_z : C_p(Z) \to \mathbb{R}$ defined by $e_z(f) = f(z)$ for all $f \in C_p(Z)$ is continuous by Problem 166 of [TFS], so $P(z) = e_z(P)$ is a continuous image of P. Thus $P(z)$ is pseudocompact; being second countable it is compact and hence Fact 1 is proved. ☐

Recall that a family $\mathcal{U} \subset \tau(X)$ is called *an open ω-cover of X* if for any finite $K \subset X$, there is $U \in \mathcal{U}$ such that $K \subset U$. We will first show that

(1) if $C_p(X)$ is a k-space, then X has the Gerlits property φ.

Assume towards a contradiction that $C_p(X)$ is a k-space and X does not have φ. It was proved in Fact 1 of T.464 that φ is equivalent to the following property φ': for any open ω-cover $\mathcal{U}$ of the space X such that $\mathcal{U} = \bigcup_{n\in\omega} \mathcal{U}_n$ and $\mathcal{U}_n \subset \mathcal{U}_{n+1}$ for all $n \in \omega$, there are sequences $\{X_n : n \in \omega\} \subset \exp(X)$ and $\{k_n : n \in \omega\} \subset \omega$ such that $X_n \to X$, $k_n < k_{n+1}$ and X_n is ω-covered by $\mathcal{U}_{k_n}$ for all $n \in \omega$. Since φ does not holds for X, the property φ' does not hold either, so there is an open ω-cover $\mathcal{U} = \bigcup_{n\in\omega} \mathcal{U}_n$ of the space X which witnesses this. In particular, X is not ω-covered by $\mathcal{U}_n$ for all $n \in \omega$.

For any $f \in C_p(X)$ and $n \in \mathbb{N}$ let $Y(f,n) = \{x \in X : f(x) < n\}$ and consider the set $A_n = \{f \in C_p(X) : Y(f,n)$ is ω-covered by $\mathcal{U}_n\}$ for any $n \in \mathbb{N}$. If $g \in C_p(X)\backslash A_n$, then there is a finite set $K \subset Y(g,n)$ such that $K \not\subset U$ for any $U \in \mathcal{U}_n$. The set $O = \{f \in C_p(X) : f(x) < n$ for all $x \in K\}$ is open in $C_p(X)$ and $g \in O \subset C_p(X)\backslash A_n$ which proves that A_n is closed in $C_p(X)$ for all $n \in \mathbb{N}$. Let $A = \bigcup\{A_n : n \in \mathbb{N}\}$.

If $u \in C_p(X)$ is the function which is identically zero on X, then the set $Y(u,n) = X$ is not ω-covered by $\mathcal{U}_n$ for each $n \in \mathbb{N}$ and hence $u \in C_p(X)\backslash A$. However, $u \in \overline{A}$; to see this take any $W \in \tau(u, C_p(X))$. There exists a finite $K \subset X$ and $\varepsilon > 0$ such that $V = \{f \in C_p(X) : |f(x)| < \varepsilon$ for all $x \in K\} \subset W$. There is $n \in \mathbb{N}$ and $U \in \mathcal{U}_n$ such that $K \subset U$. It is easy to find a function $f \in C_p(X)$

for which $f|K = u|K$ and $f(x) = n + 1$ for all $x \in X \backslash U$. It is evident that $Y(f, n) \subset U$ and therefore $f \in A \cap V \subset A \cap W$. Thus $W \cap A \neq \emptyset$ for any $W \in \tau(u, C_p(X))$ which proves that $u \in \overline{A}$.

Since $u \in \overline{A} \backslash A$, the set A is not closed in $C_p(X)$, so we can apply k-property of $C_p(X)$ to find a compact set $C \subset C_p(X)$ such that $C \cap A$ is not closed. For any $x \in X$ the set $C(x) = \{f(x) : f \in C\}$ is compact by Fact 1, so there is $n(x) \in \mathbb{N}$ such that $C(x) \subset [-n(x), n(x)]$. Let $X_n = \{x \in X : n(x) \leq n\}$ for every $n \in \mathbb{N}$; it is immediate that $X_n \subset X_{n+1}$ for any $n \in \mathbb{N}$ and $\bigcup \{X_n : n \in \mathbb{N}\} = X$. Therefore $X_n \to X$; since $\mathcal{U}$ witnesses that φ' does not hold, there is $m \in \mathbb{N}$ such that X_m is not ω-covered by $\mathcal{U}_n$ for any $n \in \omega$.

We claim that $C \cap A_n = \emptyset$ for any $n > m$. Indeed, if $f \in A_n$ for some $n > m$, then $Y(f, n)$ is ω-covered by $\mathcal{U}_n$; since $\mathcal{U}_n$ does not ω-cover X_m, there is $x \in X_m \backslash Y(f, n)$. We have $f(x) \geq n > m$ and hence $f \notin C$ because $g(x) \leq m$ for any $g \in C$ and $x \in X_m$.

As a consequence, $C \cap A = \bigcup \{C \cap A_k : k \leq m\}$ is a closed subset of $C_p(X)$. This contradiction shows that X has φ' and hence φ proving the property (1). Our next step is to establish that

(2) if $C_p(X)$ is a k-space, then $\text{ext}^*(X) = \omega$.

Assume, towards a contradiction, that there is an uncountable closed discrete set $D \subset X^n$ for some $n \in \mathbb{N}$. For every $m \in \mathbb{N}$, consider the set $P_m = \{f \in C_p(X) : |(Y(f, m))^n \cap D| \leq m\}$ for every $m \in \mathbb{N}$. The set P_m is closed in the space $C_p(X)$ for any $m \in \mathbb{N}$; indeed, if $f \in C_p(X) \backslash P_m$, then there are $y_1, \ldots, y_{m+1} \in (Y(f, m))^n \cap D$. We have $y_i = (y_1^i, \ldots, y_n^i)$ for all $i \leq m + 1$ and $f(y_j^i) < m$ for every $i \leq m + 1$ and $j \leq n$. The set $K = \{y_j^i : i \leq m + 1, \ j \leq n\}$ is finite, so $W = \{h \in C_p(X) : h(y) < m \text{ for all } y \in K\}$ is open in $C_p(X)$. Given any $h \in W$, it is immediate that $y_i \in (Y(h, m))^n \cap D$ for each $i \leq m + 1$ which shows that $h \notin P_m$. Thus $W \cap P_m = \emptyset$ and hence P_m is closed in $C_p(X)$.

Observe that $u \notin P = \bigcup \{P_m : m \in \mathbb{N}\}$ because $Y(u, m) = X$ and therefore the set $(Y(u, m))^n \cap D = D$ is uncountable for any $m \in \mathbb{N}$. However, $u \in \overline{P}$; to see it take any $O \in \tau(u, C_p(X))$. There is a finite set $L \subset X$ and $\varepsilon > 0$ such that $G = \{f \in C_p(X) : |f(x)| < \varepsilon \text{ for all } x \in L\} \subset O$. The set $L^n \subset X^n$ is finite, so it has a neighborhood in X^n which contains finitely many points of D. Therefore there is $k \in \mathbb{N}$ and $V \in \tau(L, X)$ such that $k > 1$ and $|V^n \cap D| \leq k$. Choose a function $f \in C_p(X)$ such that $f|K = u|K$ and $f(x) = k + 1$ for all $x \in X \backslash V$. Then $Y(f, k) \subset V$ which implies $|(Y(f, k))^n \cap D| \leq k$, i.e., $f \in P_k \cap G \subset P_k \cap O$. Since $P \cap O \neq \emptyset$ for any $O \in \tau(u, C_p(X))$, we proved that $u \in \overline{P} \backslash P$ and hence P is not closed in $C_p(X)$.

Now, $C_p(X)$ is a k-space, so there exists a compact $C \subset C_p(X)$ such that $C \cap P$ is not closed in C and hence in $C_p(X)$. This implies that C is not contained in $P_1 \cup \cdots \cup P_m$ for any $m \in \mathbb{N}$; as a consequence, there is an increasing sequence $\{m_i : i \in \mathbb{N}\}$ for which $P_{m_i} \cap C \neq \emptyset$ for all $i \in \mathbb{N}$. Pick a function $f_i \in P_{m_i} \cap C$ for all $i \in \mathbb{N}$ and observe that $X^n \backslash (\bigcup \{(Y(f_i, m_i))^n : i \in \mathbb{N}\}) \neq \emptyset$ because each $(Y(f_i, m_i))^n$ covers only finitely many points of D and D is uncountable.

Take any $z = (z_1, \ldots, z_n) \in X^n \backslash (\bigcup \{ (Y(f_i, m_i))^n : i \in \mathbb{N} \})$ and observe that for any $i \in \mathbb{N}$, there is $k \in \{1, \ldots, n\}$ such that $f_i(z_k) \geq m_i$ and therefore $\sup\{ f_i(z_j) : j = 1, \ldots, n \} \geq m_i$. Thus the set $S = \{ \sup\{ f_i(z_j) : j = 1, \ldots, n \} : i \in \mathbb{N} \}$ is not bounded in $\mathbb{R}$; on the other hand, $S \subset C(z_1) \cup \cdots \cup C(z_n)$ and hence S is bounded in $\mathbb{R}$ by Fact 1. This contradiction shows that $\text{ext}^*(X) \leq \omega$, i.e., (2) is proved.

Finally assume that $C_p(X)$ is a k-space. Then $\text{ext}^*(X) \leq \omega$ by (2) and hence $t(F) \leq \omega$ for any compact $F \subset C_p(X)$ (see Problem 462). Now, if A is not closed in $C_p(X)$, then choose a compact $F \subset C_p(X)$ such that $F \cap A$ is not closed; since $t(F) = \omega$, there is a countable $B \subset F \cap A$ such that $\overline{B} \backslash (F \cap A) \neq \emptyset$. The set F being compact, we have $\overline{B} \subset F$ and therefore $\overline{B} \backslash A \neq \emptyset$. We proved that for any non-closed $A \subset C_p(X)$, there is a countable $B \subset A$ such that $\overline{B} \backslash A \neq \emptyset$, so $t(C_p(X)) \leq \omega$ (see Lemma of S.162).

We finally have $t(C_p(X)) = \omega$; since also $X \vdash \varphi$ by (1), we can apply Problem 464 to conclude that $C_p(X)$ is a Fréchet–Urysohn space. This settles (iii)$\Longrightarrow$(i) and makes our solution complete.

T.466. *Suppose that X is a σ-compact space such that $C_p(X) = \bigcup_{n \in \omega} Y_n$ and Y_n is a k-space for every $n \in \omega$. Prove that $C_p(X)$ is a Fréchet–Urysohn space. In particular, if X is σ-compact and $C_p(X)$ is a countable union of sequential spaces, then $C_p(X)$ is a Fréchet–Urysohn space.*

Solution. The space $C_p(X)$ is Whyburn by Problem 216; it is evident that Whyburn property is hereditary, so every Y_n is Whyburn and hence Fréchet–Urysohn by Problem 210. Now apply Problem 450 to conclude that $C_p(X)$ is a Fréchet–Urysohn space.

T.467. *Suppose that we have arbitrary spaces X and Y and a continuous map $\varphi : X \to Y$. Let $\varphi^*(f) = f \circ \varphi$ for any function $f \in C_p(Y)$; this gives us a map $\varphi^* : C_p(Y) \to R = \varphi^*(C_p(Y)) \subset C_p(X)$. Define $r_\varphi : C_p(C_p(X)) \to C_p(C_p(Y))$ by $r_\varphi(\delta) = (\delta | R) \circ \varphi^*$ for any $\delta \in C_p(C_p(X))$. Prove that r_φ is a continuous ring homomorphism such that $r_\varphi | X = \varphi$ (here we identify the spaces X and Y with their canonical copies in $C_p(C_p(X))$ and $C_p(C_p(Y))$ respectively). Prove that the map φ is the unique continuous ring homomorphism with this property, i.e., if $s : C_p(C_p(X)) \to C_p(C_p(Y))$ is a continuous ring homomorphism such that $s | X = \varphi$, then $s = r_\varphi$.*

Solution. It is straightforward that φ^* is a ring homomorphism; besides, the map φ^* is continuous by Problem 163 of [TFS]. Applying Problem 163 of [TFS] once more, we can convince ourselves that the map $\mu : C_p(R) \to C_p(C_p(Y))$ defined by $\mu(\delta) = \delta \circ \varphi^*$ for any $\delta \in C_p(R)$ is a ring homomorphism. Since any restriction map is also a continuous ring homomorphism, this proves that the map r_φ is a continuous ring homomorphism.

Recall that the canonical identification of X with a subspace of $C_p(C_p(X))$ is determined by the map $e : X \to C_p(C_p(X))$ defined by $e(x)(f) = f(x)$ for any $x \in X$ and $f \in C_p(X)$. The mapping $e : X \to e(X)$ is a homeomorphism and $e(X)$ is a closed subspace of $C_p(C_p(X))$ (see Problem 167 of [TFS]).

Analogously, let $q(y)(f) = f(y)$ for any $y \in Y$ and $f \in C_p(Y)$. Then the map $q : Y \to q(Y) \subset C_p(C_p(Y))$ is a homeomorphism. To check that r_φ extends φ, we must show that $r_\varphi(e(X)) \subset q(Y)$ and $q^{-1} \circ r_\varphi \circ e = \varphi$; since q is a homeomorphism, it is equivalent to proving that $r_\varphi \circ e = q \circ \varphi$, so take any $x \in X$ and $f \in C_p(Y)$. Then $r_\varphi(e(x))(f) = e(x)(\varphi^*(f)) = \varphi^*(f)(x) = f(\varphi(x)) = q(\varphi(x))(f)$ which shows that $r_\varphi(e(x))(f) = q(\varphi(x))(f)$ for any $f \in C_p(Y)$ and therefore $r_\varphi \circ e = q \circ \varphi$. Thus, if we identify $e(X)$ with X and $q(Y)$ with Y, then $r_\varphi|X = \varphi$.

Now assume that $s : C_p(C_p(X)) \to C_p(C_p(Y))$ is a ring homomorphism such that $s|e(X) = r_\varphi|e(X)$. For any set $A \subset C_p(C_p(X))$, consider the sets $P(A) = \{f_1 \cdot \ldots \cdot f_n : n \in \mathbb{N}, \ f_i \in A \text{ for all } i \leq n\}$ and $R(A) = \{\lambda_0 + \lambda_1 \cdot g_1 + \cdots + \lambda_m \cdot g_m : m \in \mathbb{N}, \ \lambda_i \in \mathbb{R} \text{ and } g_i \in P(A) \text{ for all } i \leq m\}$.

It was proved in Fact 1 of S.312 that $R(A)$ is an algebra in $C_p(C_p(X))$ with $A \subset R(A)$. Now observe that $e(X)$ separates the points of $C_p(X)$, so $R(e(X))$ is an algebra which is dense in $C_p(C_p(X))$ by Problem 192 of [TFS]. Since s and r_φ are ring homomorphisms, we have $r_\varphi(\delta) = s(\delta)$ for any $\delta \in R(e(X))$ (to see it recall that $s|e(X) = r_\varphi|e(X)$ and every $\delta \in R(e(X))$ is obtained from elements of $e(X)$ applying finitely many products and multiplications; both maps s and r_φ commute with products and multiplications so $s(\delta) = r_\varphi(\delta)$).

As a consequence r_φ and s coincide on a dense subspace $R(e(X))$ of the space $C_p(C_p(X))$ and hence $r_\varphi = s$ by Fact 0 of S.351. This proves uniqueness and makes our solution complete.

T.468. *Suppose that X is an ω-monolithic compact space. Prove that for every surjective continuous map $\varphi : X \to Y$, the map $r_\varphi : C_p(C_p(X)) \to C_p(C_p(Y))$ is surjective.*

Solution. Given spaces Z, T and a continuous map $\beta : Z \to T$, let $\beta^*(f) = f \circ \beta$ for any $f \in C_p(T)$. This defines a continuous map $\beta^* : C_p(T) \to C_p(Z)$ which is an embedding if β is onto (see Problem 163 of [TFS]). As usual, for any subspace $A \subset Z$, let $C_p(A|Z) = \{f|A : f \in C_p(Z)\} \subset C_p(A)$.

Fact 1. Let K be an ω-monolithic compact space. If $\beta : K \to L$ is a continuous onto map, then $E = \beta^*(C_p(L))$ is C-embedded in $C_p(K)$.

Proof. For any $C \subset D \subset L$ let $\pi_C^D : C_p(D|L) \to C_p(C|L)$ be the restriction map and let $\pi_C = \pi_C^L$. Analogously, if $A \subset K$, then $p_A : C_p(K) \to C_p(A|K)$ is the relevant restriction map.

Since β is closed and onto, the map β^* is a homeomorphic embedding and E is closed in $C_p(K)$ (see Problem 163 of [TFS]). If $\delta \in C_p(E)$, then $\delta \circ \beta^* : C_p(L) \to \mathbb{R}$ is continuous and hence there is a countable $B \subset L$ and a continuous map $\varphi' : C_p(B|L) \to \mathbb{R}$ such that $\delta \circ \beta^* = \varphi' \circ \pi_B$ (see Problem 300 of [TFS]). Let $G = \overline{B}$; it is evident that for $\varphi = \varphi' \circ \pi_B^G$, we have $\delta \circ \beta^* = \varphi \circ \pi_G$. Choose a countable $A \subset K$ such that $\beta(A) = B$; if $F = \overline{A}$ then $\beta(F) = G$. We claim that

(1) $p_F(E) = (\beta|F)^*(C_p(G))$.

Indeed, if $f \in C_p(G)$, then by normality of L there is $g \in C_p(L)$ such that $g|G = f$. Then $h = g \circ \beta \in E$ and $p_F(h) = f \circ \beta = (\beta|F)^*(f)$ which shows that $(\beta|F)^*(C_p(G)) \subset p_F(E)$. If, on the other hand, $h \in E$, then $h = g \circ \beta$ for some $g \in C_p(L)$. It is evident that for $f = g|G$ we have $f \circ (\beta|F) = p_F(h)$ and hence (1) is proved.

The spaces F and G being compact, the map $\beta|F : F \rightarrow G$ is closed and hence $p_F(E) = (\beta|F)^*(C_p(G))$ is closed in $C_p(F)$. Since X is ω-monolithic, we have $w(F) = \omega$ and hence $nw(C_p(F)) = \omega$ which proves that $C_p(F)$ is normal. The map $(\beta|F)^* : C_p(G) \rightarrow p_F(E)$ is a homeomorphism, so its inverse $\gamma : p_F(E) \rightarrow C_p(G)$ is continuous. Since $C_p(G|L) = C_p(G)$ by normality of L, the map φ is defined on the whole $C_p(G)$, so $\varphi \circ \gamma : P_F(E) \rightarrow \mathbb{R}$ can be continuously extended over $C_p(F)$, i.e., there is $\mu \in C_p(C_p(F))$ such that $\mu|p_F(E) = \varphi \circ \gamma$. If $\lambda = \mu \circ p_F$, then $\lambda \in C_p(C_p(K))$; we claim that $\lambda|E = \delta$.

Indeed, take any function $h \in E$; there is $g \in C_p(L)$ with $g \circ \beta = h$. We have $\delta(h) = \delta(\beta^*(g)) = \varphi(\pi_G(g))$. Furthermore, if $f = \pi_G(g)$, then $p_F(h) = f \circ (\beta|F) = (\beta|F)^*(f)$. This implies that

$$\lambda(h) = \mu(p_F(h)) = \mu((\beta|F)^*(f)) = \varphi(\gamma((\beta|F)^*(f))) = \varphi(f) = \varphi(\pi_G(g)) = \delta(h).$$

Thus $\delta(h) = \lambda(h)$ for any $h \in E$ and hence λ is a continuous extension of the function δ. Since a continuous $\delta : E \rightarrow \mathbb{R}$ was chosen arbitrarily, the set E is C-embedded in $C_p(K)$ and Fact 1 is proved.   □

Returning to our solution let $R = \varphi^*(C_p(Y)) \subset C_p(X)$. Then $r_\varphi(\delta) = (\delta|R) \circ \varphi^*$ for any $\delta \in C_p(C_p(X))$ (see Problem 467). The map $\varphi^* : C_p(Y) \rightarrow R$ is a homeomorphism by Problem 163 of [TFS] so its inverse $\gamma : R \rightarrow C_p(Y)$ is continuous. The set R is C-embedded in $C_p(X)$ by Fact 1, so for any $\mu \in C_p(C_p(Y))$ the continuous map $\mu \circ \gamma : R \rightarrow \mathbb{R}$ can be continuously extended over $C_p(X)$, i.e., there is $\delta \in C_p(C_p(X))$ such that $\delta|R = \mu \circ \gamma$. We have $r_\varphi(\delta) = (\delta|R) \circ \varphi^* = (\mu \circ \gamma) \circ \varphi^* = \mu \circ (\gamma \circ \varphi^*) = \mu$, i.e., for any $\mu \in C_p(C_p(Y))$, there is $\delta \in C_p(C_p(X))$ such that $r_\varphi(\delta) = \mu$ which shows that r_φ is surjective and finishes our solution.

T.469. *Given spaces X and Y, let $\varphi : X \rightarrow Y$ be a continuous onto map. Prove that the mapping $r_\varphi : C_p(C_p(X)) \rightarrow r_\varphi(C_p(C_p(X))) \subset C_p(C_p(Y))$ is open if and only if φ is $\mathbb{R}$-quotient.*

Solution. For any $f \in C_p(Y)$ let $\varphi^*(f) = f \circ \varphi$. Then $\varphi^* : C_p(Y) \rightarrow C_p(X)$ is an embedding by Problem 163 of [TFS]. If $R = \varphi^*(C_p(Y)) \subset C_p(X)$, then $r_\varphi(\delta) = (\delta|R) \circ \varphi^*$ for any $\delta \in C_p(C_p(X))$ (see Problem 467). Consider the restriction map $\pi : C_p(C_p(X)) \rightarrow C_p(R)$ and let $\Phi(\delta) = \delta \circ \varphi^*$ for any $\delta \in C_p(R)$. Then $\Phi : C_p(R) \rightarrow C_p(C_p(Y))$ is a homeomorphism by Problem 163 of [TFS]. Therefore $r_\varphi = \Phi \circ \pi$ is open if and only if the map π is open which happens if and only if R is closed in $C_p(X)$ (see Problem 152 of [TFS]). Finally apply Problem 163 of [TFS] to conclude that R is closed in $C_p(X)$ if and only if φ is $\mathbb{R}$-quotient.

T.470. *Suppose that there exists a continuous map of $C_p(X)$ onto $C_p(Y)$. Prove that $nw(Y) \leq nw(X)$.*

Solution. Network weight is not raised by continuous maps (see Problem 157 of [TFS]), so if there is a continuous onto map $\varphi : C_p(X) \rightarrow C_p(Y)$, then $nw(C_p(Y)) \leq nw(C_p(X))$ and hence $nw(Y) = nw(C_p(Y)) \leq nw(C_p(X)) = nw(X)$.

T.471. *Suppose that there exists a continuous map of $C_p(X)$ onto $C_p(Y)$. Prove that $iw(Y) \leq iw(X)$.*

Solution. The density is not raised by continuous maps (see Problem 157 of [TFS]), so if there is a continuous onto map $\varphi : C_p(X) \rightarrow C_p(Y)$, then $d(C_p(Y)) \leq d(C_p(X))$ and hence $iw(Y) = d(C_p(Y)) \leq d(C_p(X)) = iw(X)$ (see Problem 174 of [TFS]).

T.472. *Suppose that there exists a continuous map of $C_p(X)$ onto $C_p(Y)$. Prove that $s^*(Y) \leq s^*(X)$, $hl^*(Y) \leq hl^*(X)$ and $hd^*(Y) \leq hd^*(X)$.*

Solution. We will apply three times the following evident fact.

(1) If a space T is a continuous image of a space Z, then T^n is a continuous image of the space Z^n for every $n \in \mathbb{N}$.

It follows from (1) and Problem 157 of [TFS] that s^* is not raised by continuous maps, so if there is a continuous onto map $\varphi : C_p(X) \rightarrow C_p(Y)$, then $s^*(C_p(Y)) \leq s^*(C_p(X))$ and hence $s^*(Y) = s^*(C_p(Y)) \leq s^*(C_p(X)) = s^*(X)$ (see Problem 025).

Now apply the property (1) and Problem 157 of [TFS] again to see that hd^* is not raised by continuous maps, so if $\varphi : C_p(X) \rightarrow C_p(Y)$ is continuous and onto, then we have the inequality $hd^*(C_p(Y)) \leq hd^*(C_p(X))$ and consequently, $hl^*(Y) = hd^*(C_p(Y)) \leq hd^*(C_p(X)) = hl^*(X)$ (see Problem 026).

Finally, apply (1) and Problem 157 of [TFS] to see that hl^* is not raised by continuous maps, so if $\varphi : C_p(X) \rightarrow C_p(Y)$ is continuous and onto, then $hl^*(C_p(Y)) \leq hl^*(C_p(X))$ and hence $hd^*(Y) = hl^*(C_p(Y)) \leq hl^*(C_p(X)) = hd^*(X)$ (see Problem 027).

T.473. *Suppose that there exists a continuous map of $C_p(X)$ onto $C_p(Y)$. Prove that if X is κ-monolithic, then Y is also κ-monolithic.*

Solution. If X is κ-monolithic, then $C_p(X)$ is κ-stable by Problem 152; since there exists a continuous onto map $\varphi : C_p(X) \rightarrow C_p(Y)$, the space $C_p(Y)$ is also κ-stable by Problem 123. Finally, apply Problem 152 again to conclude that Y is κ-monolithic.

T.474. *Suppose that there exists a quotient map of $C_p(X)$ onto $C_p(Y)$. Prove that $l^*(Y) \leq l^*(X)$ and $q(Y) \leq q(X)$.*

Solution. We have the inequality $t(C_p(Y)) \leq t(C_p(X))$ by Problem 162 of [TFS] which shows that $l^*(Y) = t(C_p(Y)) \leq t(C_p(X)) = l^*(X)$ (see Problem 149 of [TFS]). Analogously, we have $t_m(C_p(Y)) \leq t_m(C_p(X))$ by Problem 420 of [TFS] and hence it follows from Problem 434 of [TFS] that $q(Y) = t_m(C_p(Y)) \leq t_m(C_p(X)) = q(X)$.

T.475. *Suppose that there exists a quotient map of $C_p(X)$ onto $C_p(Y)$. Prove that if X is $l^*(\kappa)$-monolithic, then Y is also $l^*(\kappa)$-monolithic.*

Solution. Since X is $l^*(\kappa)$-monolithic, the space $C_p(X)$ is $t(\kappa)$-quotient-stable by Problem 183. It is evident that the property of being $t(\kappa)$-quotient-stable is preserved by quotient maps, so if there exists a quotient map $\varphi : C_p(X) \to C_p(Y)$, then $C_p(Y)$ is $t(\kappa)$-quotient-stable as well, and hence we can apply Problem 183 again to conclude that Y is $l^*(\kappa)$-monolithic.

T.476. *Suppose that there exists a continuous open map of $C_p(X)$ onto $C_p(Y)$. Prove that $|Y| \leq |X|$.*

Solution. We have $w(C_p(Y)) \leq w(C_p(X))$ by Problem 161 of [TFS] and hence it follows from Problem 169 of [TFS] that $|Y| = w(C_p(Y)) \leq w(C_p(X)) = |X|$.

T.477. *Suppose that there exists a continuous open map of $C_p(X)$ onto $C_p(Y)$. Prove that if X is κ-scattered, then Y is also κ-scattered.*

Solution. It follows from the fact that X is κ-scattered that the space $C_p(X)$ is $w(\kappa)$-open-stable (see Problem 187). It is evident that being $w(\kappa)$-open-stable is preserved by open maps, so if there exists an open map $\varphi : C_p(X) \to C_p(Y)$, then the space $C_p(Y)$ is also $w(\kappa)$-open-stable. Finally, apply Problem 187 again to conclude that Y is κ-scattered.

T.478. *Suppose that there exists a continuous closed map of $C_p(X)$ onto $C_p(Y)$. Prove that if X is κ-stable, then Y is also κ-stable.*

Solution. Since X is κ-stable, the space $C_p(X)$ is κ-monolithic by Problem 154. There exists a closed continuous onto map $\varphi : C_p(X) \to C_p(Y)$, so the space $C_p(Y)$ is also κ-monolithic by Problem 121. Finally, apply Problem 154 again to conclude that Y is κ-stable.

T.479. *Give an example of spaces X and Y for which there is a continuous map of $C_p(X)$ onto $C_p(Y)$ while $|Y| > |X|$.*

Solution. Let X be the set ω with the discrete topology. If $Y = \mathbb{I}$, then $C_p(Y)$ is analytic by Problem 367 and hence $\mathbb{R}^\omega$ maps continuously onto $C_p(Y)$ by Problem 360. Since $C_p(X)$ is homeomorphic to $\mathbb{R}^\omega$, we have $|X| = \omega < \mathfrak{c} = |Y|$ while $C_p(X)$ maps continuously onto $C_p(Y)$.

T.480. *Give an example of spaces X and Y for which there is a continuous map of $C_p(X)$ onto $C_p(Y)$ while $l^*(Y) > l^*(X)$.*

Solution. Let $X = \omega_1 + 1$ and $Y = \omega_1$ where both ordinals are endowed with their interval topology. Then $Y \subset X$ is countably compact and $X = \beta Y$ (see Problem 314 of [TFS]). Therefore the restriction map $\pi : C_p(X) \to C_p(Y)$ is surjective and hence $C_p(X)$ maps continuously onto $C_p(Y)$. Since Y is a countably compact non-compact space, we have $l(Y) > \omega$ and therefore $l^*(X) = \omega < l(Y) \leq l^*(Y)$.

T.481. *Give an example of spaces X and Y for which there is a continuous map of $C_p(X)$ onto $C_p(Y)$ while $q(Y) > q(X)$.*

Solution. Let $X = \omega_1 + 1$ and $Y = \omega_1$ where both ordinals are endowed with their interval topology. Then $Y \subset X$ is countably compact and $X = \beta Y$ (see Problem 314 of [TFS]). Therefore the restriction map $\pi : C_p(X) \to C_p(Y)$ is surjective and hence $C_p(X)$ maps continuously onto $C_p(Y)$. The space X is compact so $q(X) = \omega$; since Y is a countably compact non-compact space, we have $q(Y) > \omega$ (see Problems 401 and 407 of [TFS]) and therefore $q(X) < q(Y)$.

T.482. *Give an example of spaces X and Y for which there is a continuous map of $C_p(X)$ onto $C_p(Y)$ while X is compact and Y is not σ-compact.*

Solution. Let $X = \omega_1 + 1$ and $Y = \omega_1$ where both ordinals are endowed with their interval topology. Then $Y \subset X$ is countably compact and $X = \beta Y$ (see Problem 314 of [TFS]). Therefore the restriction map $\pi : C_p(X) \to C_p(Y)$ is surjective and hence $C_p(X)$ maps continuously onto $C_p(Y)$. The space X is compact; since Y is a countably compact non-compact space, it is not Lindelöf and hence not σ-compact,

T.483. *Give an example of spaces X and Y for which there is an open continuous map of $C_p(X)$ onto $C_p(Y)$ while $d(Y) > d(X)$.*

Solution. Let $X = \beta\omega$ and $Y = \beta\omega \setminus \omega \subset X$; then X is compact and Y is closed in X (see Fact 1 of S.370). Therefore the restriction map $\pi : C_p(X) \to C_p(Y)$ is open and surjective (see Problem 152 of [TFS]). This proves that there is an open continuous map of $C_p(X)$ onto $C_p(Y)$ while $d(X) = \omega$ and $d(Y) \geq c(Y) = \mathfrak{c} > d(X)$ (see Problem 371 of [TFS]).

T.484. *Give an example of spaces X and Y for which there is an open continuous map of $C_p(X)$ onto $C_p(Y)$ while $t_m(Y) > t_m(X)$.*

Solution. Let $X = \beta\omega$ and $Y = \beta\omega \setminus \omega \subset X$; then X is compact and Y is closed in X (see Fact 1 of S.370). Therefore the restriction map $\pi : C_p(X) \to C_p(Y)$ is open and surjective (see Problem 152 of [TFS]). This proves that there is an open continuous map of $C_p(X)$ onto $C_p(Y)$; we have $t_m(X) \leq d(X) = \omega$ (see Problem 418 of [TFS]) while $t_m(Y) > \omega$ by Problem 430 of [TFS] and hence $t_m(X) < t_m(Y)$.

T.485. *Give an example of spaces X and Y for which there is an open continuous map of $C_p(X)$ onto $C_p(Y)$ while $c(Y) > c(X)$ and $p(Y) > p(X)$.*

Solution. Let $X = \beta\omega$ and $Y = \beta\omega \setminus \omega \subset X$; then X is compact and Y is closed in X (see Fact 1 of S.370). Therefore the restriction map $\pi : C_p(X) \to C_p(Y)$ is open and surjective (see Problem 152 of [TFS]). This proves that there is an open continuous map of $C_p(X)$ onto $C_p(Y)$; we have $c(X) \leq d(X) = \omega$ while $c(Y) = \mathfrak{c}$ by Problem 371 of [TFS] and hence $c(X) < c(Y)$. Furthermore, we have $c(K) = p(K)$ for any compact space K by Problem 282 of [TFS] so $p(X) = c(X) < c(Y) = p(Y)$.

T.486. *Give an example of spaces X and Y for which there is a continuous map of $C_p(X)$ onto $C_p(Y)$ while X is discrete and Y is not discrete.*

Solution. Let X be the set ω with the discrete topology. If $Y = \mathbb{I}$, then $C_p(Y)$ is analytic by Problem 367 and hence $\mathbb{R}^\omega$ maps continuously onto $C_p(Y)$ by Problem 360. Since $C_p(X)$ is homeomorphic to $\mathbb{R}^\omega$, the space $C_p(X)$ maps continuously onto $C_p(Y)$ while X is discrete and Y isn't.

T.487. *Suppose that there exists a perfect map of $C_p(X)$ onto $C_p(Y)$. Prove that $d(X) = d(Y)$.*

Solution. The following easy facts will be useful for our solution.

Fact 1. If Z is a space with $w(Z) \leq \kappa$, then $\chi(K, Z) \leq \kappa$ for any compact $K \subset Z$.

Proof. Take a base $\mathcal{B}$ in Z with $|\mathcal{B}| \leq \kappa$. The family $\mathcal{U} = \{U \in \tau(K, Z) :$ there is a finite $\mathcal{B}' \subset \mathcal{B}$ with $U = \bigcup \mathcal{B}'\}$ has cardinality $\leq \kappa$. If $O \in \tau(K, Z)$, then choose for any $x \in K$ a set $U_x \in \mathcal{B}$ such that $x \in U_x \subset O$. Since K is compact, there is a finite $A \subset K$ for which $K \subset U = \bigcup \{U_x : x \in A\}$. It is clear that $U \in \mathcal{U}$ and $K \subset U \subset O$, so $\mathcal{U}$ is a base of K in Z and Fact 1 is proved. $\square$

Fact 2. For any space Z and compact $K \subset C_p(Z)$, we have $\psi(K, C_p(Z)) = \psi(C_p(Z))$.

Proof. Assume first that $\psi(K, C_p(Z)) \leq \kappa$ for some compact $K \subset C_p(Z)$. Since the space $C_p(Z)$ is homogeneous (i.e., for any $f, g \in C_p(Z)$, there is a homeomorphism $\varphi : C_p(Z) \to C_p(Z)$ such that $\varphi(f) = g$ (see Problem 079 of [TFS])), we can assume that $u \in K$ where $u \equiv 0$. Given a standard set $U = [x_1, \ldots, x_n; O_1, \ldots, O_n] = \{f \in C_p(Z) : f(x_i) \in O_i \text{ for all } i \leq n\} \in \tau(C_p(Z))$, let supp$(U) = \{x_1, \ldots, x_n\}$. Let $\mathcal{V}$ be a family of neighborhoods of K such that $|\mathcal{V}| \leq \kappa$ and $\bigcap \mathcal{V} = K$. Fix any $V \in \mathcal{V}$. For every $f \in K$ pick a standard open set U_f such that $f \in U_f \subset U$. Taking any finite subcover $\{U_{f_1}, \ldots, U_{f_m}\}$ of the open cover $\{U_f : f \in K\}$ of the compact set K, we obtain a set $W_V = U_{f_1} \cup \cdots \cup U_{f_m}$ with $K \subset W_V \subset V$ and the set $A_V = \text{supp}(U_{f_1}) \cup \cdots \cup \text{supp}(U_{f_m})$. It is evident that the set $A = \bigcup \{A_V : V \in \mathcal{V}\}$ has cardinality $\leq \kappa$, so it suffices to prove that $\overline{A} = Z$ because then $\psi(C_p(Z)) = d(Z) \leq \kappa$ (see Problem 173 of [TFS]).

Suppose that $x \in Z \backslash \overline{A}$. The map $e_x : C_p(Z) \to \mathbb{R}$ defined by $e_x(f) = f(x)$ is continuous and therefore the set $e_x(K)$ is bounded in $\mathbb{R}$. Choose any $r > 0$ such that $|f(x)| < r$ for all $f \in K$ and find some $g \in C_p(Z)$ such that $g(x) = r$ and $g(A) \subset \{0\}$. It follows from $g(x) = r$ that $g \notin K$. However $g|A = u|A$ implies $g \in \bigcap \mathcal{V}$ which contradicts the fact that $\bigcap \mathcal{V} = K$. Thus $\psi(C_p(Z)) = d(Z) \leq \kappa$ which shows that $\psi(C_p(Z)) \leq \psi(K, C_p(Z))$.

Now, if $\psi(C_p(Z)) \leq \kappa$ and $K \subset C_p(Z)$ is compact, then there is a condensation $\varphi : C_p(Z) \to M$ such that $w(M) \leq \kappa$ (see Problem 173 of [TFS]). The set $L = \varphi(K)$ is compact, so $\psi(L, M) \leq \chi(L, M) \leq \kappa$ by Fact 1. Fix a family $\mathcal{V} \subset \tau(L, M)$ such that $|\mathcal{V}| \leq \kappa$ and $\bigcap \mathcal{V} = L$. Then, for the family $\mathcal{U} = \{\varphi^{-1}(V) : V \in \mathcal{V}\}$, we have $\mathcal{U} \subset \tau(K, C_p(Z))$, $|\mathcal{U}| \leq \kappa$ and $\bigcap \mathcal{U} = K$ so $\psi(K, C_p(Z)) \leq \kappa$. This shows that $\psi(K, C_p(Z)) = \psi(C_p(Z))$ for any compact $K \subset C_p(Z)$ so Fact 2 is proved. $\square$

Returning to our solution, assume that $\varphi : C_p(X) \to C_p(Y)$ is a perfect map. If $\psi(C_p(X)) \leq \kappa$, then take any $f \in C_p(Y)$; the set $K = \varphi^{-1}(f)$ is compact, so $\psi(K, C_p(Z)) \leq \kappa$ by Fact 2. Fix a family $\mathcal{U} \subset \tau(K, C_p(X))$ such that $|\mathcal{U}| \leq \kappa$ and $\bigcap \mathcal{U} = K$. For every $U \in \mathcal{U}$, the set $O_U = C_p(Y)\backslash\varphi(C_p(X)\backslash U)$ is open in $C_p(Y)$ and contains f and $\varphi^{-1}(O_U) \subset U$ (see Fact 1 of S.226). An immediate consequence is that for the family $\mathcal{V} = \{O_U : U \in \mathcal{U}\}$, we have $|\mathcal{V}| \leq \kappa$ and $\bigcap \mathcal{V} = \{f\}$, so $\psi(f, C_p(Y)) = \psi(C_p(Y)) \leq \kappa$ which proves that $\psi(C_p(Y)) \leq \psi(C_p(X))$.

Now, if $\psi(C_p(Y)) \leq \kappa$, then take any $f \in C_p(Y)$. The set $K = \varphi^{-1}(f)$ is compact and $\psi(K, C_p(X)) \leq \kappa$, so we can apply Fact 2 to see that $\psi(C_p(X)) \leq \kappa$. Therefore $\psi(C_p(X)) \leq \psi(C_p(Y))$, i.e., $d(X) = \psi(C_p(X)) = \psi(C_p(Y)) = d(Y)$ (see Problem 173 of [TFS]) and hence our solution is complete.

T.488. *Suppose that there exists a perfect map of $C_p(X)$ onto $C_p(Y)$. Prove that $nw(X) = nw(Y)$.*

Solution. It follows from Problem 470 that $nw(Y) \leq nw(X)$, so let $\varphi : C_p(X) \to C_p(Y)$ be a perfect map and assume that $nw(Y) \leq \kappa$. Then $d(X) = d(Y) \leq nw(Y) \leq \kappa$ by Problem 487, and hence there is a condensation $r : C_p(X) \to M$ such that $w(M) \leq \kappa$ (see Problem 173 of [TFS]). For the map $\alpha = r\Delta\varphi : C_p(X) \to M \times C_p(Y)$, let $C = \alpha(C_p(X))$. Then $\alpha : C_p(X) \to C$ is a condensation because so is r; furthermore, α is perfect because φ is perfect (see Fact 1 of T.266). Consequently, α is a homeomorphism (see Problem 155 of [TFS]) and hence $nw(X) = nw(C_p(X)) = nw(C) \leq nw(M \times C_p(Y)) \leq \kappa$. This proves that $nw(X) \leq nw(Y)$ and therefore $nw(X) = nw(Y)$.

T.489. *Suppose that there exists a perfect map of $C_p(X)$ onto $C_p(Y)$. Prove that $|X| = |Y|$.*

Solution. We will need the following general fact about perfect maps.

Fact 1. Given spaces Z and T, if $f : Z \to T$ is a perfect map, then $w(T) \leq w(Z)$. In other words, perfect maps do not increase weight (recall that all perfect maps in this book are surjective).

Proof. Fix a base $\mathcal{B}$ in the space Z such that $|\mathcal{B}| \leq \kappa = w(Z)$. For any $U \in \tau(Z)$ let $f^{\#}(U) = T\backslash f(Z\backslash U)$. It is evident that $f^{\#}(U)$ is open in T for any $U \in \tau(Z)$. Therefore the family $\mathcal{C} = \{f^{\#}(U) : \text{there is a finite } \mathcal{B}' \subset \mathcal{B} \text{ such that } U = \bigcup \mathcal{B}'\}$ consists of open subsets of T; it is immediate that $|\mathcal{C}| \leq \kappa$.

To see that $\mathcal{C}$ is a base in T, take a point $t \in T$ and $W \in \tau(t, T)$. For any $z \in K = f^{-1}(t)$, choose $U_z \in \mathcal{B}$ for which $z \in U_z \subset f^{-1}(W)$. Since K is compact, there is a finite $A \subset K$ such that $K \subset \bigcup\{U_z : z \in A\}$; for the set $U = \bigcup\{U_z : z \in A\}$, we have $V = f^{\#}(U) \in \mathcal{C}$ and $t \in V \subset W$ (see Fact 1 of S.226). Therefore $\mathcal{C}$ is a base in T which shows that $w(T) \leq |\mathcal{C}| \leq \kappa$. Thus $w(T) \leq w(Z)$, so Fact 1 is proved.   □

Fact 2. For any space Z and compact $K \subset C_p(Z)$ we have $\chi(C_p(Z)) = \chi(K, C_p(Z))$.

Proof. If $\chi(C_p(Z)) \leq \kappa$, then $w(C_p(Z)) \leq \kappa$ by Problem 169 of [TFS] and hence $\chi(K, C_p(Z)) \leq \kappa$ by Fact 1 of T.487. Now assume that $\chi(K, C_p(Z)) \leq \kappa$ for some compact $K \subset C_p(Z)$.

Given a standard set $U = [x_1, \ldots, x_n; O_1, \ldots, O_n] = \{f \in C_p(Z) : f(z_i) \in O_i$ for all $i \leq n\} \in \tau(C_p(Z))$, let $\mathrm{supp}(U) = \{x_1, \ldots, x_n\}$. Fix a base $\mathcal{B}$ of neighborhoods of K in $C_p(Z)$ with $|\mathcal{B}| \leq \kappa$. If $V \in \mathcal{B}$, then pick for every $f \in K$ a standard set U_f such that $f \in U_f \subset V$. Taking any finite subcover $\{U_{f_1}, \ldots, U_{f_m}\}$ of the open cover $\{U_f : f \in K\}$ of the compact set K, we obtain a set $G_V = U_{f_1} \cup \cdots \cup U_{f_m}$ with $K \subset G_V \subset V$ and the set $A_V = \mathrm{supp}(U_{f_1}) \cup \cdots \cup \mathrm{supp}(U_{f_m})$. It is evident that the set $A = \bigcup\{A_V : V \in \mathcal{B}\}$ has cardinality $\leq \kappa$, so it suffices to prove that $A = Z$ because then $\chi(C_p(Z)) = w(C_p(Z)) = |Z| \leq \kappa$ (see Problem 169 of [TFS]).

Suppose that $x \in Z \backslash A$. The map $e_x : C_p(Z) \to \mathbb{R}$ defined by $e_x(f) = f(x)$ is continuous and therefore the set $e_x(K)$ is bounded in $\mathbb{R}$. Choose any $r > 0$ such that $|f(x)| < r$ for all $f \in K$ and observe that $W = [x; (-r, r)]$ is an open neighborhood of K. There exists $V \in \mathcal{B}$ such that $K \subset V \subset W$ and hence $G_V = U_{f_1} \cup \cdots \cup U_{f_m} \subset W$. This implies $U_{f_1} = [x_1, \ldots, x_n; O_1, \ldots, O_n] \subset W$ while $x \notin \{x_1, \ldots, x_n\}$. Apply Problem 034 of [TFS] to find $g \in C_p(Z)$ such that $g(x_i) \in O_i$ for all $i \leq n$ and $g(x) = r$. It is immediate that $g \in V \backslash W$ which is a contradiction. Thus $A = Z$ and hence $|Z| \leq \kappa$ which implies $\chi(C_p(Z)) = |Z| \leq \kappa$, i.e., $\chi(C_p(Z)) \leq \chi(K, C_p(Z))$. Thus $\chi(K, C_p(Z)) = \chi(C_p(Z))$ and Fact 2 is proved.   □

Returning to our solution observe that $w(C_p(Y)) \leq w(C_p(X))$ by Fact 1 and therefore $|Y| = w(C_p(Y)) \leq w(C_p(X)) = |X|$ (see Problem 169 of [TFS]). To prove that $|X| \leq |Y|$ assume that $\varphi : C_p(X) \to C_p(Y)$ is a perfect map and $\chi(C_p(Y)) \leq \kappa$. Take any $f \in C_p(Y)$ and let $K = \varphi^{-1}(f)$. There is a base $\mathcal{B}$ at the point f in the space $C_p(Y)$ with $|\mathcal{B}| \leq \kappa$. The family $\mathcal{C} = \{\varphi^{-1}(B) : B \in \mathcal{B}\}$ has cardinality $\leq \kappa$. To see that $\mathcal{C}$ is a base of K in the space $C_p(X)$ take any $U \in \tau(K, C_p(X))$.

The set $W = C_p(Y) \backslash f(C_p(X) \backslash U)$ is open in the space $C_p(Y)$ while $f \in W$ and $\varphi^{-1}(W) \subset U$ (see Fact 1 of S.266). Pick a set $B \in \mathcal{B}$ with $B \subset W$; then $V = \varphi^{-1}(B) \in \mathcal{C}$ and $K \subset V \subset U$. Thus $\mathcal{C}$ is a base of K in $C_p(X)$ such that $|\mathcal{C}| \leq \kappa$. This proves that $\chi(K, C_p(X)) \leq \kappa$ and therefore $w(C_p(X)) = \chi(C_p(X)) \leq \kappa$ by Fact 2. Finally, $|X| = w(C_p(X)) \leq \kappa$ (see Problem 169 of [TFS]) and hence $|X| \leq |Y|$, i.e., $|X| = |Y|$, so our solution is complete.

T.490. *Suppose that there exists a perfect map of $C_p(X)$ onto $C_p(Y)$. Prove that $hd^*(X) = hd^*(Y)$.*

Solution. Given a space P and an infinite cardinal κ, a set $A \subset P$ is an F_κ-subset of P if $A = \bigcup \mathcal{F}$ where every $F \in \mathcal{F}$ is closed in P and $|\mathcal{F}| \leq \kappa$.

Fact 1. Given spaces Z, T and an infinite cardinal κ if $hl(Z) \le \kappa$ and $w(T) \le \kappa$, then every $G \in \tau(Z \times T)$ is an F_κ-subset of $Z \times T$.

Proof. Let $\mathcal{B}$ be a base in T such that $|\mathcal{B}| \le \kappa$; for any $B \in \mathcal{B}$, consider the set $O_B = \bigcup\{U \in \tau(Z) : U \times B \subset G\}$. The set O_B is open in Z, so O_B is an F_κ-set in Z (see Problem 001). Since $hl(T) \le w(T) \le \kappa$, we can also apply Problem 001 to conclude that B is an F_κ-set in T. Thus $O_B \times B$ is an F_κ-set in $Z \times T$ for any $B \in \mathcal{B}$. Consequently, $G = \bigcup\{O_B \times B : B \in \mathcal{B}\}$ is an F_κ-subset of $Z \times T$ and Fact 1 is proved.   □

Fact 2. Given spaces Z, T and an infinite cardinal κ, if there is a perfect map $f : Z \to T$ and $l(T) \le \kappa$, then $l(Z) \le \kappa$.

Proof. Let $\mathcal{U}$ be an open cover of the space Z. For any $t \in T$ there is a finite $\mathcal{U}_t \subset \mathcal{U}$ such that $f^{-1}(t) \subset \bigcup\mathcal{U}_t$; by Fact 1 of S.226, there is $W_t \in \tau(t, T)$ such that $f^{-1}(W_t) \subset \bigcup\mathcal{U}_t$.

Since $\{W_t : t \in T\}$ is an open cover of T, there is $A \subset T$ such that $|A| \le \kappa$ and $\bigcup\{W_t : t \in A\} = T$. If $V_t = f^{-1}(W_t)$ for any $t \in A$, then $\{V_t : t \in A\}$ is a cover of Z such that $V_t \subset \bigcup\mathcal{U}_t$ for all $t \in A$. Therefore $\mathcal{U}' = \bigcup\{\mathcal{U}_t : t \in A\}$ is a subcover of $\mathcal{U}$ with $|\mathcal{U}'| \le \kappa$. This shows that $l(Z) \le \kappa$ so Fact 2 is proved.   □

Returning to our solution observe that it follows from Problem 472 that we have $hd^*(Y) \le hd^*(X)$. To prove the inverse inequality, we will need the following property:

(1) For any spaces Z and T if there exists a perfect map $\varphi : C_p(Z) \to C_p(T)$, then $hl(C_p(Z)) \le hl(C_p(T))$.

To show that (1) holds let $\kappa = hl(C_p(T))$; then $\psi(C_p(Z)) = d(Z) = d(T) = \psi(C_p(T)) = \kappa$ by Problem 487. Therefore there is a condensation $r : C_p(Z) \to M$ such that $w(M) \le \kappa$ (see Problem 173 of [TFS]). For the map $\alpha = r\Delta\varphi : C_p(Z) \to M \times C_p(T)$, let $Q = \alpha(C_p(Z))$. Then $\alpha : C_p(Z) \to Q$ is a condensation because so is r; besides, α is perfect because φ is perfect (see Fact 1 of Problem 266 of [TFS]). Consequently, α is a homeomorphism (see Problem 155 of [TFS]) and hence $C_p(Z)$ embeds in $M \times C_p(T)$. Therefore Fact 1 is applicable to conclude that every open subset of $C_p(Z)$ is an F_κ-subset of $C_p(Z)$ (Fact 1 actually says that this holds for the whole product, but it is evident that the relevant property is hereditary). Since $l(C_p(Z)) \le \kappa$ by Fact 2, we can apply Problem 001 to see that $hl(C_p(Z)) \le \kappa$; this proves the property (1).

Finally, assume that $hd^*(Y) \le \kappa$; then $hl^*(C_p(Y)) \le \kappa$ (see Problem 027). The space $(C_p(X))^n$ can be also perfectly mapped onto $(C_p(Y))^n$ for any $n \in \mathbb{N}$ (see Fact 4 of S.271). Given $n \in \mathbb{N}$, let D_n be a discrete space of cardinality n; since $(C_p(Z))^n$ is homeomorphic to $C_p(Z \times D_n)$ for any space Z (see Problem 114 of [TFS]), the space $C_p(X \times D_n)$ maps perfectly onto the space $C_p(Y \times D_n)$ for any $n \in \mathbb{N}$. This makes it possible to apply (1) to conclude that $hl(C_p(X \times D_n)) \le \kappa$ for all $n \in \mathbb{N}$, i.e., $hl^*(C_p(X)) \le \kappa$. Therefore $hd^*(X) = hl^*(C_p(X)) \le \kappa$ (see Problem 027) which implies $hd^*(X) \le hd^*(Y)$. Consequently, $hd^*(X) = hd^*(Y)$ so our solution is complete.

T.491. *Can $C_p(\beta\omega\setminus\omega)$ be mapped continuously onto $C_p(\omega_1)$?*

Solution. No, the space $C_p(\beta\omega\setminus\omega)$ cannot be mapped continuously onto $C_p(\omega_1)$. We will deduce it from the following fact.

Fact 1. For any cardinal κ with $\operatorname{cf}(\kappa) > \omega$ we have $p(C_p(\kappa)) = \kappa$. Here κ is endowed with its interval topology.

Proof. For every $\alpha < \kappa$ let $e_\alpha(f) = f(\alpha)$ for any $f \in C_p(\kappa)$. Then $e_\alpha : C_p(\kappa) \to \mathbb{R}$ is continuous by Problem 167 of [TFS] and hence $\varphi_\alpha = e_\alpha - e_{\alpha+1}$ is also continuous on $C_p(\kappa)$, so the set $U_\alpha = \{f \in C_p(\kappa) : \varphi_\alpha(f) > 1\}$ is open in $C_p(\kappa)$. It is easy to see that for every $\alpha < \kappa$ there is a function $f \in C_p(\kappa)$ such that $f(\alpha) = 2$ and $f(\alpha + 1) = 0$; as a consequence, $f \in U_\alpha$ which proves that $U_\alpha \neq \emptyset$ for any $\alpha < \kappa$. To see that the family $\mathcal{U} = \{U_\alpha : \alpha < \kappa\}$ is point-finite suppose not and fix a function $f \in C_p(\kappa)$ such that $A = \{\alpha \in \kappa : f \in U_\alpha\}$ is infinite. Choose a countably infinite $B \subset A$; since κ is not ω-cofinal, there is $\nu < \kappa$ such that $B \subset \nu$. The space $\nu + 1$ is compact by Problem 306 of [TFS], so there is a faithfully indexed sequence $S = \{\alpha_n : n \in \omega\} \subset B$ which has an accumulation point $\alpha \leq \nu$.

The function f being continuous, there is $\gamma < \alpha$ such that $|f(\beta) - f(\alpha)| < \frac{1}{3}$ for all β such that $\gamma < \beta < \alpha$. Since α is an accumulation point of the sequence S, there is $n \in \omega$ for which $\gamma < \alpha_n < \alpha$. Then $|\varphi_{\alpha_n}(f)| = |f(\alpha_n) - f(\alpha_n + 1)| \leq |f(\alpha_n) - f(\alpha)| + |f(\alpha_n + 1) - f(\alpha)| < \frac{1}{3} + \frac{1}{3} < 1$; however $f \in U_{\alpha_n}$ and hence $\varphi_{\alpha_n}(f) > 1$ which is a contradiction. Thus $\mathcal{U}$ is a point-finite family of non-empty open subsets of $C_p(\kappa)$; since $|\mathcal{U}| = \kappa$, we have $p(C_p(\kappa)) \geq \kappa$. The inverse inequality follows from $p(C_p(\kappa)) \leq w(C_p(\kappa)) = \kappa$ (see Problem 169 of [TFS]) so Fact 1 is proved. □

Now observe that the restriction map $\pi : C_p(\beta\omega) \to C_p(\beta\omega\setminus\omega)$ is continuous and onto (see Problem 152 of [TFS] and Fact 1 of S.370). We have $p(C_p(\beta\omega)) = \omega$ by Problem 382 of [TFS], so $p(C_p(\beta\omega\setminus\omega)) = \omega$ because point-finite cellularity is not raised by continuous maps. For the same reason any continuous image of $C_p(\beta\omega\setminus\omega)$ must have countable point-finite cellularity. However, $p(C_p(\omega_1)) = \omega_1$ by Fact 1 so $C_p(\omega_1)$ is not a continuous image of $C_p(\beta\omega\setminus\omega)$.

T.492. *Suppose that there exists a perfect irreducible map $\varphi : C_p(X) \to \mathbb{R}^\kappa$ for some cardinal κ. Prove that X is discrete.*

Solution. The space $\mathbb{R}^\kappa$ is pseudocomplete by Problems 465 and 470 of [TFS], so we can take a pseudocomplete sequence $S = \{\mathcal{B}_n : n \in \omega\}$ of π-bases in $\mathbb{R}^\kappa$. The family $\mathcal{C}_n = \{\varphi^{-1}(U) : U \in \mathcal{B}_n\}$ is a π-base in $C_p(X)$ for every $n \in \omega$ (see Fact 2 of S.373). To see that the sequence $\{\mathcal{C}_n : n \in \omega\}$ is pseudocomplete assume that $C_n \in \mathcal{C}_n$ and $\overline{C}_{n+1} \subset C_n$ for each $n \in \omega$. There is $B_n \in \mathcal{B}_n$ for which $C_n = \varphi^{-1}(B_n)$ for all $n \in \omega$. The map φ being closed, we have $\overline{B}_{n+1} = \varphi(\overline{C}_{n+1}) \subset \varphi(C_n) = B_n$ for every $n \in \omega$, so $A = \bigcap_{n\in\omega} B_n \neq \emptyset$ because the sequence S is pseudocomplete. If $f \in A$, then $\emptyset \neq \varphi^{-1}(f) \subset \bigcap_{n\in\omega} C_n$ and therefore $\bigcap_{n\in\omega} C_n \neq \emptyset$. This proves that $C_p(X)$ is pseudocomplete.

Fact 1. Any perfect preimage of a realcompact space is realcompact.

Proof. Let $f : S \to T$ be a perfect map such that T is realcompact. There exists a continuous map $g : \beta S \to \beta T$ such that $g|S = f$ (see Problem 257 of [TFS]). Besides, $g(\beta S \backslash S) \subset \beta T \backslash T$ (see Fact 3 of S.261) and therefore $g^{-1}(T) = S$. This makes it possible to apply Fact 1 of S.408 (for $R = \beta S$, $Z = \beta T$ and $B = T$) to see that S is realcompact and finish the proof of Fact 1.   □

To end our solution observe that $C_p(X)$ is realcompact by Fact 1 because so is $\mathbb{R}^\kappa$ (see Problem 401 of [TFS]). Thus $C_p(X)$ is pseudocomplete and realcompact so we can apply Problem 486 of [TFS] to conclude that X is discrete.

T.493. *Let $M = \prod_{t \in T} M_t$ where M_t is a metrizable space for all $t \in T$; assume that $\varphi : M \to C_p(X)$ is a closed continuous onto map. Prove that for every $t \in T$ we can choose a closed separable $N_t \subset M_t$ in such a way that $\varphi(\prod_{t \in T} N_t) = C_p(X)$. In particular, a space $C_p(X)$ is a closed continuous image of a product of (completely) metrizable spaces if and only if it is a closed continuous image of a product of separable (completely) metrizable spaces.*

Solution. Let $\pi_t : M \to M_t$ be the natural projection for all $t \in T$.

Fact 1. Let Y be a space such that every closed bounded subspace of Y is compact. Suppose that Z is a space in which any point is a limit of a nontrivial convergent sequence. Then any closed map $h : Y \to Z$ is irreducible on some closed subset of Y, i.e., there is a closed $F \subset Y$ such that $h(F) = Z$ and $h_F = h|F$ is irreducible.

Proof. For every $y \in Z$ fix a sequence $S_y = \{y_n : n \in \omega\} \subset Z \backslash \{y\}$ converging to y. We will prove first that the set $P_y = h^{-1}(y) \cap \overline{\bigcup\{h^{-1}(y_n) : n \in \omega\}}$ is compact for every $y \in Z$. Indeed, if for some $y \in Z$ the set P_y is not compact, then it is not bounded being closed in Y. Therefore there is $f \in C_p(Y)$ such that f is not bounded on P_y and hence there is a sequence $\{p_n : n \in \omega\} \subset P_y$ such that $|f(p_n)| > n$ for any $n \in \omega$. It is easy to choose a strictly increasing sequence $\{k_n : n \in \omega\} \subset \omega$ such that $f(p_{k_n}) \neq f(p_{k_m})$ whenever $n \neq m$ and $\{f(p_{k_n}) : n \in \omega\}$ is a closed discrete subset of $\mathbb{R}$. Let $r_n = f(p_{k_n})$ for all $n \in \omega$; since $\mathbb{R}$ is collectionwise normal, we can choose a discrete family $\{O_n : n \in \omega\} \subset \tau(\mathbb{R})$ such that $r_n \in O_n$ for all $n \in \omega$. If $U_n = f^{-1}(O_n)$ for all $n \in \omega$ then $\mathcal{U} = \{U_n : n \in \omega\} \subset \tau^*(Y)$ is discrete in Y; if $x_n = p_{k_n}$, then $x_n \in U_n \cap P_y$ for all $n \in \omega$.

If A is an arbitrary finite subset of ω, then for each natural number n, we have $U_n \cap (\bigcup\{h^{-1}(y_k) : k \in \omega \backslash A\}) \neq \emptyset$. This makes it possible to choose a point $z_n \in U_n \cap (\bigcup\{h^{-1}(y_k) : k \in \omega\})$ in such a way that $h(z_m) \neq h(z_n)$ if $n \neq m$.

The family $\mathcal{U}$ being discrete the set $D = \{z_n : n \in \omega\}$ is closed and discrete in Y. The set $h(D)$ is also closed because h is a closed map. Note that $h(D)$ has also to be discrete because $h(C)$ is closed for any $C \subset D$. However $h(D)$ is a nontrivial sequence converging to y, a contradiction with the fact that $h(D)$ is closed and discrete. This proves that P_y is compact for all $y \in Z$.

Claim. Suppose that H is a closed subset of Y such that $h(H) = Z$. Then $H \cap P_y \neq \emptyset$ for all $y \in Z$.

Proof of the claim. Fix any $y \in Z$. It follows from $h(H) = Z$ that it is possible to choose a point $t_n \in H \cap h^{-1}(y_n)$ for all $n \in \omega$. The map h is closed and therefore $\overline{\{t_n : n \in \omega\}} \cap h^{-1}(y) \neq \emptyset$. But $H \supset \{t_n : n \in \omega\}$ and $\overline{\{t_n : n \in \omega\}} \cap h^{-1}(y) \subset P_y$. Thus $H \cap P_y \neq \emptyset$ and the claim is proved.   □

Suppose that we have a family $\mathcal{F}$ of closed subsets of Y such that $\mathcal{F}$ is totally ordered by inclusion and $h(H) = Z$ for every $H \in \mathcal{F}$. Then $h(\bigcap \mathcal{F}) = Z$. Indeed, $H \cap P_y \neq \emptyset$ for any $y \in Z$ and $H \in \mathcal{F}$. We proved that the set P_y is compact, so $(\bigcap \mathcal{F}) \cap h^{-1}(y) \supset (\bigcap \mathcal{F}) \cap P_y \neq \emptyset$ for all $y \in Z$; consequently, $(\bigcap \mathcal{F}) \cap h^{-1}(y) \neq \emptyset$ which implies $y \in h(\bigcap \mathcal{F})$ for every $y \in Z$, i.e., $h(\bigcap \mathcal{F}) = Z$. Finally, use Zorn's lemma to find a closed $F \subset Y$ which is maximal (with respect to the inverse inclusion) in the family of all closed sets $H \subset Y$ such that $h(H) = Z$. It is evident that h_F is irreducible so Fact 1 is proved.   □

Returning to our solution observe that

(1) every closed bounded subspace B of the space M is compact.

Indeed, for every $t \in T$ the set $B_t = \pi_t(B)$ is bounded in M_t by Fact 1 of S.399. Therefore $C_t = \overline{B_t}$ is compact (see Fact 1 of S.415 and Problem 406 of [TFS]), so the set $C = \prod_{t \in T} C_t$ is also compact. The set B is closed in M and $B \subset C$ which shows that B is closed in C and hence compact so (1) is proved.

For every $f \in C_p(X)$, let $f_n = f + \frac{1}{n}$ for any $n \in \mathbb{N}$; then $\{f_n : n \in \mathbb{N}\}$ is a nontrivial sequence which converges to f. This, together with (1), shows that we can apply Fact 1 to the map φ to find a closed $F \subset M$ such that $\varphi(F) = C_p(X)$ and $\varphi|F$ is irreducible. By Fact 1 of S.228 we have $c(F) = c(C_p(X)) = \omega$; let $F_t = \pi_t(F)$ and $N_t = \overline{F_t}$ for all $t \in T$. Then $c(N_t) = c(F_t) = \omega$ and hence N_t is a closed separable subspace of M_t for any $t \in T$ (see Problem 214 of [TFS]). Since $F \subset N = \prod_{t \in T} N_t$, we have $\varphi(N) \supset \varphi(F) = C_p(X)$ and hence $\varphi(N) = C_p(X)$. The map $\varphi' = \varphi|N$ maps N continuously onto $C_p(X)$; it is evident that a restriction of a closed map to a closed subspace is a closed map onto the image of that subspace so $C_p(X)$ is a closed continuous image of N and our solution is complete.

T.494. *Let M be a product of completely metrizable spaces. Suppose that there exists a continuous closed onto map $\varphi : M \to C_p(X)$. Prove that X is discrete. In particular, X is discrete if $C_p(X)$ is a closed continuous image of $\mathbb{R}^\kappa$ for some cardinal κ.*

Solution. It follows from Problem 493 that we can assume, without loss of generality, that the space M is a product of second countable completely metrizable spaces, i.e., $M = \prod\{M_t : t \in T\}$ where $w(M_t) = \omega$ and M_t is Čech-complete for every $t \in T$. For any non-empty $S \subset T$ the map $p_S : M \to M_S = \prod\{M_t : t \in S\}$ is the natural projection; we write p_t instead of $p_{\{t\}}$ for any $t \in T$. As usual, $\pi_A : C_p(X) \to C_p(A|X) \subset C_p(A)$ is the restriction map for any $A \subset X$.

Fact 1. If $h : Y \to Z$ is a closed irreducible map and Z is Pseudocompact, then Y is also pseudocompact.

Proof. If the space Y is not pseudocompact, then there exists a discrete family $\{O_n : n \in \omega\} \subset \tau^*(Y)$. The set $W_n = Z \setminus h(Y \setminus O_n)$ is open, non-empty and $h^{-1}(W_n) \subset O_n$ for any $n \in \omega$. Choose $V_n \in \tau^*(Z)$ such that $\overline{V}_n \subset W_n$ for all $n \in \omega$. If $F_n = h^{-1}(\overline{V}_n)$, then $F_n \subset O_n$ for all $n \in \omega$, so $\{F_n : n \in \omega\}$ is a discrete family of closed subsets of Y. Consequently, $F = \bigcup_{n \in \omega} F_n$ is closed in Y which implies that $\bigcup_{n \in \omega} \overline{V}_n = h(F)$ is closed in Z.

Now take any $z \in Z$. If $z \in W_n$ for some $n \in \omega$, then $W_n \cap V_i = \emptyset$ for all $i \neq n$. If $z \in Z \setminus (\bigcup_{n \in \omega} W_n)$, then $O = Z \setminus h(F)$ is an open neighborhood of z which does not intersect any element of $\mathcal{V} = \{V_n : n \in \omega\}$. This proves that every $z \in Z$ has a neighborhood which intersects at most one element of $\mathcal{V}$. Therefore $\mathcal{V}$ is an infinite discrete family of non-empty open subsets of Z which contradicts pseudocompactness of Z. Thus Y is also pseudocompact and Fact 1 is proved.   □

Returning to our solution, observe first that $C_p(X)$ is stable because so is M (see Problem 268). Thus X is ω-monolithic, i.e., $nw(\overline{A}) = \omega$ for any countable $A \subset X$. We will establish that, in fact, $\overline{A}$ is countable for any countable $A \subset X$. The first step is to show that

(1) every closed bounded subspace B of the space M is compact.

Indeed, for every $t \in T$, the set $B_t = \pi_t(B)$ is bounded in M_t by Fact 1 of S.399. Therefore $C_t = \overline{B}_t$ is compact (see Fact 1 of S.415 and V1.406), so the set $C = \prod_{t \in T} C_t$ is also compact. The set B is closed in M and $B \subset C$ which shows that B is closed in C and hence compact so (1) is proved.

For any $f \in C_p(X)$ let $f_n = f + \frac{1}{n}$ for all $n \in \mathbb{N}$; then $\{f_n : n \in \mathbb{N}\}$ is a nontrivial sequence which converges to f. Thus

(2) every $f \in C_p(X)$ is a limit of a nontrivial convergent sequence.

The properties (1) and (2) show that we can apply Fact 1 of T.493 to find a closed $Y \subset M$ such that $\varphi(Y) = C_p(X)$ and $\varphi|Y$ is an irreducible map. Now assume that $A \subset X$ is a countable closed subset of X. The map $\pi_A : C_p(X) \to C_p(A|X)$ is open by Problem 152 of [TFS]; apply Problem 299 of [TFS] to find a countable $S \subset T$ and a continuous map $\alpha : M_S \to C_p(A|X)$ such that $\alpha \circ p_S = \pi_A \circ \varphi$. If $Y_S = p_S(Y)$ and $Z = \overline{Y}_S$, then it follows from $\varphi(Y) = C_p(X)$ that $\alpha(Y_S) = C = C_p(A|X)$ and hence $\alpha(Z) = C$. Let $p = p_S|Y : Y \to Y_S$. We claim that

(3) $\text{Int}_C(\alpha(U)) \neq \emptyset$ for any $U \in \tau^*(Z)$.

To see that the property (3) is satisfied, observe that $U' = U \cap Y_S \in \tau^*(Y_S)$ and hence $V = p^{-1}(U')$ is a non-empty open subset of Y. The map $\varphi' = \varphi|Y$ being irreducible, there is $W \in \tau^*(C_p(X))$ such that $W \subset \varphi'(V) = \varphi(V)$ (see Fact 1 of S.383). Since the map π_A is open, $\pi_A(W)$ is a non-empty open subset of C such that $\pi_A(W) \subset \alpha(U)$ so (3) is proved.

An immediate consequence of (3) is that $\alpha' = \alpha|Z : Z \to C$ is *weakly open*, i.e., $\text{Int}_C(\overline{\alpha'(U)})$ is dense in $\alpha'(U)$ for any $U \in \tau^*(Z)$. The space Z is closed in a Čech-complete space M_S, so it is Čech-complete and hence pseudocomplete. Applying Fact 1 of S.471 we can conclude that the space C is also pseudocomplete, and hence it has a dense Čech-complete subspace D by Problem 468 of [TFS]. Since D is also dense in $C_p(A)$, the space A is discrete by Problem 265 of [TFS].

Furthermore, $C = C_p(A|X)$ is an algebra and, in particular, $f - g \in C_p(A|X)$ for any $f, g \in C_p(A|X)$. Take any $h \in \mathbb{R}^A \backslash C_p(A|X)$; the map $T_h : \mathbb{R}^A \to \mathbb{R}^A$ defined by $T_h(g) = h + g$ is a homeomorphism (Problem 079 of [TFS]) and $T_h(C_p(A|X)) \cap C_p(A|X) = \emptyset$. Indeed, if $g \in C_p(A|X)$ and $f = h + g \in C_p(A|X)$, then $h = f - g \in C_p(A|X)$ which is a contradiction. Therefore $\mathbb{R}^A$ has two disjoint dense Čech-complete subspaces D and $T_h(D)$ which contradicts Problem 264 of [TFS]. Thus $C_p(A|X) = \mathbb{R}^A$ and hence we proved that

(4) every countable closed $A \subset X$ is discrete and C-embedded in X.

Now it is easy to show that $\overline{A}$ is countable for any countable $A \subset X$. Indeed, we have $nw(\overline{A}) = \omega$, so if $|\overline{A}| > \omega$, then there is a nontrivial convergent sequence $S \subset \overline{A}$ (see Fact 1 of S.497); since S is countable, closed and non-discrete, this contradicts (4) and proves that $|\overline{A}| = \omega$ for any countable $A \subset X$. Applying (4) once more we can see that $\overline{A}$ is discrete and C-embedded in X for any countable $A \subset X$. Consequently, any countable $A \subset X$ is also closed, discrete and C-embedded in X and hence $C_p(X)$ is pseudocomplete by Problem 485 of [TFS].

It follows from pseudocompleteness of $C_p(X)$ that $C_p(X, \mathbb{I})$ is pseudocompact (see Problem 476 of [TFS]). If $H = \varphi^{-1}(C_p(X, \mathbb{I}))$, then $\gamma = \varphi|H : H \to C_p(X, \mathbb{I})$ is a closed map. We leave it to the reader as an easy exercise to prove that every point of $C_p(X, \mathbb{I})$ is a limit of a nontrivial convergent sequence. Furthermore, if F is closed and bounded in H, then it is compact being closed and bounded in M [see (1)].

Thus we can apply Fact 1 of T.493 to conclude that there is a closed $G \subset H$ such that $\gamma(H) = C_p(X, \mathbb{I})$ and $\gamma|G$ is irreducible. Since H is closed in M, the set G is also closed in M. By Fact 1 the space G is pseudocompact and hence bounded in M. Therefore G is compact [see (1)], so $C_p(X, \mathbb{I})$ is also compact being a continuous image of a compact space G. Finally apply Problem 396 of [TFS] to conclude that X is discrete and finish our solution.

T.495. *Let X be a pseudocompact space. Suppose that $C_p(X)$ contains a dense subspace which is a continuous image of a product of separable spaces. Prove that X is compact and metrizable.*

Solution. There exists a family $\{Z_t : t \in T\}$ of separable spaces such that for the space $Z = \prod_{t \in T} Z_t$, there is a continuous map $\varphi : Z \to C_p(X)$ for which the set $Y = \varphi(Z)$ is dense in $C_p(X)$. Let $e(x)(f) = f(x)$ for any $x \in X$ and $f \in Y$. Then $e(x) \in C_p(Y)$ for any $x \in X$ and the map $e : X \to C_p(Y)$ is an injection (see Problem 166 of [TFS] and Fact 2 of S.351); let $X' = e(X)$. Since $\varphi : Z \to Y$ is continuous and onto, the dual map $\varphi^* : C_p(Y) \to C_p(Z)$ embeds $C_p(Y)$ in $C_p(Z)$ (see Problem 163 of [TFS]) and therefore X' is a pseudocompact space which embeds in $C_p(Z)$. This implies that X' is metrizable (and hence second countable and compact) by Problem 307 of [TFS]; any condensation of a pseudocompact space onto a second countable space is a homeomorphism (Problem 140 of [TFS]), so $e : X \to X'$ is a homeomorphism. Therefore X is compact and metrizable being homeomorphic to a compact metrizable space X'.

T.496. *Let X be a pseudocompact space. Suppose that $C_p(X)$ is a closed continuous image of a product of metrizable spaces. Prove that X is countable.*

Solution. By Problem 493 the space $C_p(X)$ is a closed continuous image of a product of *separable* metric spaces. This implies that X is compact and metrizable (see Problem 495). Fix a family $\{M_t : t \in T\}$ of separable metrizable spaces such that there is a closed continuous onto map $\varphi : M = \prod_{t \in T} M_t \to C_p(X)$. For any $S \subset T$ let $p_S : M \to M_S = \prod_{t \in S} M_t$ be the projection of M onto its face M_S.

We have $nw(C_p(X)) = \omega$, so there is a condensation $\pi : C_p(X) \to Z$ of $C_p(X)$ onto a second countable space Z. By Problem 299 of [TFS], there is a countable $S \subset T$ and a continuous map $\alpha : M_S \to Z$ such that $\alpha \circ p_S = \pi \circ \varphi$. For the map $\gamma = \pi^{-1} \circ \alpha$, we have $\gamma : M_S \to C_p(X)$ and $\gamma \circ p_S = \varphi$; therefore γ is surjective. If $U \in \tau(C_p(X))$, then $\gamma^{-1}(U) = p_S(\varphi^{-1}(U))$ is an open subset of M_S because φ is continuous and p_S is open. As a consequence, γ is a continuous map. Besides, if F is closed in M_S, then $\gamma(F) = \varphi(p_S^{-1}(F))$ is closed in $C_p(X)$ because p_S is continuous and φ is closed. Consequently, $C_p(X)$ is a closed image of the metrizable space M_S and hence X is countable by Problem 228 of [TFS].

T.497. *Let X be a pseudocompact space. Suppose that $C_p(X)$ is an open continuous image of a product of separable metrizable spaces. Prove that X is countable.*

Solution. The space X is compact and metrizable by Problem 495. By our hypothesis there is a family $\{M_t : t \in T\}$ of separable metrizable spaces such that there is an open continuous onto map $\varphi : M = \prod_{t \in T} M_t \to C_p(X)$. For any $S \subset T$ let $p_S : M \to M_S = \prod_{t \in S} M_t$ be the projection of M onto its face M_S.

We have $nw(C_p(X)) = \omega$, so there is a condensation $\pi : C_p(X) \to Z$ of $C_p(X)$ onto a second countable space Z. By Problem 299 of [TFS] there is a countable $S \subset T$ and a continuous map $\alpha : M_S \to Z$ such that $\alpha \circ p_S = \pi \circ \varphi$. For the map $\gamma = \pi^{-1} \circ \alpha$, we have $\gamma : M_S \to C_p(X)$ and $\gamma \circ p_S = \varphi$; therefore γ is surjective. If $U \in \tau(C_p(X))$, then $\gamma^{-1}(U) = p_S(\varphi^{-1}(U))$ is an open subset of M_S because φ is continuous and p_S is open. As a consequence, γ is a continuous map. Besides, if G is open in M_S, then $\gamma(G) = \varphi(p_S^{-1}(G))$ is open in $C_p(X)$ because p_S is continuous and φ is open. Consequently, $C_p(X)$ is an open image of the metrizable space M_S and hence X is countable by Problem 229 of [TFS].

T.498. *Let M be a product of separable completely metrizable spaces. Assuming that there is a finite-to-one open map $\varphi : M \to C_p(X)$, prove that X is discrete.*

Solution. Suppose that we are given spaces Y, Z and a map $f : Y \to Z$. We will say that $y \in Y$ is *a point of local homeomorphism of f* if there is $U \in \tau(y, Y)$ such that $f|U : U \to f(U)$ is a homeomorphism. As usual, $\mathbb{D} = \{0, 1\}$; we let $\mathbb{D}^0 = \{\emptyset\}$ and $\mathbb{D}^{<\alpha} = \bigcup\{\mathbb{D}^n : n < \alpha\}$ for any $\alpha \leq \omega$. If $n \in \omega$, $i \in \mathbb{D}$ and $s \in \mathbb{D}^n$, then $s^\frown i \in \mathbb{D}^{n+1}$ is defined by $(s^\frown i)|n = s$ and $s(n) = i$.

Fact 1. Given spaces Y and Z let $h : Y \to Z$ be an open finite-to-one map. If Z is pseudocomplete, then any non-empty open subset of Y contains a point of local homeomorphism of h.

Proof. Let $\{\mathcal{B}_n : n \in \omega\}$ be a pseudocomplete sequence in Z. The map h being open, it suffices to prove that for every $U \in \tau^*(Y)$ there is an open non-empty $O \subset U$ such that $h|O$ is an injection. To obtain a contradiction assume that this is not true for some $U \in \tau^*(Y)$. Then for every $O \in \tau^*(U)$ there are distinct $y_0, y_1 \in O$ with $h(y_0) = h(y_1)$. Let us construct inductively families $\{U_s : s \in \mathbb{D}^{<\omega}\} \subset \tau^*(Y)$ and $\{V_n : n \in \omega\} \subset \tau^*(Z)$ with the following properties:

(1) $V_n \in \mathcal{B}_n$ and $\overline{V}_{n+1} \subset V_n$ for all $n \in \omega$;
(2) the family $\{U_s : s \in \mathbb{D}^n\}$ is disjoint for all $n \in \omega$;
(3) if $n \in \omega$ and $s \in \mathbb{D}^n$, then $h(U_s) \supset V_n$ and $U_{s \frown i} \subset U_s$ for any $i \in \mathbb{D}$;
(4) If $n \in \omega$ and $s \in \mathbb{D}^{n+1}$, then $h(U_s) \subset V_n$.

If we let $U_\emptyset = U$ and choose any $V_0 \in \mathcal{B}_0$ such that $V_0 \subset h(U_\emptyset)$, then (1)–(4) are fulfilled for $n = 0$ (the second parts of (1) and (3) as well as the property (4) are satisfied vacuously). Now assume that $m \in \mathbb{N}$ and we have families $\{U_s : s \in \mathbb{D}^{<m}\}$ and $\{V_n : n < m\}$ such that (1)–(4) are fulfilled for all $n < m$ (to save space let us agree that the last statement actually says that (2) and the first part of (1) are satisfied for all $n < m$ while the second parts of (1) and (3) as well as the property (4) are true for all $n < m - 1$).

Let $\{s_i : 0 \le i < 2^{m-1}\}$ be an enumeration of the set $\mathbb{D}^{m-1}$. By our assumption and (3), there are distinct $y_0, y_1 \in U_{s_0} \cap h^{-1}(V_{m-1})$ such that $h(y_0) = h(y_1)$. Choose $O_0^i \in \tau(y_i, U_{s_0} \cap h^{-1}(V_{m-1}))$ for every $i \in \mathbb{D}$ such that $O_0^0 \cap O_0^1 = \emptyset$; then $W_0 = h(O_0^0) \cap h(O_0^1) \in \tau^*(V_{m-1})$.

Now assume that $k < 2^{m-1}$ and we have open non-empty sets O_j^i and W_j for all $j < k$ and $i \in \mathbb{D}$ with the following properties:

(5) $O_j^0 \cap O_j^1 = \emptyset$ and $O_j^0 \cup O_j^1 \subset U_{s_j}$ for any $j < k$;
(6) $W_j = \bigcap\{h(O_p^i) : p \le j, i \in \mathbb{D}\} \in \tau^*(V_{m-1})$ for all $j < k$.

It follows from (6) and (3) that $G = U_{s_k} \cap h^{-1}(W_{k-1}) \neq \emptyset$, so we can choose distinct $y_0, y_1 \in G$ such that $h(y_0) = h(y_1)$. Take disjoint $O_k^0 \in \tau(y_0, U_{s_k} \cap h^{-1}(V_{m-1}))$ and $O_k^1 \in \tau(y_1, U_{s_k} \cap h^{-1}(V_{m-1}))$; it is evident that (5) and (6) are fulfilled for all $j \le k$. Thus we can construct inductively a family $\{O_j^i : j < 2^{m-1}, i \in \mathbb{D}\}$ with the properties (5) and (6) fulfilled for all $j < 2^{m-1}$. Consequently, $W = W_{2^{m-1}-1} \neq \emptyset$, so we can choose $V_m \in \mathcal{B}_m$ such that $\overline{V}_m \subset W$. If $s \in \mathbb{D}^{m-1}$, then $s = s_j$ for some $j < 2^{m-1}$; let $U_{s \frown i} = O_j^i$ for all $i \in \mathbb{D}$. This defines the set U_s for every $s \in \mathbb{D}^m$ and it is immediate from the choice of O_j^i that $h(U_s) \subset V_{m-1}$ for each $s \in \mathbb{D}^m$. It also follows from (6) and the choice of V_m that $h(U_s) \supset V_m$ for all $s \in \mathbb{D}^m$. Therefore (3) is satisfied for all $n \le m$. Since (1) and (2) are clear, we proved that for the families $\{U_s : s \in \mathbb{D}^{<m+1}\}$ and $\{V_i : i < m+1\}$ the properties (1)–(4) are fulfilled for all $n \le m$.

Thus our inductive procedure gives us families $\{U_s : s \in \mathbb{D}^{<\omega}\} \subset \tau^*(Y)$ and $\{V_n : n \in \omega\} \subset \tau^*(Z)$ with (1)–(4). The sequence $\{\mathcal{B}_n : n \in \omega\}$ being pseudocomplete, we have $\bigcap_{n \in \omega} V_n \neq \emptyset$. If $x \in \bigcap_{n \in \omega} V_n$, then for any $n \in \mathbb{N}$ we have $h^{-1}(x) \cap U_s \neq \emptyset$ for any $s \in \mathbb{D}^n$ by (3). The property (2) shows that $|h^{-1}(x)| \geq 2^n$ for any $n \in \mathbb{N}$ and hence the set $h^{-1}(x)$ is infinite. This contradiction shows that U must have points of local homeomorphism and finishes the proof of Fact 1.   □

Returning to our solution observe that $C_p(X)$ is pseudocomplete by Problem 475 of [TFS]. Therefore Fact 1 can be applied to the map φ to conclude that there is a non-empty open $U \subset M$ such that $\varphi|U : U \to \varphi(U)$ is a homeomorphism. Observe that the space M is realcompact (see Problems 406 and 402 of [TFS]); we can choose $V \in \tau^*(U)$ such that V is an F_σ-set in M (see Fact 1 of T.252). The space V is realcompact by Problem 408 of [TFS] and therefore $\varphi(V)$ is a non-empty open realcompact subspace of $C_p(X)$. This implies that $C_p(X)$ is realcompact (see Problem 428 of [TFS]), so we can apply Problem 486 of [TFS] to conclude that X is discrete.

T.499. *Let M be a product of separable completely metrizable spaces. Assuming that there is a finite-to-one open map $\varphi : C_p(X) \to M$, prove that X is discrete.*

Solution. It follows from Problem 465 of [TFS] that every completely metrizable space is pseudocomplete. Therefore any product of completely metrizable spaces is pseudocomplete by Problem 470 of [TFS]. This shows that M is pseudocomplete and hence there is a non-empty open $U \subset C_p(X)$ such that $\varphi|U : U \to \varphi(U)$ is a homeomorphism (see Fact 1 of T.498). The set $\varphi(U)$ is pseudocomplete being open in M (see Problem 466 of [TFS]), so U is a non-empty open pseudocomplete subspace of $C_p(X)$. By homogeneity of $C_p(X)$, every $f \in C_p(X)$ has a pseudocomplete open neighborhood, so $C_p(X)$ is pseudocomplete by Fact 2 of S.488.

Observe that the space M is realcompact (see Problems 406 and 402 of [TFS]); we can choose $W \in \tau^*(\varphi(U))$ such that W is an F_σ-set in M (see Fact 1 of T.252). The space W is realcompact by Problem 408 of [TFS]; the set $W' = \varphi^{-1}(W) \cap U \subset U$ is open in $C_p(X)$ and $\varphi(W') = W$ which shows that W' is homeomorphic to W. Consequently, W' is a non-empty open realcompact subspace of $C_p(X)$ which implies that $C_p(X)$ is realcompact (see Problem 428 of [TFS]), so we can apply Problem 486 of [TFS] to conclude that X is discrete.

T.500. *Let H be a $G_{\delta\sigma}$-subspace of $\mathbb{R}^\kappa$ for some κ. Prove that if $C_p(X)$ is homeomorphic to a retract of H, then X is discrete. In particular, if $C_p(X)$ is homeomorphic to a retract of $\mathbb{R}^\kappa$ for some cardinal κ, then X is discrete.*

Solution. Fix a retraction $r : H \to C$ such that C is homeomorphic to $C_p(X)$; from now on we will identify C with the space $C_p(X)$. For any set $A \subset X$ the map $\pi_A : C \to C_p(A)$ is the restriction; given any $E \subset C$ denote the space $\pi_A(E)$ by $E[A]$. Furthermore, if $A \subset A' \subset X$, then $\pi_A^{A'} : C[A'] \to C[A]$ is also the restriction map. Call a set $E \subset C$ *factorizable in C* if for any continuous map $p : E \to M$

of the space E onto a second countable space M there is a countable $A \subset X$ and a continuous map $u : E[A] \to M$ such that $u \circ (\pi_A|E) = p$.

Given a set $B \subset \kappa$ the map $p_B : \mathbb{R}^\kappa \to \mathbb{R}^B$ is the natural projection of $\mathbb{R}^\kappa$ onto its face $\mathbb{R}^B$; if $G \subset \mathbb{R}^\kappa$, then $G_B = p_B(G)$. If $B \subset B' \subset \kappa$, then $p_B^{B'} :$ $\mathbb{R}^{B'} \to \mathbb{R}^B$ is the natural projection of $\mathbb{R}^{B'}$ onto its face $\mathbb{R}^B$. We will say that a set $G \subset \mathbb{R}^\kappa$ *is factorizable in* $\mathbb{R}^\kappa$ if for any continuous map $p : G \to M$ of the space G onto a second countable space M, there is a countable $B \subset \kappa$ and continuous map $v : G_B \to M$ for which $v \circ (p_B|G) = p$. If $B \subset \kappa$ and $f \in \mathbb{R}^B$, then let $G(B, f) = p_B^{-1}(f)$; a set $G \subset \mathbb{R}^\kappa$ is called *a standard G_δ-subset of* $\mathbb{R}^\kappa$ if $G = G(B, f)$ for some countable $B \subset \kappa$ and $f \in \mathbb{R}^B$. We let $\Omega_m = \{1, \dots, m\}$ for any $m \in \mathbb{N}$.

Fact 1. Suppose that Z is a space and $A \subset Z$ is countable. Let $\pi_A : C_p(Z) \to C_p(A)$ be the restriction map and assume that $\pi_A(C_p(Z, \mathbb{I}))$ has a dense Čech-complete subspace. Then A is a discrete subspace of Z.

Proof. Let $P \subset \pi_A(C_p(Z, \mathbb{I}))$ be a dense Čech-complete subspace of the space $C' = \pi_A(C_p(Z, \mathbb{I}))$. It is easy to see that C' is dense in $C_p(A, \mathbb{I})$ and hence P is a dense Čech-complete subspace of $C_p(A, \mathbb{I})$. Therefore A is discrete by Problem 287 of [TFS] and Fact 1 is proved.   □

Fact 2. Given a space Z_t for all $t \in T$ assume that $T = \bigcup\{T_n : n \in \omega\}$ and $T_n \subset T_{n+1}$ for all $n \in \omega$. For any $S \subset T$ let $Z_S = \prod\{Z_t : t \in S\}$; if $n \in \omega$ then $\pi_n : Z_{T_{n+1}} \to Z_{T_n}$ is the natural projection. Assume additionally that Y is a space and $f_n : Y \to Z_{T_n}$ is a continuous map such that $\pi_n \circ f_{n+1} = f_n$ for all $n \in \omega$. Given any $y \in Y$ we have $f_n(y) \subset f_{n+1}(y)$ for all $n \in \omega$, so $f(y) = \bigcup_{n \in \omega} f_n(y)$ is well-defined and $f(y) \in Z_T$. Then the map $f : Y \to Z_T$ is continuous.

Proof. For any $t \in T$ and $n \in \omega$ let $p_t : Z_T \to Z_t$ and $p_t^n : Z_{T_n} \to Z_t$ be the respective natural projections. It suffices to show that $p_t \circ f$ is continuous for all $t \in T$ (see Problem 102 of [TFS]), so take an arbitrary $t \in T$. There is $n \in \omega$ such that $t \in T_n$; it is immediate that $p_t \circ f = p_t^n \circ f_n$; it follows from continuity of f_n that $p_t^n \circ f_n$ and hence $p_t \circ f$ is a continuous map, so Fact 2 is proved.   □

Fact 3. Given a space Z assume that $A_n \subset Z$ and $A_n \subset A_{n+1}$ for every number $n \in \omega$. Let $\pi_n : C_p(A_{n+1}|Z) \to C_p(A_n|Z)$ be the restriction map for all $n \in \omega$. Suppose also that Y is a space and there is a continuous map $f_n : Y \to C_p(A_n|Z)$ such that $\pi_n \circ f_{n+1} = f_n$ for all $n \in \omega$. Let $A = \bigcup_{n \in \omega} A_n$; for any $y \in Y$ we have $f_n(y) \subset f_{n+1}(y)$ for all $n \in \omega$, so the function $f(y) = \bigcup_{n \in \omega} f_n(y)$ is well-defined. Then the map $f : Y \to \mathbb{R}^A$ is continuous; in particular, if $f(y) \in C_p(A|Z)$ for all $y \in Y$, then the map $f : Y \to C_p(A|Z)$ is continuous.

Proof. This is an immediate consequence of Fact 2 if we note that $f_n : Y \to \mathbb{R}^{A_n}$ for any $n \in \omega$ and every π_n is the restriction to $C_p(A_{n+1}|Z)$ of the natural projection of $\mathbb{R}^{A_{n+1}}$ onto its face $\mathbb{R}^{A_n}$. Fact 3 is proved.   □

Fact 4. Let E be a non-empty subset of $\mathbb{R}^\kappa$. For each $f \in E$, take a countable set $B_f \subset \kappa$ and let $G_f = G(B_f, f|B_f)$ Then there is a countable $E' \subset E$ such that $\bigcup\{G_f : f \in E'\}$ is dense in $G = \bigcup\{G_f : f \in E\}$.

Proof. For any $D \subset E$ let $G\langle D \rangle = \bigcup\{G_f : f \in D\}$. Take an arbitrary $f_0 \in E$ and let $T_0 = \{f_0\}$, $B_0 = B_{f_0}$. Assume that we have countable sets $T_0 \subset \cdots \subset T_n \subset E$ and $B_0 \subset \cdots \subset B_n \subset \kappa$ such that

$$(*) \quad p_{B_i}(G\langle T_{i+1}\rangle) \text{ is dense in } p_{B_i}(G) \text{ for any } i < n.$$

The space $p_{B_n}(G)$ is second countable, so there is a countable set $T_{n+1} \subset E$ such that $T_n \subset T_{n+1}$ and the set $p_{B_n}(G\langle T_{n+1}\rangle)$ is dense in the space $p_{B_n}(G)$. If we let $B_{n+1} = B_n \cup \{B_f : f \in T_{n+1}\}$, then $(*)$ will hold for all $i \leq n$, so our inductive procedure gives us sequences $\{T_i : i \in \omega\}$ and $\{B_i : i \in \omega\}$ such that $T_i \subset E$ and $B_i \subset \kappa$ are countable and $(*)$ holds for all $i \in \omega$. We claim that $E' = \bigcup_{n\in\omega} T_n$ is as promised.

Indeed, take any $g \in G$, and $O \in \tau(g, \mathbb{R}^\kappa)$; there is a finite set $L \subset \kappa$ and $O_\alpha \in \tau(g(\alpha), \mathbb{R})$ for every $\alpha \in L$ such that $O' = \{f \in \mathbb{R}^\kappa : f(\alpha) \in O_\alpha$ for all $\alpha \in L\} \subset O$. If $B = \bigcup_{n\in\omega} B_n$, then there is $m \in \omega$ such that $L' = B \cap L = B_m \cap L$. Since $p_{B_m}(G\langle T_{m+1}\rangle)$ is dense in $p_{B_m}(G)$ by $(*)$, there is $f \in T_{m+1}$ for which there is $v \in G_f$ with $v(\alpha) \in O_\alpha$ for all $\alpha \in L'$. Observe that if $L_0 = (L\backslash L')$, then $L_0 \cap B = \emptyset$ and define $\hat{v} \in \mathbb{R}^\kappa$ by $\hat{v}(\alpha) = v(\alpha)$ for all $\alpha \in \kappa\backslash L_0$ and $\hat{v}(\alpha) = g(\alpha)$ for every $\alpha \in L_0$. Then $p_B(\hat{v}) = p_B(v)$ which implies $p_{B_f}(\hat{v}) = p_{B_f}(v)$ and therefore $\hat{v} \in G_f$. Since $\hat{v} \in G_f \cap O' \subset G_f \cap O$, we proved that every neighborhood of g intersects the set $\bigcup\{G_f : f \in E'\}$, so Fact 4 is proved.   □

Fact 5. If G is a non-empty G_δ-subset of $\mathbb{R}^\kappa$ and F is a retract of G, then both G and F are factorizable in $\mathbb{R}^\kappa$.

Proof. Fix a retraction $s : G \to F$; it follows from Fact 1 of S.426 that for any $f \in G$ there is a countable set $B_f \subset \kappa$ such that $f \in G_f = G(B_f, f|B_f) \subset G$. Fact 4 guarantees that there is a countable $G' \subset G$ such that $Q = \bigcup\{G_f : f \in G'\}$ is dense in G.

Let $N = \bigcup\{B_f : f \in G'\}$; then N is a countable subset of κ; besides, we have $Q = p_N^{-1}(p_N(Q))$. Since Q is dense in G, the set $p_N(Q)$ is dense in $p_N(G)$, so if $R = \overline{p_N(Q)}$, then $G \subset p_N^{-1}(R)$. The map p_N is open, so the set G is dense in $p_N^{-1}(R) = R \times \mathbb{R}^{\kappa\backslash N}$. It turns out that G is a dense subspace of a product of second countable spaces, so G is factorizable in the relevant product by Problem 299 of [TFS]. Since all factors of $R \times \mathbb{R}^{\kappa\backslash N}$ except countably many coincide with the factors of $\mathbb{R}^\kappa$, the set G is also factorizable in $\mathbb{R}^\kappa$.

To see that F is factorizable in $\mathbb{R}^\kappa$ take a continuous onto map $\varphi : F \to M$ such that $w(M) = \omega$. Since G is factorizable, there is a countable $S \subset \kappa$ and a continuous map $v : G_S \to M$ for which $v \circ (p_S|G) = \varphi \circ s$. We have $F_S \subset G_S$; if $v_0 = v|F_S$, then $v_0 \circ (p_S|F) = \varphi$, so F is factorizable and Fact 5 is proved.   □

Fact 6. Suppose that G is a non-empty G_δ-subset of $\mathbb{R}^\kappa$. Assume that a set $F \subset C$ is factorizable in C and there is a retraction $\rho : G \to F$. Then, for any countable $A_0 \subset X$ and $B_0 \subset \kappa$, there exist countable sets $A \subset X$ and $B \subset \kappa$ such that

$A_0 \subset A$, $B_0 \subset B$ and there exists a homeomorphism $s : F_B \to F[A]$ and a retraction $\xi : G_B \to F_B$ such that $s \circ (p_B|F) = \pi_A|F$ and $(p_B|F) \circ \rho = \xi \circ (p_B|G)$, i.e., the diagram

$(\ast\ast)$

$$
\begin{array}{ccc}
G & \xrightarrow{\rho} & F \\
{\scriptstyle p_B}\downarrow & {\scriptstyle p_B}\downarrow & \searrow {\scriptstyle \pi_A} \\
G_B & \xrightarrow{\xi} & F_B \xrightarrow{s} F[A]
\end{array}
$$

is commutative.

Proof. To save space, we will use the relevant projections without mentioning that, in fact, their restrictions to some sets are considered. Those sets will be always clear from the context (we start using this convention in the diagram $(\ast\ast)$). In the expression "continuous map" we will often omit "continuous", so the word "map" actually says "continuous map".

The set F is factorizable in $\mathbb{R}^\kappa$ by Fact 5, so there exists a countable $B_1 \subset \kappa$ such that $B_0 \subset B_1$ and there is a continuous map $s_1 : F_{B_1} \to F[A_0]$ such that $s_1 \circ p_{B_1} = \pi_{A_0}$. By factorizability of G, there is a countable $D_1 \supset B_1$ and a continuous map $\xi_1 : G_{D_1} \to F_{B_1}$ for which $\xi_1 \circ p_{D_1} = p_{B_1} \circ \rho$. The set F is factorizable in C, so there is a countable $A_1 \supset A_0$ and a map $q_1 : F[A_1] \to F_{B_1}$ for which $q_1 \circ \pi_{A_1} = p_{B_1}$.

To proceed inductively, assume that $m \in \mathbb{N}$ and we have constructed families $\{A_i : i \in \Omega_m\}$, $\{B_i : i \in \Omega_m\}$ and $\{D_i : i \in \Omega_m\}$ of countable sets with the following properties:

(2) $A_0 \subset A_1 \subset \cdots \subset A_m \subset X$;
(3) $B_0 \subset B_1 \subset D_1 \subset B_2 \subset \cdots \subset D_{m-1} \subset B_m \subset D_m \subset \kappa$;
(4) for each $i \in \Omega_m$, there is a map $s_i : F_{B_i} \to F[A_{i-1}]$ such that $s_i \circ p_{B_i} = \pi_{A_{i-1}}$;
(5) for each $i \in \Omega_m$, there is a map $\xi_i : G_{D_i} \to F_{B_i}$ such that $\xi_i \circ p_{D_i} = p_{B_i} \circ \rho$;
(6) for each $i \in \Omega_m$, there is a map $q_i : F[A_i] \to F_{B_i}$ such that $q_i \circ \pi_{A_i} = p_{B_i}$.

The set F is factorizable in $\mathbb{R}^\kappa$ by Fact 5, so there exists a countable $B_{m+1} \subset \kappa$ such that $D_m \subset B_{m+1}$ and there is a map $s_{m+1} : F_{B_{m+1}} \to F[A_m]$ such that $s_{m+1} \circ p_{B_{m+1}} = \pi_{A_m}$. By factorizability of the set G, there is a countable set $D_{m+1} \supset B_{m+1}$ and a map $\xi_{m+1} : G_{D_{m+1}} \to F_{B_{m+1}}$ with $\xi_{m+1} \circ p_{D_{m+1}} = p_{B_{m+1}} \circ \rho$. The set F is factorizable in C, so there is a countable $A_{m+1} \supset A_m$ and a map $q_{m+1} : F[A_{m+1}] \to F_{B_{m+1}}$ for which $q_{m+1} \circ \pi_{A_{m+1}} = p_{B_{m+1}}$. It is evident that the families $\{A_i : i \in \Omega_{m+1}\}$, $\{B_i : i \in \Omega_{m+1}\}$ and $\{D_i : i \in \Omega_{m+1}\}$ have the properties (2)–(6), so our inductive procedure gives us families $\{A_i : i \in \mathbb{N}\}$, $\{B_i : i \in \mathbb{N}\}$ and $\{D_i : i \in \mathbb{N}\}$ for which (2)–(6) hold. We claim that $A = \bigcup_{i \in \omega} A_i$ and $B = \bigcup_{i \in \omega} B_i$ are as promised.

For any $i \in \mathbb{N}$ let $v_i = \xi_i \circ (p_{D_i}^B | G_B)$. Then $v_i : G_B \to F_{B_i}$; if $f \in G_B$, then $f = p_B(g)$ for some $g \in G$ and we have

(7)

$$p_{B_i}^{B_i+1}(v_{i+1}(f)) = p_{B_i}^{B_i+1}(\xi_{i+1}(p_{D_{i+1}}^B(p_B(g))))$$

$$= p_{B_i}^{B_i+1}(\xi_{i+1}(p_{D_{i+1}}(g))) = p_{B_i}^{B_i+1}(p_{B_{i+1}}(\rho(g))) = p_{B_i}(\rho(g))$$

$$= \xi_i(p_{D_i}(g)) = \xi_i(p_{D_i}^B(p_B(g))) = \xi_i(p_{D_i}^B(f)) = v_i(f)$$

which shows that $p_{B_i}^{B_i+1} \circ v_{i+1}(f) = v_i(f)$ for any $i \in \mathbb{N}$, so $\xi(f) = \bigcup\{v_i(f) : i \in \mathbb{N}\}$ is well-defined. Besides, it follows from (7) that $v_i(f) = p_{B_i}(\rho(g))$ for every $i \in \mathbb{N}$ and hence

(8) $\xi(f) = \bigcup\{p_{B_i}(\rho(g)) : i \in \mathbb{N}\} = p_B(\rho(g)) \in F_B$.

The elements $f \in G_B$ and $g \in G$ with $p_B(g) = f$ were chosen arbitrarily, so we proved that $p_{B_i}^{B_i+1} \circ v_{i+1} = v_i$ for all $i \in \mathbb{N}$, and hence $\xi : G_B \to F_B$ is a continuous map (see Fact 2); besides, it follows from (8) that $\xi \circ p_B = p_B \circ \rho$. Now, if $f \in F_B$, then $f = p_B(g)$ for some $g \in F$, so we can apply (8) to see that $\xi(f) = p_B(\rho(g)) = p_B(g) = f$, i.e., $\xi : G_B \to F_B$ is a retraction.

For any $i \in \mathbb{N}$, let $\mu_i = s_i \circ (p_{B_i}^B|F_B)$. Then $\mu_i : F_B \to F[A_{i-1}]$; if $f \in F_B$, then $f = p_B(g)$ for some $g \in F$ and we have

(9)

$$\pi_{A_{i-1}}^{A_i}(\mu_{i+1}(f)) = \pi_{A_{i-1}}^{A_i}(s_{i+1}(p_{B_{i+1}}^B(p_B(g))))$$

$$= \pi_{A_{i-1}}^{A_i}(s_{i+1}(p_{B_{i+1}}(g))) = \pi_{A_{i-1}}^{A_i}(\pi_{A_i}(g)) = \pi_{A_{i-1}}(g)$$

$$= s_i(p_{B_i}(g)) = s_i(p_{B_i}^B(p_B(g))) = s_i(p_{B_i}^B(f)) = \mu_i(f)$$

which shows that $\pi_{A_{i-1}}^{A_i} \circ \mu_{i+1}(f) = \mu_i(f)$ for any $i \in \mathbb{N}$, so $s(f) = \bigcup\{\mu_i(f) : i \in \mathbb{N}\}$ is well-defined. Besides, it follows from (9) that $\mu_i(f) = \pi_{A_{i-1}}(g)$ for every $i \in \mathbb{N}$ and hence

(10) $s(f) = \bigcup\{\pi_{A_{i-1}}(g) : i \in \mathbb{N}\} = \pi_A(g) \in F[A]$.

The elements $f \in F_B$ and $g \in F$ with $p_B(g) = f$ were chosen arbitrarily, so we proved that $\pi_{A_{i-1}}^{A_i} \circ \mu_{i+1} = \mu_i$ for all $i \in \mathbb{N}$ and hence $s : F_B \to F[A]$ is a continuous map (see Fact 3) with $s \circ p_B = \pi_A$. This shows that the diagram $(**)$ is commutative.

To finally prove that s is a homeomorphism, for any $i \in \mathbb{N}$, consider the map $\zeta_i = q_i \circ (\pi_{A_i}^A|F[A])$. Then $\zeta_i : F[A] \to F_{B_i}$; if $f \in F[A]$, then $f = \pi_A(g)$ for some $g \in F$ and we have

(11)

$$p_{B_i}^{B_{i+1}}(\zeta_{i+1}(f)) = \pi_{B_i}^{B_{i+1}}(q_{i+1}(\pi_{A_{i+1}}^A(\pi_A(g))))$$

$$= p_{B_i}^{B_{i+1}}(q_{i+1}(\pi_{A_{i+1}}(g))) = p_{B_i}^{B_{i+1}}(p_{B_{i+1}}(g)) = p_{B_i}(g)$$

$$= q_i(\pi_{A_i}(g)) = q_i(\pi_{A_i}^A(\pi_A(g))) = q_i(\pi_{A_i}^A(f)) = \zeta_i(f)$$

which shows that $p_{B_i}^{B_{i+1}} \circ \zeta_{i+1}(f) = \zeta_i(f)$ for any $i \in \mathbb{N}$, so $\zeta(f) = \bigcup\{\zeta_i(f) : i \in \mathbb{N}\}$ is well-defined. Besides, it follows from (11) that $\zeta_i(f) = p_{B_i}(g)$ for every $i \in \mathbb{N}$ and hence

(12) $\zeta(\pi_A(g)) = \zeta(f) = \bigcup\{p_{B_i}(g) : i \in \mathbb{N}\} = p_B(g) \in F_B.$

The elements $f \in F[A]$ and $g \in F$ with $\pi_A(g) = f$ were chosen arbitrarily, so we proved that $p_{B_i}^{B_{i+1}} \circ \zeta_{i+1} = \zeta_i$ for all $i \in \mathbb{N}$ and hence $\zeta : F[A] \to F_B$ is a continuous map (see Fact 2). Now, if $f \in F[A]$ and $\pi_A(g) = f$, then $s(\zeta(f)) = s(p_B(g)) = \pi_A(g) = f$ (we used (12) and commutativity of the diagram $(**)$); finally, if $f \in F_B$ and $p_B(g) = f$, then $\zeta(s(f)) = \zeta(\pi_A(g)) = p_B(g) = f$ [we used (10) and (12)] which shows that s is a homeomorphism and ζ is the inverse of s. Fact 6 is proved.   □

Returning to our solution let $u(x) = 0$ for all $x \in X$; we have $H = \bigcup_{n \in \omega} H_n$ where H_n is a G_δ-subset of $\mathbb{R}^\kappa$ for all $n \in \omega$. There is $m \in \omega$ for which $u \in H_m$; we claim that there exist countable sets $A \subset X$ and $B \subset \kappa$ such that

(13) there is a function $h \in \mathbb{R}^B$ for which $Q = G(B, h) \subset H_m$ and we have $r(Q) = Q \cap C = \{f \in C : f|A = u|A\}.$

Let $W(L) = \{f \in C : f|L = u|L\}$ for any $L \subset X$. Since $u \in H_m$ and $H_m \cap C$ is a G_δ-subset of C, there is a countable $A_0 \subset X$ such that $W(A_0) \subset H_m$ (see Fact 1 of S.426). Observe that $Z_0 = r^{-1}(W(A_0)) \cap H_m$ is a G_δ-subset of H_m and hence of $\mathbb{R}^\kappa$, so there is $B_0 \subset \kappa$ and $h_0 \in \mathbb{R}^{B_0}$ for which $u \in G(B_0, h_0) \subset Z_0$. Assume that $k \in \omega$ and we have families $\{A_i : i \le k\}$ and $\{B_i : i \le k\}$ of countable sets with the following properties:

(14) $A_0 \subset \cdots \subset A_k \subset X$ and $B_0 \subset \cdots \subset B_k \subset \kappa$;
(15) for each $i \le k$, there is $h_i \in \mathbb{R}^{B_i}$ such that $h_i \subset h_{i+1}$ for all $i < k$ and $u \in G(B_i, h_i) \subset r^{-1}(W(A_i)) \cap H_m$ for all $i \le k$;
(16) $W(A_{i+1}) \subset G(B_i, h_i)$ for all $i < k$.

Since $E = G(B_k, h_k) \cap C$ is a G_δ-subset of the space C with $u \in E$, there is a countable set $A_{k+1} \supset A_k$ such that $W(A_{k+1}) \subset G(B_k, h_k)$. It is easy to see that $r^{-1}(W(A_{k+1})) \cap G(B_k, h_k))$ is a G_δ-subset of $G(B_k, h_k)$ and hence of $\mathbb{R}^\kappa$, so there is a countable set $B_{k+1} \supset B_k$ and $h_{k+1} \in \mathbb{R}^{B_{k+1}}$ such that $h_k \subset h_{k+1}$ and $u \in G(B_{k+1}, h_{k+1}) \subset r^{-1}(W(A_{k+1})) \cap G(B_k, h_k)$. It is immediate that (14)–(16) are still fulfilled for $\{A_i : i \le k + 1\}$ and $\{B_i : i \le k + 1\}$, so we can construct inductively families $\{A_i : i \in \omega\}$ and $\{B_i : i \in \omega\}$ such that (14)–(16) hold for all $i \in \omega$. To see that $A = \bigcup_{i \in \omega} A_i$, $B = \bigcup_{i \in \omega} B_i$ and $h = \bigcup_{i \in \omega} h_i$ are as promised, observe that $Q = G(B, h) \subset H_m$ by (15). It follows from (16) that $W(A) \subset \bigcap_{i \in \omega} G(B_i, h_i) = Q$ and hence $W(A) \subset Q \cap C \subset r(Q)$. On the other hand, $r(Q) \subset r(G(B_i, h_i)) \subset W(A_i)$ for all $i \in \omega$ by (15), so $r(Q) \subset W(A) = \bigcap_{i \in \omega} W(A_i)$. This shows that $r(Q) = Q \cap C = W(A)$ and therefore (13) is proved.

Let $Y = \overline{A}$; then Y is a closed separable subspace of X. We claim that

(17) for any countable non-empty $L \subset X \backslash Y$, we have $\pi_L(W(A)) = \mathbb{R}^L$.

It follows from the property (13) that $r_1 = r|Q$ is a retraction of the space Q onto $W(A)$, so we apply Fact 6 to find countable sets $D \subset \kappa$ and $N \subset X$ such that $L \subset N$, $B \subset D$, and there is a retraction $\rho : p_D(Q) \to p_D(W(A))$ and a homeomorphism $s : p_D(W(A)) \to \pi_N(W(A))$. Now, $p_D(Q) = \{h\} \times \mathbb{R}^{D \backslash B}$ is a Čech-complete space and hence so is $p_D(W(A))$ being a closed subspace of $p_D(Q)$. Thus $\pi_N(W(A))$ is Čech-complete. All functions from $W(A)$ are identically zero on Y, so it is easy to see that $\pi_{N \backslash Y}^N : \pi_N(W(A)) \to \pi_{N \backslash Y}(W(A))$ is a homeomorphism. Therefore $Z = \pi_{N \backslash Y}(W(A))$ is Čech-complete and dense in $C_p(N \backslash Y | X)$ and hence in $\mathbb{R}^{N \backslash Y}$. Since Z is an algebra in $\mathbb{R}^{N \backslash Y}$, if $p \in \mathbb{R}^{N \backslash Y} \backslash Z$, then $p + Z$ is a dense Čech-complete subspace of $\mathbb{R}^{N \backslash Y}$ which does not intersect another dense Čech-complete subspace Z of $\mathbb{R}^{N \backslash Y}$. This contradiction with Problem 264 of [TFS] shows that $\pi_{N \backslash Y}(W(A)) = \mathbb{R}^{N \backslash Y}$; since $L \subset N \backslash Y$, we have $\pi_L(W(A)) = \pi_L^{N \backslash Y}(Z) = \pi_L^{N \backslash Y}(\mathbb{R}^{N \backslash Y}) = \mathbb{R}^L$ and hence (17) is proved.

The space $W(A)$ is realcompact being closed in $\mathbb{R}^\kappa$; since $\pi_{X \backslash Y} | W(A)$ is a homeomorphism and $W = \pi_{X \backslash Y}(W(A))$ is dense in a Moscow space $\mathbb{R}^{X \backslash Y}$, it must be ω-placed in $\mathbb{R}^{X \backslash Y}$ by Problem 425 of [TFS]. Thus, if $g \in \mathbb{R}^{X \backslash Y} \backslash W$, then there is a G_δ-subset R of the space $\mathbb{R}^{X \backslash Y}$ such that $g \in R \subset \mathbb{R}^{X \backslash Y} \backslash W$. This implies that there is a countable $L \subset X \backslash Y$ for which $\pi_L(g) \notin \pi_L(W) = \mathbb{R}^L$ [see (17)] which is a contradiction. As a consequence, $C_p(X \backslash Y) \supset \pi_{X \backslash Y}(W(A)) = \mathbb{R}^{X \backslash Y}$ which shows that $C_p(X \backslash Y) = \mathbb{R}^{X \backslash Y}$, i.e., $X \backslash Y$ is discrete. Another consequence of $\pi_{X \backslash Y}(W(A)) = \mathbb{R}^{X \backslash Y}$ is that the function f which is equal to 1 on $X \backslash Y$ and zero on Y is continuous on X, i.e., $Y \backslash X$ is clopen in X. This shows that

(18) $X = Y \oplus (X \backslash Y)$ and $X \backslash Y$ is discrete.

Consequently, the space $C_p(Y)$ is a retract of $C_p(X)$ and hence of H. This shows that we are in the same situation with Y as we were with X, so we can apply Facts 1–6 as if we were assuming that $Y = X$.

Let $\rho : H \to C_0 = C_p(Y)$ be a retraction. The set $H_n \cap C_0$ is a G_δ-subspace of C_0, so $\tilde{H}_n = \rho^{-1}(H_n \cap C_0) \cap H_n$ is a G_δ-subset of H_n and hence of $\mathbb{R}^\kappa$ for all $n \in \omega$. Then $\tilde{H} = \bigcup_{n \in \omega} \tilde{H}_n$ is $G_{\delta\sigma}$ in $\mathbb{R}^\kappa$ and C_0 is still a retract of $\tilde{H}$. This shows that we can assume, without loss of generality, that $H'_n = \rho(H_n) = H_n \cap C_0$ for every $n \in \omega$.

Since $\bigcup_{n \in \omega} H'_n = C_p(Y)$, there exist $m \in \omega$, $\varepsilon > 0$ and $h_0 \in C_p(Y)$ such that $(H'_m + h_0) \cap C_p(Y, (-\varepsilon, \varepsilon))$ is dense in $C_p(Y, (-\varepsilon, \varepsilon))$ and hence in $C_p(Y, [-\varepsilon, \varepsilon])$ (see Problem 431). It is easy to see that there exists a homeomorphism $\varphi : C_0 \to C_0$ such that $\varphi(C_p(Y, [-\varepsilon, \varepsilon])) = C_p(Y, \mathbb{I})$; since $H'_m + h_0$ is homeomorphic to a retract of H_m, we can conclude that

(19) there is a set $H' \subset C_0$ homeomorphic to a retract of H_m, such that the set $C' = H' \cap C_p(Y, \mathbb{I})$ is dense in $C_1 = C_p(Y, \mathbb{I})$,

so we will consider that $\rho(H_m) = H'$. The space Y being separable, we have the inequality $iw(H') \le iw(C_p(Y)) = \omega$. For any $f \in C'$, there is a countable $B_f \subset \kappa$ such that $G_f = G(B_f, f|B_f) \subset H_m$ (see Fact 1 of S.426)

It follows from Fact 4 that for some countable $C'' \subset C'$ the set $\bigcup\{G_f : f \in C''\}$ is dense in $\bigcup\{G_f : f \in C'\}$. It is clear that $G_f \simeq \mathbb{R}^\kappa$, so G_f is stable for any $f \in C''$ by Problem 268. Therefore $\tilde{Q} = \bigcup\{G_f : f \in C''\}$ is also stable by Problem 124. The set $\tilde{R} = \rho(\tilde{Q})$ is stable; since $C'' \subset \tilde{R}$, the set $\tilde{R} \cap C'$ is dense in C' which together with $iw(\tilde{R}) \le iw(H') = \omega$ implies that $nw(\tilde{R}) = \omega$ and hence C' is separable.

Let $F_0 \subset C'$ be a countable dense subset of C' (it is clear that F_0 is also dense in $C_p(Y, \mathbb{I})$). We have $H_m = \bigcap_{n \in \omega} O_n$ where the set O_n is open in the space $\mathbb{R}^\kappa$ for all $n \in \omega$. For each $f \in F_0$ and $n \in \omega$, there exists a finite set $B(n, f) \subset \kappa$ and $W(n, f) \in \tau(\mathbb{R}^{B(n,f)})$ such that $f \in \tilde{W}(n, f) = p_{B(n,f)}^{-1}(W(n, f)) \subset O_n$. If $N_0 = \bigcup\{B(n, f) : f \in F_0, n \in \omega\}$, then N_0 is a countable subset of κ. Let $W_n = \bigcup\{\tilde{W}(n, f) : f \in F_0\}$ for every $n \in \omega$; it is evident that $p_{N_0}^{-1}(p_{N_0}(W_n)) = W_n$ for all $n \in \omega$. Thus $K = \bigcap\{p_{N_0}(W_n) : n \in \omega\}$ is a Čech-complete subspace of $\mathbb{R}^{N_0}$ such that $K' = K \times \mathbb{R}^{\kappa \setminus N_0} \subset H_m$ and $F' = K' \cap C'$ is dense in C' because $F_0 \subset K' \cap C'$. Consequently, F' is dense in $C_1 = C_p(Y, \mathbb{I})$ because C' is dense in C_1. Furthermore, K' is stable being a product of Lindelöf Σ-spaces (see Problem 268); therefore $nw(\rho(K')) \le iw(\rho(K')) \le iw(C_0) = \omega$ which shows that $nw(\rho(K')) = \omega$. The set $F = C_p(Y, \mathbb{I}) \cap \rho(K')$ is closed in $\rho(K')$, so it is a G_δ-set in $\rho(K')$ (it is an easy exercise that every closed subspace of a space with a countable network is G_δ), so $G = \rho^{-1}(F) \cap K'$ is a G_δ-set in K' and hence in $\mathbb{R}^\kappa$. We have $\rho(G) = F = G \cap C_1$; besides, $F \supset K' \cap C_1 \supset K' \cap C' \supset F_0$ and hence F is dense in C_1. This proves that

(20) there exists a G_δ-set G in the space $\mathbb{R}^\kappa$ for which we have $G \subset H_m$ and $F = \rho(G) = G \cap C_1$ while $F_0 \subset F$ and hence F is dense in C_1.

We still have to find a nice subset of the set G in the same way we found the set $K' \subset H_m$. We have $G = \bigcap_{n \in \omega} E_n$ where the set E_n is open in the space $\mathbb{R}^\kappa$ for all $n \in \omega$. For each $f \in F_0$ and $n \in \omega$ there exists a finite set $D(n, f) \subset \kappa$ and $Q(n, f) \in \tau(\mathbb{R}^{D(n,f)})$ such that $f \in \tilde{Q}(n, f) = p_{D(n,f)}^{-1}(Q(n, f)) \subset E_n$. Then $N = \bigcup\{D(n, f) : f \in F_0, n \in \omega\}$ is a countable subset of κ. For every $n \in \mathbb{N}$ let $Q_n = \bigcup\{\tilde{Q}(n, f) : f \in F_0\}$; it is evident that $p_N^{-1}(p_N(Q_n)) = Q_n$. Thus $R = \bigcap\{p_N(Q_n) : n \in \omega\}$ is a Čech-complete subspace of the space $\mathbb{R}^N$ such that $K_1 = R \times \mathbb{R}^{\kappa \setminus N} \subset G$ and $K_1 \cap C_1$ is dense in C_1 because $F_0 \subset K_1 \cap C_1$.

We are going to prove next that

(21) every countable $L \subset Y$ is discrete.

To do it observe that F is factorizable in $\mathbb{R}^\kappa$ by Fact 5; since F is dense in $C_p(Y, \mathbb{I})$, it is factorizable in C_0 being dense in $\mathbb{I}^Y$ (see Problem 299 of [TFS]).

Therefore we can apply Fact 6 to the sets L, N to obtain countable sets $L_1 \subset Y$ and $N_1 \subset \kappa$ such that $L \subset L_1$, $N \subset N_1$ and there exists a retraction $\rho_1 : G_{N_1} \to F_{N_1}$ and a homeomorphism $\xi : F_{N_1} \to F[L_1]$.

The space $K_2 = p_{N_1}(K_1) = R \times \mathbb{R}^{N_1 \setminus N}$ is Čech-complete; since $K_1 \cap F$ is dense in F, the set $K_2 \cap F_{N_1} \supset p_{N_1}(K_1 \cap F)$ is dense in F_{N_1}. The set F_{N_1} is closed in $p_{N_1}(G)$, so $K_2 \cap F_{N_1}$ is closed in the Čech-complete space K_2. Therefore $K_2 \cap F_{N_1}$ is a dense Čech-complete subspace of F_{N_1}. The spaces F_{N_1} and $F[L_1]$ are homeomorphic, so $F[L_1]$ has a dense Čech-complete subspace; now, $F[L_1]$ is dense in $\pi_{L_1}(C_p(Y, \mathbb{I}))$, so $\pi_{L_1}(C_p(Y, \mathbb{I}))$ also has a dense Čech-complete subspace and therefore we can apply Fact 1 to conclude that L_1 is discrete. Thus $L \subset L_1$ is also discrete and (21) is settled.

Finally observe that a separable space is discrete if and only if all its countable subspaces are discrete, so Y is discrete by (21) and hence $X = Y \oplus (X \setminus Y)$ is discrete by (18) which shows that our solution is complete.

Chapter 3
Bonus Results: Some Hidden Statements

The reader has, evidently, noticed that an essential percentage of the problems of the main text is formed by purely topological statements, some of which are quite famous and difficult theorems. A common saying among C_p-theorists is that any result on C_p-theory contains only 20 % of C_p-theory; the rest is general topology.

It is evident that the author could not foresee all topology which would be needed for the development of C_p-theory, so a lot of material had to be dealt with in the form of auxiliary assertions. After accumulating more than seven hundred such assertions, the author decided that some deserve to be formulated together to give a "big picture" of the additional material that can be found in solutions of problems.

This section presents 100 topological statements which were proved in the solutions of problems without being formulated in the main text. In these formulations the main principle is to make them clear for an average topologist. A student could lack the knowledge of some concepts of the formulation so the index of this book can be used to find the definitions of the necessary notions.

After every statement we indicate the exact place (in this book) where it was proved. We did not include any facts from C_p-theory because more general statements are proved sooner or later in the main text.

The author considers that most of the results that follow are very useful and have many applications in topology. Some of them are folkloric statements and quite a few are published theorems, sometimes famous ones. For example, Fact 1 of T.015 is a Hajnal–Juhász theorem [see Theorem 4.9 in Hodel (1984)]; Fact 2 of T.092 is a result proved in Tkachenko (1978). Fact 5 of T.298 is a partial (but still a very nontrivial) case of Shapirovsky's theorem [see Theorem 3.18 in Juhász (1980)]. Besides, the result of Fact 9 of T.298 is published in Juhász and Szentmiklóssy (1992).

To help the reader find a result he/she might need, we have classified the material of this section according to the following topics: *standard spaces, metrizable spaces, compact spaces and their generalizations, properties of continuous maps, completeness and convergence properties, product spaces, cardinal invariants and*

V.V. Tkachuk, *A Cp-Theory Problem Book: Special Features of Function Spaces*,
Problem Books in Mathematics, DOI 10.1007/978-3-319-04747-8_3,
© Springer International Publishing Switzerland 2014

set theory. The last subsection is entitled *Raznoie* (which in Russian means "miscellaneous") and contains unclassified results. The author hopes that once we understand in which subsection a result should be, then it will be easy to find it.

3.1 Standard Spaces

By *standard spaces* we mean the real line, its subspaces and its powers, Tychonoff and Cantor cubes as well as ordinals together with the Alexandroff and Stone–Čech compactifications of discrete spaces.

T.040. Fact 2. *Under the Continuum Hypothesis there exists an HFD space* $Z = \{z_\beta : \beta < \omega_1\} \subset \mathbb{D}^{\omega_1}$.

T.040. Fact 4. *Any HFD space* $Z = \{z_\beta : \beta < \omega_1\} \subset \mathbb{D}^{\omega_1}$ *is hereditarily separable.*

T.042. Fact 1. *No non-empty open subset of* ω^* *is a union of* $\leq \omega_1$-*many of nowhere dense subsets of* ω^*.

T.063. Fact 3. *Let* I *be the space* $[0, 1]$ *with the natural topology. No Luzin space can be mapped continuously onto* I. *No additional axioms are needed for the proof of this fact.*

T.063. Fact 4. *Let* Z *be an arbitrary space. If* Z *is not zero-dimensional, then it maps continuously onto* I.

T.069. Fact 2. *Given any* $f : \omega_1 \to \omega_1$, *the set* $C_f = \{\alpha < \omega_1 : f[\alpha] \subset \alpha\}$ *is a club. Here* $f[\alpha] = \{f(\beta) : \beta < \alpha\}$.

T.097. Fact 1. *Let* A *be a countable subset of* ω^ω. *Then there exists* $f \in \mathbf{P}$ *such that* $g <^* f$ *for any* $g \in A$. *Here* $\mathbf{P} = \{f \in \omega^\omega : i < j$ *implies* $f(i) < f(j)\}$.

T.126. Fact 1. *Let* D *be a discrete space of cardinality* ω_1. *Then there exist countably compact subspaces* X *and* Y *of the space* βD *such that* $D = X \cap Y$ *and hence* $X \times Y$ *contains an uncountable clopen discrete subspace.*

T.131. Fact 2. *Every infinite closed subspace of* $\beta\omega$ *has cardinality* $2^\mathfrak{c}$ *and hence* $\beta\omega$ *has no nontrivial convergent sequences.*

T.203. Fact 2. *If* κ *is an uncountable cardinal and* $Z = A(\kappa)$, *then* $Z^2 \backslash \Delta(Z)$ *is not normal.*

T.322. Fact 4. *In the family* $\{\omega \cup \{\xi\} : \xi \in \beta\omega\backslash\omega\}$, *there exists a subfamily of* $2^\mathfrak{c}$-*many pairwise non-homeomorphic spaces. Here* $\omega \cup \{\xi\}$ *is considered with the subspace topology inherited from* $\beta\omega$ *for all* $\xi \in \beta\omega\backslash\omega$.

T.349. Fact 1. *If* P *is a countable dense subspace of the real line* $\mathbb{R}$, *then there exists a homeomorphism* $f : \mathbb{Q} \to P$ *such that* $x, y \in \mathbb{Q}$ *and* $x < y$ *implies* $f(x) < f(y)$. *In particular, if* P *and* Q *are countable dense subspaces of* $\mathbb{R}$, *then* $P \simeq Q$.

T.371. Fact 2. *Let D be an infinite discrete space. Then*

(a) *for any $A \subset D$ the set $\overline{A}$ is clopen in βD (the bar denotes the closure in the space βD);*
(b) *if $z \in \beta D \backslash D$ and $\mathcal{D}_z = \{U \cap D : U \in \tau(z, \beta Z)\}$, then a set $A \subset D$ belongs to $\mathcal{D}_z$ if and only if $z \in \overline{A}$;*
(c) *the family $\mathcal{D}_z$ is an ultrafilter on D for any $z \in \beta D \backslash D$.*

T.455. Fact 2. *Suppose that a property $\mathcal{P}$ is closed-hereditary and $\mathbb{R}^{\omega_1}$ is a countable union of its subspaces with the property $\mathcal{P}$. Then there exists a sequence $\mathcal{S} = \{Y_n : n \in \omega\}$ of subspaces of $\mathbb{R}^{\omega_1}$ such that $\mathbb{R}^{\omega_1} = \bigcup \mathcal{S}$, the set Y_i covers all countable faces of $\mathbb{R}^{\omega_1}$ and has $\mathcal{P}$ for any $i \in \omega$. The sequence $\mathcal{S}$ will be called $\mathcal{P}$-representation of the space $\mathbb{R}^{\omega_1}$.*

T.455. Fact 3. *There exists a closed discrete $D \subset \mathbb{R}^{\omega_1}$ such that $|D| = \omega_1$, while $\pi_A(D)$ is countable for any countable $A \subset \omega_1$. Here $\pi_B : \mathbb{R}^{\omega_1} \to \mathbb{R}^B$ is the natural projection for any $B \subset \omega_1$.*

3.2 Metrizable Spaces

The results of this section deal with metrics, pseudometrics or metrizable spaces in some way. We almost always assume the Tychonoff separation axiom so our second countable spaces are metrizable and hence present here too.

T.055. Fact 1. *For any second countable space Z there exists a metric d on Z and a base $\mathcal{B} = \{U_n : n \in \omega\}$ in the space Z such that $\tau(d) = \tau(Z)$ and $\text{diam}_d(U_n) \to 0$ when $n \to \infty$.*

T.131. Fact 4. *Let M be a non-compact second countable space. Then there exists a continuous onto map $f : M \to Z$ such that $\pi\chi(Z) > \omega$. Consequently, M is not $\pi\chi(\omega)$-stable.*

T.220. Fact 1. *If M is an uncountable second countable space, then there is an uncountable $M' \subset M$ such that every $U \in \tau^*(M')$ is uncountable.*

T.220. Fact 2. *If $w(M) \leq \omega < |M|$, then the space $Z = L(\omega_1) \times M$ is not weakly Whyburn.*

T.285. Fact 1. *Let Z be a metrizable space. If λ is not a countably cofinal cardinal and $w(Z) \geq \lambda$, then there is a closed discrete $D \subset Z$ such that $|D| = \lambda$.*

T.333. Fact 1. *Suppose that M is a complete metric space. Given any space Z and a continuous map $f : A \to M$ for some dense $A \subset Z$, there is a G_δ-set G of the space Z such that $A \subset G$ and there is a continuous map $g : G \to M$ with $g|A = f$.*

T.333. Fact 2. *Let M and N be complete metric spaces. If $A \subset M$, $B \subset N$ and there is a homeomorphism $f : A \to B$, then there exist $A' \subset M$, $B' \subset N$ such that $A \subset A'$, $B \subset B'$, the set A' is G_δ in M, the set B' is G_δ in N and there is a homeomorphism $h : A' \to B'$ such that $h|A = f$.*

T.348. Fact 1. *Let (M, ρ) be a complete metric space. Suppose that we have a family $\mathcal{U} = \{U_s : s \in \mathbb{D}^{<\omega}\} \subset \tau^*(M)$ with the following properties:*

(a) the family $\mathcal{U}(n) = \{U_s : s \in \mathbb{D}^n\}$ is disjoint for all $n \in \omega$;
(b) if $0 \leq k < n < \omega$, $s \in \mathbb{D}^k$, $t \in \mathbb{D}^n$ and $s \subset t$, then $\overline{U_t} \subset U_s$;
(c) $diam(\mathcal{U}(n)) \to 0$ when $n \to \infty$.

Then the space $L(\mathcal{U}) = \bigcap\{\bigcup \mathcal{U}(n) : n \in \omega\}$ is homeomorphic to $\mathbb{K}$.

T.368. Fact 3. *Given any space Z suppose that (M, d) is a complete metric space and $f : A \to M$ is a continuous map for some dense $A \subset Z$. For a point $z \in Z$ say that $osc(f, z) = 0$ if for any $\varepsilon > 0$ there is $W \in \tau(z, Z)$ such that $diam(f(W \cap A)) < \varepsilon$. If $G = \{z \in Z : osc(f, z) = 0\}$, then $A \subset G$ and there is a continuous map $g : G \to M$ with $g|A = f$.*

3.3 Compact Spaces and Their Generalizations

This section contains some statements on compact, countably compact and pseudo-compact spaces.

T.041. Fact 4. *Let X be a compact space of countable tightness. Then there exists a countable set $A \subset X$ and a non-empty G_δ-set $H \subset X$ such that $H \subset \overline{A}$.*

T.082. Fact 2. *A space Z is called a CS-space if all closed subspaces of Z are separable. If K is a compact CS-space, then K is hereditarily separable.*

T.090. Fact 2. *Any infinite compact space has a non-closed countable subspace.*

T.203. Fact 1. *A countably compact space Z is metrizable if and only if there exists a point-countable T_1-separating family $\mathcal{U} \subset \tau(Z)$.*

T.211. Fact 1. *Let Z be a compact space. If z is a non-isolated point of Z, then there exists a regular cardinal κ and a κ-sequence $S = \{z_\alpha : \alpha < \kappa\} \subset Z \backslash \{z\}$ such that $S \to z$.*

T.235. Fact 2. *Every countably compact space Z with $\Delta(Z) = \omega$ is compact and hence metrizable.*

T.298. Fact 4. *Given an infinite cardinal κ the following conditions are equivalent for any compact space X:*

(a) the space X can be mapped continuously onto $\mathbb{I}^\kappa$;
(b) there exists a continuous onto map $f : F \to \mathbb{D}^\kappa$ for some closed $F \subset X$;
(c) there exists a κ-dyadic family of closed subsets of X.

T.298. Fact 5. *If X is a compact space such that $\pi\chi(x, X) > \omega$ for each $x \in X$, then there is a closed subset $P \subset X$ for which there is a continuous onto map $f : P \to \mathbb{D}^{\omega_1}$.*

T.298. Fact 7. *Suppose that X is a compact space such that there is a continuous onto map $f : X \to \mathbb{D}^{\omega_1}$. Then X contains a convergent ω_1-sequence.*

T.298. Fact 8. *If κ is an infinite cardinal and X is a compact space with $t(X) > \kappa$ then there is a closed $Y \subset X$ which maps continuously onto $\kappa^+ + 1$.*

T.298. Fact 9. *If X is a compact space with $t(X) > \omega$, then X contains a convergent ω_1-sequence.*

T.357. Fact 2. *If K is a compact space, then $w(C_u(K)) = w(K)$. In particular, if K is metrizable, then $C_u(K)$ is a Polish space.*

T.357. Fact 3. *Every locally compact space condenses onto a compact space.*

T.494. Fact 1. *If $h : Y \to Z$ is a closed irreducible map and Z is pseudocompact, then Y is also pseudocompact.*

3.4 Properties of Continuous Maps

We consider the most common classes of continuous maps: open, closed, perfect and quotient. The respective results basically deal with preservation of topological properties by direct and inverse images.

T.132. Fact 7. *If Z is an infinite space, then there is $f \in C(Z)$ such that $f(Z)$ is infinite.*

T.139. Fact 2. *Assume that Z and T are Tychonoff spaces and $f : Z \to T$ is a continuous onto map. Then there exists a Tychonoff space T' such that for some $\mathbb{R}$-quotient continuous onto map $g : Z \to T'$ and a condensation $h : T' \to T$, we have $f = h \circ g$.*

T.201. Fact 1. *A closed continuous image of a normal space is a normal space.*

T.316. Fact 1. *If Y is a sequential space, then a continuous onto map $f : Z \to Y$ is closed if and only if $f(D)$ is not a nontrivial convergent sequence for any closed discrete $D \subset Z$.*

T.354. Fact 2. *Given spaces Z and Y assume that $\mathcal{F} = \{F_t : t \in T\}$ is a closed locally finite cover of Z and we have a family $\{f_t : t \in T\}$ of functions such that $f_t : F_t \to Y$ is continuous for all $t \in T$ and $f_t|(F_t \cap F_s) = f_s|(F_t \cap F_s)$ for any $s, t \in T$. Then the function $f = \bigcup_{t \in T} f_t : Z \to Y$ is also continuous.*

T.363. Fact 3. *Let Y and Z be second countable spaces. If $f : Y \to Z$ is a measurable onto map and Y is analytic, then the space Z is also analytic.*

T.363. Fact 4. *If Z is a cosmic space, then there exist second countable spaces M and N and condensations $f : M \to Z$ and $g : Z \to N$ such that the mappings f^{-1} and g^{-1} are measurable.*

T.245. Fact 1. *Let Z be a paracompact p-space; if $f : Z \to T$ is a perfect map (recall that our definition of a perfect map implies that f is continuous and onto), then T is also a paracompact p-space. In other words, a perfect image of a paracompact p-space is a paracompact p-space.*

T.245. Fact 2. *Let Z be a normal space and assume that F is a non-empty closed subset of Z; for any $A \subset Z$, let $A^* = (A \backslash F) \cup \{F\}$. Given $z \in Z$, let $p_F(z) = z$ if $z \in Z \backslash F$ and $p_F(z) = F$ if $z \in F$. It is clear that $p_F : Z \to Z_F = \{F\} \cup (Z \backslash F)$. Then*

(i)   the family $\tau_F = \{U \in \tau(Z) : U \subset Z \backslash F\} \cup \{U^* : U \in \tau(F, Z)\}$ is a topology on the set Z_F;
(ii) the space $Z / F = (Z_F, \tau_F)$ is T_1 and normal (and hence Tychonoff) and the map $p_F : Z \to Z / F$ is continuous, closed and onto.

The operation of obtaining the space Z / F from a space Z is called collapsing the set F to a point.

T.246. Fact 1. *Let Y be a paracompact space. Suppose that Z is a space in which any point is a limit of a nontrivial convergent sequence. Then any closed map $h : Y \to Z$ is irreducible on some closed subset of Y, i.e., there is a closed $F \subset Y$ such that $h(F) = Z$ and $h_F = h | F$ is irreducible.*

T.266. Fact 1. *Let Z be a space and suppose that $f_a : Z \to Y_a$ is a continuous map for all $a \in A$. For the map $f = \Delta\{f_a : a \in A\} : Z \to Y = \prod\{Y_a : a \in A\}$ let $Z' = f(Z)$. If the map f_b is perfect for some $b \in A$, then $f : Z \to Z'$ is also perfect.*

T.268. Fact 1. *Given spaces Y and Z, if $q : Y \to Z$ is an $\mathbb{R}$-quotient map, then, for any space M, a map $p : Z \to M$ is continuous if and only if $p \circ q$ is continuous.*

T.489. Fact 1. *Given spaces Z and T, if $f : Z \to T$ is a perfect map then $w(T) \le w(Z)$. In other words, perfect maps do not increase weight (recall that all perfect maps in this book are surjective).*

T.492. Fact 1. *Any perfect preimage of a realcompact space is realcompact.*

3.5   Completeness and Convergence Properties

This section deals mainly with Čech-complete spaces. Some results on convergence properties are presented as well.

T.041. Fact 2. *If X is a sequential space and $|A| \le \mathfrak{c}$ for some $A \subset X$, then $|\overline{A}| \le \mathfrak{c}$.*

T.041. Fact 3. *Any sequential space has countable tightness.*

T.045. Fact 1. *If Z is a Fréchet–Urysohn space without isolated points, then there exists a closed separable dense-in-itself subspace $Y \subset Z$.*

T.046. Fact 1. *Let X be a dense-in-itself space with the Baire property such that $c(X) \leq \omega$ and $w(X) \leq \omega_1$. Then, under CH, there is a dense Luzin subspace in the space X.*

T.203. Fact 3. *If Z is a locally compact paracompact space, then there is a disjoint family $\mathcal{U}$ of clopen subsets of Z such that each $U \in \mathcal{U}$ is σ-compact and $\bigcup \mathcal{U} = Z$.*

T.210. Fact 1. *Every Čech-complete space is a k-space.*

T.223. Fact 3. *Every locally compact space is p-space.*

T.272. Fact 1. *For any non-scattered Čech-complete space Z, there exists a dense-in-itself compact $K \subset Z$. In other words, if Z is Čech-complete and every compact subspace of Z is scattered, then Z is itself scattered.*

T.298. Fact 1. *If a space X has a small diagonal, then no convergent ω_1-sequence is embeddable in X.*

T.371. Fact 3. *Let Z be a homogeneous space, i.e., for every $x, y \in Z$, there is a homeomorphism $h : Z \to Z$ such that $h(x) = y$. If Z is of second category in itself, then it is a Baire space.*

T.385. Fact 4. *Given a Polish space M, suppose that $E \subset C_p(M)$ is a countable set and $K = \overline{E}$ is compact (the bar denotes the closure in $\mathbb{R}^M$). If the set K is not contained in $B_1(M)$, then $\beta\omega$ embeds in K.*

T.385. Fact 5. *Suppose that M is a Polish space and $E \subset C_p(M)$ is a countable set such that $K = \overline{E}$ is compact (the bar denotes the closure in $\mathbb{R}^M$) and $K \subset B(M)$. Then $K \subset B_1(M)$ and hence K is Rosenthal compact.*

3.6 Product Spaces

The space $C_p(X)$ being dense in $\mathbb{R}^X$, the results on topological products form a fundamental part of C_p-theory. The main line here is to classify spaces which could be embedded in (or expressed as a continuous image of) a nice subspace of a product.

T.109. Fact 1. *Suppose that X_t is a space such that $nw(X_t) = \omega$ for every $t \in T$ and consider the space $X = \prod\{X_t : t \in T\}$; given any $A \subset T$, the map $p_A : X \to X_A = \prod\{X_t : t \in A\}$ is the natural projection of X onto the face X_A defined by $p_A(x) = x|A$ for every $x \in X$. Suppose that Y is a dense subspace of the space X and $f : Y \to M$ is a continuous map of Y onto a space M such that $w(M) = \kappa \geq \omega$. Then there is a set $S \subset T$ and a continuous map $g : p_S(Y) \to M$ such that $|S| \leq \kappa$ and $f = g \circ (p_S|Y)$.*

T.110. Fact 1. *Let X_t be a space for each $t \in T$; given a point $x \in X = \prod\{X_t : t \in T\}$, let $\sigma(X, x) = \{y \in X : |\{t \in T : x(t) \neq y(t)\}| < \omega\}$. The natural projection $p_S : X \to X_S = \prod_{t \in S} X_t$ is defined by $p_S(x) = x|S$ for any $x \in X$. Suppose that $Y \subset X$ and $\sigma(X, x) \subset Y$ for any $x \in Y$. Then the map $p_S|Y : Y \to p_S(Y)$ is open for any $S \subset T$.*

T.268. Fact 2. *Assume that we have a family $\{X_t : t \in T\}$ of spaces such that the product $X_A = \prod_{t \in A} X_t$ is Lindelöf for every finite $A \subset T$. Then for any $a \in X = \prod_{t \in T} X_t$, the space $\sigma = \sigma(X, a)$ is Lindelöf.*

T.268. Fact 3. *Given a space X_t for every $t \in T$, let $X = \prod\{X_t : t \in T\}$. Suppose that Y is a Lindelöf subspace of the space X and $f : Y \to M$ is a continuous map of Y onto a space M such that $w(M) = \kappa \geq \omega$. Then there is a set $S \subset T$ and a continuous map $g : p_S(Y) \to M$ such that $|S| \leq \kappa$ and $f = g \circ (p_S|Y)$.*

T.268. Fact 4. *Assume that we have a family $\{X_t : t \in T\}$ of spaces such that the product $\prod_{t \in A} X_t$ is Lindelöf for every finite $A \subset T$. Let $X = \prod\{X_t : t \in T\}$ and suppose that $Y \in \{\sigma(X, a), \Sigma(X, a)\}$. Then, for any infinite cardinal κ and any continuous map $f : Y \to M$ of Y to a space M with $w(M) \leq \kappa$, there is a set $S \subset T$ and a continuous map $g : p_S(Y) \to M$ such that $|S| \leq \kappa$ and $f = g \circ (p_S|Y)$.*

T.268. Fact 5. *Assume that we have a family $\{X_t : t \in T\}$ of spaces such that the product $X_A = \prod_{t \in A} X_t$ is Lindelöf for every finite $A \subset T$. Let $X = \prod\{X_t : t \in T\}$ and suppose that $f : X \to M$ is a continuous map of X to a space M such that $w(M) \leq \kappa$. Then there is a set $S \subset T$ and a continuous map $g : X_S \to M$ such that $|S| \leq \kappa$ and $f = g \circ p_S$.*

T.298. Fact 6. *Let M_t be a second countable space for every $t \in T$. If $M = \prod_{t \in T} M_t$, then let $p_S : M \to M_S = \prod_{t \in S} M_t$ be the natural projection of M onto its face M_S (recall that p_S is defined by $p_S(x) = x|S$ for any $x \in M$). Then, for any open $U \subset M$, the set $\overline{U}$ depends on countably many coordinates, i.e., there exists a countable $S \subset T$ such that $p_S^{-1}(p_S(\overline{U})) = \overline{U}$.*

T.415. Fact 1. *Suppose that we have a product $Y = \prod\{Y_t : t \in T\}$ and a set $H \subset Y$ which covers all finite faces of Y. Then the map $p = p_A|H : H \to Y_A$ is open for any finite $A \subset T$.*

T.415. Fact 2. *Assume that $Z^\omega = \bigcup\{Z_n : n \in \omega\}$. Then, for some $n \in \omega$, there is $H \subset Z_n$ homeomorphic to some $G \subset Z^\omega$ which covers all finite faces of Z^ω.*

T.455. Fact 1. *If $w(M_t) \leq \omega$ for any $t \in T$ and $E \subset M = \prod_{t \in T} M_t$ covers all countable faces of M, then E is C-embedded in M.*

3.7 Cardinal Invariants and Set Theory

To classify function spaces using cardinal invariants often gives crucial information. This section includes both basic, simple results on the topic as well as very difficult classical theorems.

T.015. Fact 1. *For any space Z, we have $|Z| \leq 2^{\psi(Z) \cdot s(Z)}$.*

T.036. Fact 1. *Suppose that Y and Z are spaces such that $s(Y \times Z) \leq \kappa$ for some infinite cardinal κ. Then either $hd(Y) \leq \kappa$ or $hl(Z) \leq \kappa$.*

T.039. Fact 1. *Let Z be a space such that $c(Z) \leq \omega$ and $w(Z) \leq \omega_1$. If CH holds, then there exists a family $\mathcal{N}$ of nowhere dense closed subspaces of Z such that $|\mathcal{N}| \leq \omega_1$, and, for any nowhere dense $F \subset Z$, there is $N \in \mathcal{N}$ such that $F \subset N$. We will say that the family $\mathcal{N}$ is cofinal in the family of all nowhere dense subsets of Z.*

T.040. Fact 1. *Assume that $\mathcal{U} = \{U_n : n \in \omega\}$ is a family of infinite sets. Then there exists a disjoint family $\mathcal{V} = \{V_n : n \in \omega\}$ such that V_n is infinite and $V_n \subset U_n$ for all $n \in \omega$. We will say that the family $\mathcal{V}$ is a (disjoint) π-net for $\mathcal{U}$.*

T.050. Fact 1. *Let Z be a space with $c(Z) = \omega$. If MA$+\neg$CH holds, then ω_1 is a precaliber of the space Z.*

T.050. Fact 2. *Given an infinite cardinal κ, suppose that $c(Z_s) \leq \kappa$ for all $s \in S$. Assume additionally that $c(\prod_{s \in A} Z_s) \leq \kappa$ for each finite $A \subset S$. Then $c(\prod_{s \in S} Z_s) \leq \kappa$.*

T.082. Fact 1. *A space Z is called a CS-space if all closed subspaces of Z are separable. If Z is a CS-space, then $s(Z) = \omega$.*

T.092. Fact 2. *Let Z be any space. Assume that $w(Y) \leq \omega$ for every $Y \subset Z$ with $|Y| \leq \omega_1$. Then $w(Z) = \omega$.*

T.102. Fact 1. *Let Z be an infinite space with $w(Z) = \kappa$. Then, for any base $\mathcal{B}$ for the space Z, there is $\mathcal{B}' \subset \mathcal{B}$ such that $|\mathcal{B}'| \leq \kappa$ and $\mathcal{B}'$ is a base in Z. In other words, any base of a space contains a base of minimal cardinality.*

T.187. Fact 1. *If Z is a space and Y is dense in Z, then $\pi w(Y) = \pi w(Z)$.*

T.235. Fact 1. *A space Z has a G_δ-diagonal if and only if there is a sequence $\{\mathcal{D}_n : n \in \omega\}$ of open covers of Z such that $\bigcap\{St(z, \mathcal{D}_n) : n \in \omega\} = \{z\}$ for any $z \in Z$. The sequence $\{\mathcal{D}_n : n \in \omega\}$ is called a G_δ-diagonal sequence for Z.*

T.412. Fact 2. *Suppose that T is a space and $\mathcal{U}$ is an open cover of T. If $|\mathcal{U}| \leq \kappa$ and $w(U) \leq \kappa$ for every $U \in \mathcal{U}$, then $w(T) \leq \kappa$.*

3.8   Raznoie (Unclassified Results)

Last, but not least, we place here some interesting results which do not fit in any of the previous subsections.

T.098. Fact 2. *Let Z be any space (no axioms of separation are assumed). If, for any $z \in Z$, there exists $U \in \tau(z, Z)$ such that $\overline{U}$ is compact and Hausdorff, then Z is a Tychonoff space.*

T.139. Fact 1. *The least upper bound of any family of completely regular (not necessarily Tychonoff) topologies on a set Z is a completely regular topology on Z.*

T.219. Fact 1. *Let Z be any space; if $\mathcal{A}$ is a family of resolvable subspaces of Z, then $\bigcup \mathcal{A}$ is also resolvable.*

T.219. Fact 2. *If Z is an irresolvable space, then there is a non-empty open hereditarily irresolvable $U \subset Z$.*

T.219. Fact 3. *A space Z is ultradisconnected if and only if for any crowded set $A \subset Z$, if $Z \backslash A$ is also crowded, then A is clopen. In particular, every ultradisconnected T_1-space is irresolvable.*

T.219. Fact 5. *The conditions below are equivalent for any crowded T_1-space Z:*

(1) Z is maximal;
(2) Z is perfectly disconnected;
(3) every crowded subspace of Z is open;
(4) Z is ultradisconnected and nodec;
(5) Z is ultradisconnected and every discrete subspace of Z is closed;
(6) Z is submaximal and extremally disconnected;
(7) Z is extremally disconnected, hereditarily irresolvable and nodec.

T.219. Fact 6. *For any Tychonoff crowded space Z, there exists a maximal Tychonoff topology μ on the set Z such that $\tau(Z) \subset \mu$.*

T.219. Fact 7. *A space Z is maximal Tychonoff if and only if Z is Tychonoff and ultradisconnected.*

T.219. Fact 8. *Suppose that Z is a countable Tychonoff hereditarily irresolvable space. Then the set $T = \{z \in Z : z \in \overline{D} \backslash D$ for some discrete subspace $D \subset Z\}$ is nowhere dense in Z.*

T.219. Fact 9. *There exists a countable Tychonoff maximal space.*

T.219. Fact 10. *If Z is a Tychonoff submaximal countable space, then $Z \times A(\omega)$ is not weakly Whyburn.*

T.222. Fact 1. *If Z is a p-space, then it is of pointwise countable type, i.e., for every $z \in Z$, there is a compact $K \subset Z$ such that $z \in K$ and $\chi(K, Z) \leq \omega$.*

T.300. Fact 1. *Suppose that Z is a space in which there exists a countable closed network $\mathcal{F}$ with respect to a compact cover $\mathcal{C}$ of the space Z such that every $C \in \mathcal{C}$ is metrizable. If, additionally, the space Z has a small diagonal, then $nw(Z) = \omega$.*

T.412. Fact 1. *Suppose that T is a space such that $T = T_0 \cup \ldots \cup T_n$ and T_i has a σ-disjoint base for every $i \leq n$. Then $F = \overline{T}_0 \cap \ldots \cap \overline{T}_n$ has a σ-disjoint base.*

Chapter 4
Open Problems

The unsolved problems form an incentive for the development of any area of mathematics. Since this book has an ambitious purpose to embrace all or almost all modern C_p-theory, it was impossible to avoid dealing with open questions.

In this book, we have a wide selection of unsolved problems of C_p-theory. Of course, "unsolved" means "unsolved to the best of the knowledge of the author". I give a classification by topics, but there is no mention whatsoever of whether the given problem is difficult or not. One good parameter is the year of publication, but sometimes the problem is not solved for many years because of lack of interest or effort and not because it is too difficult.

I believe that almost all unsolved problems of importance in C_p-theory are present in this chapter. The reader understands, of course, that there is a big difference between the textbook material of the first four chapters and open questions to which an author must be assigned. I decided that it was my obligation to make this assignment and did my best to be frowned at (or hated!) by the least possible number of potential authors of open problems.

This volume contains 100 unsolved problems which are classified by topics presented in eight sections the names of which outline what the given group of problems is about. At the beginning of each subsection we define the notions *which are not defined in the main text*. Each published problem has a reference to the respective paper or book. If it is unpublished, then my opinion on who is the author is expressed. The last part of each problem is a very brief explanation of its motivation and/or comments referring to the problems of the main text or some papers for additional information. If the paper is published and the background material is presented in the main text, we mention the respective exercises. If the main text contains no background, we refer the reader to the original paper. If no paper is mentioned in the motivation part, then the reader must consult the paper/book in which the unsolved problem was published.

V.V. Tkachuk, *A Cp-Theory Problem Book: Special Features of Function Spaces*, Problem Books in Mathematics, DOI 10.1007/978-3-319-04747-8_4,
© Springer International Publishing Switzerland 2014

To do my best to assign the right author to every problem, I implemented the following simple principles:

1. If the unsolved problem is published, then I cite the publication and consider myself not to be involved in the decision about who is the author. Some problems are published many times and I have generally preferred to cite the articles in journals/books which are more available for the Western reader. Thus it may happen that I do not cite the earliest paper where the problem was formulated. Of course, I mention it explicitly, if the author of the publication attributes the problem to someone else.
2. If, to the best of my knowledge, the problem is unpublished, then I mention the author according to my personal records. The information I have is based upon my personal acquaintance and communication with practically all specialists in C_p-theory. I am aware that it is a weak point and it might happen that the problem I attributed to someone was published (or invented) by another person. However, I did an extensive work ploughing through the literature to make sure that this does not happen.

4.1   Analyticity and Similar Properties

Quite a few methods here come from descriptive set theory and functional analysis. Some problems are likely to be very difficult because they stem from the old question whether every cosmic space embeds in an analytic one.

4.1.1. Suppose that $C_p(X)$ embeds in an analytic space. Must the space X be σ-compact?

    Published in   Arhangel'skii and Calbrix (1999)

    Related to   Problems 2.365–2.368

4.1.2. Let $\mathbb{P}$ be the space of irrationals. Is it true that $C_p(\mathbb{P})$ embeds in an analytic space?

    Published in   Arhangel'skii and Calbrix (1999)

    Related to   Problems 2.365–2.368

4.1.3. Is it true that $C_p(C_p(\mathbb{P}))$ embeds in an analytic space?

    Published in   Arhangel'skii and Calbrix (1999)

    Related to   Problems 2.365–2.368

4.1.4. Given a space X suppose that there exists a K-analytic space Y such that $C_p(X) \subset Y \subset \mathbb{R}^X$. Must X be σ-bounded? Is the same conclusion true if Y is a $K_{\sigma\delta}$-space?

    Published in   Arhangel'skii and Calbrix (1999)

    Related to   Problems 2.365–2.368

4.1.5. Let X be a Lindelöf space such that $C_p(X)$ is K-analytic. Must X be σ-compact?

Published in   Arhangel'skii and Calbrix (1999)

Related to   Problems 2.365–2.368

4.1.6. Let X be a Lindelöf space for which there exists a K-analytic space Y such that $C_p(X) \subset Y \subset \mathbb{R}^X$. Must X be σ-compact?

Published in   Arhangel'skii and Calbrix (1999)

Related to   Problems 2.365–2.368

4.1.7. Suppose that $C_p(X)$ is a $K_{\sigma\delta}$-space. Must X be σ-compact?

Published in   Arhangel'skii (1988a)

Related to   Problems 2.365–2.368

4.2 Whyburn Property in Function Spaces

The nice algebraic structure of $C_p(X)$ implies very strong dependencies between local and convergence properties of $C_p(X)$. The most famous result is the coincidence of the k-property and Fréchet–Urysohn property in the spaces $C_p(X)$. Many problems below reflect the intention to generalize this result.

4.2.1. Is it true that the Whyburn property of $C_p(X)$ implies countable tightness of $C_p(X)$?

Published in   Bella and Yaschenko (1999)

Related to   Problems 2.216–2.219

4.2.2. Suppose that $C_p(C_p(X))$ is a Whyburn space. Must X be finite?

Published in   Tkachuk and Yaschenko (2001)

Motivated by   the fact that this is true when X is cosmic (Problem 2.218)

4.2.3. Is $C_p(\mathbb{R}^\omega)$ a weakly Whyburn space?

Published in   Tkachuk and Yaschenko (2001)

Motivated by   the fact that $C_p(\mathbb{R}^\omega)$ is not a Whyburn space (Problem 2.217)

4.2.4. Call a space Z *discretely generated* if, for any $A \subset Z$ and any $z \in \overline{A}$, there is a discrete $D \subset A$ such that $z \in \overline{D}$. Suppose that $C_p(X)$ is a Whyburn space. Must $C_p(X)$ be discretely generated?

Published in   Tkachuk and Yaschenko (2001)

Motivated by   the fact that this is true if $C_p(X)$ has countable tightness

4.2.5. Suppose that $C_p(X)$ is a Whyburn space for some second countable space X. Is it true that all finite powers of X are Hurewicz spaces?

Published in   Tkachuk and Yaschenko (2001)

Motivated by   the fact that X has to be a Hurewicz space (Problem 2.217)

4.2.6. Suppose that $C_p(X)$ is a Whyburn space. Is it true that $(C_p(X))^\omega$ is also a Whyburn space?

Published in   Tkachuk and Yaschenko (2001)

Motivated by   the fact that the Fréchet–Urysohn property in $C_p(X)$ is countably multiplicative (Problem 1.145)

4.3   Uniformly Dense Subspaces

It is standard in general topology to try to prove some nice properties of a space Z if it has a nice dense subspace. The same approach is valid for the spaces $C_p(X)$. For example, $C_p(X)$ is second countable if it has a dense metrizable subspace. The most interesting line of research in this direction is to find properties $\mathcal{P}$ such that every $C_p(X)$ has a dense subspace with the property $\mathcal{P}$.

4.3.1. Suppose that $C_p(X)$ has a uniformly dense Lindelöf subspace. Is it true in ZFC that $C_p(X)$ is Lindelöf? Recall that $A \subset C_p(X)$ is uniformly dense in $C_p(X)$ if, for any $f \in C_p(X)$, there is a sequence $\{f_n\}_{n \in \omega} \subset A$ which converges uniformly to f.

Published in   Tkachuk (2003)

Motivated by   Problems 2.458 and 2.459 (this is true if the uniformly dense subspace in question is Lindelöf Σ or hereditarily Lindelöf)

4.3.2. Suppose that $C_p(X)$ has a uniformly dense normal subspace. Is it true in ZFC that $C_p(X)$ is normal?

Published in   Tkachuk (2003)

Motivated by   Problems 2.458 and 2.459 (this is true if the uniformly dense subspace in question is Lindelöf Σ or hereditarily Lindelöf)

4.3.3. Suppose that $C_p(X)$ has a uniformly dense perfectly normal subspace. Must $C_p(X)$ be perfectly normal?

Published in   Tkachuk (2003)

Motivated by   Problems 2.458 and 2.459 (this is true if the uniformly dense subspace in question is hereditarily Lindelöf)

4.3.4. Suppose that $C_p(X)$ has a uniformly dense hereditarily normal subspace. Must $C_p(X)$ be hereditarily normal?

Published in   Tkachuk (2003)

Motivated by   Problems 2.458 and 2.459 (this is true if the uniformly dense subspace in question is hereditarily Lindelöf)

4.3.5. Suppose that a space $C_p(X)$ has a uniformly dense subspace Y such that $iw(Y) \leq \omega$. Must $C_p(X)$ have countable i-weight?

Published in   Tkachuk (2003)

Motivated by   Problem 2.457 (this is true if the uniformly dense subspace in question is cosmic)

Comment   It is not sufficient to require that $\psi(Y) \leq \omega$

4.3.6. Suppose that $C_p(X)$ has a uniformly dense realcompact subspace. Must $C_p(X)$ be realcompact?

Published in   Tkachuk (2003)

Motivated by   Problem 2.459 (this is true if the uniformly dense subspace in question is Lindelöf Σ)

4.3.7. Suppose that $C_p(X)$ has a uniformly dense ω-monolithic subspace. Must $C_p(X)$ be ω-monolithic?

Published in   Tkachuk (2003)

Motivated by   Problem 2.457 (this is true if the respective uniformly dense subspace is cosmic)

4.3.8. Suppose that $C_p(X)$ has a uniformly dense set Y such that $\text{ext}(Y) \leq \omega$. Is it true in ZFC that $\text{ext}(C_p(X)) = \omega$?

Published in   Tkachuk (2003)

Motivated by   Problems 2.458 and 2.459 (this is true if the uniformly dense subspace in question is Lindelöf Σ or hereditarily Lindelöf)

4.3.9. Suppose that $C_p(X)$ has a uniformly dense pseudoradial subspace. Is it true that $C_p(X)$ is pseudoradial?

Published in   Tkachuk (2003)

Motivated by   the fact that this is true if the uniformly dense subspace in question is sequential

4.4 Countable Spread and Similar Properties

Hereditary density being finitely multiplicative in spaces $C_p(X)$, a considerable effort has been made to prove the same for spread and hereditary Lindelöf number. However, in spite of a wide variety of positive results, the general question still remains open.

4.4.1. Let $C_p(X)$ be perfectly normal. Is it true in ZFC that $C_p(X) \times C_p(X)$ is perfectly normal?

Published in Tkachuk (1995)

Related to Problems 2.081–2.089

4.4.2. Suppose that every closed subspace of $C_p(X)$ is separable. Is it true that every closed subspace of $C_p(X) \times C_p(X)$ is separable?

Published in Tkachuk (1995)

Related to Problems 2.081–2.089

4.4.3. Suppose that every closed subspace of $C_p(X)$ is separable and X is zero-dimensional. Is it true that every closed subspace of $C_p(X) \times C_p(X)$ is separable?

Published in Tkachuk (1995)

Related to Problems 2.081–2.089

4.4.4. Suppose that every closed subspace of $C_p(X) \times C_p(X)$ is separable. Is it true that every closed subspace of $(C_p(X))^\omega$ is separable?

Published in Tkachuk (1995)

Related to Problems 2.081–2.089

4.4.5. Suppose that $L_p(X)$ is perfectly normal. Is it true that $C_p(C_p(X))$ is perfectly normal?

Published in Tkachuk (1995)

Related to Problems 2.081–2.089

4.4.6. Suppose that every closed subspace of $C_p(X, \mathbb{I})$ is separable. Is it true that every closed subspace of $C_p(X)$ is separable?

Published in Tkachuk (1995)

Related to Problems 2.081–2.089

4.4.7. Suppose that every closed subspace of $C_p(X)$ is separable and Y is a (closed) subspace of X. Is it true in ZFC that every closed subspace of $C_p(Y)$ is separable?

Published in Tkachuk (1995)

Related to Problems 2.081–2.089

4.4.8. Suppose that $C_pC_p(X)$ is perfectly normal. Is it true that every closed subspace of $C_pC_pC_p(X)$ is separable?

Published in   Tkachuk (1995)

Related to   Problems 2.081–2.089

4.4.9. Suppose that every closed subspace of $C_p(X)$ is separable. Is then $L_p(X)$ perfectly normal?

Published in   Tkachuk (1995)

Related to   Problems 2.081–2.089

4.4.10. Suppose that $L_p(X)$ is perfectly normal. Is it true that every closed subspace of $C_p(X)$ is separable?

Published in   Tkachuk (1995)

Related to   Problems 2.081–2.089

4.4.11. Suppose that all closed subspaces of $C_p(X)$ are separable. Is it true in ZFC that $C_p(X)$ is hereditarily separable?

Published in   Tkachuk (1995)

Related to   Problems 2.081–2.089

4.4.12. Suppose that all closed subspaces of X^ω are separable. Is it true in ZFC that all closed subspaces of $C_pC_p(X)$ are separable?

Published in   Tkachuk (1995)

Related to   Problems 2.081–2.089

4.4.13. Is it true that $hl(C_p(X) \times C_p(X)) = hl(C_p(X))$ for any space X?

Published in   Arhangel'skii (1989)

Related to   Problems 2.016–2.036

4.4.14. Is it true that $s(C_p(X) \times C_p(X)) = s(C_p(X))$ for any space X?

Published in   Arhangel'skii (1989)

Related to   Problems 2.016–2.036

4.4.15. Suppose that $C_p(X)$ is hereditarily Lindelöf. Is it true in ZFC that $C_p(X) \times C_p(X)$ is hereditarily Lindelöf?

Published in   Arhangel'skii (1989)

Related to   Problems 2.016–2.036

4.4.16. Let X be a hereditarily Lindelöf space. Is it true that every compact subspace of $C_p(X)$ is (hereditarily) separable? Must every compact subspace of $C_p(X)$ have countable spread?

Published in   Arhangel'skii (1998b)

Motivated by   the fact that $C_p(X)$ is hereditarily separable in case all finite powers of X are hereditarily Lindelöf (Problem 2.026)

4.4.17. Let X be a space of countable spread. Is it true that every compact subspace of $C_p(X)$ is (hereditarily) separable? Must every compact subspace of $C_p(X)$ have countable spread?

Published in   Arhangel'skii (1998b)

Motivated by   the fact that $C_p(X)$ has countable spread in case all finite powers of X have countable spread (Problem 2.025)

4.4.18. Let X be a hereditarily Lindelöf space. Is it true that every separable compact subspace of $C_p(X)$ is hereditarily separable?

Published in   Arhangel'skii (1998b)

Motivated by   the fact that $C_p(X)$ is hereditarily separable in case all finite powers of X are hereditarily Lindelöf (Problem 2.026)

4.4.19. Suppose that X^n is hereditarily Lindelöf for all $n \in \mathbb{N}$. Is it true in ZFC that there is a Lindelöf Z such that $C_p(X) \subset Z \subset \mathbb{R}^X$?

Published in   Yaschenko (1992a) (attributed to Arhangel'skii)

Motivated by   the fact that under MA+¬CH the space $Z = C_p(X)$ is Lindelöf (Problems 2.027 and 2.059)

4.4.20. Let X be a hereditarily Lindelöf space. Is it true that $C_p(X)$ condenses onto a hereditarily separable space?

Published in   Arhangel'skii (1998b)

Motivated by   the fact that $C_p(X)$ is hereditarily separable in case all finite powers of X are hereditarily Lindelöf (Problem 2.026)

4.4.21. If X is a space and $Y \subset X$, let $C_p(Y|X) = \{f|Y : f \in C_p(X)\}$. Is it true that $s(C_p(Y)) = s(C_p(Y|X))$ for any $Y \subset X$?

Published in   Arhangel'skii (1996b)

Motivated by   the fact that it is consistent with ZFC that the spread of $C_p(Y)$ is countable if and only if the spread of $C_p(Y|X)$ is countable

4.4.22. Assume MA+¬CH. Suppose that spread of $C_p(X)$ is countable. Is it true that $C_p(X)^\omega$ is hereditarily Lindelöf and hereditarily separable?

Published in   Arhangel'skii (1996a)

Motivated by   the fact that this is true under the axiom SA (Problem 2.036)

4.4.23. Assume MA+¬CH. Suppose that spread of $C_p(X)$ is countable. Is it true that $t(X) = \omega$?

Published in   Arhangel'skii (1996a)

Motivated by   the fact that a stronger statement is true under SA (Problem 2.036)

4.4.24. Assume MA+¬CH. Suppose that spread of $C_p(X)$ is countable. Is it true that X is separable?

Published in   Arhangel'skii (1996a)

Motivated by   the fact that a stronger statement is true under SA (Problem 2.036)

4.5 Metacompactness and Its Derivatives

The results on metacompactness in $C_p(X)$ are scarce. So far it is not known whether metacompactness of $C_p(X)$ implies its Lindelöf property. Even compactness of X does not help. This subsection contains the most evident questions about metacompactness and similar properties in $C_p(X)$.

4.5.1. Suppose that $C_p(X)$ is metacompact. Must it be Lindelöf? Is it true if X is compact?

Author   V.V. Tkachuk

Motivated by   Problem 1.219 and the fact that any normal metacompact $C_p(X)$ is Lindelöf (Problem 2.438)

4.5.2. Suppose that $C_p(X)$ is metacompact. Must it be realcompact? Is it true if X is compact?

Author   V.V. Tkachuk

Motivated by   the fact that nothing is known about metacompactness in $C_p(X)$

4.5.3. Suppose that $C_p(X)$ is metacompact. Must $C_p(X) \times C_p(X)$ be metacompact? Is this true if X is compact?

Author   V.V. Tkachuk

Motivated by   the analogous problem about Lindelöf $C_p(X)$

4.5.4. Suppose that $C_p(X)$ is Lindelöf. Must $C_p(X) \times C_p(X)$ be metacompact? Is this true if X is compact?

Author   V.V. Tkachuk

Motivated by   the fact that almost nothing is known about $C_p(X) \times C_p(X)$ in case when $C_p(X)$ is Lindelöf

4.5.5. Let X be a compact space such that $C_p(X)$ is metacompact. Is it true that $C_p(Y)$ is metacompact for any closed $Y \subset X$?

Author   V.V. Tkachuk

Motivated by   the fact that nothing is known about metacompactness in $C_p(X)$

4.5.6. Let X be a compact space such that $C_p(X)$ is metacompact. Is it true that $t(X) = \omega$?

Author   V.V. Tkachuk

Motivated by   Problem 1.189

4.5.7. Suppose that $C_{p,n}(X)$ is metacompact for all $n \in \mathbb{N}$. Must $C_p(X)$ be Lindelöf? What happens if X is compact?

Author   V.V. Tkachuk

Motivated by   the fact that nothing is known about metacompactness in $C_p(X)$

4.5.8. Suppose that $C_p(X)$ is hereditarily metacompact. Must $C_p(X)$ be (hereditarily) Lindelöf? What happens if X is compact?

Author   V.V. Tkachuk

Motivated by   the fact that a hereditarily paracompact space $C_p(X)$ is hereditarily Lindelöf (Problems 1.219, 1.292, 2.001)

4.5.9. Suppose that the space $C_p(X)$ is hereditarily metacompact. Must $C_p(X) \times C_p(X)$ be hereditarily metacompact? What happens if X is compact?

Author   V.V. Tkachuk

Motivated by   the analogous problem about hereditarily Lindelöf $C_p(X)$

4.5.10. Suppose that $C_{p,n}(X)$ is hereditarily metacompact for all $n \in \mathbb{N}$. Must $C_p(X)$ be (hereditarily) Lindelöf? What happens if X is compact?

Author   V.V. Tkachuk

Motivated by   the fact that a hereditarily paracompact space $C_p(X)$ is hereditarily Lindelöf (Problems 1.219, 1.292, 2.001)

4.5.11. Is $C_p(\omega_1 + 1)$ (hereditarily) metacompact?

Author   V.V. Tkachuk

Motivated by   the fact that nothing is known about metacompactness in $C_p(X)$

4.5.12. An arbitrary space Z is called σ-*metacompact* if any open cover of Z has an open σ-point-finite refinement. Suppose that $C_p(X)$ is σ-metacompact. Does this imply that $C_p(X)$ is metacompact? Must $C_p(X)$ be Lindelöf? What happens if X is compact?

Author   V.V. Tkachuk

Motivated by   the fact that if $C_p(X)$ is normal and σ-metacompact then it is Lindelöf (see Burke 1984)

4.5.13. Suppose that the space $C_p(X)$ is σ-metacompact. Must $C_p(X) \times C_p(X)$ be σ-metacompact? Is this true if X is compact?

Author V.V. Tkachuk

Motivated by the analogous problem about Lindelöf $C_p(X)$

4.5.14. Suppose that X is a space such that $C_p(X)$ is Lindelöf. Must the space $C_p(X) \times C_p(X)$ be σ-metacompact? Is this true if X is compact?

Author V.V. Tkachuk

Motivated by the fact that almost nothing is known about $C_p(X) \times C_p(X)$ in case when $C_p(X)$ is Lindelöf

4.5.15. Let X be a Corson (or Eberlein) compact space such that $C_p(X)$ is hereditarily σ-metacompact. Is it true that X is metrizable?

Author V.V. Tkachuk

Motivated by the fact that if $C_p(X)$ is hereditarily Lindelöf then X is metrizable

4.5.16. Suppose that $C_{p,n}(X)$ is σ-metacompact for all $n \in \mathbb{N}$. Must $C_p(X)$ be metacompact (or Lindelöf)? What happens if X is compact?

Author V.V. Tkachuk

Motivated by the fact that nothing is known about σ-metacompactness in $C_p(X)$

4.5.17. Suppose that $C_p(X)$ is hereditarily σ-metacompact. Must $C_p(X)$ be (hereditarily) Lindelöf? What happens if X is compact?

Author V.V. Tkachuk

Motivated by the fact that a hereditarily paracompact space $C_p(X)$ is hereditarily Lindelöf (Problems 1.219, 1.292, 2.001)

4.5.18. Suppose that $C_p(X)$ is hereditarily σ-metacompact. Must the space $C_p(X) \times C_p(X)$ be hereditarily σ-metacompact? What happens if X is compact?

Author V.V. Tkachuk

Motivated by the analogous problem about hereditarily Lindelöf $C_p(X)$

4.5.19. Suppose that $C_{p,n}(X)$ is hereditarily σ-metacompact for all $n \in \mathbb{N}$. Must $C_p(X)$ be (hereditarily) Lindelöf? What happens if X is compact?

Author V.V. Tkachuk

Motivated by the fact that a hereditarily paracompact space $C_p(X)$ is hereditarily Lindelöf (Problems 1.219, 1.292, 2.001)

4.5.20. Is $C_p(\omega_1 + 1)$ (hereditarily) σ-metacompact?

Author V.V. Tkachuk

Motivated by the fact that nothing is known about metacompactness in $C_p(X)$

4.5.21. A space Z is called *submetacompact* if every open cover $\mathcal{U}$ of the space Z has an open refinement $\mathcal{V} = \bigcup_{n \in \omega} \mathcal{V}_n$ such that every $\mathcal{V}_n$ covers Z and, for any $z \in Z$, there exists an $n \in \omega$ for which the set $\{V \in \mathcal{V}_n : z \in V\}$ is finite. Suppose that $C_p(X)$ is submetacompact. Must $C_p(X)$ be σ-metacompact (or metacompact or Lindelöf)? What happens when X is compact?

Author   V.V. Tkachuk

Motivated by   the fact that if $C_p(X)$ is normal and submetacompact then it is Lindelöf (see Burke 1984)

4.5.22. Suppose that $C_p(X)$ is submetacompact. Must $C_p(X) \times C_p(X)$ be submetacompact? Is this true if X is compact?

Author   V.V. Tkachuk

Motivated by   the analogous problem about Lindelöf space $C_p(X)$

4.5.23. Suppose that $C_p(X)$ is Lindelöf. Must the space $C_p(X) \times C_p(X)$ be submetacompact? Is this true if X is compact?

Author   V.V. Tkachuk

Motivated by   the fact that nothing is known on $C_p(X) \times C_p(X)$ in case when $C_p(X)$ is Lindelöf

4.5.24. Suppose that $C_{p,n}(X)$ is submetacompact for all $n \in \mathbb{N}$. Must $C_p(X)$ be σ-metacompact (or Lindelöf)? What happens if X is compact?

Author   V.V. Tkachuk

Motivated by   the fact that nothing is known about submetacompactness in $C_p(X)$

4.5.25. Suppose that $C_p(X)$ is hereditarily submetacompact. Must $C_p(X)$ be (hereditarily) Lindelöf? What happens if X is compact?

Author   V.V. Tkachuk

Motivated by   the fact that a hereditarily paracompact $C_p(X)$ is hereditarily Lindelöf (Problems 1.219, 1.292, 2.001)

4.5.26. Suppose that $C_p(X)$ is hereditarily submetacompact. Must the space $C_p(X) \times C_p(X)$ be hereditarily submetacompact? What happens if X is compact?

Author   V.V. Tkachuk

Motivated by   the analogous problem about hereditarily Lindelöf $C_p(X)$

4.5.27. Suppose that $C_{p,n}(X)$ is hereditarily submetacompact for all $n \in \mathbb{N}$. Must $C_p(X)$ be (hereditarily) metacompact? What happens if X is compact?

Author   V.V. Tkachuk

Motivated by   the fact that a hereditarily paracompact $C_p(X)$ is hereditarily Lindelöf (Problems 1.219, 1.292, 2.001)

4.5.28. Is it true that $C_p(X)$ is submetacompact for any compact X?

Published in Arhangel'skii (1997)

Motivated by the fact that nothing is known about submetacompactness in $C_p(X)$

4.6 Mappings Which Involve C_p-Spaces

The existence of an algebraic structure compatible with the topology of $C_p(X)$ radically improves its topological properties. In particular, if $C_p(X)$ is an open continuous image of some nice space Z, then $C_p(X)$ might have even better properties than Z. For example, if $C_p(X)$ is an open image of a metrizable space, then it is second countable. Also, if a space Z is a continuous image of $C_p(X)$, then we can expect very strong restrictions on Z if, say, Z is compact. A lot of research has been done in this area and this subsection contains a compilation of the respective open questions.

4.6.1. Let K be a linearly ordered compact space. Is it true that every compact continuous image of $C_p(K)$ is metrizable?

Published in Tkachenko and Tkachuk (2005)

4.6.2. Let X be a Lindelöf P-space. Is it true that every compact continuous image of $C_p(X)$ is metrizable?

Published in Tkachenko and Tkachuk (2005)

4.6.3. Let X be a hereditarily Lindelöf space. Is it true in ZFC that every compact continuous image of $C_p(X)$ is metrizable?

Published in Tkachenko and Tkachuk (2005)

4.6.4. Suppose that K is a compact space with $t(K) \leq \omega$ and $\varphi : C_p(X) \to K$ is a continuous map. Is it true that $\varphi(C_p(X))$ is cosmic or even metrizable?

Published in Tkachuk (2009)

4.6.5. Suppose that $C_p(X)$ is Lindelöf, K is a compact space of countable tightness and $\varphi : C_p(X) \to K$ is a continuous map. Is it true that $\varphi(C_p(X))$ is cosmic or even metrizable?

Published in Tkachuk (2009)

4.6.6. Suppose that X is compact, K is a compact space of countable tightness and $\varphi : C_p(X) \to K$ is a continuous map. Is it true that $\varphi(C_p(X))$ is cosmic or even metrizable?

Published in Tkachuk (2009)

4.6.7. Is it true that, for any cardinal κ and any compact space K of countable tightness, if $\varphi : \mathbb{R}^\kappa \to K$ is a continuous map, then $\varphi(\mathbb{R}^\kappa)$ is cosmic or even metrizable?

Published in   Tkachuk (2009)

4.6.8. Suppose that $C_p(X)$ is Lindelöf and there exists a condensation of $C_p(X)$ onto a σ-compact space Y. Must Y be cosmic?

Published in   Tkachuk (2009)

Motivated by   the fact that this is true if $C_p(X)$ is a Lindelöf Σ-space

4.6.9. Suppose that $C_p(X)$ is a Lindelöf Σ-space and $\varphi : C_p(X) \to Y$ is a continuous surjective map of $C_p(X)$ onto a σ-compact space Y. Must Y be cosmic?

Published in   Tkachuk (2009)

Motivated by   the fact that this is true if φ is a condensation

4.6.10. Suppose that K is a perfectly normal compact space. Is it true that every σ-compact continuous image of $C_p(X)$ has a countable network?

Published in   Tkachuk (2009)

4.6.11. Suppose that $C_p(X)$ condenses onto a space of countable π-weight. Must X be separable?

Published in   Tkachuk (2009)

Motivated by   the fact that it is true if $iw(C_p(X)) = \omega$

4.7   Additivity of Topological Properties

The existence of an algebraic structure compatible with the topology of $C_p(X)$ implies that many non-additive topological properties turn out to be additive in $C_p(X)$. For example, if $C_p(X) = \bigcup_{n\in\omega} M_n$ and every M_n is metrizable, then $C_p(X)$ is metrizable, and hence, X is countable. On the other hand, countable π-weight is not countably additive in $C_p(X)$ even when X is compact. Therefore it is an interesting task to find out what topological properties are finitely or countably additive in spaces $C_p(X)$.

4.7.1. Suppose that $C_p(X) = A \cup B$, where A and B are Whyburn spaces. Must $C_p(X)$ be a Whyburn space?

Published in   Tkachuk and Yaschenko (2001)

Motivated by   this is true if A and B are Fréchet–Urysohn spaces (Problem 2.450)

4.7.2. Suppose that $C_p(X) = A \cup B$, where A and B are metacompact. Must $C_p(X)$ be metacompact? What happens if X is compact?

Published in Casarrubias–Segura (1999)

Motivated by the fact that paracompactness is finitely additive in spaces $C_p(X)$ (Problem 2.422)

4.7.3. Suppose that $C_p(X) = A \cup B$, where A and B are normal. Must $C_p(X)$ be normal? What happens if X is compact?

Published in Casarrubias–Segura (1999)

Motivated by the fact that some non-additive properties are finitely additive in spaces $C_p(X)$ (Problems 2.422–2.430)

4.7.4. Suppose that $C_p(X) = A \cup B$, where A and B are perfectly normal. Must $C_p(X)$ be perfectly normal? What happens if X is compact?

Published in Casarrubias–Segura (1999)

Motivated by the fact that some non-additive properties are finitely additive in spaces $C_p(X)$ (Problems 2.422–2.430)

4.7.5. Suppose that $C_p(X) = A \cup B$, where A and B are hereditarily normal. Must $C_p(X)$ be perfectly normal? What happens if X is compact?

Published in Casarrubias–Segura (1999)

Motivated by the fact that some non-additive properties are finitely additive in spaces $C_p(X)$ (Problems 1.292 and 2.422–2.430)

4.7.6. Suppose that $C_p(X) = A \cup B$, where A and B are collectionwise normal. Must $C_p(X)$ be normal? What happens if X is compact?

Published in Casarrubias–Segura (1999)

Motivated by the fact that some non-additive properties are finitely additive in spaces $C_p(X)$ (Problems 2.422–2.430)

4.7.7. Suppose that $C_p(X) = \bigcup_{n \in \omega} Y_n$ and every Y_n is realcompact. Must $C_p(X)$ be realcompact?

Published in Casarrubias–Segura (1999)

Motivated by the fact that realcompactness is finitely additive in in spaces $C_p(X)$ (Problem 2.423)

4.7.8. Suppose that $C_p(X) = \bigcup_{n \in \omega} Y_n$ and every Y_n is paracompact. Must $C_p(X)$ be Lindelöf?

Published in Casarrubias–Segura (1999)

Motivated by the fact that paracompactness is finitely additive in in spaces $C_p(X)$ (Problem 2.422)

4.7.9. Suppose that $C_p(X) = \bigcup_{n \in \omega} Y_n$ and every Y_n is metacompact. Must $C_p(X)$ be metacompact?

Published in   Casarrubias–Segura (1999)

Motivated by   the fact that paracompactness is finitely additive in in spaces $C_p(X)$
(Problem 2.422)

4.7.10. Suppose that $C_p(X) = \bigcup_{n \in \omega} Y_n$ and every Y_n is ω-monolithic. Must $C_p(X)$ be ω-monolithic?

Published in   Casarrubias–Segura (1999)

Motivated by   the fact that ω-monolithity is finitely additive in in spaces $C_p(X)$
(Problem 2.426)

4.8   Raznoie (Unclassified Questions)

It is usually impossible to completely classify a complex data set such as the open problems in C_p-theory. This last group of problems contains the open questions which do not fit into any of the twenty-four previous subsections.

4.8.1. Is it consistent with ZFC that $w(X) = s(C_p(X))$ for any compact space X?

Published in   Arhangel'skii (1989)

Motivated by   the fact that under MA$+\neg$CH if the spread of $C_p(X)$ is countable then X is metrizable (Problems 2.016 and 2.062)

4.8.2. Let X be a Lindelöf ω-stable space. Is then X^n Lindelöf for all $n \in \mathbb{N}$? Must $X \times X$ be Lindelöf?

Published in   Arhangel'skii (1998b)

Motivated by   the fact that if X is Lindelöf Σ-space then X^ω is Lindelöf

4.8.3. Let X be a Lindelöf ω-stable space. Suppose that F is a compact subspace of $C_p(X)$. Must the tightness of F be countable?

Published in   Arhangel'skii (1998b)

Motivated by   the fact that this is true if X is a Lindelöf Σ-space

4.8.4. Is it true in ZFC that there exists a compact separable space X such that some $Y \subset C_p(X)$ is Lindelöf and has uncountable network weight?

Published in   Arhangel'skii (1992b)

Comment   Such a space exists consistently (Problems 2.098–2.099)

4.8.5. Suppose that X is a zero-dimensional space such that $C_p(X, \mathbb{D})$ is a k-space (sequential space). Is it true that $C_p(X, \mathbb{D})$ is Fréchet–Urysohn?

Author   V.V. Tkachuk

Related to   Problem 2.465

Bibliography

The bibliography of this book is intended to reflect the state of the art of modern C_p-theory; besides, it is obligatory to mention the work of all authors whose results, in one form or another, are cited here. The bibliographic selection for this volume has 300 items to solve the proportional part of the task.

ALAS, O.T., COMFORT, W.W., GARCIA-FERREIRA, S., HENRIKSEN, M., WILSON, R.G., WOODS, R.
   [2000]   *When is* $|C(X \times Y)| = |C(X)||C(Y)|$? Houston J. Math. **26**:1(2000), 83–115.

ALAS, O.T., GARCIA-FERREIRA, S., TOMITA, A.H.
   [1999]   *The extraresolvability of some function spaces,* Glas. Mat. Ser. III **34(54)**:1(1999), 23–35.

ALAS, O.T., TAMARIZ-MASCARÚA, A.
   [2006]   *On the Čech number of* $C_p(X)$ *II,* Questions Answers Gen. Topology **24**:1(2006), 31–49.

ARGYROS, S., MERCOURAKIS, S., NEGREPONTIS, S.
   [1983]   *Analytic properties of Corson compact spaces,* General Topology and Its Relations to Modern Analysis and Algebra, 5. Berlin, 1983, 12–24.

ARGYROS, S., NEGREPONTIS, S.
   [1983]   *On weakly K-countably determined spaces of continuous functions,* Proc. Amer. Math. Soc., **87**:4(1983), 731–736.

ARHANGEL'SKII, A.V.
   [1976]   *On some topological spaces occurring in functional analysis (in Russian),* Uspehi Mat. Nauk, **31**:5(1976), 17–32.
   [1978]   *The structure and classification of topological spaces and cardinal invariants (in Russian),* Uspehi Mat. Nauk, **33**:6(1978), 29–84.
   [1981]   *Classes of topological groups (in Russian),* Uspehi Mat. Nauk, **36**:3(1981), 127–146.
   [1984]   *Continuous mappings, factorization theorems and function spaces (in Russian),* Trudy Mosk. Mat. Obsch., **47**(1984), 3–21.
   [1986]   *Hurewicz spaces, analytic sets and fan tightness of function spaces (in Russian),* Doklady AN SSSR, **287**:3(1986), 525–528.
   [1987]   *A survey of* C_p*-theory,* Questions and Answers in General Topology, **5**(1987), 1–109.

V.V. Tkachuk, *A Cp-Theory Problem Book: Special Features of Function Spaces,*
Problem Books in Mathematics, DOI 10.1007/978-3-319-04747-8,
© Springer International Publishing Switzerland 2014

[1989]   *Topological Function Spaces (in Russian),* Moscow University P.H., Moscow, 1989.

[1989b]   *Hereditarily Lindelof spaces of continuous functions,* Moscow University Math. Bull., **44:3**(1989b), 67–69.

[1990a]   *Problems in C_p-theory,* in: Open Problems in Topology, North Holland, Amsterdam, 1990a, 603–615.

[1990b]   *On the Lindelöf degree of topological spaces of functions, and on embeddings into $C_p(X)$,* Moscow University Math. Bull., **45:5**(1990b), 43–45.

[1992a]   *Topological Function Spaces (translated from Russian),* Kluwer Academic Publishers, Dordrecht, 1992a.

[1992b]   *C_p-theory,* in: Recent Progress in General Topology, North Holland, Amsterdam, 1992, 1–56.

[1996a]   *On Lindelöf property and spread in C_p-theory,* Topol. and Its Appl., **74**: **(1–3)**(1996a), 83–90.

[1996b]   *On spread and condensations,* Proc. Amer. Math. Soc., **124:11**(1996b), 3519–3527.

[1997]   *On a theorem of Grothendieck in C_p-theory,* Topology Appl. **80**(1997), 21–41.

[1998a]   *Some observations on C_p-theory and bibliography,* Topology Appl., **89**(1998a), 203–221.

[1998b]   *Embeddings in C_p-spaces,* Topology Appl., **85**(1998b), 9–33.

[2000a]   *On condensations of C_p-spaces onto compacta,* Proc. Amer. Math. Soc., **128:6**(2000a), 1881–1883.

[2000b]   *Projective σ-compactness, ω_1-caliber and C_p-spaces,* Topology Appl., **104**(2000b), 13–26.

[1999]   Arhangel'skii, A.V., Calbrix, J. *A characterization of σ-compactness of a cosmic space X by means of subspaces of R^X,* Proc. Amer. Math. Soc., **127:8**(1999), 2497–2504.

ARHANGEL'SKII, A.V., CHOBAN M.M.

[1990]   *On the position of a subspace in the whole space,* Comptes Rendus Acad. Bulg. Sci., **43:4**(1990), 13–15.

[1996]   *On continuous mappings of C_p-spaces and extenders,* Proc. Steklov Institute Math., **212**(1996), 28–31.

ARHANGEL'SKII, A.V., OKUNEV, O.G.

[1985]   *Characterization of properties of spaces by properties of their continuous images (in Russian)* Vestnik Mosk. Univ., Math., Mech., **40:5**(1985), 28–30.

ARHANGEL'SKII, A.V., PAVLOV, O.I.

[2002]   *A note on condensations of $C_p(X)$ onto compacta,* Comment. Math. Univ. Carolinae, **43:3**(2002), 485–492.

ARHANGEL'SKII, A.V., PONOMAREV, V.I.

[1974]   *Basics of General Topology in Problems and Exercises (in Russian),* Nauka, Moscow, 1974.

ARHANGEL'SKII, A.V., TKACHUK, V.V.

[1985]   *Function Spaces and Topological Invariants (in Russian),* Moscow University P.H., Moscow, 1985.

[1986]   *Calibers and point-finite cellularity of the spaces $C_p(X)$ and some questions of S. Gul'ko and M. Hušek,* Topology Appl., **23:1**(1986), 65–74.

ARHANGEL'SKII, A.V., USPENSKIJ, V.V.

[1986]   *On the cardinality of Lindelöf subspaces of function spaces,* Comment. Math. Univ. Carolinae, **27:4**(1986), 673–676.

ASANOV, M.O., VELICHKO, M.V.
  [1981]   Compact sets in $C_p(X)$ (in Russian), Comment. Math. Univ. Carolinae,
            22:2(1981), 255–266.

BATUROV D.P.
  [1987]   On subspaces of function spaces (in Russian), Vestnik Moskovsk. Univ., Math.,
            Mech., **42:4**(1987), 66–69.
  [1988]   Normality of dense subsets of function spaces, Vestnik Moskovsk. Univ., Math.,
            Mech., **43:4**(1988), 63–65.
  [1990]   Normality in dense subspaces of products, Topology Appl., **36**(1990), 111–116.
  [1990]   Some properties of the weak topology of Banach spaces, Vestnik Mosk. Univ.,
            Math., Mech., **45:6**(1990), 68–70.

BELLA, A., YASCHENKO, I.V.
  [1999]   On AP and WAP spaces, Comment. Math. Univ. Carolinae, **40:3**(1999), 531–536.

BESSAGA, C., PELCZINSKI, A.
  [1960]   Spaces of continuous functions, 4, Studia Math., **19**(1960), 53–62.
  [1975]   Selected Topics in Infinite-Dimensional Topology, PWN, Warszawa, 1975.

BOUZIAD, A., CALBRIX, J.
  [1995]   Images usco-compactes des espaces Čech-complets de Lindelöf, C. R. Acad. Sci.
            Paris Sér. I Math. **320:7**(1995), 839–842.

BURKE, D.K.
  [1984]   Covering properties, Handbook of Set–Theoretic Topology, edited by K. Kunen
            and J.E. Vaughan, Elsevier Science Publishers B.V., 1984, 347–422.
  [2007]   Weak-bases and D-spaces, Comment. Math. Univ. Carolin. **48:2**(2007),
            281–289.

BURKE, D.K., LUTZER, D.J.
  [1976]   Recent advances in the theory of generalized metric spaces, in: Topology: Proc.
            Memphis State University Conference, Lecture Notes in Pure and Applied Math.,
            Marcel Dekker, New York, 1976, 1–70.

BUZYAKOVA, R.Z.
  [2006a]   Spaces of continuous step functions over LOTS, Fund. Math., **192**(2006a), 25–35.
  [2006b]   Spaces of continuous characteristic functions, Comment. Math. Universitatis
            Carolinae, **47:4**(2006b), 599–608.
  [2007]   Function spaces over GO spaces, Topology Appl., **154:4**(2007), 917–924.

CALBRIX, J.
  [1996]   k-spaces and Borel filters on the set of integers (in French), Trans. Amer. Math.
            Soc., **348**(1996), 2085–2090.

CASARRUBIAS–SEGURA, F.
  [1999]   Realcompactness and monolithity are finitely additive in $C_p(X)$, Topology Proc.,
            24(1999), 89–102.
  [2001]   On compact weaker topologies in function spaces, Topology and Its Applications,
            115(2001), 291–298.

CASCALES, B.
  [1987]   On K-analytic locally convex spaces, Arch. Math., **49**(1987), 232–244.

CASCALES, B., KĄKOL, J., SAXON, S.A.
  [2002]   Weight of precompact subsets and tightness, J. Math. Anal. Appl., **269**(2002),
            500–518.

CASCALES, B., ORIHUELA, J.
[1991]   *Countably determined locally convex spaces,* Portugal. Math. **48:1**(1991), 75–89.

CHOBAN, M.M.
[2005]   *On some problems of descriptive set theory in topological spaces,* Russian Math. Surveys **60:4**(2005), 699–719.

CHRISTENSEN, J.P.R.
[1974]   *Topology and Borel Structure,* North Holland P.C., Amsterdam, 1974.
[1981]   *Joint continuity of separably continuous functions,* Proceedings of Amer. Math. Soc., **82:3**(1981), 455–461.

CIESIELSKI, K.
[1993]   *Linear subspace of* $\mathbb{R}^\lambda$ *without dense totally disconnected subsets,* Fund. Math. **142** (1993), 85–88.

COMFORT, W.W., FENG, L.
[1993]   *The union of resolvable spaces is resolvable,* Math. Japon. **38:3**(1993), 413–414.

COMFORT, W.W., HAGER, A.W.
[1970a]   *Estimates for the number of real-valued continuous functions,* Trans. Amer. Math. Soc., **150**(1970a), 619–631.
[1970b]   *Dense subspaces of some spaces of continuous functions,* Math. Z. **114**(1970b), 373–389.
[1970c]   *Estimates for the number of real-valued continuous functions,* Trans. Amer. Math. Soc. **150**(1970c), 619–631.

COMFORT, W.W., NEGREPONTIS, S.A.
[1974]   *The theory of ultrafilters.* Die Grundl. Math. Wiss., **211**, Springer, New York, 1974.
[1982]   *Chain Conditions in Topology,* Cambridge Tracts in Mathematics, **79**, New York, 1982.

CONTRERAS-CARRETO, A., TAMARIZ-MASCARÚA, A.
[2003]   *On some generalizations of compactness in spaces* $C_p(X, 2)$ *and* $C_p(X, \mathbb{Z})$, Bol. Soc. Mat. Mexicana **(3)9:2**(2003), 291–308.

DEBS, G.
[1985]   *Espaces K-analytiques et espaces de Baire de fonctions continues,* Mathematika, **32**(1985), 218–228.

DIJKSTRA, J., GRILLOT, T., LUTZER, D., VAN MILL, J.
[1985]   *Function spaces of low Borel complexity,* Proc. Amer. Math. Soc., **94:4**(1985), 703–710.

DIJKSTRA, J., MOGILSKI, J.
[1991] $C_p(X)$-*representation of certain Borel absorbers,* Topology Proc., **16**(1991), 29–39.

DI MAIO, G., HOLÁ, L., HOLÝ, D., MCCOY, R.A.
[1998]   *Topologies on the space of continuous functions,* Topology Appl. **86:2**(1998), 105–122.

DIMOV, G., TIRONI, G.
[1987]   *Some remarks on almost radiality in function spaces,* Acta Univ. Carolin. Math. Phys. **28:2**(1987), 49–58.

DOUWEN, E.K. VAN
[1984]   *The integers and topology,* Handbook of Set–Theoretic Topology, K. Kunen and
           J.E. Vaughan, editors, Elsevier Science Publishers B.V., 1984, 111–167.

DOUWEN, E.K. POL, R.
[1977]   *Countable spaces without extension properties,* Bull. Polon. Acad. Sci., Math.,
           25(1977), 987–991.

DOW, A.
[2005a]   *Closures of discrete sets in compact spaces,* Studia Sci. Math. Hungar.
           42:2(2005a), 227–234.
[2005b]   *Property D and pseudonormality in first countable spaces,* Comment. Math.
           Univ. Carolin. **46:2**(2005b), 369–372.

DOW, A., PAVLOV, O.
[2006]   *More about spaces with a small diagonal,* Fund. Math. **191:1**(2006), 67–80.
[2007]   *Perfect preimages and small diagonal,* Topology Proc. **31:1**(2007), 89–95.

EFIMOV, B.A.
[1977]   *Mappings and imbeddings of dyadic spaces, I,* Math. USSR Sbornik, **32:1**(1977),
           45–57.

ENGELKING, R.
[1977]   *General Topology,* PWN, Warszawa, 1977.

FORT, M.K.
[1951]   *A note on pointwise convergence,* Proc. Amer. Math. Soc., **2**(1951), 34–35.

FOX, R.H.
[1945]   *On topologies for function spaces,* Bull. Amer. Math. Soc., **51**(1945), 429–432.

FREMLIN, D.H.
[1977]   *K-analytic spaces with metrizable compacta,* Mathematika, **24**(1977), 257–261.
[1994]   *Sequential convergence in $C_p(X)$,* Comment. Math. Univ. Carolin., **35:2**(1994),
           371–382.

GARTSIDE, P.
[1997]   *Cardinal invariants of monotonically normal spaces,* Topology Appl.
           77:3(1997), 303–314.
[1998]   *Nonstratifiability of topological vector spaces,* Topology Appl. **86:2**(1998),
           133–140.

GARTSIDE, P., FENG, Z.
[2007]   *More stratifiable function spaces,* Topology Appl. **154:12**(2007), 2457–2461.

GERLITS, J.
[1983]   *Some properties of C(X), II,* Topology Appl., **15:3**(1983), 255–262.

GERLITS, J., NAGY, ZS.
[1982]   *Some properties of C(X), I,* Topology Appl., **14:2**(1982), 151–161.

GERLITS, J., JUHÁSZ, I., SZENTMIKLÓSSY, Z.
[2005]   *Two improvements on Tkačenko's addition theorem,* Comment. Math. Univ.
           Carolin. **46:4**(2005), 705–710.

GERLITS, J., NAGY, ZS., SZENTMIKLOSSY, Z.
[1988]   *Some convergence properties in function spaces,* in: General Topology and Its
           Relation to Modern Analysis and Algebra, Heldermann, Berlin, 1988, 211–222.

GILLMAN, L., JERISON, M.
    [1960]   *Rings of Continuous Functions*, D. van Nostrand Company Inc., Princeton, 1960.

GORDIENKO, I.YU.
    [1990]   *Two theorems on relative cardinal invariants in C_p-theory*, Zb. Rad. Filozofskogo
                Fak. Nišu, Ser. Mat., **4**(1990), 5–7.

GRAEV, M.I.
    [1950]   *Theory of topological groups, I (in Russian)*, Uspehi Mat. Nauk, **5:2**(1950), 3–56.

GRUENHAGE, G.
    [1976]   *Infinite games and generalizations of first-countable spaces*, General Topology
                and Appl. **6:3**(1976), 339–352.
    [1997]   *A non-metrizable space whose countable power is σ-metrizable*, Proc. Amer.
                Math. Soc. **125:6**(1997), 1881–1883.
    [1998]   *Dugundji extenders and retracts of generalized ordered spaces*, Fundam. Math.,
                158(1998), 147–164.
    [2002]   *Spaces having a small diagonal*, Topology Appl., **122**(2002), 183–200.
    [2006]   *The story of a topological game*, Rocky Mountain J. Math. **36:6**(2006), 1885–
                1914.

GRUENHAGE, G., MA, D.K.
    [1997]   *Bairness of $C_k(X)$ for locally compact X*, Topology Appl., **80**(1997), 131–139.

GUL'KO, S.P.
    [1979]   *On the structure of spaces of continuous functions and their hereditary paracom-*
                pactness (in Russian), Uspehi Matem. Nauk, **34:6**(1979), 33–40.

GUL'KO, S.P., SOKOLOV, G.A.
    [1998]   *P-points in $\mathbb{N}^*$ and the spaces of continuous functions*, Topology Appl.,
                85(1998), 137–142.
    [2000]   *Compact spaces of separately continuous functions in two variables*, Topology
                and Its Appl. **107:1-2**(2000), 89–96.

HAGER, A.W.
    [1969]   *Approximation of real continuous functions on Lindelöf spaces*, Proc. Amer.
                Math. Soc., **22**(1969), 156–163.

HAGLER, J.
    [1975]   *On the structure of S and C(S) for S dyadic*, Trans. Amer. Math. Soc.,
                214(1975), 415–428.

HAO-XUAN, Z.
    [1982]   *On the small diagonals*, Topology Appl., **13:3**(1982), 283–293.

HAYDON, R.G.
    [1990]   *A counterexample to several questions about scattered compact spaces*, Bull.
                London Math. Soc., **22**(1990), 261–268.

HEATH, R.W., LUTZER, D.J.
    [1974a]   *The Dugundji extension theorem and collectionwise normality*, Bull. Acad. Polon.
                Sci., Ser. Math., **22**(1974a), 827–830.
    [1974b]   *Dugundji extension theorem for linearly ordered spaces*, Pacific J. Math.,
                55(1974b), 419–425.

HEATH, R.W., LUTZER, D.J., ZENOR, P.L.
    [1975]   *On continuous extenders*, Studies in Topology, Academic Press, New York, 1975,
                203–213.

HEWITT, E.
[1948]   *Rings of real-valued continuous functions, I,* Trans. Amer. Math. Soc., **64:1**(1948), 45–99.

HODEL, R.
[1984]   *Cardinal Functions I,* in: Handbook of Set-Theoretic Topology, Ed. by K. Kunen and J.E. Vaughan, Elsevier Science Publishers B.V., 1984, 1–61.

HUŠEK, M.
[1972]   *Realcompactness of function spaces and $\upsilon(P \times Q)$,* General Topology and Appl. **2**(1972), 165–179.
[1977]   *Topological spaces without κ-accessible diagonal,* Comment. Math. Univ. Carolinae, **18:4**(1977), 777–788.
[1979]   *Mappings from products. Topological structures, II,* Math. Centre Tracts, Amsterdam, **115**(1979), 131–145.
[1997a]   *Productivity of some classes of topological linear spaces,* Topology Appl. **80:1-2**(1997), 141–154.
[2005]   *$C_p(X)$ in coreflective classes of locally convex spaces,* Topology and Its Applications, **146/147**(2005), 267–278.

IVANOV, A.V.
[1978]   *On bicompacta all finite powers of which are hereditarily separable,* Soviet Math., Doklady, **19:6**(1978), 1470–1473.

JUHÁSZ, I.
[1971]   *Cardinal functions in topology,* Mathematical Centre Tracts, **34**, Amsterdam, 1971.
[1980]   *Cardinal Functions in Topology—Ten Years Later,* Mathematical Centre Tracts, North Holland P.C., Amsterdam, 1980.
[1992]   *Cardinal functions,* Recent Progress in General Topology, North-Holland, Amsterdam, 1992, 417–441.

JUHÁSZ, I., MILL, J. VAN
[1981]   *Countably compact spaces all countable subsets of which are scattered,* Comment. Math. Univ. Carolin. **22:4**(1981), 851–855.

JUHÁSZ, I., SOUKUP, L., SZENTMIKLÓSSY, Z.
[2007]   *First countable spaces without point-countable π-bases,* Fund. Math. **196:2**(2007), 139–149.

JUHÁSZ, I., SZENTMIKLÓSSY, Z.
[1992]   *Convergent free sequences in compact spaces,* Proc. Amer. Math. Soc., **116:4**(1992), 1153–1160.
[1995]   *Spaces with no smaller normal or compact topologies,* 1993), Bolyai Soc. Math. Stud., **4**(1995), 267–274.
[2002]   *Calibers, free sequences and density,* Topology Appl. **119:3**(2002), 315–324.

JUST, W., SIPACHEVA, O.V., SZEPTYCKI, P.J.
[1996]   *Non-normal spaces $C_p(X)$ with countable extent,* Proceedings Amer. Math. Soc., **124:4**(1996), 1227–1235.

KALAMIDAS, N.D.
[1985]   *Functional properties of $C(X)$ and chain conditions on X,* Bull. Soc. Math. Grèce **26**(1985), 53–64.
[1992]   *Chain condition and continuous mappings on $C_p(X)$,* Rendiconti Sem. Mat. Univ. Padova, **87**(1992), 19–27.

KALAMIDAS, N.D., SPILIOPOULOS, G.D.
[1992]   *Compact sets in $C_p(X)$ and calibers,* Canadian Math. Bull., **35:4**(1992), 497–502.

KUNEN, K.
[1980]   *Set Theory. An Introduction to Independence Proofs,* Studies Logic Found. Mathematics, **102**(1980), North Holland P.C., Amsterdam, 1980
[1981]   *A compact L-space under CH,* Topology Appl., **12**(1981), 283–287.

KUNEN, K., DE LA VEGA, R.
[2004]   *A compact homogeneous S-space,* Topology Appl., **136**(2004), 123–127.

KURATOWSKI, C.
[1966]   *Topology, vol. 1,* Academic Press Inc., London, 1966.

LELEK, A.
[1969]   *Some cover properties of spaces,* Fund. Math., **64:2**(1969), 209–218.

LINDENSTRAUSS, J., TZAFRIRI, L.
[1977]   *Classical Banach Spaces I,* Springer, Berlin, 1977.

LUTZER, D.J., MCCOY, R.A.
[1980]   *Category in function spaces I,* Pacific J. Math., **90:1**(1980), 145–168.

LUTZER, D.J., MILL, J. VAN, POL, R.
[1985]   *Descriptive complexity of function spaces,* Transactions of the Amer. Math. Soc., **291**(1985), 121–128.

LUTZER, D. J., MILL, J. VAN, TKACHUK, V.V.
[2008]   *Amsterdam properties of $C_p(X)$ imply discreteness of X,* Canadian Math. Bull. **51:4**(2008), 570–578.

MALYKHIN, V.I.
[1987]   *Spaces of continuous functions in simplest generic extensions,* Math. Notes, **41**(1987), 301–304.
[1994]   *A non-hereditarily separable space with separable closed subspaces,* Q & A in General Topology, **12**(1994), 209–214.
[1998]   *On subspaces of sequential spaces (in Russian),* Matem. Zametki, **64:3**(1998), 407–413.
[1999]   *$C_p(I)$ is not subsequential,* Comment. Math. Univ. Carolinae, **40:4**(1999), 785–788.

MALYKHIN, V.I., SHAKHMATOV, D.B.
[1992]   *Cartesian products of Fréchet topological groups and function spaces,* Acta Math. Hungarica, **60**(1992), 207–215.

MARCISZEWSKI, W.
[1983]   *A pre-Hilbert space without any continuous map onto its own square,* Bull. Acad. Polon. Sci., **31:(9–12)**(1983), 393–397.
[1988]   *A remark on the space of functions of first Baire class,* Bull. Polish Acad. Sci., Math., **36:(1–2)**(1988), 65–67.
[1993]   *On analytic and coanalytic function spaces $C_p(X)$,* Topology and Its Appl., **50**(1993), 241–248.
[1995a]   *On universal Borel and projective filters,* Bull. Acad. Polon. Sci., Math., **43:1**(1995a), 41–45.
[1995b]   *A countable X having a closed subspace A with $C_p(A)$ not a factor of $C_p(X)$,* Topology Appl., **64**(1995b), 141–147.

[1997a] *A function space $C_p(X)$ not linearly homeomorphic to $C_p(X) \times R$*, Fundamenta Math., **153:2**(1997a), 125–140.

[1997b] *On hereditary Baire products*, Bull. Polish Acad. Sci., Math., **45:3**(1997b), 247–250.

[1998a] *P-filters and hereditary Baire function spaces*, Topology Appl., **89**(1998a), 241–247.

[1998b] *Some recent results on function spaces $C_p(X)$*, Recent Progress in Function Spaces., Quad. Mat. **3**(1998b), Aracne, Rome, 221–239.

[2002] *Function Spaces*, in: Recent Progress in General Topology II, Ed. by M. Hušek and J. van Mill, Elsevier Sci. B.V., Amsterdam, 2002, 345–369.

[2003] *A function space $C_p(X)$ without a condensation onto a σ-compact space*, Proc. Amer. Math. Society, **131:6**(2003), 1965–1969.

MARCISZEWSKI, W., PELANT, J.

[1997] *Absolute Borel sets and function spaces*, Trans. Amer. Math. Soc., **349:9**(1997), 3585–3596.

McCoy, R.A.

[1975] *First category function spaces under the topology of pointwise convergence*, Proc. Amer. Math. Soc., **50**(1975), 431–434.

[1978a] *Characterization of pseudocompactness by the topology of uniform convergence on function spaces*, J. Austral. Math. Soc., **26**(1978a), 251–256.

[1978b] *Submetrizable spaces and almost σ-compact function spaces*, Proc. Amer. Math. Soc., **71**(1978b), 138–142.

[1978c] *Second countable and separable function spaces*, Amer. Math. Monthly, **85:6**(1978c), 487–489.

[1980a] *Countability properties of function spaces*, Rocky Mountain J. Math., **10**(1980a), 717–730.

[1980b] *A K-space function space*, Int. J. Math. Sci., **3**(1980b), 701–711.

[1986] *Fine topology on function spaces*, Internat. J. Math. Math. Sci. **9:3**(1986), 417–424.

McCoy, R.A., NTANTU, I.

[1986] *Completeness properties of function spaces*, Topology Appl., **22:2**(1986), 191–206.

[1988] *Topological Properties of Spaces of Continuous Functions*, Lecture Notes in Math., 1315, Springer, Berlin, 1988.

MICHAEL, E.

[1966] $\aleph_0$-*spaces*, J. Math. and Mech., **15:6**(1966), 983–1002.

[1973] *On k-spaces, k_R-spaces and $k(X)$*, Pacific J. Math., **47:2**(1973), 487–498.

[1977] $\aleph_0$-*spaces and a function space theorem of R. Pol*, Indiana Univ. Math. J., **26**(1977), 299–306.

MILL, J. VAN

[1984] *An introduction to $\beta\omega$*, Handbook of Set-Theoretic Topology, North-Holland, Amsterdam, 1984, 503–567.

[1989] *Infinite-Dimensional Topology. Prerequisites and Introduction*, North Holland, Amsterdam, 1989.

[1999] *$C_p(X)$ is not $G_{\delta\sigma}$: a simple proof*, Bull. Polon. Acad. Sci., Ser. Math., **47**(1999), 319–323.

[2002] *The Infinite-Dimensional Topology of Function Spaces*, North Holland Math. Library **64**, Elsevier, Amsterdam, 2002.

MORISHITA, K.

[2014] *The k_R-property of function spaces*, Preprint.

[1992a]   *The minimal support for a continuous functional on a function space*, Proc. Amer.
           Math. Soc. **114:2**(1992a), 585–587.
[1992b]   *The minimal support for a continuous functional on a function space. II*, Tsukuba
           J. Math. **16:2**(1992b), 495–501.

NAKHMANSON, L. B.
[1982]   *On continuous images of σ-products (in Russian)*, Topology and Set Theory,
           Udmurtia Universty P.H., Izhevsk, 1982, 11–15.
[1984]   *The Souslin number and calibers of the ring of continuous functions (in Russian)*,
           Izv. Vuzov, Matematika, 1984, N 3, 49–55.
[1985]   *On Lindelöf property of function spaces (in Russian)*, Mappings and Extensions
           of Topological Spaces, Udmurtia University P.H., Ustinov, 1985, 8–12.

NOBLE, N.
[1969a]   *Products with closed projections*, Trans Amer. Math. Soc., **140**(1969a), 381–391.
[1969b]   *Ascoli theorems and the exponential map*, Trans Amer. Math. Soc., **143**(1969b),
           393–411.
[1971]   *Products with closed projections II*, Trans Amer. Math. Soc., **160**(1971),
           169–183.
[1974]   *The density character of function spaces*, Proc. Amer. Math. Soc., **42:1**(1974),
           228–233.

NOBLE, N., ULMER, M.
[1972]   *Factoring functions on Cartesian products*, Trans Amer. Math. Soc., **163**(1972),
           329–339.

NYIKOS, P.
[1981]   *Metrizability and the Fréchet–Urysohn property in topological groups*, Proc.
           Amer. Math. Soc., **83:4**(1981), 793–801.

OKUNEV, O.G.
[1984]   *Hewitt extensions and function spaces (in Russian)*, Cardinal Invariants and
           Mappings of Topological Spaces (in Russian), Izhvsk, 1984, 77–78.
[1985]   *Spaces of functions in the topology of pointwise convergence: Hewitt extension
           and τ-continuous functions*, Moscow Univ. Math. Bull., **40:4**(1985), 84–87.
[1993a]   *On Lindelöf Σ-spaces of functions in the pointwise topology*, Topology and Its
           Appl., **49**(1993a), 149–166.
[1993b]   *On analyticity in cosmic spaces*, Comment. Math. Univ. Carolinae, **34**(1993b),
           185–190.
[1995]   *On Lindelöf sets of continuous functions*, Topology Appl., **63**(1995), 91–96.
[1996]   *A remark on the tightness of products*, Comment. Math. Univ. Carolinae,
           37:2(1996), 397–399.
[1997]   *On the Lindelöf property and tightness of products*, Topology Proc., **22**(1997),
           363–371.
[2002]   *Tightness of compact spaces is preserved by the t-equivalence relation*, Com-
           ment. Math. Univ. Carolin. **43:2**(2002), 335–342.
[2005a]   *Fréchet property in compact spaces is not preserved by M-equivalence*, Com-
           ment. Math. Univ. Carolin. **46:4**(2005a), 747–749.
[2005b]   *A σ-compact space without uncountable free sequences can have arbitrary
           tightness*, Questions Answers Gen. Topology **23:2**(2005b), 107–108.

OKUNEV, O.G., SHAKHMATOV, D.B.
[1987]   *The Baire property and linear isomorphisms of continuous function spaces (in
           Russian)*, Topological Structures and Their Maps, Latvian State University P.H.,
           Riga, 1987, 89–92.

OKUNEV, O.G., TAMANO, K.
[1996] *Lindelöf powers and products of function spaces,* Proceedings of Amer. Math. Soc., **124:9**(1996), 2905–2916.

OKUNEV, O.G., TAMARIZ-MASCARÚA, A.
[2004] *On the Čech number of* $C_p(X)$, Topology Appl., **137**(2004), 237–249.

OKUNEV, O.G., TKACHUK, V.V.
[2001] *Lindelöf* Σ-*property in* $C_p(X)$ *and* $p(C_p(X)) = \omega$ *do not imply countable network weight in* X, Acta Mathematica Hungarica, **90:(1–2)**(2001), 119–132.
[2002] *Density properties and points of uncountable order for families of open sets in function spaces,* Topology Appl., **122**(2002), 397–406.

PASYNKOV, B.A.
[1967] *On open mappings,* Soviet Math. Dokl., **8**(1967), 853–856.

PELANT, J.
[1988] *A remark on spaces of bounded continuous functions,* Indag. Math., **91**(1988), 335–338.

POL, R.
[1977] *Concerning function spaces on separable compact spaces,* Bull. Acad. Polon. Sci., Sér. Math., Astron. et Phys., **25:10**(1977), 993–997.
[1978] *The Lindelöf property and its analogue in function spaces with weak topology,* Topology. 4-th Colloq. Budapest, **2**(1978), Amsterdam, 1980, 965–969.
[1989] *Note on pointwise convergence of sequences of analytic sets,* Mathem., **36**(1989), 290–300.

PONTRIAGIN, L.S.
[1984] *Continuous Groups (in Russian),* Nauka, Moscow, 1984.

PYTKEEV, E.G.
[1976] *Upper bounds of topologies,* Math. Notes, **20:4**(1976), 831–837.
[1982a] *On the tightness of spaces of continuous functions (in Russian),* Uspehi Mat. Nauk, **37:1**(1982a), 157–158.
[1982b] *Sequentiality of spaces of continuous functions (in Russian),* Uspehi Matematicheskih Nauk, **37:5**(1982b), 197–198.
[1985] *The Baire property of spaces of continuous functions, (in Russian),* Matem. Zametki, **38:5**(1985), 726–740.
[1990] *A note on Baire isomorphism,* Comment. Math. Univ. Carolin. **31:1**(1990), 109–112.
[1992a] *On Fréchet–Urysohn property of spaces of continuous functions, (in Russian),* Trudy Math. Inst. RAN, **193**(1992a), 156–161.
[1992b] *Spaces of functions of the first Baire class over K-analytic spaces (in Russian),* Mat. Zametki **52:3**(1992b), 108–116.
[2003] *Baire functions and spaces of Baire functions,* Journal Math. Sci. (N.Y.) **136:5**(2006), 4131–4155

PYTKEEV, E.G., YAKOVLEV, N.N.
[1980] *On bicompacta which are unions of spaces defined by means of coverings,* Comment. Math. Univ. Carolinae, **21:2**(1980), 247–261.

RAJAGOPALAN, M., WHEELER, R.F.
[1976] *Sequential compactness of X implies a completeness property for* $C(X)$, Canadian J. Math., **28**(1976), 207–210.

REZNICHENKO, E.A.

[1987]   *Functional and weak functional tightness (in Russian),* Topological Structures and Their Maps, Latvian State University P.H., Riga, 1987, 105–110.

[1989]   *A pseudocompact space in which only the subsets of not full cardinality are not closed and not discrete,* Moscow Univ. Math. Bull. **44:6**(1989), 70–71.

[1990]   *Normality and collectionwise normality in function spaces,* Moscow Univ. Math. Bull., **45:6**(1990), 25–26.

[2008]   *Stratifiability of $C_k(X)$ for a class of separable metrizable X,* Topology Appl. **155:17-18**(2008), 2060–2062.

RUDIN, M.E.

[1956]   *A note on certain function spaces,* Arch. Math., **7**(1956), 469–470.

RUDIN, W.

[1973]   *Functional Analysis,* McGraw-Hill Book Company, New York, 1973.

SAKAI, M.

[1988a]   *On supertightness and function spaces,* Commentationes Math. Univ. Carolinae, **29:2**(1988a), 249–251.

[1988b]   *Property C'' and function spaces,* Proc. Amer. Math. Soc., **104:3**(1988b), 917–919.

[1992]   *On embeddings into $C_p(X)$ where X is Lindelöf,* Comment. Math. Univ. Carolinae, **33:1**(1992), 165–171.

[1995]   *Embeddings of κ-metrizable spaces into function spaces,* Topol. Appl., **65**(1995), 155–165.

[2000]   *Variations on tightness in function spaces,* Topology Appl., **101**(2000), 273–280.

[2003]   *The Pytkeev property and the Reznichenko property in function spaces,* Note di Matem. **22:2**(2003/04), 43–52

[2006]   *Special subsets of reals characterizing local properties of function spaces,* Selection Principles and Covering Properties in Topology, Dept. Math., Seconda Univ. Napoli, Caserta, Quad. Mat., **18**(2006), 195–225,

[2007]   *The sequence selection properties of $C_p(X)$,* Topology Appl., **154**(2007), 552–560.

[2008]   *Function spaces with a countable cs^*-network at a point,* Topology and Its Appl., **156:1**(2008), 117–123.

[2014]   *On calibers of function spaces,* Preprint.

SHAKHMATOV, D.B.

[1986]   *A pseudocompact Tychonoff space all countable subsets of which are closed and C^*-embedded,* Topology Appl., **22:2**(1986), 139–144.

SHAPIROVSKY, B.E.

[1974]   *Canonical sets and character. Density and weight in compact spaces,* Soviet Math. Dokl., **15**(1974), 1282–1287.

[1978]   *Special types of embeddings in Tychonoff cubes, subspaces of Σ-products and cardinal invariants,* Colloquia Mathematica Soc. Janos Bolyai, **23**(1978), 1055–1086.

[1981]   *Cardinal invariants in bicompacta (in Russian),* Seminar on General Topology, Moscow University P.H., Moscow, 1981, 162–187.

SIPACHEVA, O.V.

[1988]   *Lindelöf Σ-spaces of functions and the Souslin number,* Moscow Univ. Math. Bull., **43:3**(1988), 21–24.

[1989]   *On Lindelöf subspaces of function spaces over linearly ordered separable compacta,* General Topology, Spaces and Mappings, Mosk. Univ., Moscow, 1989, 143–148.

[1992]   *On surlindelöf compacta (in Russian),* General Topology. Spaces, Mappings and Functors, Moscow University P.H., Moscow, 1992, 132–140.

SOKOLOV, G.A.
[1986]   *On Lindelöf spaces of continuous functions (in Russian),* Mat. Zametki, **39**:6(1986), 887–894.
[1993]   *Lindelöf property and the iterated continuous function spaces,* Fundam. Math., **143**(1993), 87–95.

STONE, A.H.
[1963]   *A note on paracompactness and normality of mapping spaces,* Proc. Amer. Math. Soc., **14**(1963), 81–83.

TALAGRAND, M.
[1979]   *Sur la K-analyticité des certains espaces d'operateurs,* Israel J. Math., **32**(1979), 124–130.
[1984]   *A new countably determined Banach space,* Israel J. Math., **47**:1(1984), 75–80.
[1985]   *Espaces de Baire et espaces de Namioka,* Math. Ann., **270**(1985), 159–164.

TAMANO, K., TODORCEVIC, S.
[2005]   *Cosmic spaces which are not μ-spaces among function spaces with the topology of pointwise convergence,* Topology Appl. **146/147**(2005), 611–616.

TAMARIZ–MASCARÚA, A.
[1996]   *Countable product of function spaces having p-Frechet-Urysohn like properties,* Tsukuba J. Math. **20**:2(1996), 291–319.
[1998]   *α-pseudocompactness in C_p-spaces,* Topology Proc., **23**(1998), 349–362.
[2006]   *Continuous selections on spaces of continuous functions,* Comment. Math. Univ. Carolin. **47**:4(2006), 641–660.

TANI, T.
[1986]   *On the tightness of $C_p(X)$,* Memoirs Numazu College Technology, **21**(1986), 217–220.

TKACHENKO, M.G.
[1978]   *On the behaviour of cardinal invariants under the union of chains of spaces (in Russian),* Vestnik Mosk. Univ., Math., Mech., **33**:4(1978), 50–58.
[1979]   *On continuous images of dense subspaces of topological products (in Russian),* Uspehi Mat. Nauk, **34**:6(1979), 199–202.
[1982]   *On continuous images of dense subspaces of Σ-products of compacta (in Russian),* Sibirsk. Math. J., **23**:3(1982), 198–207.
[1985]   *On continuous images of spaces of functions,* Sibirsk. Mat. Zhurnal, **26**:5(1985), 159–167.
[2005]   Tkachenko, M.G., Tkachuk, V.V. *Dyadicity index and metrizability of compact continuous images of function spaces,* Topol. Appl., **149**:(1–3)(2005), 243–257.

TKACHUK, V.V.
[1984b]   *Characterization of the Baire property in $C_p(X)$ in terms of the properties of the space X (in Russian),* Cardinal Invariants and Mappings of Topological Spaces (in Russian), Izhevsk, 1984b, 76–77.
[1986a]   *The spaces $C_p(X)$: decomposition into a countable union of bounded subspaces and completeness properties,* Topology Appl., **22**:3(1986a), 241–254.
[1986b]   *Approximation of $\mathbf{R}^X$ with countable subsets of $C_p(X)$ and calibers of the space $C_p(X)$,* Comment. Math. Univ. Carolinae, **27**:2(1986b), 267–276.

[1987a]   *The smallest subring of the ring $C_p(C_p(X))$ which contains $X \cup \{1\}$ is dense in $C_p(C_p(X))$ (in Russian),* Vestnik Mosk. Univ., Math., Mech., **42:1**(1987a), 20–22.

[1987b]   *Spaces that are projective with respect to classes of mappings,* Trans. Moscow Math. Soc., **50**(1987b), 139–156.

[1988]   *Calibers of spaces of functions and metrization problem for compact subsets of $C_p(X)$ (in Russian),* Vestnik Mosk. Univ., Matem., Mech., **43:3**(1988), 21–24.

[1991]   *Methods of the theory of cardinal invariants and the theory of mappings applied to the spaces of functions (in Russian),* Sibirsk. Mat. Zhurnal, **32:1**(1991), 116–130.

[1994]   *Decomposition of $C_p(X)$ into a countable union of subspaces with "good" properties implies "good" properties of $C_p(X)$,* Trans. Moscow Math. Soc., **55**(1994), 239–248.

[1995]   *What if $C_p(X)$ is perfectly normal?* Topology Appl., **65**(1995), 57–67.

[1998]   *Mapping metric spaces and their products onto $C_p(X)$,* New Zealand J. Math., **27:1**(1998), 113–122.

[2000]   *Behaviour of the Lindelöf Σ-property in iterated function spaces,* Topology Appl., **107:3-4**(2000), 297–305.

[2001]   *Lindelöf Σ-property in $C_p(X)$ together with countable spread of X implies X is cosmic,* New Zealand J. Math., **30**(2001), 93–101.

[2003]   *Properties of function spaces reflected by uniformly dense subspaces,* Topology Appl., **132**(2003), 183–193.

[2005a]   *A space $C_p(X)$ is dominated by irrationals if and only if it is K-analytic,* Acta Math. Hungarica, **107:4**(2005a), 253–265.

[2005b]   *A nice class extracted from C_p-theory,* Comment. Math. Univ. Carolinae, **46:3**(2005b), 503–513.

[2007a]   *Condensing function spaces into Σ-products of real lines,* Houston Journal of Math., **33:1**(2007a), 209–228.

[2007b]   *Twenty questions on metacompactness in function spaces,* Open Problems in Topology II, ed. by E. Pearl, Elsevier B.V., Amsterdam, 2007a, 595–598.

[2007c]   *A selection of recent results and problems in C_p-theory,* Topology and Its Appl., **154:12**(2007c), 2465–2493.

[2009]   *Condensations of $C_p(X)$ onto σ-compact spaces,* Appl. Gen. Topology, **10:1**(2009), 39–48.

TKACHUK, V.V., SHAKHMATOV, D.B.
[1986]   *When the space $C_p(X)$ is σ-countably compact? (in Russian),* Vestnik Mosk. Univ., Math., Mech., **41:1**(1986), 70–72.

TKACHUK, V.V., YASCHENKO, I.V.
[2001]   *Almost closed sets and topologies they determine,* Comment. Math. Univ. Carolinae, **42:2**(2001), 395–405.

TODORCEVIC, S.
[1989]   *Partition Problems in Topology,* Contemporary Mathematics, American Mathematical Society, **84**(1989). Providence, Rhode Island, 1989.

[1993]   *Some applications of S and L combinatorics,* Ann. New York Acad. Sci., **705**(1993), 130–167.

[2000]   *Chain-condition methods in topology,* Topology Appl. **101:1**(2000), 45–82.

TYCHONOFF A.N.
[1935]   *Über einer Funktionenraum,* Math. Ann., **111**(1935), 762–766.

USPENSKIJ, V.V.

[1978] *On embeddings in functional spaces (in Russian),* Doklady Acad. Nauk SSSR, **242:3**(1978), 545–546.

[1982a] *On frequency spectrum of functional spaces (in Russian),* Vestnik Mosk. Univ., Math., Mech., **37:1**(1982a), 31–35.

[1982b] *A characterization of compactness in terms of the uniform structure in function space (in Russian),* Uspehi Mat. Nauk, **37:4**(1982b), 183–184.

[1983b] *A characterization of realcompactness in terms of the topology of pointwise convergence on the function space,* Comment. Math. Univ. Carolinae, **24:1**(1983), 121–126.

VALOV, V.M.

[1986] *Some properties of $C_p(X)$,* Comment. Math. Univ. Carolinae, **27:4**(1986), 665–672.

[1997a] *Function spaces,* Topology Appl. **81:1**(1997), 1–22.

[1999] *Spaces of bounded functions,* Houston J. Math. **25:3**(1999), 501–521.

VALOV, V., VUMA, D,

[1996] *Lindelöf degree and function spaces,* Papers in honour of Bernhard Banaschewski, Kluwer Acad. Publ., Dordrecht, 2000, 475–483.

[1998] *Function spaces and Dieudonné completeness,* Quaest. Math. **21:3-4**(1998), 303–309.

DE LA VEGA, R., KUNEN, K.

[2004] *A compact homogeneous S-space,* Topology Appl., **136**(2004), 123–127.

VELICHKO, N.V.

[1981] *On weak topology of spaces of continuous functions (in Russian),* Matematich. Zametki, **30:5**(1981), 703–712.

[1982] *Regarding the theory of spaces of continuous functions (in Russian),* Uspehi Matem. Nauk, **37:4**(1982), 149–150.

[1985] *Networks in spaces of mappings (in Russian),* Mappings and Extensions of Topological Spaces, Udmurtia University P.H., Ustinov, 1985, 3–6.

[1995b] *On normality in function spaces,* Math. Notes, **56:5-6**(1995), 1116–1124.

[2001] *On subspaces of functional spaces,* Proc. Steklov Inst. Math. **2**(2001), 234–240.

[2002] *Remarks on C_p-theory,* Proc. Steklov Inst. Math., **2**(2002), 190–192.

VIDOSSICH, G.

[1969a] *On topological spaces whose function space is of second category,* Invent. Math., **8:2**(1969a), 111–113.

[1969b] *A remark on the density character of function spaces,* Proc. Amer. Math. Soc., **22**(1969b), 618–619.

[1970] *Characterizing separability of function spaces,* Invent. Math., **10:3**(1970), 205–208.

[1972] *On compactness in function spaces,* Proc. Amer. Math. Soc., **33**(1972), 594–598.

WHITE, H.E., JR.

[1978] *First countable spaces that have special pseudo-bases,* Canadian Mathematical Bull., **21:1**(1978), 103–112.

WILDE, M. DE

[1974] *Pointwise compactness in space of functions and R.C. James theorem,* Math. Ann., **208**(1974), 33–47.

YASCHENKO, I.V.
    [1992c]   *On fixed points of mappings of topological spaces,* Vestnik Mosk. Univ., Math.,
               Mech., **47:5**(1992a), 93.
    [1992d]   *Cardinality of discrete families of open sets and one-to-one continuous mappings,*
               Questions and Answers in General Topology, **2**(1992b), 24–26.
    [1994]   *On the monotone normality of functional spaces,* Moscow University Math. Bull.,
               49:3(1994), 62–63.

ZENOR, PH.
    [1980]   *Hereditary m-separability and hereditary m-Lindelöf property in product spaces
               and function spaces,* Fund. Math., **106**(1980), 175–180.

List of special symbols

For every symbol of this list we refer the reader to a place where it was defined. There could be many such places, but we only mention one here. Note that a symbol is often defined in the first volume of this book entitled "Topology and Function Spaces"; we denote it by *TFS*. We *never* use page numbers; instead, we have the following types of references:

(a) *To an introductory part of a section*
   For example,
$\exp X$ $\cdots\cdots\cdots\cdots\cdots\cdots$ **1.1**
   says that $\exp X$ is defined in the Introductory Part of Section 1.1.
   Of course,
$C_p(X)$ $\cdots\cdots\cdots\cdots\cdots$ **TFS-1.1**
   says that $C_p(X)$ is defined in the Introductory Part of Section 1.1 of the book TFS.

(b) *To a problem*
   For example,
$C_u(X)$ $\cdots\cdots\cdots\cdots\cdots$ **TFS-084**
   says that the expression $C_u(X)$ is defined in Problem 084 of the book TFS.

(c) *To a solution*
   For example,
$O(f, K, \varepsilon)$ $\cdots\cdots\cdots\cdots$ **S.321**
   says that the definition of $O(f, K, \varepsilon)$ can be found in the Solution of Problem 321 of the book TFS.
   The expression,
HFD $\cdots\cdots\cdots\cdots\cdots\cdots$ **T.040**
   says that the definition of HFD can be found in the Solution of Problem 040 of this volume.

V.V. Tkachuk, *A Cp-Theory Problem Book: Special Features of Function Spaces*,
Problem Books in Mathematics, DOI 10.1007/978-3-319-04747-8,
© Springer International Publishing Switzerland 2014

Every problem is short, so it won't be difficult to find a reference in it. An introductory part *is never longer than two pages* so, hopefully, it is not hard to find a reference in it either. Please keep in mind that a solution of a problem can be pretty long, but its definitions *are always given in the beginning.*

The symbols are arranged in alphabetical order; this makes it easy to find the expressions $B(x, r)$ and βX, but it is not immediate what to do if we are looking for $\bigoplus_{t \in T} X_t$. I hope that the placement of the expressions which start with Greek letters or mathematical symbols is intuitive enough to be of help to the reader. Even if it is not, then there are only three pages to plough through. The alphabetic order is *by line* and not by column. For example, the first three lines contain symbols which start with "A" or something similar and lines 3–5 are for the expressions beginning with "B", "β" or "$\mathbb{B}$".

$A(\kappa)$	**TFS-1.2**	$a(X)$	**TFS-1.5**
$AD(X)$	**TFS-1.4**	$\bigwedge \mathcal{A}$	**T.300**
$\mathcal{A}\|Y$	**T.092**	$B_d(x, r)$	**TFS-1.3**
$B(x, r)$	**TFS-1.3**	(B1)–(B2)	**TFS-006**
βX	**TFS-1.3**	$\mathbb{B}(X)$	**1.4**
$\mathrm{cl}_X(A)$	**TFS-1.1**	$\mathrm{cl}_\tau(A)$	**TFS-1.1**
$C(X)$	**TFS-1.1**	$C^*(X)$	**TFS-1.1**
$C(X, Y)$	**TFS-1.1**	$C_p(X, Y)$	**TFS-1.1**
$C_u(X)$	**TFS-084**	$C_p(Y\|X)$	**TFS-1.5**
$C_p(X)$	**TFS-1.1**	$C_p^*(X)$	**TFS-1.1**
$c(X)$	**TFS-1.2**	$\mathrm{conv}(A)$	**1.2**
CH	**1.1**	$\chi(X)$	**TFS-1.2**
$\chi(A, X)$	**1.2**	$\chi(x, X)$	**1.2**
$D(\kappa)$	**TFS-1.2**	$d(X)$	**TFS-1.2**
$\mathrm{dom}(f)$	**1.4**	$\mathrm{diam}(A)$	**TFS-1.3**
$\mathbb{D}$	**1.4**	$\Delta \mathcal{F}$	**TFS-1.5**
Δ_X	**TFS-1.2**	$\Delta(X)$	**TFS-1.2**
$\Delta_n(Z)$	**T.019**	$\Delta_n^{ij}(Z)$	**T.019**
$\diamondsuit$	**1.1**	$\diamondsuit^+$	**1.1**
$\Delta_{t \in T} f_t$	**TFS-1.5**	$ext(X)$	**TFS-1.2**
$\exp X$	**TFS-1.1**	$\mathrm{Fin}(A)$	**S.326**
$f <^* g$	**1.1**	$f \subset g$	**1.4**

Index

Printed in the United States
By Bookmasters